Plants with Anti-Diabetes Mellitus Properties

Plants with Anti-Diabetes Mellitus Properties

Appian Subramoniam

CRC Press
Taylor & Francis Group
Boca Raton   London   New York

CRC Press is an imprint of the
Taylor & Francis Group, an **informa** business

CRC Press
Taylor & Francis Group
6000 Broken Sound Parkway NW, Suite 300
Boca Raton, FL 33487-2742

First issued in paperback 2020

Version Date: 20151120

ISBN 13: 978-0-367-57496-3 (pbk)
ISBN 13: 978-1-4822-4989-7 (hbk)

Library of Congress Cataloging-in-Publication Data

Names: Subramoniam, Appian, 1950- author.
Title: Plants with anti-diabetes mellitus properties / Appian Subramoniam.
Description: Boca Raton : Taylor & Francis, 2016. | Includes bibliographical references and index.
Identifiers: LCCN 2015033139 | ISBN 9781482249897 (alk. paper)
Subjects: LCSH: Diabetes--Alternative treatment. | Materia medica, Vegetable. | Herbs--Therapeutic use.
Classification: LCC RC661.H4 S83 2016 | DDC 616.4/620654--dc23
LC record available at http://lccn.loc.gov/2015033139

Visit the Taylor & Francis Web site at
http://www.taylorandfrancis.com

and the CRC Press Web site at
http://www.crcpress.com

To my mentors Professor C.V. Ramakrishnan and (late) Professor R. Rajalakshmi

Contents

Preface

Diabetes mellitus is an age-old disease; it is described in ancient medical systems. From time immemorial, plant products have been used, to a large extent, to treat diabetes. Even today, the majority of the world's population uses botanicals in the treatment of diabetes mellitus. It is one of the major diseases whose incidence and severity are increasing all over the world, and it presents a huge burden to society in terms of the economy and the well-being of people.

At present, unlike almost all other diseases, diabetes mellitus needs lifelong treatment, and the treatment, to a large extent, could provide a near-normal life. Although there are many oral medicines and injectible insulin to treat diabetes mellitus, they are inadequate and ineffective, particularly in the long run; adverse reactions and loss of efficacy occur. There are time-tested anti-diabetes mellitus plant foods that are very safe and could be effective when consumed judiciously with a concomitant change in lifestyle, which includes increase in physical activity and reduction in mental stress.

Various cultural groups around the world used and are using a large number of plants to treat diabetes mellitus; the number may be more than 2000. More than 1000 anti-diabetes mellitus ethnomedicinal plants have been subjected to varying levels of scientific studies. Out of these, about 120 plants are promising for further studies, leading to the development of medicines for diabetes mellitus.

Surprisingly, in spite of a vast and intensive traditional use of the plants to treat the disease and substantial supporting scientific studies, conventional chemical entity drugs from these plants were not successfully developed and commercialized, except metformin and sugar blockers such as acarbose. One of the major reasons is the following: In most cases, crude preparations of medicinal plants including decoctions, extracts, or fractions and poly-plant-formulations could be more successful as anti-diabetes mellitus medicines than single chemical entity drugs isolated from plants. Many such herbal medicines (crude preparations) are locally in use in alternative systems of medicine and folklore medicines. So, with this information in mind, one of the aims of this book is to facilitate the development of such medicines with modern standards of efficacy and safety and make them widely available.

One fact that emerged from the literature is that almost all of the promising anti-diabetes mellitus plants are endowed with various other pharmacological properties as well; in the majority of cases, these associated activities are beneficial. Further, more than one active principle is present in anti-diabetes mellitus plants. The search for anti-diabetes mellitus compounds brought to light the fact that already known bioactive phytochemicals with anti-diabetes mellitus activities, such as β-sitosterol, quercetin, rutin, ursolic acid, oleanolic acid, gallic acid, ellagic acid, kaempferol, ferulic acid and resveratol, are present in many plants, including certain food plants. There are certain plants with both anti-diabetes properties and toxicities. Another point noted in the literature is that a major part of scientific studies on extracts and fractions were carried out in academic institutions using a few common animal models such as alloxan-induced and streptozotocin-induced diabetic rats. A good amount of such work is repetitious and superficial in nature. The identification of active principles (anti-diabetes mellitus phytochemicals) and unraveling the mechanism of action have not been done on some of the therapeutically promising anti-diabetes mellitus plants, including a few very important anti-diabetes mellitus food plants. This gap remains to be filled. Furthermore, a long-term and detailed toxicity evaluation is lacking in most of the cases. This is needed particularly when drug preparation is different from the traditional preparation and its use in traditional medicine is limited.

There are anti-diabetic molecules that are reported in phytochemical studies on many plants. Anti-diabetes mellitus properties of these compounds are reported from different plant sources in anti-diabetes studies. Plants with very low levels of such compounds are not considered anti-diabetes mellitus plants in this book.

The occurrences of phytochemically different varieties of the same species (chemotypes), ecotypes, and variations due to soil nutrition, including water content in the soil, could affect the results obtained in various studies using crude preparations and, in particular, extracts without extract standardization. Methods of processing and subsequent preparations of the extracts of plant materials could also influence bioactivity. This is one of the limitations in interpreting published experimental results.

The book begins with a detailed introduction on diabetes mellitus (Chapter 1). This is followed by an exhaustive compilation of information on the anti-diabetes mellitus activities of 1095 plants occurring across the world. In the case of most of the important anti-diabetes mellitus plants, brief botanical description, distribution, other pharmacological properties, and phytochemicals are provided; images of 101 important plants are provided. A partial list of traditional medicinal plants used to treat diabetes, but not tested for anti-diabetic activity, is also given. Chapter 3 highlights anti-diabetes mellitus plant foods along with a list of the edible parts of plants with anti-diabetes mellitus properties.

The author fervently hopes that this book will serve as a ready source for basic information on anti-diabetes mellitus medicinal plants and help researchers, students, doctors, and diabetic patients and others concerned about plant-based treatments for diabetes mellitus. Attempts have been made to include all anti-diabetes mellitus traditional plants subjected to at least one scientific study supporting the anti-diabetes mellitus property. To facilitate ready reference, all plants are presented in alphabetical order based on their botanical names. This book is unique—the only one of its kind.

Acknowledgments

I sincerely thank all those who helped me in the process of bringing this book to its final form.

Professor M. Daniel, former professor of botany, Maharaja Sayajirao University of Baroda, Vadodara, India, and his colleague Dr. Nagar, provided me with 26 images of plant species presented in the book. Regy Yohannan, a talented young research fellow in botany (PhD student), Kollam S. N. College, and his research guide, Dr. Devipriya V., associate professor of botany, Sree Narayana College, Trivandrum, India, generously provided me with 18 images of anti-diabetes mellitus plant species. Dr. Arunkumar R., my student, and Dhanya, L., graduate field assistant, Department of Botany, University of Kerala, provided skilled assistance in taking photographs from the university's herbal garden; and Dr. A. Gangaprasad, assistant professor, Department of Botany, University of Kerala, facilitated in getting access to the herbal garden. Dr. T.S. Nayar, former head, Conservation Biology, Tropical Botanic Garden and Research Institute, Trivandrum, and his colleague Suresh provided me with the image of *Oleae europaea*.

The technical help provided by D. R. Susan (MSc computer science) in the drawing of figures is gratefully acknowledged. My deep appreciation to my son-in-law (Dr. Subeesh), daughters (Vini and Dr. Vanathi), and wife (Thankam) for their support, enthusiasm, and encouragement during the preparation of this book.

I would like to acknowledge gratefully John Sulzycki (Senior Editor, CRC Press, Taylor & Francis Group) whose involvement and high level of professionalism was instrumental in the initiation and completion of this book. I am thankful to Jill Jurgensen (Senior Project Co-ordinator, CRC Press, Taylor & Francis Group) for her understanding and valuable professional efforts in handing this book. I sincerely thank Marsha Hecht (Project Editor, CRC Press, Taylor & Francis Group) and Christine Selvan (Senior Project Manager, SPi Global) for ensuring an extremely efficient and smooth production process of this book. Besides, I would like to acknowledge the sincere efforts and hard work put in by the copy editor. Cheryl Wolf (Editorial Assistant, CRC Press, Taylor & Francis Group), Al Staropoli (CRC Press, Taylor & Francis Group), and others involved in the promotion and marketing of this book are acknowledged sincerely.

Author

Dr. A. Subramoniam completed his MSc in zoology in 1974 (Annamalai University, Tamil Nadu, India), PhD in biochemistry in 1979 (Maharaja Sayajirao University of Baroda, India), and postdoctoral research in biochemical pharmacology in the United States (Howard University, Washington DC, and Temple University, Philadelphia, Pennsylvania). He worked in many reputed national institutes (Central Food Technology Research Institute, Mysore; Industrial Toxicology Research Centre, Lucknow; and Bose Institute, Calcutta) and carried out original, very high quality multidisciplinary research work in the broad areas of biomedical sciences and plant sciences. He is a recognized PhD guide, for a few universities in India, in biochemistry, biotechnology, pharmacology, chemistry, and zoology.

He has guided 10 PhD scholars. Dr. Subramoniam is the author of more than 170 scientific publications, which include original research publications in reputed international journals, chapters in books, and review papers in journals. He has nine patents to his credit. He served as a reviewer of some scientific journals in the fields of ethnopharmacology, phytopharmacology, biochemistry, and toxicology. He joined the Tropical Botanic Garden and Research Institute (TBGRI), Trivandrum, India, as a scientist in ethnopharmacology and ethnomedicine in 1994 and was appointed director in 2009. He retired from government service in June 2011. At TBGRI, he established advanced phytopharmacological research. During his tenure, TBGRI earned national and international recognition in medicinal plant research for discovering many important leads for the development of valuable medicines from plants. For example, his research group discovered a potent aphrodisiac principle, 2,7,7-trimethyl bicyclo[2.2.1] heptane, from an orchid, *Vanda tessellata*, and his group discovered the promising anti-inflammatory property of chlorophyll-a and its degradation products. Dr. Subramoniam has received several national awards for his excellent scientific contributions, such as the Hari Om Ashram Award for Research in Indian medicinal plants, Swaminathan Research Endowment Award for outstanding contribution in the scientific evaluation of medicinal properties of plants for their therapeutic use (awarded by the Indian Association of Biomedical Scientists), Jaipur Prize from the Indian Pharmacological Society, and Dr. B. Mukherjee Prize (2006) from the Indian Pharmacological Society. He served as president, Southern Regional Indian Pharmacological Society, 2009; vice president, Indian Association of Biomedical Scientists, 2007–2010; vice president, Kerala Academy of Sciences (2011–2013). He is currently a consultant in medicinal plant research.

1

Introduction

1.1 Diabetes Mellitus

Diabetes mellitus (DM) is one of the most challenging diseases facing the society and health-care professionals today. DM and its complications are a leading cause of mortality and morbidity and are a global problem. The increasing prevalence and severity are heavy burdens to society in terms of the economy and well-being of people. The disease affects more than 370 million people, according to the International Diabetes Foundation (IDF 2012). DM is a group of common serious metabolic disorders characterized by hyperglycemia resulting mainly from defects in insulin production/secretion and/or insulin action. DM is caused by a complex interaction of genetic and environmental factors. Deregulation of carbohydrate, lipid, and protein metabolism occurs in DM. Glucagon, a peptide hormone produced by pancreatic α-cells and gut-derived hormones such as incretin and other agents have important roles in glucose homeostasis including hepatic glucose production and insulin resistance (Mingrone and Gastagneto-Gissey 2014). There are two major types of DM (type 1 and type 2). Chronic hyperglycemia in DM leads to secondary pathophysiological changes including long-term life-threatening complications in major organs. DM is one of the leading causes of end-stage renal disease, nontraumatic lower extremity amputations, blindness, and cardiovascular diseases (Powers 2008).

1.1.1 Prevalence

The estimated number of adults with DM is more than 8.3% (370 million). It is predicted to affect over 550 million people by 2030 and increase rapidly in the foreseeable future (Cooper et al. 2012; IDF 2012). Among DM patients, about 90% are affected with type 2 DM. Type 2 DM has become a major health problem in both developed and developing countries; however, its prevalence is greater in the developing countries. According to the World Health Organization (WHO), the global population is in the middle of a diabetes epidemic with people in Southeast Asia and Western Pacific being mostly at risk. The prevalence of DM worldwide has increased dramatically from an estimated 30 million cases in 1985 to 177 million in 2000. In 2009, more than 220 million people worldwide had DM (WHO 2009a). In 2010, type 2 DM was reported affecting 285 million (6.4%) adults (Shaw et al. 2010). The prevalence of type 2 DM is rising much more rapidly than type 1 because of changing food habits, increasing obesity, and reduced physical activity levels. In the United States, 20.8 million persons or 7% of the population had DM in 2005. In the same year, in the United States, the prevalence of DM was 20.9% in elderly people (more than 60 years of age). The estimation was carried out by the U.S. Centers for Disease Control and Prevention.

There are considerable geographical variations in the prevalence and severity of both type 1 and type 2 DM. According to IDF (2011), the Western Pacific region has the most people with DM (132 million). Most of these people have type 2 DM. However, the countries with the highest number of people with type 2 DM are China, India, the United States, Brazil, and the Russian Federation. The greatest increase in prevalence is predicted to occur in Africa and the Middle East. Scandinavia has the highest incidence of type 1 DM. Japan and China have relatively low incidences of type 1 DM. The prevalence of type 2 DM is the highest in certain Pacific islanders, intermediate in countries such as India and the United States, and relatively low in Russia (Powers 2008). In Basra, Iraq, one in five adults is affected by DM (Mansour et al. 2014). The prevalence of DM in Saudi Arabia is one of the highest in the world reaching 23.7% people

(30–70 years of age) in 2004 (Al-Nozha et al. 2004). The prevalence also varies among different ethnic populations within a given country. In 2005, prevalence of DM in the United States, among people older than 20 years, was 13.3% in African Americans, whereas it was 8.7% in non-Hispanic whites.

1.1.2 Glucose Homeostasis

Now it is recognized that as with most endocrine hormones in humans, a feedback loop operates to ensure integration of glucose homeostasis and maintenance of glucose concentration in a specific range. This feedback loop relies on cross talk between β-cells and insulin-sensitive tissues. These tissues feed back information to β-cells. The mediator of this process has not been identified. The brain and humeral system may be involved in this process. If insulin resistance is present, β-cells increase insulin output to maintain normal glucose tolerance. When β-cells fail to respond adequately due to decreased β-cell function, glucose levels increase beyond the normal range. The magnitude of reduction in β-cell function establishes the degree of increase in plasma glucose in a glucose tolerance test. Additionally, an aging-associated reduction in the responsiveness of β-cells to carbohydrate partly underlines the fall in glucose tolerance with aging. Human pancreas seems to be incapable of renewing β-cell loss resulting from apoptosis after 30 years of age (review: Kahn et al. 2014). In addition to insulin and glucagon, certain other hormones and cytokines have important roles in controlling glucose homeostasis (review: Kahn et al. 2014).

1.1.2.1 Insulin and Glucose Homeostasis

1.1.2.1.1 Insulin Biosynthesis

Insulin is synthesized in the β-cells of islets of Langerhans in the pancreas. The mature mRNA for insulin codes for a single-chain 110 amino acid preproinsulin. Subsequently, the amino terminal signal peptide (24 amino acids) is cleaved off from the preproinsulin by a signal peptidase in the rough endoplasmic reticulum (RER) to form proinsulin. The signal peptide is rapidly degraded. Within the cisternae of the RER, proinsulin undergoes rapid folding and disulfide bond formation to generate the native tertiary structure of proinsulin. Proinsulin is then transported from RER to Golgi apparatus where it is packed into secretory granules. Cleavage of an internal 31-residue connector fragment or C-peptide generates chain A (21 amino acids) and chain B (30 amino acids) of an insulin molecule. A and B chains are connected by two disulfide bonds. The newly synthesized insulin and C-peptide are released at the cell membrane in equimolar concentrations along with a small amount (2%–4%) of proinsulin and intermediate cleavage products. Since the C-peptide is cleared more slowly than insulin, it is a useful marker of insulin secretion in the evaluation of hypoglycemia.

A single insulin gene is present in humans and most of the mammals. But in rats and mice, there are two genes; they synthesize two molecules that differ from each other by two amino acids in the B chain. The primary sequence of porcine insulin was determined by Sanger and coworkers in 1960 and this led to the complete synthesis of the protein. The sequence of human insulin is identical to that of porcine insulin except for the 30th amino acid in the B chain, which is alanine in porcine insulin and threonine in human insulin. Human insulin is now produced by recombinant DNA technology; physical and pharmacological characteristics of insulin can be modified by structural alterations at one or more amino acid residues.

1.1.2.1.2 Secretion of Insulin

Normally, glucose is the prime stimulus for insulin secretion. Certain amino acids, ketones, various nutrients, gastrointestinal peptides, and neurotransmitters also stimulate insulin secretion. Glucose levels above a critical level (3.9 mmol/L) stimulate insulin synthesis, primarily by enhancing protein translation and processing. Glucose is transported into the β-cells by glucose transporter 2 (GLUT2). Phosphorylation of glucose by glucokinase (the rate-limiting step) produces glucose-6-phosphate. Metabolism of glucose-6-phosphate via glycolysis in mitochondria generates ATP. Increase in intracellular ATP levels causes the closure of ATP-sensitive K^+ channels. This channel consists of two separate proteins: One of them is a binding site for drugs like sulfonylurea and meglitinides; the other is a K^+ channel protein. Inhibition of this K^+ channel (closure of the channel) results in depolarization of plasma membrane and opening of voltage-gated calcium channels. The increased intracellular calcium ion concentration leads to the exocytosis of insulin-containing secretory granules. When the demand for insulin is high, the glucokinase activity in β-cells increases. High levels of intracellular glucose in β-cells also stimulate calcium-independent pathways that enhance secretion of insulin. These pathways involve enhanced glucokinase activity, increased citrate levels, increased diacylglycerol formation, and enhanced protein kinase C (PKC) signaling. Binding of glucagon-like peptide-1 (GLP-1) to its receptor in β-cells promotes insulin release via intermediates such as protein kinase B (Akt) and also increases the number of β-cells via improved cell survival and decreased apoptosis (Cooper et al. 2012).

The insulin molecules (monomers) are released from the β-cells of islets into the portal vein at a rate of about 1 U/h to achieve a concentration of 50–100 μU/mL in portal circulation and 12 μU/mL in peripheral circulation. Once insulin enters the systemic circulation, it binds to receptors in target cells. The normal half-life of insulin in plasma is about 5–6 min, and its degradation occurs primarily in the liver, kidney, and muscle (Schade and Eaton 1983).

1.1.2.1.3 Insulin Action

The major functions of insulin are the stimulation of glucose uptake from the systemic circulation and suppression of hepatic gluconeogenesis. It activates the transport systems as well as the enzymes engaged in the intracellular utilization and storage of glucose, amino acids, and fatty acids and inhibits the breakdown of glycogen, fat, and protein. Insulin stimulates glycolysis (catabolism of glucose) and glycogenesis (synthesis of glycogen from glucose) and inhibits both hepatic gluconeogenesis and glycogenolysis (Cummings 2006). Postprandially, the glucose load elicits a rise in insulin that promotes the storage of carbohydrates and fat and the synthesis of protein. Insulin reduces circulating free fatty acid levels and promotes triglyceride synthesis and storage. It reduces intracellular lipolysis of stored triglyceride in adipocytes and increases the glucose transport into adipocytes to generate glycerophosphate, which permits esterification of fatty acids.

1.1.2.1.4 Insulin Signaling

A simplified version of insulin signaling is given in Figure 1.1. Insulin mediates its multiple actions by binding to its receptors and triggering intracellular signaling. There are two types of insulin receptor (IR), namely, IR-A and IR-B. IR-B preferentially activates metabolic signals, whereas IR-A leads the predominance of growth and proliferation signals. The insulin receptors are found over the surface of cell membranes in all mammalian cells. IR-B is found in hepatocytes, adipocytes, and muscle cells in higher concentrations (300,000/cells), whereas it is present in very low concentrations in blood cells, neurons, and so on (40/cells) (Khan and White 1998). Tissues that contain low concentrations of IR-B, the brain in particular, utilize glucose, to a large extent, in an insulin-independent fashion. Both IR-A and IR-B are present in the brain, heart, kidney, pancreas, and many other tissues (Belfiore and Malaguarnera 2011).

IR is a large transmembrane glycoprotein with two α-subunits and two β-subunits linked by disulfide bridges to form a heterotetramer. The α-subunits are extracellular, bearing insulin-binding domains, while β-subunits are transmembrane, which possess tyrosine protein kinase activity. When the insulin binds to α-subunits of the IR, the internal domain of its β-subunit undergoes autophosphorylation in its several tyrosine residues, which enhances the receptor's tyrosine kinase activity toward other substrates in insulin signaling pathways. The activated IR kinase phosphorylates insulin receptor substrate (IRS) proteins IRS-1 to IRS-6 on their tyrosine residues. Most of the insulin responses are mediated through IRS-1 and IRS-2. Phosphorylation of IRSs leads to the activation of phosphatidylinositol 3-kinase (PI-3K) pathway of signal transduction. Phosphorylation of IRSs at multiple tyrosine motifs by IR kinase serves as the docking site for PI-3K. Other pathway activated by insulin is mitogen-activated protein kinase (MAPK), which mediates the mitogenic effects. The MAPK and PI-3K pathways are linear in nature, but in many places, they cross talk. PI-3K pathway controls various cytoplasmic and nuclear events. The activated PI-3K leads to the formation of plasma membrane–bound phosphatidylinositol-3,4,5-trisphosphate (PIP3) from phosphatidylinositol-4,5-bisphosphate, which in turn recruits Akt, PIP3-dependent protein kinase 1 (PDK-1), and atypical isoforms of PKC (PKCa) to the plasma membrane. There are three isoforms of Akt, of which Akt-2 is the relevant isoform that associates tightly to GLUT4 containing vesicles in the cytosol and stimulates translocation of GLUT4 to the plasma membrane (Cho et al. 2001; Balasubramanyam and

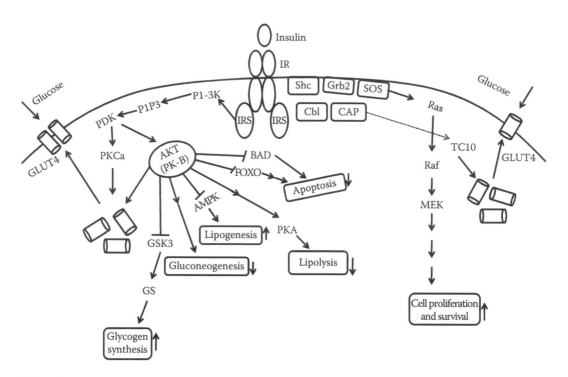

FIGURE 1.1 Simplified schematic representation of major pathways of insulin signaling. There are two types of insulin receptors (IR) (IR-A and IR-B). IR-B (abundant in hepatocytes, adipocytes, and muscle cells) preferentially activates metabolic signals, whereas IR-A predominantly activates growth and proliferation signals. AMPK-dependent mechanisms also stimulate glucose transport activity in skeletal muscle and decrease hepatic glucose production. *Abbreviations*: AMPK, adenosine monophosphate–activated kinase; BAD, Bcl₂ antagonists of cell death; CAP, an adaptor protein associated with Cbl; Cbl, an IRS family protein; FOXO, forkhead box subgroup O transcription factors; GLUT4, glucose transporter-4; Grb2, an adaptor protein that contains SH3 domains; GS, glycogen synthase; GSK3, glycogen synthase kinase-3; IR, insulin receptor; IRS, insulin receptor substrate; PDK-1, 3-phosphoinositide-dependent protein kinase-1; PKB/Akt, protein kinase B; PKCa, atypical form of protein kinase C; PIP3, phosphatidylinositol-3,4,5-trisphosphate; PI-3K, phosphatidylinositol 3-kinase; Ras, a GTP-binding protein; Raf and MEK, protein kinases in the mitogen-activated protein kinase (MAPK) pathway; Shc, a substrate for insulin receptor kinase; SOS, Son of Sevenless; TC10, a GTPase.

Mohan 2004). GLUT4 transports glucose into the cell. Defect in phosphorylation and activation of PKB/ Akt leads to insulin resistance. Akt also phosphorylates many substrates like Bcl₂ antagonists of cell death, glycogen synthase kinase-3, and forkhead transcription factor forkhead box protein 01 (FOXO1). The activation of these substrates finally leads to multifarious effects like survival and multiplication of cells, glycogen synthesis, lipogenesis, and controlling of gene expression (Mackenzie and Elliott 2014). Akt inhibits AMP-activated protein kinase (AMPK) (Kovacic et al. 2003). Glucose transport activity in skeletal muscle is also facilitated by AMPK-dependent mechanisms (Mackenzie and Elliott 2014). GLUT4 is primarily present in striated muscles and in adipose tissue; absence of insulin results in deficiency of glucose in these tissues; hyperglycemia causes excess of glucose entry in cells where it penetrates freely without insulin. Non-insulin-dependent glucose carriers are present in the liver, pancreas, kidney, intestine, erythrocytes, and so on. Among the non-insulin-dependent glucose transporters, cocarriers glucose/ Na⁺ ensure the digestive absorption of glucose and reabsorption of glucose in the renal tubules.

1.1.2.1.5 Glucose Transporters
Glucose transporters are the transmembrane proteins responsible for the transport of glucose and other substrates into the cells. There are several types of glucose transporters. They also have distinctive tissue distributions (Joost and Thorens 2001). GLUT2 is present in the β-cells to facilitate glucose entry into the cells, whereas GLUT4 is present in skeletal muscle and adipocytes. GLUT4 has 12 membrane-spanning regions with amino and carboxyl termini located intracellularly (Wood and Trayhurn 2003).

The translocation of GLUT4 from an intracellular cytosolic location to the plasma membrane is facilitated by insulin-dependent ways. Generally, plasma membrane GLUT4 determines the entry of glucose into the cells. Even though GLUT4 is expressed sufficiently in the cell, insulin resistance is associated with insufficient recruitment of GLUT4 to the plasma membrane. In addition to the major insulin-dependent GLUT4 translocation, a non-insulin-dependent GLUT4 translocation mechanism is also present, and it could be due to the combined action of AMPK and muscle contractions (Jessen and Goodyear 2005).

Insulin regulates GLUT4 recruitment in adipocytes through one minor PI-3K-independent pathway also. In this pathway, insulin mediates the formation of a complex involving three proteins. The complex in turn recruits two other proteins and activates TC10 (a small Rho family nucleotide-binding protein) facilitating GLUT4 trafficking to the plasma membrane (Figure 1.1).

In addition to insulin, several other hormones and growth factors can also activate signaling targets downstream of the IR. However, only insulin and highly related hormones such as insulin-like growth factor 1 efficiently stimulate acute glucose transport.

Termination of insulin signaling: Inositol 3'- and 5'-phosphatases attenuate PIP3 signaling. Insulin signaling is also terminated by internalization of the insulin–IR complex into endosomes and the degradation of insulin by insulin-degrading enzymes.

1.1.2.2 Glucagon, Incretins, and Other Hormones in Glucose Homeostasis

Glucagon: A peptide hormone secreted by α-cells of the pancreas and, to some extent, by intestinal tract, its plasma half-life is only a few minutes. The actions of insulin are opposed by glucagon, which is normally secreted when the blood glucose level tends to be low. Glucagon has hyperglycemic effects by stimulating the breakdown of glycogen into glucose (glycogenolysis) and gluconeogenesis and by inhibiting the synthesis of glycogen (glycogenesis) (Klover and Mooney 2003; Postic et al. 2004). Secretion of glucagon is inhibited by glucose and somatostatin (produced by δ-cells in pancreas) and is stimulated by certain amino acids.

The incretin hormones: The incretin hormones, GLP-1, and glucose-dependent insulinotropic polypeptide (GIP) play great roles in glucose homeostasis by improving β-cell differentiation, mitogenesis, survival, and insulin secretion. It also inhibits gastric emptying. GLP-1 potentiates glucose-stimulated insulin release through the G-protein-coupled receptor and the activation of protein kinase A (Drucker 2006; review: Nauck 2014). GLP-1, the most potent incretin, is released from L-cells in the small intestine and stimulates insulin secretion only when the blood glucose is above the fasting level (Powers 2008). Incretin-based drugs to treat type 2 DM include GLP-1 mimetic and dipeptidyl peptidase inhibitors that inhibit the proteolytic degradation of GLP-1.

Other hormones: In addition to incretins, many gastrointestinal hormones such as the gastrointestinal inhibitory peptide, gastrin, secretin, cholecystokinin, vasoactive intestinal peptide, gastrin-releasing peptide, and enteroglucagon promote insulin secretion. In glycemic regulation and in appetite regulation, several hormones secreted by the digestive tract, adipose tissue, and hypothalamic neurons are involved. Hormones such as adiponectin, leptin, resistin (from adipocytes), cholecystokinin, GLP-1, and ghrelin (from digestive tract) modulate the actions of insulin.

Catecholamine: Glycogen breakdown and production of glucose are induced when catecholamine binds its β-receptor.

1.1.3 Diagnosis of DM

The current criteria for the diagnosis of DM emphasize that fasting plasma glucose (FPG) is the most reliable and convenient test for identifying DM in asymptomatic individuals. FPG ≥ 7.0 mmol (126 mg/dL) warrants the diagnosis of DM. A random plasma concentration ≥11.1 mmol/L (200 mg/dL) accompanied by classical symptoms of DM (polyuria, polydipsia, and weight loss) is sufficient for the diagnosis of DM. The impairment of glucose metabolism clearly starts when the fasting glucose concentrations exceed about 7.78 mmol/L (140 mg/dL). Oral glucose tolerance testing is also a valid means for the diagnosis of DM; however, it is not recommended as a part of routine care (Powers 2008). Glycohemoglobin (HbA1c or A1C) values reflect average glycemic control over the previous period of about 3 months. Normal range of HbA1c values is between 4.0% and 6.4%. HbA1c levels between 5.7% and 6.4% indicate increased risk of diabetes and levels of 6.5% or higher indicate diabetes.

1.1.4 Different Types of DM

There are two broad categories of DM that are designated type 1 and type 2. Type 1 DM is the result of near-total insulin deficiency or the absence of insulin. Type 2 DM is a heterogeneous group of disorders characterized by variable degrees of insulin resistance, impaired insulin secretion, and increased glucose production of the liver. Gestational DM (glucose intolerance during pregnancy) is another type of DM. It may be related to the metabolic changes of late pregnancy and the increased insulin requirement. It occurs in about 4% of pregnancies in the United States. Most women revert to normal glucose tolerance postpartum but have a substantial risk of developing type 2 DM later in life. Maturity onset diabetes of the young (MODY) is a subtype of DM characterized by early onset of hyperglycemia and impairment in insulin secretion. It is inherited (autosomal dominant inheritance). Excess production of hormones such as glucagon that antagonize insulin action can also lead to DM. An extremely rare case of DM is pancreatic β-cell destruction by viral infections (Powers 2008). Mutations in the IR may cause severe insulin resistance. Pancreatitis and malnutrition-related diabetes have also been reported, particularly in tropical countries.

1.1.4.1 Type 1 DM

Among DM patients, about 10% suffer from type 1 DM. Type 1 DM (insulin-dependent DM) normally develops as a result of a chronic progressive autoimmune disorder (Pietropaolo 2001). Type 1 DM most commonly develops before the age of 30, but it can develop at any age. It results from total or near-total deficiency of insulin, commonly caused by complete destruction of β-cells in genetically susceptible individuals by a chronic autoimmune disease believed to be triggered by an infection or environmental factor (Kukreja and Maclaren 1999).

Pathologically, early in life, in most cases, islets are infiltrated with lymphocytes (insulitis). Then the islets become atrophic (complete β-cell destruction). These cells seem to be particularly susceptible to the toxic effects of tumor necrosis factor-α (TNF-α), interferon-γ, and interleukin-1 (IL-1). Most type 1 DM cases exhibit autoimmune antibodies against islet cell antigens such as glutamic acid decarboxylase, insulin, and protein tyrosine phosphatase; these antibodies are considered as the markers for the presence of autoimmune process. The presence of islet cell antibodies in nondiabetic individuals predicts a risk of developing type 1 DM.

Genetic risk of type 1 DM is conferred by polymorphism in many genes that regulate innate and adaptive immunity. Some of the gene loci identified now show strong linkage to increased occurrence of type 1 DM. The major susceptibility gene for type 1 DM is located in HLA class II gene located in the chromosome 6 (Kelly et al. 2001). There are other important gene loci that influence the expression of type 1 DM. There are additional modifying factors of genetic risk in determining development of type 1 DM. Nongenetic factors such as viral infection and vitamin D deficiency may increase risk. The infection of susceptible people with certain viruses acts as an important factor in causing type 1 DM (Lammi et al. 2005).

Environmental factors and their interaction with the immune system also give rise to the occurrence of type 1 DM. Some dietary ingredients, certain chemicals, and pollutants may cause type 1 DM.

Although in most individuals, type 1 DM is due to autoimmune destruction of β-cells (type 1A), some individuals develop type 1 DM by unknown nonimmunological mechanisms (type 1 B) (Dejkhamron et al. 2007).

At present, there is no satisfactory method for the prevention of type 1 DM. However, it has been reported that treating patients (with new-onset type 1 DM) with anti-CD3 monoclonal antibodies has slowed the decline in insulin production as judged from C-peptide levels (Powers 2008).

1.1.4.2 Type 2 DM

Type 2 DM accounts for more than 90% of diabetes cases all over the world. Type 2 DM more typically develops with increasing age (particularly after the age of 40 years). However, it occurs in obese adolescents also. Obesity is present in over 80% of type 2 diabetic patients (Powers 2008).

There are convincing data to indicate genetic components associated with insulin resistance. When both parents have type 2 DM, the risk approaches 40% (Powers 2008). Although many roles for several genes have been identified, the contribution of the genes to an individual's risk of type 2 DM is influenced by factors such as sedentary lifestyle, increased nutritional intake, and obesity. The dramatic increase in type 2 DM at present is due to changing environmental factors as well as lifestyle and dietary habits. In most cases, obesity-related insulin resistance plays a major role in the onset and progression of type 2 DM.

The contribution of maternal environment and *in utero* factors to the risk of type 2 DM in subsequent generations via epigenetic modifications is now being recognized as potentially important in explaining the very high rate of type 2 DM currently seen in many populations in the developing world.

During the early stage of type 2 DM, insulin resistance is compensated by increased production of insulin; thus, normal glucose levels are preserved (DeFronzo 2004). The glucose homeostasis appears to be normal during the early period, but when analyzed for plasma C-peptide concentration, it is evident that the insulin release is many times higher than that in normal subjects. Studies suggest that insulin resistance precedes defect in insulin secretion. Eventually, the defect in insulin secretion progresses to a level where insulin secretion is grossly inadequate. In normal adults, there is continuous renewal and loss of β-cells, and thus a dynamic relation between the apoptosis and regeneration of the cells is maintained; ductal endothelial cells of the pancreas could be the source of new β-cells (Rhodes 2005). This delicate balance between β-cell replication and apoptosis is interrupted in DM. Further, the replacement with new β-cells appears to be limited in humans after 30 years of age (review: Kahn et al. 2014).

At the onset of the pathogenesis of type 2 DM, the peripheral insulin-responsive tissues such as muscle and adipose exhibit a decreased rate of disposal of excess glucose and fatty acid from the circulatory system. At the same time, due to reduced insulin sensitivity of the liver, hepatic glucose production increases. In type 2 DM, pronounced insulin resistance is observed in the muscle, liver, and adipocytes along with defects in visceral glucose uptake. Increased hepatic glucose output predominantly accounts for increased FPG levels, whereas decreased peripheral glucose utilization results in postprandial hyperglycemia. There is a progressive decline in fasting and glucose-stimulated plasma insulin level due to the decline in β-cell function. Glucose utilization in insulin-independent tissues such as the brain is not substantially altered in type 2 DM.

1.1.4.2.1 Insulin Resistance and Type 2 DM

Insulin resistance has a central role in the development of type 2 DM. Induction of the resistance is partly by activation of various serine/threonine protein kinases that phosphorylate IRS proteins and other components of the insulin signaling pathway. This intense signaling leads to activation of negative feedback mechanisms. Normally, these feedback mechanisms are there to terminate excess insulin action. Phosphorylation of IRS proteins inhibits their function and interferes with insulin signaling in a number of ways leading to the development of an insulin-resistant state (Cooper et al. 2012).

In general, postreceptor defects in insulin signaling lead to insulin resistance. In rare cases due to gene defects, the IR gets mutated at the site of ATP binding or replaces tyrosine residues at the major sites of phosphorylation. This leads to failure of insulin signaling and cell response to insulin (Ellis et al. 1986).

The ectonucleotide pyrophosphatase/phosphodiesterase-1 (ENPP-1), a family of class II glycoprotein enzymes known to hydrolyze the 5'-phosphodiester bond in nucleotides, bears great significance in insulin signaling. When ENPP-1 interacts with IR, a decrease in insulin-dependent tyrosine phosphorylation of its α-subunit occurs. This leads to the failure of the autophosphorylation of β-subunit and switching off of insulin signaling (Abate and Chandalia 2007; Ohan et al. 2007). Inappropriate degradation of IRS-1 and IRS-2 in the insulin signaling pathway partly through the upregulation of suppressors of cytokine signaling was reported in many cases (Balasubramanyam et al. 2001).

In obese subjects, retinol-binding protein-4 interacts with PI-3K and reduces its activity. This also leads to insulin resistance in muscles and enhances the expression of phosphoenolpyruvate carboxylase in the liver. The changes in the production of adipokines are also reported in nonobese Asian Indians having direct link with obesity-independent insulin resistance. Obesity also reduces the phosphorylation of proteins involved in intracellular insulin signaling via IRS-1 and PI-3K. This results in the reduction of GLUT4-mediated influx of glucose (Ishiki and Klip 2005). The decreased insulin signaling in the skeletal muscles also promotes lipid accumulation in the muscle cells. In muscle cells, there is impairment in glycogen formation.

Insulin resistance develops in humans within an hour of acute increase in plasma nonesterified fatty acids. The increased rate of nonesterified fatty acid delivery and decreased intracellular fatty acid metabolism results in an increase in the levels of diacylglycerols, fatty acyl coenzyme A, and ceramides. These metabolites in turn activate the serine/threonine phosphorylation of IRS-1 and IRS-2 and reduce the ability of PI-3K in proper downstream regulation of insulin signaling (Khan et al. 2006). The excessive accumulation of triglycerides in the skeletal muscle cells of obese subjects is observed due to the greater mobilization of free fatty acids from insulin-resistant adipocytes. Increased free fatty acid flux from adipocytes leads to increased synthesis of very-low-density lipoprotein (VLDL) and triglycerides. This may lead to fatty liver diseases. Lipid accumulation and impaired fatty acid accumulation may generate reactive oxygen species (ROS) such as lipid peroxides. Leptin, resistin, adiponectin, and TNF-α produced by adipocytes, in addition to the contribution of fatty acid and triglycerides, have roles in the pathogenesis of type 2 DM. TNF-α, overexpressed in obese subjects, also modulates the serine/threonine phosphorylation and blocks the insulin signaling pathway (Rosen and Spiegelman 2006). Leptin in normal physiological conditions causes the accumulation of fat and reduces appetite through the hypothalamic effect, but in obese subjects, leptin resistance is developed, which leads to excessive flux of free fatty acids. This in turn leads to insulin resistance and β-cell dysfunction. Resistin and adiponectin also have roles in insulin sensitivity in obese subjects. Adiponectin, the sensitizer of insulin, stimulates fatty acid oxidation through AMPK and peroxisome proliferator-activated receptor-γ (PPAR-γ)-dependent ways (Rosen and Spiegelman 2006). PPAR-γ is an essential transcriptional mediator of adipogenesis, lipid metabolism, insulin sensitivity, and glucose homeostasis, which is increasingly recognized as a key factor in inflammatory cells as well as in cardiovascular diseases (Duan et al. 2008). Adipocyte products, adipokines, also produce an inflammatory state. Inhibition of inflammatory signaling pathways appears to reduce insulin resistance in animal models. Not all signal transduction pathways normally influenced by insulin are resistant to the effect of insulin in type 2 DM. For example, the MAPK pathway involved in the regulation of cell growth is not impaired. Hyperinsulinemia may increase the insulin action through these pathways. A supraphysiological dose of GLP-1 enhances β-cell responsiveness to glucose in patients with type 2 DM (Hojberg et al. 2008). The response of β-cells to nonglucose secretagogues such as arginine is preserved.

1.1.4.2.1.1 Role of Various Organs in Glucose Homeostasis in Type 2 Diabetes   Various organs seem to have important roles in the pathogenesis of type 2 DM (Figure 1.2). Substantial new information has emerged regarding the role of intestine and brain and feedback control of β-cell function. Certain hormones and cytokines have been identified as important in modulating communications among the various organs of the body and thereby controlling glucose homeostasis. These pathways are implicated in the pathogenesis of type 2 DM (Cooper et al. 2012). Many pathways can promote both insulin resistance and β-cell dysfunction. Oxidative stress, endoplasmic reticulum stress, and inflammation can adversely influence the pathways (review: Kahn et al. 2014).

The brain is involved in the regulation of appetite and satiety as well as in modulating the function of pancreatic α-cells and β-cells. Hypothalamus and sympathetic and parasympathetic systems are involved in this. The brain controls glucose metabolism through neuronal input. The hypothalamus is an integrator of insulin secretion and the vagus nerve is important in insulin secretion. Structural changes occur in the hypothalamus consistent with the occurrence of gliosis in obesity. Clock genes expressed in the brain are important in the establishment of the circadian rhythm. Changes in diurnal patterns and quality of sleep can have important effects on metabolic processes. Hypothalamic inflammation might also contribute to central leptin (produced by adipose tissue that acts at the level of the hypothalamus to suppress appetite) resistance and weight gain (review: Kahn et al. 2014).

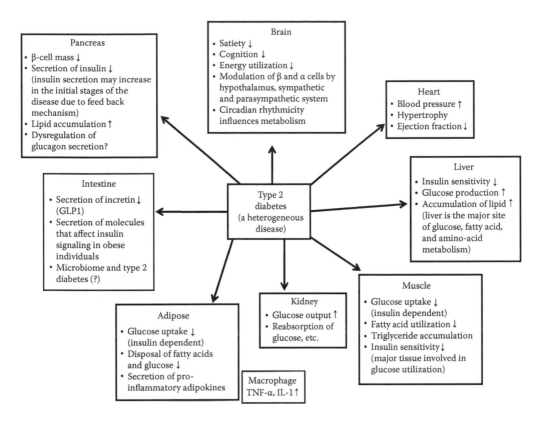

FIGURE 1.2 Involvement of major organs in type 2 diabetes mellitus.

In addition to the secretion of incretin hormones, the gastrointestinal tract has crucial roles in type 2 DM. Recent studies suggest that microbes present in the gut also have a role in development of insulin resistance. Bile acids also have important roles in glucose homeostasis. Bile acids are ligands for farnesoid X receptor. Activation of this receptor by bile results in the release of fibroblast growth factor (FGF) 1, which has insulin-like actions and insulin-sensitizing properties (Kingwell 2014; Suh et al. 2014). Further, when bile acid binds its receptor in the L-cells, GLP-1 secretion from the cells increases (review: Kahn et al. 2014). The gut may have an important role in insulin resistance in obese type 2 DM (Mingrone and Castagneto-Gissey 2014). Jejunal proteins secreted by type 2 obese diabetic mice or insulin-resistant obese humans impair insulin signaling. These proteins induce insulin resistance in normal mice and inhibit insulin signaling *in vitro* in rat skeletal muscle cells. Metabolic surgery has been shown to be effective in inducing remission of type 2 DM prior to any significant weight reduction. In metabolic surgery, the secretion of these proteins may be drastically impaired or abolished (Mingrone and Castagneto-Gissey 2013). Further, in duodenal–jejunal bypass surgery, jejunal nutrient sensing is required to rapidly lower glucose concentration (Breen et al. 2012). Stimulation of a nutrient sensor located in the proximal jejunum by glucose and/or lipid reduces hepatic glucose production.

An increase in visceral adipose tissue deposition leads to central obesity and an increase in the production of proinflammatory adipokines. Such individuals are at high risk of type 2 DM and cardiovascular diseases. Expansion of adipose tissue is associated with accumulation of macrophages that express several proinflammatory genes including TNF and IL-1, which locally impair insulin signaling (review: Kahn et al. 2014).

The liver is a major source of glucose production through glycogenolysis and gluconeogenesis (Klover and Mooney 2004). Excess accumulation of lipids in liver develops and causes insulin resistance and type 2 DM. Studies in mice and humans have elucidated an important role for hepatic diacylglycerol activation of atypical PKCε in triggering hepatic insulin resistance (Perry et al. 2014). Lipid accumulation is probably associated with secretion of proinflammatory cytokines from Kupffer cells (resident macrophages) and recruited macrophages that impair insulin signaling. Markers of systemic inflammation including C-reactive proteins and its upstream regulator IL-6 are associated with insulin sensitivity and β-cell function. Decrease in inflammation improves β-cell function in patients with type 2 DM (review: Kahn et al. 2014).

1.1.5 Complications of DM

In both type 1 and type 2 DM, uncontrolled hyperglycemia and, to some extent, hyperlipidemia lead to the development of both acute and long-term complications.

1.1.5.1 Acute Complications of DM

Diabetic ketoacidosis (DKA) and the hyperglycemic hyperosmolar state (HHS) are acute complications of DM. DKA is very common in type 1 DM patients, but it also occurs in certain type 2 DM cases. Major symptoms include nausea, thirst/polyuria, abdominal pain, and shortness of breath; in children, it is frequently associated with cerebral edema. Ketosis results from a marked increase in fatty acid release from adipocytes with a shift toward ketone body synthesis in the liver. Normally, these fatty acids are converted to triglycerides or VLDL in the liver. But hyperglucagonemia alters hepatic metabolism to favor ketone body formation. Both insulin deficiency (absolute or relative deficiency) and glucagon excess are generally required for DKA to develop. Excess catecholamines, cortisol, and/or growth hormone also contribute to the development of DKA.

HHS is primarily seen in individuals with type 2 DM with a history of polyuria, weight loss, and diminished oral intake. Clinical features include profound dehydration, hyperosmolality, hyperglycemia, tachycardia, and altered mental status. Hyperglycemia associated with DM and inadequate fluid intake induces an osmotic diuresis that leads to intravascular volume depletion. Laboratory diagnosis includes marked hyperglycemia (more than 55.5 mmol/L) and hyperosmolality (more than 3.5 mosmol/L) and prerenal azotemia.

Therapies for both DKA and HHS include appropriate intravenous replacement of fluid and insulin infusion considering glucose levels, serum electrolytes, acid base status, and renal function. Underlying or precipitating problems such as infection and cardiovascular problems should be identified and treated (Powers 2008).

1.1.5.2 Chronic Complications of DM

Chronic complications of DM can be divided into vascular and nonvascular complications. Microvascular complications lead to retinopathy, neuropathy, and nephropathy, whereas coronary artery disease, peripheral arterial disease, and cerebrovascular disease are due to macrovascular complications. The microvascular complications of both type 1 and type 2 DM result from chronic hyperglycemia. Coronary heart disease events and morbidity are two to four times greater in patients with DM. Dyslipidemia and hypertension also play important roles in macrovascular complications. Nonvascular complications include gastroparesis, diarrhea, uropathy, sexual dysfunction, infections, periodontal diseases, dermatological complications, and glaucoma. Foot ulcers and infections can lead to gangrene, which may require amputation; DM is the leading cause of nontraumatic lower extremity amputation in the United States.

Abnormality in cell-mediated immunity and phagocyte function and diminished vascularization lead to a greater frequency and severity of infection in DM. Further, hyperglycemia facilitates the growth of pathogenic fungi and other organisms. In general, infections are more common in the diabetic population.

Diabetic nephropathy is the leading cause of DM-related morbidity and mortality (Lopes 2009; Wada and Makino 2009). Diabetic nephropathy develops in 30%–40% of patients with both type 1 and type 2 DM within 20–25 years after the onset of DM (Powers 2008).

DM is the leading cause of blindness between ages 20 and 74 years in the United States. Individuals with DM are 25 times more likely to become blind than normal individuals. Blindness is primarily due to diabetic retinopathy and clinically significant macular edema. Diabetic retinopathy is classified into two stages: nonproliferative and proliferative. Nonproliferative retinopathy is marked by retinal vascular microaneurysms. The appearance of neovascularization in response to retinal hypoxia is typical of proliferative retinopathy (Powers 2008).

Diabetic neuropathy occurs in about 50% of individuals with chronic type 1 or type 2 DM. Dysfunction of isolated cranial or peripheral nerves is less frequent than distal symmetric polyneuropathy in DM. Neuropathy involving the autonomic nervous system may also occur. This may lead to genitourinary dysfunction.

The mechanisms by which hyperglycemia leads to complications of DM are not very clear. However, the four major mechanisms, namely, (1) formation of advanced glycosylation end products (AGEs), (2) increased levels of sorbitol formation, (3) sustained activation of the PKC pathway, and (4) increased glucose flux through the hexose amine pathway, have been suggested. Intracellular hyperglycemia causes the formation of AGEs by nonenzymatic glycosylation of proteins. AGEs have been shown to cross-link proteins and accelerate atherosclerosis, promote glomerular dysfunction, reduce nitric oxide synthesis, induce endothelial dysfunction, and alter extracellular matrix composition and structure. Intracellular glucose is predominantly metabolized by phosphorylation and subsequent glycolysis, whereas during hyperglycemia, some glucose is converted to sorbitol by the enzyme aldose reductase. Increased sorbitol concentration leads to increase in cellular osmolality, alterations in redox potential, and increase in ROS generation. These may facilitate retinopathy, nephropathy, neuropathy, and so on. Hyperglycemia increases the formation of diacylglycerol, which activates PKC. This enzyme, among other actions, alters the transcription of genes for fibronectin, type IV collagen, contractile proteins, and extracellular matrix proteins in endothelial cells. Hyperglycemia increases glucose flux through the hexose amine pathway, which generates fructose-6-phosphate, a substrate for O-linked glycosylation and proteoglycan production. This pathway may alter functions of proteins by glycosylation of proteins such as endothelial nitric oxide synthase.

The production of growth factors, such as transforming growth factor-β, vascular endothelial growth factor-B, platelet-derived growth factor, and epidermal growth factor, is increased by most of these proposed pathways. Excess of these growth factors may play an important role in complications of DM. Increased production of ROS in the mitochondria under hyperglycemic conditions and peroxidation of lipids may influence the aforementioned pathways. It is likely that some pathways predominate in certain organs.

Intensive glycemic control in DM can prevent the complications more effectively (Powers 2008).

1.1.6 Treatment/Management of DM

Management of DM includes optimum therapy, appropriate nutrition, and required level of exercise. Glycemic control is central to optimum DM therapy; DM care includes detection and management of DM-specific complications and modification of risk factors for DM-associated diseases.

Insulin is the major therapeutic agent used for DM, particularly for type 1 DM. Parenteral therapy includes insulin, GLP-1, and amylin. Oral glucose-lowering agents currently in use are biguanides, insulin secretagogues (sulfonylureas, repaglinide, nateglinide, etc.), thiazolidinediones, α-glucosidase inhibitors, dipeptidyl peptidase-4 (DPP-4) inhibitors, nutraceuticals, and so on.

1.1.6.1 Insulin and Other Parenteral Therapy

1.1.6.1.1 Insulin

Treatment of DM follows different regimens in accordance with the type of diabetes, age, severity of DM, meal pattern, and physiological status of the patient. Insulin is the major drug used for the treatment of type 1 DM and also in certain cases of type 2 DM. Since in type 1 DM absolute absence of insulin is observed without insulin resistance, it can be treated with insulin. Nowadays, the insulin used is produced through recombinant DNA technology using genetically engineered *Saccharomyces cerevisiae* or *Escherichia coli*. The recombinant insulin consists of the amino acid sequence of human insulin or minor variants of that. For example, in one of the short-acting insulin variants, insulin lispro, the 28th and 29th amino acids have been reversed by genetic manipulations. This insulin analogue has fewer tendencies for self-aggregation resulting in more rapid absorption and onset of action and a shorter duration of action. The different classes of insulin used for the treatment of diabetes are rapid-acting insulin, short-acting insulin, intermediate-acting insulin, and long-acting insulin. The rapid-acting insulin starts its action within 5–15 min after administration and lasts for about 3–5 h. Generally, the short-acting insulin is soluble crystalline zinc insulin; its effect starts within 30 min and lasts for 5–8 h. The intermediate-acting insulin starts its action after 2 h of administration and lasts for 18–24 h; it includes lente and neutral protamine Hagedorn insulin. Long-acting insulin includes ultralente insulin and insulin glargine with no pronounced peak of activity, and the duration of action is more than 24 h. In the treatment of type 1 DM, a combination of intermediate-acting and short-acting insulin is typically prescribed because using only one of the aforementioned insulin types would be insufficient to maintain normal glycemic condition. In addition, compounds like bis(maltolato)oxovanadium(IV) can also be used in the treatment of DM owing to its insulin-sensitizing and insulin-mimicking actions (Shinde et al. 2004). The standard mode of insulin administration is subcutaneous, and it may be done with syringes and needles, pen injectors, or via insulin pumps. In addition, powered and aerosolized insulin formulations are used for inhalation. It should be noted that the precise normal insulin secretory pattern of β-cells is not reproduced by any of the insulin regimens.

Hypoglycemia is the most common complication of insulin therapy. It mostly results from inadequate carbohydrate diet after insulin administration or very high physical exertion or relatively a high insulin dose. In certain cases, antibodies are produced against insulin, which leads to both neutralization of some quantity of administered insulin and allergic reactions. However, use of human insulin has reduced these complications to almost nil. In certain cases, atrophy of subcutaneous fatty tissue may occur at injection sites secondary to immune reactions.

1.1.6.1.2 Amylin

Amylin is a 37-amino acid peptide cosecreted with insulin in normal glucose homeostasis. An analogue of amylin (pramlintide) was found to reduce postprandial glycemic excursions in type 1 and type 2 DM patients taking insulin. Addition of pramlintide with insulin produces a modest reduction in A1C and seems to dampen meal-related glucose excursions. It slows gastric emptying and suppresses glucagon levels, but not insulin levels. The major side effects of this peptide are occasional nausea and vomiting (Powers 2008).

1.1.6.1.3 Analogues of GLP-1

Agents that act as GLP-1 (and to some extent GIP) receptor agonists or enhance endogenous GLP-1 activity amplify glucose-stimulated insulin release. They also suppress glucagon secretion, slow gastric emptying, and suppress appetite. GLP-1 receptors are found in islets, the gastrointestinal tract, and the brain. Exenatide, an analogue of GLP-1 that differs from GLP-1 in amino acid sequence, has more half-life time compared to native GLP-1 by virtue of its resistance to the enzyme that degrades GLP-1(DPP-4). Exenatide lowers glucose and suppresses appetite without weight gain in type 2 DM. The A1C reduction with exenatide is only moderate. This drug is given as an adjuvant or combination therapy with metformin or sulfonylurea. Nausea is the reported side effect especially at higher doses (Powers 2008).

1.1.6.2 Oral Hypoglycemic Agents

In type 2 DM, oral hypoglycemic agents (OHAs) are used either singly or in combination with or without insulin administration. The pathophysiology of type 2 DM is highly heterogeneous and the individual response to drugs can differ greatly. The major OHAs used in the treatment of type 2 DM are sulfonylureas, biguanides, repaglinide, nateglinide, and thiazolidinediones, agents that enhance GLP-1 receptor signaling, α-glucosidase inhibitors, and inhibitors of sodium glucose cotransporter-2 (SGLT2) (Kahn et al. 2014).

1.1.6.2.1 Sulfonylureas

Sulfonylurea drugs were initially developed in the 1950s. These drugs constitute the majority of the insulin secretagogues used in the treatment of type 2 DM. Normally, pancreatic β-cells sense and secrete an appropriate amount of insulin in response to a glucose stimulus. Sulfonylureas increase insulin release from β-cells by interacting with ATP-sensitive potassium channels on β-cells. ATP-sensitive potassium channels have two subunits: one subunit contains cytoplasmic binding sites for both sulfonylureas and ATP, which is named as the sulfonylurea receptor type 1, and the other subunit of the potassium channel acts as the pore-forming subunit. A higher rate of mitochondrial activity leads to an increase in the ATP/ADP ratio. Either this ATP or sulfonylurea interacts with the sulfonylurea receptor type 1, resulting in the closure of the K_{ATP} channel. Closure of this channel depolarizes the plasma membrane and triggers the opening of voltage-sensitive calcium channels, leading to the rapid influx of calcium. Increased intracellular calcium causes an alteration in the cytoskeleton, which stimulates a translocation of insulin-containing secretory granules to the plasma membrane, leading to the exocytotic release of insulin (Evans and Rushakoff 2007).

Sulfonylurea drugs are classified into first and second generations, of which acetohexamide, chlorpropamide, tolbutamide, and tolazamide are first-generation drugs that possess a lesser affinity to bind to sulfonylurea receptor type 1. Second-generation sulfonylureas include glibenclamide (also known as glyburide), glipizide, gliclazide, and glimepiride; these drugs are now widely used. These second-generation sulfonylureas are most effective in individuals with type 2 DM of recent onset (less than 5 years). They generally possess a more rapid onset and shorter half-life (Powers 2008). However, glimepiride (1–4 mg) is administered once a day and has a long duration of action.

1.1.6.2.2 Biguanide

Metformin is the major therapeutically useful biguanide; it is regularly advised for the treatment of type 2 DM in the United States, and it is the second most prescribed OHA in Europe. Metformin reduces hepatic glucose production, increases peripheral utilization of glucose, reduces FPG levels, and improves lipid profiles. *In vitro* and *in vivo* studies show that instead of stimulating insulin secretion, metformin activates AMPK, which is a major cellular regulator of lipid and glucose metabolism. Metformin increases peripheral glucose uptake and reduces hepatic glucose production in an AMPK-dependent manner (Mackenzie and Elliott 2014). The mechanisms of action of metformin are not fully understood, and the receptors of the compounds, if any, are yet to be identified. Metformin can reduce acetyl-CoA carboxylase activity, induce fatty acid oxidation, and increase expression of enzymes for lipogenesis (Cleasby et al. 2004; Zhang et al. 2007). Reported side effects of metformin include diarrhea, anorexia, nausea, metallic taste, and lactic acidosis (Powers 2008).

Phenformin is another OHA coming under the group biguanides. Phenformin is not prescribed extensively nowadays due to increased co-occurrence of lactic acidosis with DM and relatively less long-term benefits.

1.1.6.2.3 Glinides (Meglitinides)

Meglitinides are also insulin secretagogues. These include repaglinide and nateglinide, characterized by a very rapid onset and short duration of action. Repaglinide is a structural analogue of glyburide, while nateglinide is the derivative of the amino acid D-phenylalanine. Unlike sulfonylureas, meglitinides stimulate first-phase insulin release in a glucose-sensitive manner. Repaglinide is approximately five times more potent in stimulating insulin secretion than glyburide, and in the case of nateglinide, it is three times more rapid than repaglinide. The mechanism of action of meglitinides is binding with sulfonylurea receptor type 1, stimulating the closing of ATP-sensitive potassium (K_{ATP}) channel resulting in an influx of calcium and insulin exocytosis. This class of drugs can control postprandial glucose increase and can be used in patients with sulfonylurea allergy. These drugs have relatively short half-life and are given along with meals (Powers 2008).

1.1.6.2.4 Thiazolidinediones (PPAR-γ Agonists)

Thiazolidinediones are a class of insulin-sensitizing compounds with glucose- and lipid-lowering activity. They are the selective agonist for PPAR-γ. PPARs are ligand-activated transcription factors of the nuclear receptor family. There are three PPAR subtypes (PPAR-α, PPAR-δ or PPAR-β, and PPAR-γ) that have different ligand specificity and tissue distribution. PPAR-γ is highly expressed in adipose tissue, macrophages, and cells of the vasculature but is expressed at lower levels in many other tissues. PPAR-α is the receptor for the fibrate class of lipid-lowering drugs, and PPAR-δ orchestrates the regulation of high-density lipoprotein metabolism (Balasubramoniam and Mohan 2000; Berger and Moller 2002). Synthetic ligands of PPAR-α and PPAR-γ such as fibric acid and thiazolidinediones showed significant improvement in insulin resistance in type 2 DM and prediabetes (Jay and Ren 2007). PPAR-γ receptor signaling plays essential roles in adipogenesis, glucose, and lipid homeostasis (Auwerx 1999; Lehrke and Lazar 2005). PPAR-γ agonists such as the thiazolidinediones are efficient and clinically useful insulin sensitizers (Giaginis et al. 2009; Borniquel et al. 2010).

Pioglitazone and rosiglitazone are the major therapeutically used thiazolidinediones, which decrease insulin resistance and enhance the biological activity of both endogenous and injected insulin. Pioglitazone acts on both PPAR-γ and PPAR-α and hence has a better glucose- and triglyceride-lowering activity than that of rosiglitazone.

Thiazolidinediones may cause liver toxicity, peripheral edema, and heart failure in certain cases. An increased risk of fractures in women has been reported (Powers 2008). Rosiglitazone was withdrawn from the market in many countries due to concern about a possible increase in the risk of cardiovascular adverse effects, including congestive heart failure. Pioglitazone was withdrawn in 2011 in France considering a possible high risk of bladder cancer. The risks are not substantiated. The continued use of this drug is a subject of debate (Kahn et al. 2014).

1.1.6.2.5 α-Glucosidase Inhibitors

α-Glucosidase inhibitors in current use are acarbose, miglitol, and voglibose. These drugs inhibit an enzymatic degradation of complex carbohydrates in the small intestine and thereby reduce the entry of glucose into the blood stream. Acarbose is a nitrogen-containing pseudotetrasaccharide α-glucosidase inhibitor, while miglitol is a synthetic analogue of deoxynojirimycin. These compounds improve glycemic control in DM without increasing the risk of weight gain or hypoglycemia. The pancreatic α-amylase and membrane-bound intestinal α-glucosidase enzymes are inhibited by the inhibitors in a competitive and reversible manner. Acarbose shows little affinity for isomaltase and no affinity for lactase, while miglitol does inhibit intestinal isomaltase (Evans and Rushakoff 2002). Side effects include diarrhea, flatulence, and abdominal distention. It is contraindicated in individuals with inflammatory bowel disease, gastroparesis, and so on (Powers 2008).

1.1.6.2.6 Incretin-Based Drugs

DPP-4 is the major enzyme responsible for degrading incretin hormones *in vivo*. The inhibitor of DPP-4 increases insulin secretion, reduces glucagon secretion, improves glucose tolerance, and reduces A1C levels in type 2 DM patients (McIntosh 2008). Thus, the inhibition of DPP-4 improves type 2 DM (McIntosh et al. 2005; Ahrén 2007; Green 2007). DPP-4 inhibitors in current use are alogliptin, linagliptin, saxagliptin, sitagliptin, and vildagliptin (review: Kahn et al. 2014).

These orally active DPP-4 inhibitors show antihyperglycemic effect without severe hypoglycemia, reduce glycated hemoglobin (H1C), improve islet function, and modify the course of DM. These agents promote insulin secretion without hypoglycemia or weight gain and appear to have preferential effects on postprandial blood glucose.

DPP-4 has a wide range of substrates other than GLP-1, GIP, and peptide YY. Therefore, the inhibitors of DPP-4 not only influence the regulation of energy homeostasis but also other functions unrelated to energy homeostasis like immunity. The long-term effect of these drugs on cardiovascular and immune systems and safety is yet to be studied in detail (Fisman et al. 2008; Richter et al. 2008). Further, questions were raised regarding the safety of GLP-1 mimetics and DPP-4 inhibitors (incretin-based drugs); these drugs possibly increase the risk of pancreatitis and pancreatic cancer. However, the causal association between these drugs and pancreatic cancer, if any, is not established (review: Kahn et al. 2014).

1.1.6.2.7 Inhibitors of SGLT2 Inhibitors

The kidney not only excretes and reabsorbs glucose but also produces glucose through gluconeogenesis. Generally, the quantity of glucose filters does not exceed the kidney's threshold to reabsorb it and thus little appears in urine. The finding that SGLT2 reabsorbed glucose from urine led to the development of inhibitors of this transporter. Dapagliflozin and canagliflozin were recently introduced to the market and others are under clinical trial. SGLT2 inhibitors appear to be generally tolerated and have been used as monotherapy or in combination with other oral anti-DM agents or insulin. The risk of hypoglycemia is low with these inhibitors (review: Nauck 2014). However, the increase in urinary glucose is associated with a five times higher rate of genital mycotic infections and a 40% increase in infections of the lower urinary tract. Long-term studies are in progress to assess the cardiovascular safety of these drugs (review: Kahn et al. 2014).

1.1.6.2.8 Other Drugs

The dopamine receptor agonist bromocriptine is an approved drug to regulate glucose metabolism. The drug acts centrally and probably restores circadian rhythm. The circadian rhythm influences several organ systems associated with metabolism (review: Kahn et al. 2014).

Bile acid–binding resin colesevelam is also approved to treat type 2 DM. Bile acids are ligands of the farnesoid X receptor and activation of this receptor results in release of FGF19. FGF has insulin-like actions. Bile acids also activate G-protein coupled bile acid receptor-1 located on intestinal L-cells leading to GLP-1 secretion (review: Kahn et al. 2014).

1.2 Plant Products: A Promising Source for the Development of New Anti-DM Medicines (Anti-DM Medicines from Plants)

In spite of the introduction of various antihyperglycemic and hypoglycemic agents, DM and its associated secondary complications continue to be a major problem in the world population. The high prevalence of DM as well as its dreaded long-term complications warrants systematic search for anti-DM agents. Various types of currently used anti-DM drugs cannot fully control glucose level and cause side effects and/or insufficient response after prolonged use of these drugs. Therefore, it is necessary to look for new medicines and interventions that can be used to manage this complex and highly heterogeneous metabolic disorder.

Plants are known to be excellent sources of anti-DM medicines (Marles and Farnsworth 1995). A perusal of literature shows that there are more than 1700 recorded plants used around the world to treat or control DM by different cultural groups in traditional medicine. Varying levels of pharmacological evaluation and/or bioassays have been carried out on more than 1000 of these traditional anti-DM plants. Most of the studies on these plants are not complete enough to determine their likely therapeutic value, showing very marginal or no activity to substantial therapeutically promising antihyperglycemic/hypoglycemic/anti-DM activities. However, based on the studies conducted, more than 120 plants are very promising for further investigation for the development of medicine for DM. In some cases, the same active molecules are distributed in nature in many plants. The known mechanisms of actions and active molecules of the anti-DM plants are diverse. These plants are attractive sources for the development of novel safe and effective medicines including combination drugs and polyherbal formulations for DM and its complications (Reviews: Behara and Yadav 2013; Campbell-Tofte et al. 2013; Chauhan et al. 2010; Chang et al. 2013; Coman et al. 2012; El-Abhar and Schaalan 2014, Fatima et al. 2012; Grover et al. 2002; Gulshan and Rao 2013; Gunjan et al. 2011; Joseph and Jini 2011; Jung et al. 2006; Khan 2005; Lakshmi et al. 2012; Mitra Pandey 2014; Malvi et al. 2011; Mentreddy et al. 2005; Mukesh and Namita 2013; Noor et al. 2013; Prabhakar and Doble 2008; Raman et al. 2012; Saravanamuttu and Sudarsanan 2012; Shankar Murthy and Kiran 2012; Sidhu and Sharma 2013; Subramoniam and Babu 2003; Talaviya et al. 2014; Tripathi and Verma 2014; Yeh et al. 2003). Combination therapy can be developed focusing on several important drug targets involved in DM.

It is true that most of the traditional anti-DM plants do not have any practical utility in controlling or satisfactorily managing type 1 and type 2 DM. Further, some of these plant drugs may have adverse side effects. Scientific studies in light of present knowledge are required to select out the potential plants. Since a majority of the world population is using anti-DM plants to control or treat DM, there is a need to carry out fuller systematic studies on these plants, not only to determine their safety and efficacy, but also to develop new and improved drugs in light of modern science.

In spite of the availability of hundreds of anti-DM plants, plant-based drugs used in conventional medicine are only metformin (a derivative of galegine from *Galega officinalis*) and α-glucosidase inhibitors. Various aspects of anti-DM plant research, such as determination of efficacy and safety, isolation of active principles and lead molecules, elucidation of mechanisms of action, development of combination medicines, development of food supplements for DM patients with edible anti-DM plants (nutraceuticals), and development of agrotechnology and/or *in vitro* propagation techniques for deserving plants, should be adequately considered and integrated. The important gap areas have to be identified for giving priority in research. Facilitating focused research with proper approach on anti-DM plants will certainly lead to the development of many valuable medicines such as metformin to treat DM.

Regular intake of appropriate amounts of specific immune modulatory nutraceuticals may prevent or delay the development of type 1 DM. Food materials containing antioxidants, anti-inflammatory agents, and/or antihyperlipidemic property (nutraceuticals) could protect or delay the expression of type 2 DM and could work as antirisk factors.

1.3 The Need for This Book

There are many good reviews in journals and chapters in books on the subject. Many of these reviews are on different aspects of anti-DM plants and overlap to a considerable extent. In the recent past, the quantity of publications on medicinal plants and diabetes has multiplied along with the appearance of many new journals in the general area of medicinal plant-related research. To some extent, repetition of research work and reviews on anti-DM plants is also not uncommon in the vast literature. There is a need to provide all relevant information in one place for easy reference, among other things, for the development of anti-DM medicines and food supplements (nutraceuticals) for diabetic patients. At present, specific pieces of important information are hidden in the vast and scattered literature on diverse aspects of anti-DM medicinal plants. This is one of the major hindrances in the progresses of anti-DM medicinal plant research and development. In this context, a book with full or almost full information on the subject could serve as a ready source of information. This book is intended to serve this very important purpose.

2

Anti-Diabetes Mellitus Plants

2.1 Background

There are numerous plants used to treat diabetes in traditional medicine, which includes tribal and folk medicines practiced by various ethnic groups in remote villages and tribal pockets of the world. Varying levels of scientific studies have been done on less than a half of these plants. Out of these plants studied, most of them exhibited antidiabetic and/or hypoglycemic activities in *in vitro* and/or *in vivo* pharmacological investigations. There are two excellent sources available to get information on antidiabetic traditional medicinal plants. A detailed review on this subject was published in 1995 (Marles and Farnsworth 1995). After 10 years, updated information on traditional anti-diabetes mellitus (anti-DM) plants was published in a book chapter (Simmonds and Howes 2005). Now, another 10 years have passed. This period witnessed a rapid growth in the scientific validation of traditional medicinal plants and the emergence of new journals in this and related areas of study. There have been substantial amounts of new study reports on individual anti-DM plants and review papers on the pharmacological and phytochemical aspects of such plants. However, these reports are scattered, and the review papers on anti-DM plants are incomplete and required information is not available in one place for reference.

Most of the antidiabetic plants used in ethnomedicine are also used for various other ailments in traditional medicine. Scientific studies have also detected several other pharmacological properties in almost all of the anti-DM plants. This review is aimed at providing almost all anti-DM plants found in the world with at least one scientific study to substantiate their antidiabetes property. Attempts have been made to include, in this chapter, all those plants whose crude preparations, extracts, and/or isolated constituents showed pharmacological effects with hypoglycemic or anti-DM activity in animal models, *in vitro* assays, and/or human clinical studies.

In the development of herbal medicines for diabetes, knowledge on all pharmacological activities, (other than anti-DM activities) including toxicity, if any, of the plants is needed (Subramoniam 2003; Subramoniam et al. 2004). Brief information on traditional uses, other pharmacological properties, toxicity (if any), and phytochemicals isolated is included in the case of important and/or therapeutically promising anti-DM plants. Sufficient studies were not done in a majority of the traditional anti-DM plants, and in many cases the studies done are insufficient and incomplete to judge their likely therapeutic value, if any; therefore, new important and therapeutically promising anti-DM plants could emerge in light of future studies on these plants.

There are numerous plants with α-glucosidase or carbohydrate breakdown inhibition activities. Therefore, plants with marginal or negligible *in vitro* inhibitory activity on these enzymes without any possible therapeutic value are not included under the list of anti-DM plants in this chapter. Similarly, plants with only marginal or negligible *in vitro* protein tyrosine phosphatase 1B (PTP1B) inhibitory activity are also not included in this chapter. Some of the phytochemicals known to possess some level of beneficial effects in DM are distributed in numerous plants. For example, a data search has shown that there are 146 families, 698 genera, and 1620 species of plants reported to contain varying amounts of oleanolic acid (Fai and Tao 2009). This compound exhibits several biological activities including anti-DM activity. Plants that contain very low levels of such compounds are not included in this chapter. Lower groups of plants such as fungi, algae, and bryophytes are not covered in this book.

Limitations include likely or possible variations in the reported studies in efficacy and safety depending on ecotype and genotype and nutritional conditions of a given plant species (Subramoniam 2001). Such variations are not studied in almost all of the anti-DM medicinal plants.

The aim of this chapter is to consolidate the available knowledge on each plant with emphasis on further studies to determine their therapeutic utility and to obtain new leads. A partial list of traditional plants used to control diabetes, but not subjected to pharmacological studies, is shown separately in Table 2.1 to facilitate future work on these plants.

Plants are arranged in alphabetical order of their botanical names. In the case of important anti-DM plants, as per available literature and the author's perception, details such as ethno-medical uses, other pharmacological properties, and phytochemicals reported are given briefly. The data were collected, to a large extent, from Google scholar, PubMed, and ScienceDirect.

2.2 Plants Tested for Anti-DM Properties

Abelmoschus esculentus (L.) Moench, Malvaceae

Synonym: *Hibiscus esculentus* L.

Common Names: Lady's finger, okra, etc.

Hibiscus esculentus (*Abelmoschus esculentus*) fruit is a common vegetable. The seeds of this plant lowered blood glucose levels in normoglycemic, glucose-fed hyperglycemic, and streptozotocin-induced diabetic rats (Aslan et al. 2003). In another study, the effects of crude hydroalcoholic extract of *A. esculentus* on glucose, albumin, and total bilirubin levels of streptozotocin-induced diabetic rats were investigated. The extract (200, 400, and 800 mg/kg, p.o., 14 days) decreased blood glucose and bilirubin and increased albumin levels and body weight compared to diabetic control rats (Uraku et al. 2010). The antidiabetic activity of *A. esculentus* fruit extract was observed in rabbits also. A gradual decrease in the blood glucose levels was observed by regular feeding of the fruit (vegetable) extract for 10 days (Subrahmanyam et al. 2011).

Abelmoschus manihot (L.) Medik., Malvaceae

Synonym: *Hibiscus manihot*

Common name: Sunset muskmallow

Compounds with known antidiabetic properties such as stigmasterol, γ-sitosterol, myrecetin and anthocyanins were isolated from this plant leaf and woody stem (Jain et al. 2009).

Abelmoschus moschatus Medik., Malvaceae

Abelmoschus moschatus is an aromatic medicinal plant, distributed in many parts of Asia. The anti-hyperglycemic action of myricetin, purified from the aerial part of *A. moschatus*, was investigated in streptozotocin-induced diabetic rats. Bolus intravenous injection of myricetin decreased the plasma glucose concentrations in a dose-dependent manner in diabetic rats. Myricetin (1.0 mg/kg) attenuated the increase of plasma glucose induced by an intravenous glucose challenge test in normal rats. A concentration-dependent stimulatory effect of myricetin (0.01–10.0 μmol/L) on glucose uptake of the soleus muscles isolated from streptozotocin-induced diabetic rats was observed. The increase of glucose utilization by myricetin was further characterized by enhancing glycogen synthesis in isolated hepatocytes of streptozotocin-induced diabetic rats. These results suggest that myricetin enhances glucose utilization to lower plasma glucose in diabetic rats lacking insulin (Liu et al. 2005a). The extract of *A. moschatus* had a higher level of polyphenolic flavonoids. The extract (200 mg/kg/day) displayed the characteristics of rosiglitazone (4 mg/kg/day) in reducing the higher homeostatic model assessment of insulin resistance (HOMA-IR) index in fructose chow-fed rats after a 2-week treatment, increasing postreceptor insulin signaling mediated by enhancements in insulin receptor (IR) substrate 1–associated phosphatidylinositol 3-kinase step and glucose transporter (GLUT4) translocation in insulin-resistant soleus muscles. The authors conclude that the plant is a potentially useful adjuvant therapy for patients with insulin resistance and/or for subjects wishing to increase insulin sensitivity (Liu et al. 2010). *A. moschatus* extracts exhibited other beneficial properties such as antioxidant, free radical scavenging, antimicrobial, and antiproliferative activities in *in vitro* assays (Gul et al. 2011).

TABLE 2.1

Traditionally Used Anti-Diabetes Mellitus Plants, Not Subjected to Scientific Studies[a]

No.	Botanical Names of Plants	Family	Part Commonly Used
1	*Abutilon lignosum* (Cav.) D.Don	Malvaceae	Leaves, flowers
2	*Abutilon trisulcatum* (Jacq.) Urban	Malvaceae	Leaves
3	*Acacia bilimekii* J.F. Macbr.	Fabaceae	Leaves
4	*Acacia farnesiana* (L.) Willd.	Fabaceae	Bark
5	*Acacia pennata* Dalzell & A. Gibson	Mimosaceae	
6	*Acacia retinodes* Schltdl.	Fabaceae	Leaves
7	*Acanthopanax sessiliflorum* (Rupr. & Maxim.) Seem.	Araliaceae	
8	*Achillea fragrantissima* Sch. Bip.	Asteraceae	All parts
9	*Aconitum ferox* Wall.	Ranunculaceae	Roots (poisonous plant?)
10	*Aconitum palmatum* Don.	Ranunculaceae	Roots (poisonous plant?)
11	*Acrocomia Mexicana* Karw. ex Mart.	Leguminosae	Roots
12	*Adansonia digitata* L.	Bombacaceae	Fruits
13	*Adiantum capillus-veneris*	Polypodiaceae	Whole plant
14	*Adiantum caudatum* L.	Pteridaceae	Leaves
15	*Agapetes sikkimensis* Airy Shaw	Ericaceae	Aerial parts
16	*Agastache mexicana* (Kunth) Lint et Epling	Lamiaceae	Aerial parts
17	*Agave atrovirens* Karw. ex Salm-Dyck	Agavaceae	Leaves
18	*Agave lecheguilla* Torr.	Agavaceae	Stems, leaves
19	*Agave salmiana* Otto	Agavaceae	Leaves
20	*Ageratina petiolaris* (*Moc.* et Sesse ex DC.) R.M. King et H. Rob.	Asteraceae	Aerial parts
21	*Ajuga bracteosa* Benth.	Lamiaceae	All parts
22	*Ajuga remota* Benth.	Lamiaceae	Leaves
23	*Albizia amara* Boivin	Mimosoideae	Leaves
24	*Alchemilla vulgaris* L.	Rosaceae	
25	*Alisma orientale* (Sam.) Juzep.	Alismataceae	
26	*Allionia choisyi* Standl.	Nyctaginaceae	Whole plant
27	*Alloispermum integrifolium* (DC.) H.Rob	Asteraceae	
28	*Aloe arborescens* Mill.	Aloaceae	Leaves
29	*Aloe excels* Berger	Aloaceae	Leaves
30	*Aloe succotrina* Lam.	Aloaceae or Asphodelaceae	
31	*Alpinia galanga* (L.) Sw. A	Zingiberaceae	Rhizome
32	*Alternanthera sessilis* (L.) R.Br. ex DC.	Amaranthaceae	Whole plant
33	*Amaranthus esculantus* Guil. et. Per.	Amaranthaceae	Whole plant
34	*Ambrosia artemisiifolia* L.	Asteraceae	
35	*Amomum aromaticum* Roxb.	Zingiberaceae	Root
36	*Amomum subulatum* Roxb.	Zingiberaceae	Root
37	*Amorphophallus konjac* K. Koch	Araceae	Rhizome
38	*Amphipterygium adstringens* Diederich F.L. von S.	Anacardiaceae	Bark
39	*Andrographis lineate* Nees	Acanthaceae	Leaves
40	*Andropogon citratus* (DC.) Stapf.	Poaceae	Ariel parts
41	*Ananas comosus* (L.) Merr	Bromeliaceae	Leaf, fruit
42	*Andansonia digitata* L.	Bombacaceae	
43	*Anemopsis californica* Hook. & Arn.	Saururaceae	Root and leaves
44	*Annona cherimola* Mill.	Annonaceae	Leaf, bark, etc.
45	*Annona glabra* L.	Annonaceae	Leaf

(Continued)

TABLE 2.1 (*Continued*)

Traditionally Used Anti-Diabetes Mellitus Plants, Not Subjected to Scientific Studies[a]

No.	Botanical Names of Plants	Family	Part Commonly Used
46	*Anthemis herba-alba* Asso.	Compositae	Ariel parts
47	*Anthocephalus indicus* Achille Richard	Rubiaceae	Bark
48	*Anthocleista nobilis* G.Don.	Loganiaceae	Bark
49	*Anthocleista rhizophoroides* Baker	Loganiaceae	Bark
50	*Apodanthera buraeavi* Cogn.	Cucurbitaceae	Aerial parts
51	*Aporocactus flagelliformis* (L.) Lem.	Cactaceae	Stem
52	*Aporosa lanceolata* Thwaites	Euphorbiaceae	Leaves
53	*Aporocactus flagelliformis* (L.) Lem.	Cactaceae	
54	*Aquilaria agallocha* Roxb.	Thymelaeaceae	Stem
55	*Aralia nudicaulis* L.	Araliaceae	
56	*Aralia racemosa* L.	Araliaceae	
57	*Arceuthobium vaginatum* (Humb. & Bonpl. ex Willd.) Presl.	Santalaceae/Viscaceae	Whole plant
58	*Arctostaphylos uva-ursi* (L.) Spreng	Ericaceae	Fruits
59	*Argemone mexicana* L.	Papaveraceae	Fruit and leaf
60	*Argemone ochroleuca* Sweet.	Papaveraceae	Root
61	*Argemone ochroleuca* Sweet ssp. *ochroleuca*	Papaveraceae	Root and stem
62	*Argemone platyceras* Link & Otto	Papaveraceae	
63	*Argyreia nervosa* (N. Burman.) Bojer	Convolvulaceae	Roots
64	*Arisaema triphyllum* (L.) Schott	Araceae	
65	*Aristolochia anguicida* Jacq.	Aristolochiaceae	Root
66	*Aristolochia brevipes* Benth.	Aristolochiaceae	Root
67	*Aristolochia maxima* Jacq.	Aristolochiaceae	Root
68	*Aristolochia odoratissima* L.	Aristolochiaceae	Root
69	*Aristolochia pilosa* H.B.K.	Aristolochiaceae	Root
70	*Aristolochia sericea* Benth.	Aristolochiaceae	
71	*Aristolochia trilobata* L.	Aristolochiaceae	Leaf
72	*Arracacia brandegei* J.M.Coult & Rose	Umbelliferae	Roots
73	*Artemisia absinthium* L.	Asteraceae	
74	*Artemisia abyssinica* Sch.Bip.	Asteraceae	Leaves, aerial parts
75	*Artemisia ludoviciana* Nutt.	Compositae/Asteraceae	Leaves
76	*Artocarpus altilis* (Parkinson) Fosberg A	Moraceae	Leaves
77	*Asarum canadense* var. *acuminatum* Ashe	Aristolochiaceae	
78	*Asclepias linaria* Cav.	Asclepiadaceae	
79	*Asparagus gonoclados* Baker	Asparagaceae	Bulbs
80	*Asparagus officinalis* L.	Asparagaceae	
81	*Astragalus crotalariae* A. Gray	Leguminosae	Roots
82	*Atractylodes lancea* (Thunb.) DC.	Asteraceae	
83	*Atractylodes ovata* DC.	Asteraceae	Aerial part
84	*Atropa belladonna* L.	Solanaceae	Leaf
85	*Auricularia auricula-judae* (Fr.) Quel.	Primulaceae	Fruits
86	*Avena sativa* L., oats	Poaceae	Whole plant
87	*Averrhoa carambola* L.	Oxalidaceae	Leaves
88	*Baccharis salicifolia* Lam.	Asteraceae	Leaves
89	*Baccharis trimera* (Less.) DC.	Myrtaceae	Leaves
90	*Barleria noctiflora* Nees	Acanthaceae	Whole plant
91	*Barringtonia acutangula* (L.) Gaertn	Lecythidaceae	Stem bark

(Continued)

TABLE 2.1 (*Continued*)

Traditionally Used Anti-Diabetes Mellitus Plants, Not Subjected to Scientific Studies[a]

No.	Botanical Names of Plants	Family	Part Commonly Used
92	*Bauhinia divaricata* Lam.	Leguminosae	Leaves
93	*Bergenia stracheyi* Engl.	Saxifragaceae	Roots
94	*Bergia capensis* L.	Elatinaceae	
95	*Billia hippocastanum* Peys.	Hippocastanaceae	Aerial parts
96	*Barosma betulina* Bartl. & H.L. Wendl. [*Agathosma betulina* (P.J.Bergius) Pillans]	Rutaceae	
97	*Begonia heracleifolia* Schltdl. & Cham.	Begoniaceae	
98	*Berberis moranensis* Schult. & Schult. f.	Berberidaceae	
99	*Bidens aurea* (Aiton) Sherff	Asteraceae	Leaves
100	*Bidens leucantha* (L.) Willd.	Asteraceae	Aerial parts
101	*Bidens odorata* Cav.	Asteraceae	Leaves
102	*Bocconia arborea* S. Watson	Papaveraceae	
103	*Borreria verticillata* (L.) G.F.W. Mey	Rubiaceae	Whole plant
104	*Brachystegia nigerica* Hoyle & A.P.D. Jones	Leguminosae	Seed flour
105	*Brassica napiformis* (Pailleux & Bois) L.H. Bailey	Brassicaceae	Roots, leaves
106	*Brassica rapa* L.	Brassicaceae	Roots
107	*Brickellia cavanillesii* (Cass.) A. Gray	Asteraceae	Aerial parts
108	*Brickellia squarrosa* B.L. Rob. & Seaton.	Asteraceae	Aerial parts
109	*Bridelia ndellensis* Beille	Euphorbiaceae	Leaves
110	*Bryonia cretica* L.	Cucurbitaceae	Aerial parts
111	*Buchanania axillaris* (Desr.) Ramamoorthy	Anacardiaceae	Leaves, bark
112	*Buddleja stachyoides* Cham. & Schltd.	Scrophulariaceae	Whole plant (all parts)
113	*Bursera simaruba* (L.) Sarg.	Burseraceae	Bark
114	*Cacalia peltata* Kunth	Asteraceae	
115	*Caesalpinia crista* L.	Fabaceae	Seed
116	*Calamintha macrostema* Benth.	Lamiaceae	Root, stem
117	*Calamintha umbrosa* (Mbieb.) Rchb	Lamiaceae	Root, stem
118	*Calamus rotang* L.	Arecaceae	Root
119	*Calea integrifolia* (DC.) Hemsl.	Compositae	Stem and leaf
120	*Calea urticifolia* (Mill.) DC.	Compositae	
121	*Calliandra anomala* (Kunth) J.F. Macbr.	Fabaceae	
122	*Calystegia japonica* Choisy	Convolvulaceae	Flower
123	*Canarium zeylanicum* (Retz.) Blume	Burseraceae	Bark
124	*Capparis iberica* Trevir. ex Spreng.	Capparaceae	Leaf, etc.
125	*Capparis incana* Kunth	Capparaceae	Leaves, bark
126	*Capraria biflora* L.	Scrophulariaceae	Leaves
127	*Cardiospermum halicacabum* L.	Sapindaceae	Leaves
128	*Carissa lanceolata* R. Br.	Apocynaceae	Whole
129	*Carmona retusa* (Vahl) Masamune	Boraginaceae	Leaves, root
130	*Casearia glauca* Lam.	Salicaceae	Bark
131	*Casearia zeylanica* (Gaertn.) Thwaites	Flacourtiaceae	Stem bark, root bark
132	*Cassia alata* L. [syn: *Senna alata* L. Roxb]	Fabaceae/Caesalpiniaceae	Leaves
133	*Cassia fruticosa* Mill.	Leguminosae	Flower
134	*Cassia glauca* L.	Caesalpiniaceae	Leaves
135	*Cassia siamea* (Lam.) Irwin & Barneby	Fabaceae	Leaves
136	*Cassia sieberiana* DC.	Fabaceae	Leaves, flower
137	*Cassia skinneri* Benth.	Fabaceae	Leaves

(*Continued*)

TABLE 2.1 (*Continued*)

Traditionally Used Anti-Diabetes Mellitus Plants, Not Subjected to Scientific Studies[a]

No.	Botanical Names of Plants	Family	Part Commonly Used
138	*Cassia tomentosa* L.	Fabaceae	Leaves
139	*Castela texana* (T. & G.) Rose	Simaroubaceae	Leaves
140	*Cecropia peltata* L.	Moraceae	Leaves
141	*Cedronella canariensis* Webb & Berthel	Lamiaceae	
142	*Celastrus scandens* L.	Celastraceae	
143	*Centaurea salmantica* L.	Compositae	Aerial parts
144	*Cephalanthus glabratus* K.Schm	Rubiaceae	Wood
145	*Cephalaria syriaca* (L.) Scrad. ex Roem. & Schult	Dipsacaceae	Aerial parts
146	*Ceratonia siliqua* L.	Fabaceae	Leaves, fruits
147	*Chamaecrista hispidula* (Vahl) H.S. Irwin & Barneby	Fabaceae	Leaves, fruits
148	*Chaptalia nutans* Polak.	Compositae	Whole plant
149	*Chasmanthera cordifolia* (DC.) Baill [syn: *Cocculus cordifolius* DC.]	Menispermaceae	Stem, leaves, root
150	*Cheirolophus arbutifolius* (Sevent) G. Kundel	Asteraceae/Compositae	Aerial parts
151	*Cheirolophus canariensis* (Willd.) Holob	Asteraceae	Aerial parts
152	*Cichorium endivia* L.	Asteraceae	Leaf, root
153	*Cichorium pumilium* L.	Asteraceae	
154	*Cirsium dipsacolips* Matsum	Asteraceae	Root
155	*Cirsium mexicanum* DC.	Asteraceae	Root
156	*Cirsium ochrocentrum* A. Gray	Asteraceae	Root
157	*Cirsium rhaphiolepis* (Hemsl.) Petr.	Asteraceae	Flower
158	*Cissampelos pareira* L.	Menispermaceae	Root
159	*Cissus populnea* Guill & Perr	Vitaceae	Stem
160	*Citrus aurantifolia* (Christm.) Swingle Lim.	Rutaceae	
161	*Citrus bigaradia* Riss.	Rutaceae	
162	*Citrus medica* L.	Rutaceae	
163	*Citrus sinensis* (L.) Osbeck	Rutaceae	Peels
164	*Clematis chinensis* Osbeck	Ranunculaceae	
165	*Cleome aspera* L.	Capparidaceae	Whole plant
166	*Clerodendranthus spicatus* (Thunb.) C.Y.Wu.	Lamiaceae	
167	*Cnidoscolus aconitifolius* (Mill.) I.M. Johnst	Euphorbiaceae	Leaf
168	*Cnidoscolus aconitifolius* ssp. *aconitifolius* Breckon	Euphorbiaceae	Leaf
169	*Cocculus villosus* Diels	Menispermaceae	Leaf, root
170	*Cola gigantea* A. Chev.	Sterculiaceae	Seed
171	*Colubrina glomerata* (Benth.) Hemsl.	Rhamnaceae	Bark
172	*Convallaria majalis* L.	Asparagaceae	Bulb
173	*Convolvulus microphyllus* Sieber ex Spreng	Convolvulaceae	Stem bark
174	*Conyza filaginoides* (DC.) Hieron.	Asteraceae	Whole plant
175	*Coptis japonica* Makino	Ranunculaceae	
176	*Cordia alliodora* (Ruiz et Pav.) Cham.	Boraginaceae	Bark
177	*Cordia dichotoma* Forst.	Boraginaceae	Stem bark
178	*Cordia myxa* L.	Boraginaceae	Stem bark
179	*Cordia morelosana* Standl.	Boraginaceae	Leaves
180	*Cordia tinifolia* Willd.	Boraginaceae	Leaves
181	*Coridothymus capitatus* (L.) Rchb. F.	Lamiaceae	
182	*Cornus officinalis* Sieb. et Zucc (*Corni fructus*)	Cornaceae	Whole plant

(Continued)

TABLE 2.1 (*Continued*)

Traditionally Used Anti-Diabetes Mellitus Plants, Not Subjected to Scientific Studies[a]

No.	Botanical Names of Plants	Family	Part Commonly Used
183	*Cornus stolonifera* Michx.	Cornaceae	
184	*Corylus cornuta* Marsh.	Betulaceae	
185	*Costus mexicanus* Liebm.	Costaceae	Leaves, roots
186	*Costus schlechteri* H.J.P. Winkl.	Costaceae	Leaves
187	*Cotoneaster aitchisonii* C.K.Schneid	Rosaceae	Aerial parts
188	*Coutarea hexandra* K Schum	Rubiaceae	
189	*Coutarea latiflora* Moc & Sess. [syn: *Hintonia latiflora* Moc & Sess.]	Rubiaceae	Whole plant
190	*Crataegus azarolus* L.	Rosaceae	
191	*Crataegus mexicana* Moc. & Sesse ex DC.	Rosaceae	Root
192	*Crataegus pubescens* f. stipulacea Stapf	Rosaceae	Whole plant
193	*Crescentia alata* (H.B.K.)	Bignoniaceae	
194	*Crescentia cujete* L.	Bignoniaceae	
195	*Crotalaria medicaginea* Lam.	Fabaceae	Seeds
196	*Croton draco* Schltdl.	Euphorbiaceae	Bark and leaf
197	*Croton niveus* Jacq.	Euphorbiaceae	Bark
198	*Croton tonduzii* Pax.	Euphorbiaceae	Bark
199	*Croton xalapensis* H.B.K.	Euphorbiaceae	Bark
200	*Cryptostegia grandiflora* R. Br.	Apocynaceae	
201	*Cucumeropsis mannii* Naudin	Cucurbitaceae	Leaf, fruit
202	*Cucumis callosus* (Rottl.) Rogn.	Cucurbitaceae	Seed
203	*Cucumis metuliferus* E. Mey. ex Naud.	Cucurbitaceae	Fruit
204	*Cuscuta chinensis* Lam.	Convolvulaceae	Leaves
205	*Cuscuta jalapensis* Schltdl.	Convolvulaceae	
206	*Cyathea divergens* Kunze	Cyatheaceae	Bark
207	*Cyclanthera pedata* (L.) Schrad.	Cucurbitaceae	Shoot, leaves
208	*Cymbalaria muralis* Gaertn	Scrophulariaceae	Whole plant
209	*Cynanchum schlechtendalii* (Deche.) Standl. et Steyerm.	Asclepiadaceae	Resin
210	*Cynara scolymus* L.	Asteraceae	
211	*Cynometra ramiflora* L.	Fabaceae	
212	*Cyperus iria* L.	Cyperaceae	Root
213	*Dahlia pinnata* Cav	Asteraceae	
214	*Datura quercifolia* HB & K	Solanaceae	All parts
215	*Decalepis hamiltonii* Wight et Arn.	Apocynaceae	Tuber
216	*Decatropis bicolor* (Zucc.) Radlk.	Rutaceae	Leaf
217	*Delonix regia* (Bojer ex Hook.) Raf.	Fabaceae	Leaf
218	*Dendrobium loddigesii* Rolfe	Orchidaceae	Stem
219	*Dendrobium nobile* Lindl.	Orchidaceae	Stem
220	*Descurainia sophia* (L.) Prantl	Brassicaceae	Whole plant
221	*Desmodium motorium* DC.	Fabaceae	Leaf
222	*Dicliptera squarrosa* Nees [syn: *Jacobinia suberecta* Andre]	Acanthaceae	
223	*Dillenia indica* L.	Dilleniaceae	Leaf
224	*Dioscorea asclepiadea* Prain & Burkill	Dioscoreaceae	Tuber
225	*Dioscorea batatas* (L.) Lam.	Dioscoreaceae	Tuber
226	*Dioscorea gracillima* Miq.	Dioscoreaceae	Bulb
227	*Dioscorea hispida* Dennst	Dioscoreaceae	Tuber

(*Continued*)

TABLE 2.1 (*Continued*)

Traditionally Used Anti-Diabetes Mellitus Plants, Not Subjected to Scientific Studies[a]

No.	Botanical Names of Plants	Family	Part Commonly Used
228	*Dioscorea transversa* R. Br. [syn: *Dioscorea punctata R.Br.*]	Dioscoreaceae	Whole
229	*Diospyros ebenaster* Retz. [syn: *D. ebenum* J. König ex Retz.]	Ebenaceae	Seed
230	*Diospyros lotus* Standl.	Ebenaceae	Fruit
231	*Dipteracanthus prostrates* Nees	Acanthaceae	Whole plant
232	*Dirca palustris* L.	Thymelaeaceae	
233	*Dracaena arborea* Wild	Dracaenaceae	
234	*Eichhornia crassipes* (C. Martius) Solms-Laub	Pontederiaceae	
235	*Elaeocarpus serratus* L.	Elaeocarpaceae	Fruits
236	*Elytropappus rhinocerotis* Lees	Asteraceae	Roots
237	*Embelia madagascariensis* A.DC.	Myrsinaceae	Leaves
238	*Ensete superbum* (Roxb.) Cheesman	Musaceae	
239	*Ephedra alata* Decne	Ephedraceae	Aerial parts
240	*Ephedra sinica* Stapf.	Ephedraceae	
241	*Epimedium sagittatum* (Sieb. et Zucc.) Maxim.	Berberidaceae	
242	*Equisetum robustum* A. Br.	Equisetaceae	Leaf
243	*Eragrostis bipinnata* (L.) Schum [syn: Desmostachya bipinnata (L.) Stapf.]	Poaceae	
244	*Eriodendron anfractuosum* ar. *africanum* DC. [syn: *Ceiba pentandra* (L.) Gaertn.]	Bombacaceae	Gum, all parts
245	*Erythrina variegata* (L.) Merr.	Fabaceae	Leaves
246	*Eupatorium odoratum* L. or = *Chromolaena odoratum* (L.) King and Robinson	Asteraceae	Aerial parts
247	*Euphorbia antiquorum* L.	Euphorbiaceae	Leaves, fruit
248	*Euphorbia maculata* L.	Euphorbiaceae	Leaves
249	*Evolvulus alsinoides* (L.) L.	Convolvulaceae	Whole plant
250	*Exostema mexicanum* A. Gray	Rubiaceae	
251	*Eysenhardtia polystachya* (Ortega) Sarg	Fabaceae	Roots
252	*Fagopyrum cymosum* (Trev.) Meisn.	Polygonaceae	
253	*Feronia elephantum* Correa	Rutaceae	Fruits
254	*Ferula hermonis* Boiss.	Umbelliferae (Apiaceae)	
255	*Ferula persica* Willd.	Umbelliferae (Apiaceae)	
256	*Ficus capensis* Thunb.	Moraceae	Leaves
257	*Ficus retusa* L.	Moraceae	Leaves
258	*Foeniculum vulgare* Gaertn.	Umbelliferae	
259	*Fouquieria splendens* Engelm.	Fouquieriaceae	Stem
260	*Fraxinus alba* Marshall	Oleaceae	Leaves
261	*Fumaria officinalis* L.	Fumariaceae	Aerial parts
262	*Gaultheria procumbens* L.	Ericaceae	
263	*Gelsemium sempervirens* (L.) J St Hill	Loganiaceae	
264	*Gekko gecko* L.	Gekkonidae	
265	*Geranium maculatum* L.	Geraniaceae	Roots
266	*Glossostemon bruguieri* Desf.	Sterculiaceae	Root mucilage
267	*Gnetum africanum* Welw	Gnetaceae	Leaves
268	*Grewia flavescens* Juss.	Tiliaceae	Leaves
269	*Grifola frondosa* (Fr.) S. F. Gray	Meripilaceae	Fruits
270	*Guaiacum sanctum* L.	Zygophyllaceae	

(*Continued*)

TABLE 2.1 (*Continued*)

Traditionally Used Anti-Diabetes Mellitus Plants, Not Subjected to Scientific Studies[a]

No.	Botanical Names of Plants	Family	Part Commonly Used
271	*Guazuma ulmifolia* Lam.	Sterculiaceae	Barks
272	*Guiera senegalensis* J.F.Gmel.	Combretaceae	
273	*Gundelia tournefortii* L.	Asteraceae	
274	*Gynostemma pentaphyllum* (Thunb.) Mak.	Cucurbitaceae	Stem, leaves
275	*Gynura procumbens* (Lour.) Merr.	Asteraceae	Leaves
276	*Haematoxylum brasiletto* H.Karst.	Fabaceae	Aerial parts
277	*Haloxylon salicornicum* Bunge	Chenopodiaceae	Whole plant
278	*Hamada salicornica* (Moq.) Iljin	Hamamelidaceae	Whole plant
279	*Hamiltonia suaveolens* D. Don	Rubiaceae	Roots
280	*Harpagophytum procumbens* DC. ex Meissner	Pedaliaceae	
281	*Hedychium gardnerianum* Sheppard ex Ker Gawl	Zingiberaceae	Leaves
282	*Heracleum lanatum* Michx. (*Heracleum maximum* Bartram)	Apiaceae	
283	*Heterophragma quadriloculare* K. Schum.	Bignoniaceae	
284	*Hexachlamys edulis* (O.Berg) Kausel & D. Legrand	Myrtaceae	Leaves
285	*Hodgsonia heteroclita* (Roxb.) Hook. f. & Thomson	Cucurbitaceae	Fruits
286	*Holostemma annularis* (Roxb.) K. Schum	Apocynaceae	Seeds
287	*Hoodia currorii* (Hook) Descaine	Apocynaceae	
288	*Hybanthus enneaspermus* (L.) F. Muell.	Violaceae	Leaves
289	*Hydrolea zeylanica* (L.) J. Vahl	Hydrophyllaceae	
290	*Hygrophila longifolia* (Linnaeus) Kurz.	Acanthaceae	Whole plant
291	*Hyssopus officinalis* L.	Lamiaceae	Leaf
292	*Ipomoea crassicaulis* Rob.	Convolvulaceae	Bark
293	*Ipomoea nil* (L.) Roth	Convolvulaceae	Leaf
294	*Ipomoea stans* Cav.	Convolvulaceae	
295	*Isoplexis canariensis* (L.) Lindl.	Scrophulariaceae	Leaf
296	*Isoplexis isabelliana* Morris	Scrophulariaceae	Leaf
297	*Jatropha dioica* Cerv.	Euphorbiaceae	Root
298	*Jatropha elbae* J.	Euphorbiaceae	
299	*Jatropha glandulifera* Roxb.	Euphorbiaceae	Tuber
300	*Jatropha gossypiifolia* L.	Euphorbiaceae	Leaf
301	*Juglans mandshurica* Maxim.	Juglandaceae	Leaf
302	*Juniperus phoenicea* L.	Cupressaceae	
303	*Juniperus virginiana* L.	Cupressaceae	
304	*Justicia spicigera* Schlecht.	Acanthaceae	Bark and leaf
305	*Kalanchoe laciniata* (L.) DC.	Crassulaceae	Leaf
306	*Kalanchoe verticillata* Scott-Elliot.	Crassulaceae	Leaf
307	*Kalmia angustifolia* L.	Ericaceae	
308	*Kickxia ramosissima* (Wall) Lanchen	Scrophulariaceae	Aerial parts
309	*Krameria triandra* Ruiz.	Krameriaceae	Roots
310	*Kyllinga triceps* Rottb.	Cyperaceae	Roots
311	*Lagerstroemia parviflora* (L.) Pers	Lythraceae	Aerial parts
312	*Landolphia dulcis* (Sabine) Pichon	Apocynaceae	Leaves, roots
313	*Landolphia heudelotii* DC.	Apocynaceae	
314	*Lasia spinosa* (L.) Thwaites	Araceae	Rhizomes
315	*Laurus nobilis* L.	Lauraceae	Leaves
316	*Lavandula dentata* L.	Lamiaceae	

(*Continued*)

TABLE 2.1 (*Continued*)

Traditionally Used Anti-Diabetes Mellitus Plants, Not Subjected to Scientific Studies[a]

No.	Botanical Names of Plants	Family	Part Commonly Used
317	*Lavandula multifida* Burm. [present name: *Lavandula bipinnata* (Roth) Kuntze]	Lamiaceae	
318	*Ledum groenlandicum* Oeder.	Ericaceae	
319	*Leea crispa* M. Laws	Vitaceae	Aerial parts
320	*Leea indica* Merr.	Vitaceae	Leaves
321	*Lemaireocereus dumortieri* (Scheidweiler) P.V. Heath	Cactaceae	Stem
322	*Lepidium ruderale* L.	Cruciferae	Aerial part
323	*Leucaena leucocephala* (Lam.) de Wit	Leguminosae	Seeds
324	*Ligusticum porteri* Coult. & Rose	Umbelliferae	Root
325	*Lilium auratum* Lindl.	Liliaceae	Bulbs
326	*Lilium speciosum* Thunb.	Liliaceae	Bulbs
327	*Lepidium virginicum* L.	Cruciferae	Whole plant
328	*Litchi chinensis* Sonn	Sapindaceae	
329	*Litsea coreana* Levl.	Lauraceae	Leaves
330	*Lobelia laxiflora* H.B.K.	Lobeliaceae	Bark
331	*Lodoicea sechellarum* Labill.	Palmae	Fruit
332	*Lophophora williamsii* (Lem.) J.M.Coult.	Cactaceae	Aerial parts
333	*Loranthus quandang* Auct. non Lindl	Loranthaceae	Fruit
334	*Luffa echinata* Roxb.	Cucurbitaceae	Aerial parts
335	*Luffa tuberosa* (Roxb.) [syn]; *Momordica cymbalaria* Hook.f [botanical name]	Cucurbitaceae	Fruits
336	*Lupinus albus* L.	Papilionaceae/Fabaceae	Seeds
337	*Lupinus varius* Gaertn.	Fabaceae	Seeds
338	*Lycopersicon esculentum* Mill.	Solanaceae	Seeds
339	*Lycopus virginicus* L.	Lamiaceae	
340	*Lycoris radiata* (L'Hér.) Herb.	Amaryllidaceae	Bulbs
341	*Lycoris squamigera* Maxim.	Amaryllidaceae	Bulbs
342	*Lygodium flexuosum* (L.) Sw.	Lycopodiaceae	Whole plant
343	*Lyophyllum decastes* (Fr.) Sing.	Lyophyllaceae	Fruits
344	*Lysiloma acapulcense* (Kunth) Benth.	Fabaceae	Leaves
345	*Malachra alceifolia* Jacq.	Malvaceae	Leaf
346	*Mallotus philippensis* (Lam.) Müll.-Arg.	Euphorbiaceae	Stems
347	*Malvastrum coromandelianum* (L.) Garcke	Malvaceae	Leaf
348	*Malva verticillata* L.	Malvaceae	Seeds
349	*Maprounea africana* Müll. Arg.	Euphorbiaceae	Roots
350	*Matricaria aurea* Sch.Bip.		
351	*Matricaria frigidum* (H.B.K.) Kunth	Asteraceae	Whole plant
352	*Mazus surculosus* D. Don	Scrophulariaceae	Whole plant
353	*Megacarpaea polyandra* Benth. ex Madden	Brassicaceae	Whole plant
354	*Melhania incana* B.Heyne	Malvaceae	Leaves
355	*Melia azedarach* L.	Meliaceae	Leaves
356	*Melissa annua* L.	Lamiaceae	All parts
357	*Mentha longifolia* (L.) Huds.	Lamiaceae	Whole plant
358	*Mentha pulegium* L.	Lamiaceae	Aerial parts
359	*Mentha rotundifolia* (L.) Huds.	Lamiaceae	
360	*Mentha suaveolens* Ehrh.	Lamiaceae	Aerial parts
361	*Momordica balsamina* L.	Meliaceae	Fruit

(Continued)

TABLE 2.1 (*Continued*)

Traditionally Used Anti-Diabetes Mellitus Plants, Not Subjected to Scientific Studies[a]

No.	Botanical Names of Plants	Family	Part Commonly Used
362	*Momordica foetida* Schumach.	Meliaceae	Aerial parts
363	*Morus australis* Poiret	Moraceae	Leaves
364	*Murraya paniculata* (L.) Jack	Rutaceae	Leaves
365	*Musa balbisiana* Colla	Musaceae	Flowers
366	*Musanga cecropioides* R.Br.&Tedlie	Urticaceae	
367	*Nopalea indica* L.	Cactaceae	Stem
368	*Nardostachys jatamansi* DC.	Valerianaceae	
369	*Narcissus tazetta* (Loisel.) Baker.	Amaryllidaceae	Bulbs
370	*Nauclea orientalis* (L.) L.	Rubiaceae	Stem bark
371	*Nauclea pobeguinii* Petit	Rubiaceae	Barks
372	*Nerium oleander* L.	Apocynaceae	
373	*Nicotiana tobacum* L.	Solanaceae	Leaves
374	*Nuphar variegatum* Durand	Nymphaeaceae	Stem, leaf
375	*Ocimum album* Blanco	Lamiaceae	Leaf
376	*Ocimum campechianum* Miller. [syn: *O. micranthum* Willd.]	Lamiaceae	Leaf
377	*Ocimum viride* Willd.	Lamiaceae	Leaf
378	*Opuntia atropes* Rose	Cactaceae	Stem
379	*Opuntia ficus-indica* (L.) Mill.	Cactaceae	Stem
380	*Opuntia imbricata* (Haw) D.C	Cactaceae	Fruit
381	*Opuntia lindheimeri* var. *tricolor* (Griffiths) L. Benson	Cactaceae	Aerial parts
382	*Opuntia leucotricha* DC.	Cactaceae	Stem and root
383	*Opuntia vulgaris* Mill.	Cactaceae	Aerial part
384	*Orchis latifolia* L.	Orchidaceae	Root
385	*Orchis mascula* L.	Orchidaceae	Root
386	*Origanum compactum* Benth.	Lamiaceae	Leaf
387	*Origanum syriacum* L.	Lamiaceae	Leaf
388	*Orthosiphon spiralis* Mers	Lamiaceae	Leaf
389	*Pachira aquatica* Aubl.	Bombacaceae	Root
390	*Pachycereus marginatus* (DC.) Britton & Rose	Cactaceae	Root, pulp
391	*Paeonia emodi* Wall. ex Royle	Ranunculaceae	Whole plant
392	*Paeonia moutan* Sims	Ranunculaceae	Whole plant
393	*Paeonia veitchii* Lynch	Ranunculaceae	Roots
394	*Pandanus spiralis* R. Br.	Pandanaceae	Wood
395	*Panicum miliaceum* L.	Poaceae	Whole plant
396	*Pappea capensis* Eckl. & Zeyh	Sapindaceae	
397	*Parinari microphylla* Sabine	Chrysobalanaceae	
398	*Paronychia argentea* L.	Caryophyllaceae	Aerial part
399	*Passiflora foetida* L.	Passifloraceae	Whole plant
400	*Passiflora quadrangularis* L.	Passifloraceae	Leaf
401	*Paullinia cupana* Kunth.	Sapindaceae	Leaf
402	*Pedicularis rhinanthoides* Larranagas	Scrophulariaceae	Whole plant
403	*Pedilanthus palmeri* Millsp.	Euphorbiaceae	Stem bark
404	*Pelargonium graveolens* L'Her. ex Ait.	Geraniaceae	
405	*Periploca laevigata* Aiton	Apocynaceae	
406	*Persea gratissima* C.F. Gaertn.	Lauraceae	Seed, bark
407	*Petiveria alleaceae* L.	Phytolacaceae	Leaf, stem powder

(*Continued*)

TABLE 2.1 (*Continued*)

Traditionally Used Anti-Diabetes Mellitus Plants, Not Subjected to Scientific Studies[a]

No.	Botanical Names of Plants	Family	Part Commonly Used
408	*Peumus boldus* Molina	Monimiaceae	Leaf
409	*Phalaris canariensis* L.	Poaceae	Seed
410	*Phlomis persica* Boiss	Lamiaceae	Aerial part
411	*Phoebe wightii* Meisner	Lauraceae	Aerial part
412	*Phoradendron tomentosum* (DC.) Engelm. ex A. Gray	Viscaceae	Flower
413	*Phragmites communis* Trin.	Poaceae	
414	*Phyllanthus carolinensis* Walt.	Euphorbiaceae	Leaf, stem
415	*Phyllanthus swartzii* Kostel [syn: *P. microphyllus* Kth.]	Euphorbiaceae	Aerial part
416	*Phyllanthus urinaria* L.	Euphorbiaceae	Leaf
417	*Phyllostachys nigra* (Lodd.) Munro	Poaceae	
418	*Physalis alkekengi* L.	Solanaceae	Fruit
419	*Physalis philadelphica* Lam.	Solanaceae	Bark, leaf
420	*Picea mariana* (Miller.) Britton	Pinaceae	
421	*Pileostigma thonningii* Milne-Redh.	Fabaceae	
422	*Pimenta officinalis* Lindl.	Myrtaceae	Leaf, fruit
423	*Pinellia ternata* (Thunb.) Breit	Araceae	
424	*Pinus halepensis* Mill.	Pinaceae	
425	*Pinus maritime* Mill.	Piperaceae	Bark
426	*Pinus teocote* Schiede ex Schltdl.	Piperaceae	Aerial part
427	*Piper aduncum* L.	Piperaceae	Leaf
428	*Piper auritum* Kunth.	Piperaceae	Leaves
429	*Piper cubela* L.	Piperaceae	Whole plant
430	*Piper hispidum* Sw.	Piperaceae	Leaf
431	*Piper nigrum* L.	Piperaceae	Whole plant
432	*Piper retrofractum* Vahl.	Piperaceae	Fruits
433	*Piper sanctum* (Miq.) Schltdl.	Piperaceae	Leaf
434	*Pisum sativum* L.	Fabaceae	Fruit
435	*Plantago australis* Lam.	Plantaginaceae	All parts
436	*Plantago himalacia* Pilg	Plantaginaceae	Whole plant
437	*Plantago major* L.	Plantaginaceae	Whole plant
438	*Plantago psyllium* L.	Plantaginaceae	Leaf
439	*Pleurostylia opposite* (Wall.) Alston.	Celastraceae	
440	*Pluchea symphytifolia* (Mill.) Gillis	Compositae	Root
441	*Plumbago scandens* L.	Plumbaginaceae	
442	*Plumbago zeylanica* L.	Plumbaginaceae	Leaves
443	*Plumeria rubra* L.	Apocynaceae	Stems
444	*Poa pratensis* L.	Poaceae	Whole plant
445	*Polygala erioptera* DC.	Polygalaceae	
446	*Polygala javana* DC.	Polygalaceae	Leaves
447	*Polygala tenuifolia* Willd.	Polygalaceae	Leaves
448	*Polygonatum sibiricum* Red.	Asparagaceae	
449	*Polygonum cuspidatum* Siebold & Zucc	Polygonaceae	Rhizome
450	*Polygonum odoratum* Lour. [= *Persicaria odorata* (Lour.) Sojak]	Polygonaceae	Rhizome
451	*Populus tremuloides* Michx.	Salicaceae	Leaves
452	*Poria cocos* (Schw.) Wolf.	Polyporaceae	
453	*Potamogeton crispus* L.	Potamogetonaceae	Whole plant

(*Continued*)

TABLE 2.1 (*Continued*)

Traditionally Used Anti-Diabetes Mellitus Plants, Not Subjected to Scientific Studies[a]

No.	Botanical Names of Plants	Family	Part Commonly Used
454	*Pouteria tomentosa* (Roxb.) Baehni	Sapotaceae	Aerial part
455	*Prunus armeniaca* L.	Rosaceae	Root
456	*Prunus serotina* Ehrh.	Rosaceae	
457	*Psacalium palmeri* (Greene) Rob. et Brett. [syn: *Senecio palmeri* (Greene) Rydb.]	Asteraceae	Root
458	*Psacalium sinuatum* (Cerv.) Rob et Brett.	Asteraceae	Root
459	*Psidium yucatanense* Lundell	Myrtaceae	Leaf
460	*Psychotria elata* (Sw.) Hammel.	Rubiaceae	Flower, stem and leaves
461	*Psychotria poeppigiana* Muell. Arg.	Rubiaceae	Flower, stem and leaves
462	*Pterocarpus erinaceus* Poir.	Leguminosae	
463	*Puemus boldus* Molina J.	Monimiaceae	
464	*Pyrus communis* L.	Rosaceae	Aerial parts
465	*Quercus alba* L.	Fagaceae	
466	*Quercus acutifolia* Nees	Fagaceae	Barks
467	*Quercus coccifera* L.	Fagaceae	
468	*Quercus lanata* Sm.	Fagaceae	
469	*Quercus lineata* Blume	Fagaceae	Barks
470	*Quercus robur* L.	Fagaceae	Gall
471	*Quercus rubra* L.	Fagaceae	
472	*Quercus spicata* Bonpl.	Fagaceae	Barks
473	*Randia dumetorum* Lam.	Rubiaceae	Bark, fruit
474	*Raphanus sativus*	Brassicaceae	Whole plant
475	*Ravenala madagascariensis*	Strelitziaceae	Flower, root
476	*Rehmannia lute* Maxm	Scrophulariaceae	
477	*Rhamnus purshiana* DC.	Rhamnaceae	Bark
478	*Rheum palmatum* L.	Polygonaceae	
479	*Rheum tanguticum* (Maxim. ex Regel) Maxim. ex Balf.	Polygonaceae	Root
480	*Rhodiola sachalinensis* Boriss	Crassulaceae	Root
481	*Rhus hirta* L.	Anacardiaceae	
482	*Rhus virens* Lindh. Ex A. Gray.	Anacardiaceae	Peel
483	*Rivea ornate* Choisy	Convolvulaceae	Juice
484	*Rosa brunonii* Lindl.	Rosaceae	Aerial parts
485	*Rosa canina* L.	Rosaceae	Fruit
486	*Rosa centifolia* L.	Rosaceae	Aerial part
487	*Rouvolfia vomitoria* Afzel. [*Rauvolfia senegambiae* A. DC.; *Hylacium owariense* Afzel.]	Apocynaceae	Leaves, roots
488	*Rumex abyssinicus* Jacq.	Polygonaceae	Root
489	*Russelia equisetiformis* Schltdl. & Cham	Scrophulariaceae	Roots
490	*Sageretia parviflora* (Klein) G.Don Small	Rhamnaceae	Roots
491	*Salpianthus arenarius* Humb & Bonpl	Nyctaginaceae	Flowers
492	*Salvia canariensis* L.	Lamiaceae	Aerial parts
493	*Salvia leucantha* Cav.	Lamiaceae	Leaves
494	*Samanea saman* Jacq.	Fabaceae	Bark
495	*Samyda yucatanensis* Standl.	Flacourtiaceae	Root
496	*Sansevieria roxburghiana* Schult.	Ruscaceae	Rhizome
497	*Sapindus laurifolia* Vahl.	Sapindaceae	Fruit
498	*Sapindus mukorossi* Gaertn.	Sapindaceae	Fruit

(*Continued*)

TABLE 2.1 (Continued)

Traditionally Used Anti-Diabetes Mellitus Plants, Not Subjected to Scientific Studies[a]

No.	Botanical Names of Plants	Family	Part Commonly Used
499	*Sassafras albidum* (Nutt.) Nees.	Lauraceae	
500	*Schleichera oleosa* (Lour.) Oken	Sapindaceae	Stem bark
501	*Scilla siberica* Andr.	Hyacinthaceae	Bulb
502	*Scindapsus officinalis* L.	Araceae	Fruit and root
503	*Scrophularia buergeriana* Miq	Scrophulariaceae	Root
504	*Scutia myrtina* (Burm. f.) Kurz	Rhamnaceae	Whole plant
505	*Senecio peltiferus* Hemsl.	Asteraceae	
506	*Senna multiglandulosa* (Jacq.) H.S. Irwin & Barneby	Fabaceae	Leaf
507	*Senna obtusifolia* L.	Fabaceae	Leaf, seed
508	*Serenoa serrulata* (Michx.) Hook.	Arecaceae	Fruit
509	*Serjania racemosa* Schumach.	Sapindaceae	Stem
510	*Sesbania grandifolia* (L.) Pers.	Fabaceae	Bark
511	*Setaria italica* (L.) P.	Poaceae	Seeds
512	*Sida cordifolia* L. var. mutica Benth.	Malvaceae	Aerial part and root
513	Sigesbeckia *orientalis* L.	Asteraceae	All parts
514	*Smilax aristolochiifolia* Mill.	Smilacaceae	Root
515	*Smilax australis* R. Br.	Smilacaceae	Leaf
516	*Smilax canariensis* Willd.	Smilacaceae	Aerial parts
517	*Smilax china* L.	Smilacaceae	Rhizome
518	*Smilax glyciphylla* Sm.	Smilacaceae	Leaf
519	*Solanum diversifolium* Dunal	Solanaceae	Leaves
520	*Solanum lycocarpum* A.St.-Hil.	Solanaceae	Fruits
521	*Solanum seaforthianum* Andr.	Solanaceae	Leaf
522	*Solanum verbascifolium* L.	Solanaceae	Leaves
523	*Solenostemon rotundifolius* JK Mortin	Lamiaceae	Tubers
524	*Solidago canadensis* L.	Asteraceae	
525	*Sorbus americana* Marsh	Rosaceae	
526	*Soymida febrifuga* (Roxb.) A. Juss	Meliaceae	Barks
527	*Sphaeranthus indicus* L.	Asteraceae	Whole plant
528	*Spondias dulcis* Sol. ex Parkinson	Anacardiaceae	Stem bark
529	*Stachytarpheta indica* Vahl.	Verbenaceace	Leaf
530	*Stachytarpheta jamaicensis* Vahl.	Verbenaceace	Leaf
531	*Stachyurus himalaicus* J. D. Hooker & Thomson ex Bentham	Stachyuraceae	
532	*Stellaria media* (L.) Vill	Caryophyllaceae	Whole plant
533	*Stenocereus marginatus* (DC.) Britton & Rose	Cactaceae	Aerial parts
534	*Striga gesnerioides* Vatke ex Engl.	Scrophulariaceae	All parts
535	*Strobilanthes crispa* Blume	Acanthaceae	Leaf
536	*Struthanthus densiflorus* (Benth.) Standl.	Loranthaceae	Branch
537	*Sweetia panamensis* Benth.	Leguminosae	Stem bark
538	*Swertia japonica* Makino	Gentianaceae	Whole plant
539	*Swietenia humilis* Zucc.	Meliaceae	Seed
540	*Symphytum officinale* L.	Boraginaceae	Extract (toxic)
541	*Symplocos cochinchinensis* (Lour.) S.Moore	Symplocaceae	Leaves
542	*Symplocos paniculata* (Thunb.) Miq	Symplocaceae	Leaves, stem
543	*Syringa vulgaris* L.	Lamiaceae	Buds
544	*Syzygium jambolanum* DC.	Myrtaceae	Fruits
545	*Syzygium suborbiculare* T. Hartley & Perry	Myrtaceae	Bark, root and fruit

(Continued)

TABLE 2.1 (*Continued*)

Traditionally Used Anti-Diabetes Mellitus Plants, Not Subjected to Scientific Studies[a]

No.	Botanical Names of Plants	Family	Part Commonly Used
546	*Tagetes erecta* L.	Asteraceae	Leaves, flower
547	*Talinum portulacifolium* (Forssk.) Asch. ex Schweinf	Portulacaceae	Leaves
548	*Tapinanthus bangwensis* (Engl. & K. Krause) Danser	Loranthaceae	Leaves
549	*Taxus canadensis* Marsh.	Taxaceae	
550	*Taxus yunnanensis* W.C. Cheng & L.K. Fu.	Taxaceae	Wood
551	*Tecoma mollis* H.B.K.	Bignoniaceae	Whole plant
552	*Tetraclinis articulata* Benth.	Cucurbitaceae	
553	*Tetrapanax papyrifer* K. Koch	Araliaceae	Roots
554	*Tetrastigma leucostaphylum* (Dennst.) Alston ex Mabb.	Vitaceae	
555	*Teucrium royleanum* Wall	Lamiaceae	Whole plant
556	*Thuja occidentalis* L.	Cupressaceae	
557	*Tillandsia usneoides* L.	Bromeliaceae	Whole plant
558	*Tinospora smilacina* Benth.	Menispermaceae	Seeds
559	*Torenia asiatica* Mardr.ex Wall	Scrophulariaceae	All parts
560	*Tournefortia petiolaris* DC.	Boraginaceae	Barks
561	*Trema orientalis* (L.) Blume	Ulmaceae	Barks
562	*Tremella mesenterica* Retz.	Combretaceae	Fruits
563	*Trixis radialis* (L.) Kuntze	Asteraceae	Leaves, flowers
564	*Tsuga canadensis* (L.) Carr.	Pinaceae	
565	*Turbina corymbosa* (L.) Raf.	Convolvulaceae	Leaf
566	*Urginea indica* Kunth	Liliaceae	All parts
567	*Uvaria chamae* P. Beauv.	Annonaceae	
568	*Vaccinium leschenaultii* Wight	Ericaceae	Aerial parts
569	*Vaccinium vitis-idaea* L.	Ericaceae	Leaves
570	*Varthemia iphionoides* Boiss	Asteraceae	
571	*Vateria copallifera* (Retz.) Alston.	Dipterocarpaceae	Fruit
572	*Verbascum thapsus* L.	Scrophulariaceae	
573	*Verbena bonariensis* L.	Verbenaceae	
574	*Verbesina encelioides* (Cav.) A. Gray	Asteraceae	Flower, root
575	*Viburnum opulus* L.	Caprifoliaceae	Fruit
576	*Vigna mungo* L.	Fabaceae	Seed
577	*Vigna subterranea* (L.) Verdc	Fabaceae	Seed
578	*Vinca major* L.	Apocynaceae	Whole plant
579	*Vitex mollis* H.B.K.	Verbenaceae	Root
580	*Xanthoxalis corniculata* (L.) Small.	Oxalidaceae	
581	*Xanthium sibiricum* Patr.	Asteraceae	Leaf
582	*Xylopia aethiopica* (Dun.) A. Rich.	Annonaceae	Leaf and fruit
583	*Yucca guatemalensis* Baker. [syn: *Y. elephantipes*]	Agavaceae	Leaf and flower
584	*Ziziphus acuminate* Benth.	Rhamnaceae	
585	*Ziziphus oenoplia* Mill.	Rhamnaceae	Bark and fruit

Sources: Marles, R.J. and Farnsworth, N.R., *Phytomedicine*, 2, 137, 1995; Swargiary et al. 2001; Alarcon-Aguilar, F.J. and Roman-Ramos, R., Antidiabetic plants in Mexico and Central America, in *Traditional Medicines for Modern Times: Antidiabetic Plants*, Soumyanath, A. (ed.), CRC Press, Boca Raton, FL, 2005, pp. 221–224; Ghisalberti, E.L., Australian and New Zealand plants with antidiabetic properties, in *Traditional Medicines for Modern Times: Antidiabetic Plants*, Soumyanath, A. (ed.), CRC Press, Boca Raton, FL, 2005, pp. 241–256; Simmonds and Howes 2005; Seaforth, C.E., Antidiabetic plants in the Caribbean, in *Traditional Medicines for Modern Times: Antidiabetic Plants*, Soumyanath, A. (ed.), CRC Press, Boca Raton, FL, 2005, pp. 221–224; Jarald, E. et al., *Indian J. Exp. Biol.*, 46, 660, 2008; Jarald, E. et al., *Iran. J. Pharmacol. Ther.*, 7, 97, 2008; Jaya Preethy 2013; Vijayalakshmi, N. et al., *Int. J. Curr. Microbiol. Appl. Sci.*, 3, 405, 2014; Unrecorded information.

[a] This is not a complete list.

Abies balsamea (L.) Mill, Pinaceae

Common Name: Balsam fir

Abies balsamea is widely distributed in Boreal regions of Canada and northern parts of the United States. *A. balsamea* has been traditionally used to combat colds, fever, rheumatism, and respiratory tract infections. In ethnomedicine, the plant has been used for the management of symptoms and complications of DM. The antidiabetic activity of this plant has been reviewed recently (Eid and Haddad 2014). The ethanol extract of the inner bark of this plant significantly enhanced basal and insulin-stimulated glucose uptake in cultured skeletal muscle cells and adipocytes and exerts its action on muscle cells by moderately uncoupling mitochondrial energy production, which activated adenosine monophosphate–activated protein kinase (AMPK), but had mild effects on cell pH or ATP levels. However, more prominent effects were observed on glucose production and storage in hepatocyte cultures. Further, mechanistic studies revealed enhanced phosphorylation of Akt (protein kinase B) and AMPK. The plant extract may stimulate glycogen synthesis by increasing the phosphorylation of glycogen synthase kinase-3. The phosphorylation of this enzyme by the plant extract relieves its inhibitory effect on glycogen synthase and promotes glycogen synthesis. The plant extract influences the activity of cytochrome P450 in the cell-free system. Therefore, the plant should be used cautiously, notably in combination with other drugs. Thus, balsam fir preparations appear to exert their antidiabetic potential in part by improving muscle and adipose tissue glucose uptake but mostly by reducing hepatic glucose production through both insulin-dependent and insulin-independent mechanisms (Eid and Haddad 2014). Follow-up studies needed include identification of active principles, toxicity evaluation, and clinical trials.

Abies pindrow Royle, Pinaceae

Abies pindrow, an evergreen tree, has many ethnopharmacological applications. It is reported to have hypoglycemic effects (Rahman and Zaman 1989) and is used to treat diabetes and other diseases in traditional medicine. Under *in vitro* conditions, the ethanol extract of *A. pindrow* showed insulin secretagogue activity in rat insulinoma (INS-1) cells at 10 μg/mL level (Hussein et al. 2004). Phytochemical and pharmacological values of this plant have been reviewed recently; the plant contains triterpenoids, pinitol, maltol, etc. (review: Majeed et al. 2013). Pinitol is an important anti-DM compound.

Abroma augusta L. f., Sterculiaceae

Synonyms: *Abro. fastuosum* Jacq., *Abro. mollis* DC, *Theobroma augusta* L.

Common Names: Ulatkambal (Hindi), flame of the forest, devil's cotton (Eng.)

Description: *Abroma augusta* is a shrub or a small tree. The leaves are alternate and polymorphous about 10–30 cm. long and 6–18 cm broad, base 3–7 lobed caudate nerved, upper nonlobed, ovate-lanceolate; smaller, entire, glabrescent above, tomentose below and stipules linear. The flowers are about 5 cm in diameter, dark red, purple, or yellow in color occurring on a few flowered cymes. Sepals are lanceolate, free nearly to the base. Petals scarcely exceeding sepals imbricate in bud and are deciduous. Stamens are present on a short stamen tube; capsules are almost 4 cm long, membranous, and finally pubescent and truncate at the apex. Each carpel has a triangular wing behind it. Seeds are abundant, small, and blackish and are covered with silky hairs. The roots have 0.5–1.0 mm thick highly fibrous brown barks. The bark is slimy, odorless, tough, and not brittle (review: Gupta et al. 2011a).

Distribution: *A. augusta* is of Indo-Malayan origin and occurs in the tropical forests of Asia, Africa, and Australia. It grows in the wild. It is planted for its showy deep scarlet flowers (review: Gupta et al. 2011).

Traditional Medicinal Uses: *A. augusta* is used in traditional medicine to treat gynecological disorders, dermatitis, dysmenorrhea, scabies, diabetes, rheumatic pain, headache with sinusitis, etc. It is also used as an abortifacient, antifertility, and anti-inflammatory agent. The leaves and bark of roots are mainly used in traditional medicine (review: Gupta et al. 2011).

Antidiabetes: The methanol extract of the leaves of this plant at a dose of 300 mg/kg, when administered daily for 7 days, showed promising antidiabetic activity against alloxan-induced diabetic rats. The ethanol extract (100 mg/kg) of this plant's roots also exhibited glucose-lowering activity in alloxan-induced DM in rats. The water extract of the plant showed efficacy in combination therapy along with the water extract of other plants with antidiabetic properties such as *Curcuma longa* or *Coccinia indica* in streptozotocin-induced diabetic rats (review: Gupta et al. 2011). Further, the combined water extracts of this plant and *Azadirachta indica* (1:1) was active against alloxan-induced diabetic rats. Oral administration of the aqueous extract of curcumin from *C. longa* and partially purified preparation from *A. augusta*, when administered for 8 weeks at a dose of 300 mg/kg, markedly reduced the levels of lipids and creatinine in streptozotocin-induced diabetic rats compared to untreated diabetic rats (review: Gupta et al. 2011). The water extract of fresh leaves of *A. augusta* reduced the absorption of glucose administered orally in fasted rats. But the extract reduced the absorption of metformin hydrochloride in alloxan-induced diabetic rats. This study suggests that the extract may be beneficial in diabetic patients to improve glycemic control but should not be coadministered with metformin for the management of type 2 DM (Islam et al. 2012).

The active principles involved and their mechanisms of action remain to be studied. This plant is a promising material for further studies.

Other Activities: Other reported pharmacological properties of this plant include as an antioxidant, anti-inflammatory, antibacterial and antifungal, and insecticidal and for wound healing (review: Gupta et al. 2011).

Phytochemicals: The major reported constituents of the plants are alkaloids, abromine, sterol, friedelin, abromasterol, taraxeryl acetate, taraxerol, β-sitosterol, and α-amyrin (review: Gupta et al. 2011). The anti-DM properties of β-sitosterol and α-amyrin from other sources are known.

Abrus precatorius L., Fabaceae

Synonym: *Abrus abrus* (L.) W. F. Wight.

Common Name: Rosary pea

The antidiabetic effect of a crude glycoside fraction from the *A. precatorius* seed was studied in alloxan-induced diabetic rabbits. Blood glucose levels were reduced in the glycoside-treated (50 mg/kg) diabetic rabbits. The peak glucose reduction occurred after 30 h of oral administration (Monago and Akhidue 2002). In another study, the chloroform–methanol extract (50 mg/kg) of *A. precatorius* seed markedly decreased blood glucose levels in alloxan-induced diabetic rabbit (Monago and Alumanah 2005). Moreover, treatment with water extract (100 and 200 mg/kg; p.o.) of the seeds of this plant to streptozotocin-induced diabetic rats resulted in significant reduction in blood glucose levels, serum total cholesterol, low-density lipoprotein cholesterol (LDL-C), and triacylglycerol levels compared to untreated diabetic control rats. The treatment also increased body weight (Nwanjo 2008). The seed is toxic.

Abutilon indicum (L.) Sweet, Malvaceae

Alcohol and water extracts of *Abutilon indicum* leaves (400 mg/kg, p.o.) showed significant hypoglycemic effect in normal rats 4 h after administration (Seetharam et al. 2002). In another study, the water extract of *A. indicum* (whole plant) inhibited glucose absorption and stimulated insulin secretion in rodents (Krisanapun et al. 2009).

Acacia albida Del., Mimosaceae

The water extracts of the stem bark of *A. albida* (250 mg/kg) exhibited antihyperglycemic and anti-hyperlipidemic effects in alloxan-induced diabetic rats. The treatment also improved body weight and liver function in diabetic rats (Umar et al. 2014). The methanol extract (200 mg/kg) of the root bark of *A. albida* also showed hypoglycemic effects in alloxan-induced diabetic rats (Agunu et al. 2011). *A. albida* root bark has been used in Nigeria traditional medicine for the treatment of numerous diseases including DM (Agunu et al. 2011).

Acacia arabica (Lam) Wild., Fabaceae/Mimosaceae

Acacia arabica plants are used in traditional Indian medicine for the treatment of DM. Oral administration of a cold water extract of *A. arabica* bark to diabetic and normal rats at a dose of 400 mg/kg resulted in the reduction of blood glucose, cholesterol, and triglycerides (TGs). The cold water extract was found to reduce blood glucose level to its normal level within 7 days. Histological studies of the β-cells showed the plant's favorable action on the pancreas (Yasir et al. 2010). Phytochemical investigations found that phenolic compounds are present in the extracts. Acacia polyphenols present in *Acacia mearnsii* showed antiobesity and antidiabetic effects in obese diabetic KK-Ay mice fed with a high-fat diet (HFD) (Ikarashi et al. 2011). Administration of *A. arabica* extract orally for 21 days (100 and 200 mg/kg) to streptozotocin-induced diabetic rats resulted in a significant decrease in the levels of serum glucose, insulin resistance, total cholesterol, TG, LDL, and malondialdehyde (MDA) and a significant increase in high-density lipoprotein (HDL) in the treated diabetic groups when compared to the untreated diabetic group. The changes were dose dependent (Hagazy et al. 2013).

Acacia auriculiformis A. Cunn. ex Benth. Babool, Fabaceae/Leguminosae

The bark and empty pod extract of *Acacia auriculiformis* showed protective effects against alloxan-induced type 2 diabetes in animals. The treatment (100–400 mg/kg) reduced blood glucose levels, improved the lipid profile, and decreased blood creatinine and urea levels (Arumugam and Siddhuraju 2013).

Acacia catechu (L.f.) Willd., Fabaceae (Subfamily Mimosoideae)

Synonyms: *Acacia catechu* (L.f.) Willd. var. *catechuoides* (Roxb.) Prain, *A. wallichiana* DC, *A. catechuoides* Benth, *A. sundra* (Roxb.) Bedd, *Mimosa catechu* L.f., *Mimosa catechuoides* Roxb. (Figure 2.1)

Local/Common Names: Cutch tree or black catechu or catechu (Eng.), katha, catechu, kachu, khadiram, khair, etc.

Description: It is about 15 m high. Its bark is dark gray or grayish-brown. The leaves are bipinnately compound, with 9–30 pairs of pinnae and a glandular rachis. Flowers are 5–10 cm long, white to pale yellow, and with a campanulate calyx and corolla 2.5–3 mm long. Stamens are numerous and far exerted from the corolla (Hashmat and Hussain 2013).

Distribution: It is widely distributed throughout Central Asia especially Pakistan, India, and Bangladesh.

Ethnomedical Uses: The stem bark, wood, flowering tops, and gum are used in traditional medicine. In local health traditions and traditional systems of medicine, it is used as an astringent, coolant, and digestive; it is also used to treat skin diseases, cough, diarrhea, etc. It is an ingredient of many formulas in Ayurveda (Lakshmi et al. 2011).

Antidiabetes Activities: Various extracts and crude water suspension of barks of *A. catechu* and fractions of active ethanol extract were tested for antihyperglycemic activity in glucose-loaded hyperglycemic rats. The effective extract and fraction of *A. catechu* were subjected to antidiabetic study in alloxan-induced diabetic rats (200 and 400 mg/kg). The ethanol extract of *A. catechu* and the water-insoluble fraction of

FIGURE 2.1 *Acacia catechu* (twig with flower).

the ethanol extract exhibited significant antihyperglycemic activity and produced dose-dependent hypoglycemia in fasted normal rats. Treatment of diabetic rats with ethanol extract and the water-insoluble fraction of *A. catechu* restored the elevated biochemical parameters (glucose, urea, creatinine, serum cholesterol, serum TG, HDL, LDL, hemoglobin, and glycosylated hemoglobin) almost to the normal level. Comparatively, the water-insoluble fraction of the ethanol extract was more effective than the ethanol extract and the activity was comparable to that of the standard, glibenclamide (5 mg/kg). The active ethanol extract and the water-insoluble fraction of the ethanol extract contain alkaloids and flavonoids (Jarald et al. 2009b). In another study, the ethyl acetate fraction of this plant showed hypoglycemic activity in albino rats. In alloxan-induced diabetic rats, *A. catechu* (250 and 500 mg/kg), when administered for 7 days, exerted marked hypoglycemic activity compared to untreated diabetic control animals (Lakshmi et al. 2011). The active principles and mechanisms of action remain to be studied. However, compounds with anti-DM activities such as quercetin and catechin are reported from this plant.

Other Pharmacological Activities: Catechu (one of the products of this plant) has several biological properties, including hepatoprotection, antipyretic, and digestive properties. Catechin, an active principle, has antioxidant and anticancer activity and promotes fat oxidation. Condensed tannins from this plant inhibited fatty acid synthase (Zhang et al. 2008). Cyanidanol, an active principle of *A. catechu*, is claimed to be effective in treating liver diseases. Further, the plant possesses antibacterial, antimycotic, immunomodulatory, antipyretic, antidiarrheal, and antiulcer activities. The tannin and polyphenols present in it impart an astringent activity (Lakshmi et al. 2011).

Phytochemicals: The main chemical constituents of this plant are catechin, catechuic acid, pyrocatechin, phloroglucin, protocatechuic acid, quercetin, cyanidanol, and gum.

Acacia leucophloea (Roxb.) Willd., Fabaceae/Mimosaceae

Synonyms: *Mimosa leucophloea* Roxb.; *Vachellia leucophloea* (Roxb.) Maslin, Seigler & Ebinger

Common Names: Distiller's acacia, white babul, white barked acacia, etc.
Acacia leucophloea is reported to contain myricetin, β-sitosterol, β-amyrin, etc. (Ahmed et al. 2014). These compounds are known to have antidiabetes properties. Detailed studies remain to be done.

Acacia mearnsii De Wild., Fabaceae/Mimosaceae
Catechin-like flavan-3-ols, such as robinetinidol- and fisetinidol-rich extract from the bark of *A. mearnsii* (black wattle tree), exhibited antiobesity and antidiabetic effects in obese diabetic KK-Ay mice (Ikarashi et al. 2011).

Acacia nilotica (L.) Willd ex Delile, Fabaceae/Mimosaceae

Synonym: *Acacia nilotica* Lam
The methanol extract of *Acacia nilotica* showed antidiabetes activity in alloxan-induced diabetic rats (Tanko et al. 2013). The aqueous methanolic extract of pods of this plant prevented diabetic nephropathy in streptozotocin-induced diabetic rats (Omara et al. 2012).

Acacia tortilis (Forssk.) Hayne, Fabaceae/Mimosaceae
The plant seed extract lowered serum glucose in normal and alloxan-induced diabetic rats. Further, treatment increased glucose tolerance in the diabetic rats (Agrawal et al. 2014).

Acalypha indica L., Euphorbiaceae
Acalypha indica is a traditional medicinal plant used to treat various ailments including diabetes. The anti-DM effects of the methanol and acetone extract of *A. indica* was evaluated in normal and alloxan-induced diabetic rats. The extracts exhibited dose-dependent blood glucose-lowering effects compared to control rats (Masih et al. 2011). In another study, the ethanol extract of the plant (200 and 400 mg/kg daily for 14 days) lowered blood glucose levels in streptozotocin–nicotinamide-induced type 2 diabetic rats. Further, treatment improved the lipid profile in diabetic rats compared to diabetic control rats. The β-cell regeneration effect of the extract was observed in a histopathological study (Raghuram Reddy et al. 2012).

Acalypha wilkesiana Mull., Euphorbiaceae

The methanol extracts of the root of *A. wilkesiana* (200 and 400 mg/kg) showed anti-DM activity in alloxan-induced diabetic rats. A 74% significant reduction in the fasting blood glucose level was observed for the 400 mg/kg dose in alloxan-induced diabetic rats. A reduction in serum total cholesterol and TG levels was also observed for the same dose. Treatment improved body weight and glucose tolerance and decreased the levels of serum glutamate pyruvate transaminase (SGPT) and glutamate oxaloacetate transaminase (SGOT) (Odoh et al. 2013). In another study, the methanol extract showed improvement in the regeneration of the β-cells of the pancreas in diabetic rats. A histopathological study showed healing of the pancreas by the extract, which was a possible mechanism of their antidiabetic activity (Odoh et al. 2013).

Acanthopanax koreanum Nakai, Araliaceae

The roots and stem barks of *Acanthopanax koreanum* have been traditionally used as a tonic and to treat rheumatism, hepatitis, and diabetes in Korea. A diterpene isolated from the CH_2Cl_2 extract of the *A. koreanum* root inhibited PTP1B at a low concentration (IC_{50} = 7.1 µmol/L) (Jiang et al. 2012).

Acanthopanax senticosus Rupr. & Maxim, Araliaceae

Synonym: *Eleutherococcus senticosus* (Rupr. & Maxim) Maxim

Common Name: Siberian ginseng

Saponin from *Acanthopanax senticosus* was reported to decrease various cases of experimental hyperglycemias induced by injection of adrenaline, glucose, and alloxan, without affecting the levels of blood sugar in normal mice (Sui et al. 1994). Syringin (a compound isolated from this plant) injection to streptozotocin-induced diabetic rats resulted in an increase in plasma glucose utilization accompanied by an increase of plasma insulin and C-peptide levels in anesthetized diabetic rats (Niu et al. 2008). This result suggests that syringin may be useful in the treatment of human diabetes (Niu et al. 2008). Further, investigation revealed the likely mechanisms of action. Syringin enhances the release of acetylcholine from the nerve terminals that stimulates muscarinic M3 receptors in pancreatic cells thereby augmenting insulin release (Liu et al. 2008). The plant is reported to have many pharmacological activities (review: Yan-Lin et al. 2011).

Acer saccharum Marshall, Sapindaceae

Common Names: Maple, sugar maple, etc.

Extracts of the leaves of *Acer saccharum* showed a potent inhibitory effect on α-glucosidase in both *in vivo* and *in vitro* experiments. *In vitro* enzyme-inhibitory assay-guided fractionation of the crude extract gave the active compound acertannin (2,6-di-*O*-galloyl-1,5-anhydro-D-glucitol) (Honma et al. 2010).

Achillea millefolium L., Yarrow, Asteraceae

Synonyms: *A. borealis* Bong., *A. lanulosa* Nutt., *A. millefolium* ssp. *borealis* (Bong.) Breitung., *A. millefolium* ssp. *lanulosa* (Nutt.) Piper, *A. millefolium* var. *occidentale* DC. (monograph: WHO 2009)

Achillea millefolium is a traditional medicinal plant with many biological activities. A decoction of the plant (flowering tops) was documented to reduce experimental epinephrine and alloxan-induced diabetic hyperglycemia by 12.7%–45.2% but not elevate the low blood insulin level in the epinephrine hyperglycemia model and alloxan-induced diabetes in mice and rats (Molokovskij et al. 2002).

Achillea santolina L., Asteraceae

Achillea santolina is used in Jordan as an ethnomedicine to control diabetes. The effect of *A. santolina* (hydroalcoholic extract) on the blood glucose level, serum nitric oxide (NO) concentration, and oxidative stress in rat pancreatic tissue has been evaluated. This herbal treatment could reduce blood glucose serum NO, pancreatic MDA, protein carbonyls, and advanced oxidation protein products. In addition, the GSH content (reduced glutathione) was restored to the normal level in the control group. Furthermore, catalase (CAT) and superoxide dismutase (SOD) activities in the treated rats were increased significantly. In conclusion, *A. santolina* has a high hypoglycemic activity that may be due to its antioxidative potential (Yazdanparast et al. 2007). *In vitro* studies using MIN6 β-cells in culture have shown that the water extract of this plant enhances β-cell proliferation as judged from a dose-dependent augmentation of bromodeoxyuridine incorporation (Kasabri et al. 2012a).

Achyranthes aspera L., Amaranthaceae

Common Names: Prickly chaff flower (Eng.), latjiva (Hindi), nayurivi (Tamil)

Description: *Achyranthes aspera* is an erect herb of 3 ft height, bearing velvety tomentose, orbicular, obovate, or elliptic, usually obtuse, thick leaves that are 4 in. long and 3 in. broad. The flowers are pink or greenish-white with slender spikes, often reaching 18 in. long, and the fruit is deflexed.

Distribution: It is commonly found as a weed throughout India up to an elevation of 3000 ft.

Traditional Medicinal Uses: It is used for diabetes, chicken pox, dysentery, conjunctivitis, cuts and wounds, measles, deafness, headache, piles, rheumatism, ulcers, earache, tooth infection, migraine, cough, tumors, urinary troubles, tuberculosis, gonorrhea, gastric trouble, night blindness, swelling of the eyes, asthma with cough, snake bite, scorpion sting, uterine bleeding, dog bite, leprosy, leukoderma, and cholera and as an antifertility drug (Kirtikar and Basu 1975).

Antidiabetes: The whole plant powder and its extracts (water and methanol) produced a significant hypoglycemic effect in normal as well as in alloxan-induced diabetic rabbits without any acute toxicity (Akhtar and Iqbal 1991). The water extract of the whole *A. aspera* plant (250 mg/kg, p.o., daily, for 45 days) exhibited promising antidiabetic activity in alloxan-induced diabetic animals as judged from blood glucose, serum insulin, glycosylated hemoglobin, and liver glycogen levels (Vidhya et al. 2012). In another study, the ethanol extract of *A. aspera* leaves (100, 200, and 400 mg/kg) showed dose-dependent hypoglycemic effects in normal rabbits (Ahmad et al. 2013). Mechanisms of action and active principles involved are not known.

Other Activities: The methanol extract of leaves and the nonalkaloid fraction of *A. aspera* possessed pronounced anticarcinogenic effects on *in vivo* two-stage mouse skin carcinogenesis test and *in vitro* assay. Ethanol extracts of *A. aspera* exhibited anti-inflammatory and antiarthritic activity in mice and rats. The aqueous leaf extract of *A. aspera* has both antioxidative and prothyroidic activities. The whole plant (benzene extract) showed an abortifacient effect. Its ethanol extract caused reproductive toxicity in male rats (review: Ajikumaran and Subramoniam 2005).

Phytochemicals: The phytochemicals present in *A. aspera* include the following: alkaloid, glycosides, saponins, ecdysone hormone, oleanolic acid, aliphatic dihydroxyketone achyranthine, betaine, and hentriacontane (*The Wealth of India* 1992).

Achyranthes rubrofusca L., Amaranthaceae

The water and ethanol extracts of *Achyranthes rubrofusca* leaves, when administered for 28 days (120 mg/kg; p.o.), showed anti-DM activity in alloxan-induced diabetic rats as judged from improvement in body weight and decrease in the levels of blood glucose, lipids, and lipid peroxides. Further, histopathological observations showed regenerative effect of the extract in the pancreas (Geetha et al. 2011). Further, the whole plant extract exhibited hypolipidemic activity in HFD-induced hyperlipidemic rats (Geetha et. al. 2008).

Achyrocline satureioides (Lam) DC, Asteraceae

Achyrocline satureioides is a South American medicinal plant. A new prenylated dibenzofuran, achyrofuran, a compound derived from *A. satureioides* significantly lowered blood glucose levels when administered orally (20 mg/kg) to type 2 diabetic *db/db* mouse (Carney et al. 2002). The plant is also known to have hepatoprotective properties (Simoes et al. 1998). Inflorescences of the plant have been used as remedies in folk medicine for the treatment of a variety of human ailments, particularly those of the gastrointestinal tract (Simoes et al. 1988).

Aconitum carmichaelii Debeaux, Ranunculaceae

Glycans (aconitans A, B, C, and D) of *Aconitum carmichaelii* roots lowered blood glucose (Konno et al. 1985b). All parts of this plant are extremely toxic, and it has, historically, been used as a poison on arrows.

Acorus calamus L., Acoraceae
The methanol extract of the rhizome of this plant showed potent antihyperglycemic activity in normal and streptozotocin-induced diabetic rats (Prisilla et al. 2012). A new sesquiterpenoid, 1β,5α-guaiane-4β,10α-diol-6-one, was isolated from 70% ethanol extract of the rhizomes of *Acorus calamus*. This compound showed antidiabetic activity *in vitro* in HepG2 cells as judged from the increase in insulin-mediated glucose consumption (Zhou et al. 2012).

Acosmium panamense Benth., Fabaceae
Synonym: Sweetia panamensis Benth.
Acosmium panamense is a traditional medicinal tree occurring in tropical rain forests. Oral administration of the water extract (20 and 200 mg/kg) and butanol extract (20 and 100 mg/kg) of the bark of *A. panamense* lowered the glucose levels in streptozotocin-induced diabetic rats within 3 h. Three new compounds isolated from the plant were the main constituents in both extracts (Andrade-Cetto and Wiedenfeld 2004). In another study, the butanol extract of *A. panamense* produced marked reduction in blood glucose in maltose-loaded streptozotocin-induced diabetic rats. *In vitro* assays of α-glucosidase activity showed an IC_{50} of 109 µg/mL for the butanol extract of *A. panamense*, which was lower than that of acarbose (128 µg/mL) (Andrade-Cetto et al. 2008).

Acourtia thurberi (A. Gray) Reveal & R.M. King, Asteraceae
The root extract of *Acourtia thurberi* reduced blood glucose levels in normal healthy mice and rabbits (Aguilar et al. 1997).

Acrocomia aculeata (Jacq.) Lodd. ex R.Keith, Arecaceae
Synonym: *Acrocomia mexicana* Karw. ex Mart.
The methanol extract of the *Acrocomia mexicana* (*Acrocomia aculeata*) root (2.5–40 mg/kg, i.p.) showed a significant dose-dependent blood glucose-lowering effect on normal and alloxan-induced diabetic mice. A new tetrahydropyran compound was isolated from the methanol extract of the root (Perez et al. 1997).

Adansonia digitata L., Malvaceae
Treatment of streptozotocin-induced diabetic rats with the stem bark extract of *Adansonia digitata* caused a significant reduction in the blood glucose levels when compared with diabetic control rats. The extract exhibited a significant decrease in blood glucose after 1, 3, 5, and 7 h of administration when compared to control. The highest activity was observed at 100 mg/kg with 51% glycemic change after 7 h, while 200 and 400 mg/kg showed a glycemic change of 39% and 31%, respectively, after 7 h of extract administration (Tanko et al. 2008b). In a recent study, administration of methanol extract of the fruit pulp (100, 200, and 300 mg/kg for 4 weeks) to alloxan-induced diabetic rats resulted in significant reduction of serum total cholesterol, LDL, and TGs when compared with the diabetic control rats (Bako et al. 2014).

Adenia lobata (Jacq.) Engl., Passifloraceae
The stem extract of *Adenia lobata* (ethanol extract) reduced the blood glucose levels in streptozotocin-induced diabetic rats (Sarkodie et al. 2013).

Adhatoda zeylanica Medicus, Acanthaceae
Synonym: *Adhatoda vasica* Nees, *Justicia adhatoda* L.
Adhatoda zeylanica leaf exhibited significant reduction in blood glucose levels in alloxan-induced diabetic rats (Bhatt et al. 2011).

Adiantum capillus-veneris L., Adiantaceae
Common Name: Maidenhair fern
The alcohol extract of the dried fronds of *Adiantum capillus-veneris* increased oral glucose tolerance in normal mice. The extract contained triterpenoidal and flavonoid compounds. The identified compounds

include quercetin, quercetin-3-*O*-glucoside, and quercetin-3-*O*-rutinoside (rutin). The extract also exhibited anti-inflammatory activity (Ibraheim et al. 2011). The water and methanol extracts of the whole *A. capillus-veneris* plant were investigated for their antidiabetic activity in streptozotocin-induced diabetic rats. The water and methanol extracts (200 and 400 mg/kg, orally, for 21 days) of the fern markedly decreased fasting blood glucose levels in diabetic rats. The major chemical constituents of the extract were found to be flavonoids and tannins (Ranjan et al. 2014).

Adiantum philippense L., Adiantaceae

Adiantum philippense is a traditional medicinal fern in the Philippines and elsewhere. The ethanol and water extracts of *A. philippense* (250 and 500 mg/kg, for 14 days) exhibited significant hypoglycemic effect in alloxan-induced diabetic rats. Further, the treatment increased liver glycogen, decreased oxidative stress, and improved the lipid profile in diabetic rats (Paul et al. 2012).

Aegiceras corniculatum (L.) Blanco, Myrsinaceae

Aegiceras corniculatum is a mangrove plant species that showed anti-DM activity. Oral administration of 80% alcohol extract of the *A. corniculatum* leaf (100 mg/kg daily for 60 days) to alloxan-induced diabetic rats resulted in marked reduction in blood glucose, glycosylated hemoglobin, and the activities of glucose-6-phosphatase and fructose 1,6-bisphosphatase and an increase in the activity of liver hexokinase (Gurudeeban et al. 2012). Further, falcarindiol isolated from this plant inhibited PTP1B at a low concentration ($IC_{50} = 9.1$ μmol/L) (Jiang et al. 2013).

Aegle marmelos (L.) Correa, Rutaceae (Figure 2.2)

Common Names: Holy fruit tree, bael or stone apple (Eng.), bilva (Sanskrit), bel (Hindi)

Description: A medium-sized, deciduous tree, armed with spines; the alternate, trifoliate but rarely 5-foliate, pinnately trifoliate, leaflets are glabrous but gray pubescent in arid localities and profoundly glandular. The flowers are greenish-white, produced in axillary or terminal raceme or cymes; the fruits are large, oblate, or subglobose berry.

Distribution: This tree originated in India. It is found in moist deciduous forests in subtropical Western Himalayas and Central and South India up to an elevation of 1200 m. It is also cultivated for its fruits.

Traditional Medicinal Uses: Its fruits (both ripe and unripe), roots, bark, leaves, rind of ripe fruits, and flowers are used in Ayurveda, Siddha, and Unani systems of traditional medicine. In folklore medicine and local health traditions, it is used for various ailments and diseases: purgative, abortifacient, astringent, digestive, stomachic, refrigerant, sterilant, febrifuge, antidiarrheal, laxative, diuretic, antifungal, and antibacterial and for sleeplessness, cardiac depression, intermittent fever, high blood pressure,

(a)

(b)

FIGURE 2.2 *Aegle marmelos.* (a) Crown with fruits. (b) Twig with flower.

dandruff, snake bite, hemorrhoids, dysentery, gastritis, stomach pain, diabetes, asthma, eczema, cholera, jaundice, ulcers in the throat, cold, peptic ulcers, chest pain, cataract, ophthalmia, leukoderma, leprosy, hiccups, deafness, joint pains, constipation, convolution, itch, nausea, palpitation, dyspepsia, proctitis, piles, catarrhs, hepatitis, tuberculosis, and chronic enteritis (Kirtikar and Basu 1975; Johnson 1999).

Pharmacology

Antidiabetes: A number of studies on experimental animals have found that the methanol, ethanol, and water extracts of the plant leaf have anti-DM activities. These extracts decreased serum glucose levels, improved the utilization of the external glucose load, and increased plasma insulin levels in experimental diabetic animals (review: Maity et al. 2009). The extract of the plant leaf exhibited hypoglycemic activity in male mice without altering the serum cortisol concentration. The alcohol extract of the leaf has been reported to show glucose tolerance in glucose-loaded hyperglycemic rats, and the mechanism of action is either by direct stimulation of glucose uptake or via the mediation of enhanced insulin secretion. Oral administration of water and ethanol extracts of leaves of this plant (500 mg/kg) also induced hypoglycemia in normal fasted rabbits (Maity et al. 2009). Treatment with the leaf extract to streptozotocin-induced diabetic rats resulted in improvement in the functional state of β-cells and improved regeneration of parts of the pancreas damaged by streptozotocin (Maity et al. 2009). Biochemical evaluation of the tissue antioxidant system and histological studies on β-cells confirmed the protective effect of *Aegle marmelos* leaf extract against streptozotocin-induced diabetes in rats (Narendhirakannan and Subramanian 2010). The leaf extract of this plant exhibited a protective effect against streptozotocin challenge in rats by augmenting the antioxidant defense system and improving β-cell function (Narendhirakannan and Subramanian 2010). This plant may also help ameliorate diabetic complications. The ethanol extract of the plant leaves ameliorated early-stage diabetic cardiomyopathy in alloxan-induced diabetic rats (Bhatti et al. 2011).

From the leaves of *A. marmelos*, an alkaloid amide, aegeline 2, was isolated and found to have antihyperglycemic activity in streptozotocin-induced diabetic rats. The compound also decreased the levels of cholesterol, TGs, and free fatty acid and increased HDL cholesterol (HDL-C) levels (Narender et al. 2007). Based on pharmacophoric hypothesis and 3D quantitative structure–activity relationship (QSAR) model, the authors suggest that the compound might be a β-3-AR agonist. Scopoletin (7-hydroxy-6-methoxy coumarin), isolated from the leaves of *A. marmelos*, when administered to levothyroxine-induced hyperthyroid rats, decreased the levels of serum thyroid hormones, glucose, and liver glucose-6-phosphatase activity. It also inhibited hepatic lipid peroxidation (LPO) (Panda and Kar 2006).

The water extract of *A. marmelos* fruits also showed hypoglycemic and antidiabetes actions in streptozotocin-induced diabetic rats (Kamalakkanan et al. 2003). In alloxan-induced diabetic rats, the ethanol extracts of the fruits significantly lowered blood glucose levels. The methanol extract effectively reduced the oxidative stress induced by alloxan and caused a reduction in blood sugar in alloxan-induced diabetic rats (Sabu and Kuttan 2004).

There may be more active principles in addition to aegeline involved in the anti-DM action; in ameliorating the disease, among other things, aegeline and its interaction with other bioactive molecules remain to be studied.

Clinical Trials: In a preliminary clinical trial, administration of the leaf extract for 15 days reduced blood cholesterol levels and, to a small extent, blood glucose levels in patients with type 2 DM (review: Ajikumaran and Subramoniam 2005). In another study, the decoction of *A. marmelos* leaf powder (5 g daily for 16 weeks) decreased postprandial blood glucose levels from 11.2 mmol/L (mean of 20 type 2 DM patients) to 8.8 mmol/L. In another study, the decoction, when administered as combination therapy along with a standard hypoglycemic agent to type 2 DM patients, showed improved efficacy in patients. The same finding was also reported in another double-blind placebo trial on type 2 DM patients. In this study, administration of *A. marmelos* (2 g/twice a day) plus the sulfonylurea drug for 8 weeks showed improvement in glycemic control compared to the sulfonylurea plus placebo group (review: Ghorbani 2013).

Other Activities: Experiments using animal models showed antiulcer, antioxidant, anti-LPO, anticancer, antiasthma, antimicrobial, antiviral, and radioprotective properties of this plant (review: Maity et al. 2009). The water extract of the leaves showed antihyperlipidemic effect in rats (Rajadurai and Prince 2005).

The leaf extracts showed anti-inflammatory, antinociceptive, and antipyretic activities in rats. The water extract of fruits exhibited anti-lipid peroxidative activity in streptozotocin-induced diabetic rats (review: Kamalakkanan et al. 2003; Ajikumaran and Subramoniam 2005). Under *in vitro* conditions, extract from the fruit exhibited a dose-dependent nitric oxide scavenging activity (Jagetia and Baliga 2004).

Toxicity: Higher doses of the hydroalcoholic extract of the plant leaf showed toxicity to mice (Jagetia et al. 2005).

Phytochemicals: Although more than 100 compounds have been isolated from different parts (leaf, fruit, stem, and root) of this plant, only some of them have been studied in detail for their utility as therapeutic agents against specific disease conditions. The biologically active compounds include aegelin, aurapten, cineol, citral, cuminaldehyde, eugenol, fagarine, lupeol, luvangetin, marmelide, marmelosin, marmin, marmesinin, psoralen, rutin, skimmianine, and tannin (Maity et al. 2009). Marmenol, umbelliferone, aegelinoside A, trans-cinnamic acid, valencic acid, 4-methoxy benzoic acid, betulinic acid, N-p-cis- and trans-coumaroyltyramine, montanine, phellandrene, β-sitosterol and rutaretin are some of the other compounds isolated in pure form (review: Maity et al. 2009; Ajikumaran and Subramoniam 2005).

Aerva lanata (L.) Juss ex Schultes, Amaranthaceae

Synonym: *Achyranthes lanata L.*

Aerva lanata, a herb, commonly known as sunny khur, is widely used in Indian folk medicine for the treatment of DM. The alcohol extract of *A. lanata* (375 and 500 mg/kg daily for 2 weeks) reduced blood glucose levels, improved body weight, and decreased the levels of lipid peroxides in alloxan-induced diabetic rats (Vetrichelvan and Jegadeesan 2002). These effects on body weight and blood glucose were confirmed by another study. In this study, the alcohol extract of *A. lanata* (400 mg/kg) improved oral glucose tolerance in alloxan-induced diabetic rats; subacute treatment for 28 days to the diabetic rats caused a marked reduction in blood glucose levels and prevented the decrease in body weight (Deshmukh et al. 2008). Aerial parts of this plant (water and methanol extracts) showed antihyperglycemic and antihyperlipidemic activity in streptozotocin-induced diabetic rats. Treatment with the extracts (200 and 400 mg/kg; for 14 days) resulted in increases in body weight and decreases in the levels of blood glucose, lipid peroxides, urea, creatinine, GPT, GOT, and lipids in treated diabetic animals (Rajesh et al. 2012).

Aesculus hippocastanum L., Hippocastanaceae

Common Name: Horse chestnut

Aesculus hippocastanum is a large tree indigenous to Asia Minor and Greece. In traditional medicine, it is used to treat many ailments.

Escins (triterpene oligoglycosides) present in the seeds of *A. hippocastanum* have been shown to have hypoglycemic activity in glucose tolerance tests in rats (Yoshikawa et al. 1994a, 1996c). The steroidal constituents from the plant bark have anti-inflammatory activities. Saponins and sapogenins from *A. hippocastanum* have antielastase and antihyaluronidase activities. Fruits contain cytotoxic sapogenals (review: Ajikumaran and Subramoniam 2005).

Aframomum melegueta (Roscoe) K. Schum., Zingiberaceae

Common Name: Grains of paradise

The water extract of *A. melegueta* leaf (50, 100, and 200 mg/kg) reduced blood glucose levels and improved body weight in alloxan-induced diabetic rats. The extract induced hypoglycemia in normal rats (Mojekwu et al. 2011). This is an important traditional medicinal plant with hepatoprotective, antioxidant, and other properties.

Afzelia africana Smith/*Afzelia africana* Sm. ex Pers., Fabaceae

The water extract of the stem bark of *Afzelia africana* (100 and 200 mg/kg daily for 10 days) reduced blood glucose levels and improved body weight in streptozotocin-induced diabetic rats compared to untreated diabetic control. Further, the levels of red blood cells and white blood cells and their functional indices were significantly improved after administration of the extract (Oyedemi et al. 2011).

Agarista mexicana (Hemsl) Judd, Ericaceae

Oral administration of 100 and 150 mg/kg of chloroform extract of *Agarista mexicana* produced a significant hypoglycemic effect in normal as well as in diabetic mice and rats. In addition, the extracts altered glucose tolerance in alloxan-induced diabetic rats, enhanced glucose uptake in skeletal muscles, and significantly inhibited glycogenolysis in the liver. These results indicate that the hypoglycemic effect may be similar to tolbutamide (Perez et al. 1996). Triterpenes from the stem extract of *A. mexicana* showed potent glucose-lowering effect in alloxan-induced diabetic mice (Perez and Vargas 2002).

Agave tequilana Gto., Asparagaceae

Diet supplemented with fructons (10%) from *Agave tequilana* for 5 weeks influenced glucose and lipid metabolism in mice (Urias-Silvas et al. 2007). The body weight gain and food intake in mice fed with fructon-containing diet were lower than those fed with standard diet. Serum glucose and cholesterol levels also decreased in the fructan-containing diet-fed mice. The supplementation with fructons induced a higher concentration of glucagon-like peptide-1 (GLP-1) and its precursor mRNA in the different colonic segments, suggesting that fermentable fructons from a different botanical origin are able to promote the production of incretin peptides in the lower parts of the gut (Urias-Silvas et al. 2007).

Ageratum conyzoides L., Asteraceae

Synonyms: *Ageratum album* Willd. ex Steud., *Ageratum caeruleum* Hort. ex. Poir., *Ageratum coeruleum* Desf., *Ageratum cordifolium* Roxb.

Common Names: Billy-goat weed, Chick weed

Ageratum conyzoides is a plant used in traditional medicine against many human infectious diseases and also in cases of diabetes. The water extract of leaves of *A. conyzoides* (300 mg/kg) reduced blood glucose levels in streptozotocin-induced diabetic rats, 1 h (10%) and 4 h (21%) after (i.p.) administration (Nyunai et al. 2006). In another study, the water extracts of leaves of *A. conyzoides* (100, 200, and 300 mg/kg) showed antihyperglycemic activity in streptozotocin-induced diabetic rats and hypoglycemic activity in normal rats. In the oral glucose tolerance test, 100 mg/kg dose significantly attenuated the rise of blood glucose in normal fasted rats (Nyunai et al. 2009). In another study, different fractions of the water extract were tested for their glucose-lowering effect in rats loaded with glucose and in streptozotocin-induced diabetic rats. The water extract showed a glucose-lowering effect in both glucose-loaded normal rats and diabetic rats. However, one of the fractions was more potent in glucose-loaded normal rats, while another fraction was more potent in streptozotocin-induced diabetic rats. The authors suggested that *A. conyzoides* contains more than one antihyperglycemic compound with different chemical characteristics and mechanisms of action (Nyunai et al. 2010). In another study, oral administration of the water extract (500 mg/kg) of *A. conyzoides* to alloxan-induced diabetic rats reduced fasting blood glucose by 39%. The hypoglycemic effect of *A. conyzoides* was less than that of the reference hypoglycemic agent, glibenclamide. Alkaloids, cardenolides, saponins, and tannins were detected in the extract (Agunbiade et al. 2012). Recently, it has been reported that the water extract of *A. conyzoides* leaves also exhibited blood glucose-lowering activity in glucose-loaded hyperglycemic rabbits (Stephane et al. 2013).

Agrimonia eupatoria L., Rosaceae

Agrimonia eupatoria is used to treat DM in traditional medicine. Incorporation of *A. eupatoria* into the diet (62.5 g/kg) and drinking water (2.5 g/L) countered the weight loss, polydipsia, hyperphagia, and hyperglycemia of streptozotocin-induced diabetic mice. Moreover, its water extract stimulated 2-deoxyglucose transport, glucose oxidation, and incorporation of glucose into glycogen in mouse abdominal muscle and evoked stimulation of insulin secretion from the BRIN-BD11 pancreatic β-cell line. The effect of the extract on insulin secretion was glucose independent and was not evident in the cells exposed to a depolarizing concentration of KCl (Gray and Flatt 1988).

Agrimonia pilosa Ledeb., Rosaceae

In Chinese traditional medicine, *Agrimonia pilosa* has a very important place in the treatment of type 2 DM. Further, *A. pilosa* is a medicinal plant with antitumor, antioxidant, anti-inflammatory, and anti-hyperglycemic activities. A recent study shows the α-glucosidase inhibitory activity and the antioxidant activity of the flavonoid and triterpenoid compounds from *A. pilosa* (Liu et al. 2014). An *in vitro* study showed the mechanisms of the antidiabetic effect of *A. pilosa*. The water extract of this plant inhibited α-amylase activity *in vitro*. Moreover, the 10 μg/mL extract reversed free fatty acid-induced insulin resistance in C2C12 myotubes and the 100 μg/mL extract significantly increased the utilization of glucose in C2C12 myotubes cultured in normal glucose (7 mM). Treatment increased the relative mRNA and protein expression levels of Akt in the myotubes. In particular, the effect of *A. pilosa* on the insulin signaling system is associated with the upregulation of Akt genes and glucose uptake in C2C12 myotubes. These results suggest that *A. pilosa* is useful in the prevention of diabetes and the treatment of hyperglycemic disorders. The plant drug has more than one mechanisms of action (Sang-Mi et al. 2013).

Ainsliaea latifolia (D. Don) Sch. Bip., Compositae/Asteraceae

Extracts of this plant showed anti-DM activity in rodents (Marles and Farnsworth 1995).

Ajuga iva L., **Lamiaceae**

Synonym: *Ajuga iva* (L.) Schreb (Figure 2.3)

Local/Common Names: Herb ivy, musky bugle, bugle weed, carpet bugle, chendgoura

Distribution: It is widely distributed in the Mediterranean region: southern Europe and northern Africa, particularly in Algeria, Morocco, Tunisia, and Egypt (Sabah 2010).

Description: *Ajuga iva* is a small, aromatic perennial bitter taste herb, with green stems creeping and hairy. Leaves are narrowly oblong to linear, pubescent, 14–35 mm long. *A. iva* grows in rocky slopes up to 2700 m of altitude. The flowers are purple, pink, or yellow, 20 mm long; the upper lip of the corolla is absent or reduced and the lower lip is divided into three lobes and is hairy. The side lobes are small, while the central lobe is relatively larger, usually in purple, and is decorated at the base with a central yellowish spot the same color as the flower. Within the flower there are four stamens related to four black carpels. The seeds are brown (review: Sabah 2010).

FIGURE 2.3 *Ajuga iva.*

Traditional Medicinal Uses: It is used in folk medicine as anthelmintic and for a variety of ailments, including diabetes. It is ingested for its useful action against stomach and intestinal pains, enteritis, fever, sinusitis, and headache. It has been used to treat dysuria and painful joints of the limbs (review: Sabah 2010).

Pharmacology

Antidiabetes: *A. iva* is used in traditional medicine in the treatment of diabetes. The hypoglycemic effect of the lyophilized aqueous extract of the whole *A. iva* plant was studied in normal and streptozotocin-induced diabetic rats. Single and repeated oral administration of the extract (10 mg/kg) produced a slight and significant decrease in plasma glucose levels in normal rats 6 h after administration and after 3 weeks of treatment. The extract reduced plasma glucose levels of streptozotocin-induced diabetic rats from 18.7 to 5.7 mmol/L after 6 h of oral administration. Repeated oral administration of the extract to streptozotocin-induced diabetic rats decreased the plasma glucose levels at 1 week of treatment (mean 6.2 mmol/L on day 8 vs. 18.7 mmol/L at the baseline value). It continuously decreased thereafter and showed a rapid normalization after 1 week of treatment (El-Hilaly and Lyouss 2002).

The hypoglycemic and hypolipidemic effect of continuous intravenous infusion for 4 h at a dose of 4.2 μg/min/100 g body weight of a lyophilized water extract of the whole *A. iva* plant was investigated in anesthetized normal and streptozotocin-induced diabetic rats. In normal rats, the extract infusion had no effect on plasma glucose or TGs, but plasma cholesterol levels were significantly decreased. In streptozotocin-induced diabetic rats, the infusion reduced plasma levels of glucose by 24%, cholesterol by 35%, and TGs by 13%. The authors conclude that intravenously administered, the extract exerts hypoglycemic and hypolipidemic effects in diabetic rats by mechanism(s) that appear to be similar to that of taurine, which involves insulin sensitization or an insulin-like effect (El-Hilaly et al. 2007). The single and repeated administration of the water extract of *A. iva* orally for 3 weeks reduced the blood glucose levels significantly in diabetic rats compared to untreated diabetic rats (Rouibi et al.2013). The hot water extract of aerial parts of *A. iva* also showed hypoglycemic potential in alloxan-induced diabetic rats (Chabane et al. 2013).

The studies have established the anti-DM activities of this plant. However, the mechanisms of actions and active principles involved remain to be studied.

Other Activities: Pharmacological studies have shown that *A. iva* has antiulcer and anti-inflammatory activities. *A. iva* extract decreased plasma cholesterol and TGs. In addition, *A. iva* induced an inhibition of calcium oxalate monohydrate crystal growth (Beghalia et al. 2008; Sabah 2010).

Toxicity: No apparent toxicity was observed for this plant (El-Hilaly et al. 2004).

Phytochemicals: *A. iva* was found to contain a large number of compounds such as 8-*O*-acetylharpagide, ajugarine, apigenin-7-*O* neohesperidoside, barpagide, caffeine, clorogenes, cyasterone, diglycerides, 14,15-dihydroajugapitin, ecdysones, ecdysterones, flavonoids, iridoides, makisterone A, neohesperidoside, phenylcarboxylic acids, and tannin polyphenols. It was also reported that *A. iva* contains tannins, phytoecdysteroids, polyhydroxylated sterols, essential oil, and ecdysteroids (makisterone A, 20-hydroxyecdysone, and cyasterone) with several minor compounds including 24,28-dehydromakisterone A and two new phytoecdysteroids (22-oxocyasterone and 24,25-dehydroprecyasterone). In addition, it contained polypodine B and 2-deoxy-20-hydroxyecdysone. The occurrence of the antifeedant 14,15-dihydroajugapitin in the aerial parts of *A. iva* from Algeria was also shown (review: Sabah 2010).

Alangium salviifolium (L.f.) Wang., Alangiaceae

Alangium salviifolium is commonly known as sage-leaved Alangium. It is a tall thorny tree native to India. The bark is ash colored, rough, and faintly fissured. The leaves are elliptic oblong, elliptic lanceolate, or oblong lanceolate. In Ayurveda, the roots and fruits are used for the treatment of rheumatism and hemorrhoids. Externally it is used for the treatment of rabbit bites. The leaves are used for the treatment of diabetes and the fruits as an astringent. The seeds of *A. salviifolium* have also been traditionally used for a variety of biological activities, including antidiabetic activity.

The effect of water extracts of the stem and leaves of *A. salviifolium* on blood glucose levels in normal and alloxan-induced diabetic rats was studied. Oral administration of the extract (200, 400, and 800 mg/kg) resulted in a significant reduction in blood glucose level. The effect was comparable with that of 0.5 mg/kg (i.p.) glibenclamide. The results support the traditional usage of the plant for the control of diabetes (Hepcy-Kalarani et al. 2011).

From the acute toxicity study, it was observed that chloroform, ethanol, and water extracts of the seeds are nontoxic at 2000 mg/kg. The ethanol extract of the seeds exhibited significant antidiabetic, antiepileptic, analgesic, and anti-inflammatory activities. In rats, the antiarthritic activity of the bark extracts and the antifertility activity of the stem bark of *A. salviifolium* have been reported. Phytochemical analysis revealed the presence of alkaloids, glycosides, terpenoids, steroids, and tannins in this extract (Sharma et al. 2011a).

Alchornea cordifollia Mull. Arg., Euphorbiaceae

Administration of the *n*-butanol fraction of *Alchornea cordifolia* (200 and 400 mg/kg) for 28 days to streptozotocin-induced diabetic rats resulted in marked reduction in blood glucose levels and increase in body weight. Further, the treatment showed erythropoietic effects (Mohammed et al. 2012).

Alhagi camelorum Fisch., Fabaceae

The methanol extract of the aerial parts of *Alhagi camelorum* showed about 92% inhibition of α-glucosidase activity under *in vitro* conditions (Gholamhoseinian et al. 2008).

Alisma orientale (Sam.) Juzepcz., Alismataceae

Alisma rhizome is widely used in the therapy of diabetes in the traditional folk medicine of China. Alisma rhizome is the rhizome of the perennial marsh plant *Alisma orientale*. Eight protostane-type triterpenes were isolated from the alcohol extract of Alisma rhizome. Two of the compounds, alisol F and alisol B, displayed *in vitro* inhibition of α-glucosidase activity. The alcohol extract also inhibited α-glucosidase activity. Further, the extract increased glucose uptake in the 3T3-L1 adipocyte model without an increase in adipogenesis (Li and Qu 2012).

Alisma plantago-aquatica L., Alismataceae

Common Name: European water-plantain

A powder prepared from the dried roots of *Alisma plantago-aquatica* is used in popular medicine as a cure for rabies and crushed leaves are used against mammary congestion. It is also used to treat DM. The ethanol extract of *A. plantago-aquatica* activated the peroxisome proliferator–activated receptor-γ (PPAR-γ) in a concentration-dependent manner. This could be a mechanism of the likely anti-DM activity of this plant (Rau et al. 2006).

Allium cepa L., Liliaceae (Figure.2.4)

Common Names: Onion (Eng.), piyaj (Hindi), vengayam (Tamil), etc.

Description: *Allium cepa* is a bulbous biennial herb with linear, hollow, fleshy, cylindrical leaves arising from an underground bulb and umbellate inflorescent flowers. Bulbs constitute the crop and vary in size, color, and shape from cultivar to cultivar and are often depressed globose; the outer tunics

(a)   (b)

FIGURE 2.4 *Allium cepa*. (a) Bulb. (b) Flower.

are membranous. The stems are up to 100 cm tall and 30 mm in diameter, tapering from the inflated lower part. The leaves are up to 40 cm in height and 20 mm in diameter, usually almost semicircular and slightly flattened on the upper side; the umbel is subglobose or hemispherical, dense, and many-flowered; the pedicels are up to 40 mm and are almost equal; the perianth is stellate with white segments and green stripes and is slightly unequal. The stamens are exerted with filaments 4–5 mm long; the outer part is subulate and the inner part has an expanded base up to 2 mm wide bearing short teeth on each side. The ovary is whitish with a capsule measuring about 5 mm (monograph: WHO 1999).

Distribution: Onion is extensively cultivated in India and is grown as a garden crop. Onion is probably indigenous to Western Asia but is commercially cultivated worldwide, especially in regions of moderate climate (monograph: WHO 1999).

Traditional Medicinal Uses: Bulbs of onions are primarily used in ethnomedicine. It is an antiseptic, aperitif, carminative, digestive, disinfectant, stimulant, diuretic, expectorant, stomachic and vermifuge useful in flatulence, antibacterial, antiperiodic, emmenagogue, and aphrodisiac. In addition, the plant is used to treat dysentery, dyspepsia, abscesses, albuminuria, bronchitis, cough, diphtheria, dropsy, diabetes, snake bite, tumor, warts, hepatopathy, hepatosis, obalgia, skin diseases, cholera, arteriosclerosis, cataracts, common cold, hyperglycemia, sore throat, ear ailment, piles, rheumatism, asthma, bruises, colic, earache, fevers, high blood pressure, pimples, ulcers, wounds, and inflammation (monograph: WHO 1999).

Pharmacology

Antidiabetes: The hypoglycemic effects of onion bulb have been demonstrated *in vivo*. Intragastric administration of the juice, a chloroform, ethanol, petroleum ether (0.25 g/kg), or water extract (0.5 mL), suppressed alloxan-, glucose- and epinephrine-induced hyperglycemia in rabbits and mice (monograph: WHO 1999). The hypoglycemic and hypolipidemic actions of *A. cepa* extracts were associated with antioxidant activity. Tissue extracts from callus cultures of onion exhibited much higher antidiabetic activity in diabetic rats as compared to natural bulbs; traditional preparation of *A. cepa* decreased the hyperglycemic peak in experimental rabbits (Roman-Ramos et al. 1995). The onion juices showed antihyperglycemic and antioxidant effects in alloxan-induced diabetic rats and may alleviate liver and renal damage caused by alloxan (El-Demerdash et al. 2005). S-Methyl cysteine sulfoxide isolated from onions showed antioxidant and anti-DM properties in alloxan-induced diabetic rats (Augusti and Sheela 1996; Kumari et al. 1995; Kumari and Augusti 2002). Diphenyl amine is an antihyperglycemic agent present in onion, tea, etc. (Karawya et al. 1984). Interestingly, a study suggests that a high-fat onion diet may increase insulin secretion and consequently insulin resistance in a dose-dependent manner, resulting in a worsened hyperglycemic and hyperlipidemic diabetic state. The authors conclude that higher dietary fat may impair the antidiabetic effects of dietary onion intake. Onion is reported to contain oleanolic acid, an antidiabetic compound.

Clinical Trials: The antihyperglycemic activity of onion has been demonstrated in clinical studies. Administration of the water extract (100 mg) of onion decreased glucose-induced hyperglycemia in human adults. Onion juice (50 mg), administered orally to diabetic patients, reduced blood glucose levels. Addition of raw onion to the diet of type 2 diabetic subjects decreased the dose of antidiabetic medication required to control the disease. However, a water extract of onion (200 mg/kg) was not active (monograph: WHO 1999). An acute hypoglycemic effect of onion was also observed in a self-controlled study on 20 patients with type 2 DM. The treatment also attenuated rice in plasma glucose 2 h after glucose ingestion. In another study, it was shown that intake of 100 g onion decreased fasting blood glucose levels and improved glucose tolerance in both type 1 and type 2 DM patients (review: Ghorbani 2013). Thus, onion appears to be useful in combination therapy and/or as a food supplement to DM patients.

Other Activities: The antihyperlipidemic and anticholesterolemic activities of onion were observed in rabbits and rats. Inhibition of platelet aggregation by onion has been demonstrated both *in vitro* and *in vivo*. Both raw onions and its essential oil increased fibrinolysis in *ex vivo* studies on rabbits and humans. Other reported activities include antibacterial, antifungal, antifibroblast proliferation, diuretic, antiallergic, and anti-inflammatory. Topical application of a 45% ethanol extract of onion inhibited allergic skin reactions induced by anti-IgE in humans. The butanol extract of onion (200 mg) to subjects given a high-fat meal prior to testing suppressed platelet aggregation associated with an HFD (monograph: WHO 1999).

Toxicity: It is a vegetable and there is no recorded toxicity.

Phytochemicals: Essential oil (chief constituent is allyl-propyl disulfide), cycloallin, oleanolic acid, arabinose, quercetin 3-glucoside, quercetin 4'-glucoside, quercetin 3,4'-diglucoside, quercetin 7,4'-diglucoside, quercetin 3,7,4'-triglucoside, and isorhamnetin 4'-glucoside occur in onion. The organic sulfur compounds isolated from onion included thiosulfinates, thiosulfonates, cepaenes, *S*-oxides, *S,S*-dioxides, monosulfides, disulfides, and trisulfides. When the onion bulb is crushed, degradation of naturally occurring cysteine sulfoxide occurs. Thiosulfonates occur in freshly chopped onions in low concentrations, whereas sulfides accumulate in stored extracts or steam distilled oils (monograph: WHO 1999).

Allium porrum L., Liliaceae

The effect of *Allium porrum* on glucose uptake was studied using the rat everted intestinal sac technique. *All. porrum* (2.5 and 5 mg/mL) significantly reduced glucose uptake and it was found to be marginally better than *Allium sativum* in the reduction of glucose uptake (Belemkar et al. 2013). The inhibitory effects of six selected *Allium* species (*A. akaka, A. ampeloprasum* ssp. *iranicum, A. cepa, A. hirtifolium, A. porrum*, and *A. sativum*) on α-amylase enzyme activity were investigated using an *in vitro* model. According to the results, the ethanol extracts of *A. akaka, A. sativum, A. porrum*, and *A. cepa* were found to have a favorable α-amylase inhibitory activity, revealing no significant differences in their IC_{50} values (Nickavar and Yousefian 2009).

The ethanol extract of bulbs of *A. porrum* (200 and 500 mg/kg, p.o., daily for 5 days) decreased blood glucose levels and kidney oxidative stress in streptozotocin-induced diabetic rats (Aslan et al. 2010). In another study, the extract showed hypolipidemic effect in normal and streptozotocin-induced diabetic rats (Eydi et al. 2007).

Allium sativum L., Alliaceae

Synonym: *Porvium sativum* Rchb. (Figure 2.5)

Common Names: Garlic (Eng.), lasan (Hindi), vellaipundu (Tamil), etc.

Description: *Allium sativum* is a perennial herb with narrow flat leaves and short bulb and strong smelling when crushed. The underground portion consists of a compound bulb (bulblets enclosed in a pink or white envelope) with numerous fibrous rootlets; the bulb gives rise aboveground to a number of narrow, keeled, grasslike leaves. The leaf blade is linear, flat, solid, 1.0–2.5 cm wide, and 30–60 cm long and has an acute apex. The leaf sheaths form a pseudostem. Inflorescences are umbellate; the scape is smooth, round, solid, and coiled at first, subtended by membranous, long-beaked spathe, splitting on one side and remaining attached to the umbel. Small bulbils are produced in inflorescences; the flowers are variable in number and sometimes absent, seldom open, and may wither in bud. Flowers are on slender pedicels; perianth is about 4–6 mm long and pinkish; stamens 6, ovary superior, 3-locular. The fruit is a small loculicidal capsule; seeds are seldom if ever produced (monograph: WHO 1999).

(a)　　　　　　　　　　　　　　　(b)

FIGURE 2.5　*Allium sativum.* (a) Bulb. (b) Flower head.

Distribution: Garlic is cultivated throughout India, especially at an elevation of 3000–4000 ft. Garlic is probably indigenous to Asia, but it is commercially cultivated in most countries (WHO 1999).

Traditional Medicinal Uses: Fresh or dried bulbs of *A. sativum* are used in traditional medicine. It is used for the treatment of respiratory and urinary tract infections, ringworm infections, and rheumatic conditions. The herb has been used to promote hair growth and as a carminative in the treatment of dyspepsia. It is used as an aphrodisiac, antipyretic, diuretic, emmenagogue, expectorant, ascaricide, and sedative. It is also used to treat diabetes, kidney ailments, arteriosclerosis, hypertension, inflammation, viral infections, asthma, bronchitis, and senescence (monograph: WHO 1999).

Pharmacology

Antidiabetes: Several studies have demonstrated hypoglycemic effects of the garlic bulb in experimental animals. Oral administration of the water, ethanol, petroleum ether, or chloroform extract, or the essential oil of garlic lowered blood glucose levels in rabbits and rats (review: Ajikumaran and Subramoniam 2005). In alloxan-induced diabetic rats, garlic juice exerted antihyperglycemic and antioxidant activities (El-Demerdash et al. 2005). However, three similar studies reported negative results (monograph: WHO 1999). In one study, the garlic bulb, administered in feed to normal or streptozotocin-induced diabetic mice, reduced hyperphagia and polydipsia but had no effect on hyperglycemia or hypoinsulinemia (Swanston-Flatt et al. 1990). The protective effects of garlic extract on alloxan-induced diabetic rats were confirmed in another study. In this study, garlic extract was administered for 7 days before alloxan challenge and the daily administration was continued for 14 more days. The results show that preadministration of garlic extract before alloxan challenge prevents hyperglycemia in alloxan-administered rats (Oja et al. 2012).

The hypoglycemic effect of garlic juice (1 mL/100 g) feeding for 6 weeks was studied in streptozotocin-induced diabetic rats. In one group of animals, garlic juice was administered 3 weeks before streptozotocin challenge and the juice treatment was continued for 3 more weeks. The elevated levels of glucose, cholesterol, and TGs increased activities of alanine aminotransferase (ALT) and aspartate aminotransferase (AST), and increased food and water consumption observed in the diabetic control groups were prevented, to a large extent, in the garlic juice treated (garlic juice was administered both before and after streptozotocin administration) streptozotocin challenged rats. In the diabetic control group, decrease of pancreatic islet numbers and diameter and atrophy of pancreatic islets were observed. These abnormalities in histology were ameliorated in the garlic treated group. Thus, garlic juice may help in the prevention of certain toxic chemical-induced DM (Masjedi et al. 2013). In a recent study, the hypoglycemic effect and increase in body weight in streptozotocin-induced diabetic rats with the combination of glibenclamide and *A. sativum* extract was greater than any one of the drugs given alone. The drugs may show additive effect. This may be important in reducing the dose of glibenclamide and achieving an enhanced therapeutic effect with minimal side effects (Poonum et al. 2013).

Allicin, a compound from garlic, administered orally to alloxan-induced diabetic rats lowered blood glucose levels and increased insulin activity in a dose-dependent manner (Mathew and Augusti 1973). Garlic extract's hypoglycemic action may be due to enhanced insulin production, and allicin has also been shown to protect insulin against inactivation (monograph: WHO 1999). The bulb of *A. sativum*, its oil, and S-allyl cysteine also prevented diabetes complications such as oxidative stress, diabetic cardiomyopathy, and diabetic nephropathy in experimental diabetic rats (Mariee et al. 2009; Ou et al. 2010; Chang et al. 2011; Saravanan and Ponmurugan 2011). Garlic oil alleviated MAPK- and interleukin-6 (IL-6)-mediated diabetes-related cardiac hypertrophy in streptozotocin-induced diabetic rats (Chang et al. 2011). Cardiac contractile dysfunction and apoptosis in streptozotocin-induced diabetic rats were ameliorated by garlic supplementation (Ou et al. 2010).

Clinical Studies: In a double-blind, placebo-controlled study, oral administration of garlic powder (800 mg/day) to 120 patients for 4 weeks decreased the average blood glucose by 11.6%. Another study found no such activity after dosing non-insulin-dependent DM (NIDDM) patients with 700 mg/day of a spray-dried garlic preparation for 1 month (monograph: WHO 1999).

Other Activities: The antihypercholesterolemic and antihyperlipidemic effects were observed in various animal models (WHO 1999). Oral administration of allicin to rats during a 2-month period lowered serum and liver levels of total lipids, phospholipids, TGs, and total cholesterol. Total plasma lipids and cholesterol in rats were reduced after intraperitoneal injection of a mixture of diallyl disulfide and diallyl trisulfide. At low concentrations, garlic extracts inhibited the activity of hepatic 3-hydroxy-3-methylglutaric acid (HMG)-CoA reductase, but at higher concentrations, cholesterol biosynthesis was inhibited (Brosche and Platt 1991; Gebhardt 1993; Beck and Wagnerk 1994). Allicin, ajoene, allyl mercaptan, nicotinic acid, and adenosine may inhibit cholesterol biosynthesis. The subjects receiving garlic supplementation (powder) showed a reduction in total cholesterol and TGs. Meta-analysis of the clinical studies confirmed the lipid-lowering action of garlic. A systematic review of the lipid-lowering potential of a dried garlic powder preparation in eight studies with 500 subjects had similar findings (monograph: WHO 1999).

Garlic bulb, essential oil of garlic, etc., have a wide range of antibacterial and antifungal activities. Allicin, ajoene, and diallyl trisulfide constituents of the bulb have antibacterial and antifungal activities. Garlic has been used in the treatment of roundworm and hookworm infections (monograph: WHO 1999). Allicin exhibited antiproliferative and antigenotoxic activity in mammalian cells in culture. Allicin showed antioxidant activity by inhibiting induced nitric oxide synthase (iNOS) and the arginine transporter CAT-2 activity in cultured cardiac myocytes (Dirsch et al. 1998; Schwartz et al. 2002). Garlic lowered blood pressure in experimental animals. The drug appeared to decrease vascular resistance by directly relaxing smooth muscles. Garlic may increase production of nitric oxide, which is associated with a decrease in blood pressure. Garlic inhibited collagen-, ADP-, arachidonic acid-, epinephrine-, and thrombin-induced platelet aggregation *in vitro*. Prolonged administration of the essential oil or a chloroform extract of the garlic bulb inhibited platelet aggregation in rabbits. Adenosine, alliin, allicin, the ajoenes, the vinyldithiins, and the dialkyloligosulfides are responsible for the inhibition of platelet adhesion and aggregation. In addition, methyl allyl trisulfide, a minor constituent of garlic oil, inhibited platelet aggregation at least 10 times as effectively as allicin. The mechanism of action involves the inhibition of the metabolism of arachidonic acid by both cyclooxygenase and lipoxygenase. A meta-analysis of the effect of garlic on blood pressure concluded that garlic may have some clinical usefulness in mild hypertension in humans. Intragastric administration of an ethanol extract of garlic decreased carrageenin-induced rat paw edema at a dose of 100 mg/kg. The anti-inflammatory activity of the drug appears to be due to its antiprostaglandin activity (monograph: WHO 1999).

Phytochemicals: Essential oil contains allyl-propyl disulfide, diallyl disulfide, cycloallin, methylallyl thiosulfinate, allicin (diallylthiosulfinate), alliin, *S*-allyl cysteine sulfoxide, ajoene, etc. Other reported compounds include dithiane, allisatin, allisatin-1, allisatin-2, allium fructans, allium lectin, allixin, allyl trisulfide, allyl methyl di- and trisulfide, caffeic acid, chlorogenic acid, cis-ajoene, cycloallin, desgalactotigonin, diallyl heptasulfide, diallyl hexasulfide, diallyl pentasulfide, eruboside, geraniol, kaempferol, linalool, myrosinase, phloroglucinol, *p*-coumaric acid, phytic acid, prostaglandins, protoeruboside, pseudoscoridinine-a and pseudoscoridinine-b, quercetin, saponin, rutin, sativosides, scordinines, scorodose, stigmasterol, taurine, thiamocornine, and thiamamidine. Cysteine sulfoxides and the nonvolatile γ-glutamylcysteine peptides make up more than 82% of the total sulfur content of garlic. The thiosulfinates (e.g., allicin), ajoenes, vinyldithiins, and sulfides are not naturally occurring compounds. Rather, they are degradation products from the naturally occurring cysteine sulfoxide alliin. Allicin itself is an unstable product and will undergo additional reactions to form other derivatives depending on environmental and processing conditions. Extraction of garlic cloves with ethanol at 0°C gave alliin; extraction with ethanol and water at 25°C led to allicin and no alliin; and steam distillation (100°C) converted the alliin totally to diallyl sulfides. Whole garlic cloves (fresh) contained 0.25%–1.15% alliin, while material carefully dried under mild conditions contained 0.7%–1.7% alliin (monograph: WHO 1999).

Toxicity or Adverse Reactions: Garlic may develop allergic reactions occasionally. Consumption of a large amount of garlic may increase the risk of postoperative bleeding. Average daily use of garlic up to 5 g (wet weight) or 1.2 (dried powder) is safe for this nutraceutical (monograph: WHO 1999).

Alnus incana ssp. *rugosa* (Du Roi) R.T. Clausen, Betulaceae

Common Name: Speckled alder

Alnus incana is a deciduous tree growing in North America. The bark is used in ethnomedicine to treat cutaneous disorders, mouth and gum diseases, and syphilis and to alleviate symptoms of type 2 DM. The antidiabetic activity of this plant has been reviewed recently (Eid and Haddad 2014). It was one of the more potent plants to stimulate glucose transport in cultured myocytes and to stimulate AMPK in myocytes. It also inhibited glucose-6-phosphatase activity in cultured hepatocytes, albeit without significantly affecting either Akt or AMPK phosphorylation. Besides, it moderately inhibited intestinal glucose absorption. However, in a screening assay for adipogenic glitazone-like activity, the alcohol extract of *A. incana* bark unexpectedly and strongly prevented the intracellular accumulation of lipid droplets. Subsequent detailed mechanistic studies demonstrated that the plant blocked the differentiation and the maturation of 3T3-L1 preadipocytes. This activity was attributed to oregonin, a diarylheptanoid glycoside and a minor constituent of the alcohol extract. *In vivo* follow-up studies are required to establish the antidiabetic activity of this plant extract. Further, it should be noted that the plant extract possesses some level of inhibitory effect on drug-metabolizing enzymes.

Aloe ferox Mill, Aloaceae
The ethanol extract of the *Aloe ferox* leaf (300 mg/kg) showed mild antidiabetic activity in streptozotocin-induced diabetic rats; the treatment moderately increased insulin levels and decreased lipids (Loots et al. 2011).

Aloe greatheadii var. *davyana*, Aloaceae
Ethanol extract of *Aloe greatheadii* leaf (300 mg/kg) moderately increased insulin levels and decreased blood glucose and lipids in streptozotocin-induced diabetic rats (Loots et al. 2011).

Aloe perryis Baker, Liliaceae
Aloe perryis is a traditional medicinal plant. The ethanol extract of this plant showed hypoglycemic effects in alloxan-induced diabetic rats (Ibegbulem and Chikezie 2013).

Aloe vera (L.) Burm. f., Aloaceae/Liliaceae

Synonyms: *A. barbadensis* Mill., *A. chinensis* Bak., *A. elongata* Murray, *A. indica* Royle, *A. officinalis* Forsk., *A. perfoliata* L., *A. rubescens* DC, *A. vera* L. var. *littoralis* König ex Bak., *A. vera* L. var. *chinensis* Berger, *A. vulgaris* Lam. (a closely related species is *Aloe ferox* Mill) (monograph: WHO 1999) (Figure 2.6)

Common Names: West Indian aloe, aloes, Barbados aloe, Indian aloe, zambila (Eng.), Ghee-kunwar (Hindi), chirukattalai (Tamil) (monograph: WHO 1999)

Description: A dwarf succulent and perennial herb, with fleshy leaves 30–50 cm long and 10 cm broad at the base and bright yellow tubular flowers 25–35 cm in length arranged in a slender loose spike; stamens frequently project beyond the perianth tube; and its fruit is an ellipsoidal capsule (monograph: WHO 1999).

FIGURE 2.6 *Aloe vera*.

Distribution: *Aloe vera* is native to Southern and Eastern Africa and subsequently introduced into Northern Africa, the Arabian Peninsula, China, Gibraltar, the Mediterranean countries, and the West Indies. It is commercially cultivated in Aruba, Bonaire, Haiti, India, South Africa, the United States, and Venezuela (WHO 1999).

Traditional Medicinal Uses: The juice of the leaves (dried or fresh) of this plant is generally used in traditional medicine.

It is stomachic, carminative, aperient, emollient, expellant, purgative, emmenagogue, and vulnerary; it is used in piles, rectal fissures, poultice inflammation, asthma, burns, cough, cuts, eczema, fever, gonorrhea, headache, mouth sores, skin disease, pyorrhea, fever, cancer, congestion, hemorrhoids, insect bite, intestinal worms, pregnancy, skin irritation, nausea, swelling, tuberculosis, wounds, and wrinkles. It is also used to treat seborrheic dermatitis, peptic ulcers, tuberculosis, fungal infections, and diabetes (WHO 1999).

Pharmacology

Antidiabetes: Experiments on animals have shown that *A. vera* has antidiabetic properties (Ghannam et al. 1986; Koo 1994). *A. vera* leaf pulp extract showed hypoglycemic activity on type 1 and type 2 diabetic rats, the effectiveness being more for type 2 diabetes in comparison with glibenclamide. On the contrary, *A. vera* leaf gel extract showed hyperglycemic activity on type 2 diabetic rats. It may, therefore, be concluded that the pulps of *A. vera* leaves devoid of the gel could be useful in the treatment of type 2 DM (Okyar et al. 2001). However, in alloxan-induced diabetic rats, the gel extract of this plant showed antidiabetes, antihypercholesterolemic, and antioxidative effects (Mohamed 2011a). Extracts of the *Aloe* gum increased glucose tolerance in normal as well as diabetic rats. *A. vera* sap, when taken daily for 4–14 weeks, showed hypoglycemic effect both clinically and experimentally (review: Sahu et al. 2013).

Five phytosterols isolated from *A. vera*, namely, lophenol, 24-methyl-lophenol, 24-ethyl-lophenol, cycloartanol, and 24-methylenecycloartanol exhibited antidiabetic effects in type 2 diabetic mice (Tanaka et al. 2006). The plant also contains sitosterol, quercetin, and rutin, known hypoglycemic compounds, occurring in certain plants. Moreover, the plant contains polysaccharides that increase the insulin level and show hypoglycemic properties (review: Sahu et al. 2013).

Clinical Studies: *A. vera* sap showed hypoglycemic effect in one clinical study. *A. vera*'s high-molecular-weight fractions containing less than 10 ppm of barbaloin and polysaccharide (MW 1000 kDa) with glycoprotein, verectin (MW 29 kDa), produced significant decrease in blood glucose level sustained for 6 weeks of the start of the study. Significant decrease in TGs was only observed 4 weeks after treatment and continued thereafter. No deleterious effects on kidney and liver functions were apparent (Yagi et al. 2009). Further, *A. vera* is used in reducing sugar in DM (review: Sahu et al. 2013). However, further controlled clinical trials using isolated compounds, mixers of compounds, and specified, standardized extracts are urgently needed.

Other Activities: Proven pharmacological activities of this plant include wound healing, antiinflammatory activity, antitumor activity, immunomodulatory property, laxative effects, antimicrobial activities, and antiviral properties. *A. vera*'s laxative activity stimulates colonic motility reducing fluid absorption from the fecal mass; it also increases water in the large intestine (review: Sahu et al. 2013).

Anthraquinone compounds are the active antibacterial components in aloe and aloin is the main active compound. There is experimental evidence that *A. vera* can prevent the development of, at least, certain types of cancer (Furukawa et al. 2002). *A. vera* extract has antilipoxygenase activity, and this may be involved in the inhibition of the acute inflammation process, especially in the topical application for healing of minor burns and skin ulcers. *A. vera* can abrogate the toxicity of certain environmental toxins and protect from radiation and oxidative stress (Rajasekaran et al. 2005). *A. vera* polysaccharides has a radioprotective effect to different cell lines (Wang et al. 2005b). The antioxidative stress of this plant can help from developing cardiovascular diseases under hyperlipidemic conditions.

Toxicity: The major symptoms of overdose are severe diarrhea, abdominal spasms, and abdominal pain. Chronic abuse of anthraquinone stimulant laxatives should be avoided (monograph: WHO 1999).

Phytochemicals: Compounds isolated from this plant include aloin, emodin, chrysophanic acid, isobarbaloin, b-barbaloin, isoemodin, quercetin, rutin, free and cinnamoylated 8-methyl-7-hydroxyaloins, acemannan, alliinase, alocutin, aloe emodin, aloeferon, aloesin, aloesone, aloetic acid, aloetin, aloetinic acid, aloinose, anthrol, apoise, barbaloin, behenic acid, campesterol, casantranol I and II, chrysamminic acid, chrysazin, chrysophanic acid, cinnamic acid, coniferyl alcohol, coumarin, cycloecosane, glucitol, dehydro abietal, emodin, hecogenin, heneicosanoic acid methyl ester, homonataloin, isobarbaloin, lauric acid, lupeol, margaric acid, salicylic acid, myristic acid, nataloin, niacinnamide, *p*-coumaric acid, pectic acid, phytosterols, quinine, resin, resitannol, rhein, sapogenin, saponins, threitol, lupeol, sitosterol, and polysaccharides. Aloe contains as its major active principles hydroxyanthrone derivatives, mainly of the aloe-emodin-anthrone 10-*C*-glucoside type. The major constituent is known as barbaloin (aloin) (15%–40%). It also contains hydroxyaloin (about 3%). Barbaloin (aloin) is in fact a mixture of aloin A and B. Aloin A and B interconvert through the anthranol form as do aloinoside A and B (reviews: Ajikumaran and Subramoniam 2005; Sahu et al. 2013).

Alpinia officinarum Hance, Zingiberaceae

The extract from *Alpinia officinarum* inhibited α-glucosidase activity (Gholamhoseinian et al. 2008).

Alstonia macrophylla Wall & G. Don, Apocynaceae

Alstonia macrophylla is a tree used in traditional medicine to treat various diseases. Phytochemicals from the leaves of *A. macrophylla* inhibited the activity of Na (+)-glucose cotransporter that is involved in the reabsorption of glucose in the kidney. This transporter is a therapeutic target in the control of DM. Three new picraline-type alkaloids, alstiphyllanines E–G, and a new ajmaline-type alkaloid, alstiphyllanine H, were isolated from the leaves of *A. macrophylla* together with 16 related alkaloids. Alstiphyllanines E and F showed moderate Na (+)-glucose cotransporter (SGLT1 and SGLT2) inhibitory activity. A series of hydroxy-substituted derivatives of the picraline-type alkaloids have been derived as having potent SGLT inhibitory activity. 10-Methoxy-*N*(1)-methylburnamine-17-*O*-veratrate exhibited potent inhibitory activity, suggesting that the presence of an ester side chain at C-17 may be important to show SGLT inhibitory activity (Arai et al. 2010).

Alstonia scholaris L., Apocynaceae

Synonyms: *Echites scholaris* L., *Echites pala* Buch.-Ham. ex Spreng. (Figure 2.7)

Local Names: Devil tree, white cheese wood, verbal, milk wood pines, black board tree, mill wood, etc.

Distribution: *Alstonia scholaris* is widely distributed in dried forests of India especially in Western Himalayas, Western Ghats, and the Southern region. It is occurring in many countries such as Thailand, Malaysia, Philippines, China, Africa, and Australia.

Description: *A. scholaris* is a large, buttressed evergreen tree, 6–10 m in height; the bark is rough, gray-white in color and yellowish inside and exudes a bitter latex when injured. The leaves are thick, dark green, in whorls of 4–8, obovate to oblanceolate, narrow at the base, entire, coriaceous, rounded

FIGURE 2.7   *Alstonia scholaris.*

or bluntly acuminate at apex, shining above and pale beneath; the petioles are 6–12 mm long. The flowers are compact, umbellately branched and pubescent, with panicled cymes; the bracts are leafy; bractioles are minute; pedicels are very short; calyx is 1.5–2 mm long and pubescent. The corolla is greenish-white and pubescent outside; the lobes are spreadingly ovate–obtuse, 2–3 mm long. The ovaries are distinct, with slender 20–50 cm long follicles and are terrate and pendulous. The seeds are oblong, 6–8 mm long, and flattened with a tuft of brownish hair at either end (review: Khyade et al. 2014).

Traditional Medicinal Uses: *A. scholaris* is extensively used in traditional medicine in Thailand, Malaysia, Philippines, China, Africa, Australia, etc. Different organs of the *A. scholaris* are used in traditional medicine for the treatment of malaria, jaundice, gastrointestinal troubles, cancer, etc. (review: Khyade et al. 2014).

Pharmacology

Antidiabetes: Several *in vitro* and *in vivo* studies have shown the anti-DM potential of the leaves and stem bark of *A. scholaris* in traditional and folklore medicine. The plant also has several other beneficial pharmacological properties (review: Khyade et al. 2014). The ethanol extract of the leaves of *A. scholaris* (100, 200, and 499 mg/kg, p.o., daily for 6 weeks) reduced the blood glucose level, glycosylated hemoglobin, and LPO, whereas the extract increased body weight, liver and muscle glycogen, and antioxidant status in streptozotocin-induced diabetic rats. The antidiabetic effect was sustained from 1 week onward till the end of the study (6 weeks). The histopathology of the pancreas revealed that the pancreatic β-cell damage with streptozotocin did not reverse in any of the treatment groups (Arulmozhi et al. 2010). The water extract of the *A. scholaris* bark reduced elevated blood glucose level in streptozotocin-induced diabetic rats without showing any hypoglycemic effect in normal rats. The antidiabetic effect of the extract could be due to increased utilization of glucose by peripheral tissues, improved sensitivity of target tissues for insulin, or improved metabolic regulation of glucose. The bark (150 and 300 mg/kg, daily for weeks) reduced fasting blood glucose, glycosylated hemoglobin, and TG levels in streptozotocin-induced diabetic rats. Thus, the bark of *A. scholaris* possesses antidiabetic and antihyperlipidemic effects in diabetic rats (Bandawane et al. 2011). The antidiabetic potential of *A. scholaris* was partly due to its promising α-glucosidase inhibitory effects (Anurakkun et al. 2007).

Clinical Human Study: The powder of *A. scholaris* leaves exerted a consistent hypoglycemic effect in patients suffering from NIDDM. The hypoglycemic effect *A. scholaris* leaf powder in patients suffering from NIDDM was attributed to their insulin triggering and direct insulin-like actions (Akhtar and Bano 2002).

Other Activities: Other reported activities include antimalarial activity, antimicrobial activity, antioxidant activity, analgesic action, anti-inflammatory activity, anticancer and cytotoxic activities, radioprotective effect, effect on central nervous system, immune-stimulating activity, and antifertility activity (review: Khyade et al. 2014).

Toxicity: Although this plant is used in traditional medicine, caution is required when utilizing this plant for health care in view of the reported toxicity including reproductive toxicity and teratogenicity. The hydroalcoholic extract of this plant bark, above 240 mg/kg, damaged all major organs in rats. The ethanol extract of the leaf was also found to be teratogenic. The bark extract prepared in the summer was found to be more toxic compared to that obtained in the rainy season (Khyade et al. 2014). Detailed toxicity studies are urgently warranted for the safe traditional use of this plant.

Phytochemicals: A wide range of phytochemicals have been reported in different parts of *A. scholaris*. Alkaloids, phlobatannins, simple phenolics, steroids, saponins, and tannins have been reported in all parts of *A. scholaris*. Iridoids, coumarins, and flavonoids in the leaf and terpenoids in the bark and root have also been reported. Alkaloids are the most important class of compounds found in this plant species. Over 70 different types of alkaloids have been reported in different parts such as the root, stem bark, leaves, fruit, and flower. A majority of the alkaloids are found in the leaves (review: Khyade et al. 2014).

Althaea officinalis L., Malvaceae

Common Name: Marshmallow

Intraperitoneal administration of a polysaccharide (althaeamucilage-*O*) isolated from the root of the African plant *Althaea officinalis* to nondiabetic mice resulted in significant reduction in the levels of blood glucose (Tomoda et al. 1987). *A. officinalis* is reported to contain scopoletin and many other phytochemicals (review: Al-Snafi 2013). Scopoletin (7-hydroxy-6-methoxy coumarin) is therapeutically evaluated in rats for hyperthyroidism, LPO, and hyperglycemia. Scopoletin (1.00 mg/kg, p.o.) administered daily for 7 days decreased the levels of serum thyroid hormones and glucose as well as hepatic glucose-6-phosphatase activity. Scopoletin also inhibited hepatic LPO and promoted antioxidant activity, superoxide dismutase, and catalase (Panda and Kar 2006).

Amaranthus caudatus L., Amaranthaceae

The methanol extracts of *Amaranthus caudatus*, *A. spinosus*, and *A. viridis* showed significant antidiabetic and anticholesterolemic activity in streptozotocin-induced diabetic rats, which provides the scientific proof for their traditional claim (Girija et al. 2011).

Amaranthus spinosus L., Amaranthaceae

Amaranthus spinosus is used in the Indian traditional system of medicine for the treatment of diabetes.

The methanol extract of the stem of *A. spinosus*, when administered for 15 days, significantly controlled blood glucose levels in streptozotocin-induced diabetic rats. Further, the methanol extract showed significant antihyperlipidemic and spermatogenic effects in streptozotocin-induced diabetic rats. The extract increased the sperm count and accessory sex organ weights (Sangameswaran and Jayakar 2008). In alloxan-induced diabetic rats also, the methanol stem extract of *A. spinosus* at doses of 250 and 500 mg/kg daily for 15 days showed antihyperglycemic and antihyperlipidemic activities (Balakrishnan and Pandhare 2010). The methanol extract of *A. spinosus* inhibited α-amylase activity under *in vitro* conditions (IC_{50} 46.02 μg/mL). *In vivo*, in alloxan-induced diabetic rats, the extract (200 and 400 mg/kg for 15 days) showed promising antioxidant activity and decreased blood glucose levels. This study provides evidence that the methanol extract of *A. spinosus* has potent α-amylase inhibitory, antidiabetic, and antioxidant activities (Ashokkumar et al. 2011).

The effect of the leaf extract of *A. spinosus* on the histoarchitecture of the pancreas, kidney, and liver of streptozotocin–nicotinamide-induced diabetic rats was studied. Administration of the ethanol extract (250 and 500 mg/kg for 21 days) to diabetic rats caused a significant reduction in blood glucose and improvement in antioxidant parameters compared to diabetic control rats. Degenerative changes of pancreatic cells in diabetic rats were minimized to near normal morphology by the treatment (Mishra et al. 2012).

In another study, the ethanol extract of *A. spinosus* leaves was administered (150, 300, and 450 mg/kg) to type 1 and type 2 diabetic rats. Standard drugs, glibenclamide and metformin, were used as a positive control for comparison. Changes in carbohydrate and lipid metabolism and antioxidants were assessed and compared with control and standard drug-treated animals. Higher doses of the extract significantly decreased plasma glucose levels and hepatic glucose-6-phophatase activity and increased the hepatic glycogen content with a concurrent increase in hexokinase activity in both type 1 and 2 diabetic rats. It also significantly lowered the plasma and hepatic lipids, urea, creatinine, and LPO with an improvement in the antioxidant profiles of both type 1 and type 2 diabetic rats. It is concluded that *A. spinosus* has potential antidiabetic activity and significantly improves disrupted metabolisms and antioxidant defense in type 1 and type 2 diabetic rats (Bavarva and Narasimhacharya 2013).

Amaranthus viridis L., Amaranthaceae

The methanol extract of whole *Amaranthus viridis* plant (200 and 400 mg/kg, p.o., daily, for 15 days) showed significant reduction in blood glucose and lipid profiles and significant improvement in antioxidant parameters in alloxan-induced diabetic rats when compared to diabetic control group. *In vitro* α-amylase inhibition activity of the extract was also observed (Ashok Kumar et al. 2012). In another study, the methanol extract of the levels of *A. viridis* (200 and 400 mg/kg, p.o., daily, for 21 days) showed significant reduction in blood glucose, total cholesterol, and TGs and increase in body weight and HDL levels in streptozotocin-induced diabetic rats when compared to diabetic control rats. Further, histologically, focal necrosis observed in the diabetic rat pancreas was less obvious in treated groups (Krishnamurthy et al. 2011). The water extract of the *A. viridis* stem (100, 200, and 400 mg/kg, p.o., for 30 days) also exhibited dose-dependent antihyperglycemic and antihyperlipidemic effects in streptozotocin-induced diabetic rats. The significant control of serum lipids levels in the extract treated diabetic rats may be directly attributed to improvement in glycemic control (Pandhare et al. 2012). The water extract of the aerial parts of *A. viridis* also showed hypoglycemic and hypolipidemic effects in streptozotocin-induced diabetic rats (Patel et al. 2012b). In a recent study, oral administration of the ethanol extract of *A. viridis* (400 and 500 mg/kg daily for 10 days) resulted in a significant decrease in serum glucose, TG, total cholesterol, LDL, and very-low-density lipoprotein (VLDL) and significant increase in body weight, HDL, liver glycogen, and muscle glycogen levels in dexamethasone-induced type 2 diabetic rats (Paramesha et al. 2014).

The studies indicate that water, methanol, and ethanol extracts of the whole plant, stem, and leaves show antidiabetic properties. This suggests the involvement of more than one active principle and mechanism of action. Further studies are warranted to address these points.

Ambrosia maritime L., Compositae

The aerial parts of *Ambrosia maritime* are used in Egypt in traditional medicine to control diabetes and are reported to possess antidiabetic properties (review: Haddad et al. 2005). In a recent study, the antidiabetic, hypolipidemic, and antioxidant effects of the water extract of *A. maritime* were shown in the alloxan-induced diabetic male albino rats (Abu-amra et al. 2014).

Ammi visnaga (L.) Lam., Apiaceae

Synonym: *Daucus visnaga* L.

Common Name: Khella, bishop's weed

The effect of the aqueous extract of *Ammi visnaga* on blood glucose levels was investigated in fasting normal and streptozotocin-induced diabetic rats after single and repeated oral administration. The water extract (20 mg/kg) significantly reduced blood glucose in normal rats 6 h after a single oral administration and 9 days after repeated oral administration. This hypoglycemic effect was more pronounced in streptozotocin-induced diabetic rats (Jouad et al. 2002b).

Ammoides pusilla (Brot.) Breistr., Apiaceae

Ammoides pusilla is used as an antidiabetic medication in Eastern Morocco. The water extract (250 mg/kg) of *A. pusilla* improved both oral and intravenous glucose tolerance in rats. Further, the water extract induced a significant inhibition of jejunal glucose absorption in rats. Preliminary toxicity evaluation did not show any adverse effects (Bnouham et al. 2007).

Amomum villosum var. *xanthioides* (Wall. ex Baker) T.L.Wu & S.J.Chen., Zingiberaceae

Synonym: *Amomum xanthioides* Wall.

Common Name: Black cardamom

Amomi semen (seeds of *Amomum xanthioides*/*Amomum villosum*) has been used as a folk remedy for the treatment of diabetes in Korea. *A. xanthioides* extract exerted anti-DM activity against alloxan-induced DM through the suppression of NF-κB activation (Park and Park 2001). The aqueous ethanol extract (0.5 mg/mL) of *A. xanthioides* stimulated glucose uptake in 3T3-L1 adipocytes. The extract (0.02–0.5 mg/mL) potentiated insulin-stimulated glucose uptake with a dose-dependent manner. The results suggest that the antidiabetic action of Amomi may be mediated through the stimulation of glucose uptake and the potentiation of insulin action (Kang and Kim 2004).

Amorpha fruticosa L., Fabaceae

Amorpha fruticosa (false indigo bush) is a traditional medicinal shrub used to treat hypertension, contusions, etc., in certain parts of Asia. Amorfrutins, present in the fruits of this plant, activated PPAR-γ and are potent natural antidiabetic dietary products (Weidner et al. 2012; Wang et al. 2014).

Amorphophallus konjac K. Koch., Araceae

Amorphophallus konjac has long been used in China, Japan, and Southeast Asia as a food source and as a traditional medicine. It is commonly known as elephant yam, voodoo lily, snake plant, dragon plant, etc. Glucomannan is a sugar made from the root of the konjac plant. A randomized, controlled metabolic trial concluded that konjac-mannan fiber added to conventional treatment may ameliorate glycemic control, blood lipid profile, and systolic blood pressure in high-risk diabetic individuals, possibly improving the effectiveness of conventional treatment in type 2 DM (Vuksan et al. 1999). Three oligosaccharide fractions from the root of *A. konjac* were isolated and studied using streptozotocin-treated diabetic rats. At concentrations less than 1.5 mM, one of the fractions, KOS-A, positively decreased streptozotocin-induced NO(*) level of islets, but normal NO(*) release for non-streptozotocin-treated islets was not affected within the range. At 15 mM, KOS-A played a contrary role and increased NO(*) level for islets both with and without streptozotocin treatment. Islet insulin secretion changed corresponding to the NO(*) level in the assay. Increased insulin secretion appeared parallel to the decrease of NO(*), and normal insulin release was not affected by KOS-A less than 1.5 mM. KOS-A is a tetrasaccharide with MW of 666 Da and a reductive end of α-D-mannose. The NO(*) attenuation function of KOS-A on the diabetes model mainly resulted from environmental free radical scavenging by the oligosaccharide. Present results also imply that the mechanism of clinical *A. konjac* hypoglycemic function may be connected with free radical attenuation and lower risks of islet damage from the NO(*) radical (Lu et al. 2002).

Amorphophallus paeoniifolius (Dennst.) Nicolson, Araceae

Amorphophallus paeoniifolius is cultivated for its stem, which is a tuber and is edible. It is commonly known as elephant foot yam or whitespot giant arum. It is also used in traditional medicine in India. Administration of the acetone extract of elephant foot yam (*A. paeoniifolius*) in the diet (0.1% and 0.25% for 12 weeks) to streptozotocin-induced diabetic rats resulted in concentration-dependent amelioration of the diabetes status, which was observed with respect to water intake, diet, urine output, body weight gain, fasting blood glucose, and glomerular filtration rate. The study indicated that the acetone extract of elephant foot yam is an effective anti-DM agent in streptozotocin-induced diabetic rats (Arva et al. 2013).

Ampelodesma mauritanica Durand & Schinz, Poaceae

The methanol extract of *Ampelodesma mauritanica* roots significantly reduced blood glucose levels in normal glucose-fed hyperglycemic mice. The root extract contained antioxidant phenolics and flavonoids (Djilani et al. 2011).

Anabasis articulata Forssk. Moq., Chenopodiaceae

A decoction of *Anabasis articulata* leaves is widely used by Algerian traditional medicine practitioners as a remedy for the treatment of diabetes. Experiments were performed in nondiabetic mice and in hyperglycemic mice (glucose-treated and alloxan-treated mice) to confirm the antidiabetic potential of the *A. articulata* water extract. Administration of the water extract (400 mg/kg, p.o.) in normoglycemic mice decreased the glycemia by 29.89% after 6 h. This dose also lowered blood glucose concentrations in diabetic mice revealing the antihyperglycemic effect of *A. articulata* leaves. Phytochemical screening showed that the water extract contained alkaloids (1.25%) and saponin (1.30%). Saponin (5 mg/kg) fraction restored the normal blood glucose levels after 21 days of treatment. The alkaloid fraction did not significantly reduce the blood glucose level. The present study confirms the antidiabetic proprieties of *A. articulata* leaves previously reported by Algerian healers (Kambouche et al. 2009).

The saponin fraction of ethanol extract of the aerial parts of *A. articulata* (400 mg/kg) was studied for its effects against diabetes complication-induced tissue injury in streptozotocin-induced diabetic rats. Oral administration of the fraction controlled the increase in blood glucose and cortisol levels. Further, the treatment increased blood insulin and α-fetoprotein levels and decreased the levels of tumor necrosis factor-α (TNF-α) and fructose amine. Besides, the plant extract effectively modulated hepatic damage induced by oxidative stress. The study suggests that *A. articulata* exerts multibeneficial actions in controlling diabetes and its complications in the pancreas and liver (Metwally et al. 2012).

Anacardium occidentale L., Anacardiaceae (Figure 2.8)

Common Names: Cashew nut tree (Eng.), kaju (Hindi), mundiri (Tamil), etc.

Description: *Anacardium occidentale* is an erect, spreading evergreen tree; its leaves are simple, petiolated, alternate with an entire leaf margin. Small flowers are produced in terminal panicles, and a greenish-gray nut is produced in the yellow or scarlet fleshy peduncle.

Distribution: *A. occidentale* is a native of tropical America and is introduced to India and elsewhere from Brazil and naturalized.

Traditional Medicinal Uses: *A. occidentale* is an astringent, diuretic, caustic, purgative, intoxicant, aphrodisiac, anthelmintic, rubefacient, vermifuge, demulcent, and emollient used in treating diabetes, asthma, diarrhea, burns, ascites, vomiting, catarrh, common cold, constipation, cough, dyspepsia, dysentery, inflammation, hypertension, leprosy, sore throat, piles, ringworms, snake bite, toothache, sore gums, ulcers, leukoderma, skin diseases, stomachache, warts, tumor, congestion, caries, fever, freckles, nausea, stomatitis, and rash (Kirtikar and Basu 1975; review: Ajikumaran and Subramoniam 2005).

Pharmacology

Antidiabetes: The ethanol extract of dried nuts and stem bark extracts of the plant showed hypoglycemic activity in both normal and streptozotocin-induced diabetic rats (Arul et al. 2004). The methanol extract of the stem bark of *A. occidentale* showed antihyperglycemic and antioxidant effects and improvement in plasma lipids in fructose-induced diabetic rats (Olatunji et al. 2005). The plant bark water extract protected streptozotocin-induced diabetic rats (Kamtchouing et al. 1998). Stigmast-4-en-3-one, an isolate from the bark of the plant, showed glucose-lowering activity in normal dogs (Alexander-lindo et al. 2004). Hydroethanol extracts of cashew seed and its active component, anacardic acid, stimulated glucose uptake into C2C12 myotubes in a concentration-dependent manner. Extracts of the leaves, bark, and apple of the plant were inactive (Tedong et al. 2010). The extract of the seed as well as anacardic acid activated AMPK in the myotubes after 6 h of incubation. No significant effect was observed on Akt and IR phosphorylation. Both the extract and the compound exerted uncoupling of succinate-stimulated respiration in rat liver mitochondria. Activation of AMPK is likely to increase the number of plasma membrane glucose transporters resulting in elevated glucose uptake. The authors suggest that the extract could be a potential antidiabetic nutraceutical (Tedong et al. 2010). The leaf extract also reduced blood glucose levels in a dose-dependent manner in streptozotocin-induced diabetic rats (Sokeng et al. 2007a).

Other Activities: Tannins from the bark of *A. occidentale* showed anti-inflammatory actions. The methanol extract of the stem bark inhibited lipopolysaccharide-induced microvascular permeability (Olajide et al. 2004). The kernel oil of the cashew nut has an antioxidant property. The plant bark as well as

FIGURE 2.8 *Anacardium occidentale* (twig with fruit).

cashew apple has antibacterial activity (Akinpelu 2001). The bark extract also inhibited *Leishmania braziliensis* infection in an *in vitro* model (Franca et al. 1993).

Phytochemicals: Phytochemicals isolated from this plant include anacardic acid, cardol, cardanol, terpenes, carboxylic acids, lactones, norisophenols, methional, (Z)-1,5-octadin-3-one, nonenal, damascenone, decalactone tannins, and stigmast-4-en-3-one (Bicalho and Rezende 2001; Alexander-lindo et al. 2004).

Andrographis paniculata (Burm. f.) Nees, Acanthaceae

Common Names: Creat (Eng.), charayetah (Hindi), nilavembu (Tamil), etc.

Andrographis paniculata is present throughout India. In traditional medicine, it is utilized as a tonic alternative, anthelmintic, anodyne, astringent, laxative, and stomachic and is used in the treatment of snake venom, bug bite, cholera, diabetes, colic, diarrhea, dog bite, dysentery, fever, malaria, and rabies (Kirtikar and Basu 1975; review: Ajikumaran and Subramoniam 2005).

Antidiabetes: The antihyperglycemic and antidiabetic properties of this plant (alcohol extract) and the isolate andrographolide have been shown in streptozotocin-induced diabetic rats (Yu et al. 2003). Mediation of β-endorphin in andrographolide-induced plasma glucose-lowering action in type 1 diabetes-like animals has been reported (Yu et al. 2008). *A. paniculata* or its active compound andrographolide showed hypoglycemic and hypolipidemic effects in high-fat, high-fructose fed rats also (Nugroho et al. 2012). In a recent study, andrographolide significantly decreased the levels of blood glucose and improved diabetic rat islet β-cells in streptozotocin-induced type 2 diabetic rats. The treatment could restore the decrease of pancreatic insulin content in diabetic rats (Nugroho et al. 2014).

Other Activities: *A. paniculata* is a well-known hepatoprotective agent. This plant extract has an antioxidant property. Andrographolide inhibited interferon-γ (IFN-γ) and IL-2 cytokine production in mouse T-cells and protected against cell apoptosis. Andrographolide interfered with T-cell activation and reduced experimental autoimmune encephalomyelitis in mouse. *A. paniculata* has anti-inflammatory and immunomodulatory activities (Ji et al. 2005). Neoandrographolide suppressed NO production in activated macrophages *in vitro* and *ex vivo*. Andrographolide showed anticancer activity on diverse cancer cells. Xanthones isolated from the roots of this plant have antimalarial activity. The plant has antimicrobial and psychopharmacological activities too. The plant extract induced relaxation of the uterus by blocking voltage-operated calcium channels and inhibiting Ca^{+2} influx. The water extract of the plant has cardiovascular activities including hypotensive activity. Andrographolide has male reproductive toxicity/male antifertility property in albino rats. It has testicular toxicity. Further, the plant extract influenced progesterone levels in the blood plasma of pregnant rats (review: Ajikumaran and Subramoniam 2005).

Isolated phytochemicals include andrograpanin, andrographolide (main labdane diterpene), 14-deoxy-11,12-didehydroandrographolide and neoandrographolide, bis-andrographolide ether xanthones, and flavonoids (review: Ajikumaran and Subramoniam 2005).

Andrographis stenophylla C.B. Clarke, Acanthaceae

The leaf extract of *Andrographis stenophylla* exhibited hypoglycemic activity in rats (Parasuraman et al. 2010).

Anemarrhena asphodeloides Bunge, Liliaceae (Figure 2.9)

Common Names: Anemarrhena (Eng.), zhi mu (Chinese) (the dried rhizome of the plant is known as Rhizoma Anemarrhenae)

Description: Anemarrhena is a glabrous perennial herb. Leaves grow out from the base and are tufted and slender lanceolate and 33–66 cm long. Stems grow out from the foliage, are upright, cylindrical, and racemose with lavender flowers. Fruit are oblong with many black seeds inside. Roots grow horizontally in the ground and are slightly flat round; the upper surface densely grows golden color long hair. The rhizome grows many brown fibrous residues at the leaf base, and the lower part grows many fleshy fibrous roots. Leaves grow from the base and are linear, and the base often expands into

FIGURE 2.9 *Anemarrhena asphodeloides.*

a sheathlike form, with multiple parallel veins, and has no obvious midrib. Scapes grow erect and unbranched, on which pointed caudate bracts grow out, with sparse and narrow spikes; the flowers often grow in clusters of 2–3 and grow on the top and become a spike; the pedicels are sessile or very short, about 3 mm; there are six persistent, oblong perianth segments, arranged in two rings, three purple vertical veins, and pinkish-red, lilac, or white lower petals; stamens 3, shorter than the perianth, adnate to the middle of the inner wheel tepals, filaments very short, with a T-shape; ovary superior, three rooms, stylus long 2 mm. Capsule oblong, six vertical edges, 10–15 mm long, cracking above the ventral slit when mature, each room contains 1–2 seeds, seeds prismatoidal shape, pointed at both ends, color black.

Distribution: Mongolia to Korea

Traditional Medicinal Uses: The rhizome of this plant is used as a traditional oriental medicine for diabetes, inflammation, fever, depression, etc.

Pharmacology

Antidiabetes: The antidiabetes activity of the rhizome of this plant has been reported (Suzuki and Kimura 1984). A hot water extract of the rhizomes of this plant lowered the blood glucose levels in alloxan-induced diabetic rats. Hypoglycemic activity-guided isolation of active principle resulted in the identification of a new glycoside, pseudoprototimosaponin AIII. This compound exhibited hypoglycemic effects in a dose-dependent manner in streptozotocin-induced diabetic rats. It did not show any effect on glucose uptake and insulin release. The authors suggest that the hypoglycemic mechanism may be due to the inhibition of hepatic gluconeogenesis and/or glycogenolysis (Nakashima et al. 1993).

In another study, the water extract of the rhizome decreased blood glucose levels in KK-Ay mice, a model of obese genetic type 2 diabetes. The water extract (90 mg/kg) reduced blood glucose levels from about 31.6 to 22.3 mmol/L (570–401 mg/dL) 7 h after oral administration in KK-Ay mice. The treatment also showed marginal decrease in serum insulin levels in KK-Ay mice. The anti-DM mechanism of the rhizome extract appears to be reducing insulin resistance. Interestingly, the active components of the extract were confirmed to be mangiferin and its glucoside (Miura et al. 2001a). Mangiferin is an anti-DM compound present in many plants (see *Mangifera indica* L.).

The ethanol extract of this plant stimulated insulin secretion from isolated islet of healthy rats and diabetic Goto-Kakizaki rats (Hoa et al. 2004). A recent study suggests that the plant extract may prevent complications of diabetes also. Administration of the rhizome of *Anemarrhena asphodeloides* (60% ethanol extract) to streptozotocin-induced diabetic rats resulted in prevention of diabetic ophthalmopathy. The treatment increased the activities of serum SOD and GSH-Px, whereas it decreased the levels of MDA and advanced glycosylation end products (AGEs) in serum and sorbitol in the lens in diabetic rats.

The pathological changes of the lens and retina were alleviated by the extract treatment. Moreover, the subnormal growth of retinal pericytes induced by high glucose was ameliorated by mangiferin and neo-mangiferin, the two major constituents of the extract (Li et al. 2013).

Interestingly, the water extract of the rhizome of *A. asphodeloides* stimulated GLP-1 secretion in human enteroendocrine L-cell line (NCI-H716) in culture. This study demonstrated that the extract possibly stimulates GLP-1 secretion under *in vivo* conditions (Kim et al. 2013). Follow-up *in vivo* study is needed. From the aforementioned reports, it may be concluded that the plant rhizome has a few anti-DM compounds acting via different mechanisms.

Other Activities: Timosaponin AIII isolated from *A. asphodeloides* ameliorated learning and memory deficits in scopolamine-treated mice (Lee et al. 2010a). Treatment with the compound counteracted the increase in TNF-α and IL-1β induced by scopolamine administration. It also inhibited the activation of NF-κB (*nuclear factor* kappa-light-chain-enhancer of activated B cells) signaling in BV-2 microglia and in SK-N-SH neuroblastoma cells induced by TNF-α or scopolamine (Lee et al. 2009).

Phytochemicals: The rhizomes mainly contain saponins (timosaponin A-I, timosaponin A-II, timosaponin A-III, timosaponin A-IV, timosaponin B-I, timosaponin B-II); timosaponin B-I is prototimosaponin A-III. The rhizome also contains amemarsaponin A2, or marlogenin-3-O-β-D-glucopyranosyl (1 $\rightarrow$ 2)-β-D-galactopyranoside B, desgalactotigonin, F-gitonin, pseudoprototimosaponin A-III, and smilageninoside. The sapogenin found in the rhizome is known as sarsasapogenins; its concentration in the dry rhizome is about 0.5%; in addition it contains markogenin and neogitogenin. The rhizomes also contain a large amount of reducing sugars, mucilage, tannic acids, fatty acids, etc.; it also contains four kinds of polysaccharides known as anemaran A, B, C, and D; other compounds reported include cis-hinokiresinol, monomethyl-cis-hinokiresinol, oxy-cis-himokiresinol, 2,6,4′-trihydroxy-4-methoxy benzophenone, p-hydroxyphenyl crotonic acid, pentacosyl vinyl ester, β-sitosterol, and mangiferin (0.5%). The aboveground part contains mangiferin and isomangiferin, the leaves contain mangiferin 0.7%, and the flowers contain saponins.

Anethum graveolens L., Umbelliferae

Anethum graveolens is an Indian traditional herb. The leaves and seeds of this plant are known for their medicinal properties. The hydroalcoholic extract of this plant seed showed a glucose-lowering effect in diabetic rats (Madani et al. 2005). An investigation was made to evaluate the role of *A. graveolens* leaf extract in the regulation of corticosteroid-induced type 2 DM in female rats. In dexamethasone-treated animals (1 mg/kg for 22 days), an increase in serum concentration of insulin and glucose and in hepatic LPO was observed. However, there was a decrease in serum concentration of thyroid hormones and in the endogenous antioxidant enzymes, such as superoxide dismutase and catalase, and reduced GSH in the liver. The leaf extract (100 mg/kg/day) treatment for 15 days to dexamethasone-treated diabetic animals resulted in a decrease in the concentration of both serum glucose and insulin, indicating the potential of the plant extract to regulate corticosteroid-induced diabetes. Dexamethasone-induced alterations in the levels of thyroid hormones as well as in hepatic antioxidant enzymes and GSH were also reversed by the plant extract (Panda 2008). In another study, the hypoglycemic and hematopoietic potential of the seed extracts of *A. graveolens* was investigated. Control and alloxan-induced diabetic mice were orally treated once per day with water and ethanol extract, respectively, for 15 days, and the effect of both extracts on body weight, organ weight, and blood glucose level was determined. A significant decrease in blood glucose level after administration of the water extract was observed along with a significant increase in body and organ weights in alloxan-induced diabetic mice. The water extract, but not the ethanol extract, increased red blood cell count, hemoglobin, and mean corpuscular hemoglobin content when compared with control. The water and ethanol extracts showed beneficial effects on hematological parameters (Mishra 2013).

Clinical Studies: In a randomized, double-blind, placebo-controlled clinical trial, the effects of *A. graveolens* supplementation on proinflammatory biomarkers (IL-6, TNF-α) and acute phase protein high-sensitivity C-reactive protein (hs-CRP) concentrations in type 2 diabetic patients were assessed.

The intervention group received 3.3 g/day *A. graveolens* powder for 8 weeks, and the placebo group received the same amounts of starch. At the onset and end of the study, 5 mL blood samples were collected. The serum concentration of IL-6, TNF-α, and hs-CRP decreased at the end of the study in the intervention group. Further clinical studies are needed to confirm the results (Payahoo et al. 2014). A double-blind, randomized, placebo-controlled clinical study was designed to investigate whether *A. graveolens* extract could improve metabolic components in patients with metabolic syndrome. In this small clinical study, 12 weeks of the extract treatment had a beneficial effect in terms of reducing TG from baseline. However, the treatment was not associated with a significant improvement in metabolic syndrome-related markers compared to the control group. Larger studies might be required to prove the efficacy and safety of long-term administration of *A. graveolens* to resolve metabolic syndrome components (Mansouri et al. 2012).

Angelica acutiloba (Siebold and Zucc.) Kitag., Apiaceae
Administration of this plant root (flavonoid fraction) at a dose of 50–200 mg/kg (p.o.) to streptozotocin-induced diabetic rats ameliorated glycation-mediated renal damage (Liu et al. 2011).

Angelica sinensis (Oliv.) Diels, Apiaceae
Synonym: *Angelica polymorpha* Maxim. var. *sinensis* Oliv.
Angelica sinensis is a traditional anti-DM medicinal plant. The polysaccharide purified from the fresh roots of *A. sinensis* (oral administration, daily for 4 weeks) reduced fasting blood glucose levels and improved blood insulin levels in prediabetic and streptozotocin-induced diabetic Balb/c mice. Further, the treatment improved insulin resistance in the prediabetic mice. Besides, hepatic and muscle glycogen concentrations were increased, and insulin resistance-related inflammatory factors, IL-6 and TNF-α, in serum were reduced in streptozotocin-induced diabetic mice by the treatment. Histopathological examination indicated that the impaired pancreatic, hepatic, and adipose tissues were effectively restored in the diabetic and prediabetic mice (Wang et al. 2015).

Angylocalyx pynaertii De Wild. Fabaceae
Common Name: Oko aja
Alkaloids from the pods of *Angylocalyx pynaertii* (West African plant) inhibited α-glucosidase activity (Yasuda et al. 2002).

Anisodus tanguticus Pascher, Solanaceae
Synonym: *Scopolia tangutica* Maximowicz
Anisodus tanguticus is used for the treatment of type 2 diabetes by Chinese doctors. It is said to be quiet effective in improving complications while lowering blood glucose (review: Chauhan et al. 2010). However, controlled scientific studies remain to be done to establish its utility.

Anisopus mannii N.E.Br., Apocynaceae
Anisopus mannii is a familiar herb in traditional medicinal preparations in northern Nigeria, where a decoction of the whole plant is used as a remedy for diabetes, diarrhea, and pile. The water extract of the *A. mannii* leaf (100, 200, and 400 mg/kg) showed potent hypoglycemic activity in alloxan-induced diabetic rats. This study confirmed the traditional use of these Nigerian medicinal plants in diabetes treatment. These plants showed high potential for further investigation to novel antidiabetic drugs (Manosroi et al. 2011). The hypoglycemic activities of different fractions from the methanolic leaf crude extract of *A. mannii* were investigated in normoglycemic and alloxan-induced diabetic mice. One of the fractions (which contains saponin) at 250 mg/kg significantly reduced the PBG (postprandial blood glucose) levels in a postprandial study. The hypoglycemic effect of this plant was eliminated when coadministered with isosorbide dinitrate or nifedipine indicating the induction of insulin secretion via K+ ATP-dependent channels. This study confirmed the traditional use of *A. mannii* for DM and the potential for the further development as a novel hypoglycemic drug (Zaruwa et al. 2013).

FIGURE 2.10   *Annona muricata* (twig with fruit).

Annona muricata L., Annonaceae

Synonym: *Annona muricata* Vell. (Figure 2.10)

Local Names: Soursop, graviola, guanábana, sirsak, paw-paw, etc.

Distribution: *Annona muricata* is native to tropical Central and South America and the Caribbean but is now widely cultivated in tropical areas worldwide. The fruit of this plant is found to be edible and its fruits are used commercially for the production of juice, candy, and sherbets. It has occasionally escaped cultivation and became naturalized (Morton 1987; review: Gajalakshmi et al. 2012).

Description: *A. muricata* is a slender, small, and cold-intolerant tree, generally reaching heights of 4–6 m (13–20 ft); it flowers and bears fruit 3–5 years after planting. Leaves are glossy, dark green, and generally evergreen, with a distinctive odor. Its fruits are the largest of the *Annona* species, 20–30 cm long, with exceptional fruits as large as 7 kg (Flora of Pakistan 2011). The fruits have green inedible skins, with many soft, curved spines, but the white, juicy pulp is edible. The fruit gives a flavor of custard when it is in ripened condition and hence has a pleasant, distinctive aroma and fibrous pulp that can be consumed. It has a sour or musky flavor and is eaten fresh or used for making juices and other beverages, as well as custards and sherbets and other dishes in South America, the Caribbean, and Indonesia (Morton 1987; Gajalakshmi et al. 2012).

Traditional Medicinal Uses: Extracts from various morphological parts of *A. muricata* are widely used medicinally in many parts of the world for the management, control and/or treatment of a plethora of human ailments. The leaves of *A. muricata* are used in Cameroon and elsewhere to manage diabetes and its complications. It is used as a nutritional medicinal supplement. The leaves are used as a strong diuretic, tonic, antiparasitic, sedative, and antispasmodic; it is also used for liver problems, high blood pressure, neuralgia, rheumatism, arthritis pain, dysentery, intestinal colic, difficult child birth, asthma, catarrh, internal ulcers, malaria, and diabetes. The seeds are used to kill parasites and as a carminative. The fruits are used for dysentery, mouth sores, fever, parasites, and diarrhea. Its bark is used as a sedative and antispasmodic and is used to treat diabetes and hypertension (Gajalakshmi et al. 2012; review: Kedari and Khan 2014).

Pharmacology

Antidiabetes: Many studies have confirmed the effects of the methanol extract of *A. muricata* leaves on glycemic control in streptozotocin-induced diabetic rat (review: Sawant and Gogle 2014). The methanol extract of *A. muricata* leaf (100 mg/kg, i.p., daily for 2 weeks) reduced the blood glucose level of streptozotocin-induced diabetic rats. The treatment increased body weight compared to untreated diabetic control rats (Adeyemi et al. 2009). *A. muricata* leaf (water extract) treatment has beneficial effects on pancreatic tissues subjected to streptozotocin-induced oxidative stress by directly quenching lipid peroxides and indirectly enhancing the production of endogenous antioxidants. The extract protected and preserved pancreatic β-cell integrity as judged from histological observations (Adewole and Caxton-Martins 2006). In another study, the effects of the methanol extract of *A. muricata* leaves on the pancreatic islet cells of streptozotocin-induced diabetic rats were evaluated by histopathological studies. DM was experimentally induced in rats by a single intraperitoneal injection of 80 mg/kg streptozotocin. Daily intraperitoneal injection of 100 mg/kg methanol extract to diabetic rats for 2 weeks resulted in restoration of the histopathological changes to the normal state. The regeneration of the β-cells of the streptozotocin-destructed islets is probably due to the fact that the pancreas contains stable (quiescent) cells that have the capacity to regenerate. Therefore, the surviving cells can proliferate to replace the lost cells. Treatment with the extract also showed a significant antihyperglycemic activity in diabetic rats at the end of the experiment (Adeyemi et al. 2007). Treatment with the water extract of the leaf of this plant (100 mg/kg/day, p.o.) for 4 consecutive weeks resulted in improvement in liver antioxidant enzymes and lipid profiles and decrease in the levels of blood glucose and lipid peroxides. The treatment also increased serum insulin levels. These findings suggest that the *A. muricata* extract has a beneficial effect on hepatic tissues subjected to streptozotocin-induced oxidative stress, possibly by decreasing LPO and indirectly enhancing production of insulin and endogenous antioxidants (Adewole and Ojewole 2009).

Soursop leaf extract treatment decreased the blood glucose concentration of diabetic rats due to the regeneration/proliferation of pancreatic β-cells, making these leaves beneficial for diabetics (review: Sawant and Gogle 2014). In another study, the methanol extract of this plant seed also improved liver function in alloxan-induced diabetic rats (Agbai and Nwanegwo 2013). In a recent study, the hypoglycemic and antidiabetic activity of the ethanol extract of the *A. muricata* stem bark was demonstrated in alloxan-induced diabetic rats. The ethanol extract of *A. muricata* bark was administered daily at single doses of 150 and 300 mg/kg, p.o. to the diabetic rats for a period of 14 days. The treatment resulted in reduction in blood glucose levels, serum lipid profiles, total cholesterol, TGs, phospholipids, and LDL. Thus, ethanol extract of *A. muricata* stem bark is an antidiabetic agent (Ahalya et al. 2014). In another study, the effect of oral administration of *A. muricata* water extract (100 or 200 mg/kg) was studied in normal and streptozotocin-induced diabetic rats. The plant extract was not effective in normal rats. In diabetic rats, single administration of the extract reduced blood glucose levels by 75% and 58%, respectively, at the dose of 100 and 200 mg/kg as compared to the initial value. Treatment with the extract at the dose of 100 mg/kg (daily, p.o., for 2 weeks) significantly reduced blood glucose levels as compared with diabetic control rats. Immunohistochemical staining of pancreatic β-cells of diabetic rats treated with the repeated dose of 100 mg/kg expressed strong staining for β-cell compared to diabetic control. Daily administration of the extract for 28 days to diabetic rats reduced blood glucose levels, serum creatinine, MDA, AST, ALT activity, nitrite levels, and LDL-C. Further, total cholesterol and TG levels and SOD and CAT activities were restored. Thus, the extract exerted antioxidant, hypolipidemic, and β-cell protective actions (Florence et al. 2014).

The nutrients in soursop (*A. muricata*) leaves are believed to stabilize blood sugar levels in the normal range in humans. Moreover, the extracts of soursop leaves may be used as one of the natural diabetes remedies. In view of the aforementioned points, controlled clinical trials and safety evaluation on humans are warranted. The active principles involved remain to be identified.

Other Activities: Research on this plant has documented activities such as hepatoprotective, antinociceptive, anti-inflammatory, antirheumatic, anticancer, antiherpes simplex virus, antihyperlipidemia, antidepression, antioxidant, antihypertensive, and antimicrobial (review: Kedari and Khan 2014).

Phytochemicals: Phytochemical screening of the plants showed the presence of alkaloids, coumarins, flavonoids, glycosides, phenolic compounds, phytosterols, quinones, saponins, steroids, and terpenoids. Annonaceous acetogenins are a series of polyethers. Most acetogenins are white waxy derivatives of long-chain fatty acids (C32 or C34), and the terminal carboxylic acid is combined with a 2-propanol unit at the C-2 position to form a methyl-substituted α,β-unsaturated-γ-lactone. One of their interesting structural features is a single, adjacent, or nonadjacent tetrahydrofuran or tetrahydropyran system with one or two flanking hydroxyl group(s) at the center of a long hydrocarbon chain (review: Kedari and Khan 2014).

Annona squamosa L., Annonaceae

Common Names: Custard apple (Eng.), agrimakhya (Sanskrit), ati (Tamil), etc.

Annona squamosa is a small tree with oblong leaves. Its fruit is fleshy, tubercled, and areolate bearing many brownish, black seeds; the ripe fruit is edible. It occurs in naturalized condition in the Western Indian Peninsula and also in tropical America. In traditional medicine, it is a sedative, cathartic, stimulant, expectorant, abortifacient, purgative, astringent, diuretic, pectoral, stomachic, vermifuge, etc. It is also used to treat vomiting, dysentery, spinal diseases, diarrhea, ulcers, tumor, common cold, and rheumatism (Kirtikar and Basu 1975).

The water extract of the plant leaf showed antidiabetic activity in streptozotocin–nicotinamide-induced type 2 diabetic rats (Shirwaikar et al. 2004a). This anti-DM effect of the water extract was shown in streptozotocin-induced hyperglycemic rats also (Mujeeb et al. 2009). The hypoglycemic and antidiabetic effects of the ethanol extract of its leaves in experimental animals have also been reported (Watal et al. 2005). The alcohol extract of the leaves of *A. squamosa* has free radical scavenging potential (Shirwaikar et al. 2004b).

Ent-kauranes from the stems inhibited superoxide anion generation by human neutrophils and exhibited antiplatelet aggregation activity (Yang et al. 2002, 2004). Other reported activities include cytotoxicity to human cancer cell lines, antimicrobial activity, and anti-HIV activity (Wu et al. 1996; Hopp et al. 1997; Kotkar et al. 2002; Pardhasaradhi et al. 2005). A preparation from the seed extract is reported to have anti-head lice activity (Tiangda et al. 2000). The leaves contain neurotoxic benzyltetrahydroisoquinoline alkaloids (Caparros-Lefebvre and Elbaz 1999).

Phytochemicals: The phytochemicals found in *A. squamosa* include the following: acetogenins, liriodenine, moupinamide, -(–)-kauran-16α-ol-19-oic acid, 16β, 17-dihydroxy-(–)-kauran-19-oic acid, anonaine, 16α, 17-dihydroxy-(–)-kauran-19-oic acid, squamosamide, 16α-methoxy-(–)-kauran-19-oic acid, sachanoic acid, (–)-kauran-19-al-17-oic acid, daucosterol benzoquinazoline alkaloid, and *bis*-tetrahydrofuran acetogenins (Yang et al. 1992; Araya et al. 2002; Yu et al. 2005).

Anoectochilus formosanus Hayata, Orchidaceae

The water extract of *Anoectochilus formosanus* showed antihyperglycemic and antioxidant effects in streptozotocin-induced diabetic rats. The extract treatment at a dose of 1 and 2 g/kg (daily for 21 days) reduced fasting blood glucose, serum fructosamine, TGs, and total cholesterol compared with vehicle-treated diabetic rats. Further, the treatment reduced renal lipid peroxide levels and increased reduced GSH levels in diabetic rats (Shih et al. 2002).

Anoectochilus roxburghii (Wall.) Lindl. (*Anoectochilus setaceus* Lindl.), Orchidaceae

Kinsenoside isolated from *Anoectochilus roxburghii* (15 mg/kg) exerted antihyperglycemic and antioxidant activities in streptozotocin-induced diabetic rats. Histopathological observations revealed the presence of much more intact β-cells in the pancreas of kinsenoside-treated diabetic rats. Further, the compound improved oral glucose tolerance in both normal and streptozotocin-induced diabetic rats (Zhang et al. 2007).

Anogeissus acuminata Roxburgh ex Candolle, Combretaceae

A widely used traditional medicinal plant for many ailments including diabetes in Thailand, oral administration of methanol extracts of *Anogeissus acuminata* leaves (300 mg/kg) for 7 days resulted in a significant reduction in blood glucose levels in alloxan-induced diabetic rats. The extract also prevented body weight loss in diabetic rats (Hemamalini and Vijusha 2012). In another study also, the methanol bark extract and an active fraction obtained from the extract showed potent anti-DM activity in alloxan-induced diabetic rats (Zaruwa et al. 2012). Further, the water extract (100 mg/kg) of this plant exhibited marked glucose-lowering activity in alloxan-induced diabetic mice at the 4th h after oral administration (Manosroi et al. 2011a).

Anthemis nobilis L., Compositae

In Saudi Arabia, *Anthemis nobilis* is one of the commonly used medicinal plants to treat diabetes. This plant contains HMG-containing flavonoids, glucoside hamaemeloside. This compound has been shown to have *in vivo* hypoglycemic activity. In alloxan-challenged rabbits and mice, the water extract of this plant exhibited an initial hyperglycemia that was followed by hypoglycemia (review: Narayan et al. 2012).

Anthocephalus indicus A. Rich., Rubiaceae

Synonym: *Anthocephalus cadamba* (Roxb.) Miq., *Neolamarckia cadamba* (Roxb.) Bosser
Anthocephalus indicus is an important traditional medicinal plant used to treat a variety of illnesses. Oral administration of the ethanol extract of the *Ant. indicus* root (500 mg/kg) resulted in decrease in the levels of blood glucose, TGs, total cholesterol, phospholipids, and free fatty acids in alloxan-induced diabetic rats. Further, the root extract showed antioxidant activity *in vitro* (Kumar et al. 2009c). In another study, the root extract of *A. cadamba* (*Ant. indica*) caused significant dose-dependent reduction in the blood glucose level in both normoglycemic and alloxan-induced diabetic rats at the tested dose levels (100, 200, and 400 mg/kg, p.o.). In glucose-loaded animals, the extract reduced the elevated blood glucose concentration (Achrayya et al. 2010). In addition to roots, the leaves and fruits of this plant also showed antidiabetic activity. The water extract of *A. indicus* leaves (400 mg/kg, p.o., daily for 21 days) decreased the levels of serum glucose, total cholesterol, TG, and LDL and increased the levels of HDL in alloxan-induced diabetic rats. Further, administration of the water extract of the leaves caused an improvement in damaged liver cells and degenerated pancreatic islet cells (Sanadhya et al. 2012). In another study, the water extract of *A. indicus* fruits (400 mg/kg, p.o., daily for 21 days) decreased the levels of serum glucose, total cholesterol, TG, and LDL and increased the levels of HDL in alloxan-induced diabetic rats. Further, the treatment almost normalized the histopathological changes in the liver and pancreas of alloxan-induced diabetic rats (Sanadhya et al. 2013). This is a promising plant for further studies.

Anthocleista djalonensis A.Chev., Gentianaceae

Anthocleista djalonensis is traditionally used in West Africa to treat various diseases such as malaria, hernia, hypertension, stomachaches, hemorrhoids, syphilis, and diabetes.

The antidiabetic activities of the ethanol extract/fraction of *A. djalonensis* root were evaluated in alloxan-induced diabetic rats. Administration of the root extract (37–111 mg/kg) to diabetic rats resulted in a marked reduction in fasting blood glucose levels in both the acute study and prolonged treatment (2 weeks). The activities of the extract and fractions were more than that of the reference drug, glibenclamide (Okokon et al. 2012). The leaves and stem bark (methanol extracts) of *A. djalonensis* strongly inhibited α-amylase activity (Olubomehin et al. 2013). The methanol extract of the stem bark, leaves, and root of *A. djalonensis* (1 g/kg, p.o., daily for 7 days) also decreased the levels of blood glucose by 72%, 45%, and 47%, respectively, in alloxan-induced diabetic rats (Olubomehin et al. 2013).

Anthocleista schweinfurthii Gilg., Gentianaceae

A new steroid schweinfurthiin and known compounds, bauerenone and bauerenol, isolated from *Anthocleista schweinfurthii* were found to be highly promising α-glucosidase inhibitors (Mbouangouere et al. 2007).

Anthocleista vogelii Planch, Gentianaceae

Anthocleista vogelii is used in West Africa traditionally to treat various diseases such as malaria, hernia, hypertension, stomachaches, hemorrhoids, syphilis, and diabetes. The aqueous extract of the roots of *A. vogelii* possesses hypoglycemic activity in both normal and alloxan-induced hyperglycemic in mice, rats, and rabbits (Abuh et al. 2006). The leaves and stem bark (methanol extracts) of *A. vogelii* inhibited α-amylase activity (Olubomehin et al. 2013). Administration of *A. vogelii* root extract (500 mg/kg) resulted in significant reduction in glucose levels in glucose tolerance test in rats. The extract also decreased blood glucose levels in diabetic rats. There were significant reductions in LDL-C levels and significant increase in HDL-C in the treated diabetic group compared to the negative control (Ogbonnia et al. 2011).

Apium graveolens L., Apiaceae

Apium graveolens is a biennial medicinal plant with a ridged, shiny stem and can grow to a height of 45 cm. In a recent study, the antidiabetic and antiglycation effects of *A. graveolens* chloroform extract was evaluated in streptozotocin-induced type 1 and type 2 diabetic rats. Oral administration of the extract (400 mg/kg) for 30 days resulted in improvement in oral glucose tolerance in both normal and diabetic rats. Further, the treatment reduced oxidative stress, blood glucose levels, and glycosylated hemoglobin levels. Gluconeogenic enzymes were significantly increased while hexokinase was decreased in the liver along with increase in glycogen in treated diabetic animals (Gutierrez et al. 2014). Further, the treatment reduced the levels of blood lipids (except HDL, which increased). The serum insulin levels also increased. The chloroform extract also showed antiglycation activity *in vitro* in RIN-5F cells. The study supports that *A. graveolens* improves glucose metabolism by reducing insulin resistance and stimulating insulin production by protecting pancreatic β-cells from oxidative stress (Gutierrez et al. 2014). Thus, the chloroform extract appears to be promising for further studies to determine its utility for drug development. In another study, the effects of administration of *A. graveolens* water extract (200 mg/kg, i.p., and alternate days for 4 weeks) on the serum glucose, TG, total cholesterol, and HDL-C and LDL-C levels of diabetic rats were evaluated. The extract-treated diabetic rats did not show any significant effect on blood glucose levels compared to untreated diabetic control. However, there was a significant lower level of TG and cholesterol in the treated diabetic rats (Roghani et al. 2007). Thus, it appears that the water extract has only lipid-lowering properties while the chloroform extract has anti-DM activity.

Aporosa lindleyana Baill., Phyllanthaceae
Aporosa lindleyana, an evergreen tree, is used in traditional medicine to treat skin diseases, as a diuretic and as an ingredient of folk medicines used in the treatment of diabetes, eye diseases, etc. The hypoglycemic effect of the water and alcohol extracts of the *Aporosa lindleyana* root was investigated in both normal and alloxan-induced diabetic rats. The blood glucose levels were measured at 0, 1, 2, and 3 h after the treatment. The extracts of *A. lindleyana* (100 mg/kg) reduced the blood glucose levels of normal rats 3 h after oral administration and also significantly lowered blood glucose levels in alloxan-induced diabetic rat from (mean, mmol/L) 17.0–8.9 (water extract) and 18.2–8.4 (alcohol extract), respectively (Jayakar and Suresh 2003).

Aquilaria sinensis (Lour.) Gilg, Thymalaeaceae
Common Name: Agarwood
The leaves of *Aquilaria sinensis* have been used in folk medicine for the treatment of diabetes in Guangdong Province, China. In a scientific validation, administration of the 95% ethanol extract of *A. sinensis* leaves (600 mg/kg) for 4 weeks resulted in activation of AMPK and reduced fasting blood glucose and glycosylated hemoglobin levels in diabetic *db/db* mice. In addition, the oral glucose tolerance test showed that the extract could remarkably improve insulin resistance. Compared to thiazolidinediones, no weight gain was observed after the administration, which is a severe side effect of thiazolidinediones. The authors suggest that the extract could act as an activator of AMPK and might be used as an alternative to thiazolidinediones in the management of obesity-related diabetes (Jiang et al. 2011). In another study, the effect of hexane, methanol, and water extracts of *A. sinensis* leaf on hyperglycemia in streptozotocin-induced diabetic rats was investigated. The methanol and water extracts (1 g/kg), but not the hexane extract, lowered the fasting blood glucose levels by 54% and 40%, respectively. In an *in vitro* experiment, these extracts (10 μg/mL) enhanced glucose uptake in rat adipocytes. The glucose uptake enhancement activity of the extracts was comparable to that of 1.5 nM insulin (Pranakhon et al. 2011).

Arachis hypogaea L., Fabaceae
Common Names: Earth nut, groundnut, peanut (Eng.), bhuchanaka (Sanskrit), mungphali (Hindi), nilakkadalai (Tamil), etc.

Description: It is an annual, branched herb bearing compound leaves, with the base of the petiole clasping and the sheath producing two linear-lanceolate stipules. The leaflets are oblong to obovate. The flowers are yellow and are produced axillarilly. Its fruit lies underground and is an oblong, leathery, reticulate pod, bearing 1–3 seeds. The seeds yield a valuable edible oil.

Distribution: It is cultivated in the tropical and subtropical regions of India and is a native of tropical America.

Traditional Medicinal Uses: *Arachis hypogaea* is used as an astringent, lactagogue, aphrodisiac, cyanogenetic, decoagulant, and emollient and is used in the treatment of bronchitis, flatulence, acute abdominal pain, gonorrhea, inflammation, nephritis, rheumatism, and warts (Kirtikar and Basu 1975).

Antidiabetes: The hypoglycemic and hypolipidemic effects of the water extract of the *A. hypogaea* nut have been demonstrated in normal and alloxan-induced diabetic rats (Bilbis et al. 2002). *A. hypogaea* seed oil (500 and 1000 mg/kg daily for 28 days) decreased the blood glucose and glycosylated hemoglobin levels in streptozotocin–nicotinamide-induced diabetic rats as compared to untreated diabetic control rats. *In vivo* antioxidant studies on the diabetic rats revealed decreased MDA and increased reduced GSH levels in the treated rat. Thus, the investigations show that oil of seeds of *A. hypogaea* has significant antihyperglycemic and antioxidant activity (Kumar et al. 2013d). Moreover, consumption of peanut (*A. hypogaea*) reduced oxidative stress and improved HDL-C levels in streptozotocin-induced diabetic rats (Emekli-Alturfan et al. 2008).

Other Activities: Biologically active lectins have been found in ground nuts (Sanford et al. 1990). β-galactoside-specific lectins from peanut stimulate vascular cell proliferation (Sanford et al. 1990).

Toxicity: Raw ground nut contains neurotoxic agglutinins (Aletor et al. 1986).

Phytochemicals: Many alkaloids and flavonoids have been isolated from peanut (Lou et al. 2001). It also contains lectins; the seed coat contains bioactive resveratrol.

Aralia chinensis L. var. *glabrescens* Harms & Rehder, Araliaceae
Common Name: Chinese angelica tree
The *Aralia chinensis* fruit is rich in oleanolic acid, which lowers blood glucose in normal and diabetic mice. Further, it inhibits α-amylase activity (cited from Simmonds and Howes 2005).

Aralia decaisneana Hance
This plant also contains oleanolic acid. Total saponins from this plant exhibited hypoglycemic activity in animal models (Dayuan et al. 1995).

Aralia elata (Miq.) Seem, Araliaceae
Common Name: Japanese angelica tree
A new α-glucosidase inhibitor named elatosides E (which was shown to affect the elevation of plasma glucose level in oral sugar tolerance test in rats) was isolated from the root cortex of *Aralia elata* (Yoshikawa et al. 1994b). Five new saponins (elatosides G, H, I, J, and K) and hederagenin-3-*O*-glucuronopyranoside and elatoside C were isolated from the young shoot of *A. elata*. Elatosides G, H, and I were found to exhibit potent hypoglycemic activity in oral glucose tolerance test (Yoshikawa et al. 1995). The water extract (250 mg/kg) from the root cortex of *A. elata* inhibited the increase in plasma glucose concentration induced by oral administration of maltose and trehalose, but not glucose, in mice. In *in vitro* intestinal α-glucosidase assay, the extract inhibited the activities of intestinal maltase and trehalase with IC_{50} values of 0.05 and 0.65 mg/mL, respectively. The observation indicates that the reduced glucose response to disaccharides in mice with the water extract was, at least in part, due to the inhibition of intestinal α-glucosidase activity (Ohno et al. 2012). Araliosides isolated from the root bark of this plant species prevented diabetic cardiomyopathy in streptozotocin-induced diabetic rats during the early stages (Xi et al. 2009).

Aralia taibaiensis Z. Z. Wang & H. C. Zheng, Araliaceae
Total saponins isolated from *A. taibaiensis* (320 mg/kg, p.o., daily for 28 days) showed excellent antihyperglycemic, hypolipidemic, and antioxidant activities in streptozotocin–nicotinamide-induced type 2 DM rats. Compared with the diabetic control group, administration of the saponins resulted in the fall in the levels of fasting blood glucose, glycosylated hemoglobin, creatinine, urea, alanine transaminase, AST, total cholesterol, TGs, LDL-C, and MDA, but significant increase in the levels of serum insulin, superoxide dismutase, and reduced GSH. However, the saponins did not have any effect on normal rats (Weng et al. 2014).

Arbutus unedo L., Ericaceae
Arbutus unedo (strawberry) is an evergreen shrub. The leaves and roots of *A. unedo* are used in traditional medicine to treat diabetes in Morocco. The water extract (500 mg/kg) of *A. unedo* improved oral glucose tolerance in rats. Further, the water extract induced a significant inhibition of jejunal glucose absorption in rats. Preliminary toxicity evaluation did not show any adverse effects (Bnouham et al. 2007).

Arctium lappa L. (*A. majus* (Gaerth) Bernh), Asteraceae

Common Name: Burdock

Traditional Drug Name: Fructus arctii

Burdock (*Arctium lappa*) is a traditional edible and medicinal plant. This plant is used for the treatment of DM in Europe in traditional medicine. The ethanol extract of the root of *A. lappa* (400 mg/kg, p.o.) decreased blood glucose and increased insulin levels in streptozotocin-induced diabetic rats. Further, the treatment decreased the levels of total cholesterol, TG, and LDL and increased the levels of HDL in the serum in diabetic rats compared to untreated diabetic control rats. There was improvement in body weight gain and glycogen content in the liver in the treated diabetic animals; the treatment also decreased lipid peroxides in the liver and kidney and urea and creatinine in the serum. No hypoglycemic effect was observed in normal rats; the ethanol extract (400 mg/kg) improved oral glucose tolerance in normal rats (Cao et al. 2012). Thus the root extract showed promising antidiabetic activity. In another study, the alloxan-induced diabetic animals showed significant reductions in plasma glucose, TGs, and total cholesterol after treatment with the total lignan fraction from *A. lappa* (Xu et al. 2008). As per one report, the plant drug (traditional form) aggravated the diabetic condition in streptozotocin-induced diabetic rats (Swanston-Flatt et al. 1989). This negative result could, possibly, be due to the differences in the chemical composition of the drug preparations.

Ardisia japonica (Thunb.) Blume, Myrsinaceae

Ardisia japonica is an evergreen shrub. Benzoquinones isolated from *A. japonica* inhibited human PTP1B with IC_{50} values ranging from 3.0 to 19.1 μmol/L (Jiang et al. 2012).

Areca catechu L., Arecaceae (Figure 2.11)

Common Names: Betel nut palm, areca nut, supari palm, Pinang palm, kamugu, pakku tree, etc.

Description: *Areca catechu* is a slender, single-trunked, monoecious palm with a prominent crown shaft. Areca nut is the seed or endosperm (nut) of *A. catechu*.

Distribution: *A. catechu* is a species of nut palm which grows in much of the tropical Pacific, Asia, and parts of east Africa. It is cultivated for the kernel obtained from its fruit that is chewed in its tender, ripe, or processed form.

Traditional Medicinal Uses: In traditional medicine, betel nut has been used as a therapeutic agent for leukoderma, leprosy, and obesity; it is also used as vermifuge and for deworming (review: Amudhan et al. 2012).

(a)   (b)

FIGURE 2.11 *Areca catechu.* (a) Trees (crowns). (b) Areca nut (fruit).

Pharmacology

Antidiabetes: The hypoglycemic, hypolipidemic, and antioxidant properties of the areca nut extract was evaluated in alloxan-induced diabetic rats. Oral administration of the nut extract (25 mg/kg) to diabetic animals decreased the elevated levels of serum glucose and lipids to near normal levels in alloxan-induced diabetic rats. Further, the level of liver glycogen content was improved upon the extract treatment. The elevated levels of lipid peroxides found in the plasma and pancreatic tissue of diabetic rats were normalized by the extract treatment. The study indicates that the nut extract possesses anti-DM, antioxidant, and antilipidemic activity in diabetic rats (Kavitha et al. 2013). Subcutaneous administration of an alkaloid fraction of the nut to alloxan-induced diabetic rabbits showed significant hypoglycemic effect lasting up to 6 h. Arecoline, an alkaloid, has been reported to have hypoglycemic activity in the animal model of DM upon subcutaneous administration (review: Amudhan et al. 2012; Chempakam 1993).

Ethanol extract of areca nut showed *in vitro* inhibitory activity of intestinal α-glucosidase enzymes maltase and surcease with an IC_{50} value of 12 μg/mL for maltase and 30 μg/mL for surcease. The ethanol extract (250 and 500 mg/kg) reduced postprandial elevation in blood glucose levels at 30 and 60 min after administration of maltose. Thus, areca nut extract inhibits α-glucosidase and is effective in the suppression of blood glucose level elevation after oral maltose loading to rats (review: Amudhan et al. 2012).

Detailed studies including all the active principles involved and mechanisms of action other than α-glucosidase inhibition remain to be studied.

Other Activities: The ethanol extract of areca nut showed potent antioxidant activities including radical scavenging; it also inhibited the protein enzyme hyaluronidase. The extract showed 1,1-diphenyl 2-picryl (DPPH) radical scavenging activity and superoxide scavenging activity. It also inhibited hydrogen peroxide-induced RBC hemolysis *in vitro* (review: Amudhan et al. 2012).

Areca nut extract exhibited cholesterol absorption inhibition activity in high-cholesterol-fed rats. Further, supplementation of the extract resulted in significant reduction in the absorption of TGs and the levels of plasma lipids. In addition, the extract inhibited the activity of pancreatic cholesterol esterase and lowered the absorption of dietary cholesterol (Amudhan et al. 2012). Other reported activities of the nut include anti-inflammatory activity, antihypertensive activity, antidepressant property, wound healing activity, inhibition of platelet aggregation, CNS stimulation, anticonvulsant activity, inhibition of proteasome activity, antimicrobial action, and anti-HIV activity (review: Amudhan et al. 2012).

Phytochemicals: Phytochemical analysis revealed the presence of alkaloids, flavonoids, saponins, tannins, phytosterols, terpenoids, and phenols (Kavitha et al. 2013). The polyphenols include (+) catechin, epicatechin, and leucocyanidin. The four major alkaloids isolated from areca nut are arecoline, arecaidine, guvacoline, and guvacine (review: Amudhan et al. 2012).

Argyreia cuneata (Willd.) Ker-Gawl., Convolvulaceae
Extracts of *Argyreia cuneata* prevented the onset of hyperglycemia in alloxan-induced diabetic rats (Birader et al. 2010).

Aronia melanocarpa (Michx.) Ell., Rosaceae
Aronia melanocarpa (aroma berry) is cultivated for its edible fruits. The fruit juice of *A. melanocarpa* (10 and 20 mg/kg, p.o., daily for 6 weeks) did not influence plasma glucose and lipid levels in normal rats. But in streptozotocin-induced diabetic rats, the treatment with the fruit juice markedly reduced plasma glucose and TG to levels that did not significantly differ from those of the normal control rats. Further, the treatment almost normalized the levels of total cholesterol, LDL, and HDL in diabetic rats. The fruit juice may be useful in the prevention and control of diabetes and its complications (Valcheva-Kuzmanova et al. 2007). Anthocyanins and procyanidins from the *A. melanocarpa* fruit (or methanol extract) exhibited potent α-amylase inhibitory activity and free radical scavenging activity (Bräunlich et al. 2013).

Artemisia afra Jacq., Asteraceae
The water extract of *Artemisia afra* (leaf and stem) at a dose of 50 and 100 mg/kg decreased blood glucose levels and ameliorated oxidative stress in the pancreas of streptozotocin-induced diabetic rats (Afolayan and Sunmonu 2011).

Artemisia amygdalina Decne

The hydroethanolic and methanol extracts (250 or 500 mg/kg b.w.) reduced glucose levels in streptozotocin-induced diabetic rats. These extracts also reduced the levels of serum cholesterol, TG, and LDL. The extract-treated diabetic animals exhibited a reduction in the feed and water consumption compared to diabetic control animals. The histopathological analysis showed the regenerative/protective effects of the extracts on pancreatic β-cells of the rats. Thus, the study validated the traditional use of this plant as an anti-DM medicine (Ghazanfar et al. 2014).

Artemisia campestris L., Asteraceae

Common name: Field wormwood

The leaf extract of *Artemisia campestris* effectively ameliorated diabetic renal dysfunction by reducing oxidative and nitrosative stress in alloxan-induced diabetic rats. Histological studies also supported the experimental findings (Sefi et al. 2012).

Artemisia capillaris Thunb, Asteraceae

Synonym: *A. scoparia*, Waldst. and Kit.

In alloxan-induced diabetic rats, *Artemisia capillaris* exhibited antihyperglycemic activity. The treatment showed beneficial lipid profiles in hyperlipidemic and alloxan-induced diabetic mice. These results indicated that *A. capillaris* would have a similar hypoglycemic effect as biguanide drugs that improve endogenous and exogenous metabolic derangement in blood lipid. These observations suggest that this plant could be useful in protecting and treating DM and its chronic complications (Pan et al. 1998).

The effects of *A. capillaris* extract on cytokine-induced β-cell damage were examined. Treatment of RINm5F rat insulinoma cells with IL-1β- and IFN-γ-induced cell damage. The *A. capillaris* extract completely protected IL-1β- and IFN-γ-mediated cytotoxicity in a concentration-dependent manner. Incubation with the extract resulted in a significant reduction in IL-1β- and IFN-γ-induced NO production, a finding that correlated well with reduced levels of the iNOS mRNA and protein. The molecular mechanism by which the extract inhibited iNOS gene expression appeared to involve the inhibition of NF-κB activation. Furthermore, the plant extract restored the cytokine-induced inhibition of insulin release from isolated islets. These results suggest that the extract protects β-cells by suppressing NF-κB activation (Kim et al. 2007).

Artemisia dracunculus L., Asteraceae/Compositae

Two well-described cultivars are Russian tarragon and French tarragon. Synonyms for French tarragon include *A. dracunculus f.* L., *A. dracunculus* var. *sativa* Besser, and *A. dracunculus* L., while those for Russian tarragon include *A. dracunculus f.* Redowskii Hort, *A. dracunculus* var. *dracunculus*, *A. dracunculus* var. *dracunculus*, *A. dracunculus* var. *pratorum* Krasch., *A. dracunculus* var. *turkestanica* Krasch. Krasc, *A. dracunculus* var. *pilosa* Krasch., *A. dracunculus* var. *humilis* Kryl., *A. dracunculus* var. *redovskyi* Ldb., and other numerous varieties. Other synonyms for *A. dracunculus* L. (without distinction between Russian and French origin) include *A. aromatica* A. Nelson, *A. dracunculina* S. Watson, *A. dracunculoides* Pursh, *A. dracunculoides* ssp. *dracunculina* (S. Watson) H. M. Hall & Clements, *A. glauca* Pallas ex Willdenow, *A. glauca* var. *megacephala* B. Boivin, *A. redowskyi* Ledeb, *Oligosporus condimentarius* Cass., *Oligosporus dracunculus* L. Polj., and *A. inodora* Willd (review: Obolskiy et al. 2011) (Figure 2.12).

Common Names: Tarragon, silky wormwood, little dragon, dragon herb, etc.

Description: *Artemisia dracunculus* L. is a perennial herb, 60–80 cm in height formed by short rhizomes forming dense mats and numerous herbaceous stems. Tarragon stalks are thin and hairless, erect or prostrate at 40 cm. The leaves are linear, lanceolate, entire, sessile, dark green, and hairless. The basal leaves have three possible lobulations at the apex. The inflorescence is a globose floral chapter with small, greenish-yellow, pendant, pistillate, and bisexual flowers measuring about 3 mm. The fruit is an achene. However, the flowers are sterile in warm weather. Multiplication occurs from the stem cuttings (www.botanical-online.come).

FIGURE 2.12 *Artemisia dracunculus.*

Distribution: The plant is native to Central Asia, Siberia, Mongolia, and the Pamir mountain range. It was introduced in Europe in the middle ages. It is cultivated in Asia, central Russia, central Europe, North Africa, and the United States.

Traditional Medicinal Uses: Fresh leaves, floral chapters, etc., are used in traditional medicine. It has a long history of use in culinary traditions and as an herbal medicine. The main therapeutic uses in traditional medicine are for the nervous (mitigative, antiepileptic), digestive (appetite stimulation, spasmolytic, and laxative), renal (diuretic action), and liver (choleretic) functions. It is also used as an anti-inflammatory (wound healing, antiulcer), anticancer, and antibacterial agent. Arabic cultures have used *A. dracunculus* to treat insomnia and to dull the taste of medicines. It is also used as an anesthetic for aching teeth, sores, and cuts; it is used to treat skin wounds, irritations, allergic rashes, dermatitis, and fevers. It is also used as an antiepileptic, laxative, antispasmodic, vermifuge, and gynecological aid to reduce excessive flow during the menstrual cycle and to aid in difficult labor (review: Obolskiy et al. 2011).

Pharmacology

Antidiabetes: Studies have shown the blood glucose lowering properties of *A. dracunculus* (tarragon) extract and its mechanism of action in improving insulin resistance. The ethanol extract of *A. dracunculus* enhanced IR signaling and modulated gene expression in skeletal muscle in KK-A(y) mice (Wang et al. 2010a). The extract enhanced cellular insulin signaling in primary human skeletal muscle culture also (Wang et al. 2008a). Further, the antihyperglycemic activity of the extract in animal models was reported to decrease phosphoenolpyruvate carboxykinase (PEPCK) mRNA expression in streptozotocin-induced diabetic rats. Two polyphenolic compounds that inhibited PEPCK mRNA levels were isolated and identified as 6-demethoxycapillarisin and 2,4-dihydroxy-4-methoxy dihydrochalcone with IC_{50} values of 43 and 61 μM, respectively. The phosphoinositide-3 kinase (PI3K) inhibitor LY-294002 showed that 6-demethoxycapillarisin exerts its effect through the activation of the PI3K pathway, similarly to insulin. The effect of 2,4-dihydroxy-4-methoxydihydrochalcone is not regulated by PI3K and dependent on the activation of the AMPK pathway (Govorko et al. 2007).

A standardized ethanol extract of *A. dracunculus*, called PMI-5011, was shown to be hypoglycemic in animal models for type 2 diabetes and contains at least six bioactive compounds responsible for its antidiabetic properties and was shown to reduce hyperglycemia by improving insulin resistance. At doses of 50–500 mg/kg/day, the hypoglycemic activity of the extract was enhanced three- to fivefold with the bioenhancer Labrasol, making it comparable to the activity of the antidiabetic drug metformin. When combined with Labrasol, one of the active compounds, 2′, 4′-dihydroxy-4-methoxydihydrochalcone, was at least as effective as metformin at doses of 200–300 mg/kg/day (Ribnicky et al. 2009). PMI-5011 treatment to KK-Ay mice decreased insulin levels and improved insulin sensitivity from week 2 of the treatment when compared with control. Animals treated with PMI-5011 had greater phosphorylation of Akt-1 and insulin stimulated activities of PI3K and Akt-1 in skeletal muscles versus control. In addition, higher abundances of intracellular proteins such as AS160 were noted. There was a trend for decreased PTP1B activity in PMI-5011-treated mice (Wang et al. 2007). In another study, C2C12 myotubes and the KK-A(y) mice model of type 2 diabetes were used to evaluate the effect of PMI-5011 on steady-state levels of ubiquitylation, proteasome activity, and expression of atrogin-1 and MuRF-1, muscle-specific ubiquitin ligases that are upregulated with impaired insulin signaling. PMI-5011 inhibited proteasome activity and steady-state ubiquitylation levels *in vitro* in the myocytes and *in vivo* in the diabetic mice. The effect of PMI-5011 is mediated by PI3K/Akt signaling and correlates with decreased expression of atrogin-1 and MuRF-1. Under *in vitro* conditions of hormonal or fatty acid–induced insulin resistance, PMI-5011 improves insulin signaling and reduces atrogin-1 and MuRF-1 protein levels. In the KK-A(y) murine model of type 2 diabetes, skeletal muscle ubiquitylation and proteasome activity are inhibited and atrogin-1 and MuRF-1 expression is decreased by PMI-5011. PMI-5011-mediated changes in the ubiquitin–proteasome system *in vivo* correlate with increased phosphorylation of Akt and FoxO3a and increased myofiber size (Kirk-Ballard et al. 2013).

Furthermore, PMI-5011 treatment to human myocytes in culture led to an inhibition of cytokine (MCP-1 and TNF-α)-induced activation of inflammatory signaling pathways such as Erk1/2, and NF-κB, possibly by preventing further propagation of the inflammatory response within muscle tissue. Thus, PMI-5011 improved insulin sensitivity in diabetic obese human myotubes to the level of normal lean myotubes despite the presence of proinflammatory cytokines (Vandanmagsar et al. 2014). Ectopic lipids in peripheral tissues have been implicated in attenuating insulin action *in vivo*. The antidiabetes extract of *A. dracunculus* (PMI-5011) improved insulin action, yet the precise role of these lipids is not known. Free fatty acids (FFA) treatment resulted in increased levels of total ceramides and ceramide species in L6 myotubes. Saturated FFAs and ceramide C2 inhibited insulin-stimulated phosphorylation of protein kinase B/Akt and reduced glycogen content. PMI-5011 had no effect on ceramide formation or accumulation but increased insulin sensitivity via restoration of Akt phosphorylation. PMI-5011 also attenuated the FFA-induced upregulation of an inhibitor of insulin signaling (PTP1B) (Obanda et al. 2012). The effect of PMI-5011 on neuropathy in HFD-fed mice (a model of prediabetes and obesity developing oxidative stress and proinflammatory changes in the peripheral nervous system) was evaluated. C57Bl6/J mice fed an HFD for 16 weeks developed obesity, moderate nonfasting hyperglycemia, nerve conduction deficit, thermal and mechanical hypoalgesia, and tactile allodynia. They displayed 12/15-lipoxygenase overexpression, 12(*S*)-hydroxy eicosatetraenoic acid accumulation, and nitrosative stress in peripheral nerves and the spinal cord. PMI-5011 (500 mg/kg/day, 7 weeks) normalized glycemia, alleviated nerve conduction slowing and sensory neuropathy, and reduced 12/15-lipoxygenase upregulation and nitrated protein expression in the peripheral nervous system. The authors conclude that PMI-5011 may find use in the treatment of neuropathic changes at the earliest stage of the disease (Watcho et al. 2010).

The extract or its isolated active compounds appear to be an attractive material for drug development; detailed toxicity study and subsequent clinical studies are warranted.

Other Activities: Other validated pharmacological properties include antibacterial activity, antifungal activity, antiplatelet aggregation activity, anti-inflammatory activity, hepatoprotective activity, hypolipidemic action, antioxidant activity, antihypoxic activity, gastroprotective properties, neurotrophic activity, analgesic action, and anticonvulsant activity (review: Obolskiy et al. 2011).

Phytochemicals: The compounds isolated and identified from this plant include flavonoids (5,6,7,8,40-pentahydroxymethoflavone, estragoniside, 7-*O*-β-D-glycopyranoside 5,7-dihydroxyflavavone, pinocembrin, 7-*O*-β-D-glucopyranoside, luteolin, quercetin, rutin, kaempferol, annangenin, 5,7-dihydroxyflavone, naringenin, 3,5,40-trihydroxy-7-methoxyflavanone, 3,5,4-trihydroxy-7,30-dimethoxyflavanone, davidigenin, 20,40-dihydroxy-4-methoxydihydrochalcone, and sakuranetin), phenylpropanoids (chicoric acid, hydroxybenzoic acid, (*E*)-2-hydroxy-4-methoxycinnamic, chlorogenic acid, caffeic acid, 5-*O*-caffeoylquinic acid, and 4,5-di-*O*-caffeoylquinic acid), chromones/

coumarins ((–)-(*R*)-20-methoxydihydro-artemidin, (+)-(*S*,*R*)-epoxyartemidin, dracumerin, (+)-(*R*)-(*E*)-30-hydroxyartemidin, capillarin isovalerate, 7,8-methylenedioxy-6-methoxycoumarin, γ,γ-dimethylallyl ether of esculetin, scopoletin, scoparone, daphnetin methylene ether, daphnetin 7-methyl ether, artemidiol, 4-hydroxycoumarin, artemidin, artemidinal, artemidinol, esculin, capillarin, 8-hydroxycapillarin, 8-hydroxyartemidin, and 6-demethoxycapillarisin), alkamides (pellitorine, neopellitorine A, and neopellitorine B), and benzodiazepines (trace amounts). The quantity and quality of the phytochemicals vary widely among different varieties and geographical and soil conditions (review: Obolskiy et al. 2011). Medicinally important secondary metabolites include essential oils that vary widely (0.15%–3.1%) among various varieties and cultivars. Major essential oil constituents include methyleugenol, methylisoeugenol, terpinen-4-ol, sabinene, elemicin, β-ocimene, allo-ocimene, cis-ocimene, cis-allo-ocimene, trans-ocimene, estragole, 7-methoxy coumarin, 3,4,7-trimethoxycoumarin, terpinolene, elemicin, cis-ocimene, α-trans-ocimene, limonene, trans-anethole, α-pinene, azarone, phytol, tetradecanoic acid, *n*-hexadecanoic acid, 9-octadecenoic acid, germacrene D, germacrene D-4-ol, squalene, 3,7-dimethyl-1,3,7-octatriene, 1-methoxy-4-(2-propenyl) benzene, capillarin, capillene, spathulenol, α-chimachalene, geranyl acetate, 5-phenyl-1,3-pentadiyne, phellandrene, myrcene, γ-terpinene, terpinen-4-ol, citronellyl formate, isoelemicin, and citronellol (review: Obolskiy et al. 2011).

Toxicity: The herb may contain estragole and methyleugenol that are potential carcinogens. The levels of these compounds differ in different cultivars; and the aqueous extract does not contain these compounds (Weinoehrl et al. 2012).

Artemisia herba-alba Asso, Asteraceae

Synonym: *Artemisia sieberi* Bess. (Figure 2.13)

Local Names: Wormwood, chim, armoise, etc.

Description: *Artemisia herba-alba* is a greenish-silver perennial herb and grows 20–40 cm in height. *A. herba-alba* is a chamaeophyte (i.e., the buds giving rise to new growth each year are borne close to the ground). The stems are rigid and erect. The gray leaves of sterile shoots are petiolate and ovate to orbicular in outline, whereas the leaves of flowering stems are much smaller. The flowering heads are sessile, oblong, and tapering at the base (review: Mohamed et al. 2010).

Distribution: *A. herba-alba* grows wild in arid areas of the Mediterranean basin, extending into northwestern Himalayas. This plant is abundant in the Iberian Peninsula and reaches highest population in the center of Spain spreading over eastern, southeastern, and southern Spain and West Irano-Turanian, but it extends until the Atlantic coast in North Africa.

FIGURE 2.13 *Artemisia herba-alba.*

Traditional Medicinal Uses: The plant is used in traditional medicine for skin trouble. It is also used as vermifuge, emmenagogue, diuretic, stomachic, intestinal antiseptic, tonic, cholagogue, depurative, and anti-DM. *A. herba-alba* has been used in folk medicine by many cultures since ancient times and is used in Moroccan folk medicine to treat arterial hypertension and/or diabetes and stomach disorders. Herbal tea from this species has been used as analgesic, antibacterial, antispasmodic, and hemostatic agents (Mohamed et al. 2010).

Pharmacology

Antidiabetes: The extract of aerial parts of *A. herba-alba* (85 mg/kg, p.o.) was administered to normal and streptozotocin-induced diabetic rabbits. The extract treatment resulted in a significant hypoglycemic effect in normal and diabetic rabbits (Iriadam et al. 2006). In another study, oral administration of 390 mg/kg of the water extract of *A. herba-alba* exhibited a marked reduction in blood glucose levels. It was shown that the antidiabetic effect of this plant extract was similar to that of repaglinide and insulin as judged from blood glucose and other biochemical parameters in alloxan-induced diabetic rats (Tastekin et al. 2006). *In vitro* and *in vivo* studies on *A. herba-alba* (alcohol extract) confirmed its hypoglycemic activity. The alcohol extract (70%) of this plant produced more hypoglycemic activity than any of the fractions obtained from the extract. Four hypoglycemic compounds were isolated and identified from the 70% alcohol extract (Awad et al. 2012). Thus, it appears that several hypoglycemic agents are present in this plant and their combined effects are better than any single compound present in the extract.

In another study, the preventive effect of the hydroalcoholic extract of *A. herba-alba* was evaluated in a type 2 diabetic mouse model induced with a standardized HFD. The extract was administered orally (2 g/kg) daily for 20 weeks to male C57BL/6J mice fed an HFD. After 6 weeks, blood glucose levels increased in HFD control mice. At the end of the study (20 weeks) in the extract-treated group, there was a significant reduction in fasting blood glucose levels, TG concentrations, and serum insulin levels. The extract also markedly reduced insulin resistance as measured by the homeostasis model assessment compared to HFD controls. The plant extract had no effect on calorie intake or body weight. Thus, the *A. herba-alba* extract prevented HFD-induced diabetes in mice. In a follow-up study, the hydroalcoholic extract of *A. herba-alba* was tested in established type 2 diabetes mice induced with a standardized HFD. After confirmation of diabetes, the plant extract (2 g/kg) was administered daily for 18 weeks. The diabetic mice that received the extract showed marked reduction in blood glucose, TG, total cholesterol, and serum insulin levels compared to untreated diabetic mice. The treatment also reduced insulin resistance. The plant extract decreased calorie intake and had little effect on body weight or HDL-C (Hamza et al. 2010). This is an important anti-DM plant for further studies, which include identification of active principles and elucidation of mechanisms of action. The plant is reported to contain chlorogenic acid, an anti-DM compound.

Clinical Trial: In a human trial, 15 patients with type 2 DM were treated with *A. herba-alba* extract. Results showed that the extract caused considerable lowering of elevated blood sugar, and 14 out of 15 patients had good remission of diabetic symptoms with the use of the extract. It is concluded that *A. herba-alba* extract contains material capable of reducing raised blood sugar in DM. No side effects were recorded during or after the treatment (Al-Waili 1986).

Other Activities: The water extract of this plant has antihyperlipidemic activity (Abass 2012). Other reported biological activities of this plant include antioxidant activity, antivenom action, nematicidal activity, anthelmintic activity, antileishmanial activity, antibacterial activity, antispasmodic effect, neurological effects, cytotoxic effects, and gene induction effects (review: Mohamed et al. 2010).

Toxicity: Studies showed that long-term exposure of female rats to *A. herba-alba* caused adverse effects on the reproductive system and fertility. In treated rats, testicular cell population showed a decrease in number of spermatocytes and spermatids when compared to controls (review: Mohamed et al. 2010).

Phytochemicals: Various secondary metabolites have been isolated from *A. herba-alba*, perhaps the most important being the sesquiterpene lactones that occur with great structural diversity within the genus *Artemisia*. Further, studies have reported the presence of flavonoids and essential oils. Several polyphenolics and related constituents were isolated. These included chlorogenic acid, 4,5-*O*-dicaffeoylquinic acid, isofraxidin 7-*O*-β-D-glucopyranoside, 4-*O*-β-D rutin, schaftoside, isoschaftoside, and vicenin-2 (review: Mohamed et al. 2010).

Artemisia indica Willd., Asteraceae

Synonyms: *A. indi*ca Willd. var *maximowiczii* (Nokia) H. Hara, *A. princeps* Pamp, *A. vulgaris* L. var. *maximowiczii*

Common Name: Japanese mugwort

Artemisia species have been extensively used for the management of diabetes in folklore medicine. In a study, hydromethanolic extracts and its various fractions of aerial parts of *Artemisia indica* were tested for their antidiabetic potential in streptozotocin-induced diabetic rats. The extracts were further subjected to preliminary phytochemical analysis. A daily oral dose of hydromethanolic crude extract (200 and 400 mg/kg) and chloroform fraction (200 mg/kg) of the extract for 15 days showed a significant reduction in blood glucose level that was comparable to that of the standard antidiabetic drug gliben-clamide (500 μg/kg, p.o.). The extract and fraction also showed reduction in total cholesterol, TGs, and LDLs as well as serum creatinine level, serum GPT, GOT, and alkaline phosphatase (ALP) in diabetic rats. Thus, *A. indica* possesses hypoglycemic, antihyperlipidemic, and protective effects on the liver and renal functions in diabetic rats (Ahmad et al. 2014).

Artemisia judaica L., Asteraceae

The whole plant extract of *Artemisia judaica* reduced blood glucose levels in diabetic rats (Nofal et al. 2009).

Artemisia minor Jacq. ex Besser, Asteraceae

Artemisia minor is a widely used medicinal plant in South Africa. It is used in traditional medicine to treat diabetes. Caffeic acid (a widespread phenolic acid that occurs naturally in many agriculture products) isolated from the aerial parts of *A. minor* was found to be a PTP1B inhibitor with an IC_{50} value of 3.06 μmol/L (Jiang et al. 2013). Additionally, 1,4-benzodioxane lignan, isolated from the methanol extract of *A. minor*, inhibited PTP1B with an IC_{50} of 1.62 μg/mL (Jiang et al. 2012).

Artemisia pallens Wall. ex D.C., Asteraceae

Common Names: Davana, kozhunthu, kozhunnu, etc.

Artemisia pallens is about 50–60 cm in height with much divided leaves. It is distributed in Northeast and South India and also in Thailand. It is used in Indian folk medicine for the treatment of DM. The plant has good aroma and in folk medicine is used in aromatherapy.

Antidiabetes: Oral administration of the methanol extract of the aerial parts of this plant resulted in significant blood glucose lowering in glucose-fed hyperglycemic and alloxan-induced diabetic rats. However, the extract is unstable upon storage for a few days (Subramoniam et al. 1996). There is no reported toxicity. Follow-up studies are warranted to identify the active principles and find out conditions for its stability.

Reported phytochemicals include 2,5-divinyl-2-methyltetrahydrofuran, 2-acetyl-5-methyl-furan, 2-methyl-2-vinyl-5-ethyl-tetrahydrofuran, 5-acetyl-2-methyl-2-vinyl-tetrahydrofuran alcohols, allo-aromadendrene, β-eudesmol, β-maaliene, bicycloelemene, bicyclogermacrene, cadinene, caryophyllene, cinnamyl cinnamate, cis-(erythro)-davanafuran, cis-(threo), davanafuran, cis-linalool-oxide, davana acid, davana ether, davanafurans, davanone, dehydro-α-linalool, fenchyl alcohol, sodavanone, lavender lactone, ledene, lilac alcohol, lilac aldehyde, linalool, nor-davanol, spathulenol, T-cadinol, terpinen-4-ol, trans-(erythro)-davanafuran, trans-(threo)-davanafuran, and trans-linalool-oxide (*The Wealth of India* 1992; review: Ajikumaran and Subramoniam 2005).

Artemisia roxburghiana Besser, Asteraceae

Under *in vitro* conditions, the ethanol extract of this plant showed insulin secretagogue activity in INS-1 cells at the 1 μg/mL level (Hussain et al. 2004).

Artemisia santonicum L., Asteraceae
The plant panicle extract showed blood glucose-lowering effect in normal and alloxan-induced diabetic rabbits (Korkmaz and Gurdal 2002).

Artemisia sphaerocephala Krasch, Asteraceae
Artemisia sphaerocephala is a perennial shrub and is widely distributed in desert areas of Gansu and Nei Monggol in China. Since ancient times, *A. sphaerocephala* seed powder has been used in Chinese traditional medicine for the treatment of diabetes. In Chinese medicine, the plant seed is used to treat diseases such as parotitis, abdominal distention, and diabetes. A new polysaccharide isolated from this plant seed, when administered to alloxan-induced diabetic rats at a dose of 200 mg/kg (p.o., daily for 4 weeks), produced a significant decrease in blood glucose levels and plasma cholesterol and TG levels (Zhang et al. 2006). A recent study has also shown that *A. sphaerocephala* seed polysaccharide (400 and 800 mg/kg) can ameliorate the high fructose-induced hyperglycemia, hepatic steatosis, and oxidative stress in mice (Ren et al. 2014). Moreover, addition of *A. sphaerocephala* gum (extracted from seed powder) to the rats' food significantly lowered fasting blood glucose, glycated serum protein, serum cholesterol, and TG in HFD and low-dose streptozotocin-induced type 2 diabetic rats. Further, the gum reduced insulin resistance and liver fat accumulation in diabetic rats (Xing et al. 2009). This gum also showed antioxidant activity in HFD streptozotocin-induced diabetic rats (Hu et al. 2011b).

Artemisia vulgaris L., Asteraceae
A. vulgararis is commonly known as mugwort. Extracts of this plant lowered blood glucose in experimental animals, but the plant is toxic (Marles and Farnsworth 1995).

Arthrocnemum glaucum (Del.) Ung.-Sternb., Chenopodiaceae
The aerial parts of *Arthrocnemum glaucum* are used in Egypt and elsewhere to treat DM in traditional medicine. *A. glaucum* extract showed potent hypoglycemic effects in alloxan-induced diabetic rats (Shabana et al. 1990).

Artocarpus heterophyllus Lam., Moraceae
Artocarpus heterophyllus is commonly known as jackfruit tree, panasa, kathal, palapalam, chakka, etc.

It is a large tree reaching a height of 50 ft or more with a dense crown with entire leaves, coriaceous, round at the base, and obtuse at the apex; its flowers are produced on fleshy peduncle and are pendulous; the fruiting receptacle attains large-sizes bearing tubercles. The plant is indigenous to India, especially in Western Ghats, Bengal, Bihar, and the Deccan. It is also present in Burma, Sri Lanka, Malaya, and Brazil. In traditional medicine, it is utilized as an astringent, laxative, and demulcent and is used in carbuncles, diabetes, leprosy, diarrhea, small pox, and the puerperium.

Antidiabetes: The hot water extract of the *A. heterophyllus* leaf significantly improved glucose tolerance in normal subjects and diabetic patients (Fernando et al. 1991). The water extracts of *A. heterophyllus* showed highly promising hypoglycemic effects in rats loaded with glucose (Fernando 1990). In normoglycemic rats the leaf extract reduced fasting blood glucose levels and improved glucose tolerance (Chackrewarthy et al. 2010). The unripe as well as ripe fruit is edible. It is of interest to test the antidiabetic activity of the different parts of the fruit in view of its use as a vegetable in South India.

Other Activities: Lectins from the seed of *A. heterophyllus* have various biological properties. Prenyl flavones from *A. heterophyllus* possess antioxidant properties. The extracts from this plant have potent antibacterial flavones. The methanol extracts of the leaves, bark (stem and root), fruit, and seeds show antibacterial activity. *A. heterophyllus* contains anti-inflammatory flavones. Artocarpanone (a flavonoid) inhibits nitric oxide production in lipopolysaccharide-activated macrophages. The *A. heterophyllus* seed inhibits sexual competence of male rats. This is a temporary effect and does not affect fertility.

The phytochemicals reported include caoutchouc, resins, steroketone, artostenone, artocarpetin, artocarpin, isoartocarpin, cyanomaclurin, dihydromorin, artocarpesin, artocarpanone (flavonoid), and lectins (review: Ajikumaran and Subramoniam 2005).

Aspalathus linearis (Burm.f.) R. Dahlgren, Fabaceae

Aspalathus linearis is a medicinal plant that is endemic to South Africa. Rooibos is a slightly sweet and mildly astringent fragrant tea produced by fermentation of the commercially cultivated leaves and twigs of *A. linearis*. It is a popular tea in view of its low tannin content and potential health-promoting properties including antioxidant activity. Rooibos extracts are also used by the beverage, food, and nutraceutical industries for its flavor and antioxidant properties.

Aspalathin, a green rooibos tea component, dose dependently increased glucose uptake by L6 myotubes at 1–100 mM concentrations. It also increased insulin secretion from cultured RIN-5F β-cells at 100 mM. Inclusion of aspalathin in the diet (0.1% and 0.2%) suppressed the increase in fasting blood glucose levels of *db/db* type 2 diabetic mice. In i.p. glucose tolerance test, aspalathin improved impaired glucose tolerance at 30, 60, 90, and 120 min in diabetic *db/db* mice (Kawano et al. 2009). Another study carried out to confirm the antidiabetes activity of aspalathin-rich rooibos extract showed the synergic action of a mixture of compounds present in the extract. An extract was selected after screening for high aspalathin content and α-glucosidase inhibitory activity. Under *in vitro* conditions, the extract induced a dose-dependent increase in glucose uptake on C2C12 myocytes. Aspalathin was effective at 1, 10, and 100 μM, whereas rutin was effective at 100 μM. *In vivo* the extract sustained a glucose-lowering effect comparable to metformin over a 6 h period after administration (25 mg/kg) to streptozotocin-induced diabetic rats. In an oral glucose tolerance test, the extract (30 mg/kg) was more effective than vildagliptin (10 mg/kg), a dipeptidyl peptidase IV (DPP-IV) inhibitor. A mixture of aspalathin–rutin (1:1) at a low dose (1.4 mg/kg), but not the single compounds separately, reduced blood glucose concentrations over a 6 h monitoring period in streptozotocin-induced diabetic rats. The improved hypoglycemic activity of the mixture and the extract showed synergic actions of the polyphenols in mixture (Muller et al. 2012).

High glucose markedly increased vascular permeability, monocyte adhesion, expression of cell adhesion molecules, formation of reactive oxygen species (ROS), and activation of NF-κB in human umbilical vein endothelial cells *in vitro*. Remarkably, treatment with aspalathin or nothofagin (two major antioxidant dihydrochalcones found in green rooibos) inhibited high-glucose-mediated vascular hyperpermeability, adhesion of monocytes toward human umbilical vein endothelial cells, and expression of cell adhesion molecules *in vitro*. In addition, these compounds suppressed the formation of ROS and the activation of NF-κB and also suppressed vascular inflammation *in vivo* in mice. Since vascular inflammation induced by high levels of glucose is critical in the development of diabetic complications, the authors suggest that these compounds may have significant benefits in the treatment of diabetic complications (Ku et al. 2015).

Asparagus adscendens Buch. Ham. ex Roxb., Liliaceae, Shweta musali

Asparagus adscendens is a suberect, prickly medicinal plant. The water extract of *A. adscendens* was found to have antidiabetic potentials as it stimulated both the secretion and action of insulin as well as inhibition of starch digestion in clonal pancreatic β-cell line (Mathews et al. 2006).

Asparagus racemosus Willd., Liliaceae

Synonym: *Asparagus fasciculatus* R.Br., *Asparagus acerosus* Roxb. (Figure 2.14)

Common Names: Shatavari, satavar

Description: *Asparagus fasciculatus* is a woody climbing herb with small and uniform leaves like pine needles and white flowers in small spikes; it contains an adventitious root system with tuberous roots. The branches contain spines on them. Inflorescences develop after axillary cladodes, each a many-flowered raceme or 1–4 cm panicle; the flowers are white and unisexual in nature, with equal stamens; the fruits are globular or obscurely three lobed, with pulpy berries that are purplish-black when ripe; the seeds are hard and brittle.

Distribution: *A. fasciculatus* is native to India including the Andamans; the herb is distributed in tropical and subtropical forests and in central parts of India.

Traditional Medicinal Uses: In Ayurvedic medicine, the root of shatavari is used in the form of a juice, paste, decoction, and powder to treat intrinsic hemorrhage, peptic ulcer, diarrhea, piles, hoarseness of voice, cough, arthritis, diseases of the female genital tract, erysipelas, and fever and as an aphrodisiac

FIGURE 2.14   *Asparagus racemosus.* (a) Whole plant. (b) Flower.

and rejuvenative. The herb is also used to promote milk secretion and as a demulcent, diuretic, anti-spasmodic, and tonic to the nerve. Further, it is used in the treatment of stomach ulcers, lung abscess, menopause, herpes, chronic fevers, dyspepsia, diarrhea, dysentery, tumors, inflammations, hyperdip-sia, neuropathy, hepatopathy, cough, bronchitis, hyperacidity, and certain infectious diseases. The root has been referred as bittersweet, emollient, cooling, rejuvenating, antiseptic, and carminative (review: Chawla et al. 2011; Sharma and Sharma 2013).

Pharmacology

Antidiabetes: The *A. racemosus* root is known to reduce blood glucose in rats and rabbits and has been shown to enhance insulin secretion in perfused pancreas and isolated islets. The ethanol extract and each of the hexane, chloroform, and ethyl acetate partition fractions concentration-dependently stimulated insulin secretion in isolated perfused rat pancreas, rat islets, and clonal β-cells (Hannan et al. 2007). Inhibition of *A. racemosus*-induced insulin release was observed with diazoxide and verapamil. The extract and fractions increased intracellular calcium levels, consistent with the observed abolition of insulin secretion under calcium-free conditions (Hannan et al. 2007). These studies suggest that there may be active molecules that stimulate physiological insulinotropic pathways.

The anti-DM activity of the alcohol extract of the root (150 mg/kg) was also shown in alloxan-induced diabetic rats (Dheeba et al. 2012). Further, the extract exhibited antihyperlipidemic prop-erty in alloxan-induced diabetic rats as evidenced by improvement in lipid profiles (Dheeba et al. 2012). When administered orally along with glucose, the ethanol extract of the plant root improved glucose tolerance in normal as well as diabetic rats. The extract suppressed postprandial hyper-glycemia after sucrose ingestion and reversibly increased unabsorbed sucrose content throughout the gut. The extract also inhibited the absorption of glucose during *in situ* gut perfusion with glucose. Further, the plant root extracts enhanced glucose transport and insulin action in 3T3-L1 adipocytes in culture conditions. Daily administration of the extract to type 2 diabetic rats for 28 days decreased the levels of serum glucose and plasma insulin and increased liver glycogen content, pancreatic insulin levels, and total oxidant status. Thus, the mechanism of action of the plant extract is inhibition of carbohydrate digestion and absorption as well as enhancement of insulin secretion (Hannan et al. 2012). The water extract of *A. racemosus* leaves ameliorated early diabetic nephropathy in streptozotocin-induced diabetic rats (Somani et al. 2012). There is an urgent need to identify the active principles involved in the anti-DM action of this promising plant. *A. racemosus* may also be used as a dietary adjunct for the management of DM. Phytochemical studies reported the presence of kaempferol, quercetin, and rutin (compounds with anti-DM prop-erties) in this plant.

Other Activities: This plant possesses a variety of biological properties. The roots and rhizomes of *A. racemosus* have potent antioxidant, antitussive, antidyspepsia, immunomodulatory, antiulcer, anticancer, and aphrodisiac activities. The roots were found to possess antioxidant, antitumor, antihepatotoxic, anticancer, antiulcerogenic, anti-inflammatory, and antimicrobial activities (review: Chawla et al. 2011; Sharma and Sharma 2013).

Phytochemicals: Recent chemical analysis indicates that the following active constituents are present in *A. racemosus*: steroidal saponins, known as shatavarins (I, IV), sarsasapogenin, adscendin (A, B), and asparanin (A, B, C). Shatavarin 1 is the major glycoside with 3 glucose and rhamnose moieties attached to sarsasapogenin. Shatavarin IV is a glycoside of sarsasapogenin having two molecules of rhamnose and one molecule of glucose. Sarsasapogenin and shatavarin I–IV are present in the roots, leaves, and fruits. A new isoflavone, 8-methoxy-5, 6, 4′-trihydroxyisoflavone-7-*O*-β-D-glucopyranoside, was also reported. Polycyclic alkaloid called asparagamine, a new 9,10-dihydrophenanthrene derivative named racemosol, kaempferol, oligofurostanosides (curillins G and H), and spirostanosides (curilloside G and H) have been isolated from the roots. Other chemical constituents of *A. racemosus* include essential oils, asparagine, arginine, tyrosine, flavonoids (kaempferol, quercetin, and rutin), resin, and tannin. Its leaves mainly contain rutin, diosgenin, and flavonoid as quercetin 3-glucuronide. The flowers contain quercetin hyperoside and rutin (review: Chawla et al. 2011; Sharma and Sharma 2013).

Aspidosperma macrocarpon Mart., Apocynaceae
The extract from *Aspidosperma macrocarpon* inhibited mammalian α-glucosidase activity (Silva et al. 2009).

Aspilia pluriseta Schweinf., Compositae
The antidiabetic activity of *Aspilia pluriseta* was assessed by intraperitoneally injecting varying doses of the water extract of the plant into alloxan-induced diabetic mice. The extract showed hypoglycemic activity (Piero et al. 2011).

Aster koraiensis Nakai, Asteraceae
The ethanol extract of the aerial parts of *Aster koraiensis* reduced the development of diabetic nephropathy in streptozotocin-induced diabetic rats (Eunjin et al. 2010).

Asteracantha longifolia (L.) Nees, Acanthaceae
Asteracantha longifolia is a medicinal herb used to treat various common ailments in traditional medicine. The ethanol extract of *A. longifolia* at a dose of 300 mg/kg markedly lowered blood glucose levels and increased serum insulin levels and liver glycogen content in alloxan-induced diabetic rats (Ramesh and Ragavan 2007). In another study, the water extract of *A. longifolia* leaf (100–400 mg/kg) decreased blood glucose, glycosylated hemoglobin, and total cholesterol levels and increased the levels of insulin and antioxidant enzymes in alloxan-induced diabetic male rats; the treatment also improved the status of β-cells in diabetic rats as seen from histological observations (Muthulingam 2010). In a recent study also, the antidiabetic activity of the alcohol extract of this plant in alloxan-induced diabetic rats was reported. Oral administration of the extract (250 mg/kg) for 21 days reduced the levels of blood glucose levels; acute treatment also reduced the blood glucose levels (Doss and Anand 2014).The plant appears to improve β-cell function. Further studies are needed to identify active principles and determine details of mechanisms of action and safety.

Astianthus viminalis (H.B.K.) Baill., Bignoniaceae
Astianthus viminalis has been used in traditional medicine in Mexico for the treatment of diabetes, inflammatory diseases, and more (Alarcon-Aguilar and Roman-Ramos 2005). The hypoglycemic ingredients of *A. viminalis* such as oleanolic acid, ursolic acid, and a new tetracyclic triterpenoid were isolated from the chloroform extract of their leaves. The new tetracyclic triterpenoid showed potent hypoglycemic and hypolipidemic effects in streptozotocin-induced diabetic mice. Continuous administration of the compound (30 mg/kg, orally) led to significant decrease in the level of glucose, TGs, total cholesterol, LDL, and VLDL in diabetic mice (Perez-Gutierrez et al. 2009).

Astilbe grandis Stapf ex E.H.Wilson, Saxifragaceae

Synonym: *Astilbe koreana* (Kom.) Nakai

Five oleanane-type triterpene PTP1B inhibitors, including 3α,24-dihydroxyolean-12-en-27-oic acid, were isolated from the methanol extract of the rhizomes of *Astilbe koreana*. Two of the isolates exhibited strong inhibitory activity with IC_{50} values of 4.9 and 5.2 μmol/L, respectively (Jiang et al. 2013).

Astragalus membranaceus Bunge, Fabaceae

A polysaccharide-enriched extract from *Astragalus membranaceus* showed hypoglycemic effect in diet-induced insulin-resistant C57BL/6j mice (Mao et al. 2009). Extracts of *A. membranaceus* significantly activated PPAR-α and PPAR-γ. Bioassay-guided fractionation resulted in the isolation of the isoflavones, formononetin, and calycosin from *A. membranaceus* as the PPAR-activating compound (Shen et al. 2006). Formononetin, an isoflavone isolated from the ethanol extract of this plant, activated PPAR-γ (review: Wang et al. 2014). *A. membranaceus* root exhibited protective effect against diabetic nephropathy in animal models (Zhang et al. 2009b).

Asystasia gangetica (L.) T. Anderson, Acanthaceae, a Weed

Asystasia gangetica showed anti-DM and antioxidant activity in alloxan-induced diabetic rats. The treatment almost normalized blood glucose levels and liver glycogen content in diabetic rats. Further, the treatment increased the activity of antioxidant enzymes and decreased lipid peroxide levels compared to untreated diabetic rats (Kumar et al. 2010d).

Atractylodes japonica Koidz., Asteraceae

A water extract of the oriental crude drug byaku-jutsu, from *Atractylodes japonica* rhizomes, showed hypoglycemic activity in mice. The extract was fractionated by monitoring the pharmacological activity to obtain three glycans, atractans A, B, and C. These constituents exerted significant hypoglycemic actions in normal and alloxan-induced hyperglycemic mice (Konno et al. 1985c).

Atractylodes lancea (Thunb.) DC or *Atractylodes chinensis* DC, Compositae

Atractylodes lancea is a traditional medicinal plant in China. Eudesmol, a sesquiterpenoid alcohol isolated from *A. lancea*, potentiated succinylcholine-induced neuromuscular blockade, and this effect was greater in diabetic muscles than normal ones (review: Chauhan et al. 2010). Further studies are required to validate its traditional use.

Atractylodes macrocephala Koidz, Asteraceae

A complex polysaccharide (AMP-B) isolated from the root of *Atractylodes macrocephala* showed hypoglycemic activity in alloxan-induced diabetic rats. AMP-B was found to markedly reduce blood glucose level in diabetic rats at doses of 50, 100, and 200 mg/kg (i.g.), but no effect in normal rats. AMP-B was found to recover pancreas damage in alloxan-induced diabetic rats (Shan and Tian 2003).

Atriplex halimus L., Chenopodiaceae

A shrubby saltbush found all over the Mediterranean region and North Africa. The water extract of the *Atriplex halimus* leaf decreased blood glucose levels in streptozotocin-induced diabetic rats. The treatment improved glucose tolerance and also increased body weight and total protein levels of diabetic rats compared to untreated diabetic control rats (Chikhi et al. 2014). Several *in vitro* studies and studies on alloxan-induced diabetic rats suggest that *A. halimus* leaf ash containing trace amounts of chromium potentiates the effect of insulin on glucose uptake (review: Haddad et al. 2005).

Averrhoa bilimbi L., Oxalidaceae

Common Names: Belambu, bilimpi, bilimbi, etc.

Averrhoa bilimbi is a small tree with alternate exstipulate and pinnate leaves; its fruit is a berry, oblong with rounded lobes carrying nonarillate seeds. It is cultivated as a garden plant throughout the hotter parts of India. In traditional medicine, the plant is used in the treatment of hemorrhage, stomach troubles, piles, scurvy, inflammation, hepatitis, fever, diarrhea, biliousness, beriberi, cough, itch, mumps, oph-thalmia, osteosis, pimples, rheumatism, swelling, syphilis, thrush, and many other ailments (Kirtikar and Basu 1975).

A. bilimbi leaf extract lowered blood glucose and lipids in streptozotocin-induced diabetic rats (Pushparaj et al. 2000). An active fraction from the alcohol extract of the leaves ameliorated hyper-glycemia and associated complications in streptozotocin-induced diabetic rats (Pushparaj et al. 2001, 2005). The leaf extract also has antilipid peroxidative, antiatherogenic, and hypotriglyceridemic proper-ties (Pushparaj et al. 2000).

Axonopus compressus (Sw.) P. Beauv., Poaceae

Synonyms: *Axonopus compressus* (Sw.) P. Beauv. var. *australis* G. A. Black, *Milium compressum* Sw., *Paspalum compressum* (Sw.) Nees, *Paspalum platycaule* Willd. ex Steud., *Paspalum platycaulon* Poir.

Common Name: Blanket grass

Axonopus compressus is commonly used by the people of Southern Nigeria to treat different ail-ments such as common cold and diabetes. The methanol extract of the leaf of this plant (250, 500, and 1000 mg/kg, p.o.) at all the doses studied caused a time-dependent reduction of the blood glucose levels in diabetic rats when compared to the diabetic control group (Ibeh and Ezeaja 2011).

Azadirachta indica A. Juss. Meliaceae (Subfamily Melioideae)

Synonyms: *Melia azadirachta* L., *M. indica* (A. Juss.) Brand, *M. indica* Brand. (monograph: WHO 2007) (Figure 2.15)

Common Names: Neem tree, Indian lilac tree, margosa tree (Eng.), nimba, arista, nim, vepa, vepu, etc. (Indian names)

Description: Neem is a large tree of 40–60 ft height with alternate and imparipinnately compound leaves, with subopposite, serrate leaflets that are very unequal at base; its flowers are small born in axil-lary panicles. It produces yellow drupes that are ellipsoid, glabrous, and 12–20 cm long. The fruits are green turning yellow when ripe and are aromatic.

Distribution: Neem is native of India and Burma and naturalized in most of the tropical and subtropical countries and distributed widely in the world. It grows in Southeast Asia, West Africa, Caribbean, and South and Central America.

Traditional Medicinal Uses: It is used as an astringent, antiperiodic, stomachic, purgative, and emollient; it is used in skin diseases like eczema, septic sores, ulcers due to burns, bleeding gum, pyorrhea, diabetes, scrofula, indolent ulcers, sores and ringworm, rheumatism, cholera, carbun-cles, fever, measles, malaria, smallpox, snake bite, gingivitis, and nausea. The leaf juice is given in gonorrhea and leukorrhea and is also used as nasal drop to treat worm infections in nose; the leaves are applied as poultice to relieve boils. Infusion of the flower is given for dyspepsia and general debility. The seed oil is used for the treatment of leprosy, syphilis, eczema, and chronic ulcer (Hashmat et al. 2012).

Pharmacology

Antidiabetes: The leaf water extract and seed oil administration resulted in a hypoglycemic effect in normal as well as alloxan-induced diabetic rats (Khosla et al. 2000; Kar et al. 2003). The water extract of root and leaves also lowered blood glucose in alloxan-induced diabetic rats (Chattopadhyay 1999b). Petroleum ether extract of the kernel and husk of neem seeds protected rats from streptozotocin-induced diabetes (Gupta et al. 2004). Administration of neem kernel powder (500 mg/kg) as well as a combination of the neem (250 mg/kg) and glibenclamide (0.25 mg/kg) decreased the concentration of serum lipids, blood glucose, and activities of serum alkaline phosphatase, acid phosphatase, lactic dehydrogenase, liver glucose-6-phosphatase, and HMG-CoA reductase activity of the liver and intes-tine of alloxan-induced diabetic rabbits. All the treatments produced an increase in liver hexokinase activity. The effects observed were greater when the treatment was given in combination than with neem kernel powder (Bopanna et al. 1997). The leaf extract blocked the inhibitory effect of serotonin

(a)

(b)

FIGURE 2.15   *Azadirachta indica.* (a) Twig with flower. (b) Twig with fruit.

and epinephrine on insulin secretion mediated by glucose (Chattopadhyay 1999a). The water extract of *Azadirachta indica* leaf produced a marked decrease in blood glucose level in normal rats. The glycogen content of the liver, skeletal muscle, and heart was increased after 1 h of extract administration. The authors conclude that increased glycogen synthesis is one of the mechanisms responsible for its hypoglycemic action (Das et al. 2014). In another study, neem leaf extract (250 mg/kg, single dose) administration to diabetic animals reduced glucose, cholesterol, lipids, creatinine, and urea, 24 h after treatment, compared to untreated diabetic control animals. Multiple daily dose administration for 15 days also reduced the aforementioned parameters in diabetic animals. Besides, neem extract (250 mg/kg) administration for 15 days increased glucose tolerance in diabetic rats (Hashmat et al. 2012).

The chloroform extract of *A. indica* also showed improved glucose tolerance in diabetic mice. Further, the extract inhibited intestinal α-glucosidase activity. Besides, the chloroform extract of the plant leaves increased the activity of glucose-6-phosphate dehydrogenase and the glycogen content in the liver and skeletal muscle after 21 days of treatment. Immunohistochemical analysis revealed regeneration of β-cells in the extract-treated diabetic mice pancreas and a corresponding increase in plasma insulin and C-peptide levels compared to untreated diabetic control animals (Bhat et al. 2011a). The water extract of *A. indica* fruit showed hypoglycemic activity in normoglycemic rabbits (Rao et al. 2012).

Whole and fractionated *A. indica* seed oil decreased blood glucose levels in alloxan-induced diabetes in New Zealand white rabbits (Koffuor et al. 2011).

The phytochemicals reported from this plant include the antidiabetes molecules β-sitosterol and quercetin.

Other Activities: This plant has been reported to possess antioxidant, antihypercholesterolemic, anti-inflammatory, and anxiolytic properties (Yanpallewar et al. 2005). Administration of the mature leaf extract of neem to rats resulted in a decrease in serum cholesterol levels without significant change in serum proteins, urea, and uric acid levels (Hashmat et al. 2012). The ethanol extract of neem leaves inhibited buccal pouch carcinogenesis in hamsters (Subapriya et al. 2004). Neem leaf ethanol extract protected rats against *N*-methyl-*N'*-nitro-*N*-nitrosoguanidine-induced gastric carcinogenesis (Subapriya and Nagini 2003). The neem bark extract has therapeutic potential for controlling gastric hypersecretion and gastroesophageal and gastroduodenal ulcers (Bandyopdhyay et al. 2004). The neem leaf extract offers antiulcer activity by blocking acid secretion through inhibition of H^+-K^+-ATPase and by preventing oxidative damage and apoptosis (Chattopadhyay et al. 2004). The stem bark extract of *A. indica* possesses antiulcer agents, which probably act via histamine H2 receptors (Raji et al. 2004). The extracts of the leaves of this plant showed antibacterial and antiviral properties (Hashmat et al. 2012). The fresh juice of the leaves showed hepatoprotective activity against paracetamol toxicity in rats (Yanpallewar et al. 2003). The leaf-mediated immune activation causes prophylactic growth inhibition of Ehrlich carcinoma and B16 melanoma in mice (Baral and Chattopadhyay 2004). It also has antifertility activity (Raji et al. 2003). The chloroform extracts from dried fresh leaves of *A. indica* showed marked inhibitory activity on epimastigote growth of *Trypanosoma cruzi*. The leaf and seed extracts of the plant have antidermatophytic activity (review: Ajikumaran and Subramoniam 2005).

Toxicity: *A. indica* leaves adversely affect sperm parameters and fructose levels in vas deferens fluid of albino rats (Ghosesawar et al. 2003). The leaf extract has genotoxic activity to mouse germ cells (Khan and Awasthy 2003).

Phytochemicals: The compounds reported to be present in the neem tree include alkaloids, flavonoids, triterpenoids, phenolic compounds, azadirachtin, carotenoids, steroids, and ketones (Hashmat et al. 2012). The seed also contains tignic acid responsible for the distinctive odor of the oil. Other compounds isolated from neem include nimbidin, flavin solannin, meliacin-solannolide, azadirachtannin, tetranorterpenoid isoazaditolide, nimbocinolide, limonoids, quercetin, triterpenoid nimocinol, β-sitosterol, nimbolin A and B, nimbene triterpenoid, miliantriol, 3-a-acetoxy-1-hydroxy-azadirachtol, and azadirachtol (a tetranortriterpenoid) (Malathi et al. 2002; Siddiqui et al. 2003).

Azorella compacta Phil, Umbelliferae

Local Name: Llareta

Aqueous and ethyl alcohol infusions of *Azorella compacta* have been used as a medicine to treat DM in the folk medicine in Chile. Diterpenoids isolated from *A. compacta* exhibited antihyperglycemic activity in streptozotocin-induced diabetic rats. The diterpenoids isolated were mulenic acid, azorellanol, and mullin-11,13-dien-20-oic acids. Administration of mulenic acid or azorellanol to diabetic rats markedly reduced hyperglycemia, and the effect was comparable to that of chlorpropamide. Azorellanol treatment resulted in elevation of serum insulin levels in diabetic rats, but mulenic acid did not influence serum insulin levels (Fuentes et al. 2005). Thus, the studies indicated the presence of two anti-DM compounds that act through different mechanisms. Follow-up studies are needed to determine the mechanism of action, toxicity, if any, etc.

Baccharis articulata (Lam.) Pers., Asteraceae

Baccharis species, commonly known as carqueja, have many traditional uses in folk medicine. Infusions or decoctions of *Baccharis articulata* are traditionally used as antidiabetic remedies in local folk medicine in Southern Brazil. Oral administration of the crude ethanol extract and *n*-butanol and water fractions of *B. articulata* reduced glycemia in hyperglycemic rats. Additionally, the *n*-butanol fraction stimulated insulin secretion and increased glycogen content in both the liver and muscles. *In vitro* incubation with the crude extract and *n*-butanol and water fractions inhibited maltase activity and the formation of AGEs. Thus, *B. articulata* exhibited antihyperglycemic and insulin secretagogue effects (Kappel et al. 2012).

Baccharis trimera (Less.) DC, Asteraceae

Baccharis trimera is a widespread South American plant known as carqueja. This plant is used as an anti-DM medicine in popular medicine in Brazil and elsewhere. In diabetic mice, the water fraction of the aerial parts of *B. trimera* (2000 mg/kg, twice daily, p.o.) reduced glycemia after a 7-day treatment. This effect was not associated with a body weight reduction. The fraction did not exhibit any acute effect on glucose levels (Oliveira et al. 2005).

Bacopa monnieri (L.) Wettst., Scrophulariaceae

Bacopa monnieri is an important traditional medicinal herb known for its memory-enhancing property. The ethanol extract of *B. monnieri* produced a significant decrease in the blood glucose level when compared with the controls in alloxan-induced diabetic rats in both single dose and multiple dose experiments. The ethanol extract reversed the weight loss of diabetic rats. Administration of the extract decreased the thiobarbituric acid reactive substances (TBARS) levels, increased the GSH content, and increased the SOD and catalase activity in the liver of diabetic rats. The extract prevented elevation of glycosylated hemoglobin *in vitro*, with IC_{50} value being 11. 25 μg/mL. The extract increased glucose uptake in the diaphragm of diabetic rats *in vitro*, which is comparable with the action of insulin. Thus, the antihyperglycemic effect of the extract might be due to an increase in peripheral glucose consumption as well as protection against oxidative damage in alloxan-induced diabetes (Ghosh et al. 2008). The aqueous ethanol extract of *B. monnieri* (whole plant) modulated oxidative stress in the brain and kidney of streptozotocin-induced diabetic rats (Kapoor et al. 2009).

Bacosine, a triterpene isolated from the ethyl acetate fraction of the ethanol extract of *B. monnieri*, produced a significant decrease in the blood glucose level in alloxan-induced diabetic rats when compared with the diabetic control rats in both the single administration and multiple administration study. It was observed that the compound reversed the weight loss of diabetic rats, returning the values to near normal, and increased the glycogen content in the liver. Bacosine also prevented elevation of glycosylated hemoglobin *in vitro* with an IC_{50} value of 7.44 μg/mL. Administration of bacosine decreased the levels of MDA (TBARS) and increased the levels of reduced glutathione and the activities of SOD and catalase in the liver of diabetic rats. Bacosine increased glucose utilization in the diaphragm of diabetic rats *in vitro*, which is comparable with the action of insulin. Thus, bacosine might have insulin-like activity, and its antihyperglycemic effect might be due to an increase in peripheral glucose consumption as well as protection against oxidative damage in alloxanized diabetes (Ghosh et al. 2011). Bacosides from this plant are known to increase memory and learning.

Balanites roxburghii Plunch, Balanitaceae

Synonym: *Balanites aegyptiaca* (L.) Del., *Ximenia aegyptiaca* L.

Common Name: Desert date

Balanites aegyptiaca is an evergreen xerophytic medicinal tree distributed in the tropical regions of Asia, Africa, etc. In ethnomedicine including folk medicine, *B. aegyptiaca* is used to treat jaundice, intestinal worm infection, wounds, malaria, syphilis, epilepsy, dysentery, diarrhea, stomachaches, and fever (Daya et al. 2011). Fruits of *B. aegyptiaca* are used in Egyptian folk medicine to treat DM.

Antidiabetes Activity: The water extract of the mesocarp of the fruits of this plant exhibited antidiabetic activity against streptozotocin-induced diabetic mice; the extract also ameliorated impaired renal function of diabetic rats (Mansour and Newairy 2000). In another study, *B. aegyptiaca* (fruit extract) showed hypoglycemic, hypolipidemic, and liver protective properties in senile diabetic rats. The fruit flesh was found to increase serum insulin levels and stimulate glucose metabolism (review: Gajalakshmi et al. 2013). Two new steroidal saponins were isolated from an active fraction and their structures were determined. In addition, two known saponins and their methyl ether were isolated. Interestingly, the individual saponins did not show antidiabetic activity, but the combination of these saponins showed significant antidiabetic activity (Kamel et al. 1991). The known antidiabetic compound β-sitosterol and its glucoside have been reported from this plant (Saboo et al. 2014).

Other reported activities of this plant include antioxidant, hypocholesterolemic (Abdel-Rahim et al. 1986), antibacterial, anthelmintic, antivenin, anticancer, anti-inflammatory, analgesic, antinociceptive, antiviral, and wound healing activities (Chothani and Vaghasiya 2011; Saboo et al. 2014). Phytochemicals such as flavonoids, saponins, β-sitosterol, and its glucoside are reported from this plant (Gajalakshmi et al. 2013).

Bambusa arundinacea (Retz.) Willd, Poaceae

Synonym: *B. bambos* Druce

Bambusa arundinacea is a spiny bamboo used in traditional medicine. The water extract (500 mg/kg, p.o.) of the leaves showed a glucose-lowering effect in both normal rats and streptozotocin-induced diabetic rats (Joshi et al. 2009). In another study, the hypoglycemic activities of the ethanol extract and its fractions of *B. arundinacea* leaves were evaluated in streptozotocin-induced diabetic rats. The ethyl acetate fraction of the ethanol extract (150 mg/kg) of this plant's leaves lowered blood glucose levels in diabetic rats, comparable to that of glibenclamide. Further, the treatment elevated antioxidant enzymes and decreased oxidative stress in diabetic rats. Phytochemical analysis showed the presence of β-sitosterol glucoside and stigmasterol in the leaves (Nazreen et al. 2011). The aqueous ethanol extracts of *B. arundinacea* seeds also lowered blood glucose levels in alloxan-induced diabetic rats, and the effect of the extract was comparable to that of glibenclamide (Marcharla 2011).

In another study, the ethanol extract of the root of *B. arundinacea* showed hypoglycemic potential in alloxan-induced diabetic rats and glucose-loaded normal hyperglycemic rats. Preliminary phytochemical study revealed the presence of flavonoids, tannins, and phenolic compounds in the extract (Sundeepkumar et al. 2012). Thus, the root, leaves, and seeds of this plant have hypoglycemic effects. The presence of α-amyrin and its derivatives in *B. arundinacea* (*B. bambos*) has been reported (review: Aakruti et al. 2013). α-Amyrin and its derivatives and β-sitosterol glucoside are known antidiabetic principles.

Bambusa vulgaris Schrad. ex Wendl., Poaceae

Common Names: Yellow bamboo, manjamula, etc.

Bambusa vulgaris is a large bamboo with green, yellow, or striped hollow and polished stem. The nodes are hardly raised, usually with a ring of brown hairs. The leaves are petiolated with a twisted and scabrid tip. This plant is cultivated throughout India; it originated in the old world, probably in tropical Asia. The leaves of *B. vulgaris* possess several bioactivities and are used in traditional medicinal systems.

The water extract of *B. vulgaris* leaves significantly lowered the fasting blood glucose level and markedly improved glucose tolerance in rats. A maximum hypoglycemic activity was observed at 3 h and the activity was better than that of tolbutamide (Fernando et al. 1990). In another study, the antidiabetic activity of petroleum ether extract of *B. vulgaris* leaves was evaluated in streptozotocin-induced diabetic rats. Administration of the extract (200 and 400 mg/kg daily for 15 days, p.o.) resulted in a dose-dependent reduction in the blood glucose levels in diabetic rats. This effect was comparable to that of glibenclamide. The preliminary phytochemical study showed the presence of phytosterols and tannins in the extract. Based on preliminary toxicity evaluation in rats, petroleum ether extract was nontoxic up to the dose of 2000 mg/kg (Senthilkumar et al. 2011). Reported phytochemicals include hydrocyanic acid, silicic acid, β-sitosterol, campesterol, stigmasterol, 4-hydroxy benzaldehyde, and taxiphyllin (*The Wealth of India* 1988). The anti-DM properties of water as well as petroleum ether extract indicate the presence of more than one active molecules. Further studies are needed on this plant to determine its utility as an anti-DM medicine.

Barleria lupulina Lindl., Acanthaceae

The extract of the aerial part of this plant reduced blood glucose levels in streptozotocin-induced diabetic rats (Sabu et al. 2004).

Barleria prionitis L. (*Barleria prionitis* L.), Acanthaceae

The alcohol extract of *Barleria prionitis* leaves (200 mg/kg daily for 14 days) decreased blood glucose and glycosylated hemoglobin levels in alloxan-induced diabetic rats. Further, the treatment resulted in improvement in body weight and increase in serum insulin level and liver glycogen content in diabetic rats compared to diabetic control animals (Dheer and Bhatnagar 2010).

Barringtonia racemosa (L.) Roxb., Lecythidaceae

Synonym: *B. racemosa* (L.) Spreng

Bartogenic acid isolated from *Barringtonia racemosa* seeds inhibited α-glucosidase and amylase (Gowri et al. 2007).

Basella rubra L., Basellaceae (= *Bas. alba* L.)

Synonym: *B. lucida*

Common Name: Malabar spinach

Basella lucida is a heat-tolerant fast-growing perennial vine. It is native to tropical Southern Asia. Considered to be one of the best tropical spinach, it is an important leafy vegetable grown for its nutritive value. The leaf is used in traditional medicine to cure digestive disorders, skin diseases, bleeding piles, pimples, urticaria, irritation, whooping cough, leprosy, aphthae, insomnia, cancer, gonorrhea, burns, liver disorder, ulcers, diarrhea, etc. (Deshmukh and Gaikwad 2014).

Treatment with the water extract of *Basella alba* leaves brought back the increased blood glucose level to almost normal in streptozotocin-induced diabetic rats. Further, the extract of leaves markedly increased the levels of antioxidant enzymes in the liver compared to untreated diabetic control rats. Thus, the plant has hypoglycemic and antioxidant properties (Nirmala et al. 2009, 2011; review: Deshmukh and Gaikwad 2014).

Other pharmacological activities reported include androgenic, anti-inflammatory, antimicrobial, antiviral, hepatoprotective, and wound healing activities. The plant is rich in flavonoids, saponins, carotenoids, many amino acids, and organic acids (Deshmukh and Gaikwad 2014).

Bauhinia divaricata L., Leguminosae

Bauhinia divaricata is used in traditional medicine in Mexico to treat DM. A decoction of *B. divaricata* leaf significantly decreased hyperglycemia induced by the subcutaneous injection of 50% dextrose solution in normal healthy rabbits (Roman-Ramos et al. 1992).

Bauhinia forficata Link, Fabaceae

Synonym: *Bauhinia candicans* Benth., *Bauhinia breviloba* Benth. (Figure 2.16)

(a)   (b)

FIGURE 2.16   *Bauhinia forficata.* (a) Twig. (b) Dried fruits.

Common Names: Orchid tree, cow paw, cow's foot (Eng.); pata de vaca, casco de vaca (Spanish)

Description: *Bauhinia forficata* is a perennial shrub or small tree that can be found in the rain forests. In morphology, *B. forficata* has thorns, white flowers, and leaves that are divided by two lobes, and in adult trees, the leaves can reach 7–12 cm.

Distribution: *B. forficata* occurs in Argentina, Bolivia, Brazil, Paraguay, Peru, Suriname, Uruguay, India, and Tonga. It is a native of Argentina, Bolivia, Brazil Paraguay, Peru, Suriname, and Uruguay (Contu 2012).

Traditional Medicinal Uses: The leaves of this plant are used in traditional medicine to treat DM. *B. forficata* is frequently used as an antidiabetic herbal medicine. It is also used as a diuretic for kidney and urinary disorders (including polyuria, cystitis, and kidney stones), as a blood cleanser, to build blood cells, and for high cholesterol (Contu 2012).

Pharmacology

Antidiabetes: Several studies on experimental animals have shown that the water extract, decoction, and flavonoid fraction of the leaves of *B. forficata* possess hypoglycemic properties. Decoction of the leaves (when administered to streptozotocin-induced diabetic rats [150 g leaf/L] for 1 month as a drinking water substitute) decreased serum and urinary glucose levels and urinary urea levels as compared to untreated diabetic rats. The authors think that the decoction may act in a way similar to bigunaides (Pepato et al. 2002). In a follow-up study, the authors have shown that the decoction is not toxic to diabetic and normal rats when administered for 33 days, as judged from the activity of serum toxicity markers lactate dehydrogenase, creatine kinase, amylase, angiotensin-converting enzyme, and the level of bilirubin (Pepato et al. 2004). The leaf extract reduced both hyperglycemia and hyperlipidemia in alloxan-induced diabetic rats (Viana et al. 2004a). The dried extracts (200 mg/kg, p.o., for 7 days) of *B. forficata* leaves exhibited hypoglycemic activity in streptozotocin-induced diabetic rats. The spray-drying or oven-drying processes applied to the preparation *B. forficata* extracts did not significantly alter its flavonoid profile or its hypoglycemic activity (da Cunha et al. 2010).

In normal and alloxan-induced diabetic rats, the butanol fraction (500 mg/kg) of the plant leaves exhibited an acute glucose-lowering effect. The maximum effect of the butanol fraction at 800 mg/kg was detected in diabetic animals 1 h after administration, and this profile was maintained for the next 3 h. Treatment of normal and alloxan-induced diabetic rats with butanol fraction decreased glucose levels, while this fraction was devoid of a hypoglycemic effect in glucose-fed hyperglycemic normal rats (Silva et al. 2002). The butanol fraction of the methanol extract of this plant's leaves reduced plasma glucose levels and urinary glucose excretion in induced diabetic rabbits. Further, the fraction reduced plasma glucose levels in normal as well as glucose-loaded rabbits (Fuentes et al. 2004).

The water extract of the leaves improved body weight and decreased glucose levels in nonobese diabetic mice. In diabetic mice, the salivary glands were characterized by involution of the secretory epithelium, presence of an inflammatory infiltrate, and an increase of extracellular fibrillar components. These changes were, however, not reversed by the treatment (Curcio et al. 2012).

The active principles involved are found in the flavonoid fraction and flavonoid glycosides. Kaempferitrin (kaempferol-3,7-O-(α)-L-dirhamnoside) is the predominant flavonol glycoside found in *B. forficata* leaves. Kaempferitrin was found to have an acute blood glucose-lowering effect in diabetic rats and to stimulate the glucose uptake percentile as efficiently as insulin in muscles of normal rats. This compound did not have any effect on glucosuria or on protein synthesis in muscles of normal and diabetic animals. Thus, the hypoglycemic effect and the prompt efficiency of the kaempferitrin in stimulating [U-14C]-2-deoxi-D-glucose uptake in muscles constitute the first evidence to indicate that the acute effect of this compound on blood glucose lowering may occur as a consequence of the altered intrinsic activity of the glucose transporter (Jorge et al. 2004). The hypoglycemic activity of glycosylated flavonoids, kaempferol-3,7-O-(α)-dirhamnoside and kaempferol-3-neohesperidoside, in experimental diabetic rats are enhanced substantively when complexes were prepared with vanadium (IV) (Cazarolli et al. 2006). Flavonoids from *Bauhinia megalandra* leaves inhibited intact microsomal glucose-6-phosphatase. The highest inhibitory activity was exhibited by quercetin 3-O-α-(2″-galloyl)rhamnoside and kaempferol 3-O-α-(2″galloyl)rhamnoside; astilbin, quercetin 3-O-α-rhamnoside, kaempferol 3-O-α-rhamnoside, and quercetin 3-*O*-α-arabinoside an intermediate effect; and quercetin and kaempferol the lowest effect (Estrada et al. 2005). Kaempferitrin, a major constituent of this plant leaf's phytochemicals, activates the insulin signaling pathway and stimulates secretion of adiponectin in 3T3-L1 adipocytes (Tzeng et al. 2009). Kaempferitrin treatment resulted in an upregulated level of phosphorylation on IR β and IR substrate 1 and ser473 site in PKB/Akt. PI3K acted upstream of PKB/Akt phosphorylation and GLUT4 translocation, as the inhibitor of PI3K wortmannin abolished both. GLUT4 translocated to the membrane and the GLUT4 protein level increased upon kaempferitrin stimulation. Kaempferitrin also stimulated more sustained adiponectin secretion than insulin did (Tzeng et al. 2009). The flavonoids also inhibit the activity of α-glucosidase (Ferreres et al. 2012). Thus, it appears that the flavonoid glycosides of *B. forficata* act through multiple mechanisms to improve glycemia in diabetic animals.

Clinical Trial: In a clinical trial, administration of aqueous extract from the leaves (daily for 75 days) to human patients with type 2 DM resulted in substantial reduction in glycemia (Moraes et al. 2010). In another clinical trial, infusions of leaves of *B. forficata* (3g/day for 56 days) had no hypoglycemic effect on type 2 DM patients (Russo et al. 1990).

Other Activities: Other reported activities include antioxidant properties, anticoagulant effect, antifibrinogenolytic activity, anxiolytic effect, and antihypertensive effect (dos Anjos et al. 2013).

Phytochemicals: Anti-DM compounds, β-sitosterol and kaempferol-3,7-dirhamnoside (kaempferitrin), are present in *B. forficata* leaves. Several chemical constituents, including lactones, terpenoids, glycosides, saponins, glycosyl steroids, β-sitosterol, alkaloids, flavonoids, mucilage, essential oils, tannins, and quinines, have been isolated and identified from this species (Souza et al. 2009).

Bauhinia multinervia (Kunth) DC, Caesalpiniaceae

Synonym: *Bauhinia megalandra* Griseb.

From the methanol extract of *B. megalandra* (*Bauhinia multinervia*) leaves, eight flavonoids were isolated and evaluated by rat liver microsomal glucose-6-phosphatase bioassay, which might be a useful methodology for screening antihyperglycemic substances. All the flavonoids assayed showed an inhibitory effect on the intact microsomal glucose-6-phosphatase: quercetin and kaempferol exhibited the lowest effect and astilbin, quercetin 3-*O*-α-rhamnoside, kaempferol 3-*O*-α-rhamnoside, and quercetin 3-*O*-α-arabinoside an intermediate effect. The highest inhibitory activity was shown by quercetin 3-*O*-α-(2″-galloyl)rhamnoside and kaempferol 3-*O*-α-(2″galloyl)rhamnoside. None of the flavonoids mentioned earlier showed an inhibitory effect on the disrupted microsomal glucose-6-phosphatase (Estrada et al. 2005). Follow-up studies including *in vivo* studies are warranted to establish its antidiabetic property.

Bauhinia purpurea L., Caesalpiniaceae/Fabaceae

Common Names: Pink bauhinia (Eng.), khairwal (Hindi), mandari (Tamil)

Bauhinia purpurea is an evergreen, moderate-sized tree distributed in the sub-Himalayan tract at an elevation of 4000 ft and also in the Western Indian Peninsula. In traditional medicine, it is used as an astringent, anthelmintic, and anodyne and used in insect sting, convulsion, dropsy, diabetes, epilepsy, intoxication, rheumatism, hemorrhage, snake bite, tumor, septicemia, diarrhea, etc. (*The Wealth of India* 1988).

Antidiabetes: An active antidiabetic principle has been isolated from the areal part of the plant (Abdel et al. 1987). The hypoglycemic activity of the ethanol extract and the purified fraction-1 of the *B. purpurea* stem were studied. The extract (100 mg/kg, i.p.) reduced serum glucose levels of alloxan-induced diabetic rats. The mechanism could be inhibition of cyclooxygenase and promotion of β-cell regeneration (Muralikrishna et al. 2008).

B. purpurea is reported to regulate the levels of circulating thyroid hormone in female mice. The plant is known to contain bioactive lectins (review: Ajikumaran and Subramoniam 2005). Reported phytochemicals include astragalin, quercetin, isoquercitrin, flavone glycoside, halcon glycosides, phytohemagglutinin, and pelargonidin-3-glucoside and its triglucoside (Yadava and Tripathi 2000).

Bauhinia rufescens Lam., Fabaceae

The methanol extract of *Bauhinia rufescens* (200, 300, and 400 mg/kg, p.o., daily for 4 weeks) showed marked reduction in blood glucose levels in alloxan-induced diabetic rats compared to diabetic control rats. Further, the treatment attenuated the elevated serum concentrations of urea and creatinine suggesting the nephroprotective effect of the extract (Aguh et al. 2013).

Bauhinia semla Wunderlin. Fabaceae

Synonym: *Bauhinia retusa* Roxb.

Proteins from *Bauhinia retusa* seed showed hypoglycemic and hypocholesterolemic effects in albino rats (Singh and Chandra 1977).

Bauhinia tomentosa L., Fabaceae

Common Name: St. Thomas tree

Bauhinia tomentosa is a very small tree. The water extract of the *B. tomentosa* leaf (100–400 mg/kg, p.o.) improved oral glucose tolerance in normal and alloxan-induced diabetic rats; the most effective dose reported was 300 mg/kg (Devaki et al. 2011a). Oral administration of the water extract of the *B. tomentosa* leaf (300 mg/kg daily for 30 days) to alloxan-induced diabetic rats resulted in increases in total protein and glycogen levels in the liver. Further, the treatment reduced glycemic parameters and lipid parameters and increased the level of HDL-C. Furthermore, the liver carbohydrate metabolizing enzymes were normalized by the administration of the extracts. Histopathological examination results of the liver, pancreas and kidney of the extract-treated diabetic rats were normal in general, thus indicating the antidiabetic efficacy of the *B. tomentosa* leaf extract (Devaki et al. 2011b). The roots of *B. tomentosa* also showed anti-DM activity. The ethanol extract of the root (250 and 500 mg/kg, i.p., daily for 14 days) decreased blood glucose levels in a concentration-dependent manner (Kaur et al. 2011). Follow-up studies are needed.

Bauhinia variegata L., Leguminosae/Fabaceae

Synonym: *Bauhinia alba* Wall. (Figure 2.17)

Common Names: Mountain ebony, Buddhist bauhinia, orchid tree, variegated bauhinia (Eng.), kanchanara, kovidara kanchan, kachnar, barial, sigappu-mandarai or sigappu-mandharai, etc. (Indian languages)

Description: *Bauhinia variegata* is a moderate-sized, deciduous tree with gray bark and prominent lenticels with short pedicelled and deeply cordate leaves. The flowers are large, light pink, sessile, or short pedicelled and born in a corymb. The pods are oblong at the base and narrowed and horned at the apex. Unlike *B. forficata*, *B. variegata* has no thorns, the flowers can be white or pink, and the leaves also have two lobes but reach a smaller size.

(a) (b)

FIGURE 2.17 *Bauhinia variegata* (flowers). (a) Flower [white variety]. (b) Flower [pink variety].

Distribution: The kachnar tree grows throughout India. It is distributed in the sub-Himalayan tract from Indus eastward and in dry forests of Eastern, Central, and South India and also in Burma, Iraq, Java, and Nepal (review: Sahu and Gupta 2012).

Traditional Medicinal Uses: In Ayurveda, the different parts, bark and leaves in particular, of this tree are used to treat various diseases. It is an astringent, alterative, tonic, carminative, and anthelmintic and is used in scrofula, skin diseases, ulcers, diabetes, hematuria, menorrhagia, leprosy, cough, piles, dermatosis, parturition, scrofula, asthma, diarrhea, snake bite, arthritis, anthelmintic, dysentery, and tumors. In addition, it is used to treat mouth ulcer, bad breath, diarrhea, jaundice, liver-related problems, loss of appetite, burning sensation upon urinating, weakness, cysts in the uterus, fibroids, tumors, etc. (*The Wealth of India* 1988). The leaves of many *Bauhinia* species are used in the treatment of diabetes by many populations of the world. In India, the stem bark is used as an antidiabetic in the Ayurvedic system of medicine.

Pharmacology

Antidiabetes: In an *in vitro* study, the ethanol extracts of *B. variegata* (20 µg/mL) leaves showed insulin-releasing effects from INS-1 cells in the presence of 5.5 mM glucose (Hussain et al. 2004). In another *in vitro* study, the ethanol extract of *B. variegata* and its major constituent, roseoside, have demonstrated enhanced insulin release from the β-cell INS-1 lines (Frankish et al. 2010). Further, isolation and intracellular localization of insulin-like proteins from the leaves of *B. variegata* have been reported (Azevedo et al. 2006). The flavonoid constituents of the ethanol extract of this plant was reported to reduce blood glucose levels (Sahu and Gupta 2012).

The ethanol (95%) extract of the *B. variegata* bark showed antidiabetic activity by reducing the blood glucose level, improving body weight, attenuating the altered lipid profile toward normal, and regenerating the islets of Langerhans in alloxan-induced diabetic rats (Koti et al. 2009). In streptozotocin-induced diabetic rats also, the water and ethanol extracts of *B. variegata* at a dose of 200 mg/kg showed both antihyperglycemic and antihyperlipidemic activities (Thiruvenkatasubramoniam and Jayakar 2010). In another study, the hydroalcoholic extract of *B. variegata* (stem hark) at a dose of 200 and 400 mg/kg (p.o., daily for 7 days) showed antihyperglycemic effects that may be attributed to increased glucose metabolism. The extract and metformin (500 mg/kg) treatment significantly reduced the blood glucose levels in alloxan-induced diabetic animals. The glucose levels reduced with a single dose on day 1 and decreased further after subsequent doses. The extract and metformin did not influence blood glucose in normal rats, suggesting that the extract has no hypoglycemic effect. Further, the extract normalized the impaired glucose tolerance with observable reduction in glucose levels from 60 to 120 min after glucose load (Kumar et al. 2012c). Although this study may suggest that insulin is not involved in the antihyperglycemic activity, former studies (Koti et al. 2009; Frankish et al. 2010) suggest improved insulin secretion. Further studies are required to clear the mechanisms of action. It appears that this plant possesses more than one antidiabetic principle and their levels may differ in different parts of the plant and they may have different mechanisms of action. There is an urgent need to identify the anti-DM principles. The stem is reported to have the antidiabetic compound β-sitosterol (Sahu and Gupta 2012).

Other Activities: The reported biological activities other than anti-DM activity include anti-inflammatory activity, immunomodulatory activity, antitumor activity, hepatoprotective activity, antibacterial activity, hemagglutinating activity, hematinic activity, antimicrobial activity, antiulcer activity, and anticarcinogenic activity (review: Sahu and Gupta 2012).

Phytochemicals

Root: Flavonol glycosides, 5,7,3,4 tetrahydroxy-3-methoxy-7-*O*-α-L-rhamnopyranosyl $(1 \rightarrow 3)$–*O*-β-D-galactopyranoside, flavanone (2S)-5, 7-dimethoxy-3,4 methylenedioxyflavanone, dihydrodibenzoxepin, and three other flavonoids have been isolated.

Stem: Quercitroside, isoquercitroside, rutoside, myricetol glycoside and kaempferol glycoside; β-sitosterol, lupeol and naringenin 5,7 dimethyl ether 4-rhamnoglucoside, and a new phenanthraquinone (bauhinione).

Leaves: Catechol, tannins, ellagic acid, sterol, and two long-chain compounds (heptatriacontane-12, 13-diol 7 dotetracont-15-en-9-ol).

Flowers: Cyanidin-3-glucoside, malvidin-3-glucoside, malvidin-3-diglucoside, peonidin-3-glucoside, and peonidin-3-diglucoside, etc. The white flowers contain kaempferol-3-galactoside and kaempferol-3-rhamnoglucoside.

Seeds: Yellow fatty oil (myristic, palmitic, stearic, lignoceric, behenic, arachidic, oleic, and linoleic acid) (review: Sahu and Gupta 2012).

Begonia malabarica Lam., Begoniaceae
The stem bark extract of *Begonia malabarica* reduced fasting and postprandial plasma glucose levels and increased the levels of blood insulin and liver glycogen content in streptozotocin-induced diabetic rats. In normal rats it exhibited hypoglycemic activity (Pandikumar et al. 2009).

Belamcanda chinensis (L.) DC, Iridaceae
The leaf extract of this plant showed a hypoglycemic effect in normal rats and streptozotocin-induced diabetic rats. The serum insulin concentration in normal rats is also enhanced by the treatment. Further, the oral glucose tolerance of streptozotocin-induced diabetic rats is largely improved by the extract treatment. Coadministration of the extract with nifedipine, a Ca(2+) ion channel blocker, or nicorandil, an ATP-sensitive K(+) ion channel opener, thoroughly abolishes the hypoglycemic effect of the extract (Wu et al. 2011).

Benincasa hispida, Cucurbitaceae
The fruit of this plant improved blood glucose and lipid levels in alloxan-induced diabetic rats (review: Behera and Yadav 2013).

Berberis aristata DC., Berberidaceae (Figure 2.18)
Local or Common Names: Indian barberry, tree turmeric (Eng.), chitra, dar-hald, rasaut (Hindi), mara-manjal (Malayalam)

Description: This is a shrub that grows up to 1.5–2 m in height with a thick woody root covered with a thin brittle bark. The leaves are cylindrical, straight, tapering, and hard with smooth spine. Flowers are yellow and arranged in drooping racemes; fruits are ovoid and smooth berry (Tamilselvi et al. 2014).

Distribution: The plant is a native to Himalayas at an elevation of 2000–3500 m. It is a red-listed endemic medicinal plant that warrants conservation and propagation.

Traditional Medicinal Uses: *Berberis aristata* is used to treat infections of the eye, nose, and throat, dysentery, indigestion, and uterine and vaginal disorders. It is also used as a tonic and to cure ulcers and wounds. This plant is used in traditional systems of medicine as an ingredient of several polyherbal formulations for treating eye diseases, diarrhea, and cholera. Tender leaf buds are used to treat dental caries.

FIGURE 2.18 *Berberis aristata.*

Pharmacology

Antidiabetes: The ethanol extracts of the stem bark of this plant reduced blood glucose levels in alloxan-induced diabetic rats (Gupta et al. 2010). *B. aristata* root (50% aqueous ethanol extract; 250 mg/kg) also showed antihyperglycemic and antioxidant activities in alloxan-induced diabetic rats. The extract increased glucokinase and glucose-6-phosphate dehydrogenase activities and decreased glucose-6-phosphatase activity in diabetic rats. Thus, it plays an important role in glucose homeostasis (Singh and Kakkar 2009). Further, the root and its water and methanol extracts exhibited hypoglycemic activity in normal and alloxan-induced diabetic rabbits (Akhtar et al. 2008). In another study also, the antihyperglycemic activity of the plant root was shown in alloxan-induced diabetic rats (Semwal et al. 2009). Several studies have shown the promising anti-DM properties of berberine present in *Berberis* sp. The plant also contains β-sitosterol, an anti-DM compound. Berberine, an alkaloid, possesses anti-DM and hypoglycemic properties. The compound has been studied for its mechanism of action. It improved insulin action by activating AMPK that helps in regulating the cellular uptake of glucose, the oxidation of free fatty acids, and the synthesis of GLUT4 (review: Arif et al. 2014; Lee et al. 2006b; Turner 2008). Berberine mimicked insulin action by increasing glucose uptake by 3T3-L1 adipocytes and L6 myocytes in an insulin-independent manner, inhibiting PTP1B activity, and increasing phosphorylation of IR, IRS1, and Akt in 3T3 L1 adipocytes. It did not increase insulin synthesis and release (Chen et al. 2010; Zhang et al. 2010b). It increased GLP-1 secretion in streptozotocin-induced diabetic rats; *in vivo* 5-week treatment with berberine enhanced GLP-1 secretion induced by glucose load–promoted L-cell (GLP-1 secreting cell) proliferation in the intestine (Yu et al. 2010). Inhibition of protein kinase C (PKC) or AMPK inhibited berberine-mediated GLP-1 secretion. Some signaling pathways including PKC-dependent pathways are involved in promoting GLP-1 secretion and biosynthesis (Yu et al. 2010). Experimentally, berberine was found to inhibit human recombinant DPP-4 *in vitro*. This inhibition is one of the mechanisms that explain the antihyperglycemic activity of berberine (Almasri et al. 2009).

Other Activities: Pharmacological studies demonstrated several cardiovascular effects of berberine and its derivatives such as positive inotropic on isolated guinea pig atria (Lau and Yao 2001), negative chronotropic activity (Shaffer 1985), antiarrhythmic activity (Chi and Hu 1997), heart failure improvement in rats (Ying et al. 2002) and humans (Zeng et al. 2003), vasodilator and antihypertensive (Liu and Chan 1999), lowering the resistance of peripheral vessels in humans (Marin 1988), and lowering cholesterol levels in the blood (Kong et al. 2004). Other pharmacological activities of this plant include anti-inflammatory activity, antimicrobial activity, cytotoxic and antitumor activity, antidiarrheal activity, antiosteoporotic activity, and hypolipidemic activity (Mokhber-Dezfuli et al. 2014; review: Tamilselvi et al. 2014).

Phytochemicals: The roots of *B. aristata* yielded berbamine, dihydroberlamine, and noroxyhydrastine. Studies on the stem identified the presence of berberine, oxyberberine, oxycanthine, ceryl alcohol, hentriacontane, sitosterol, and saponin. Phytochemical screening of the stems of this plant revealed the presence of alkaloids, phenolic compounds, tannins, flavonoids, phytosterols, saponins, and glycosides. Aromoline is also present in this plant (review: Tamilselvi et al. 2014).

Berberis brevissima Jafri, Berberidaceae
Methanol extracts of the roots of *Berberis brevissima* showed antidiabetic activities. Berberine and 8-oxo-berberine were isolated for the first time from this plant (Ali et al. 2013). The anti-DM activity was determined against PTP1B, a negative insulin regulator. 8-Oxo-berberine showed activity more than that of berberine against PTP1B (Ali et al. 2013).

Berberis parkeriana C.K.Schneid., Berberidaceae
Berberis parkeriana (methanol extract of the root) showed antidiabetic activities. Berberine and 8-oxo-berberine were isolated from *B. parkeriana* root (Ali et al. 2013). The antidiabetes activity was determined against PTP1B, a negative insulin regulator. 8-Oxo-berberine showed activity more than that of berberine against PTP1B (Ali et al. 2013).

Berberis vulgaris L., Berberidaceae
Synonyms: *B. dumetorum* (Figure 2.19); related species include *Berberis brevissima* Jafri and *Berberis parkeriana* C.K.Schneid.

FIGURE 2.19 *Berberis vulgaris.* (Photo courtesy of Arnstein Ronning.)

Local/Common Names: Barberry, rocky mountain grape (Eng.), agracejo, aigtet, kattuvally, etc.

Traditional Medicinal Uses: The fruits, leaves, and stem of this plant have been used for medical purposes including hepatoprotection, cardiotonic, and antimicrobial activity.

Distribution and Description: *B. vulgaris* is a shrub 1–3 m in height that grows in many areas of world, including Iran.

Antidiabetes: The water extract as well as saponins from *B. vulgaris* showed hypoglycemic activity in normal rats and antidiabetic activity against streptozotocin-induced diabetic rats (Meliani et al. 2011). High concentrations of berberine have been reported in this plant. Barberry crude extract (ethanol extract of defatted root) containing 0.6 mg berberine/mg crude extract showed potent antioxidative capacity. The inhibitory effect of *B. vulgaris* crude extract on α-glucosidase was more potent than that of berberine chloride, while both had the same acetylcholinesterase inhibitory effect (El-Wahab et al. 2013). Berberine is an important antidiabetic agent with other bioactivities (see under *B. aristata*). Moreover, the plant is reported to contain lupeol and oleanolic acid, known to have, among other things, anti-DM activities.

Other Activities: An *in vitro* study showed that barberry crude ethanol extract has potent antioxidant activity, antidiabetic property and anticancer effect (Abeer et al. 2013), and that different concentrations of both berberine chloride and barberry ethanol extract possess no growth inhibitory effect on normal blood cells. Otherwise, both berberine chloride and barberry ethanol extract inhibited the growth of breast, liver, and colon cancer cell lines, and the inhibitory effect increased with time in a dose-dependent manner (El-Wahab et al. 2013). Other reported activities include antihistaminic, anticholinergic, anti-inflammatory, and antinociceptive activities (review: Mokhber-Dezfuli et al. 2014).

Phytochemicals: The major isolated compounds reported from *B. vulgaris* include lupeol, oleanolic acid, stigmasterol, berberamine, stigmasterolglucoside, berberine, columbamine, palmatine, oxyberberine, isocorydine, berbamine, lambertine, mangiflorine, oxycanthine, *N*-(*p*-trans-coumaroyl) tyramine, cannabisin G, lyoniresinol, jatrorrhizine, and berberubin (Mokhber-Dezfuli et al. 2014).

Bergenia ciliata (Haw.) Sternb., Saxifragaceae

The Nepalese herb pakhanbhed (*Bergenia ciliate*) is one of the traditional remedies used for diabetes since prehistoric times. In an *in vitro* study, two compounds isolated from this plant demonstrated significant dose-dependent enzyme inhibitory activities against rat intestinal α-glucosidase and porcine pancreatic α-amylase. The active compounds are (−)-3-*O*-galloylepicatechin and (−)-3-*O*-galloylcatechin, which are reported from this plant species for the first time (Hong et al. 2008).

Bergenia himalaica Boriss., or *Bergenia pacumbis* (Buch.-Ham. ex D.Don) C.Y.Wu & J.T.Pan., Saxifragaceae (*B. himalaica* Boriss is considered a synonym)
Bergenia himalaica is mainly distributed in the temperate Himalayas from the southeastern regions in Central Asia and northern regions in South Asia. The plant has a long history of use in traditional medicine for the treatment of various diseases such as diabetes, urinary complaints, kidney stones, hemorrhagic diseases, and epilepsy. Under *in vitro* conditions, the ethanol extract of *B. himalaica* showed insulin secretagogue activity in INS-1 cells at the 1 µg/mL level (Hussain et al. 2004). In a recent study, two new compounds, bergenicin and bergelin, were isolated from the methanol extract of aerial parts of *B. himalaica*. Significant decrease of blood glucose was observed 1 h (1.0 mg/kg) and 2 h (0.5 mg/kg) after bergenicin administration to streptozotocin–nicotinamide-induced diabetic rats and 2 h (1.0 mg/kg) and 3 h (0.5mg/kg) after bergelin administration. Bergenicin, but not bergelin, enhanced glucose-stimulated insulin secretion in isolated pancreatic islets (Siddiqui et al. 2014).

Bersama engleriana Gurke, Melianthaceae
The water extract of the leaves of *Bersama engleriana* exhibited hypoglycemic activity in normal rats at a dose of 300 mg/kg 8 h after oral administration. The methanol extract was active only at a high dose (600 mg/kg) (Njike et al. 2005). In another study, administration of the water and methanol extracts from *B. engleriana* leaves (300 and 600 mg/kg daily for 4 weeks) to streptozotocin–nicotinamide-induced diabetic rats resulted in a significant decrease in the levels of blood glucose, total cholesterol, TG, and LDL; the treatment increased body weight and blood HDL levels in diabetic rats compared to the diabetic control. The methanol extract (600 mg/kg) was found to be better than the water extract and glibenclamide in this repeated dose treatment to diabetic rats (Pierre et al. 2012).

Beta vulgaris L. ssp. *maritima* (L.) Arc., Amaranthaceae
Common Name: Chard
Chard (*Beta vulgaris*) is used as a hypoglycemic agent by diabetic patients in Turkey. Chard extract treatment to streptozotocin-induced diabetic rats resulted in an increase in the number of β-cells of Langerhans islets and in the secretory granules, together with many hypertrophic Golgi apparatus and granules of low densities. The extract, while having no effect on blood glucose in the normal group, reduced the blood glucose values in streptozotocin-induced hyperglycemic animals. The authors concluded that the extract of this plant may reduce blood glucose levels by regeneration of the β-cells (Bolkent et al. 2000). *Beta vulgaris* extract ameliorated oxidative injury in the aorta and heart of streptozotocin-induced diabetic rats (Sener et al. 2002). Betavulgaroside (glucuronide saponin) was isolated from the root and leaves of this plant. The compound showed hypoglycemic effects in rats (Yoshikawa et al. 1996a).

Bidens leucantha Willd, Asteraceae
Extracts from this plant lowered blood glucose in experimental animals (Marles and Farnsworth 1995).

Bidens pilosa L., Asteraceae
Synonym: *Bidens leucantha* (L.) Willd. (Figure 2.20)
Common Names: Vernacular names include Spanish needles, beggar's ticks, devil's needles, cobbler's pegs, broom stick, pitchforks, and farmers' friends (Eng.). There are many more names in other languages.
Description: *Bidens pilosa* is an erect, perennial herb. It is either glabrous or hairy, with green opposite leaves that are serrate, lobed, or dissected. It has white or yellow flowers and long narrow ribbed black achenes (seeds). It grows to an average height of 60 cm and a maximum of 150 cm in favorable environments. *B. pilosa* prefers full sun and moderately dry soil. *B. pilosa* propagates via seeds. A single plant can produce 3000–6000 seeds that are viable for at least 3 years. Minimal agricultural techniques are required for *B. pilosa* cultivation. Due to its invasive tendencies, *B. pilosa* is generally considered to be a weed. *B. pilosa* has several varieties including *B. pilosa* var. *radiata*, var. *minor*, var. *pilosa*, and var. *bisetosa*. Alongside examination of morphological traits, authentication of *B. pilosa* can be aided by chemotaxonomy and molecular characterization (review: Bartolome et al. 2013).

(a) (b)

FIGURE 2.20 *Bidens pilosa.* (a) Flowers. (b) Immature stellate infructescence.

Distribution: *B. pilosa* is believed to have originated in South America and subsequently spread almost everywhere on earth. It is an easy-to-grow herb that is widely distributed all over the world. It is mainly distributed across temperate and tropical regions of the world. It is considered to be a rich source of food and medicine for humans and animals (review: Bartolome et al. 2013).

Traditional Medicinal Uses: *B. pilosa* is used as an herb and as an ingredient in sauces, teas, and herbal medicines.

In the 1970s, the United Nations Food and Agriculture Organization (FAO) promoted the cultivation of *B. pilosa* in Africa because it is easy to grow, edible, palatable, and safe. All parts of the *B. pilosa* plant are used as ingredients in folk medicines. In traditional medicine, it is frequently prepared as a dry powder, decoction, maceration, and tincture. *B. pilosa*, either as a whole plant or different parts, has been reported to be useful in the treatment of more than 40 disorders such as inflammation, immunological disorders, digestive disorders, infectious diseases, cancers, metabolic syndrome, and wounds. *B. pilosa* is usually ingested; however, it is also utilized externally and as an ingredient in medicinal mixtures together with other medicinal plants such as *Aloe vera*, *Plectranthus mollis*, *Valeriana officinalis*, and *Cissus sicyoides*. *B. pilosa* varieties share similar phytochemical compositions to a large extent and may be substituted for each other (review: Bartolome et al. 2013).

Pharmacology

Antidiabetes: Several studies have shown that *B. pilosa* exerted an anti-DM property. It prevented type 1 DM and controlled type 2 DM in animal experimental models (review: Yang et al. 2014). Although there are many plants active against type 2 DM, drugs active against type 1 are rare. Immune modulators may prevent or delay type 1 DM development in type 1 DM–prone individuals. Helper T (Th) cell differentiation regulates type 1 DM development. Further, Th1 cell differentiation promotes type 1 DM, whereas Th2 cell differentiation alleviates type 1 DM (review: Yang 2014).

Prevention of Type 1 DM: One study showed that *B. pilosa* extract and its butanol fraction could decrease Th1 cells and increase Th2 cells (Chang et al. 2005). This inhibition was reported to be partially attributed to selective cytotoxicity, because the butanol fraction at 180 μg/mL could cause 50% death of Th1 cells. Using the bioactivity-directed isolation and identification approach, three active polyynes (3-β-D-glucopyranosyl-1-hydroxy-6(*E*)-tetradecene-8,10,12-triyne, 2-β-D-glucopyranosyloxy-1-hydroxy-5(*E*)-tridecene-7,9,11-triyne, and 2-β-D-glucopyranosyloxy-1-hydroxytrideca-5,7,9,11-tetrayne [cytopiloyne]) were isolated from the butanol extract (Chang et al. 2004; Chiang et al. 2007). These compounds showed similar effects on Th cell differentiation as the *B. pilosa* butanol fraction. Moreover, cytopiloyne showed greater activity than the other two compounds in terms of enhancement of differentiation of Th0 to Th2 and inhibition of differentiation to Th1 (Chang et al. 2004; Chiang et al. 2007; Chang et al. 2013a,b). Further, the butanol fraction of *B. pilosa* effectively prevented type 1 DM in nonobese diabetic mice (Chang et al. 2004, 2007). Consistently, this prevention involved downregulation of Th1 cells or upregulation of Th2 cells. This was proven by intraperitoneal injection of the butanol fraction at a dose of 3 mg/kg, three times a week, to nonobese diabetic mice from 4 to 27 weeks (Chang et al. 2004). This treatment resulted in lower incidence of diabetes, whereas at a higher dose (10 mg/kg), the butanol fraction totally stopped the initiation of the disease. Th1 cytokine IFNγ and Th2 cytokine IL-4 favor the production of IgG2a and IgE, respectively. To further confirm whether this butanol fraction *in vivo* regulated Th cell differentiation and Th cytokine profiling, IgG2a and IgE production was measured in the serum of nonobese diabetic mice. As expected, high levels of IgE and some decline in the levels of IgG2a were observed in the serum (Chang et al. 2004). In another study, the nonobese mice received intraperitoneal or intramuscular injection of the most active cytopiloyne at 25 μg/kg, 3 times per week. Remarkably, 12–30-week-old mice treated with cytopiloyne showed normal levels of blood glucose (200 mg/dL) and insulin (1–2 ng/mL), whereas untreated control 12-week-old nonobese diabetic mice started to develop type 1 DM, and 70% of these mice aged 23 weeks and over developed type 1 DM. Consistent with type 1 DM incidence, cytopiloyne delayed and reduced the invasion of CD4+ T-cells into the pancreatic islets (Chang et al. 2007).

The underlying mechanism by which cytopiloyne and its derivatives inhibited type 1 DM covered inactivation of T-cells, polarization of Th cell differentiation, and Th cell depletion, leading to islet protection (Chang et al. 2007). First, [³H] thymidine incorporation assay showed that cytopiloyne inhibited ConA/IL-2- and CD3 antibody-mediated T-cell proliferation, implying that cytopiloyne could inhibit T-cell activation. Second, *in vitro* study showed that cytopiloyne inhibited the differentiation of CD4+ T-cells into Th1 cells and promoted differentiation of Th0 cells into Th2 cells (Chiang et al. 2007). The *in vitro* data are consistent with the *in vivo* results, indicating that cytopiloyne reduced Th1 differentiation and increased Th2 differentiation as shown by intracellular cytokine staining and fluorescence-activated cell sorting (FACS) analysis (Chang et al. 2007). Cytopiloyne also enhanced the expression of GATA-3, a master gene for Th2 cell differentiation, but not the expression of T-bet, a master gene for Th1 cell differentiation, further supporting its role in skewing Th differentiation. In line with the skewing of Th differentiation, the level of serum IFN-γ and IgG2c decreased, while that of serum IL-4 and serum IgE increased compared to the negative controls. Third, cytopiloyne partially depleted CD4+ rather than CD8+ T-cells in non-obese diabetic (NOD) mice (Chang et al. 2007). Coculture assays showed that the depletion of CD4+ T-cells was mediated through the induction of Fas ligand expression on pancreatic islet cells by cytopiloyne, leading to apoptosis of infiltrating CD4+ T-cells in the pancreas via the Fas and Fas ligand pathway. However, cytopiloyne did not induce the expression of TNF-α in pancreatic islet cells and, thus, had no effect on CD8+ T-cells (Chang et al. 2007). Cytopiloyne is an immunomodulatory compound rather than an immunosuppressive compound (Chang et al. 2004, 2007). The mechanism of action of cytopiloyne in type 1 DM includes inhibition of T-cell proliferation, skewing of Th cell differentiation and partial depletion of Th cells, and protection of β-cells of pancreatic islets (review: Yang 2014).

Action on Type 2 DM: *B. pilosa* has been scientifically investigated for its antidiabetic activity, which was assessed by intraperitoneally injecting varying doses of the plant water extract into alloxan-induced diabetic mice. The extract showed hypoglycemic activity (Piero et al. 2011). The aqueous ethanol extracts of the aerial part of *B. pilosa* at 1 g/kg lowered blood glucose in *db/db* type 2 diabetic mice (Ubillas et al. 2000). Bioactivity-guided identification led to the identification of two polyynes, 3-β-D-glucopyranosyl-1-hydroxy-6(E)-tetradecene-8,10,12-triyne and 2-β-D-glucopyranosyloxy-1-hydroxy-5(E)-tridecene-7,9,11-triyne, two of the three compounds discussed earlier. Moreover, the mixture of the compounds in a 2:3 ratio significantly reduced blood glucose level and food intake on the second day when administered at doses of 250 mg/kg twice a day to C5BL/Ks *db/db* diabetic mice. When evaluated at 500 mg/kg, a more substantial decrease in the blood glucose level as well as a stronger anorexia was noticed. This study suggested that these two compounds were active ingredients of *B. pilosa* for diabetes (Ubillas et al. 2000). The antidiabetic effect of both polyynes was partially caused by the hunger suppressing effect. However, the anoxic effect of the ethanol extract of *B. pilosa* was not observed in the studies described later. In this study (Hsu et al. 2009), water extracts of *B. pilosa* were tested in diabetic *db/db* mice. Like oral antidiabetic glimepiride, which stimulates insulin release, one single dose of the water extract reduced blood glucose levels from 20.8 to 8.0 mmol/L (374–144 mg/dL).

The antihyperglycemic effect of the extract was relevant to an increase in serum insulin levels, implying that the extract drops blood glucose concentration through an upregulation of insulin production (Hsu et al. 2009). In a long-term study also, the extract lowered blood glucose, boosted blood insulin, improved glucose tolerance, and reduced the percentage of HbA1c. Both one-time and long-term experiments strongly support the action of the extract on type 2 DM. Unlike glimepiride, which failed to preserve pancreatic islets, the extract significantly protected against islet atrophy in mouse pancreas (Hsu et al. 2009).

In another study, the antidiabetic effect of three *B. pilosa* variants, *B. pilosa* var. *radiata*, *B. pilosa*, var. *pilosa*, and *B. pilosa* var. *minor*, was evaluated in *db/db* mice (Chien et al. 2009). A single oral dose (10, 50, and 250 mg/kg) of methanol extracts of the three variants decreased postprandial blood glucose levels in *db/db* mice for up to 4 h, and this reduction was dose dependent. The extract of variety *radiata* exhibited a higher reduction in blood glucose levels and higher increase in insulin levels when administered at the same dose as the other two variants. The three polyynes were identified from the three variants though their contents varied. Cytopiloyne at 0.5 mg/kg exerted a better stimulation for insulin production in *db/db* mice than the other two compounds. In the 28-day treatment, the positive control (glimepiride) as well as the crude extracts of the three varieties lowered the blood glucose levels in *db/db* mice. However, only the extract of *radiata* containing a higher content of cytopiloyne reduced blood glucose levels and augmented blood insulin levels more than the two other varieties.

Cytopiloyne reduced postmeal blood glucose levels, increased blood insulin, improved glucose tolerance, suppressed HbA1c level, and protected pancreatic islets in *db/db* mice. Nevertheless, cytopiloyne never managed to decrease blood glucose in streptozocin-induced diabetic mice whose β-cells were already almost completely destroyed. In addition, cytopiloyne dose-dependently promoted insulin secretion and expression in β-cells as well as calcium influx and activation of diacylglycerol and PKCα. Taken together, the mechanistic studies suggest that cytopiloyne acts via regulation of β-cell functions (insulin production and β-cell preservation) involving the calcium/diacylglycerol/PKCα cascade. Intriguingly, 36 polyynes have been found in *B. pilosa* so far. Multiple biological activities reflect the phytochemical complexities of this herb. It remains to be investigated whether all the polyynes present in this plant have antidiabetic activities (review: Yang 2014).

Other Activities: Scientific studies, although not extensive, have demonstrated that *B. pilosa* extracts and/or compounds have, in addition to anti-DM activity, antitumor, anti-inflammatory, antioxidant, immunomodulatory, antimalarial, antibacterial, antifungal, antihypertensive, vasodilatory, and antiulcerative activities (review: Bartolome et al. 2013).

Toxicity: Preliminary studies suggest that consumption of *B. pilosa* water extract up to 1 g/kg/day is highly safe in rats. However, detailed toxicological studies and drug interactions remain to be investigated (review: Yang 2014).

Phytochemicals: More than 100 publications have documented the medical use of *B. pilosa*. To date, more than 200 compounds comprising aliphatics, flavonoids, terpenoids, phenylpropanoids, aromatics, porphyrins, and other compounds have been identified from this plant. However, the association between *B. pilosa* phytochemicals and their bioactivities is not yet fully established and should become a future research focus. *B. pilosa* is an extraordinary source of phytochemicals, particularly flavonoids and polyynes. However, the bioactivities of only 7 of the more than 60 flavonoids present in *B. pilosa* have been studied (review: Bartolome et al. 2013).

Biophytum sensitivum (L.) DC, Oxalidaceae
Biophytum sensitivum is used as a remedy to control diabetes in the Philippines. The leaf extract of *B. sensitivum* reduced the blood glucose and glycosylated hemoglobin levels in streptozotocin–nicotinamide-induced diabetic rats (Ananda Prabu et al. 2012). The whole plant water and ethanol extracts exhibit aldolase reductase inhibitory activity and anticataract activity under *in vitro* conditions (Gacche and Dhole 2011). In another study, 80% aqueous ethanol extracts of this plant were examined for its α-glucosidase enzyme inhibitory activity using yeast α-glucosidase enzyme. *B. sensitivum* extract showed yeast α-glucosidase inhibitory activity with IC_{50} of 2.24 μg/mL (Lawag et al. 2012).

Bixa orellana L., Bixaceae

Synonyms: *Bixa platycarpa* Ruiz & Pav. ex G. Don, *Bixa purpurea* Hort, *Bixa tinctoria* Salixb, *Bixa urucurana* Willd, etc. (Figure 2.21)

Local Names: Achiote, chiot, annatto tree, etc. (annatto is a pigment produced from the seed)

Description: *Bixa orellana* is a shrub or bushy tree that ranges from 3 to 10 m in height. Its glossy, ovate leaves are evergreen with reddish veins; they have a round, heart-shaped base and a pointed tip. With a thin, long stem, the leaves are between 8 and 20 cm long and 5 and 14 cm wide. The twigs are covered with rust-colored scales when young and bare when older. The flowers are pink, white, or some combination and are 4–6 cm in diameter. From the flower protrudes a striking two-valved fruit, covered either with dense soft bristles or a smooth surface. These round fruits, approximately 4 cm wide, appear in a variety of colors: scarlet, yellow, brownish-green, maroon, and most commonly bright red. When ripe, they split open and reveal numerous small, fleshy seeds, about 5 mm in diameter and covered with red-orange pulp, the embryo of which is poisonous (review: Ulbricht et al. 2012).

Distribution: *Bixa orellana* is native to the tropics of North and South America, the Caribbean, and the East Indies. It is most abundant from Mexico to Ecuador, as well as Brazil, Bolivia, Venezuela, and several other South American countries. *B. orellana* is cultivated in South America, Southeastern Asia, etc. (Ulbricht et al. 2012).

Traditional Medicinal Uses: In traditional medicine it is used to treat a variety of ailments. These include healing of minor wounds and burns and ailments of the womb or uterus; it is used as a female aphrodisiac agent and is also used to treat inflammation, asthma, pleurisy, labored breathing, fever, diabetes, flu, venereal diseases, gonorrhea, jaundice, kidney diseases, diarrhea, hemorrhoids, nausea, and vomiting; it is considered as an astringent, antiperiodic, diuretic, and purgative (review: Ulbricht et al. 2012).

Pharmacology

Antidiabetes: In West Indian folklore, *B. orellana* (annatto) is commonly used in the treatment of DM. Preliminary studies have reported that a crude annatto seed extract exhibited either a glucose-lowering or hyperglycemia-inducing effect. There is a potential danger in using the extract in any form without further detailed studies. The effect of the extract may differ depending on the nutritional status and, possibly, gut microbes. Species differences have also been reported for the effect of the extract. Therefore, caution is needed in extrapolating results of studies on animals to humans. Further, the differences in the phytochemical composition of the extracts and preparations from them may have influences on DM and toxicity. Therefore, the human use of the extract and preparations from them may be prohibited until further studies establish the safety in view of the following reports.

FIGURE 2.21 *Bixa orellana.*

The red powdery extract from the seeds of the annatto, *B. orellana*, is used as a food coloring agent. In an oil suspension, the extract of *B. orellana* is used as a folk remedy in the West Indies for DM. Detailed investigations on this extract (bush teas) yielded a methyl ester, trans-bixin, which caused hyperglycemia in anesthetized mongrel dogs. Concomitant electron microscopy of tissue biopsies revealed damage to the mitochondria and endoplasmic reticulum mainly in the liver and pancreas. When dogs were fed on a diet fortified with riboflavin, there was neither demonstrable tissue damage nor associated hyperglycemia. These findings point to (1) the potential dangers of informal medications such as "bush teas" and (2) the possible role of plant extracts/food additives in the development of DM especially in the undernourished state (Morrison et al. 1991). In contrast, an oil-soluble, partially purified annatto extract caused hypoglycemia in dogs, which was mediated by an increase in plasma insulin concentration as well as an increase in insulin binding on the IR due to the elevated affinity of the ligand for the receptor (Russell et al. 2005). In another study, annatto extract was found to decrease blood glucose levels in fasting normoglycemic and streptozotocin-induced diabetic dogs. In addition, in normal dogs, it suppressed the postprandial rise in blood glucose after an oral glucose load. The extract also caused an increase in insulin-to-glucose ratio in normal dogs. Increased insulin levels were not due to increased insulin synthesis as after 1 h residence time and half-hour postprandial, decreased C-peptide levels were observed. The authors concluded that *B. orellana* (annatto) lowered blood glucose by stimulating peripheral utilization of glucose (Russell et al. 2008).

The carotenoids bixin and norbixin are present in the seeds of *B. orellana*. In a study, the toxicity of norbixin or annatto extract containing 50% norbixin was investigated in mice and rats after 21 days of ingestion through drinking water. Mice were exposed to doses of 56 and 351 mg/kg of annatto extract and 0.8, 7.6, 66, and 274 mg/kg of norbixin. In rats, no toxicity was detected. In mice, norbixin induced an increase in plasma alanine aminotransferase activity, while both norbixin and annatto extract induced a decrease in plasma total protein and globulins. However, no signs of toxicity were detected in the liver by histopathological analysis. No enhancement in DNA breakage was detected in the liver or kidney from mice treated with annatto pigments, as evaluated by the comet assay. Nevertheless, there was a remarkable hyperglycemic effect of norbixin on both rodent species. Annatto pigment treatment showed hyperinsulinemia in rats and hypoinsulinemia in mice. More studies should be performed to fully understand how species-related differences influence the biological fate of norbixin (Fernandes et al. 2002).

Bixin and norbixin from *B. orellana* regulated mRNA expression involved in adipogenesis and enhanced insulin sensitivity in 3T3-L1 adipocytes through PPAR-γ activation (Wang et al. 2014). The hot water extract of *B. orellana* was found to have potent inhibitory activity toward lens aldose reductase. Isoscutellarein from the extract was identified as the potent inhibitor compound (Terashima et al. 1991).

Other Activities: The leaves, seeds, and fruits were subjected to pharmacological studies. Other reported activities include analgesic, anticarcinogenic, anticlastogenic, anticoagulant, antidiarrheal, antimicrobial, antioxidant, and antivenom activity. Bixin isolated from this plant inhibited cyclooxygenase; annatto exhibited diuretic and hypotensive effects. The leaves exhibited gastrointestinal effects, neurological effects, etc. (review: Ulbricht et al. 2012).

Toxicity: In animal studies, annatto at 2 g/kg daily was not toxic. However, high doses can influence the mutagenic effect of cyclophosphamide in mice (Ulbricht et al. 2012).

Phytochemicals: More than 35 compounds have been identified from *B. orellana*, of which (*Z,E*)-farnesyl acetate, occidentalol acetate, spathulenol, ishwarane, bixin, and norbixin are major constituents. The floral scent of the seed is caused by a tricyclic sesquiterpene hydrocarbon, ishwarane. Other constituents include acetone, geranylgeraniol, achiotin, tomentosic acid, carotenoids, a methyl ester, trans-bixin, apocarotenoids, and orellin (Ulbricht et al. 2012).

Blighia sapida Koenig, Sapindaceae

Common Names: Aki, arbe fricassee (French), arbol de seso, seso vegetal (Spanish); aki, akee, ackee, etc.

Description: The tree is densely branched and symmetrical with smooth gray bark. The tree reaching about 12 m possesses evergreen (rarely deciduous) alternate leaves with up to five pairs of oblong or elliptic leaflets, bright green and glossy on the upper surface, dull and paler and finely hairy on the veins on the underside. It produces bisexual male flowers, borne together in simple racemes, with five fragrant, white, and hairy petals. The fruit is a leathery, pear-shaped, more or less distinctly 3-lobed capsule. When the fruit is fully mature, it splits open revealing three glossy arils attached to the black, smooth hard seeds, normally three in number. The base of each aril is attached to the inside of the stem end of the jacket (Olubunmi et al. 2009). The fruit aril is edible when fully ripe.

Distribution: The plant has its origin in West Africa but has traversed the Atlantic making the Caribbean its home. It is cultivated in Trinidad, Haiti, West Indies, and the Bahamas. There are scattered trees in many countries including India (Olubunmi et al. 2009).

Traditional Medicinal Uses: The pulverized bark is mixed with ground hot peppers and rubbed on the body as a stimulant. The water extract of the seed is used as a parasite expellant from the intestinal tract. The crushed new foliage is applied on the forehead to relieve headache. The leaf juice is employed as eye drops in ophthalmia and conjunctivitis. Various preparation and combinations of the leaf extract have been made for the treatment of dysentery, epilepsy, yellow fever, and diabetes. The plant has been reported to be effective against colds and pain (Olubunmi et al. 2009).

Antidiabetes: Studies have shown the presence of two extremely hypoglycemic compounds (unusual amino acids). One of them, hypoglycin A (α-amino-β-(2-methylene cyclopropyl) propionic acid), is present in unripe fruit, and the ripe fruit contains only traces of the compound. The other compound, hypoglycin B, is present only in the seeds. Hypoglycin A is more potent than hypoglycin B in inducing hypoglycemia; hypoglycin A causes Jamaica vomiting sickness. Depletion of glucose reserves and the inability of cells to regenerate glucose lead to hypoglycemia (Olubunmi et al. 2009). The injection of hypoglycin A forms a metabolite called methylenecyclopropane acetyl CoA that inhibits several enzymes that are essential for the metabolism of lipids, gluconeogenesis, etc. However, the fully ripe fruit is considered edible in Africa (review: Atolani et al. 2009).

Although these hypoglycins do not appear to be promising antidiabetic compounds, the effect of different doses including extreme low doses on diabetic animal models is not studied. They may show anti-DM effects at very low levels. Further, the effects of these compounds on carbohydrate metabolism and insulin action could potentially shed new light in the complex metabolism of carbohydrates in the body in health and DM conditions. Above all, these two compounds can serve as lead molecules for synthetic transformation into useful compounds including beneficial anti-DM drugs.

The water extract of the air-dried root of *Blighia sapida* (when administered at intervals of 48 for 21 days, at 100 and 200 mg/kg doses) exerted hypoglycemic effects in normal rats. These findings support the traditional use of the extract for controlling diabetes (Saidu et al. 2012). It is not known whether or not the root extract contains any toxic compounds. Nontoxic hypoglycemic agents, if any, remain to be identified.

Toxicity: Subacute intraperitoneal administration of the lipid portion of the unripe ackee oil to rats resulted in neutropenia and increase in platelets without anemia. Hypoglycin A is considered to be responsible for the vomiting sickness, an acute condition characterized by persistent vomiting and coma and death in severe cases. This toxic hypoglycemic syndrome is associated with severe disturbances in carbohydrate and lipid metabolism (Olubunmi et al. 2009). It appears that these effects are brought about by metabolites of hypoglycin A. The metabolite exerts its effects by inhibiting several coenzyme A dehydrogenases that are essential for gluconeogenesis. Experiments on animals show that hypoglycin A also causes fatty degeneration of the liver (Olubunmi et al. 2009).

Other Activities: Platelet and neutrophil counts were significantly lowered in mice treated with water and lipid extracts of the unripe fruits of this plant. These extracts may be useful in disease conditions where these two blood parameters are elevated, for example, myeloid leukemia, essential thrombocythemia, and polycythemia (review: Atolani et al. 2009).

Phytochemicals: The ackee fruit contains a number of other compounds other than hypoglycin A and hypoglycin B. These include glighinone, quinine from the fruit, and vomifoliol from the leaves and stems. Another nonprotein amino acid, (2S, 1′S, 2S) 2 (2′ carboxycycloprophyl) glycine, has been isolated from the fruit.

FIGURE 2.22 *Boerhaavia diffusa.*

Boerhaavia diffusa L., Nyctaginaceae

Synonyms: *Boerhaavia repens* L.; *Boerhavia coccinea* Mill. (Figure 2.22)

Local and Vernacular Names: Hogweed, horse purslane (Eng.); local names in different parts of India include mukaratee-kirei, punarnava, snathikari, kommegida, tambadivasu, and varshabhu.

Distribution: Widely distributed throughout the world, this is a persistent herbaceous plant growing in moist areas in India, South America, Antilles, Africa, etc. (review: Riaz et al. 2014).

Description: This is a diffusely branched, young, and straight herb. It shows a mode of vertical growth and covered axis that grows loose and the renewable shoots rise on its upper surface. The roots are firm, tuberous, plump, rod-shaped, thick, fusiform with woody stock and bearing rootlets. The leaves are ovate-oblong, fleshy, or subcordate at the base and smooth above, having flat margins. The upper surface of the leaves is green, smooth, and glabrous, though it is hairy, white, and pinkish below. The stem is greenish-purple, inflexible, trim, tube-shaped, and swollen at the nodes. The flowers are subcapitate, tiny, fascicled, and pink. It is corymb and axillary, and in terminal panicles, bracteoles and minor, acute, perianth tube are limited above the ovary. The fruit is glandular and narrowly oblong obovoid. The nut is one seeded, and the clavate is 6 mm in length, curved, broadly and bluntly 5 ribbed, and viscidly glandular (review: Riaz et al. 2014).

Traditional Medicinal Uses: Most of the traditional uses were for the reproductive system, jaundice, kidney problems, skin troubles, eye diseases, wounds, and inflammation. The Unani and Ayurvedic preparations made from *Boerhaavia diffusa* are used for the management of several illnesses. The roots are used as a diuretic, laxative, and expectorant and the leaves are used as an appetizer and alexiteric preparation. The seeds are used as an energizer and helps in digestion. The juice of fresh leaves is also used for the treatment of pain in general and of sore throat. In Brazil, the plant is used in traditional medicine as a diuretic (roots) and against snake venom. The leaves of *B. diffusa* are eaten as a green vegetable in many parts of India. It is believed to cure ulcers of the cornea and night blindness and helps bring back virility in men. People in tribal areas use it to accelerate childbirth. The plant is also consumed as a vegetable as it is thought to be a rich source of vitamins, minerals, protein, and carbohydrate (review: Riaz et al. 2014).

Pharmacology

Antidiabetes: The chloroform extract of the *B. diffusa* leaf produced dose-dependent reduction in blood glucose in streptozotocin-induced type 2 diabetic rats comparable to that of glibenclamide. The results indicate that the reduction in blood glucose produced by the extract is probably through rejuvenation of pancreatic β-cells or through extrapancreatic action (Nalamolu et al. 2004).

Treatment with the water extract (200 mg/kg) for 4 weeks resulted in a decrease in blood glucose and increase in plasma insulin levels in normal and alloxan-induced diabetic rats. Further, the treatment reduced the levels of glycosylated hemoglobin. The activities of liver hexokinase were increased and that of glucose-6-phosphatase and fructose-1,6-bisphosphatase were decreased by the administration in normal and diabetic rats. The effect of the extract was more prominent when compared to glibenclamide (Pari and Satheesh 2004a). The main chemical constituents of the extract were flavonoid aglycones, flavonoid-*O*-glycosides, and acylated flavonoid-*O*-glycoside derivatives of quercetin and myricetin (Pari and Satheesh 2004a). Another study also confirmed the antidiabetic effects of the water extract (200 mg/kg) in alloxan-induced diabetic rats. The study further noted a reduction of HbA1c similar to glibenclamide and reduction in TGs and total cholesterol was also similar (Pari and Satheesh 2004b). In another study, administration of a hot water extract of the leaf (200 mg/kg) for 4 weeks in alloxan-induced diabetes was able to reduce circulating glucose and increase plasma insulin (relative to alloxan control) and oxidative stress, which has a potency similar to that of glibenclamide; slight reduction of blood glucose and insulin was noted in healthy control rats (Satheesh and Pari 2004).

Another study provided evidence for the renoprotective and antihyperglycemic effect of the ethanol extract in diabetic animals. The ethanol extract of *B. diffusa* administered orally at a dose of 500 mg/kg for a period of 30 days to alloxanized diabetic rats not only maintained the ionic balance and renal Na$^+$–K$^+$ ATPase activity but also significantly minimized diabetic hyperglycemia. The renal antioxidant status (GPx, catalase, SOD, and GSH) remained in the near normal range, and lipid peroxide levels were lower than the nondiabetic level. These effects are comparable to the changes brought about by metformin treatment and even better (Singh et al. 2011a). The alcohol extract of *B. diffusa* aerial parts (200 mg/kg, p.o.) showed antidiabetic activity in alloxan-induced as well as streptozotocin-induced diabetic rats. The extract improved glucose tolerance and body weight and decreased total cholesterol and TGs in diabetic rats compared to the diabetic control. The methanol extract showed more activity than the ethanol extract (Bhatia et al. 2011). In another study, the antihyperglycemic effects of oral daily administration of *B. diffusa* extract (100 and 200 mg/kg) for 10 weeks to normal and streptozotocin-induced type 2 rats were evaluated. A significant reduction in blood glucose levels was observed in normal and type 2 diabetic rats treated with the extract. The effect of glibenclamide was more prominent compared to that of *B. diffusa* extract (Nim et al. 2013). In addition to improving β-cell function and carbohydrate and lipid metabolism, the extract may also influence sugar absorption to a small extent. It was reported that *B. diffusa* extract inhibited α-glucosidase enzyme (IC$_{50}$ 1.72 μg/mL) with no apparent effect on α-amylase (Gulati et al. 2012). In another test on α-amylase, the plant outright failed to inhibit this enzyme (Prashanth et al. 2001).

Thus, several studies have established the anti-DM effects of *B. diffusa* extracts. Water, hot water, ethanol, and chloroform extracts have showed activity. The best extract and the active principles involved remain to be studied. It appears that several active principles and mechanisms of actions are involved in exerting the antidiabetic property of this plant. Further studies are warranted on this promising plant.

Other Activities: Other reported pharmacological and biological activities of *B. diffusa* include antibacterial activity, immunomodulatory activity, adaptogenic/antistress activity, hepatoprotective effect, analgesic effect, anti-inflammatory activity, antitumor activity, anticonvulsant activity, antiproliferative effect, antiestrogenic activity, antifibrinolytic activity, antioxidant activity, antiviral activity, spasmolytic activity, angiotensin-converting enzyme (ACE) inhibition, and antiasthma activity (Mishra et al. 2014; review: Riaz et al. 2014).

Toxicity: Toxicological studies done on *B. diffusa* confirmed the absence of teratogenic and mutagenic effects and ingestion of bigger doses is linked with nausea (review: Riaz et al. 2014). Since it has many pharmacological properties, interaction with conventional drugs cannot be ruled out.

Phytochemicals: Chemical analysis of *B. diffusa* gives a wide variety of chemical constituents, namely, flavonoids, alkaloids, steroids, triterpenoids, lipids, lignins, carbohydrates, proteins, and glycoproteins. Punarnavine, boeravinone A–F, hypoxanthine 9-L-arabinofuranoside, punarnavoside, liriodendron, ursolic acid, and a glycoprotein having a molecular weight of 16–20 kDa have been isolated. *B. diffusa* also contains many boeravinones, syringaresinol mono-D-glycoside, purine nucleoside hypoxanthine 9-larabinose, dihydroisofuroxanthone-borhavine, phytosterols, and alkaloids known as punarnavine and punernavoside (review: Riaz et al. 2014).

Bombax ceiba L., Bombacaceae

Shamimin, a C-flavonol glucoside from *Bombax ceiba* leaves, showed significant potency as a hypotensive agent at doses of 1–25 mg/kg, whereas it showed significant hypoglycemic activity at 500 mg/kg in Sprague-Dawley rats (Saleem et al. 1999).

Boswellia serrata Roxb. ex Colebr., Burseraceae

Synonyms: *B. glabra* Roxb., *B. thurifera* (Colebr.) Roxb. (monograph: WHO 2009)

Boswellia serrata is a deciduous medicinal tree. Under *in vitro* conditions, boswellic acid isolated from the gum resin of this plant inhibited aldose reductase in rat lens homogenate. Boswellic acid (10 mg/kg) inhibited the formation of AGEs in diabetic rats (Rao et al. 2013). The water extract of *B. serrata* reduced blood glucose, oxidative stress, and the complications of diabetes in the liver and kidneys of fertile female diabetic rats (Azemi et al. 2012).

In a recent clinical trial, treatment of the diabetic patient with *B. serrata* gum resin (900 mg daily) for 6 weeks resulted in an increase in blood HDL levels as well as a remarkable decrease in cholesterol, LDL, fructosamine, SGPT, and SGOT levels (Ahangarpour et al. 2014). The study was carried out with one dose of gum resin.

Bougainvillea glabra Choisy, Nyctaginaceae

Bougainvillea glabra is closely related to *Bougainvillea spectabilis*. The water extract of the leaves of *B. glabra* (100 and 400 mg/kg daily for 3 weeks) produced antidiabetes activity as judged from the marked reduction in blood glucose levels and improvement in the lipid profile in treated alloxan-induced diabetic rats (Grace et al. 2009a). In another study, the hydroalcoholic extract of *B. glabra* (150 and 300 mg/kg daily for 21 days) showed antidiabetic and antihyperlipidemic activity in streptozotocin-induced diabetic rats (Tomar and Sisodia 2013). The water extract of *B. glabra* leaves (100 mg/kg daily for 10 days) showed marked cutaneous wound healing effect in normal and diabetic rats (Soni et al. 2013).

Bougainvillea spectabilis Willd., Nyctaginaceae

Synonyms: *B. spectabilis* Pers., *B. spectabilis* Raeusch., *B. spectabilis* Lund ex Choisy (invalid), *B. spectabilis* Nees & Mart. (illegitimate), *B. spectabilis* Schnize, *B. spectabilis* var. *hirsutissima* J. A. Schimidt, *B. spectabilis* var. *parviflora* Mart ex. J.A. Schimidt, *B. spectabilis* var. *virescens* (Choisy) Schimidt, *B. virescens* Choisy (book: Lim 2014) (Figure 2.23)

Common Names: Bougainvillea, great bougainvillea, bouganvila, paper flower, purple bougainvillea (Eng.), etc.

Description: *B. spectabilis* is a woody evergreen rambling vine; the stems and branches have stout recurved spines; the leaves are alternate, petiolate, simple ovate or elliptic, up to 10 cm long, covering an entire margin. Flowers in axillary cymose inflorance of 3 small, trumped flowers; bracts are showy and

FIGURE 2.23 *Bougainvillea spectabilis.*

of various colors; perianth tube is tubular, rounded, and densely pubescent, with five spreading creamy or white lobes at the apex. From the base of the tube arises a single carpel and encircling the ovary is a cup-shaped nectar, and from its rim eight stamens usually arise. The fruits are small, dry, elongate, one-seeded ribbed achene (Lim 2014).

Distribution: *B. spectabilis* is native to South America from Brazil to Peru and to South Argentina. It has been introduced pantropically and is a popular ornamental plant in the warm areas of Asia, Southeast Asia, Australia, the Pacific Islands, the Mediterranean region, the Caribbean, Mexico, South Africa, and the United States (Lim 2014).

Traditional Medicinal Uses: The stem bark, root bark, leaves, and flowers are used in traditional medicine to treat a variety of disorders such as diarrhea and stomach acidity.

Pharmacology

Antidiabetes: The hypoglycemic activity of the alcohol extract of *B. spectabilis* leaf was first reported in 1984 (Narayanan et al. 1984). One of the hypoglycemic principles of this plant leaf was isolated and identified as D-pinitol (Narayanan et al. 1987). This compound is a methyl ester of chiro-inositol (3-*O*-methyl-1,2,3-*trans*-hexahydroxycyclo hexanol). This compound also occurs in other plants. Studies confirmed that D-pinitol could exert an insulin-like effect in improving glycemia in streptozotocin-induced diabetic mice (Bates et al. 2000). In these diabetic rats, D-pinitol at a dose of 100 mg/kg decreased hypoglycemia by 22% in 6 h. Intraperitoneal and oral administration showed almost similar effects. The insulin concentration and the rate of insulin-induced glucose disappearance were not altered by 100 mg/kg pinitol. Chronic administration of the compound (100 mg/kg twice a day for 11 days) to streptozotocin-induced diabetic rats maintained a reduction in plasma glucose concentration. However, in normal nondiabetic and severely insulin-resistant *ob/ob* mice, this dose of pinitol did not significantly affect plasma glucose or insulin during acute studies. This showed that pinitol may act via a postreceptor pathway of insulin action affecting glucose uptake. In another study, oral administration of pinitol from *B. spectabilis* to streptozotocin-induced diabetic rats elicited decrease in the elevated levels of blood glucose, total cholesterol, TG, and free fatty acids in the serum, liver, kidney, heart, and brain (Geetham and Prince 2008).

Administration of water and ethanol extracts of *B. spectabilis* leaves showed improved glucose tolerance in diabetic mice. Further, the extract inhibited intestinal glucosidase activity. Besides, the extract of the plant leaves increased the activity of glucose-6-phosphate dehydrogenase and the glycogen content in the liver and skeletal muscle after 21 days of treatment. Immunohistochemical analysis revealed regeneration of β-cells in the extract-treated diabetic mice pancreas and a corresponding increase in plasma insulin and C-peptide levels compared to untreated diabetic control animals (Bhat et al. 2011a). Additionally, the chloroform extract of this plant's leaves was found to have significant α-amylase inhibitory activity and could be useful in managing postprandial hypoglycemia (Bhat et al. 2011b).

The bark extract of the plant exhibited anti-DM activity against streptozotocin-diabetic rats (Purohit and Sharma 2006). In another study, the ethanol extract of *B. spectabilis* root bark (100 mg/kg daily for 7 days) reversed hyperglycemia in alloxan-induced diabetic rats. The hypoglycemic activity of the extract was found to be more potent than glibenclamide (Jawla et al. 2011). Recently, from the stem bark, pinitol, β-sitosterol, quercetin, and quercetin-3-*O*-α-L-rhamnopyranoside were isolated (Jawla et al. 2013). All these compounds possess anti-DM properties.

Thus, the plant contains active principles other than pinitol that improves insulin levels and blocks sugar absorption in the intestine. There is a need to identify all active principles and determine the distribution of these active principles in various parts of the plant. An appropriate combination of these photochemicals may work better than an individual compound.

Clinical Trial: In a randomized placebo, 7-week study on 15 elderly (66 ± 8 years) nondiabetic individuals, pinitol supplementation was found not to influence whole body insulin-mediated glucose metabolism and muscle IR content and phosphorylation (Campbell et al. 2004). Clinical studies on diabetic patients are not available. In acute studies, in normal rats, pinitol also did not significantly influence plasma insulin levels (Geetham and Prince 2008).

Other Activities: Other pharmacological properties include anti-inflammatory, antiviral, antitumor, antibacterial, antihypolipidemic, antihypercholesterolemic, and antifertility activity (Lim 2014).

Phytochemicals: Phytochemical analysis revealed the presence of alkaloid, flavonoids, glycosides, phlobatannins, saponins, steroids, tannins, and terpenoids in distilled water, acetone chloroform, ethanol, and methanol extracts of the *B. spectabilis* stem, flowers, and leaves (Kumaraswamy et al. 2012; Rashid et al. 2013).

Bouvardia terniflora, Rubiaceae

The chloroform extract of *Bouvardia terniflora* (300 mg/kg, i.p.) lowered blood glucose levels in alloxan-induced diabetic mice. Further, it showed hypoglycemic activity in normal mice (Perez et al. 1998).

Brachylaena discolor D.C., Asteraceae

An *in vitro* glucose utilization study was done using 3T3-L1 adipose cells, C2C12 muscle cells, and Chang liver cells to determine the antidiabetic effect of the *B. discolor* extract. All parts of this plant stimulated glucose utilization in the three cell types with negligible cytotoxicity (van de Venter et al. 2008).

Brachylaena elliptica Less, Asteraceae

The leaves are believed to be used as a treatment for diabetes in traditional medicine. Extracts from this plant root and stem lowered blood glucose in experimental animals (Marles and Farnsworth 1995).

Brassica juncea Czern. & Coss., Brassicaceae

Synonyms: *Brassica alba* (L.) Rabenh, *Brassica nigra* (L.) Koch., *Sinapis alba* (L.) Hooker filius & Thompson (Figure 2.24)

Common Names: Indian mustard (Eng.), rajika, rai, kadugu, kaduka (Indian languages)

Description: It is an economically important plant well known for its nutritive and medicinal values (Kumar et al. 2011g). It is a tall, erect, branching annual plant, rarely glaucous or hispid at the base. The upper leaves are narrow, while the lower leaves are oblong-lanceolate and toothed. The flowers are bright yellow. Siliqua are linear-lanceolate with a straight, flattened beak bearing dark, small, rugose seeds. Edible oil is obtained from the seeds.

(a)   (b)

FIGURE 2.24 *Brassica juncea*. (a) Plant with flower. (b) Seeds.

Distribution: Its primary center of origin is Central Asia with secondary centers in China, Burma, and through Iran to the Near East. It is cultivated in India, Bangladesh, China, Central Africa, Japan, Nepal, Pakistan, etc. It is also found in Egypt, Europe, and Afghanistan.

Traditional Medicinal Uses: The leaves and seeds of this plant are edible and possess diverse medicinal uses. The leaves are used in a range of folk medicines as stimulants, diuretics, and expectorants as well as a spice. In Korea, it is used for both food itself and a major ingredient of kimchi, a traditional fermented vegetable food with medicinal value. The plant is an anodyne, rubefacient and emmenagogue and is used for headache, diabetes, gastrointestinal disturbance, rheumatism, and syphilis (review: Ajikumaran and Subramoniam 2005).

Pharmacology

Antidiabetes: *B. juncea* showed hypoglycemic action in rats. Hepatic glycogen and glycogenesis increased as a result of increased activity of glycogen synthase and hepatic glycogenolysis and gluconeogenesis decreased (Khan et al. 1995). *B. juncea* showed both hypoglycemic and antihyperglycemic activities in rats (Grover et al. 2002). The ethyl acetate fraction from mustard leaf showed inhibitory effects, which were concentration dependent, on the formation of AGEs and free radical-mediated protein damage in an *in vitro* system. Further, oral administration of the ethyl acetate fraction (50 and 200 mg/kg, for 10 days) reduced the serum levels of glucose, glycosylated protein, and oxidative stress (Yokozawa et al. 2003).

However, no definitive statements on the nature of phytoconstituents involved in their observed anti-DM effects can yet be made. This is partly because of the diverse types of extracts and experimental designs and conditions used in different studies. There are apparent discrepancies in the studies also. For example, one study described the dose-dependent (250, 350, and 450 mg/kg/day) beneficial effects of mustard seed (water extract) against hyperglycemia and insulin deficiency in streptozotocin-induced diabetic rats (Thirumalai et al. 2011), whereas in an earlier study, no hypoglycemic effect of the seeds was observed in an analogous rat diabetes model (Grover et al. 2002). Since *B. juncea* caused the reduction in glucose levels in moderately diabetic but not in severely diabetic rats, it seems that the antihyperglycemic activity of *B. juncea* depends upon the presence of the functional β-cells to release the insulin and the degree of β-cell damage. *B. juncea* seeds probably prevented the destruction of the β-cells of the islets in the pancreas by its antioxidant effects (Grover et al. 2002). In normal rats, *B. juncea* increased glucose utilization (increase in glycogenesis as evidenced by the increased activity of glycogen synthase) and decreased glycogenolysis and gluconeogenic enzymes (Khan et al. 1995). However, in diabetic rats, *B. juncea* did not influence enzyme activity, indicating a positive role only in the prediabetic state.

B. juncea significantly prevented the development of insulin resistance in rats fed a fructose-enriched diet (Yadav et al. 2004a). The study suggests that *B. juncea* can play a role in the management of the prediabetic state of insulin resistance and its use in higher quantity as a food ingredient may help those patients prone to diabetes (Yadav et al. 2004a). Since *B. juncea* significantly prevented a rise in creatinine levels, it may delay the development of diabetic nephropathy. Depending on the seed powder content of the animals' food (5%–15%), antihyperglycemic, antihyperlipidemic, and other beneficial effects of the seeds against diverse metabolic disorder-associated pathologies have also been observed in earlier reports. Feeding rats with a 10% *B. juncea* diet for 60 days had no adverse effect on food intake and various hematological parameters. These reports could indicate that fairly high doses of its bioactive components are well tolerated by experimental animals. The broad spectrum of beneficial effects of the seeds observed in these studies warrant further exploration of *B. juncea* seeds as a potential source for obtaining pharmacologically standardized phytotherapeutics that are potentially useful for combating insulin resistance and obesity (Farag and Gaballa 2011; Kumar et al. 2011g).

Isolation of active compounds and their actions individually and in combination under various degrees of DM conditions remain to be studied to establish its benefit as a therapeutic agent and/or nutraceutical.

Clinical Studies: Controlled clinical studies remain to be done on this nutraceutical to establish its therapeutic value.

Other Activities: Isorhamnetin 3,7-di-*O*-β-D-glucopyranoside from this plant leaf shows antioxidant activity (Yokozawa et al. 2002). Four different active fractions from the plant leaf also show *in vitro* and *in vivo* antioxidant activities. In addition to antioxidant activities, several other activities such as antifungal, antibacterial, antitumor, cognition-improvement activity, astrocyte developing activity, and other health benefits are reported (review: Kumar et al. 2011).

Phytochemicals: Several pharmacologically active compounds have been isolated. These include allyl isothiocyanate, phenyl isothiocyanate crotonyl isothiocyanate, essential oils, and isorhamnetin 3,7-di-*O*-β-D-glucopyranoside (isorhamnetindiglucoside) (Yokozawa et al. 2002). Phenolic compounds, sinapic acid and sinapine, were also isolated. Kaempferol glycosides (kaempferol 7-*O*-β-D-glucopyranosyl-(1 → 3)-[β-D-glucopyranosyl-(1 → 6)]-glucopyranoside, kaempferol-3-*O*-(2-*O*-feruloyl-β-D-glucopyranosyl-(1 → 2)-β-D-glucopyranoside)-7-*O*-β-D-glucopyranoside, kaempferol-3-*O*-β-D-glucopyranosyl-(1 → 2)-*O*-β-D-glucopyranoside-7-*O*-β-D-glucopyranoside, and 1-*O*-sinapoyl glucopyranoside) with antioxidant activities were present (Kumar et al. 2011). Methyl cinnamate, glucosinolate, sinigrin, 3-butenyl isothiocyanate, isorhamnetin 3,7-di-*O*-β-D-glucopyranoside (isorhamnetin diglucoside), etc., were also isolated from this plant (review: Kumar et al. 2011). However, specific anti-DM molecules were not identified.

Brassica nigra (L.) Koch, Brassicaceae

In the streptozotocin-induced diabetic rats treated separately with water, ethanol, acetone, and chloroform extracts of the seeds of *Brassica nigra*, although all the extracts improved glucose tolerance, the water extract was found to be better than the other tested extracts. In further studies, the effective dose was found to be 200 mg/kg water extract in glucose tolerance test. Administration of 200 mg/kg of the water extract to diabetic animals (once daily for 1 month) brought down fasting serum glucose (FSG) levels, while in the untreated group, FSG remained at a higher level. In the treated animals, the increase in glycosylated hemoglobin and serum lipids was much less when compared with the levels in untreated diabetic controls (Anand et al. 2007). In a follow-up study, to understand the mechanism of the antidiabetic action of the water extract, the effect of oral administration of the extract for 2 months on glycolytic and gluconeogenic enzymes was studied in the liver and kidney of streptozotocin-induced diabetic rats. The activities of gluconeogenic enzymes were higher and of glycolytic enzymes were decreased in both the liver and kidney tissues during diabetes. However, in diabetic rats treated with the extract for 2 months, decrease of serum glucose, increase of serum insulin, and release of insulin from the pancreas (shown *in vitro* from isolated pancreas) along with the restoration of key regulatory enzyme activities of carbohydrate metabolism and glycogen content were observed. The therapeutic role of the extract in diabetic rats can be attributed to the release of insulin from the pancreas and the change of glucose metabolizing enzyme activities to normal levels (Anand et al. 2009).

Brassica oleracea L., var. *botrytis*, Brassicaceae

Common names of *Brassica oleracea* include cauliflower, phulgobi, kohlrabi, and gospoovu. Cauliflower is an important vegetable. It is a low herb with a stout, short stem; the leaves are long-oblong or elliptic and ascending; the flowers are abortive on closely crowded short fleshy stalk forming a dense terminal head or curd, topped by leaves. It is cultivated in most parts of India, especially in the northwest in higher altitudes. In traditional medicine, it is an analgesic and vermifuge and used in cancer, skin diseases, infections, tumor, scurvy, sores, and warts (review: Ajikumaran and Subramoniam 2005).

Antidiabetes: In glucose tolerance test in rats, *B. oleracea* var. *botrytis* showed antihyperglycemic activity (Roman-Ramos et al. 1995). Crude petroleum ether extract of the stem of *B. oleracea* reduced blood glucose levels in alloxan-induced diabetic rats (Koti et al. 2004). In another study, petroleum ether extracts of *B. oleracea*, when treated for 60 days, exhibited antihyperglycemic and antioxidant activities in streptozotocin-induced diabetic rats (Rasal et al. 2005). In a recent study, antidiabetic and anti-inflammatory activities of green and red cauliflowers were evaluated via PTP1B and rat lens aldose reductase inhibitory assays and cell-based lipopolysaccharide-induced nitric oxide inhibitory assays in RAW 264.7 murine macrophages. The methanol extracts of green and red cauliflowers exhibited potent inhibitory activities against PTP1B with the corresponding IC_{50} values of 207 ± 3.48 and 287 ± 3.22 μg/mL, respectively. Interestingly, the red kohlrabi (cauliflower) extract exhibited significantly stronger anti-inflammatory, antidiabetic, and antioxidant effects than that of green kohlrabi. The total phenolic content was found to be increased in the red variety (Jung et al. 2014).

Other Activities: Cauliflower and the plant extract have cancer preventive properties. The cancer chemopreventive effects of *Brassica* vegetables may be partly due to the presence of organosulfur phytochemicals. In an aspirin-induced rat ulcer model, *B. oleracea* (leaf) water extract lowered the ulcer index and increased hexosamine levels, suggesting gastric mucosal protection. Fresh aqueous *B. oleracea* has antifungal activity against *Candida albicans* and other pathogenic fungi. The water extract of the plant has antitrypanosomal activity (review: Ajikumaran and Subramoniam 2005).

Phytochemicals: Phenolic compounds, flavonols, and organosulfur phytochemicals have been reported. Reported compounds include allantoin, allantoic acid, ascorbic acid, allyl isothiocyanate, 3-methyl-thiopropylcynide, 4-methyl-thyobutylcynide, 2-phenyl ethylcyanide, and leucoanthocyanidin (review: Ajikumaran and Subramoniam 2005).

Brickellia grandiflora (Hook.) Nutt., Compositae

Common Names: Hamula, bricklebush, tassel flower, etc.
The aboveground parts and the flower of the plant are used dry or fresh in traditional medicine. Generally, 1–3 teaspoons of the leaves are steeped in one cup hot water for 15–20 min, and then one cup is consumed in the morning and one cup in the evening to control diabetes. Alternately, a tincture can be used at a dose of 1–3 mL twice a day (again, morning and afternoon). It is said that *Brickellia grandiflora* helps lower high blood sugar in insulin-resistant diabetics; it stimulated fat digestion in the gallbladder and improved stomach lining and digestion (book: Yarnell 2009).

Brickellia veronicaefolia (Kunth) a. Gray, Compositae
The hexane extract of *Brickellia veronicaefolia* (300 mg/kg, i.p.) lowered blood glucose levels in alloxan-induced diabetic mice. Further, it exhibited hypoglycemic activity in normal mice (Perez et al. 1998).

Bridelia ferruginea Benth., Euphorbiaceae
Bridelia ferruginea is used in traditional medicine in Africa to treat many diseases. The methanol extract of *B. ferruginea* bark lowered blood glucose levels in both normoglycemic rats and glucose-induced hyperglycemic rats (Mathew et al. 2006). In another study, the hypoglycemic activity of the water extract of *B. ferruginea* (stem bark) was investigated in both normal and alloxan-induced diabetic rats. The extract (100 mg/kg) significantly reduced the fasting blood glucose level of normal and alloxan-induced diabetic rats. Glucose tolerance in normal rats was enhanced, which shows that there is marked improvement in glucose utilization after administration of this extract (Adewale and Oloyede 2012). The possible therapeutic potential of *B. ferruginea* (water extract) in gestational diabetes was evaluated in pregnancy-induced glucose intolerance in rats. *B. ferruginea* caused a reduction in glycemic response to glucose challenge and an increased glucose tolerance in rats that had pregnancy-induced glucose intolerance. Thus, the diabetogenic effect of pregnancy was ameliorated by oral administration of the water extract of *B. ferruginea* to pregnant albino rats (Taiwo et al. 2012). Another study confirmed the glucose-lowering property of the water extract of this plant. In alloxan-induced diabetic rats, there was significant reduction in blood glucose levels in rats treated with the water extract (200, 400, and 800 mg/kg) of *B. ferruginea* leaves (Aja et al. 2013). Although the aforementioned studies show moderate glucose-lowering property of the extracts of this plant's bark and leaves, the mechanisms of action and active principles involved remain to be studied.

Broussonetia papyrifera L., Moraceae
Broussonetia papyrifera is a traditional medicinal plant with anti-inflammatory and other properties. Five natural flavonoids, namely, 8-(1,1-dimethylallyl)-5'-(3-methylbut-2-enyl)-3',4',5,7-tetrahydroxyflanvonol, 3'-(3-methylbut-2-enyl)-3',4',7-trihydroxyflavane, quercetin, uralenol, and broussochalcone A isolated from the roots of *B. papyrifera* showed inhibitory activity against PTP1B with IC_{50} values ranging from 4.3 to 36.8 μmol/L (Jiany et al. 2012).

Brucea javanica (L.) Merr, Simaroubaceae

Brucea javanica is an evergreen shrub. In China, the seeds of *B. javanica* have been used as traditional herbal medicine due to its multifaceted activities. To date, 153 compounds have been reported from the seeds and aerial parts of this plant. The seeds of *B. javanica* are also recommended by traditional practitioners for the treatment of DM. Hypoglycemic activity-guided fractionation led to the isolation of bruceines E and D from the seeds of *B. javanica*. Normoglycemic mice administered with 1 mg/kg of these compounds exhibited hypoglycemic activity. In streptozotocin-induced diabetic rats, these compounds also reduced blood glucose levels, which was comparable to that of glibenclamide (Noorshahida et al. 2009).

In a recent study, the ethyl acetate fraction of the ethanol extract of *B. javanica* seed, rich in tannin, exhibited the strongest antioxidant activities. The fraction also exerted a dose-depended inhibition of glycogen phosphorylase-α *in vitro*. Further evaluation of hypoglycemic effect on oral glucose tolerance test indicated that rats treated with the fraction (125 mg/kg) showed marked reduction in glucose at 30, 60, and 90 min after glucose loading compared to the normal control. When the tannins were separated using column chromatography, the glycogen phosphorylase-α inhibitory activity was retained and the tannins showed only an antioxidant activity (Ablat et al. 2014).

The acute and subchronic toxicity of the seed extracts with hypoglycemic activity was examined. Methanol and butanol extracts reduced blood glucose levels of mice at 32 mg/kg with no mortality. The LD_{50} values of methanol and butanol extracts were 281.71 and 438.43 mg/kg, respectively. The inactive hydrophilic and toxic hydrophobic constituents of the butanolic extract were removed via aqueous residual and hydrophobic solvent partitioning during extraction. Histology examination and blood tests of AST, ALT, ALP, urea, bilirubin, and creatinine indicated that the butanolic extract did not induce liver and kidney toxicity upon 9-week consumption. The butanol extract contained blood glucose-lowering quassinoids bruceine D (10.3%) and E (0.4%) and is safe for treatment of DM (Shahida et al. 2011). Detailed long-term toxicity evaluation may be required to establish the safety of the isolated compounds and active extracts.

Bruguiera gymnorrhiza (L.) Sav., Rhizophoraceae

The seeds of *Bruguiera gymnorrhiza*, a mangrove tree, are used in traditional medicine as a contraceptive and to treat diabetes. Gymnorrhizol, a novel, unusual 15-membered macrocyclic polydisulfide, isolated from *B. gymnorrhiza* showed inhibitory activity against PTP1B (Jiany et al. 2012).

Bryonia alba L., Cucurbitaceae

Unsaturated fatty trihydroxy acids isolated from *Bryonia alba* showed hypoglycemic activity in alloxan-induced diabetic rats (Panosian et al. 1981). In a follow-up study, the effects of the aforementioned indicated compounds on the activity of glycogen phosphorylase (a and b forms), phosphoprotein phosphatase and hexokinase, in the liver and muscle tissues of white rats with alloxan-induced diabetes were investigated. One of the possible mechanisms of the hypoglycemic action of trihydroxyoctadecadiene acids is discussed (Panosian et al. 1984, article in Russian). The alcohol extract of the dried roots of *B. alba* (200 mg/kg, oral) reduced blood sugar levels in rats (Singh et al. 2012). Further studies are needed to get fuller information on the mechanism of action and all active principles present in the plant.

Bryophyllum pinnatum Lam., Crassulaceae

The plant *Bryophyllum pinnatum* is ethnobotanically used in the treatment of various diseases including diabetes. The water extract of this plant's leaves showed antidiabetic activity in streptozotocin-induced diabetic rats (Ojewole et al. 2005). In another study, in normal rats, the aqueous ethanol (80%) extract of *B. pinnatum* (500 mg/kg p.o.) improved glucose tolerance upon testing. Administration of the extract (500 mg/kg) daily for 21 days to streptozotocin-induced diabetic rats resulted in a significant reduction in blood glucose levels, TG levels, and LDL level and increase in HDL level in diabetic rats. Thus, the plant extract showed anti-DM activity (Ogbonnia et al. 2008). A recent study also demonstrated the antidiabetic property of the water extract of *B. pinnatum* leaves. The water extract showed a significant drop in the blood glucose levels of diabetic rats to a value close to the normal blood glucose level within

120 min. When a sample containing a mixture of the water extract and glibenclamide was administered, a further drop in blood glucose level was observed. Therefore, the results reveal that the performance of the existing drugs (glibenclamide) could be enhanced with the judicial use of the aqueous extract (Aransiola et al. 2014).

Buddleia americana L., Loganiaceae (Buddleia = Buddleja)
In Mexico, decoction of *Buddleia americana* leaves is used to treat diabetes in traditional medicine (Alarcon-Aguilar and Roman-Ramos 2005). Administration of the leaf decoction of this plant resulted in hypoglycemic effects in normal healthy rabbits (Roman-Ramos et al. 1992).

Buddleja officinalis Maxim., Loganiaceae
In a screening study to detect the DPP-IV inhibitory activity in natural products, a promising inhibitory activity was found in *Buddleja officinalis*. Further, studies showed that the plant had a hypoglycemic effect *in vivo* in alloxan-induced diabetic rats, which was related to the inhibition of the enzyme activity (Lei 2008). Additionally, the extract of this plant flower inhibited the activity of aldose reductase 2 under *in vitro* conditions (Matsuda et al. 1995).

Bumelia sartorum Mart., Sapotaceae
The reputed use of *Bumelia sartorum* in the treatment of DM and inflammatory disorders has been mentioned in Brazilian folklore. An ethanol extract of the root bark elicited a hypoglycemic effect in normal and alloxan-induced diabetic rats. In addition, the extract altered glucose tolerance in alloxan-induced diabetic rats, enhanced glucose uptake in skeletal muscle, and significantly inhibited glycogenolysis in the liver. These results indicate that the hypoglycemic effect may be similar to chlorpropamide and possibly due to an enhanced secretion of insulin from the islets of Langerhans or an increased utilization of glucose by peripheral tissues. Besides, the ethanol extract elicited significant anti-inflammatory activity (Almeida et al. 1985). Bassic acid, an unsaturated triterpene acid isolated from the ethanol extract of the root bark, exhibited hypoglycemic effect in alloxan-induced diabetic rats. The hypoglycemic potential was also demonstrated *in vitro* using isolated rat diaphragm (Naik et al. 1991). In another study, assessment of the hypoglycemic effect of the plant extract rich in polyphenols and the possible mechanism of action was also made. Hypoglycemia is produced by the polyphenol-rich extract and sarco-/endoplasmic reticulum Ca^{2+}-ATPase inhibition may be one of the possible mechanisms involved in the hypoglycemic action (Ruela et al. 2013).

Bunium persicum (Boiss) Fedts, Apiaceae or Umbelliferae

Synonym: *Carum bulbocastanum* Koch., *Carum bulbocastanum* Clarke
Carum bulbocastanum is a spice commonly known as black cumin, black caraway, etc. The ethyl acetate extract of this plant seed inhibited α-amylase activity with an IC_{50} value of 28 µg/mL (Giancarlo et al. 2006).

Butea monosperma (Lam) Taub., Fabaceae

Common Name: Flame of the forest or flame tree
Butea monosperma is a traditional medicinal plant; it is used as an anthelmintic, appetizer, aphrodisiac, laxative, etc. Studies have reported its anti-DM activities. A 50% ethanol extract of *B. monosperma* flower showed antidiabetic activity against alloxan-induced diabetic rats. Daily oral administration of the extract for 45 days to alloxan-induced diabetic rats resulted in the maintenance of blood glucose levels and body weight close to the values observed in normal control and glibenclamide-treated diabetic rats. Further, the levels of total cholesterol, TG, LDL, and VLDL were decreased in the treated diabetic rats (Somani et al. 2006). In another study, the ethanol extract of this plant leaves showed hypoglycemic and antioxidant activities in alloxan-induced diabetic mice (Sharma and Garg 2009). The antioxidant and anti-DM activities of the alcohol extract (80%) of this plant flower in alloxan-induced diabetic rats were confirmed by another study (Talubmook and Buddhakala 2012).

Caesalpinia bonduc (L.) Roxb., Fabaceae

Synonym: *Caesalpinia bonducella* (L.) Flem. (Figure 2.25)

Common Names: Yellow nicker nut, fever nut, bonduc nut (Eng.), putikaranja, kantkarej, nata karanja, kat-karanj, lalarci, kalichikai, kaka-moullou, etc. (Indian languages) (review: Singh and Raghav 2012)

Description: *Caesalpinia bonduc* is a large, straggling, thorny evergreen shrub (height 10–12 m) bearing pinnate leaves and large foliaceous stipules, with elliptic oblong, obtuse, mucronate leaflets, lanceolate, acuminate bracts; deep taproots; hard woody stem; yellow flowers, producing pods covered with wiry prickles; and hard, globose, and gray seeds. It is a perennial shrub vine. All parts of this plant have medicinal properties.

Distribution: It is widely distributed all over the world. It occurs in most plain districts of India especially near the coastal area (in hedges and wasteland). It is also grown as an ornamental plant.

Traditional Medicinal Uses: In traditional medicine, it is an emmenagogue, febrifuge, anthelmintic, emollient, cosmetic, tonic, and vermifuge used in stopping discharge from the ear, inflammation, swelling, cirrhosis, cough, diabetes, fever, piles, sclerosis, spasms, diarrhea, hematochezia, freckles, malaria, snake bite, fractures, and tumor (*The Wealth of India* 1992).

Pharmacology

Antidiabetes: *C. bonducella* seed has a hypoglycemic effect in normal as well as type 1 and 2 diabetic rats (Sharma et al. 1997; Chakrabarti et al. 2003). The seed fractions have an insulin secretagogue property on isolated islets, which could be one of its mechanisms of action (Chakrabarthi et al. 2005).

The seed kernel was reported to have hypoglycemic activity in rats (Parameshwar et al. 2007). In diabetic rats, both ethyl acetate and water extracts of the seed showed significant hypoglycemic action and reversed the diabetes-induced increases in blood lipid levels and liver glycogen content. The petroleum extract was inactive. In *in vitro* antioxidant studies, the water extract was devoid of any free radical scavenging activity, whereas the ethyl acetate extract showed free radical scavenging activity. Thus, although the antioxidant potential of the ethyl acetate extract may contribute to reduce diabetes-linked oxidative stress, it may not contribute to its hypoglycemic activity (Parameshwar et al. 2002). Further, the hypoglycemic activity and antioxidant activity may be mediated by different active molecules. In another study, the oral administration of the seed extracts (300 mg/kg) resulted in significant antihyperglycemic activity and a reduction in blood urea nitrogen and LDL levels in alloxan-induced diabetic rats (Singh and Raghav 2012).

The hydromethanolic extract of the seeds, when administered, at a dose of 250 mg/kg for 21 days, showed antihyperglycemic and antioxidant effects in streptozotocin-induced diabetic rats (Jana et al. 2012).

FIGURE 2.25   *Caesalpinia bonduc.*

Treatment with the hydroethanolic extract resulted in significant recovery in the activities of carbohydrate metabolizing enzymes (hexokinase, glucose-6-phosphatase, and glucose-6-phosphate dehydrogenase), antioxidant enzymes, and lipid peroxide levels in the blood along with correction in fasting blood glucose and liver glycogen levels as compared to untreated diabetic animals (Jana et al. 2012).

The aqueous ethanol and chloroform extracts of the roots of this plant lowered the blood glucose levels in the glucose tolerance test in both normal and alloxan-induced diabetic rats. In alloxan-induced diabetic rats, the maximum reduction in blood glucose was observed after 3 h at a dose level of 250 mg/kg. In the long-term treatment of alloxan-induced diabetic rats, the antidiabetic activity was assessed by measuring the blood glucose, TGs, cholesterol, and urea levels on the 0th, 3rd, 5th, 7th, and 10th day. Both the extracts showed marked antidiabetic activity that is comparable with that of glibenclamide (Patil et al. 2010). The water extract of this plant seed shell at a dose of 250 mg/kg produced marked blood glucose-lowering effect in glucose-loaded, streptozotocin-induced diabetic and alloxan-induced diabetic rats. Contrary to a previous report, a suspension of the powdered seed kernel (in 0.5% carboxy methyl cellulose) of *C. bonducella* (growing in Dar es Salaam, Tanzania) did not exhibit hypoglycemic activity in fasted and glucose-fed normal albino rabbits. The treatment did not influence oral glucose tolerance also. Administration of the seed kernel (0.2, 0.4, and 0.8 g/kg daily) for 7 days also did not result in any significant change in the blood glucose levels in albino rabbits (Moshi and Naga 2000). This could be due to the difference in the preparation of the seed drug or the difference in ecotype, if any. Pintol and sitosterol, known anti-DM compounds, are reported to be present in this plant (review: Singh and Raghav 2012).

Other Activities: This plant has antitumor activity against Ehrlich's ascites carcinoma in mice and anti-LPO (Gupta et al. 2004). The seeds have an antibacterial activity (Saeed and Sabir 2001). Other known pharmacological activities of this plant include antioxidant, adaptogenic, anti-inflammatory, anthelmintic, antifilarial, antimalarial, antimicrobial, antiestrogenic, antiproliferative and antitumor, immunomodulatory, hepatoprotective, antipyretic and analgesic, antipsoriatic, anxiolytic, and larvicidal (review: Singh and Raghav 2012).

Phytochemicals: The plant contains chemical constituents such as alkaloids, flavonoids, tannins, glycosides, triterpenoids, steroidal saponins, fatty acids, hydrocarbons, phytosterols, isoflavones, and phenolics. The presence of bounducin, triterpenoids, sitosterol, heptacosane, saponin, different types of caesalpinins, bonducellin, homoisoflavone, and pinitol has been reported (review: Singh and Raghav 2012).

Caesalpinia digyna Rottler, Leguminosae/Caesalpiniaceae

Caesalpinia digyna is a large scandent, sparingly prickly shrub. The methanol extracts of *C. digyna* root (500 mg/kg daily for 15 days) reduced blood glucose, glycosylated hemoglobin, and lipid levels in streptozotocin-induced diabetic rats. The treatment also improved body weight compared to diabetic control rats (Sarathchandran et al. 2008). In another study, the standardized alcohol extract of *C. digyna* (root) (250, 500, and 750 mg/kg) showed anti-DM activity in streptozotocin–nicotinamide-induced diabetic rat. The treatment resulted in a dose-dependent reduction in fasting blood glucose levels, improvement in lipid profiles, and reduction in oxidative stress. In normoglycemic rats, the extract did not reduce blood glucose levels, whereas oral glucose tolerance test showed better tolerance of glucose in the treated rats (Kumar et al. 2012d). In a follow-up study, bergenin, a major constituent of *C. digyna*, was isolated from its roots. Isolated bergenin was then evaluated for efficacy against streptozotocin–nicotinamide-induced type 2 DM in rats, which exhibited significant antidiabetic, hypolipidemic, and antioxidant activity. Histopathological studies suggest the regenerative effect of bergenin on pancreatic β-cells. However, bergenin showed no significant effect on liver glycogen at all doses studied (Kumar et al. 2012e).

Further, the methanol extract of the plant root inhibited α-glucosidase and α-amylase activities in a dose-dependent manner under *in vitro* assay conditions (Narkhede et al. 2012).

Caesalpinia ferrea C. Martius ex Tul., Caesalpiniaceae

The tea from the stem bark of *Caesalpinia ferrea* has been popularly used in the treatment of diabetes in Brazil. The water extract of the stem bark of *C. ferrea* (300 and 450 mg/kg daily for 4 weeks) reduced blood glucose levels and improved the metabolic status of streptozotocin-induced diabetic rats. Phosphorylated

Akt was increased in the liver and skeletal muscles of treated diabetic animals; phosphorylated AMPK was reduced only in the skeletal muscles of these animals and phosphorylated acetyl-CoA carboxylase was reduced in both when compared with untreated rats. The authors conclude that the extract possibly acts to regulate glucose uptake in the liver and muscles by way of Akt activation, restoring the intracellular energy balance confirmed by inhibition of AMPK activation (Vasconcelosa et al. 2011).

Caesalpinia sappan L., Caesalpiniaceae

Common Names: Brazil wood, pattavanjaka, bakam, patunagam, sappannam

Caesalpinia sappan is a small woody, spreading tree with a few prickles. It is a commonly cultivated hedge plant in India especially in Bengal and South India and also in Sri Lanka, Burma, and Malaysia. In traditional medicine, it is an astringent, sedative, antiseptic, depurative emmenagogue, etc. It is used in biliousness, fever, delirium, ulcers, blood complaints, internal bleeding, wounds, rheumatism, skin diseases, hemorrhage, menstrual disorders, dysentery, diarrhea, syphilis, sapremia, hematemesis, and mastalgia (Kirtikar and Basu 1975).

Antidiabetes: The wood extract of this plant showed hypoglycemic activity. Brazilin (7,11b-dihydrobenz[b]indeno-[1,2-d]pyran-3,6a,9,10(6H)-tetrol) from *C. sappan* was found to have hypoglycemic action and increase glucose metabolism in experimental diabetic animals. Brazilin increased basal glucose transport in 3T3-L1 fibroblasts and adipocytes. However, insulin-stimulated glucose transport was not influenced by the compound. Autophosphorylation of the partially purified IR was not affected by brazilin treatment in 3T3-L1 fibroblasts. However, brazilin decreased the PKC activity in 3T3-L1 fibroblasts and adipocytes (Kim et al. 1995). Brazilin isolated from *C. sappan* exhibited antioxidant activity and inhibited adipocyte differentiation (Liang et al. 2013). Anti-DM compounds β-sitosterol and quercetin were reported from this plant.

Phytochemicals reported include sappanchalcone, campesterol, stigmasterol, β-sitosterol, brazilein, protosappanin E, tetraacetylbrazilin, protosappanin-B, homoisoflavonoids, sappanol, episappanol, 3′-0-methyl brazilein, sappanon-B, 8-methoxy-bonucellin, quercetin, rhamnetin, and ombuin (review: Ajikumaran and Subramoniam 2005).

Cajanus cajan (L.) Millsp., Fabaceae

Common Names: Pigeon pea, red gram, Congo pea (Eng.), arhar, thovaray, etc.

Description: *Cajanus cajan* is an erect shrub of 1.5–3 m height and branches that are provided with silky hairs. Its leaves are trifoliately compound and pulvinate; the leaflets are oblong-lanceolate, entire, and densely silky beneath. Yellow flowers are produced in terminal panicles or in corymbose racemes. The fruit is a pod with 2–7 seeds.

Distribution: The shrub's center of origin is most likely Asia. It has been cultivated in Egypt, Africa, and Asia since prehistoric times. Now it is distributed in several tropical countries. It is cultivated extensively in India, both as a food crop and forage crop.

Traditional Medicinal Uses: It is an astringent, diuretic, and laxative and is used for bronchitis, colic, convolution, diarrhea, ciguatera, common cold, cough, earache, dermatosis, dysentery, jaundice, sore, urticaria, leprosy, stroke, swelling, and vertigo. It is also used for the treatment of diabetes and as an energy stimulant. The leaves are used in colic and constipation; in Chinese folk medicine, the leaves are used as an analgesic and to kill parasites. The leaves are used in India to cure gingivitis and as a laxative; the leaves and seeds are applied as poultice over the breast to induce lactation (Pal et al. 2011).

Antidiabetes: In normal as well as alloxanized mice, single doses of unroasted seeds of *C. cajan* significantly reduced the serum glucose level at 1–2 h, but at 3 h a rise of glucose level was observed (Amalraj and Ignacimuthu 1998a). In another study, the water extract of the stem and leaves in normoglycemic mice showed hyperglycemia in doses of 500 and 1000 mg/kg, but at 300 mg/kg dose, a short-lived decrease in the glycemia is observed (Esposito et al. 1991). The antidiabetic activity of the methanol extract of leaves of *C. cajan* was studied in alloxan-induced diabetic and oral glucose-loaded rats. The results showed that the extract significantly reduced the fasting blood sugar of alloxan-induced diabetic rats in a dose-dependent manner with a maximum hypoglycemic effect at 4–6 h (Ezike et al. 2010). In contrast, the water extract of *C. cajan* leaves showed significant increment in fasting blood glucose levels in streptozotocin-induced type 2 diabetic rats (Jaiswal et al. 2008). Thus, further studies are required to establish the antidiabetic property of this plant. However, the phytochemical profile of the plant shows the presence of α-amyrin and β-sitosterol, compounds known to have anti-DM activity.

Other Activities: In high-fat, high-cholesterol diet-fed rats, significant lipid-lowering effect was exhibited by the administration of a protein from *C. cajan*. The plant has antioxidant activity; two isoflavonoids, genistein and genistin, isolated from the root were found to possess antioxidant activity. Other reported activities of this plant include antimicrobial activity, anticancer activity, hepatoprotective effects, anthelmintic activity, and neuroactive properties. In chloroquine-sensitive *Plasmodium falciparum* strain, 3D7 pure compounds such as longistylin A and C, and betulinic acid showed a moderately high degree of *in vitro* activity. The extract of this plant was shown to be active in the reduction of painful crises and may ameliorate the unwanted effects of sickle cell anemia on the liver (Akinsulie et al. 2005; review: Pal et al. 2008, 2011).

Phytochemicals: The leaves are rich in flavonoids and stilbenes; they also contain saponins, tannins, resins, and terpenoids. Chemical studies showed the presence of betulinic acid, biochanin A, cajanol, genistein, 2'-hydroxygenistein, longistylin A and C, pinostrobin (Duker-Eshun et al. 2004), vitexin, salicylic acid, hentriacontane, 2-carboxy-3-hydroxy-4-isoprenyl-5-methoxystilbene, laccerol, niringenin-7,4'-dimethyl ether, β- and α-amyrin, β-sitosterol, C_{26-27} alkanes, lupeol, stigmasterol, campesterol, cholesterol, cajaflavanone, cajanol, 2'-*O*-methyl cajanone, isoflavone glucoside, sitosterol-β-D-glucoside, cajaquinone, etc., in this plant (Pal et al. 2011).

Calamintha officinalis Moench, Lamiaceae
The aerial parts of this plant showed blood glucose-lowering activity in normal and streptozotocin-induced diabetic rats without influencing the levels of serum insulin levels (Lemhadri et al. 2004).

Calea zacatechichi Schlecht, Compositae
Calea zacatechichi is one of the plants used in Mexico to treat diabetes. The extract of the areal part of this plant showed hypoglycemic effect *in vivo* (Roman Ramos et al. 1992).

Callistemon lanceolatus DC., Myrtaceae
Callistemon lanceolatus is a traditional medicinal shrub. The plant leaf (ethyl acetate fraction) possessed antihyperglycemic property; the fraction improved body weight, liver profile, renal profile, and total lipid levels in alloxan-induced diabetic rats (Kumar et al. 2011d). In another study, the chloroform fraction of the ethanol extract of this plant leaf (150 mg/kg) showed hypoglycemic activity comparable to that of glibenclamide in streptozotocin-induced diabetic rats (Nazreem et al. 2011). Two new flavones characterized as 5,7-dihydroxy-6,8-dimethyl-4'-methoxy flavone and 8-(2-hydroxypropan-2-yl)-5-hydroxy-7-methoxy-6-methyl-4'-methoxy flavone were isolated from the leaves of *C. lanceolatus*. The isolated flavones exhibited blood glucose lowering effect in streptozotocin-induced diabetic rats (Nazreen et al. 2012). *C. lanceolatus* fruits (methanol extract) also exhibited prominent antidiabetic effect at a dose of 400 mg/kg (p.o., daily for 21 days) in alloxan-induced diabetic rats, which may be due to its antioxidant property. The treatment resulted in significant reduction in blood glucose, serum cholesterol, TG, AST, and ALT levels, whereas the HDL level was found to be improved as compared to the diabetic control group. The pancreas and liver histology indicated significant recovery with the extract administration (Sanjita and Uttam 2014). Further studies are needed to establish its utility for therapy.

Callistemon rigidus R. Br., Myrtaceae
Common Names: Bitter orange, seville orange
Piceatannol and scirpusin B were isolated from the stem bark of *Callistemon rigidus* (Kobayashi et al. 2006). These compounds inhibited α-amylase activity in isolated mouse plasma and gastrointestinal tract (Kobayashi et al. 2006).

Calotropis gigantea L., Asclepiadaceae

The chloroform extract of the flowers of *Calotropis gigantea* (200 mg/kg, p.o., daily for 21 days) showed moderate antidiabetic potential in terms of reduction of fasting glucose level in alloxan-induced diabetic rats. It also improved the level of serum antioxidant enzymes (Choudhary et al. 2011). In another study, *C. gigantea* leaves and flowers (chloroform extract, 50 mg/kg) were effective in lowering serum glucose levels in normal rats. Improvement in oral glucose tolerance was also registered by the treatment. The administration of the leaf and flower (chloroform extracts) (50 mg/kg, p.o., for 27 days) to streptozotocin-induced diabetic rats showed a reduction in serum glucose levels (Rathod et al. 2011).

Calotropis procera (Ait.) R. Br., Asclepiadaceae

The water suspension of the dried latex of *Calotropis procera* showed antioxidant and antihyperglycemic properties in alloxan-induced diabetic rats (Kumar and Padhy 2011). In another study also, the latex of *C. procera* exhibited antioxidant and antidiabetic effects in alloxan-induced diabetic rats (Jaya Preethi et al. 2012). The extracts (methanol, petroleum ether, and water) of *C. procera* leaves (250 mg/kg for 15 days) showed antihyperglycemic and antihyperlipidemic activities in streptozotocin-induced diabetic rats (Bhaskar and Ajay 2009). In another study, the antihyperglycemic activity of the hydroalcoholic extract of the leaves of *C. procera* was evaluated. Administration of the extract (300 and 600 mg/kg/day) orally to streptozotocin-induced diabetic rats resulted in the reduction in the level of blood glucose throughout the evaluation period, improvement in the metabolic status of the animals, and amelioration of the oral glucose tolerance (Neto et al. 2013).

Camellia sinensis (Linn.) Kuntze, Theaceae

Common Name: Tea plant

Camellia sinensis is an evergreen tree growing to a height of 9–15 m with simple, alternate, elliptic, obovate, and lanceolate leaves with a serrate margin. The white and fragrant flowers may be either solitary or may occur in clusters of 2–4. One to three seeds are produced in a brownish-green capsule. It is used worldwide for the preparation of tea.

Distribution: It is considered as a native of Southern Yunnan and Upper Indo-China and now it is cultivated over a circular range in tropical and subtropical regions.

Traditional Medicinal Uses: It is an analgesic, astringent, demulcent, diuretic, lactagogue, and narcotic and is used in inflammation, abdominal disorders, fever, strangury, fatigue and hemicrania, chest pain, eye troubles, etc. (Kirtikar and Basu 1975).

Antidiabetes: *C. sinensis* (black tea) showed an antihyperglycemic effect in rats (Gomes et al. 1995). The tea showed *in vitro* insulin-enhancing activity, and the predominant active ingredient is epigallocatechin gallate (Anderson and Polansky 2002). The water extract of the *C. sinensis* leaf was found to be effective in reducing most of the diabetes-associated abnormalities in streptozotocin-induced diabetic rats (Islam 2011). The water extract of the leaves of *C. sinensis* (200 mg/kg) containing catechins reduced hyperglycemia-induced renal oxidative stress and inflammation in streptozotocin-induced male diabetic rats (Kumar et al. 2012a). In an *in vitro* study, catechin promoted adipocyte differentiation in human bone marrow mesenchymal stem cells through PPAR-γ activation (Wang et al. 2014).

Other Activities: *C. sinensis* (black tea) is a chemopreventive agent. Inhibition of lung carcinogenesis by oral administration of green tea is likely to be mediated by inhibition of angiogenesis and induction of apoptosis in A/J mice. Further, it has a potent antioxidant activity. Green tea protected against ethanol-induced LPO in rat organs. Green tea polyphenols exhibited antimutagenic activity against tobacco-induced mutagenicity. It has hypolipidemic and anti-inflammatory activities (review: Ajikumaran and Subramoniam 2005).

Phytochemicals: Reported phytochemicals include (–)-epigallocatechin gallate (the major flavonoid of green tea), polysaccharide conjugates, purine alkaloids, theaflavins, thearubigins (pigments), caffeine, theophylline, xanthine, hypoxanthine, gums, dextrins, inositol, kaempferol, quercetin, barringtogenol, cinnamic acid, angelic acid, glucuronic acid, sapogenol, putrescine, theanine, typhasterol, teasterol, umbelliferone, skimming, scopoletin, and α-spinasterol-β-D-gentiobioside (review: Ajikumaran and Subramoniam 2005).

Campomanesia xanthocarpa Berg, Myrtaceae

Administration of a leaf decoction of *Campomanesia xanthocarpa* (20 g/L) for 3 weeks to streptozoto-cin-induced diabetic rats resulted in decrease in blood glucose levels, inhibition of hepatic glycogen loss, and prevention of potential histopathological alterations in the pancreas and kidneys. No differences were found between the control rats treated with the decoction and the control rats maintained on water only. In conclusion, these results suggest that *C. xanthocarpa* leaf decoction (20 g/L) might be useful for DM management (Vinagre et al. 2010).

Campsis grandiflora (Thunb.) K.Schum, Bignoniaceae

Synonym: *Tecoma grandiflora*

Common Name: Chinese trumpet vine

Pentacyclic triterpenoid isolated from *Campsis grandiflora* showed insulin-mimetic and insulin-sensitiz-ing activities. The compounds enhanced the activity of insulin on phosphorylation of the IR in Chinese hamster ovary cells expressing the human IR. Out of the five triterpenoids tested, ursolic acid showed the maximum activity. The authors conclude that enhancement of insulin activity by ursolic acid may be useful for developing specific IR activators for the treatment of type 1 and type 2 DM (Jung et al. 2007).

Canarium schweinfurthii Engl., Burseraceae

Stem bark extracts of *Canarium schweinfurthii* are used in Africa for the treatment of various ailments, including DM. In a study, the antidiabetic effects of the methanol/methylene chloride extract of the stem bark were evaluated on male streptozotocin-induced diabetic rats. Administration of the extract (300 mg/kg daily for 14 days) resulted in marked reduction in blood glucose levels in diabetic rats compared to diabetic control rats. The treatment also improved body weight (Kamtchouing et al. 2005). In another study, the antidiabetic effects of the hexane extract of the *C. schweinfurthii* stem bark was evaluated in streptozotocin-induced diabetic rats. In acute treatment, the extract (75 and 150 mg/kg) pro-duced a significant hypoglycemic effect 2 h postdosing in both normal and diabetic rats. In subchronic (14 days daily) treatment to the diabetic rats, the hexane extract (38, 75, and 150 mg/kg, p.o.) significantly decreased blood glucose levels by 72%, 80%, and 74%, respectively, while insulin (10 UI/kg) given sub-cutaneously and once daily had 71% reduction as compared to initial values. Further, the extract treat-ment increased body weight in diabetic rats (Kouambou et al. 2007).

Canavalia ensiformis (L.) DC, Fabaceae

Common Name: Jack bean

Canavalia ensiformis or (common) jack bean is a legume that is used for animal fodder and human nutri-tion. Seeds of *C. ensiformis* were investigated for hypoglycemic activity in normal and streptozotocin-induced diabetic albino rats. The seeds exhibited potent hypoglycemic effects in normal animals but not in streptozotocin-induced diabetic ones. The authors suggest that the hypoglycemic effect of *C. ensiformis* is probably due to either direct or indirect stimulation of β-cells of the islets of Langerhans to secrete more insulin, thereby producing hypoglycemia in normal albino rats (Envikwola et al. 1991). In another study, the effect of *C. ensiformis* seeds on alloxan-induced diabetic rats was studied. One-week oral administration of the water extract of the seeds of *C. ensiformis* significantly reduced hyperlipidemia and hyperketonemia in diabetic rats. These results suggest that *C. ensiformis* seeds can be useful for the treatment of DM (Nimenibo-Uadia 2003).

Cannabis sativa L., Cannabaceae

Related Species: *Cannabis indica* Lam., *Cannabis ruderalis* Janisch (Figure 2.26)

Common Names: The plant is colloquially known as marijuana and hemp; the common names for this and related species include ganja, Mary Jane, Indian hemp, etc.

FIGURE 2.26 *Cannabis sativa.*

Description: *Cannabis sativa* is an annual herb that can reach up to 3 m high or more with suitable humidity and soil. The stem is covered by rigid hairs and is rough to the touch. Species are male and female; the latter has more leaves. The leaves are long-stalked, palmate, with 3–7 narrow and toothed leaflets. The upper leaves are alternate and the low ones opposite. The flowers are very small and green and have axillary branches. The fruit is oval, flat, 5–6 mm long and 4 mm wide, with a light green color. It has an herbaceous strong smell and taste.

Distribution: The plant is originally from the Caspian and Black Sea area and was taken to Persia and India eight centuries ago. *C. sativa* (native to Central Asia) has a worldwide distribution. It is cultivated in North, Central, and South America, Asia, Europe, and North and Central Africa. Major producers include Mexico, Brazil, Paraguay, Colombia, Peru, New Zealand, and Saudi Arabia.

Traditional Use and Marijuana Abuse: The flowering tops of the female cannabis plant, consumed by humans since the beginning of recorded human history, are known as marijuana. This plant is cultivated and used as a medicinal and mood-altering agent. But there is a worldwide ban on the plant's cultivation and use. Marijuana as well as the biologically active compounds, cannabinoids, present in *C. sativa* and related species are considered as highly potential materials for abuse and dangerous to the public. In spite of this, cannabis is one of the most investigated, therapeutically active substances. However, the ban on its use and cultivation is a hindrance for clinical trials and cannabis-based medicine development.

Antidiabetes Activities: Preclinical studies indicated that cannabinoids modify diabetes progression and that they may also provide symptomatic relief to those suffering from the complications of diabetes.

The flowers of *C. sativa* (15 and 30 mg/kg extract, p.o.) prevented diabetic neuropathy and oxidative stress in streptozotocin-induced diabetic male rats (Comelli et al. 2009). Administration of 5 mg/kg/day of the nonpsychoactive cannabinoid reduced the incidence of diabetes in obese mice (Weiss et al. 2006). In another study, using genetically modified obese mice, administration of tetrahydrocannabivarin resulted in several metabolically beneficial effects such as improved glucose tolerance, improved insulin sensitivity, and decreased lipid accumulation in the liver. The authors conclude that tetrahydrocannabivarin may be useful in the treatment of type 2 diabetes, either alone or in combination with existing treatments (Wargent et al. 2013). Streptozotocin-induced diabetic rats treated with cannabidiol (nonpsychotropic cannabinoid) for 1–4 weeks experienced significant protection from diabetic retinopathy; the treatment reduced neurotoxicity, inflammation, and blood–retinal barrier breakdown in diabetic rats through actions that may involve inhibition of p38 MAPK (El-Remessy et al. 2006). Cannabinoids alleviate neuropathic pain associated with diabetes in animal models. Administration of a cannabis receptor agonist to diabetic rats resulted in a reduction in diabetes-related pain resulting from noninjurious stimulus to the skin compared to nontreated diabetic control mice (Dogrul et al. 2004). Besides, administration of cannabidiol reduces various symptoms of diabetic cardiomyopathy in a mouse model of type 1 diabetes. Furthermore, cannabidiol also attenuated the high glucose–induced ROS generation, nuclear factor-κB activation, and cell death in primary human cardiomyocytes in culture. The authors conclude that it may have great therapeutic potential in the treatment of diabetic complications (Rajesh et al. 2010). Δ-9-Tetrahydrocannabinol isolated from the leaves activated PPAR-γ *in vitro* (Wang et al. 2014).

Human Clinical Trials: In recent years, observation trials have reported that those who consume cannabis possess a lower risk of developing type 2 diabetes than do nonusers. Further studies are needed to show a direct effect of marijuana on DM. The study did not find an association between cannabis use and other chronic diseases (Rajavashisth et al. 2012). In another trial, U.S. subjects who reported using marijuana the previous month had lower levels of fasting insulin, improvement in insulin resistance (homeostasis model assessment of insulin resistance), smaller waist circumference, and high levels of HDL-C, suggesting that the cannabinoid system may play a role in insulin resistance and obesity (Penner et al. 2013).

Other Activities: Several clinical observations shed light on the therapeutic value of marijuana and cannabinoids. It can alleviate severe neuropathic pain associated with cancer, diabetes, certain viral diseases, etc.; studies also suggest that it could be useful for patients suffering from multiple sclerosis, rheumatoid arthritis, and inflammatory bowl disorders. It also has a therapeutic value in certain neurological disorders. A growing body of preclinical and clinical data concludes that cannabinoids can reduce the spread of certain cancer cells via apoptosis and antiangiogenesis.

Toxicity: In normal circumstances the consumption of marijuana cannot result in death even in overdose cases. It has a good safety record. Nonpsychotropic cannabinoid has excellent safety and tolerability profile in humans (Rajesh et al. 2010).

Phytochemicals: This plant contains volatile terpenes and sesquiterpenes that give a characteristic odor to this plant. This plant contains more than 60 cannabinoids including cannabinol, cannabidiol, cannabinolic acid, cannabigerol, cannabicyclol, and various tetrahydrocannabinol isomers, the most important being Δ-9-tetrahydrocannabinol (Penner et al. 2013).

Canscora decussata (Roxb) Roem & Schult, Gentianaceae

The hypolipidemic and renoprotective effects of the methanol extract of *Canscora decussata* (whole plant) in alloxan-induced diabetic rabbits was evaluated. Oral administration with the extract (400 and 600 mg/kg) for 30 days to diabetic rabbits decreased the raised parameters including TG, total cholesterol, and LDL-C, atherogenic index, and coronary risk index up to normal values compared to diabetic control rabbits. Further, the treatment increased HDL-C, albumin, globulin, and total protein levels. Therefore, it is suggested that the methanol extract of *C. decussata* exerted hypolipidemic and renoprotective effects in alloxan-induced diabetic rabbits (Irshad et al. 2013). In a follow-up study, 0.5, 1, and 2 g/kg of the methanol extract decreased blood glucose levels in normal rabbits and diabetic rabbits. The extract (600 mg/kg), when administered along with insulin (2 and 3 unit/kg, s/c), further reduced blood glucose levels of alloxan-induced diabetic rabbits. The oral glucose tolerance test revealed lowered area under curve values in the extract-treated rabbits. Treatment with the extract (400 and 600 mg/kg) for 30 days showed highly significant decrease in blood glucose level by augmenting insulin secretion; in the treated diabetic rabbits, HbA1c decreased and body weight increased; histopathology study showed regeneration of β-cells (Irshad et al. 2014).

Capparis decidua (Forsk.) Edgew., Capparidaceae

Capparis decidua is one of the traditional remedies used for various medicinal treatments including diabetes. Extracts from different aerial parts of *C. decidua* showed potent antidiabetic and antihemolytic activity. All examined extracts were prominently rich in phenolics and glucosinolates (Zia-Ul-Haq et al. 2011). The antidiabetic activity of the water and ethanol extracts of the *C. decidua* stem in alloxan-induced diabetic rats was evaluated. Oral administration of the water and ethanol extracts (250 and 500 mg/kg daily for 21 days) to diabetic rats decreased fasting blood glucose levels; the alcohol extract was found to be slightly better than the water extract (Rathee et al. 2010). Fruits of *C. decidua* also exhibited antidiabetic activity; it decreased blood glucose levels and reduced LPO and oxidative stress in erythrocytes, liver, kidney, and heart in aged alloxan-induced diabetic rats (Yadav et al. 1997).

Capparis sepiaria L., Capparidaceae

Capparis sepiaria, a profusely branched hedge plant, is used in Indian traditional medicine. The ethanol extract of *C. sepiaria* was investigated for possible hypoglycemic effect in the streptozotocin-induced diabetic rats. The extract (100, 200, and 300 mg/kg) caused maximum fall of plasma glucose level after 12 h of treatment in diabetic rats (Selvamani et al. 2008).

Capparis spinosa L., Capparidaceae

Common Names: Caper bush, kakadani, kabra, etc.

Description: *Capparis spinosa* is a diffuse, prostrate shrub with terete branches with variable, entire, glaucescent, orbicular, or from broadly ovate to obovate, retuse or sometimes acute, mucronate, solitary leaves and axillary pedicellate flowers. The fruit is ovoid or obovoid, ellipsoid berry, red when ripe, bearing smooth brown seeds.

Distribution: It occurs in the rocky and hilly areas of the Deccan Peninsula, Rajputana, and Northwestern India.

Traditional Medicinal Uses: *C. spinosa* is an anthelmintic, alternative, analgesic, astringent, expectorant, emmenagogue, and tonic and used in rheumatism, paralysis, toothache, tumors, infections of the liver and spleen, tubercular gland, carcinomas, gout, scurvy, common cold, spleen sclerosis, sore, and snake bite (Kirtikar and Basu 1975; *The Wealth of India* 1992).

Antidiabetes: The water extract of the fruit of *C. spinosa* showed a significant glucose-lowering effect on streptozotocin-induced diabetic rats without any pronounced hypoglycemia in normal rats (Eddouks et al. 2004a). Besides, the water extract of the fruits showed hypolipidemic activity in streptozotocin-induced diabetic rats (Eddouks et al. 2005a). In a study, the possible antihyperglycemic and body weight–reducing activities of the water extract of *C. spinosa* (20 mg/kg) were evaluated: The repeated oral administration of the water extract evoked a potent antihyperglycemic activity in HFD obese mice. Postprandial hyperglycemic peaks were significantly lower in the extract-treated experimental groups. The body weight was also significantly lower in the treated group when compared to the control one after daily oral administration of the extract for 15 days (Lemhadri et al. 2007).

Other Activities: The water extract of *C. spinosa* showed hypolipidemic activity in normal and diabetic rats. The methanol extract from flowering buds possessed a chondroprotective effect, which might be useful in the management of cartilage damage. Antiallergic and antihistaminic effects have been shown in the extracts of *C. spinosa* flowering buds. Polyprenol cappaprenol-13 from the plant showed anti-inflammatory activity in carrageenan-induced paw edema in rats. The flower bud extract has *in vitro* antioxidant and *in vivo* hepatoprotective effects. The water extract of this plant has antifungal activity. The extract completely prevented the growth of *Microsporum canis* and *Trichophyton violaceum* (review: Ajikumaran and Subramoniam 2005).

Phytochemicals: The phytochemicals found in this plant include the following: rutin, rutic acid, pectic acid, resin, tannins, quercetin-3-rutinoside, kampferol-3-rutinoside, quercetin triglycoside, alkaloid cadabicine, β-sitosterylglucoside-6′-octadecanoate, 3-methyl-2-butenyl-β-glucoside, (6S)-hydroxy-3-oxo-α-ionol glucosides, quercetin-7–O-glucorhamnoside, stachydrine, and isothiocyanate glucoside (*The Wealth of India* 1992; Ajikumaran and Subramoniam 2005).

Capsicum annuum L., Solanaceae

Synonyms: *Capsicum fastigiatum* Blume (morphological features of *Capsicum frutescens* [tabasco pepper] varieties can overlap with those of *C. annuum*; there is no consistently recognizable *C. frutescens* species (Zhigila et al. 2014))

Common Names: Capsicum, chili (Eng.); lal mirich, gachmarich, milagay, mulaku, etc. (Indian languages) (there are several varieties)

Description: *Capsicum annuum* is a herbaceous or suffrutescent annual or biennial with entire, oblong, and glabrous leaves and pedicellate, solitary, rarely in pairs, and corolla white flowers. The fruit is a berry becoming yellow or orange to red grading to brown to purple when mature and bearing discoid, smooth, or subscabrous seeds.

Distribution: It is introduced from South America and cultivated throughout India for dry chili of commerce.

Traditional Medicinal Uses: It is bitter and is used as an expectorant, analgesic, carminative, tonic, and aphrodisiac and for the treatment of lumbago, neuralgia, rheumatic complications, dyspepsia, gastroenteritis, pharyngitis, relaxed sore throat, asthma, gonorrhea, ciguatera, sores, scrofula, yellow fever, diarrhea, piles, and scarlatina (Kirtikar and Basu 1975).

Antidiabetes: The effect of dried red chili pepper at 1% and 2% and the pure capsaicin at 0.015% in the diet was studied in HFD-fed alloxan-induced diabetic rats. Blood serum glucose levels were markedly reduced in the diabetic group fed an HFD + 0.015% capsaicin compared to diabetic control animals. The decrease was less in the red chili–fed groups compared to that in capsaicin-fed diabetic rats. Moreover, serum cholesterol as well as serum TGs for the diabetic groups fed an HFD + 2% red dried chili pepper was significantly low. The HDL concentration for the groups fed with red chili or capsaicin were significantly higher than that in diabetic control rats fed with an HFD. Feeding the groups of diabetic rats with an HFD + 1% or 2% dried red chili pepper or 0.015% capsaicin resulted in a decrease in the levels of LDL and VLDL as well as total lipids compared with diabetic control rats (Magied et al. 2014).

Clinical Trials: Daily intake of 30 g of freshly chopped chili for 4 weeks reduced postprandial hyperinsulinemia without reducing blood glucose levels in humans (Ahuja et al. 2006). In a crossover study on 12 healthy volunteers, daily oral intake of 5 g of capsicum improved oral glucose tolerance. Furthermore, plasma insulin levels were significantly higher at 60, 75, 105, and 120 min (Kamon et al. 2009).

The studies on animals and humans suggest that chili and its major active ingredient capsaicin are beneficial to diabetic patients. However, detailed studies are warranted for its therapeutic use considering the capsaicin content of capsicum, the method of processing chili, the dose response and, possibly, the drug interaction, and other adverse effects.

Other Activities: Oleoresin of *C. annum* showed a hypocholesterolemic effect in gerbils. Other reported effects include weight-reducing effect and hypolipidemic properties in diabetics (review: Frhirhie et al. 2013).

Phytochemicals: The major active ingredients present in *Capsicum* include capsaicin, homocapsaicin, norhydrocapsaicin, dihydrocapsaicin, homodihydrocapsaicin, apsorubin, zeaxanthin, lutein, cryptoxanthin, sesquiterpenes capsinide, oleoresin, fucosterol, citrostadienol, isocitrostadienol, α-methoxy-3-isobutylpyrazine,trans-β-ocimene, limonene, methyl salicylate, capsorubin, polyphenols, fatty acids, carotene, vitamins A and C, thiamine, and citric, tartaric, malic, and tannic acids (review: Frhirhie et al. 2013).

Capsicum frutescens L., Solanaceae

Synonym: *Capsicum annum* L. var. *frutescens* (*L.*) Kuntze

A study was conducted to clarify whether a low or a high, but tolerable, dietary dose of red chili can ameliorate the diabetes-related complications in an HFD-fed streptozotocin-induced type 2 diabetes model of rats. After 4 weeks feeding of experimental diets, the fasting blood glucose concentrations in both red chili-fed groups were not significantly different. The serum insulin concentration was significantly increased in the 2.0% red chili group compared with the diabetic control and 0.5% red chili groups. Blood HbA1c, liver weight, liver glycogen, and serum lipids were not influenced by the feeding of red chili-containing diets. The data of this study suggest that 2% dietary red chili is insulinotropic rather than hypoglycemic at least in this experimental condition (Islam and Choi 2008). In another study, *C. frutescens* fruit supplementation (1 and 2 g) improved the biochemical parameters, blood glucose levels, and body weight of alloxan-induced diabetic rats. This may suggest cardioprotective and antidiabetic properties of the *C. frutescens* fruit (Anthony et al. 2013). The major active ingredient capsaicin is present in both *C. annum* and *C. frutescens* fruits. They may share many of the pharmacological properties. There could be variations in phytochemical profiles within the species due to the ecotype and other variations such as varieties.

Caralluma adscendens var. *attenuata* (Wight.) Grav. & Mayur., Asclepiadaceae

Synonym: *Caralluma attenuata* Wight.

Common Name: Yugmaphallottama

Caralluma adscendens is an erect, branched succulent herb that is distributed from the Mediterranean region to the East Indies; in India, it occurs in the rocky areas of the Deccan Peninsula, Punjab, etc. This plant is used for the treatment of diabetes in certain remote villages in Kerala, India.

Antidiabetes: The butanol extract of *C. adscendens* showed antihyperglycemic activity in both glucose-loaded normal rats and alloxan-induced diabetic rats (Venkatesh et al. 2003). In another study, the water and alcohol extracts of the whole plant significantly lowered blood glucose level in normal and alloxan-induced diabetic rats (Jayakar et al. 2004). These effects were confirmed by another study wherein the whole plant extract exhibited a blood glucose-lowering effect in both normal and alloxan-induced diabetic animals (Suresh et al. 2004). The butanol extract of *C. adscendens* exhibited a promising blood glucose-lowering effect in alloxan-induced diabetic rats. Further, the extract improved glucose tolerance and lipid profile in the treated diabetic rats (Tatiya et al. 2010). Thus, the anti-DM activity of the plant has been established in rodents. Follow-up studies are warranted.

A flavonoid isolated from *C. attenuata* exhibited antinociceptive and anti-inflammatory activities. Phytochemicals reported include special glycosides and n-pentatriacontane (review: Ajikumaran and Subramoniam 2005).

Caralluma sinaica L., Asclepiadaceae

The ethanol extract of the leaves of *Caralluma sinaica* lowered plasma glucose levels in streptozotocin-induced diabetic rabbits (Mohammed et al. 2008).

Caralluma tuberculata N. E. Br., Asclepiadaceae

Caralluma tuberculata is a leafless succulent plant used in traditional medicine in India and elsewhere. The antihyperglycemic and hypolipidemic effects of the methanol extract of *C. tuberculata* were investigated in streptozotocin-induced diabetic rats. The extract showed reduction in the fasting blood glucose levels and improvement in oral glucose tolerance in diabetic rats. Further, the tested extract also increased plasma insulin by 207%. The hypolipidemic action of the extract was evident by the significant decrease in the levels of total cholesterol, TGs, and LDL compared to diabetic control rat values. The extract increased the levels of HDL (Abdel-Sattar et al. 2011). In a follow-up study, the methanol extract and the water fraction of *C. tuberculata* showed high antidiabetic activity in streptozotocin-induced diabetic rats. The methanol extract was superior to the fraction. The antidiabetic activity was evaluated through assessing fasting blood glucose levels, insulin levels, oral glucose tolerance test, glucose utilization by isolated rat psoas muscle, gut glucose absorption, and G-6-Pase activity in isolated rat liver microsomes. The mechanism underlying the observed antihyperglycemic activity of the methanol extract could be attributed, at least in part, to enhanced skeletal muscle utilization of glucose, inhibition of hepatic gluconeogenesis, and stimulation of insulin secretion. The extract was standardized through liquid chromatography-mass spectroscopy (LC-MS) analysis for its major pregnanes (Abdel-Sattar et al. 2013).

Carica papaya L., Caricaceae

Carica papaya leaf extract exerted hypoglycemic and antioxidant effects in streptozotocin-induced diabetic rats; further, the extract improved the lipid profile in diabetic rats (Juarez-Rojop et al. 2012). The unripe fruits of *C. papaya* showed hypoglycemic and hypolipidemic activity and improvement in body weight in alloxan-induced diabetic rats; the treatment reduced glycated hemoglobin and increased HDL levels (Sunday et al. 2014). This unripe fruit may be useful as nutraceuticals for diabetic patients.

Carissa carandas L., or *Carissa congesta* Wight, Apocynaceae
Carissa carandas, commonly known as karanda, has a long history of use in traditional medicine. The unripe fruit of this plant was evaluated for its antidiabetic activity. The methanol extract and its fractions were screened for their antidiabetic activity in alloxan-induced diabetic rats. The methanol extract and its ethyl acetate-soluble fraction lowered the elevated blood glucose levels at a dose level of 400 mg/kg, p.o., after 24 h as compared to diabetic control. The polyphenolic and flavonoid compounds were present in the methanol extract and its ethyl acetate-soluble fraction (Itankar et al. 2011).

Carissa edulis Vahl, Apocynaceae
Consumption of the ethanol extract of *Carissa edulis* leaves resulted in significant decreases in the blood glucose level in streptozotocin-induced diabetic rats during the first 3 h of treatment. On the other hand, in normal rats, the ethanol extract treatment produced insignificant changes in the blood glucose levels compared to glibenclamide treatment (El-Fiky 1996). Further, chlorogenic acid (5-caffeoylquinic acid) is present in this plant; this compound is known to have antioxidant, anti-DM, and other activities (review: Al-Youssef and Hassan 2014).

Carthamus tinctorius L., Asteraceae

Common Name: Safflower
The hydroalcoholic extract (200 mg/kg) of *Carthamus tinctorius* significantly decreased the blood glucose level in alloxan-induced diabetic rats. Furthermore, in the extract-treated rats, there was a significant decrease in serum contents of total cholesterol with a significant increase in the insulin level. Histological morphology examination showed that the extract restored pancreas tissue damage in diabetic rats (Parivash et al. 2009). Another study also reported essentially the same results. The hydroalcoholic extract of this plant (flower) improved blood glucose levels and lipid profile in alloxan-induced diabetic rats (Asgary et al. 2012). A recent study confirmed these observations in rabbits. In this study, *C. tinctorius* extract (200 and 300 mg/kg, p.o., daily for 30 days) showed significant hypoglycemic effect in alloxan-induced diabetic rabbits as compared to diabetic control rabbits. Insulin levels were also significantly increased in *C. tinctorius*-treated groups as compared to diabetic control (Qazi et al. 2014). Further studies are needed to determine its utility as a therapeutic agent.

Carum carvi L., Apiaceae

Common Names: Caraway (Eng.), sushavi, shia-jira, shimai-shembu, etc. (in India)
Carum carvi is an annual or biennial, erect, stemmed, herbaceous plant forming a taproot occasionally acting as a perennial stock. It is distributed in the temperate regions of the world including Europe, Holland, Central Asia and the Northern Himalayas. In India, it is cultivated as a cold crop in the hills of Kumaon, Garhwal, Kashmir, and Chambal.

In traditional medicine, *C. carvi* is used as an aphrodisiac, antiperiodic, anthelmintic, astringent, lactagogue, stomachic, carminative, diuretic, emmenagogue, and laxative. Further, it is used in the treatment of scabies, hookworm, halitosis, fistula, cholera, flatulence, cancer, headache, stomachache, indigestion, inflammation, leukoderma, dysentery, hiccups, piles, hysteria, etc. (*The Wealth of India* 1992).

The plant exhibited potent antihyperglycemic activity in streptozotocin-induced diabetic rats without affecting plasma insulin levels. The treatment did not show highly significant hypoglycemic activity in normal rats (Eddouks et al. 2004). Oral administration of the seed extract of *C. carvi* resulted in both antihyperglycemic and hypolipidemic activity in streptozotocin-induced diabetic rats (Haidari et al. 2011). The water extract of *C. carvi* seeds exhibited renoprotective activity in streptozotocin-induced diabetic nephropathy in rats (Sadiq et al. 2010).

Essential oil from *C. carvi* (carvone, limonene, germacrene D, and trans-dihydrocarvone) showed antibacterial activity (Iacobellis et al. 2005). The water extract of the fruits has antioxidant activity. *C. carvi* oil has chemopreventive properties (review: Ajikumaran and Subramoniam 2005). Phytochemicals reported include carvone, dl-limonene, carvacrol, flavon (ol)-*O*-glycosides, essential oil, and monoterpenes (*The Wealth of India* 1992; Iacobellis et al. 2005).

Casearia esculenta Roxb., Flacourtiaceae/Samydaceae (Figure 2.27)

Common Names: Carilla fruit (Eng.), chilla, kakkaipilai, malampavatta, etc.

Description: *Casearia esculenta* is a shrub or tree of 6 m height with pale, yellow, smooth bark with elliptic-lanceolate, coriaceous, shining and glabrous, entire or obscurely crenate leaves. The flowers are born in axillary clusters and are light yellow in color; the fruit is a capsule of 1–2 cm long with bright red aril; there are many seeds with arils.

Distribution: It is distributed in the Indian Peninsula in Eastern and Western Ghats up to 1200 m and in the hills of Northeastern India.

Traditional Medicinal Uses: The plant root is widely used in traditional of medicine to treat diabetes (*The Wealth of India* 1992).

Pharmacology

Antidiabetes: The plant root extract has antihyperglycemic and antidiabetic properties in rats (Joglegar et al. 1959; Choudhury and Basu 1967; Gupta et al. 1967; Prakasam et al. 2003a,b). The hypoglycemic effect of the ethanol extract of *C. esculenta* was investigated on alloxan-induced diabetic rats. The blood glucose levels were measured at 0, 1, 2, and 3 h after the treatment. The ethanol extract (250 mg/kg) reduced the blood glucose of normal rats 3 h after oral administration. It also markedly lowered blood glucose level in alloxan-induced diabetic rats from (mean ± S.E.) 18.4 ± 0.3 to 7.2 ± 0.4 mmol/L, 3 h after oral administration (Arul et al. 2006). The root extract also showed lipid-lowering activity in streptozotocin-induced diabetic rats. Administration of the water extract of the root (200 and 300 mg/kg) for 45 days resulted in a significant reduction in serum and tissue cholesterol, phospholipids, free fatty acids, and TGs in streptozotocin-induced diabetic rats. In addition to that, decrease in HDL and significant increase in LDL and VLDL were observed in diabetic rats, which were normalized after 45 days of the extract treatment. The root extract at a 300 mg/kg body weight dose showed significant hypolipidemic effects than the 200 mg/kg body weight dose (Prakasam et al. 2003b).

An active compound, 3-hydroxymethyl xylitol, has been isolated from the root extract and its optimum dose has been determined in a short duration study (40 mg/kg). The long-term effect of the compound in type 2 diabetic rats has been investigated. An optimum dose of 40 mg/kg was orally administered for 45 days to streptozotocin-induced diabetic rats for the assessment of glucose, insulin, hemoglobin, glycated hemoglobin (HbA1c), hepatic glycogen, and activities of carbohydrate metabolizing enzymes, such as glucokinase, glucose-6-phosphatase, fructose 1,6-bisphosphatase, and glucose-6-phosphate dehydrogenase, and hepatic marker enzymes, such as AST, ALT, ALP, and γ-glutamyl transferase (GGT) in normal and streptozotocin-induced diabetic rats. 3-Hydroxymethyl xylitol (40 mg/kg) produced similar effects on all biochemical parameters studied as that of glibenclamide, a standard drug. Histological study of the pancreas also confirmed the biochemical findings (Chandramohan et al. 2008). Further, the effect of this compound on plasma and tissue lipid profiles in streptozotocin-induced diabetic rats

FIGURE 2.27 *Casearia esculenta.*

was studied. Total cholesterol, TG, free fatty acid, and phospholipid levels significantly decreased in plasma and tissues (liver, kidney, heart, and brain), whereas plasma HDL significantly increased in the compound-treated (40 mg/kg daily for 45 days) diabetic rats compared to diabetic control rats. Besides, the treatment decreased plasma levels of LDL and VLDL. Histological study of the liver also confirmed the biochemical findings (Chandramohan et al. 2010).

In another study, the compound showed favorable effects in type 2 diabetic rats as evidenced by antihyperglycemic, antihyperlipidemic, and antioxidant properties coupled with improvement in kidney function (Govindasamy et al. 2011). The significant increase observed in the levels of hexose, hexosamine, and fucose in the liver and kidney of diabetic rats was reversed to near normal levels by the treatment with the compound (Govindasamy et al. 2011). Thus, 3-hydroxymethyl xylitol appears to be promising for drug development. Besides, the presence of anti-DM compounds, β-sitosterol and leucopelargonidin, in the plant has been reported. This appears to be a promising plant for further studies.

Other Activities: The root extracts possess antioxidant and hypolipidemic actions in normal as well as diabetic rats (Prakasam et al. 2003a,b).

Phytochemicals: The phytochemicals reported include arabinose, dulcitol, tannin, β-sitosterol, and leucopelargonidin (*The Wealth of India* 1992).

Cassia alata L.

Synonym: *Senna alata* (L.) Roxb., Caesalpiniaceae
Cassia alata (*Senna alata*) is used for treating various disease conditions including diabetes in traditional medicine. It was found that the α-glucosidase inhibitory effect of the crude extract of *C. alata* leaves was far better than the standard clinically used drug, acarbose. The ethyl acetate (IC_{50}, 2.95 ± 0.47 µg/mL) and n-butanol (IC_{50}, 25.80 ± 2.01 µg/mL) fractions that contained predominantly kaempferol and kaempferol 3-*O*-gentiobioside, respectively, displayed the highest α-glucosidase inhibitory effect (Varghese et al. 2013b). In a recent study, the ethanol extract of the plant markedly reduced the blood glucose levels in oral glucose tolerance test in streptozotocin-induced diabetic mice. In acute toxicity test in mice, the extract up to 3 g/kg did not cause mortality and the mice appeared to be normal (Priyadarshini et al. 2014).

Cassia auriculata L., Caesalpiniaceae (Figure 2.28)

Common or Local Names: Tanner's cassia, avarttaki, awal, taroda, avaram, aviram, etc.

Description: *Cassia auriculata* is a tall shrub with a smooth bark and variegate branches. The leaves are alternate, pinnately compound, very numerous, closely placed, slender, and pubescent with an erect linear gland between the leaflets of each pair; leaflets are 16–24, slightly overlapping, oval oblong and dull

FIGURE 2.28   *Cassia auriculata* (twig with flower).

green; stipules very long, reniform, persistent, and produced at base. The flowers are large, bright yellow, and showy in copious corymbose racemes. The flowers are irregular and bisexual and the pedicles are glabrous and about 2.5 cm long. The fruit is a short legume, 7–11 cm long and about 1.5 cm broad; the pod is straight, ligulate, dark brown, and compressed tapering toward the base. There are 12–20 seeds present in each fruit (review: Joy et al. 2012).

Distribution: It is present in Asian countries including India and Sri Lanka. It is distributed in the dry zone of Southern, Western, and Central India extended up to Rajasthan. The plant is cultivated in the dry parts of Punjab, Haryana, Uttar Pradesh, and West Bengal.

Traditional Medicinal Uses: It is one of the important constituents of "avaarai panchaga chooranam," an Indian traditional herbal formulation used in the treatment of diabetes. It is also a component of various herbal formulations used to control DM. Various parts, including the flowers and leaves, of the plant are used in traditional medicine. It is considered as alexeteric, astringent, and alexipharmic in Ayurveda. In traditional medicine, it is used in sore throat, enemas, rheumatism, eye disease, diabetes, stomachache, dysentery, antiviral, conjunctivitis, migraine, insect attack, inflammation, ulcers, asthma, menorrhea, skin diseases, fractures, burns, and tumors (Kirtikar and Basu 1975; *The Wealth of India* 1992; review: Joy et al. 2012).

Pharmacology

Antidiabetes: The *C. auriculata* flower (water extract) suppressed the elevated blood glucose and lipid levels in streptozotocin-induced diabetic rats (Pari and Latha 2002; Latha and Pari 2003b). It exhibited hypoglycemic effects in normal rats also (Sabu and Subburaju 2002).

Oral administration of 450 mg/kg of the water extract of the flowers to streptozotocin-induced diabetic rats resulted in a significant reduction in blood glucose and an increase in plasma insulin levels, but at lower concentrations (150 and 300 mg/kg), the changes were not significant. The treatment also reduced serum lipid levels (Pari and Latha 2002).

The effect of *C. auriculata* flowers (water extract) on glucose as well as hepatic glycolytic and gluconeogenic enzymes was evaluated in streptozotocin-induced diabetic rats, which was given orally for 30 days. The treatment resulted in an antihyperglycemic effect in animals and almost normalized gluconeogenesis in diabetic animals. Further, the extract enhanced the utilization of glucose through increased glycolysis. The effect of the extract was more prominent than that of glibenclamide (Latha and Pari 2003a).

In another experiment on alloxan-induced diabetic male rats, the antihyperglycemic properties of the extracts of the root, stem, stem, leaves, and flower were compared. The extracts were orally administered for 28 days at a daily dose of 250 mg/kg in each case. There was a significant reduction in the serum glucose, TGs, and cholesterol and increase in the plasma insulin levels in the flower or leaf extract-treated diabetic animals. However, the root and stem extracts did not exhibit significant activity. Thus, it is confirmed that the leaves and flowers of *C. auriculata* have antidiabetic and antilipidemic effect, and these could be used in diabetes and coronary heart disease (CHD) management (Umadevi et al. 2006).

In another study, the antidiabetic potential of the water and ethanol extract of *C. auriculata* flowers was assessed in alloxan-induced diabetic rats. Phytochemical screening revealed that flavonoids and phenolic acid and free radical scavenging activity were higher in the water-soluble fraction of the ethanol extract compared to those of the water extract. Diabetic rats showed increase in blood glucose and decrease in plasma insulin levels after 48 h of alloxan administration. The oral administration of the water-soluble fraction of the ethanol extract (250 and 500 mg/kg) for 30 days exhibited a marked reduction in the blood glucose level and a remarkable increase in plasma insulin level compared to the water extract-treated rats and diabetic control. The level of serum TGs and total cholesterol was increased in diabetic rats. The marker enzymes of liver toxicity serum ALT, AST, acid phosphatase, and ALP were also elevated in diabetic control animals. The liver glycogen content and body weight were also decreased in alloxan-induced diabetic rats. Treatment with the water-soluble fraction of the ethanol extract and water extract of *C. auriculata* flowers significantly restored the aforementioned altered parameters in diabetic animals. The water-soluble fraction of the ethanol extract showed a more efficient antihyperglycemic effect compared to the water extract (Lukmanul et al. 2007).

Oral administration of the water extract (100, 200, 400, and 600 mg/kg daily for 21 days) of *C. auriculata* leaf to alloxan-induced mild diabetic (MD) and severe diabetic (SD) rabbits produced a dose-dependent fall in fasting blood glucose up to 400 mg/kg dose from day 3 to day 21. Further, a significant elevation in the levels of insulin and reduction in glycosylated hemoglobin (HbA1c) was observed in both MD and SD rabbits when treated with 400 mg/kg dose of the extract. The significant decrease in serum levels of TGs, total cholesterol, and LDL-C with a concomitant increase in HDL-C was exhibited by MD as well as SD rabbits following treatment with the extract. Atherogenic indices were also significantly reduced in both diabetic models of rabbits fed with the extract. The effect of the extract (400 mg/kg) was comparable to that of glibenclamide (600 µg/kg). Thus, the present study demonstrated that the water extract of the leaf can be a possible candidate for anti-DM drug (Gupta et al. 2009a). *C. auriculata* exerted a strong antihyperglycemic effect in rats comparable to the therapeutic drug acarbose (Abesundara et al. 2004).

Thus, the aforementioned studies show that the flower and leaf water extracts or the water fraction of the alcohol extract has promising antidiabetic and hypolipidemic activities. There is an urgent need to identify the active principles and all mechanisms of actions involved. The plant is known to contain anti-DM compounds such as quercetin, β-sitosterol, and kaempferol (review: Majumder and Paridhavi 2010; Nageswara Rao et al. 2010). Detailed toxicity evaluation is also needed.

Other Activities: The flower extract as well as the leaf extract has antioxidant activities (Latha and Pari 2003b). The leaf extracts of *C. auriculata* exhibit significant broad-spectrum antibacterial activity (Samy and Ignacimuthu 2000). Studies have shown pharmacological activities such as hepatoprotection, antipyretic activity, anthelmintic activity, diuretic activity, and antiulcer activity of this plant's flowers and leaves (review: Joy et al. 2012).

Toxicity: Occurrence of hepatotoxic pyrrolizidine alkaloids in *C. auriculata* has been reported (Arseculeratne et al. 1981).

Phytochemicals: The following phytochemicals are found in this plant: emodin, chrysophanol, rubiadin, nonacosane, nonacosane-6-one, auricassidin, kaempferol, kaempferol 3-*O*-rutinoside, luteolin, quercetin, β-sitosterol, β-sitosterol-β-D-glucoside, catechol-type tannins, di-(2-ethylhexyl)phthalate, polysaccharides, anthracene, dimeric procyanidins, myristyl alcohol, 3–0-methyl-glucose, α-tocopherol-β-D-mannoside, resorcinol, ethyl caprylate, capric acid ethyl ester, linoleic acid, oleic acid, etc. (review: Majumder and Paridhavi 2010).

Cassia fistula L., Caesalpiniaceae (Figure 2.29)

Local Names: Indian laburnum, purging fistula, the golden shower (Eng.), amulthus, bundaralati, garmala, sonhali, kakkemara, shrakkonnai, bahava, kondrakayi, nopadruma, amaltas, etc. (Indian languages)

FIGURE 2.29 *Cassia fistula.*

Description: *Cassia fistula* is a deciduous tree with a greenish-gray bark, compound leaves, and 5–12 cm long leaflet pairs. It is a semiwild tree known for its beautiful bunches of yellow flowers. The leaves are 23–40 cm long with main pubescent rhachis and minute, linear-oblong, obtuse, pubescent stipules. Leaflets are 4–8 pairs, ovate or ovate-oblong, acute, bright green above, and silvery-pubescent beneath when young; main nerves are numerous, close, and conspicuous beneath; petiolules of leaflets are 6–10 mm long and pubescent or glaborous. The flowers are in lax racemes 30–50 cm long, with 3.8–5.7 cm long, slender, pubescent, and glabrous pedicels. The calyx is 1 cm long divided to the base and pubescent; the segments are oblong and obtuse; the corolla is 3.8 cm across and yellow; the stamens are all antheriferous. The fruit is a cylindrical pod with many seeds that are black, with a sweet pulp separated by transverse partitions. The long pods are green when unripe and turn black on ripening. The fruit pods are 40–70 cm long and 20–27 mm in diameter, straight or slightly curved, smooth but finely striated transversely; the striations appear as fine fissures. The rounded distal ends bear a small point marking the position of the style. The seed contains a whitish endosperm in which the yellowish embryo is embedded (review: Danish et al. 2011).

Distribution: *C. fistula* is a deciduous plant occurring in mixed-monsoon forests throughout greater parts of India, ascending to 1300 m in outer Himalaya. In Maharashtra, it occurs as a scattered tree throughout the Deccan Plateau and Konkan. The plant is cultivated as an ornamental throughout India. The plant is also seen in Bangladesh, Sri Lanka, Myanmar, etc. (review: Rajagopal et al. 2013).

Traditional Medicinal Uses: The root is prescribed as a tonic, astringent, febrifuge, strong purgative, and anti-inflammatory medicine. The root is used in cardiac disorders, biliousness, rheumatic conditions, hemorrhages, wounds, ulcers and boils, and various skin diseases. The roots are also used for chest pain, joint pain, migraine, and bloody dysentery. Both leaves and flowers are purgative like the pulp. The leaves are laxative and used externally as an emollient and for insect bites, swelling, rheumatism, and facial paralysis. The leaves are used in jaundice, piles, and ulcers and also externally for skin eruptions, ringworms, and eczema. The fruits are used as a cathartic and in snake bite. The flowers and pods are used as purgative, febrifugal, biliousness, and astringent. The fruits are reported to be useful for asthma. The pulp is given in disorders of the liver. It is also used for blood poisoning, anthrax, dysentery, leprosy, and DM. The fruits are used in the treatment of diabetes, antipyretic, abortifacient, demulcent, inflammation, and fever and are useful in chest complaints, throat troubles, constipation, colic, chlorosis, urinary disorders, liver complaints, diseases of eye, and gripping. The pith is particularly useful if there is swelling in the stomach, liver, or intestine. The seeds are emetic, used in constipation, and have cathartic properties. The seeds are used as a laxative, carminative, and coolant, improves the appetite, and has antipyretic activity. The bark possesses tonic properties and is also used for skin complaints; the powder or decoction of the bark is administered in leprosy, jaundice, syphilis, and heart diseases. The stem bark is used against amenorrhea, chest pain, and swellings (review: Danish et al. 2011).

Pharmacology

Antidiabetes: The water extract of the stem of *Cassia fistula* reduced the levels of serum glucose and serum transaminase activity and improved body weight in diabetic rats (Daisy and Saipriya 2012). The hexane extract of the *C. fistula* bark (150, 300, and 450 mg/kg for 30 days) suppressed the elevated blood glucose in streptozotocin-induced diabetic rats. The extract at 450 mg/kg was found to be comparable with glibenclamide, the reference drug. The lipid profile in the extract-treated (450 mg/kg) diabetic rats showed remarkable improvement compared to the diabetic control animals (Nirmala et al. 2008).

The hydroalcoholic extract of the *C. fistula* leaf was found to be hypoglycemic and antidiabetic in alloxan-induced diabetic rats; it also improved oral glucose tolerance. The extract increased the uptake of glucose in psoas muscle, decreased the absorption of glucose in rat intestine, and increased the glycogen level in the liver of rats. Further, the extract exhibited antioxidant and free radical scavenging activity *in vitro* and *in vivo* in rats (Silawat et al. 2009).

The dose-dependent antihyperglycemic activity of the methanol extract of the plant leaves was shown in oral glucose tolerance test in glucose-loaded mice. The efficacy of the extract (400 mg/kg) was equivalent to that of 10 mg/kg glibenclamide (Khan et al. 2010). In another study, different extracts of the flower were compared for their anti-DM activity and the methanol extract was found to be the best extract. The extracts of flowers were tested for antihyperglycemic activity in glucose-overloaded hyperglycemic rats. The petroleum ether and ethanol extracts and the water-soluble fraction of the ethanol extract (200 and 400 mg/kg) were found to exhibit significant antihyperglycemic activity. However, the extracts, at the given doses, did not produce hypoglycemia in fasted normal rats. Treatment of alloxan-induced diabetic rats with ethanol extract and water-soluble fraction of this extract (200 and 400 mg/kg) restored the elevated biochemical parameters (glucose, urea, creatinine, serum cholesterol, serum TG, HDL, LDL, hemoglobin, and glycosylated hemoglobin) of diabetic rats to the normal level. Comparatively, the water-soluble fraction of the ethanol extract was found to be more effective than the ethanol extract, and the activity was comparable with that of glibenclamide (Jarald et al. 2013).

In another study, the methanol and water extracts of the different parts of the plant were compared for their antidiabetic activity in both normoglycemic and streptozotocin–nicotinamide-induced type 2 diabetic rats. Oral glucose tolerance test and glucose uptake in isolated rat hemidiaphragm were performed in normal rats. The methanol extract of the bark and leaves were found to be more effective in causing hypoglycemia in normoglycemic rats. Glucose uptake studies in isolated rat hemidiaphragm have shown enhanced peripheral utilization of glucose. Different extracts of the plant were administered to diabetic rats at 250 and 500 mg/kg doses for 21 days. Diabetic rats showed increased levels of glycosylated hemoglobin and reduced levels of plasma insulin; these changes were significantly reverted to near normal levels after oral administration of the bark and leaf methanol extracts. Further, the treatment remarkably restored the normal status of the histopathological changes observed in the diabetic rat pancreas. The methanol extract of the bark showed the best effect followed by the leaf methanol extract (Einstein et al. 2013). The antihyperglycemic activity of the methanol extract of this plant's stem bark was also reported in 2012. The extract (250 and 500 mg/kg) reduced blood sugar level in alloxan-induced diabetic (hyperglycemic) and glucose-induced hyperglycemic (normo-/hyperglycemic) rats (Ali et al. 2012). The ethanol extract of this plant's flower showed potent antioxidant activity in streptozotocin-induced diabetic rats. The observed antioxidant potential of *C. fistula* flowers may be partially responsible for its antidiabetogenic properties (Jeyanthi 2012).

Further, the ethyl acetate fraction of the ethanol extract of *C. fistula* bark inhibited α-amylase activity *in vitro* (Malpani and Manjunath 2013).

Thus, several studies suggest that this is an important antidiabetic medicinal plant containing several active principles and mechanisms of action. The major active compounds involved are not isolated and identified; the actions of appropriate combinations of the active principles could, possibly, turn out to be a good medicine for type 2 DM.

Other Activities: Other reported pharmacological activities include analgesic activity, antitussive activity, CNS activities, leukotriene inhibition activity, clastogenic effect, antipyretic activity, antioxidant activity, laxative activity, anti-inflammatory activity, wound healing activity, hepatoprotective activity, antifungal activity, antibacterial activity, larvicidal and ovicidal activity, hypocholesterolemic activity, hypolipidemic activity, antitumor activity, antiparasitic activity, and antileishmaniatic activity (Danish et al. 2011; Ali et al. 2012).

Toxicity: The petroleum ether extract of *C. fistula* seeds possessed a pregnancy-terminating effect by virtue of the anti-implantation activity. A 50% ethanol extract of pods shows antifertility activity in female albino rats (review: Danish et al. 2011).

Phytochemicals: A major anthraquinone derivative called rhein and proanthocyanidins like catechin, epicatechin, procyanidin B-2, and epiafzelechin were isolated from fruit. The anthraquinones like rhein, chrysophanol, and physcion, 9-(–)-epiafzelechin, 3-*O*-β-D-glucopyranoside, 7 bioflavonoids, and two triflavonoids together with (–)-epiafzelechin, (–)-epicatechin, and procyanidin B-2 were isolated from the leaves. An isoprenoid compound, a new diterpene, 3-β-hydroxy-17-norpimar-8(9)-en-15-one, 3-formyl-1-hydroxy-8-methoxy anthraquinone, etc., were isolated from the pods. Fistucacidin, an optically inactive leucoanthocyanidin (phenolic compound), was isolated from the heartwood. A bianthoquinone glycoside, fistulin, together with kaempferol and rhein has been isolated from the ethanol extracts of the flowers. Lupeol, β-sitosterol, and hexacosanol were isolated from the stem bark. A new bioactive flavone glycoside 5,3′,4′-trihydroxy-6-methoxy-7-*O*-α-L-rhamnopyranosyl-(1–2)-*O*-β-D-galactopyranoside was isolated from the seeds (review: Rajagopal et al. 2013).

Cassia grandis L., Leguminosae

Common Name: Bukut tree

The water and ethanol extracts of *Cassia grandis* improved glucose tolerance in normal rats in the glucose tolerance test. In alloxan-induced diabetic rats, the maximum reduction in blood glucose was observed after 3 h at a dose level of 150 mg/kg of body weight. In the long-term treatment of alloxan-induced diabetic rats, both of the extracts lowered the levels of blood glucose, cholesterol, and TGs; this effect was comparable to that of glibenclamide (Lodha et al. 2010).

Cassia javanica L., Leguminosae

Synonym: *Bactyrilobium fistula* Willd, *Cassia bonplandiana* DC

Cassia javanica is one of the ornamentals and medicinally less known plant. In a preliminary study, the dried leaf powder of this plant exhibited hypoglycemic activity in streptozotocin-induced diabetic rats. Single and multiple doses of the leaf powder (0.5 g/kg/day) were given to diabetic rats. Unlike acute treatment, subacute treatment of the leaf powder for 10 days showed highly significant reduction in blood glucose levels of diabetic rats. This effect was considerably good in comparison with glibenclamide (Kumavat et al. 2012).

Cassia kleinii W. & A., Fabaceae/Caesalpiniaceae

Local Names: Mullillathottalvadi, mullillathottavadi, nela tangedu, etc. (Indian languages)

Description: *Cassia kleinii* is a diffuse undershrub with pinnately compound leaves with 10–20 pairs of usually glabrous leaflets and a crispate-villous rachis. The flowers are rather larger and yellow, usually with long pedicels; the pods are small, ligulate, and dehiscent.

Distribution: It is distributed in the Deccan, Cuddapah, and West Coast in Southern Canara, Malabar, and Travancore, India.

Traditional Medicinal Uses: The plant is used for the treatment of diabetes in certain remote villages in Tamil Nadu and Kerala.

Antidiabetes: The *C. kleinii* leaf as well as its alcohol extract (but not the other extracts) exhibited a concentration-dependent antihyperglycemic effect in glucose-loaded rats. However, the extract did not show a hypoglycemic effect in fasted normal rats. In alloxan-induced diabetic rats, the extract (200 mg/kg) showed remarkable efficacy. The *C. kleinii* leaf (alcohol extract) showed antihyperglycemic activity in both glucose-fed hyperglycemic and alloxan-induced diabetic rats (Babu et al. 2002). In a follow-up study, the extract (200 mg/kg) showed significant antidiabetic property in streptozotocin-induced diabetic rats as judged from the body weight, serum glucose, lipids, cholesterol and urea, and liver glycogen levels. However, the extract did not significantly influence the levels of serum insulin in both diabetic and normoglycemic rats. The alcohol extract was devoid of any conspicuous acute and short-term general toxicity in mice. The antihyperglycemic activity was found predominantly in the chloroform fraction of the alcohol extract, which contained terpenoids, coumarins, and saponins. The active fraction of the *C. kleinii* leaf extract is very promising for further studies leading to the development of standardized phytomedicine for DM (Babu et al. 2003).

Toxicity: The extracts of the plant are nontoxic based on preliminary short-term toxicity evaluation in mice (Babu et al. 2003).

Phytochemicals: Oxanthrone esters have been isolated from the roots and aerial parts of *C. kleinii* (Anu and Rao 2002).

Cassia occidentalis L., Leguminosae (*Senna occidentalis* (L.) Link is the present accepted botanical name)

Synonyms: *Cassia occidentalis* L., *Senna occidentalis* Roxb., *Cassia foetida* Pers., (*Cass. occidentalis* L. is used in most of the publications) (Figure 2.30)

Local Names: Coffee senna, Negro coffee, stinking weed, fetid cassia, etc. (Eng.)

Description: *Cassia occidentalis* is a common weed; it is an annual herb or undershrub and reaches 60–150 cm in height. It grows at altitudes up to 1500 m. It bears compound leaves that are lanceolate or ovate lanceolate in shape. Each compound leaf has three pairs of leaflets. The leaflets are membranous and ovate-lanceolate. The plant bears yellow flowers in short racemes. The plants produce characteristic pods. The pods are glabrous and recurved. They are 10–13 cm long and 0.8 cm wide. The pods carry numerous dark olive green-colored seeds. The seeds are up to 6 mm long and 4 mm wide. They are hard in texture and have a lustrous appearance (review: Kaur et al. 2014).

Distribution: *C. occidentalis* occurs throughout the tropical and subtropical regions of America, Africa, Asia, and Australia. It is a common weed found throughout India (review: Kaur et al. 2014).

Traditional Medicinal Uses: *C. occidentalis* (the leaves, roots, and entire plant) has been used in China, Brazil, Sri Lanka, India, etc., in the treatment of a variety of ailments including diabetes. All plant parts have purgative, febrifugal, expectorant, and diuretic actions. The plant is used for the treatment of sore eyes, hematuria, rheumatism, typhoid, and leprosy. Decoction of the whole plant is used for treating hysteria, dysentery, and gastrointestinal problems including inflammation of the rectum. An infusion of the stem is used in treating DM. The roots are bitter and have purgative, anthelmintic, and diuretic properties. The roots are given with lime to treat dysentery and diarrhea associated with malaria. They are used for relief in cramps, itches, and sore throat. The root bark is also used to cure malaria. The root bark decoction is a remedy against gonorrhea and hepatic malfunction. The leaves possess purgative, febrifugal, diuretic, and stomachic properties and are also used in the management of cough, hysteria, and skin problems. The seeds have febrifugal and purgative properties. They are used as blood tonic and diuretics. The seed powder is externally applied for treating skin problems (review: Kaur et al. 2014).

Pharmacology

Antidiabetes: Several studies have established the antidiabetic property of the plant extracts in experimental animals. The methanol extract of *C. occidentalis* leaf showed anti-DM activity against streptozotocin-induced diabetic rats. Oral administration of the extract exhibited anti-DM activity as judged from the dose-dependent reduction in blood glucose levels in diabetic rats and reduction in glycosylated hemoglobin, lipid peroxides, and hepatic marker enzymes for liver damage and increase in the levels of liver glycogen, antioxidant enzymes, and GSH. Further, histopathological

FIGURE 2.30   *Cassia occidentalis.*

examination showed that the extract protected the pancreatic tissue from streptozotocin-induced damage (Emmanuel et al. 2010). In another investigation, the whole plant extract (200 mg/kg of different extracts, p.o.) exhibited anti-DM activity in alloxan-induced diabetic rats. The treatment also improved lipid profiles (Verma et al. 2010a). In another study, treatment with the butanol and water extracts of the plant leaf brought down the plasma glucose values to normal in alloxan-induced diabetic rats, with the butanol extract having a more profound effect. The treatment also improved body weight and lipid profiles in diabetic rats (Singh et al. 2011c). In another study, the methanol extract of *C. occidentalis* leaves was tested for anti-DM activity against alloxan-induced diabetic mice. Treatment with the leaf extract to diabetic mice resulted in reduction of the blood glucose levels to normal (Onakpa and Ajagbonna 2012).

The antidiabetic activities of *C. occidentalis* aerial parts were evaluated in alloxan-induced diabetic rats. The water and methanol extracts of the aerial parts possessed antihyperglycemic/antidiabetic activity against the alloxan-induced animal model. Among all the extracts, a potent antidiabetic activity was observed in the water extract of the leaves followed by the water extracts of seeds and stem. In normal animals, significant reduction in the blood glucose level was observed with the water extract (Arya et al. 2013). In a recent study, the ethanol extract of the *C. occidentalis* root was administered orally (250 and 500 mg/kg) to normal and streptozotocin-induced type 2 diabetic rats. Both doses of the extract caused a marked decrease in fasting blood glucose levels in the streptozotocin-induced diabetic rats. The extract decreased the blood glucose, food intake, water intake, organ weight, serum cholesterol, TG, creatinine, SGOT, and SGPT levels and increased the levels of HDL and total protein. The decrease in body weight in diabetic rats was also restored by the treatment (Sharma et al. 2014).

Thus, the anti-DM activity was detected in various extracts (water, butanol, methanol, ethanol, etc.) of the leaf, aerial parts, and whole plant in several studies. Nothing is known regarding the active principles and mechanisms of action. More than one active principle and mechanisms of action may be involved. Further studies are warranted to determine the utility of this plant for medicine development.

Other Activities: Other reported activities of *C. occidentalis* include hepatoprotective, antioxidant, antimicrobial, anti-inflammatory, antiallergic, muscle relaxant, wound healing effect, immunosuppressant, cholinergic, antidepressant, and antimalarial (review: Kaur et al. 2014).

Toxicity: Consumption of the seeds results in weight loss and toxicity in animals.

Phytochemicals: The leaves contain chrysophanol, emodin and their glycosides, physcion, metteucinol-7-rhamnoside, jaceidin-7-rhamnoside and 4,4,5,5-tetrahydroxy-2,2-dimethyl-1,1-bianthraquinone4, and flavonoid glycosides. The roots contain pinselin (cassiallin), rhein, aloe emodin and their glycosides, chrysophanol, physcion, emodin, islandicin, helminthosporin, xanthorin, sitosterol, campesterol, stigmasterol, 1,8-dihydroxyanthraquinone, 1,7-dihydroxy-3-methoxyxanthone, α-hydroxyanthraquinone and quercetin, questin, germichrysone, methylgermitorosone, singueanol I, pinselin, bis(tetrahydro) anthracene derivatives, occidentalol-1 and occidentalol-II, C-glycosidic flavonoids, and cassiaoccidentalins A, B, and C flavonoids. The seeds contain anthraquinones, 1,8-dihydroxy-2-methylanthraquinone, 1,4,5-trihydroxy-7-methoxy-3-methylanthraquinone, physcion and its glucoside, rhein, aloe emodin, chrysophanol and its glycoside, *N*-methylmorpholine, α-glucosides of campesterol and β-sitosterol, and a galactomannan. The flowers contain the phytoconstituents: β-sitosterol, emodin, physcion, and physcion-1-β-D-glucoside. The fruits (pods) contain 3,5,3,4-tetrahydroxy-7-methoxyflavone-3-*O*-(2-rhamnosyl glucoside) and 5,7,4-trihydroxy-3,6,3-trimethoxyflavone-7-*O*-(2″-rhamnosylglucoside) (review: Kaur et al. 2014).

Cassia sophera L., Caesalpiniaceae/Leguminosae

The water extract of *Cassia sophera* showed a blood glucose-reducing effect in oral glucose tolerance in normal rats. Treatment with the extract (200 mg/kg) reduced the elevated levels of blood glucose and glycosylated hemoglobin and increased the levels of insulin in the blood in streptozotocin-induced diabetic rats. Further, the treatment improved the lipid profile in diabetic rats. Under *in vitro* conditions, the extract stimulated insulin secretion from isolated pancreatic islets. The extract-induced enhancement in insulin secretion appears to be not dependent on K-ATP channels of β-cells (Sharma et al. 2013a).

Castanospermum australe A. Cunn. ex Mudie, Fabaceae

Common Name: Moreton Bay chestnut

Castanospermum australe is a traditional Australian medicinal tree. The toxic seeds are processed and used. The seed contains the alkaloid castanospermine, which exhibits hypoglycemic activity and inhibits intestinal sucrase in normal and streptozotocin-treated rats (review: Ghisalberti 2005).

Catha edulis (Vahl.) Forsskål ex Endl., Celastraceae

The antidiabetic activity of *Catha edulis* was assessed by intraperitoneally injecting varying doses of the water extract of the plant into alloxan-induced diabetic mice. The extract showed hypoglycemic activity (Piero et al. 2011). However, in another study, oral administration of a single dose of the hydroethanol extract of *C. edulis* did not cause any significant change in blood glucose levels in normal and alloxan-induced diabetic rats with or without glucose loading (Dallak et al. 2010).

Catharanthus roseus (L.) G. Don f., Apocynaceae

Synonym: *Vinca rosea* (Figure 2.31)

Common Names: Madagascar periwinkle, vinca, catharanthus (Eng.); nityakalyani, sadabahar, sudukattu mallikai, savanari, etc. (Indian languages)

Description: *Catharanthus roseus* is a herbaceous, erect, beautiful annual herb growing to 80 cm high with deep green leaves that are simple, opposite, oval, oblong or obovate, and glossy and white or deep rose-colored flowers arising from the axillary clusters of cymes; the fruits are paired follicles and flower all summer long (review: Retna and Ethalsha 2013).

Distribution: *C. roseus* is native to the Indian Ocean island of Madagascar. This herb is now common in many tropical and subtropical regions of the world including the Southern United States. It is distributed throughout India (in wastelands and gardens).

Traditional Medicinal Uses: *C. roseus* is a traditional medicinal plant used to control diabetes in various regions of the world. In India, wasp sting is treated with the juice from the leaves. In Hawaii, an extract of the boiled plant is used to arrest bleeding. In Central America and parts of South America, a gargle is made to ease sore throats and chest ailments and laryngitis. In Cuba, Puerto Rico, Jamaica, and other islands, an extract of the flower is commonly administered as eyewash for the eyes of infants. In Africa, the leaves are used for menorrhagia and rheumatism. Surinamese boil 10 leaves and 10 flowers together for diabetes. Bahamians take the flower decoction for asthma and flatulence and the entire plant for tuberculosis. In Mauritius, the leaves' infusions are given for dyspepsia and indigestion. In Vietnam,

FIGURE 2.31 *Catharanthus roseus.*

it is taken for diabetes and malaria. Curaçao and Bermuda natives take the plant for high blood pressure. Indochinese uses the stalks and leaves for dysmenorrhea (Duke 1985). It is an emetic, emmenagogue, euphoriant, hemostat, purgative, vermifuge, sedative, and depurative and used in cancer, asthma, chest cold, catarrh, common cold, diabetes, hypertension, ophthalmia, tuberculosis, and toothache (review: Retna and Ethalsha 2013).

Pharmacology

Antidiabetes: The leaves of the plant exhibited hypoglycemic and antihyperglycemic activity in experimental animals. Normal and streptozotocin-induced diabetic rats treated with the plant leaf extract exhibited potent blood sugar-lowering activity (Chattopadhyay 1999). In streptozotocin-induced diabetic rats, the dichloromethane/methanol extract (1:1) of leaves and twigs stimulated carbohydrate metabolism and reduced oxidative stress. Prior treatment with the extract (500 mg/kg) for 30 days completely protected the rats from streptozotocin challenge (Singh et al. 2001a).

The leaf juice of *C. roseus* also showed significant hypoglycemic and antidiabetic activity in alloxan-induced diabetic rabbits, and the mechanism may be enhancing the secretion of insulin from the β-cells or through the extrapancreatic mechanism (Nammi et al. 2003). The whole plant methanol extract (300 and 500 mg/kg) showed anti-DM activity in alloxan-induced diabetic rats. The methanol extract also showed improvement in parameters like body weight and lipid profile as well as regeneration of β-cells of the pancreas in diabetic rats. Histopathological studies reinforced the healing of the pancreas by the methanol extracts as a possible mechanism of the anti-DM activity (Ahmed et al. 2010).

In another study, the antidiabetic and hypolipidemic effect of *C. roseus* leaf powder was evaluated in streptozotocin-induced diabetic rats. In the untreated diabetic control rats, the plasma glucose was increased and the plasma insulin was decreased gradually. In diabetic rats treated with the leaf powder (100 mg/kg/day/60 days), reduction of plasma glucose and increase in plasma insulin were observed after 15 days of treatment, and by the end of the experimental period (60 days), the plasma glucose had almost reached the normal level, but insulin had not. The enhancement in plasma total cholesterol, TGs, LDL-C, and VLDL-C and the atherogenic index of diabetic rats were normalized in diabetic rats treated with the leaf powder. Decreased hepatic and muscle glycogen content and alterations in the activities of enzymes of glucose metabolism (glycogen phosphorylase, hexokinase, phosphofructokinase, pyruvate kinase, and glucose-6-phosphate dehydrogenase), observed in diabetic control rats, were prevented with the leaf administration. Thus, the leaf powder has both antidiabetes and lipid-lowering properties (Rasineni et al. 2010).

In another study, oral administration of the water extract of the flower (250, 350, and 450 mg/kg daily for 30 days) to diabetic rats resulted in significant reduction in blood glucose and lipids. The treatment prevented a decrease in body weight. Histological observation demonstrated significant fatty changes and inflammatory cell infiltration in the pancreas of diabetic rats. But supplementation with the extract significantly reduced the fatty changes and inflammatory cell infiltration in diabetic rats (Natarajan et al. 2012).

Effects of daily oral administration of the *C. roseus* leaf dichloromethane/methanol (1:1) extract (500 mg/kg) for 20 days on blood glucose and hepatic carbohydrate metabolizing enzymes in alloxan-induced diabetic rats were studied. Blood glucose, urea, and cholesterol levels decreased and body weight and liver glycogen content increased in treated diabetic animals. The activity of hepatic enzymes such as hexokinase increased while that of glucose-6-phosphatase and fructose 1,6-bisphosphatase decreased in treated diabetic animals compared to untreated diabetic control animals (Jayanthi et al. 2010).

In a study, different extracts of various parts (flower, leaf, stem, and root) of *C. roseus* were compared for their hypoglycemic property in normal and alloxan-induced diabetic rats. The water extracts reduced the blood glucose of both healthy and diabetic mice. The stem water extract (250 mg/kg) and its alkaloid-free fraction (300 mg/kg) containing polyphenol compounds reduced blood glucose in diabetic mice. The best hypoglycemic activity was presented by this stem's extract and fraction (Vega-Avila et al. 2012). Methyl (18β)-3,4-didehydroibogamine-18-carboxylate, an alkaloid isolated from this plant, lowered blood glucose levels (Chattopadhyay 1999). There could be more active principles. More focused research is needed to determine its therapeutic utility.

Other Activities: *C. roseus* is a well-known anticancer agent and the alkaloids vincristine and vinblastine isolated from this plant is used against blood cancer. The water extract elicited significant inhibition of angiogenesis on chick embryo chorioallantoic membrane model and cell proliferation of cultured bovine aortic endothelial cells. Bisindole alkaloids are widely used in the chemotherapy of malignant diseases. The total phenolic contents of the plant show significant oxygen radical scavenging activity. Other reported activities include hypotensive effect, wound healing activity, anthelmintic activity, and antimicrobial activity (review: Retna and Ethalsha 2013).

Phytochemicals

More than 100 alkaloids were reported from this plant: (–)tabersonine, 3-epiajmalicine, 3-hydroxy voafrine, ios ajmalicine, 3′-4′-anhydro-vincoleucoblastine, 4-deacetoxy vincoleucoblastine, 10-geraniol hydroxylase, 11-methoxy tabersonine, 16-epi vindolinine-*N*-oxide and its derivatives, tabersonine derivatives, vindolinines, 21-hydroxycyclochnerine, ajmalicine, akuamicine, akuammicine, akuammigine, akuammiline, akuammine, α-amyrin acetate, alstonine, ammorosine, ammocalline, antirhine, aparicine, bannucine, β-sitosterol, calmodulin, campesterol, cantharanthine, carosidine, cathalanceine, catharanthine, catharanthamine, catharanthine, catharicine, catharine, catharosin, cathasterone, cathenamine, cathindine, cathovaline, cavincidine, clevamine, coronaridine, camphor, vincaleukoblastins, vindolines, loganins, pseudo-vincaleukoblastin, epi-vindolinine, extension, fluorocarpamine, geraniol, geissoschizines, horhammericine, isositsirikine, isovallesiachotamine, isovincoside, isoleurodine, isositsirikine, kaempferol, bornesitol, lanceine, leurocolombine, leurocristine, leurosidines, leurosine, leurosinone, leurosivine, lochnerallol, lochnericine, lochneridine, lochnerivine, lochnerol, lochnovine, lochrovidine, lochrovine, locnovidine, loganic acid, maandrosie, malvidin, minovincinine, mitraphylline, neoleurocristine, pericalline, pericyclivine, perividine, perosine, petunidin, phytochelatin, pleiocarpamine, quercetin, reserpine, resin, rhazimol, posamine, roseadine, roseoside, rosicine, rovidine, serpentine, sitsirikine, stigmasterol, sweroside, srictosidine, syringic acid, tannin, tetrahydroalstonine, tetrahydroserpentine, tubotaiwine, ursolic acid, vallessiachotamine, vanillic acid, vinamide, vinaspine, vinblastine, vincandioline, vincaleukoplastine, vincalines, vincamicine, vincamine, vincarodine, vincathicine, vinceine, vincolidine, vincoline, vincristine, vindolicine, vindilidine, vincubine, vindolidine, vindoline, vindolinines, vindorosine, vinseine, vinosidine, vinsedine, vinsedicine, virosine, vivaspine, and yohimbine (review: Retna and Ethalsha 2013).

Catunaregam tomentosa (Blume ex DC.) Tirveng, Rubiaceae

Catunaregam tomentosa is one of the traditional Thai medicinal plants. The water extract (100 mg/kg) of this plant showed hypoglycemic activity in alloxan-induced diabetic mice (Manosroi et al. 2011a).

Caylusea abyssinica Fresen. Fisch. & Mey, Resedaceae

The leaves of this plant are used in Ethiopia as a cooked vegetable and also for the management of DM in folk medicine. *Caylusea abyssinica* leaf (hydroethanol extract, 200 mg/kg) exhibited a blood glucose-lowering effect in diabetic mice and improved oral glucose tolerance in normal mice. This study supports the traditional use of the leaf extract for the management of DM (Tamiru et al. 2012). In a recent study, successive extraction of the leaf powder with chloroform, methanol, and water was done. The methanol extract (300 mg/kg) produced significant hypoglycemic effect 4 h after treatment in mice. The methanol extract and, to a lesser extent, water extract reduced blood glucose levels in glucose-loaded normal animals and streptozotocin-induced diabetic animals (Gebreyohannis et al. 2014). Further studies are warranted to determine the utility of this plant for treating DM patients.

Cecropia obtusifolia Bertol., Cecropiaceae

Cecropia obtusifolia is extensively used for the treatment of type 2 diabetes in ethnomedicine in Mexico. The butanol (9 and 15 mg/kg) and water (90 and 150 mg/kg) extracts of the leaves of this plant showed a glucose-lowering effect 3 h after administration in streptozotocin-induced diabetic rats. The flavones isoorientin and chlorogenic acid (3-caffeoylquinic acid) were identified as the major constituents in both of the extracts (Andrade-Cetto and Wiedenfeld 2001). In another study, the hypoglycemic effect of the methanol extract of *C. obtusifolia* leaf was evaluated in healthy mice. A significant decrease in plasma glucose levels was recorded 2 and 4 h after a single oral administration of the methanol extract (1 g/kg). This effect was correlated with the chlorogenic acid contents in the leaves (Nicasio et al. 2005). The water extract of *C. obtusifolia* and chlorogenic acid isolated from *C. obtusifolia* stimulated 2-NBD-glucose (2-[N-(7-nitrobenz-2-oxa-1,3-diazol-4-yl)amino]-2-deoxyglucose) uptake in both insulin-sensitive and insulin-resistant 3T3 adipocytes in culture. The potency of chlorogenic acid to stimulate 2-NBD-glucose uptake was comparable to that of the antidiabetic drug rosiglitazone (Alonso-Castro et al. 2008). In another study, the butanol extract of *C. obtusifolia* leaf produced marked reduction in blood glucose in maltose-loaded streptozotocin-induced diabetic rats. The blood glucose in the treated diabetic rats at 90 min was lower than the fasting level suggesting the mechanisms of action other than inhibition of α-glucosidase. *In vitro* assays of α-glucosidase activity showed an IC_{50} of 14 µg/mL for the butanol extract that was lower than that of acarbose (128 µg/mL) (Andrade-Cetto et al. 2008b). In a study, the effects of *C. obtusifolia* leaf extracts on gluconeogenesis (*in vivo*) and the activity of the enzyme (*in vitro*) were investigated. The study confirmed that the main components of *C. obtusifolia* water and butanol extracts are chlorogenic acid and isoorientin. Diabetic rats treated with the extracts (150 mg/kg) showed a lower glucose curve. The extracts were able to reduce the increase in the blood glucose levels *in vivo*, and *in vitro* they inhibited the glucose-6-phosphatase activity with IC_{50} of 224 µg/mL for the water extract and 160 µg/mL for the butanol extract. The results of these experiments suggest that the administration of *C. obtusifolia* can improve glycemic control by blocking the hepatic glucose output, especially in the fasting state (Andrade-Cetto and Vazquez 2010).

Clinical Trials: The efficacy of the leaf (water extract) of *C. obtusifolia* was tested on type 2 noncontrolled diabetes patients. Patients with poor response to the conventional treatment were selected for the studies. The patients (22 in number) received the leaf extract along with their regular medical treatment for 21 days (about 3 mg of chlorogenic acid was present in 1 g of the dried plant leaf). The treatment resulted in decreases in the levels of fasting blood glucose (15.3%), total cholesterol (14.6%), and TG (42%). Thus, it appears to be useful as an adjunct on patients with type 2 DM with poor response to conventional treatment (Herrera-Arellano et al. 2004). However, the lack of a control group without the extract treatment is a hindrance for the interpretation of the data. In another clinical trial, the water extract of the leaves of this plant was administered daily for 32 weeks to 12 recently diagnosed type 2 diabetes patients controlled only with diet and exercise. A significant reduction of glucose was observed after 4 weeks of administration. The HbA1c was also reduced after 6 weeks of treatment. The levels of insulin, cholesterol, and TGs were not changed by the treatment. It is concluded that the water extract of the plant possesses a blood glucose-lowering effect without any adverse effects (Revilla-Monsalve et al. 2007).

Cecropia pachystachya Trecul, Cecropiaceae

Cecropia pachystachya is a medicinal plant with several pharmacological properties. It is used in folk medicine to treat DM. The methanol extract of *C. pachystachya* leaf exhibited hypoglycemic and antioxidant activity in normal and alloxan-induced diabetic rats (Aragao et al. 2010). The plant is reported to contain antidiabetic or glucose-lowering compounds such as chlorogenic acid, β-sitosterol, α-amyrin, and ursolic and oleanolic acids (Hikawczuk et al. 1998).

Cecropia peltata L., Cecropiaceae

Cecropia peltata is widely used in Mexico and elsewhere to treat diabetes in traditional medicine. The hypoglycemic effect of the methanol leaf extracts from *C. peltata* was evaluated in healthy mice. A significant decrease in plasma glucose levels was recorded 2 and 4 h after a single oral administration of the methanol extracts (1 g/kg). This effect correlated with the chlorogenic acid content in the leaves. The hypoglycemic effect produced by different doses (0.1, 0.25, 0.50, 0.75, and 1 g/kg body wt, p.o.) of *C. peltata* showed a linear relationship with chlorogenic acid content, reaching an $ED_{50} = 0.540$ g/kg for the extract and an $ED_{50} = 10.8$ mg/kg for chlorogenic acid (Nicasio et al. 2005). In another study, the effects of *C. peltata* leaf extracts on gluconeogenesis (*in vivo*) and the activity of the enzyme (*in vitro*) were investigated. The study showed that the main components of *C. peltata* water and butanol extracts are chlorogenic acid and isoorientin. Diabetic rats treated with the extracts (150 mg/kg) showed a lower glucose curve. The extracts were able to reduce the increase in the blood glucose levels *in vivo*, and *in vitro* they inhibited the glucose-6-phosphatase activity with IC_{50} of 146 μg/mL for the water extract and 150 μg/mL for the butanol extract. The results of these experiments suggest that the administration of *C. peltata* can improve glycemic control by blocking the hepatic glucose output, especially in the fasting state (Andrade-Cetto et al. 2010).

Cedrus libani A. Rich., Pinaceae
Cedrus libani is an evergreen coniferous tree. Essential oil from the woods of *C. libani* strongly inhibited α-amylase activity (review: El-Abhar and Schaalan 2014).

Ceiba pentandra (L.) Gaertn., Bombacaceae
Ceiba pentandra, commonly called silk-cotton tree, has been extensively used by traditional medicine practitioners in Northern and Eastern Nigeria in the control and management of diabetes.

The effects of daily oral administration of the root bark methylene chloride/methanol extract of *C. pentandra* for 28 days (40 and 75 mg/kg) to streptozotocin-induced type 2 diabetic rats were investigated. The extract treatment significantly reduced the intake of both food and water as well as the levels of blood glucose, serum cholesterol, TG, creatinine, and urea in diabetic rats, in comparison with diabetic controls. The treatment also improved glucose tolerance but no effect was observed in the level of hepatic glycogen. The effect of the extract (40 mg/kg) was more prominent when compared to glibenclamide in lowering blood glucose (Dzeufiet et al. 2006). In another study, *C. pentandra* bark ethanol extract (200 and 400 mg/kg) produced no significant hypoglycemic effect in normal rats but the extract decreased blood glucose level in streptozotocin-induced diabetic rats. In oral glucose tolerance tests, the extract (200 mg/kg) significantly reduced elevated glucose level in normal and diabetic rats. In a long-term (21 days) study, *C. pentandra* (200 mg/kg) significantly decreased blood glucose, total cholesterol, and TG level, prevented degeneration of the liver and pancreas, and increased serum insulin level and liver glycogen content in diabetic rats. Acute toxicity studies in rats did not show any signs of toxicity up to the dose of 2 g/kg (Satyaprakash et al. 2013).

Celastrus vulcanicola J.D. Smith, Celastraceae
Celastrus vulcanicola is a subtropical woody vine with medicinal properties. Two friedelane-type triterpenes isolated from the root barks of *C. vulcanicola* exhibited increased insulin-mediated signaling *in vitro*. In human hepatic cells, these compounds mediated phosphorylation of the IR in the absence of insulin, suggesting that these friedelane triterpenes have potential therapeutic use in insulin-resistant states. In 3T3-L1 adipocytes, astragaloside IV from this plant stimulated glucose uptake and antagonized TNF-induced insulin resistance (review: Nazaruk and Borzym-Kluczyk 2014).

Celosia argentea L., Amaranthaceae
Common Names: Quil grass (Eng.), vitunna, sarwari, pannakeerai (Indian languages)

Description: *Celosia argentea* is an erect, stout, or slender glabrous annual with alternate, linear, or lanceolate leaves; solitary, few or many, peduncle, slender spikes; and white, glistening flowers. The top of the spike sometimes branches out in a cock's comb form. The capsules contain 4–8 black, polished shining seeds.

Distribution: Typical to Asia, Africa, and America, *C. argentea* was introduced to India and Ceylon and is growing throughout at altitudes up to 4000 ft.

Traditional Medicinal Uses: It is galactogogue, cooling agent, aphrodisiac, antidiabetic, and antiscorbutic and is used to treat wounds, scorpion stings, blood dysentery, dysuria, otosis, pellagra, sore, spitting of blood, menorrhagia, diarrhea, tumors, mouth sores, eye disease, uterine bleeding, and hemorrhoids (Kirtikar and Basu 1975; *The Wealth of India* 1992).

Antidiabetes: *C. argentea* is widely used in India as a traditional treatment for DM. The alcohol extract of *C. argentea* seeds exhibited antidiabetic activity in alloxan-induced diabetic rats (Vetrichelvan et al. 2002). An ethanol extract of the *C. argentea* root was found to lower blood glucose in basal conditions and after a heavy glucose load in normal rats. Maximum reduction in serum glucose was observed after 90 min at a dose of 500 mg/kg (63.28%) of body weight, but petroleum ether and chloroform extracts did not reduce the serum glucose. The ethanol extract (500 mg/kg daily for 15 days) was also found to markedly reduce blood sugar levels in streptozotocin-induced diabetic rats compared to diabetic control rats. It was further found to reduce the levels of cholesterol, TGs, and urea and to increase the levels of serum proteins and liver glycogen in streptozotocin-induced diabetic animals. The treatment also improved body weight gain (Santhosh et al. 2010). These results are encouraging to carry on follow-up studies.

Other Activities: *C. argentea* plant extract has antimicrobial activity. The leaf extract improved wound healing in a rat burn wound model. The water extract from *C. argentea* seeds showed anti-metastatic and immunomodulating properties. Celosian, an acidic polysaccharide from the plant, protected rats from chemically and immunologically induced liver injuries (review: Ajikumaran and Subramoniam 2005).

Phytochemicals: The phytochemicals isolated include celosianin, isocelosianin, betalains, celogenamide A, a new cyclic peptide, hydrocyanic acid, and steroidal sapogenin (*The Wealth of India* 1992).

Centaurea aspera L., Compositae
Water and hot water extracts of *Centaurea aspera* flowers were tested for hypoglycemic activity in normal and alloxan-induced diabetic rats. Both extracts (p.o., chronic administration) showed significant hypoglycemic activity in diabetic rats, but only the extract obtained by exhaustion with hot water displayed an acute hypoglycemic activity in normal animals (Masso and Adzet 1976).

Centaurea alexandrina Delile, Compositae
In Egyptian traditional medicine, the flower, leaf, and root of *Centaurea alexandrina* are used to treat diabetic patients (Haddad et al. 2005). The 80% methanol extract of the leaves of *C. alexandrina* showed significant hypoglycemic and anti-DM activity. The extract also showed anticancer, anti-inflammatory, and analgesic activities. The extract contained compounds with antidiabetes and other biological activities such as kaempferol 3-*O*-rutinosise, rutin, kaempferol, quercetin, naringenin, and vitexin (Kubacey et al. 2012).

Centaurea bruguierana (DC.) Hand.-Mazz. ssp. *belangerana* (DC.) Bornm., Asteraceae
The methanol and ethyl acetate extracts (200 mg/kg) and dichloromethane and water extracts (400 mg/kg) of dried areal fruiting parts of *Centaurea bruguierana* spp. *belangerana* showed hypoglycemic activity after a single administration in streptozotocin–alloxan-induced diabetic rats. When the liver from treated diabetic rats was removed 3 h after treatment and examined for the activities of carbohydrate metabolizing enzymes, it was found that the extracts reduced phosphoenolpyruvate carboxykinase activity and increased the activity of glycogen phosphorylase. None of the extracts influenced blood insulin levels. The authors conclude that the plant is able to lower blood glucose via stimulation of hepatic glycogenolysis and inhibition of gluconeogenesis (Khanavi et al. 2012).

Centaurea calcitrapa L., Compositae/Asteraceae
Common Name: Red star thistle
The methanol extract of the aerial parts of *Centaurea calcitrapa* inhibited the activity of α-glucosidase under *in vitro* conditions. The IC_{50} value was about 4.4 mg/mL (Kaskoos 2013).

Centaurea corcubionensis Lainz, Compositae
The extracts of the leaves and flowers of *Centaurea corcubionensis* were assayed for their hypoglycemic activity in normal and hyperglycemic rats. Infusion of an active dose of 5 g/kg to normal rats lowered blood glucose levels by 16%–19% and increased circulating insulin by 27%–50%. Infusions induced a significant fall in glycemia levels in rats with glucose-induced hyperglycemia, but had no effect on alloxan-induced diabetic animals. Heavy release of insulin from islets of Langerhans isolated from rat pancreas was induced by the extracts (25, 50, and 100 μg/mL) *in vitro* (Chucla et al. 1998).

Centaurea iberica Spreng., Compositae

Common Name: Centaury thistle (Spanish)
Under *in vitro* conditions, the ethanol extract of *Centaurea iberica* showed insulin secretagogue activity in INS-1 cells at the 10 μg/mL level (Hussain et al. 2004).

Centaurea melitensis L., Compositae
Extracts from the aerial part of *Centaurea melitensis* lowered blood glucose in experimental animals (Marles and Farnsworth 1995).

Centaurea seridis L., Compositae
The effect of β-sitosterol 3-β-glucoside (antihyperglycemic principle isolated from the aerial part of *Centaurea seridis* var. *maritima*) and its aglycone on plasma insulin and glucose levels in normo- and hyperglycemic rats was investigated. The results indicated that oral treatment with the glycoside or with the β-sitosterol increased the fasting plasma insulin levels. There was a corresponding decrease in fasting glycemia when β-sitosterol was administered orally. In addition, these compounds improved the oral glucose tolerance with an increase in glucose-induced insulin secretion. But when these products were administered orally, the effect of glycoside, either on fasting insulinemia or on glucose-induced insulin secretion, lasted longer than that of aglycone. It can be assumed that β-sitosterol 3-β-D-glucoside acts by increasing circulating insulin levels and that this effect is due to their aglycone, β-sitosterol (Ivorra et al. 1998). In a follow-up study, the effect of oral administration of this compound on plasma insulin and glucose levels in streptozotocin-induced diabetic rats was studied. The compound did not change insulin and glucose levels in rats with severe diabetes. However, it stimulated insulin release from isolated rat islets in the presence of a nonstimulatory glucose concentration, but did not increase the insulin-releasing capacity of glucose (16 mmol/L). These data suggest that the compound exerts its action on intact pancreatic β-cells by stimulating insulin secretion (Ivorra et al. 1990).

Centaurea solstitialis L., Compositae
Extracts from flowers of *Centaurea solstitialis* lowered blood glucose in normal experimental animals (Marles and Farnsworth 1995).

Centaurium erythraea Rafn., Gentianaceae

Synonyms: *Cent. umbellatum* Gilib., *Cent. minus* Moench., *Erythraea centaurium* var. *sublitoralis* Wheldon & Salmon
The preventive effect of the hydroalcoholic extract of *Centaurium erythraea*, used in the traditional treatment of diabetes in Northeastern Algeria, was evaluated in a type 2 diabetic mouse model induced with a standardized HFD. The extract was administered orally (2 g/kg) daily for 20 weeks to male C57BL/6J mice fed HFD. After 6 weeks, blood glucose levels increased in HFD control mice. At the end of the study (20 weeks) in the extract-treated group, there was a significant reduction in fasting blood glucose levels, TG concentrations, and serum insulin levels. The extract also markedly reduced insulin resistance as measured by the homeostasis model assessment compared to HFD controls. The plant extract had no effect on calorie intake or body weight. Thus, *C. erythraea* extract prevented HFD-induced diabetes in mice. In a follow-up study, the hydroalcoholic extract of *C. erythraea* was tested in established type 2 diabetes induced with a standardized HFD in mice. After confirmation of diabetes, the plant extract (2 g/kg) was administered daily for 18 weeks to the diabetic mice fed an HFD. The diabetic mice that received the extract treatment showed marked reduction in blood glucose, TG, total cholesterol, and serum insulin levels compared to untreated diabetic mice. The treatment also reduced insulin resistance. The plant extract decreased calorie intake and had little effect on body weight or HDL-C (Hamza et al. 2011). Administration of 20% (w/v) water extract (0.66 mL/100 g body weight) and butanol extract (0.015 mL/100 g) of *Erythraea centaurium* var. *sublitoralis* (*C. erythraea*) resulted in significant reduction in blood glucose levels in oral glucose tolerance test in normal mice. The glucose-lowering effect of the extracts was comparable to that of glibenclamide (0.25 mg/100 g). But in the medium term, administration of this plant

in rats showed adverse effects in the liver and kidney (Mansar-Benhamza et al. 2013). In view of the reported adverse effect, detailed toxicological studies of standardized extracts are to be performed to proceed further.

Centella asiatica (L.) Urban, Apiaceae

Common Name: Vallarai

Centella asiatica, an herb traditionally used to improve memory, was reported to have antihyperglycemic effects in rodent diabetic models. The ethanol extract (200 mg/kg) of *C. asiatica* showed significant anti-DM activity in streptozotocin-induced diabetic rats as judged from body weight, serum glucose, lipids, cholesterol, and urea and liver glycogen levels (Gayathri et al. 2011). In subacute and acute toxicity evaluation in mice, the ethanol extract did not show any conspicuous toxic symptoms. Among the various fractions of this extract tested, antihyperglycemic activity (glucose tolerance test) was found predominantly in the chloroform fraction (12.5 mg/kg). This fraction showed positive reaction to terpenoids, coumarins, and saponins (Gayathri et al. 2011).

In a recent study, the effect of *C. asiatica* extract and insoluble fiber on carbohydrate absorption, insulin secretion, insulin sensitivity, and glucose utilization was studied. *C. asiatica* showed no significant change in insulin secretion *in vivo* and in isolated rat islets. Additionally, no effect of the extract was seen on liver glycogen deposition. Retarded glucose absorption was seen in the *in situ* perfused rat intestinal model. The extract was also found to inhibit the action of both intestinal disaccharidase and α-amylase. This was confirmed, yet again, via the Six-Segment study, in which sucrose digestion was found to be inhibited throughout the length of the gastrointestinal tract. Significant glucose–fiber binding was demonstrated in the *in vitro* models. During the chronic study, the body mass of *C. asiatica*-treated type 2 diabetic rats returned to normal and their polydipsic and polyphagic conditions were also improved. Chronic treatment with *C. asiatica* also improved the lipid profile of diabetic rats. Thus, the antihyperglycemic activity of *C. asiatica* is partly mediated by the inhibition of carbohydrate absorption and glucose–fiber binding (Kabir et al. 2014).

C. asiatica prevented diabetes-related hippocampal dysfunction. The water extract of the plant leaf was administered (100 and 200 mg/kg) for 4 consecutive weeks to streptozotocin-induced adult male diabetic rats. Administration of the leaf extract to diabetic rats maintained near normal ATPase activity levels and prevented the increase in the levels of inflammatory and oxidative stress markers in the hippocampus. Lesser signs of histopathological changes were observed in the hippocampus of the extract-treated diabetic rats. Thus, *C. asiatica* leaf protected the hippocampus against diabetes-induced dysfunction that could help preserve memory (Giribaba et al. 2014). High doses of *C. asiatica* extract (1 and 2 g/kg) decreased plasma glucose, TG, and total cholesterol levels in lipid emulsion-induced hyperlipidemic rats in 3 h. *In vitro* study showed the pancreatic lipase inhibitory activity of the alcohol extract of *C. asiatica*, which also inhibited the activity of α-amylase and α-glucosidase (Supkamonseni et al. 2014). Centellsapogenol A, a triterpene, from this plant, inhibited aldolase reductase (Matsuda et al. 2001). Asiatic acid, isolated from this herb, exhibited anti-DM activity in rats; besides, the compound improved the lipid profile in diabetic rats (Ramachandrana et al. 2014).

Centratherum anthelminticum Kuntz, Asteraceae

Synonym: *Vernonia anthelmintica*

Common Names: Bitter cumin, black cumin, etc.

Administration of the water extract of *Centratherum anthelminticum* (200 and 500 mg/kg, p.o., daily for 1 week) to alloxan-induced diabetic rats resulted in a promising dose-dependent reduction in the levels of blood glucose (Bhatia et al. 2008).

Ceratonia siliqua L., Caesalpiniaceae

Ceratonia siliqua seed is used as a traditional medicinal plant to treat diabetes in Israel. Administration of the hydroalcoholic seed extract of *C. siliqua* reduced blood glucose and lipid levels in diabetic male rats. The decrease in blood glucose level is probably due to the presence of fiber, phytosterols, and tocopherol in the extract (Mokhtari et al. 2011).

Ceriops decandra (Griff.) Ding Hou, Rhizophoraceae

The extract of *Ceriops decandra* showed antidiabetic activity in alloxan-induced diabetic rats (Nabeel et al. 2010).

Ceriops tagal (Perr.) C.B. Rob., Rhizophoraceae

Ceriops tagal is used as a remedy to control diabetes in the Philippines. The 80% aqueous ethanolic extracts of this plant were examined for their α-glucosidase enzyme inhibitory activity using yeast α-glucosidase enzyme. *C. tagal* extract manifested a very promising α-glucosidase inhibitory activity with $IC_{50} = 0.85 \pm 1.46$ μg/mL, which could be one of the mechanisms of its antidiabetes activity (Lawag et al. 2012).

Chaenomeles sinensis (Thouin) Koehne, Rosaceae

The *Chaenomeles sinensis* fruit extract inhibited the progression of diabetes induced by streptozotocin, which might be associated with its hypoglycemic effect, modulation of lipid metabolism, and its ability to scavenge free radicals (Sancheti et al. 2010). Administration of active α-glucosidase, α-amylase, and lipase inhibitory ethyl acetate fraction (50 and 100 mg/kg daily for 14 days) from the fruits of *C. sinensis* to streptozotocin-induced diabetic rats resulted in marked reduction in the levels of blood glucose, lipids, and transaminases (liver function markers) and improvement in antioxidant levels (Sancheti et al. 2013).

Chamaemelum nobile (L.) All., Compositae

Synonym: *Anthemis nobilis* L.

Common Name: Chamomile

This plant is used to treat DM in traditional medicine in Morocco. The effect of both a single dose and daily oral administration of the water extract of the aerial part of *Chamaemelum nobile* (20 mg/kg for 15 days) on blood glucose levels and basal insulin levels in normal and streptozotocin-induced diabetic rats was investigated (Eddouks et al. 2005b). A single dose of the extract reduced blood glucose levels in both normal and diabetic rats 6 h after administration. Furthermore, blood glucose levels were decreased markedly after 15 days of treatment in both normal and streptozotocin-induced diabetic animals. Basal plasma insulin concentrations remained unchanged after treatment in both normal and streptozotocin-induced diabetic rats; so the mechanism of this pharmacological activity seems to be independent of insulin secretion (Eddouks et al. 2005b).

In another study, the possible antihyperglycemic and body weight–reducing activities of the water extract of *C. nobile* (20 mg/kg) were evaluated. The repeated oral administration of the water extract evoked a potent antihyperglycemic activity in HFD obese mice. Postprandial hyperglycemic peaks were significantly lower in the extract-treated experimental groups. Body weight was significantly lower in the treated group when compared to the control mice after daily oral administration of the water extract for 15 days (Lemhadri et al. 2007). The 3-hydroxy-3-methylglutaric acid (HMG) containing flavonoid glucoside chamaemeloside, isolated from this plant, showed *in vivo* hypoglycemic activity comparable to that of free HMG (Konig et al. 1998).

Chelidonium majus L., Papaveraceae

Berberine, an isoquinoline alkaloid obtained from *Chelidonium majus*, is used widely in China to reduce blood glucose in type 2 diabetes. Berberine inhibited mitochondria function and decreased intracellular ATP in streptozotocin-induced diabetes in rats (Xuan et al. 2011). This led to a reduction in transcription factors such as FoxO1, SREBP1, and ChREBP. As a result, expression of gluconeogenic genes (PEPCK and G-6-Pase) and lipogenic gene (FAS) decreased. These molecular changes represent stimulation of a signaling pathway for improvement of fasting glucose in berberine-treated diabetic rats (Xia et al. 2011). Berberine is an important anti-DM compound present in many other plants also (see *Berberis aristata*). The pharmacological properties of this plant have been reviewed (Jyoti 2013).

Chiliadenus iphionoides (Boiss. & Blanche) Brullo, Asteraceae

Synonym: *Varthemia iphionoides* Boiss. & Blanche.

Chiliadenus iphionoides, a small aromatic shrub found throughout Israel, Jordan, and elsewhere, is used traditionally in the treatment of DM. The water extract of *Varthemia iphionoides* reduced blood glucose levels in normoglycemic rats from (mean) 5.3 to 4.2 and to 3.5 mmol/L at 4 and 24 h, respectively, after administration of the extract by gastric intubation. No changes in the blood glucose concentration of normoglycemic rats were noticed 1 h after administration. Significant decreases in the blood glucose levels of hyperglycemic rats were observed 1, 4, and 24 h after administration of the plant extract (from 29.8 to 8.2, 13.8, and 19.2 mmol/L, respectively). However, no significant dose–response relationship was observed in hyperglycemic rats on administration of various doses of the extract (Afifi et al. 1997). The ethanol extract of this plant increased insulin secretion in β-cells as well as glucose uptake in adipocytes and skeletal myotubes *in vitro*. The extract also displayed hypoglycemic activity in diabetic rats (Gorelick et al. 2011). In another study, the water extract of *V. iphionoides* exerted significant dose-dependent inhibition of α-amylase and α-glucosidase activities *in vitro*. Under *in vivo* conditions, the extract showed acute postprandial antihyperglycemic efficacies in starch-fed rats. However, the extract did not improve glucose tolerance in fasted rats on glucose loading. The authors conclude that the extract can be considered as potential candidates for therapeutic modulation of impaired fasting glycemia, impaired glucose tolerance, and type 2 diabetes (Kasabri et al. 2011).

Chonemorpha fragrans (Moon) Alston, Apocynaceae

Chonemorpha fragrans is used in traditional medicine to treat DM. In a scientific validation study, the alcohol extract of *C. fragrans* root (200 mg/kg, p.o.) produced marked reduction in blood glucose levels in fasted normal rats, glucose-overloaded normal rats, and alloxan-induced diabetic rats at a single dose as well as a 12-day treatment. Histopathology studies detected inflammatory changes and selective destruction of β-cells in the pancreas of alloxan-induced diabetic rats. These changes were inhibited by the extract treatment (Shende et al. 2009).

Chromolaena odorata (L.) R.M. King &H. Rob., Asteraceae

In traditional medicine, *Chromolaena odorata* is used for dysentery, cough, chest pain, diabetes, wounds, etc., in Thailand and elsewhere. (9S,13R)-12-oxo-phytodienoic acid and odoratin isolated from this plant were shown to activate PPAR-γ receptors (review: Wang et al. 2014).

Chrysophyllum cainito L., Sapotaceae

Common Name: Star apple

Chrysophyllum cainito is used to treat DM by traditional healers. Although at lower doses the aqueous decoction of the leaf of this plant lowered blood glucose levels in alloxan-induced diabetic rabbits, at higher doses, after 6 weeks of treatment, the treated diabetic rabbits stopped eating and succumbed (Koffi et al. 2009). Thus, the plant leaf is toxic, the healers may be alerted, and detailed studies may be carried out regarding the toxicity.

Cichorium glandulosum Boiss. & A. Huet, Compositae

Lactucin isolated from the root of the Chinese medicinal plant *Cichorium glandulosum* was reported to inhibit PTP1B with an IC_{50} value of about 1 μmol/L. A patent has been filed for the same (review: Jiany et al. 2012).

Cichorium intybus L., Asteraceae

Common Name: Chicory

Cichorium intybus is a traditional medicinal plant widely used in India as a treatment for DM. The whole plant extract (80% alcohol extract; 125 mg/kg) showed hypoglycemic activity in oral glucose tolerance test in rats. Repeated administration of the extract (125 mg/kg, p.o.) daily for 14 days to streptozotocin-induced diabetic rats resulted in reduction in blood glucose and lipids. However, there was no change in serum insulin levels. In addition, hepatic glucose-6-phosphatase activity was markedly reduced by the extract treatment when compared with the diabetic control group. The reduction in the activity of this enzyme could decrease hepatic glucose production (Pushparaj et al. 2007). Feeding with a standard diet supplemented with 10% chicory herb to alloxan-induced diabetic rats resulted in significant decrease in plasma glucose and total cholesterol compared to diabetic control rats. HDL-C showed a significant increase in the supplemented diabetic rats after 50 days compared with the diabetic control. AST and ALT enzyme activity showed a significant reduction after 10 and 50 days compared with diabetic control animals. Thus, chicory herb has good effects as a hypoglycemic and hypolipidemic agent in diabetic animals (Abozid 2010). In another study, the effect of oral administration of the infusion of *C. intybus* (100 mg/kg) and its active principle esculetin (6 mg/kg) for 4 weeks on oral glucose tolerance, insulin secretory response, serum lipid profile, and oxidative stress in streptozotocin-induced diabetic rats was evaluated. The treatment with *C. intybus* infusion and esculetin resulted in a marked amelioration of the impaired glucose tolerance in diabetic rats; the treatment lowered insulin and C-peptide levels. The impoverished liver glycogen content and elevated liver glucose-6-phosphatase and serum AST and ALT activities of fasting diabetic rats were profoundly corrected as a result of treatments with the plant infusion and esculetin. Also, these treatments lead to decrease in serum total lipid, total cholesterol, TG, LDL, and VLDL levels and increase in HDL level. The antioxidant defense system was potentially improved in diabetic rats as a result of the treatments. The plant seems to be very effective. However, further clinical studies are required to assess the efficacy and safety of these treatments in diabetic human beings (Hozayen et al. 2011). These results were almost confirmed by another study, wherein the ethanol extract of *C. intybus* (125 mg/kg daily for 4 weeks) was administered (i.p.) to streptozotocin-induced diabetic rats. The treatment resulted in significant reduction in blood glucose, TG, total cholesterol, and LDL levels and significant elevation of HDL levels as well as increase in the body weight as compared with the untreated diabetic rats. In the treated diabetic group, the levels of lipid peroxides decreased and the levels of antioxidant enzymes reduced and GSH increased. These results suggest that the *C. intybus* extract has antioxidant properties and prevents diabetes complications by modulation of the oxidative stress system (Samarghandian et al. 2013). In an *in vitro* study, tannins present in *C. intybus* enhanced glucose uptake and inhibited adipogenesis in 3T3-L1 adipocytes through PTP1B inhibition (Muthusamy et al. 2008).

Cimicifuga dahurica (Turcz. ex Fisch. & C. A. Mey.) Maxim, Ranunculaceae
Isoferulic acid extracted from the rhizome of *Cimicifuga dahurica* showed *in vivo* antihyperglycemic activity. An antihyperglycemic action of isoferulic acid in spontaneously diabetic rats, similar to type 1 diabetes, was observed (Liu et al. 1999).

Cimicifuga racemosa (L.) Nutt., Ranunculaceae

New Name: *Actaea racemosa* L.
Cimicifuga racemosa is used in Chinese traditional medicine to treat diabetes. Isoferulic acid isolated from *C. racemosa* showed anti-DM activity. Studies suggest that isoferulic acid can inhibit hepatic gluconeogenesis and/or increase the glucose utilization in peripheral tissue to lower plasma glucose in streptozotocin-induced diabetic rats lacking insulin (Liu et al. 2000).

Cinchona calisaya Wedd., Rubiaceae
Cinchona calisaya is a tropical medicinal plant endowed with important pharmacological properties. The effect of the water extract of the *C. calisaya* bark in alloxan-induced DM in rats was evaluated. The bark extract (50 and 100 mg/kg, i.p.) caused remarkable decreases in blood sugar levels; at 50 mg/kg the bark extract caused a 64.4% decrease. The bark extract exhibited higher activity at a lower dose (50 mg/kg) compared to a higher dose (100 mg/kg). The extract exhibited antioxidant activity in diabetic rats. Quantitative phytochemical analyses of the extract showed saponins, flavonoids, alkaloids, and cardiac glycosides in the extract (Ezekwesili et al. 2012).

Cinnamomum burmannii (Nees & Th. Nees) Nees ex Blume, Lauraceae

Common Name: Indonesian cinnamon

Cinnamomum burmannii is an evergreen tree growing up to 7 m in height with an aromatic bark and smooth, angular branches. In mouse adipocytes, the *C. burmannii* (bark) water extract increased GLUT1 mRNA levels sevenfold after a 16 h treatment. It decreased the expression of genes encoding insulin signaling pathway proteins including GSK3B, IGF1R, IGF2R, and PIK3R1. Thus, cinnamon regulates the expression of multiple genes in adipocytes (Cao et al. 2010). Water-soluble polyphenol type A polymers isolated from *C. burmannii* increased insulin-dependent *in vitro* glucose metabolism roughly 20-fold and displayed antioxidant activity (Anderson et al. 2004).

Cinnamomum cassia Blume, Lauraceae

Synonym: *Cinnamomum aromaticum* Nees; a closely related species: *Cinnamomum verum* J.S. Presl. (synonym: *Cinnamomum zylanicum* Nees, *Laurus cinnamomum* L.) (monograph: WHO 1999) (Figure 2.32)

Common Names: Common names for *Cinnamomum cassia* include cinnamon, Chinese cinnamon, and cassia cinnamon; common names for *Cinnamomum verum* include Ceylon cinnamon and true cinnamon.

Description: *C. cassia* is an evergreen tree, up to 10 m high. Leaves are alternate, petiolate, oblong (or elliptical-oval), 8–15 cm long, 3–4 cm wide, entire and three-nerved; leaf base is rounded and tip is acuminate; petiole is about 10 mm long and lightly pubescent. Inflorescence is a densely hairy panicle as long as the leaves; the panicles are cymose, terminal, and axillary. Flowers are yellowish white, small, in cymes of 2–5; the perianth is six lobed. There are no petals. There are six pubescent stamens; the ovary is free and one-celled. The fruit is a globular drupe, 8 mm long, and red in color. The bark is used in channelled pieces or simple quills, 30–40 cm long by 3–10 cm wide and 0.2–0.8 cm in thickness. The surface is grayish-brown, slightly coarse, with irregularly fine wrinkles and transverse lenticels. Here and there are found scars of holes, indicating the insertion of leaves or lateral shoots; the inner surface is rather darker than the outer. The fracture is short, the section of the thicker pieces showing a faint white line (pericyclic sclerenchyma) sometimes near the center and sometimes near and parallel to the outer margin (monograph: WHO 1999).

    C. verum is a small evergreen tree with a smooth, pale bark; the young parts are glabrous except the buds that are finely silky; the leaves are opposite or subopposite (rarely alternate), hard and coriaceous,

(a)   (b)

FIGURE 2.32 (a) *Cinnamomum verum*. (b) *Cinnamomum cassia*.

7–18 cm long, ovate or ovate lanceolate, subacute or shortly acuminate, glabrous and shining above, slightly paler beneath; the base is acute or rounded; there are 3–5 strong, main nerves from the base or nearly so, with fine reticulate venation between; the petioles are 1.3–2.5 cm long and flattened above. The flowers are numerous, in silky pubescent, lax panicles usually longer than the leaves; the peduncles are long, often clustered, glabrous, or pubescent. The perianth is 5–6 mm long; the tube is 2.5 mm long; the segments are pubescent on both sides, oblong or somewhat obovate, and usually obtuse. The fruit is purple generally about 1 cm drupe containing a single seed (monograph: WHO 1999). Ceylon cinnamon sticks (quills) have many thin layers and can easily be made into powder, whereas Chinese cinnamon is generally hard and woody in texture and thicker (2–3 mm thick), as all of the layers of the bark are used.

Distribution: *C. zylanicum* Blume is indigenous to Sri Lanka and southern parts of India and *C. cassia* J. Presl. originated in southern China and widely cultivated there and elsewhere in Southern and Eastern Asia (Ranasinghe et al. 2013).

Traditional Medicinal Uses: Cinnamon is used as a common spice by different cultures around the world for several centuries. It is obtained from the inner bark of trees from the genus *Cinnamomum*. In Ayurvedic medicine, cinnamon is considered a remedy for respiratory, digestive, and gynecological ailments. The cinnamon tree, including the bark, leaves, flowers, fruits, and roots, has some medicinal use in traditional medicine.

Pharmacology

Antidiabetes: *Cinnamomum verum*
The effects of the inner bark of *C. zylanicum* on DM were reviewed in the recent past. *C. zylanicum* possesses numerous beneficial effects both *in vitro* and *in vivo* (Bandara et al. 2012; Ranasinghe et al. 2012). *In vitro* studies suggest a reduction in postprandial intestinal glucose absorption by inhibiting the activity of enzymes involved in carbohydrate metabolism (pancreatic α-amylase and α-glucosidase), stimulation of cellular glucose uptake by translocation of GLUT4 to the membrane, stimulation of glucose metabolism and glycogen synthesis, inhibition of gluconeogenesis by influencing key regulatory enzymes, and stimulation of insulin release and potentiation of insulin signaling. Cinnamtannin B1 was identified as the potential active compound responsible for these effects (Bandara et al. 2012).

The *in vivo* studies using experimental animals revealed beneficial effects of *C. zylanicum*. These effects include attenuation of weight loss associated with diabetes, reduction of fasting blood glucose levels in diabetic animals, reduction in LDL levels and increase in HDL-C, and reduction in the levels of HbA1c and increase in the levels of insulin levels in the blood. In addition, cinnamon also showed beneficial effects against diabetic neuropathy and nephropathy.

Besides, cinnamon reduced total cholesterol, LDL-C, and TGs while increasing HDL-C in diabetic rats (Hassan et al. 2012b).

The methylhydroxychalcone polymer from *Cinnamomum* sp. was found to be an effective mimetic of insulin in 3T3-LI adipocytes. This compound may be useful in the treatment of insulin resistance and in the study of the pathways leading to glucose utilization in cells (Jarvill-Taylor et al. 2001).

Water-soluble polyphenol type A polymers from cinnamon that increase insulin-dependent *in vitro* glucose metabolism roughly 20-fold and display antioxidant activity were isolated and characterized from *Cinnamomum* sp. (Anderson et al. 2004).

It is predicted that cinnamtannin B, methylhydroxychalcone, polyphenol type A molecules, and other compounds may be involved in bringing about these multiple effects observed in the aforementioned experimental studies. Further studies involving the effect of individual active compound as well as the effect of active principles in various combinations may shed more light.

Clinical Trials: In a clinical trial, supplementation of a single high-fructose breakfast with 3 g of cinnamon (*C. zylanicum*) did not result in a significant change in gastric emptying parameters, postprandial triacylglycerol, or glucose concentrations after a single administration (Markey et al. 2011). This is a single dose study on non-DM patients; therefore, detailed controlled studies on DM patients with various doses remain to be studied for a meaningful conclusion.

Cinnamomum cassia

In an *in vitro* study, *C. cassia* has been shown to potentiate the hypoglycemic effect of insulin through upregulation of the glucose uptake in cultured adipocytes of rats. The water extract of extract of cinnamon containing polyphenols purified by high-performance liquid chromatography showed insulin-like activity *in vitro*. The water extract of cinnamon markedly decreased the absorption of alanine in the rat intestine. Alanine plays a vital role in gluconeogenesis and is altered back to pyruvate in the liver and is utilized as a substrate for gluconeogenesis (review: Rao and Gan 2014). In an *in vivo* study, cinnamon extract (200 mg/kg daily for 6 weeks) showed a significant decrease of blood glucose concentration in alloxan-induced diabetic rats, but there was slight or no change in the level of lipid parameters (Mohmood et al. 2011). The antidiabetic effect of the cinnamon bark (*C. cassia*) was investigated in a type 2 diabetic animal model (C57BIKsj *db/db* mice). The cinnamon extract (50, 100, 150, and 200 mg/kg for 6 weeks) administration to diabetic mice resulted in a marked dose-dependent decrease in blood glucose concentration compared with the diabetic control group. In addition, serum insulin levels and HDL-C levels were significantly higher, and the concentration of TG, total cholesterol, and intestinal α-glycosidase activity was significantly lower in the treated diabetic animals (Kim et al. 2006).

Clinical Trials: Although there were conflicting reports regarding the clinical efficacy of cassia cinnamon, it appears that cinnamon has beneficial effects on type 2 diabetic patients. In a clinical trial, different doses of cinnamon (1, 3, and 6 g daily) were found to be equally effective at lowering fasting glucose, total cholesterol, LDL-C, and TG levels in subjects with type 2 diabetes. However, in another study, cassia cinnamon taken at a dose of 1 g daily for 3 months produced no significant change in fasting glucose, lipid, A1C, or insulin levels. The authors concluded that the effects of cinnamon differ by population. Studies should be conducted to determine how specific variables (diet, ethnicity, BMI, glucose levels, cinnamon dose, and concurrent medication) affect cinnamon responsiveness. In another study, the cinnamon doses (1–3 g/day) significantly reduced the mean fasting serum glucose levels while the placebo doses did not affect the serum glucose levels. In light of this research, it is recommended that type 2 diabetic individuals should use 1–3 g cinnamon in their food preparations on a regular basis.

In contrast, in another clinical trial, cinnamon supplementation (1.5 g/day) did not improve whole body insulin sensitivity or oral glucose tolerance and did not modulate the blood lipid profile in postmenopausal patients with type 2 diabetes. It is plausible that differences in the dose of cinnamon or content of active principle used, as well as baseline glucose and lipid levels, have led to these variations. However, in another randomized, double-blind clinical study on type 2 DM patients, not on insulin therapy, but treated with oral antidiabetic drugs, the water extract of cinnamon (336 mg/day for 4 months) showed a moderate effect in reducing fasting plasma glucose concentrations in diabetic patients with poor glycemic control (review: Rao and Gan 2014). In another randomized, controlled trial, cassia cinnamon (1g daily for 90 days) lowered HbA1C level in type 2 DM patients compared with usual care alone patients. The study concluded that taking cinnamon could be useful for lowering serum HbA1C in type 2 diabetics with HbA1C > 7.0 in addition to the usual care (Crawford 2009).

Other Activities: In hyperlipidemic albino rabbits, cinnamon reduced total cholesterol, LDL-C, and TGs while increasing HDL-C (Javed et al. 2012). The essential oils obtained from the bark and eugenol have shown powerful antioxidant activity. The dried fruit extracts also exhibited antioxidant activity; the water extract was found to be the most active extract (Ranasinghe et al. 2013). Other reported important pharmacological properties of *Cinnamomum* sp. include antimicrobial properties, antiparasitic activities, vasorelaxant activity, hepatoprotective activity, antiulcer activity, antidiarrhea activity, analgesic action, anti-inflammatory activity, and anticancer activity (Ranasinghe et al. 2013).

Toxicity: *In vivo* studies on animals showed lack of significant toxic effects on the liver, kidney, etc. However, there are contradicting reports regarding the possible abortive or embryo toxicity of cinnamon to mice (Ranasinghe et al. 2013).

Phytochemicals: The different parts of the plant possess the same array of hydrocarbons in varying proportions. Cinnamaldehyde is rich in the bark, whereas eugenol concentration is high in the leaf, and camphor level is more in the root. The root that has camphor as the main constitute has minimal commercial value unlike the leaf and bark (Ranasinghe et al. 2013). Three of the main components of the essential oils obtained from the bark of *C. zeylanicum* (Ceylon cinnamon) are trans-cinnamaldehyde, eugenol, and linalool, which represent 82.5% of the total composition. One important difference between Chinese cinnamon (*C. cassia*) and Ceylon cinnamon is their coumarin (1, 2-benzopyrone) content. The levels of coumarins in Chinese cinnamon appear to be very high and, according to the German Federal Institute for Risk Assessment, pose health risks if consumed regularly in higher quantities (Ranasinghe et al. 2013). Other reported compounds in *Cinnamomum* sp. include rutin, catechin, quercetin, kaempferol, and isorhamnetin (review: Rao and Gan 2014).

Cinnamomum osmophloeum Kaneh., Lauraceae
Cinnamaldehyde isolated from the leaves of *Cinnamomum osmophloeum* inhibited high-glucose-induced hypertrophy in renal interstitial fibroblasts (Chao et al. 2010). In another study, different doses of cinnamon (5, 10, and 20 mg/kg) of *C. osmophloeum* were found to help with glycemic control in diabetics due to enhanced insulin secretion. It is plausible that the amelioration of oxidative stress and the proinflammatory environment in the pancreas may confer protection to pancreatic β-cells, which should be further investigated (Lee et al. 2013).

Cinnamomum parthenoxylon (Jack) Nees, Lauraceae
Cinnamon bark has been reported to be effective in the alleviation of diabetes through its antioxidant and insulin-potentiating activities. The hypoglycemic activity of a polyphenolic oligomer-rich extract from the barks of this plant was studied in normal, transiently hyperglycemic, and streptozotocin-induced diabetic rats. In streptozotocin-induced diabetic rats, oral administration of the polyphenolic oligomer-rich extract from the barks of this plant at doses of 100, 200, and 300 mg/kg, daily for 14 days, caused a dose-dependent decrease in the blood glucose level; the plasma insulin levels were significantly increased over pretreatment levels. In an oral glucose tolerance test, the extract produced a significant decrease in glycemia 90 min after the glucose pulse. The authors suggest that the extract could be potentially useful for postprandial hyperglycemia treatment (Jia et al. 2009).

Cinnamomum tamala (F. Hamilt.) Nees. & Eberm., Lauraceae
Common Names: Indian cassia lignea; tejpat; Malabar leaf (Eng.); Indian bay leaf, tamalaka, tejpatra, tejpat, talishapattiri, talisapatri, etc. (Indian languages)

Description: *Cinnamomum tamala* is a moderate-sized tree with opposite or subopposite triple-nerved leaves, shining above and rarely elliptic or obtuse with flowers borne on panicles and sparingly silky-pubescent perianth. The fruit is sealed in elongated perianth and the perianth lobs are deciduous.

Distribution: It is distributed in tropical and subtropical Himalayas in an altitude between 3000 and 8000 ft and also in Khasi, Jaintia Hills, and Eastern Bengal. It is native to India, Nepal, Bhutan, and China.

Traditional Medicinal Uses: It is carminative and an astringent and is used for diabetes, colic, diarrhea, rheumatism, gonorrhea scabies, diseases of the anus and rectum, piles, heart troubles, ozena, bad taste, etc. In Unani medicine, the leaf is used as a tonic to the brain, anthelmintic, and diuretic. It is also used in inflammation, sore eyes, etc. The bark is given for gonorrhea and given in decoction or powder in suppression of lochia after child birth (review: Pravin et al. 2013).

Antidiabetes: The leaves of this plant are reported to have significant antidiabetic and antioxidant properties (review: Chaurasia 2013). *C. tamala* leaves showed hypoglycemic and hypolipidemic effects in rats (Sharma et al. 1996). *C. tamala* (95% ethanolic extract) showed some level of blood glucose-lowering effect within 2 weeks of treatment in alloxan-induced diabetic rats (Kar et al. 2003). In streptozotocin-induced diabetic rats, the water extract of this plant's leaves at a dose of 125 and 250 mg/kg (p.o., daily for 3 weeks) markedly decreased the levels of fasting blood glucose and urine sugar with a concomitant increase in body weight; the extract also caused a significant decrease in peroxidation products (Chakraborty and Das 2010). *C. tamala* oil also showed antidiabetic, hypolipidemic, and antioxidant activities in streptozotocin-induced diabetic rats. It also reduced the levels of blood glycosylated hemoglobin (Kumar et al. 2012g,h). It may have more than one mechanism of action. The methanol and successive water extracts of this plant bark were found to inhibit α-amylase activity. The methanol extract of the bark showed a higher activity than the water extract (Kumanan et al. 2010).

Other Activities: The experimentally proven pharmacological activities include lipid-lowering activity, free radical scavenging effect, renoprotective properties against gentamicin-induced kidney toxicity, immunomodulatory property, gastroprotective property, antidiarrheal activity, and antimicrobial activity. The essential oils of this plant have an antidermatomycosis action and antifungal activities (review: Pravin et al. 2013).

Phytochemicals: The phytochemicals present in this plat include the following: cinnamicaldehyde, linalool, eugenol, eugenolacetate, β-caryophyllin, benzaldehyde, sabiene, furanosesquiterpenoids, sesquiterpenoids, curcuminol, camphor, cardinene, α-terpinol, α- and β-pinene, p-cymene, limonene, geraniol, ocimene, γ-terpinene, β-phyllandrene, benzyl cinnamate, benzyl acetate, flavone, kaempferol derivatives, quercetin derivatives, etc. (review: Pravin et al. 2013).

Cinnamomum verum J. Presl, Lauraceae

Synonym: *Cinnamomum zylanicus* Blume

Common Names: Ceylon cinnamon, a species related to *Cinnamomum cassia* (Nees & T. Nees) J. Presl. (see *Cinn. cassia* J. Presl)

Cirsium pascuarense (HBK) Spreng, Compositae

The plant *Cirsium pascuarense* is a common herb that grows wild and abundantly in the fields of Mexico. In traditional medicine, a water extract of the fresh leaves has long been used for treating DM. Hexane extracts of this plant (100, 200, and 400 mg/kg, i.p.) exhibited significant blood glucose-lowering effect in normal as well as alloxan-induced diabetic mice (Perez et al. 2001).

Cissampelos mucronata A. Rich., Menispermaceae/Leguminosae

Synonyms: *Cissampelos macrostachya* Klotzsch, *Cissampelos pareira* L. var. *mucronata* (A. Rich.) Engl., *Cissampelos sinensis* Klotzsch

Single intraperitoneal administration of the ethanol extract of *Cissampelos mucronata* (200, 400, and 800 mg/kg) to streptozotocin-induced diabetic rats resulted in significant reduction in the blood glucose levels. The 200 mg/kg dose was more effective with the highest glycemic change of 67% after 24 h of extract administration than the other two higher doses with a glycemic change of 60%. The LD_{50} dose in rats was greater than 5 g/kg of the extract (Tanko et al. 2007).

Cissampelos owariensis P. Beauv. ex DC., Menispermaceae

Cissampelos owariensis is a medicinal plant used in traditional medicine for treating diseases including DM. In alloxan-induced diabetic rats, the ethanol leaf extract of this plant (100 and 200 mg/kg orally for 14 days) significantly increased body weight gain and decreased blood glucose when compared with glibenclamide. Preliminary phytochemical screening of the ethanol extract of *C. owariensis* revealed the presence of tannins, flavonoids, alkaloids, and saponins. The study supported the traditional usage of the herbal preparations for the therapy of diabetes (Ekeanyanwu et al. 2012).

Cissus cornifolia (Baker) Planch, Vitaceae

Cissus cornifolia is used in traditional medicine to treat diabetes. The effect of the methanol extract of this plant's leaf on blood glucose levels and histology of the pancreas was studied using alloxan-induced diabetic rats. The extract (50, 100, and 200 mg/kg for 7 days) significantly decreased fasting blood glucose levels in diabetic rats and protected pancreatic islet cells from degeneration, to a large extent. The lowest dose tested (50 mg/kg) showed more activity than higher doses (100 and 200 mg/kg) (Jimoh et al. 2013).

Cissus sicyoides L., Vitaceae

Common Name: Princess vine

C. sicyoides is a runner plant found abundantly in Brazil, especially in the Amazon. In ethnomedicine it is used as a diuretic, anti-influenza, anti-inflammatory, hypoglycemic, and anticonvulsant agent. Studies have shown that the water extract of the plant leaves (100 and 200 mg/kg) and stem has hypoglycemic and antilipidemic effects on alloxan-induced diabetic rats (Viana et al. 2004b). The water extract of *C. sicyoides* leaves and stem exhibited a hypoglycemic effect (45% decrease after 60 days of administration) in alloxan-induced diabetic rats. Furthermore, indices of hepatic glycogen, blood glucose, C-reactive peptide, and fructosamine were found to be efficient biomarkers to monitor diabetes in rats (Salgado et al. 2009).

Cistanche tubulosa (Schenk) Wight, Orobanchaceae

The dried succulent stem of *Cistanche tubulosa* is one component of traditional Chinese medicine prescriptions for diabetes. Different doses of *C. tubulosa* powder (equivalent to 120.9, 72.6, or 24.2 mg verbascoside/kg) were administered orally once daily for 45 days to male *db/db* diabetic mice. Age-matched *db/+* mice were used as normal controls. The *C. tubulosa* treatment significantly suppressed the elevated fasting blood glucose and postprandial blood glucose levels, improved insulin resistance and dyslipidemia, and suppressed body weight loss in *db/db* mice. However, *C. tubulosa* did not significantly affect serum insulin levels or hepatic and muscle glycogen levels. The authors suggest that *C. tubulosa* has the potential for development into a functional food ingredient or drug to prevent hyperglycemia and treat hyperlipidemia (Xiong et al. 2013).

Cistus laurifolius L., Cistaceae

The water and ethanol extracts of this plant inhibited α-amylase and α-glucosidase activities *in vitro* and exhibited antidiabetic activity in streptozotocin-induced diabetic rats (Orhan et al. 2013).

Citrullus colocynthis (L.) Schrd, Cucurbitaceae (Old Name: *Colocynthis citrullus*) (Figure 2.33)

Common Names: Bitter cucumber, bitter apple (Eng.); mahendravaruni, indrayan, peykkumutti, peykommutti, coloquint, etc. (Indian languages)

Description: *Citrullus colocynthis* is a slender, stemmed, diffuse or creeping monoecious plant with perennial root. The tendrils are simple with bifid bearing hairs. The leaves are variable in nature, usually deltoid in outline, pale green above and ashy beneath, and scabrid on both surfaces, and the lamina is deeply lobed. The flowers are unisexual. The fruit is globular, variegated, green and white when young, while glabrous when ripe.

Distribution: Native to the warmer regions of Africa and India, *C. colocynthis* is distributed throughout India, and is growing wild in arid, warm sandy tracts of the northwest, central, and south Coromandel

FIGURE 2.33 *Citrullus colocynthis.*

coast, Gujarat, and Western India. It also grows in Saudi Arabia, Syria, Egypt, Spain, Sicily, Morocco, and the Mediterranean.

Traditional Medicinal Uses: *C. colocynthis* is used by various tribes in India for the treatment of many diseases. The plant parts are used for bowel complaints and epilepsy (seed oil); malaria (seed pulp); acute stomachache, dropsy, antibacterial, hepatitis, constipation, and to cause abortion (fruits in various forms); inflammation of the breasts and joints, rheumatism, enlarged abdomen, and to ease child birth (root); and diabetes (whole plant extract) (Meena et al. 2014). It is purgative, antidiabetic, cathartic, emmenagogue, parasiticide, anthelmintic, antipyretic, and carminative and is used in ascites, jaundice, urinary diseases, cancer, fever, dropsy, intermittent fever, leukoderma, asthma, bronchitis, ailment of the spleen, dyspepsia, constipation, throat disease, elephantiasis, joint pain, ophthalmia, uterine pain, pimples, migraine, neuralgia, snake bite, and amenorrhea (Kirtikar and Basu 1975).

Pharmacology

Antidiabetes: The water extract of *C. colocynthis* seeds on oral administration could ameliorate some toxic effects of streptozotocin in streptozotocin-induced diabetic rats (Al-Ghaithi et al. 2004). The water extract of *C. colocynthis* fruit exhibited hypoglycemic and antihyperglycemic effects in normal and alloxan-induced diabetic rabbits (Abdel-Hassan et al. 2000). Different *C. colocynthis* seed extracts have insulinotropic effect that could at least partially account for the anti-DM activity of the seed (Nmila et al. 2000). The *C. colocynthis* root extract also exhibited hypoglycemic activity in animals; further, the extract also reduced serum creatinine and urea levels (Agarwal et al. 2012).

Clinical Trial: To determine the efficacy and toxicity of *C. colocynthis*, a 2-month clinical trial was conducted in 50 type 2 diabetic patients. Two groups of 25 each (under standard antidiabetic therapy) received 100 mg *C. colocynthis* fruit capsules or placebos three times a day, respectively. Glycosylated hemoglobin (HbA1c), fasting blood glucose, total cholesterol, LDL, HDL, TG, AST, ALT, ALP, urea, and creatinine levels were determined at the beginning and after 2 months. The results showed a significant decrease in HbA1c and fasting blood glucose levels in *C. colocynthis*-treated patients. Other serological parameters in both groups did not change significantly. No notable gastrointestinal side effect was observed in either group. In conclusion, *C. colocynthis* fruit treatment had a beneficial effect on improving the glycemic profile without severe adverse effects in type 2 diabetic patients. Further clinical studies are recommended to evaluate the long-term efficacy and toxicity of *C. colocynthis* fruit in diabetic patients (Huseini et al. 2009).

Other Activities: The immunostimulating activity of the hot water-soluble polysaccharide extracts of *C. colocynthis* has been reported (Bendjeddou et al. 2003). A crude ethanol extract of *C. colocynthis* fruit induced reversible antifertility in male rats (Chaturvedi et al. 2003).

Toxicity: Saponin extracted from *C. colocynthis* induced pathological changes in the small intestine, liver, and kidney (Diwan et al. 2000). The toxicity of the diet containing 10% *C. colocynthis* fruits to rats is manifested as diarrhea, enterohepatic nephropathy, weight loss, and behavioral changes (Al-Yahya et al. 2000). The carcinogenic activity of the condensate from *C. colocynthis* (seeds) after chronic epicutaneous administration to mice has been reported (Habs et al. 1984).

Phytochemicals: The phytochemicals isolated from *C. colocynthis* are as follows: citrullin, α-elaterin, elaterin A and B, hentriacontane, saponins, ipuranol, tannin, phytosterols, citbittol, α-spinasterol, glycosidic components, cucurbitacin and its derivatives, citrulline, citrullic acid, citrullol, citronellal, methylheptenone, methyl eugenol, phenylethyl alcohol, docosan-1-ol acetate, 10,13,dimethylpentadec-13-en-1-ol,hepatocosan-1-ol, hydroxyketo tetracyclic triterpene, and linoleic acid (*The Wealth of India* 1992).

Citrullus lanatus (Thunb.) Matsumara & Makai, Cucurbitaceae

Synonyms: *Citrullus vulgaris* Eckl., *Citrullus vulgaris Schrad*, *Citrullus vulgaris Schrad* ex Ecklon & Zeyher (Figure 2.34)

Local Names: Watermelon, wild watermelon, egusi melon, tsamma melon, cooking melon, citron melon, etc. (review: Erhirhie and Ekene 2013)

FIGURE 2.34   *Citrullus lanatus.*

Distribution: Watermelon is thought to have originated in Southern Africa because it is found growing wild throughout the area and reaches a maximum diversity of forms there. It has been cultivated in Africa for over 4000 years. Now it is a very popular crop in many countries. Cultivation of the watermelon as a fruit crop began in ancient Egypt and Asia Minor and spread from there. The cultivars of this popular summer fruit grown today bear little resemblance to their ancestral African forms (review: Erhirhie and Ekene 2013).

Description: *Citrullus lanatus* is a prostrate or climbing annual with several herbaceous, rather firm and stout stems; the young parts are densely woolly with yellowish to brownish hairs, while the older parts become hairless. The leaves are herbaceous but rigid, becoming rough on both sides, 60–200 mm long and 40–150 mm broad, usually deeply three lobed with the segments again lobed or doubly lobed; the central lobe is the largest. The leaf stalks are somewhat hairy and up to 150 mm long. The tendrils are rather robust and usually divided in the upper part (Ajithadas et al. 2014). Male and female flowers occur on the same plant (monoecious) with the flower stalk up to 40 mm long and hairy. The receptacle is up to 4 mm long, broadly campanulate and hairy. The corolla is usually green or green-veined outside and white to pale or bright yellow inside and up to 30 mm in diameter. The fruit in the wild form is subglobose, indehiscent, and up to 200 mm in diameter. The fruits in the cultivated forms are globose to ellipsoid or oblong and up to 600 mm long and 300 mm in diameter. The rind in the ripe fruit is not woody. The flesh in the wild form and some cultivated forms (citron watermelon) is firm and rather hard, white, green-white, or yellowish. In cultivated forms the flesh is somewhat spongy in texture but very juicy and soft, pink to bright red-pink. The seeds are numerous, ovate in outline, sometimes bordered; in wild forms they are usually black or dark brown; in cultivated forms they are also white or mottled, mostly 6–12 mm long. Wild *C. lanatus* ssp. *lanatus* var. *caffer* occurs in two biochemically different forms, both with whitish flesh: one with the bitter elaterinid (karkoer, bitterboela) and the other without (*tsamma*).

Traditional Medicinal Uses: The leaves of *C. lanatus* is used as an analgesic, anti-inflammatory, mosquitocidal, antigonorrheal, and antimicrobial property. Traditionally, *C. lanatus* had been reportedly used as purgative and emetic in high dose, vermifuge, demulcent, diuretic, pectoral, and tonic. The seed is used in the treatment of urinary tract infections, bed wetting, dropsy and renal stones, alcohol poisoning, hypertension, diabetic, diarrhea, and gonorrhea. The fruits are eaten as a febrifuge when fully ripe. The rind of the fruit is used in diabetes (Erhirhie and Ekene 2013).

Pharmacology

Antidiabetes: The ethanol extract of *C. lanatus* seed showed very good reduction in the blood glucose level in normoglycemic animals. In oral glucose tolerance test (400 mg/kg), the ethanol extract also showed better glycemic control. In diabetic rats, daily administration of the extract (400 mg/kg) for

30 days controlled the blood glucose levels. Further, the ethanol extract of *C. lanatus* seed attenuated oxidative stress in various organs in diabetic rats. In an acute toxicity study, there was no mortality observed up to the maximum dose level of 2000 mg/kg, p.o., of the extract (Varghese et al. 2013a). In another study, daily oral administration of the watermelon seed (methanol extract; 200 mg/kg) for 28 days to streptozotocin-induced diabetic rats resulted in marked decrease in plasma glucose concentrations. However, the treatment did not influence Na^+, K^+, Cl^-, and HCO_3^- concentrations in the blood (Omigie and Agoreyo 2014).

The effect of supplementation with 10% watermelon flesh powder or 1% watermelon rind extract for 4 weeks on the antidiabetic potential in mice was evaluated. ICR mice fed with 1% watermelon rind ethanol extract for 4 weeks, when challenged with streptozotocin (i.p., 40 mg/kg, for 5 consecutive days), showed decreased blood glucose levels and increased serum insulin levels compared to untreated streptozotocin-challenged diabetic rats. Further, the ethanol extract effectively protected pancreatic cell death. However, mice fed with 10% watermelon flesh powder did not show significant influence on these parameters (Ahn et al. 2011).

In a recent study, the antidiabetes activity of ethanol and petroleum ether extracts of *C. lanatus* (watermelon) rind in alloxan-induced diabetic mice was investigated. Treatment of diabetic rats with the extracts (150, 200, and 250 mg/kg) decreased the blood glucose levels in a dose-dependent manner. The ethanol extract was more potent than the petroleum ether extract (Sani 2014). This observation is in agreement with the previous report that the ethanol extract supplementation with diet has antidiabetic activity in mice (Ahn et al. 2011).

In another study, antidiabetic screening of the methanol extract of *C. lanatus* leaves was done by various *in vitro* methods. Preliminary phytochemical screening of the methanol extract of the leaves showed the presence of carbohydrates, triterpenoids, proteins, alkaloids, tannins, saponins, flavonoids, and sterols and the absence of glycosides, volatile oil, and fixed oil. The extract showed the presence of quercetin, gallic acid, and catechin. In nonenzymatic glycosylation of hemoglobin assay, the IC_{50} value was found to be 65.648 and 59.762 µg/mL for the extract and α-tocopherol, respectively. The extract exhibited significant inhibition of glycosylation as compared with the standard α-tocopherol. In glucose uptake assay in yeast cells, the IC_{50} value was found to be 77.031 and 67.408 µg/mL for the extract and acarbose, respectively. In the inhibition of α-amylase enzyme assay, the IC_{50} value was found to be 58.6 and 47.9 µg/mL for the extract and acarbose, respectively. In α-glucosidase inhibition assay, the IC_{50} value was found to be 627.3 and 482.2 µg/mL for the extract and acarbose, respectively. Overall the extract showed significant *in vitro* antidiabetic activity. Follow-up *in vivo* studies are required to determine the possible use of this leaf extract for diabetic treatment (Aruna et al. 2014). The active principles involved in this important food medicine remain to be studied. However, gallic acid and quercetin reported in the methanol extract of leaves are compounds known to have antidiabetic activities and present in many other anti-DM plants.

Other Activities: Biological activities reported include antimicrobial, antioxidant, antiplasmodial, anti-inflammatory, antiprostatic hyperplasia, antigiardial, and analgesic properties. Further, it has laxative, antiulcerogenic, and hepatoprotective activities (Erhirhie and Ekene 2013).

Phytochemicals: Watermelon is a rich natural source of lycopene, a carotenoid of great interest because of its antioxidant capacity and potential health benefits. Cucurbitaceae plants are known to contain bioactive compounds such as cucurbitacin, triterpenes, sterols, and alkaloids. The amino acid citrulline had been extracted from watermelon (Erhirhie and Ekene 2013). Quercetin, gallic acid, and catechin were detected in the leaf (Aruna et al. 2014).

Citrus aurantium L., Rutaceae, sour orange

The alcohol extract of *Citrus aurantium* fruit peel was evaluated for its hypoglycemic and hypolipidemic activity in normal and alloxan-induced diabetic rats. The extract treatment daily for 21 days produced significant reduction in blood glucose and also had beneficial effects on the lipid profile in normal as well as alloxan-induced diabetic rats (Sharma et al. 2008). The antidiabetes fraction (hexane fraction) from *C. aurantium* stimulated NCI-H716 cells to secrete GLP-1. Through microarray analysis, it was found that voltage-gated potassium channels drove membrane depolarization and then influenced calcium current in NCI-H716 cells (Choi et al. 2012). Thus, at least one of the mechanisms of the antidiabetic action of *C. aurantium* is the release of GLP-1. Pentamethylquercetin from citrus fruits showed anti-DM activity in neonatally streptozotocin-induced diabetic rats (Wang et al. 2011c).

Clinical Trial: In a human clinical trial, the effect of short-term consumption of sour orange juice on the blood glucose and lipid profile of dyslipidemic diabetic patients was evaluated. Each patient consumed 240 mL of sour orange juice daily for 4 weeks. The patients maintained their usual diet, physical activity, and consumption of their oral hypoglycemic agent during the entire experimental period. Body weight was measured at baseline and after consumption of 240 mL of sour orange juice daily for 4 weeks. Fasting blood glucose, lipids, and ascorbic acid level were measured. There was no statistically significant change in body weight, energy, and macronutrient intakes before and after 4-week consumption of sour orange juice, but vitamin C intake increased. Fasting blood sugar level significantly decreased after consumption of the juice. Plasma ascorbic acid level was significantly increased; but glycosylated hemoglobin and lipid profile did not change. Thus, short-term incorporation of sour orange juice in the diet of diabetic patients had fasting blood glucose-lowering effect (Ravanshad et al. 2006).

Citrus lemon (L.) Burm.f., Rutaceae

Common Names: Lemon (Eng.), nimbu, jambira, jambirah, paharinimbu, paharikaghzi, yelumichai (Indian languages)

Description: *Citrus lemon* is a spinous shrub or small tree with green barks; its leaves are unifoliate and glandular with narrowly winged petiole. Inflorescence is a condensed raceme, bearing 5–7 flowers; the flowers are bisexual, pedicellate, and purplish in bud condition. The fruit is ovoid-oblong or oblate with a medium-sized hesperidium and becomes yellowish when mature bearing many seeds.

Distribution: It is native of India and cultivated throughout up to an elevation of 4000 ft. In Southeast Asia, it is grown in some gardens as well as in orchards, especially in Uttar Pradesh, Bombay, Madras, and Mysore.

Traditional Medicinal Uses: It is carminative, stimulant, stomachic, and rubefacient and used to treat diarrhea, dysentery, hypertrophy of the spleen, inflammatory affections, pneumonia, rheumatism, scurvy, epilepsy, tumor, and warts (Kirtikar and Basu 1975; *The Wealth of India* 1992).

Antidiabetes: Lemon peel pectin is reported to have an antidiabetic property. The effect of the citrus flavonoids hesperidin and naringin on glucose and lipid regulation in C57BL/KsJ-*db/db* mice was investigated. Hesperidin and naringin both significantly increased the glucokinase mRNA level, while naringin also lowered the mRNA expression of phosphoenolpyruvate carboxykinase and glucose-6-phosphatase in the liver. In addition, the hepatic glucose transporter 2 protein expression was significantly reduced, while the expression of adipocyte glucose transporter 4 and hepatic and adipocyte PPAR-γ was elevated in the hesperidin and naringin groups when compared with the control group. The two flavonoids also led to a decrease in the plasma and hepatic cholesterol levels that may have been partly due to the decreased hepatic HMG-CoA reductase and acyl CoA/cholesterol acyltransferase activities and increased fecal cholesterol excretion (Jung et al. 2006c). *C. lemon* juice decreased blood glucose levels and increased blood insulin levels in a dose-dependent manner in alloxan-induced diabetic rats (Riaz et al. 2013).

Phytochemicals: The phytochemicals present in *C. lemon* include the following: citric acid, lemon oil, eriocitrin, eriodictyol, apigenin, luteolin, quercetin, isorhamnetin, limocitrin, sinapic acid, scoletin, umbelliferone, hesperidin, and naringin (*The Wealth of India* 1992; Jung et al. 2006c).

Citrus limetta Risso, Rutaceae

Common Names: Pomona sweet lemon, mosambi
Oral administration of the methanol extract of *Citrus limetta* fruit peel (200 and 400 mg/kg daily for 15 days) to streptozotocin-induced diabetic rats resulted in dose-dependent normalization of blood glucose levels and serum biochemical parameters. The treatment also decreased LPO (Kundusen et al. 2011).

Citrus maxima (Burm.) Merr., Rutaceae

Common Name: Pummelo

In vivo hypoglycemic activity of the *Citrus maxima* fruits and fruit peel has been reported. The ethanol extract of the stem bark of *C. maxima* exhibited antidiabetic activity in alloxan- and streptozotocin-induced diabetic rats. The bark extract (200 and 400 mg/kg) showed a dose-dependent glucose-lowering effect in diabetic rats (Abdul Muneer et al. 2014).

The methanol extract of *C. maxima* leaf (200 and 400 mg/kg, daily for 15 days) showed antihyperglycemic and antioxidant effects in streptozotocin-induced diabetic rats (Sen et al. 2011).

Citrus paradisi Macfad., Rutaceae

Common Name: Grapefruit

Alcohol decoction of *Citrus paradisi* seed is reputed for the local management of an array of human diseases including, anemia, DM, and obesity by Yoruba herbalists (Southwest Nigeria). In a scientific validation study, the methanol extract of *C. paradisi* seed (100, 300, and 600 mg/kg, daily for 30 days) showed glucose- and lipid-lowering effects in alloxan-induced diabetic rats. Phytochemical analysis showed the presence of alkaloids, flavonoids, cardiac glycosides, tannins, and saponin in the extract (Adeneye 2008).

Citrus reticulata Blanco, Rutaceae

Essential oil (500–2000 mg/kg) from *Citrus reticulata* fruit showed anti-DM activity in alloxan-induced diabetic rats. The treatment reduced the levels of glucose and lipids in the blood of diabetic rats (Gavimath et al. 2010). Citrus fruits, tangor Elendale (*C. reticulata* × *C. sinensis*) and tangelo Minneola (*C. reticulata* × *C. paradisi*), reduced AGE- and H_2O_2-induced oxidative stress in human adipocytes (Ramful et al. 2010).

Citrus sinensis (L.) Osbeck, Rutaceae

Synonym: *C. aurantium* var. *sinensis L., C. macracantha* Hassk.

Common name: Sweet orange

The peel extract of *Citrus sinensis* showed antithyroidal, hypoglycemic, and insulin stimulatory properties, which suggest its potential to ameliorate both hyperthyroidism and diabetes mellitus (Parmar and Kar 2008a).

Clausena anisata (Willd) Hook., Rutaceae

An indigenous Southern African medicinal plant used to treat DM, the methanol extract of *Clausena anisata* root exhibited hypoglycemic activity in rats (Ojewole 2002). The extract produced dose-dependent substantial reduction in the levels of blood glucose in fasted streptozotocin-induced diabetic rats. The results of the study indicate that the herb possesses hypoglycemic activity (Ojewole 2002). The water and methanol extracts of *C. anisata* leaves strongly inhibited α-amylase and moderately inhibited rabbit hepatic glucose-6-phosphatase. These two extracts were less potent inhibitors of α-amylase than acarbose and significantly more potent inhibitors of G-6-Pase than sodium vanadate. The *in vitro* inhibition of glucose-6-phosphatase by the water and methanol extracts of *C. anisata* needs to be confirmed *in vivo* (Leshweni et al. 2012). Further studies are warranted to identify the active principles and their specific mechanisms of actions.

Clausena lansium (Lour.) Skeels, Rutaceae

Clausena lansium (fool's curry leaf) is used for various ailments in ethnomedicine in some countries. This plant grows widely in South China. The methanol extract of the stem bark (100 mg/kg) induced maximum and significant antihyperglycemic activity at 30 min and a 39% increase in plasma insulin at 60 min compared to control animals. The increase in plasma insulin after 60 min, compared to 0 min, was 62.0% (Adebajo et al. 2008). Under *in vitro* conditions, the extract also stimulated insulin release from INS-1 cells at 0.1 mg/mL (174% increase). Imperatorin and chalepin were the major active constituents increasing *in vitro* insulin release to 170.3% and 137.9%, respectively. The methanol extract also exhibited antioxidant activities. *C. lansium* stem bark may be useful in diabetes and its mechanism of action is the stimulation of insulin secretion (Adebajo et al. 2008). Clausenacoumarine is a compound isolated from the leaves of *C. lansium*. This compound lowered blood glucose level in normal mice and alloxan-induced diabetic mice at 200 mg/kg for 3 days orally and antagonized the elevation of blood glucose caused by injecting adrenaline in normal mice. No effect on blood lactic acid was observed (Shen et al. 1989).

Cleistocalyx operculatus (Roxb.) Merr & Perry, Myrtaceae

Cleistocalyx operculatus flower bud extract (used for making drinks in Vietnam) showed promising inhibitory activity against the α-glucosidase enzyme. The antihyperglycemic effects of a water extract from flower buds of *C. operculatus*, a commonly used material for drink preparation in Vietnam, were investigated. *In vitro*, the extract inhibited the rat intestinal maltase and sucrase activities, with IC_{50} values of 0.70 and 0.47 mg/mL, respectively. Postprandial blood glucose testing of normal mice and streptozotocin-induced diabetic rats by maltose loading (2 g/kg body weight) showed that the blood glucose reduction with the extract (500 mg/kg) was slightly less than that with acarbose (25 mg/kg). In an 8-week experiment, the blood glucose level of streptozotocin-induced diabetic rats treated with the extract (500 mg/kg) markedly decreased in comparison with that of nontreated diabetic rats (Mai and Chuyen 2007). In another study, *in vitro*, the water extract of *C. operculatus* flower buds (which have the highest phenolic and flavonoid contents) showed a strong antioxidant effect and highest pancreatic lipase inhibitory activity when compared with green tea and guava leaf extracts. Oral administration of the water extract (500 mg/kg/day) to streptozotocin-induced diabetic rats for 8 weeks resulted in significant reduction in the levels of glucose, total cholesterol, and TG in plasma as well as the concentration of glucose and sorbitol in the lens. In addition, the extract showed significant recovery in the activities of antioxidant enzymes (superoxide dismutase, GSH S-transferase) and GSH level in the liver with a markedly decrease lipid peroxide level in the liver and lens of treated diabetic rats. Thus, the water extract of *C. operculatus* showed antioxidant activities, prevention of sorbitol accumulation in the lens, and hypolipidemic effects in addition to its antidiabetic effects and may be considered as a promising material for the prevention of diabetic complications and metabolic syndrome (Mai et al. 2009).

Clematis pickeringii A. Gray, Ranunculaceae

Clematis pickeringii is used as an anti-inflammatory agent in traditional medicine in Australia. The stem ethanol extract of *C. pickeringii* activated the expression of PPAR-α and PPAR-γ proteins in HepG2 cells. This could be the anti-DM mechanism of this plant species. *Clematis glycinoides* and *Clematis microphylla* also showed activity to a lesser extent (review: El-Abhar and Schaalan 2014).

Cleome droserifolia (Forssk.) Del, Cleomaceae (= *Roridula droserifolia* (Delile) Forsk.)

Cleome droserifolia is used in Egyptian traditional medicine for the treatment of DM. *C. droserifolia* leaf (ethanol extract) lowered blood glucose levels in normal and alloxan-induced diabetic mice. Further, the extract (310 mg/kg, i.p., daily for 30 days) treatment resulted in not only an antihyperglycemic effect and increase in liver glycogen content but also reduction in oxidative stress and enhancement of blood insulin levels in the alloxan-induced diabetic mice. The authors conclude that the increase in insulin levels caused by the plant extract could be secondary to its antioxidant properties (El-Shenawy and Abdel-Nabi 2006). A recent study almost confirmed these observations. In this study, the possible protective effects of *C. droserifolia* methanol extract against the damage of pancreas β-cells and antioxidant defense systems in alloxan-induced diabetic rats were evaluated. The increase in blood glucose and MDA levels with the decrease in GSH content and the decrease in activities of antioxidant enzymes were the salient features observed in alloxan-induced diabetic rats. Administration of the extract (310 mg/kg/day, orally) for 30 days caused a significant reduction in blood glucose and MDA levels in alloxan-treated diabetic rats when compared with untreated diabetic control rats. Furthermore, diabetic rats treated with the extract showed a significant increase in the activities of both enzymatic and nonenzymatic antioxidants. Degenerative changes of pancreatic β-cells in alloxan-induced diabetic rats were minimized to near normal morphology by administration of the extract as evidenced by histopathological examination (Nagy and Mohamed 2014).

Cleome viscosa L., Cleomaceae

Cleome viscosa is an annual herb occurring in Australia and New Zealand. It is reported to contain hypoglycemic activity (Wang and Ng 1999).

Clerodendrum infortunatum L., Verbenaceae

Clerodendrum infortunatum, commonly known as *bhant* in Hindi, is a small shrub occurring throughout the plains of India, which is traditionally used for several medicinal purposes. The antihyperglycemic activity of the methanol extract of the leaves of *C. infortunatum* was evaluated in rats. Streptozotocin-induced diabetic rats were treated with the methanol extract of the leaves (250 and 500 mg/kg, i.p., daily) for 15 days. The treatment significantly and dose-dependently reduced and normalized blood glucose levels as compared to that of the diabetic control rats. Serum biochemical parameters were restored toward normal levels in the extract-treated rats as compared to the diabetic control rats. The treatment also decreased LPO and recovered GSH levels and catalase (CAT) activity toward normal values (Kumar et al. 2011e). In another study, the leaf extract of *C. infortunatum* also reduced blood glucose levels in streptozotocin-induced diabetic rats (Das et al. 2011a).

Clerodendrum phlomidis Linn.f., Verbenaceae

Synonyms: *C. multiflorum* (Burm. f.) O. Kuntze, *Volkameria multiflorum* Burm. f. (Figure 2.35)

Common Names: Wind killer, clerodendrum (Eng.), agnimantha, agnimanthini, arni, takkari, taludalai, irutali, etc. (Indian languages)

Description: *Clerodendrum phlomidis* is a large bush sometimes reaching up to 30 ft height with pubescent shoots. The leaves are simple, petiolated, and opposite. The root is 7–15 cm long, occasionally branched, cylindrical, tough, and yellowish-brown with a thin bark. The stem is straight, unbranched, and cylindrical with uneven surface. The flowers produced from an axillary or terminal panicle with a puberulous calyx. The fruit is rigid, slightly enlarged, and glabrescent in nature.

Distribution: It is a common shrub of arid plains, low hills, and tropical deserts. The plant occurs in almost all parts of India, Sri Lanka, Myanmar, and Southeast Asia.

Traditional Medicinal Uses: In India, this plant is traditionally recommended in formulations as well as a single drug for various ailments, including digestive disorders, acidity, gas, diarrhea, liver tonic, and general health tonic; it is also used to treat gonorrhea, anthrax, colic, dyspepsia, fever, stomachache, cholera, and dysentery. The root and leaves are used in traditional and folk medicine. Whole plant decoction is used to treat diabetes. Leaf decoction is used for inflammation, bronchitis, headache, weakness, drowsiness, and digestive problems. The root is used as a bitter tonic, analgesic, antiasthmatic, and antirheumatism (*The Wealth of India* 1992; review: Mohan and Mishra 2010).

(a)   (b)

FIGURE 2.35 *Clerodendrum phlomidis.* (a) Plant with flowers. (b) Flowers.

Pharmacology

Antidiabetes: The plant extract showed hypoglycemic activity. The ethanol extract of the leaves of this plant (100 and 200 mg/kg) showed hypoglycemic and hypolipidemic activities in alloxan-induced diabetic rats (Dhanabal et al. 2008). In histopathological studies, more prominent islet cells were seen in the ethanol extract-treated group. In another report, decoction and alcohol extract of this plant brought down the blood sugar levels effectively in both adrenaline-induced hyperglycemia and alloxan-induced diabetes in rats. The decoction in normal rats produced comparable fall in blood glucose both on immediate (hourly basis) and on long-term (30-day) treatment (review: Mohan and Mishra 2010). In another study, the leaf extract showed promising anti-DM activity. Oral administration of *C. phlomidis* leaf extract to alloxan-induced diabetic rats showed a marked decrease in the levels of biochemical parameters such as blood glucose and glycosylated hemoglobin. The levels of blood urea and serum creatinine were also brought to normal levels in diabetic rats treated with the extract. The lipid profile of the treated diabetic rats was also brought back to normal. SGOT, SGPT, and ALP activities were increased in diabetic rats. Treatment with the extract also brought these enzymes back to normalcy (Gopinathan and Naveenraj 2014). A decoction of the entire *C. phlomidis* plant has been reported to have anti-DM activity. A dose of 1 g/kg showed antidiabetic effects in epinephrine- and alloxan-induced hyperglycemia in rats and also showed antihyperglycemic activity in human adults at a dose of 15–30 g/day (Chaturvedi et al. 1984).

Other Pharmacological Properties: Many pharmacological studies using experimental animals have showed the diversified biological properties of this plant, including analgesic, antiamnesic, antidiarrheal, anti-inflammatory, antimicrobial, antiplasmodial, immunomodulatory, psychopharmacological, and nematicidal activity (review: Mohan and Mishra 2010).

Phytochemicals: Essential oils containing D-α-phellandrene, eugenol, cinnamic aldehyde, and chalcone glycoside were reported. Other isolated compounds include β- and γ-sitosterol, sterol, ceryl alcohol, cerotic acid, D-mannitol, sctellarein, petolinarigenin, (2 4 5) ethyl-cholesta-5,22,25-trien-3β-ol, unidentified sterols, apigenin, luteolin, pectolinarigenin, hispidulin, and clerodin (review: Mohan and Mishra 2010).

Clitoria ternatea L., Fabaceae

Common Name: Butterfly pea
The pancreatic regeneration potential of different fractions of the ethanol extract of *Clitoria ternatea* aerial parts was investigated. The antidiabetic and antihyperlipidemic potential was evaluated in streptozotocin-induced diabetic rats and correlated with its *in vivo* and *in vitro* antioxidant activity. The most significant pancreatic regeneration activity, antidiabetic activity, and antihyperlipidemic activity were shown by the ethanol extract and butanol-soluble fraction of the ethanol extract at a dose level of 200 mg/kg. It is suggested that the factors causing regeneration are present within the pancreas. The newly generated islets may have formed from the ductal precursor cells and reduced oxidative stress helps in the restoration of β-cell function (Verma et al. 2013).

Cluytia richardiana L., Euphorbiaceae
Cluytia richardiana is a traditional medicinal herb used medicinally in Saudi Arabia with demonstrated hypoglycemic effects. Saudin, a novel diterpene, was isolated from the leaves of this plant. This compound (40 mg/kg, i.p.) showed hypoglycemic activity in alloxan-induced diabetic rats (Mossa et al. 1985). Further, it has been shown to be effective in treating diabetes (U.S. Patent 1988, 4740521).

Cnestis ferruginea DC, Connaraceae
The methanol and ethyl acetate extracts of *Cnestis ferruginea* showed promising antidiabetic activities in streptozotocin-induced diabetic rats (Adewoye and Olorunsogo 2010). The methanol and ethyl acetate extracts (250 mg/kg), when administered orally for 10 consecutive days to streptozotocin-induced diabetic rats, lowered blood glucose levels by 68% and 74%, respectively, whereas glibenclamide reduced the same by 60%. The extracts reduced the elevated levels of serum ALT

and AST in diabetic rats. Similarly, both extracts significantly lowered the levels of serum creatinine, urea, total cholesterol, TG, and thiobarbituric acid reactive substance. A single dose of the extracts also reduced blood glucose levels 4 h after administration in diabetic rats (Adewoye and Olorunsogo 2010). Further studies are warranted to identify the active principles and determine its therapeutic value.

Cnidoscolus chayamansa McVaugh, Euphorbiaceae

Oral administration of *Cnidoscolus chayamansa* extracts resulted in marked hypoglycemic effects in a dose-dependent manner in diabetic rats. In addition, the plant extract induced dose-dependent beneficial changes in lipid levels (Figueroa-Valverde et al. 2009).

Coccinia indica Wight & Arn., Cucurbitaceae

Synonyms: *Coccinia cordifolia* (L.) Cogn., *Coccinia grandis* (L.) Voigt, *Cephalandra indica* Naud., *Bryonia grandis* L. (Figure 2.36)

Common Names: Scarlet vine or ivy gourd (Eng.), bimba, kunduri, kovakkai, koval (Indian languages)

Description: *Coccinia indica* is a climbing, branched perennial tropical vine that grows several meters long. It is a scandent or prostrate, much branched herb with grooved stem. The roots are thick, simple, succulent, and slender; the stems are succulent with striated tendrils. The leaves are arranged alternatively, larger, and bright green above while pale beneath. The flowers are large, white, and star-shaped; unisexual, male flowers are solitary and subfiliform with three stamens. Female flowers are bearing fusiform, glabrous, slightly ribbed ovary; the ovary is inferior. The fruit is fusiform-ellipsoid and slightly beaked, bearing compressed, yellowish-gray seeds. The seeds are 6–7 mm long and tan-colored, with thickened margins (review: Pekamwar et al. 2013).

Distribution: *C. indica* is native of East Africa. It has been introduced to tropical regions of Asia, America, and the Pacific and is present throughout India in the wild as well as in cultivated land. It also occurs in the Philippines, China, Indonesia, Malaysia, Thailand, Vietnam, Papua New Guinea, and Australia. The plant is used by humans mostly as a food crop in several countries (review: Pekamwar et al. 2013).

Traditional Medicinal Uses: It is cooling, aphrodisiacal, sweet, acrid, antispasmodic, cathartic, astringent, pungent, laxative, emetic, galactagogue, and antipyretic and used for vomiting, urinary loss, burning of hands and feet, uterine discharge, flatulence, itching, biliousness, jaundice, leprosy, fever, bronchitis, asthma, disease of the blood, inflammation, anemia, diabetes, sore tongue, sores, fever, eruption of skin, gonorrhea, earache, ringworm attack, psoriasis, itch, ulcers, chronic sinuses,

(a)   (b)

FIGURE 2.36 *Coccinia indica*. (a) Vine with flower. (b) Fruits (vegetable).

snake bite, and catarrh. The plant has been a part of Indian traditional medicine and has been used, among other things, for reducing blood glucose. The leaf, fruit, stem, and roots are used in traditional medicine (review: Pekamwar et al. 2013).

Pharmacology

Antidiabetes: The promising antidiabetic activity of this plant's parts was shown in dogs in 1985 (Singh et al. 1985). *C. indica* (95% ethanol extract) showed marked blood glucose-lowering effect in alloxan-induced diabetic rats (Kar et al. 2003). In normal rats, administration of the pectin from *C. indica* resulted in a significant reduction in blood glucose levels and an increase in the liver glycogen content (Kumar et al. 1993). The protective effect of *C. indica* on changes in the fatty acid composition in streptozotocin-induced diabetic rats has been reported (Pari and Venkateswaran 2003b). Reports suggest that the antidiabetes activity of the plant can be attributed to the presence of some compounds that inhibit the enzyme glucose-6-phosphatase, which is one of the key liver enzymes involved in regulating glucose metabolism (Gosh and Roy 2013). The effectiveness of 60% methanol extract (150 and 300 mg/kg) of this plant's leaves in the maintenance of blood glucose levels in both normal and streptozotocin-induced diabetic rats is indicated by significant reduction in elevated blood sugar levels after 10 days of treatment that was comparable to that of glibenclamide under similar conditions (Gosh and Roy 2013). The water extract of *C. indica* leaves (daily administration for 21 days, p.o.) exerted hypoglycemic and hypolipidemic activities in alloxan-induced diabetic rats. It lowered the levels of blood glucose and lipids (cholesterol, TG, LDL, and VLDL) in diabetic rats (Manjula and Ragavan 2007). In another study also, the water extract of *C. indica* leaves (daily administration for 10 days, p.o.) reduced the blood glucose, cholesterol, protein, and urea in alloxan-induced diabetic and normal rats; serum insulin levels were not increased in the rats treated with the extract (Doss and Dhanabalan 2008). The hypoglycemic activity of *C. indica* fruits in alloxan-induced diabetic rats has been shown (Ramakrishnan et al. 2011). Recently also, 70% ethanol extract of the root of *Cephalandra indica* (*C. indica*) was reported to have antidiabetic activity against alloxan-induced diabetic rats (Ravichandran et al. 2014). The fruit can be used as a medicinal food. The active principle involved and the details of the mechanism of action remain to be studied.

Clinical Trial: In a double-blind controlled study, the leaves of the plant showed favorable results to untreated, but uncomplicated, maturity-onset DM patients. There were no adverse effects during the 6-week treatment. The data showed that the active principle could be slow acting, since the maximum effect was obtained only after 3 weeks of treatment (Khan et al. 1979, 1980). Another double-blind, placebo-controlled, randomized clinical trial was carried out on 60 type 2 DM patients requiring only dietary or lifestyle modifications to determine the efficacy of the alcohol extract of the aerial parts (leaves and fruits) of *C. indica* (1 g/day). There was a significant decrease in the fasting and postprandial blood glucose and blood glycosylated hemoglobin levels of the herbal drug–treated group compared with that of the placebo group. There was no significant change in the serum lipid levels. The study suggests that the *C. indica* extract has a potential hypoglycemic action in patients with mild DM (Kuriyan et al. 2008). In another study, the dried extract of *C. indica* in doses of 500 mg/kg body weight was administered orally to 30 diabetic patients for 6 weeks. Mild diabetes had no effect on lipoprotein lipase (LPL), lactic dehydrogenase (LDH), and glucose-6-phosphatase (G-6-Pase). But reduced activities of LPL enzymes and raised levels of G-6-Pase and LDH in plasma of severe diabetics were found. The alterations in these parameters in untreated diabetic patients were restored after treatment with *C. indica*. The authors postulate that the ingredients present in the extract of *C. indica* act like insulin, correcting the elevated enzymes G-6-Pase and LDH in the glycolytic pathway and restoring the LPL activity in the lipolytic pathway with the control of hyperglycemia in diabetes (Kamble et al. 1998).

Other Activities: *C. indica* leaf extract has antioxidant and hypolipidemic activities in experimental streptozotocin-induced diabetic rats (Venkateswaran and Pari 2003). Other reported activities of this plant include antibacterial, antifungal, antimalarial, antioxidant, antiulcer, anti-inflammatory, antipyretic, analgesic, hepatoprotective, anticancer, and antitussive activity (review: Pekamwar et al. 2013).

Phytochemicals: Reported phytochemicals include pectin, alkaloid, β-sitosterol, taraxerol, palmitic acid, linoleic acid, β-carotene, lycopene, terasocrone, cryptoxanthin, apo-6'-lycopenal, β-amyrin, cucurbitacin-B, cephalandrol, tritriacontane, cephalandrin A and B, and stigmast-7-en-3-one. Ivory gourd is rich in β-carotene (Kumar et al. 1993; Ajikumaran and Subramoniam 2005).

Cocculus hirsutus (L.) Diels, Menispermaceae

Cocculus hirsutus is used in traditional medicine to treat diabetes. In alloxan-induced diabetic mice, the water extract of leaves of *C. hirsutus* (250, 500, and 100 mg/kg, p.o.) lowered blood glucose levels and improved oral glucose tolerance (Badole et al. 2006). In another study, the hypoglycemic and hypolipidemic activity of the methanol extract of the *C. hirsutus* leaf was evaluated in streptozotocin-induced diabetic rats, which showed increase in the levels of glycosylated hemoglobin, glucose, glucose-6-phosphatase, cholesterol, phospholipids, TGs, and LPO and decrease in insulin and hemoglobin; these changes were considerably restored to near normal levels in the extract-treated (500 mg/kg) diabetic rats. The plasma lipoproteins (HDL and LDL) were altered significantly in streptozotocin-induced diabetic rats; these levels were also restored back to near normal by the plant extract treatment (Palsamy and Malathi 2007). A repetition of the same study was also reported wherein the methanol extract of the aerial part of *C. hirsutus* showed promising blood glucose-lowering effect in streptozotocin-induced diabetic rats (Sangameswaran and Jayakar 2007).

Cocos nucifera L., Arecaceae

Common Name: Coconut tree

Cocos nucifera have been used in traditional medicine around the world to treat numerous ailments. The hypoglycemic activity of the alcohol extract of the flower and fruit oil of *C. nucifera* was studied in alloxan-induced diabetic rats. The ethanol extract of *C. nucifera* flower and fruit oil (300 mg/kg, p.o., daily for 15 days) suppressed the elevated blood glucose and lipid levels in alloxan-induced diabetic rats (Kabra et al. 2012). In another study, the antidiabetic activities and phytochemical composition of the immature inflorescence of *C. nucifera* were evaluated. Phytochemical analyses on inflorescence showed the presence of phenolic compounds, flavonoids, resins, and alkaloids. Administration of the methanol extract of coconut inflorescence to diabetic rats showed dose-dependent reduction in hyperglycemia. The levels of serum AST, ALT, and ALP were significantly decreased in treated rats (Renjith et al. 2013). In a recent study, the antidiabetic and antioxidant nature of *C. nucifera* flower extract in streptozotocin-induced diabetic rats was evaluated. Oral administration of *C. nucifera* flower extract (300 mg/kg/day) to diabetic rats for 30 days significantly reduced the levels of blood glucose, glycosylated hemoglobin, urea, uric acid, and creatinine. The altered levels of serum aminotransferases and ALP were normalized upon treatment with the fruit extract. The observed decrease in the levels of plasma protein in diabetic rats was elevated to near normal by the extract treatment. The level of glycogen content and the altered activities of glycogen synthase and glycogen phosphorylase were improved upon treatment with the extract. The antioxidant competence was also improved upon extract treatment. The results of the study indicated that the flower extract is nontoxic and possesses antidiabetic and antioxidant potential (Saranya et al. 2014).

Interestingly, treatment with the lyophilized form of mature coconut water and glibenclamide in alloxan-induced diabetic rats reduced the blood glucose and glycated hemoglobin along with improvement in plasma insulin levels. Elevated levels of liver function enzyme markers like ALP, SGOT, and SGPT in diabetic rats were significantly reduced on treatment with mature coconut water. In addition to this, diabetic rats showed altered levels of blood urea, serum creatinine, albumin, and albumin/globulin ratio that were significantly improved by treatment with mature coconut water and glibenclamide. Activities of nitric oxide synthase in the liver and plasma L-arginine were reduced significantly in alloxan-induced diabetic rats, while treatment with mature coconut water reversed these changes. The overall results show that mature coconut water has significant beneficial effects in diabetic rats and its effects were comparable to that of glibenclamide, a well-known antidiabetic drug (Preetha et al. 2013). There is an urgent need to identify active principles and carry on clinical trials.

Codonopsis pilosula (Franch.) Nannf, Campanulaceae

Administration of the plant extract once a day for 4 weeks to streptozotocin-induced diabetic mice resulted in reduction in blood glucose levels and oxidative stress (He et al. 2011).

Coffea arabica L., Rubiaceae

Synonyms: *Coffea vulgaris* Moench., *Coffea laurifolis* Salisb., *Coffea moka* Hort. ex Heynh. (Figure 2.37)

Common Names: Arabica coffee, arabica, coffee, etc.

Description: Coffee is a medium-sized tree of the Rubiaceae family living up to 25 years. Coffee is an upright, evergreen shrub or small tree up to 5 m in height and 7 cm in diameter. The plant may grow with a single stem but often develops multiple stems by branching at the base or on the lower stem. The bark is light gray and thin and becomes fissured and rough when old. The root system consists of a short, stout central root, with secondary roots radiating at all angles. The glabrous, shiny, dark green, opposite leaves have petioles 4–12 mm long and ovate to elliptic blades 7–20 cm long, with entire edges, and pointed at both ends. The fragrant, white flowers are in axillary clusters of two to nine. The 1.0–1.8 cm drupes are ovoid, fleshy, green turning red and finally blue-black. The fruits usually contain two seeds, 8–12 mm long, that are rounded and flattened on one site with a medial groove.

Coffee is made from the roasted seeds of the genus *Coffea* and is brought from plant to cup via a complex process. The glossy red fruits are picked, the fleshy outer part of the berry is removed, and then their pale-colored seeds (beans) are sent to mills to remove the hard outer layer encasing the seeds. At this stage the beans are either exported or are blended with others from the region before exportation. The complicated process of roasting usually occurs in the country where the beverage is consumed.

Distribution: The original native population of coffee was in the highlands of Ethiopia and, possibly, nearby highland areas of Sudan and Kenya. Coffee was first cultivated by Arabs during the fourteenth century and introduced into the New World. In the first century, it was cultivated in Arabic countries and then in Iran and India. Currently, the main producers of this plant are Brazil and Columbia. Coffee is one of the world's favorite drinks, one of the most important commercial crop plants, and the second most valuable international commodity.

Traditional Medicinal Uses: Coffee, prepared from the coffee bean, is the most frequently consumed drink around the world; coffee has medicinal properties. In traditional medicine, different parts of coffee plants are used for influenza, anemia, edema, rage, hepatitis, and externally for nervous shock and as a stimulant for sleepiness, an antitussive in flu and lung ailments, a cardiotonic and neurotonic, for asthma, etc.

FIGURE 2.37 *Coffea arabica.*

Pharmacology

Antidiabetes: Most of the data on the effects of coffee phytochemicals on glucose metabolism are based on animal and *in vitro* studies. Studies in mice showed that caffeine promotes the release of catecholamines and stimulates an increased metabolic rate and thermogenesis of brown adipose tissue, which is expected to reduce obesity (Yoshioka et al. 1990). Caffeine also upregulates the expression of uncoupling protein 3. This protein is lower in patients with type 2 diabetes compared with healthy control subjects and is inversely related to the body mass index (review: Bisht and Sisodia 2010). A recent study supports this. The influence of integral and decaffeinated coffee brews (*Coffea arabica* and *Coffea canephora*) on the metabolic parameters of rats fed with hyperlipidemic diet was evaluated. Integral coffee brew treatment resulted in a reduction in weight gain and accumulation of lipids, lower diameter of adipocytes, and a lower relative weight of the liver and kidneys of rats fed with a hyperlipidemic diet. Integral coffee brew reduced the obesity in rats receiving a hyperlipidemic diet, but the same effect did not occur with the decaffeinated types (Gomes et al. 2013).

In a recent animal experiment, oral administration of the water extract of *C. arabica* green grain (63 and 93 mg/kg/day) to alloxan-induced diabetic rats for 15 days significantly reduced the levels of blood glucose after 8 and 15 days of treatment compared to untreated diabetic rats (Florian et al. 2013).

Chlorogenic acid and diabetes: Chlorogenic acid is a major component of coffee and is present in many other plants and is being recognized as an important antidiabetic compound. Chlorogenic acid has been shown to delay intestinal glucose absorption and inhibit gluconeogenesis. It inhibits glucose-6-phosphate translocase 1 and reduces the sodium gradient–driven apical glucose transport. Besides, *in vitro* and animal studies on chlorogenic acid and its derivates showed that these substances can decrease the hepatic glucose output through inhibition of glucose-6-phospatase. Further, chlorogenic acid also exerts antioxidant effects. This may have beneficial effects on glucose metabolism as oxidative stress plays a role in the development of insulin resistance and type 2 diabetes. Chlorogenic acid can act as a metal chelator, and chlorogenic acid changed the soft tissue mineral composition (e.g., increased magnesium concentrations in the liver) in rats. This change in mineral composition may have improved glucose tolerance (review: Bisht and Sisodia 2010). In *db/db* mice, a significant decrease in fasting blood sugar was observed 10 min after the intraperitoneal administration of 250 mg/kg chlorogenic acid and the effect persisted for another 30 min after the glucose challenge. Besides, chlorogenic acid stimulated and enhanced both basal and insulin-mediated glucose transports in soleus muscle. In L6 myotubes, chlorogenic acid caused a dose- and time-dependent increase in glucose transport. Chlorogenic acid was found to phosphorylate AMPK, consistent with the result of increased AMPK activities. Chlorogenic acid did not appear to enhance association of IRS-1 with p85. However, activation of Akt by chlorogenic acid was observed. These parallel activations in turn increased translocation of GLUT 4 to the plasma membrane. At 2 mmol/L, chlorogenic acid did not cause any significant changes in viability or proliferation of L6 myotubes. Chlorogenic acid stimulated glucose transport in skeletal muscle via the activation of AMPK. It appears that chlorogenic acid may contribute to the beneficial effects of coffee on type 2 DM (Ong et al. 2012).

Coffee contains substantial amounts of several lignans. Lignans may affect glucose metabolism through its antioxidant and (anti-)estrogenic properties. Intake of the lignan secoisolariciresinol resulted in a lower incidence of diabetes in Zucker diabetic rats, an animal model of type 2 diabetes (review: Bisht and Sisodia 2010). Coffee also contains substantial amounts of magnesium, potassium, trigonelline (*N*-methylnicotinic acid), and niacin. Trigonelline lowered the glucose concentrations in diabetic rats. Higher magnesium intake was associated with a lower risk of type 2 diabetes in several cohort studies, and pharmacological doses were associated with improved insulin sensitivity and insulin secretion in some intervention studies. However, adjustment for magnesium intake did not explain the association of coffee consumption with glucose tolerance and a lower risk of type 2 diabetes (review: Bisht and Sisodia 2010).

Clinical Studies: Several studies suggest a significant reduction in the risk of type 2 diabetes in coffee drinkers (review: Bisht and Sisodia 2010). As per one study, current or past coffee drinkers who did not have diabetes at baseline had a 60% reduced risk of type 2 diabetes during the next 8 years when compared with those who never drank coffee. Additionally, those without diabetes, who had impaired glucose at baseline, were similarly protected against diabetes. A significant reduced risk of diabetes among coffee drinkers is consistent with many studies (review: Bisht and Sisodia 2010).

A number of human population studies associated coffee consumption with a reduced risk for the development of type 2 diabetes (review: van Dam and Hu 2005). An optimum level of coffee consumption may be useful for the prevention or delay of metabolic disease.

Coffee consumption was not appreciably associated with early insulin secretion assessed during an oral glucose tolerance test. In several, but not all, studies, higher habitual coffee consumption was associated with higher insulin sensitivity. Results from studies that included an oral glucose tolerance test suggested that coffee consumption affected the postprandial glucose metabolism rather than the fasting glucose concentrations. In an intervention study, 14 days of higher decaffeinated coffee consumption was associated with a decrease in the plasma glucose concentrations. In a randomized crossover study, 4 weeks of very high coffee consumption did not affect the fasting glucose concentrations and increased the fasting insulin concentration. This increase may reflect the effects on hepatic extraction of insulin or the direct effects on insulin secretion (review: Bisht and Sisodia 2010). As similar results have been found with decaffeinated coffee, compounds in coffee other than caffeine have been proposed as being potentially responsible for the reduced risk (Bisht and Sisodia 2010).

Human studies indicated that acute caffeine ingestion impairs glucose tolerance while regular consumption of caffeinated or decaffeinated coffee beverage exerts a protective effect against type 2 diabetes (Battram et al. 2006). Caffeine intake was associated with an acute reduction of insulin sensitivity in short-term metabolic studies in humans. This effect reflects decreased glucose storage, probably due to increased epinephrine release. However, the acute effects of caffeine on the epinephrine levels wane after continued intake (review: Bisht and Sisodia 2010).

It should be noted that coffee contains, in addition to caffeine, several other bioactive compounds such as chlorogenic acid, trigonelline, quinides, and diterpenes such as cafestol and kahweol. These compounds differ in their mechanisms of actions. They may even have opposing biological effects. The relative ratios of these bioactive phytochemicals could differ depending on the source of the coffee bean and methods of brewing/preparation of coffee. Further, coffee is relatively rich in magnesium that is known to have several health benefits; so the effect of coffee can differ in people who have magnesium deficiency or borderline deficiency. Further interaction of these phytochemicals with dietary phytochemicals and gut microbial products could possibly influence the antidiabetic activity of coffee.

The diterpene content is lower in coffee that has been drip-filtered. Diterpenes, in particular cafestol and kahweol, have been reported to increase the serum total cholesterol levels and to be associated with higher rates of coronary heart disease in coffee drinkers from Norway (review: Bisht and Sisodia 2010).

Other Activities: Pharmacological studies have shown coffee bean's antibacterial activity; antioxidant property; effect on asthma and bronchitis; beneficial effects on cardiovascular diseases, cognition, and mood; effect on the gastrointestinal tract as a laxative and digestive aid; hepatoprotective property; beneficial effects in Parkinson's disease and other neurologic conditions; etc. (review: Bisht and Sisodia 2010). Integral coffee brew reduced the obesity in rats receiving a hyperlipidemic diet, but the same effect did not occur with the decaffeinated types (Gomes et al. 2013).

Phytochemicals: The biologically active phytochemicals include caffeine, chlorogenic acid, coumaric acid, ferulic acid, sinapic acid, flavonoids, polyphenols, lignans, trigonelline (*N*-methylnicotinic acid), diterpenes such as cafestol and kahweol, and niacin (Bisht and Sisodia 2010; Gomes et al. 2013).

Coffea canephora Pierre ex A. Froehner, Rubiaceae

Common Name: Robusta coffee

Chlorogenic acids from green coffee beans inhibited pancreatic α-amylase isozyme 1 (Narita and Inouye 2011), and this compound has more than one mechanism of antidiabetic action (see *Coffea arabica* L.). Further a lipid transfer protein from *C. canephora* seeds inhibited α-amylase (Zottich et al. 2011).

Cogniauxia podoleana Baillon, Cucurbitaceae

Cogniauxia podoleana leaves are used in Congolese traditional medicine for the treatment of DM. The flavonoid fraction (diethyl ether fraction) separated from the water extract of the leaves (100 mg/kg) of this plant reduced the blood glucose levels in 40.0% 3 h after oral administration to normal rats. The diethyl ether fraction (50 mg/kg) was found to reduce the increase of blood glucose levels in 29.4% and 44.5%, respectively, 3 and 4 h after oral administration to alloxan-induced diabetic rats, whereas at a dose of 100 mg/kg, it decreased the levels of hyperglycemia in 41.4% and 70.4%, respectively, after 3 and 4 h. Thus, the fraction showed a hypoglycemic effect in normal rats and an antihyperglycemic effect in diabetic rats (Diatewa et al. 2004).

Coix lacryma-jobi L., Gramineae or Poaceae

Common Names: Coix seed, adlay, Job's tears (Eng.), jargadi, sankru, netpavalam, poochakkuru, etc.

Description: *Coix lacryma-jobi* is an annual herb with culm up to 1.5 m high and linear lanceolate leaves up to 1.5 ft long that are cordate at the base. The flowers are in axillary and terminal spiciform racemes. The fruits are subglobose and stony and enclosed in the bract.

Distribution: It is native of Southeast Asia and distributed extensively in tropical and subtropical parts of the world. In India, it is present in warmer slopes of hills up to 5000 ft, as well as in the plains.

Traditional Medicinal Uses: It is an anodyne, antiphlogistic, diuretic, pectoral tonic, vermifuge, cathartic, and depurative and used in cancer, headache, catarrhal infection of the air passage, urinary inflammation, rheumatism, diabetes, metroxenia, small pox, toothache, dysentery, urinary infection, anthrax, diarrhea, fever, goiter, metroxenia, and tenesmus (*The Wealth of India* 1950; Kirtikar and Basu 1975).

Antidiabetes: *C. lacryma-jobi* var. *ma-yuen* seeds have hypoglycemic activity and active glycans have been isolated (Takahashi et al. 1986). Diabetic *db/db* mice fed with a diet containing adlay protein concentrates for 21 days showed lower levels of total cholesterol, TGs, lipid peroxides, and arteriosclerotic index compared to those in the diabetic control group (Watanabe et al. 2012). Administration of adlay seed water extract reduced the expression of neuropeptide Y and leptin receptor levels in the hypothalamus of HFD-fed rats; the treatment also reduced obesity (Kim et al. 2007). In another study, PPAR-γ ligands (hydroxy unsaturated fatty acids) were isolated from adlay seeds (Wang et al. 2014).

Other Activities: The plant has anticarcinogenic activity and inhibited COX-2 expression. The antiproliferative and chemopreventive effects of the adlay seed have been shown on lung cancer *in vitro* and *in vivo*. The anti-inflammatory activity of benzoxazinoids from roots has been shown. The plant seed has hypolipidemic activity and inhibited nitric oxide production from macrophages (review: Ajikumaran and Subramoniam 2005).

Phytochemicals: The phytochemicals found in *C. lacryma-jobi* include glycans (coixans A, B, and C), palmitic acid, stearic, oleic, linoleic acids, benzoxazinoids, coixenolide, 2-hydroxy-7-methoxy-2*H*-1,4-benzoxazin-3-one, 6-methoxybenzoxazolinone, coixenolide, 2,4-dihydroxy-7-methoxy-2*H*-1,4-benzoxazin-3(4H)-one (Ajikumaran and Subramoniam 2005).

Colocasia esculenta (L.) Schott, Araceae

Colocasia esculenta leaves are used in folk medicines of Bangladesh for the treatment of pain, inflammation, and lowering of blood sugar. This plant is grown for its edible corms (root vegetable). The ethanol extract of *C. esculenta* leaves exhibited antidiabetic activity in alloxan-induced diabetic rats (Kumawat et al. 2010). In another study, the methanol extract of the leaves of *C. esculenta* dose-dependently reduced blood glucose levels in glucose tolerance test in mice. The results demonstrate that the methanol extract possesses antihyperglycemic potential (Akter et al. 2013). In a recent study, several compounds were isolated from the ethanol extract of this plant's leaves and tested for their effect against lens aldose reductase activity. Among tested compounds, two of the flavonoids significantly inhibited rat lens aldose reductase, with IC_{50} values of 1.65 and 1.92 μM, respectively. These results suggest that flavonoid derivatives from this plant represent potential compounds for the prevention and/or treatment of diabetic complications (Li et al. 2014). Administration of cocoyam (*C. esculenta*) incorporated feed for 21 days to streptozotocin-induced diabetic rats resulted in 58.75% decrease in hyperglycemia and amelioration of their elevated urinary protein, glucose, urine specific gravity, and relative kidney weights. It is concluded that cocoyam could be useful in the management of diabetic nephropathy (Eleazu et al. 2013).

Combretum lanceolatum Pohl ex Eichler, Combretaceae

Oral administration of the ethanol extract of *Combretum lanceolatum* flowers (250 and 500 mg/kg/day) to streptozotocin-induced diabetic rats for 21 days significantly increased body weight and the weight of the liver, white adipose tissue, and skeletal muscle. Further, treatment with the extract resulted in a reduction in glycemia, glycosuria, and urinary urea levels and increase in the liver glycogen content. Quercetin was the major compound in the extract. The phosphorylation levels of AMPK were increased in liver slices incubated *in vitro* with 50 µg/mL of the extract, similarly to the incubation with metformin (50 µg/mL) or quercetin (10 µg/mL). The antihyperglycemic effect of the extract was similar to that of metformin and appears to be through the inhibition of gluconeogenesis, since urinary urea was reduced and skeletal muscle mass was increased. These data indicate that the antidiabetic activity of the *C. lanceolatum* extract could be mediated, at least in part, through the activation of AMPK by quercetin (Dechandt et al. 2013).

Combretum micranthum G. Don, Combretaceae

Common Name: Cocoyam

Combretum micranthum, a medicinal plant, is used for treating diabetes in Northwestern Nigeria. The water extracts of *C. micranthum* leaf (100 mg/kg) lowered blood glucose levels in normal glucose-loaded rats and alloxan-induced diabetic rats. The effect was comparable to that of glibenclamide (Chika and Bello 2010).

Commelina africana L., Commelinaceae

Oral administration of the water extract (500 mg/kg) of *Commelina africana* to alloxan-induced diabetic rats reduced fasting blood glucose by 78%. The hypoglycemic effect of *C. africana* was comparable with the reference hypoglycemic agent glibenclamide. Alkaloids, cardenolides, saponins, and tannins were detected in the extract (Agunbiade et al. 2012).

Commelina communis L., Commelinaceae

Common Name: Asiatic day flower

Commelina communis is a herb and considered to be a weed. A decoction of this plant has been traditionally used for the treatment of diabetes in Korea. The effect of the water extract of *C. communis* on the activity of α-glucosidase was evaluated *in vitro* and *in vivo*. The extract showed inhibitory activity of the α-glucosidase in a dose-dependent manner *in vitro*. Further, the extract alleviated hyperglycemia caused by maltose or starch loading in normal and streptozotocin-induced diabetic mice with better efficacy than that of acarbose. In addition, prolonged administration of the extract tends to normalize hyperglycemia in streptozotocin-induced diabetic mice. Thus, the extract has potential for use in the management of NIDDM (Youn et al. 2004).

Commiphora mukul (Stocks) Hook./*Commiphora wightii* (Arn.) Bhandari, Burseraceae

The alcohol extract of *Commiphora mukul* gum resin exhibited antihyperglycemic and antioxidant activities in streptozotocin-induced diabetic rats (Bellamkonda et al. 2010). Feeding the normal rats with fat-rich diet resulted in increased serum glucose, cholesterol, and TG levels along with the increase in insulin resistance in comparison to control animals. Different biochemical parameters like glucose tolerance, glycogen content, glucose homeostatic enzymes (like glucose-6-phosphatase, hexokinase), insulin release *in vivo*, and expression profiles of various genes involved in carbohydrate and lipid metabolism clearly demonstrated the antidiabetic effect of the guggulsterone extract isolated from *C. mukul* in HFD-fed rats. Guggulsterone demonstrated improved PPAR-γ expression and activity in *in vivo* and *in vitro* conditions. However, it inhibited 3T3-L1 preadipocyte differentiation *in vitro*. The results suggest that guggulsterone has both hypoglycemic and hypolipidemic effects that can help cure type 2 DM

(Sharma et al. 2009). Commipheric acid, isolated from the ethyl acetate extract of the gum of the tree, activated PPAR-α and PPAR-γ that may contribute to the antidiabetic activity of guggulipid in Lep(ob)/Lep(ob) mice (review: Wang et al. 2014).

Convolvulus althaeoides L., Convolvulaceae
Convolvulus althaeoides is an Egyptian traditional medicinal plant used to control DM (Haddad et al. 2005). The plant extract exhibited hypoglycemic effects in normal rats (Shabana et al. 1990).

Convolvulus pluricaulis Chois., Convolvulaceae
Common Name: Shankhapushpi
The ethanol extract of the leaves of *Convolvulus pluricaulis* (800 mg/kg) exhibited antihyperglycemic activity in normal and streptozotocin-induced diabetic rats. The treatment improved body weight and serum lipid profiles in diabetic rats (Agarwal et al. 2014).

Conyza dioscoridis (L.) Desf. DC, Asteraceae
The aerial parts of *Conyza dioscoridis* are used to control diabetes in Egypt. A 70% ethanol extract of the root, 50 mg/kg in streptozotocin-induced diabetic rats, reduced serum glucose, increased insulin levels, and decreased oxidative stress. Compounds present in the extract include known antidiabetes compounds such as β-sitosterol, α-amyrin, lupeol acetate, β-sitosterol glucoside, gallic acid, syringic acid, rutin, quercetin, and kaempferol (El-Zalabani et al. 2012). This is an attractive plant for further studies.

Coptis chinensis Franch, Ranunculaceae
Common Name: Chinese goldthread
Coptis chinensis is used to treat DM in ethnomedical practices in Asian countries. Oral administration of the extract of *C. chinensis* inflorescence (125, 250, and 500 mg/kg) to alloxan-induced diabetic rats for 4 weeks resulted in a concentration-dependent reduction in the levels of blood sugar and total cholesterol and LDL-C (Youn et al. 2006). The 100 g dried water extract of the inflorescence contained 8.11 g total alkaloid, 3.34 g berberine, 1.08 g palmatine, and 0.66 g jatrorrhizine, which had long been identified as active compounds in *C. chinensis* root (Youn et al. 2006). Berberine is the main alkaloid of *C. chinensis* (Ma et al. 2010). Berberine is known to have multiple antidiabetic mechanisms of action (see under *Berberis aristata* DC). It activated AMPK and glucose transport in rat skeletal muscle (Ma et al. 2010). Further, berberine inhibited PTP1B in L-6 myocytes and mimics insulin action (Chen et al. 2010). *In vitro* and *in vivo* studies have shown that berberine modulates the release of GLP-1 (Yu et al. 2010). It increased the levels of GLP-1 by inhibiting DPP-IV (Al-masri et al. 2009). Protoberberine-type alkaloids present in the rhizome of *C. chinensis* inhibited rat lens aldose reductase activity (Jung et al. 2008). Therefore, *C. chinensis* and the protoberberine alkaloids contained therein may have beneficial uses in the development of therapeutic and preventive agents for diabetic complications and DM (Jung et al. 2008).

Coptis deltoides C.Y. Cheng et Hsiao, Ranunculaceae
Berberine isolated from the *Coptis deltoides* root exhibited hypoglycemic effect *in vivo* (Chen and Xie 1986). Berberine is a known antidiabetic compound.

Coptis japonica (Thunb.) Makino, Ranunculaceae
Isoquinoline alkaloids present in the root of *Coptis japonica* inhibited rat lens aldose reductase activity (Lee 2002). Therefore, *C. japonica* alkaloids may have beneficial uses in the development of therapeutic and preventive agents for diabetic complications and DM.

Corallocarpus epigaeus (Rottl.) C.B.Clarke, Cucurbitaceae
The root and rhizome of *Corallocarpus epigaeus* are used in traditional medicine to treat syphilis, venereal complaints, diabetes, etc. The ethanol extracts of this plant rhizome (400 mg/kg) effectively inhibited elevation of blood glucose levels (1–4 h after administration) in hyperglycemic rats. This effect was comparable to that of glibenclamide (Gnananath et al. 2013).

Corchorus olitorius L., Tiliaceae

The ethanol extract of *Corchorus olitorius* seed (100–1000 mg/kg, p.o.) reduced blood sugar levels in normoglycemic rats, normal glucose-loaded rats, and alloxan-induced diabetic rats. In a repeated dose study (500 mg/kg daily for 14 days), the ethanol extract reduced glycosylated hemoglobin levels and increased the levels of insulin in the blood. The ethanol extract was found to contain alkaloids, tannins, flavanoids, glycosides, saponin, cardiac glycosides, anthraquinones, steroids, and volatile oil (Egua et al. 2013). In a follow-up study, various fractions of the alcohol extract of *C. olitorius* were tested for their antidiabetic activity in alloxan-induced diabetic rats. All the fractions (250 and 500 mg/kg) showed varying levels of glucose-lowering activity in diabetic rats. Significant glucose-lowering activity was seen only with the water fraction (1 h posttreatment), chloroform fraction (1, 2, and 4 h posttreatment), and ethyl acetate fraction (2 and 3 h posttreatment) (Egua et al. 2014). The observations may suggest the presence of more than one active principle.

Coriandrum sativum L., Apiaceae (Figure 2.38)

Local Names: *Coriander*, cilantro, Chinese parsley (Eng.), dhanyaka, dhania (Hindi), yuan sui (Chinese)

Distribution: *Coriander* could be indigenous to the Mediterranean region and the Caucasus. It is cultivated in many parts of the world. South Asia is the world's biggest producer of coriander (review: Nadeem et al. 2013). Coriander is grown as a spice crop all over the world.

Description: *Coriandrum sativum* is an annual, branched, smooth herb, growing about 30 cm in height. The leaves are pinnately or ternately decompounds; the ultimate segment of the lower leaves is ovate or lanceolate and deeply cut; the upper leaves are more finely dissected into narrow linear segments. The flowers are small white or pinkish-purple in compound terminal umbels. The fruits are yellowish-brown, globose, 4–5 mm in diameter, and ribbed, separating into two halves. The seeds are convex–concave, about thrice as broad as they are thick.

Traditional Medicinal Uses: *C. sativum* (coriander) has been documented as a traditional treatment of diabetes. In addition, the seeds have been used to treat indigestion, rheumatism, and pain in the joints.

Infusion of the fruit is used for dyspepsia; coriander oil is used for flatulence, colic, rheumatism, and neuralgia. The seeds are chewed for halitosis and are used in lotions or bruised for poultice in rheumatic pains; the paste of the seeds are applied for headaches; the juice of the fresh plant is applied for erythema; decoction of the plant in milk is used for bleeding piles; cold infusion of the seeds or the powder of dried seeds with a little sugar is used for colic in children (Paarakh 2009; Nadeem et al. 2013).

(a)

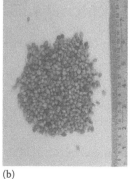
(b)

FIGURE 2.38 *Coriandrum sativum*. (a) Young plant. (b) *Coriander* seeds.

Pharmacology

Antidiabetes: Coriander incorporated into the diet (62.5 g/kg) and drinking water (2.5 g/L, decoction) reduced hyperglycemia in streptozotocin-induced diabetic mice. The water extracts of coriander (1 mg/mL) increased 2-deoxyglucose transport (1.6-fold), glucose oxidation (1.4-fold), and incorporation of glucose into glycogen (1.7-fold) in isolated murine abdominal muscle. In acute 20 min tests, 0.25–10 mg/mL water extract of coriander evoked a stepwise 1.3–5.7-fold stimulation of insulin secretion from a clonal β-cell line. This effect was abolished by 0.5 mM diazoxide and prior exposure to the extract did not alter subsequent stimulation of insulin secretion by 10 mM-L alanine, thereby negating an effect due to detrimental cell damage. The effect of the extract was potentiated by 16.7 mM glucose and 10 mM-L alanine but not by 1 mM 3-isobutyl-1-methylxanthine. Insulin secretion by hyperpolarized β-cells (16.7 mM glucose, 25 mM KCl) was further enhanced by the presence of the extract. Activity of the extract was found to be heat stable, acetone soluble, and unaltered by overnight exposure to acid (0.1 M HCl) or dialysis to remove components with molecular mass <2000 Da. Activity was reduced by overnight exposure to alkali (0.1 M NaOH). Sequential extraction with solvents revealed insulin-releasing activity in hexane and water fractions indicating a possible cumulative effect of more than one extract constituent. These results demonstrate the presence of antihyperglycemic, insulin-releasing, and insulin-like activity in *C. sativum* (Gray and Flatt 1999).

An ethanol extract of the seeds of this plant (200 and 250 mg/kg, i.p.) stimulated the release of insulin from the pancreatic β-cells in streptozotocin-induced diabetic rats. Treatment with the seed extract (200 mg/kg) increased significantly the activity of the β-cells in comparison with the diabetic control rats. Further, administration of the ethanol extract (200 and 250 mg/kg, i.p.) exhibited a significant reduction in serum glucose levels (Eidi et al. 2008). *C. sativum* (fruit) reduced plasma glucose, total cholesterol, and LDL in obese, hyperlipidemic diabetic rats (Aissaoui et al. 2011). In another study, the water extract of *C. sativum* fruits (250 and 500 mg/kg) decreased blood glucose levels in streptozotocin-induced diabetic rats. The extract also decreased total cholesterol levels and increased HDL significantly (Maquvi et al. 2012). The active principles involved remain to be identified. However, it should be noted that the seed is reported to contain compounds with known anti-DM properties such as quercetin, rutin, β-sitosterol, and chlorogenic acid.

Clinical Trial: In a clinical trial, the hypoglycemic effect of *C. sativum* was investigated in type 2 DM patients. Ten patients of type 2 DM with no previous medication, 10 patients of type 2 DM taking oral hypoglycemic agents with a history of inadequate control, and 6 control subjects were given low (2.5 g) and high (4.5 g) doses of the powdered part, water extract, and alcohol extract of *C. sativum* for 14 days. The study showed that the high dose of *C. sativum* has significant hypoglycemic activity (Waheed et al. 2006).

Other Activities: Other reported pharmacological activities include hypolipidemic activity, antioxidant activity, anti-inflammatory activity, antibacterial action, immunostimulant effect, anthelmintic activity, cardiovascular activity, anxiolytic activity, hepatoprotective activity, diuretic effect, and cytotoxicity of essential oils.

Toxicity: The water extract of fresh coriander seeds showed anti-implantation effect in rats. Further, the extract produced a significant decrease in serum progesterone levels on day 5 of pregnancy, which could be responsible for the anti-implantation effect observed (review: Paarakh 2009).

Phytochemicals: Seeds contain β-sitosterol; D-mannitol; flavonoid glycoside; rutin; umbelliferone; scopoletin; coriandrinonediol; chlorogenic acid, caffeic acid, palmitic, petroselinic, oleic, linolenic, lauric, myristic, myristoleic, palmitoleic and octadecenoic acids; quercetin-3-*O*-caffeyl glycoside; and kaempferol-3-glucoside. Seed oil components include α-pinene, limonene, β-phellandrene, 1, 8-cineole, linalool, borneol, β-caryophyllene, citronellol, geraniol, thymol, linalyl acetate, geranyl acetate, caryophyllene oxide, elemol, methyl heptenone, and petroselinic acid. β-Sitosterol, triacontane, triacontanol, tricosanol, psoralen, angelicin, coriandrinol, β-sitosterol glucoside, butylphthalide-neoenidilide, Z-ligustilide, coriandrin, dihydrocoriandrin28, coriandrone A, coriandrone B, and coriandrone C to E have been reported from the whole plant. Compounds isolated from the leaves include nonane, C9–16 alkenals, C7–17 alkanals, C10–12 primary alkenols, and alkanols. Fruits contain gnaphaloside A, gnaphaloside B, quercetin, isorhamnetin, rutin, and luteolin. Linalool, furfural, geranio, etc., are components of the essential oil of this plant (review: Paarakh 2009).

Cornus alternifolia L.f., Cornaceae
Cornus alternifolia is a traditional medicinal plant used as a tonic, analgesic, diuretic, etc. Kaempferol-3-*O*-β-glucopyranoside isolated from the dried leaves of *C. alternifolia* activated liver PPAR-γ and X receptors (Wang et al. 2014). This study suggests the antidiabetic activity of this plant.

Cornus kousa F. Buerger ex Miquel, Cornaceae
Cornus kousa, an oriental medicinal plant, has been traditionally used for the treatment of hyperglycemia. The leaf extract of this plant increased PPAR-γ ligand-binding activity in a dose-dependent manner in cultured 3T3-L1 cells. In addition, the extract enhanced adipogenesis and the expression of PPAR-γ target proteins, including glucose transporter 4 (GLUT4) and adiponectin, as well as proteins involved in adipogenesis in 3T3-L1 adipocytes. Furthermore, the extract led to significant induction of glucose uptake and stimulated insulin signaling, but not to activation of AMPK signaling (Kim et al. 2011).

Cornus officinalis Sieb., Cornaceae
Common Names: Dogwood, Japanese cornel or Japanese cornelian cherry
Several compounds with anti-DM activity have been isolated from this plant. Triterpene acids (50 mg/kg) from the fruit of *Cornus officinalis* alleviated diabetic cardiomyopathy in streptozotocin-induced diabetic rats (Qi et al. 2008). Triterpene fraction from the fruit of this plant at a dose of 80 mg/kg also alleviated diabetic cardiomyopathy in alloxan-induced diabetic rats (Gong et al. 2012). Ursolic acid from this plant inhibited PTP1B (Nazaruk and Borzym-Kluczyk 2014). Another triterpenoid, loganin, isolated from the fruits at a dose of 5 and 10 mg/kg showed renoprotective activity in streptozotocin-induced diabetic rats (Jiang et al. 2012). In a recent study, loganin and its derivatives isolated from the *C. officinalis* fruit (shan zhu yu) attenuated diabetic nephropathy *in vitro* in high-glucose-induced mesangial cells as judged from the inhibition of expression of collagen IV, fibronectin, and IL-6 (Ma et al. 2014).

Coscinium fenestratum (Gartn.) Colebr., Menispermaceae
Coscinium fenestratum is widely used in Ayurveda and Siddha systems of medicine for the treatment of DM. Oral administration of different doses of the ethanol extract of *C. fenestratum* stem for 12 days significantly reduced the levels of blood glucose in normal and streptozotocin–nicotinamide-induced diabetic rats. In addition, the extract treatment improved body weight, serum lipid profiles, thiobarbituric acid reactive substance levels, glycosylated hemoglobin, and liver glycogen levels in treated diabetic rats compared to control diabetic rats. Serum insulin levels were not stimulated in the animals treated with the extract (Shirwaikar et al. 2005). The water extract of *C. fenestratum* stem also showed antihyperglycemic activity in streptozotocin–nicotinamide-induced type 2 diabetic rats. In the extract-treated diabetic rats, an insulin-independent action with significant reduction in blood glucose, serum TG, and cholesterol levels was observed (Shirwaikar et al. 2008). In another study, the crude dichloromethane and ethyl acetate extracts (250 mg/kg, p.o., daily for 4 weeks) of *C. fenestratum* stem showed a strong hypoglycemic effect by lowering the blood glucose levels and increasing the body weight in streptozotocin-induced diabetic rats. These extracts contained phenolics and antioxidant activity (Malarvili et al. 2011). The anti-DM activity of the ethanol extract of *C. fenestratum* stem was also confirmed in alloxan-induced diabetic rats. The ethanol extract showed potent antihyperglycemic effect in alloxan-induced diabetic rats. The antihyperglycemic efficacy was comparable to that of glibenclamide (Manoharan et.al. 2011). Among other things, the active principles involved and details of the mechanism of action remain to be investigated.

Costus pictus D. Don., = *Costus igneus* Nak., Costaceae
Synonyms: *Costus mexicanus* Liebm ex Petersen; *Cst. congestus* Rowle (Figure 2.39)
Common Names: Spiral ginger, spiral flag, fiery costus, insulin plant (Eng.), cana agria or cana de jabali (in Mexico)

(a)

(b)

FIGURE 2.39 *Costus pictus.* (a) Young plant. (b) Flower.

Description: Formerly it was a subfamily under Zingiberaceae. It is a perennial, upright, spreading plant reaching about 60 cm tall, with the tallest stems falling over and lying on the ground. The leaves are simple, alternate, entire, oblong, evergreen, 10–20 cm in length with parallel venation. The large, smooth, dark green leaves of this tropical evergreen have light purple undersides and are spirally arranged around stems, forming attractive, arching clumps arising from underground rootstocks. Beautiful, 1.5 in. diameter, orange flowers are produced in the warm months, appearing on cone-like heads at the tips of branches. The fruits are inconspicuous, not showy, about 1 cm, and green-colored (review: Hedge et al. 2014).

Distribution: The plant occurs in Mexico and elsewhere. It is native to Central and South America and has been introduced from Mexico or America to India. It is widely grown in gardens as an ornamental plant in South India and also runs wild in many places.

Traditional Medicinal Uses: The leaves of this plant are used in India and elsewhere for antihyperglycemic activity. In Mexican folk medicine, the aerial parts of *C. pictus* is used as an infusion in the treatment of renal disorders (Melendez-Camargo et al. 2006).

Pharmacology

Antidiabetes: In the recent past, a lot of research work has been carried out on the hypoglycemic and antidiabetic property of the leaf extracts of *C. pictus* (*C. igneus*) using diabetic rodent models (review: Hedge et al. 2014). Although the leaf powder and extracts (water, methanol, and ethanol) showed varying levels of hypoglycemic activity, the methanol extract showed better activity compared to others (Mani et al. 2010; Krishnan et al. 2011; Kumudhavalli and Jaykar 2012). The methanol extract of both leaf and rhizome showed almost the same level of antidiabetic activity (Kalailingam et al. 2011; Krishnan et al. 2011). The hypoglycemic activity of the plant (shown in normal glucose-loaded mice) is present in the stem, leaf, and rhizome extracts (hexane, ethyl acetate, methanol, and water extracts) collected from different parts of South India. Although all the extracts tested showed activity, the methanol extract of the leaf showed maximum yield of phytochemicals (Shiny et al. 2013). The leaf showed antidiabetic activity in alloxan-induced as well as streptozotocin-induced and dexamethasone-induced diabetic rats (Jothivel et al. 2007; Shetty et al. 2010a; Krishnan et al. 2011). In addition to lowering glucose levels, the extract increased liver glycogen content, increased insulin level in the blood, improved lipid profiles, and reduced oxidative stress (Jothivel et al. 2007; Sethumathi et al. 2009; Devi and Urooj 2011; Kumudhavalli and Jaykar 2012; Suganya et al. 2012).

The methanol extracts of the leaf stimulated insulin release under *in vitro* conditions from mouse and human islets of Langerhans (Al-Romaiyan et al. 2010). In another *in vitro* study, the water extract of the leaf induced glucose-stimulated insulin secretion from the islets (Gireesh et al. 2009). However, an *in vitro* study of the ethanol extract of the *C. pictus* leaf was analyzed to study GLUT4 translocation and glucose uptake activity, which showed no direct peripheral action at 300 µg/mL dose comparable with insulin and metformin (Pareek et al. 2010).

Recently, an insulin-like protein was purified from *C. igneus*. The protein was found to be structurally different but functionally similar to insulin. It significantly decreased blood glucose in oral glucose tolerance tests (OGTT) when given orally to normal and diabetic mice. This protein purified from *C. igneus* is a novel protein having oral hypoglycemic activity (Joshi et al. 2013).

Clinical Trial: In a cross-sectional clinical study, diabetic patients consuming either one fresh leaf or 1 teaspoon of shade-dried powder/day of *C. igneus* in conjunction with other modalities of treatment had effectively produced glycemic control (Shetty et al. 2010b).

Other Activities: Other reported activities include hypolipidemic activity, diuretic activity, antioxidant activity, improvement of mitochondrial activities during alcohol-induced free radical stress, antimicrobial activity, anticancer activity, and inhibitory effect on calcium oxalate urolithiasis (review: Hedge et al. 2014).

Toxicity: Folklore information available in the district of Kanyakumari, Tamil Nadu, suggests that it may have renal toxicity. *C. pictus* leaf-fed normal rats showed an increase in the levels of serum urea and creatinine (Subramoniam et al., unpublished observation). Therefore, a detailed toxicity evaluation is required. A study carried out on the methanol extract of *C. igneus* indicated toxicity at 250 mg/kg body weight (Krishnan et al. 2011). Further, in another investigation, palmitic acid was found to be the major component in the stem, leaf, and rhizome oils of *C. pictus*. Excess palmitic acid is found to induce degeneration of myofibrils in healthy adult rat cardiomyocytes and enhance LDL-C to HDL-C ratio. Excess palmitic acid intake may lead to the development of coronary heart diseases. Hence, the constant use of *C. pictus* leaves for diabetic treatment may cause serious cardiac diseases and is not recommended for the treatment (Jose and Reddy 2010).

Phytochemicals: The leaves of *C. igneus* are rich in ascorbic acid, α-tocopherol, β-carotene, carbohydrates, terpenoids, steroids (e.g., ergastanol), flavonoids, glycosides, quinines, coumarins, alkaloids, tannins, saponins, bis(2′-ethylhexyl)-1,2-benzenedicarboxylate, etc. The stem showed the presence of a terpenoid compound lupeol and a steroid compound stigmasterol. Bioactive compounds quercetin and diosgenin and a steroidal sapogenin were isolated from *C. igneus* rhizome. Steam distillation of stems, leaves, and rhizomes of *C. pictus* yielded clear and yellowish essential oils. The major constituents identified in the essential oil are hexadecanoic acid, 9,12-octadecanoic acid, dodecanoic acid, linalyl propanoate, tetradecanoic acid, α-eudesmol, γ-eudesmol, and 4-ethoxy phenol (Shiny et al. 2013; review: Hedge et al. 2014).

Costus speciosus (Koenig) Sm., Costaceae

Synonyms: *Cheilocostus speciosus* (J. Koenig) C. Specht, Costaceae (formerly it was under Zingiberaceae) (Figure 2.40)

Common Names: Spiral flag (Eng.); kembuka, kembu, keu, keukand, and channakoova (India)

Description: *Costus speciosus* is an ornamental, rhizomatous, perennial, erect, succulent herb, up to 2.7 m in height, arising from a horizontal rhizome. Rhizomes are clothed with sheaths in the lower parts, leafy upward, with elliptic to oblong or oblong-lanceolate, thick, spirally arranged, 15–35 cm × 6–10 cm, leaves that are silky beneath, with stem clasping sheaths up to 4 cm; the flowers are large and white, in thick, cone-like terminal spikes, with bright red bracts, lip with yellowish throat; the fruits are globosely trigonous, red capsules, 2 cm in diameter, with black seeds black, five in number, that have white fleshy aril. The dried rhizome is a curved or somewhat straight, cylindrical, branched piece, 10–30 cm in length and 3–5 cm in diameter in dried condition; the upper surface is marked with circular nodal scars with remnants of leaf bases; the lower and lateral surfaces exhibit small circular scars of roots. There is no characteristic taste or odor. The flowers are white in color, 5–6 cm long with cup-shaped labellum and crest yellow stamens (review: Srivastava et al. 2011; Sabitha Rani et al. 2012).

FIGURE 2.40 *Costus speciosus.*

Distribution: *C. speciosus* is native to the Malay Peninsula of Southeast Asia. The plant is widely distributed in India in the tropical or subtropical climate from the sea level to the Himalayas. It is common along roadsides, streams, and wastelands (review: Srivastava et al. 2011).

Traditional Medicinal Uses: The rhizomes and roots are ascribed to be a bitter astringent, acrid, cooling, aphrodisiac, purgative, anthelmintic, depurative, febrifuge, expectorant, and tonic. It is a stimulant herb that improves digestion and clears toxins. The juice of the rhizome is applied to the head for cooling and relief from headache. The rhizomes are given for pneumonia, rheumatism, dropsy, urinary diseases, and jaundice and the leaves are given for mental disorders. Bruised leaves are applied in fever; decoction of the stem is used in fever and dysentery. Leaf infusion or decoction is utilized as a sudorific or in a bath for patients with high fever. The rhizome juice is given with sugar internally to treat leprosy and used for abortion. The plant possesses purgative, anti-inflammatory and antiarthritic effects, as well as antifungal activities and is used in gout rheumatism and bronchial asthma. The plant is used internally for eye and ear infections, diarrhea (sap from leaves, young stems), cold, cough, dyspepsia, skin diseases (rhizome), and snake bites. Rhizomes are used as a cardiotonic, diuretic, and CNS depressant and were formerly used in Malaysia for smallpox (review: Srivastava et al. 2011; Sabitha Rani et al. 2012).

Pharmacology

Antidiabetes: The rhizome of *C. speciosus* showed hypoglycemic activity in normal and diabetic rats (Mosihuzzaman et al. 1994). Possible protective effects of *C. speciosus* rhizome extracts on the biochemical parameters in streptozotocin-induced male diabetic rats were investigated (Daisy et al. 2009). Administration of hexane, ethyl acetate, and methanol extracts of the rhizome (250, 400, and 400 mg/kg for 60 days) to streptozotocin-induced hypoglycemic and normoglycemic rats resulted in a marked decrease in plasma glucose levels; all three extracts decreased glucose levels compared with untreated controls.

The hexane extract of the plant showed antihyperglycemic and hypolipidemic activity. The hexane extract of the *C. specious* rhizome was effective in decreasing the serum glucose level and normalizing other biochemical parameters in diabetic rats (Eliza et al. 2008, 2009a,b). The water and methanol extracts of *C. speciosus* were also highly effective in bringing down the blood glucose level (Rajesh et al. 2009).

The antihyperglycemic, antihyperlipidemic, and antioxidant potencies of an ethanol extract of *C. speciosus* root were evaluated in alloxan-induced diabetic male rats. It is concluded that the *C. speciosus* root extract possesses antihyperglycemic, antihyperlipidemic, and antioxidative effects, which may prove to be of clinical importance in the management of diabetes and its complications (Bavarva and Narasimhacharya 2008). Eremanthin and costunolide isolated from the *C. speciosus* root showed anti-DM and antihyperlipidemic activities in streptozotocin-induced diabetic rats (Eliza et al. 2009a,b).

Other Activities: An antifungal principle, methyl ester of *p*-coumaric acid, has been isolated from the rhizome of *C. speciosus*. The estrogenic activity of saponins from *C. speciosus* has been demonstrated. Other activities reported include anticholinesterase activity, hepatoprotective activity, antioxidant activity, adaptogenic property, and antifungal and antibacterial activities. The saponin diosgenin from this plant showed antifertility and estrogenic activities (review: Srivastava et al. 2011).

Toxicity: In view of the suspected renal toxicity to a much related species, *C. pictus*, a detailed toxicity study is required.

Phytochemicals: Phytochemicals isolated include diosgenin; saponin A, B, and C; lupeol palmitate; tigogenin; α-amyrin stearate; β-amyrin; prosapogenin B of dioscin; diosgenone; cycloartanol, 25-encycloartenol; octacosanoic acid; methyl ester of *p*-coumaric acid; and dihydrophytyl plastoquinone and its 6-methyl derivatives. Diosgenin is the major constituent isolated from the stem of this plant. Other constituents isolated are tigogenin, dioscin, gracillin, and β-sitosterol glucoside; the underground parts contain steroid glycosides. Several oxy acids and branched fatty acids are present in the plant; the rhizomes yield essential oils that contain pinocarveol as the major constituent (review: Srivastava et al. 2011).

Costus spicatus (Jacq.) Sw., Costaceae

This plant is commonly known as spiked spiral flag ginger or Indian head ginger. An *in vivo* study evaluated the ability of a tea made from the leaves of *Costus spicatus* to alter glucose homeostasis in C57BLKS/J (KS) *db/db* mice, a model of obesity-induced hyperglycemia, with progressive β-cell depletion. Intraperitoneal insulin tolerance testing after the 10-week study period showed that *C. spicatus* tea consumption did not alter insulin sensitivity, which suggested that at the dose given, the tea made from *C. spicatus* leaves had no efficacy in the treatment of obesity-induced hyperglycemia (Keller et al. 2009).

Crinum asiaticum L., Amaryllidaceae

The ethanol extract of this plant's leaves (200 mg/kg, p.o.) attenuated hyperglycemia-mediated oxidative stress and protects hepatocytes in alloxan-induced diabetic male rats (Indradevi et al. 2012).

Crotalaria aegyptiaca Benth., Labiatae

Crotalaria aegyptiaca is a traditional medicinal plant and the aerial parts of this plant are used in Egypt to treat DM (Haddad et al. 2005). The plant extract exhibited persistent hypoglycemic effects in normal rats (Shabana et al. 1990).

Croton cajucara Benth., Euphorbiaceae

Croton cajucara is a plant found in Amazonia, Brazil, and elsewhere; the bark and leaf infusion of this plant have been popularly used to treat diabetes and hepatic disorders.

Trans-dehydrocrotonin, a nor-clerodane diterpene isolated from *C. cajucara* exhibited hypoglycemic activity in alloxan-induced diabetic rats and glucose-fed normal rats (Farias et al. 1997). Trans-dehydrocrotonin (50 mg/kg) showed reduction of hyperglycemia in streptozotocin-induced diabetic rats and reduction of TG in ethanol-induced hypertriglyceridemic rats. Thus, the compound has an anti-DM potential (Silva et al. 2001). Besides, the water extract of *C. cajucara* treatment effectively reduced the oxidative stress in streptozotocin-induced diabetic rats. The extract has the potential to attenuate diabetic complications (Rodrigues et al. 2012).

Croton klozchianus L., Euphorbiaceae

Croton klozchianus is a relatively uninvestigated traditional medicinal plant with no pharmacological or phytochemical reports available; the plant has been used clinically by Ayurvedic physicians to treat diabetes. Treatment of diabetic rats with the extract of *C. klozchianus* aerial parts (100 and 300 mg/kg) for 3 weeks showed significant reduction in blood glucose and improvement in the lipid profile. *In vitro*, the extract caused a concentration-dependent increase in insulin secretion (eightfold at 2 mg/mL for cells challenged with 20 mM glucose) from MIN6 cells grown as monolayers and as pseudo-islets (Govindarajan et al. 2008).

Croton macrostachyus Hochst. ex Delile, Euphorbiaceae

Common Names: Woodland croton, forest fever tree, broad-leaved croton (Eng.)

Croton macrostachyus is a traditional medicinal tree. In a recent study, the hydroalcoholic root extract (300 mg/kg. p.o.) of *C. macrostachyus* exhibited a promising hypoglycemic effect in streptozotocin-induced diabetic rats; further, the extract improved glucose tolerance in the diabetic rats (Bantie and Gebeyehu 2015).

Croton zambesicus Mull. Arg., Euphorbiaceae

Synonyms: *C. amabilis* Mull. Arg., *C. gratissimus* Buch.

Croton zambesicus leaves are used in Africa to treat diabetes in folk medicine. The ethanol extract of *C. zambesicus* leaves showed anti-DM activity in alloxan-induced diabetic rats. The extract produced a significant reduction in blood glucose levels after a single dose of the extract and in repeated treatment for 7 days. The antidiabetic activity was comparable to that of the reference drug—chlorpropamide (Okokon et al. 2006). Further, the extract exhibited antihyperlipidemic activity in diabetic rats. Administration of the ethanol extract of this plant leaf lowered elevated levels of lipids in streptozotocin-induced diabetic rats; the treatment increased the levels of HDL in diabetic rats (Ofusori et al. 2012).

Cryptolepis sanguinolenta (Lindl.) Schltr., Asclepiadaceae

Extracts from various parts of the plant are used traditionally in many parts of the world for the treatment of many ailments including DM. Experiments on animals have shown hypoglycemic and hypolipidemic properties of the ethanol extracts of this plant. Oral administration of the extract (50–250 mg/kg) for 21 days resulted in significant reduction in the levels of plasma glucose, total cholesterol, TGs, and LDL-C in normoglycemic rats. Further, histological observations showed increased size of β-cells in the pancreas of treated rats (Ajayi et al. 2012).

Cucumis melo L., Cucurbitaceae

Common Names: Musk melon, oriental melon, etc.

Cucumis melo fruits have been used in the Indian traditional system of medicine for the treatment of various disorders. The fruit is used as a liver tonic, cardioprotective, antidiabetic, antiobesity, etc. The fruit peel extracts of *C. melo* (musk melon) reversed atherogenic diet-induced increase in the levels of blood glucose, tissue LPO, and serum lipids. Further, musk melon increased the levels of thyroid hormone and insulin indicating their potential to ameliorate the diet-induced alterations in serum lipids, thyroid dysfunction, and hyperglycemia. These beneficial effects could be due to the rich content of polyphenols and ascorbic acid in the peel extracts (Parmar and Kar 2008b). Hexane extract from *C. melo* seed exhibited promising inhibitory activities against α-glucosidase and α-amylase (Chen and Kang 2013). The methanol and water extracts of *C. melo* fruit peel exhibited antihyperlipidemic activity in high-cholesterol diet–induced hyperlipidemia in rats. The fecal excretion of bile acids and sterols increased upon treatment with the methanol and water extracts. The methanol extract (500 mg/kg) was found to be better than the same dose of water extract in improving the lipid profile in the high-cholesterol diet–fed rats (Bidkar et al. 2012).

Cucumis metuliferus E. Mey. ex Naudin, Cucurbitaceae

Cucumis metuliferus is a traditional medicinal plant in Africa. Glycoside fraction extracted from the fruit pulp of *C. metuliferus* (25–100 mg/kg) showed dose-dependent antihyperglycemic activity in alloxan-induced diabetic rats (Jurbe 2011). In another study, the fruit pulp extract of *C. metuliferus* (1000–1500 mg/kg) showed hypoglycemic activity in alloxan-induced diabetic rats. The extract did not lower glucose levels in normal rats (Jimam et al. 2010).

Cucumis sativus L., Cucurbitaceae (Figure 2.41)

Common Names: Cucumber (Eng.), bahuphala, khira, kakrikai, vellarikhai, vellarikka (Indian languages)

Description: *Cucumis sativus* is a hispidly hairy climber with membranous deeply cordate leaves, angled or shallowly 3–5 lobed, petiolated, hispidulous on both surface and often with soft hairs. Unisexual yellow flowers are present, and males are clustered having tubular or campanulate hyphanthium with white hairs, while female flowers are solitary. The fruit is a fleshy pepo and common vegetable.

Distribution: *C. sativus* is indigenous to North India and is cultivated extensively throughout India as well as in tropical and subtropical regions of the world.

Traditional Medicinal Uses: Used as a diuretic, refrigerant, taeniafuge, anthelmintic, and astringent, *C. sativus* is also used for biliousness, inflammation, throat infection, fever, spruce, dysentery, cancer, headache, and intestinal flu (*The Wealth of India* 1950; Kirtikar and Basu 1975).

Pharmacology

Antidiabetes: The seed of *C. sativus* is one of the traditional remedies used for DM treatment. Administration of the traditional preparation of the plant resulted in moderate glucose-lowering effect in glucose tolerance tests in rabbits (Roman-Ramos et al. 1995). In another study, the alcohol extract of *C. sativus* (250 mg/kg) did not influence blood glucose levels in rats (Chandrasekar et al. 1989). However, the hydroalcoholic and butanol extracts obtained from *C. sativus* seeds (200, 400, and 800 mg/kg/day) reduced blood glucose levels in normal and streptozotocin-induced type 1 diabetic rats, when treated daily for 9 days. The treatment improved body weight of diabetic animals. But the seed extracts were not effective on reducing blood glucose levels in normal and diabetic rats in acute treatments (Minaiyan et al. 2011). Interestingly, the ethanol extract of *C. sativus* fruits (cucumber) showed excellent hypoglycemic effects in alloxan-induced diabetic rats. The cucumber extracts reduced the blood glucose level by 67% 12 h after a single intraperitoneal injection. The treatment also reduced LDL, total cholesterol, and TG levels (Sharmin et al. 2013). In a recent study, methanol extracts of the fruit pulp of cucumber (*C. sativus*) (500 mg/kg, p.o.) showed a potent hypoglycemic activity in alloxan-induced diabetic rats. It reduced the fasting glucose levels from 12.8 to 4.6 mmol/L (231–82 mg/dL). Phytochemical analysis of the extract indicated the presence of saponins, glycosides, terpenes, phenolics, flavonoids, and tannins (Saidu et al. 2014). The mechanism of action and active principles involved remain to be studied. Cucumber is a valued vegetable and appears to be an attractive material to develop an anti-DM food.

Phytochemicals: *C*-glycosyl flavonoid phytoalexins, 14α-methyl sterol, 24(*R*)14α-methyl-24-ethyl-5α-cholest-9(11)-en-3β-ol,24-methylene-pollinastanol, codisterol, 25(27)-dehydrochondrillasterol,

FIGURE 2.41 *Cucumis sativus.*

clerosterol, avenasterol, isofucosterol, stigmasterol, campesterol, sitosterol, spinasterol, 24β-ethyl-25(27)-dihydrolanosterol, and 22-dihydrospinasterol were isolated from this plant (Rastogi and Mehrotra 1999).

Cucumis trigonus Roxb., Cucurbitaceae

The *Cucumis trigonus* fruit is used in Indian traditional medicine for the treatment of diabetes. The water extract of the fruit, when administered at a dose of 500 mg/kg, p.o., for 21 days, increased body weight, liver glycogen, and serum insulin levels and decreased the levels of blood glucose and glycosylated hemoglobin, total cholesterol, and TG in streptozotocin-induced diabetic rats (Salahuddin and Jalalpure 2010). The ethanol extract of this plant fruit exhibited antioxidant and anti-LPO activities in urolithiasis-induced rats (Balakrishnan et al. 2011).

Cucurbita ficifolia Bouche, Cucurbitaceae

Synonym: *Cucurbita melanosperma* A.Braun ex Gasp

Common Names: Malabar gourd, fig leaf gourd

Cucurbita ficifolia is cultivated in Mexico and elsewhere for its edible fruit (vegetable), which is also used in traditional medicine. *C. ficifolia* (fruit) is one of the widely used herbal medications in the treatment of DM. The hypoglycemic activity of this plant fruit has been shown in experimental animals and type 2 DM patients. The mechanism of action appears to be stimulation of insulin synthesis and insulin secretion (Acosta-Patino et al. 2001; Xia and Wang 2006a,b). The water extract of the fruit of *C. ficifolia* exhibited hypoglycemic and antioxidant properties in streptozotocin-induced diabetic mice (Xia and Wang 2006a; Diaz-Flores et al. 2012). In RINm5F cells, the water extract of the fruit and D-chiro-inositol, the hypoglycemic agent isolated from the fruit, increased the levels of insulin and insulin mRNA expression (Xia and Wang 2006b; Miranda-Perez et al. 2013).

Clinical Studies: In a clinical trial, oral administration of the raw extract of this plant fruit (4 mL/kg) to type 2 DM patients with moderate hyperglycemia resulted in a significant reduction in blood glucose levels 3 h after administration; the decrease was more pronounced after 5 h. (The patients had stopped their regular pharmacological medication 24 h prior to the study) (Acosta-Patino et al. 2001). Recently, available literature concerning the effects of *C. ficifolia* on type 2 DM was reviewed and concluded that *C. ficifolia* intake might have useful effects on the prevention and treatment of DM. *C. ficifolia* has beneficial effects on insulin sensitivity and the risk factors of DM; however, due to the small number of available studies, more researches are needed in this field (review: Bayat et al. 2014).

Cucurbita maxima Duchense, Cucurbitaceae

Common Names: Pumpkin, autumn squash

The plant seed exhibited antidiabetes and antihypoglycemic activity in streptozotocin-induced diabetic rats (Sharma et al. 2013c). The fruit juice and hydroalcoholic extract of *C. maxima* showed hypoglycemic activity in streptozotocin-induced diabetic rats. Although the juice and extract caused significant decrease in hyperglycemia, the extract showed more hypoglycemic effect than the fruit juice (Lal et al. 2011). A daily supplement of pumpkin in fruit powder was found to reduce blood glucose levels in NIDDM patients (Chen et al. 1994). The aerial part of this plant also showed anti-DM activity. The methanol extract of the aerial parts of *C. maxima* showed anti-DM activity in rats against streptozotocin-induced diabetes. The treatment reduced fasting blood glucose levels in a treatment duration-dependent manner (Saha et al. 2011a). Protein-bound polysaccharides from pumpkin increased serum insulin levels and lowered blood glucose in alloxan diabetic rats (Li et al. 2005a).

Cucurbita moschata Duchesne ex Poiret, Cucurbitaceae

Synonym: *Cucurbita moschata Duchesne* ex Poir., a type of elongated pumpkin

Cucurbita moschata fruit contains polysaccharides that inhibit α-glucosidase activity. Polysaccharides of the *C. moschata* fruit noncompetitively inhibited α-glucosidase catalyzed hydrolysis of p-nitrophenyl-D-galactopyranoside (pNPG) (Song et al. 2012c). In another study, a protein-bound polysaccharide was isolated from the water extract of fruits by activity-guided isolation. The protein-bound polysaccharide increased the levels of serum insulin, improved glucose tolerance and decreased the levels of glucose in alloxan-induced diabetic rats. It showed excellent antidiabetic activity at a higher dose (1000 mg/kg) (Quanhong et al. 2005). It appears to be an attractive material to be developed as an anti-DM nutraceutical.

Cucurbita pepo L., Cucurbitaceae

Synonyms: *C. aurantia* Willd., *C. courgero* Ser., *C. esculenta* Gray, *C. fastuosa* Salisb., *C. melopepo* L., *C. ovifera* L., *C. subverrucosus* Willd., *C. verrucosus* L., *Pepo melopepo* Moench., *P. verrucosus* Moench., *P. vulgaris* Moench. (monograph: WHO 2009)

Common Names: Abobora, bitter bottle gourd, etc.

Dried seeds of *C. pepo* are mainly used in traditional medicine. It is an annual running, monoecious herb with dark green leaves. The fruit powder exhibited hypoglycemic activity in alloxan-induced diabetic rats. Histological analysis showed improvement in the number and size of the islets in the fruit-powder-treated alloxan-induced diabetic rats (Sedigheh et al. 2011).

Cuminum cyminum L., Apiaceae/Umbelliferae (Figure 2.42)

Common Names: Cumin, huminum (Eng. and French); cumino (Spanish); jiraka, jira, jeera, siragam, jeerakam, etc. (Indian languages)

Description: *Cuminum cyminum* is a small, slender annual, glabrous except the fruit; it has blue-green linear partite leaves and an ultimate filiform segment and inflorescence a compound umbel, bearing white flowers with several bracts and bracteoles. The fruits are cylindrical with primary filiform ridges. secondary are usually hispidulous. The seeds are somewhat dorsally compressed. Cumin is grown from the seed. A hot climate is generally ideal for its growth.

Distribution: *C. cyminum* is a native of eastern Mediterranean countries; it is distributed throughout the world, especially in Southeastern Europe, North Africa, Morocco, Iran, Turkey, India, China, and America. In India, it is cultivated in all states except Assam and Bengal.

Traditional Medicinal Uses: In traditional medicine, *C. cyminum* is a stimulant, emmenagogue, lactagogue, carminative, stomachic, astringent, pungent, hot, sweet, cooling, aphrodisiac, antipyretic, and tonic and is used in treating dyspepsia, diabetes, colic, flatulence, dermatosis, tumor, enteritis, leukorrhea, skin disease, leprosy, anorexia, leukoderma, asthma, cough, ulcers, hiccups, dysentery, eye disease, fever, biliousness, enlargement of the spleen, boils, snake bite, flatulence, and puerperium (Kirtikar and Basu 1975; *The Wealth of India* 1992). It is also used to treat swellings of the breast and testicles; cumin stimulates menstruation (Kaur and Sharma 2012).

(a)

(b)

FIGURE 2.42 *Cuminum cyminum.* (a) Plant with flowers. (b) Seeds.

Pharmacology

Antidiabetes: Oral administration of cumin seeds for 6 weeks to diabetic rats resulted in significant reduction in blood glucose levels and body weight compared to untreated diabetic animals. Further, in a glucose tolerance test conducted in rabbits, cumin significantly improved glucose tolerance. In another study, the methanol extract of cumin seeds reduced the levels of blood glucose, glycosylated hemoglobin, creatinine, and blood urea nitrogen and increased the levels of serum insulin and glycogen content of the liver and skeletal muscle in alloxan- as well as streptozotocin-induced diabetic rats (Gagtap and Patil 2010; review: Kaur and Sharma 2012). An 8-week subacute administration of cumin to streptozotocin-induced diabetic rats reduced hyperglycemia and glycosuria along with an improvement in body weight and blood urea compared to untreated control diabetic animals. In normal rabbits also cumin administration resulted in a decrease in glucose (hypoglycemia) (Kaur and Sharma 2012). One of the active constituents of cumin seed oil, cuminaldehyde, inhibited aldose reductase and α-glucosidase isolated from rat (Lee 2005). It appears that more than one active principle may be involved in the antidiabetic action of cumin because the enzyme inhibition alone may not result in improvement in body weight in diabetic animals.

Other Activities: The hypolipidemic effect of *C. cyminum* on alloxan-induced diabetic rats has been reported (Dhandapani et al. 2002). Ethanol extracts of *C. cyminum* showed *in vitro* antibacterial activity against *Helicobacter pylori* (Nostro et al. 2005). The ethanol extract of cumin inhibited arachidonate-induced platelet aggregation (Srivastava 1989). Essential oils from the plant have antibacterial activity (Iacobellis et al. 2005). Cuminaldehyde has *in vitro* platelet aggregation activity (Subramoniam and Satyanarayana 1989). The plant seed extract has chemopreventive effects in mouse tumor models (Gagandeep et al. 2003). Cumin increased acid secretion in injured stomachs of rats and is good for digestion and related gastrointestinal problems. Estrogenic activity of the seed extract has been shown in rats (Malini and Vanithakumari 1987). The seed has galactagogue activity (Agarwala et al. 1968). Cumin seed possesses antimicrobial, anticancer, antioxidant, antiosteoporotic, immunomodulatory, antiasthmatic, and antiepileptic activities (Kaur and Sharma 2012).

Toxicity: Cumin contains safrole, a mutagen that is degraded by cooking or heating (Kaur and Sharma 2012).

Phytochemicals: It has been found to contain essential oils such as cuminaldehyde (4-isoprophylbenz-aldehyde), pyrazines, 2-methoxy-3-sec-butylpyrazine, 2-ethoxy-3-isopropylpyrazine, and 2-methoxy-3-methylpyrazine. The seeds are also rich in many flavonoid antioxidants such as carotene, zeaxanthin, and lutein; it also contains flavonoid glycosides including derivatives of apigenin and luteolin. Reported compounds include p-cymene, pinene, dipentene, cumene, cuminic alcohol, β-phellandrene, α-terpineol, apigenin-7-*O*-glucoside, luteolin 7-*O*-glucoside, sesquiterpene lactone glucosides, and alkyl glycosides. Monoterpene hydrocarbons are another major component of the oil and sesquiterpenes are minor constituents. Cuminaldehyde and p-menthen-7al in combination with related aldehydes provide a characteristic aroma (review: Kaur and Sharma 2012).

Cuminum nigrum L., Apiaceae/Umbelliferae

Common Names: Kala zeera, zira siga, bitter cumin

The seeds of *Cuminum nigrum* are used in traditional medicine, among other things, to treat DM. Oral administration of various doses of *C. nigrum* seeds (1, 2, 3, and 4 g/kg) produced a significant hypoglycemic effect in normal as well as in diabetic rabbits. The water and methanol extracts also decreased the blood glucose level in normal and alloxan-induced diabetic rabbits. However, the total blood lipids were not influenced by the seeds in either normal or diabetic rabbits. Acute toxicity studies carried out on rabbits could not reveal any adverse or side effects of this folk medicine (Akhtar and Ali 1985).

In another study, oral administration of the flavonoid contents of seeds of *C. nigrum* caused a hypoglycemic effect at a dose range of 0.5–1.5 g/kg, in both normal and alloxan-induced diabetic rabbits. The hypoglycemic effect started 2 h after drug administration, reaching a maximum within 4–8 h, and the blood glucose levels returned close to normal within 24 h of drug administration. Glibenclamide (5 mg/kg) produced a hypoglycemic effect in normal rabbits, whereas it had no effect on the blood glucose levels of alloxan-induced diabetic rabbits. The alkaloids isolated from *C. nigrum* seeds, however, failed to exert any significant hypoglycemic effect in either normal or diabetic rabbits. A 7-day acute toxicity study in rabbits did not produce any apparent adverse effect at oral doses as high as 5 g/kg (Ahmad et al. 2000).

Curcuma amada Roxb., Zingiberaceae

Common Name: Mango ginger

Curcuma amada (commonly known as mango ginger) is used in traditional medicine. Crude water–ethanol (1:4) extracts of this plant rhizome (250–650 mg/kg) exhibited hypoglycemic activity in normal mice and antihyperglycemic activity in alloxan-induced diabetic mice. The treatment also improved oral glucose tolerance in normal and diabetic rats (Syiem et al. 2010).

Curcuma longa L., Zingiberaceae (Figure 2.43)

Common Names: Turmeric (Eng.), haridra, haldi, manjal, manja, etc. (Indian languages)

Description: *Curcuma longa* is a perennial rhizomatous herb with orange-yellow rhizomes; lanceolate or oblong-lanceolate leaves that are rounded or attenuate at the base and acuminate at the apex; white flowers appearing with leaves and central to the leaf tuft; and a labellum with a yellow band.

Distribution: *C. longa* is considered native to South Asia, particularly India, and is cultivated extensively in almost all states of India as well as in the warmer parts of the world, particularly in China and East Indies.

Traditional Medicinal Uses: *C. longa* is a tonic, blood purifier, antiperiodic, stomachic, appetizer, antispasmodic, antiseptic, antacid, carminative, and alternative and used in common cold, skin diseases, indolent ulcers, purulent ophthalmia, bacterial diseases, arthritis, bone fractures, cough, inflammation, whooping cough, abscesses, conjunctivitis, dysentery, gonorrhea, jaundice, hepatosis, parturition, pyuria, skin sores, wounds, scabies, swellings, burns, and vertigo (*The Wealth of India* 1950; Kirtikar and Basu 1975).

Pharmacology

Antidiabetes: The hypoglycemic and antidiabetic properties of the rhizome of this plant are known. It is recommended for the prevention and control of type 2 diabetes. Active principles such as curcumin, demethoxycurcumin, sesquiterpenoids, bisdemethoxycurcumin, and ar-turmerone, act via PPAR-γ ligand-binding activity (Arun and Nalini 2002; Kuroda et al. 2005; Nishiyama et al. 2005). Curcumin, the principal constituent of the rhizomes of *C. longa*, was found to inhibit PTP1B.

(a)   (b)

FIGURE 2.43   *Curcuma longa*. (a) Plant with flowers. (b) Rhizome.

The compound improved insulin and leptin sensitivity in the liver of rats; it prevented hypertriglyc-eridemia and hepatic steatosis in fructose-fed rats (Li et al. 2010a). The antidiabetic capacity of *C. longa* volatile oil in terms of its ability to inhibit α-glucosidase activity was evaluated. Turmeric volatile oils inhibited α-glucosidase enzymes more effectively than the reference standard drug acar-bose. Drying of rhizomes was found to enhance α-glucosidase and α-amylase inhibitory capacities of volatile oils. Ar-turmerone, the major volatile component in the rhizome, also showed potent α-glucosidase (IC_{50} 0.28 µg/mL) and α-amylase (IC_{50} 24.5 µg/mL) inhibition (Lekshmi et al. 2012).

Further, curcumin can prevent some diabetic complications primarily due to its antioxidant and anti-inflammatory properties. Curcumin (150 mg/kg, p.o.) prevented diabetic nephropathy in streptozotocin-induced diabetic rats by inhibiting the activation of Sphk1-S1P signaling pathway (Huang et al. 2013). Besides, curcumin (100 mg/kg, p.o.) prevented diabetic cardiomyopathy in streptozotocin-induced dia-betic rats (Soetikno et al. 2012).

Studies on Healthy Volunteers: A study was carried out on healthy volunteers to determine whether the ingestion of 6 g of *C. longa* in a single meal lowered postprandial plasma glucose and insulin levels in healthy subjects. The ingestion of 6 g *C. longa* had no significant effect on the glucose response. But the ingestion resulted in a significantly higher serum insulin response 30 and 60 min after the OGTT, com-pared with the reference OGTT. The change in insulin level (Δ insulin) was also significantly different from that seen after the reference meal 30 and 60 min postprandially. The results indicate that *C. longa* may have an effect on insulin secretion (Wickenberg et al. 2010).

Other Activities: The biological properties of curcumin have been reviewed (review: Labban 2014). The plant rhizomes have several pharmacological properties such as inhibition of cellular ROS generation, anti-inflammatory activity, hypolipidemic activity, antiviral activity, wound healing, hepatoprotection from toxic chemicals, protective property on the cardiovascular system, and other bioprotective proper-ties. Therapeutic benefits have been demonstrated for a variety of gastrointestinal disorders including dyspepsia, peptic ulcer, and *Helicobacter pylori* infection. It is also beneficial to irritable bowel syn-drome. Turmeric and curcumin isolated from it have potentiality to prevent carcinogenesis and sup-press certain types of cancer growth. Curcumin, ar-turmerone, etc., are known to induce apoptotic cell death. However it can prevent oxidative stress-mediated apoptosis due to its antioxidant property. Contraceptive effect of turmeric in male albino rat has been shown (Ashok and Meenakshi 2004; review: Labban 2014).

Toxicity: Due to its cytotoxic and apoptosis induction potential, a high concentration could be toxic.

Phytochemicals: Phytochemicals include curcumin, essential oil containing D-α-phellandrene, b-turmerone, D-sabinene, cineol, borneol, zingiberene, dimethoxy curcumin, bis-demethoxy curcumin, turmerones, *p*-tolylmethyl carbinol, campesterol, stigmasterol, β-sitosterol, cholesterol, monoenoic acid, dienoic acid, 4-hydroxycinnamoyl(feruloyl) methane, *bis*-(4-hydroxycinnamoyl)methane, and *bis*-demethoxy curcumin (*The Wealth of India* 1992).

Curcuma xanthorrhiza Roxb., Zingiberaceae

Curcuma xanthorrhiza is used in traditional medicine to treat DM. *C. xanthorrhiza* (ethanol extract) has an antidiabetic activity in alloxan-induced diabetic rats, probably by increasing insulin production, thereby justifying its traditional claim (Adnyana et al. 2013).

Cuscuta chinensis Lam., Convolvulaceae

Seeds of *Cuscuta chinensis* are used in China to treat DM. Oral administration of 300 and 600 mg/kg *C. chinensis* polysaccharide decreased fasting blood glucose level and increased weight of the body and immune organs in alloxan-induced diabetic mice. In another study, oral administration of 200 and 400 mg/kg of *C. chinensis* polysaccharide for 15 days to diabetic rats could reduce fasting blood glucose and glycosylated serum protein and increase body weight without any obvious effect on insulin levels. Thus, *C. chinensis* polysaccharides show potential effect on reducing blood sugar in type 2 diabetes, but the mechanism of this effect is unclear (review: Donnapee et al. 2014). It is used in prescriptions for diabetic complications because of the actions to invigorate the kidney and supplement essence (review: Chauhan et al. 2010).

Cuscuta reflexa Roxb., Cuscutaceae

Cuscuta reflexa is a parasitic vine used in folk medicine as remedy for prostate cancer, impotency, scabies, fevers, diarrhea, diabetes, and throat pains.

The hypoglycemic effects of the methanol and chloroform extracts of whole plants of *C. reflexa* were investigated in oral glucose tolerance tests in rats and mice. Both methanol and chloroform extracts of *C. reflexa* (whole plant) demonstrated significant oral hypoglycemic activity in glucose-loaded rats at doses of 50, 100, and 200 mg/kg. The chloroform extract was found to be better than the methanol extract in lowering blood glucose levels (Mollik et al. 2009).

C. reflexa is known to contain α-glucosidase inhibitory compounds and compounds such as flavonoids (kaempferol, quercetin), coumarins, and flavonoid glycosides. Earlier studies have shown that both kaempferol and quercetin could significantly improve insulin-stimulated glucose uptake in mature 3T3-L1 adipocytes. It was further reported that these two compounds act at multiple targets to ameliorate hyperglycemia. A phenolic compound (6,7-dimethoxy-2*H*-1-benzopyran-2-one) isolated from this plant inhibited α-glucosidase (Anis et al. 2002).

Cyamopsis tetragonoloba (L.) Taub., Fabaceae

Synonyms: *C. psoralioides* (Lam.) DC., *Dolichos fabaeformis* L'Her., *Dolichos psoraloides* Lam., *Lopinus trifoliolatus* Cav., *Psoralea tetragonoloba* L. (review: Sharma et al. 2011b) (Figure 2.44)

Common Names: Cluster bean, gowar plant (Eng.), bakuchi, guar, kothaveray, etc. (Indian languages)

Description: *Cyamopsis tetragonoloba* is a moderate-sized erect annual herb with a grooved stem, clothed more or less with appressed medifixed, grayish hairs. The leaves are trifoliate and stipulate. The leaflets are elliptic, acute, sharply dentate, clothed on both sides with appressed, medifixed hairs; the flowers are small and purplish and produced in axillary racemes. The fruit is a pod, which is thick, fleshy, subtetragonal, and slightly pubescent, bearing 5–6 seeds.

Distribution: Considered as native to India, *C. tetragonoloba* is not found in the wild. It is found throughout the country as a cultivar for its pods, which are used as a vegetable.

FIGURE 2.44 *Cyamopsis tetragonoloba* fruits (vegetable).

Traditional Medicinal Uses: The leaves are used in asthma and to cure night blindness, whereas the pods and seeds are used to cure inflammation, sprains, and arthritis; as an antioxidant, antibilious, and laxative; and in flatulence, polluting boiling, and biliousness. As per Ayurveda, the plant can be used as cooling, digestive, tonic, galactagogue, useful in constipation, dyspepsia, anorexia, agalactia, hyetalopia, etc. In folklore the decoction of the bean is used to control diabetes (Kirtikar and Basu 1975; review: Sharma et al. 2011).

Pharmacology

Antidiabetes: The water extract of the beans showed remarkable blood glucose-lowering effect in alloxan-induced diabetic rats, but marginal activity was seen in normal and glucose-loaded rats (Mukhtar et al. 2004a). The ethanol extract of the beans (250 mg/kg) significantly reduced the blood glucose levels in alloxan-induced diabetic rats within 3 h of oral administration. Repeated daily administration of the extract (250 mg/kg) for 10 days produced substantial reduction in blood glucose levels, whereas marginal activity was seen in normal and glucose-loaded rats (Mukhtar et al. 2006). The hypoglycemic property of the seeds has been shown in normal and diabetic guinea pigs also (Shrivastava et al. 1987). The effect of guar gum obtained from the endosperm of the seeds was studied for its anti-DM and antihyperlipidemic potential. The gum drastically reduced blood glucose levels and improved the lipid profile. Further, it promoted a general improvement in the condition of diabetic rats including body weight gain and improvement in the indexes of protein absorption and utilization (Frias and Sgarbieri 1998).

Recently, the antidiabetic activity of *C. tetragonoloba* beans was evaluated in HFD-fed streptozotocin-induced type 2 diabetic rats. A dose-dependent response of the oral treatment of the methanol extract (200 and 400 mg/kg) of the beans was assessed by measuring fasting blood glucose, changes in body weight, plasma insulin, homeostasis model assessment of insulin resistance, total cholesterol, TGs, oral glucose tolerance, intraperitoneal insulin tolerance, hepatic glycogen, and marker enzymes of carbohydrate metabolism in HFD-fed streptozotocin-induced type 2 diabetic rats. Histology and immunohistochemical analysis of pancreatic islets were also performed. High-performance liquid chromatography (HPLC) analysis of the methanol extract showed the presence of polyphenols such as gallic acid and caffeic acid at 2.46% (W/W) and 0.32% (W/W) concentrations, respectively. The extract reverted the altered biochemical parameters to near normal levels in diabetic rats. Furthermore, the extract showed the protective effect on the β-cells of pancreatic tissues in diabetic rats. These findings indicate that *C. tetragonoloba* beans have a therapeutic potential in HFD-fed streptozotocin-induced hyperglycemia (Gandhi et al. 2014). The bean is a vegetable that can be used in the management of type 2 diabetes. It is of interest to note that the plant contains compounds with anti-DM properties such as quercetin, gallic acid, genistein, ellagic acid, kaempferol, and chlorogenic acid (review: Sharma et al. 2011).

Other Activities: The bean and its guar gum are shown to have hypolipidemic and hypocholesterolemic properties (Srivastava et al. 1987; Frias and Sgarbieri 1998). Other reported activities include antiulcer, anticholinergic, anticoagulant, hemolytic, antimicrobial, antiasthma, and anti-inflammatory (review: Sharma et al. 2011).

Phytochemicals: The leaves and the pods contain fibers, galactomannans, ascorbic acid, and condensed tannins together with caffeic acid, gallic acid, gentisic acid, *p*-coumaric acid, astragalin, P-hydroxycinnamyl, and coniferyl alcohol. Its flavonoid content in seeds includes quercetin, daidzein, genistein, kaempferol, and its 3-arabinosides. Other polyphenol composition of the plant includes gallotannins, gallic acid and gallic acid derivatives, myricetin-7-glucoside-3-glycoside, chlorogenic acid, ellagic acid, 2,4,3-trihydroxy benzoic acid, texasin-7-*O*-glucoside, and *p*-coumaryl quinic acid. Its sterol includes campesterol, avenasterol, stigmasterol, sitosterol, and traces of Δ-7-avenasterol, stigmast-7-enol, brassicasterol, and cholesterol (review: Sharma et al. 2011).

Cyclocarya paliurus (Batal.) Ijinskaja, Cyclocaryaceae
Cyclocarya paliurus has been used as a traditional tonic, and its leaves have been processed as tea products. In ICR mice, *C. paliurus* showed a glucose-lowering effect. The blood glucose level was lower in the *C. paliurus* extract-treated group than in the control group after animals were given sucrose.

This difference was not observed following the administration of glucose. Further, the chronological change in the level of blood glucose in genetically hyperglycemic obese KK-Ay mice was lower when the *C. paliurus* extract was administered daily for 3 weeks. *C. paliurus* inhibited α-glucosidase activity *in vitro* (Kurihara et al. 2003). Phenolic compounds from the leaves of *C. paliurus* showed inhibitory activity against PTP1B. Quercetin-3-*O*-β-D-glucuronide isolated from the leaves was active against PTP1B. Further, a naphthoquinone derivative, cyclonoside A, from this plant inhibited the enzyme (review: Jiany et al. 2012).

Cydonia oblonga Miller, Rosaceae

Common Name: Quince

Cydonia oblonga leaves are used as a folk remedy for the treatment of diabetes and they are also consumed as food in Turkey. The ethanol extract of *C. oblonga* leaves (250 and 500 mg/kg, p.o., daily for 5 days) showed antidiabetic and antioxidant activities in normal and streptozotocin-induced diabetic rats (Aslam et al. 2010). Moreover, *C. oblonga* extracts induced significant alleviation on heart tissue TBARS levels and not GSH levels (Aslam et al. 2010).

Cymbopogon citratus (DC.) Stapf, Poaceae

Common Name: Lemongrass

In traditional medicine, lemongrass is used to treat various ailments. In Nigeria, this plant has been utilized for treating diabetes, obesity, and cardiovascular problems. The hypoglycemic and hypolipidemic effects of the fresh leaf (water extract) of *C. citratus* were shown in rats (Adeneye and Agbaje 2007). *C. citratus* (leaf sheath) essential oil and its components (geraniol and myrcene) showed good antihyperglycemic action. The whole essential oil, geraniol and myrcene, showed insulin secretagogue action. Among the three, geraniol demonstrated the most potent insulin secretagogue action in both *in vivo* and *in vitro* studies (Bharti et al. 2013). When compared to diabetic control rats, the essential oil-treated poloxamer-407-induced type 2 diabetic rats presented significant amelioration of glycemia, insulinemia, and lipid dysmetabolism, accompanied by increased GLP-1 content in the cecum and remarkable reduction of oxidative markers. Histopathological analysis of the pancreas showed increase in β-cell mass, islet number, and quality of insulitis. The phytoconstituents of this plant like myrcenol, linalool, α-elemol, and β-eudesmol showed significant interaction with PPAR-γ and DPP-IV, while only pimeloyl dihydrazide showed interaction with PTP-1B. Citral isolated from lemongrass exhibited a PPAR-γ agonist action. Thus, studies provided a pharmacological evidence of the essential oil as antidiabetic mediated by interaction of various phytoconstituents with multiple targets operating in DM (Adeneye and Agbaje 2007; Bharti et al. 2013; review: Wang et al. 2014). This is a promising plant for further studies.

Cymbopogon proximus Stapf, Gramineae

Common Name: Lemongrass

Cymbopogon proximus is a traditional medicinal herb with hypoglycemic activity in alloxan-induced diabetic rats (review: Haddad et al. 2005). The plant contains phytoconstituents such as flavonoids and phenolic compounds, which consist of luteolin, isoorientin 2′-*O*-rhamnoside, quercetin, kaempferol, and apigenin. This plant possesses various pharmacological properties including hypoglycemic activity (Shah et al. 2011).

Cynara cornigera L., Asteraceae

Common Name: Choke

Administration of the water extract of *Cynara cornigera* (1.5 g/kg, p.o.) to alloxan-induced diabetic rats resulted in a significant reduction in blood glucose levels. In addition, the treatment significantly recovered serum insulin levels and liver glycogen content and prevented the decrease in body weight. Further, the treatment with the extract significantly reversed the hyperlipidemic condition observed in untreated diabetic rats (Mohamed 2011b).

Cynara scolymus L., Asteraceae

Cynara scolymus (artichoke) is a vegetable with medicinal value. The leaves have long been used effectively for treating a variety of diseases including DM. Oral administration of the water extract of *C. scolymus* aerial parts (200 and 400 mg/kg) for 21 days to streptozotocin-induced diabetic rats resulted in a significant reduction in hyperglycemia and hyperlipidemia (total cholesterol, TG, LDL, and VLDL) compared to untreated diabetic rats. The treatment also reduced oxidative stress in diabetic rats (Heidarian and Soofiniya 2011).

In a clinical study on type 2 diabetic patients, supplementation with artichoke-incorporated wheat biscuit to diabetic patients for a period of 90 days resulted in a highly significant reduction in postprandial blood glucose levels compared to diabetic control patients. The treatment also reduced total cholesterol, LDL, VLDL, and TG levels. Thus, the results of the study recommended that the type 2 diabetic subjects should use the globe artichoke vegetable in their food preparation on a regular basis (Nazni et al. 2006). A randomized double-blind placebo-controlled clinical trial concluded that a fiber-free artichoke extract may be a safe antihypercholesterolemic agent but does not improve glycemic control in hypercholesterolemic type 2 diabetic patients, suggesting the involvement of fibers in the antihyperglycemic effect of artichoke (Falla et al. 2012).

Cynodon dactylon Pers, Poaceae

When various extracts (petroleum ether, chloroform, acetone, ethanol, water, and crude hot water extracts) of the whole plant of *Cynodon dactylon* and the two fractions of the water extract were tested for antihyperglycemic activity in glucose-overloaded hyperglycemic rats and in alloxan-induced diabetic rats (200 and 400 mg/kg, p.o.), the water extract and the nonpolysaccharide fraction of the water extract were found to exhibit significant antihyperglycemic activity. Treatment of diabetic rats with the water extract and nonpolysaccharide fraction of the plant decreased the elevated biochemical parameters, glucose, urea, creatinine, serum cholesterol, serum TG, HDL, LDL, hemoglobin, and glycosylated hemoglobin, significantly. Comparatively, the nonpolysaccharide fraction of the water extract was found to be more effective than the water extract (Jarald et al. 2008a). In another study, the antidiabetic activity of the ethanol extract of *C. dactylon* root stalks in streptozotocin-induced diabetic rats was reported to be comparable to that of glibenclamide. Phytochemical investigation showed the presence of alkaloids, carbohydrates, phytosterols, glycosides, saponins, phenolic compounds, flavonoids, and triterpenoids in the extract (Sanjeeva et al. 2011). In a recent study, the methanol extract of *C. dactylon*, when administered orally for 21 days to alloxan-induced diabetic rats, markedly reduced the levels of blood glucose, cholesterol, and TGs (Ramya et al. 2014).

Cyperus rotundus L., Cyperaceae

Common Names: Nut grass (Eng.), mastaka, musta, mutha, korai, arencole, etc. (Indian languages)

Description: *Cyperus rotundus* is an erect glabrous perennial herb with subsolitary stem. Bearing hard ovoid tunicate black fragrant tubers, the flowers are reddish brown in simple or compound umbels with 2–8 rays that often reach up to 3 in. long, and the nuts broadly ovoid and trigonous grayish-black in color.

Distribution: It is distributed throughout India up to an elevation of 6000 ft.

Traditional Medicinal Uses: *C. rotundus* is a diaphoretic, alternative, emmenagogue, astringent, demulcent, carminative, diuretic, emollient, lactagogue, vermifuge, and stomachic and used in alleviating disorders of the stomach, edema, common cold, dyspepsia, leukorrhea, insect bite, ulcers, kidney stone, intermittent fever, vomiting, nausea, traumatic injuries, dysmenorrhea, gravel, leukorrhea, sores, and wounds (*The Wealth of India* 1992).

Antidiabetes: The *C. rotundus* root is used in the treatment of diabetes in traditional systems of medicine. Herbal mixtures containing *C. rotundus* are reported to have anti-DM activity in juvenile diabetes. The hydroethanol extract of *C. rotundus* exhibited an antidiabetic activity in alloxan-induced diabetic rats (Raut and Gaikwad 2006). In a model of fructose-mediated protein glycoxidation, *C. rotundus* suppressed AGE formation and protein oxidation (Ardestani and Yazdanparast 2007). Oral daily administration of (500 mg/kg) ethanol extract of *C. rotundus* (once a day for 7 consecutive days) significantly lowered the blood glucose levels in alloxan-induced diabetic rats. The extract also showed *in vitro* antioxidant activity (Raut and Gaikwad 2006). In a follow-up study, the antidiabetic activity of the fractions of the hydroethanol extract of *C. rotundus* was evaluated in alloxan-induced diabetic rats. Significant antidiabetic activity was found at a dose of 300 mg/kg in the acetone fraction and residue left after successive fractionation. The activities of the active extracts were comparable to those of metformin (450 mg/kg, p.o.). The results suggested that the antidiabetic activity was due to the presence of polyphenols and flavonoids (Raut 2012). Compounds with anti-DM properties such as β-sitosterol and ferulic acid have been reported from this plant.

Other Activities: The plant roots contain antimalarial substances. It is useful in the treatment of intestinal metaplasia.

Phytochemicals: The essential oil contains L-α-pinene, cineol, sesguiterpenes, secondary dicyclic alcohol, isocyperol, α-cyperone, cyperene, cyperol, isocyperol, patchoulenone, copadiene, rotundone, epoxy-guaiene, a-rotanol, b-rotunol, isokobusone, 2-norsesquiterpenoid-kobusone, β-selinene, sesquiterpene alkaloids (Jeong et al. 2000), (–)-isorotundene, (–)-cypera-2,4(15)-diene, (–)-norrotundene and ketone (+)-cyperadione (Sonwa and Konig 2001), patchoulenone, 10,12-peroxycalamenene, and 4,7-dimethyl-1-tetralone. (+)Copadiene,2-carboxy arabinitol, 1,8-cineole, α-copaene, α-humulene, α-rotundol, α-rotunol, α-selinene, β-caryophyllene, β-cyperone, β-elemene, β-pinene, β-rotundol, β-rotunol, β-santalene, β-selinene, camphene, caryophyllene and its derivatives, cineol, copaene, cyperenone, cyperolone, cyperotundone, ferulic acid, flavanoids, γ-cymene, humulene, isokobusone, isocyperol, kobusone, limonene, luteolin, mustakone, myristic acid, *p*-cymol, *p*-coumaric acids, patchouylenone, pinene, rotundene, rotunenol, rotundone, sitosterol, and sugenol and its derivatives (review: Ajikumaran and Subramoniam 2005).

Daemonorops sp. (*D. verticillaris* (Griff.) Martius; *D. macrophylla* Bece; *D. draco*, etc.), Palmae
Some of the species of *Daemonorops* yield a resin known as dragon's blood resin (it is also known as sanguis draconis). This resin is used in ethnomedicine, among other things, to treat DM. The resin from *D. draco* attenuated high-glucose-induced oxidative stress and endothelial dysfunction in human umbilical vein endothelial cells (Chang et al. 2014). Hyperglycemia and hypoinsulinemia found in streptozotocin-induced diabetic mice were completely prevented when mice were orally administered with the ethanol extract of sanguis draconis (dragon's blood resin) for 3 weeks. In addition the extract can inhibit streptozotocin-induced iNOS expression, pancreatic injury, and LPO (Hu et al. 2011a). Further in cultured RIN-m5F cells, the extract (below 100 µg/mL) prevented streptozotocin (5 mM)-induced cell death and apoptosis (Hu et al. 2011). Detailed studies using the resin from different species are needed to select the best species and determine its utility as a therapeutic agent for diabetes.

Dalbergia sissoo Roxb. ex DC., Fabaceae
The antidiabetic activity of the ethanol extract of *Dalbergia sissoo* leaves in alloxan-induced diabetic rats was studied. The ethanol extract of the *D. sissoo* leaves was found to be 12% more effective in reducing blood glucose levels compared to standard glibenclamide (Niranjan et al. 2010). In another study, the ethanol extract of the bark of this plant showed reduction in blood glucose and increase in liver glycogen content in alloxan-induced diabetic rats (Pund et al. 2012).

Daniella oliveri Bull. Misc., Caesalpiniaceae
Daniella oliveri is a traditional Nigerian anti-DM plant. The water extracts of *D. oliveri* roots (250 mg/kg) lowered blood sugar levels in alloxan-induced diabetic rats (Iwueke and Nwodo 2008). The extract (250 mg/kg) caused a marked lowering of blood sugar level in alloxan-induced diabetic rats within 6 h, from (mean, mmol/L) 16.7 to 6.6 after 3 weeks. The same dose did not show blood sugar-lowering effect in normoglycemic rats. The extract significantly restored both hepatic glycogen content and the activities of hexokinase, glucokinase, and phosphofructokinase in diabetic rats (Iwueke and Nwodo 2008). Besides, the water extract of *D. oliveri* showed a hypoglycemic effect in alloxan-induced diabetic mice (Manosroi et al. 2011).

Daphniphyllum macropodum Miq., Daphniphyllaceae

Daphniphyllum macropodum is a small tree. 5,7-dihydroxychromone isolated from this plant fruit showed promising PPARγ agonist activity. The compound potently induced the differentiation of mouse 3T3-L1 preadipocytes. Further, it increased PPARγ and liver X receptor-α mRNA expression levels. Furthermore, the comound rich fraction from *D. macropodum* reduced serum glucose, total cholesterol and triglyceride levels in streptozotocin-high-fat diet-induced type 2 DM mice (Koo et al. 2014).

Dasylirion spp., Nolinaceae

Diet supplemented with fructons (10%) from *Dasylirion* spp. for 5 weeks influenced glucose and lipid metabolism in mice (Urias-Silvas et al. 2007). The body weight gain and food intake in mice fed with fructon-containing diet were lower than those fed with standard diet. Serum glucose and cholesterol levels also decreased in the fructon-containing diet–fed mice. The supplementation with fructons induced a higher concentration of GLP-1 and its precursor mRNA in the different colonic segments, suggesting that permentable fructons are able to promote the production of incretin peptides in the lower parts of the gut (Urias-Silvas et al. 2007). Different species of *Dasylirion* have to be studied to select the best species for further studies.

Datura metel L., Solanaceae

Datura metel is commonly known as downy datura or thorn apple (Eng.). It is an erect herbaceous plant, clothed with grayish-green tomentum. It is supposed to have originated from South America. In traditional medicine the plant is used to treat numerous ailments (Kirtikar and Basu 1975). The seed powder of this plant (25, 50, and 75 mg/kg, p.o.) showed promising dose-dependent hypoglycemic activity in normal and alloxan-induced diabetic rats (Krishnamurthy et al. 2004). The chloroform fraction of *D. metel* is reported to be endowed with antifungal activity against pathogenic *Aspergillus*. Acute intoxication with the dried flowers and seeds has been reported (review: Ajikumaran and Subramoniam 2005).

Daucus carota L., Apiaceae

Common Name: Carrot

In a recent study, falcarinol (polyacetylene) isolated from carrot stimulated glucose uptake in normal and insulin-resistant primary porcine myotubes. This effect was attenuated in the presence of indinavir (GLUT4 inhibitor) and wortmannin (PI3K and MAPK inhibitor), indicating a dependence on GLUT4 activity as well as PI3K and/or p38MAPK activity. Further, falcarinol-stimulated glucose uptake was independent of AMPK activity. Besides, falcarinol enhanced phosphorylation of TBC1D1 suggesting that this compound enhanced translocation of GLUT4-containing vesicles and thereby glucose uptake via a TBC1D1-dependent mechanism (Bhattacharya et al. 2014).

Dietary carrot (10% and 20% in the diet for 6 weeks) exhibited antihypercholesterolemic and antioxidant activities in hypercholesterolemic rats. The weight of organs and body weight gain significantly decreased in the carrot-fed rats. In addition, dietary carrot powder improved the blood picture and reduced the blood glucose level in hypercholesterolemic rats (Afify et al. 2013).

Dendrobium huoshanense C.Z.Tang & S.J.Cheng, Orchidaceae

Dendrobium huoshanense is a traditional medicinal orchid. In a recent study, *D. nobile* polysaccharides exhibited hypoglycemic activity (Pan et al. 2014). In an earlier study, a polysaccharide with a molecular weight of approximately 150 kDa, isolated from the stem of *D. chrysotoxum*, exhibited marked reduction in the blood glucose level in alloxan-induced diabetic mice. Further, the polysaccharide showed a strong *in vitro* antioxidant activity (Zhao et al. 2007). Future studies may clear the situation.

Dendrobium moniliforme (L.) Sw., Orchidaceae

Polysaccharides from *Dendrobium moniliforme* (100 and 200 mg/kg) showed hypoglycemic effects in both adrenalin and alloxan-induced diabetic mice (Yunlong et al. 2003). A phenanthraquinone-type metabolite was isolated from *D. moniliforme*, which inhibited PTP1B activity with an IC_{50} value of 38.0 μmol/L (Jiany et al. 2012).

Dendrobium nobile Lindl, Orchidaceae

D. nobile is a traditional medicinal orchid. Syringic acid extracted from *D. nobile* provided protection from D-gal-induced damage to rat lens by consistently maintaining lens transparency and delaying lens turbidity development. Syringic acid prevented diabetic cataract in rat lenses by inhibiting aldose reductase activity and gene expression, which has potential to be developed into a novel drug for the therapeutic management of diabetic cataract (Wei et al. 2012). In a recent study also, *D. nobile* polysaccharides exhibited hypoglycemic activity in alloxan-induced diabetic rats (Pan et al. 2014).

Derris scandens Benth, Fabaceae

Free radical scavenging and α-glucosidase inhibitory activities were detected in some of the chemical isolates of *Derris scandens* (Rao et al. 2007).

Desmodium gangeticum (L.) DC, Fabaceae

The aerial parts of *Desmodium gangeticum* showed a reduction in blood glucose levels in rats and an increase in insulin secretion from MIN6 cells grown as monolayers and pseudo-islets (Govindarajan et al. 2007).

Detarium microcarpum Guill. & Perr., Fabaceae

Detarium microcarpum is a traditional anti-DM plant in Nigeria. Various fractions of the methanol extract of *D. microcarpum* stem bark were subjected to antidiabetic study in alloxan-induced diabetic rats at two dose levels, 200 and 400 mg/kg. The results indicated that intraperitoneal injection of various fractions reversed the effect of alloxan in rats by different degrees. The methanol fraction of the methanol extract showed maximum activity and the activity of this fraction was comparable to that of glibenclamide (5 mg/kg). Treatment of diabetic rats with the methanol fraction restored the biochemical parameters (glucose, urea, creatinine, serum cholesterol, serum TG, HDL, LDL, hemoglobin, and glycosylated hemoglobin) to the normal level in alloxan-induced diabetic rats. Histological studies showed a restorative effect on the islet cells by the fraction. Acute toxicity test showed that the fractions were safe at doses up to 5 g/kg (Odoh et al. 2013). The water extract of this plant also showed hypoglycemic activity in alloxan-induced diabetic mice (Manosroi et al. 2011). In a recent study, oral administration of the methanol extract of *D. microcarpum* leaf (200 and 400 mg/kg) resulted in a potent dose-dependent reduction of blood sugar level after 4 h of administration. Preliminary phytochemical screening of the extract revealed the presence of terpenoids, flavonoids, saponins, glycosides, proteins, and resins (Uchenna et al. 2014).

Dichrostachys glomerata Chiov., Fabaceae/Mimosoideae

In ethnomedicine *Dichrostachys glomerata* is used, among other things, to control obesity and diabetes. In a clinical trial, the effects of a standardized water extract of *D. glomerata* on certain anthropometric, biochemical (including proinflammatory and prothrombotic states), and hemodynamic parameters were evaluated in obese patients with metabolic syndrome. Various parameters were measured at baseline and after 4 and 8 weeks of treatment. At the end of the study, the *D. glomerata*-treated group showed significant differences in the parameters studied compared to baseline values. Changes in BMI and waist circumference were accompanied by changes in biochemical parameters with the exception of adiponectin levels that were not correlated to waist circumference and PAI-1values. The results confirm the hypothesis that the water extract has anti-inflammatory properties and is effective in reducing cardiovascular disease risk factors associated with the metabolic syndrome in obese human subjects (Kuate et al. 2013).

Dillenia indica L., Dilleniaceae

The leaf extracts of this plant reduced blood glucose levels and increased insulin levels in alloxan-induced diabetic rats. Further, the extract showed hypolipidemic activity (Kumar et al. 2011c). From the leaves of *Dillenia indica*, betulinic acid, n-heptacosan-7-one, n-nonatriacontan-18-one, quercetin, β-sitosterol, stigmasterol, and stigmasteryl palmitate were isolated. Among these isolates, β-sitosterol, stigmasterol, and stigmasteryl palmitate were evaluated for their antidiabetic activity in streptozotocin–nicotinamide-induced diabetic mice, and the other four compounds were evaluated for *in vitro* enzyme inhibition. These four compounds (50 µg/kg) showed α-amylase and α-glucosidase inhibition. β-Sitosterol, stigmasterol, and stigmasteryl palmitate showed antidiabetic activity in streptozotocin–nicotinamide-induced diabetic mice at the dose of 10 mg/kg. Quercetin, β-sitosterol, and stigmasterol are known anti-DM compounds. Thus, *D. indica* might be a potential source of antidiabetic agents (Kumar et al. 2013c).

Dioecrescis erythroclada (Kurz) Tirveng., Rubiaceae
Dioecrescis erythroclada (extract) occurring in Thai/Lanna showed hypoglycemic activity in experimental animals (Manosroi et al. 2011).

Dioscorea alata L., Dioscoreaceae
The tuber (ethanol extract) *Dioscorea alata* reduced blood glucose levels, serum lipids, and creatine levels to near normal in alloxan-induced diabetic rats (Maithili et al. 2011). The hot water extract of the edible portion of *D. alata* tuber (100, 200, and 300 mg/kg daily for 21 days) reduced the food intake, fasting blood glucose level, and body weight of normal male rats. These observations suggest that this tuber could serve as a therapeutic diet in the management of DM (Helen et al. 2013).

Dioscorea bulbifera L., Dioscoreaceae

Synonyms: *Dioscorea anthropophagum* Chev., *Dioscorea hoffa* Cordemoy, *Dioscorea sativa* Thunb., *Dioscorea sylvestris* Will., *Helmia bulbifera* Kunth., (very related *Dioscorea* spp. include *Dioscorea dumetorum*, *Dioscorea japonica*, and *Dioscorea batatas*) (Figure 2.45)

Common or Local Names: Air potato, potato yam, air yam, bitter yam (Eng.); varahi, zimikand, mekachil (Indian languages)

Description: *Dioscorea bulbifera* (air potato) is a yam species. It is a perennial vine with broad leaves and two types of storage organs. The plant forms bulbils in the leaf axils of the twining stems and tubers beneath the ground. (These tubers are like small, oblong potatoes and they are edible and cultivated as a food crop, especially in West Africa. The tubers often have a bitter taste, which can be removed by boiling. They can then be prepared in the same way as other yams, potatoes, and sweet potatoes. The air potato is one of the most widely consumed yam species.) Leaves have long petioles and alternate; flowers are small, male and female flowers arising from leaf axils on separate plants. The fruit is a capsule with partially winged seeds (Okan and Ofeni 2013).

Distribution: Air potato plant is native to tropical Asia and Africa. In some places, such as Florida, USA, it is an invasive plant species because of its quick-growing, large-leafed vine that spreads tenaciously and shades out plants growing beneath it. (The bulbils on the vines sprout and become new vines,

(a) (b)

FIGURE 2.45 *Dioscorea bulbifera*. (a) Vine with leaves. (b) Tubers.

twisting around each other to form a thick mat. If the plant is cut to the ground, the tubers can survive for extended periods and produce shoots subsequently.)

Traditional Medicinal Uses: Air potato has been used as a folk medicine to treat diarrhea, dysentery, conjunctivitis, etc. (Duke and Judith 1993). It is a medicinal plant commonly used in Cameroonian traditional medicine to treat pain and inflammation. *D. bulbifera*, the air potato, has been used in the Chinese system of medicine to treat diseases of the lungs, kidneys, and spleen and many types of diarrhea. Commonly known as yams, these plants have been traditionally used to lower the glycemic index, thus providing a more sustained form of energy and better protection against obesity and diabetes. This herb was also found to have a beneficial effect in improving digestion and metabolism. In Asia, this herb has been highly recommended for treating diabetes. It has been traditionally used to lower the glycemic index, providing a more sustained form of energy and better protection against obesity and diabetes. The leaf of aerial yam is used as a poultice for pimples and in bath water to soothe skin irritations and stings. In Northern Bangladesh, air yams are used to treat leprosy and tumor (Okan and Ofeni 2013).

Pharmacology

Antidiabetes: The water extract of *D. bulbifera* tubers was investigated for its antihyperglycemic effect in glucose-primed and streptozotocin-treated rats and for its antidyslipidemic effect in HFD-fed C57BL/6J mice. The water extract (250, 500, and 1000 mg/kg) was administered for 3 weeks to streptozotocin-treated rats and for 4 weeks to HFD-fed C57BL/6J mice. The extract treatment to the streptozotocin-induced diabetic rats resulted in marked reduction in blood glucose level and increase in body weight. In the extract-treated HFD-fed mice, serum glucose and lipid levels were reversed toward normal (Ahmed et al. 2009). *D. bulbifera* extract administration resulted in activation of β-cells and granulation returned to normal, showing an insulinogenic effect. Another mechanism of action of the extract for its antihyperglycemic effect might be due to the stimulation of insulin secretion from the remnant β-cells and/or regenerated β-cells (Ahmed et al. 2009). Antidiabetic activity of the ethanol extract of the tuber of this plant (380–1140 mg/kg) in alloxan-induced diabetic rats has also been reported. The extract showed maximum acute glucose-lowering effect 5 h after intra peritoneal administration (Okon and Ofeni 2013).

   D. bulbifera bulb was tested for its efficiency to inhibit α-amylase and α-glucosidase. The bulb extracts showed excellent inhibitory properties against crude murine pancreatic, small intestinal, and liver glucosidase enzymes (Ghosh et al. 2012). Among petroleum ether, ethyl acetate, methanol, and 70% ethanol (v/v) extracts of bulbs, the ethyl acetate extract showed the highest inhibition up to 72% and 83% against α-amylase and α-glucosidase, respectively. Diosgenin, the active principle, was isolated from the ethyl acetate extract. Kinetic studies confirmed the uncompetitive mode of binding of diosgenin to α-amylase (Ghosh et al. 2014b). In another study also, the potent α-amylase inhibitory activity of the bulb of *D. bulbifera* was reported (Chopade et al. 2012).

Other Activities: The plant has anticancer activity (Yu et al. 2004). The methanol extract of the bulbs of this plant showed antinociceptive activities (Nguelefack et al. 2010).

Toxicity: Hepatotoxicity of *D. bulbifera* has been reported. Studies on mice indicate that the ethyl acetate fraction of the ethanol extract contains the main hepatotoxic toxic ingredients of the *D. bulbifera* rhizome, and the mechanism of hepatotoxicity may be due to liver oxidative stress (Wang et al. 2010d). Uncultivated forms, such as those growing wild, can be poisonous. These varieties contain the steroid diosgenin. There have been claims that even the wild forms are rendered edible after drying and boiling.

Chemical Constituents: The plant tubers contain ennogenin glycoside, furanoid norditerpene, norditerpene glucosides, clerodane diterpenoids such as bafoudiosbulbins F and G, etc. (Teponno et al. 2008).

Dioscorea dumetorum (Kunth) Pax, Dioscoreaceae
Dioscorea dumetorum tuber is used in traditional medicine in Nigeria for the management of diabetes. From the tubers a compound was isolated, which was named dioscoretine. This compound isolated from

the water fraction of the methanol extract (20 mg/kg, i.p.) exhibited hypoglycemic activity in normal and alloxan-induced diabetic rabbits (Iwu et al. 1990a, b). The fasting blood glucose in normoglycemic rabbits was reduced from 5.9 to 3.2 mmol/L after 4 h by the compound (20 mg/kg). In alloxan-induced diabetic rabbits, the blood glucose was lowered from 28.9 to 15.6 mmol/L at the same time interval. The water fraction of the methanol extract produced comparable effects at 100 mg/kg. In contrast, the chloroform fraction of the same extract raised the fasting blood sugar of normal rabbits to 196 mg/100 mL after 6 h. Acute toxicity studies gave LD_{50} values of 1.4 g/kg for the aqueous fraction and 0.58 g/kg for dioscoretine when tested on mice (Iwu et al. 1990b). The ethanol extract of the tuber may contain compounds with both pro- and antidiabetic activities.

Dioscorea japonica Thunb., Dioscoreaceae
The antihyperglycemic activity of the ethanol extract of the leaves of *Dioscorea japonica* was evaluated in streptozotocin-induced diabetic rats. Administration of the ethanol extract (350 mg/kg, p.o., daily) to diabetic rats for 14 days resulted in 73.3% fall in fasting blood glucose levels and no sugar was observed in fasting urine. Further, the treatment brought about the fall in the level of total cholesterol by 49.3% with an increase of 30.3% in HDL and decrease of 71.9% and 28.7% in LDL and TG levels, respectively, in treated diabetic rats (Budhwani et al. 2012). In another study, glycans with hypoglycemic activity were isolated from this plant. Aqueous methanol extracts of the Oriental crude drug, sanyaku (*D. japonica* and *D. batatas* rhizophors) notably lowered blood glucose concentration in mice. Activity-guided fractionation of the extract from *D. japonica* afforded six glycans, dioscorans A, B, C, D, E, and F, which exhibited remarkable hypoglycemic effects in normal and alloxan-induced hyperglycemic mice (Hikino et al. 1986a).

Dioscorea nipponica Makino, Dioscoreaceae
Diosgenin (a phytosterol) isolated from *Dioscorea nipponica* rhizome ameliorated diabetic neuropathy in alloxan-induced mice and streptozotocin-induced diabetic rats by inducing the nerve growth factor (Kang et al. 2011). A mixture of extracts of the rhizomes of *D. nipponica* and *D. japonica* Thunb (DA-9801) showed neurotrophic activity under *in vitro* conditions in PC12 cells and dorsal root ganglion neurons (Kim et al. 2011c).

Dioscorea opposita Thunb., Dioscoreaceae
Dioscorea opposita extract reduced the blood insulin and glucose levels in dexamethasone-induced diabetic rats. *In vitro*, the extract significantly enhanced insulin-stimulated glucose uptake in 3T3-L1 adipocytes. Moreover, the extract increased the mRNA expression of GLUT4 glucose transporter in 3T3-L1 adipocytes (Gao et al. 2007).

Diospyros lotus L., Ebenaceae
Common Name: Date plum
The fruits of this plant are used traditionally to treat diabetes in Iran and elsewhere. The water extract of *Diospyros lotus* fruit (500 and 1000 mg/kg daily for 16 days) showed reduction in blood glucose levels and improvement in body weight in streptozotocin-induced diabetic rats (Sharma and Arya 2011).

Diospyros melanoxylon Roxb., Ebenaceae
Synonym: *Diospyros dubia* Wall. ex A.DC
The ethanol extract of *Diospyros melanoxylon* bark powder was tested for its efficacy in alloxan-induced diabetic rats. The extract showed significant antihyperglycemic activity as compared to standard drug. Further, the extract reversed the diabetes-induced hyperlipidemia. Histopathological studies of the pancreas revealed its significant effects on β-cell count. Thus, the ethanol extract could serve as a good adjuvant to other oral hypoglycemic agents and seems to be promising for the development of phytomedicines for DM (Gupta et al. 2009b). *D. melanoxylon* leaf extracts (ethanol and water, 200 mg/kg) reduced the fasting serum glucose of normal rats and alloxan-induced diabetic rats, which was comparable with the reduction caused by glibenclamide. All the alloxan-induced diabetic rats showed significant elevation

of serum cholesterol, TG, urea, and creatinine levels compared to the normal control during the 21 days of the study period. These changes were attenuated by the extract treatment (Vangapelli et al. 2014).

Diospyros peregrine (Gaertn) Garke, Ebenaceae

The fruit extract of *Diospyros peregrine* exhibited dose-dependent hypoglycemic, hypolipidemic, and antioxidant activities in type 2 diabetic animals (Dewanjee et al. 2009a). In type 1 diabetic rats, the methanol extract of the fruit showed antidiabetes and antioxidant activities (Dewanjee et al. 2009b).

Dodecadenia grandiflora Nees, Lauraceae

Dodecadenia grandiflora is used in traditional medicine to treat diabetes. Bioactivity-guided separation of an antihyperglycemic extract from the leaves of *D. grandiflora* afforded two phenylpropanoyl esters of catechol glycosides and two lignane bis(catechol glycoside)esters. These compounds showed significant antihyperglycemic activity in streptozotocin-induced diabetic rats, which was comparable to the standard drug metformin (Kumar et al. 2009b). In another study, phytochemical investigation of *D. grandiflora* leaves led to the isolation and identification of three new phenolic glycosides, along with nine known compounds. Two new glycosides and two already known compounds isolated from the leaves exhibited significant glucose-6-phosphatase inhibitory activity (63%–85%; IC_{50} values 50–88 μM). On the basis of biological results, a structure–activity relationship has been discussed (Kumar et al. 2010c).

Dodonaea viscosa (L.) Jacq., Sapindaceae

Dodonaea viscosa is a traditional medicinal plant. *D. viscosa* leaves exhibited antidiabetic and antioxidant activities in rats. The methanol extract of *D. viscosa* leaves (200 and 400 mg/kg, p.o.) produced significant hypoglycemic effect in normal rats after 6 h of administration. The extract also showed improvement in glucose tolerance. Oral administration of the methanol extract to streptozotocin-induced diabetic rats (200 and 400 mg/kg daily for 28 days) resulted in reduction in the levels of blood glucose, HbA1c, and total cholesterol and increase in liver glycogen content. The treatment also reduced oxidative stress (Veerapur et al. 2010a). The anti-DM and antioxidant effects of the methanol extract in sytreptozotocin-induced diabetic rats were confirmed by another report (Meenu et al. 2011). In another study, the aqueous ethanol and butanol extracts of *D. viscosa* improved glucose tolerance in normal rats. In alloxan-induced diabetic rats, the extracts (250 mg/kg) exhibited the maximum reduction in blood glucose levels at 3 h after administration. In chronic treatment (10 days), both extracts showed significant antidiabetic activity in alloxan-induced diabetic rats comparable with that of glibenclamide (Muthukumran et al. 2011). Administration of the water extract and polar fraction of the ethanol extract of *D. viscosa* aerial parts daily for 4 weeks to HFD and streptozotocin-induced type 2 diabetic rats resulted in a dose-dependent reduction in blood glucose levels, serum insulin levels, improvement in glucose tolerance, improvement in lipid profiles, and marked reduction in oxidative stress (Veerapur et al. 2010b). In *in vitro* bioassays, the extract and fraction showed inhibition of PTP1B, and at a concentration of 10 μg/mL, the extract and fraction showed 60% and 54.2% binding to PPAR-γ, respectively. Both extract and fraction exhibited stimulation of glucose uptake by skeletal muscles (Veerapur et al. 2010b). In fructose-fed insulin-resistant rats, *D. viscosa* also showed antidiabetic activity (Veerapur et al. 2010c).

Dracaena cochinchinensis (Lour) S.C. Chen, Dracaenaceae

Dracaena cochinchinensis is a Chinese traditional medicinal herb used to treat diabetic disorders. Sanguis draconis is a red resin obtained from the wood of this plant. Several studies have shown its antidiabetic effects in experimental animals. In insulin-resistant type 2 diabetic rats, sanguis draconis administration (300 and 500 mg/kg once a day for 14 days, p.o.) resulted in an increase in the response to exogenous short-acting porcine insulin (Hou et al. 2005). In another study, anti-DM effects of total flavonoids of sanguis draconis were evaluated in streptozotocin-induced diabetic rats. The flavonoids not only exhibited marked hypoglycemic activity but also alleviated dyslipidemia, tissue steatosis, and oxidative stress associated with type 2 DM. Moreover, pancreatic islet protecting effects of the flavonoids could be observed. Further investigations revealed a potential anti-inflammatory action of the flavonoids as judged from the decrease in the levels of IL-6, TNF-α, and C-reactive protein in the flavonoid-treated diabetic animals (Chen et al. 2013).

Dregea volubilis (L.f.) Benth. ex Hook.f., Apocynaceae

Dregea volubilis leaf is used in ethnomedicine to treat diabetes. The ethanol extract of *D. volubilis* leaf (200 mg/kg, p.o.) was found to lower the fasting blood glucose level in streptozotocin-induced diabetic rats. An active fraction was separated from the ethanol extract. This fraction (100 mg/kg) reduced blood glucose levels in diabetic rats. Additionally, it caused reduction in cholesterol and TG levels and improvement in HDL level in diabetic rats (Natarajan and Arul Gnana Dhas 2013).

Ducrosia anethifolia Boiss., Apiaceae

Ducrosia anethifolia is used as a flavoring additive. The antidiabetic property of the plant was studied recently (Shalaby et al. 2014). *D. anethifolia* extract showed hypoglycemic, hypolipidemic, and antioxidant activities; further, it improved kidney function. Some furanocoumarins isolated from the extract inhibited activities of α-amylase, α-glucosidase, and β-galactosidase (Shalaby et al. 2014).

Echinacea purpurea (L.) Moench, Asteraceae

Common Name: Cone flower

In traditional medicine, *Echinacea purpurea* is used as a remedy for the treatment and prevention of upper respiratory tract infections and the common cold. *E. purpurea* has been shown to have antidiabetic activities; for example, it activated PPAR-γ and increased insulin-stimulated glucose uptake. When adipocyte differentiation was induced with insulin plus 3-isobutyl-1-methylxanthine and dexamethasone, the accumulation of lipid droplets and the cellular TG content were significantly increased by the ethanol extract of *E. purpurea*. The expressions of PPAR-γ and C/EBPα in adipocytes treated with the extract were gradually increased as compared with control cells (Shin et al. 2014). Alkamides from the n-hexane extract of the flowers activated PPAR-γ (Wang et al. 2014). Two novel isomeric dodeca-2*E*,4*E*,8*Z*,10*E*/*Z*-tetraenoic acid 2-methylbutylamides, isolated from the active fraction of *E. purpurea*, were found to activate PPAR-γ to increase basal and insulin-dependent glucose uptake in adipocytes in a dose-dependent manner and to exhibit characteristics of a PPAR-γ partial agonist (Kotowska et al. 2014).

Eclipta alba (L.) Hassk., Asteraceae

Synonym: *Eclipta prostrata* L.

Eclipta alba, commonly known as false daisy, is a traditional medicinal herb used to treat various ailments including liver diseases. In alloxan-induced diabetic rats, oral administration of the leaf suspension of *E. alba* (2 and 4 g/kg) for 60 days resulted in significant reduction in blood glucose (from 20.7 to 6.5 mmol/L, mean) and glycosylated hemoglobin, a decrease in the activities of glucose-6-phosphatase and fructose 1,6-bisphosphatase, and an increase in the activity of liver hexokinase (Ananthi et al. 2003). In another study, the antidiabetic effect of *E. alba* ethanol extract has been investigated for possible beneficial effects against hyperglycemia and diabetic nephropathy in streptozotocin-induced diabetic rats. Single dose treatment of the extract (250 mg/kg) was found to significantly lower blood glucose level after 5 h of oral administration. Treatment of streptozotocin-induced diabetic rats for 10 weeks with the aforementioned dose level of the extract reduced the elevated levels of blood glucose, glycosylated hemoglobin, urea, uric acid, and creatinine and significantly increased the depressed serum insulin level. The extract exerted a significant inhibitory effect on α-glucosidase in a noncompetitive manner with an IC_{50} value of around 54 μg/mL and was found to inhibit eye lens aldose reductase with an IC_{50} value of about 4.5 μg/mL. Inhibition of α-glucosidase and aldose reductase was postulated to be the reason behind the other observed antidiabetic effects (Jaiswal et al. 2012). A bioactivity-guided isolation approach based on α-glucosidase inhibition led to the isolation of four echinocystic acid glycosides of which eclalbasaponin VI was found to be the most potent (IC_{50} 54.2 ± 1.3 μM) (Kumar et al. 2012b).

Elaeis guineensis Jacq., Arecaceae

This African oil palm is used to treat various disease conditions in African traditional medicine. Tocotrienols (in palm oil) of this palm was reported to improve insulin sensitivity through activating PPAR-γ (review: Wang et al. 2014).

Elaeocarpus ganitrus Roxb., Elaeocarpaceae

Local Name: Rudraksha (India)

The water extract of *Elaeocarpus ganitrus* has been traditionally used to treat diabetic patients. Studies have shown that this plant possesses sedative, tranquilizing, anticonvulsive, antiepileptic, and antihypertensive properties. The water extract of *E. ganitrus* seeds has potential antidiabetic activities. In normoglycemic rats the extract (250, 500, and 1000 mg/kg) showed a hypoglycemic effect at 2 h. In streptozotocin-induced diabetic rats, the extract (daily oral administration for 30 days) decreased blood glucose levels and improved lipid profiles in a dose-dependent manner (Hule et al. 2011).

Elaeodendron glaucum (Rottb.) Pers., or *Cassine glauca* (Rottb.) Kuntze, Celastraceae

Common Name: Ceylon tea

The methanol extract of *Elaeodendron glaucum* stem bark was evaluated for its antidiabetic potential in normal and alloxan-induced diabetic rats. The methanol extract showed positive response to alkaloids, saponins and triterpenes, tannins, flavonoids, carbohydrates, and sterols. Oral administration of the extract daily for 21 days resulted in a hypoglycemic effect in oral glucose tolerance test in normal rats and antidiabetic activity in alloxan-induced diabetic rats (Lanjhiyana et al. 2011).

Elaeodendron transvaalense (Burtt Davy) R.H. Archer, Celastraceae

The acetone extract of *Elaeodendron transvaalense* inhibited the activities of α-glucosidase and α-amylase. However, a cytotoxicity study revealed that the acetone extract of *E. transvaalense* is toxic and raises concern for chronic use (Deutschlander et al. 2009).

Elephantopus mollis Kunth, Asteraceae

An antioxidant compound (3,4-di-*O*-caffeoyl quinic acid), isolated from the polyphenolic-rich extract of *Elephantopus mollis*, exhibited anti-α-glucosidase activities. It also showed an apoptosis-inducing property and cytotoxicity (Ooi et al. 2011).

Elephantopus scaber L., Compositae (Figure 2.46)

Common Names: Prickly leaved elephant foot (Eng.), dila-dila

Description: *Elephantopus scaber* is an erect and hairy herb with a height of 30–60 cm. The leaves are mostly in basal rosette and 15–20 cm in length. The stem is usually dichotomously branched and the flowers are borne in clusters at the tip of the branches. The flowering heads are numerous and sessile forming a large terminal inflorescence comprising about four violet flowers.

Distribution: It is a tropical perennial herb that grows widely in many Asian countries including China, India, Vietnam, and Malaysia. It is popular as a medicinal herb in many countries of Southeast Asia, Latin America, and Africa for a long time (Kabeer and Prathapan 2014).

(a)

(b)

FIGURE 2.46 *Elephantopus scaber.* (a) Young plant. (b) Plant with flowers.

Traditional Medicinal Uses: The plant is an ingredient of several medicinal formulations to treat diarrhea, dysentery, neoplasm, and liver diseases. This plant is traditionally used to treat skin diseases and wounds, neoplasm, diarrhea, edema, rheumatism, snake bite, fever, scabies, stomach troubles, hepatitis, jaundice, bronchitis, nephritis, leukorrhea, menstrual complaints, and cardiac problems. It is also used as an anthelmintic, antifebrile, diuretic, etc. (Kabeer and Prathapan 2014).

Pharmacology

Antidiabetes: The hexane, methanol, and water extracts of the plant showed a significant dose-dependent decrease in the levels of total cholesterol, TG, and LDL-C with an increase in the levels of HDL-C in diabetic rats compared to untreated diabetic control animals. Further, the hexane extract of the plant showed antihyperlipidemic activity and it also improved kidney function (Daisy and Priya 2010). The acetone extract of *E. scaber* reduced blood glucose levels in streptozotocin-induced diabetic rats. Fractionation of the extract yielded a new steroid, 28nor-22(R) with a 2,6,23-trienolide. The compound exhibited significant antidiabetic activity against streptozotocin-induced diabetic rats by reducing the elevated blood glucose levels and restoring the insulin levels (Daisy et al. 2009a; Murugan 2009). This compound may warrant further studies to establish its utility as a medicine to treat type 2 DM.

The ethyl acetate extract of the root and the methanol extract of the leaf of this plant, when orally administered to diabetic animals, resulted in a reduction in the levels of blood glucose, glycosylated hemoglobin, total cholesterol, LDL, and TG; the HDL level increased. The treatment also increased serum insulin level and glycogen content in the liver and skeletal muscle. The histopathological studies showed regeneration of β-cells of the pancreas in treated diabetic animals (Daisy et al. 2011). Oral administration of the water extracts of the leaf and root to alloxan-induced diabetic rats showed antidiabetic activity including a reduction in the activities of the gluconeogenic enzyme glucose-6-phosphatase and an increase in the activity of the glycolytic enzyme glucokinase (Modilal and daisy 2011). Deoxyelephantopin has been reported to function as a selective partial agonist against PPAR-γ (review: Wang et al. 2014). Thus, this plant is likely to have more than one active principles and mechanisms of action.

Other Activities: *E. scaber* exhibited a vast range of important pharmacological activities. These include anticancer and antitumor; antioxidant and free radical scavenging; anti-inflammatory and antiasthma; antibacterial, antifungal; and antiviral; antidiarrheal; hepatoprotective; anticoagulant; wound healing; antiulcer; anthelmintic; anti-snake venom; antimutagenic; and nephroprotective activities (Kabeer and Prathapan 2014).

Toxicity: The active acetone extract did not exhibit any toxicity in acute toxicity studies in rats (Daisy et al. 2009).

Phytochemicals: The major phytochemicals of *E. scaber* are sesquiterpene lactones and phenolic acids and flavonoids. The major sesquiterpene lactones isolated from the ethanol extracts of the plant are deoxyelephantopin, isodeoxyelephantopin, scabertopin, isoscabertopin, scabertopinol, 17,19-dihydrodeoxyelephantopin, iso-17,19-dihydrodeoxyelephantopin, 11,13-dihydrodeoxyelephantopin, molephantinin, elescaberin, deacylcyanopicrin, deacylcyanopicrin 3, glucozaluzanin-C, and β-glucopyranoside crepiside E. The phenolic compounds include 3,4-dihydroxy benzaldehyde, *p*-coumaric acid, vanillic acid, syringic acid, isovanillic acid, *p*-hydroxybenzoic acid, ferulic acid, 3-methoxy-4-hydroxyl cinnamic aldehyde, tricin, E-3-(3-ethoxy-4-hydroxyphenyl) acrylic acid, and 2-hydroxybenzolate (review: Kabeer and Prathapan 2014). The flavonoid glycoside luteolin and flavonoid glycosides luteolin-7-*O*-glucuronide 6″-methyl ester and luteolin-4-*O*-β-D-glucoside were identified along with three polyphenols, namely, trans-*p*-coumaric acid, methyl trans-caffeate, and trans-caffeic acid, from the methanol extract of the aerial parts (Chang et al. 2012). Anti-viral-assay-guided isolation of the ethanol extract of the rhizome led to the identification of several dicaffeoyl derivatives (Geng et al. 2011).

Eleusine coracana (L.) Gaertner, Poaceae

Synonym: *Eleusine indica* (L.) Gaertn. subspp. *coracana* (L.) Lye

Common Name: Finger millet

Finger millet (*Eleusine coracana*) is extensively cultivated and consumed in India and Africa. The millet seed coat is a rich source of dietary fiber and phenolic compounds. The effect of feeding a diet containing 20% finger millet seed coat matter was examined in streptozotocin-induced diabetic rats. Diabetic rats maintained on the millet diet for 6 weeks exhibited a lesser degree of fasting hyperglycemia and partial reversal of abnormalities in serum albumin, urea, and creatinine compared with the diabetic control group. The millet diet-fed group of diabetic rats excreted comparatively lesser amounts of glucose, protein, urea, and creatinine and was accompanied by improved body weights compared with their corresponding controls. Hypercholesterolemia and hypertriacylglycerolemia associated with diabetes were also notably reversed in the millet-fed group. Slit lamp examination of the eye lens revealed an immature subcapsular cataract with mild lenticular opacity in the millet-fed group of rats compared to the mature cataract with significant lenticular opacity and corneal vascularization in the diabetes control group. Lower activity of lens aldose reductase, serum AGEs, and blood glycosylated Hb levels were observed in the millet-fed group. Further, the millet diet showed pronounced ameliorating effects on kidney pathology as reflected by near normal glomerular and tubular structures and lower glomerular filtration rate compared with the shrunken glomerulus and tubular vacuolations in the diabetes control group. Thus, the animal study suggests the utility of the finger millet seed coat as a functional ingredient in diets for diabetics (Shobana et al. 2010).

Eleutherine americana (Aubl.) Merr. ex K.Heyle, Iridaceae

Synonym: *Eleutherine bulbosa* (Mill.) Urb

A 50% aqueous methanol extract of the bulbs of *Eleutherine americana* inhibited α-glucosidase activity. The active principle eleutherinoside A was isolated from the extract (Leyama et al. 2011).

Eleutherine palmifolia (L.) Merr., Iridaceae (This name is considered as a synonym of *Sisyrinchium palmifolium* L., Iridaceae.)

Bawang Dayak (*Eleutherine palmifolia*) is traditionally used to cure DM and other diseases by Dayak tribes in Kalimantan Island, Indonesia. Treatment of alloxan-induced diabetic rats with the water and ethanol extracts of *E. palmifolia* bulbs (100 mg/kg daily for 28 days) maintained the body weight of diabetic rats similar to those of nondiabetic rats, reduced the blood serum glucose level, increased the serum insulin level, and lowered the serum total cholesterol and LDL levels compared to untreated diabetic rats (Febrinda et al. 2014). The suggested mechanisms of antidiabetic actions of the plant are inhibition of α-glucosidase, which could decrease postprandial blood glucose level, and also improvement of pancreatic β-cells, thus enhancing the insulin secretion directly (Febrinda et al. 2014).

Eleutherococcus senticosus (Rupr. & Maxim) Maxim., Araliaceae

Common Names: Devil's bush, Siberian ginseng

Eleutherococcus senticosus is an oriental traditional medicinal plant. Eleutheroside E, a principal component of *E. senticosus*, has anti-inflammatory and protective effects in the ischemic heart. The effect of eleutheroside E containing *E. senticosus* extracts as well as eleutheroside E on hyperglycemia and insulin resistance in *db/db* mice was studied. When 5-week-old *db/db* mice were fed a diet consisting of the plant extract or eleutheroside E for 5 weeks, serum lipid profiles improved and blood glucose levels decreased. The extract and the compound effectively attenuated HOMA-IR. Glucose tolerance and insulin tolerance tests showed that the compound increased insulin sensitivity. Immunohistochemical staining indicated that eleutheroside E and the extract protected pancreatic α- and β-cells from diabetic damage. In addition, the extract and the compound improved hepatic glucose metabolism by upregulating glycolysis and downregulating gluconeogenesis in obese type 2 diabetic mice (Ahn et al. 2013). Eleutheroside E increased the insulin-provoked glucose uptake in C2C12 myotubes. Moreover, the compound improved TNF-α-induced suppression of glucose uptake in 3T3-L1 adipocytes (Ahn et al. 2013). Syringin (an active principle from *E. senticosus*) injection resulted in an increase in plasma glucose utilization accompanied with the increase of plasma insulin and C-peptide in anesthetized rats (Liu et al. 2008). Further investigations showed that syringin has an ability to raise the release of acetylcholine from nerve terminals to stimulate muscarinic M3 receptors in pancreatic cells and augment the insulin release (Liu et al. 2008). In addition, *E. senticosus* extracts showed promising hypolipidemic activities (Choi et al. 2008).

Embelia ribes Burm.f., Myrsinaceae

The ethanol extract of *Embelia ribes* (fruits) attenuated isoproterenol-induced cardiotoxicity in streptozotocin-induced diabetic rats (Bhandari and Ansari 2009). In another study, *E. ribes* berry extract exhibited hypoglycemic activity in alloxan-induced diabetic rats (Purohit et al. 2008).

Enicostemma littorale Blume, Gentianaceae (= *Enicostemma hyssopifolium* (Willd.) Verd.)

Synonym: *Enicostemma axillare* (Lam.) Raynal. (Figure 2.47)

Common Names: Chota chirayata, vellarugu, vellari, nahi (local names in India), white head (Eng.) (Saranya et al. 2013)

Description: *Enicostemma littorale* is an erect perennial glabrous herb (5–30 cm tall) bearing sessile leaves that are variable in size, linear or linear-oblong or elliptic-oblong or lanceolate, obtuse or acute, and glabrous. Sessile flowers are produced in clusters from the auxiliary position. The flowers are white with green lines with long bracts that are shorter than the calyx, a corolla tube 3.5–6.0 mm long, stamens inserted below the sinuses, just above the middle of the tube, and filaments with a double hood at the insertion point. The fruit is an ellipsoid capsule with a narrow base and round apex (Saranya et al. 2013).

Distribution: The plant is widely distributed in South America, Africa, and Asia. It grows in many diverse habitats from savannas, grasslands, and forests to beaches, from wet to dry land. It is distributed in greater parts of India, especially in the coastal area, up to an altitude of 1500 ft.

Traditional Medicinal Uses: It is a laxative, tonic, and stomachic and used as a blood purifier, in dropsy, rheumatism, hernia, swelling, abdominal ulcers, itches, insect poisoning, malaria, snake bite, skin diseases, leprosy, and diabetes (Kirtikar and Basu 1975). In Ayurvedic medicine this plant is taken in combination with other herbs especially for diabetes (Saranya et al. 2013).

Pharmacology

Antidiabetes: A dose-dependent hypoglycemic effect of the water extract of *E. littorale* in alloxan-induced diabetic rats has been observed (Vijayvargia et al. 2000; Maroo et al. 2003), and the suggested mechanisms include potentiation of glucose-mediated insulin release and increase in insulin sensitivity (Maroo et al. 2003; Murali et al. 2002). The water extract of the whole plant (2 g/kg) almost normalized some key carbohydrate metabolic enzymes and antioxidant defense in alloxan-induced diabetes in rats (Srinivasan et al. 2005). Oral administration of the water extract to diabetic rats for 45 days resulted in marked reduction in the levels of blood glucose and tissue (liver, kidney, and pancreas) TBARS; the treatment also improved the status of antioxidant enzymes. The efficacy of the extract (2 g/kg) was better than 6 units/kg insulin in alloxan-induced diabetic rats (Prince and Sreenivasan 2005). In streptozotocin-induced type 1 diabetic rat, treatment with the hot water extract of the whole plant (500 mg/kg) for

FIGURE 2.47 *Enicostemma littorale.*

3 weeks reversed the diabetes-induced body weight loss and polyuria, increased level of blood glucose, and decreased insulin levels. Besides, the treatment decreased serum lipid profiles and glycosylated hemoglobulin levels (Vishwakarma et al. 2010). Swertiamarin was found to be a major component in the hot water extract, while it was almost absent in the cold extract. This compound decreased glycosylated hemoglobulin and glucose-6-phaspatase levels (Vishwakarma et al. 2010). However, activity-guided isolation of all the active principles remains to be done.

The extract prevented diabetes-induced neuropathy and oxidative stress in a rat model of diabetic neuropathy (Bhatt et al. 2009). Besides, administration of the water extract of *A. marmelos* and *E. littorale* for 15 days prevented hyperinsulinemia and hyperglycemia induced by a diet high in fructose (Gohil et al. 2008). Interestingly, an isolated compound (SGL-1) from the plant extract stimulated islet neogenesis in cell culture conditions. The authors conclude that it is a therapeutic agent for islet neogenesis (Gupta et al. 2010).

Clinical Trial: Administration of pills prepared from this plant to type 2 diabetes patients for 3 months results in a reduction in blood glucose as well as serum insulin levels and prevention of the progression of complications in diabetic patients (Upadhyay and Goyal 2004).

Other Activities: Antioxidant and hypolipidemic effects of the plant in diabetic rats are reported. This plant exhibits several pharmacological activities. These include antitumor activity of the methanol extract (Kavimani and Manisenthlkumar 2000), anti-inflammatory activity (Sadique et al. 1987), antimicrobial activity, antioxidant activity, anthelmintic property, antinociceptive activity, antiulcer activity, and hepatoprotective activity against toxic chemical-induced liver damage (Saranya et al. 2013).

Phytochemicals: Ophelic acid, bitter glycoside, gentiocrucine, enicoflavine, erythrocentaurin, flavonoids, and swertiamarin are present in the plant (Ghosal and Jaiswal 1980; Rastogi and Mehrotra 1999). Further, compounds isolated include volatile oils, betulin, sapogenin, monoterpene alkaloids like enicoflavin, gentiocrucine, flavonoids (apigenin, genkwanin, isovitexin, swertisin, saponarin, 5-*O*-glucosylswertisin, and 5-*O*-glucosylisoswertisin), catechins, steroids, sapogenin, triterpenoids, flavonoids, xanthones, verticilliside (a new flavone C-glucoside), phenolic acids (vanillic acid, syringic acid, p-hydroxyl benzoic acid, protocatechuic acid, *p*-coumaric acid, and ferulic acid), and amino acids (review: Saranya et al. 2013).

Ephedra alata Decne., Ephedraceae

Aerial parts of *Ephedra alata* are used in folk medicine to control diabetes in Egypt. This plant is reported to have antidiabetic activity and is predicted (*in silico*) to contain DPP-IV inhibitors (Guasch et al. 2012). The plant extract exhibited persistent hypoglycemic effects in normal rats (Shabana et al. 1990).

Ephedra distachya L., Ephedraceae

Synonym: *Ephedra vulgaris* Rich.

Ephedrines A, B, C, D, and L (glycans) from *Ephedra distachya* showed antihyperglycemic activity in streptozotocin-induced diabetic mice. Further, the glycans caused pancreatic islet regeneration in diabetic mice (Xiu et al. 2001).

Epimedium brevicornum Maxim, Berberidaceae

Icariin isolated from the *Epimedium brevicornum* leaf reduced mitochondrial oxidative stress and improved cardiac function in streptozotocin-induced diabetic rats (Bao and Chen 2011). Further, icariin (5 mg/kg/day, p.o.) ameliorated streptozotocin-induced diabetic retinopathy in rats (Xin et al. 2012). The compound (80 mg/kg, i.g.) protected streptozotocin-induced diabetic rats from renal damage in the early stage of nephropathy via modulating transforming growth factor β1 and type IV collagen expression (Qi et al. 2011). *E. brevicornum* leaves (water extract) exerted a potent and specific inhibition against yeast α-glucosidase, with 50% inhibition of activity (IC_{50}) at 11.8 μmol/L. Baohuoside I (a flavonol from the leaves) exhibited strong inhibition against the α-glucosidase (Phan et al. 2013).

Epimedium elatum C. Morren & Decne., Berberidaceae
Epimedium elatum is used in traditional medicine to treat rheumatic conditions, to improve kidney and bone functions, etc. This plant is endemic to the Western Himalayas. Acylated flavonol glycosides isolated from the ethanol extract of *E. elatum* (whole plant) activated PPAR-γ suggesting the likely anti-diabetes activity of this plant (review: Wang et al. 2014).

Equisetum myriochaetum Schlecht and Cham, Equisetaceae
The water as well as butanol extract of the aerial parts of this plant exhibited hypoglycemic activity in streptozotocin-induced diabetic rats. A single oral administration of the water extract (13 mg/kg) or butanol extract (16 mg/kg) lowered blood glucose levels within 3 h in diabetic rats; the effect was comparable to that of glibenclamide. Three kaempferol glycosides and one caffeoyl glucoside were found to be the main constituents in both extracts (Cetto et al. 2000).

Clinical Trial: Oral administration of a single dose of the water extract (330 mg/kg) of the aerial parts of the plant to recently diagnosed type 2 diabetic patients resulted in a significant reduction in blood glucose levels within 90, 120, and 180 min. There were no significant changes in the insulin levels (Revilla et al. 2002). The plant appears to be attractive for further studies.

Eremophila alternifolia R. Br., Myoporaceae
Eremophila alternifolia is a medicinal shrub occurring in Australia. This plant was valued for medicinal purposes by aboriginal people in coastal and central Australia. This plant yields phenylethanoid verbascoside (acteoside) and the iridoid geniposidic acid. Verbascoside has a number of pharmacological activities. This compound is a potent lens aldose reductase inhibitor. Some of the chemical isolates from this plant such as deacetyl-asperulosidic acid methyl ester showed hypoglycemic activity in normal and alloxan-induced diabetic rats (review: Ghisalberti 2005).

Eremophila longifolia (R.Br.) F.Muell., Myoporaceae
Eremophila longifolia is an Australian traditional medicinal tree. This plant produces the phenylethanoid verbascoside (acteoside) and the iridoid geniposidic acid. Verbascoside has a number of pharmacological activities. This compound is a potent lens aldose reductase inhibitor. Some of the chemical isolates from this plant such as deacetyl-asperulosidic acid methyl ester showed hypoglycemic activity in normal and alloxan-induced diabetic rats (review: Ghisalberti 2005).

Erigeron annuus (L.) Pers, Compositae
A novel 2,3-dioxygenated flavanone, erigeroflavanone, isolated from the ethyl acetate–soluble extract of the flowers of *Erigeron annuus* inhibited AGE formation and rat lens aldose reductase activity. This property may be useful in attenuating complications of DM (Yoo et al. 2008). A composition (compositions and functional food for the prevention and treatment of diabetic complications) containing *E. annuus* as an active ingredient for preventing and treating diabetes was developed to suppress generation of AGEs and abnormal activation of aldose reductase and patented by the Korea Institute of Oriental Medicine (IPC: A61P35/00 A61P3/10 A61K36/515).

Erigeron breviscapus (Vaniot) Hand., Asteraceae
Erigeron breviscapus is a small Chinese herb used in herbal medicine. The effect of *E. breviscapus* supplementation on the endothelium-dependent relaxation response to acetylcholine in streptozotocin-induced diabetic rat aorta *in vitro* was studied. At the stage of 6-week duration, diabetes caused an approximately 60% deficit in maximum relaxation, which was significantly prevented by a dietary supplement with 0.5% *E. breviscapus* from 1 week of diabetes induction (Zhu et al. 1999). Breviscapine, a flavonoid extracted from *E. breviscapus*, at a dose of 10 and 25 mg/kg/day (p.o.) exerted protective effects in the pathogenesis of cardiomyopathy induced by high glucose levels in streptozotocin-induced diabetic rats via the PKC/NF κ-B/c-fos signal transduction pathway (Wang et al. 2009a). Further, the compound (10 and 25 mg/kg/day, p.o.) ameliorates cardiac dysfunction and regulated myocardial calcium cycling proteins in streptozotocin-induced diabetic rats (Wang et al. 2010b). It was shown that treatment with

breviscapine attenuated renal injury in diabetic rats. Further, the combination of enalapril and breviscapine conferred superiority over monotherapies on renoprotection in diabetic rats. The mechanism of action may be, at least partly, correlated with synergetic suppression on increased oxidative stress and PKC activities as well as overexpression of TGFβ1 in renal tissue (Xu et al. 2013).

Clinical Trial: In a prospective, randomized, double-blind, placebo-controlled clinical trial, *E. breviscapus* treatment resulted in improvement of the mean sensitivity of the visual field defects for diabetic patients with intraocular pressure-controlled glaucoma. *E. breviscapus* showed a favorable neuroprotective effect for patients with middle- or late-stage glaucoma (Ye and Jiang 2003).

Erigeron canadensis L., Asteraceae
Extracts from this plant lowered blood glucose in experimental animals (Marles and Farnsworth 1995).

Eriobotrya japonica Lindl., Rosaceae

Common/Local Names: Logat, ilakotta, nokkotta, Japanese medlar, loquat, etc.

Description: *Eriobotrya japonica* is a small evergreen branched tree with simple, narrowly oblanceolate, acuminate, remotely serrate, thick, coriaceous leaves that is narrowed at the base into a very short stout woody petiole. The flowers are stipulate, white, and fragrant and produced in terminal panicles. The fruit is succulent, formed from the calyx tube, yellowish, and pyriform, bearing 2–5 large angular seeds.

Distribution: *Eriobotrya japonica* is a native of China and Japan and is naturalized in India; it is grown nearly throughout India up to an altitude of 5000 ft.

Traditional Medicinal Uses: *E. japonica* is a sedative, antipyretic, and expectorant and used in diarrhea, diabetes, cancer, mental depression, swelling, vomiting, indigestion, asthma, and apoplexy (Kirtikar and Basu 1975).

Antidiabetes: *E. japonica* leaves had been used traditionally for the treatment of diabetes by immersing the dried leaves in a hot water drink. It has been shown to lower blood glucose levels of normal and alloxan-induced diabetic rabbits (Noreen et al. 1988). Sesquiterpene glycosides and polyhydroxylated triterpenoids from the plant have hypoglycemic effects (De Tommasi et al. 1991). A sesquiterpene glycoside, isolated from the dried leaves of loquat (*E. japonica*), showed hypoglycemic effects in normal and alloxan-induced diabetic mice. Administration of the sesquiterpene glycoside (25 and 75 mg/kg, p.o.) resulted in a marked glucose-lowering effect in alloxan-induced diabetic mice and a slight effect in normal mice (Chen et al. 2008). The water extract of *E. japonica* leaves increased insulin secretion from INS-1 cells in a dose-dependent manner. Oral administration on the extract (230 mg/kg), however, decreased plasma insulin level for 240 min postadministration and caused a transient drop of blood glucose at 15 and 30 min. Cinchonain Ib, epicatechin, procyanidin B-2, and chlorogenic acid were isolated from the water extract of *E. japonica* leaves. Cinchonain Ib significantly enhanced insulin secretion from INS-1 cells, whereas epicatechin inhibited the insulin secretion. Cinchonain Ib at a dose of 108 mg/kg enhanced plasma insulin levels in rats for as long as 240 min after oral administration, but it did not induce any change in blood glucose levels (Qadan et al. 2009). The extract containing more than 50% of triterpene acids (tormentic, corosolic, maslinic, oleanolic, and ursolic acids) increased serum insulin levels and decreased glycosylated serum protein, total cholesterol, and TGs in alloxan- and streptozotocin-induced diabetic mice (review: Nazaruk and Borzym-Kluczyk 2014). It is known that chlorogenic acid, ursolic acid, oleanolic acid, and β-sitosterol have antidiabetic properties and occur in many plants. It is of interest to study the effect of specific combinations of different bioactive compounds in different diabetic animals.

Other Activities: Antitumor, antioxidant, liver function improvement, and anti-inflammatory properties of this plant have been reported.

Phytochemicals: Amygdalin, triterpenoidal saponins, flavonol glycoside, megastigmane glycosides, polyphenols, sesquiterpene glycosides, flavonol glycoside, malic acid, citric acid, tartaric acid, cryptoxanthin, neo-β-carotene, β-sitosterol, 4-methyleneproline, 4-methylene-dl-proline, trans-4-hydroxy methyl-D-proline, loquatifolin, tormentic, corosolic, maslinic, oleanolic, and ursolic acids have been isolated in this plant (Ajikumara and Subramoniam 2005; review: Nazaruk and Borzym-Kluczyk 2014).

Eruca sativa Mill., Brassicaceae

Synonym: *Eruca vesicaria* ssp. *sativa* (Miller) Thell., *Brassica eruca* L.

Common Name: Rocket (roquette)

Eruca sativa is used as a vegetable and spice especially among Europeans. Daily oral administration of *E. sativa* seed oil 2 weeks before or after diabetes induction ameliorated hyperglycemia, improved lipid profile, blunted the increase in MDA and 4-hydroxynonenal, and stimulated GSH production in the liver of alloxan-treated rats (El-Missiry and El-Gindy 2000). In a recent study, the water and ethanol extracts of *E. sativa* leaves (edible part) inhibited, in a concentration-dependent manner, the activities of α-amylase, α-glucosidase, and β-galactosidase. The ethanol extract showed greater inhibitory effect than the water extract (Hetta et al. 2014).

Ervatamia microphylla (Pit.) Kerr (Botanical Name: *Tabernaemontana bufalina* Lour.), Apocynaceae
Canophylline, an alkaloid compound, isolated from the leaves of *Ervatamia microphylla* (*Tabernaemontana bufalina* Lour.), a small tropical tree, was effective in stimulating the differentiation of progenitor cells into β-cells. The compound promoted β-cell differentiation in fetal and neonatal rat pancreas. It suppressed pancreatic stellate cells and improved islet fibrosis in Goto-Kakizaki rats (review: Chang et al. 2013b).

Eryngium creticum Lam., Umbelliferae

Common Name: Field eryngo

Eryngium creticum is used in Jordan as an ethnomedicine to control diabetes. *In vitro* studies using MIN6 β-cells in culture have shown that the water extract of this plant potentiates insulin secretion from the β-cells. Further, comparable to GLP-1, the extract enhanced β-cell proliferation (Kasabri et al. 2012a).

Erythrina abyssinica Lam. ex DC, Fabaceae
The antidiabetic activity of *Erythrina abyssinica* was assessed by intraperitoneally injecting varying doses of the water extract of the plant into alloxan-induced diabetic mice. The extract showed hypoglycemic activity (Piero et al. 2011). Studies showed a hypoglycemic activity of the methanol extract of *E. abyssinica* tuber in normal rats. This activity was comparable to that of glibenclamide (Kihara 2012). Several abyssinoflavanones and pterocarpan derivatives isolated from the stem bark of *E. abyssinica* exhibited PTP1B inhibitory activity (Jiany et al. 2012).

Erythrina addisoniae Hutch & Dalziel, Fabaceae
Prenylated isoflavonoids isolated from the stem bark of *Erythrina addisoniae* inhibited PTP1B activity *in vitro* (review: Jiany et al. 2012).

Erythrina lysistemon Hutch., Fabaceae
Several pterocarpans isolated from the methanol extract of the stem bark of *Erythrina lysistemon* inhibited PTP1B activity with IC_{50} values ranging from 1.0 to 18.1 μg/mL (review: Jiany et al. 2012).

Erythrina mildbraedii Harms, Fabaceae
Bioassay-guided fractionation of the ethyl acetate extract of the root bark of *Erythrina mildbraedii* yielded several flavonoids that inhibited PTP1B activity with IC_{50} values ranging from with 5.5 to 42.6 μmol/L (review: Jiany et al. 2012).

Erythrina variegata var. *orientalis* (L.) Merr., Leguminosae

Synonym: *Erythrina indica* Lam.

Erythrina variegata (*Erythrina indica*), commonly known as tiger's claw, is a thorny deciduous tree grown in tropical and subtropical regions of Eastern Africa, Asia, Australia, etc. Its leaves, root, and bark are traditionally used for DM. The methanol extract of *E. variegata* (*E. indica*) leaf (300, 600, and 900 mg/kg, p.o., daily for 21 days) exhibited a dose-dependent hypoglycemic activity in streptozotocin-induced diabetic rats. Further, serum biochemical parameters including the lipid profile were restored toward normal levels in the extract-treated rats as compared to diabetic control animals (Kumar et al. 2011a). In a follow-up study, the water and alcohol extract of the stem bark of *E. indica* was evaluated for their antidiabetic activity in alloxan-induced and dexamethasone-induced diabetic rats. In normal rats, the extracts decreased the blood glucose level in a dose-dependent manner after repeated administration for 7 days. In alloxan-induced diabetic rats, both the extracts decreased blood sugar levels with significant improvement in glucose tolerance and body weight at the end of the first, second, and third week after test extract treatment. In case of dexamethasone-induced insulin-resistant diabetic rats, the extracts exhibited increases in blood glucose levels and improved glucose tolerance as compared to untreated dexamethasone control rats (Kumar et al. 2011b). In another study, the water extract of the *E. variegata* bark (400 mg/kg daily for 30 days) exhibited antidiabetic activity in streptozotocin-induced diabetic rats. The hypoglycemic effect of the extract was comparable to that of glibenclamide. Further, the treatment improved serum biochemical parameters. Histological studies suggest regeneration of pancreatic β-cells in treated diabetic animals (Anupama et al. 2012).

Eucalyptus camaldulensis Dehnh, Myrtaceae
The essential oil of *Eucalyptus camaldulensis* inhibited both α-amylase and α-glucosidase in a noncompetitive manner and also exhibited greater antioxidant potential than butylated hydroxyl toluene and curcumin (Basak and Candan 2010).

Eucalyptus citriodora Hook, Myrtaceae
Current Botanical Name: *Corymbia citriodora* (Hook) K.D.Hill & L.A.S.Johnson
Synonym: *Eucalyptus rostrata* Schltdl.
Eucalyptus citriodora is an evergreen tree common in Australia. The water extract of *E. citriodora* leaf showed a dose-dependent blood glucose-lowering activity in glucose-loaded rats and alloxan-induced diabetic rats. On oral administration of the extract at a dose of 500 mg/kg of body weight, the reduction of the blood glucose level was 22.9% after 4 h, and on continuous daily administration, the reduction in the blood glucose level after 21 days was 49.9% and 56.8% with doses of 250 and 500 mg/kg of body weight, respectively (Patra et al. 2009).

Eucalyptus globulus Labill, Myrtaceae
Eucalyptus globulus is used in the traditional treatment of diabetes. *E. globulus* exhibited various pharmacological actions including anti-DM activity (review: Dey and Mitra 2013). Incorporation of *E. globulus* in the diet (62.5 g/kg) and drinking water (2.5 g/L) reduced the hyperglycemia and associated weight loss of streptozotocin-treated mice. The water extracts of *E. globulus* (0.5 g/L) enhanced 2-deoxy-glucose transport by 50%, glucose oxidation by 60%, and incorporation of glucose into glycogen by 90% in the abdominal muscle of mice. Acute, 20 min incubation with the extract evoked a stepwise 70%–160% enhancement of insulin secretion from the clonal pancreatic β-cell line. Thus, *E. globulus* represents an effective antihyperglycemic dietary adjunct for the treatment of diabetes (Gray and Flatt 1998). In another study, *E. globulus* (20 g/kg in the diet and 2.5 g/L in the drinking water) exhibited antidiabetic and antioxidant properties. The treatment reduced the levels of blood glucose, glycated hemoglobin, and lipid peroxides in streptozotocin-induced diabetic rats (Nakhaee et al. 2009). In another study, oral administration of *E. globules* leaf extract reduced oxidative stress in alloxan-induced diabetic rats and lowered blood glucose but did not improve liver glycogen content (Ahlema et al. 2009). Further, when examined using an *in vitro* method to assess the possible effects on glucose diffusion across the gastrointestinal tract, the water extract of *E. globulus*, at a concentration of 50 g/L, decreased glucose movement *in vitro* (Gallagher et al. 2003). *E. globulus* leaf extract (10 g/kg in the diet) inhibited intestinal fructose absorption and suppressed adiposity due to dietary sucrose in rats. The suggested mechanism is that fructose is transported across the intestinal brush border membrane by the specific transporter GLUT5 and inhibiting intestinal fructose absorption prevents adiposity in subjects consuming large amounts of sucrose and fructose. Besides, *E. globulus* leaf extract is found to inhibit the activities of fructokinase and G6PDH, preventing the activation of fructose metabolism and

fatty acid synthesis induced by dietary sucrose. *E. globulus* leaf extract simultaneously inhibited intestinal fructose and sucrose absorption and shows enough potential to be used as a natural food additive in fructose-/sucrose-rich junk foods (review: Dey and Mitra 2013). The alcohol extract of *E. globulus* leaves caused a significant reduction of serum TG and cholesterol in diabetic rats, whereas it did not change the levels of serum TG and cholesterol in normal rats (Eidi et al. 2009a). Available studies suggest that the antidiabetic action of the *E. globulus* water extract is due to reduction in carbohydrate absorption from the intestine and reduction in oxidative stress. Due to its antioxidant property, it may protect β-cells to some extent. However, caution is required due to its possible adverse effect on the liver function in view of the following report. Eucalyptus water extract, when consumed in drinking water (2.5 mg/mL) for 4 weeks, decreased blood glucose level but increased liver function marker enzyme activities in streptozotocin-induced diabetic male rats (Shahraki and Shahraki 2013). The species of *Eucalyptus* studied was not indicated. The report suggests that prolonged use of a high dose of eucalyptus water extract can lead to liver damage.

Eucalyptus tereticornis Sm., Myrtaceae
Validation of the ethnobotanical use of *Eucalyptus tereticornis* leaves as an antidiabetic agent using the oral glucose tolerance test showed antihyperglycemic activity. At a dosage of 5 mg/20 g mouse, *E. tereticornis* leaf showed a significant decrease in blood glucose levels in 60 min (Villasenor and Lamadrid 2006).

Euclea undulata Thunb, Ebenaceae

Synonym: *Euclea myrtina* Burch (Figure 2.48)

Common Names: Ghwarrie, common guarri, guarrie, kholi, etc.

Description: *Euclea undulata* is a very sturdy, evergreen, dense, and twiggy shrub to a small tree reaching an approximate size of 7 m in height and very often with an equal width. The leaves are opposite to subopposite or may be arranged in whorls of 3 or 4, mostly toward the ends of branchlets. The leaves are small, obovate to narrowly elliptical, stiff, leathery, dark green or blue-green above, paler below, and sometimes rusty brown. The leaf apex is broadly tapering, or rounded to abruptly attenuate, base tapering, margin entire, finely rolled under, conspicuously wavy, or almost flat. The bark is gray, scaly longitudinally fissured or cracked, particularly with age. The branchlets are twiggy, angular, and densely leafy with short internodes. The flowers are small, whitish to cream, carried on unbranched 5- to 7-flowered spikes, with rust-colored stalked glands. The spikes are solitary in the leaf axils and are somewhat

FIGURE 2.48 *Euclea undulata.*

scented. The fruits appear from February and are single-seeded rounded berries, thinly fleshy, reddish brown becoming dark when ripe, in short sprays (van Wyk et al. 2009).

Distribution: The plant is widespread on rocky slopes in all provinces of South Africa.

Traditional Medicinal Uses: The plant is used by traditional healers in the treatment of diabetes in Limpopo province. Leaf preparations are taken orally in the Western Cape to treat diarrhea and disorders of the stomach and as a gargle to relieve tonsillitis. Root infusions are used as enemata or as an ingredient of *inembe* (medication taken regularly to ensure a trouble-free pregnancy). The use of root preparations to induce emesis or purgation is also recorded and of bark preparations for the treatment of headache, toothache, and other pains. Infusions of the roots have been traditionally used for heart diseases (van Wyk et al. 2009).

Pharmacology

Antidiabetes: The acetone extract of *E. undulata* root bark stimulated glucose uptake (162.2%) by Chang liver cells at 50 μg/mL. Further, the extract inhibited the activities of α-glucosidase and α-amylase (Deutschlander et al. 2009). The extract was devoid of cytotoxicity. In a follow-up study, the hypoglycemic activity of a crude acetone extract of the root bark of *E. undulata* var. *myrtina* was evaluated in a streptozotocin–nicotinamide-induced type 2 diabetes rats. The extract was administered for 21 days orally at a concentration of 50 mg/kg and 100 mg/kg, respectively. The acetone extract (100 mg/kg) significantly lowered fasting blood glucose levels as well as elevated cholesterol and TG levels to near normal levels without any weight gain. The results indicate that the crude acetone root bark extract of *E. undulata* exhibits antidiabetic activity in type 2 streptozotocin–nicotinamide-induced diabetic rats (Deutschlander et al. 2012).

Phytochemical studies on crude acetone extract of the root bark of *E. undulata* var. *myrtina* produced a new compound, α-amyrin-3-*O*-β-(5-hydroxy) ferulic acid, and three known compounds (betulin, lupeol, and epicatechin). *In vitro* assays on cultured C2C12 myocytes revealed that epicatechin (166.3%) and betulin (21.4%) were active in lowering glucose levels, whereas α-amyrin-3-*O*-β-(5-hydroxy) ferulic acid was active to a lesser extent only. Epicatechin (IC$_{50}$ 5.86 μg/mL) and lupeol (IC$_{50}$ 6.27 μg/mL) inhibited α-glucosidase activity. *E. undulata* acetone extract also inhibited α-amylase (IC$_{50}$ 2.80 μg/mL). These results indicated that the crude *E. undulata* acetone extract does contain compounds that display hypoglycemic activity (Deutschlander 2010; Deutschlander et al. 2011). Thus, the plant appears to be promising for further studies.

Other Activities: The water extract of dried leaf showed *in vitro* antimicrobial activity against *Staphylococcus aureus*. Except for the antidiabetes studies, no other information is available regarding the bioactivity of this species (Deutschlander 2010).

Phytochemicals: Studies on the phytochemistry of *Euclea* species have identified triterpenoids and aliphatics in the branches and leaf and naphthoquinones in the root, stem, and fruit. Naphthoquinones 7-methyljuglone and diospyrin were isolated from the roots and isodiospyrin from the fruits of *E. undulata* var. *myrtina*. Stems appeared devoid of naphthoquinones. The presence of tannins, saponins, and reducing sugars in the leaf and stem was observed. The bark is reported to contain 3.26% of tannin. The acetone extract of the root bark of *E. undulata* var. *myrtina* produced a new α-amyrin-3-*O*-β-(5-hydroxy)ferulic acid compound and three known compounds (betulin, lupeol, and epicatechin) (Costa et al. 1978; Deutschlander 2010).

Eucommia ulmoides Oliv., Eucommiaceae

Eucommia ulmoides leaves have been used as a folk remedy for the treatment of DM in Korea.

Flavonol glycosides (a new flavonol glycoside [quercetin 3-*O*-α-L-arabinopyranosyl-(1 → 2)-β-D-glucopyranoside], astragalin, and quercetin) isolated from the leaves of *E. ulmoides* inhibited AGE formation, which is comparable to that of aminoguanidine, a known glycation inhibitor (Kim et al. 2004). Accumulation of AGEs is one of the main molecular mechanisms implicated in diabetic complications (Hy et al. 2004). The water extract of *E. ulmoides* leaves exerted hypoglycemic and hypolipidemic actions in C57BL/KsJ *db/db* mice (Park et al. 2006). The extracts of this plant stimulated glucose uptake in L6 rat skeletal muscle cells. The extract stimulated the activity of phosphatidylinositol 3-kinase and its downstream effectors, protein kinase B and the atypical form of protein kinase C. But it did not influence AMPK (Hong et al. 2008). Supplementation with powdered du-zhong (*E. ulmoides*) leaves (1% in the diet) or its water extract (0.187%) for 3 weeks resulted in a significant reduction in the blood glucose levels and increase in the levels of insulin and C-peptide in streptozotocin-induced diabetic rats (Lee et al. 2005).

Eugenia floccose Bedd., Myrtaceae
The ethanol extract of *Eugenia floccosa* leaf (150 and 300 mg/kg daily for 14 days) elicited significant reductions in blood glucose, lipid parameters except HDL, and serum enzymes (GPT, GOT, ALP) and significantly increased HDL and antioxidant enzymes in alloxan-induced diabetic rats. The extracts also caused significant increase in plasma insulin in diabetic rats. Thus, the ethanol extract of *E. floccosa* showed significant antidiabetic, antihyperlipidemic, and antioxidant effects in alloxan-induced diabetic rats (Jelastin et al. 2012).

Eugenia uniflora L., Myrtaceae
The leaf extract of *Eugenia uniflora* improved oral glucose tolerance in animals and α-glucosidase inhibitory activity has been detected in *E. uniflora* (Matsumura et al. 2000).

Euonymus alatus (Thunb.) Sieb., Celastraceae
Euonymus alatus has been used as a folk medicine for diabetes in China for more than 1000 years. The ethyl acetate fraction of *E. alatus* displayed significant effects on reducing plasma glucose in normal mice and alloxan-induced diabetic mice. In oral glucose tolerance, the fraction at 17.2 mg/kg could significantly decrease the blood glucose of both normal mice and diabetic mice. After 4 weeks of administration of the fraction, when compared with the diabetic control, there were significant differences in biochemical parameters, such as glycosylated serum protein, superoxide dismutase and MDA, TG, and total cholesterol, between the treated diabetic mice and the diabetic control mice. Besides, histopathological studies of pancreatic islets also showed beneficial effects of the fraction on diabetic mice. The major components in the extract were flavonoids and phenolic acids, which had antioxidative effects on scavenging DPPH-radical *in vitro*. The mechanism of action of *E. alatus* for glucose control could be stimulation of insulin release, improvement in glucose uptake, and suppression of oxidative stress (Fang et al. 2008). Kaempferol and quercetin isolated from *E. alatus* improved glucose uptake of 3T3-L1 cells without adipogenesis activity and activated PPAR-γ (Wang et al. 2014). The methanol extract of *E. alatus* inhibited yeast α-glucosidase activity in a concentration-dependent manner. A single oral dose of *E. alatus* extract (500 mg/kg) significantly inhibited increases in blood glucose levels at 60 and 90 min and decreased incremental response areas under the glycemic response curve in streptozotocin-induced diabetic rats (Lee et al. 2007).

Eupatorium purpureum L., Compositae
Extracts from *Eupatorium purpureum* root and flower lowered blood glucose in experimental animals (Marles and Farnsworth 1995).

Euphorbia helioscopia L., Euphorbiaceae
Under *in vitro* conditions, the ethanol extract of *Euphorbia helioscopia* showed insulin secretagogue activity in INS-1 cells at the 10 μg/mL level (Hussain et al. 2004).

Euphorbia hirta L., Euphorbiaceae
Euphorbia hirta is a traditional Australian medicinal herb. The plant reportedly possesses hypoglycemic activity. The flavonoid quercetin has been isolated from the plant. Quercetin exhibits hypoglycemic activity and is an inhibitor of aldose reductase (review: Ghisalberti 2005).

Euphorbia prostrata Ait, Euphorbiaceae
Euphorbia prostrata is used in traditional medicine to treat diabetes in Pakistan and elsewhere. The antihyperglycemic and hypolipidemic effects of the methanol extract of the whole *E. prostrata* plant (250 and 500 mg/kg/day, orally, for 14 days) was evaluated in alloxan-induced diabetic rabbits. The extract exhibited an antidiabetic effect as judged from a significant reduction in fasting blood glucose levels and

serum cholesterol and TG levels. A 500 mg/kg dose of the extract produced more significant results as compared to the 250 mg/kg dose (Shamin et al. 2014).

Euphrasia officinalis L., Scrophulariaceae
Common Name: Tree of light
The water extract of the leaves of *Euphrasia officinalis* (600 mg/kg) reduced hyperglycemia in alloxan-induced diabetic rats 3 and 6 h after the oral administration. The extract was devoid of hypoglycemic effects in normal rats (Porchezhian et al. 2000).

Exostema caribaeum (Jacq.) Schultes, Rubiaceae
The extract prepared from the stem barks of *Exostema caribaeum* showed hypoglycemic and antihyperglycemic effects in normal and streptozotocin-induced diabetic rats, respectively (Guerrero-Analco et al. 2007).

Eysenhardtia platycarpa Pennell. & Saff., Fabaceae
The methanol extracts from the branches and leaves of *Eysenhardtia platycarpa* markedly decreased the blood glucose levels in normal and streptozotocin-induced diabetic rats. 3-O-acetyloleanolic acid, the major constituent of the methanol extract of branches (31 mg/kg), showed a significant decrease in the glucose levels of streptozotocin-induced diabetic rats (Narvaez-Mastache et al. 2006).

Eysenhardtia polystachya (Ortega) Sarg., Leguminosae
Eysenhardtia polystachya is used in Mexico and Central America to treat DM. In streptozotocin-induced diabetic rats, the *E. polystachya* bark methanol–water extract reduced the blood glucose and increased serum insulin, body weight, marker enzymes of hepatic function, glycogen, and HDL, while there was reduction in the levels of TG, cholesterol, TBARS, LDL, and G-6-Pase. Further, the extract inhibited glycation of hemoglobin *in vitro* and exhibited antioxidant activity (Gutierrez and Baez 2014).

Fagophyrum tataricum Geartn, Polygonaceae
Common Name: Buckwheat
Buckwheat bran contains D-chiroinositol. D-chiroinositol-enriched tartary buckwheat bran extract lowered the blood glucose level in KK-Ay diabetic mice (review: El-Abhar and Schaalan 2014).

Feronia elephantum Correa, Rutaceae
Common Name: Wood apple
Feronia elephantum is a traditional medicinal tree occurring in India and elsewhere. The water extract (500 mg/kg) of *F. elephantum* fruit showed anti-DM activity in alloxan-induced diabetic rats (Sharma and Arya 2011). In another study, the ethanol extract of *F. elephantum* leaf and bark (400 mg/kg daily for 14 days) caused a decrease in blood glucose and lipids (except HDL), increase in insulin levels, and decrease in oxidative stress in alloxan-induced diabetic rats (Muthulakshmi et al. 2013). In a recent study, administration of the water extract of *F. elephantum* leaf (1000 mg/kg, p.o.) to alloxan-induced diabetic rats resulted in a decrease in blood glucose levels and improvement in glucose tolerance (Kamalakannan and Balakrishnan 2014).

Ferula assafoetida L., Apiaceae
Synonym: *Ferula asafoetida* Regel
Asafetida, an oleo-gum-resin obtained from the roots of *Ferula assafoetida* is used in traditional medicine to treat various diseases. Streptozotocin-induced diabetic rats that received the water extract of asafetida daily in the drinking water (50 mg/kg) for 4 weeks showed a decrease in the levels of serum glucose. However, the treatment did not influence the lipid profile compared to diabetic control rats (Akhlaghi et al. 2012). In another study, *F. assafoetida* extract exhibited antidiabetic activity in normal and alloxan-induced diabetic rats; it appears that it protects the β-cells from alloxan toxicity (Abu-Zaiton 2010).

Ficus amplissima Smith, Moraceae

Ficus amplissima, commonly known as kal-itchchi, has been used in folklore medicine for the treatment of diabetes. The methanol extract of the plant bark showed antidiabetic activity in streptozotocin-induced diabetic rats. The extract showed a regenerative effect on the β-cells of diabetic rats (Karuppusamy and Thangaraj 2013). These observations were confirmed by another study wherein the methanol extract of the bark of *F. amplissima* (50, 100, and 150 mg/kg) exhibited a glucose-lowering effect in normal, glucose-loaded, and streptozotocin-induced diabetic rats. Further, the extract treatment increased body weight and serum insulin levels, decreased LPO, and improved the lipid profiles in diabetic rats. Histological analysis showed the regenerative effect of the extract on the β-cells of diabetic rats (Arunachalam and Parimelazhagan 2013).

Ficus arnottiana Miq., Moraceae

Ficus arnottiana is commonly known as paras papal. It is used in traditional medicine to treat inflammation, diarrhea, diabetes, burning sensation, leprosy, scabies, wounds, skin diseases, etc. *F. arnottiana* bark (acetone extract, 100 mg/kg for 21 days) showed hypoglycemic effects in streptozotocin-induced diabetes in rats (Mazumdar et al. 2009).

Ficus asperifolia L., Moraceae

The water extract of the stem of *Ficus asperifolia* (400 and 800 mg/kg daily for 21 days) showed a significant decrease in blood glucose concentration, serum TG, total cholesterol, and LDL and a significant increase in the levels of HDL in alloxan-induced diabetic rats compared to diabetic control rats. But the serum levels of markers of liver function (GPT, GOT, and bilirubin) increased, suggesting likely liver damage (Omoniwa and Luka 2012).

Ficus benghalensis L., Moraceae (Figure 2.49)

Common/Local Names: Banyan tree, bahupada, vata, avaroha, bar, bargad, al, alu, alamaram, etc.

Description: *Ficus benghalensis* is a very large tree of 25–25 m height with branches bearing aerial roots forming accessory trunks and leaves up to 10–30 cm long and 7–20 cm wide, simple, stipulate, glabrescent above and glabrous or minute pubescent below. Fruits are 1–2 cm in diameter, without stalks, in pairs in leaf axils, and when ripe are bright red; a globular receptacle with hundreds of small fleshy flowers inside develops into a ripe fruit; bracts 4–5, cupular, 6 mm, shortly connate, obtuse persistent, tepals 3–5, shortly connate, glabrous. Male flowers are rather numerous near the mouth of the receptacle. Stem bark is smooth, light gray-white and about 1.7 cm thick; wood is moderately hard and gray or grayish-white (review: Patil and Patil 2010). It is epiphytic when young. It develops from seeds dropped by birds on old walls or on other trees.

(a)

(b)

FIGURE 2.49 *Ficus benghalensis.* (a) Young plant. (b) Twig with fruit.

Distribution: The tree is commonly found all over India from sea level to an elevation of about 3000 ft. It is also reported from Sri Lanka, Pakistan, etc.; it occurs in the wild in sub-Himalayan forests and in the slopes of the Deccan and South India. It is also planted in the plains of India.

Traditional Medicinal Uses: In Ayurveda, the water extract of leaf buds is mixed with sugar and honey for checking diarrhea; the aerial roots or leaf buds are used in hemorrhages and bleeding piles; a decoction of leaf buds and aerial roots mixed with honey was given for checking vomiting and thirst and also during fevers with a burning sensation. The aerial roots are useful in obstinate vomiting and leukorrhea and are said to be used in osteomalacia of the limbs. The bark is useful in hemoptysis, hemorrhages, diarrhea, dysentery, diabetes, enuresis, ulcers, skin diseases, gonorrhea, leukorrhea, and hyperpiesia. The leaves are good for ulcers, leprosy, allergic conditions of the skin, and abscesses. The buds are useful in diarrhea and dysentery. The fruits are refrigerant and tonic. The latex is useful in neuralgia, rheumatism and lumbago bruises, nasitis, ulorrhagia, ulitis, odontopathy, hemorrhoids, gonorrhea, inflammations, cracks of the sole, and skin diseases. The milky juice and seeds are beneficial as local application to sores and ulcers, soles of the feet when cracked or inflamed, and in rheumatism. The leaves are heated and applied as a poultice to abscesses. The tender ends of the hanging (aerial) roots are antiemetic. The seeds are cooling and tonic (review: Patil and Patil 2010).

Pharmacology

Antidiabetes: Hypoglycemic and antidiabetic compounds (bengalenoside, glycoside of pelargonidin, leucopelargonin, leucocyanidin derivative) have been isolated from this plant bark. The antidiabetes properties of leucopelargonin derivatives have been shown in rats and dogs (Cherian et al. 1992; Geetha et al. 1994; Kumar and Augusti 1994). The glucoside of leucopelargonidin from *F. benghalensis* showed significant hypoglycemic, hypolipidemic, and serum insulin-raising effects in moderately diabetic rats (Cherian and Augusti 1993). Further, this compound significantly enhanced fecal excretion of sterols and bile acids. A leucodelphinidin derivative isolated from the bark of this plant demonstrated hypoglycemic action (250 mg/kg) in both normal and alloxan-induced diabetic rats (Geetha et al. 1994).

The glycosides of leucopelargonidin from the bark exhibited hypoglycemic and hypolipidemic activities in diabetic rats and dogs (Cherian and Augusti 1993; Augusti et al. 1994); further, it increased the levels of serum insulin (Cherian and Augusti 1993). A leucocyanidin derivative from this plant also showed antidiabetes activity; it showed an insulin-sparing action (Kumar and Augusti 1994). A glycoside of pelargonin isolated from the plant bark also showed anti-DM activity (Cherian et al. 1992).

A leucodelphinidin glycoside showed hypoglycemic property in normal as well as in alloxan-induced diabetic rats (Geetha et al. 1994). The 5, 7, 3′ trimethyl ether of leucodelphinidin 3-O-α-L rhamnoside is one of the active compounds (Geetha et al. 1994). A leucodelphinidin derivative from *F. benghalensis* showed a hypolipidemic effect also in cholesterol-fed rats (Mathew et al. 2012). Another study also showed an antidiabetic activity of this plant's aerial roots in different diabetic animal models (Singh et al. 2009a). The hypoglycemic effect of the *F. benghalensis* water extract was also shown in hypercholesterolemic rabbits (Shukla et al. 2004). The anti-DM and ameliorative potential of the *F. benghalensis* bark (water extract) was shown in streptozotocin-induced diabetic rats (Gayathri and Kannabiran 2008). Anti-DM compounds, α-amyrin, β-amyrin, α-amyrin acetate, and β-amyrin acetate, were also isolated from this plant bark (Singh et al. 2009; Santos et al. 2012; Karan et al. 2013). In spite of many experimental studies on animals, detailed toxicity evaluation and human clinical trials are lacking.

Other Activities: The major pharmacological properties other than anti-DM activities include antioxidant activity (water extract of the bark), antiatherogenic activity of the bark, hypolipidemic activity of the stem bark, anti-inflammatory activity of the bark, antitumor activity of the fruit extract, analgesic activity, antiallergic action, antidiarrheal activity of hanging roots, immunomodulatory activity of areal roots, and wound healing property (review: Patil and Patil 2010). Compounds with antiatherogenic (flavonoids) and antioxidant activities have been isolated from the bark extract of this plant (Daniel et al. 1998, 2003; Shukla et al. 2004).

Phytochemicals: Caoutchouc, resin, fucosterol, flavonoids, glycoside of pelargonidin, leucodelphinidin (Geetha et al. 1994), ketones, β-sitosterol-α-D-glucose and meso-inositol (Subramanian and Misra 1978), taraxasteroltigalate, quercetin-3-galactoside, rutin, methyl ethers of leucoanthocyanins, 20-tetratriaconten-2-one, pentatriacontan-5-one, 6-heptatriacontane-10-one, and friedelin have been isolated from this plant. The leaves of *F. benghalensis* contain quercetin-3-galactoside, rutin, friedelin, taraxasterol, lupeol, and β-amyrin along with psoralen, bergapten, and β-sitosterol. The bark showed the presence of 5,7 dimethyl ether of leucopelargonidin-3-*O*-α-L rhamnoside, 5,3 dimethyl ether of leucocyanidin 3-*O*-α-D galactosyl cellobioside, glucoside, 2-*O*-tetratriaconthene-2-one, leucodelphinidin derivative, bengalenoside (a glucoside), leucopelargonin derivative, leucocyanidin derivative, and the glycoside of leucopelargonidin (review: Patil and Patil 2010).

Ficus carica L. (*Fi. sycomorus* L.), Moraceae (Figure 2.50)

Local Names: Fig tree, edible fig, common fig (this plant is commonly known as edible fig)

Distribution: This plant is thought to be native to Western Asia, Northern Africa, and/or Southern Europe. However, the plant has been cultivated since very ancient times and its exact native range is not clear. Common fig is widely naturalized in the temperate regions of Southern Australia, parts of Europe, Southern Africa, New Zealand, and parts of the United States. *F. carica* has been cultivated for a long time in various places worldwide for its edible fruit. Places with typically mild winters and hot dry summers are the major producers of edible figs (fruits can be eaten raw, dried, canned, or in other preserved forms such as jam; currently it is an important crop worldwide) (review: Mawa et al. 2013).

Description: *F. carica* is a deciduous tree growing at a medium rate. It flowers from June to September, and the seeds ripen from August to September. The flowers are monoecious (individual flowers are either male or female, but both sexes can be found on the same plant). The plant is self-fertile. Morphological data propose that the fig is gynodioecious, whereas from a functional standing point, the fig is considered dioecious with two tree morphs: capri fig and edible fig (Mawa et al. 2013).

Traditional Medicinal Uses: In traditional medicine different parts of this plant are used to treat various diseases and disorders such as gastrointestinal (colic, indigestion, loss of appetite, and diarrhea), respiratory (sore throats, cough, and bronchial problems), inflammatory, and cardiovascular disorders. The roots are used in the treatment of leukoderma, ringworms, etc. The fruits are used as an antipyretic, purgative, and aphrodisiac. It is also reported that figs have been used as laxative, cardiovascular, respiratory, antispasmodic, and anti-inflammatory remedies. In India, the fruits are used as a mild laxative, expectorant, and diuretic. It is used in liver and spleen diseases. The dry fruit of *F. carica* is

FIGURE 2.50 *Ficus carica.*

a supplement food for diabetics. The leaves are also added to boiling water and used as a steam bath for painful or swollen piles. The latex from the stems is used to treat warts and piles. It also has an analgesic effect against insect stings and bites. The unripe green fruits are cooked with other foods as a galactagogue and tonic. The roasted fruit is used as a poultice in the treatment of gumboils, dental abscesses, etc. A decoction of the young branches is an excellent pectoral in traditional medicine (review: Mawa et al. 2013).

Pharmacology

Antidiabetes: A few studies have confirmed the hypoglycemic property of *F. carica*. The stem, bark, leaf, and fruit of this plant showed a hypoglycemic property. Administration of a decoction of *F. carica* leaf to streptozotocin-induced diabetic rats resulted in a marked decrease in blood glucose levels. However, the decoction treatment did not significantly influence blood glucose levels in normal rats. Plasma insulin levels were decreased by treatment in nondiabetic rats. Thus, *F. carica* extract showed a clear hypoglycemic effect in diabetic rats. Such an effect cannot be mediated by an enhanced insulin secretion, so a yet undefined insulin-like peripheral effect may be suggested. In another study, the organic extract of the leaves showed a hypoglycemic activity with concomitant improvement in the lipid profile in streptozotocin-induced diabetic rats (review: Khan et al. 2011). Further, it has been shown that both water and chloroform extracts of the leaves exhibited similar effects in reducing hyperglycemia in streptozotocin-induced diabetic rats, while the chloroform extract had a greater effect in reducing fatty acid levels (Perez et al. 2000a).

The effect of supplementation of the *F. carica* fruit (5%, 10%, and 20% in the diet) and leaves (4%, 6%, and 8% in the diet) on hyperglycemia in alloxan-induced diabetic mice was studied. The supplementation with the fruit as well as leaves improved glycemia. In addition, hyperlipidemia and hypercholesterolemia were significantly corrected by feeding the diabetic rats with the fruit or leaf containing feed. Phytochemical analysis showed the presence of pyrogallic acid, ferulic acid, coumaric acid, galangin, cinnamic acid, quercetin, and others (El-Shobaki et al. 2010).

Another study has shown that oral consumption of the water extract (aromatic water) of *F. carica* decreased blood glucose level in normal rats and streptozotocin-induced diabetic rats (Rashidi and Noureddini 2011). The methanol extract of *F. carica* (200 mg/kg, p.o., for 21 days) showed prominent anti-DM activity in alloxan-induced diabetic rats. The treatment also reduced the levels of serum TG levels (Stalin et al. 2012).

The methanol extract of the *F. carica* stem bark lowered the blood glucose levels to normal in oral glucose tolerance test. Long-term treatment of streptozotocin-induced diabetic rats resulted in significant protection as judged from the levels of blood glucose, TGs, total cholesterol, and serum insulin levels. Phytochemical investigation of the stem bark resulted in the isolation of two new flavonol esters characterized as 3,5-dihydroxy-7,4′-dimethoxy-flavonol-3-octadec-9″-en-oxy-5-hexadecanoate and 3,5,3′-trihydroxy-7,4′-dimethoxyflavonol-3-octadec-9″-en-oxy-5-hexadecanoate, along with known compounds of β-amyrin acetate and β-sitosterol acetate (Bhat et al. 2013). β-Amyrin acetate and β-sitosterol are known hypoglycemic compounds. The plant could contain several bioactive compounds. Studies on the mechanism of action and studies using isolated active principles are needed. The edible fruit is an attractive material for further studies leading to the development of dietary supplements for diabetic patients.

Other Activities: Pharmacological studies have revealed many activities of *F. carica*. These include anticancer activity, antioxidant properties, hypolipidemic activity, hepatoprotective action, antibacterial and antifungal activities, antipyretic activity, antituberculosis property, nematicidal activity, antispasmodic and antiplatelet activity, anthelmintic activity, antimutagenic effect, and antiviral property (review: Mawa et al. 2013).

Toxicity: Detailed toxicity studies including the effect of long-term feeding with *F. caria* on the liver function have to be carried out to establish safety.

Phytochemicals: Numerous bioactive compounds such as phenolic compounds, phytosterols, organic acids, anthocyanins, triterpenoids, coumarins, and volatile compounds such as hydrocarbons, aliphatic alcohols and few other classes of secondary metabolites have been isolated from different parts of *F. carica*. Phenolic acids such as 3-*O*- and 5-*O*-caffeoylquinic acids, ferulic acid, quercetin-3-*O*-glucoside, quercetin-3-*O*-rutinoside, psoralen, bergapten, and organic acids have been isolated from the water extract of the leaves. Coumarin has been isolated from the methanol extract of the leaves. Four triterpenoids, bauerenol, lupeol acetate, methyl maslinate, and oleanolic acid, have been isolated from *F. carica* leaves. Fifteen anthocyanin pigments were isolated from the fig fruit and bark. Most of them contain cyanidin, an aglycone, and some pelargonidin derivatives. Total and individual phenolic compounds, phenolic acid, chlorogenic acid, flavones, and flavonols, have been isolated from fresh and dried fig skins of *F. carica*, and dried figs contained total higher amounts of phenolics than the pulp of fresh fruits (review: Mawa et al. 2013).

Ficus exasperata Vahl., Moraceae

Ficus exasperata is used in ethnomedicine to control diabetes in Western Nigeria. In a scientific validation study, 4 weeks of daily oral administration of the water extract from the *F. exasperata* leaf (100 mg/kg/day) to streptozotocin-induced type 1 and obese Zucker rats with type 2 diabetes decreased blood glucose and lipid profiles. Administration of the water extract also restored the microanatomy of the blood vessels to almost normal levels. This study suggests that *F. exasperata* leaf aqueous extract possesses hypoglycemic and hypolipidemic properties. The extract also reduced blood pressure in spontaneously hypertensive rats (Adewole et al. 2011).

The different extracts of the *F. exasperata* leaf were evaluated for their effect on the activities of α-amylase and α-glucosidase *in vitro*. The water extract was found to be the most active extract in inhibiting the enzymes with an IC_{50} value of 3.7 and 1.7 µg/mL against α-amylase and α-glucosidase, respectively (Kazeem et al. 2013b). Inhibition of these enzymes by the extract may explain, at least partly, its anti-DM mechanism. Further, the water extract of *F. exasperata* leaves (100 mg/kg) ameliorated streptozotocin-induced nephropathy in streptozotocin-induced diabetic rats (Adewole et al. 2012).

Ficus hispida L., Moraceae

The water-soluble portion of the ethanol extract of *F. hispida* bark showed significant reduction of blood glucose level in both normal and alloxan-induced diabetic rats. However, the reduction in the blood glucose level was less than that of glibenclamide. There was a significant increase in the glycogen content of the liver, skeletal muscle, and cardiac muscle. The herbal drug also increased the uptake of glucose by rat hemidiaphragm (Ghosh et al. 2004).

Ficus lutea Vahl, Moraceae

Common Name: Dahomey rubber tree

Ficus lutea is used traditionally to treat diabetes in Africa. It is commonly known as giant-leaved fig. Acetone extracts of *F. lutea* leaf showed promising α-amylase and α-glucosidase inhibitory activity *in vitro*. The EC_{50} value for α-amylase inhibition was 9.4 µg/mL (Olaokun et al. 2013). In the hepatoma cell line, the acetone extract of the dried leaf was as effective as metformin in decreasing extracellular glucose concentration by approximately 20%. In the pancreatic insulin secretory assay, the extract was four times greater in its secretory activity than commercial glibenclamide. *F. lutea* extract significantly increased glucose uptake in the primary muscle cells, primary fat cells, C2C12 muscle, and H-4-II-E liver cells; the extract may act by increasing the activity of cell surface glucose transporters. *F. lutea* possessed substantial *in vitro* activity related to glucose metabolism. The authors conclude that while the clinical effectiveness of *F. lutea* is not known, this plant species does possess the ability to modify glucose metabolism (Olaokun et al. 2014).

Ficus microcarpa L., Moraceae, Chinese banyan

The ethanol extract of *Ficus microcarpa* leaves showed potential hypoglycemic action against alloxan-induced diabetic rats; this effect may be attributed to the increased levels of antioxidant enzymes and protection of pancreatic β-cell integrity as evidenced by histopathological observations (Ashok Kumar et al. 2007).

Ficus pumila L., Moraceae

Common Name: Creeping fig

The ethanol extract of *Ficus pumila* showed hypoglycemic and hypolipidemic effects in streptozotocin-induced diabetic rats, which is comparable to that of glibenclamide (Thamotharang et al. 2013).

Ficus racemosa L., Moraceae

Synonym: *Ficus glomerata* Roxb., *Ficus lucescens* Blume (Figure 2.51)

Common/Local Names: Cluster fig, apushpaphalasambandha, gular, umar, athi, etc.

Description: *Ficus racemosa* is a moderate-sized to large spreading evergreen tree with glabrous, pubescent young shoots. The leaves are membranous, petiolate, stipulate, obovate-oblong or elliptic-lanceolate, subacute, glabrous, and rarely softly pubescent above and asperous beneath. The receptacles are subglobose or pyriform, reddish when ripe, and produced in short leafless branches in large clusters arising from the main branch.

Distribution: It is distributed in outer Himalayas and plains and low hills of India, extending from Khasia Mountains to Rajputana. It is also found in Burma, Deccan Peninsula, and Sri Lanka.

Traditional Medicinal Uses: It is an astringent, cooling, acrid, galactagogue, aphrodisiac, carminative, anorectic, expectorant, and tonic and used to allay thirst and sweat, styptic, biliousness, burning sensations, fatigue, disease of the vagina, hydrophobia, fatigue, urinary discharge, leprosy, menorrhagia, piles, intestinal worms, chronic bronchitis, disease of kidney and spleen, inflammation, stomachic, diarrhea, dysentery, small pox, cancer, diabetes, urinary diseases, gonorrhea, toothache, and backache (Kirtikar and Basu 1975).

Pharmacology

Antidiabetes: *F. racemosa* is used in traditional systems of medicine for the treatment of diabetes. *F. glomerata* has been claimed to possess antidiabetic properties by many investigators. The effects of powdered *F. racemosa* (*F. glomerata*) fruits on blood glucose levels in groups of normal and alloxan-induced diabetic rabbits were studied. In normal rabbits, administration of 1, 2, 3, and 4 g/kg body weight of the *F. glomerata* fruit powder lowered the blood glucose levels significantly. The methanol extract of the fruit powder also produced significant hypoglycemia but the water extract could not produce this effect. In alloxan-induced diabetic rabbits, treatment with the fruit powder (2, 3, and 4 g/kg) produced a significant fall in blood glucose levels. The methanol extract of the fruit also produced a

FIGURE 2.51 *Ficus racemosa.*

significant decrease in blood glucose in diabetic rabbits, but the water extract could produce only a slight fall in glucose levels in these rabbits. Therefore, it is conceivable that the indigenous plant contains more than one type of hypoglycemic principles that seems to act by producing an organotropic effect on the β-cells resulting in an increased secretion of insulin. In addition, it is also possible that the drug acts by providing certain necessary elements to the β-cells, especially in alloxan-induced diabetic rabbits (Akhtar and Qureshi 1998). In another study, the hypoglycemic activity of the ethanol extract of the leaves of *F. glomerata* was tested. The results showed that it has a significant antihyperglycemic effect in the alloxan-induced rat model of DM (Sharma et al. 2010d). The hypoglycemic effect of the leaf extract was confirmed by another study. In this study, graded doses (100, 200, and 300 mg/kg) of the ethanol extract of leaves of *F. racemosa* produced significant dose-dependent reductions in blood glucose levels 6 h after administration of the extract in both normal and alloxan-induced diabetic rats (Patil et al. 2010).

The glucose-lowering activity of *F. racemosa* (bark) was also shown in normal and alloxan-induced diabetic rats (Bhaskara et al. 2002). The bark (95% v/v ethanol extract) of this plant showed a hypoglycemic activity in alloxan-induced diabetic rats (Kar et al. 2003). Further, the antihyperglycemic and hypolipidemic activities of the ethanol extract of the bark of this plant were shown in alloxan-induced diabetic rats. Oral administration of the extract (300 mg/kg) to alloxan-induced diabetic rats restored the levels of blood glucose, lipids, and lipoproteins. These effects were comparable to that of glibenclamide (Sophia and Manoharan 2007). In normal rats, the water and alcohol extracts (400 mg/kg) of *F. racemosa* bark showed almost the same level of hypoglycemic activity, while in alloxan-induced diabetic rats, the decrease in blood glucose level by the water and alcohol extracts was found to be about 27% and 47%, respectively (Sachan et al. 2009). This activity was confirmed by another study where the paper has been published with the synonym (*F. glomerata*). In this study bark extract also showed antidiabetic activity in alloxan-induced diabetic rats (Sharma et al. 2010).

The effects of the *F. racemosa* fruit extract and fraction on fasting serum glucose levels of normal, type 1, and type 2 diabetic model rats were studied. The 80% ethanol (20% water) extract and its water-soluble fraction of *F. racemosa* fruit did not show any serum glucose-lowering effect on nondiabetic and type 2 diabetic rats at the fasting condition, whereas the extract showed significant hypoglycemic effect on type 1 diabetic model rats. Both the extract and fraction were consistently active in nondiabetic and types 1 and 2 diabetic model rats when fed simultaneously with glucose load. On the contrary, they were ineffective in lowering blood glucose levels when fed 30 min prior to glucose load. The butanol-soluble part of the ethanol extract exhibited significant antioxidant activity in DPPH free radical scavenging assay. 3-*O*-(*E*)-Caffeoyl quinate was isolated for the first time from this plant, which also showed significant antioxidant activity (Jahan et al. 2009). Although many studies were carried out, they were not focused to identify active principles and details of the mechanism of action and therapeutic utility.

Clinical Trial: The hypoglycemic effect of *F. racemosa* was studied in a group of diabetic subjects in Lahore taking an oral hypoglycemic drug. The extract (5 mL) of the bark of *F. racemosa* (about 100 mg) was given orally to diabetic patients twice per day for 15 days. Blood samples for the estimation of blood glucose and parameters of liver and renal functions were estimated. It was observed that after taking the herb in combination with the drug, the blood glucose level (fasting and after breakfast) was markedly decreased in both male and female patients. To rule out herb toxicity, liver and renal function tests of patients were also performed, which were observed to be in the normal range (Gul-e-Rana et al. 2013). This is a preliminary study; only one dose was selected arbitrarily for the study, and liver toxicity at higher doses for a longer duration cannot be ruled out.

Other Activities: The plant extract possesses antifungal activity against cutaneous pathogens (Vonshak et al. 2003).

Phytochemicals: Phytochemicals isolated include tannins, caoutchouc, β-sitosterol glucoside, friedelin, lupeol, tetracyclic triterpene-glucanol acetate, leucoanthocyanins, α-amyrin acetate, and glauanol (*The Wealth of India* 1992).

Ficus religiosa L., Moraceae

Common Names: Peepal, ashvattha, pipul, arasu, arayal, etc.

Description: *Ficus religiosa* is a large, deciduous tree, epiphytic when young; leaves are simple and cordate in shape with long petiole, entire margin and a distinctive extended tip. The receptacle is in axillary pairs, sessile, smooth, depressed, spheroidal, and dark purple when ripe.

Distribution: It is distributed in sub-Himalayan forests, Bengal, and Central India in the wild. It is commonly planted throughout India and Sri Lanka, less frequently in Burma, and rarely in Malaya region.

Traditional Medicinal Uses: It is bitter, sweet, acrid, antibacterial, refrigerant, cooling, alexipharmic, astringent, aphrodisiac, alternative, purgative, and laxative and used in the treatment of diseases of the blood, vagina, uterus, and heart leukorrhea, burning sensation, foul taste, thirst, biliousness, gout, stomatitis, ulcers, bone fracture, urinary discharge, inflammation, vomiting, gonorrhea, scabies, asthma, obstinate hiccup, cracked foot, fistula, toothache, snake venom, skin disease, atrophy, cachexia, carbuncles, cholera, dysuria, fever, gravel, otitis, pimples, rheumatism, skin diseases, small pox, and sores (Kirtikar and Basu 1975).

Antidiabetes: The bark extract of this plant showed antidiabetic property (Brahmachari and Augusti 1961).

The water extract of *F. religiosa* bark (25, 50, and 100 mg/kg) showed a glucose-lowering effect in streptozotocin-induced diabetic rats and improved glucose tolerance in normal glucose-loaded rats. The extract also showed a significant increase in serum insulin, body weight, and glycogen content in the liver and skeletal muscle of streptozotocin-induced diabetic rats while there was a significant reduction in the levels of serum TG and total cholesterol. Besides, the extract showed significant anti-LPO effect in the pancreas of diabetic rats (Pandit et al. 2010). In another study, the methanol extract of *F. religiosa* bark also exhibited significant antihyperglycemic activity in streptozotocin-induced diabetic rats. Preliminary phytochemical investigation revealed the presence of phenolic and flavonoid compounds (Verma et al. 2012b).

Other Activities: The fruit extract has antibacterial activity and anticancer activity in the potato disc bioassay.

Phytochemicals: Tannins, caoutchouc, etc. (*The Wealth of India* 1992)

Ficus sycomorus L., Moraceae

The water extract of the stem of *Ficus sycomorus* showed remarkable antidiabetic potential in alloxan-induced diabetic rats. It lowered blood glucose levels like insulin in a dose-dependent manner (Njagi et al. 2012).

Filipendula ulmaria (L.) Maxim. Rosaceae

Filipendula ulmaria is used as a substitute to tea without toxic properties. The plant infusion exhibited antihyperglycemic activity in alloxan-induced diabetic rats and mice and improved glucose tolerance in epinephrine-induced hyperglycemia in rats (Барнаулов and Поспелова 2005).

Fraxinus excelsior L., Oleaceae

Synonyms: *Fraxinus americana* L., *Fraxinus elatior*, *Fraxinus ornus* L., etc. (Figure 2.52)

Local Names: Common ash, English ash, European ash, golden ash, Venus of the forest, etc.

Distribution: *Fraxinus excelsior* is the most widely distributed ash species in Europe. Its distribution extends across Europe from the Atlantic Coast in the west into continental Russia (*F. excelsior* seeds are consumed as a food and condiment and used in folk medicine).

Description: *F. excelsior* is a deciduous tree up to 40 m in height. The twigs are greenish-gray, with firm and ridged bark. The plant grows in moist areas with deep soil, generally in hollows and gullies at the medium and subalpine mountain levels. The leaves are large, measuring between 20 and 35 cm long, divided into 9–13 folioles. These have a lanceolate form, serrated margins, and with lamina base touching the principal vein; the leaves are glabrous except for tufts of hair at the base of the underside of the midrib of each leaflet. The flowers are small, polygamous in bunches, lacking petals, flower stigmas long appearing before leaves.

FIGURE 2.52 *Fraxinus excelsior.* Twig with immature fruit. (From Wikimedia Foundation, Inc.)

The fruits are slightly twisted ellipsoid samaras 3–5 cm long, with a wide wing aiding wind dispersal. There are variants of *F. excelsior* with different morphological features (review: Kostava 2001).

Traditional Medicinal Uses: The seeds of *F. excelsior* are traditionally used as a potent hypoglycemic agent. Before the introduction of quinine, a bitter tonic made from the bark of *F. excelsior* was used as a remedy for malaria. An infusion from the leaves has purgative and laxative properties. The seeds have long been considered an aphrodisiac. A bark decoction served to control kidney function and as a febrifuge and diuretic. The leaves were used for their cathartic properties. Other medicinal uses of the ash are for vomit induction, peritonitis, as a substitute for cinchona bark, constipation, arthritis, adder bite, intestinal worms, insect bites, nervous disease, diabetes, syphilis, healing wounds, tuberculosis, and dropsy. Ash bark has been employed as a bitter tonic and astringent and is said to be valuable as an antiperiodic. On account of its astringency, it has been used, in decoction, extensively in the treatment of intermittent fever and ague. It has been considered useful to remove obstructions of the liver and spleen.

Pharmacology

Antidiabetes: The hypoglycemic effect of the water extracts from the *F. excelsior* seed was investigated in normal and streptozotocin-induced diabetic rats. Oral administration of the water extract (20 mg/kg, single dose or 12 daily doses) produced a significant decrease of blood glucose levels in both normal and streptozotocin-induced diabetic rats. In addition, no changes were observed in basal plasma insulin concentrations after the extract treatments in both normal and streptozotocin-induced diabetic rats indicating that the extract exerted its pharmacological activity without affecting insulin secretion (Maghrani et al. 2004a).

To understand the underlying mechanism of the hypoglycemic activity of the water extract of *F. excelsior* in normal and streptozotocin-induced diabetic rats, it was administered intravenously and the blood glucose changes, plasma insulin, and glucose in the urine were determined within 4 h after starting the treatment. The water extract at a dose of 10 mg/kg/h produced a significant decrease in blood glucose levels in normal rats and even more in diabetic rats. This hypoglycemic effect might be due to an extrapancreatic action of the water extract, since the basal plasma insulin concentrations were unchanged after the treatment. A potent increase of glycosuria was observed in both normal and diabetic rats. The authors concluded that the water extract caused a potent inhibition of renal glucose reabsorption. This renal effect might be at least one mechanism explaining the observed hypoglycemic activity of this plant in normal and diabetic rats (Eddouks and Maghrani 2004).

Clinical Trial: The clinical efficacy of the seed extract (FraxiPure, Naturex), containing 6.8% of nuzhenide and 5.8% of GI3 (w/w), was assessed on plasma glucose and insulin levels against glucose (50 g) induced postprandial glycemia in healthy volunteers. The extract lowered the incremental postprandial plasma glucose concentration as compared to placebo at 45 and 120 min. It reduced the glycemic area under the blood glucose curve. The seed, also, induced a significant secretion of insulin 90 min after glucose administration. However, the insulinemic area under the blood insulin curve was not different than the placebo. No adverse events were reported (Visen et al. 2009).

In a recent longitudinal, randomized, crossover, double-blind, placebo-controlled 7-week nutritional intervention study, the potential benefit of the seed extracts on glucose homeostasis and associated metabolic markers in nondiabetic overweight/obese subjects was studied. Compared to baseline, administration of 1 g of Glucevia® (seed extract) for 3 weeks resulted in a significantly lower incremental glucose area under the curve and significantly lower 2 h blood glucose values following an oral glucose tolerance test. No significant changes were found in the control group. Furthermore, significant differences were found between responses in the control and Glucevia groups with respect to serum fructosamine and plasma glucagon levels. Interestingly, administration of Glucevia significantly increased the adiponectin/leptin ratio and decreased fat mass compared to control. The study concluded that the administration of an extract from *F. excelsior* seeds/fruits in combination with a moderate hypocaloric diet may be beneficial in metabolic disturbances linked to impaired glucose tolerance, obesity, insulin resistance, and inflammatory status, specifically in older adults (Zulet et al. 2014). Long-term toxicity studies, details of mechanisms of action, clinical studies, and identification of active principles are warranted.

Other Activities: Other reported pharmacological properties include antioxidant activity, antiviral effect, antimicrobial action, immunomodulatory property, wound healing, and photodynamic damage prevention (review: Kostava 2001).

Toxicity: The leaves and fruits are reportedly poisonous to cattle.

Phytochemicals: Ash contains flavonoids, hydroxycoumarins, phenylethanoids, and secoiridoid glucosides; the bark contains the bitter glucoside fraxin, fraxetin, tannin, quercetin, mannite, a little volatile oil, and malic acid (review: Kostava 2001).

Fumaria parviflora Lam., Fumariaceae/Papaveraceae

Common Names: Araka, pitpapra, thuva, etc.

Description: *Fumaria parviflora* is a much branched annual with divided and narrow segmented leaves and small, white or rose, or purplish flowers in terminal or opposed racemes. The fruits are globose and 1-seeded.

Distribution: It is seen in the Indo-Gangetic plains, Lower Himalayas, and Nilgiri mountains up to 9000 ft. It is also found as a weed among cultivated crops.

Traditional Medicinal Uses: The plant is bitter, slightly acrid, anodyne, depurative, astringent, diuretic, alternative, and laxative and is used in the treatment of dyspepsia, scrofulous skin affections, body pain, boils, carcinoma, fever, and sweating.

Antidiabetes: *F. parviflora* leaf extracts lowered blood glucose in normoglycemic and alloxan-treated hyperglycemic rabbits (Akhtar et al. 1984). In another study, the methanol extract of this plant (150 and 250 mg/kg, i.p., daily for 7 days) showed a potent glucose-lowering effect in streptozotocin-induced diabetic rats. However, no significant effect in the glucose levels was observed in normal rats (Fathiazad et al. 2013). In contrast, in a recent report, oral administration of *F. parviflora* for 6 weeks to streptozotocin-induced diabetic rats improved TG, total cholesterol, and HDL serum levels but no significant effects on glucose and LDL (Ghosian-Moghadam et al. 2014). Further studies are required to get a clear understanding of its action considering the route of administration, dose levels, type of extract, etc.

Other Activities: Significant oral antipyretic activity in rabbits has been reported for the extracts of *F. parviflora*. An aqueous methanolic extract of the plant exhibits a selective protective effect against paracetamol-induced hepatotoxicity. The ethanol extract of *F. parviflora* is active against gastrointestinal trichostrongylids in artificially infected lambs (review: Ajikumaran and Subramoniam 2005).

Phytochemicals: Alkaloids, methylhydrastine and *N*-methylhydrasteine (Forgacs et al. 1974), protopine, cryptopine, D-bicuculline, l-adlumine, fumaramine, D-α-hydrastine, villantine, stylopine, synactine, nor-pallidine, and β-sitosterol are isolated from this plant (review: Ajikumaran and Subramoniam 2005).

Galega officinalis L., Leguminosae/Fabaceae

Common Names: Goat's rue, Italian fitch

Galega officinalis has been used as a traditional botanical (tea infusion) for over 3000 years to relieve polyuria and halitosis. The leaves and flowering tops of *G. officinalis* are diaphoretic, diuretic, galactagogue, and hypoglycemic. The herbal extract contains guanidine and galegine as the major chemical components. Studies demonstrated that galegine and other guanidine derivatives reduced blood sugar levels. These compounds, although they have an antidiabetic effect, are too toxic for clinical use. Studies on guanidine and galegine analogues for antidiabetic activity culminated in the discovery of metformin, introduced as Glucophage, by Jean Sterne, in 1957. The hypoglycemic drug metformin is now one of the leading therapies for DM (Patade and Marita 2014).

Garcinia kola Heckel, *Guttiferae*

Garcinia kola is a popular medicinal plant in Nigeria. Oral administration of the water extract of *G. kola* seed at a concentration of 200 mg/kg, over a period of 21 days, significantly decreased the levels of blood glucose and increased the activity of superoxide dismutase. The treatment, however, had no significant effect on the activity of the catalase (Kingsley et al. 2010). The saponin extract from the root of *G. kola* (100, 200, and 400 mg/kg daily for 7 days) was evaluated for antidiabetic activity in alloxan-induced diabetic rats. The saponin extract (100 mg/kg) produced a reduction of 36% in blood glucose after the third day treatment as compared to the 31% observed for metformin. A 200 mg/kg dose of saponin produced a maximum reduction of about 73% after the seventh day treatment as compared to 36% observed for metformin. Thus, the saponin extract from the root of *G. kola* demonstrated a remarkable glucose-lowering activity (Smith et al. 2012). Kolaviron, a bioflavonoid complex isolated from the seeds of *G. kola*, possessed significant hypoglycemic effect in alloxan-induced diabetic rats. Kolaviron also inhibited rat lens aldolase activity *in vitro*. Further, kolaviron showed remarkable protective effects on renal, cardiac, and hepatic tissues of streptozotocin-induced diabetic rats (review: Akinmoladun et al. 2014).

Garuga pinnata Roxb., Burseraceae

The water extract of *Garuga pinnata* bark showed increase in the liver glycogen and serum insulin and a decrease in the fasting blood glucose levels and glycated hemoglobin levels in streptozotocin–nicotinamide-induced type 2 DM (Shirwaikar et al. 2006).

Gastrodia elata Blume, Orchidaceae

Gastrodia elata, a widely used traditional herbal medicine, was reported with anti-inflammatory and antidiabetes activities. Administration of the water extract of *G. elata* (1 g/kg daily for 8 weeks) to HFD-fed rats reduced insulin resistance by decreasing fat accumulation in adipocytes by activating fat oxidation and potentiating leptin signaling in obese rats. These effects are mainly a result of the action of 4-hydroxybenzaldehyde and vanillin present in the extract (Park et al. 2011).

Fermented *G. elata* was prepared by fermentation with *Saccharomyces cerevisiae*. Administration of fermented *G. elata* (100 or 20 mg/kg daily for 3 weeks, p.o.) to streptozotocin-induced type 1 diabetic rats resulted in significant reduction in blood glucose in oral glucose tolerance test, fasting blood glucose, and glycosylated hemoglobin. In addition, the treatment improved the lipid profile and increased insulin levels (Kwon et al. 2013).

The ethanol extract of *G. elata* ameliorated obesity, insulin resistance, dyslipidemia, hypertension, and fatty liver in HFD-fed rats (Kho et al. 2013). A recent study has shown that the ethanol extract of *G. elata* (100 mg/kg/day for 8 weeks) attenuated lipid metabolism and endothelial dysfunction in a high-fructose diet-fed rats. Treatment with the extract significantly suppressed the increments of epididymal fat weight, blood pressure, plasma TG, total cholesterol levels, and oral glucose tolerance. In addition, the treatment markedly prevented increase of adipocyte size and hepatic accumulation of TGs. The plant extract ameliorated endothelial dysfunction by downregulation of endothelin-1 and adhesion molecules in the aorta. Moreover, the extract recovered the impairment of vasorelaxation to acetylcholine and levels of endothelial nitric oxide synthase expression and markedly induced upregulation of phosphorylation of

AMPK in the liver, muscle, and fat (Kho et al. 2014). Both water and ethanol extracts appear to contain anti-DM compounds and they may have more than one mechanism of action. Fermentation may yield a different set of compounds with different mechanisms of action.

Genista tenera (Jacq. ex Murr.) O. Kuntze, Leguminosae

Common Name: Madeira broom

The butanol extract of the plant decreased blood glucose levels in streptozotocin-induced diabetic rats. In the butanol extract, 21 monoglycosyl and 12 diglycosyl flavonoids were detected. *In vitro* toxicity studies showed no evidence for acute cytotoxicity or genotoxicity of the extract. This is the first report on antidiabetic activity of the genus *Genista* (Rauter et al. 2009).

Gentiana olivieri Griseb., Gentianaceae

Gentiana olivieri showed a glucose-lowering effect in streptozotocin-induced diabetic rats. Through *in vivo* bioassay-guided fractionation, isoorientin, a known C-glycosylflavone, was isolated from the ethyl acetate fraction by silica gel column chromatography as the main active ingredient from the plant. Isoorientin exhibited hypoglycemic and antihyperlipidemic effects at a 15 mg/kg dose (Sezik et al. 2005).

Geranium robertianum L., Geraniaceae

Geranium robertianum is a traditional medicinal herb and its leaves are used in Egypt to treat DM. Oral administration of the decoction of *G. robertianum* leaf to Goto-Kakizaki type 2 diabetic rats, daily for weeks, lowered the plasma glucose levels. Furthermore, the treatment improved liver mitochondrial respiratory parameters and increased oxidative phosphorylation efficiency (Ferreira et al. 2010).

Ginkgo biloba L., Ginkgoaceae

Synonyms: *Salisburia adiantifolia* Sm., *Salisburia macrophylla* Reyn., *Pterophylla salisburiensis* (Figure 2.53)

Common Names: Kew tree, ginkgo, maidenhair tree, etc.

Description: The ginkgo tree has flourished in forests for over 150 million years and hence is called a "living fossil." It is the oldest living tree species. It is a dioecious tree with male and female reproductive

(a)

(b)

FIGURE 2.53 *Ginkgo biloba.* (a) Trees. (b) Leaves.

organs on separate trees. They have a large trunk with a girth of about 7 m and a height of about 30 m. Young *Ginkgo biloba* trees are like conifer and exhibit branching dimorphism. Leaves that grow in clusters are golden yellow in fall during senescence. The leathery leaves are uniquely shaped with two lobes and resemble the maidenhair fern in shape and venation. The pollination process involves the male microstrobili bearing loosely distributed sporangiophores containing microspores with male gameto-phytes and the female pendulous pairs of ovules borne on the shoots. These trees begin to reproduce after about 20 years by developing naked seeds (nuts) with an outer fleshy layer (fruits) (review: Mahadevan and Park 2008).

Distribution: The ginkgo tree is the only surviving member of the Ginkgoaceae family, class of Ginkpoosida (Gingoatae), rediscovered in Asian graced temple gardens by Kaempfer in 1670. It occurs in many parts of the world such as Japan, China, Korea, and Malaysia. The tree is also grown in many parts of Europe and the United States for mainly its ornamental value. It grows well in most places (McKenna et al. 2001).

Traditional Medicinal Uses: In traditional Chinese medicine, the fruits and seeds of *G. biloba* have been used for the treatment of asthma, cough, and enuresis (Cong et al. 2011). Both the leaves and nuts of this tree have been in use for the past several centuries in traditional Chinese medicine. For over 5000 years, the seeds (nuts) have been known to treat pulmonary disorders (like asthma, cough, and enuresis), alcohol abuse, and bladder inflammation, while the leaves have been mainly used to treat heart and lung dysfunctions and skin infections. However, it was only in the last 20–30 years that the use of the ginkgo leaf and its standardized extract formulation, EGb 761, originated in Germany and now is the most used form of supplement for cognitive ailments in the United States (review: Mahadevan and Park 2008).

Pharmacology

Antidiabetes: The root extract of *G. biloba* showed antihyperglycemic, antioxidant, and antihyperlipid-emic activities in streptozotocin-induced diabetic rats (Cheng et al. 2013). It was reported that EGb761 (a standardized and well-defined product extract of *G. biloba* leaves) could protect pancreatic β-cells against HFD-induced apoptosis in rats (Dong et al. 2009). In addition, it has been observed that EGb761 could increase insulin secretion from INS-1 cells (Choi et al. 2007).

EGb761 has beneficial effects on the treatment of multiple diseases, including diabetes and dys-lipidemia. The effects of EGb761 on insulin sensitivity were evaluated in an obese and insulin-resistant mouse model, established through chronic feeding of C57BL/6J mice with an HFD. Mice fed with an HFD for 18 weeks developed obesity, dyslipidemia, and insulin resistance compared to control mice fed with a standard laboratory chow. Oral treatment of the HFD-fed mice with EGb761 (100, 200, and 400 mg/kg) dose-dependently enhanced glucose tolerance in oral glucose tolerance test and decreased both the insulin levels and the HOMA-IR index values. EGb761 treatment also ameliorated HFD-induced obesity, dyslipidemia, and liver injury, as indicated by decreases in body weight, blood lipid levels, and liver function test (Cong et al. 2011). In further mechanism studies, EGb761 was found to protect hepatic IR β and IR substrate 1 from HFD-induced degradation and to keep the AMPK, which plays a crucial role in reducing lipotoxicity, from HFD-induced inactivation (Cong et al. 2011). The ethanol extract of the plant leaf exerted a protective effect against experi-mental hypoxia of streptozotocin-induced diabetic rat myocardium (Welt et al. 2001) and showed neuroprotective activity on the enteric nervous system of streptozotocin-induced diabetic rats (Perez da Silva et al. 2011).

Clinical Trial: Recently, 16 randomized controlled trials were reviewed to evaluate the effectiveness and safety of a *G. biloba* extract for patients with early diabetic nephropathy. *G. biloba* extract decreased the urinary albumin excretion rate, fasting blood glucose, serum creatinine, and blood urea nitrogen. The extract also improved hemorheology. The methodological quality in the included studies was low. Three studies had observed adverse events. It is concluded that *G. biloba* extract is a valuable drug that has prospect in treating early diabetic nephropathy, especially with high urinary albumin excretion rate and baseline level. Long-term, double-blinded, randomized, controlled trials with large sample sizes are still needed to provide stronger evidence (Zhang et al. 2013).

In another randomized, double-blind, placebo-controlled trial, the effect of standardized *G. biloba* leaf, dry extract (160 mg/day for 9 months and then 240 mg/day for another 9 months), was compared with the placebo on the microcirculation lesions of the eye in type 2 DM patients. Although the values of total conjunctival index decreased during the study in the extract-treated group, there were no significant differences between this group and the placebo group. The authors conclude that further studies are needed to determine the beneficial effects of the ginkgo extract (Spadiene et al. 2013).

Other Activities: Other validated pharmacological activities include antioxidant activity, prevention of neurodegenerative diseases, cardioprotective effects, anticancer activity, antidepressant activity, beneficial effects in the treatment of tinnitus, and beneficial effects in the treatment of old age-related disorders (review: Mahadevan and Park 2008).

Phytochemicals: The two main pharmacologically active groups of compounds present in the *Ginkgo* leaf extract are the flavonoids and the terpenoids. Flavonoids present in the ginkgo leaf extract are flavones, flavonols, tannins, biflavones (amentoflavone, bilobetol, 5-methoxybilobetol, ginkgetin, isoginkgetin and sciadopitysin), and associated glycosides of quercetin and kaempferol attached to 3-rhamnosides, 3-rutinosides, or *p*-coumaric esters. Two types of terpenoids are present in ginkgo as lactones (nonsaponifiable lipids present as cyclic esters): ginkgolides and the bilobalide. Ginkgolides are diterpenes with five types, A, B, C, J, and M, where types A, B, and C account for around 3.1% of the total ginkgo leaf extract. Bilobalide, a sesquiterpene trilactone, accounts for the remaining 2.9% of the total standardized ginkgo leaf extract (review: Mahadevan and Park 2008).

Girardinia heterophylla Decne, Urticaceae
The plant is known as Himalayan nettle. An epimer of β-sitosterol, γ-sitosterol, has been isolated from the plant leaves (Tripathi et al. 2013). The leaves of the plant are edible and γ-sitosterol is known to have antidiabetes properties (Balamurugan et al. 2011).

Globularia alypum L., Globulariaceae
The effect of oral administration of the water extract of *Globularia alypum* leaves on blood glucose levels in normal and streptozotocin-induced diabetic rats was studied. In normal rats, single and repeated oral administration of the water extract did not change blood glucose levels. However, in streptozotocin-induced diabetic rats, single and repeated administration of the water extract produced significant decrease in blood glucose levels. The extract treatments did not affect insulin secretion in both normal and streptozotocin-induced diabetic rats, indicating that mechanism(s) by which the extract decreased blood glucose levels was extrapancreatic (Jouad et al. 2002).

The effect of administration (i.p.) of globularin on blood glucose levels in normal and streptozotocin-induced diabetic rats was evaluated. Globularin, an iridoid glucoside, was isolated from the leaves of *G. alypum*. In normal and streptozotocin-induced diabetic rats, single (i.p.) administration of globularin (100 mg/kg) produced significant decrease of blood glucose levels. In prolonged treatment study, the repeated (i.p.) administration of globularin significantly decreased the blood glucose levels when compared to diabetic control rats. In addition, daily injection of globularin reduced serum levels of total cholesterol and TGs in diabetic rats. The acute toxicity test demonstrated that globularin is not lethal up to an i.p. dose of 1 g/kg injection (Merghache et al. 2013).

Glycine max (L.) Merr., Fabaceae

Common Name: Soybean
Glycine max (soybean) is being used as a fresh, fermented, or dried food. The petroleum ether, alcohol, and water extracts of *G. max* seeds were evaluated for their antihyperglycemic activity in alloxan-induced diabetic mice. The petroleum ether (100 mg/kg) and ethanol extracts (100 mg/kg) did not reduce serum glucose level. The water extracts (100, 200, and 400 mg/kg) significantly reduced serum glucose levels (antihyperglycemic effect) in acute and chronic studies. The antihyperglycemic effect of the water extract showed onset at 2 h and peak effect at 4 h, and the antihyperglycemic effect was sustained for 24 h. In chronic study (28 days) also, reduction in serum glucose level was observed and the effect remained after withdrawal also for the next 7 days. In oral glucose tolerance test, increased

glucose tolerance was observed in water extract-treated nondiabetic as well as diabetic mice (Badole and Bodhankar 2009). When alloxan-induced diabetic rats were fed with diet containing 20% of soybean seeds for 4 weeks, blood glucose concentration decreased, body weight increased, and lipid profiles improved in treated diabetic rats. The treatment also minimized histopathological changes in the liver (Amer 2012). Soy phytoestrogen genistein activated PPAR-γ (review: Wang et al. 2014). Thus, genistein may show anti-DM activity.

Glycyrrhiza glabra L., Fabaceae

The dried root and rhizome of *G. glabra* are known as radix glycyrrhizae.

Common Names: Liquorice, licorice, sweet wood, aslussiera, madhuka, klitaka, yashtimadhu, jethi-madh, mulhatti, irattimadhuram, etc.

Description: It is a perennial herb/subshrub for subtropical and temperate zones. The underground stem grows horizontally up to 2 m length and is highly branched consisting of a short taproot with a large number of rhizomes. The diameter of the root varies from 0.75 to 2.5 cm. It has a characteristic pleasant sweet taste. The leaves are alternate and pinnate, with yellow-green leaflets of 4–7 pairs that are covered with soft hairs on the underside. The flowers appear in the axil of terminal and axillary leaves in raceme and are pealike and lavender to purple in color. The seed pod is 2–2.5 cm long containing 2–5 seeds (Vispute and Khopade 2011).

Distribution: *G. glabra* is distributed in Southern Europe, Syria, Iran, Afghanistan, Russia, China, Pakistan, and Northern India. Large-scale commercial cultivation is seen in Spain, Sicily, and England (review: Vispute and Khopade 2011).

Traditional Medicinal Uses: Traditionally the plant has been recommended for gastric and duodenal ulcers, dyspepsia, and allergenic reactions. In folk medicine, it is used as a laxative, emmenagogue, contraceptive, galactagogue, antiasthmatic drug, and antiviral agent. *Glycyrrhiza* roots are used for their demulcent and expectorant property. It is useful in treating anemia, gout, sore throat, tonsillitis, flatulence, sexual debility, hyperdipsia, fever, coughs, skin diseases, swellings, acidity, leukorrhea, bleeding, jaundice, hiccup, hoarseness, and bronchitis (review: Sharma and Agrawal 2013).

Antidiabetes: Glycerin exhibited a potent PPAR-γ ligand-binding activity and therefore reduced the blood glucose level in knockout diabetic mice (KK-Ay). This finding is of much significance as licorice has also been traditionally used as an artificial sweetening agent and could be helpful in insulin resistance syndrome prevalent in the modern society (Kuroda et al. 2004). Glycyrrhizin has also exhibited antidiabetic activity in an NIDDM model (Takii et al. 2000). 5-*O*-Formylglabridin, (2R,3R)-3,40,7-trihydroxy-30-prenylflavane, echinatin, (3*R*)-20,30,7-trihydroxy-40-methoxyisoflavan, kanzonol X, kanzonol W, shinpterocarpin, licoflavanone A, glabrol, shinflavanone, gancaonin L, and glabrone (isolated from the ethanol extract of this plant roots) showed PPAR-γ ligand-binding activity (review: Wang et al. 2014).

Other Activities: Glycyrrhizin has a potential hepatoprotective activity. It has a capacity to regenerate the liver cell and inhibit fibrosis. The water extract of *G. glabra* decreased acetylcholinesterase activity in the brain. Glabridin isolated from the roots of *G. glabra* improved memory and cognitive functions and reduced the brain cholinesterase activity. Other reported activities include antitussive, expectorant, antibacterial, antioxidant, anticoagulant, antiviral, antitumor, antiulcer, and immunomodulatory activity (review: Sharma and Agrawal 2013).

Phytochemicals of *G. glabra* include triterpene saponin, flavonoids, polysaccharides, pectins, asparagines, essential oil, female hormone estrogen, gums, resins, sterols, volatile oils, tannins, and glycosides. Glycyrrhizin, a triterpenoid compound, accounts for the sweet taste of licorice root (review: Sharma and Agrawal 2013).

Glycyrrhiza foetida Desf., Fabaceae

Glycyrrhiza foetida is a traditional medicinal plant of Morocco. It is mainly used in the treatment of stomach and throat problems in traditional medicine. Amorfrutins (present in the edible roots) are likely antidiabetic dietary natural products (Weidner et al. 2012). Amorfrutins are shown to activate PPAR-γ (review: Wang et al. 2014).

Glycyrrhiza inflata Batalin, Fabaceae

Licochalcone E is a retrochalcone isolated from the root of *Glycyrrhiza inflata*. This compound induced 3T3-L1 preadipocyte differentiation. Further, licochalcone E evidenced weak, but significant, PPAR-γ ligand-binding activity. Two weeks of licochalcone E treatment to diet-induced diabetic mice lowered blood glucose and serum TG levels. Additionally, treatment with this compound resulted in marked reductions in adipocyte size and increases in the mRNA expression levels of PPAR-γ in white adipose tissue. Licochalcone E was also shown to significantly stimulate Akt signaling in epididymal white adipose tissue (Park et al. 2012). Chalcones and their derivatives isolated from the chloroform extract of *G. inflata* inhibited PTP1B (review: Jiany et al. 2012).

Glycyrrhiza uralensis Fisch., Fabaceae

The antidiabetic effect and mechanism of raw *Glycyrrhiza uralensis* and roasted *G. uralensis* extracts and their major respective components, glycyrrhizin and glycyrrhetinic acid, were examined. In partial pancreatectomized diabetic mice, both raw and roasted *G. uralensis* extracts improved glucose tolerance, but only the roasted extract enhanced glucose-stimulated insulin secretion. Extracts from both roasted and raw *G. uralensis* enhanced insulin-stimulated glucose uptake through PPAR-γ activation in 3T3-L1 adipocytes. Consistently, only the extract from roasted *G. uralensis* and glycyrrhetinic acid enhanced glucose-stimulated insulin secretion in isolated islets. In addition, they induced mRNA levels of IR substrate 2, pancreas duodenum homeobox-1, and glucokinase in the islets, which contributed to improving β-cell viability. In conclusion, roasted *G. uralensis* extract containing glycyrrhetinic acid improved glucose tolerance better than raw *G. uralensis* extract by enhancing insulinotropic action (Ko et al. 2007). The inhibitory effects of 10 components from the root of *G. uralensis* on aldose reductase and sorbitol formation in rat lenses with high levels of glucose were investigated. Of the compounds tested, semilicoisoflavone B showed the most potent inhibition, with the IC_{50} values of 1.8 and 10.6 μM for rat lens aldose reductase and human recombinant aldose reductase, respectively. It showed noncompetitive inhibition against rat lens aldose reductase. Further, semilicoisoflavone B inhibited sorbitol formation of rat lens incubated with a high concentration of glucose, indicating that this compound may be effective for preventing osmotic stress in hyperglycemia (Lee et al. 2010b). PTP1B inhibitors, glycyrrhisoflavone, glisoflavone, and licoflavone A, were isolated from the roots of *G. uralensis*. Besides, 2-arylbenzofuran glycybenzofuran and licocoumaron isolated from the roots also inhibited PTP1B (review: Jiany et al. 2012).

Gmelina arborea Roxb., Verbenaceae

The water extract of *Gmelina arborea* bark did not significantly decrease the plasma glucose level in normoglycemic rats in the acute and subacute (28 days) treatments. However, the water extract (250 and 500 mg/kg, p.o.) decreased plasma glucose levels in streptozotocin-induced diabetic rats in both the acute and long-term administration. Additionally, the water extract also improved body weight and decreased food and water intake in diabetic animals. Moreover, it did not show significant glucose-lowering effect in normoglycemic rats (Kulkarni and Veeranjaneyulu 2013).

Recently, based on folkloric use, the antidiabetes potential of *G. arborea* leaf powder (100 mg/kg for 21 days) was evaluated in male alloxan-induced diabetic rats. As judged from the changes in the levels of glycosylated hemoglobulin, glucose, urea, creatinine, protein, and liver function marker enzymes, the leaf powder showed anti-DM activity and it appeared to be nontoxic (Kumaresan et al. 2014). The antidiabetes activity was also reported for the fruit extract (Nayak et al. 2012). All the extracts (ethanol, ethyl acetate, butanol, and petroleum ether) of the fruit reduced blood glucose levels in alloxan-induced diabetic rats. However, the ethanol extract was found to have maximum glucose-lowering activity and antioxidant activity in comparison with other extracts (Nayak et al. 2012).

Gomphrena globosa L., Amaranthaceae

Gomphrena globosa is a traditional Caribbean anti-DM medicinal plant. The ethanol extract of the leaves of *G. globosa* exhibited hypoglycemic activity in alloxan-induced diabetic albino rat. It may be safe to conclude that the plants may be useful in the management of diabetes (Omodamiro and Jimoh 2014).

Gongronema latifolium Benth., Asclepiadaceae (Figure 2.54)

Common Names: Utalis, arokeke (in Nigeria), akan-asante, aborode-aborode (West Africa)

Description: *Gongronema latifolium* has simple and opposite leaves and dehiscent seed pod (follicle) that opens along a single seam. The seeds are flat with white hairy pappus; the flowers are bisexual, regular with pale yellow-colored petals and superior ovary (Osuagwu et al. 2013).

Distribution: *G. latifolium* is a spice plant growing in the humid forest vegetation of Southeastern Nigeria. It is now widespread in tropical Africa and can be found from Senegal east to Chad and south to Congo. It occurs in rainforest, deciduous and secondary forests, and also mangrove and disturbed roadside forest, from sea level up to 900 m altitude. The leafy vegetable can be propagated by seed or softwood, semihardwood, and hardwood cuttings.

Traditional Medicinal Uses: *G. latifolium* is a tropical rainforest plant primarily used in traditional folk medicine in the treatment of malaria, diabetes, and hypertension and utilized as a laxative (Chime et al. 2014). The leaves of *G. latifolium* are used as a traditional anti-DM medicine.

Pharmacology

Antidiabetes: Scientific studies have established the hypoglycemic effects of the ethanol extracts of the *G. latifolium* leaf. The antihyperglycemic effects of the water and ethanol extracts from *G. latifolium* leaves (100 mg/kg for 2 weeks) on glucose and glycogen metabolism in livers of nondiabetic and streptozotocin-induced diabetic rats were investigated. The ethanol extract significantly increased the activities of hexokinase, phosphofructokinase, and glucose-6-phosphate dehydrogenase in diabetic rats, while it markedly decreased the activity of glucokinase and the levels of blood glucose. The water extract of *G. latifolium* was only able to significantly increase the activities of hexokinase and decrease the activities of glucokinase but did not produce any significant change in the hepatic glycogen and both hepatic and blood glucose content of diabetic rats. Thus, the studies show that the ethanol extract from *G. latifolium* leaves has antihyperglycemic potency (Ugochukwu and Babady 2002).

The antidiabetic properties of the methanol extract of *G. latifolium* leaves loaded in solid lipid microparticles (the lipid matrix composed of fat from *Capra hircus* and Phospholipon® 90H) were evaluated in alloxan-induced diabetic rats. *G. latifolium*-loaded microparticles had a mean percentage reduction in blood glucose of 76% at 2 h, 42.3% at 8 h, and 24.4% at 12 h, while the group that received the reference glibenclamide had 82.6%, 61.7%, and 46.7% at 2, 8, and 12 h, respectively, after oral administration. *G. latifolium*-loaded microparticles had blood glucose reduction effect higher than the pure extract (Chime et al. 2014).

FIGURE 2.54 *Gongronema latifolium*. (From Brunken, U. et al., *Romanian Biotechnol. Lett.*, 19, 9649, 2008.)

Histological studies showed improvement in β-cells and kidneys in alloxan-induced diabetic rats treated with the ethanol extract of this plant leaf. Investigations were carried out on the effect of the ethanol extract of this plant leaf on the histology of pancreas, kidney, heart, and liver tissues of diabetic and nondiabetic male rats. The pancreas of alloxan-induced diabetic rats showed reduction in β-cell density and size and distorted reticular support and infiltration of inflamed cells.

The extract (400 mg/kg) administration resulted in substantial recovery of these histological changes. The renal tubules of diabetic rat kidneys indicated inflammation and obscured borders between convoluted tubules. However, gavaging with the extract caused a restoration and regeneration of these hitherto inflamed cells. Hyperglycemia as well as extract administration caused no significant change in the heart tissues, except that the myocytes of the extract-treated rats were multinucleated and hypertrophied. No significant change was also observed in the histology of the liver (Edet et al. 2013). The water and ethanol extracts of *G. latifolium* leaves ameliorated renal oxidative stress and LPO in streptozotocin-induced diabetic rats (Ugochukwu and Cobourne 2003). In another study, the water and methanol extracts and fractions of the methanol extract of *G. latifolium* leaf (200–800 mg/kg) exhibited an antidiabetic effect by ameliorating alloxan-induced increase in blood glucose levels in diabetic rats (Akah et al. 2011).

Administration of a combination of *G. latifolium* leaf extract and selenium to alloxan-induced diabetic rats for 28 days significantly reduced aspartate transaminase activity and glucose levels and increased superoxide dismutase activity, GSH-S-transferase activity, and albumin and protein levels. Treatment with selenium alone significantly increased catalase activity (Iweala et al. 2013). Follow-up studies should include identification of active principles, toxicity evaluation, and, if required, clinical trials.

Other Activities: Scientific studies have established antibacterial, cardioprotective, hypolipidemic, anti-inflammatory, and antioxidative effects of water and ethanol extracts of the *G. latifolium* leaf (review: Edim et al. 2012).

Phytochemicals: Phytochemical analysis of the tender fruits and mature leaves of *G. latifolium* revealed the presence of alkaloids, tannins, saponins, flavonoids, phenols, phytic acid, and hydrocyanic acid. The phytochemicals were higher in the fruits compared to the leaves except the flavonoids that were higher in the leaves than in the fruits (Osuagwu et al. 2013). Phytochemical screening of the leaf extract revealed the presence of alkaloids, saponins, tannins, phlobatannins, cardiac glycosides, reducing sugars, and polyphenols in both ethanol and water extracts. However, polyphenols and saponins were relatively higher in the ethanol extract compared to the water extract (Edet et al. 2013).

Goodenia ovata Sm., Goodeniaceae
Goodenia ovata is a sticky erect shrub used in Australian traditional medicine by early settlers. This plant was reported to have antidiabetic properties. The leaves have been shown to contain ursolic acid (1%–2% dry wt), a triterpene that exhibits hypoglycemic activity in streptozotocin-induced diabetic rats (review: Ghisalberti 2005).

Gossypium herbaceum L., Malvaceae
Gossypium herbaceum is a traditional medicinal plant with a hypoglycemic property. The hypoglycemic and hypolipidemic effects of ethyl ether and ethanol extracts (200 mg/kg) of *G. herbaceum* leaves were evaluated in alloxan-induced diabetic rats. The extracts showed significant antihyperglycemic and hypolipidemic activity as compared to diabetic control animals (Velmurugan and Bhargava 2014).

Grewia asiatica L., Tiliaceae
Local Name: Phalsa or fasla
Grewia asiatica is native to South Asia. It is cultivated for its edible fruit, which are a rich source of nutrients and phytochemicals such as anthocyanins, tannins, phenolics, and flavonoids. In folk medicine the edible *G. asiatica* fruit is used in a number of pathological conditions. In a study, the antihyperglycemic activities of the methanol extract and its fractions of the fruit, stem bark, and leaves of *G. asiatica* were compared in alloxan-induced diabetic rabbits. Oral administration of the extracts and fractions (100 and 200 mg/kg) reduced serum glucose levels in diabetic rabbits. All three parts showed antihyperglycemic

activity (Parveen et al. 2012). In another study, the extracts (chloroform, ethanol, and water) of leaves of *G. asiatica* (200 mg/kg) showed significant reduction in blood glucose levels in alloxan-induced diabetic rats (Priyanka et al. 2010).

In a human study, the *G. asiatica* fruit showed a low glycemic index value of 5.34 with modest hypoglycemic activity in healthy nondiabetic human subjects (Mesaik et al. 2013). Thus, the studies have confirmed the glucose-lowering effect of *G. asiatica*. Follow-up studies on the mechanisms of action, identification of active principles, etc., remain to be done.

Guaiacum coulteri A. Gray., Zygophyllaceae

Guaiacum coulteri is a traditional antidiabetes medicinal plant occurring in Mexico and Central America. The plant extract exhibited promising antihyperglycemic activity in subcutaneously dextrose-injected normal healthy rabbits (Roman-Ramas et al. 1992).

Guazuma ulmifolia Lam, Sterculiaceae

Common Names: Bastard cedar, guácima, guácimo

Guazuma ulmifolia is a traditional medicinal plant extensively used in México for the empirical treatment of type 2 diabetes. *G. ulmifolia* (water extract) exerted antidiabetic effects by stimulating glucose uptake in both insulin-sensitive (3T3-F442A adipocytes) and insulin-resistant adipocytes without inducing adipogenesis (Alonso-Castro and Salazar-Olivo 2008). The water extract of *G. ulmifolia* leaf showed antidiabetic activity in alloxan-induced diabetic mice (Adnyana et al. 2013).

Gymnema montanum Hook. f., Asclepiadaceae

Gymnema montanum is an endangered medicinal plant endemic to India. The molecular mechanism of the antidiabetic property of *G. montanum* leaf extract against alloxan-induced apoptotic cell death in rat insulinoma cells was evaluated. In alloxan-treated (7 mM/mL) rat insulinoma cells, pretreatment with *G. montanum* leaf extract (5 μg and 10 μg/mL) resulted in a significant decrease in intracellular Ca^{2+} concentration and nitric oxide production along with increase in mitochondrial membrane potential. The extract treatment reduced oxidative stress and apoptosis. The extract significantly increased the cellular antioxidant levels and decreased the lipid peroxides in alloxan-treated RINm5F cells (Ramkumar et al. 2009a).

In another study, *G. montanum* (leaf) ethanol extract (200 mg/kg) prevented renal damage associated with diabetic oxidative stress in alloxan-induced diabetic rats (Ramkumar et al. 2009b).

G. montanum stem showed antidiabetic activity in streptozotocin-induced diabetic rats. Oral administration of the stem alcohol extract (25–200 mg/kg daily for 4 weeks) to diabetic rats resulted in a dose-dependent reduction in blood glucose levels and improvement in plasma insulin levels. The extract showed significant increase in hexokinase, glucose-6-phosphate dehydrogenase, and glycogen content in the liver of diabetic rats while there was significant reduction in the levels of glucose-6-phosphatase and fructose 1,6-bisphosphatase (Ramkumar et al. 2011).

Gymnema sylvestre R. Br., Asclepiadaceae (Figure 2.55)

Common Names: Periploca of the wood, small Indian ipecacuanha (Eng.), meshashringi, madhunashini, gurmar, adigam, cheru kurinja, chakkarakolli (Indian languages), Australian cowplant, waldschlinge (German)

Description: *G. sylvestre* is a slow-growing vulnerable species and a large, more or less pubescent, woody, perennial climber with opposite, elliptic or obovate, acute, rarely cordate leaves, glabrous or puberulose beneath and small, yellow flowers having slender pedicels in umbellate subglobose cymes. The follicle is slender bearing pale brown seeds with broad, thin wings.

Distribution: *G. sylvestre* is present in the Deccan peninsula, extending to Northern and Western India and is cultivated as a medicinal plant in different parts of the world.

Traditional Medicinal Uses: The leaves are mostly used in traditional medicine for diabetes, arthritis, anemia, osteoporosis, hypercholesterolemia, cardiomyopathy, asthma, constipation, microbial infections,

(a)   (b)

FIGURE 2.55   *Gymnema sylvestre.* (a) Plant. (b) Stem with flower.

and indigestion and as a diuretic and anti-inflammatory (review: Tiwari et al. 2014). It is bitter, acrid, stomachic, cooling, tonic, alternative, antihelmintic, expectorant, stimulant and used to treat snake bite, eye complaints, piles, leucoderma, biliousness, bronchitis, asthma, ulcers, and cough (Kirtikar and Basu 1975).

Pharmacology

Antidiabetes: *G. sylvestre* (leaf powder, ethanol extract, and hydroethanolic extract of the leaves, saponin fraction) has shown hypoglycemic and antihyperglycemic activities in animal experiments (review: Porchezhian and Dobriyal 2003; Tiwari et al. 2014). The stimulatory effects of *G. sylvestre* on insulin release have been reported (Persaud et al. 1999). It does not improve insulin resistance (Tominaga et al. 1995). The leaf extract of this plant showed anti-DM and hypolipidemic activity in alloxan-induced diabetic rats (Mall et al. 2009). A novel dihydroxy gymnemic triacetate (5–20 mg/kg) isolated from the leaves of *G. sylvestre* showed normoglycemic and hypolipidemic activities in streptozotocin-induced diabetic rats (Daisy et al. 2009b).

Gymnemic acid, a mixture of triterpene glycosides extracted from the leaves of *G. sylvestre*, inhibited the intestinal absorption of glucose in human and rats (review: Tiwari et al. 2014). A crude saponin fraction derived from the methanol extract of leaves of *G. sylvestre* reduced blood glucose levels after intraperitoneal administration to streptozotocin-induced diabetic mice. Gymnemic acid IV, a triterpene glycoside, was found to be the active principle, which at doses of 3.4–13.4 mg/kg reduced the blood glucose levels by 13.5%–60% 6 h after the administration comparable to the potency of glibenclamide and did not change the blood glucose levels of normal rats. This compound (13.4 mg/kg) increased the levels of plasma insulin in diabetic mice. The compound (1 mg/mL) did not inhibit α-glycosidase activity in the brush border membrane vesicles of rat small intestine (Sugihara et al. 2000). In another study, gymnemoside b and gymnemic acids III, V, and VII were found to inhibit glucose absorption from the intestine (Yoshikawa et al. 1997a).

The following are the likely mechanisms of action of *G. sylvestre* leaf: It promotes regeneration of islet cells and increases secretion of insulin from β-cells. It increases the activation of enzymes responsible for the utilization of glucose by insulin-dependent pathways. It also inhibits absorption of glucose from intestine. It modulates incretin activity that triggers insulin secretion and release.

Clinical Trials: In clinical trials *G. sylvestre* showed promise; it decreased fasting and postprandial blood glucose levels and glycated hemoglobulin level (Baskaran et al. 1990; Shanmugasundaram et al. 1990; review: Tiwari et al. 2014). Studies suggest that the β-cells may be regenerated/repaired in type 2 diabetic patients on *G. sylvestre* supplementation (Bhaskaran et al. 1995). In another study diabetic patients were given a product containing the *G. sylvestre* leaf extract (400 mg/kg twice a day) for 3 months. The treatment decreased fasting as well as postprandial blood glucose levels. In another trial, treatment of type 2 DM patients with a *G. sylvestre*-based product (1 g/day for 2 months) resulted in significant decreases in fasting and postprandial blood glucose levels with an associated increase in the levels of insulin and C-peptide. Further, it stimulated insulin secretion from isolated human islets of Langerhans (review: Ghorbani et al. 2013). *G. sylvestre* improves blood sugar homeostasis and promotes regeneration of the pancreas. This plant is very promising for further clinical trials, among other things, to fix optimum dose and long-term toxicity studies to establish safety.

Other Activities: *G. sylvestre* leaf extract improved serum cholesterol and TG levels (Shigematsu et al. 2001). Gymnema preparations have a profound action on the modulation of taste, particularly suppressing sweet taste sensations due to gurmarin (a protein), sapogenin, and triterpenoid (Yoshikawa et al. 1989). Antiallergic, antiviral, lipid-lowering, and other effects of the plant extracts are also reported (review: Porchezhian and Dobriyal 2003). Gymnemic acid inhibits intestinal absorption of oleic acid in rats (Wang et al. 1998). The ethanolic extract of *G. sylvestre* leaves exhibited antimicrobial activity (Satdive et al. 2003). Other known pharmacological actions of this plant include antiarthritic activity, anti-inflammatory activity, anticancer and cytotoxic property, immunomodulatory activity, and wound healing activity. The hepatoprotective activity of this plant extract has also been reported (review: Tiwari et al. 2014).

Toxicity: Administration of 1.0% leaf powder in the diet in rats for 52 weeks has shown no toxic effects. However, treatment of diabetic patients with *G. sylvestre* has been reported to cause hepatitis or drug-induced liver injury (review: Tiwari et al. 2014).

Phytochemicals: Hentriacontane, pentatriacontane, phytin, resins, tartaric acid, formic acid, butyric acid, anthraquinone derivatives, inositol, D-quercitol, triterpene glycosides (gymnemic acid, saponin, gymnestrogenin, gymnemagenin (Puratchimani and Jha 2004)), nonacosane, hexahydroxy-olean-12-ene, gymnamine, gymnemic acid (I, II, III, IV, V, VI and VII), gypenosides (II, V, XLIII, XLV, XLVII, and LXXIV), and gynosaponin TN-2 have been isolated in this plant. From the glycosidic fraction of the dried leaves of the plant, six triterpene glycosides, gymnemosides a, b, c, d, e, and f, were isolated together with nine known triterpene glycosides (Yoshikawa et al. 1997; review: Tiwari et al. 2014). The leaves of *G. sylvestre* contain triterpene saponins belonging to oleanane and dammarene classes. Oleanane saponins are gymnemic acids and gymnemasaponins, while dammarene saponins are gymnemasides. Besides this, flavones, anthraquinones, hentriacontane, pentatriacontane, phytin, resins, D-quercitol, tartaric acid, formic acid, butyric acid, lupeol, β-amyrin-related glycosides, and stigmasterol are present. The plant extract also tests positive for alkaloids. The leaves of this species yield acidic glycosides and anthroquinones and their derivatives (Kanetkar et al. 2007; review: Tiwari et al. 2014).

Gymnema yunnanense Tsiang, Asclepiadaceae
The antihyperglycemic activity of the plant extract was demonstrated in obese *ob/ob* and diabetic *db/db* mice (Xie et al. 2003). In diabetic mice, after treatment with 100 mg/kg of the extract for 12 days, the fasting glucose levels decreased from about 247 to 190 mg/dL. The body weight was also slightly decreased by the treatment (Xie et al. 2003).

Gynandropis gynandra (L.) Brig., Capparidaceae

Synonym: *Cleome gynandra* L.

Common Name: Cat whiskers
In some part of Western Orissa, India, the leaves and roots are used by some tribal and traditional healers as an anti-DM drug. In a recent study, the ethanol extract of *Cleome gynandra* (*Gynandropis gynandra*) lowered blood glucose levels and improved the lipid profiles in alloxan-induced diabetic rats (Ravichandiran et al. 2014).

Gynura divaricata (L.) DC, Asteraceae
A polysaccharide from *Gynura divaricata* decreased the activities of intestinal disaccharidases in streptozotocin-induced diabetic rats. The specific activities of intestinal disaccharidases were increased significantly during diabetes, and the polysaccharide attenuated this increase (Deng et al. 2011). In another study, the ethyl acetate fraction of the water extract of the aerial parts showed α-amylase and α-glycosidase inhibition, and the active compounds were identified as flavonoids and alkaloids (Wu et al. 2011).

Gynostemma pentaphyllum (Thunb.) Makino, Cucurbitaceae

Gynostemma pentaphyllum is traditionally used as in Vietnamese folk medicine for the treatment of diabetes. In a study, supplementation of the ethanol extract of *G. pentaphyllum* leaf containing standardized concentrations of gypenosides in the diet (0.01%) lowered the blood glucose level by altering the hepatic glucose metabolic enzyme activities in C57BL/KSJ *db/db* mice. The plasma insulin concentrations of the extract-supplemented mice were elevated compared to the control group. The histology of the pancreatic islets revealed that the insulin-positive β-cell numbers were higher in the high-dose extract of *G. pentaphyllum*-treated *db/db* mice (Yeo et al. 2008). In another study, dammarane derivatives isolated from *G. pentaphyllum* inhibited PTP1B (Jiany et al. 2012).

Gynura procumbens (Lour.) Merr., Asteraceae (Compositae)

Gynura procumbens is an annual evergreen shrub and a traditional medicinal plant used to treat diabetes. The extract of the leaf of this plant significantly lowered fasting blood glucose, in acute dose (1 g/kg), in streptozotocin-induced diabetic rats. Furthermore, the extracts suppressed peak glucose levels in subcutaneous glucose tolerance test. Moreover, in the 14-day study, the 25% ethanol extract exerted fasting blood glucose-lowering effect. *G. procumbens* contains antidiabetic principles mostly extracted in 25% ethanol. Interaction among active components appears to determine the antidiabetic efficacy achieved likely by a metformin-like mechanism (Alariri et al. 2013).

Haloxylon salicornicum (Moq.) Bunge ex Boiss., Chenopodiaceae

Haloxylon salicornium (aerial parts and whole plant) is used in traditional medicine to treat diabetes in Egypt. This plant is reported to have antidiabetic activity and is predicted (*in silico*) to contain DPP-IV inhibitors (Guasch et al. 2012). The plant extract exhibited persistent hypoglycemic effects in normal rats (Shabana et al. 1990).

Hedychium spicatum Ham. ex Smith, Zingiberaceae

Phytochemical investigation of the antihyperglycemic extract of the rhizomes of *Hedychium spicatum* led to the isolation of two new labdane-type diterpenes along with seven known compounds. One of the new compounds displayed strong intestinal α-glucosidase inhibitory activity. Other compounds also displayed varying degrees of intestinal α-glucosidase inhibitory potential (Reddy et al. 2009; Tiwari et al. 2009).

Hedyotis biflora (L.) Lamk., Rubiaceae

Synonym: *Oldenlandia paniculata*

The methanol extract of *Hedyotis biflora* showed potent α-glucosidase inhibition. At a concentration of about 480 μg/mL, the methanol extract showed 50% inhibition of α-glucosidase activity (Nirmal Christhudas et al. 2013). The methanol extract also showed free radical scavenging activity. The active compound was identified as ursolic acid using various spectroscopical studies (Nirmal Christhudas et al. 2013).

Heinsia crinita (Afzel) G. Taylor, Rubiaceae

Treatment of alloxan-induced diabetic rats with the leaf extract caused a significant reduction in fasting blood glucose levels of normal and alloxan-induced diabetic rats both in an acute study and with prolonged treatment (2 weeks) (Okokon et al. 2009).

Helianthus annuus L., Asteraceae

Common Name: Sunflower

Administration of the ethanol extract of *Helianthus annuus* seed (250 and 500 mg/kg) showed only marginal changes in blood glucose level of normoglycemic rats, whereas an oral glucose tolerance test in streptozotocin–nicotinamide-induced type 2 diabetic rats showed marked reduction in blood glucose levels. Administration of the extract to streptozotocin–nicotinamide-induced type 2 diabetic rats resulted in a marked decrease in blood glucose levels comparable to that of glibenclamide. Further, the treatment restored the lipid profile and increased the body weight, liver glycogen content, glycosylated hemoglobin, serum insulin level, and GSH level in diabetic rats (Saini and Sharma 2013).

Helianthus tuberosus L., Asteraceae

Synonym: *Helianthus tomentosus* Michx.

Tubers of *Helianthus tuberosus* are consumed as food in Turkey; they are also used as a folk remedy for the treatment of diabetes. The antidiabetic and antioxidant activities of the ethanol extract of the tubers were studied in normal and streptozotocin-induced diabetic rats. Oral administration of the extract (250 and 500 mg/kg daily for 5 days) decreased thiobarbituric acid reactive substance levels in the kidney. The extract did not restore GSH levels in kidney, liver, and heart tissues of diabetic rats (Aslan et al. 2010). In an *in vitro* study, *H. tuberosus* extract increased cell viability, had a protective effect on β-cells, and increased insulin secretion level and NAD+/NADH ratio in HIT-T15 cells (Jeong-Lan et al. 2010).

Helichrysum graveolens (M. Bieb.) Sweet, Asteraceae

Helichrysum graveolens is used as traditional medicine for diabetes in different regions of Anatolia.

The ethanol extract of *H. graveolens* fruit exhibited promising inhibitory activity (55.7% ± 2.2%) on α-amylase enzyme (Orhan et al. 2014).

Helicteres isora L., Apocynaceae/Sterculiaceae (Figure 2.56)

Common Names: Indian screw tree, East Indian screw tree, vishanika, murva, avartani, marorphali, valampiri, edampiri valampiri

Description: *Helicteres isora* is a subdecidous shrub or small tree with gray bark, reaching a height of 5–15 ft with obovate or obliquely cordate leaves with serrate margin, scabrous above and pubescent below; the flowers may be solitary or in spare clusters with red, long, reflexed petals, turning pale blue when old. The fruit is cylindrical with spirally twisted carpels; the seeds are turbercled (review: Kumar and Singh 2014).

Distribution: *H. isora* is distributed throughout India especially Bihar, Bengal, Western and Southern India, and Andaman and also in Nepal.

Traditional Medicinal Uses: The root, bark, and fruit are used in traditional medicine. It is an expectorant, demulcent, astringent, antigalactagogue, aphrodisiac, and tonic and used in diabetes, empyema, stomach affection, diarrhea, dysentery, biliousness, colic, flatulence, carbuncle, parturition, and tumor (Kirtikar and Basu 1975).

Pharmacology

Antidiabetes: The water and alcohol extracts of the fruit, stem bark, and root of *H. isora* showed glucose-lowering properties in diabetic animals. *H. isora* root extracts improved glucose tolerance in glucose-induced hyperglycemic rats (Venkatesh et al. 2004). The water extract of the *H. isora* root

(a)   (b)

FIGURE 2.56 *Helicteres isora.* (a) Stem with fruit and leaves. (b) Flower.

showed antidiabetic and hypolipidemic activity in streptozotocin-induced diabetic rats (Kumar and Murugesan 2008). Another study suggests that the extract of *H. isora* has insulin-sensitizing and hypolipidemic activity and has the potential for use in the treatment of type 2 diabetes (Chakrabarti et al. 2002). The fruit of *H. isora* exhibited high antioxidant activity and moderate antidiabetes activity in experimental animals (Suthar et al. 2009). However, the ethanol extract of the *H. isora* fruit (300 mg/kg), when administered orally, also showed potent antihyperglycemic effect and improved the glycemic control mechanism in streptozotocin-induced diabetic rats. The antidiabetic effect of the extract (300 mg/kg) was comparable to that of 0.6 mg/kg glibenclamide (Boopathy et al. 2009). The hot water extract of the fruits of this plant activated glucose uptake in the L-6 cell line of mouse skeletal muscle (Gupta et al. 2009a). Further, the fruit extract also has antihyperlipidemic activity. Treatment with *H. isora* fruit extract restored to almost near normal levels and altered the levels of total cholesterol, LDL, VHDL, TGs, and HDL in streptozotocin-induced diabetic rats (Boopathy et al. 2010). The water extract of the bark of *H. isora* also exhibited significant antihyperglycemic and *in vivo* antioxidant activity in streptozotocin-induced diabetic rats, and the results were found to be dose dependent (Kumar et al. 2007).

Thus, the studies have established the antidiabetic properties of the extracts of the root, bark, and fruit of this plant. The active principle involved remains to be studied; the mechanism of action is also not clear. Further studies are warranted to determine the utility of this plant for the development of medicine to treat diabetes.

Other Activities: Other reported activities based on experimental pharmacological studies include hypolipidemic activity, antispasmodic activity of the fruits, antibacterial activity of the root (Tambekar et al. 2008), antioxidant activity, anticancer activity, hepatoprotective activity, antidiarrheal activity, and antinociceptive activity (review: Kumar and Singh 2014; Qu et al. 1991).

Toxicity: Cytotoxic compounds cucurbitacin B and isocucurbitacin B have been reported (Bean et al. 1985).

Phytochemicals: Flavonoid glucuronides, cucurbitacin B and isocucurbitacin B, phytosterols, saponins, phlobatannins and hydroxy carboxylic acid are reported. The roots and barks of *H. isora* showed the presence of flavones, triterpenoids, cucurbitacin, phytosterols, saponins, sugars, and phlobatannins. Saponins, tannins, anthraquinones, alkaloids, triterpenes, flavonoids, glycosides, reducing sugar, and phlobatannins were detected from fruits. Cucurbitacin B and isocucurbitacin B are isolated from the root. Betulinic acid, daucosterol, sitosterols, and isorin were isolated from the roots and barks of *H. isora* (review: Kumar and Singh 2014).

Heliotropium subulatum Hoclist. ex DC., Boraginaceae
Heliotropium subulatum is a traditional anti-DM plant. Extracts from this plant exhibited antidiabetes activity *in vivo*; but the plant is toxic (Marles and Farnsworth 1995).

Helleborus purpurascens Waldst. & Kit., Ranunculaceae
MCS-18, a novel natural product isolated from the plant *Helleborus purpurascens* (Christmas rose), was able to increase diabetes-free survival in the NOD mouse model, which was accompanied with a diminished IFN-γ secretion of splenocytes. MCS-18 exerted an efficient immunosuppressive activity with potential for therapies to prevent or treat type I diabetes (Seifarth et al. 2011).

Hemidesmus indicus (L.) R. Br., Periplocaceae/Asclepiadaceae
Recently, it was transferred to Periplocaceae (Figure 2.57).

Common Names: Indian sarsaparilla, anantamula, sariva, magrabu, salsa, nannari, narunindi, zaiyana, etc.

Description: *Hemidesmus indicus* is a diffusely twining, sometimes prostrate, undershrub having numerous slender wiry laticiferous branches with a purplish-brown bar. The leaves are simple, petioled, exstipulate, entire, dark green above but paler, and sometimes pubescent below. The flowers are greenish-yellow to greenish-purple outside; the calyx is deeply five lobed with a gamopetalous corolla;

(a)   (b)

FIGURE 2.57 *Hemidesmus indicus.* (a) Vine on the ground. (b) Flowers.

there are five stamens inserted near the base of the corolla with distinct filaments and small anthers ending in inflexed appendages; the pistil is bicarpellary, with free ovaries and many ovules; the fruit has two straight, slender, narrow follicles with many, flat, oblong seeds, with a long tuft of white silky hairs (review: Austin 2008). The transverse section of the fresh tuberous root is circular. It shows a slightly compact porous strand of wood at the center enveloped by a massive cream-colored starchy tissue and peripheral strip of rind (review: George et al. 2008).

Distribution: This plant is found throughout India growing under mesophytic to semidry conditions in the plains and up to an altitude of 600 m. It is also found in Sri Lanka, Pakistan, Iran, Bangladesh, and Moluccas (Nayar et al. 2006). It is quite common in open scrub jungles, hedges, and uncultivated soil.

Traditional Medicinal Uses: The roots are used as a tonic, antipyretic, antidiarrheal, astringent, blood purifier, diaphoretic, diuretic, and refrigerant. The plant is used to treat biliousness, blood diseases, dysentery, respiratory disorders, skin diseases, syphilis, fever, leprosy, leukoderma, leukorrhea, itching, bronchitis, asthma, eye diseases, epileptic fits, kidney and urinary disorders, loss of appetite, burning sensation, dyspepsia, ulcer, syphilis, uterine complaints, cough, and rheumatism (review: Austin 2008). Root, root bark, and stem are used in ethnomedicine.

Pharmacology

Antidiabetes: Oral administration of a compound (2-hydroxy 4-methoxy benzoic acid) isolated from the root of *H. indicus* showed anti-DM activity against streptozotocin-induced diabetic rats. At an oral dose of 500 µg/kg body weight, the compound normalized the elevated levels of glycosylated hemoglobin, total cholesterol, and LDL-C found in diabetic animals. The levels of plasma insulin and liver glycogen content were also restored in drug-treated diabetic rats (Gayathri and Kannabiran 2009). The hypoglycemic activity of *H. indicus* root in streptozotocin-induced diabetic rats has been reported (Gayathri and Kannabiran 2008).

While evaluating the toxicity of the tuberous root extracts of *H. indicus*, the glucose-lowering property of the root was observed by Subramoniam and coworkers (Ajikumaran et al. 2014). Follow-up studies resulted in the isolation of the antihyperglycemic principle from the root and establishment of its anti-DM property. The active principle was isolated by antihyperglycemic activity-guided chromatographic techniques. A glucose tolerance test in rats was used to evaluate the antihyperglycemic property. The anti-DM property was evaluated in alloxan-induced diabetic rats as well as streptozotocin-induced (type 2 model) diabetic rats. The active principle was isolated and identified with spectral data as β-amyrin palmitate. Although it is a known compound, its presence in *H. indicus* was not known previously. Interestingly, it was observed for the first time that β-amyrin palmitate has remarkable antihyperglycemic activity in orally glucose-loaded rats. Further, it exhibited excellent anti-DM activity in both alloxan-induced diabetic and streptozotocin-induced diabetic rats at a very low concentration (50 µg/kg). One of the mechanisms of action of β-amyrin palmitate appears to be blocking the entry of glucose from the intestine. Since the drug restored the weight of body and liver and liver glycogen content in diabetic rats, it is likely to have more than one mechanisms of action. In addition to the inhibition of glucose absorption, it may have delayed actions such as insulin-like (partial or complete) action and/or sensitiza-tion of insulin action. It is also possible that the compound could protect β-cells and/or stimulate the cells to produce more insulin. α- and β-Amyrin (10–100 mg/kg) have been shown to have beneficial effects in the preservation of β-cell integrity in streptozotocin-challenged mice (Santos et al. 2012). The authors suggest that one or more of these delayed actions may occur, at least, a few hours after the administra-tion of β-amyrin palmitate. Further studies are required to elucidate its multiple mechanisms of action. β-Amyrin palmitate is very promising to develop a medicine for diabetes for combination therapy and/or monotherapy (Ajikumaran et al. 2014).

The anti-DM activities of relatively high doses of some compounds related to β-amyrin palmitate have been reported. Interestingly, these compounds, α- and β-amyrins, β-amyrin acetate, and β-sitosterol, are also present in the root of this plant (review: Austin et al. 2008). α-Amyrin acetate isolated from *Ficus benghalensis* has been reported to have anti-DM activity at the 50 mg/kg dose level (Singh et al. 2009). α-Amyrin acetate (25–75 mg/kg) from *Streblus asper* showed anti-DM activity in streptozotocin-induced diabetic rats (Karan et al. 2013). Mice treated with the α-/β-amyrin mixture (10–100 mg/kg) from *Protium heptaphyllum* showed significant reduction in streptozotocin-induced increase in blood glucose, total cholesterol, and serum TGs (Santos et al. 2012). But β-amyrin palmitate is about 1000 times more active than these compounds (Ajikumaran et al. 2014).

Other Activities: The reported pharmacological properties include anti-inflammatory, antiulcer, diuretic, antinociception, antidiarrheal, anti-snake venom, anticancer, antileprotic, chemopreventive, antioxidant, and free radical scavenging properties. The hepatoprotective property of this plant has been reported by a few researchers. However, *H. indicus* var. *pubescens* produced hepatomegaly in one study (review: Austin 2008). The difference in the extraction procedure, variety used, plant part used, and maturity stage of the plant may account this discrepancy (review: Austin 2008). The water extract of this plant root, but not the alcohol extract, stimulated absorption of water and electrolyte from the intestinal tract of the rat (Evans et al. 2004).

Phytochemicals: Numerous phytochemicals have been reported from this plant, root, stem, flower, and leaf. From the root resin, hemidesmol, tannin, glycosides, lupeol, α- and β-amyrins, β-sitosterol, lupeol acetate, β-amyrin acetate, hexatriacontane acid, lupeol octacosonate, a coumarinolignoid like hemidesmine, hemi-desmin-1, hemidesmin-2, hemidesmol, hemidesterol, 2-hydroxy-4-methoxybenzadehyde, etc., were reported (2-hydroxy-4-methoxybenzadehyde is responsible for its aromatic nature). From the stem, two novel preg-nane glycosides were isolated. Isoquercetin and rutin were reported from the flower. Coumarinolignoids, hyperoside, and rutin were also reported from the leaves (review: Austin 2008). Recently the presence of β-amyrin palmitate in the root of this plant has been brought to light (Ajikumaran et al. 2014).

Hemigraphis colorata (Blume) H.G. Hallier, Acanthaceae

Hemigraphis colorata is a small herb with medicinal properties. The *n*-hexane and, to some extent, the ethanol extract of *H. colorata* (whole plant) lowered the levels of blood glucose in glucose-fed rats. The hypoglycemic activity may be due to one or more of the steroids and coumarins present in the extract (Gayathri et al. 2012). In folk medicine the leaves are ground into a paste and applied on fresh cut wounds to promote wound healing (Deviprya 2013). The wound healing property of this plant has been scientifi-cally validated (Subramoniam et al. 2001).

Hemionitis arifolia (Burn.) Moore, Hemionitidaceae

Synonyms: *Asplenium arifolium* Burm., *Hemionitis cordifolia* Roxb. ex Beddome (Figure 2.58)

Common Name: Heart fern

Description: *Hemionitis arifolia* is a small herb with short creeping rhizome, 2 cm thick, densely covered by scales; the scales are ovate-lanceolate, dark at the center, whitish at the periphery, margin entire. The stipes are compact, numerous, black or dark brown, polished, brittle, up to 33 cm long in fertile fronds, up to 23 cm in sterile ones, up to 2 mm thick, densely scaly all over when young, sparsely when mature. The lamina is simple, dimorphic, dark green, chartaceous, cordiform, up to 9 × 6 cm, fertile ones 7 × 10 cm, deltoid, trilobed, entire; the costa is raised below, grooved above, and densely scaly below; the veins are obscured both above and below, anastomosing closely; long soft, pale brown scales are distributed all over

FIGURE 2.58 *Hemionitis arifolia.*

the lower surface of the sterile laminae and very rare on the adaxial surface of the fertile and sterile laminae. Sori are continuous along the veins filling the entire surface of the lamina when mature intermixed with hairs and scales; the spores are trilete, spherical, 45 μm in diameter (Manickam and Irudayaraj 1992). Vegetative propagation of *H. arifolia* is carried out by dividing its root balls with rhizome, which is the most effective method. Sexual reproduction takes place naturally by spore formation.

Distribution: *H. arifolia* is found in Western Ghats (South India) and at low altitudes, frequently along road sides and clearings in fully exposed dry places between 20 and 1400 m. It is also present in Bangladesh, Burma, Sri Lanka, and the Philippines (Manickam and Irudayaraj 1992).

Traditional Medicinal Uses: Tribes of Kerala, India, are using the whole plant decoction of *H. arifolia* to treat fever and the whole plant juice as an ingredient of herbal mixture to treat vomiting in children. The tribal people of Tirunelveli District, Tamil Nadu, India, use the leaf paste of *H. arifolia* to get relief from joint pain. *H. arifolia* is used in the Philippines to treat burns and the crushed juice of fronds to treat aches (Dixit 1984; Ajikumaran 2008). The fronds are used in ethnomedicine in the treatment of aches, diabetes, burns, antifertility, and menstrual disorders, and as vermifuge and antiflatulence; the rhizome is used for antibacterial activity (Ajikumaran et al. 2006; Hima Bindhu et al. 2012).

Pharmacology

Antidiabetes: Antidiabetic activity against alloxan-induced diabetic rats and hypoglycemic property in glucose-loaded normal rats of the ethanol extract of *H. arifolia* were reported (Ajikumaran et al. 2006). The active ethyl acetate fraction of the alcohol extract exhibited promising antidiabetic activity in streptozotocin-induced diabetic rats at a dose of 50 mg/kg (per oral) as judged from the levels of blood glucose, liver glycogen, and serum lipids. Intraperitoneal as well as oral administration of the active fraction was found to be effective in glucose tolerance test. The fraction also showed antihyperglycemic effect in intraperitoneally glucose-loaded rats. The fraction did not influence insulin release into the medium from cultured β-cells of rat islets. However, the fraction stimulated glucose uptake in isolated rat hemidiaphragm suggesting a direct effect of the drug (without the involvement of insulin) on glucose uptake. A steroid positive active principle was isolated from the active ethyl acetate fraction by chromatography (Ajikumaran 2008, PhD thesis). The antihyperglycemic properties of *H. arifolia* were confirmed by other studies. The ethanol and aqueous extracts lowered the levels of blood glucose in glucose-fed rats (Kumudhavalli and Jaykar 2012).

Other Activities: The plant showed antibacterial activity. Gram-negative bacteria such as *Salmonella typhi*, *S. paratyphi* A, and *Enterobacter aerogenes* were more susceptible to the crude extracts than Gram-positive bacteria (Hima Bindhu et al. 2012). The plant has antioxidant activity with flavonoids and, to some extent, phenols (Udayabhanu et al. 2014).

Toxicity: Daily administration of the anti-DM fraction (up to four times higher than the therapeutic dose) to normal mice for 30 days did not show any toxicity (Ajikumaran 2008).

Phytochemicals: Phytochemical screening showed the presence of flavonoids, steroids, and glycosides along with reducing sugar in all the extracts investigated (Hema et al. 2012).

Heritiera minor Lam., Sterculiaceae

Synonyms: *Heritiera littoralis* Dryand, *Heritiera minor* (Gaertn.) Lam.
Extracts from the aerial parts of *Heritiera minor* lowered the levels of blood glucose in experimental animals (Marles and Farnsworth 1995).

Hibiscus cannabinus L., Malvaceae

The methanol extract of *Hibiscus cannabinus* leaves (400 mg/kg, p.o., daily for 15 days) significantly lowered the blood glucose level in streptozotocin-induced diabetic rats. The methanol extract was non-toxic up to 5 g/kg; phytosterols, flavonoids, and glycosides were present in the extract (Sundarrajan et al. 2011). Further, the hydroalcoholic extract of *H. cannabinus* leaves exhibited a potent lipid-lowering activity in diet-induced hyperlipidemia (Mahadevan and Kamboj 2010).

Hibiscus platanifolius L., Malvaceae

Administration of the ethanol and water (hot) extracts of *Hibiscus platanifolius* leaves (100 and 150 mg/kg) decreased blood glucose, TG, total cholesterol, and LDL levels to near normal levels. The extracts also exhibited antioxidant activity. Thus, the plant leaf had good antioxidant, hypoglycemic, and hypolipidemic effects (Saravanan et al. 2011).

Hibiscus rosa-sinensis L., Malvaceae

Synonym: *Hibiscus boryanus* DC.

Shoe Flower: The ethanol extract of *Hibiscus rosa-sinensis* flower reduced blood glucose levels and lipid profile in streptozotocin-induced diabetic rats (Sachdewa and Khemani 2003). This effect was confirmed by another study wherein *H. rosa-sinensis* flower extract (125, 250, and 500 mg/kg) exhibited antioxidant, hypoglycemic, and hypolipidemic activities against streptozotocin-induced diabetic rats. Among the three doses of the extract studied, 250 mg/kg showed the best result when compared to the two other doses (Sankaran and Vadivel 2011). In another study, water extracts of the flowers of *H. rosa-sinensis* (500 mg/kg) exhibited significant hypoglycemic and hypolipidemic activities. A marked reduction in glycosylated hemoglobin was also observed, while insulin levels did not show any significant change in streptozotocin-induced diabetic rats (Bhaskar and Vidhya 2012). Oral administration of *H. rosa-sinensis* (500 mg/kg) water extract to streptozotocin-induced diabetic rats for 4 weeks significantly reduced blood glucose, urea, uric acid, and creatinine but increased the levels of insulin, C-peptide, albumin, and albumin/globulin ratio and restored all marker enzymes for liver function to near control levels in streptozotocin-induced diabetic rats (Mandade and Sreenivas 2011). In another study, on fractionation of the ethanol extract of *H. rosa-sinensis* leaves, five fractions were obtained. Of these, two fractions were more active than others. Both fractions on oral feeding (100 and 200 mg/kg) demonstrated an insulinotropic nature and protective effect in nonobese diabetic mice as judged from the levels of serum glucose, glycosylated hemoglobin, TG, cholesterol, blood urea, insulin, LDL, VLDL, and HDL. These fractions may contain potential oral hypoglycemic agents (Moqbel et al. 2011). Further studies are required to determine its utility as a therapeutic agent. Among other things, the mechanisms of actions and active principles involved remain to be studied.

Hibiscus sabdariffa L., Malvaceae

Common Name: Roselle

Polyphenol extracts from *Hibiscus sabdariffa* (flowers) attenuated nephropathy in experimental type 1 DM in rats (Lee et al. 2006b). The water extract of this plant attenuated diabetic nephropathy by improving oxidative stress and via Akt/Bad in streptozotocin-induced diabetic rats (Wang et al. 2011b). A report suggested that *H. sabdariffa* has a potential protective role against diabetes-induced sperm damage. In streptozotocin-induced diabetic rats, treatment with the extract of *H. sabdariffa* calyx resulted in decrease in blood glucose levels, increase in serum insulin levels, and improvement in sperm quality and count (Idris et al. 2012). In a recent study, different extracts of roselle calyx (*n*-hexane, ethyl acetate, and ethanol) were tested for their antidiabetic activity against streptozotocin-induced diabetic mice. The ethanol extract of roselle calyx (200, 400, and 600 mg/kg, p.o.) reduced blood glucose levels in diabetic mice; this effect was comparable to that of glibenclamide (Rosemary et al. 2014). Recently, the flavonoid-rich water fraction of the methanol extract of *H. sabdariffa* calyx was evaluated for its anti-hepatotoxic activities in streptozotocin-induced diabetic rats. The ameliorative effects of the fraction on streptozotocin-induced diabetes liver damage were evident from the histopathological analysis and the biochemical parameters evaluated in the serum and liver homogenates (Adeyemi et al. 2014).

Hibiscus tiliaceus L., Malvaceae

The methanol extract of *Hibiscus tiliaceus* flower (250 and 500 mg/kg, p.o., daily for 21 days) improved body weight, decreased blood glucose levels, and lowered serum cholesterol and TG levels in strepto-zotocin-induced diabetic rats (Kumar et al. 2010e). The ethyl acetate fraction of the methanol extract of *H. tiliaceus* leaves (200 and 400 mg/kg, p.o., daily for 21 days) reduced blood glucose levels and improved the serum lipid profile in alloxan-induced diabetic rats. The fraction also improved liver and kidney functions and histopathological changes of the pancreas, liver, and kidney (Kumar 2014).

Hibiscus vitifolius L., Malvaceae

Gossypin, a pentahydroxyflavone glucoside found in the flowers of *Hibiscus vitifolius*, has many biological properties, including antioxidant, anti-inflammatory, and anticancer activities. Oral administration of gos-sypin (20 mg/kg, p.o., daily for 30 days) to streptozotocin-induced diabetic rats improved glucose tolerance. Further, the treatment increased blood glucose and HbA1c levels, and the reduced plasma insulin and hemo-globin levels in diabetic rats were significantly reversed to near normal levels after oral administration of gos-sypin. Furthermore, the glycogen content of the liver and muscles was significantly improved after gossypin treatment of diabetic rats. The data obtained in gossypin-treated rats were comparable with those obtained following gliclazide treatment of rats, a standard reference drug for diabetes (Venkatesan and Pillai 2012).

Hintonia latiflora (Sesse & Moc.) Bullock., Rubiaceae

Synonyms: *Coutarea latiflora* Sesse & Moc., *Coutarea pterosperma* (S. Watson) Standl., *Hintonia standleyana* Bullock (Figure 2.59)

Common Name: Copalchi

Description: *Hintonia latiflora* is a slender tree 5–10 m height, with a profusion of pendent white flowers over the crown. The fruits persist on the plant for a few months.

Distribution: *H. latiflora* (*H. standleyana*) is a Mexican traditional medicinal plant.

FIGURE 2.59 *Hintonia latiflora.*

Ethnomedicinal Use: The bark is used to treat fever and as a purgative. The bark is highly regarded as a treatment for diabetes. It is brewed into a bitter tea and used as a tonic. The bark is also used to treat malaria.

Antidiabetes: The antidiabetic effects of the copalchi bark extract in various animal species with hyperglycemia induced by different methods have been reported (Korec et al. 2000). In a study, a blood sugar-lowering effect of oral or intragastric administration of a copalchi native extract (*H. latiflora* cortex) or intragastric administration of coutareagenin (neoflavonoid) (5-hydroxy-7-methoxy-4-(3,4-dihydroxyphenyl)-2H-benzo-1-pyran-2-on) present in the extract was demonstrated (Korec et al. 2000). An extract (100 mg/kg) of the stem bark of *H. standleyana* caused a significant decrease in blood glucose levels in both normal and streptozotocin-induced diabetic rats when compared with vehicle-treated groups. From the active extract, 3-*O*-β-D-glucopyranosyl-23,24-dihydrocucurbitacin F and 5-*O*-[β-D-apiofuranosyl-(1 → 6)-β-D-glucopyranosyl]-7-methoxy-3′,4′-dihydroxy-4-phenylcoumarin were isolated. The coumarin is a new natural product. These compounds did not decrease blood glucose levels in normal rats. However, in two different long-term subacute experiments, using animals with a developing diabetes condition and with streptozotocin-induced diabetes, both compounds at daily doses of 10 mg/kg (developing diabetes condition) or 30 mg/kg (streptozotocin-induced diabetes condition) provoked a significant antihyperglycemic activity. Furthermore, the new coumarin restored normal blood glucose levels in streptozotocin-induced diabetic rats. Treatment with the compounds also improved body weight (Guerrero-Analco et al. 2005).

In another subacute study, the stem bark extract of *H. latiflora* showed significant hypoglycemic and antihyperglycemic effects in normal and streptozotocin-induced diabetic rats, respectively. From the active extract of this plant, 25-*O*-acetyl-3-*O*-β-D-glucopyranosyl-23,24-dihydrocucurbitacin F, an analogue of 23,24-dihydrocucurbitacin F, and several known compounds were isolated. Oral administration of *H. latiflora* extract (100 mg/kg) or 5-*O*-β-D-glucopyranosyl-7,3′,4′-trihydroxy-4-phenylcoumarin (30 mg/kg) to streptozotocin-induced diabetic rats for 30 days restored blood glucose levels to normal values. The treatment also improved body weight. Further, the extract improved hepatic glycogen levels and plasma insulin levels (Guerrero-Analco et al. 2007).

The chloroform–methanol (1:1) extract of the leaves of *H. latiflora* caused significant decrease in blood glucose levels in both normal and streptozotocin-induced diabetic rats when compared with vehicle-treated groups. The extract was not toxic to mice according to the Lorke criteria. The extract of *H. latiflora* yielded the new 5-*O*-[β-D-xylopyranosyl-(1 → 6)-β-D-glucopyranosyl]-7,4′-dimethoxy-4-phenylcoumarin along with several known compounds, including ursolic acid and desoxycordifolinic acid. The new phenylcoumarin showed hypoglycemic activity. HPLC profiles of the leaf extracts of the plants revealed the presence of known hypoglycemic phenylcoumarins as well as chlorogenic acid (Cristians et al. 2009). It is of interest to note that the plant contains compounds such as chlorogenic acid, ursolic acid, quercetin, and new phenylcoumarin with anti-DM activities.

Clinical Trial: A concentrated extract from the bark of *H. latiflora* in capsule form was tested in an open, prospective clinical study in 41 dietetically stabilized subjects with type 2 DM.

Fasting and postprandial glucose and the HbA1c values declined significantly in the extract-treated patients. Furthermore, cholesterol and TGs were slightly reduced and no negative effects on other laboratory parameters were observed. Thus, the study confirmed the positive effect of the extracts from the bark of *H. latiflora* on blood glucose values (Korecova and Hladikova 2014).

Further studies, including long-term toxicity evaluation, may be carried out for likely medicine development from the extract of the stem bark and leaves of this plant.

Other Activities: Other established pharmacological properties include antimalarial and anticancer activities.

Phytochemicals: From the leaf extract, the new 5-*O*-[β-D-xylopyranosyl-(1 → 6)-β-D-glucopyranosyl]-7,4′-dimethoxy-4-phenylcoumarin along with several known compounds, including ursolic acid and desoxycordifolinic acid, was isolated. *H. latiflora* contains a high amount of polyphenols, mainly neoflavone and their glycosides (e.g., coutareagenin), quercetin, ursolic acid, and phenolic acids. Phenylcoumarins as well as chlorogenic acid were also reported from this plant.

Hippophae rhamnoides L., Elaeagnaceae

Common name: Seabuckthorn

Seabuckthorn is an important medicinal plant with many pharmacological properties. Water extract of sea-buckthorn seed residues has been shown to possess hypoglycemic and hypolipidemic properties in normal mice. Besides, the water extract of seeds significantly lowered the serum glucose, triglyceride and nitric oxide levels in streptozotocin-induced diabetic rats (Zhang et al. 2010).

Holarrhena antidysenterica (Linn.) Wall., Apocynaceae

Synonym: *Holarrhena pubescens* (Buch. Ham.) Wall. ex G. Don.

Common Names: Kutaja, kalinga, kurchi, veppalei, kodagapalei, kodagapala (Indian languages)

Description: *Holarrhena antidysenterica* is a deciduous laticiferous shrub or small tree of 30–40 ft height with pale bark; the leaves are opposite and subsessile, elliptic or ovate-oblong; the flowers are white, arranged in corymbose cyme. The follicles are very slender usually with small, long, white spots, bearing linear oblong seeds with spreading deciduous coma of brown hairs.

Distribution: It is distributed from the Chenab westward and throughout the drier forests of India to Travancore up to an elevation of 3500 ft.

Traditional Medicinal Uses: It is a carminative, stimulant, rubefacient, bitter, acrid, anthelmintic, galactagogue, tonic, and aphrodisiac and used in amoebic dysentery, diarrhea, piles, colic, dyspepsia, chest affections, spleen diseases, skin disease, toothache, dropsy, chronic bronchitis, boils, ulcers, blood diseases, fever, leprosy, biliousness, leukoderma, pains, fatigue, hallucination, headache, excessive menstrual flow, lumbago, urinary discharge, asthma, intestinal worms, liguatera, gonorrhea, and snake bite (Kirtikar and Basu 1975).

Antidiabetes: The seed of *H. antidysenterica* is reported to have antidiabetic activity (Gopal and Chauhan 1994). The water extract of *H. antidysenterica* seeds exhibited both antidiabetic and antihyperlipidemic activities in experimentally induced diabetic rats. Treatment with the extract resulted in a significant recovery of the activities of carbohydrate metabolic enzymes like glucose-6-phosphatase, glucose-6-phosphate dehydrogenase, and hexokinase in the liver and glycogen content in the liver and muscle of diabetic rats. The treatment also improved the lipid profile and liver and kidney function (Ali et al. 2009, 2010). In another study, almost the same type of effects was reported for the methanol extract of the seed. The methanol extract of the seeds, when administered orally at a dose of 400 mg/kg daily for 21 days, exhibited promising antidiabetic efficacy in streptozotocin-induced diabetic rats (Kazi et al. 2010). Alcohol, butanol, chloroform, water, and butanone extracts of *H. antidysenterica* stems (250 mg/kg) showed significant blood glucose-lowering activity in acute as well as prolonged treatment in alloxan-induced diabetic rats compared to diabetic control. Among all the extracts, the alcohol extract had significantly reduced the blood glucose level after a single dose and nearly equal to standard glibenclamide after prolonged treatment (Jalalpure et al. 2006).

Other Activities: The methanol extracts of *H. antidysenterica* have strong antibacterial activity. Alkaloids have strong antibacterial and antidiarrheal activities. *H. antidysenterica* was also reported to stimulate phagocytic function while inhibiting the humoral component of the immune system (review: Ajikumaran and Subramoniam 2005).

Toxicity: The presence of hepatotoxic pyrrolizidine alkaloids in *H. antidysenterica* has been reported (Arseculeratne et al. 1981).

Phytochemicals: Alkaloids like conessine, nor-conessimine, iso-conessimine, kurchine, comine, con-amine, conarrhimine, conkurchine, conessidine, trimethyl conkurchine, holarrhimine, holarrhemine, holarrhessimine, lettocine, conkurchinine, kurchinicine, resin, tannin, β-sitosterol, α-triterpene alcohol, lupeol, caoutchoui, lettoresinol-A, lettoresinol-B, 3β-dimethylaminocon-5-enin-18-one (antidysentericine), concuressine, 3-epihetero-conessine, kurcholessine, kurchamine, holadysone, holarrifine, holarrhetine, kurchiphyllamine, kurchiphylline, regholarrhenines A, B, D, C, holantosines A and B, holarosine B, holantosine E and F, and holacetine were isolated in this plant (review: Ajikumaran and Subramoniam 2005).

Holarrhena floribunda G. Don., Apocynaceae

Holarrhena floribunda is a common plant that has traditionally been used in Africa to treat many diseases such as fever, dysentery, sterility, and diabetes. Effects of the ethanol extract of the *H. floribunda* leaf and various fractions of this extract were evaluated in normal fasted and fed hyperglycemic rats. The extract showed a remarkable downregulation of blood glucose levels in fasted rats at 1000 mg/kg and significantly reduced or totally prevented the induction of hyperglycemia at 500 or 1000 mg/kg, respectively. This activity of the extract was found to be present in the dichloromethane and ethyl acetate fractions of the plant (Gnangoran et al. 2012). Further studies are warranted to establish the anti-DM property of this plant.

Holarrhena pubescens Wall. ex G. Don., Apocynaceae

Synonym: *Holarrhena antidysenterica* (G. Don) Wall.ex A. DC.

Recently, the antioxidant and antidiabetic potential of the stem bark of *Holarrhena pubescens* (*Holarrhena antidysenterica* (G. Don.) Wall. ex A. DC) was evaluated. The methanol and water extracts showed strong antioxidant activity with inhibition of more than 90% DPPH free radicals at the 100 µg/mL concentration. The methanol extracts (250 and 500 mg/kg) showed hypoglycemic activity in glucose tolerance test in mice (Bhusal et al. 2014).

Holostemma ada-kodien Schult, Apocynaceae

Synonyms: *Asclepias annularis* Roxb., *Holostemma annulare* (Roxb.) K. Schum.

The ethanol extract of *Holostemma ada-kodien* (200 and 400 mg/kg, p.o., daily for 7 days) significantly lowered the blood levels of glucose in fed hyperglycemic and alloxan-induced diabetic rats. The extract was found to be nontoxic in the preliminary toxicity study. Phytochemical analysis showed the presence of alkaloids, flavonoids, flavanones, tannins, terpenoids, amino acids, and carbohydrates in the extract (Janapathy et al. 2008c).

Homalium letestui Pellegr., Flacourtiaceae

The ethanol extract of *Homalium letestui* root (500–1000 mg/kg) caused a significant reduction in fasting blood glucose levels of streptozotocin-induced diabetic rats both in acute study and prolonged treatment (2 weeks). The activity of the extract was comparable to that of the reference drug glibenclamide (Okokon et al. 2007b).

Hordeum vulgare L. ssp. *spontaneum* (K.Koch) Korn, Poaceae

Synonym: *Hordeum spontaneum* Was.

Common Name: Barley

Barley is the world's fourth most important cereal crop. The hydroalcoholic extract of barley seeds and a protein-enriched fraction on the blood glucose of normal and streptozotocin-induced diabetic rats were investigated. In the acute treatment, the hydroalcoholic extract or the fraction did not influence blood glucose levels in normal and diabetic rats. Nevertheless, the extract (250 and 500 mg/kg) was only effective in reducing blood glucose levels of diabetic rats after 11 days of continued daily administration. The treatment also restored body weight at the end of the treatment. Thus, the hydroalcoholic extract of barley seeds has a role in the control of DM in long-term consumption. The mechanism of action and active principles involved remain to be studied (Minaiyan et al. 2013). In another study, administration of the ethanol extract of the seeds of *Hordeum vulgare* for 28 days to streptozotocin-induced diabetic rats resulted in improvement in body weight and decrease in food and water intake, urine volume, and blood glucose levels. The treatment also increased insulin levels; it also showed prevention in increase in the TG, cholesterol, LDL-C, and VLDL-C levels in the serum of diabetic rats. Chronic treatments with the extract significantly prevented the increase of MDA and the reduction of superoxide dismutase and reduced GSH and catalase levels when compared with the diabetic rats. Thus, the extract attenuated oxidative stress, hypercholesterolemia, and hyperglycemia (Shah et al. 2012). Histological and biochemical studies showed that feeding barley for 4 weeks could protect the rat liver from HFD-induced liver damages in DM (Khalaf and Mohamed 2008).

Clinical Study: In a human clinical study, barley showed a low glycemic index in both normal and type 2 diabetic subjects (Shukla et al. 1991).

Hovenia dulcis Thunb., Rhamnaceae
Hovenia dulcis is well known as a treatment for liver disease. Several studies have demonstrated that extracts of *H. dulcis* or its purified compounds can serve as detoxifying agents for alcohol poisoning. When alloxan-induced diabetic mice were treated with *H. dulcis* for 7 days, the blood glucose significantly decreased and the liver glycogen content increased compared to untreated diabetic mice (Ji et al. 2002). In a recent study, the antiobesity effect of the water extracts from the fruits or stems of *H. dulcis* was examined in 3T3-L1 preadipocytes. The cellular lipid contents in 3T3-L1 adipocytes were assessed by Oil Red O staining. The fruit extract, but not the stems, significantly inhibited lipid accumulation during adipogenesis in a dose-dependent manner. The fruit extract (100 μg/mL) downregulated the expression of the PPAR-γ, CCAAT/enhancer-binding protein-α, adipocyte fatty acid–binding protein 2, adiponectin, and resistin, and the inhibition rates were 29.3%, 54.3%, 34.5%, 55.7%, and 60.4%, respectively. In addition, the fruit extract upregulated the phosphorylation of AMPK-α, liver kinase B1 as a major AMPK, and the downstream substrate acetyl CoA carboxylase, and the inhibition rates were 43.52%, 38.25%, and 20.39%, respectively. These results indicate that the water extracts of the fruits have a significant antiobesity effect through the modulation of the AMPK pathway, suggesting that the fruit has a potential benefit in preventing obesity (Kim et al. 2014).

Humulus lupulus L., Cannabaceae
Older taxonomists included this species in the mulberry family, Moraceae.

Common Name: Hops, hop

Description: *Humulus lupulus* is a dioecious perennial climbing vine. It grows vigorously from the end of April to the beginning of July in the temperate climate zone. It is found in shrubbery and at the edge of forests with access to sufficient water, and it reaches a height of up to 7–8 m. Many female flowers form an inflorescence, called strobiles, which consist of membranous stipules and bracts that are attached to a zigzag, hairy axis. Each small branch of the axis bears a bract, represented only by its pair of stipules, which subtends either four or six bracts, each enclosing a flower or fruit. The stipules and bracts constitute the drug lupulin. The dried inflorescences have limited stability, are nonhomogeneous, and have a low bulk density.

Distribution: *H. lupulus* is native to the Northern Hemisphere. Generally growing at 38° to 51° latitude, it is cultivated in many parts of the world including the United States, Germany, Great Britain, Czech Republic, China, Japan, Korea, and India.

Traditional Medicinal Uses: *H. lupulus* has been used in the beer-making process for centuries. It is also a traditional medicinal plant. Inflorescences (strobiles) are the major part of hop plants used in beer making and in traditional medicine. In traditional medicine, it is used as a diuretic, bile-increasing agent, sleep promoter, a remedy against hair loss, antirheumatic, analgesic, and tonic. It is also used to treat anorexia (due to gastritis and sleeplessness), inflammation, breast and womb problems, etc. In Ayurvedic medicine, it has been recommended for restlessness associated with nervous tension, headache, and indigestion. Young hop shoots are used for cleaning the blood, liver, and spleen (review: Koetter and Biendl 2010).

Antidiabetic: Hop and hop-derived compounds have antidiabetic and antihyperlipidemic properties. Isohumulones, bitter acids from the flower, activated both PPAR-α and PPAR-γ and reduced insulin resistance (Yajima et al. 2004). Dietary isohumulones from hops raised plasma HDL-C and reduced liver cholesterol and triacylglycerol accumulation similar to PPAR-α activation in C57BL/6 mice (Miura et al. 2005b). In another study, dietary isomerized hop extract containing isohumulones prevented diet-induced obesity in rodents (Yajima et al. 2005b). Isohumulones modulate the blood lipid state favorably through the activation of PPAR-α (Shimura et al. 2005). In 3T3-L1 adipocytes and *db/db* mice, hop and *Acacia* phytochemicals decreased lipotoxicity (Minich et al. 2010).

Clinical Studies: In a double-blind clinical study, isohumulones (isomerized hop extract) improved hyperglycemia and decreased glycated hemoglobulin levels and body fat after 8 weeks of treatment in Japanese subjects with prediabetes (Obara et al. 2009). Hop and acacia phytochemicals decreased lipotoxicity in individuals with metabolic syndrome (Minich et al. 2010).

Other Activities: One of the important medicinal properties of hops is its ability to induce sleep. It has a sedative-like action in experimental animals. In a human trial, a valerian–hops combination reduced sleep latency (Koetter et al. 2007). Oral administration to rats resulted in increased gastric secretion. Gel with hops showed estrogenic activity. *H. lupulus* has antibiotic, antiseptic, antituberculosis, and antioxidant properties, and *in vitro* tests have mostly shown anticarcinogenic, antigenotoxic, and anti-inflammatory activities. The drug also has antihyperlipidemic activity (review: Koetter and Biendl 2010).

Phytochemicals: The bracts and stipules of the hop contain polyphenols. Fresh lupulin contains bitter acids, resins (such as humulon, lupulon and its derivatives 2-methyl-3-butenol), tannins, flavonoids (xanthohumol), and essential oils (with its main constituent myrcene, α-humulene and β-caryophyllene, and farnesene) (review: Koetter and Biendl 2010).

Hunteria umbellata K. Schum. Hallier f., Apocynaceae

Synonym: *Picralima elliotii* (Stapf) Stapf, *Hunteria eburnea* Pichon, *Hunteria elliotii* (Stapf) Pichon, *Hunteria mayumbensis* Pichon (Figure 2.60)

Local/Vernacular Names: Demouain (Fr), akan-asante kanwini (Ghana), *edo* òsù (Nigeria)

Distribution: *Hunteria umbellata* occurs throughout West and Central Africa, east to central DR Congo, and south to Cabinda (Angola).

Description: *H. umbellata* is a tropical rainforest glabrous tree having smaller flowers and fruits and broad and elongated leaves, measuring about 10–20 cm × 3.5–10 cm and ubiquitous to the West African and Central African rainforests. Its fruit measures up to 5–25 cm in diameter and consists of two separate globose mericaps that are about 3–6 cm long, yellow, smooth, 8–25 seeded, and embedded in a gelatinous pulp (Adeyemi et al. 2011).

Traditional Medicinal Uses: In Sierra Leone, the bark of *H. umbellata* is used as a stomachic and as a lotion to treat fever. In Ghana and Nigeria, the root and stem bark are used as an anthelmintic. The aqueous and alcoholic extracts of the seeds are used as a cure for piles, yaws, diabetes, and stomach ulcers in Nigeria. The bark and the root are used as a bitter tonic in Nigeria, and the powdered root and root decoctions are used to prevent miscarriage and to treat menorrhagia. In Cameroon, a bark or fruit decoction

FIGURE 2.60   *Hunteria umbellata.*

is taken to treat stomachache, liver problems, and hernia. The plant is also used in the treatment of geriatric problems. *H. umbellata* extracts are used in Germany to reduce the heart rate, as an aphrodisiac, to decrease blood pressure, and to reduce the blood lipid content. In African traditional medicine, water decoction made from the dry seeds of *H. umbellata* is highly valued in the management of DM.

Pharmacology

Antidiabetes: The antihyperglycemic activity of the water extract of the seed of *H. umbellata* was investigated in alloxan-induced, high fructose-induced, and dexamethasone-induced diabetic rats. The effects of the water extract and glibenclamide on FBG, free plasma insulin levels, HbA(1c), serum TG and TC, and insulin resistance indices were investigated. Glibenclamide and graded oral doses of the water extract caused significant dose-related reductions in FBG when compared to the values obtained for the model control rats. Administration of 50, 100, and 200 mg/kg of the extract significantly and dose dependently attenuated the development of hyperglycemia and decreased the levels of plasma HbA(1c), insulin, and serum TG and cholesterol in high fructose-induced as well as dexamethasone-induced diabetic rats compared to untreated diabetic control rats. The authors suggest that the hypoglycemic and antihyperlipidemic effects of the extract are mediated via enhanced peripheral glucose uptake and improvements in hyperinsulinemia (Adeneye and Adeyemi 2009a). In another study by the same authors, the water extract of *H. umbellata* seeds (50–200 mg/kg) caused a dose-dependent decrease in blood glucose levels in normal and glucose- and nicotine-induced diabetic rats. This effect was more than that of glibenclamide and mediated via inhibition of intestinal glucose uptake and adrenergic homeostatic mechanisms. The extract also caused reduction in body weight. The acute toxicity study showed that the plant extract had an LD_{50} of about 1 g/kg and as such is slightly toxic (Adeneye and Adeyemi 2009b).

Chloroform, ethyl acetate, and butan-1-ol fractions (200 mg/kg each) of the seed water extract were investigated for their acute oral hypoglycemic effects in normal rats over 6 h, while their antihyperglycemic effects were evaluated in alloxan-induced hyperglycemic rats over 5 days (daily oral administration). In addition, 50 mg/kg of the crude alkaloid fraction extracted from the water extract was also evaluated for its possible antihyperglycemic activity. Oral pretreatment with 200 mg/kg of different fractions resulted in a significant time-dependent hypoglycemic effect in normal rats, with the butan-1-ol fraction causing the most significant hypoglycemic effect. In the alloxan-induced hyperglycemic rats, repeated oral treatment with the same fractions for 5 days resulted in significant decreases in the fasting blood glucose concentrations with the most significant antihyperglycemic effect also recorded for the butan-1-ol fraction. Similarly, oral pretreatment with 50 mg/kg of alkaloid fraction significantly attenuated an increase in the postabsorptive glucose concentration in the oral glucose tolerance test in alloxan-induced hyperglycemic rats. In conclusion, the water extract contains a relatively high amount of alkaloids that could have accounted for the antihyperglycemic action of the water extract that was mediated via intestinal glucose uptake inhibition (Adeneye et al. 2011).

An *in vitro* study showed the antihyperglycemic action of erinidine, isolated from this plant; this effect may be, to some extent, mediated via the α-glucosidase inhibition mechanism as the results for other *in vitro* tests such as DPP-IV, glycogen phosphorylase, HIT-T15 cell insulin secretion, glucose uptake activity, and aldose reductase assays were all negative. However, erinidine (50 mg/kg, p.o.) attenuated the increase in blood glucose levels in oral glucose tolerance test and in normal and alloxan-induced hyperglycemic rats. Its antiglycemic mechanism may be the inhibition of intestinal glucose absorption; biotransformation in the gut may enhance its inhibitory activity (Adejuwon et al. 2013). However, other *in vivo* mechanisms of action may also be involved.

Other Activities: Alkaloids extracted from the stem bark exhibited antinociceptive and antipyretic effects. Fresh leaves showed oxytoxic action mediated via the muscarinic acetylcholinergic mechanism; the water extract of the seed showed analgesic activity in rodents (Adeyemi et al. 2011).

Toxicity: The LD_{50} values for the acute oral and intraperitoneal toxicity for mice for the water extract of *H. umbellata* seed were 1000 mg/kg and 459 mg/kg, respectively. Visible signs of immediate and delayed toxicities including starry hair coat, respiratory distress, and dyskinesia were observed (Adeneye et al. 2010). A fresh root bark extract is applied in Côte d'Ivoire to sores caused by leprosy.

This medication is toxic and fatalities have been recorded. The fruits are toxic; the fruit is rich in latex that is an ingredient of arrow poison in Côte d'Ivoire.

Phytochemicals: The phytochemical analysis of the crude medicinal plant extract revealed the presence of saponin, saponin glycosides, steroid, tannins, volatile oils, phenols, and copious amount of alkaloids (Falodun et al. 2006).

Hydnocarpus wightiana Blume, Achariaceae/Flacourtiaceae

Synonym: *Hydnocarpus laurifolia* (Dennst.) Slemmur
Hydnocarpus wightiana is advocated in traditional Indian medicine to possess strong antidiabetic activity. The acetone extract of the seed hulls of *H. wightiana* showed strong free radical scavenging activity, inhibition of α-glucosidase activity, and moderate *N*-acetyl-β-D-glucosaminidase inhibitory activities. Hydnocarpin, luteolin, and isohydnocarpin were isolated from the acetone extract. All the compounds showed varying levels of antioxidant activity. Furthermore, all the three compounds also showed varying degrees of α-glucosidase and *N*-acetyl-β-D-glucosaminidase inhibitory activity, luteolin being the superior (Reddy et al. 2005). Oral administration of the ethanol extract of *H. wightiana* seed hull for 28 days to streptozotocin-induced diabetic rats resulted in a decrease in blood glucose levels (Reddy et al. 2013). The antihyperglycemic activity of different extracts of *H. laurifolia* (*H. wightiana*) seeds was evaluated in both normal and diabetic rats. Administration of the petroleum ether and ethyl acetate extracts of *H. laurifolia* seeds at different doses to diabetic rats resulted in a highly significant reduction in blood glucose levels at 1, 2, and 4 h when compared to the diabetic control group (Rao et al. 2014). In a recent study, the antihyperglycemic and antihyperlipidemic effects of the chloroform extract of *H. laurifolia* seeds were evaluated in streptozotocin-induced diabetic rats. The chloroform extract (100 and 300 mg/kg, p.o.) of *H. laurifolia* seed markedly decreased blood glucose levels in diabetic rats. The serum levels of cholesterol, TGs, LDL, and VLDL were found to be lowered in the extract-treated diabetic rats (Rao and Krishnamohan 2014).

Hydrangea macrophylla (Thunb.) Ser., Hydrangeaceae

Hydrangeic acid (3–100 μM), a stilbene constituent of the processed leaves of *Hydrangea macrophylla* (Hydrangeae Dulcis Folium), promoted adipogenesis of 3T3-L1 cells. Hydrangeic acid significantly increased the amount of adiponectin released into the medium, the uptake of 2-deoxyglucose into the cells, and the translocation of glucose transporter 4 (GLUT4). Hydrangeic acid also increased mRNA levels of adiponectin, PPAR-γ2, GLUT4, and fatty acid–binding protein (aP2) while it decreased the expression of TNF-α mRNA. However, it did not activate PPAR-γ in a nuclear receptor cofactor assay system. Furthermore, hydrangeic acid significantly lowered blood glucose, TG, and free fatty acid levels after its administration for 2 weeks at a dose of 200 mg/kg/day (p.o.) in KK-A(y) mice (Zhang et al. 2009a).

Hydrangea paniculata Sieb., Hydrangeaceae

Hydrangea paniculata is a traditional medicinal plant. Skimmin, a coumarin isolated from *H. paniculata*, suppressed diabetic nephropathy in streptozotocin-induced diabetic rats (Zhang et al. 2012).

Hydrastis canadensis L., Ranunculaceae

Synonym: Goldenseal
Hydrastis canadensis (goldenseal) is rich in berberine (Brown and Roman 2008), a bioactive compound with antidiabetes and other activities (see under *Berberis aristata* DC).

Hygrophila auriculata (Schum.) Hiene, Acanthaceae

Synonym: *Hygrophila spinosa* T. Ander
A fraction was isolated from the methanol extract of the whole *Hygrophila auriculata* plant. This fraction at a dose of 80 mg/kg produced maximum fall in blood glucose levels in streptozotocin-induced diabetic rats 5 h after treatment. The hypoglycemic action of the methanol extract of the plant could be due to terpenoids in the fraction. Another fraction of the methanol extract showed spasmolytic activity (Ramesh et al. 2014).

Hymenaea courbaril L., Leguminosae

Synonym: *H. resinifera* Salisb.

Common Name: Jatobá

Hymenaea courbaril is a tree usually 30–40 m high. In traditional medicine, in Mexico and Central America, the plant is used to treat diabetes and other diseases. A water extract of jatobá leaves has demonstrated significant hypoglycemic activity, producing a significant reduction in blood sugar levels. Antidiabetic compounds sitosterol and astilbin were isolated from this plant (Lopez 1976; review: Alarcon-Aguilar and Roman-Ramos 2005).

Hymenocardia acida Tul., Euphorbiaceae

Effects of the methanol extract of *Hymenocardia acida* leaves on diabetes and associated lipidemia were investigated on alloxan-induced diabetic rats. The extract exhibited dose-dependent blood glucose- and lipid-lowering properties. The extract did not demonstrate any acutely toxic effect in rats within the dose range (250–2000 mg/kg) employed in the study (Ezeigbo and Asuzu 2011).

Hypericum perforatum L., Hypericaceae

Common Name: St. John's Wort

Hypericum perforatum (ethyl acetate extract) showed potent antihyperglycemic activity in streptozotocin-induced diabetic rats. In normal rats, the ethyl acetate extract showed dose-dependent fall in fasting blood glucose after 30 min of extract administration. *H. perforatum* ethyl acetate extract (50, 100, and 200 mg/kg, p.o. for 15 days) produced significant reduction in plasma glucose level, serum total cholesterol, TGs, and glucose-6-phosphatase levels in streptozotocin-induced diabetic rats. Further, the treatment decreased tissue glycogen content, serum HDL-C level, and glucose-6-phosphate dehydrogenase activity compared with untreated diabetic control (Arokiyaraj et al. 2011). *H. perforatum* extract (25 mg/kg daily for 30 days) prevented the deleterious effects of diabetes on passive avoidance learning and memory in streptozotocin-induced diabetic rats (Hasanein and Shahidi 2011).

Hyphaene thebaica (L.) Mart, Arecaceae

Common Name: Doum palm

Oral administration of the water extract of *Hyphaene thebaica* mesocarp produced a significant decrease in blood glucose levels in normoglycemic rats 12–18 h postadministration (Auwal et al. 2012). In another study, the acetone extract of the epicarp of *H. thebaica* fruits, when fractionated with methanol and ethyl acetate, a residual water-soluble active fraction, was obtained. Phytochemical analysis of this active fraction revealed the presence of 10 flavonoids. This fraction improved glucose tolerance and decreased blood levels of glycosylated hemoglobulin in alloxan-induced diabetic adult male rats. One of the new compounds isolated from the fraction (chrysoeriol 7-*O*-β-D-galactopyranosyl-(1–2)-α-L-arabinofuranoside) significantly reduced serum AST, ALT, urea, and creatinine levels in diabetic rats, indicating improvement in liver and kidney functions (Salib et al. 2013).

Hypolepis punctata (Thunb.) Mett., Dennstaedtiaceae

Hypolepis punctata is a fern. A low molecular weight antidiabetic compound, pterosin A, was isolated from this fern. Pterosin A, when administered orally for 4 weeks, effectively improved hyperglycemia and glucose intolerance in streptozotocin, HFD-fed, and *db/db* diabetic mice. There were no adverse effects in normal or diabetic mice treated with pterosin A for 4 weeks. The compound reversed the increased serum insulin and insulin resistance in dexamethasone- and insulin-resistant mice and in *db/db* mice. Further, the compound significantly reversed the reduced muscle GLUT-4 translocation and the increased liver phosphoenoylpyruvate carboxyl kinase expression in diabetic mice. It also increased phosphorylation of AMPK and Akt in muscles of diabetic mice. The decreased AMPK phosphorylation and increased p38 phosphorylation in livers of *db/db* mice were effectively reversed by pterosin A. The compound also enhanced AMPK phosphorylation in cultured human muscle cells. In cultured cells, it inhibited inducer-enhanced PEPCK expression; triggered the phosphorylations of AMPK, acetyl CoA carboxylase, and glycogen synthase kinase-3; and increased the intracellular glycogen level (Hsu et al. 2013). Thus, it appears to be an attractive phytomolecule for studies of drug development.

Hypoxis hemerocallidea Fisch. & C.A. Mey, Hypoxidaceae

Common Names: African potato, African star grass, etc.

Hypoxis hemerocallidea is an important ethnomedical herb. The water extract of the corm exhibited dose-dependent hypoglycemic activity in both normoglycemic and streptozotocin-induced diabetic rats. The extract also showed anti-inflammatory and antinociceptive activities (Ojewole 2002).

Hyptis suaveolens Poit., Lamiaceae

Traditionally, *Hyptis suaveolens* is used in the treatment of DM, fever, eczema, flatulence, cancers, and headache. *H. suaveolens* is abundantly found in farmlands. The methanol extract of the *H. suaveolens* leaf exhibited a blood glucose-lowering effect in alloxan-induced diabetic rats. An acute oral toxicity study of the methanol extract showed an LD_{50} value of more than 2 g/kg in rats (Danmalam et al. 2009). The anti-DM activity was confirmed by another study wherein the extract of the *H. suaveolens* leaf (250 and 500 mg/kg daily for 21 days) reduced blood glucose levels and serum lipids (TGs, total cholesterol, LDL, and VLDL) in streptozotocin-induced diabetic rats (Mishra et al. 2011). The extract of the aerial parts of this plant (250 and 500 mg/kg daily for 11 days) showed marginal decrease in blood glucose levels in normal rats, while in the alloxan-induced diabetic rats, the decrease in blood glucose was substantial in both chronic study (11 days) and acute glucose tolerance test. The treatment improved the lipid profile and parameters studied for liver function and kidney function in diabetic rats. The extract also exhibited *in vitro* antioxidant activity (Nayak et al. 2013).

Ibervillea sonorae (*S.*Watson) Greene, Cucurbitaceae

Ibervillea sonorae is widely used for the control of diabetes in Mexican traditional medicine. The plant exerted antidiabetic effects on animal models. Oral administration of freeze-dried decoction of *I. sonorae* roots did not exhibit hypoglycemic activity in normal fasted mice. However, the decoction reduced the blood glucose of normal mice in a dose-dependent manner after intraperitoneal injection. Also, this extract significantly lowered the glycemia of mild alloxan-induced diabetic mice and rats, but did not in severe alloxan-induced diabetic rats, so it seems that this antidiabetic plant needs the presence of insulin to show its hypoglycemic activity (Alarcon-Aguilar et al. 2002a). In a recent study, the antidiabetes mechanism of the water extract of this plant was investigated. The extract stimulated the 2-NBD-glucose uptake by mature adipocytes in a concentration-dependent manner. The extract (50 µg/mL) induced the 2-NBDG uptake in insulin-sensitive 3T3-F442A, 3T3-L1, and human adipocytes by 100%, 63%, and 33%, respectively, compared to insulin control. Inhibitors for the IR, PI3K, AKT, and GLUT4 blocked the 2-NBDG uptake in murine cells, but human adipocytes were insensitive to the PI3K inhibitor Wortmannin. The extract also stimulated the 2-NBDG uptake in insulin-resistant adipocytes by 117% (3T3-F442A), 83% (3T3-L1), and 48% (human). The extract induced 3T3-F442A adipogenesis but lacked proadipogenic effects on 3T3-L1 and human preadipocytes. Chemical analyses showed the presence of phenolics in the extract, mainly an appreciable concentration of gallic acid. Thus, the water extract of *I. sonorae* exerted its antidiabetic properties by stimulating the glucose uptake in human preadipocytes by a PI3K-independent pathway and without proadipogenic effects (Zapata-Bustos et al. 2014).

Ichnocarpus frutescens L., Apocynaceae

Ichnocarpus frutescens is used in Indian traditional medicine for centuries. The plant has anticancer and antioxidant activities. Decoction prepared from the leaves of *I. frutescens* is used to alleviate the symptoms of DM in folk medicine.

Administrations of the polyphenolic extract (150 and 300 mg/kg) of *I. frutescens* leaves resulted in a significant reduction of fasting blood glucose levels in alloxan-induced diabetic rats. Administration of the extract (300 mg/kg for 21 days) showed significant decrease in hepatic HMG-CoA reductase activity of alloxan-induced diabetic rats. No significant effects were found in normoglycemic rats. Further, the extract exhibited significant hypolipidemic effect as evident from the correction of hyperlipidemic indicators in diabetic rats. Oral administration of the polyphenolic extract (100 mg/kg) enhanced the release of the lipoprotein lipase enzyme. Histopathological studies of aorta in polyphenolic extract-treated alloxan-induced diabetic rats revealed almost recovery to normal appearance. Thus, the polyphenolic extract showed the therapeutic potential against diabetes and associated hyperlipidemia and atherosclerosis (Kumarappan et al. 2007). Another study showed that the polyphenolic extracts of *I. frutescens* exhibited antioxidant and antihyperglycemic effects in streptozotocin-induced diabetic rats (Kumarappan et al. 2012). In another study, antidiabetic activity of the water extract of the roots of *I. frutescens* was evaluated in streptozotocin–nicotinamide-induced type 2 diabetes in rats. Administration of the water extract (250 and 500 mg/kg, p.o.) for 15 days resulted in significant reduction of fasting blood glucose levels and increase in body weight in type 2 diabetic rats on the 10th and 15th days. In the oral glucose tolerance test, the extract increased the glucose tolerance in normal rats (Barik et al. 2008). The ethyl acetate extract of the plant root showed α-glucosidase inhibitory activity (Singha et al. 2013).

Ilex guayusa Loes., Aquifoliaceae

Common Name: Aguayusa

In traditional medicine it is used to treat DM. It showed hypoglycemic activity. Guanidine is believed to be the main active component (Marles and Farnsworth 1995).

Ilex paraguariensis A. St-Hill., Aquifoliaceae

Ilex paraguariensis is rich in polyphenols, especially chlorogenic acid. This compound is known to have antidiabetic properties (see under *Berberis aristata* DC). The water extract of *I. paraguariensis* decreased intestinal SGLT1 gene expression, indicating that the extract could decrease glucose absorption from the intestine. However, administration of a high dose (1g/kg) of the extract for 28 days did not influence blood glucose, insulin, and liver glucose-6-phosphatase activity in alloxan-induced diabetic rats (Oliveira et al. 2008). There is a need for further studies to clear the situation. Possibly, low doses may have a different effect; besides, it is also possible that interaction of different phytochemicals present in the extract, especially at very high doses, could alter the expression of the antidiabetic property of compounds like chlorogenic acid.

In another study, the acute *in vivo* effect and *in vitro* effects of samples from native and commercial *I. paraguariensis* on glucose homeostasis was examined. The results indicated that both infusions from green and roasted *I. paraguariensis* were able to improve significantly the oral glucose tolerance curve in rats. Additionally, both the infusions and the ethyl acetate and n-butanol fractions of *I. paraguariensis* induced insulin secretion and increased liver glycogen content in rats. Also, these fractions inhibited *in vitro* disaccharidase activities. Besides, these fractions inhibited protein glycation *in vitro*. Thus, this study showed that *I. paraguariensis* has antihyperglycemic activity and is probably a source of multiple hypoglycemic compounds (Pereiraa et al. 2012).

Indigofera arrecta Hochst. ex A.Rich, Fabaceae

Indigofera arrecta is used for the treatment of diabetes. The freeze-dried extract of *I. arrecta* decreased the plasma glucose levels of fasting normoglycemic rats, but did not prevent the rise in plasma glucose after an oral glucose load in these rats. The extract increased plasma insulin levels. In diabetic rats, the rise in blood glucose after an oral glucose load was not affected when the extract was administered 17 days after the induction of streptozotocin-induced diabetes. When administered 7 days after induction of diabetes, the rise in blood glucose was decreased and was stabilized after 30 min. The results indicate that *I. arrecta* is insulinotropic, requiring functional β-cells to express its effect (Nyarko et al. 1993). *I. arrecta* was investigated in ddY mice to determine its acute and subchronic effects and whether it modulated hepatic cytochrome P450 (CYP) isozymes and GSH. No mortality was observed in the acute (up to 10 g/kg, p.o.) and subchronic (2 g/kg, p.o., daily for 30 days) treatment with *I. arrecta*. The extract was devoid of overt acute and subchronic toxic effects and did not affect CYPs and GSH whose modulation may cause interactions of components in a multiple drug therapy (Nyarko et al. 1999).

Indigofera pulchra L., Fabaceae

Intraperitoneal administration of the hydromethanolic extract of *Indigofera pulchra* leaves at a high dose (1g/kg) did not change blood glucose levels in alloxan-induced diabetic rats 4, 8, and 24 h after administration. But a lower dose of the extract (250 mg/kg) lowered blood glucose levels significantly at 4, 8, and 24 h after administration. In normoglycemic rats, the high dose (1 g/kg) decreased blood glucose levels at 8 and 24 h after administration, whereas the lower dose (250 mg/kg) decreased at 4, 8, and 24 h (Tanko et al. 2009b). The ethyl acetate portion of the hydromethanolic extract (50 mg/kg) decreased blood glucose levels in normal and alloxan-induced diabetic rats after 24 h of treatment, whereas higher doses (100 and 200 mg/kg) did not influence the levels of blood glucose in alloxan-induced diabetic rats; in normal rats 100 mg/kg, but not 200 mg/kg, also showed a decrease in blood glucose levels (Tanko et al. 2009b). The unusual effect of the extract at lower concentration, but not at high concentrations, is of interest; the effect of further low concentrations of the fractions and dose-response study using isolated individual active principles may give light on the dose effect of this plant extract and fractions.

In another study, the antidiabetic and some hematological effects of ethyl acetate and *n*-butanol fractions of *I. pulchra* were evaluated. Preliminary phytochemical screening of ethyl acetate and n-butanol fractions revealed the presence of alkaloids, flavonoids, and saponins. The LD_{50} values were 775 and 2154 mg/kg for ethyl acetate and n-butanol fractions, respectively. There was a significant reduction in the fasting blood glucose levels of alloxan-induced diabetic rats after 1 and 2 weeks of treatment with the ethyl acetate fraction (50 mg/kg, i.p.). With regard to the n-butanol fraction (250 mg/kg), there was a significant reduction in the glucose levels of the diabetic groups after 2 weeks of treatment, not after 1 week treatment. Also, in relation to the erythrocyte indices, there was no significant change in rats treated with the two fractions when compared to control. However, in both the fraction-treated groups, there were significant increases in the white blood cell, neutrophil, and eosinophil counts after 2 weeks of treatment (Tanko et al. 2011).

Indigofera tinctoria L., Fabaceae

Indigofera tinctoria leaf reduced blood glucose levels in diabetic rabbits (Verma et al. 2010b). In streptozotocin-induced diabetic rats, the alcohol extract of *I. tinctoria* leaves (40, 80, 160, and 200 mg/kg) showed a significant blood glucose-lowering effect. Further, the leaf extract improved renal creatinine clearance and reduced renal total protein loss demonstrating nephroprotective properties. The organ to body weight ratios showed pancreas- and liver-specific beneficial effects of the leaf extract. The authors conclude that the alcohol extract of this plant may be beneficial to type 1 and type 2 DM patients for long-term treatment (Bangar and Saralaya 2011).

Inula britannica L., Asteraceae

The water extract from the flowers of *Inula britannica* prevented immunologically induced experimental hepatitis in mice and suggested that the antihepatitic effect of the flower extract is due to the inhibition of IFN-γ production. Diabetes in mice induced by multiple low doses of streptozotocin is a mouse model for IFN-γ-dependent autoimmune diabetes. C57BL/KsJ mice (male, 7 weeks) were provided with the flower extract (500 mg/kg/day) in drinking water *ad libitum*, starting 7 days before the first streptozotocin injection. Autoimmune diabetes was induced by streptozotocin (40 mg/kg/day daily for 5 days, i.p.). The extract treatment significantly suppressed the increase of blood glucose levels. Histological analysis of the pancreas showed that the degree of insulitis and destruction of β-cells were reduced by the extract treatment. The IFN-γ production from stimulated splenic T lymphocytes was inhibited by the treatment. Moreover, the proportion of IFN-γ-producing cells in the CD4(+) population, which was increased by multiple low doses of streptozotocin, was significantly decreased by the treatment. These results suggest that the plant extract has a preventative effect on autoimmune diabetes by regulating cytokine production (Kobayashi et al. 2002).

Inula helenium L., Asteraceae

Alantolactone from *Inula helenium* is reported to lower blood glucose (Marles and Farnsworth 1995).

Inula racemosa Hook., Asteraceae

Synonym: *Inula royleana* C.B. Clarke

Common Names: Inula, pushkaram, pohakarmul, puskkaramulam, pushkkaramulam, etc.

Description: *Inula racemosa* is a tall, stout herb of 1–5 ft height with grooved stem; coriaceous, radical leaves, narrowed into a long petiole; and cauline often deeply lobed at the base. The head is very large, having outer involucre of broad bracts with recurved triangular tips.

Distribution: It is found in the Western Himalayas, in Kashmir, up to an elevation of 7000 ft, especially in the borders of fields.

Traditional Medicinal Uses: It is a stomachic, alexiteric, tonic, carminative, analgesic, aphrodisiac, expectorant, and antiseptic and is used in the treatment of pain of the heart, spleen, liver, and joints, hemicrania, eruptions, inflammation, earache, cough, and boils (Kirtikar and Basu 1975).

Antidiabetes: *I. racemosa* (root extract) exhibited antiperoxidative, hypoglycemic, and cortisol-lowering activities in humans; it is suggested that the hypoglycemic effect of this plant is mediated through inhibition in corticosteroid concentration (Gholap and Kar 2005). The extract acted mainly by potentiating insulin sensitivity in an animal model (Tripathi and Chaturvedi 1995). The methanol extract of *I. racemosa* root (300 mg/kg daily for 28 days) treatment markedly decreased blood glucose levels and reduced oxidative stress in alloxan-induced diabetic rats (Ajani et al. 2009).

Other Activities: The plant extract possesses antiallergic activity in rats. One of the constituents of the plant extract may have been its β-adrenergic receptor blocking property (review: Ajikumaran and Subramoniam 2005).

Phytochemicals: Alantolactone, isoalantolactone, dihydroalantolactone, germacranolide inunolide, β-sitosterol, octadecanoic acid, D-mannitol, inulal, aplotaxene, phenylacetonitril, inulolide, dihydroinulolide, oxygenated derivatives of alantolides, alantolactones, and telekin were isolated (review: Ajikumaran and Subramoniam 2005).

Inula viscosa L., Asteraceae/Compositae

The water extract of *Inula viscosa* showed hypoglycemic, but not hypolipidemic, effects in normal and diabetic mice without influencing insulin levels in the blood. In normal rats, a significant reduction in blood glucose levels at 2 h was observed after a single oral administration. Repeated daily oral administration significantly reduced blood glucose levels after 4 days of treatment. In diabetic rats, a significant reduction in blood glucose levels was observed 1 h after a single oral administration. Repeated oral administration reduced blood glucose levels on the fourth day. No change in total plasma cholesterol and TG levels was observed after both a single and repeated oral administration in both normal and diabetic rats. In addition, plasma insulin levels and body weight remained unchanged after 15 days of repeated oral administration in normal and diabetic rats (Zeggwagh and Ouahidi 2006).

Ipomoea aquatica Forsk., Convolvulaceae

Common Names: Swamp cabbage, kadambi, kalmisag, vellaikeerai, etc.

Description: *Ipomoea aquatica* is an aquatic, trailing, or floating, herbaceous perennial, sometimes annual plant with a long, hollow stem producing adventitious roots from the nodes. The leaves are simple, pedicellate, ovate-oblong or elliptic, cordate or hastate at the base. The flowers are produced as solitary or five-flowered cyme, and the corolla is infundibuliform, white or pale, purple colored with a dark purple eye. The fruit is a capsule bearing 2–4 seeds.

Distribution: *I. aquatica* is distributed throughout India and also in Sri Lanka, tropical Asia, Africa, and Australia.

Ethnomedical Uses: The juice of the plant is an emetic and purgative and used in the treatment of nervous and general debility, piles, as a poultice, febrile delirium, fever, and ringworm (*The Wealth of India* 1959).

Antidiabetes: The water extract of *I. aquatica* (leaf) was reported as effective as the oral hypoglycemic drug tolbutamide in reducing the blood sugar levels of normal rats (Malalavidhane et al. 2000,

2001). It was also effective as an oral hypoglycemic agent in streptozotocin-induced type 2 diabetic rats (Malalavidhane et al. 2003). The methanol extract of the leaves of *I. aquatica* exhibited hypoglycemic and antioxidant activities. The methanol extract (200 and 400 mg/kg) of leaves showed a dose-dependent hypoglycemic activity in mice. It showed potent free radical scavenging activity with an IC_{50} value of 4.4 μg/mL (Hamid et al. 2011). One of the mechanisms of action could be inhibition of glucose absorption from the intestine. The effect of the water extract and dichloromethane/methanol extracts of *I. aquatica* on the glucose absorption using a rat intestinal preparation *in situ* was evaluated. The extracts (160 mg/kg, p.o.) exerted a significant inhibitory effect on glucose absorption when compared with control animals. The most pronounced effect was observed with the water extract. Ouabain used as a reference inhibitor strongly inhibited glucose absorption. On the other hand, both plant extracts inhibited the gastrointestinal motility suggesting that the inhibition of glucose absorption is not due to the acceleration of intestinal transit (Sokeng et al. 2007b).

Other Activities: Prostaglandin and leukotriene biosyntheses *in vitro* are inhibited by cinnamoyl-β-phenethylamine and *N*-acyl dopamine derivatives from the plant leaves.

Phytochemicals: Cinnamoyl-β-phenethylamine and *N*-acyl dopamine derivatives were isolated (review: Ajikumaran and Subramoniam 2005).

Ipomoea batatas L., Convolvulaceae

Synonyms: *Convolvulus batatas* L., *Ipomoea apiculata* M. Martens & Galeotti (Figure 2.61)

Common Names: Sweet potato, patate douce, batata da terra, batata doce, etc.

Description: Sweet potato, *Ipomoea batatas*, is a tuberous-rooted perennial mainly grown as an annual. The roots are adventitious, mostly located within the top 25 cm of the soil. Some of the roots produce elongated starchy tubers. Tuber flesh colors can be white, yellow, orange, and purple, while the skin color can be red, purple, brown, or white. The stems are creeping slender vines, up to 4 m long. The leaves are green or purplish, cordate, palmately veined, and borne on long petioles. The flowers are white or pale violet, axillary, sympetalous, solitary, or in cymes. The fruits are round, 1–4-seeded pods. The seeds are flattened (Duke 1983).

Distribution: The center of origin of the sweet potato is thought to be located between the Yucatán Peninsula of Mexico and the mouth of the Orinoco River, in Venezuela. Sweet potatoes as old as 8000 years have been found in Peru. It has then spread to the Caribbean and Polynesia. It is now widely

(a)   (b)

FIGURE 2.61   *Ipomoea batatas.* (a) Vine with flowers. (b) Tubers.

cultivated between 40° N and 32° S, up to 2000 m (and up to 2800 m in equatorial regions). Sweet potato is cultivated for food in more than 100 countries on the continents of Asia, North and South America, and Africa (Ramirez 1992).

Traditional Medicinal Uses: White-skinned sweet potato has been used in Shikoku, Japan, as a folk medicine for the treatment of diabetes and other diseases. The leaf decoction is used in folk remedies for tumors of the mouth and throat. Sweet potato is considered as alterative, aphrodisiac, astringent, bactericide, demulcent, fungicide, laxative, and tonic and is a folk remedy for asthma, bugbites, burns, catarrh, ciguatera, convalescence, diarrhea, dyslactea, fever, nausea, renosis, splenosis, stomach distress, and tumors (Duke and Wain 1981).

Pharmacology

Antidiabetes: White-skinned sweet potato showed remarkable anti-DM activity in Zucker fatty rats and improved the abnormality of glucose and lipid metabolism by reducing insulin resistance (Kusano and Abe 2000). In another study, the potato has been shown to have hypoglycemic activity in streptozotocin-induced diabetic rats and increased blood insulin levels. The active component was a high-molecular-weight glycol protein found mainly in the cortex of the tuber (Kusano et al. 2001). Follow-up studies were carried out to determine the effect of sweet potato on the number of pancreatic β-cells and on insulin expression by immunohistochemical staining with an anti-insulin antibody. The sweet potato treatment (flour suspension, 100–800 mg/kg) to streptozotocin-induced diabetic rats resulted in a dose-dependent marked decrease in blood glucose levels, an increase in the number of pancreatic β-cells, and an increase in the expression of insulin (Royhan et al. 2009). Thus, sweet potato induces regeneration of pancreatic β-cells and increases insulin expression.

The potato alleviates oxidative stress in diabetes. Administration of the methanol extract of white-skinned sweet potato (200 mg/kg for 2 weeks) to streptozotocin-induced diabetic rats caused a significant reduction in blood glucose, lipid peroxides, transaminases, TG, and cholesterol in diabetic rats. Furthermore, administration of the extract showed significant improvement in the activities of antioxidant enzymes. The authors conclude that the protective effects of the extract could be used to benefit diabetic patients (Bachri et al. 2010). Another study suggests that the hypoglycemic effects of *I. batatas* result from the suppression of oxidative stress and proinflammatory cytokine production followed by improvement of pancreatic β-cell mass. When streptozotocin-induced diabetic rats were treated with powdered *I. batatas* (5 g/kg/day, oral feeding) for 2 months, there was significant suppression in the increases of fasting plasma glucose and hemoglobin A1c levels. The treatment increased serum insulin levels and improved oral glucose tolerance and body weight in diabetic rats. Moreover, the treatment reduced superoxide production from leukocytes and vascular homogenates, serum 8-oxo-2′-deoxyguanosine, and vascular nitrotyrosine formation of diabetic rats to comparable levels of normal control animals. Stress- and inflammation-related p38 mitogen-activated protein kinase activity and TNF-α production of diabetic rats were significantly depressed by *I. batatas* administration. Histological examination also exhibited improvement of pancreatic β-cell mass after the treatment (Niwa et al. 2011). The ethanol extract of purple sweet potato tubers decreased blood glucose and increased total antioxidant levels in rats with high glucose intake (Yasa et al. 2013).

An arabinogalactan protein isolated from white-skinned sweet potato decreased plasma glucose levels and improved glucose tolerance and insulin sensitivity in spontaneously diabetic *db/db* mice. This suggests that amelioration of insulin resistance by the arabinogalactan protein leads to its hypoglycemic effects (Oki et al. 2011). In this connection it should be noted that *I. batatas* peel proteins are susceptible to digestive enzymes to a considerable extent (Maloney et al. 2014).

The water extract of the leaves also showed antidiabetic and antihyperlipidemic activities in alloxan-induced diabetic mice (Fenglin et al. 2009a). Hypolipidemic and hypoglycemic effects of the ethyl acetate fraction from the leaves of sweet potato in obese and streptozotocin-induced type 2 diabetic mice were demonstrated. Administration of the ethyl acetate fraction (orally, daily, 1000 mg lyophilized powder/kg for 21 days) to obese diabetic mice resulted in marked hypoglycemic and hypolipidemic effects compared to the control (untreated diabetic mice). The ethyl acetate fraction accelerated hexokinase activity, stimulated insulin secretion, and inhibited gluconeogenesis enzymatic activity (glucose-6-phosphatase) (Lien et al. 2011). Administration of the flavone extract from

I. batatas leaf (50 mg/kg daily for 2 weeks) to rats with NIDDM resulted in a significant decrease in the concentration of plasma TG, cholesterol, LDL, fasting glucose levels, and MDA levels in diabetic rats; the treatment increased the insulin sensitivity index and superoxide dismutase level in diabetic rats. In addition, the flavone extract did not show any physical or behavioral signs of toxicity. The aforementioned results suggest that flavone extracted from *I. batatas* leaf could control blood glucose and modulate the metabolism of glucose and blood lipid and decrease oxidative stress in diabetic rats (Zhao et al. 2007). In a recent report, the hypoglycemic effects of different doses (200, 300, and 400 mg/kg daily for 14 days) of *I. batatas* leaf (hot water extract) on the blood glucose level of rats whose glucose level exceeded 200 mg/dL after alloxan induction were studied. The extract at a dose of 300 mg/kg produced the best hypoglycemic effect (69.67%) in diabetic rats (Ijaola et al. 2014). Interestingly, a recent report indicated that sweet potato leaf (edible) extract powder attenuated hyperglycemia by enhancing the secretion of GLP-1. Administration of dietary sweet potato leaf for 5 weeks significantly lowered glycemia in type 2 diabetic mice. In a follow-up *in vitro* study, the extract and polyphenols such as caffeoylquinic acid present in the leaf enhanced GLP-1 secretion. Further, administration of the extract to rats resulted in stimulation of GLP-1 secretion and enhanced insulin secretion (Nagamine et al. 2014).

Clinical Trial: In a placebo-controlled, randomized, and double-blind clinical study, the tolerability, efficacy, and mode of action of Caiapo, an extract of white sweet potatoes, on metabolic control in type 2 diabetic patients was investigated. After treatment with Caiapo (4 g/day for 12 weeks), HbA1c decreased significantly, whereas it remained unchanged in subjects given placebo. Further, fasting blood glucose levels decreased; the mean cholesterol at the end of the treatment was significantly lower in the Caiapo group than in the placebo group. Thus, this study confirms the beneficial effects of Caiapo on plasma glucose as well as cholesterol levels in patients with type 2 diabetes (Luvik et al. 2004).

Other Activities: Other activities reported include potent inhibitory activity toward rat lysosomal β-glucosidase (alkaloids), antispasmodic, collagenase inhibitory (phenolics), hepatoprotective, spasmolytic, inhibition of prostate cancer proliferation, acetylcholinesterase inhibitory, anticoagulant, anti-HIV (coumarins), antioxidant, antimutagenic (anthocyanosides), antinociceptive, antibacterial, and antifungal (triterpenes) (review: Mohanraj and Sivasankar 2014).

Phytochemicals: The sweet potato leaves contain polyphenols such as caffeoylquinic acid derivatives. The phytochemical groups present in the leaf are tannins, saponin, flavonoid, alkaloids, steroids, phenol, anthraquinone, phlobatannin, glycosides, and terpenoids (Ijaola et al. 2014). Phenolics and anthocyanosides are involved in exerting antidiabetes activity (review: Mohanraj and Sivasankar 2014).

Ipomoea digitata L., Convolvulaceae

Common Name: Wild yam
Ipomoea digitata tuber/root exhibited considerable hypoglycemic activities, which could be due to the presence of flavonoids and β-sitosterol as active principles, although the mechanism of action remains to be determined. The hypoglycemic activity of *I. digitata* tuber (hydroalcoholic extract, 100 and 200 mg/kg, p.o.) was demonstrated in streptozotocin-induced diabetic rats in acute (2–24 h) and chronic (28 days) studies. The extract treatment also prevented body weight loss in diabetic rats (Pandey et al. 2013). A solid pharmaceutical dosage formulation using a dry plant extract of *I. digitata* tuberous roots using various excipients was prepared and reported to be effective as an antidiabetic medicine (Chandira and Jayakar 2010).

Ipomoea reniformis Chios, Convolvulaceae
The ethanol and water extracts of the stem of *Ipomoea reniformis* exhibited antihyperglycemic and antihyperlipidemic activities in alloxan-induced diabetic rats (Sangeswaran et al. 2010).

Irvingia gabonensis (Aubry-Lecomte ex O'Rorke.) Baill., Irvingiaceae
In a study, the hypoglycemic effect of the methanol extract of *Irvingia gabonensis* seeds was examined in streptozotocin-induced diabetic rats. A single oral administration of the methanol extract (150 and 250 mg/kg) significantly lowered the plasma glucose levels in diabetic rats 2 h after treatment

(Ngondi et al. 2006). In another study, the seed extract (10% in the diet) exhibited hypoglycemic and hypolipidemic effects in streptozotocin-induced diabetic rats (Desire et al. 2009). Incorporation of *I. gabonensis* seed powder (10%) or oil-free seed powder (5%) in the diet lowered blood glucose levels in type 2 diabetic Long–Evans rats without increasing liver glycogen content and body weight. The treatments also decreased the level of TGs (Hossain et al. 2012b). The water extract of *I. gabonensis* bark also decreased blood glucose levels and improved liver function in normal rabbits (Omonkhua and Onoagbe 2012).

Human Clinical Study: In a double-blind, randomized study involving 40 obese subjects, *I. gabonensis* seed (1.05 g, three times a day for 1 month) treatment decreased body weight by 5%. Further, the treatment resulted in a significant decrease of total cholesterol, LDL-C, and TGs and an increase of HDL-C in obese subjects (Ngondi et al. 2005).

Jasonia montana (Vahl) Botsch., Asteraceae
The aerial parts of *Jasonia montana* are used in traditional medicine to treat diabetes in Egypt. The ethanol and water extracts (150 mg/kg daily for 30 days, p.o.) of the aerial parts of *J. montana* showed anti-DM activity in streptozotocin-induced diabetic rats. Oral administration of both the extracts resulted in a significant decrease in fasting blood glucose, hepatic and renal thiobarbituric acid reactive substances, and hydroperoxides. The treatment also resulted in a significant increase in reduced GSH, superoxide dismutase, catalase, GSH peroxidase, and GSH-s-transferase in the liver and kidney of diabetic rats. The extracts exerted rapid protective effects against LPO by scavenging free radicals by reducing the risk of diabetic complications. The effect was more pronounced in the ethanol extract compared to the water extract (Hussain 2008).

Jatropha curcas L., Euphorbiaceae
Jatropha curcas is a traditional medicinal plant. The leaf extract of *J. curcas* reduced blood glucose levels in normal and alloxan-induced diabetic rats (Mishra et al. 2010). In a screening study, the ethanolic extract of Barbados nut (*J. curcas*) activated PPAR-γ (review: El-Abhar and Schaalan 2014).

Juglans regia L., Juglandaceae
Common Names: Walnut, common walnut, Persian walnut, English walnut (Eng.), aksotah, akrotu, akrottu, etc. (local names in India)

Description: Walnuts are the oldest tree food known to man (review: Shah et al. 2014). *J. regia* is a large tree with a trunk up to 2 m diameter, commonly with a short trunk and broad crown. The bark is smooth, olive brown when young and silvery gray on older branches, and features scattered broad fissures with a rougher texture. The pith of the twigs contains air spaces. The leaves are alternately arranged, 25–40 cm long, with 5–9 leaflets, paired alternately with one terminal leaflet; the margins of the leaflets are entire. The male flowers are in drooping catkins and the female flowers terminal, in clusters of two to five. The fruit is a green, semifleshy husk and a brown corrugated nut. The seed is large with a relatively thin shell and edible, with a rich flavor (review: Shah et al. 2014).

Distribution: *J. regia* is native to the mountain range of Central Asia. Early history shows that English walnuts came from ancient Persia. It is cultivated in the Himalayas and the Khasi Hills. It occurs in many countries, including China, Kazakhstan, Uzbekistan, Kirghizia, Nepal, Bhutan, Pakistan, Sri Lanka, Afghanistan, Turket, and the United States (review: Shah et al. 2014).

Traditional Medicinal Uses: The nut, leaves, and bark are used in traditional medicine. Many traditional systems of medicine have recommended this plant leaf for the treatment of DM. The leaves are used as an antimicrobial, anthelmintic, astringent, keratolytic, antidiarrheal, hypoglycemic, depurative, and carminative and for the treatment of sinusitis, cold, stomachache, and swelling in the joint. The kernel of this plant is used for the treatment of inflammatory bowel disease, diabetes, asthma, and prostate disturbance. The plant is also used to treat chronic eczema and scrofula. The leaves are used to treat scalp itching, dandruff, superficial burns, etc. The bark, branches, and exocarp of the immature fruit have been used to treat gastric, liver, and lung cancer. The bark is also used in arthritis, skin diseases, toothache, and hair growth; the seed coat is used for wound healing (review: Shah et al. 2014).

Antidiabetes: The extracts of the plant leaf showed anti-DM activity in animals (Dzhafarova et al. 2009). Treatment to streptozotocin-induced diabetic rats with the plant leaf (90% ethanol extract) at a dose of 200 or 400 mg/kg for 28 days resulted in a marked decrease in blood glucose, glycosylated hemoglobin, LDL, TG, and total cholesterol, whereas the levels of insulin and HDL in the blood increased. Thus, the plant extract is capable of ameliorating hyperglycemia and hyperlipidemia in streptozotocin type 1 diabetic rats (Mohammadi et al. 2011, 2012). The walnut polyphenol fraction (200 mg/kg) attenuated oxidative stress in type 2 diabetic mice. Further, the polyphenols inhibited glycosidase, sucrose, maltase, and amylase (Fukuda et al. 2004). Consumption of walnut leaf pellets (185 mg/kg) reduced fasting blood sugar and improved regeneration of β-cells in alloxan-induced diabetic rats (Jelodar et al. 2007). In another study, oral administration of the hydroalcoholic extract of the *J. regia* leaf for 4 weeks resulted in a significant reduction of glucose, HbA1c, total cholesterol, and serum TGs in streptozotocin–nicotinamide-induced diabetic rats (Jamshid et al. 2012). Detailed studies including identification of active principles from the leaves remain to be done. Walnut, an ingredient of the diet, needs more attention and studies on its likely role in controlling DM-associated complications.

Clinical Trials: Recently, a human trial was carried out for the first time to investigate the hypoglycemic activity of the plant leaves (water extract) in type 2 DM patients. It is a placebo-controlled, randomized trial on 58 patients (placebo control, 28; herbal drug group, 30); the extract administration for 2 months resulted in decrease in the levels of serum HbAIC and fasting blood glucose; the insulin levels were increased in the extract-treated patients. Thus, the plant leaf (water extract) showed favorable anti-DM activity against type 2 DM patients (Hosseini et al. 2014). In another study, a daily intake of 43–57 walnuts incorporated into the Japanese diet for 4 weeks to 40 healthy Japanese men and women lowered blood cholesterol. Further, walnut-enriched meals effectively prevented postprandial lipidemia where triacylglycerol was markedly reduced (review: Shah et al. 2014).

Other Pharmacological Activities: Allergic sensitization, anti-inflammation (Comstock et al. 2010; Wills et al. 2010), antifungal (Noumi et al. 2010), and acaricidal (Wang et al. 2009b) activities have been reported for this plant. Other pharmacological properties reported include antibacterial, antiviral, antioxidant, anthelmintic, antidepressant, inhibition of melanin formation, hepatoprotection, hypotriglyceridemic, and anticancer (review: Shah et al. 2014).

Phytochemicals: The walnut oil contains a great concentration of linolenic acid including omega-6 and omega-3 polyunsaturated fatty acids; flavonoids, saponins, glycosides, alkaloids, and steroids are present in the plant. Phytochemicals isolated include methyl palmitate, juglone, juglanin B, ellagic acid, glansreginins A and B, and glansrin D. Polyphenolic compounds present in the nut include casuarictin, tellimagrandin I, and tellimagrandin II; the leaf contains gallic acid and caffeoylquinic acids (review: Shah et al. 2014).

Juniperus communis L., Cupressaceae

Juniperus communis is used as traditional medicine for diabetes in different regions of Anatolia. *J. communis* was reported to have antidiabetic and antihyperlipidemic activities in streptozotocin–nicotinamide-induced diabetic rats. The methanol extract of *J. communis* (100 and 200 mg/kg, p.o., daily for 21 days) caused significant reduction in blood glucose levels and lipid levels (except HDL that increased) in diabetic rats (Banerjee et al. 2013). The hydroalcoholic extract of *J. communis* fruit exhibited potent inhibitory activity on the α-glucosidase enzyme (Orhan et al. 2014).

Juniperus oxycedrus L., Cupressaceae

Juniperus oxycedrus fruits and leaves are used internally and pounded fruits are eaten for diabetes in Turkey. Treatment of streptozotocin-induced diabetic rats with the fruit and leaf extracts of this plant (500 mg/kg daily, p.o., for 10 days) decreased the blood glucose levels and the levels of LPO in liver and kidney tissues. Further, the extracts have augmented Zn concentrations in the liver of diabetic rats (Orhan et al. 2011). The hydroalcoholic extract of *J. oxycedrus* leaf exhibited potent inhibitory activity on the α-amylase enzyme (Orhan et al. 2014).

Justicia adhatoda L., Acanthaceae

The alcohol extracts of *Justicia adhatoda* (roots and leaves) showed antidiabetes and hypolipidemic activities in alloxan-induced diabetic rats (Gulfraz et al. 2011).

Justicia beddomei (C.B.Clarke) Bennet., Acanthaceae

Justicia beddomei is a shrub that grows in the shadow and moist areas. The leaves of the plant are reported to be useful in the treatment of diabetes. The ethanol extract of leaves (100 mg/kg, i.p.) reduced the serum glucose level in alloxan-induced diabetic rats (Srinivasa et al. 2008).

Kaempferia parviflora Wall. ex Baker, Zingiberaceae

Kaempferia parviflora has been used in folklore medicine in Thailand and elsewhere to lower blood glucose levels, improve blood flow, and increase vitality. The ethanol extract (100 mg/kg) of the *K. parviflora* rhizome prevented vascular complications of diabetes in streptozotocin-induced diabetic rats (Malakul et al. 2011). When feed containing 1% or 3% of *K. parviflora* was given *ad libitum* to Tsumura Suzuki obese diabetic mice (a multifactorial genetic disease mouse model wherein metabolic diseases develop spontaneously) and nonobese mice (corresponding control mice) for 8 weeks, body weight increase, visceral fat accumulation, abnormal lipid metabolism, hyperinsulinemia, glucose intolerance, insulin resistance, and peripheral neuropathy were suppressed in obese diabetic mice, whereas no marked differences were found in nonobese mice. Thus, the plant had prevented the effect of the metabolic disease with an antiobesity activity only in obese animals (Akase et al. 2011). The plant ameliorated complications of diabetes also. The ethanol extract of *K. parviflora* reduced oxidative stress and preserved endothelial function in the aorta of streptozotocin-induced diabetic rats. When aortic rings were acutely exposed to the *K. parviflora* extract (1, 10, and 100 µg/mL), there was a significant reduction in the detection of superoxide anion and enhanced relaxation to acetylcholine. Two separate groups of rats (control and diabetic) were orally administered daily with the extract (100 mg/kg) for 4 weeks. The treatment reduced superoxide generation and increased the nitrite levels in diabetic aorta and enhanced acetylcholine-induced relaxation (Malakul et al. 2011).

Kalanchoe crenata Andr. Haw., Crassulaceae

Kalanchoe crenata is a vegetable widely used in Cameroon and largely efficient in the treatment of DM. The effects of the water–ethanol extract of this plant on blood glucose levels were investigated in fasting normal and hypercaloric sucrose diet-induced diabetic rats after a short- and medium-term treatment. Six hours after a single oral administration of the extract (135 and 200 mg/kg), blood glucose levels decreased in both normal and diabetic rats. During the medium-term treatment (200 mg/kg for 4 weeks), blood glucose levels decreased within week 1, with a maximum effect at week 4 (52%); the treatment maintained glycemia within the normal range. All the extract-treated diabetic rats exhibited significant increase in the insulin sensitivity index compared with the initial time and to the untreated diabetic animals. Animals treated for 4 weeks exhibited decrease in food and water intake. Qualitative phytochemical screening revealed that the extract contained terpenoids, tannins, polysaccharides, saponins, flavonoids, and alkaloids (Kamgang et al. 2008). The methanol extract (50 and 68 mg/kg, p.o., for 6 weeks) of *K. crenata* (whole plant) treatment to streptozotocin-induced diabetic, nephropathic rats decreased glycemia, glycosuria, and proteinuria. The treatment decreased MDA levels and increased the activities of catalase and superoxide dismutase in the blood, liver, and kidney. The treatment improved the lipid profile and decreased the atherogenic index. Thus, the extract holds promise for the development of a phytomedicine for DM (Fondjo et al. 2012).

Kalanchoe pinnata (Lam.) Pers., Crassulaceae

The ethanol extract of the leaves of *Kalanchoe pinnata* inhibited α-glucosidase activity *in vitro* (IC_{50} 16–41 ppm). The IC_{50} values for the leaves collected from plants occurring in various areas differed (Dewiyanti et al. 2012). The ethanol and water extracts of dried stem bark of *K. pinnata* showed hypoglycemic activity in normal fasted rats and antihyperglycemic effect in alloxan-induced diabetic rats. The alcohol extract, but not the water extract, inhibited α-amylase activity (Mathew et al. 2013). Thus, the stem bark contains more than one antidiabetic principle and mechanisms of action.

Kalopanax pictus (Thunb.) Nakai (= *Kalopanax septemlobus* ex A.Murr.) Koidz., Araliaceae
Kalopanax pictus is known as castor-aralia or prickly castor-oil tree. *K. pictus* extracts have been used for dietary health supplements and as medicine for a variety of ailments in East Asia. *K. pictus* extracts, isolated compounds and metabolites of compounds (formed by the actions of intestinal microflora) exhibited anti-diabetes activity in experimental diabetic animals.

The hot water extract from *K. pictus* bark showed insulin-like action and glucose uptake in 3T3-L1 cells. Treatment of 3T3-L1 fibroblasts with 1 and 10 μg/mL of *K. pictus* total hot water extract increased the differentiation of the cells. When co-treated with inducers such a dexamethasone, 1-methyl-3-isobutylxanthine and insulin, the differentiation was increased at 1 μg /mL of total extract, but not at 10 μg /mL. In 3T3-L1 adipocytes, glucose uptake was increased by 3.3 times with addition of 0.3 μg/mL of an active fraction from the water extract at 3 ng/mL insulin. Thus, the extract contains such compounds that play a role of insulin-like action and insulin sensitizer (Ko et al. 2002). The antidiabetic evaluation of chemical isolates from the stem bark of *K. pictus* in the streptozotocin-induced diabetic rats showed that kalopanaxsaponin A has a potent antidiabetic activity in contrast to a mild activity of hedearagenin. In addition, significant hypocholesterolemic and hypolipidemic activities of kalopanaxsaponin A and hedearagenin were observed (Park et al. 1998). Human intestinal microflora metabolized kalopanaxsaponin B to kalopanaxsaponin A, hederagenin 3-*O*-α-L-arabinopyranoside and hederagenin. Kalopanaxsaponin H was metabolized to kalopanaxsaponin A and I, hederagenin 3-*O*-α-L-arabinopyranoside and hederagenin. Among kalopanaxsaponin B, H, and their metabolites, kalopanaxsaponin A showed the most potent antidiabetic activity, followed by hederagenin. Kalopanaxsaponin A (25 mg/kg, i.p.) significantly reduced blood levels of glucose, cholesterol, total lipids and triglycerides in streptozotocin-induced diabetic rats. However, the main components, kalopanaxsaponin B and H in *K. pictus* were inactive (Kim et al. 1998).

Khaya senegalensis Desr. A.Juss., Meliaceae

Common Name: African mahogany
The bark of *Khaya senegalensis* is used to treat diabetes in folklore medicine. The extract of *K. senegalensis* bark (2–20 μg/mL) reduced the release of glucose in a concentration- and time-dependent manner from fragments of mice liver under *in vitro* conditions (Takin et al. 2014).

Kielmeyera coriacea Mart, Calophyllaceae
The hydroethanolic extract of *Kielmeyera coriacea* stem bark presented a strong inhibition of α-amylase activity (Silva et al. 2009).

Kigelia pinnata, Bignoniaceae
Kigelia pinnata flower significantly reduced blood glucose, serum cholesterol, and TG levels in streptozotocin-induced diabetic rats (Kumar et al. 2012f).

Kochia scoparia (L.) Schrad., Chenopodiaceae
The methanol extract of a food garnish (tonburi), the fruit of Japanese *Kochia scoparia*, inhibited the increase in serum glucose in glucose-loaded rats. Through bioassay-guided separation, momordin Ic and its 2'-*O*-β-D-glucopyranoside were isolated as the active principles from this medicinal foodstuff. These are the principal saponin constituents of this medicinal foodstuff that inhibit glucose and ethanol absorption in rats (Yoshikawa et al. 1997b).

Lactuca indica L., Compositae/Asteraceae
Three novel sesquiterpene lactones (lactucain A, B, and C) and a new furofuran lignin, lactucaside, were isolated from *Lactuca indica* along with nine known compounds. The known compounds include

quercetin, quercetin 3-*O*-glucoside, rutin, apigenin, luteolin, and chlorogenic acid. Among the new compounds, lactucain C and lactucaside showed antidiabetic activity (Hou et al. 2003). Quercetin, rutin, and chlorogenic acid (from other sources) are known anti-DM compounds. *L. indica* could be a promising plant for further studies.

Lactuca sativa L., Asteraceae, Lettuce

Lactuca sativa leaf extract lowered blood glucose *in vivo* (Marles and Farnsworth 1995). This plant is also an ingredient in a traditional polyherbal formulation for diabetes.

Lagenaria siceraria L. (*La. vulgaris*), Cucurbitaceae

Lagenaria siceraria fruit (bottle gourd) may have a potential therapeutic role in the treatment of type 2 DM. The methanol extract of *L. siceraria* aerial parts was evaluated for its antidiabetic activity using streptozotocin-induced diabetic rats. The methanol extract showed potent antihyperglycemic activity in diabetic rats. Besides, the extract treatment improved lipid metabolism and represented a protective mechanism against the development of atherosclerosis and prevented diabetic complications from LPO by improving the antioxidant status in diabetic rats. The authors concluded that the methanol extract of *L. siceraria* supplementation is quite beneficial in controlling the blood glucose level, without producing hypoglycemia (Saha et al. 2011b).

Human Clinical Study: The nutraceutical potential of bottle gourd juice in the management of human DM and associated dyslipidemia was evaluated. A notable reduction in the blood glucose levels and improvement in the lipid profile were observed in diabetic subjects on bottle gourd juice (200 mL/day) for 90 consecutive days. The juice administration increased the activities of superoxide dismutase and catalase and GSH levels in both diabetic and healthy subjects. Considerable improvement in liver and kidney function was also observed in the juice-treated diabetic subjects. This may indicate that the juice may be useful as an adjunctive treatment in disorders associated with carbohydrate and lipid metabolism (Katare et al. 2013).

Lagerstroemia speciosa L., Lythraceae (Figure 2.62)

Common Names: Banaba, pride-of-India, queen's flower, etc.

Description: *Lagerstroemia speciosa* is a tree that can grow as tall as 20 m. It has multiple trunks or stems diverging from just aboveground level and has a wide spreading crown. The bark is distinctive as it is light brown in color and often peels from the trunk in large regions revealing the smooth new bark that is forming underneath. The leaves of pride-of-India are oblong, are up to 30 cm long and 13 cm wide, and are quite leathery. The flowers of this species are variable in color, with white, purple, and lavender. The flowers are produced in clusters at the tips of the branches and are up to 7.5 cm wide and have five

FIGURE 2.62 *Lagerstroemia speciosa.*

petals that appear crumpled. The yellow color in the center of the flower is the stamens that produce the pollen with 130–200 stamens found in a single flower. The fruit is a capsule, which is dry at maturity and splits open to release the winged seeds.

Distribution: *L. speciosa* is a tropical plant found in many parts of Asia including the Philippines, Vietnam, Malaysia, China, and Australia. It is native to China, Cambodia, Myanmar, Thailand, Vietnam, Indonesia, Malaysia, and the Philippines. This species is widespread in cultivation in tropical regions.

Traditional Medicinal Uses: It is traditionally used in the Philippines for the treatment of diabetes and kidney-related diseases.

Pharmacology

Antidiabetes: The leaves of *L. speciosa* have been subjected to numerous *in vitro* and *in vivo* studies that consistently confirmed the anti-DM activity of this plant. Scientists have identified different components of banaba to be responsible for its activity. Using tumor cells as a cell model, corosolic acid was isolated from the methanol extract of banaba and shown to be an active compound. Later on, the focus on the water-soluble fraction of the leaf extract led to the discovery of other active compounds. The ellagitannin lagerstroemin was identified as an effective component of the banaba extract responsible for the antidiabetic activity. In a different approach, using 3T3-L1 adipocytes as a cell model and a glucose uptake assay as the functional screening method, it was shown that the banaba water extract exhibited insulin-like glucose transport inducing activity. Coupling HPLC fractionation with a glucose uptake assay, the gallotannin penta-*O*-galloyl-glucopyranose (PGG) was identified as the most potent compound. A comparison of published data with results obtained for PGG indicates that PGG has a significantly higher glucose transport stimulatory activity than lagerstroemin. Further, PGG exhibits antiadipogenic properties in addition to stimulating the glucose uptake in adipocytes. The combination of glucose uptake and antiadipogenesis activity is not found in the current insulin mimetic drugs and may indicate a great therapeutic potential of PGG (review: Klein et al. 2007; Yamada et al. 2008; Takagi et al. 2010).

The hypoglycemic effects of banaba have been attributed to both corosolic acid and ellagitannins (review: Miura et al. 2012). Studies have been conducted in various animal models, human subjects, and *in vitro* systems using water-soluble banaba leaf extracts, corosolic acid, and ellagitannins. Corosolic acid has been reported to decrease blood sugar levels within 60 min in human subjects. Corosolic acid also exhibits antihyperlipidemic and antioxidant activities. The beneficial effects of banaba extract, ellagitannins, tannic acid, and corosolic acid with respect to various aspects of glucose and lipid metabolism appear to involve multiple mechanisms, including stimulation of insulin action, enhanced cellular uptake of glucose, inhibition of hydrolysis of sucrose and starches, decreased gluconeogenesis, and regulation of lipid metabolism. These effects may be mediated by PPAR-γ, MAPK, NF-κB, and other signal transduction factors. Banaba extract, corosolic acid, and other constituents may be beneficial in addressing the symptoms associated with metabolic syndrome, as well as offering other health benefits (review: Klein et al. 2007; Shi et al. 2008; Yamada et al. 2008a,b; Saha et al. 2009; Tanquilut et al. 2009; Thuppia et al. 2009; Musabayane et al. 2010; Saumya and Basha 2011; review: Stohs et al. 2012).

Major problems associated with the interpretation of and generalizing the effects reported in various *in vivo* studies are due to the differences in the extract preparations, differences in the doses used, and the animal models used.

In vitro studies reported include stimulation of glucose uptake by methanol and water extracts of the plant in 3T3 adipocytes. The ellagitannins extracted from the leaf stimulated glucose uptake by isolated rat adipocytes. Lagerstroemin (an ellagitannin) stimulated glucose transport into adipocytes with a 50% effective concentration of 80 μM. Penta-*O*-galloyl-D-glucopyranose binds to IR and activates insulin-mediated glucose transport; further, seven ellagitannins isolated from this plant exhibited the ability to stimulate insulin-like glucose uptake as well as to inhibit adipocyte differentiation in 3T3-L1 adipocytes in culture. Six pentacyclic triterpene acids including corosolic acid and oleanolic acid from the leaves inhibit α-amylase and α-glucosidase activities (Li et al. 2005; review: Klein et al. 2007; Bai et al. 2008;

Shi et al. 2008; Hou et al. 2009). Since several biologically active compounds beneficial to DM patients are present in the plant, their individual effect and various combination effects in various doses remain to be studied in *in vivo* models to derive more insights.

Clinical Trial: Several preparations containing the leaf extract of banaba were tested for their efficacy and safety on DM patients. Although these studies showed varying levels of efficacy, the constituents in the product responsible for the anti-DM effect were not determined and the preparations were not standardized. In a 1-year open label safety and efficacy study on 15 subjects, daily administration of the water extract of the plant leaf (100 mg tablet/day) resulted in a decrease (16.6%) in fasting glucose in individuals with fasting glucose levels higher than 110 mg/dL. Significant improvements in glucose tolerance and glycated albumin levels were also observed in the treated patients (Ikeda et al. 2002).

In another study, the antidiabetic activity of a banaba leaf extract standardized to 1% corosolic acid in a soft gel capsule formulation has been examined. Ten type 2 diabetic subjects were given 32 or 48 mg of the product daily for 2 weeks. A 30% decrease in blood glucose levels was reported after 2 weeks. It is not clear whether the observed effect was due to corosolic acid, the tannin components, or a combination thereof (Judy et al. 2003). The plant contains oleanolic acid; this compound from other sources has been reported to have antidiabetic property.

In a double-blind, crossover design, DM patients ($n = 31$) were given a capsule containing 10 mg corosolic acid or a placebo 5 min before a 75 g oral glucose tolerance test. Blood glucose levels were measured at 30 min intervals for 2 h. The corosolic acid treatment resulted in lower blood glucose levels from 60 to 120 min compared with controls. According to the authors, the corosolic acid used was 99% pure, thus indicating that the blood sugar-lowering effect was specifically due to the corosolic acid (Fukushima et al. 2006). The clinical studies indicate that both the water extract of the banaba leaf and corosolic acid decrease fasting as well as postprandial blood glucose levels in humans. No adverse effect has been observed in any of the studies.

Other Activities: The plant leaf extract exhibited free radical scavenging activity and also contains a known antioxidant compound. Corosolic acid, isolated from this plant, stimulated osteoblast differentiation by activating transcription factors and MAPKs (Shim et al. 2009). *L. speciosa* water extract inhibited TNF-induced activation of nuclear factor-κB in rat cardiomyocyte in a time- and dose-dependent manner (Ichikawa et al. 2010).

Phytochemicals: The plant contains ellagitannins (lagerstroemin; lagerstannins A, B, and C; flosin B; reginin; etc.), methyl ellagic acid derivatives, gallotannins such as PGG/tannic acid, pentacyclic triterpene acids (oleanolic acid, corosolic acid, arjunolic acid, asiatic acid, maslinic acid, and 23-hydroxyursolic acid), and polyphenolic valoneic acid lactone (review: Klein et al. 2007; Bai et al. 2008; Hou et al. 2009).

Lantana camara L., Verbenaceae

Synonyms: *Lantana aculeata* L., *Lantana antillana* Raf., *Lantana camara* L. var. *aculeata* (L.) Mold., *Lantana scabrida* Soland

The ethanol extract of the roots of *Lantana aculeata* decreased the levels of blood glucose, total cholesterol, and TG and increased blood insulin levels and liver glycogen content in alloxan-induced diabetic rats (Kumar et al. 2010b). Another independent study also showed the anti-DM activity of this plant against alloxan-induced diabetes. Oral administration of the methanol extract of *L. camara* leaves (200 and 400 mg/kg) to alloxan-induced diabetic rats resulted in significant dose-dependent reduction in blood glucose levels. Further, the treatment increased body weight and improved glucose tolerance and lipid profile in the diabetic rats. Thus, the methanol extract showed promising anti-DM activity against alloxan-induced diabetes (Ganesh et al. 2010).

Laportea ovalifolia (Schum.) Chew., Urticaceae

Laportea ovalifolia is used in ethnomedicine to treat diabetes in Cameroon and elsewhere. A decoction of the leaves of *L. ovalifolia* is widely used in Cameroon for the treatment of several illnesses, including DM. The anti-DM and hypolipidemic effects of a methanol/methylene chloride extract of the aerial parts of this plant have been investigated in normal rats and alloxan-induced diabetic rats. In diabetic rats, 2 weeks of daily, intragastric treatment with the *L. ovalifolia* extract not only produced a significant reduction in the fasting serum glucose concentrations but also lowered the serum concentrations of total cholesterol, TGs, and LDL-C, lowered the ratio of total cholesterol to HDL-C, and increased the serum concentration of HDL-C (Momo et al. 2006a). In another study, the same authors evaluated the antidiabetic and hypolipidemic effects of the water extract of *L. ovalifolia* aerial parts in normal and alloxan-induced diabetic rats. Administration of the water extracts of *L. ovalifolia* for 2 weeks resulted in a significant reduction in fasting serum glucose levels in treated diabetic rats. Further, the treatment showed considerable lowering of serum total cholesterol, TGs, and LDL-C and an increase in HDL-C in the treated diabetic group (Momo et al. 2006b). Another report confirmed the antidiabetic activity of the water extract. The water extract (200 mg/kg) of the aerial parts of *L. ovalifolia* exhibited antihyperglycemic activity 5 h after administration to alloxan-induced diabetic rats. The same dose did not cause any hypoglycemic activity in normal rats. This observation supports the application of the water extract of *L. ovalifolia* as an antihyperglycemic agent (Claudia et al. 2007).

Larix laricina (Du Roi) K. Koch, Pinaceae

Common Name: American larch

Larix laricina has been part of the healing practices of North American for thousands of years. Inner bark preparations are applied as poultice for infected wounds, burns, and ulcers or used in the form of a herbal tea as a tonic, diuretic, or laxative. The antidiabetic activity of this plant has been reviewed recently (Eid and Haddad 2014). Ethno-botanico-medical studies revealed that this is used by Cree of Eeyou Istchee (CEI) communities of Canada to treat symptoms of diabetes.

The crude extract of *L. laricina* bark stimulated glucose uptake in cultured skeletal muscle cells while uncoupling and inhibiting mitochondrial function, thereby activating AMPK. Further, the extract inhibited liver cell glucose-6-phosphatase by a mechanism involving AMPK phosphorylation. In CaCo-2 cells, *L. laricina* extract exerted a mild to moderate inhibition of intestinal glucose transport. More importantly, it promoted adipogenesis. In fact, the plant's extract was more powerful than the rosiglitazone control used. A new triterpenoid and the diterpene labdane derivative 13-epitorulosol, isolated from this plant, had the most prominent adipogenic effect (3.7- and 2.7-fold increase in accumulation of lipids, when compared to vehicle control). The antidiabetic potential of *L. laricina* was validated in a mouse model of diet-induced obesity/hyperglycemia. The plant extract was either administered concomitantly with the HFD (prevention study) or after the development of obesity and insulin resistance (treatment study). When administered as treatment, the plant was effective at attenuating hyperglycemia, an effect that started at the onset of the treatment but stayed statistically significant with the higher dose only. Furthermore, the higher dose remarkably reduced hyperinsulinemia, which suggests an improvement of insulin resistance. Of note, the leptin/adiponectin ratio, another marker of the degree of insulin resistance, was reduced in both prevention and treatment studies. *L. laricina* exhibited antidiabetic properties both *in vitro* and *in vivo*. These actions involve muscle, liver, and adipose tissue to enhance insulin sensitivity through the activation of liver and muscle AMPK and enhancement of adipogenesis-like activities (Eid and Haddad 2014).

Larrea tridentata (DC) Coville, Zygophyllaceae

Common Name: Creosote bush

An ethnomedically driven approach was used to evaluate the ability of a pure compound isolated from the creosote bush (*Larrea tridentata*) to lower plasma glucose concentration in two mouse models of type 2 diabetes. Masoprocol (nordihydroguaiaretic acid) isolated from *L. tridentata* lowered blood glucose levels, without influencing the levels of insulin, in two mouse models (C57BL/ks *db/db* mice and C57BL/6J *ob/ob* mice) of type 2 diabetes. Further, masoprocol improved glucose tolerance and decreased insulin resistance (Luo et al. 1998). Masoprocol is a known lipoxygenase inhibitor, and the authors suggest that lipoxygenase inhibitors represent a new approach for the treatment of type 2 DM.

Lathyrus sativus L., Fabaceae

Common Names: Lanka, kasari, white vetch, grass pea

Description: *Lathyrus sativus* is an annual herb having a winged stem. The leaves are equally pinnate with broad stipule and winged petiole and the upper leaflet modified into a tendril. The flowers are solitary, reddish-purple or white; the pods are oblong and winged on the back bearing 4–5 seeds.

Distribution: It is spread from the plains of Bengal to Kumaon and often cultivated in India. It is also distributed in Europe, tropical Africa, and oriental lands.

Traditional Medicinal Uses: It is an astringent, cathartic, and tonic and used in biliousness, heart trouble, pain, inflammation, burning, piles, dysentery, and leg ache (Kirtikar and Basu 1975; *The Wealth of India* 1992).

Antidiabetes: Signal transduction through the hydrolysis of glycosylphosphatidylinositol leading to the release of the water-soluble inositol phosphoglycan (IPG) molecules has been demonstrated to be important for mediating some of the actions of insulin and insulin-like growth factor-I (IGF-I). IPG from grass pea (*L. sativus*) seeds has been purified and partially characterized on the basis of its chromatographic properties and its compositional analysis. IPG was generated from *L. sativus* seed glycosylphosphatidylinositol by hydrolysis with a GPI-specific phospholipase D. This IPG inhibited protein kinase A in an *in vitro* assay, caused cell proliferation in explanted cochleovestibular ganglia, and decreased 8-Br-cAMP-induced phosphoenolpyruvate carboxykinase mRNA expression in cultured hepatoma cells. Thus, *L. sativus* seed IPG possesses insulin-mimetic activities (Paneda et al. 2001).

Toxicity: *L. sativus* seed contains neurotoxic substances and the seed intake causes behavioral changes in rats (Jain et al. 1998).

Phytochemicals: 3-*N*-oxalyl-L-2, 3-diaminopropionic acid (neurotoxin), phytin, gluten, globulin, rhamnoside of flavone derivatives, lycopene, saponins, alkaloids, guanidinoamine-homoagnatine, cyanin, and pelargonin were isolated (*The Wealth of India* 1962; De Bruyn et al. 1993; review: Ajikumaran and Subramoniam 2005).

Launaea nudicaulis (L.) J.D.Hooker, Asteraceae

Launaea nudicaulis is a traditional antidiabetic plant in Egypt; the aerial parts of this plant are used in traditional medicine. The aerial parts showed hypoglycemic activity in normal fasting and alloxan-induced diabetic rats (Shabana et al. 1990).

Lavandula stoechas L., Lamiaceae

Administration of essential oil samples (50 mg/kg, i.p., daily for 15 days) obtained from the aerial parts of the plant by hydrodistillation showed antidiabetic and antioxidant activities in alloxan-induced diabetic rats. The essential oils protected against the increase of blood glucose as well as the decrease of antioxidant enzyme activities and LPO induced by alloxan treatment. The principal compounds detected in the oil are D-fenchone, α-pinene, camphor, camphene, eucapur, limonene, linalool, and endobornyl acetate (Sebai et al. 2013).

Lawsonia inermis L., Lythraceae

Common Name: The henna tree

Lawsonia inermis is endowed with both antidiabetic and anti-AGE formation activities. Feeding alloxan-induced diabetic mice with 0.8 g/kg *L. inermis* leaf extract (70% alcohol) for 14 days resulted in normalization of glucose levels in diabetic mice. Further, the treatment brought back the increased lipid levels to normal levels in the diabetic mice (Abdillah et al. 2008). The protein glycation inhibitory activity of the ethanol extract of *L. inermis* (henna) was evaluated *in vitro* using the model system of bovine serum albumin and glucose. Protein oxidation and glycation are posttranslational modifications that are implicated in the pathological development of many age-related disease processes. The alcohol extract of *L. inermis* could effectively protect against protein damage and showed that its action is mainly due to lawsone. In addition, the presence of gallic acid also played a role in the protective activity against protein oxidation and glycation. Lawsone and gallic acid previously isolated from this plant were subjected to glycation bioassay for the first time. It was found that the alcohol extract, lawsone, and gallic acid showed significant inhibition of AGE formation (Sultana et al. 2009).

Leandra lacunosa Cogn., Melastomataceae

Leandra lacunosa is used in Brazilian folkloric medicine for the treatment of DM. The plant showed blood glucose-lowering effects in normal and alloxan-induced diabetic rats. Administration of the hydroalcoholic extract of *L. lacunosa* aerial parts (500 mg/kg, p.o.) to normal rats caused a reduction in blood glucose levels after 2 h of treatment. The treatment also improved oral glucose tolerance. In alloxan-induced diabetic rats, the treatment caused a marked reduction in blood glucose levels after 4 h of treatment (Cunha et al. 2008). It is of interest to note that the extract contained ursolic acid, kaempferol, luteolin, and quercetin (Cunha et al. 2008). These compounds from other sources are reported as antidiabetic compounds.

Leonotis leonurus (L.) R. Br., Lamiaceae

Common Names: Wild dagga, lion's tail, etc.

The water extract of the flowers of *Leonotis leonurus* showed a glucose-lowering effect after intraperitoneal administration to fasting healthy mice. A fraction obtained from the water extract showed hypoglycemic activity in healthy and mild alloxan-induced diabetic mice, but not in severe alloxan-induced diabetic mice (Roman-Ramos et al. 2001). Marrubiin, a constituent of *L. leonurus*, increased insulin levels and glucose transporter-2 gene expression in INS-1 cells. Marrubiin increased insulin secretion and HDL-C level, while it normalized total cholesterol, LDL-C, atherogenic index, IL-1β, and IL-6 levels in an obese rat model (Mnonopi et al. 2012).

Lepechinia caulescens (Ortega) Epling., Lamiaceae

Lepechinia caulescens is a traditional medicinal plant. *L. caulescens* significantly decreased the hyperglycemic peak and the area under the glucose tolerance curve in hyperglycemic rabbits (Bnouham et al. 2006).

Lepidium sativum L., Brassicaceae or Cruciferae (Figure 2.63)

Common Names: Garden cress, ashalika, chandrashura, etc. (*Lepidium sativum* is edible and the leaves are used as salad; they are also cooked with vegetable curries and used as garnish.)

FIGURE 2.63   *Lepidium sativum.*

Description: *Lepidium sativum* is a cool season annual plant cultivated in India and elsewhere. It has long leaves at the bottom of the stem and small bright green feather like the ones arranged on the opposite side of its stalk at the top. There are plain broad leaf and curly leaf varieties that differ in texture but not taste. It can be grown indoor or outdoor. The flowers are bisexual, regular 4-merous; the sepals are ovate; the petals are spatulate with short claw, white or pale pink. The fruits are globose and purple black. The seeds are small, oval-shaped, pointed, and triangular at one end, about 2–3 mm long (review: Sharma and Agrawal 2011).

Distribution: *L. sativum* is a native plant of Southwest Asia that spread many centuries ago to Western Europe (Sharma and Agarwal 2011). Some scientists say its origin started from Ethiopia.

Traditional Medicinal Uses: In Ayurveda, it is described as galactagogue and aphrodisiac; the whole plant is used in traditional medicine for asthma, cough, expectorant, bleeding piles, etc. The seeds are used for abortion; the leaves of this plant are diuretic; root powder is used to treat syphilis (Falana et al. 2014; Manohar et al. 2012).

Pharmacology

Antidiabetes: The water extract of *L. sativum* seed (20 mg/kg), when administered orally, exerted a blood glucose-lowering effect in both normal and diabetic rats. The effect was observed in both chronic and acute treatment and the effect was independent of insulin secretion (Eddouks et al. 2005c). The water extract of this plant exhibited hypoglycemic activity in normal as well as streptozotocin-induced diabetic rats. Intravenous administration of the extract at a dose of 10 mg/kg/h reduced blood glucose levels in both normal and diabetic rats. At the same time a marked increase of glycosuria was observed in the treated animals. Besides, oral administration of the seed extract for 15 days to the diabetic rats normalized glycemia, enhanced glycosuria, and decreased the amount of urinary TGF-β. It is concluded that the extract caused a potent inhibition of renal glucose reabsorption that is, at least, one mechanism of its hypoglycemic action (Eddouks and Maghrani 2008). In another study, the total alkaloids from the seeds of *L. sativum* (250 mg/kg, p.o., for 21 days) exhibited antidiabetic activity in alloxan-induced diabetic rats. The treatment decreased blood glucose and lipid levels (Shukla et al. 2012a). Isolation of active principles remains to be done on this important anti-DM plant.

Clinical Trial: The anti-DM activity of *L. sativum* seeds (15 g/day for 21 days) was tested on NIDDM as well as normal healthy subjects. The hypoglycemic activity was reported in both normal and diabetic patients (Patole et al. 1998).

Toxicity: The plant is edible and is well tolerated. One study suggested that in order to show toxicity, if any, it should be injected in very high amounts that normal people would not take in normal circumstances. At very high doses, it can have antiovulatory and teratogenic properties in rats (review: Falana et al. 2014).

Other Activities: The plant possesses antihemagglutinating, antihypertensive, fracture healing, and bronchodilatory properties. The flavonoid groups of compounds have been observed to exhibit anti-inflammatory activity. The volatile products of crushed leaves show antibacterial activity (review: Sharma and Agrawal 2011). Other properties include hepatoprotection, antihypertensive effect, chemoprotective effect, laxative effect, increase in milk production, etc. (review: Falana et al. 2014).

Phytochemicals: In addition to its nutritional value, the major secondary compounds of this plant are glucosinolates. The plant yields, on steam distillation, a volatile oil with a characteristic pungent aroma containing benzyl isothiocyanates, benzyl cyanide, etc. Seven imidazole alkaloids, lipidine B, C, D, E, and F, two new monomeric alkaloids, semilipideneside A and B, flavonoids, tannins, glucosinolates, sterols, sitosterol, triterpenes, and 4-methoxyglucobrassicin were found in seeds. The ethanolic extract revealed the presence of alkaloids, tannins, flavonoids, steroids, etc. The main volatile constituents of the seeds are benzyl isothiocyanate and 1,8-cincole while those of roots and nonflowering aerial parts are benzyl isothiocyanate, α-pinene, and hexadecanoic acid. Flavonoids including quercetin and kaempferol glycosides and glucosinolates are reported in this plant (review: Sharma and Agrawal 2011).

Leptadenia hastata (Pers.) Decne., Asclepiadaceae

Leptadenia hastata is widely used as a vegetable and traditionally in the management of diabetes, treatment of wounds, and stomachache in Nigeria and elsewhere. *L. hastata* leaf was found to be rich in polyphenols and possessed an α-glucosidase inhibition potential. *L. hastata* leaves exhibited hypoglycemic and hypolipidemic effects in alloxan-induced diabetic rats. Oral administration of the methanol and water extracts (300 mg/kg daily for 7 days) of the leaves showed a significant decrease in the blood glucose levels and increase in glycogen content of the liver and muscle. Further, the treatment reduced the levels of serum TG and VLDL-C and increased HDL-C levels (Bello et al. 2011a). Both the methanol and water extracts of *L. hastata* leaf significantly inhibited the activity of α-glucosidase. The methanol extract was found to be better than the water extract in inhibiting α-glucosidase activity (Bello et al. 2011b). In another study, oral administration of the water extract of *L. hastata* root (600 and 800 mg/kg) to normal rats resulted in a significant decrease in blood glucose levels (Sanda et al. 2013).

Leucas indica Linn., Labiatae

Leucas indica is a traditional medicinal herb occurring in the wild in India. The hypoglycemic activity of the water extract of *L. indica* on streptozotocin-induced diabetic rats was evaluated.

The extract showed a significant dose-dependent (200 and 400 mg/kg orally) reduction in fasting blood glucose level. In addition, the treatment improved body weight and biochemical parameters like the lipid profile, glutamate oxaloacetate transaminase, glutamate pyruvate transaminase, total proteins, bilirubin, urea, and ALP compared to untreated diabetic control rats. Besides, histopathological investigation of the pancreas, liver, and kidney showed a significant protection and regeneration of streptozotocin-induced cellular necrosis in the treated diabetic animals (Sarkar et al. 2013).

Leucas lavandulaefolia Willd., Lamiaceae

Leucas lavandulaefolia is mainly used in Indian folk medicine for the treatment of DM. The oral administration of 0.15, 0.20, and 0.25 g/kg of the chloroform extract of the *L. lavandulaefolia* flowers for 30 days to alloxan-induced diabetic rats resulted in a significant reduction in blood glucose and glycosylated hemoglobin and an increase in total hemoglobin. It also prevented decrease in body weight. There was improvement in oral glucose tolerance in animals treated with the extract (Chandrashekar and Prasanna 2009).

Levisticum officinale Koch, Apiaceae

The methanol and water extracts of this plant showed promising *in vitro* inhibition of α-amylase activity (Gholamhoseinian et al. 2008).

Ligularia fischeri (Ledeb.) Trucz., Compositae

The plant *Ligularia fischeri* is common in Southwestern China and has been used as a traditional Chinese medicine since ancient time. From the roots of *L. fischeri*, an eremophilane sesquiterpene was isolated; this sesquiterpene exhibited PTP1B inhibitory activity with IC_{50} value of 1.3 μmol/L (Jiany et al. 2012).

Ligusticum chuanxiong Hort, Apiaceae

Tetramethylpyrazine isolated from *Ligusticum chuanxiong* (200 mg/kg, p.o. for 8 weeks) reduced blood glucose levels and improved renal function in streptozotocin-induced diabetic rats compared to untreated diabetic rats. Diabetic nephropathy resulted in an increase in the expression of vascular endothelial growth factor, while *tetramethylpyrazine* administration greatly decreased the expression. This protective effect may be mediated, in part, by downregulated expression of vascular endothelial growth factor in the kidney (Yang et al. 2011).

Ligustrum lucidum Ait., Oleaceae

Oleanolic acid from *Ligustrum lucidum* showed antidiabetes and antioxidant effects in alloxan-induced diabetic rats. Oleanolic acid (60 and 100 mg/kg for 40 days) showed significant hypoglycemic activity by lowering blood glucose in the alloxan-induced diabetic rats. The levels of serum total cholesterol, TGs, and LDL-C in the oleanolic acid-treated diabetic rats were lower, and the HDL-C level was higher than those in the diabetic control rats. A significant reduction in the serum AST, ALT, and ALP levels of diabetic rats following oleanolic acid treatment was also observed. Furthermore, the treatment decreased the MDA level but increased superoxide dismutase and GSH peroxidase activities of the liver and kidney in diabetic rats. Thus, oleanolic acid had hypoglycemic, hypolipidemic, and antioxidant efficacy in diabetic rats (Gao et al. 2009).

Limnocitrus littoralis (Miq.) Swingle, Rutaceae
Limnocitrus littoralis is used in traditional Vietnamese medicine as an expectorant and antitussive; it is also used to treat colds and fevers. Meranzin isolated from the leaves of *L. littoralis* activated PPAR-γ. This observation suggests the likely antidiabetic activity of this plant (review: Wang et al. 2014).

Limonia acidissima L., Rutaceae
The methanol extracts of *Limonia acidissima* stem bark (200 and 400 mg/kg daily for 21 days, p.o.) showed antidiabetic and antioxidant properties in alloxan-induced diabetic rats (Ilango and Chitra 2009).

Limonium tubiflorum (Delile) Kuntze., Plumbaginaceae
Limonium tubiflorum is an Egyptian traditional medicinal plant used, among other things, to treat DM. Aerial parts of the plant extract showed hypoglycemic activity in a screening study in rats (Shabana et al. 1990).

Linum usitatissimum L., Linaceae
Common Names: Flax, linseed, and common flax
Secoisolariciresinol diglucoside isolated from *Linum usitatissimum* delayed the development of type 2 DM in Zucker rats (Prasad 2001). Linseed oil showed antidiabetic and antihyperlipidemic activities in streptozotocin-induced diabetic rats. Further, the linseed oil showed *in vitro* antioxidant activity (Kaithwas and Majumdar 2012).

Lippia nodiflora L., Verbenaceae
Synonym: *Phyla nodiflora* (L.) Greene
Common Names: Frog fruit, lippaincise, lippie, bukkan, vasir vasuka, jalpali, poduthalai, etc.
Description: It is a creeping perennial herb. *Lippia* is well adapted to moist clay soils in riverine and floodplain environments and has been observed to grow most prolifically on sites that experience periodic flooding of short duration. It can grow to a 20–30 cm and dominate other plants. The plant is green to purple in color when young and can become somewhat gray and woody with age. Roots are produced from leaf axils along stems and consist of a central taproot. Leaves arise in pairs at stem nodes and are rounded, entire, or bluntly toothed at the tip and narrow toward the petiole. White to purple flowers are produced in heads (10 mm in diameter) on long peduncles arising from leaf axils. The fruits (1–1.5 mm in diameter) release two tiny, brown, oval, flattened seeds at maturity. The seeds are barely visible to the naked eye (review: Sharma and Singh 2013).
Distribution: *L. nodiflora* is distributed in India, Sri Lanka, Ceylon, Baluchistan, South and Central America, and tropical Africa. It is native to California (review: Sharma and Singh 2013).
Traditional Medicinal Uses: It is widely used in traditional systems on medicine to treat ulcers, bronchitis, diabetes, and heart diseases.
Antidiabetes: In streptozotocin-induced diabetic rats, treatment with the methanol extract of *Lippia nodiflora* resulted in a dose-dependent decrease in the levels of blood glucose and glycosylated hemoglobulin, while there was an increase in serum insulin and glycogen in the liver and muscle. Serum cholesterol (total) and TG decreased, whereas HDL-C level increased in the methanol extract–treated

diabetic rats (Balamurugan and Ignacimuthu 2011). Oral administration of γ-sitosterol (20 mg/kg) isolated from *L. nodiflora* for 21 days to streptozotocin-induced diabetic rats resulted in a significant decrease in blood glucose and glycosylated hemoglobin with an increase in plasma insulin level and body weight. Furthermore, the treatment showed antihyperlipidemic activity as evidenced from marked decrease in serum total cholesterol, TGs, and LDL levels coupled with an elevation of HDL levels. In isolated rat islets, γ-sitosterol increased insulin secretion in response to glucose. An immunohistochemical study of the pancreas confirmed the biochemical findings (Balamurugan et al. 2011).

Other Activities: Based on experiments on animals, the plant has several pharmacological properties such as antimicrobial, hepatoprotective, antioxidant, anti-inflammatory, antitumor, diuretic, antiurolithiatic, anticonvulsant, and anxiolytic effects (review: Sharma and Singh 2013).

Phytochemicals: The plant contains a variety of constituents such as triterpenoids, flavonoids, phenols, and steroids. Nodifloretin, β-sitosterol glycoside, and stigmasterol glycoside were isolated from the leaves of *L. nodiflora*. Nodifloridin A and nodifloridin B and two new flavone glycosides (lippiflorin A and lippiflorin B) along with the known compound nepetin and batalilfolin were isolated from the ethanol extract of the plant. From the methanol extract of the aerial parts of *L. nodiflora*, a new triterpenoid lippiacin, a new steroid 4′, 5′-dimethoxybenzoloxy stigmasterol along with the known stigmasterol and β-sitosterol was isolated (review: Sharma and Singh 2013).

Liriope spicata var. *prolifera* (Thunb.) Lour., Convallariaceae

Common Name: Lily turf

Studies have shown that the water extract containing crude polysaccharides from the tuberous roots of *Liriope spicata* var. *prolifera* possesses hypoglycemic and hypolipidemic activities. Two new water-soluble polysaccharides were isolated from the active crude polysaccharides by cellulose and AB-8 macroporous resin chromatography. These are two fructans with the molecular weights 3.2 and 4.29 kDa, respectively. Both of these polysaccharides caused a decrease of the fasting blood glucose and an improvement on glucose tolerance in type 2 DM mice (Chen et al. 2009).

Litchi chinensis Sonn., Sapindaceae

Litchi chinensis is an evergreen tree and the seeds of this plant are used in Chinese medicine to treat diabetes. Litchi nucleus extract improved blood biochemical parameters in alloxan diabetic mice (Li et al. 2006a). The water extracts of the seeds of this plant (5 g/kg) lowered blood glucose levels in normal and alloxan-induced diabetic mice. The nut of this plant has been developed into a medicine (tablet) to treat diabetic complications, especially pregnancy diabetes in a clinic in China (review: Chauhan et al. 2010).

Lithocarpus polystachyus Rehd., Fagaceae

Administration of the flavonoid-rich fraction from the leaves of *Lithocarpus polystachyus* for 4 weeks showed promising antidiabetes activity in type 2 diabetic rats as judged from blood glucose, glycosylated serum protein, cholesterol, TG, MDA, superoxide dismutase, and attenuation of liver injury in type 2 diabetic rats. Administration of the fraction also significantly reduced the fasting serum insulin and C-peptide level and improved the insulin tolerance. In type 1 diabetic rats, administration of the fraction for 3 weeks caused a significant reduction in fasting blood glucose, total cholesterol, TG, urea nitrogen, creatinine, and liver mass, along with significantly inhibiting the decline of insulin level compared to diabetic control (Hou et al. 2011). The sweet compound from *L. polystachyus*, trilobatin, inhibited α-glucosidase strongly and α-amylase mildly (Dong et al. 2012).

Lithospermum erythrorhizon Sieb. et Zucc., Boraginaceae

Lithospermum erythrorhizon, the purple gromwell, redroot gromwell, is a perennial herb growing up to 0.7 m. *L. erythrorhizon* is a Chinese medicinal plant with various antiviral and biological activities.

A water extract of the crude drug shikon, *L. erythrorhizon* roots, remarkably diminished the plasma sugar level in mice. Fractionation of the extract by monitoring the activity yielded three glycans, lithospermans A, B, and C. These glycans exerted marked hypoglycemic effects in normal and alloxan-induced hyperglycemic mice (Konno et al. 1985a). Shikonin, a naphthoquinone isolated from *L. erythrorhizon*, when administered intraperitoneally (10 mg/kg) once daily for 4 days improved plasma glucose levels in diabetic Goto-Kakizaki rats. Further, it increased the glucose uptake in skeletal muscle cells (Oberg et al. 2011). Shikonin increased glucose uptake in L-6 skeletal muscle myotubes without phosphorylating Akt indicating that in these cells its effect is mediated via a pathway distinct from that used for insulin-stimulated uptake. The compound increases intracellular levels of calcium in these cells and this increase is necessary for shikonin-stimulated glucose uptake. Furthermore, the compound stimulated the translocation of GLUT4 from intracellular vesicles to the cell surface in L6 myocytes. The authors conclude that shikonin increases glucose uptake in muscle cells via an insulin-independent pathway dependent on intracellular free calcium (Oberg et al. 2011).

Lobelia chinensis Lour., Campanulaceae

Common Names: Azemushiro, Chinese lobelia

Lobelia chinensis is a perennial herb and is one of the 50 fundamental herbs used in traditional Chinese medicine. Two new pyrrolidine alkaloids, radicamines A and B, isolated from *L. chinensis* exhibited promising α-glucosidase inhibitory activity *in vitro* (Shibano et al. 2001).

Lodoicea maldivica (J.F.Gmel.) Pers., Palmae

Synonym: *Lodoicea sechellarum* Labill.

Common Name: Double coconut

Lodoicea maldivica is endemic to Praslin and Curieuse Islands, Seychelles, and listed as an endangered species. It was previously used as a medicinal plant. The extract of *L. sechellarum* fruit (2, 3, and 4 g) showed anti-DM activity in type 2 diabetic patients (Sharma and Arya 2011).

Lonicera japonica Thunb., Caprifoliaceae

Common Name: Honeysuckle flower

The flower bud of *Lonicera japonica* (methanol extract) inhibited α-glucosidase activity. An active fraction from the methanol extracts downregulated α-glucosidase activity in Caco-2 monolayer cells. In maltose-loaded SD rats, the fraction reduced postprandial blood glucose level. 3,5-Dicaffeoylquinic acid was found to be the potent maltase inhibitor, whereas chlorogenic acid and rutin showed a weak inhibitory activity against maltase (Zhang et al. 2013). The ethanol extract of the flowering aerial parts of *L. japonica* downregulated the protein expression of p38 mitogen-activated protein kinase in the kidney of diabetic rats. The results suggest that it has the property to inhibit the activity of p38 MAPK-mediated inflammatory response to halt the progression of diabetic nephropathy (Tzeng et al. 2014). Hyperoside, chlorogenic acid, and luteolin and caffeic acid are the major important active ingredients present in *L. japonica*. Chlorogenic acid is a known antidiabetic compound (Peng et al. 2005).

Loranthus begwensis L., Loranthaceae

The African mistletoe, *Loranthus begwensis*, has been widely used in Nigerian folk medicine to treat DM. The water extract or infusion (1.32 g/kg/day) of the leaves of *L. begwensis* was supplied *ad libitum* to both nondiabetic and streptozotocin-induced diabetic rats, as their only source of fluid for a period of 28 days. The infusions of mistletoe parasite on both lemon and guava trees significantly decreased serum glucose levels in nondiabetic and diabetic rats, whereas that prepared from mistletoe parasitic on jatropha did not. The data indicate that African mistletoe possesses significant antidiabetic activity in streptozotocin-induced diabetic rats; its antidiabetic activity appears to be highly dependent on the host plant species (Obatomi et al. 1994).

Loranthus micranthus L., Loranthaceae

The antidiabetic activity of the methanol extracts of the leaves of *Loranthus micranthus* parasitic on *Persea americana, Baphia nitida, Kola acuminata, Pentaclethra macrophylla,* and *Azadirachta indica* was evaluated. The extracts were found to possess significant dose-dependent antihyperglycemic effects in alloxan-induced diabetic and normoglycemic rats. The maximum activity of the methanol extract of *L. micranthus* (400 mg/kg) harvested from *P. americana* on alloxan-induced diabetic rats showed 83% reduction of blood sugar level at 24 h after administration. The methanol extracts of *L. micranthus* from five different host trees did not show any toxicity according the acute toxicity tests in mice. The leaves of *L. micranthus* parasitic on *K. acuminata, A. indica,* and *B. nitida* showed more antihyperglycemic activity compared to other host trees investigated. The results demonstrated that the antidiabetic effect of the extract was found to be dependent on the host plant species (Osadebe et al. 2004).

The water and methanol extracts of *L. micranthus* leaves caused a significant reduction in fasting lipid levels in diabetic and nondiabetic rats with the effect of the methanol extract being significantly higher than that of the water extract (Peter and Obi 2010). An active fraction was separated; however, the active principles are not identified. The weakly acidic fraction of the water–methanol extract of the leaves of *L. micranthus* parasitic on *P. americana* revealed a glucose-lowering activity in alloxan-induced diabetic rats (Osadebe et al. 2010a). The seasonal variation for the antidiabetic effect of the aqueous methanolic extract of the leaves of *L. micranthus* parasitic on *P. americana* in alloxan-induced diabetic rats was studied, demonstrating that the antidiabetic effect of the extract is seasonal and dose dependent with the highest activity being at the peak of the rainy season (Osadebe et al. 2010b). The active principles involved in the antidiabetic action are not known. Although the extract showed hypoglycemic and hypolipidemic activities, the mechanisms of action are not known. This is an important plant for further studies.

Luffa acutangula (L.) Roxb., Cucurbitaceae

Common Names: Vegetable gourd, ridge gourd, peerkankai, etc.

Luffa acutangula is a perennial climber native to Southern and Western India. It is a traditional medicinal plant used for jaundice, spleen enlargement, and as a laxative. The plant is reported to have an α-glucosidase inhibitory effect. The methanolic extract of *Luffa acutangula* fruits (100, 200, and 400 mg/kg) showed a concentration-dependent antidiabetic activity in streptozotocin–nicotinamide-induced diabetic rats. The water extract of the fruit showed moderate activity compared to the methanol extract. Further, the methanol extract exhibited antihyperlipidemic activity in diabetic rats (Pimple et al. 2011). The antidiabetic activity of the fruit extract was further confirmed by histological studies of organs in streptozotocin-induced diabetic rats (Mohan et al. 2012).

Luffa cylindrica L., Cucurbitaceae

Synonym: *Luffa aegyptiaca* Mill

The methanol extract (200 and 400 mg/kg, p.o.) of *Luffa cylindrica,* sponge gourd, fruit exhibited remarkable antihyperglycemic activity in oral glucose tolerance test in normal rats. The extract exhibited promising antidiabetic activity against alloxan-induced diabetic rats as evidenced from glucose levels and serum biochemical parameters (Hazra et al. 2011).

Lupinus mutabilis Sweet, Fabaceae

Synonym: *Lupinus albus* L. (Figure 2.64)

Local Names: Hanchcoly, white lupin, white lupine, lupine, etc.

Description: *Lupinus mutabilis* is a cultivated legume; it is an annual, white flowered, more or less pubescent herb, 30–120 cm in height. It has wide intraspecies variability in physiological plant properties: growth rate, photoperiodic sensitivity, shade tolerance, and drought resistance. There are winter and spring forms of white lupin.

Distribution: *Lupinus mutabilis* is a traditional pulse cultivated and widely distributed in the Mediterranean region. It is widely spread as wild plants throughout the southern Balkans, the islands of

FIGURE 2.64 *Lupinus mutabilis.*

Sicily, Corsica, Sardinia, the Aegean Sea, Palestine, and western Turkey. The plant occurs in meadows, pastures, and grassy slopes. It is also cultivated in Egypt, Sudan, Ethiopia, Syria, Europe, the United States, Australia, South America, Africa, Russia, and Ukraine.

Traditional Medicinal Uses: Lupin seed is known as an anti-DM product in traditional medicine. It is an edible pulse.

Pharmacology

Antidiabetes: Lupine, an edible pulse, is a medicinal food plant with potential value in the management of DM. *L. mutabilis* seed contains an unusual antidiabetic glycoprotein and glucose-lowering quinolizidine alkaloids that are promising for medicine development.

A lupin seed glycoprotein known as conglutin-γ was found to cause significant plasma glucose reduction when orally administered to rats in glucose tolerance test. The glucose-lowering effect of this protein is comparable to that of metformin (Magni et al. 2004). Conglutin-γ in its native conformation is unusually resistant to proteolysis by trypsin. *In vitro* studies in cultured myoblastic C2C12 cells have shown that conglutin-γ stimulation of cells results in the persistent activation of protein synthetic pathway kinases and increase in glucose transport and GLUT4 translocation to the membrane. It appears to stimulate the same insulin signaling pathways (Terruzzi et al. 2011).

Further, lupin seed conglutin-γ lowered blood glucose in hyperglycemic rats and increased glucose consumption of HepG2 cells. When the glycoprotein (conglutin-γ) was administered for 28 days at a daily oral dose of 28 mg/kg to rats in which hyperglycemia was induced by providing drinking water with 10% glucose, the treatment attenuated a rise in the levels of plasma glucose and insulin. Besides, fasting insulin levels and insulin resistance decreased in conglutin-γ-treated rats (Lovati et al. 2012). Moreover, conglutin-γ increased glucose consumption by HepG2 cells under cell culture conditions. In this *in vitro* model, conglutin-γ potentiated the activity of insulin and metformin (Lovati et al. 2012).

Studies have also demonstrated the hypoglycemic activity of lupin quinolizidine alkaloids. Lupanine (a quinolizidine alkaloid from the seed) and its derivatives enhanced glucose-induced insulin secretion in *in vitro* conditions. Multiflorine, another lupin alkaloid, and its derivatives exerted hypoglycemic activity in normal as well as streptozotocin-induced diabetic mice (the most plausible action mechanism for the hypoglycemic activity of quinolizidine alkaloids is similar to that of sulfonylurea drugs) (Garcia Lopez et al. 2004; Gurrola-Diaz et al. 2008). In another study, the effect of multiflorine and the derivatives of sparteine and lupanine on insulin secretion by pancreatic islets under *in vitro* conditions was investigated. Dioxosparteine, hydroxyl lupanine, and multiflorine at 500 μM enhanced insulin secretion from the islets incubated with 16.7 mM glucose, while thionosparteine enhanced insulin secretion at 8.3 mM glucose. Enhancing insulin secretion at hyperglycemic conditions is one of the most desirable characteristics of glucose-lowering drugs (Gurrola-Diaz et al. 2008). A recent study also showed that the water extract of white lupine seeds has hypoglycemic and hypolipidemic effects by increasing insulin levels in alloxan-induced diabetic rats (Helal et al. 2013).

Clinical Trial: The hypoglycemic effects of raw *L. mutabilis* seed was evaluated in healthy volunteers and subjects with dysglycemia. Consumption of the seed by normal healthy individuals did not influence their blood glucose and insulin levels, but consumption of similar doses of lupinus by dysglycemic individuals significantly decreased blood glucose. The effects were greater in those subjects with higher basal glucose levels. A reduction in insulin levels was also observed in the lupinus-treated group after 60 min of treatment. Further, the treatment improved insulin resistance in dysglycemic subjects. The authors conclude that lupinus consumption could be a feasible and low-cost alternative to treat DM (Fornasini et al. 2012).

In another clinical trial (phase II), the hypoglycemic effects of cooked *L. mutabilis* seeds and its purified alkaloids were evaluated in subjects with type 2 DM. Consumption of cooked *L. mutabilis* or its purified alkaloids decreased blood glucose and insulin levels almost to the same level. The reduction in serum glucose concentration from basal line to 90 min after treatment was significant in both the treated groups. None of the volunteers in either group presented any side effects (Baldeon et al. 2012). This clinical trial shows that the plant seed contains anti-type 2 DM alkaloids in addition to the already studied insulin mimetic glycoprotein (conglutin-γ). This pulse could be an interesting nutraceutical for the development of food supplements to DM patients. The lupin quinolizidine alkaloid sparteine has been shown to induce hypoglycemia when administered to diabetic subjects (cited from Gurrola-Diaz et al. 2008).

Phytochemicals: Lupanine and sparteine are the main quinolizidine alkaloids present in *L. mutabilis*. Alkaloids isolated from this plant include 13-acetoxylupanine, ammodendrine, anagyrine, 13-angeloyloxylupanine, angustifoline, 13-benzoyloxylupanine, *cis*-13-cinnamoyloxylupanine, *trans*-13-cinnamoyloxylupanine, 4,13-dihydroxylupanine, α-isolupanine, *N*-methylcysteine, multiflorine, and 17-oxysparteine. Flavonoids identified include genistein, 2-hydroxygenistein, and luteone; β-sitosterol and campesterol are reported from the seed. Other bioactive compounds isolated include conglutin-γ (glycoprotein) (Bisby 1994).

Lycium barbarum L., Solanaceae

Common Names: Wolfberry, goji berries (Eng.), gouqi or ningxiagouqi (Chinese), chirchitta (Hindi)

Description: *Lycium barbarum* is a spinous, glabrous shrub with white or gray branches, armed with sharp conical spines. The leaves are small, entire, and linear-oblong; the flowers are short solitary or fascicled; the calyx is five lobed, irregular, with corolla lobes more than half as long as the tube. The fruit is a small berry, globose or oblong with several seeds.

Distribution: The original habitat of *L. barbarum* is not definitely established; it could probably be the Mediterranean Basin. Now the plant is widely distributed in warm regions of the world, in particular the Mediterranean area and Southwest and Central Asia. It is also cultivated in North America and Australia as a hedge plant (review: Potterat 2009).

Traditional Medicinal Uses: In China and other Asian countries, *L. barbarum* has a long tradition as a food and medicinal plant. The fruit, root, and bark are used in traditional medicine. In Indian traditional medicine it is bitter, emmenagogue, aphrodisiac, diuretic, and laxative and used in the treatment of toothache, spasm, piles, ascites, scabies, and eye diseases (Kirtikar and Basu 1975). Goji berries are used in Chinese traditional medicine as tonic and for the improvement of kidney, liver, and lung functions as well as to treat blurry vision, infertility, abdominal pain, dry cough, fatigue, and headache; it is also used to increase longevity and against premature gray hair. The root bark is prescribed for night sweating and chronic mild fever. It is also indicated for the treatment of diabetes and hypertension (review: Potterat 2009).

Antidiabetes: Hypoglycemic and hypolipidemic activities of *L. barbarum* fruit extracts have been shown in alloxan-induced diabetic rabbits. Polysaccharides (glycol conjugates) containing several mono-saccharides and 17 amino acids were the major bioactive constituents responsible for the hypoglycemic effects (Luo et al. 2004). The protective effects of the polysaccharide against alloxan-induced damage of isolated rat islet cells has been shown (Xu et al. 2002). A polysaccharide extracted from *L. barbarum* (40 mg/kg) significantly decreased the levels of blood glucose, total cholesterol, and TG in DM mice (Jing et al. 2009). Further, oral administration of the polysaccharide exhibited protective effects against oxidative stress and DNA damage in NIDDM rats (Wu et al. 2006, 105). The protective effects of the polysaccharides against streptozotocin-induced oxidative stress and kidney damage were confirmed in another study (Li 2007). The polysaccharide decreased insulin resistance in NIDDM rats as evidenced from a marked decrease in the fasting insulin level, postprandial glucose level, and plasma cholesterol (Zhao et al. 2005). Further, the polysaccharide fraction from this plant fruit also showed renoprotective activity in streptozotocin-induced diabetic rats (Zhao et al. 2009). The methanol extract of *L. barbarum* fruit as well as taurine isolated from the fruit reversed the caspase-dependent apoptotic cytotoxicity pathway in human retinal pigment epithelial cells (Song et al. 2012b). Thus, the studies showed that the polysaccharides reduced insulin resistance and exhibited hypoglycemic, hypolipidemic, and antioxidant properties; it may have β-cell protective effects also. Clinical studies, pharmacokinetics, and detailed mechanisms of action remain to be investigated for the possible development of a medicine from the plant polysaccharides to treat type 2 DM.

Other Activities: The fruit extracts exhibited hypolipidemic and antioxidant activities in rabbits. Polysaccharides from the fruit promoted the peripheral blood recovery of irradiation- or chemotherapy-induced myelosuppression in mice. The polysaccharide–protein complex has immunomodulation and antitumor activities. The polysaccharide inhibited proliferation of hepatoma QGY7703 cells and induced apoptosis. The fruit extract is a good source of antioxidant agent in the daily dietary supplement. A purified component of *L. barbarum* polysaccharide induces a remarkable adaptability to exercise load, enhance resistance, and accelerate elimination of fatigue. It could enhance the storage of muscle and liver glycogen, increase the activity of LDH before and after swimming, and moderate increase of blood urea nitrogen (Gan et al. 2004; Luo et al. 2004; Gong et al. 2005; review: Potterat 2009).

Toxicity: The plant berries and bark are safe to use within reasonable doses (within 1 g/kg). The oral LD_{50} for mice was more than 10 g/kg. However, confusion with morphologically similar toxic Solanaceae species cannot be ruled out with samples collected from the wild. A likely drug interaction with warfarin has been reported (review: Potterat 2009).

Phytochemicals: The roots contain cyclooctapeptides lyciumins A and B, betain, choline, linoleic acid, and β-sitosterol. The fruit contains taurine, γ-aminobutyric acid, betaine, β-cryptoxanthin palmitate, zeaxanthin monopalmitate, zeaxanthin, dopamine derivative lyciumide, L-monomenthyl succinate, and β-carotene. Other phytochemicals isolated from this plant include 2-*O*-(β-D-glucopyranosyl)ascorbic acid, polysaccharides (glycol conjugates), zeaxanthin esters, flavonoids, scopoletin, *p*-coumaric acid, daucosterol, betaine, hydrocyanic acid, zeaxanthin, vanillic acid, salicylic acid, quercetin, kaempferol, atropine, and hyoscyamine (Xie et al. 2001; Luo et al. 2004; Potterat 2010).

Lycium chinense Mill., Solanaceae
Lycium chinense is used in Chinese traditional medicine to treat various diseases including diabetes and inflammation. NF-κB inhibitors and PPAR-γ agonists were isolated from the root bark of *L. chinense*. Fatty acids present in the root bark of dichloromethane (DCM) extract of this plant activated PPAR-γ (review: Wang et al. 2014).

Lycium shawii Roem. and Schult., Solanaceae
Lycium shawii is a traditional medicinal plant. The powder and decoction of *L. shawii* aerial parts are used as a folklore remedy in the treatment of diabetes by the local community in various parts of Saudi Arabia (review: Haddad et al. 2005). The extract of *L. shawii* exhibited persistent hypoglycemic effects in normal rats (Shabana et al. 1990). In another study, a blood glucose-lowering effect was noticed in Long–Evans rats treated orally with 250 and 500 mg/kg of 80% ethanol extract of *L. shawii*

aerial parts. Further, the treatment improved glucose tolerance. The antidiabetic effect was also observed in streptozotocin-induced diabetic rats. In a chronic (90 days) toxicity evaluation in mice, the extract (100 mg/kg) induced changes in body weight and biochemical and hematological parameters and was found to possess significant spermatatoxic potential (Sher and Alyemeni 2011).

Lygos raetam (Forssk.) Heyw., Labiatae
Lygos raetam (aerial part) is used in Egyptian traditional medicine to treat DM (Haddad et al. 2005). The plant extract exhibited transient hypoglycemic effects in normal rats; the effect appeared 1 h after administration (Shabana et al. 1990).

Lythrum salicaria L., Lythraceae
Lythrum salicaria is an important medicinal plant used by Chinese to treat diabetes. The ether extract of *L. salicaria* stems and flowers exhibited significant hypoglycemic activity in rats with glucose- and epinephrine-induced hyperglycemia. This extract was also found to be active in alloxan- and streptozotocin-induced diabetic rats and alloxan-induced diabetic mice (Lamela et al. 1986). Studies carried out by Chinese investigators showed that litchi (*L. salicaria*) seed water extract increased insulin sensitivity and reduced the concentrations of blood fasting glucose, TG, leptin, and tumor necrosis factor in a type 2 DM rat model (Li et al. 2006).

Macaranga adenantha Gagnep., Euphorbiaceae
Triterpene metabolites with PTP1B inhibitory activity, identified as oleanolic acid, 3β, 28-dihydroxy-12-en-olean, maslinic acid, and 3β-*O*-acetyl aleuritolic acid, were obtained from the plant *Macaranga adenantha* (Jiany et al. 2012).

Macaranga tanarius (L.) Mull. Arg., Euphorbiaceae
Synonym: *Ricinus tanarius* L.
The ethyl acetate extract of *Macaranga tanarius* leaves showed potent α-glucosidase inhibitory activity. Five ellagitannins were successfully isolated and identified as active principles (Gunawan-Puteri and Kawabata 2012).

Machilus thunbergii Sieb. et Zucc., Lauraceae
Machilus thunbergii is an important medicinal resource distributed widely in China. The methanol extract of the plant exhibited α-glucose inhibitory activity and potent antioxidant activity *in vitro*. The EC_{50} value for the enzyme inhibition by a water fraction from the methanol extract was 1.1 µg/mL (review: El-Abhar and Schaalan 2014).

Madhuca longifolia (Koenig) Macbride, Sapotaceae
Synonym: *Madhuca indica* J.F. Gmel
Madhuca longifolia (*Madhuca indica*), commonly known as the butter nut tree, is used traditionally in Indian folk medicine for the treatment of DM. The hydroethanolic extract of the leaves of *M. longifolia* was investigated for its antidiabetic properties in alloxan-induced diabetic rats. Oral administration of the extract (150 and 300 mg/kg, once a day, for 30 consecutive days) significantly lowered blood glucose levels. Furthermore, the activity of glucose-6-phosphate dehydrogenase, serum TGs, HDL, and total cholesterol levels showed marked improvement that indicated that the hydroethanolic extract possessed antihyperglycemic activity (Ghosh et al. 2008). In another study, the ethanol extract of the bark of *M. longifolia* exhibited a dose-dependent hypoglycemic activity in normal, glucose-loaded, and streptozotocin-induced diabetic rats; the effect was comparable with that of the standard antidiabetic agent glibenclamide (Prashanth et al. 2010). In another study, the ethanol extract of seeds of *M. indica* was effective in reducing the plasma glucose level in a dose-dependent manner in normal albino rats. The hypoglycemic effect could be due to stimulation of insulin release from the β-cells and/or increase in the uptake of glucose from the plasma (Seshagiri et al. 2007). The methanol extract (100 and 200 mg/kg) of the bark of *M. longifolia* showed potent antihyperglycemic activity in normal and streptozotocin-induced diabetic rats (Dahake et al. 2010).

When different extracts (petroleum ether, methanol, and water) of the bark of *M. indica* were screened, the methanol extract showed significant antidiabetic activity against streptozotocin- and streptozotocin–nicotinamide-induced diabetic models in rats (Pavankumar et al. 2011).

Thus, the methanol extract of different parts of this plant such as the leaves, bark, and seeds shows antidiabetic activity. A comparative study is required to select the best part. More importantly, the active principles involved and their precise mode of action are not known. This is an important plant for further studies.

Magnolia officinalis Rehder & E.H.Wilson, Magnoliaceae

Synonym: *Magnolia hypoleuca* Diels (monograph: WHO 2009)

It is a large deciduous tree native to China. The stem, trunk, and root bark are generally used in traditional medicine. Magnolol (5,5-diallyl-2,2-dihydroxybiphenyl) isolated from the cortex of *Magnolia officinalis* retarded diabetic nephropathy in nonobese type 2 diabetic Goto-Kakizaki rats (Eunjin et al. 2007). In a recent study, the preventive effect on HFD-induced obesity and insulin resistance in mice by 4-*O*-methylhonokiol isolated from *M. officinalis* was compared with the *M. officinalis* extract. Administration of the extract or the compound for 24 weeks with HFD slightly reduced body weight gain, body fat mass, and the epididymis adipose tissue without any effect in food intake. Moreover, the compound significantly lowered HFD-induced plasma TG, cholesterol levels and activity of ALT, liver weight and hepatic TG level, and ameliorated hepatic steatosis. The extract reduced ALT and liver TG level only. Concurrently, low-dose 4-*O*-methylhonokiol improved HFD-induced hyperinsulinemia and insulin resistance. Furthermore, the infiltration of mast cells in adipose tissue was decreased in the compound- or extract-treated animals. These results suggested that 4-*O*-methylhonokiol might exhibit potential benefits for HFD-induced obesity by improvement of lipid metabolism and insulin resistance (Zhang et al. 2014).

Magnolol is a small polyphenolic molecule with low toxicity that is isolated from the genus *Magnolia*. In preclinical experiments, magnolol was found to have antidiabetic and several other pharmacological activities (antioxidative, anti-inflammatory, antitumorigenic, antimicrobial, antineurodegenerative, and antidepressant properties). Further, magnolol can effectively regulate pain control, hormonal signaling, and gastrointestinal and uterus modulation as well as provide cardiovascular and liver protective effects (Chen et al. 2011b). Magnolol enhanced adipocyte differentiation and glucose uptake in 3T3-L1 cells and activated PPAR-γ (review: Wang et al. 2014). Honokiol, a naturally occurring rexinoid present in this plant, could serve as a regulator of various retinoid x receptor heterodimers (Kotani et al. 2012).

Majorana hortensis Moench, Lamiaceae

Synonym: *Origanum majorana* L.

When different extracts and the volatile oil of *Origanum majorana* (*Majorana hortensis*) leaves were tested, the volatile oil (200 and 400 mg/kg, p.o.) and methanol extract (200 and 400 mg/kg, p.o.) exhibited potent dose-dependent antihyperglycemic activity. The water extract (200 and 400 mg/kg, p.o.) showed moderate effect on the blood sugar level. The volatile oil (100 mg/kg, p.o.) marginally decreased the elevated TG and cholesterol levels, whereas the methanol (200 and 400 mg/kg, p.o.) and water (200 and 400 mg/kg, p.o.) extracts showed potent antihyperlipidemic effect. From this study, it was concluded that the volatile oil and methanol extract of *O. majorana* leaves could prove to be beneficial in the management of DM (Pimple et al. 2012).

Malmea depressa (Baill) R.E. Fries., Annonaceae

Malmea depressa is traditionally used by Mayan communities of southeastern Mexico to treat type 2 diabetes. The roots of the plant lowered blood glucose levels in streptozotocin-induced diabetic rats (Andrade-Cetto et al. 2005). In another report, the butanol extract (10 mg/kg daily for 30 days) of *M. depressa* root reduced the levels of blood glucose as well as HbA1c in n5-streptozotocin-induced diabetic rats. This effect was also observed after 45 days of treatment. In acute experiments, a single administration of the butanol extracts (50 mg/kg) stimulated insulin release that was similar to the result seen in the case of tolbutamide. The new compound, 3-(3-hydroxy-2,4,5-trimethoxyphenyl) propane-1,2 diol, was isolated from the active fraction of the butanol extract. The results presented here support the utility of the traditional use of the root and suggest that the active fraction of the plant extract could be developed as a phytomedicine (Andrade-Cetto et al. 2008a). In another study, the butanol extract of *M. depressa* produced marked reduction in blood glucose in maltose-loaded streptozotocin-induced diabetic rats. *In vitro* assays of α-glucosidase activity showed an IC_{50} of 21 μg/mL for the butanol extract, which was lower than that of acarbose (128 μg/mL) (Andrade-Cetto et al. 2008b). The ethanol extract of the root bark of *M. depressa* dose-dependently inhibited gluconeogenesis *in vivo* as judged from pyruvate tolerance test in n5-strep-tozotocin-induced diabetic rats after an 18 h fasting period. Furthermore, the gluconeogenesis inhibition was confirmed by *in vitro* assay with intact rat liver microsomes. The study suggests that administration of *M. depressa* can improve glycemic control by blocking hepatic glucose production, especially in the fasting state (Andrade-Cetto 2011). In an *in vitro* study, isoorientin attenuated insulin resistance in adipocytes activating the insulin signaling pathway (Alonoso-Castro et al. 2012). From the pharmacological active fractions, two phenylbutane derivatives (2-hydroxy-3,4,5-trimethoxy-1-(2′,4′-hydroxy-3′-dihydroxy)butyl-benzene and 2-hydroxy-3,4,5-trimethoxy-1-(2′,3′,4′-hydroxy)butyl-benzene) as well as a phenylpropane derivative, 3-(3-hydroxy-2,4,5-trimethoxyphenyl)propane-1,2 diol, were isolated (Andrade-Cetto 2011).

Mammea africana Sabine, Guttiferae

The stem bark of *Mammea africana* is used in African rain forest to treat various diseases, including DM. The effects of acute (5 h) and subacute (21 days) oral administrations of the chloroform–methanol extract (19–300 mg/kg) of *M. africana* stem bark on blood glucose levels of normal and streptozotocin-induced type 1 diabetic rats were studied. Acute administration of the extract reduced blood glucose in diabetic rats (only 34%). Subacute treatment for 21 days markedly reduced blood glucose level in diabetic rats (73%). A reduction or stabilization in total serum protein, TG, cholesterol, and ALT levels was also observed. No effect was observed on body weight loss but food and water intakes were significantly reduced in diabetic rats. The maximal antidiabetic effect was obtained with the dose of 75 mg/kg; this dose was more effective than that of glibenclamide (Tchamadeu et al. 2010). In another study, treatment of streptozotocin-induced diabetic rats with the ethanol extract of *M. africana* (stem bark) caused a significant reduction in fasting blood glucose levels of diabetic rats in both acute study and prolonged treatment (2 weeks). Further, the treatment showed considerable lowering of serum total cholesterol, TGs, LDL-C, and VLDL-C and an increase in HDL-C in diabetic rats (Okokon et al. 2007a). This study confirms the hypoglycemic and hypolipidemic properties of this plant extract. Further studies are warranted to determine its therapeutic utility and identify active compounds.

Mangifera indica L., Anacardiaceae (Figure 2.65)

Common Names: Mango tree, amara, aam, mamaram, mavu, etc.

Description: *Mangifera indica* is a large tree with widely spreading branches. The leaves are simple, petiolated, alternate, crowded at the end of the branches, quite entire margin often undulate. The petioles

(a)

(b)

FIGURE 2.65 *Mangifera indica*. (a) Branch with flowers. (b) Unripe mangoes.

are swollen at the base; the flowers are small, yellow, and odorous, with male and female flowers borne on the same panicle; the fruit is a drupe, fleshy with fibers, and compressed stone.

Distribution: *M. indica* is a native from tropical Asia. It is cultivated in many tropical and subtropical regions, southward toward peninsular India.

Traditional Medicinal Uses: *M. indica* is an astringent, diuretic, parasiticide, aperitif, ascaricide, deutrifice, stomachic, taeniafuge, vermifuge, laxative, styptic, aphrodisiac, depurative, diaphoretic, and antipyretic and used in asthma, anasarca, anemia, cholera, cough, diarrhea, glossitis, hemato-chezia, hemoptysis, jaundice, hypertension, insomnia, hemiplegia, hemorrhage, leukorrhea, malaria, melena, metrorrhagia, ophthalmia, piles, rheumatism, rickets, stomatitis, warts, toothache, sore throat, bronchitis, vaginal troubles, urinary discharge, ulcers, typhoid, blood impurities, eye sore, vomiting, eruption, syphilis, snake bite, scorpion sting, diphtheria, and malignant throat diseases (Kirtikar and Basu 1962).

Pharmacology

Antidiabetes: The water extract of the plant leaf produced a reduction of blood glucose level in nor-moglycemic and glucose-induced hyperglycemic mice (Aderibigbe et al. 2001). The ethanol extracts of stem barks reduced glucose absorption in type 2 diabetic rats. Potent therapeutically promising anti-DM activity of the methanol extract of bark and leaves was shown in type 1 and type 2 diabetic rats (Bhoumik et al. 2009). Besides, *M. indica* exhibited DPP-IV inhibitory activity (Yogisha and Raveesha 2010).

The predominant antidiabetes constituent of the extract of the mango plant is mangiferin. Experiments demonstrate that mangiferin isolated from the plant leaves possesses significant antidiabetic properties in streptozotocin-induced diabetic rats (Muruganandan et al. 2005). Mangiferin from the leaves prevented diabetic nephropathy progression in streptozotocin-induced diabetic rats (Li et al. 2010b). Recent studies shed light on the likely emergence of this compound as a very important molecule in mediating insulin sensitivity and modulating lipid metabolism (Mirza et al. 2013). Extracts of the kernel of *M. indica* seeds also exhibited antidiabetic effect (Petchi et al. 2011). 1,2,3,4,6-Penta-O-galloyl-β-D-glucose isolated from *M. indica* inhibited 11-β-HSD-1 and ameliorated HFD-induced diabetes in C57BL/6 mice (Mohan et al. 2013a). The antidiabetic compounds 6-O-galloyl-5-hydroxy mangiferin, mangiferin, 5-hydroxy mangiferin, and methyl gallate were isolated from the kernel of *M. indica*. These compounds reduced the blood glucose levels in alloxan-induced diabetic rats (review: Firdous 2014).

Mango fruit peel supplementation resulted in remarkable antidiabetic effects in streptozotocin-induced diabetic rats. In a recent study, mango fruit peel (5% and 10% levels in basal diet) ameliorated streptozotocin-induced increase in urine sugar, urine volume, fasting blood glucose, total cholesterol, LDL, and TGs and decrease in HDL in rats. Besides, the treatment increased antioxidant enzyme activities and decreased LPO in the plasma, kidney, and liver in streptozotocin-induced diabetic rats compared to untreated diabetic rats (Gondi et al. 2015). Phenolic compounds identified in the raw and ripe mango peel include gallic acid, syringic acid, mangiferin, ellagic acid, gentisyl-protocatechuic acid, and quercetin (Ajila et al. 2010). It is of interest to note that these compounds are known antidiabetic agents. Mangiferin exerts its antidiabetic activity through multiple mechanisms including modulation of insulin sensitivity and lipid metabolism. Syringic acid from *Dendrobium nobile* prevented diabetic cataract pathogenesis by inhibiting aldose reductase (Wei et al. 2012); syringin from *Eleutherococcus senticosus* was reported to augment insulin release (Liu et al. 2008). Quercetin and quercetin glycosides from *Eucommia ulmoides* inhibited AGE formation (Hy et al. 2004). Quercetin from *Myrcia multiflora* inhibited aldose reductase (Varma et al. 1975). In addition, quercetin effectively blocked polyol accumulation in intact rat lenses incubated in medium containing high concentration of sugars (Varma et al. 1975). Quercetin glycosides from *Bauhinia forficata* were reported to inhibit intact microsomal glucose-6-phosphatase and activate insulin signaling pathways (Estrada et al. 2005). Ellagic acid from *Myrciaria dubia* inhibited aldose reductase (Udea et al. 2004). Gallic acid isolated from *Terminalia* species stimulated insulin secretion (insulin secretagogue) (Latha and Daisy 2011). Thus, the mango peel is an ideal food medicine (nutraceuticals) for diabetic patients.

Other Activities: Mangiferin has documented antioxidant, cardioprotective and anti-inflammatory effects (review: Telang et al. 2013). *M. indica* has been reported to possess antiviral, antibacterial, and anti-inflammatory activities. Mangiferin from the leaves of *M. indica* has antihyperlipidemic and antiatherogenic properties in rats (Muruganandan et al. 2005). Mangiferin exhibited antibacterial activity *in vivo* against specific periodontal pathogens such as *Prevotella intermedia* and *P. gingivalis* (Bairy et al. 2002). *In vivo* and *in vitro* anti-inflammatory activity of the *M. indica* extract has been reported (Garrido et al. 2004). The anti-inflammatory action of the bark extract would be related with the inhibition of iNOS and cyclooxygenase-2 expression (Beltran et al. 2004). The alcoholic extract of the stem bark of *M. indica* increased humoral antibody titer and delayed-type hyper-sensitivity reactions in mice (Makare et al. 2001). The water extract of the stem bark of the plant inhibited T-cell proliferation and TNF-induced activation of nuclear transcription factor NF-κB (Garrido et al. 2005). Anthelmintic and antiallergic activities of stem bark components Vimang and mangiferin have been reported (Garcia et al. 2003). The *M. indica* extract protected the injury associated with hepatic ischemia reperfusion (Sanchez et al. 2003). Flavonoids from *M. indica* are effective for dyslipidemia (Anila and Vijayalakshmi 2002).

Phytochemicals: Phytochemicals reported in *M. indica* include 2-octene, alanine, α-phellandrene, α-pinene, ambolic acid, ambonic acid, arginine, ascorbic acid, β-carotene β-pinene, carotenoids, furfurol, GABA, gallic acid, gallotannic acid, geraniol, histidine, isoleucine, isomangiferolic acid, kaempferol, limonene, linoleic acid, mangiferic acid, mangiferin, mangiferol, mangiferolic acid, myristic acid, neo-β-carotene-b, neo-β-carotene-u, neoxanthophyll, nerol, neryl acetate, oleic acid, oxalic acid, *p*-coumaric acid, palmitic acid, palmitoleic acid, pantothenic acid, peroxidase, phenylalanine, phytin, proline, quercetin, and xanthophylls (review: Ajikumaran and Subramoniam 2005).

Marrubium cuneatum Banks & Sol., Lamiaceae

Synonym: *Marrubium radiatum* Delile ex Benth

The methanol extract of *Marrubium radiatum* (*Marrubium cuneatum*) exerted promising *in vitro* activity against α-amylase and α-glucosidase (review: El-Abhar and Schaalan 2014)

Marrubium vulgare L., Lamiaceae

It is a traditional medicinal plant used for the treatment of diabetes in Algeria. The water extract of this plant (100–300 mg/kg daily oral administration for 21 days) showed dose-dependent antidiabetic and antihyperglycemic activities in alloxan-induced diabetic rats (Boudjelal et al. 2012). In another study, the methanol extract (500 mg/kg) of *Marrubium vulgare* ameliorated hyperglycemia and dyslipidemia in streptozotocin-induced diabetic rats. In diabetic rats, the treatment improved glucose tolerance and fasting glucose levels along with an increase in plasma insulin levels and liver glycogen content (Elberry et al. 2011).

Clinical Trial: In a preliminary clinical trial, the water extract of the plant did not show promising anti-DM activity in type 2 diabetic patients (Herrera-Arellano et al. 2004). A dose response study and efficacy evaluation with the most active extract is needed.

Matricaria chamomilla L., Asteraceae

Synonym: *Chamomilla recutita* (L.) Rauschert

In traditional medicine, *Matricaria chamomilla* is used in Egypt to treat diabetic patients. The antihyperglycemic and antioxidative activities of the aerial part of the *M. chamomilla* ethanol extract were evaluated in streptozotocin-induced diabetic rats. Treatment with different doses (20, 50, and 100 mg/kg) of the ethanol extract for 14 days significantly reduced postprandial hyperglycemia and oxidative stress and augmented the antioxidant system in streptozotocin-induced diabetic rats in a dose-dependent manner. In histological investigations, the treatment protected the majority of the pancreatic islet cells, compared to the diabetic control group (Cemek et al. 2008). In another study, the water extract of *M. chamomilla* leaves showed a strong hypoglycemic effect in streptozotocin-induced diabetic rats (Najla et al. 2012).

Maytenus jelskii Zahlbr., Celastraceae

Friedelane-type triterpenes isolated from the root barks of *Maytenus jelskii* exhibited increased insulin-mediated signaling *in vitro*, suggesting that these friedelane triterpenes have potential therapeutic use in insulin-resistant states (review: Nazaruk and Borzym-Kluczyk 2014).

Matthiola livida D.C., Cruciferae

Matthiola livida (aerial part) is used in Egyptian traditional medicine to treat DM (Haddad et al. 2005). The plant extract exhibited potent hypoglycemic effects in alloxan-induced diabetic rats (Shabana et al. 1990).

Medicago sativa L., Fabaceae

Common names: Alfalfa, lucerine, forage crop

Medicago sativa (lucerne) is used in traditional medicine to treat DM. Administration of lucerne in the diet (62.5 g/kg) and drinking water (2.5 g/L) reduced the hyperglycaemia of streptozotocin-diabetic mice. Water extracts of lucerne (1 mg/mL) stimulated 2-deoxy-glucose transport (1.8-fold), glucose oxidation (1.7-fold) and incorporation of glucose into glycogen (1.6-fold) in mouse abdominal muscle. In acute 20 min tests, 0.25–1 mg/mL the extract of lucerne evoked a stepwise 2.5–6.3-fold stimulation of insulin secretion from the BRIN-BD11 pancreatic β-cell line. This effect was abolished by 0.5 mM diazoxide, and prior exposure to extract did not affect subsequent stimulation of insulin secretion by 10 mM L-Alanine, thereby negating a detrimental effect on cell viability. The effect of extract was potentiated by 16.7 mM glucose and by 1 mM 3-isobutyl-1-methylxanthine. L-Alanine (10 mM) and a depolarizing concentration of KCl (25 mM) did not augment the insulin-releasing activity of lucerne. Activity of the extract was found to be heat stable and largely acetone insoluble, and was enhanced by exposure to acid and alkali (0.1 M HCl and NaOH) but decreased 25% with dialysis to remove components with molecular mass <2000 Da. Sequential extraction with solvents revealed insulin-releasing activity in both methanol and water fractions indicating a cumulative effect of more than one extract constituent. The results demonstrate the presence of antihyperglycaemic, insulin-releasing, and insulin-like activity in the traditional antidiabetic plant (Gray and Flatt, 1997). Histological studies on kidney have shown that although the water–ethanol extract of *M. sativa* treatment for 4 weeks (550 mg/kg) resulted reduction of blood sugar in the diabetic rats, the treatment had no distinct effect on nephropathy side effects in the 4-week study (Mehranjani et al. 2007). In another study, administration of ethanol extract of *M. sativa* leaf daily for 30 days to alloxan-induced diabetic rats exerted anti-hyperglycemic action as effective as metformin and corrected diabetes induced dyslipidemia, oxidative stress, hepatic function, and renal function (Baxi et al. 2010). Administration of the water extract of *M. sativa* seed to alloxan induced diabetic rats also resulted in decrease in blood glucose levels and serum lipids and increase in the levels of insulin (Helal et al. 2013).

Melampyrum pratense L., Orobanchaceae

Melampyrum pratense is used in traditional Austrian medicine for the treatment of different inflammation-related conditions including arthritis. The extracts of *M. pratense* stimulated PPAR-α and PPAR-γ that are well recognized for their anti-inflammatory and antidiabetes activities. Besides, the extract inhibited the activation of the transcription factor NF-κB and induction of its target genes IL-8 and E-selectin *in vitro*. Apigenin and luteolin from this plant effectively inhibited TNF-α-induced NF-κB-mediated transactivation of a luciferase reporter gene (Vogl et al. 2013). In view of these observations, this plant is likely to have anti-DM property.

Melanthera scandens (Schumach. & Thonn.) Roberty, Asteraceae

Melanthera scandens is used in ethnomedicine to treat diabetes. Antidiabetic and hypolipidemic activities of the ethanol extract and fractions of *M. scandens* leaf were evaluated in alloxan-induced diabetic rats. Administration of the extract/fractions (37–111 mg/kg) to alloxan-induced diabetic rats for 14 days caused a significant reduction in fasting blood glucose levels of diabetic rats in both the acute study (7 h) and prolonged treatment (14 days). The activities of the extract and fractions were more than those of the reference drug glibenclamide. Besides, the extract/fractions exerted a significant reduction in the levels of serum total cholesterol, TGs, LDL, and VLDL and an increase in HDL levels of diabetic rats (Akpan et al. 2012).

Melia dubia Cav., Meliaceae

The leaves of *Melia dubia* are used for the treatment of various ailments including DM. The plant was screened for α-amylase inhibition and antioxidant activities. The ethanol extract of the plant leaves inhibited α-amylase at a lower concentration than the standard acarbose. The ethanol extract contained the highest amount of phenolic and flavonoid compounds and exhibited excellent antioxidant activity (Valentina et al. 2013).

In an *in vivo* study, the ethanol extract of the *M. dubia* fruit exhibited dose-dependent hypoglycemic activity in normal mice. There was a gradual decrease in the blood sugar level from the second hour onward in streptozotocin-induced diabetic mice, and the low glucose level was maintained up to 8 h. Oral glucose tolerance test showed reduction of blood glucose level in the extract-treated diabetic mice. The therapeutic index value of the extract suggested that the extract is safe (Susheela et al. 2008). Thus, the plant is promising for follow-up studies.

Melissa officinalis L, Lamiaceae

Synonym: Lemon balm

The antioxidant activity of lemon balm (*Melissa officinalis*) essential oil on 2,2-diphenyl-1-picrylhydrazyl (DPPH) radicals and its hypoglycemic effect in *db/db* mice were investigated. The essential oil scavenged 97% of DPPH radicals at a 270-fold dilution. Administration of the oil (0.015 mg/day) for 6 weeks to mice showed significantly reduced blood glucose (65%) and TG concentrations; improved glucose tolerance, as assessed by an oral glucose tolerance test; and significantly increased serum insulin levels, compared with the control group. Among all glucose metabolism-related genes studied, hepatic glucokinase and GLUT4, as well as adipocyte GLUT4, PPAR-γ, PPAR-α, and SREBP-1c expression, were significantly upregulated, whereas glucose-6-phosphatase and phosphoenolpyruvate carboxykinase expression was downregulated in the livers of the essential oil-treated group. The results further suggest that the essential oil administered at low concentrations is an efficient hypoglycemic agent, probably due to enhanced glucose uptake and metabolism in the liver and adipose tissue and the inhibition of gluconeogenesis in the liver (Chung et al. 2010). The water extract of *M. officinalis* could relax the phenylephrine-preconstricted rings of the aorta in streptozotocin-induced diabetic rats through a nitric oxide pathway, and this extract could not affect the release of calcium from intracellular stores. This may have a role in the amelioration of vascular dysfunction in diabetes (Dehkordi and Enteshari 2013).

Melothria maderaspatana (L.) Cogn. or *Mukia maderaspatana* (L.) M. Roemer, Cucurbitaceae

Synonyms: *Cucumis maderaspatanus* L., *Bryonia cordifolia* L., *Coccinia cordifolia* (L.) Cogn., *Bryonia scabrella* L. f., *Mukia scabrella* (L.f.) Arnott., *Mukia maderaspatana* (L.) M. Roem. var. *scabrella* (L.) Kurz., *Melothria maderaspatana* var. *gracilis* Kurz., *Bryonia rottleri* Spreng., *Mukia rottleri* (Spreng.) M. Roem., *Bryonia althaeoides* Seringe, *Mukia althaeoides* (Seringe) M. Roemer, *Melothria althaeoides* (Ser.) Nakai, *M. celebica* Cogn. var. *villosior* Cogn., *Mukia leiosperma* Auct. (Wight & Arn.), *Cucumis maderaspatana* L., *C. pubescens* Willd. (review: Petrus 2012) (Figure 2.66)

Local Names: Madras pea pumpkin, bristly bryony, rough bryony, wild cucurbit, etc.

Distribution: *Melothria maderaspatana* occurs widely throughout the tropics and subtropics of the Old World, extending from the sub-Saharan Africa and Madagascar through Southwest Asia (including Yemen, Pakistan, India, Bangladesh, Sri Lanka Andaman and Nicobar Islands, and Southern China) and Southeast Asia to New Guinea and Australia (review: Petrus 2012).

Description: *M. maderaspatana* is a prostrate or climbing, many branched, annual herb with spreading bristly hairs and simple tendrils. The leaves are alternate and broadly triangular in outline. These leaves are 3–5 lobed, 3–11 cm in length and breadth. Their apices are acute, deeply cordate base, irregularly dentate, dark green colored and scabrid above but pale green and hispid beneath. The petioles are hairy and 0.6–2.5 cm long. The flowers are small and pale yellow in color. The male flowers are fascicled on very short peduncles, while the female flowers are usually solitary and sessile. The calyx is hairy and bears a tube of 2 mm long and narrowly campanulate. The corolla is pubescent,

FIGURE 2.66 *Melothria maderaspatana.*

ovate-oblong segments, rounded at the apex. The fruits are berry and are globose-ellipsoid, up to 1.5 cm in diameter and are pale green in color with longitudinal cream stripes and turn reddish when ripe. The seeds are about 4 mm long and 2 mm broad and are present in numerous numbers (review: Petrus 2012).

Traditional Medicinal Uses: *Mukia maderaspatana* (*Melothria maderaspatana*) is extensively used in folklore medicine as an antidiabetic plant. According to traditional medicinal use, the plant is expected to possess expectorant, refrigerant, carminative, aperients, vulnerary, sudorific, anodyne, and tonic values. It is used to treat asthma, cough, burning sensation, dyspepsia, flatulence, colic, constipation, ulcers, neuralgia, nostalgia, and vertigo. The fruits are used in the treatment of dysuria, piles, polyuria, and tuberculosis. In Siddha, the root and leaf are used to treat fever, dyspnea, abdominal disorders, hepatic disorders, cough, and vomiting; decoction of the leaf is used to treat hypertension and nasobronchial diseases (review: Petrus 2013).

Pharmacology

Antidiabetes: The antihyperglycemic and hypolipidemic effects of the ethanol extract of the aerial parts of *M. maderaspatana* (100 and 200 mg/kg, p.o., daily for 14 days) were shown in streptozotocin-induced diabetic rats. The extract was also found to be very effective on the recovery of altered biochemical parameters and decreased body weight in treated diabetic animals (Balaraman et al. 2010). Administration of the ethanol extract (100 and 200 mg/kg, p.o.) of *M. maderaspatana* to alloxan-induced diabetic rats decreased the blood glucose levels after 5 h of treatment (Vadivelan et al. 2010). The ethanol extract of the whole plant inhibited the activities of α-glucosidase and α-amylase *in vitro* (Vadivelan et al. 2012). The ethanol extract of leaves of *M. maderaspatana* demonstrated a significant reduction in glucose absorption in a dose-dependent manner. The ethanol extract (1–1000 μg/mL) stimulated insulin secretion *in vitro* from the β-cells of pancreatic islets and INS-1E insulinoma cells. Administration of the extract (100 and 200 mg/kg, p.o.) in an oral glucose tolerance test led to a significant fall in plasma glucose level 30 min after the administration in normal C57BL/6 mice compared to untreated mice. Oral administration of the water extract of *M. maderaspatana* resulted in a significant reduction in blood glucose levels of normal and streptozotocin-induced diabetic rats. The treatment also decreased serum cholesterol, LPO, and liver lipid peroxide levels in streptozotocin-induced diabetic rats. The extract also showed *in vitro* antioxidant activity in the DPPH assay (Hemalatha et al. 2010). The methanol extract of the roots of *M. maderaspatana* also exhibited a blood glucose-lowering activity in alloxan-induced diabetic rats (Wani et al. 2011).

In a recent study, the methanol extract of the whole plant and phenolics such as quercetin and phloroglucinol were investigated for their *in vitro* antidiabetic activity. Quercetin at 0.25 and 0.5 mg/mL caused 65% and 89% inhibition of glucose production in rat liver slices. Addition of insulin did not increase the inhibition. Phloroglucinol inhibited 100% glucose production with or without insulin. The methanol extract of this plant (0.25 mg/mL) inhibited gluconeogenesis by 45%, and with insulin, inhibition increased to 50%. The extract had no effect on glucose uptake but potentiated the action of insulin-mediated glucose uptake in isolated rat hemidiaphragm. HPLC analysis revealed the presence of phenolics in the extract. The study supports the use of *M. maderaspatana* in folk medicine as an antidiabetic nutraceutical (Srilatha and Ananda 2014).

It should be noted that the leafy vegetable (*M. maderaspatana*) is reported to exhibit potent antioxidant capacity *in vitro* and *in vivo* and to possess antihypertensive, vasodilatory, antihyperglycemic, antihyperlipidemic, and hepatoprotective activities.

Other Activities: Other pharmacological properties include antihypertensive, antioxidant, anti-inflammatory, anti-platelet aggregation, antihyperlipidemic, hepatoprotective, immunomodulatory, *in vitro* anti-platelet aggregation, antimicrobial, antiulcer, anxiolytic, and local anesthetic activities. Clinical studies on human subjects showed antihypertensive activity and anti-inflammatory activity in rheumatoid arthritis (review: Petrus 2013).

Phytochemicals: Reported phytochemicals include steroids, triterpenes, alkaloids, phenols, flavones, catechins, saponins, glycosides, and reducing sugars. The presence of spinasterol, 22,23-dihydrospinasterol, its 3-*O*-β-D-glucoside, β-sitosterol, decosanoic acid, and triterpenes has also been reported from the leaf extract. Six C-glycoflavones (isovitexin, homoorientin, vitexin, orientin, saponarin, and lutonarin) have been isolated from the water extract of leaves. Columbin has been isolated from its roots (review: Petrus 2012).

Memecylon umbellatum Burm.f., Melastomaceae

Local Names: Anjani, nijasa, alli, kasai, kasavu.

Description: *Memecylon umbellatum* is a shrub to small tree with ovate-oblong or lanceolate leaves that are coriaceous and glabrous with an entire margin. Small subsessile flowers are produced in a compact umbel in a reduced peduncle. The fruit is a globose, yellow berry with obovoid seeds.

Distribution: It is distributed in Peninsular India and Sri Lanka.

Traditional Medicinal Uses: It is an astringent and used to treat conjunctivitis, leukorrhea, gonorrhea, and excessive menstrual discharge (Kirtikar and Basu 1975).

Antidiabetes: The alcohol extract of the leaves *of M. umbellatum* (p.o.) exhibited pronounced lowering of serum glucose levels in normal and alloxan-induced diabetic mice (Amalraj and Ignacimuthu 1998b). The *in vitro* antihyperglycemic potential of *M. umbellatum* was evaluated recently. The methanol extract of the plant leaf showed higher antidiabetic activity compared to ethyl acetate and n-hexane extracts. The methanol extract (250–1000 µg/mL) inhibited substantial α-amylase activity, inhibited nonenzymatic glycosylation of hemoglobin, and enhanced glucose uptake by yeast cells (Rajesh et al. 2014). The glucose diffusion assay was performed and a significant level of inhibition of glucose movement at various time intervals was found when compared to control. Phytochemical analysis of the methanol extract revealed the presence of phenols, flavonoids, terpenoids, tannins, and saponins (Rajesh et al. 2014). These results are encouraging for further studies to determine its utility for anti-DM drug development. The anti-DM compounds β-amyrin, sitosterol, oleanolic, acid and ursolic acids were reported from this plant.

Phytochemicals: Umbelactone, β-amyrin, sitosterol, sitosterol glucoside, oleanolic acid, ursolic acid, etc., were isolated (review: Ajikumaran and Subramoniam 2005).

Mentha piperita (Ehrhart) Briquet, Labiatae
Common Name: Mint
The juice of *Mentha piperita* leaves (100 and 250 mg/kg, p.o., daily for 3 weeks) lowered blood glucose and serum lipid levels in fructose-fed hyperglycemic rats. The water extract of the plant leaves/juice produced a significant decrease in elevated levels of glucose, cholesterol, TGs, LDL, and VLDL and an

increase in HDL levels without affecting serum insulin levels (Badal et al. 2011). The offspring from *M. piperita* juice-treated streptozotocin-induced diabetic dams showed significantly reduced levels of blood glucose, serum cholesterol, LDL, and TGs and increased levels of HDL. Thus, the juice may aid in the prevention of DM and dyslipidemia (Barbalho et al. 2011). In another study, oral administration of *M. piperita* leaf juice (290 mg/kg) to alloxan-induced diabetic rats for 21 days resulted in a marked decrease in blood glucose levels (Angel et al. 2013). Further studies are needed to identify the active principles and their mechanism of action. *M. piperita* essential oil in the form of ointment showed promising wound healing activity in streptozotocin-induced diabetic rats (Umasankar et al. 2013). The plant also has an antioxidant property (Barbalho et al. 2011).

Merremia emarginata Burm. F., Convolvulaceae
Merremia emarginata is a procumbent herb that is traditionally used in Sri Lanka and India to treat diabetes. The methanol extract (100, 200, and 400 mg/kg daily for 28 days) of *M. emarginata* (whole plant) showed promising antidiabetes activity in streptozotocin-induced diabetic rats (Gandhi and Sasikumar 2012).

Merremia tridentata (L.) Hall. f., Convolvulaceae
Merremia tridentate root increased serum insulin levels, body weight, and glycogen content of the liver and skeletal muscle in streptozotocin-induced diabetic rats (Arunachalam and Parimelazhagan 2012).

Mimosa invisa Mart., Leguminosae/Mimosoideae
Mimosa invisa is traditionally prescribed for DM in Nigeria. The water extract of this plant showed potent hypoglycemic activity in alloxan-induced diabetic mice. Saponins, xanthones, tannins, and glycosides were detected in the extract (Manosroi et al. 2011). However, caution is required. Toxicity has been reported, and cyanide and nitrite have been identified as toxic elements in *M. invisa* plants. Twenty-two swamp buffaloes died 18–36 h after eating *M. invisa* Mart var. *inermis* Adelbert. The symptoms included salivation, stiffness, lack of mastication, muscular tremor, etc. The lesions revealed congestion and petechial hemorrhage of the heart, liver, kidneys, lungs, rumen, and intestines. The water extract was toxic to mice with LD_{50} 0.6 gm/kg, while an extract with alcohol showed an LD_{50} of 0.22 gm/kg (Tungtrakanpoung and Rhienpanish 1992).

Mimosa pudica L., Fabaceae
Common Names: Touch-me-not, thotta sinungi, etc.
In India, different parts of *Mimosa pudica* have long been popular for treating various ailments. A few studies have shown the blood glucose-lowering property of *M. pudica*. The ethanol extract showed significant decrease of blood glucose level in alloxan-induced diabetic rats (Sutar et al. 2009). The whole plant powder of *M. pudica* (100 and 200 mg/kg daily for 4 weeks) reduced blood glucose levels and increased body weight in alloxan-induced diabetic rats compared to untreated diabetic rats (Viswanathan et al. 2013). In another study, the blood glucose-lowering effect of *M. pudica* root powder was determined in diabetic albino rabbits. Oral administration of the root powder (6 mg/kg) for 20 days resulted in a significant reduction in blood glucose levels in diabetic rabbits on day 5, 10, and 20 (Bashir et al. 2013). Full-fledged follow-up studies are required to determine its therapeutic utility.

Mimusops elengi L., Sapotaceae
Considering the traditional claim of *Mimusops elengi* in the management of diabetes, the *in vitro* antioxidant and *in vivo* antihyperglycemic properties of the methanol extract of the bark of *M. elengi* were evaluated. The methanol extract offered significant *in vitro* reducing power capacity and radical scavenging activity. In an acute study, in alloxan-induced diabetes, the methanol extracts (400 mg/kg) exhibited a significant antihyperglycemic effect. The onset of the antihyperglycemic effect was observed 2 h after administration; the peak activity was demonstrated at 6 h. Thus, the methanol extract produced significant antihyperglycemic activity in both diabetic and normal rats (Ganu et al. 2010). In another report, these findings were confirmed. The water extract of *M. elengi* exhibited a reducing power as well as

DPPH and OH radical scavenging activity *in vitro*. The extract showed antihyperglycemic activity in alloxan-induced diabetic rats. The onset of action of antihyperglycemic activity of the extract was at the second hour and the duration of action was about 24 h in diabetic rats (Ganu and Jadhav 2010). The plant exhibited antihyperlipidemic and many other pharmacological properties (Singh et al. 2014).

Mirabilis jalapa L., Nyctaginaceae
The ethanol extract of *Mirabilis jalapa* roots showed hypoglycemic and hypolipidemic activities in streptozotocin-induced diabetic rats (Sarkar et al. 2011).

Momordica balsamina L., Cucurbitaceae
Common Names: Balsam apple, African pumpkin
The methanol extract of *Momordica balsamina* fruit (200 and 500 mg/kg) showed anti-DM activity in streptozotocin-induced diabetic rats (review: Sharma and Arya 2011).

Momordica charantia L., Cucurbitaceae
Synonyms: *M. balsamina* Blanco, *M. chinensis* Spreng., *M. elegans* Salisb., *M. indica* L. (monograph: WHO 2009) (Figure 2.67)

Common/Local Names: Bitter melon, bitter gourd, balsam pear, carilla fruit, karela, wasabella, ambu-vallika, karela, pavakkachedi, pantipavel, etc.

Description: *Momordica charantia* is a medicinal as well as vegetable plant. It is a climbing monoecious annual with simple tendril and petiolated, orbicular, glabrous or slightly pubescent or subpinnately lobed leaves. The flowers are yellow and monoecious; the male peduncle is one flowered with bract in the middle, while the female flower bract is present near the base of the peduncle. The fruit is ovoid, elongated with a knobby surface, narrowed to both ends, many ribbed, and covered with triangular tubercles (monograph: WHO 2009).

Distribution: *M. charantia* is a tropical flowering vine widely cultivated in Asia (including India, China, and Malaya), East Africa, and South America for its bitter fruits. It is a common vegetable cultivated extensively all over India and is the most bitter of all vegetables.

Traditional Medicinal Uses: More than 100 products from this plant or preparations containing this plant have been used by diverse cultures of the world to treat hyperglycemia in traditional medicine. It is used to treat DM and related conditions by the indigenous populations of Asia, South America, India, the Caribbean, and East Africa (review: Joseph and Jini 2013). Additionally, the plant (fruit, leaf, and seed) is used as refrigerant, vermifuge, abortifacient, aperitif, aphrodisiac, canicide, digestive, emmenagogue,

(a)   (b)

FIGURE 2.67 *Momordica charantia*. (a) Vine with flowers. (b) Fruit vegetable.

and contraceptive. In addition to DM, it is used to treat dysmenorrhea, eczema, skin diseases, malaria, dysentery, tumor, gout, jaundice, abdominal pain, kidney stone, spruce, diarrhea, leprosy, headache, rhinitis, leukorrhea, piles, pneumonia, snake bite, dyspepsia, hypertension, rheumatism, fever, scabies, peptic ulcer, biliousness, anemia, asthma, bronchitis, cholera, syphilis, ophthalmia, night blindness, sores, and gonorrhea as well as blood boils, scabies, psoriasis, ringworm, and other fungal diseases. It is useful for appetite and as a stomachic, laxative, bitter, cooling, antipyretic, anthelmintic, carminative, tonic, astringent, emmenagogue, galactagogue, and purgative (Kirtikar and Basu 1975; monograph: WHO 2009).

Pharmacology

Antidiabetes: The fruits, leaves, and extracts of *M. charantia* possess pharmacological properties. There are about 150 studies on the anti-DM properties of this plant in animal models of DM (Jung et al. 2006b). Hypoglycemic and antidiabetic properties of this plant fruit and seeds have been shown by several studies (review: Basch et al. 2003). The fruit juice has exhibited antidiabetic activity in streptozotocin-induced diabetic rats (Karunanayake et al. 1990; Sitasawad et al. 2000; Ahmed et al. 2004). The water extract of the plant seed also has antidiabetic activity (Sathishsekar and Subramanian 2005a; Sathishsekar et al. 2005b). The water extract of fruit lowered blood glucose and prevented cataractogenesis in alloxan-induced diabetic rats (Srivastava et al. 1988). The hypoglycemic action of the fruit and fruit pulp has been shown in a few animal diabetic models (Ali et al. 1993; Sarkar et al. 1996; Miura et al. 2001b; Nerurkar et al. 2004), and the water extract of the fruit is most effective compared to other extracts. The water extract powder (20 mg/kg) of fresh unripe whole fruits was found to reduce fasting blood glucose by 48%, an effect comparable to that of glibenclamide. The extract did not show any signs of nephrotoxicity and hepatotoxicity as judged from histological and biochemical parameters (Virdi et al. 2003). A hypoglycemic polypeptide has been isolated from the fruit and is described as plant insulin (Khanna et al. 1981; Sheng et al. 2004). However, the protein's action in the oral route is not explained and its heat sensitivity is not known. The gene for the hypoglycemic peptide was cloned and expressed (Wang et al. 2011a).

In one study, dietary inclusion of 0.5% freeze-dried bitter gourd does not significantly influence glucose levels in diabetic rats (Platel and Srinivasan 1995). The level of bitter gourd in the diet may be insufficient to exert an effect. Further, the processing can, possibly, inactivate certain active proteins.

More than one active principle and mechanisms of action are involved in the antidiabetic property of *M. charantia* (Subramoniam and Babu 2003). Like sulfonylurea drugs, the water extract of the fruit stimulated β-cells to release or secrete more insulin, and viable β-cells may be required for its action (Karunanayake et al. 1990; Ahmed et al. 1998). It is known to protect and stimulate β-cells in streptozotocin-challenged rats (Ahmed et al. 1998; Sitasawad et al. 2000). The presence of molecules with insulin-like bioactivity in *M. charantia* seeds has been reported (Ng et al. 1986, 1987a). The insulin-like peptide did not affect steroidogenesis, but ginsenoside inhibited the same (Ng et al. 1987b). The plant insulin may act in the absence of β-cells. In diabetic animals the fruit normalized the activities of enzymes involved in glucose metabolism (Shibib et al. 1993). It is also reported that the fruit inhibited glucose absorption from the intestine. This activity is attributed to some other active principles. The fruit juice (10 mg/kg, p.o., for 30 days) produced significant lowering of blood sugar level in alloxan-induced rabbits. Alkaloids like charantin (50 mg/kg, p.o.) produced 42% lowering of blood glucose on normal fasting rabbits. Saponins from the plant stimulated insulin secretion *in vitro* (Keller et al. 2011).

Other effects observed include suppression of key gluconeogenic enzymes, stimulation of key enzymes of the hexose monophosphate shunt (HMP) pathway, and preservation of islet cells and their functions (review: Joseph and Jini 2013). Studies also indicate the involvement of mechanisms such as activation of the AMPK pathway (Tan et al. 2008) and stimulation of PPAR-α and PPAR-γ. Cucurbitane-type triterpene glycosides activated PPAR-γ. These receptors are known to mitigate insulin resistance (review: Joseph and Jini 2013; Lee et al. 2009a; Wang et al. 2014).

Active Principles: The major compounds that have been isolated from bitter melon and identified as hypoglycemic agents include charantin, polypeptide-*p*, and vicine. Charantin is a mixture of two compounds, namely, sitosteryl glucoside and stigmasteryl glucoside. Polypeptide-*p* or plant insulin is a hypoglycemic protein. The protein works by mimicking the action of human insulin in the body. The gene has been cloned and a recombinant polypeptide with hypoglycemic activity in diabetic mice has been produced (Wang et al. 2011a). The glycol alkaloid vicine (a pyrimidine nucleoside) isolated from the seed has been shown to induce hypoglycemia in nondiabetic rats (review: Joseph and Jini 2013). In addition to charantin, several triterpenoids (cucurbitane-type triterpenoids in fruits, including momordicine and momordicosides) and conjugated linolenic acid (a fatty acid found in high concentrations in the seeds) that improved insulin resistance have been isolated from the fruit and stem of this plant, and some of them stimulated AMPK activity, which also contributes to the hypoglycemic activity (review: Joseph and Jini 2013). Momordicine 1 and momordicine 2 stimulated insulin secretion in MIN6 β-cells (review: Firdous 2014). Recently, a new cucurbitacin, 5β,19-epoxycucurbit-23-en-7-on-3β,25-diol, and three already known cucurbitacins showed a concentration-dependent inhibition of glucose production from liver cells (Chan et al. 2015). There could be many more molecules contributing to the anti-DM activity of the plant. Compounds like β-sitosterol and β-amyrin present in this plant are known to have anti-DM activity. Further studies are required to identify them and the mechanisms of action of individual compounds and their various combinations remain to be studied. Specific combinations may prove better medicine to treat DM.

Clinical Studies: Although there are some clinical studies, the data with human subjects are flawed by poor study design, improper doses, and low statistical power. Besides, there are only a few double-blind, randomized controlled studies (monograph: WHO 2009). The seed, fruit, fruit juice, water, and alcohol extracts of the fruit and fried fruit showed varying levels of efficacy in clinical trials. The insulin-like compound obtained from the fruit of this plant reduced blood glucose levels in DM patients. In another study, fresh fruit juice improved glucose tolerance in type 2 DM subjects. In addition, the fruit extract reduced postprandial serum glucose levels in type 2 diabetic patients (Ahmad et al. 1999). The seeds of the plant also showed hypoglycemic activity in type 2 DM patients. A preliminary dietary supplementation study showed improvement in the metabolic syndrome in the bitter gourd supplemented group (Chen et al. 2012; review: Joseph and Jini 2013).

In another study, when different groups of diabetic patients were treated with bitter gourd tablets for 12 weeks, there were significant improvements in the levels of glucose, cholesterol, HDL, LDL, and TG in bitter gourd-treated groups compared to untreated diabetic control groups. From the study, it is concluded that bitter gourd tablets have beneficial effects on glucose tolerance (Hasan and Khatoon 2012). In newly diagnosed type 2 patients, bitter melon showed hypoglycemic effect (Fuangchana et al. 2011).

Drinking a water suspension of the vegetable pulp resulted in remarkable reduction of blood glucose levels and improvement in glucose tolerance in 86 out of 100 cases with moderate type 2 DM. Similarly the fruit juice was found to improve glucose tolerance in 73% of maturity-onset diabetic patients. In another study, a decrease in HbA1c was observed in nine type 2 DM patients who consumed fried *M. charantia* fruits (230 g/day for 8–11 weeks). Another study showed that *M. charantia* (55 mL fruit juice/day for 5 months) was more effective in the management of type 2 diabetes and its complications than rosiglitazone. However, one report showed no significant decrease in fasting blood glucose and cholesterol levels of patients treated with a *M. charantia* product (two capsules/three times daily) for 3 months (review: Ghorbani 2013). The quality of *M. charantia* product, dose level, and method of preparation may influence the efficacy.

The United Kingdom has released a warning with regard to the bitter gourd capsules, because it is not yet known what doses are safe when taken with other antidiabetic agents. Thus, although the clinical studies suggest efficacy and safety, further controlled and randomized studies with different doses and defined sample preparations are required to utilize this important vegetable-cum-anti-DM plant to the fuller extent for human welfare. Since it is a dietary ingredient, a dietary approach can be developed to control diabetes using this plant's fruit (Krawinkel and Keding 2006).

Other Activities: The plant extract (leaf or fruit) has antioxidant property and has modulated the drug metabolizing enzymes in rat liver (Raza et al. 2000). The antioxidant activity of the seed is also shown in diabetic rats (Sathishsekar and Subramoniam 2005b). The fruit has antilipolytic activity (Wong et al. 1985) and the fruit extract is hypolipidemic in diabetic rats (Ahmed et al. 2001). It lowered glucose and lipid parameters in rats fed a high-cholesterol diet (Jayasooriya et al. 2000). The acidic ethanol extract of fruits and seeds influenced lipid metabolism in isolated adipocytes. The fruit juice inhibits TG transfer protein gene expression in cultured cells (Chaturvedi et al. 2005). *M. charantia* has *in vivo* antitumor (Jilka et al. 1983) and antimutagenic activities (Guevara et al. 1990). A protein with antitumor activity has been isolated (Ng et al. 1992).

An inhibitor protein of HIV-1 infection and replication has been isolated from the fruit and the enzyme integrase of HIV is inhibited by the protein (Lee-Huang et al. 1995). Bitter melon can stimulate the immune system and activate the body's natural killer cells that help fight against viral and other infections. The fruit has broad-spectrum antibacterial activity (review: Joseph and Jini 2013). The antimalarial activity of this plant has also been reported (Amorim et al. 1991). Chloroform and petroleum ether extracts of the fruits exhibited antispasmodic and antioxytocic properties. The ethanol extract of the root in doses of 10–100 µg/mL produced uterine contractions on isolated preparation in rat. The essential oil recovered from crushed unripe fruits of the plant (4 mL/100 g, i.p.) exhibited anorectic and antipyretic effects in rats.

Toxicity: Feeding normal adult rats with the fruit-containing diet (0.5%, dry weight) for 8 weeks had no adverse influence on the food intake, growth, organ weights, and hematological parameters of rats. However, serum cholesterol levels of rats receiving the fruit-containing diet were lower than those of the control rats (monograph: WHO 2009). The alcohol extract of the seed has antispermatogenic activity in rats (Naseem et al. 1998). The seeds of the plant also contain abortifacient and immunosuppressive proteins (Leung et al. 1987; Ng et al. 1992). However, the proteins are likely to be inactivated during cooking. The heat sensitivity of these proteins remains to be studied.

Phytochemicals: Many bioactive proteins including insulin-like polypeptide, anti-HIV proteins, and lectins have been isolated from the fruit. Several glycosides have been isolated from the plant and are grouped under cucurbitane-type triterpenoids. Low-molecular-weight phenolic compounds isolated include gallic acid, gentisic acid, catechin, chlorogenic acid, and epicatechin. Its other isolates include flavonoids, ginsenosides, oleanolic acid 3-*O*-monodesmoside, oleanolic acid 3-*O*-glucuronide, and oleanane-type triterpene saponins, and chemical isolates from seeds are vaccine, mycose, 3-*O*-(β-D-glucopyranosyl)-24β-ethyl-5α-cholesta-7-trans-22E,25(27)trien-3β-ol, momorcharaside A, momorcharaside B and bioactive proteins (Zhu et al. 1990), ascorbigenin, pipecolic acid, luteolin, glucoside, resin, β-sitosterol, charantin, β-sitosterol glucoside, stigmast-2,25-diene-3β-*O*-glucoside, β- and α-momorcharins, paraffins, p-cymene, hexadecanol, (–)menthol, nerolidol, pentadecanol, squalene, 10α-cucurbita-5,24-dien-3β-nerolidol, cycloartenol, taxaxerol, β-amyrin, campesterol, cycloeucalenol, 24β-ethyl-5α-cholesta-7-trans-22-dien-3β-ol, 24β-ethyl-5α-cholesta-7-trans-22,25(27)trine-3β-ol, lophenol, 4α-methylzymosterol, obtusifoliol, spinosterol, stigmasterol, stigmast-7,25-dienol, stigmast-7,22,25-trienol, benzyl alcohol, myrtenol, cis-hexenol, trans-2-hexenal, 1-penten-3-ol, and cis-2-penten-1-ol. The fruits, leaves, and roots contain vitamins (vitamins C, B2, B1, and A and calcium), amino acids (aspartic acid, alanine, glutamic acid, carotenes, charantin steroid glycoside, D-galacturonic acid), sterol (stigmasterol, β-sitosterol), lactin, pectin, fatty acids (linolenic acid) trypsin inhibitors, alkaloids (momordicine), lycopene, cryptoxanthin, terpenoid derivatives, and peptides (monograph: WHO 2009; review: Joseph and Jini 2013).

Momordica cymbalaria Hook., Fenzl ex Naud, Cucurbitaceae

Synonyms: *Momordica tuberosa* Roxb., *Luffa tuberosa* Roxb
Momordica cymbalaria is a monoecious climber with simple tendril, having a large woody root tuber. The plant is distributed in Mysore, Konkan, in India and tropical Africa. It is used to procure abortion in traditional medicine.

M. cymbalaria fruit (water extract) showed antihyperglycemic and antidiabetic effect in alloxan-induced diabetic rats without significant hypoglycemic activity in normal rats (Kameswararao et al. 2003). *Luffa tuberosa* fruit (250 and 500 mg/kg, water extract) exhibited antidiabetic activity in streptozotocin-induced diabetic rats (Sharma and Arya 2011). Glucose uptake by isolated diaphragms of both streptozotocin-induced diabetic and nondiabetic animals increased in the presence of an oleanane-type triterpenoid saponin isolated from the roots of *M. cymbalaria*. Insulin release was augmented by the presence of the saponin (1 mg/mL) in rat insulinoma cell line (RIN-5F) preexposed to adrenaline (5 µM) and nifedipine (50 µM). Pancreatic histology also indicated considerable quantitative increase in β-cells (75%) when treated with the saponin in diabetic rats. The study suggests that the saponin of *M. cymbalaria* possesses potential antidiabetic activity with respect to insulin secretion, which may be attributed to the modulation of the calcium channel and β-cell rejuvenation (Koneri et al. 2014a).

In another recent study, the steroidal saponin of *M. cymbalaria* showed a potential neuroprotective effect in diabetic peripheral neuropathy with respect to neuropathic analgesia, improvement in neuronal degenerative changes, and significant antioxidant activity in streptozotocin-induced diabetic rats (Koneri et al. 2014b). The fruit powder also showed hypolipidemic effects. Phytochemicals reported include iridoid glycosides (review: Ajikumaran and Subramoniam 2005).

Momordica dioica Roxb. ex Willd., Cucurbitaceae

Synonym: *Momordica cymbalaria* Fenzl ex Naud

The oral hypoglycemic effects of *Momordica dioica* (a wild relative of *M. charantia*) was shown in the rat in 1994 (Fernandopulle et al. 1994). After this, the antihyperglycemic activity of *M. dioica* fruits was shown in alloxan-induced diabetic rats (Reddy et al. 2005). In another study also, the antidiabetic effect of *M. dioica* fruit was shown in alloxan-induced diabetic rats. The water extract of *M. dioica* showed a maximum effect in glucose tolerance test compared to hexane, chloroform, and ethanol extracts in normal rats. The oral effective dose of the water extract was 200 mg/kg, which produced a fall of blood glucose (57.5%) in diabetic rats. Administration of this extract (200 mg/kg) once daily for 21 days to alloxan-induced diabetic rats resulted in marked reduction of elevated blood glucose and glycosylated hemoglobin levels. Blood urea, creatinine, total protein, AST, ALT, ALP, bilirubin, and urinary sugar levels were also reduced after the extract treatment in diabetic rats. Above 3 g/kg was the lethal dose of the extract, indicating a high margin of safety (Singh et al. 2011a). The antidiabetic and renal protective effect of the methanol extract of *M. dioica* was also reported in streptozotocin-treated diabetic rats. The extract treatment markedly reduced serum glucose and urea levels and increased serum insulin levels. Furthermore, histological observation of kidney of diabetic rats showed degenerative changes in the glomerulus and renal tubules. These changes were reversed, to a large extent, by the treatment (Gupta et al. 2011b).

Monstera deliciosa Liebm., Araceae

Under *in vitro* conditions, the ethanol extract of *Monstera deliciosa* showed an insulin secretagogue activity in INS-1 cells at the 1 μg/mL level (Hussain et al. 2004).

Morinda citrifolia, Rubiaceae

Common Name: Noni

Morinda citrifolia is native to Southeast Asia. The plant has been widely used to treat a number of diseases including diabetes. The young leaves are eaten as a vegetable. Iridoid glycosides and the known hypoglycemic agents β-sitosterol and ursolic acid have been obtained from the leaves (review: Ghisalberti 2005). The fruit of *M. citrifolia* regulated glucose metabolism via forkhead transcription factor 1 (FOXO1) in HFD-induced obese diabetic mice (Nerurkar et al. 2012). A number of anthraquinones and iridoid asperuloside have been isolated from the fruit (review: Ghisalberti 2005).

Morinda lucida Benth., Rubiaceae

The stem bark and leaves of *Morinda lucida* have traditionally been used in Ghana and elsewhere for the management of DM. The hypoglycemic and antihyperglycemic activities of the methanol extract of *M. lucida* leaves were studied in normal and streptozotocin-induced diabetic rats. In normal rats, the methanol extract (50–400 mg/kg) demonstrated a dose-dependent hypoglycemic activity within 4 h after oral administration. In streptozotocin-induced diabetic rats, the extract produced an antidiabetic effect from day 3 after oral administration, with 400 mg/kg extract-treated animals having a plasma glucose level of 13.8 mmol/L compared with glibenclamide (10 mg/kg)-treated animals with a plasma glucose level of 13.9 mmol/L. These results suggest that the leaves of *M. lucida* have a strong glucose-lowering property in streptozotocin-induced diabetic rats (Olajide et al. 1999). The water extracts of *M. lucida* root (140 and 280 mg/kg, p.o.) exhibited potent dose-dependent hypoglycemic effects in both normoglycemic and alloxan-induced diabetic mice. In alloxan-induced diabetic mice, the reaction time of the extract was longer with a single dose of the extract producing a significant hypoglycemic effect 4 h after administration (Kamanyi et al. 2006). The hypoglycemic activity of petroleum ether and ethanol extracts (100, 200, and 400 mg/kg, p.o., daily for 19 days) of the stem bark was evaluated in streptozotocin-induced diabetic rats. Both of the extracts exhibited significant hypoglycemic and substantial antioxidant activities throughout the study period. While the ethanol extract demonstrated the strongest hypoglycemic activity 6 h after administration, the petroleum ether extract exhibited the highest activity 24 h after administration (Fleischer et al. 2011). The composition of active molecules in the two extracts could differ. Interestingly, the leaf, root, and stem bark of this plant showed a glucose-lowering property in both normal and diabetic animals. Elucidation of the mechanism of action and identification of active principles should follow. One of the mechanisms could be inhibition of carbohydrate metabolizing enzymes. The water extract of this plant leaf inhibited the α-amylase competitively but displayed a mixed noncompetitive mode of inhibition toward α-glucosidase (Kazeem et al. 2013a).

Toxicity evaluation has shown that the leaf extract was safe. Acute oral administration of the leaf extract was nonlethal at 6400 mg/kg. The results obtained for liver and kidney function parameters indicated that ingestion of *M. lucida* leaf extract has no toxic effect on liver and kidney functions (Oduola et al. 2010).

Moringa oleifera Lam., Moringaceae

The leaf (water extract) of *Moringa oleifera* showed antihyperglycemic activity. The treatment reduced the levels of glucose and increased body weight and hemoglobulin content in rats (Jaiswal et al. 2009).

Moringa stenopetala Baker f., Moringaceae

Moringa stenopetala leaves are traditionally employed for the treatment of DM in Ethiopia. Studies confirmed that the crude water extract of the leaves of this plant and fractions isolated from these extracts have hypoglycemic and antihyperglycemic effects. Similarly, the ethanol extract and three fractions of the ethanol extract (chloroform, butanol, and water residue) (300 mg/kg, single dose or repeated doses) significantly reduced blood glucose levels in both normal (normoglycemic) and alloxan-induced diabetic rats (Nardos et al. 2011). The hydroalcoholic extract of leaves of *M. stenopetala* inhibited the activities of intestinal α-glucosidase and some pancreatic enzymes (Toma et al. 2014). This could be one of the mechanisms of action. However, other mechanisms of action and active principles involved remain to be studied.

Morus alba L., Moraceae (Figure 2.68)

Common Names: White mulberry, tutam, tut, musukette, malbari, etc.

Description: *Morus alba* is a shrub or tree 3.10 m tall with gray, shallowly furrowed bark; fine hairy branches; reddish brown, ovoid, finely hairy buds; and lanceolate stipules, 2.3–5 cm long, and densely covered with short pubescence. The petiole is pubescent with ovate to broadly ovate, irregularly lobed,

(a)   (b)

FIGURE 2.68 *Morus alba*. (a) Twig with fruits (wild variety). (b) Twig with fruits [cultivated].

abaxially sparsely pubescent leaf blade; primary lateral veins, coarsely serrate to crenate margin; acute, acuminate, or obtuse apex; pendulous, 2.3–5 cm, densely white hairy male catkins; 1.2 cm, pubescent female catkins; and 5.10 mm pubescent peduncle. In male flowers, the calyx lobes are pale green and broadly elliptic, the filaments are inflexed in a bud, and the anthers are 2-loculed. In female flowers, sessile, calyx lobes are ovoid with marginal hairs, ovary is sessile and ovoid, style is absent; stigmas have mastoid-like protuberance, and the branches are divergent and papillose. The syncarp is red when immature and blackish-purple, purple, or greenish-white when mature and is ovoid, ellipsoid, or cylindrical (review: Devi et al. 2013).

Distribution: *M. alba* (white mulberry) is native of India, China, and Japan. The plant is cultivated throughout the world wherever silkworms are raised; these include Europe, North America, and Africa (review: Devi et al. 2013).

Traditional Medicinal Uses: *M. alba* plant is used in ethnomedicine for ailments such as asthma, cough, bronchitis, edema, insomnia, wound healing, diabetes, influenza, eye infections, and nosebleeds; the fruit has been used as a medicinal agent to nourish the blood, benefit the kidneys, and treat weakness, fatigue, anemia, and premature graying of hair. It is also used to treat urinary incontinence, tinnitus, dizziness, and constipation in the elderly patient. It has been used in the indigenous system of Chinese medicine for cooling, acrid, purgative, diuretic, laxative, anthelmintic, brain tonic, antibacterial, and hepatopathy properties (review: Devi et al. 2013).

Pharmacology

Antidiabetes: *M. alba* extract inhibited the breakdown of starch by α-amylase (IC_{50} 17.60 µg/mL) in a concentration-dependent manner *in vitro* (Bahman and Golboo 2009). This suggests inhibition of starch absorption as monosaccharides from the intestine. Chalcomoracin, moracin C, moracin D, and moracin M isolated from *M. alba* inhibited α-glucosidase activity (review: Firdous 2014). Moran A, a glycoprotein, isolated from the plant exhibited hypoglycemic activity (Hikino et al. 1985).

Egyptian *M. alba* root bark extract exhibited antidiabetic and anti-LPO effects in streptozotocin-induced diabetic rats (Singab et al. 2005). Moracin M, steppogenin-4-*O*-β-glucoside, and mullberroside were isolated from the root of *M. alba*. All the three phenolic compounds showed hypoglycemic effects in alloxan-induced diabetic rats. Steppogenin-4-*O*-β-glucoside (50 mg/kg) was found to be more potent compared to the other compounds (Zhang et al. 2009c). The leaf extract of this plant also improved body weight, diameter of islets, and the number of β-cells in treated (35-day treatment) diabetic rats compared to untreated control animals as evidenced by histological observations (Jamshid and Prakash 2012). It appears that the extract exhibits hypoglycemic activity by helping the regeneration of β-cells also. It is not known whether or not this effect is brought about by the phenolic compounds that showed hypoglycemic activity (Zhang et al. 2009). Recent studies also confirmed the promising antidiabetic and antioxidant activities of this plant. The ethanol extract of the branch bark of this plant showed *in vivo* antihyperglycemic and antioxidant activity (Wang et al. 2014). A polysaccharide isolated from *M. alba* showed antidiabetic activity in streptozotocin-induced diabetic mice by blocking the inflammatory response and attenuating oxidative stress (Guo et al. 2013). Repeated administration of the leaf extract of this plant reduced insulin resistance in KK-Ay mice (Tanabe et al. 2011). A fraction from the 70% alcohol extract of this plant's root bark showed antihypercholesterolemic and antioxidant activity in HFD-fed rats (review: Devi et al. 2013). Thus, the plant appears to be very promising for further studies to determine its utility for medicine development.

Other Activities: *M. alba* possesses several other pharmacological activities such as antioxidant, antihyperlipidemic activity, antistress, anticancer, antimutagenic activity, antibacterial, anthelmintic, immunomodulatory activity, antidopaminergic effect, anxiolytic activity, nephroprotective action, hepatoprotective activity, inhibition of melanin biosynthesis, and protection of gut disorder (review: Devi et al. 2013).

Phytochemicals: The plant contains ascorbic acid, carotene, vitamin B1, folic acid, folinic acid, isoquercetin, quercetin, tannins, flavonoids, and saponins. White mulberry leaf contains triterpenes (lupeol) sterols (β-sitosterol), bioflavonoids (rutin, moracetin, quercetin-3-triglucoside, and isoquercitrin), apigenin, coumarins, volatile oil, alkaloids, amino acids, and organic acids. Many flavones were isolated from the root bark as active principles. Many biochemical compounds such as moranoline, albafuran, albanol, morusin, kuwanol, calystegin, and hydroxymoricin are isolated from mulberry plants that play an important role in the pharmaceutical industry. The plant is reported to contain the phytoconstituents 1-deoxynojirimycin, phytosterols, saponins, flavonoids, benzofuran derivatives, morusimic acid, anthocyanins, anthraquinones, glycosides, and oleanolic acid (review: Devi et al. 2013).

Morus bombycis Koidzumi, Moraceae

Morus bombycis is used in traditional medicine to treat obesity and diabetes. The compound 2,5-dihydroxy-4,3-di(β-D-glucopyranosyloxy)-trans-stilbene, isolated from *M. bombycis*, exhibited antidiabetic activity in streptozotocin-induced diabetic rats. At doses of 200–800 mg/kg, the compound improved hyperglycemia in rats, which was comparable to that of tolbutamide. The histological observations showed that the compound prevented atrophy of pancreatic β-cells and vascular degenerative changes in the islets. The compound also showed antioxidant activity (Heo et al. 2007). Bioassay-guided fractionation of the chloroform fraction of *M. bombycis* led to the isolation of kuwanons J, R, and V (chalcone-derived products). These compounds inhibited PTP1B with IC_{50} values ranging from 4.3 to 13.8 µmol/L (review: Jiany et al. 2012).

Morus indica L., Moraceae

Morus australis Poir. is a closely related species.

Powdered leaves (500 mg/kg) of *Morus indica* reduced blood glucose levels in streptozotocin-induced diabetic rats by 38% after 15 days of treatment (Devi and Urooj 2008). In another study, *M. indica* showed anti-DM and antioxidant activity in alloxan-induced diabetic rats. The treatment almost normalized blood glucose levels and liver glycogen content in diabetic rats. Further, the treatment increased the activity of antioxidant enzymes and decreased lipid peroxide levels compared to untreated diabetic rats (Kumar et al. 2010d). The flavonoid fraction from *M. indica* lowered the levels of blood glucose and lipids in hyperlipidemic diabetic rats (Ma et al. 2012). Mulberry leaves exhibited antioxidant and antihyperglycemic properties in streptozotocin-induced diabetic rats, retarding cataract formation in diabetic rats (Andallu and Varadacharyulu 2003). Mulberry leaf therapy exhibited potential hypoglycemic and hypolipidemic effects in type 2 diabetic patients (Andallu et al. 2001). Active principles were not identified.

In a recent human clinical trial, the effect of mulberry tea in the reduction of abnormally high postprandial blood glucose (PPG) levels in type 2 DM patient was evaluated. Twenty diabetic patients were given plain tea (control) and 28 diabetic patients were given mulberry tea (test subject) to measure the effect of mulberry tea on fasting blood glucose and PPG levels. After the consumption of plain tea and mulberry tea, the PPG values were recorded as 15.9 and 11.7 mmol/L, respectively, in the control and test group. A significant change in the PPG level was observed in response to mulberry tea in all the test patients compared with control (Banu et al. 2014).

Morus insignis Bureau, Moraceae

Ethyl acetate and n-butanol-soluble fractions of the leaves of *Morus insignis* showed a significant hypoglycemic activity on streptozotocin-induced hyperglycemic rats. From these hypoglycemic activity–showing fractions, two new compounds, mulberrofuran U and moracin M-3-*O*-β-D-glucopyranoside, were isolated, along with six known compounds (β-sitosterol, β-sitosterol-3-*O*-β-glucopyranoside, ursolic acid, moracin M, kaempferol-3-*O*-β-glucopyranoside, and quercetin-3-*O*-β-glucopyranoside) (Basnet et al. 1993).

Morus nigra L., Moraceae

Morus nigra, commonly known as black mulberry, is used in folk medicine for diabetes treatment. Several reports are claiming a hypoglycemic effect for black mulberry (*M. nigra*) extract. The alcohol extract of *M. nigra* leaves exhibited a hypoglycemic effect in normal and diabetic rats (Oryan et al. 2003). The hydroalcoholic extract of the *M. nigra* leaf (100 mg/kg) decreased blood glucose, serum cholesterol, and TG contents of alloxan-induced diabetic rats and increased HDL content (Baratil et al. 2012). The root bark extract and leaves contain deoxynojirimycin, an alkaloid. An infusion of leaves caused a drop in blood glucose, sometimes diuresis, and a reduction of arterial pressure (Kumar and Chauhan 2008). Deoxynojirimycin is a potent α-glycosidase inhibitor and helps establish greater glycemic control in type 2 diabetes. Young mulberry leaves, taken from the top part of branches in the summer, contained the highest amount of deoxynojirimycin. In a human study, deoxynojirimycin-enriched powder of mulberry leaves significantly suppressed the elevation of post-prandial glucose. The authors conclude that newly developed deoxynojirimycin-enriched powder can be used as a dietary supplement for preventing DM (Kumar and Chauhan 2008). Oral administration of the water extract (prepared with the maceration method) of *M. nigra* (100 mg/kg twice daily for 42 days) to streptozotocin-induced diabetic rats resulted in the reduction of blood glucose levels, but the treatment increased the levels of serum creatinine (a marker of kidney function) and GOT and GPT (liver function markers). Prolonged high dose use of mulberry leaf may damage kidney and liver function (Hemmati et al. 2010). In another report, administration of the water extract of *M. nigra* (400 mg/kg) leaves from day 0 to 20 pregnancy to streptozotocin-induced diabetic rats resulted in decreased levels of MDA, cholesterol, TGs, and VLDL and decreased placental index and weight as compared to the diabetic group. However, the *M. nigra* leaf water extract failed to control hyperglycemia in pregnant diabetic rats (Volpato et al. 2011).

Morus rubra L., Moraceae

Common Name: Egyptian mulberry

The *Morus rubra* leaf extract showed a dose-dependent fall in fasting blood glucose in streptozotocin-induced diabetic rats. Treatment with 400 mg/kg extract produced a significant reduction in glycosylated hemoglobin with a concomitant elevation in plasma insulin and C-peptide levels. The altered serum lipids in diabetic rats were significantly restored following treatment with the extract. In erythrocytes, as well as the liver, the activity of antioxidant enzymes and content of reduced GSH were found to be significantly enhanced, while the levels of serum and hepatic lipid peroxides were suppressed in extract-fed diabetic rats. Histopathological examination of pancreatic tissue revealed an increased number of islets and β-cells in extract-treated diabetic rats (Sharma et al. 2010a). In another study, the water extract (100 and 200 mg/kg) of the leaves of this plant showed hypoglycemic and antiatherosclerotic activities in diet-induced atherosclerosis in diabetic rats (Sharma et al. 2010b).

Mucuna pruriens (Linn.) DC., Fabaceae

Common Names: Cowage, ajarha, goncha, amudari, naykkurana, etc.

Mucuna pruriens is an herbaceous twining annual with slender terete branches, becoming glabrescent when mature. It is a cosmopolitan species in the tropics.

In traditional medicine, *Mucuna pruriens* is attributed with numerous pharmacological properties, including as an aphrodisiac, tonic, emmenagogue, anthelmintic, purgative, diuretic, and vermifuge (*The Wealth of India* 1962; Kirtikar and Basu 1975).

M. pruriens seeds (powder) reduced the blood glucose levels in normal and alloxan-induced diabetic rabbits (Akhtar et al. 1990). The water extract of *M. pruriens* provided only some level of protection against experimental diabetic cataract formation in rats (Rathi et al. 2002a). The alcohol extract decreased blood glucose levels in alloxan-induced diabetic rats, while it showed no influence on the same in streptozotocin-induced diabetic rats (Rathi et al. 2002b). However, another study showed the hypoglycemic effect of *M. pruriens* seed extract on normal and streptozotocin-induced diabetic rats (Bhaskar et al. 2008). Extracts (toluene, chloroform, ethyl acetate, and butanol fractions) of *M. pruriens* leaves showed hypoglycemic and hypolipidemic activities in alloxan-induced diabetic rats (Murugan et al. 2009).

The plant is known to have aphrodisiac activity. The seed powder is a natural source of L-dopa that may be of use in Parkinson's disease. The seed glycoprotein provided protection against *Echis carinatus* venom. The alcohol extract of the seeds has an anti-LPO property.

Phytochemicals reported include nicotine, purieninine, purienidine, β-sitosterol, serotonin (5-hydroxy-tryptamine), tetrahydroisoquinoline alkaloids, and bioactive glycoprotein (review: Ajikumaran and Subramoniam 2005).

Murraya koenigii (Linn.) Spreng., Rutaceae (Figure 2.69)

Common/Local Names: Curry leaf, kari patta, kathinim, kariveppilai, kariveppu, etc.

Description: *Murraya koenigii* is a strong-smelling umbrageous small tree or shrub 4–6 m height. *M. koenigii* has a gray color bark with longitudinal striations, and beneath it is a white bark. The leaves are bipinnately compound, 15–30 cm long each bearing 11–25 leaflets alternate on rachis, 2.5–3.5 cm long ovate lanceolate with an oblique base. The flowers are bisexual, white, funnel shaped, sweetly scented, stalked, complete, ebracteate, regular with an average diameter of a fully opened flower being 1.12 cm inflorescence, and terminal cymes each bearing 60–90 flowers. The fruits are ovoid to subglobose, wrinkled or rough with glands, which are 2.5 cm long and 0.3 cm in diameter and gets purplish black when ripe; they are generally biseeded. The seeds typically occur in spinach green color, 11 mm long, 8 mm in diameter (review: Handral et al. 2012).

Distribution: *M. koenigii* is native to tropical Asia and is distributed from the Himalayan foothills of India to Sri Lanka eastward through Myanmar, Indonesia, South China, and Hainan. It is cultivated throughout India and elsewhere in the world. (The leaf is used in South India as a natural flavoring agent in various curries.)

Traditional Medicinal Uses: The leaves of this plant are used traditionally in the Indian Ayurvedic system to treat diabetes. It is acrid, analgesic, bitter, cooling, alexiteric, anthelmintic, carminative, purgative, and stimulant and used to allay the heat of the body, bites of poisonous animals, blood disorders, diarrhea, dysentery, eruption, inflammation, itching, kidney pain, leukoderma, piles, snakebite, thirst, and vomiting. The leaves are applied externally to bruises and eruption; the leaves and roots are an analgesic. An infusion of the toasted leaves is used to stop vomiting. The juice of the root is good for pain associated with kidney infection. The fruits are also considered as an astringent in Indo-China. Crushed leaves are applied externally to cure skin eruption and to relieve burns while pastes of leaves are applied externally to treat poisonous animal bites. The plant is credited with tonic and stomachic properties. The branches of *M. koenigii* are very popular for cleaning the teeth (Kirtikar and Basu 1975; review: Handral et al. 2012).

FIGURE 2.69 *Murraya koenigii.*

Pharmacology

Antidiabetes: The leaf showed significant hypoglycemic action in rats (Khan et al. 1995). In another study, administration of the leaf (water or alcohol extract) significantly reduced blood glucose levels and increased insulin levels in alloxan-induced diabetic rats (Vinuthan et al. 2004). A single oral administration of variable dose levels (200, 300, and 400 mg/kg) of the water extract of leaves led to lowering of blood glucose level in normal as well as in diabetic rabbits. The maximum fall of about 15% in normal and 28% in MD rabbits was observed after 4 h of oral administration of 300 mg/kg dose. The same dose also showed a marked improvement in glucose tolerance (Kesari et al. 2005). However, as per a report, feeding curry leaf up to 15% in the diet caused only mild reduction in blood glucose in mild alloxan-induced diabetic rats without a significant effect on glucose levels in moderate streptozotocin-induced diabetic rats (Yadav et al. 2003). The dose used in this study is very high.

The water extract of the leaves of *M. koenigii* delayed diabetic nephropathy and showed hypoglycemic activity in streptozotocin-induced diabetic rats (Yankuzo et al. 2011). In another study, oral administration of the water extract of *M. koenigii* leaves (100 and 200 mg/kg) in streptozotocin-induced diabetic rats resulted in a significant decrease in cholesterol, TG, and serum glucose levels (range 55.6%–64.6%) compared to metformin (62.7%). However, there was no significant effect on body weight and serum creatinine (El-Amin et al. 2013). In another study, leaves of *M. Koenigii* showed antioxidant activity and improved pancreatic β-cells structure in diabetic rats (Arulselvan and Subramanian 2007).

Mahanimbine (carbazole alkaloid) isolated from the leaves of *M. koenigii* showed anti-DM activity when administered at a single dose of 50 or 100 mg/kg/week for 30 days in streptozotocin-induced diabetic rats. The elevated fasting blood sugar, TGs, LDL, and VLDL found in diabetic rats were decreased and the HDL level was increased (Kumar et al. 2010a). Mahanimbine showed a strong α-amylase inhibitory effect. The study indicates that mahanimbine possesses both antihyperglycemic and antihyperlipidemic activities (Kumar et al. 2010a). However, the mechanism of action of this compound remains to be studied in detail. However, there was no significant effect on body weight and serum creatinine (El-Amin et al. 2013).

Clinical Trial: Curry leaf supplementation to type 2 diabetic patients caused a transient decrease in blood glucose levels without significant influence on lipid parameters (Iyer and Mani 1990). More clinical trials with different doses and duration of treatments are needed.

Other Activities: The levels of cholesterol and phospholipids decrease in the curry leaf-treated, 1,2-dimethyl hydrazine (carcinogen)-challenged rats, and colon carcinogenesis is also inhibited (Khan et al. 1996). Curry leaf, in HFD-fed rats, altered peroxidation (thiobarbituric acid-reactive substances) to a beneficial level. Carbazole alkaloids from curry leaf showed antimicrobial and antioxidant activities (Ramsewak et al. 1999; Tachibana et al. 2001). Other reported pharmacological actions include anthelmintic activity, vasodilating activity, antiulcer activity, hypocholesterolemic activity, antidiarrheal activity, analgesic activity, wound healing property, memory enhancing property, and antiamnesic activity (review: Handral et al. 2012; Nayak et al. 2010).

Phytochemicals: The leaves are aromatic and contain carotene, nicotinic acid, vitamin C, high amount of oxalic acid, crystalline glycosides, carbazole alkaloids, koenigin, resin, girinimbin, iso-mahanimbin, koenine, koenigine, koenidine and koenimbine, mahanimbicine, bicyclomahanimbicine and phebalosin, triterpenoid alkaloids like cyclomahanimbine, tetrahydromahanmbine, murrayastine, murrayaline, pypayafolinecarbazole alkaloids, and many other chemical compounds. The fresh leaves contain yellow-colored volatile oil. The bark mainly contains the carbazole alkaloids as murrayacine, murrayazolidine, murrayazoline, mahanimbine, girinimbine, koenioline, and xynthyletin (review: Handral et al. 2012).

Musa paradisiaca L., Musaceae

Common Name: Banana

Musa paradisiaca is an important fruit (ripe fruit) and vegetable (unripe fruit) crop. Administration of the unripe plantain (*M. paradisiaca*)-incorporated feed for 21 days to streptozotocin-induced diabetic rats resulted in 38.13% decrease in hyperglycemia and amelioration of their elevated urinary protein, glucose, SPGR, and relative kidney weights. It is concluded that unripe plantain could be useful in the management of diabetic nephropathy (Eleazu et al. 2013).

M. paradisiaca flowers (alcohol extract) also showed potent antihyperglycemic activities. Oral administration of the alcohol and alcohol–water (1:1) extracts (250 and 500 mg/kg) to alloxan-induced diabetic rats resulted in reversal of hyperglycemia within a week. The ethanol extract (259 mg/kg) exerted more hypoglycemic effect than glibenclamide (Jawla et al. 2012). In contrast to flowers and fruit, the stem juice of *M. paradisiaca* (500 mg/kg) produced significant rise (28%) of blood glucose level 6 h after oral administration in normal rats. In subdiabetic rats, the same dose produced a rise of 16% in blood glucose levels within 1 h during glucose tolerance test and a rise of 16% after 4 h in fasting blood glucose levels of SD rats (Singh et al. 2007). In a recent study, different parts (leaves, fruit peels, stem, and roots) of *M. paradisiaca* were evaluated for its antidiabetic effects. The ethanol extract of leaves and fruit peels showed promising antidiabetic activity in streptozotocin-induced diabetic rats. The active molecules involved are not identified (Lakshmi et al. 2014). The mechanism of action and active principles present in the fruit, flowers, and leaves remain to be studied. The stem may contain compounds with pro-DM activity.

Musa sapientum L., Musaceae

Common Names: Banana plant, alabu, amrit, ambanam, vazha, etc.

Musa sapiens is a subarborescent, stoloniferous plant with a cylindrical stem and convolute leaf sheath. The fruit is fleshy, indehiscent, oblong or fusiform with many angled seeds. It is commercially cultivated for its edible fruits (ripe and unripe). The plant is indigenous to the Eastern Himalayas and Bihar and is cultivated throughout India and the tropics. The stem, root, leaves, and fruits are used in traditional medicine to treat many ailments.

Oral administration of the chloroform extract of the flowers resulted in a significant reduction in blood glucose and glycosylated hemoglobin and an increase in total hemoglobin in alloxan-induced diabetic rats. The extract prevented a decrease in body weight and also resulted in a decrease in free radical formation in the tissues (Pari and Maheswari 1999, 2000). The antihyperglycemic and antioxidant effects of the ethanol extract of *M. sapientum* flowers in alloxan-induced diabetic rats were also reported by another study. Oral administration of the ethanol extract showed a blood glucose-lowering effect at 200 mg/kg in alloxan-induced diabetic rats (120 mg/kg, i.p.), and the extract was also found to significantly scavenge oxygen free radicals, namely, superoxide dismutase, catalase, and also MDA. *M. sapientum*-induced blood sugar reduction may be due to the possible inhibition of free radicals and subsequent inhibition of tissue damage induced by alloxan (Dhanabal et al. 2005).

The flower also showed hypoglycemic activity in hyperglycemic rabbits (Alarcon-Aguilara et al. 1998). The *M. sapientum* fruit extract showed antidiabetic and gastric ulcer healing effects in streptozotocin-induced diabetic rats and could be more effective in diabetes with concurrent gastric ulcer (Kumar et al. 2013b). In another study, the juice of *M. sapientum* "bark" (leaf base) significantly lowered the blood glucose level and reduced the contractility of the fundus and pylorus in alloxan-induced diabetic rats. Further, gastric emptying time, intestinal transit time, and gastric acid secretion were elevated in treated diabetic rats. The treatment also reduced oxidative stress (Darvhekar et al. 2013). Phytochemical studies with reference to active principles are lacking. However, a few known anti-DM compounds are reported from this plant.

The methanol extract of plantain banana pulp has antioxidant properties. Flavonoid leucocyanidin from unripe banana has antiulcer activity in rats. Freeze-dried banana pulp shows a marked cholesterol-lowering effect when incorporated into a diet, while the banana pulp dried in a hot-air current (65°) does not. Further, both soluble and insoluble fibers fractionated from banana pulp have a cholesterol-lowering effect. Plantain banana stimulates gastric and colonic mucosal eicosanoid synthesis and has antiulcerogenic effects.

The muscle paralyzing effect of the juice from the trunk of the banana tree has been reported (Singh and Dryden 1985).

Phytochemicals reported include volatile branched chain esters, flavonoid leucocyanidin in unripe banana, dopamine, dopa, noradrenalin, serotonin, caffeic acid, cinnamic acid, *p*-coumaric acid, ferulic acid, gallic acid, protocatechuic acid, campesterol, β-sitosterol, stigmasterol, cyclomusalenol, cyclomusalenon, sitoindoside III and IV, sitosterol-3-*O*-gentiobioside, and sitosterol-3-*O*-myoinosityl(1–6)-β-D-glucose (review: Ajikumaran and Subramoniam 2005).

Myrcia bella Cambess, Myrtaceae

Myrcia species are used by the indigenous people of Brazil in traditional medicine for the treatment of DM. A 70% ethanol extract of *M. bella* leaf (600 mg/kg) reduced the fasting blood glucose levels on the seventh day in diabetic mice; the treatment increased liver glycogen content and decreased the levels of total cholesterol and TG in the blood of diabetic mice. In addition, the treatment increased the expression of IRS-1, PI3K, and Akt in the liver of diabetic mice (Vareda et al. 2014). Thus, it positively influenced the insulin signal transduction pathway.

Myrcia multiflora (Lam.) D.C., Myrtaceae

Synonym: *Aubmyrcia salicifolia* O. Berg.

Local Names: Pedra hume caá, pedra-ume-caá, pedra hume, insulina vegetal, etc. In Brazil, the common name *pedra hume caá* refers to three species of *Myrcia* plants (*Myrcia salicifolia*, *Myrcia uniflorus*, and *Myrcia sphaerocarpa*).

M. *multiflora* has small, green leaves and large, orange-red flowers. Anatomically, the leaf is hypostomatic, with compact and dorsiventral mesophyll, containing three layers of palisade parenchyma (Donato and Morretes 2011). It is one of more than 150 species of *Myrcia* indigenous to tropical South America and the West Indies. It has a deep-rooted traditional use to treat diabetes. A simple leaf tea of this plant is a very popular natural remedy for diabetes throughout South America. It is locally known as "vegetable insulin." This plant is used by indigenous tribes for diabetes, diarrhea, and dysentery. The Taiwanos tribe considers the leaves to be an astringent and use it for persistent diarrhea. It is also used for hypertension, enteritis, hemorrhages, and mouth ulcers.

Antidiabetes: Flavonoids such as quercitrin and myricitrin isolated from this plant are potent inhibitors of aldose reductase. The inhibitory activity is of the noncompetitive type. In addition, quercitrin has effectively blocked polyol accumulation in intact rat lenses incubated in medium containing high concentration of sugars (Varma et al. 1975). The methanol extract and ethyl acetate-soluble portion from the leaves of *M. multiflora* were found to show inhibitory activities on aldose reductase and α-glucosidase. These extracts reduced the increase of serum glucose level in sucrose-loaded rats and in alloxan-induced diabetic mice. From the ethyl acetate-soluble portion, new flavanone glucosides, myrciacitrins I and II, and new acetophenone glucosides, myrciaphenones A and B, were isolated together with several known compounds such as five flavonol glycosides, myricitrin, mearnsitrin, quercitrin, desmanthin-1, and guaijaverin. The principal components of this natural medicine, including new glucosides, myrciacitrin I and myrciaphenone B, were found to show potent inhibitory activities on aldose reductase and α-glucosidase (Yoshikawa et al. 1998). The plant is reported to obtain anti-DM compounds such as gallic acid and β-amyrin. Following the characterization of myrciacitrins I and II and myrciaphenones A and B, three new flavanone glucosides, myrciacitrins III, IV, and V, were isolated from the leaves of *M. multiflora* (Matsuda et al. 2002).

Other Activities: The 2′,4′,6′-trihydroxyacetophenone isolated from *M. multiflora* has antiobesity and mixed hypolipidemic effects with the reduction of intestinal absorption of the lipid. Phloroacetophenone from this plant showed potent hepatoprotective effect against CCl_4-induced hepatic injury in mice. Other reported activities include hypotensive activity (lowered blood pressure) and appetite suppressant action (Ferreira et al. 2010, 2011).

Phytochemical analysis showed high content of flavonoids, flavonols, and flavanones. Flavanone glucosides isolated from this plant were named myrciacitrins I and II; the new acetophenone glucosides were named myrciaphenones A and B. Other flavonoids found in this plant include quercitrin, myricitrin, guaijaverin, and desmanthin. The main plant chemicals documented in *M. salicifolia* include β-amyrin, catechin, desmanthin, gallic acid, ginkgoic acid, and guaijaverin.

Myrcia uniflora Barb. Rodr., Myrtaceae

Myrcia uniflora (hot water extract) showed antidiabetic activity against streptozotocin-induced diabetes in rats (Pepato et al. 1993). The extract (7.5 mg/kg twice daily for 3 weeks, p.o.) reduced the hyperglycemia, polyphagia, polydipsia, urine volume, and urinary excretion of glucose in diabetic rats fed a normal balanced diet; the extract administration has no effect on body weight, the weight of epididymal and retroperitoneal adipose tissue, and the concentration of insulin. The intestinal absorption of glucose measured with a perfusion technique *in situ* was inhibited by the extract (7.5 mg/mL of perfusion solution) (Pepato et al. 1993). The mechanism of action of the extract is not clear. The active principle involved is not identified. However, in a randomized clinical trial, infusions of leaves of *Myrcia uniflora* (3g/day for 56 days) had no hypoglycemic effect on type 2 DM patients (Russo et al. 1990).

Myrciaria dubia Mc Vaugh, Myrtaceae

Synonym: *Eugenia divaricata* Benth

Myrciaria dubia, commonly known as camu camu, is a small medicinal tree. Its fruit is used in traditional medicine in Peru and elsewhere. Ellagic acid and its two derivatives, 4-*O*-methylellagic acid and 4-(α-rhamnopyranosyl)ellagic acid, were isolated from *M. dubia*. These compounds inhibited aldose reductase. 4-(α-rhamnopyranosyl)ellagic acid showed the strongest inhibition against human recombinant aldose reductase and rat lens aldose reductase. The inhibitory activity of this compound against human recombinant aldose reductase (IC$_{50}$ value = 4.1×10^{-8} M) was 60 times more than that of quercetin. The type of inhibition was uncompetitive (Ueda et al. 2004).

Myristica fragrans Houtt., Myristicaceae, nutmeg

The petroleum ether extract of *Myristica fragrans* decreased blood glucose levels in normal, glucose-fed, and alloxan-induced diabetic rats. The hypoglycemic effect may be due to the potentiation of insulin release from β-cells. Oral administration of the extract also suppressed the increase in glucose level induced by glucose loading. This effect might be due to the decrease in the rate of intestinal glucose absorption or potentiation of pancreatic secretions or increase of glucose uptake. The extract increased body weight in diabetic animals, which might be due to increased insulin secretion and better glycemic control (Somani and Singhai 2008). Administration of the ethanol extract of *M. fragrans* fruits to streptozotocin-induced diabetic rats also resulted in moderate lowering in blood glucose. This effect was found from 3 to 7 h posttreatment; peak lowering in blood glucose was observed at 7 h posttreatment (Ahmad et al. 2008). Nutmeg has shown insulin-like activity *in vitro* (Broadhurst et al. 2000). Inhibitory effects on protein tyrosine phosphate 1B, involved in insulin cellular signaling, have been demonstrated (Yang et al. 2006). Bioassay-guided fractionation of the methanol extract of *M. fragrans* led to the isolation of meso-dihydroguaiaretic acid and otobaphenol. Both compounds inhibited PTP1B with IC$_{50}$ values of 19.6 and 48.9 μmol/L, respectively (Jiany et al. 2012). Besides, the methanol extract of nutmeg seeds showed significant PPAR-γ agonist activity in cell-based *in vitro* assays. Oral administration of the extract to type 2 diabetic rats resulted in dose-dependent glucose-lowering activity. Thus, the seed might be a potential anti-DM agent (Lestari et al. 2012). The gap to be filled includes detailed toxicity studies and search for more active principles.

Myrtus communis L., Myrtaceae

Myrtus communis is commonly known as common myrtle; it is one of the important medicinal plants being used in Unani system of medicine. Its berries, leaves, and essential oil are used to treat diseases like gastric ulcer, diarrhea, vomiting, rheumatism, hemorrhages, diabetes, and leukorrhea.

Antidiabetes Activity: Administration of this plant extract exhibited antihyperglycemic activity in streptozotocin-induced diabetic mice. The extract had no effect on the glucose levels of normal mice. These observations suggest the possible utility of this plant in the treatment of DM. The leaves and the volatile oil obtained from the leaves are used in traditional medicine to lower the glucose level in type 2 DM patients. Experiments on rabbits have shown that the leaf oil possesses hypoglycemic activity in normal and alloxan-induced diabetic rabbits. Further, the oil exhibited mild TG-lowering activity and reduced intestinal absorption of glucose (Dineel et al. 2007; review: Sumbul et al. 2011). Administration of phenolic compounds (800 mg/kg) from *M. communis* to streptozotocin-induced diabetic rats resulted in marked antihyperglycemic response and restoration of other relevant biochemical parameters to baseline levels in normal control rats (Benkhayal et al. 2009). Other reported activities of this plant include antimicrobial, antifungal, anti-inflammatory, antioxidant, and LDL-C decrease. A wide range of biologically active compounds like tannins, flavonoids, coumarins, essential oil, foxed oil, and antioxidants are present in these plants (review: Sumbul et al. 2011).

Nasturtium officinale R. Br., Cruciferae

Common Name: Watercress

Nasturtium officinale is a traditional medicinal herb used to treat DM in Morocco and Egypt (Haddad et al. 2005). Ethyl acetate extracts (100 mg/kg daily for 2 months) of *N. officinale* decreased blood glucose levels in streptozotocin-induced diabetic rats (Hoseini et al. 2009).

Nauclea latifolia S.M., Rubiaceae

The butanol extract of *Nauclea latifolia* (root and stem) decreased hyperglycemia in pregnant streptozotocin-induced diabetic rats. The plant extract also showed substantial antioxidant activity and T-cell proliferation inhibition activity (Yessoufou et al. 2013).

Nelumbo nucifera Gaertn., Nymphaeaceae

Common/Local Names: Sacred lotus, Indian lotus, ambuja, ambuj, ambal, tamara, etc.

Description: *Nelumbo nucifera* is a large, aquatic herb with creeping branches, stem producing adventitious roots from the nodes, and membranous, orbicular, exactly pellate, glaucous, glabrous and entire leaves. The petioles are long and vary in length, bearing small, distant prickles. Large, solitary flowers arise from the nodes of the stem and are rose or white in color. The torus is spongy, bearing long, ovoid, glabrous ripened carpels.

Distribution: *Nelumbo nucifera* is present throughout the warmer parts of India.

Traditional Medicinal Uses: *N. nucifera* is a sedative, astringent, aphrodisiac, pungent, diuretic, demulcent, and refrigerant and used in the treatment of fever, biliousness, blood complaints, vomiting, leprosy, piles, strangury, cough, skin eruptions, eye disease, diarrhea, inflammation, ulcers, mouth sores, dysentery, snake bite, chest pain, spermatorrhea, leukoderma, small pox, bronchitis, internal injury, menorrhagia, leukorrhea, cholera, cutaneous diseases, hemoptysis, and sclerosis (Kirtikar and Basu 1975).

Antidiabetes: The ethanol extract of the rhizomes markedly reduced the blood sugar level of normal, glucose-fed hyperglycemic, and streptozotocin-induced diabetic rats (Mukherjee et al. 1997). In another study, the effect of *N. nucifera* rhizome and flower extracts on serum glucose level in normal and streptozotocin-induced diabetic rats was evaluated. In streptozotocin-induced diabetic rats, the various extracts showed significant antidiabetic property. *N. nucifera* rhizome and flower extracts are promising antidiabetic agents (Rakesh et al. 2011). The leaves also showed anti-DM activity. Oral administration of the methanol extract of *N. nucifera* leaves for 15 days to streptozotocin-induced diabetic rats resulted in highly significant antidiabetic activity. The treatment also improved body weight of diabetic rats. The extract did not show any toxic symptoms up to a dose of 2 g/kg in acute oral toxicity study in rats. There is a need for further investigation to isolate and identify the active principles involved (Venkata et al. 2013).

Another report shed some light on its mechanism of action. The *in vitro* and *in vivo* effects of the lotus leaf methanol extract on insulin secretion and hyperglycemia were investigated. The methanol extracts increased insulin secretion from β-cells (HIT-T15) and human islets. It enhanced the intracellular calcium levels in β-cells. The extract could also enhance phosphorylation of extracellular signal-regulated protein kinases (ERK)1/2 and PKC, which could be reversed by a PKC inhibitor. *In vivo* studies showed that the extract possesses the ability to regulate blood glucose levels in fasted normal mice and HFD-induced diabetic mice. Furthermore, the *in vitro* and *in vivo* effects of the active constituents of the extract, quercetin and catechin, on glucose-induced insulin secretion and blood glucose regulation were evaluated. Quercetin did not affect insulin secretion, but catechin significantly and dose-dependently enhanced insulin secretion. Orally administered catechin significantly reversed the glucose intolerance in HFD-induced diabetic mice. These findings suggest that the *N. nucifera* methanol extract and its active constituent catechin are useful in the control of hyperglycemia in NIDDM patients through their action as insulin secretagogues (Huang et al. 2011).

Other Activities: Oligomeric procyanidins from the seed pod as well as the methanol extract of the leaf of *N. nucifera* show antioxidative activity (Wu et al. 2003; Ling et al. 2005). Further, antioxidant flavonoids have been isolated from the stamen of the plant (Jung et al. 2003).

A bisbenzylisoquinoline alkaloid from the seed embryo inhibited bleomycin-induced pulmonary fibrosis in mice (Xiao et al. 2005). The extracts from the plant leaves suppressed cell cycle progression, cytokine gene expression, and cell proliferation in human peripheral blood mononuclear cells (Liu et al. 2004). Anti-HIV principles have been isolated from the leaves of this plant (Kashiwada et al. 2005). The ethanol extract from the seeds showed hepatoprotective and free radical scavenging effects (Sohn et al. 2003). Neferine (Nef), a dibenzyl isoquinoline alkaloid isolated from *N. nucifera*, exhibited hypotensive, antiarrhythmic, and anti-platelet aggregation activities (Yu and Hu 1997). Linesinine, an alkaloid extracted from the green seed embryo, has an effect on cardiac function (Wang et al. 1993). The ethanol extract of the stalks of this plant showed antipyretic action (Sinha et al. 2000). The rhizome of the plant exhibited antidiarrheal and anti-inflammatory activities (Talukder and Nessa 1998).

Phytochemicals: Oligomeric procyanidins, bisbenzylisoquinoline alkaloid, liensinine and its analogs, flavonoids β-sitosterol glucopyranoside, neferine, linesinine (alkaloids) were isolated (Yu and Hu 1997; Kashiwada et al. 2005; Ling et al. 2005; Xiao et al. 2005).

Nepeta ciliaris L., Lamiaceae
Screening of Indian medicinal plants resulted in detection of hypoglycemic activity of *Nepeta ciliaris* extract in rats (Abraham et al. 1986).

Nepeta cataria L., Lamiaceae
The anti-DM activity of *Nepeta cataria* was demonstrated by *in vitro* and *in vivo* studies.

Extracts of *N. cataria* exhibited antioxidant and carbohydrate metabolizing enzyme (α-amylase, α-glucosidase, and β-galactosidase) inhibitory activity *in vitro*. In streptozotocin-induced diabetic rats, the crude methanol, chloroform, and petroleum ether extracts reduced blood glucose, improved lipid profiles, reduced oxidative stress, and improved liver and kidney functions (as judged from the levels of serum transaminases, ALP, urea, and creatinine). Histological studies showed normalization of the histoarchitecture of the liver and pancreas in the extract-treated diabetic mice (Aly et al. 2010).

Nephelium lappaceum L., Sapindaceae
Geraniin isolated from the rind waste of *Nephelium lappaceum* showed antihyperglycemic activity in experimental animal (Palanisamy et al. 2011).

Nerium indicum Mill., Apocynaceae

Synonyms: *N. oleander* L. and *N. odorum* Aiton.
Oral administration of *Nerium oleander* (*N. indicum*) extract (250 mg/kg) daily for 4 weeks to streptozotocin-induced diabetic rats resulted in improvement of hypoinsulinemia and hyperglycemia in diabetic rats compared to diabetic control rats. However, glibenclamide was found to be better than the extract. Further, the treatment decreased the activities of serum AST, ALT, and ALP (markers of liver function) in diabetic rats (Mwafy and Yassin 2011).

Nervilia plicata Schltr., Orchidaceae
Nervilia plicata has long been used in antidiabetic medicinal preparations of traditional healers of Wayanad (Kerala), India. Administration of the alcohol extract (5 mg/kg) of *N. plicata* stem resulted in marked decrease in blood glucose levels of streptozotocin-induced type 2 diabetic rats. Damages caused to the kidney tissue were negligible or not seen in the treated diabetic rats. Further, the treatment lowered serum urea and creatinine levels; it also reduced LPO products in the kidney and pancreas of diabetic rats. The regenerative potential of the plant extract on the kidney affected by type 2 DM was shown (Kumar and Janardhana 2011).

Neurolaena lobata R. Br., Asteraceae
Neurolaena lobata is an ethnomedicinal shrub. The extract of *N. lobata* showed hypoglycemic activity in rats (Gupta et al. 1984).

Newbouldia laevis P. Beauv., Bignoniaceae

The water extract of *Newbouldia laevis* root significantly reduced serum glucose levels in both alloxan-induced diabetic rats and normal rats at the two doses administered (500 and 1000 mg/kg). In both the normal and diabetic rats, maximal blood glucose-lowering effect following extract administration was observed after 4 and 6 h, respectively. Phytochemical analysis of the extract revealed a strong presence of alkaloids, flavonoids, and saponins (Okonkwo and Okoye 2009). In another study, the ethanol leaf extract of *N. laevis* (100, 200, and 400 mg/kg, p.o., for 14 days) reduced blood glucose levels markedly in alloxan-induced diabetic rats. Acute toxicity evaluation revealed that for the oral route, mortality was at 8 g/kg, while LD_{50} was about 6 g/kg, indicating the high safety status of the extract (Owolabi et al. 2011). The ethanol extract of the flower of *N. laevis* (250–1000 mg/kg, i.p.) also showed blood glucose-lowering activity in streptozotocin-induced diabetic rats. After 4, 8, and 24 h, but not after 2 h, of administration, there was a significant decrease in the blood glucose levels (Tanko et al. 2008a). In another study, the effects of the ethanol extract of *N. laevis* leaves on LPO and glycosylation of hemoglobin in streptozotocin-induced diabetic rats were evaluated. Oral administration of the extract (300 and 500 mg/kg) to diabetic rats for 28 days resulted in a dose-dependent marked reduction of fasting blood glucose levels. The treatment also improved oral glucose tolerance. The percentage of total glycated hemoglobulin significantly reduced after an 8-week treatment. Further, MDA levels in the kidney, liver, and pancreas decreased in the treated diabetic rats. The effects of the extract were comparable to those of glibenclamide (Timothy et al. 2013). It is of interest to study the mechanisms of action (which is delayed at least by 2 h) and active principle(s) involved in the antidiabetic activity of these plant parts.

Nigella sativa L., Ranunculaceae

Nigella sativa is commonly known as kalonji, black cumin, or simply Nigella. It is a small, annual flowering plant. The seeds of *N. sativa* have been used for centuries in traditional medicine to treat various ailments. Thymoquinone isolated from the seeds of *N. sativa* showed anti-DM (anti-β-cell damage), antioxidative, and neuroprotective activities in streptozotocin–nicotinamide-induced diabetic rats (Sankaranarayanan and Pari 2011). The hypoglycemic, antioxidant, hypolipidemic, and hypotensive effects of *N. sativa* have been reported in experimental animals (review: Ghorbani et al. 2013).

Clinical Studies: Administration of *N. sativa* oil (2–5 mL thrice a day for 6 weeks) to patients with metabolic syndrome decreased fasting blood glucose levels and LDL and increased HDL levels. In another study, *N. sativa* seeds (1, 2, and 3 g/day) were included with the anti-DM drugs of 94 type 2 DM patients. After 6 weeks, a significant reduction occurred in fasting and PPG levels and HbA1c levels in diabetic patients. In another study also, *N. sativa* seed administration decreased blood glucose levels, serum lipids, and blood pressure in type 2 DM patients (review: Ghorbani et al. 2013).

Nitraria retusa (Forssk.) Asch., Nitrariaceae

Nitraria retusa is a salt-tolerant shrub or bush and an edible halophyte. The aerial parts of this plant are used as an anti-DM medicine in Egypt. *N. retusa* extract showed hypoglycemic effects in rats (Shabana et al. 1990). In a recent study, the ethanol extract of this plant exhibited potential antiobesity effects in *db/db* model diabetic mice and may relieve obesity-related symptoms including hyperlipidemia through modulating the lipolysis–lipogenesis balance (Zar-Kalai et al. 2014).

Notopterygium incisum C.T. Ting ex H.T. Chang, Apiaceae

Notopterygium incisum is used in traditional medicine for the treatment of rheumatism, cold, and headache. Polyacetylenes, isolated from root and rhizomes of *N. incisum*, showed activities similar to selective partial agonists of PPAR-γ (review: Wang et al. 2014). In view of these findings, the plant is likely to have anti-DM activity.

Nyctanthes arbor-tristis L., Oleaceae

The methanol extracts of *Nyctanthes arbor-tristis* root exhibited hypoglycemic activity in alloxan-induced diabetic rats. Oral administration of the extract (500 mg/kg) to diabetic rats resulted in marked reduction in blood glucose levels (Sharma et al. 2011c).

Nymphaea nouchali Burm.f., Nymphaeaceae

Synonym: *Nymphaea stellata* Willd.

Common Name: Star lotus

The hydroalcoholic extract of *Nymphaea nouchali* seed exhibited antihyperglycemic activity in streptozotocin-induced diabetic rats. Oral administration of the extract (100 and 200 mg/kg) of the seed for 21 days to diabetic rats resulted in almost normalization of blood glucose levels. Further, the treatment improved the lipid profiles and the serum markers of kidney and liver function (Parimala and Shoba 2014).

Nymphaea pubescens Willd., Nymphaeaceae

The ethanol extract of this plant showed β-cell regeneration potential in alloxan-induced diabetic rats (Sreenathkumar and Arcot 2010). In another study, the ethanol extract of the *N. pubescens* tuber (20,000 mg/kg daily for 14 days) exhibited a blood glucose-lowering effect, hypolipidemic effect, and antioxidant activity in alloxan-induced diabetic rats. The extract also caused significant increase in plasma insulin levels in diabetic rats (Shajeela et al. 2012).

Nymphaea stellata OW, Nymphaeaceae

Nymphaea stellata is known as dwarf lily. It is used in traditional medicine for a variety of illnesses; it is also grown as an aquarium plant.

The defatted ethanol extract of *N. stellata* leaf (100 and 200 mg/kg, p.o.) significantly and dose-dependently reduced hyperglycemia in alloxan-induced diabetic rats. Moreover, it decreased the levels of cholesterol and TGs in diabetic rats. On the contrary, no effect was seen in normal rats, in both the glucose and lipid levels (Dhanabal et al. 2007). In another study, the hydroalcoholic extract of flowers of *N. stellata* (200, 300, and 400 mg/kg, p.o.) was evaluated in normoglycemic and alloxan-induced diabetic rats. It showed no hypoglycemic effect in normoglycemic animals but showed antihyperglycemic activity in diabetic rats. It also improved oral glucose tolerance in diabetic rats. A 300 mg/kg dose of the extract showed significant blood glucose level reduction (45%) 4 h after administration (Rajagopal et al. 2008). The mechanisms of the antihyperglycemic action are not clear. One of the mechanisms may be due to the potentiation of pancreatic secretion of insulin, which was evident from the increased level of insulin in treated rats. In another study, administration of the hydroalcoholic extract of the flower resulted in an improvement in the lipid profile in diabetic rats. The treatment reduced TC, TG, LDL, and VLDL levels in the serum and increased the HDL level. Serum phospholipids were elevated, whereas the phospholipids in the liver and kidney were decreased. Oral administration of the *N. stellata* flower extract for 30 consecutive days to diabetic rats decreased their food consumption and improved body weight (Rajagopal and Sasikala 2008a,b). Nymphayol (25, 26-dinorcholest-5-en-3b-ol), a new sterol isolated from the chloroform extract of this plant flower, has been reported for its antidiabetic activity at 20 mg/kg in streptozotocin-induced diabetic rats. Oral administration of nymphayol for 45 days significantly restored the plasma glucose levels and increased the plasma insulin levels to near normal in streptozotocin-induced diabetic rats. Light microscopy and immunocytochemical staining of nymphayol-treated diabetic pancreas revealed an increased number of insulin-positive β-cells.

Other Activities: Nymphayol enhanced the antioxidant defense against the ROS produced under hyperglycemic conditions (Subashbabu et al. 2009). *N. stellata* extract also exhibited hepatoprotective property in carbon tetrachloride toxicity in rat, suggesting an improvement in liver function in treated diabetic rats (Bhandarkar and Khan 2004).

Nypa fruticans Wurmb., Arecaceae

Nypa fruticans is a mangrove palm traditionally used to treat different ailments in Bangladesh. The methanol extract of the leaf and stem of *N. fruticans* showed antihyperglycemic activity in orally glucose-loaded hyperglycemic mice. Maximum antihyperglycemic activity was observed at 500 mg/kg, which was more than that for 10 mg/kg glibenclamide (Reza et al. 2011). The extract also had antinociceptive activity (Reza et al. 2011).

Ochradenus baccatus Del., Resdaceae

Aerial parts of *Ochradenus baccatus* are used in traditional medicine to treat diabetes in Egypt (Haddad et al. 2005). In a screening study, the plant extract produced slow hypoglycemic activity and the effect appeared 3 h after administration (Shabana et al. 1990).

Ocimum americanum L., Labiatae, syn: *O. canum* Sims

In Ghana, *Ocimum canum* is used in ethnomedicine to manage DM. Administration of the water extract of *O. canum* (whole plant) resulted in decrease in blood glucose and body weight in both genetically type 2 diabetic mice and nondiabetic mice. Under *in vitro* conditions, the extract (30 µg/mL) enhanced insulin release from isolated rat pancreatic β-cells. Insulin release was dose dependent and glucose concentration dependent (Nyarko et al. 2002a). In another study, the water extract of *O. canum* decreased fasting blood glucose levels and increased antiatherogenic lipid levels in diabetic mice. The extract treatment (p.o., daily for 13 weeks) to diabetic mice decreased body weight, blood glucose, serum total cholesterol, and LDL-C, while serum HDL-C increased. Further, the extract decreased oxidative stress in diabetic mice (Nyarko et al. 2002b).

Ocimum basilicum L., Labiatae

Synonym: *Ocimum minimum* L.

After a single oral administration, the water extract of the whole *O. basilicum* plant showed hypoglycemic activity in normal as well as streptozotocin-induced diabetic rats. Repeated oral administration of the extract daily for 15 days led to a marked reduction in blood glucose levels in diabetic rats and a moderate reduction in normal rats. Total plasma cholesterol and TG levels were reduced after repeated administration in diabetic rats. However, plasma insulin levels and body weight remained unchanged in the treated normal as well as diabetic rats compared to untreated respective control animals (Zeggwagh et al. 2007). The mechanisms of actions and the active principles involved remain to be studied.

Ocimum gratissimum L., Lamiaceae/Labiatae

Synonyms: *Ocimum suave* Willd., *Ocimum viride* Willd.

Ursolic acid isolated from the leaves of *Ocimum gratissimum* (10 mg/kg) inhibited formation of AGEs in diabetic rats. Under *in vitro* conditions, it inhibited aldose reductase in rat lens homogenate (Rao et al. 2013). Chicoric acid (phenolic compound) isolated from this plant leaf reduced the glycemic level of streptozotocin-induced diabetic mice at 2 mg/kg level (Casanova et al. 2014).

Ocimum sanctum L., Labiatae

This plant species is also known as *O. tenuiflorum* L. (Figure 2.70).

Common and Local Names: Holy basil, tulasi, sacred tulasi, ajaka, baranda, kalatulsi, karuthulasi, krishnathulasi, etc.

Description: *Ocimum sanctum* is a strongly scented, herbaceous plant with soft, patently haired leaves that are opposite, exstipulate, oblong, obtuse or acute, entire or subserrate. Small flowers are borne on a slender raceme. The fruiting calyx contains pale red-brown, slightly compressed nutlets. There are two different varieties known as Krishna tulasi (light brown or violet in color) and Rama tulasi (green) occurring in India (Figure 2.70).

Distribution: *O. sanctum* is present throughout India in the Himalayas up to 6000 ft and also in Ceylon, West Asia, Australia, Pacific, Arabia, and Malay Island.

Traditional Medicinal Uses: It is pungent, bitter, stomachic, heating, cholagogue, anthelmintic, alexiteric, antiperiodic, antipyretic, and expectorant; it is used to treat leukoderma, strangury, bronchitis, vomiting, lumbago, asthma, hiccup, eye diseases, ear pain, body pains, hepatic infection, ringworm attack, urogenital infections, insect attack, mosquito bite, snake bite, malarial fever, cholera, collapse, collyrium, common cold, conjunctivitis, cough, dog bite, dropsy, earache, gastrosis, hemiplegia, nausea, parturition, puerperium, rabies, rheumatism, septicemia, and tumor. It is also used in tooth powder preparation and antiworm activity (Kirtikar and Basu 1975).

(a)   (b)

FIGURE 2.70 *Ocimum sanctum*. (a) Variant: Krishna. (b) Variant: Rama.

Pharmacology

Antidiabetes: The ethanol extract of *O. sanctum* leaves partially attenuated streptozotocin-induced alterations in blood glucose, liver glycogen content, and carbohydrate metabolism in rats (Chattopadhyay 1993; Vats et al. 2004). *O. sanctum* was found to be one of the most effective inhibitors of lens aldose reductase that has a role in sugar-induced cataract (Halder et al. 2003). The alcohol extract of the leaves of this plant lowered blood glucose levels in normal and alloxan-induced diabetic rats (Vats et al. 2002). The leaf powder supplementation resulted in significant reduction in the levels of blood sugar, serum lipids, and tissue lipids in diabetic rats (Rai et al. 1997). The ethanol extract of *O. sanction* (leaf) showed hypoglycemic activity in normal rats and anti-DM activity in alloxan-induced diabetic rats, possibly by increasing the levels of insulin (Khan et al. 2010). The aerial part of *O. sanctum* exhibited antidiabetes activity by ameliorating glucose and lipid parameters; a tetracyclic triterpenoid, the antidiabetic compound, was isolated from the aerial part of *O. sanctum* (Patil et al. 2011). *O. sanctum* seed oil exhibited antidiabetes and cholesterol-lowering properties (Gupta et al. 2006). There could be more anti-DM molecules in this plant. Intense research is needed to determine its utility as a therapeutic agent for DM.

Clinical Trial: In a clinical trial, the plant leaf administration to type 2 diabetic patients resulted in significant improvement in blood glucose and cholesterol levels (Agarwal et al. 1996). In another study, a significant decrease in diabetic symptoms (polydipsia, polyphagia, and tiredness) has been reported in 30 type 2 DM patients consuming (2 g/day) leaf powder of *O. sanctum* for 3 months. Dietary supplementation with *O. sanctum* leaf powder for 1 month resulted in decrease in the levels of fasting blood glucose, glycated protein, total cholesterol, LDL, VLDL, and TG in 27 type 2 diabetic patients (review: Ghorbani 2013).

Other Activities: It is an adaptogenic plant with several pharmacological properties. Treatment of animals with the ethanol extract of the plant prevented the changes in plasma level of corticosterone induced by exposure to both acute and chronic noise stress (Sembulingam et al. 1997). The water extract of the plant leaf has immunomodulatory and immunotherapeutic potential (Godhwani et al. 1988; Mukherjee et al. 2005). *Neisseria gonorrhoeae* clinical isolates and WHO strains are sensitive to extracts of *O. sanctum* (Shokeen et al. 2005). Chronic oral administration of the plant homogenate augmented cardiac endogenous antioxidants and prevented isoproterenol-induced myocardial necrosis in rats (Sood et al. 2006). Both water and alcohol extracts and their fractions from this plant have marked *in vitro* and *in vivo* anti-LPO activity (Geetha and Vasudevan 2004). The leaves showed hypolipidemic activity in normal albino rabbits (Sarkar et al. 1994) and contain the antioxidant ursolic acid (Balanehru and Nagarajan1991).

With its antioxidant property, *O. sanctum* may be useful in the treatment of cerebral reperfusion injury (Yanpallewar et al. 2004). The plant leaf extract protects rats from antitubercular drug-induced hepatotoxicity (Ubaid et al. 2003). The hepatoprotective activity of *O. sanctum* leaf extract against paracetamol-induced hepatic damage in rats has been reported (Chattopadhyay et al. 1992). The plant decreased the incidence of gastric ulcers and also enhanced the healing of ulcers induced by stress and chemicals (Dharmani et al. 2004). The antinociceptive action of the alcohol extract of the leaf has been reported (Khanna and Bhatia 2003). The water extract of leaves provided protection against mercury-induced toxicity in mice (Sharma et al. 2002). The plant has radioprotective, anticarcinogenic, and anti-oxidant properties (Uma Devi 2001). The plant extract prevented DMBA-induced hamster buccal pouch carcinogenesis (Karthikeyan et al. 1999). The seed oil has immunomodulatory potential (Mediratta et al. 2002). The hydroalcoholic extract of the plant has cardioprotective action. *O. sanctum* fixed oil produced a hypotensive effect in anesthetized dog (Singh et al. 2001). Orientin and vicenin, flavonoids from the plant, protected lymphocytes from radiation (Virinda and Uma Devi 2001). The essential oil of the plant and eugenol have anthelmintic activity (Asha et al. 2001). The plant has phenolic compounds with cyclo-oxygenase inhibitory and antioxidant properties (Kelm et al. 2000). The fixed oil of the plant has anti-inflammatory and antiulcer property (Singh 1998; Singh and Majumdar 1999). The leaf extract could regulate thyroid function in the male mouse (Panda and Kar 1998).

Toxicity: The benzene extract of the plant leaf extract has an antifertility effect in male rats (Ahmed et al. 2002).

Phytochemicals: Ursolic acid, eugenol, nerol, caryophyllene, terpinen-4-ol, decyl aldehyde, γ-selinene, camphene, α-pinene, cardinene, 1–8-cineole, limonene, methyl chavicol, β-caryophyllene, humulene, bornyl acetate, camphor, carvacrol, 1,8-cineole, chavicol, delinene, vitamin C, fatty acids, β-carotene, luteolin, apigenene, bergamotene, linalool, methyl chavicol, methyl cinnamate, phenyl propanoids, and trans-β-ocimene are present in the plant (review: Ajikumaran and Subramoniam 2005).

Olea europaea L. ssp. *europaea*, Oleaceae

Synonym: *Olea officinarum* Crantz (Figure 2.71)

Local Names: Common olive, European olive, olive, olive tree, small-fruited olive, wild olive

Distribution: Common olive (*Olea europaea*) is native to southern Europe, the Madeira Islands, the Canary Islands, Northern Africa, and Western Asia (Cyprus, Israel, Jordan, and Turkey). *O. europaea* has been naturalized in southern and eastern Australia, particularly in temperate regions. It is cultivated as a garden ornamental and also grown commercially for its fruit and for the production of olive oil.

Description: *O. europaea* is an evergreen tree usually growing 2–10 m tall with many branched upright stems. The leaves are oppositely arranged and elongated in shape (3–7 cm long and 0.8–2 cm wide) with

FIGURE 2.71 *Olea europaea* (twig with flower).

pointed tips; the leaves have glossy dark green upper surfaces and silvery or whitish undersides. The flowers are small and creamy white with four petals that are joined into a very short tube at the base. The fruits are purplish-black and oval-shaped (15–30 mm long and 6–20 mm wide). The fruit contains a single hard seed (10–15 mm long) surrounded by oily flesh.

Traditional Medicinal Uses: *O. europaea* leaves have been used for centuries in folk medicine to treat diabetes. The leaves are taken orally for stomach and intestinal diseases and used as mouth cleanser. Decoctions of the dried fruit and dried leaf are taken orally for diarrhea and for treating respiratory and urinary tract infections. The hot water extract of the fresh leaves is taken orally to treat hypertension and to induce diuresis. The seed oil is taken orally as a cholagogue, to remove gallstones, in nephritis associated with lead intoxication. To prevent hair loss, oil is applied every night on the scalp. The seed oil is taken orally as a laxative and applied externally as an emollient and pectoral. Decoction of dried leaves is taken orally for diabetes. Tincture of leaves is taken orally as a febrifuge. The fruit is applied externally to fractured limb and as a skin cleanser. The hot water extract of the dried plant is taken orally for bronchial asthma. Infusion of the fresh leaf is taken orally as an anti-inflammatory medicine (Khan et al. 2007).

Pharmacology

Antidiabetes: The oral administration of the olive leaf extract (0.1, 0.25, and 0.5 g/kg) for 14 days significantly decreased the serum glucose, total cholesterol, TGs, urea, uric acid, creatinine, AST, and ALT, while it increased the serum insulin in streptozotocin-induced diabetic rats, but not in normal rats. The antidiabetic effect of the extract was more effective than that observed with glibenclamide (Eidi et al. 2009b). In another study, oral administration of the water extract of *O. europaea* leaves (100 and 200 mg/kg) in streptozotocin-induced diabetic rats resulted in a significant decrease in cholesterol, TG, and serum glucose levels (range 55.6%–64.6%) compared to the metformin (62.7%). However, there was no significant effect on body weight and serum creatinine (El-Amin et al. 2013).

The olive leaf extract inhibited high glucose-induced neural damage and suppressed diabetes-induced thermal hyperalgesia. Incubation of cells with the olive leaf extract (200, 400, and 600 µg/mL) decreased cell damage in NGF-treated PC12 cells. The streptozotocin-induced diabetic rats developed neuropathic pain that was evident from decreased tail-flick latency (thermal hyperalgesia). Activated caspase 3 and the Bax/Bcl2 ratio were significantly increased in the spinal cord of diabetic animals. The extract treatment (300 and 500 mg/kg/day) ameliorated hyperalgesia, inhibited caspase 3 activation, and decreased the Bax/Bcl2 ratio. Furthermore, the leaf extract exhibited potent DPPH free radical scavenging capacity. The authors suggest that the mechanisms of these effects may be due, at least in part, to the reduction in neuronal apoptosis (Kaeidi et al. 2011).

Human Clinical Trial: In a randomized, double-blinded, placebo-controlled, crossover trial in New Zealand, effects of supplementation with olive leaf polyphenols (51.1 mg oleuropein, 9.7 mg hydroxytyrosol/day) on insulin action and cardiovascular risk factors in middle-aged overweight men were assessed.

The patients received capsules with olive leaf extract or placebo for 12 weeks. Supplementation with olive leaf polyphenols for 12 weeks significantly improved insulin sensitivity and pancreatic β-cell secretory capacity in overweight middle-aged men at risk of developing the metabolic syndrome.

The supplementation also led to increased fasting IL-6, IGFBP-1, and IGFBP-2 concentrations. There were, however, no effects on IL-8, TNF-α, ultrasensitive CRP, lipid profile, ambulatory blood pressure, body composition, carotid intima-media thickness, or liver function (de Bock et al. 2013). In spite of the promising anti-DM activity, sufficient investigations were not carried out to establish its therapeutic value including long-term toxicity evaluation.

Other Activities: Reported pharmacological activities include antimicrobial activities, antiviral activities, antioxidant activities, anti-inflammatory activities, gastroprotective activities, hypolipidemic effect, antihypertensive effect, and increasing thyroid hormone levels (Khan et al. 2007).

Toxicity: Olive leaves should be used with care, especially when using at higher doses for longer periods of time as it may have undesirable effects on the liver and kidneys (Omer et al. 2012).

Phytochemicals: Several phytoconstituents have been isolated and identified from different parts of the plant, including flavonoids, flavone glycosides, flavanones, iridoids, iridane glycosides, secoiridoids, secoiridoid glycosides, triterpenes, biophenols, benzoic acid derivatives, xylitol, sterols, isochromans, sugars, and a few other types of secondary metabolites. Phenolic compounds, flavonoids, secoiridoids, and secoiridoid glycosides are present in almost all the parts of *O. europaea* (Khan et al. 2007).

Olneya tesota A. Gray, Fabaceae

Common Names: Palo fierro, desert iron wood tree
Palo fierro seeds contain a potent α-amylase inhibitor, a lectin (Guzman-Partida et al. 2007).

Ophiopogon japonicus (L.f.) Ker-Gawl., Liliaceae

Common Names: Snake's weed, mondo grass, fountain plant, monkey grass, etc.
Ophiopogon japonicus is widely distributed in Southeast Asia; it is a traditional Chinese medicinal plant used to treat cardiovascular and chronic inflammatory diseases (Xio 2002). A hypoglycemic polysaccharide has been isolated and characterized from the root of *O. japonicas* (Chen et al. 2011c). The polysaccharide reduced blood glucose levels and increased the insulin levels in streptozotocin-induced diabetic rats. Further, it remediated destruction of pancreatic islets in diabetic rats compared to diabetic control animals (Chen et al. 2011).

Opuntia dillenii (Ker-Gawl.) Haw., Cactaceae

Synonyms: *Cactus dillenii* Ker-Gawl., *Opuntia stricta* (Haw.) Haw. var. *dillenii* (Ker-Gawl.) L. Benson. *Opuntia dillenii* (Ker-Gawl) Haw. was an independent species; however, due to its close similarity, it is considered a variety of *Opuntia stricta* (Haw.) Haw. (*Opuntia stricta* (Haw.) Haw. var. *dillenii* (Ker-Gawl.) L. Benson). But the Flora of North America (since 2003) accepts only the species *O. stricta* (Haw.) Haw., without subspecies (Bohn 2008). Specialists differ in their opinion. Publications, with the name *Opuntia dillenii* (Ker-Gawl.) Haw, are included under this name in this book.

O. dillenii fruit is used in folk medicine as an antidiabetic agent. The fruit is a rich source of fiber, carbohydrates, and vitamins B1, B2, and C, in addition to the minerals, Fe, Zn, Cu, Cr, Mn, Ca, and Mg. Oral administration of *O. dillenii* fruit juice had no effect on normal rats. In streptozotocin-induced diabetic rats, administration of the fruit juice improved body weight gain and lipid profile and significantly reduced blood glucose and MDA levels as compared with nontreated diabetic group. Histopathological investigation of the pancreatic tissue of diabetic rats showed the presence of necrosis, edema, and congested blood vessels in the islets of Langerhans cells. The fruit juice treatment suppressed these changes; the majority of the cells tend to be normal. It could be concluded that *O. dillenii* fruit juice has a potent hypoglycemic activity. This effect may be attributed to its antioxidant activity. Therefore, it could be recommended that *O. dillenii* should be ingested as a fresh fruit to diabetic and hypercholesterolemic patients besides the usual therapy (Abdulla 2008). In another study, three kinds of *O. dillenii* polysaccharides were isolated. One of the polysaccharides (daily oral administration for 3 weeks) significantly increased the body weights, hepatic glycogen levels, HDL-C levels, and the hepatic superoxide dismutase and GSH peroxidase activity in streptozotocin-induced diabetic mice. However, the treatment did not significantly increase insulin levels in diabetic mice (Zhao et al. 2011a,b). *Opuntia* fruits may cause reproductive toxicity (Ramyashree et al. 2013).

Opuntia ficus-indica (L.) Mill., Cactaceae

Synonym: *Opuntia vulgaris* Mill.
Common names of *Opuntia ficus-indica* include Indian fig and tuna cactus. Cladodes of *O. ficus-indica* are recommended for their therapeutic properties; the fruit and stem are edible and used to prepare value-added products such as jam, pickle, and squash (review: El-Mostafa et al. 2014). The known pharmacological properties of this plant include hypoglycemic, antioxidant, anti-inflammatory, neuroprotective, and antimicrobial properties (El-Mostafa et al. 2014).

In a study, cladodes of three maturity stages, from the same crop and location, were evaluated for some of their physicochemical and nutritional characteristics and antidiabetic properties. The flours of small and medium cladodes (SCF and MCF, respectively) had higher contents of dietary fiber, water absorption, swelling, and viscosity compared to those of the large cladode flour (LCF). Streptozotocin-induced diabetic rats treated with MCF and SCF (50 mg/kg) showed 46.0% and 23.6% reduction, respectively, of PPG levels compared to the diabetic control rats; LCF had no significant effect. *In vitro*, glucose diffusion tests showed similar ranking by the two former samples, whereas the latter was close to the control. Cladode maturity stages showed different fiber content and produced suspensions with differences in viscosity, which may affect *in vitro* and *in vivo* glucose responses (Nunez-Lopez et al. 2013). Two polysaccharides isolated from *O. ficus-indica* (and *O. streptacantha*) showed hypoglycemic activity in glucose-loaded mice (Alarcon-Aguilar et al. 2003). *O. ficus-indica* fruit juice (5 mL/rat daily, once, twice, or thrice, for 5 weeks) administration to alloxan-induced diabetic rats could restore changes in blood glucose, cholesterol, creatine, urea, AST, ALT, MDA, and reduced GSH observed to their normal levels in diabetic rats (Hassan et al. 2012a). Further, this plant contains several compounds known to have beneficial effects in diabetic animals. The fruit peel contains bioactive flavonoid derivatives such as kaempferol and quercetin, the contents of which are 0.22 and 4.32 mg/100 g, respectively. The flower contains gallic acid, glycosides of kaempferol and quercetin; the pulp also contains kaempferol and quercetin; the cladode contains gallic acid, ferulic acid, and rutin (El-Mostafa et al. 2014).

It should be noted that *Opuntia* fruits may cause reproductive toxicity (Ramyashree et al. 2013).

Opuntia fuliginosa Griffiths, Cactaceae

Opuntia species have several important biological actions including anti-DM activity (review: Chauhan et al. 2010). The hypoglycemic activity of a purified extract from the stems of *Opuntia fuliginosa* was evaluated in streptozotocin-induced diabetic rats. Blood glucose and glycated hemoglobin levels were reduced to normal values by a combined treatment of insulin and *Opuntia* extract (1 mg/kg/day). When insulin was withdrawn from the combined treatment, the prickly pear extract alone maintained the normoglycemic state in diabetic rats. Although the mechanism of action is unknown, the magnitude of the glucose control by the small amount of *Opuntia* extract required (1 mg/kg/day) precludes a predominant role for dietary fiber (Gonzfilez et al. 1996).

Opuntia humifusa (Raf.) Raf., Cactaceae

Opuntia humifusa stem powder (150, 250, and 500 mg/kg) lowered blood glucose, total cholesterol, and LDL levels in streptozotocin-induced diabetic rats; the treatment increased HDL levels in diabetic rats. Besides, a significant increase in relative β-cell volume of the pancreas was observed in rats treated with 500 mg/kg of the stem powder, when compared with the untreated DM rats (Hahm et al. 2011). The serum glucose level, fasting insulin level, and homeostasis model assessment of insulin resistance were lowered in 5% *O. humifusa*-supplemented HFD sedentary group compared to the HFD sedentary control group. In addition, PPAR-γ protein expression in the *O. humifusa*-treated group was significantly higher than that of the untreated group (Kang et al. 2013).

Opuntia lindheimeri Engelm., Cactaceae (*Opuntia engelmannii* ssp. *lindheimeri* (Engelm))

Administration of *Opuntia lindheimeri* extract (0, 250, or 500 mg/kg, p.o.) to streptozotocin-induced diabetic pigs resulted in both a dose- and time-dependent decrease in blood glucose concentrations. The hypoglycemic effect of the extract was apparent within 1 h of administration, with maximal effects occurring 4 h after administration. However, blood glucose concentrations in normoglycemic pigs were not significantly and consistently lowered by the treatment (Laurenz et al. 2003).

Opuntia megacantha Salm-Dyck, Cactaceae

Opuntia megacantha leaf extracts reduced blood glucose levels in diabetic rats. Administration of the leaf extract (20 mg/kg, p.o., daily for 4 weeks) to streptozotocin-induced diabetic and nondiabetic rats resulted in marked decrease in blood glucose levels in both diabetic and normal rats. However, the leaf extracts increased plasma creatine and urea concentrations in diabetic and normal rats (Bwititi et al. 2000).

Further, the leaf extract increased urinary sodium output in streptozotocin-induced diabetic and non-diabetic rats resulting in low plasma concentration of sodium compared to untreated animals. The leaf extract did not alter plasma aldosterone levels in diabetic and nondiabetic rats (Bwititi et al. 2001). Caution is required in its use to control DM in view of its likely renal toxicity.

Opuntia robusta J.C. Wendl., Cactaceae

Opuntia robusta is a traditional medicinal plant. In a pilot study on nondiabetic and nonobese males (37–55 years) with hypercholesterolemia, modification of the diet with prickly pear edible pulp (250 g/day) for 8 weeks resulted in a decrease in blood glucose, insulin, total cholesterol, LDL, TG, apolipoprotein, fibrinogen, and uric acid while body weight and HDL level remained unchanged compared to the untreated control group (Wolfram et al. 2002).

Opuntia streptacantha Lemaire, Cactaceae

Closely related species used for diabetes in traditional medicine include *O. fuliginosa*, *O. lasiacantha* Pfeiffer, *O. velutina* Weber, and *O. macrocentra* Engelman (Figure 2.72).

Common/Local Names: Prickly pear cactus, nopal, chumbera (used generally to refer *Opuntia* sp.)

Description: This is a xerophytic plant with a large bushy cactus and a definite trunk. The flattened stems or cladodes are popularly known as pencas in Mexico. (This plant has been employed both as a medicine and a source of nourishment, since prehistoric times, and was traded by various ethnic groups in Mexico and other parts of tropical America, long before the arrival of the Europeans (Lozoya 1999).)

Distribution: This plant can be found throughout the dry regions of the Western Hemisphere.

Traditional Medicinal Uses (Medicinal Food): The sliced or diced tender young pads or "paddles" (cladodes) of some species of the genus *Opuntia* are commonly known as nopalitos, which have been a traditional vegetable in the Mexican diet for centuries and, more recently, a specialty vegetable in the United States. Usually, prickly pear cactus is consumed as a fresh or cooked green vegetable (Pimienta 1993). The cactus pads or stems are sliced, diced, and cooked (boiled or broiled) much like string beans and consumed as a salad or as part of a meal (Nobel 2002). Principally, the stems (pads or cladodes) are used as medicine or food. The roots are also sometimes used, especially as a source of dietary fiber. The fruit is very sweet but has an astringent action.

FIGURE 2.72 *Opuntia streptacantha.*

Pharmacology

Antidiabetes: The raw plant contains abundant mucilage, which is a complex carbohydrate that may delay the absorption of glucose. The cactus also contains fiber, which is known to delay glucose absorption (review: Lopez 2007). Several reports confirmed the hypoglycemic effect of high doses of *Opuntia* (*O. streptacantha*) preparations up to 3 h after its ingestion (Frati-Munari et al. 1989). Studies performed in Mexico with *O. streptacantha* in three different animal species (rabbits, rats, and dogs) induced hypoglycemic effects when orally administered to animals under induced states of moderate increase of blood sugar. In animals with normal blood glucose levels, as well as in pancreatectomized animals, the effect of the product was not detected (Frati-Munari et al. 1989). The traditional liquefied extract preparation of the cladode of *O. streptacantha* produced an antihyperglycemic effect when administered before a glucose challenge. Administration of the extract exerted glycemic control by blocking the hepatic glucose output, especially in the fasting state (Andrade-Cetto and Wiedenfeld 2011). Another study confirmed the antihyperglycemic effect of the *O. streptacantha* extract and suggests that this effect is not due to the action on α-glucosidases or related to the intestinal hydrolysis of disaccharides (Becerra-Jimenez and Andrade-Cetto 2012).

The hypoglycemic activity of the cactus (*O. lindheimeri*) was investigated in streptozotocin-induced diabetic pigs, employing an enteral route of administration. The results showed that the hypoglycemic activity of the cactus was evident 1 h after ingestion, reaching its maximum effect 4 h after ingestion (Laurenz et al. 2003).

The purified *O. fuliginosa* administration (1 mL, p.o., daily for 4 weeks) to diabetic rats resulted in normal blood glucose in treated diabetic rats, whereas untreated diabetic control animals exhibited hyperglycemia (Trejo-Gonzalez et al. 1996). In another study, polysaccharides of *O. ficus-indica* (POF) and *O. streptacantha* (POS) were isolated and evaluated for their hypoglycemic properties. When each of the two polysaccharides was injected intraperitoneally in healthy mice, no hypoglycemic activity was observed. However, oral administration of POF and a hypoglycemia-inducing agent resulted in a significant hypoglycemic effect in mice. The authors attribute this effect to a reduction in intestinal absorption of glucose. But POS produced a significant decrease in serum glucose levels in mice with subcutaneously induced hyperglycemia suggesting that the polysaccharide may be a hypoglycemic agent possibly working by increasing insulin sensitivity (Alarcon-Aguilar et al. 2003). The decrease in serum glucose levels observed upon ingestion of *Opuntia* may be attributable to both dietary fibers that inhibit absorption of glucose and specific hypoglycemic agents present in the *Opuntia* that may possibly sensitize insulin action (review: Lopez 2007).

The hypoglycemic activity of a purified extract from prickly pear cactus (*O. fuliginosa*) was evaluated on streptozotocin-induced diabetic rats. Blood glucose and glycated hemoglobin levels were reduced to normal values by a combined treatment of insulin and *Opuntia* extract. When insulin was withdrawn from the combined treatment, the prickly pear extract alone maintained a normoglycemic state in diabetic rats. The blood glucose response to administered glucose also showed that the rats receiving the combination treatment of insulin and *Opuntia* extract for 7 weeks followed by the *Opuntia* extract alone were capable of rapidly returning blood glucose to the levels of nondiabetic rats. Although the mechanism of action is unknown, the magnitude of the glucose control by the small amount of *Opuntia* extract required (1 mg/kg/day) precludes a predominant role for dietary fiber. These very encouraging results for diabetes control by the purified extract of this *Opuntia* make the need evident for clinical studies in humans (Trejo-Gonzalez et al. 1996).

Further studies are required to determine the role of polysaccharides and other active compounds, if any, in lowering blood glucose levels.

Clinical Trial: The methodology employed in the clinical studies carried out in Mexico and the United States is not always uniform or does not always meet the adequate standards for a clinical trial (Veronin and Ramirez 2002). In a study undertaken in Mexico, the daily intake of 30 *Opuntia* capsules by patients with DM had a discrete beneficial effect on glucose and cholesterol. However, this dose was considered impractical and, therefore, was not recommended in the management of DM (Frati et al. 1992). In another study, the hypoglycemic effect of nopal (*O. streptacantha*) was assessed in patients with NIDDM. The study showed that *O. streptacantha* (500 g boiled stem) has a hypoglycemic effect in patients with NIDDM. The mechanism of this effect is unknown, but an increased sensitivity to insulin was suggested (Frati-Munari et al. 1989).

The relationship between the doses of *O. streptacantha* and its acute hypoglycemic action was evaluated in type 2 diabetic patients. Four tests were performed to each patient with the intake of 0, 100, 300, and 500 g of broiled stems of *O. streptacantha*. Serum glucose was measured at 0, 60, 120, and 180 min. Maximal dose-dependent decrease of serum glucose was noticed at 180 min, with a mean of 2.3, 10.0, 30.1, and 46.7 mg/dL less than the basal value with 0, 100, 300, and 500 g, respectively (Frati et al. 1989). In order to know the extent of the hypoglycemic effect of the crude extracts of *O. streptacantha*, eight patients with type 2 DM were studied. Five tests were performed to each patient with the intake of (1) supernatant, (2) precipitate, (3) complete homogenate of 500 g of crude *O. streptacantha* stem, (4) 400 mL of water, and (5) 500 g of broiled *O. streptacantha* stems. Serum glucose levels were measured at 0, 30, 60, 120, and 180 min. Crude extracts did not cause a significant decrease of glycemia, and the results were similar to the water control test (P greater than 0.05). The intake of broiled stems caused a significant decrease of serum glucose level that reached about 2.67 mmol/L (48 mg/dL) lower than basal values at 180 min. Perhaps heating of *O. streptacantha* is necessary to obtain the hypoglycemic effect (Frati et al.1990).

To evaluate if the acute hypoglycemic effect of nopal that occurs in diabetic patients also appears in healthy individuals, 500 g of nopal stems was given orally to 14 healthy volunteers and to 14 NIDDM patients. The intake of nopal by the diabetic group was followed by a significant reduction of serum glucose and insulin concentration at 180 min. No significant changes were noticed in the healthy group as compared with the control. An acute hypoglycemic effect of nopal was observed in diabetic patients, but not in healthy subjects. The researchers conclude that the mechanisms of this effect differ from those exhibited by currently available hypoglycemic agents (Frati et al. 1991).

In a clinical trial involving 45 human subjects and 3 species of prickly pear cactus, *O. lasiacantha* Pfeiffer, *O. velutina* Weber, and *O. macrocentra* Engelman, the results showed no acute hypoglycemic effects on Hispanic type 2 diabetic patients. This study compared the possible hypoglycemic activity of these cacti to boiled zucchini and water. The study concluded that prickly pear cacti did possess a mild hypoglycemic effect, but not in a statistically significant manner. The authors did mention that ingestion of prickly pear cactus might help lower serum cholesterol levels and perhaps augment the patients' sensitivity to insulin, as well as improve glucose tolerance curves (Rayburn et al. 1998).

Activity-guided isolation of the active principle from the active preparation remains to be done. Variations in the efficacy of different species of prickly pear cactus need to be addressed in future studies.

Other Activities: In addition to its antidiabetes property, the prickly pear cactus has been shown to decrease hypercholesterolemia, optimize platelet function, and decrease oxidative stress in guinea pigs.

The ingestion of prickly pear cactus was shown to reverse the LDL receptor suppression induced by a hypercholesterolemic diet in guinea pigs. The influence of *Opuntia* was also examined in patients suffering from familial heterozygous isolated hypercholesterolemia. A daily consumption of 250 g of broiled edible pulp of prickly pear over 4 weeks decreased both total cholesterol and LDL-C (review: Lopez 2007).

Toxicity: *Opuntia* fruits and stems may cause reproductive toxicity (Ramyashree et al. 2013).

Phytochemicals: The active principles involved in this plant are currently unknown. Polysaccharides have been isolated; they may be responsible to some extent for the hypoglycemic activity.

Opuntia stricta (Haw.) Haw., Cactaceae

Synonym: *Opuntia inermis* (DC.) DC.

Opuntia stricta is also used as a food and medicine. (According to the "Flora of North America," *Opuntia dillenii* (Ker-Gawl) Haw. is a synonym of *Opuntia stricta* (Haw.) Haw.) The extracts of *O. stricta* showed strong inhibition of α-glucosidase activity as compared to the α-amylase inhibition *in vitro* (Kunyanga et al. 2014). Flavonoid glycosides isolated from the flowers of *O. stricta* exhibited antihyperglycemic activity in diabetic rats (Merina et al. 2011).

Origanum majorana L., Labiatae or Lamiaceae
A methanol extract of *Origanum majorana* leaves strongly inhibited rat intestinal α-glucosidase. Five 6-hydroxy flavonoids including two novel compounds were isolated from the extract as active principles and related compounds. One of them, 6-hydroxyapigenin (scutellarein), exhibited an IC_{50} value of 12 μM for sucrose hydrolysis by rat intestinal α-glucosidase (Kawabata et al. 2003). The methanol extract of the aerial parts (200 mg/kg) inhibited the formation of AGEs and reduced oxidative stress in streptozotocin-induced diabetic rats. Under *in vitro* conditions, it also inhibited AGE formation (Perez Gutierrez 2012).

Origanum vulgare L., Lamiaceae
Origanum vulgare is used as a culinary herb worldwide; it is also used in traditional medicine as emmenagogue, antispasmodic, carminative, expectorant, etc. The water extract of *O. vulgare* (leaf) exhibited antihyperglycemic activity in streptozotocin-induced diabetic rats without affecting insulin secretion (Eddouks et al. 2004b). Biochanin A, isolated from the leaves of this plant, activated PPAR-γ *in vitro*. However, *O. vulgare* extracts bind but do not transactivate PPAR-γ, and binding affinity differed among different oregano extracts. The extracts contain PPAR-γ antagonists, selective PPAR-γ modulators (e.g., naringenin and apigenin), and PPAR-γ agonists (e.g., biochanin A) (Mueller et al. 2008). Thus, it appears that the beneficial effect of *O. vulgare* to diabetic patients depends on the extract or fraction used (active principles). Further studies are required to specify the effect of active principles (individually and in various combinations) on diabetes.

Ornithogalum longibracteatum Jacq., Asparagaceae
Ornithogalum longibracteatum is used in South Africa in traditional medicine to treat DM. The water extracts of *O. longibracteatum* showed significant increase in glucose utilization activity *in vitro* in Chang liver cells. However, further *in vivo* studies are required to confirm its antidiabetic property (Huyssteen et al. 2011).

Orthosiphon aristatus (Blume) Miq., Lamiaceae
Synonyms: *Orthosiphon stamineus* Benth; *Clerodendranthus spicatus* (Thunb.) C. Y. Wu., *Ocimum grandiflorum* Blume
Orthosiphon stamineus (*Orthosiphon aristatus*) is a popular folk medicine widely used to treat many diseases including diabetes. Studies have shown that the subfraction of the chloroform extract of *O. stamineus* was able to inhibit the rise of blood glucose levels in a glucose tolerance test. Administration of this chloroform extract subfraction (1 g/kg twice daily for 14 days) to streptozotocin-induced diabetic rats resulted in significant reduction in blood glucose levels compared to pretreatment levels; however, the treatment did not significantly influence the serum insulin levels. *In vitro*, at a concentration of 2 mg/mL, the subfraction significantly increased the glucose uptake by the rat diaphragm muscle. The increase in glucose uptake was also shown when the muscle was incubated in a solution containing 1 IU/mL of insulin or 1 mg/mL of metformin. Furthermore, the subfraction significantly reduced glucose absorption in the everted rat jejunum. These results suggest that the antidiabetes effect of the subfraction may be due to extrapancreatic mechanisms (Mohamed et al. 2013). In a recent study, the ethanol extract of roots of *O. stamineus* showed anti-DM activity in streptozotocin- induced type 2 diabetic rats. In a short-term study, the extract (200–800 mg/kg, single dose) showed a dose-dependent reduction in blood glucose levels, after 2 h of administration, in streptozotocin-induced diabetic rats. In a long-term study, streptozotocin-induced type 2 diabetic rats were treated for 4 weeks with various doses of the methanol extract. The treatment resulted in a dose-dependent reduction in blood glucose levels. The treatment did not change the levels of insulin. However, the treatment resulted in significant reduction in glucose-6-phosphatase and significant increase in glucose-6-phosphate dehydrogenase and glycogen levels in the liver of diabetic rats (Rao et al. 2014).

Oryza sativa L., Poaceae

Common Names: Rice, dhanya, nivara, chaval, arisi, ari, nellu

Description: *Oryza sativa* is a leafy annual grass with long leaves, flat, striated, nerved, scaberulous with a smooth sheath, long ligules, and two partite. The fruit is a caryopsis, and the grain narrow adnate to glumes and palea.

Distribution: *O. sativa* is cultivated extensively all over the world for food; it is indigenous in marshes of Rajpootana, Sikkim, Bengal, and Central India.

Traditional Medicinal Uses: Brown rice bran is medicinally important. It has refrigerant, acrid, sweet, tonic, aphrodisiac, and diuretic properties and is used to treat carbuncles, diarrhea, biliousness, spitting of blood, leprous ulcers, dysentery, cataplasm, fever, dyspepsia, ophthalmia, indigestion, inflammation, and tumor (Kirtikar and Basu 1975).

Pregerminated brown rice intake results in lowering of blood glucose levels in streptozotocin-induced diabetic rats (Hagiwara et al. 2004). Rice bran extract prevented the elevation of plasma peroxylipid in KK-Ay diabetic mice (Kanaya et al. 2004). Resistant starch from rice improved glucose control, colonic events, and blood lipid concentrations in streptozotocin-induced diabetic rats. Parboiled varieties of red raw rice had a significantly lower glycemic index than white raw rice (Hettiarachchi et al. 2001). The roots contain glycans with hypoglycemic activity (Hikino et al. 1986b). The biological activity of rice bran, root, etc., may differ depending on varieties.

Rice bran has lipid-lowering effects and prevented cholesterogenesis in animals. Rice bran extracts showed antioxidative, antimutagenic, and anticarcinogenic activities. Triterpene alcohol and sterol ferulates from rice bran possess anti-inflammatory activity.

Phytochemicals reported include bran oil, tocopherol, sitosterol, stigmasterol, myricyl alcohol, myricyl cerotate, ceryl cerotate, isoceryl isocerotate, squalene, phosphatides, tocopherone, momilactone A and B, carlinoside, neocarlinoside, isoscoparine-2-glucoside, 6-*p*-coumaric acid ester, ferulic acid ester, schaftoside, neoschaftoside, isoorientin-2-glucoside, glucotricin, and oryzarans (Ajikumaran and Subramoniam 2005).

Osbeckia octandra DC., Melastomataceae

Osbeckia octandra is a woody, small leaved shrub with subquadrangular branches. The fruit a capsule, obscurely ribbed or smooth, sometimes with scattered stellate hairs. It is found in the South Deccan Peninsula in and near mountains in India and also in Ceylon. It is used in traditional medicine in the treatment of liver diseases and diabetes.

The water extract of *O. octandra* leaves lowered the fasting blood glucose level and markedly improved glucose tolerance in rats, and the efficacy is comparable with that of tolbutamide (Fernando et al. 1990). Further, the water extract of *O. octandra* had significant *in vitro* antiglycation activity and antioxidant activity. The extract was rich in phenolic content (Perera et al. 2013).

The hepatoprotective effects of the leaf extract against carbon tetrachloride toxicity in rats have also been shown. This plant (water extract) has immunomodulatory properties as evidenced by strong anti-complement effects (review: Ajikumaran and Subramoniam 2005).

Otholobium pubescens (Poir.) J.W. Grimes, Fabaceae

This is a Peruvian medicinal plant used for the treatment of diabetes. Anti-DM activity-guided isolation of the active principle from this plant resulted in the isolation of a known compound, bakuchiol. Oral administration of bakuchiol reduced blood glucose levels in a dose-dependent fashion in obese *db/db* mice. However, in lean mice it did not exhibit hypoglycemic activity. In fat-fed streptozotocin-induced diabetic animals, an oral dose of 150 mg/kg significantly lowered plasma glucose and TG levels (Krenisky et al. 1999).

Otostegia persica (Burm.) Boiss, Labiatae

The whole plant extract exhibited anti-DM activity in streptozotocin-induced diabetic rats (Ebrahimpur et al. 2009).

Ougeinia oojeinensis (Roxb.) Hochr., Leguminosae

The ethanol extract of *Ougeinia oojeinensis* bark (200 mg/kg, p.o.) showed anti-DM and hypolipidemic activity in alloxan-induced diabetic rats (Velmurugan et al. 2011).

Oxytenanthera abyssinica (A. Rich) Munro, Gramineae

Oxytenanthera abyssinica is an African bamboo. The ethanol extract of *O. abyssinica* leaves decreased hyperglycemia in pregnant streptozotocin-induced diabetic rats. The plant extract also showed substantial antioxidant activity and T-cell proliferation inhibition activity (Yessoufou et al. 2013).

Paeonia lactiflora Pall., Paeoniaceae

Common Name: Peony

Paeoniflorin from the root of *Paeonia lactiflora* prevented diabetic nephropathy in streptozotocin-induced diabetic rats (Jianfang et al. 2009). 1,2,3,4,6-penta-O-galloyl-D-glucopyranose, a PTP1B inhibitor (IC_{50} = 4.8 μmol/L), was isolated from the roots of *P. lactiflora*, which acted as an insulin sensitizer in human hepatoma cells (HCC-1.2) at 10 μmol/L (review: Jiany et al. 2012).

Paeonia suffruticosa Andrews, Paeoniaceae

Common Name: Moutan cortex

Paeonia suffruticosa is a common traditional herb used as an ingredient in formulae for treating DM in Korean, Chinese, and Japanese traditional medicines. One of the triterpenes, palbinone, isolated from this plant stimulated glucose uptake and glycogen synthesis via activation of AMPK, glycogen synthase kinase-3β (GSK-3β), and acetyl-CoA carboxylase (ACC) phosphorylation in insulin-resistant human HepG2 cells (Ha et al. 2009).

Panax ginseng C.A. Meyer, Araliaceae (Figure 2.73)

Common Names: Asian ginseng, Chinese ginseng, Korean ginseng, etc. (Asian ginseng is not a generic name). When one uses the name Asian ginseng, it includes all ginsengs originating from Asian countries, that is, *Panax ginseng* C.A. Meyer, *Panax japonicus*, *Panax notoginseng* [Sanchi ginseng], and *Panax sinensis* J. Wien. (Yun 2001))

(a)　　　　　　　　　　　　　　(b)

FIGURE 2.73 *Panax ginseng.* (a) Plant with fruits. (b) Root.

Panax ginseng C.A. Meyer, cultivated in Korea (Korean ginseng), is harvested after 4–6 years of cultivation and is classified into three types depending on how it is processed: (1) fresh ginseng (less than 4 years old; can be consumed in its fresh state), (2) white ginseng (4–6 years old; dried after peeling), and (3) red ginseng (harvested when 6 years old and then steamed and dried). Each type of ginseng is further subcategorized as ginseng products: fresh sliced, juice, extract (tincture or boiled extract), powder, tea, tablet, capsule, etc. (Yun 2001).

Description: Ginseng is a slow-growing perennial herb with characteristic branched roots extending from the middle of the main root in the form of a human figure. The stem is erect, simple, and not branching. The leaves are verticillate, compound, digitate, with five leaflets, the three terminal leaflets larger than the lateral ones, elliptical or slightly obovate, 4–15 cm long, 2–6.5 cm wide; acuminate apex; cuneate base; and a serrulate or finely bidentate margin. Inflorescence is a small terminal umbel. The flowers are polygamous and pink; the calyx is vaguely five toothed; it has five petals and five stamens. The fruit is a small berry, nearly drupaceous, and red when ripe (monograph: WHO 1999).

Distribution: Asian ginseng is a native to the mountain forests of Northeastern China, Korea, and the far eastern regions of the Russian Federation. It is cultivated extensively in China, Japan, Korea, and Russia. The Changbai mountain range is reportedly the only area in China where wild ginseng still occurs naturally (monograph: WHO 1999).

Traditional Medicinal Uses: In traditional Chinese medicine, it is usually prescribed in combination with other herbs and taken in an aqueous decoction dosage form. The Pharmacopoeia of the People's Republic of China indicates its use for prostration with impending collapse marked by cold limbs and faint pulse, diminished function of the digestive system with loss of appetite, diabetes caused by "internal heat," general weakness with irritability and insomnia in chronic diseases, impotence or frigidity, and heart failure and cardiogenic shock (monograph: WHO 1999).

Pharmacology

Antidiabetes: Several *in vitro* and *in vivo* studies, including clinical trials, show that the ginseng root and berry have glucose-lowering effects, antidiabetes activity, and antiobesity effects. Multiple mechanisms of action and several compounds are involved in the action of ginseng on the metabolic syndrome. Some compounds even have opposing effects. Further, the optimum dose required and the combinations of compounds present in the preparation or extract or fraction are important in exerting its specific antidiabetic effects.

The antidiabetic effect of the ginseng berry extract was shown in obese diabetic mice. Administration of the extract (150 mg/kg, 12 days) resulted in decrease in blood glucose and insulin levels and improvement in glucose (i.p.) tolerance. A hyperinsulinemic–euglycemic clamp study revealed a more than two fold increase in the rate of insulin-stimulated glucose disposal in treated *ob/ob* mice. Treatment with the extract also significantly reduced plasma cholesterol levels and body weight in *ob/ob* mice. Additional studies demonstrated that ginsenoside Re plays a significant role in antihyperglycemic action. This antidiabetic effect of ginsenoside Re was not associated with body weight changes, suggesting that other constituents in the extract have distinct pharmacological mechanisms on energy metabolism (Attele et al. 2002). In another study, the antioxidant and antihyperlipidemic efficacies of ginsenoside Re were shown in streptozotocin-induced diabetic rats. In addition to lowering glucose and lipid levels, ginsenoside Re decreased the levels of TNF and IL-6 involved in inflammation (El-Khayat et al. 2011). A study suggests that the compound can protect diabetic rats from oxidative stress-mediated microvasculopathy in the eye, kidney, etc. (Cho et al. 2006). Ginsenoside Re exhibited antidiabetic activity by reducing insulin resistance through activation of the PPAR-γ pathway by increasing the expression of PPAR-γ and its responsive genes and inhibition of TNF-α production in 3T3-L1 adipocytes (Gao et al. 2013). Malonyl ginsenosides, from the root of this plant, lowered fasting blood glucose level, improved insulin sensitivity, and improved the lipid profile in type 2 diabetic rats induced by HFD and streptozotocin (Liu et al. 2013a).

Wild ginseng leaf extract markedly reduced lipid peroxide levels and elevated antioxidant enzyme activities in streptozotocin-induced diabetic rats (Chang-Hwa et al. 2005). Improvement of insulin resistance by *P. ginseng* in fructose-rich chow-fed rats has been reported (Liu et al. 2005b). Panacene (a peptidoglycan from ginseng) exhibited hypoglycemic activity (Konno et al. 1984); a peptide with insulinomimetic properties has also been isolated (Ando et al. 1980).

The antidiabetic effects of ginseng root and berry were compared in *ob/ob* obese diabetic mice. When treated with the extract (150 mg/kg) for 12 days, the berry extract showed better glycemia and tolerance to i.p. glucose loading compared to the root extract, although on day 5 the reduction in blood glucose level was almost the same in both groups. In addition, the body weight did not change significantly after ginseng root extract (150 mg/kg) treatment, but the same concentration of ginseng berry extracts decreased body weight. These data suggest that, compared to ginseng root, ginseng berry exhibits more potent antihyperglycemic activity, and only ginseng berry shows marked antiobesity effects in *ob/ob* mice (Dey et al. 2003). Another study suggests that white ginseng (ginseng radix alba) can improve hyperglycemia in KK-Ay mice, possibly by blocking intestinal glucose absorption and inhibiting hepatic glucose-6-phosphatase, and the rootlet (ginseng radix palva) through the upregulation of adipocytic PPAR-γ protein expression as well as inhibiting intestinal glucose absorption (Chung and Choi 2001). Another study reported that the extract of the dried root of *P. ginseng* improved glucose-stimulated insulin secretion and β-cell proliferation through IRS2 induction (Park et al. 2008c).

Some fractions extracted from ginseng radix caused a hypoglycemic effect on alloxan-induced diabetic mice. The effect was abolished by the IV injection of antisera against bovine insulin. The same doses of the ginseng fraction (10–50 mg/kg) produced an increase in the blood insulin level in alloxan-induced diabetic mice. Insulin release from perfused rat pancreases was stimulated by the ginseng fraction (0.2 mg/mL), but the potency was not stronger than that of the sulfonylureas. The study indicates that some ginseng fractions stimulated insulin release, especially glucose-induced insulin release from pancreatic islets (Kimura et al. 1981). Subsequent studies also confirmed modulation of insulin secretion by ginseng. Increase of insulin secretion by ginsenoside Rh2 to lower plasma glucose in Wister rats has been reported (Lee et al. 2006a). Compound K (one of the ginsenosides) enhanced insulin secretion with beneficial metabolic effects in *db/db* mice (Kan et al. 2007). Korean red ginseng stimulated insulin release from isolated rat pancreatic islets (Kim and Kim 2008). Increase in acetylcholine release by *P. ginseng* root enhanced insulin secretion in rats (Su et al. 2007). In an *in vitro* study, 20(S)-ginsenoside Rg3 enhanced glucose-stimulated insulin secretion and activated AMPK (Park et al. 2008a).

Ginsenoside Rh2 is one of the ginsenosides that exerts antidiabetes, anti-inflammatory, and anticancer effects. In cell culture system, ginsenoside Rh2 effectively inhibited adipocyte differentiation via PPAR-γ inhibition. Interestingly, ginsenoside Rh2 significantly activated AMPK in 3T3-L1 adipocytes. Further, ginsenoside Rh2 effectively induced lipolysis and this induction was abolished by AMPK inhibitor treatment (Hwang et al. 2007). Another study suggests that the antiobesity effect of a red ginseng-rich constituent, ginsenoside Rg3, also involves the AMPK signaling pathway and PPAR-γ inhibition (Hwang et al. 2009). In a recent review, the ginseng extracts and ginsenosides that activate AMPK and the various likely mechanisms of their action are discussed (Jeong et al. 2014).

In an *in vitro* study, ginsenoside Rb1 stimulated glucose uptake through insulin-like signaling pathway in 3T3-L1 adipocytes. Rb1 stimulated basal and insulin-mediated glucose uptake in a time- and dose-dependent manner in 3T3-L1 adipocytes and C2C12 myotubes; in adipocytes, Rb1 promoted GLUT1 and GLUT4 translocations to the cell surface. Rb1 increased the phosphorylation of IR substrate 1 and PKB and stimulated PI3K activity in the absence of the activation of the IR. Rb1-induced glucose uptake as well as GLUT1 and GLUT4 translocations was inhibited by the PI3K inhibitor (Shang et al. 2008).

Protein kinase A (PKA) may also be involved in the antidiabetes actions of ginseng. Ginsenoside Rb1 and Rg1 suppressed TG accumulation in 3T3-L1 adipocytes and enhanced β-cell insulin secretion and viability in Min6 cells via PKA-dependent pathways (Park et al. 2008b).

Clinical Trial: Korean red ginseng (steam-treated *P. ginseng*) has been clinically shown to have beneficial effects in type 2 DM and improved cardiovascular disease and other risk factors.

The effect of *P. ginseng* on newly diagnosed NIDDM patients was investigated in a double-blind placebo-controlled study (Sotaniemi et al. 1995). Patients were randomized to ingest one tablet daily containing 0 (placebo), 100, or 200 mg ginseng, presumably an extract, but the authors did not state the type of preparation used in the study (manufactured by Dansk Droge, Copenhagen) for 8 weeks. Effects on psychophysical tests, measurements of glucose balance, serum lipids, aminoterminal propeptide concentration, and body weight were tested. Ginseng therapy elevated mood, improved psychophysical performance, and reduced fasting blood glucose and body weight. The 200 mg dose of ginseng improved glycated hemoglobin, serum aminoterminal propeptide, and physical activity. The authors concluded that ginseng may be a useful therapeutic adjunct in the management of NIDDM (Sotaniemi et al. 1995).

In another randomized, double-blind, crossover design clinical study, the antidiabetic efficacy and safety of 12 weeks of supplementation with a Korean red ginseng (2 g/meal = 6 g/day) to well-controlled type 2 diabetic patients was assessed. The treatment was given as an adjuvant to their usual antidiabetes therapy. The participants remained well controlled (HbA1c = 6.5%) throughout. The selected Korean red ginseng treatment improved glucose tolerance and decreased plasma glucose and fasting plasma insulin levels; further, the treatment increased insulin sensitivity compared to placebo group. Safety and compliance outcomes remained unchanged. However, the levels of HbA1c remained unchanged in the treated group compared to well-controlled control patients (Vuksan et al. 2008). It is possible that a lower dose may prove better.

Another clinical study reported that ginseng and ginsenoside Re do not improve β-cell function or insulin sensitivity in overweight and obese subjects with impaired glucose tolerance or diabetes (Reeds et al. 2011).

In a recent randomized, placebo-controlled, double-blind, crossover design clinical trial, the efficacy and safety of acute escalating doses of the white ginseng (dried nonsteamed Korean ginseng) on vascular and glycemic parameters in well-controlled type 2 diabetes were investigated. In this acute study, for the duration of 240 min after administration of (1, 3, and 6 g) the white ginseng compared to 3 g wheat-bran control, the white ginseng appeared to be safe but did not affect any postprandial, vascular, or glycemic parameters except the augmentation index, a cumulative indicator of arterial health. The white ginseng might have a beneficial effect on microarteries/microvessels. The authors conclude that these results are preliminary and highlight the need for long-term investigation with a focus on its active components (Shishtar et al. 2014). The dose level and relative concentrations of bioactive phytochemicals in the ginseng preparations may account for some of the variations in the anti-DM activities of ginseng.

Other Activities: Ginseng's pharmacological activities are multiple due to ginsenosides (more than 30 ginsenosides are present) and other bioactive molecules such as polysaccharides, salicylate, and vanillic acid. The pharmacological actions of individual ginsenosides may work in opposition. For example, the two main ginsenosides, Rb1 and Rg1, respectively suppress and stimulate the central nervous system. Water extracts, saponin fraction, and polysaccharide fractions of ginseng root showed immune-modulatory actions. Immune-modulatory actions include modulation of cytokine production, potentiation of humoral immune response, enhancement of CD4(+) T-cell activation, upregulation of adjuvant effects, and restoration of T lymphocyte function. People who consume ginseng preparations are at lower risk of cancers in several organs. Ginsenosides Rg3 and Rh2 are the major anticancer saponins. Ginseng influences CNS function and ameliorates certain degenerative diseases. Ginsenosides Rg3 and Rh2 are very effective in protecting the CNS, preventing neurodegenerative diseases (review: Wee et al. 2011).

Ginseng is reported to promote vasodilatation and act as an anxiolytic and antidepressant. Many studies on animals have found ginseng extracts and ginsenosides to be effective in supporting radioprotection, providing resistance to infection, providing antioxidant and antifatigue effects, enhancing energy metabolism, and reducing plasma total cholesterol and TGs while elevating HDL levels (monograph: WHO 1999).

Phytochemicals: The biologically active constituents in *P. ginseng* are a complex mixture of triterpene saponins known as ginsenosides. More than 30 ginsenosides have been isolated and characterized. More than 30 are based on the dammarane structure, and one (ginsenoside Ro) is derived from oleanolic acid. The root contains 2%–3% ginsenosides of which Rb1, Rb2, Rc, Rd, Rf, Rg1, and Rg2 are considered the most important; Rb1, Rb2, and Rg1 are the most abundant (monograph: WHO 1999).

Panax japonicus C.A. Meyer, Araliaceae
The rhizomes of *Panax japonicus* are used as a folk medicine for the treatment of lifestyle-related diseases such as arteriosclerosis, hyperlipidemia, hypertension, and NIDDM as a substitute for ginseng roots in China and Japan. Potent α-glucosidase inhibitors were reported in the roots of *P. japonicus*

(Chan et al. 2010). Chikusetsusaponins isolated from the rhizome of *P. japonicus* (1% and 3% in the diet) exhibited an antiobesity effect in HFD-fed mice. Further, total chikusetsusaponins, chikusetsusaponin III, 28-deglucosyl-chikusetsusaponin IV, and 28-deglucosyl-chikusetsusaponin V inhibited the pancreatic lipase activity *in vitro* (Han et al. 2005).

Panax notoginseng (Burk) F.H. Chen, Araliaceae

Panax notoginseng, known as tiánqī or san qi, is a well-known medicinal herb in Asia for its long history of use in Chinese medicine. The *P. notoginseng* plant looks similar to Siberian ginseng and contains 12 saponins (key active components) that are similar to the active ingredients in *P. ginseng*. The saponins have been widely used in China for the treatment of cardiovascular diseases. However, scientific studies have shown a wide range of other pharmacological applications including anticancer, neuroprotective, and anti-inflammatory agents, immunologic adjuvant, and prevention of diabetes complications. Recently, the hypoglycemic and antiobesity properties of the saponins have also been demonstrated. The saponins exhibited effects on glucose production and absorption and on inflammatory processes that seem to play an important role in the development of diabetes (review: Uzayisenga et al. 2014). In KK-Ay mice, *P. notoginseng* saponins exhibited antihyperglycemic and antiobese activities by improving insulin and leptin sensitivity. The saponin Rb1 is responsible for the antihyperglycemic effect among the various saponins in KK-Ay mice (Yang et al. 2010a).

Panax quinquefolius L., Araliaceae

Synonyms: *Panax quinquefolium* (L.) Alph., *Panax americanum* Raf., *Ginseng quinquefolium* Wool, *Aralia quinquefolia* Dec & Planch. (monograph: WHO 2009) (Figure 2.74)

Common Vernacular Names: American ginseng, American white ginseng, North American ginseng, Canadian white ginseng, wild American ginseng, Ameerika zensen, ginseng americano, etc.

Description: *Panax quinquefolius* is an herb, up to 1 m high, with fusiform rootstock, up to 2 cm in diameter and straight, slender, subterete, often striate stem. The leaves are compound, and each leaf is palmate with five widely spreading leaflets; the leaves are generally in groups of 3–5, with a slender petiole, up to 10 cm long; the leaflets are thin, elliptic to obovate, acute to rounded at the base, acuminate at the apex, and conspicuously and often doubly serrate; the teeth are deltoid, acute, and the petiolules are up to 4.5 cm long; the principal veins are slightly raised on both surfaces, there are 5–9 lateral veins and a slender and straight peduncle; the bractlets are deltoid to lanceolate, acute, 2–5 mm long; the pedicels are 15–40 per umbel, up to 12 mm long, swollen distally; a short tubular calyx with five tiny teeth, the

(a) (b)

FIGURE 2.74 *Panax quinquefolius*. (a) Whole plant. (b) Root.

lobes deltoid, acute, about 0.5 mm long; petals greenish white, membranous, slightly granular papillose distally, oblong, about 1.5 mm long and 1 mm broad, subacute and slightly incurved at the apex; filaments carnose, narrowed distally, the anthers oblong, obtuse at both ends; summit of the ovary flattened or concave, styles 2, carnose, slightly curved; locules 2; fruit laterally flattened, transversely oblong, up to 7 mm long and 10 mm broad, longitudinally sulcate, the wall at length dry, seeds 2, oblong, 4–5 mm long (monograph: WHO 2009).

Distribution: *Panax quinquefolius* is native to Canada and the United States and cultivated in France, China, and elsewhere (monograph: WHO 2009).

Traditional Medicinal Uses: *P. quinquefolius* is used orally as diuretic, digestive, tonic, and stimulant. It is also used to enhance resistance to stress and to treat cough, loss of appetite, colic, vomiting, insomnia, neuralgia, rheumatism, and headache (monograph: WHO 2009).

Pharmacology

Antidiabetes: Intraperitoneal administration of the water extract (10 mg/kg) of the root exhibited significant hypoglycemic activity in mice with alloxan-induced hyperglycemia. Activity-guided fractionation of the extract led to isolation of three glycans, the quinquefolans A, B, and C, which displayed significant hypoglycemic activity in normal mice and in mice with alloxan-induced hyperglycemia at a dose of 10, 100 and 10 mg/kg, respectively (Oshima et al. 1987).

P. quinquefolius root exhibited anti-DM activity and improved glucose tolerance in the *ob/ob* diabetic mice model; the antidiabetes effect of the root of *P. quinquefolius* may not be linked to its antioxidant activity (Xie et al. 2009). The alcohol extract of the plant root prevented diabetic nephropathy in streptozotocin-induced diabetic rats (Sen et al. 2012).

Clinical Trial: Various extracts of the roots, with a specific ginsenoside profile, are reported to decrease postprandial glycemia. A preliminary short-term clinical study evaluated the efficacy of the powdered crude drug on postprandial glycemia in humans. On 4 separate occasions, 10 subjects who did not have diabetes and 9 subjects who had type 2 DM were randomly allocated to receive 3.0 g of the powdered root or placebo capsules, either 40 min before or together with a 25.0 g oral glucose challenge. A capillary blood sample was taken during fasting and then at 15, 30, 45, 60, 90, and 120 min after the glucose challenge. The results of this study demonstrated that ginseng reduces postprandial glycemia to some extent in nondiabetic subjects and subjects with type 2 DM (Vuksan et al. 2000a).

In a follow-up study, the effect of escalation of the dose and dosage interval of a root product (500 mg powdered root per capsule) was assessed in individuals who did not have diabetes to determine whether further improvements in glucose tolerance (seen previously when 3 g of the crude drug was taken 40 min before a 25.0 g glucose challenge) could be attained. Ten healthy volunteers were randomly assigned to receive 0 (placebo), 3, 6, or 9 g of the crude drug prepared from the ground root at 40, 80, or 120 min before a 25.0 g oral glucose challenge. Capillary blood glucose was measured prior to ingestion of the crude drug or placebo capsules and at 0, 15, 30, 45, 60, and 90 min from the start of the challenge. As compared with the placebo, 3, 6, and 9 g of the crude drug reduced postprandial incremental glucose at 30, 45, and 60 min from the start of the glucose challenge; 3 g and 9 g of the crude drug also did so at 90 min (Vuksan et al. 2000b).

In a subsequent investigation, 10 patients with type 2 diabetes were randomly administered 0 g (placebo) or 3, 6, or 9 g of the ground crude drug in capsules at 120, 80, 40, or 0 min before a 25 g oral glucose challenge. Capillary blood glucose was measured before ingestion of the crude drug or placebo and at 0, 15, 30, 45, 60, 90, and 120 min from the start of the glucose challenge. Treatment (0, 3, 6, and 9 g of the crude drug), but not time of administration (120, 80, 40, or 0 min before the challenge), significantly affected postprandial glucose. Compared with administration of the placebo (0 g), doses of 3, 6, or 9 g of the crude drug reduced glucose levels and incremental glycemia at 30 min (16.3%, 18.4%, and 18.4%, respectively), 45 min (12.5%, 14.3%, and 14.3%, respectively), and 120 min (59.1%, 40.9%, and 45.5%, respectively) (Vuksan et al. 2000c).

A randomized, crossover study to assess whether a dose of the powdered root of below 3.0 g, administered 40 min prior to an oral glucose challenge, would reduce postprandial glycemia in subjects without diabetes was performed. Twelve healthy volunteers received treatment, 0 (placebo), 1, 2, or 3.0 g of the crude drug, at 40, 20, 10, or 0 min prior to a 25.0 g oral glucose challenge. Capillary blood was collected before administration and at 0, 15, 30, 45, 60, and 90 min after the start of the glucose challenge.

Glycemia was lower over the last 45 min of the test after doses of 1, 2, or 3 g than after placebo; there were no significant differences between the three doses. Glycemia in the last hour of the test was lower when the crude drug was administered 40 min before the challenge than when it was administered 20, 10, or 0 min before the challenge. Thus, doses of the crude drug within the range of 1–3 g were equally effective (Vuksan et al. 2001). However, an extract of the roots (6 g/patient, p.o.) containing low levels of ginsenosides does not affect postprandial glycemia in oral glucose tolerance test or insulin levels in the blood (Sievenpiper et al. 2003).

In the aforementioned clinical trials, the effects of a single administration of the extract, before a glucose challenge, were studied in normal and diabetic individuals. The effects of repeated dose treatments for weeks or months with very low doses (100–1000 mg) of the herbal drug are to be studied in type 2 DM patients to get better insights regarding the utility of the drug to treat DM patients.

Other Activities: Other experimental pharmacological activities include antioxidant activity, antithrombotic activity, estrogen-like activity, aphrodisiac activity in male rats, immune stimulation effect, neurological effects, and memory enhancing property (monograph: WHO 2009).

Toxicity: Ginsenoside C was more toxic than ginsenoside A2 after intraperitoneal administration to mice. Toxicity was not observed after oral administration of any of the ginsenosides. The genins panaxadiol and panaxatriol were more toxic and had larger volumes of distribution than the ginsenosides. The extracts of the root, which were not treated with calf serum inhibited CYP1 catalytic activity under *in vitro* conditions, but the effects were not due to ginsenosides (Rb1, Rb2, Rc, Rd, Re, Rf. or Rg1) (monograph: WHO 2009).

Phytochemicals: The major constituents of the root are the dammarane triterpene saponins collectively known as ginsenosides. The ginsenosides of *P. quinquefolius* are derivatives of protopanaxadiol or protopanaxatriol. The total ginsenoside content of *P. quinquefolius* is higher than that of *P. ginseng*. In cultivated *P. quinquefolius*, the dominant ginsenosides are malonyl (m)-Rb1, Rb1, and Re with the percentages of m-Rb1 and Rb1 being almost identical. Rg1 levels and total ginsenosides are much higher in wild than in cultivated *P. quinquefolius*. The combined amount of Rb1 and m-Rb1 often exceeds half of the total ginsenoside content with the total malonyl ginsenoside being approximately 40%. In a study of wild American ginseng, total ginsenosides range from 1% to 16%, with the majority being in the range of 4%–5%. Polysaccharides of biological significance include quinquefolans A, B, and C (monograph: WHO 2009).

Pandanus amaryllifolius Roxb., Pandanaceae

Synonyms: *Pandanus latifolius* Hassk., *Pandanus latifolius* var. *minor*, *Pandanus odorus* Ridl., *Pandanus hasskarlii* Merr.

Administration of the extract (0.5g/kg, p.o.) of *Pandanus amaryllifolius* (*Pandanus odorus*) root did not significantly affect the plasma glucose level in healthy rats, whereas the extract significantly lowered the plasma glucose level in streptozotocin-induced diabetic rats. In an oral glucose tolerance test, administration of the extract at (0.5 and 1.0 g/kg, p.o.) significantly lowered the plasma glucose level in diabetic rats (Peungvicha et al. 1996). Hypoglycemic activity-guided fractionation led to the isolation of the known compound, 4-hydroxybenzoic acid, from this plant. The compound (0.5g/kg, p.o.) lowered blood glucose and increased serum insulin levels and liver glycogen content in healthy rats (Peungvicha et al. 1998c). Moreover, oral administration of 4-hydroxybenzoic acid caused a dose-dependent decrease in plasma glucose levels in streptozotocin-induced diabetic rats. The component did not affect serum insulin level and liver glycogen content in the diabetic model but increased glucose consumption in healthy and diabetic rat diaphragms. These results suggest that 4-hydroxybenzoic acid produced a hypoglycemic effect mediated by an increase in the peripheral glucose consumption (Peungvicha et al. 1998a).

Pandanus fascicularis Lamk., Pandanaceae

Synonym: *Pandanus odoratissimus* L.

Common Names: Umbrella tree, screw pine, screw tree

The aerial roots of *Pandanus fascicularis* are used traditionally in the treatment of diabetes. The alcohol and water extracts of *P. fascicularis* were found to lower blood glucose level in alloxan-induced diabetic rats. Glycosylated hemoglobin level significantly reduced in the treated diabetic animals when compared to control. The alcohol and water extracts (500 and 750 mg/kg) restored serum glucose level to near normal values. The results of this study substantiated the traditional use of this drug (Madhavan et al. 2008). Administration of the methanol extract of the aerial roots of *P. fascicularis* (250, 500, and 1000 mg/kg, p.o., for 7 days) caused significant dose-dependent reduction in serum glucose in both normoglycemic and alloxan-induced hyperglycemic rats; it also improved glucose tolerance. The treatment also enhanced β-cell function. Thus, this study suggests that the methanol extract of the aerial roots of *P. fascicularis* exhibits antidiabetic activity possibly through increased secretion of insulin, and the effect may be due to the presence of flavonoids and phenolic compounds (Kumari et al. 2012). The hypoglycemic effect of this plant is confirmed by another study. The water and methanol extracts of roots of *P. fascicularis* (250 mg/kg) produced a significant fall in blood glucose levels in alloxan-induced diabetic rats (Savitha et al. 2012). In another study, the antidiabetic activity and chemical characterization of the water and ethanol extracts of prop roots of *P. fascicularis* was determined in streptozotocin-induced diabetic rats. The ethanol and water extracts of *P. fascicularis* (250 mg/kg) reduced the blood glucose level in streptozotocin-induced diabetic rats, when compared with the diabetic control. The blood glucose level of diabetic control animals after 3 h was about 12.6 mmol/L (226 mg/100 mL), whereas it was 5.7 and 7.3 mmol/L for the groups treated with water extract and ethanol extract, respectively (Rajeswari et al. 2012). Thus, this plant is promising for further studies.

Papaver somniferum L., Papaveraceae

Synonyms: *Papaver album* Mill., *Papaver setigerum* DC
Papaver somniferum yields opium and poppy seeds. Opium is the source of morphine and other drugs. This plant is grown on a large scale for its seeds and oil. Papaverine from *P. somniferum* was found to readily dock within the binding pocket of human protein tyrosine phosphatase 1B (h-PTP1B) in a low-energy orientation. Follow-up experimental studies showed the potent *in vitro* inhibitory effect of papaverine against recombinant h-PTP1B (IC_{50} = 1.20 μM). *In vivo*, papaverine significantly decreased the fasting blood glucose level of Balb/c mice. Thus, papaverine exhibited antidiabetic activity (Bustanji et al. 2009).

Parinari curatellifolia Planch ex Benth., Chrysobalanaceae
Parinari curatellifolia is used in traditional medicine, among other things, to treat diabetes in Africa. Experiments have shown that the ethanol extracts of the seeds of *P. curatellifolia* exerted a significant reduction in the plasma glucose and the level of LDL in alloxan-induced diabetic rats; the extract did not show any obvious toxicity at the pharmacological dose tested (Ogbonnia et al. 2008). In a follow-up study, the ethanol extract of seeds of *P. curatellifolia* exhibited potent antioxidant properties. The authors conclude that since type 2 diabetes is intrinsically linked with oxidative stress, *P. curatellifolia* possibly exerts its antidiabetic action using a combination of mechanisms and its antioxidant potency possibly plays a major role in ameliorating secondary complications resulting from oxidative damage in diabetes (Ogunbolude et al. 2009). In another study, administration of *P. curatellifolia* seed extract (500 mg/kg) resulted in significant reduction in glucose levels in glucose tolerance test in rats. The extract also decreased blood glucose levels in diabetic rats. There were significant reductions in LDL levels and significant increases in HDL levels in the treated diabetic group compared to diabetic control rats (Ogbonnia et al. 2011).

Parinari excelsa Sabine, Chrysobalanaceae, guinea plum
The water extracts of *Parinari excelsa* (100 or 300 mg/kg) showed antidiabetic property in alloxan-induced diabetic rats. Further, the extract lowered glucose in glucose-loaded normoglycemic rats (Ndiaye et al. 2008).

Parkia biglobosa Jacq., Mimosoideae

Common Name: African locust bean
The hypoglycemic effects of the water and alcohol extracts of fermented seeds of *Parkia biglobosa*, a natural nutritional condiment that features frequently in some African diets as a spice, were investigated in

alloxan-induced diabetic rats. The effects of the seed on lipid profiles were also examined. Administration of a single dose (120 mg/kg, IV) of the extracts to alloxan-induced diabetic rats resulted in significant increases in the blood glucose levels of test animals compared with controls. But dietary supplementation (6% w/w) with the fermented seed extracts (6 g/kg for 4 weeks, p.o.) ameliorated alloxan-induced diabetes in a manner comparable with that of the reference antidiabetic drug glibenclamide. Animals treated with the water extract of the fermented seed gained weights, whereas animals given the methanol extract and glibenclamide lost weight. In addition, high levels of HDL and low levels of LDL were observed in animals treated with the water extract, a pattern similar to that seen in normal controls. In contrast, low levels of HDL and high levels of LDL were observed in animals treated with the methanol extract, a pattern similar to that seen in nontreated diabetic controls. Thus, the water extract of the fermented seed ameliorated the loss of body weight and dyslipidemia usually associated with diabetes (Odetola et al. 2006). In contrast to IV administration, oral administration may result in changes in the phytochemicals in the gastrointestinal tract by the action of microflora and digestive enzymes. Further, certain compounds may not be observed from the intestine. This could possibly explain the acute IV dose effect on blood glucose.

In another study, the antidiabetic potentials of the methanol extract of *P. biglobosa* seeds and its chloroform, hexane, and mother liquor fractions were evaluated in glucose-loaded and alloxan-induced diabetic rats. The methanol extract of the seed exhibited a peak percentage decrease of 64% and 44.1% in blood glucose levels of the glucose-loaded and alloxan-induced diabetic rats, respectively. The blood glucose-lowering effect of the chloroform fraction of methanol extract was significant and more than that exhibited by the reference drug glibenclamide in the alloxan-induced diabetic rats (Fred-Jaiyesimi and Abo 2009).

Parkinsonia aculeata L., Caesalpiniaceae

The water extracts of the leaves and flowers of *Parkinsonia aculeata* showed antihyperglycemic and antihyperlipidemic activities in alloxan-induced diabetic rats (Leite et al. 2007). The extracts also exhibited renoprotective effects in alloxan-induced diabetic rats (Leite et al. 2011).

Parmentiera edulis C. Am., Bignoniaceae

Synonym: *Parmentiera aculeate* (Kunth) Williams

The chloroform extract of the *Parmentiera edulis* fruit (300 mg/kg, i.p.) lowered blood glucose levels in alloxan-induced diabetic mice. It showed hypoglycemic activity in normal mice (Perez et al. 1998). Hypoglycemic activity-guided fractionation led to the isolation of guaianolide (lactucin-8-*O*-methylacrylate) from the chloroform extract of the dried fruits of *P. edulis*. The compound lowered blood glucose levels in alloxan-induced diabetic mice (Perez et al. 2000b).

Paronychia argentea Lam., Caryophyllaceae

Paronychia argentea is one of the plants widely recommended by the herbalists and used for their hypoglycemic activity in Jordan; the leaves of this plant are used to treat diabetic patients in Israel also. A report suggests that the water extract of this plant is devoid of significant *in vivo* hypoglycemic activity in normal as well as streptozotocin-induced diabetic rats. However, *P. argentea* extract showed *in vitro* α-amylase inhibitory activity (Hamdan and Afifi 2004). In another *in vitro* study, the water extract of this plant augmented pancreatic MIN6 β-cell expansion and inhibited carbohydrate absorption (Kasabri et al. 2012b). Further studies are required to determine its *in vivo* efficacy, if any.

Parquetina nigrescens Afzel. Bullock, Apocynaceae/Asclepiadaceae

The plant *Parquetina nigrescens* is used in folklore medicine to treat DM and its complications in several parts of West Africa. Diabetic rats treated with the extract of *P. nigrescens* showed significant reduction of the blood glucose to levels comparable to that of the nondiabetic control and those treated with chlorpropamide (standard drug). The mechanism underlying the antidiabetic properties of *P. nigrescens* remains to be elucidated (Saba et al. 2010).

Parthenium hysterophorus L., Asteraceae

The water extract of *Parthenium hysterophorus* flower (100 mg/kg) showed significant reduction in blood glucose level in diabetic rats 2 h after administration. However, the reduction in blood glucose level with the water extract was less than that with the standard drug glibenclamide. The extract showed less hypoglycemic effect in fasted normal rats (Patel et al. 2008).

Paspalum scrobiculatum L., Poaceae, millet crop

Paspalum scrobiculatum (millet) is an important cereal cultivated in Asia, Europe, etc. *P. scrobiculatum* grain is used in traditional medicine in the management of DM. The plant grain increased the levels of serum insulin and liver glycogen content, whereas it decreased the levels of glycated hemoglobin levels in alloxan-induced diabetic rats (Jain et al. 2010). Among 2 types of millet (Kodo millet and finger millet), Kodo millet-fed rats showed a greater reduction in blood glucose and cholesterol levels than those fed the finger millet. The water and ethanol extracts of *P. scrobiculatum* (250 and 500 mg/kg, p.o., daily for 15 days) exhibited blood glucose- and lipid-lowering effects in alloxan-induced diabetic rats (Radhika et al. 2013).

Peganum harmala L., Zygophyllaceae

Common Names: Esfand, wild rue, Syrian rue, African rue, etc.

Description: *Peganum harmala* is an erect, bushy, perennial, herbaceous plant with short creeping roots. The flowers are white and about 2.5 cm in diameter. The round seed capsules carry more than 50 seeds.

Distribution: *P. harmala* grows spontaneously in semiarid conditions and is widely distributed and used as a medicinal plant in Central Asia, North Africa, and Middle East. It has also been introduced in America and Australia.

Traditional Medicinal Uses: Various parts of this plant including seeds, bark, and roots have been used in folk medicine. It is used for the treatment of depression, leishmaniasis, hallucination, diabetes, etc.

Antidiabetes: The ethanol extract of the seeds of this plant (150 and 250 mg/kg, oral administration) showed hypoglycemic activity in sucrose-challenged normal rats and streptozotocin-induced diabetic rats. The efficacy of the extract was comparable to that of metformin in both normoglycemic and streptozotocin-induced diabetic rats (Singh et al. 2008). The anti-DM activity of this plant was confirmed in several studies (review: Moloudizargari et al. 2013) and is dependent on the concentration of the extract used; at very high concentrations, the plant loses its hypoglycemic activity (Nafisi et al. 2011). Harmine is the major alkaloid of this plant species involved in its antidiabetes activity. Harmine regulated the expression of PPAR-γ and is reported that it mimics the effect of PPAR-γ ligands on insulin sensitivity and adipocyte gene expression without showing the side effects of thiazolidinedione drugs such as weight gain (Waki et al. 2007; review: Moloudizargari et al. 2013).

Other Activities: *P. harmala* has cardiovascular, vasorelaxant, and antihypertensive effects, angiogenic inhibitory effect, inhibitory effect on platelet aggregation, effect on the nervous system, monoamine oxidase inhibition, antidepressant effect, analgesic and antinociceptive effects, antineoplasm and antiproliferative activity, antioxidant effects, antifungal and antibacterial activity, antispasmodic effect, emetic effects, osteogenic activity, and immune-modulatory activity (review: Moloudizargari et al. 2013).

Phytochemicals: The most important constituents of this plant are β-carboline alkaloids such as harmalol, harmaline, and harmine. Harmine is the most studied among these alkaloids.

Toxicity: In addition to all therapeutic effects of *P. harmala*, there have been several reports of human and animal intoxications induced by this plant. There are also experimental studies indicating *P. harmala* toxicity (review: Moloudizargari et al. 2013).

Peltophorum pterocarpum (DC) K. Hyne, Fabaceae

The anti-DM activity of the methanol/ethyl acetate (1:9) extract of flower buds of *Peltophorum pterocarpum* on alloxan-induced diabetic mouse model was evaluated. The extract (200 mg/kg, i.p.) reduced the blood glucose level in alloxan-induced diabetic mice by 60% (metformin, 65%) at 12 h in a 24 h treatment period. Preliminary phytochemical screening of the extract revealed the presence of different types of compounds including flavonoids and steroids (Isalm et al. 2011).

Pentas schimperiana A. Rich, Rubiaceae

Pentas schimperiana is widely used for the treatment of DM and various other ailments in the traditional medical practices of Ethiopia. A single dose of (500 mg/kg, each) of water and hydroalcoholic extracts of fresh leaves of *P. schimperiana* did not show significant antidiabetic effects in alloxan-induced diabetic mice. However, at a dose of 1000 mg/kg, the extracts lowered blood glucose levels. At a dose of 500 mg/kg, the water and methanol fractions prepared from the dried plant material lowered blood glucose level, as compared with diabetic control mice. The antidiabetic activity of the dried plant material was found to be better than that of the fresh plant material. Both the hydroalcoholic extracts and the methanol fraction of the dried leaves displayed a very good DPPH scavenging activity with IC_{50} values of 7.83 and 8.84 μg/mL, respectively. An acute toxicity study of the water and hydroalcoholic extract of *P. schimperiana* performed on albino mice indicated that the median lethal dose (LD_{50}) of the extract is above 4000 mg/kg. Phytochemical screening carried out on the total leaf extracts of the plant confirmed the presence of flavonoids, saponins, steroids, and tannins (Dinka et al. 2010).

Peperomia pellucida (L.) HBK, Piperaceae

Alloxan-induced diabetic rats on diets supplemented with 10% and 20% of *Peperomia pellucida* for 28 days showed a reduction in blood glucose levels. Further, the levels of serum cholesterol, TGs, and LDL decreased in the supplemented group compared to untreated diabetic rats. The treatment increased the activity of antioxidant enzymes and decreased oxidative stress in diabetic rats (Hamzah et al. 2012).

Pergularia tomentosa L., Asclepiadaceae

Pergularia tomentosa (aerial part) is used in Egyptian traditional medicine to treat DM (Haddad et al. 2005). The plant extract exhibited persistent hypoglycemic effects in normal rats (Shabana et al. 1990).

Perilla frutescens (L.) Britton, Lamiaceae

The ethyl acetate-soluble fraction of the methanol extract of *Perilla frutescens* inhibited aldose reductase, the key enzyme in the polyol pathway. The main aldose reductase-inhibiting compounds were tentatively identified as chlorogenic acid (IC50, 3.16 μM), rosmarinic acid (IC50, 2.77 μM), luteolin (IC50, 6.34 μM), and methyl rosmarinic acid (IC50, 4.03 μM) (Paek et al. 2013). In another study, the rosmarinic acid–rich fraction from *P. frutescens* inhibited α-glucosidase activity and glucose transport activity *in vitro* and *in vivo* in rats (Higashino et al. 2011).

Persea americana Mill., Lauraceae

Common Name: Alligator pear

The leaves of *Persea americana* have been popularly used in the treatment of diabetes in countries in Latin America and Africa. The hydroalcoholic extract of the leaves of *P. americana* (0.15 and 0.3 g/kg/day for 4 weeks) exhibited antidiabetic activity in streptozotocin-induced diabetic rats. The hydroalcoholic extract of the leaves reduced blood glucose levels and improved the metabolic state of the animals. Additionally, PKB activation was observed in the liver and skeletal muscle of treated rats when compared with untreated rats (Lima et al. 2012).

The hypoglycemic and renal function effects of *P. americana* leaf ethanol extract in streptozotocin-induced diabetic rats were evaluated. The extract induced dose-dependent hypoglycemic responses in diabetic rats, while subchronic treatment additionally increased hepatic glycogen concentrations. Acute administration of the extract decreased urine flow and electrolyte excretion rates, while subchronic treatment reduced plasma creatinine and urea concentrations (Gondwe et al. 2008). In another study, administration of the water extract of *P. americana* leaf (100–200 mg/kg) to alloxan-induced diabetic rats produced a significant reduction in blood glucose levels in a dose-dependent fashion after a single dose of the extract, as well as following treatment for 7 days compared to the diabetic control group. Maximum antidiabetic activity was reached at 6 h after a single dose of the extract (Antia et al. 2005). The ethanol extract of *P. americana* seed exhibited antihyperlipidemic activities in alloxan-induced diabetic rats (Edem 2010). In another study, the hypoglycemic and tissue-protective effects of the hot-water extract of *P. americana* seed on alloxan-induced albino rats were investigated. The extract possessed significant hypoglycemic effect and reversed the histopathological damage that occurred in alloxan-induced diabetic rats, comparable to the effects of glibenclamide. The seed extract also had antidiabetic and protective effects on the pancreas, kidneys, and liver (Ezejiofor et al. 2013).

Petroselium crispum (Miller) Nyman ex A.W. Hill, Apiaceae (Umbelliferae)

Synonym: *Petroselinum crispum* (Mill) Nyman, Apiaceae

Common Name: Parsley

Petroselinum crispum (parsley) is one of the medicinal herbs used by diabetic patients in Turkey and Egypt. It has been reported to reduce blood glucose levels (review: Haddad et al. 2005). Treatment of streptozotocin-induced diabetic rats with parsley (2 g/kg for 28 days) reversed the effects of diabetes on blood glucose and tissue LPO and GSH levels. However, the treatment resulted in increase in LPO and decrease in GSH levels in both aorta and heart tissue of diabetic rats (Sener et al. 2003).

In another study of the streptozotocin-induced diabetic group given the *Petroselinum crispum* extract, the number of secretory granules and cells in islets and other morphologic changes were not different from the diabetic control group. However, the blood glucose levels in the diabetic group given the plant extract were reduced. It is suggested that the plant therapy can provide blood glucose homeostasis and cannot regenerate β-cells of the pancreas (Yanardag et al. 2003).

Peucedanum pastinacifolium Boiss and Hohen, Apiaceae

The hydroalcoholic extracts of the aerial parts of *Peucedanum pastinacifoliurn* exhibited antihyperlipidemic activities in streptozotocin-induced diabetic rats (Movahedian et al. 2010).

Phaseolus mungo L. (this is a synonym that is widely used)

Botanical Name: *Vigna mungo* (L.) Hepper, Fabaceae

Phaseolus mungo is a highly prized pulse and is widely cultivated. Common names of this plant include black gram, black lentil, urad bean, and urad payara. *Vigna mungo* is a suberect stemmed annual, more or less densely clothed with loose deflexed hairs; leaves are compound, with ovate stipules and membranous leaflets, with scattered and hairs on both sides. Medium-sized short pedicellate flowers are borne on capitate racemes. The pods are clothed with long spreading deciduous silky hair. In traditional medicine, it has acrid, sweet, cooling, laxative, antipyretic, astringent, etc., properties; it is used in treating biliousness, enriching blood, fever, eye troubles, headache, nose complaints, throat inflammation, bronchitis, kidney troubles, tumors, and abscesses (Kritikar and Basu 1975).

Black gram reduced serum glucose and lipid levels in alloxan-induced diabetic guinea pigs (Srivastava and Joshi 1990). Black gram fiber has hypoglycemic action in experimental animals (Boby and Leelamma 2003). Hypolipidemic action of the polysaccharide from black gram has also been reported (Menon and Kurup 1976). Black gram reduced serum lipids in normal and alloxan-induced diabetic guinea pigs. Its hypolipidemic activity has also been reported in rats. Biochanin A and formononetin (isoflavones) isolated from this plant showed hypolipidemic activity in rats.

Phytochemicals reported include phaselic acid, phaseolosides (A, B, C, D and E), phaseollidin, phaseollin isoflavin, rutin, quercetin derivatives, delphidin-3-monoglucoside, petunidin-3-monoglucoside, malvidin-3-monoglucoside, kaempferol-3-monoglucuronide, nicotiflorin, p-coumaryl glucose, feruloyl glucose, caffeoyl glucose, p-3-coumaryl-3-ferulylquinic acid, kievitone, vomifoliol, dehydrovomifoliol, 4-dihydrophaseic acid, 17-β-sitosterol, glycinoeclepins A and B, kaempferol, robinin, brassinosteroids, brassinolide, castasterone, soyasaponin V, and soyasapogenol B (review: Ajikumaran and Subramoniam 2005).

Phaseolus vulgaris L., Fabaceae

Common Names: Kidney beans, dwarf bean, haricot beans, and bakla
Phaseolus vulgaris is a suberect or twining, subglabrous annual with trifoliate leaves. The flowers are few in number in racemose inflorescence. The pods are linear or slightly curved containing 4–6 seeds. It is the native of tropical America and now cultivated throughout the tropical and temperate region. In traditional medicine *P. vulgaris* is carminative and a diuretic; it is used in diabetes and kidney and heart ailments (review: Ajikumaran and Subramoniam 2005).

Oral administration of 200 mg/kg of the water extract of the plant pods to diabetic animals led to a significant decrease in blood glucose and glycosylated hemoglobin and significant increase in total hemoglobin and plasma insulin (Pari and Venkateswaran 2003a). The seed also exhibited hypoglycemic and hypolipidemic effects in streptozotocin-induced diabetic rats (Pari and Venkateswaran 2004). The water extracts of *P. vulgaris* seed and pod (200 mg/kg) showed antioxidant and hypolipidemic activities in streptozotocin-induced diabetic rats (Venkateswaran and Pari 2002). Antioxidant activities of the bean as well as antimutagenic activities of natural phenolic compounds present in the bean have been reported. A broad-spectrum antifungal peptide has been isolated from the beans (review: Ajikumaran and Subramoniam 2005).

Rats fed raw kidney bean lost body weight rapidly and the majority die by 9 days, whereas pretreatment of the beans by extrusion cooking resulted in body weight gain.

Phytochemicals reported include phaseothione, hydrocyanic acid, allantoin, allantoic acid, phaseolin, conphaseolin, phaselin, phaseolotoxin A, quercetin glycoside, cyanogenic glycosides, flavonol glycosides, and polyphenolic in the seed coats.

Phellodendron amurense Ruprecht, Rutaceae

Phellodendri Cortex, the dried trunk bark of *Phellodendron amurense*, contains a number of alkaloids, for example, berberine, palmatine, and jatrorrhizine. Oral administration of Phellodendri Cortex extract (379 mg/kg/day) to streptozotocin-induced diabetic rats for 4 weeks resulted in marked reduction in blood glucose levels and prevention or retardation of the development of diabetic nephropathy in streptozotocin-induced diabetic rats (Kim et al. 2008).

Phlogacanthus thyrsiflorus Nees, Acanthaceae

Phlogacanthus thyrsiflorus is a traditional medicinal plant used to treat diabetes and other ailments. The crude methanol extracts of the leaf and stem bark and the chloroform and petroleum ether fractions of the methanol extract improved oral glucose tolerance in mice. Further, the methanol extract of the stem bark and leaf showed significant hypoglycemic activity at a dose of 200 mg/kg in mice (Ilham et al. 2012). In another study, administration of the water extract of the flower of *P. thyrsiflorus* (100 and 200 mg/kg daily for 21 days) to streptozotocin-induced diabetic rats resulted in a significant decrease in the levels of blood glucose and cholesterol and an increase in the levels of liver glycogen in diabetic rats compared to diabetic control animals (Chakravarty and Kalita 2012). In a follow-up study, administration of the flower extract (100 and 200 mg/kg for 7 days) showed reduction in blood glucose and lipid levels in streptozotocin-induced diabetic rats. Further, the treatment improved liver and kidney functions as judged from serum biochemical parameters (Chakravarty and Kalita 2014).

Phoenix dactylifera L., Arecaceae

Common Name: Date fruit
Subacute administration of the hydroalcoholic extract of *Phoenix dactylifera* leaf or its fractions to alloxan-induced diabetic rats significantly reduced blood glucose. Water intake, serum TG, and cholesterol also decreased while plasma insulin level increased in treated diabetic animals compared with the diabetic control group (Mard et al. 2010). In another study, the effects of a date fruit water extract diet on diabetic neuropathy in streptozotocin-induced diabetic rats were evaluated. In streptozotocin-induced diabetic neuropathy, chronic treatment for 6 weeks with the date fruit extract counteracted the impairment of the explorative activity of rats in an open field behavioral test and of the conduction velocity

of the sciatic nerve. In addition, pretreatment with the fruit significantly reversed each nerve diameter reduction in diabetic rats. Thus, the fruit extract is a neuroprotective agent in diabetic peripheral neuropathy (Zangiabadi et al. 2011).

Phoradendron reichenbachianum (Seem.) Oliv., Viscaceae/Santalaceae

Morolic and moronic acids are the main constituents of the acetone extract from *Phoradendron reichenbachianum*. Daily administration of morolic and moronic acids (50 mg/kg) to NIDDM rats significantly lowered the blood glucose levels at 60% on the 1st day and the low level was maintained until the 10th day after treatment compared to the untreated group. Moreover, both compounds diminished concentrations of cholesterol and TGs in plasma. Also, pretreatment with 50 mg/kg of each compound induced significant antihyperglycemic effect after glucose and sucrose loading (2 g/kg) compared with the control group. *In vitro* studies showed that these compounds induced inhibition of 11β-hydroxysteroid dehydrogenase type 1 (11β-HSD 1) activity at 10 µM (Ramirez-Espinosa et al. 2013). Oleanolic acid and ursolic acid isolated from this plant inhibited PTP1B activity (review: Nazaruk and Borzym-Kluczyk 2014). Oleanolic acid and ursolic acid are known to have anti-DM properties.

Phragmites vallatoria (Pluk. ex L.) Veldk., Poaceae

Administration of the ethanol extract of leaves of *Phragmites vallatoria* (500 mg/kg) to streptozotocin-induced diabetic rats resulted in a reduction in blood glucose levels and increase in body weight and liver and muscle glycogen content compared to diabetic control rats (Vamsikrishna et al. 2012).

Phyllanthus amarus Schum. & Thonn., Phyllanthaceae (Formerly Euphorbiaceae)

Phyllanthus niruri, *Phyllanthus amarus*, and *Phyllanthus sellowianus* are very similar but different species; in pharmacological studies, confusion would have occurred; they possess many similar pharmacological effects (Figure 2.75).

Common/Local Names: Hurricane weed, gale-o-wind, bahupatra, kilanelli, etc.

Description: *Phyllanthus amarus* is a branching annual glabrous herb that is 30–60 cm high and has slender, leaf-bearing branchlets and distichous leaves that are subsessile elliptic-oblong, obtuse, and rounded at the base. The flowers are yellowish, whitish, or greenish and axillary; male flowers are in groups of 1–3, whereas females are solitary. The fruits are depressed globose-like smooth capsules present underneath the branches, and the seeds are trigonous and pale brown with longitudinal parallel ribs on the back (review: Verma et al. 2014).

FIGURE 2.75 *Phyllanthus amarus.*

Distribution: This plant is distributed in tropical and subtropical regions of the world.

Traditional Medicinal Uses: The plant has bitter, astringent, sweet, cooling, diuretic, deobstruent, stomachic, febrifuge, and antiseptic properties. It is useful in treating gastropathy, dropsy, jaundice, diarrhea, dysentery, intermittent fever, ophthalmopathy, diseases of the urinogenital system, scabies, ulcers, and wounds (Warrier 2002).

Pharmacology

Antidiabetes: Several studies have shown that the water, methanol, and hydroalcoholic extracts of *P. amarus* have antidiabetic potential (review: Patel et al. 2012c). *P. amarus* methanol extract was reported to reduce the blood sugar in alloxan-induced diabetic rats by 6% at a dose level of 200 mg/kg and by 18.7% at 1000 mg/kg. It has also been observed that extract (1 g/kg) administration for 18 days to alloxan-induced diabetic rats normalized blood sugar levels (Raphael et al. 2002). In another study, the methanol extracts of this plant (200 and 400 mg/kg) exhibited marked reduction in fasting blood glucose levels and improvement in glucose tolerance in alloxan-induced diabetic rats. Further, histopathological observations showed no visible lesion in the liver, kidney, and pancreas of the extract-treated groups (Adedapo et al. 2013). Two flavanoids isolated from the ethanol extract of *P. amarus* exhibited hypoglycemic action in alloxan-treated rats. The water extract of this whole plant also showed antidiabetic properties in experimental animals. The water extract of the leaf and seed of *P. amarus* produced a dose-dependent decrease in the fasting plasma glucose and cholesterol and also reduction in weights in the treated mice (Adeneye 2006). Recently, it has been shown that the water extract decreased blood glucose in a dose-dependent manner and a maximum effect was observed 2 h after administration (400 mg/kg) in normal rats; it suppressed postprandial rise in blood glucose also. Chronic administration of the extract (200 and 400 mg/kg) for 28 days to alloxan-induced diabetic rats decreased blood glucose levels. At 400 mg/kg dose, the effect was better than that of glibenclamide. Further, the treatment improved body weight and decreased total cholesterol and TG levels in the blood (Adedapo et al. 2014). In another study also, the water extract (260 mg/kg) decreased blood glucose levels in alloxan-induced diabetic rats (Mbagwu et al. 2011). The water extracts (200 mg/kg) showed anti-DM activity in streptozotocin-induced diabetic rats also. The extract, when administered daily for 8 weeks, decreased blood glucose and lipid levels with an increase in body weight in diabetic animals. Further, the treatment decreased the lipid peroxide content, protein oxidation, and oxidative stress in the kidney suggesting nephroprotective activity of the extract (Karuna et al. 2011). *P. amarus* influenced the level of insulin in the blood also. The water and hydroalcoholic extracts of *P. amarus* (when administered daily for 15 days) decreased the blood glucose and MDA levels and increased the levels of insulin in alloxan-induced diabetic rats (Lawson-Evi and Degbeku 2011).

Thus, several studies show the antidiabetic potential of *P. amarus*. However, fuller information on the active principles involved, comparative studies to determine the best extract, and details of the mechanism of action remain to be studied.

Other Activities: *Phyllanthus amarus* is a well-known hepatoprotective plant with antihepatitis viral activity. Other validated pharmacological properties include antiviral, anticancer, antiamnesic, antioxidative, antimicrobial, antinociceptive, antileptospiral, anti-inflammatory, anticonvulsant, nephroprotective, cardioprotective, diuretic, and antifertility activities. Further, the extract of this plant decreased the erythrocyte sedimentation rate and packed cell volume; circulating leukocytes and neutrophils count were significantly increased in rats treated with the extract of this plant (review: Verma et al. 2014).

Phytochemicals: The herb contains alkaloids (isobubbialine and epibubbialine; pyrrolizidine types of alkaloids such as securinine, dihydrosecurinine, tetrahydrosecurine, securinol B, phyllanthine, allosecurine, norsecurinine, 4-methoxy dihydrosecurinine, 4-methoxytetrahydrosecurinine, and 4-hydrosecurinine), tannins (geraniin, corilagin, 1,6-digalloylglucopyranoside rutin, quercetin-3-*O*-glucopyranoside, amarulone, phyllanthusiin D, and amariin), ellagitannins (amariin, 1-galloyl-2,3-dehydrohexahydroxydiphenyl-glucose, repandusinic acid, geraniin, corilagin, phyllanthusiin D), flavonoids (rutin, quercetins, quercetin 3-*O*-glucoside, 1-*O*-galloyl-2,4-dehydrohexahydroxydiphenoyl-glucopyranose elaeocarpusin, repandusinic acid A, and geraniinic acid), lignans (niranthin, nirtetralin, phyltetralin, hypophyllanthin, phyllanthin, demethylenedioxy niranthin, 5-demethoxy-niranthin, isolintetralin), volatile oil (linalool and phytol), and triterpene (phyllanthenol, phyllanthenone, and phyllantheol) (review: Verma et al. 2014).

Phyllanthus emblica Linn., Euphorbiaceae

Synonym: *Emblica officinalis* Gaertn. (Figure 2.76)

Common Names: Indian gooseberry, amala, adiphala, amloke, nelli, emblic myrobalan, etc.

Description: *Phyllanthus emblica* is a medium-sized, deciduous tree with pubescent, slender branchlets and crooked trunk. The leaves are equally sized and symmetrically arranged like the leaflets of a pinnate leaf with scarious and lacerative stipules. The flowers are small, with yellowish racemes on branches. The fruits are nearly spherical or globular, wider than long and with a small and slight conic depression on both apexes; the fruit is 18–25 mm wide and 15–20 mm long; the surface is smooth with 6 obscure vertically pointed furrows; the mesocarp is yellow and the endocarp is yellowish-brown in ripened condition; in the fresh fruit, the mesocarp is acidulous, and in the dried fruit, it is acidulous astringent. It has four to six, smooth, and dark brown seeds (review: Khan 2009).

Distribution: *P. emblica* is present in the wild as well as in plantations throughout tropical India. It also occurs in China, Pakistan, Uzbekistan, Sri Lanka, Malaysia, etc. The fruits are edible and the plant is cultivated.

Traditional Medicinal Uses: The fruits are acrid, cooling, sour, bitter, sweetish, anodyne, astringent, refrigerant, diuretic, laxative, carminative, alexiteric, tonic, and antipyretic and are used in treating hemorrhage, diarrhea, dysentery, asthma, inflammation of the eye, bronchitis, jaundice, dyspepsia, cough, anthrax, burning sensation, vomiting, biliousness, urinary discharge, thirst, carbuncles, leprosy, piles, erysipelas, anuria, cholera, constipation, leukorrhea, nasal hemorrhage, convulsion, gonorrhea, sores, myalgia, and snakebite. The fruits are also said to be beneficial in insomnia, skin problems, gall pain, and tympanites (Kirtikar and Basu 1952; Hossen et al. 2014).

Pharmacology

Antidiabetes: The leaf extract of *P. emblica* showed promising anti-DM and antioxidant activities in streptozotocin-induced type 2 diabetic rats. The treatment resulted in an increase in serum insulin levels compared to untreated diabetic control animals (Nain et al. 2012).

Oral administration of the fruit extracts (100 mg/kg body weight) reduced the blood sugar levels in normal and in alloxan-induced (120 mg/kg) diabetic rats (Sabu and Kuttan 2002b). In another study, the water extract of the fruit (200 mg/kg, i.p.) decreased the blood glucose level in alloxan-induced diabetic rats. Further, the treatment decreased TG levels in diabetic rats. In addition, the extract improved liver

(a)   (b)

FIGURE 2.76 *Phyllanthus emblica.* (a) Stems with fruits. (b) Flowers.

functions by normalizing the activity of the liver-specific marker enzyme alanine transaminase (Shamim et al. 2009). The ethanol extract of the plant fruit caused a significant dose-dependent inhibition of intestinal disaccharidase activity. In a recent study, the methanol extract of the fruit, in a dose-dependent manner, increased residual sucrose content throughout the gut after sucrose ingestion. Further, the methanol extract of the fruit showed significant inhibition of intestinal disaccharidase activity in rats. The gut perfusion analysis also showed that the extract significantly reduced intestinal glucose absorption (Sultana et al. 2014). Hydrolyzable tannoids from *E. officinalis* extract inhibited rat lens aldose reductase that is involved in the development of some of the complications of diabetes (Suryanarayana et al. 2004). Further, *E. officinalis* and its enriched tannoids delay streptozotocin-induced diabetic cataract in rats (Suryanarayana et al. 2007).

Thus, the *P. emblica* fruit possesses promising antidiabetes properties that include increase in insulin levels, inhibition of intestinal disaccharidase activity, and inhibition of lens aldose reductase activity. However, the phytochemicals involved in producing the anti-DM properties and their interactions are not known. The fruit is an attractive nutraceutical for medicine and/or nutritional supplement development.

Clinical Trial: In a randomized, double-blind, controlled clinical study, a standardized water extract of *P. emblica* (containing low-molecular-weight hydrolyzable tannins: emblicanin A, emblicanin B, pedunculagin, and punigluconin) significantly improved endothelial function and reduced biomarkers of oxidative stress and systemic inflammation in patients with type 2 DM, without any significant changes in laboratory safety parameters (Usharani et al. 2013). Further, the treatments significantly improved the lipid profile and HbA1c levels compared with baseline and placebo.

Other Activities: The major pharmacological properties of *E. officinalis* include hepatoprotection, antioxidant activity, anti-inflammation activity, anticancer activity, antimicrobial activity, and antiulcer activity (review: Khan 2009). *E. officinalis* fruit extract exhibited antitussive, antiphlogistic and antispasmolytic, and antioxidant activities (Nosal'ova et al. 2003). The extract showed marked reduction in complete Freund's adjuvant-induced inflammation and edema and caused immunosuppression to adjuvant-induced arthritic rats (Ganju et al. 2003). The fruit caused myocardial adaptation and protects against oxidative stress in ischemic reperfusion injury in rats (Rajak et al. 2004). Norsesquiterpenoid glycosides from the roots exhibited significant antiproliferative activities against cultured human cancer cells. *E. officinalis* reversed thioacetamide-induced oxidative stress and early promotional events of primary hepatocarcinogenesis (Sultana et al. 2004). The ethanol extract of the fruit protected mice from 7,12-dimethylbenz(a)anthracene (DMBA)-induced genotoxicity (Banu et al. 2004). The polyphenols from the fruit have strong antioxidant properties (Sabu and Kuttan 2002b). The polyphenols induced apoptosis in mouse and human carcinoma cell lines (Rajeshkumar et al. 2003). The fruit also provided protection against gamma radiation-induced damages in mice (Hari Kumar et al. 2004). The fruit possessed anti-inflammatory, antimicrobial, and antidiarrheal activities (Hossen et al. 2014).

Phytochemicals: Phytochemicals isolated from this plant include myristic acid, linoleic acid, vitamin C, tannins, alkaloids, flavanoid, phyllanthin, hypophyllanthin, proanthocyanidin polymers, 3-*O*-gallated prodelphinidin and procyanidin, gallic acid, ellagic acid, 1-*O*-galloyl-β-D-glucose, 3,6-di-*O*-galloyl-D-glucose, chebulinic acid, quercetin, chebulagic acid, corilagin, 3-ethylgallic acid (3-ethoxy-4,5-dihydroxy-benzoic acid), norsesquiterpenoids, isostrictiniin, 1,6-di-*O*-galloyl-β-D-glucose, emblicanin A, emblicanin B, punigluconin, pedunculagin, ellagitannin, trigallayl glucose, and pectin (reviews: Khan 2009).

Phyllanthus fraternus GL Webster, Phyllanthaceae

Flavonoid fractions from *Phyllanthus fraternus* exhibited hypoglycemic activity in rats (Hukeri et al. 1998). In another study, administration of standardized alcohol extract of *P. fraternus* (whole plant) at a dose of 500 mg/kg for 21 days to alloxan-induced diabetic rats resulted in improvement in the lipid profile, kidney function (decrease in serum urea and creatinine), and liver function (decrease in GPT, GOT, and ALP). The extract contained tannins and flavonoids as major constituents; rutin and quercetin are the two major flavonoids (Garg et al. 2010). In another study, the water extract of *P. fraternus* (250 mg/kg daily for 3 weeks) was tested for its beneficial effects in fructose-induced insulin-resistant rats. The extract attenuated fructose-induced hyperinsulinemia, glucose intolerance, insulin resistance, hypertriglyceridemia, and hypertension (Kushwah et al. 2010).

Phyllanthus niruri L., Phyllanthaceae/Euphorbiaceae

Synonyms: *Phyllanthus carolinianus* Blanco, *Phyllanthus sellowianus* Mull., *Phyllanthus sellowianus* Mull. Arg.

Administrations of the methanol extract of *Phyllanthus niruri* (whole plant) at doses of 125 and 250 mg/kg for 14 days to streptozotocin-induced diabetic rats resulted in significant reduction in blood glucose levels and improvement in body weight; further, the treatment reduced oxidative stress in diabetic rats (Mazunder et al. 2005). The ethanol extract of the leaves of this plant exhibited antidiabetic and antihyperlipidemic effects in experimental animals (Bavarva and Narasimhacharya 2005). In another study, administration of the water extract of *P. niruri* (120 and 240 mg/kg daily for 14 days) to streptozotocin-induced diabetic rats caused a significant decrease in blood glucose and prevented the loss of body weight, to a large extent. Alkaloids, flavonoids, and saponins were found to be present in the water extract (Nwanjo 2006).

Phyllanthus sellowianus Müller Arg., Euphorbiaceae

Common Name: Sarandí blanco

Phyllanthus sellowianus is widely used in popular medicine as a hypoglycemic and diuretic. The hypoglycemic effect of a water extract of *P. sellowianus* was shown in C57BL/Ks mice. In normal animals, a slight reduction in the glucose levels was observed, and in those with glucose overload, the capacity to normalize the glucose levels was potentiated by the extract. In animals treated with low doses of streptozotocin, the blood glucose could also be maintained at normal levels. This study provided evidence for its potential use in the prevention and treatment of diabetic disorders (Navarro et al. 2004). In another study, two fractions of the water extract (200 mg/kg, p.o.) caused a significant reduction in blood glucose levels in streptozotocin-induced diabetic rats (Hnatyszyn et al. 2002). The mechanisms of action and active principles involved and long-term safety remain to be investigated.

Phyllanthus simplex Retz., Euphorbiaceae

Various fractions of *Phyllanthus simplex* (whole herb) showed antidiabetic and antioxidant activities in alloxan-induced diabetic rats (Shabeer et al. 2009).

Physalis alkekengi L., Solanaceae

Common Name: Strawberry tomato

Physalis alkekengi is a traditional medicinal plant used to treat various diseases including liver disease, cancer, and arthritis. The *P. alkekengi* fruit (50 and 100 mg/kg, water extract) exhibited antidiabetic activity in alloxan-induced diabetic rats (Sharma and Arya 2011). A purified polysaccharide fraction from the fruits of *P. alkekengi* exhibited antidiabetic activity in alloxan-induced diabetic mice. The polysaccharide significantly reduced blood glucose levels and water intake and increased the body weight of diabetic mice compared with the diabetic control group (Tong et al. 2008). The ethanol extract of *P. alkekengi* leaves and fruits (25, 50, and 100 mg/kg, p.o., daily for 30 days) reduced the level of glucose, total cholesterol, and TGs to near normal levels with an increase in insulin and glycogen concentration to near normal levels in alloxan-induced diabetic rats (Sanchooli 2011). Administration of *P. alkekengi* fruit powder (10% in diet) or methanol extract (500 mg/kg, p.o.) for 2 months to streptozotocin-induced diabetic rats resulted in amelioration of the diabetic condition to a large extant. The treatment along with chromium showed better effects (El-Mehiry et al. 2012). *P. alkekengi* appears to be an attractive plant for detailed studies to determine its therapeutic value.

Picea glauca (Moench) Voss, Pinaceae

Common Name: White spruce

Native American populations have historically used *Picea glauca* as food, traditional medicine, and firewood. The antidiabetic activity of this plant has been reviewed recently (Eid and Haddad 2014). The alcohol extract of *P. glauca* needles showed marked inhibitory activity on glucose-6-phosphatase in cultured hepatocytes. Nonetheless, the weak correlation with activation of either AMPK or Akt suggested that other pathways might be engaged in modulating the activity of glucose-6-phosphatase. Besides, this plant extract exerted a powerful dose-dependent inhibition of intestinal glucose transport in CaCa-2 cells. However, this effect may be short-lasting since it waned in longer-term experiments. Importantly, *P. glauca* showed a prominent cytoprotective effect on neuronal precursor cells against both glucose toxicity and glucose deprivation. The effect was also found to be dose dependent and organ specific as bark extracts had negligible activity, while cone extracts further enhanced the effects of high or low glucose. Compounds uniquely detected in *P. glauca* needles warrant further investigation as adjuvant therapy in diabetes neuropathic complications. *P. glauca* significantly inhibited the activity of both recombinant CYP2C9 and CYP2C19 drug metabolizing enzymes *in vitro*. Thus, white spruce preparations inhibited intestinal glucose transport and liver glucose-6-phosphatase activity while protecting preneuronal cells against high glucose toxicity (Eid and Haddad 2014).

Picralima nitida (Stapf) T. Durand & H. Durand, Apocynaceae
The ethanol and butanol extracts of *Picralima nitida* seeds decreased hyperglycemia in pregnant streptozotocin-induced diabetic rats. The plant extract also showed substantial antioxidant activity and T-cell proliferation inhibition activity (Yessoufou et al. 2013). In another study, the antioxidant and antidiabetic potential of the methanol extract of the stem bark and leaves of *P. nitida* was evaluated in streptozotocin-induced diabetic rats. The methanol extract (300 mg/kg) of leaves exhibited significant antidiabetic activities. Further, the methanol extract showed significant reduction in the levels of MDA and hydrogen peroxides and a substantial increase in catalase activity in diabetic rats (Teugwa et al. 2013).

Picrorhiza kurroa Royle., Scrophulariaceae
Synonyms: *P. kurroa* Royle ex Benth., *P. kurroa* Benth (monograph: WHO 2009)
Vernacular names of *Picrorhiza kurroa* include balakadu len, kadu, honglen, katuka, katuki, katurohini, kuri, kutki, and kadugurohini. *P. kurroa* is a hairy herb with perennial, woody, bitter root stock. The plant is found commonly in the Alpine Himalayas from Kashmir to Sikkim. This is a well-known hepatoprotective plant. In traditional medicine, it is used to treat various ailments.

The antidiabetic activity of the plant extract has been shown in alloxan-induced diabetic rats (Joy and Kuttan 1999). Administration of the water extract of *P. kurroa* (100 and 200 mg/kg, p.o.) for 14 days to streptozotocin–nicotinamide-induced diabetic rats resulted in significant reduction in elevated fasting glucose levels; further, the treatment improved oral glucose tolerance and body weight (Husain et al. 2009). The antidiabetic activity, if any, of the hepatoprotective principle picroliv remains to be studied.

Picroliv, isolated from the root and rhizome of *P. kurroa*, is known to have significant hepatoprotective activity against toxic chemical-induced liver damages (review: Subramoniam and Pushpangadan 1999). Picroliv modulated the expression of insulin-like growth factor receptor during hypoxia in rats. The rhizome of the plant contains bioactive molecules such as picroliv, iridoid glycosides, apocynin, vanillic acid and cucurbitacin glycosides, picrorhizin, kutkin, kurrin, vanillic acid, kutkiol, and kutki-sterol (monograph: WHO 2009).

Pilea microphylla (L.) Liebm., Urticaceae
The flavonoid-rich fraction of *Pilea microphylla* (100 mg/kg/day, i.p., for 28 days) produced significant reduction in body weight, plasma glucose, TGs, and total cholesterol content in HFD-fed streptozotocin-induced diabetic mice. The fraction also improved oral glucose tolerance and enhanced the endogenous antioxidant status in mice liver compared to diabetic control. The extract preserved islet architecture and prevented hypertrophy of hepatocytes as evident from the histopathology of the pancreas and liver (Bansal et al. 2012).

Pimpinella tirupatiensis Bal. & Subr., Apiaceae/Umbelliferae
Administration of the ethyl acetate extract of *Pimpinella tirupatiensis* (750 mg/kg for 30 days) reduced fasting blood glucose level and MDA, whereas body weight, antioxidant enzymes, and nonenzymatic antioxidants (reduced GSH, vitamin C and vitamin E) were increased in streptozotocin-induced diabetic rats compared to untreated diabetic control rats (Narasimhulu et al. 2012).

Pinellia ternata (Thunb.) Ten. ex Breitenb., Araceae
Pinellia ternata is a traditional medicinal plant used to treat symptoms such as nausea and vomiting. Fatty acids present in different apolar extracts of this plant rhizome activated PPAR-γ (Wang et al. 2014).

Pinus banksiana Lamb., Pinaceae

Common Name: Scrub pine
Pinus banksiana is a North American pine. The plant extract caused significant increase in the phosphorylation of AMPK *in vitro* (review: El-Abhar and Schaalan 2014).

Pinus pinaster Aiton, Pinaceae

Previous Name: *P. maritima* Mill.

Common Name: French maritime pine
The standardized maritime pine (*Pinus pinaster*) bark extract (Pycnogenol®) was reported to exert clinical antidiabetic effects after personal intake. However, an increase in insulin secretion was not observed after administration of the extract to patients. Pycnogenol showed very potent α-glucosidase inhibitory effect compared to acarbose. The inhibitory action was found to be higher in extract fractions containing higher procyanidin oligomers (Schafer and Hogger 2007).

Pinus roxburghii Sarg., Pinaceae

Common Names: Chir pine, chir, etc.
Pinus roxburghii is a large tree usually attaining a height of 100 ft with thick and furrowed bark. The leaves are dark or light green and serrulate. The male spike is cylindrical, with solitary or clustered cones bearing oblong seeds with wings. *Pinus roxburghii* is distributed in the Outer Himalayan ranges and is planted in many gardens in the plains. In traditional medicine, it is a stimulant, stomachic, and antiseptic and is used in the treatment of chronic bronchitis, gangrene of the lungs, flatulent colic, obstinate constipation, tympanitis, worm infection, diabetes, and rheumatic infections.

The antidiabetic property of this plant has been reported (Jain and Sharma 1967a). Recent molecular docking studies have showed that three molecules (secoisoresinol, pinoresinol, and cedeodarin) present in *P. roxburghii* are likely to bind with target molecules involved in diabetes (receptors PTP1β, DPP-IV, aldose reductase, and IR). The most significant molecular docking results were observed in the case of aldose reductase. Thus, the docking results have given insights into the development of better aldose reductase inhibitors from this plant (Kaushik et al. 2014).

Phytochemicals reported include turpentine oil, pine oil, rosin, phenols, rosin spirit, pine needle oil, α-longipinene, abietic acid, isopimaric acid, friedelin, ceryl alcohol, β-sitosterol, and hexacosylferulate (review: Ajikumaran and Subramoniam 2005).

Piper betle L., Piperaceae
Piper betle is a medicinal plant with a folk reputation in rural India. The water and ethanol extracts of the leaves of *P. betle* reduced glucose and glycosylated hemoglobin levels in streptozotocin-induced diabetic rats (Arambewela et al. 2005). In another study, oral administration of the leaf suspension of *P. betle* (75 and 150 mg/kg) for 30 days resulted in a significant reduction in blood glucose and glycosylated hemoglobin and decreased activities of liver glucose-6-phosphatase and fructose-1,6-bisphosphatase, while liver hexokinase increased in streptozotocin-induced diabetic rats when compared with untreated diabetic rats. *P. betle* (75 mg/kg) exhibited better sugar reduction than 150 mg/kg. In addition, protection against body weight loss of diabetic animals was also observed. The effects produced by *P. betle* were comparable with those of the standard drug glibenclamide (Santhakumari et al. 2006). Further studies are needed to determine the therapeutic utility of *P. betle*.

Piper guineense Schum and Thonn, Piperaceae
Piper guineense is used as a spice from the ancient time onward. Oral administration of the crude water extract of *P. guineense* (500 mg/kg) to alloxan-induced diabetic rats resulted in marked reduction in the levels of fasting blood glucose, total cholesterol, LDL, VLDL, and TGs compared to diabetic control rats (Ekoh et al. 2014).

Piper longum L., Piperaceae

Synonym: *Piper aromaticum* Lam.

Common Names: Long pepper, thippali

Piper longum is a native of the Indo-Malayan region. The fruit of *P. longum* is a valuable spice and is also used as a therapeutic agent in the treatment of various pathological conditions in ethnomedicine. The ethyl acetate and ethanol extracts of *P. longum* fruits showed antihyperglycemic activity and attenuated oxidative stress in streptozotocin-induced diabetic rats (Kumar et al. 2011f). The water extract of *P. longum* root (200 mg/kg) was found to possess significant antidiabetic activity after 6 h of treatment in streptozotocin-induced diabetic rats. Administration of the same dose of the extract for 30 days to streptozotocin-induced diabetic rats resulted in a significant decrease in fasting blood glucose levels with the corrections of diabetic dyslipidemia compared to untreated diabetic rats. There was a significant improvement in the activities of liver and renal functional markers in treated diabetic rats compared to untreated diabetic rats indicating the protective role of the water extract of *P. longum* against liver and kidney damages (Nabi et al. 2013).

Piper retrofractum Vahl., Piperaceae

Synonyms: *Piper officinarum* (Miquel) C.D.C., *Piper chaba* Hunter, *Piper longum* Blume, *Chavica maritema* Miquel. (a closely related species is *Piper longum* L.)

The fruits of *Piper retrofractum* have been used for their antiflatulent, expectorant, antitussive, antifungal, and appetizing properties in traditional medicine and are reported to possess gastroprotective and cholesterol-lowering properties. Piperidine alkaloids (piperine, pipermonaline, and dehydropipermonaline) from *P. retrofractum* attenuated HFD-induced obesity by activating AMPK and PPAR-δ and regulated lipid metabolism, suggesting their potential antiobesity effects. Oral administration of the alkaloids (50, 100, or 300 mg/kg/day for 8 weeks) to obese mice significantly reduced HFD-induced body weight gain without altering the amount of food intake. Fat pad mass was reduced in the treatment groups, as evidenced by reduced adipocyte size. In addition, elevated serum levels of total cholesterol, LDL-C, total lipid, leptin, and lipase were suppressed by the treatment. Piperidine alkaloids also protected against the development of nonalcoholic fatty liver by decreasing hepatic TG accumulation (Kim et al. 2011b). In view of its role in influencing AMPK and PPAR-δ, further studies are warranted to determine its likely antidiabetic property.

Piper sarmentosum Roxb., Piperaceae

Thai Name: Chaplu

Piper sarmentosum is used in Asian countries as food and in traditional medicine. The plant is abundant in Asia and used as a remedy for DM in local health traditions.

The water extract of *P. sarmentosum* (whole plant) at doses of 125 and 250 mg/kg (single oral dose) improved glucose tolerance in normal rats. Repeated oral administration of the water extract (125 mg/kg) for 7 days produced a significant reduction in blood glucose levels in diabetic rats (Peungvicha et al. 1998b). The water extract of *P. sarmentosum* leaves possessed anti-DM activity and showed signs of β-cell regeneration in streptozotocin-induced diabetic rats. Besides, the water extract of the leaves attenuated hyperglycemia and associated diabetic nephropathy (as judged from histological observations) in streptozotocin-induced diabetic rats (Hussain et al. 2013). Besides, the water extract of the plant (125 mg/kg for 28 days) ameliorated the oxidative stress in streptozotocin-induced diabetic rats. The extract-treated diabetic rats showed a marked decrease in MDA levels and an increase in erythrocyte superoxide dismutase activity (Rahman et al. 2010).

In vitro studies have also showed the anti-DM potential of the water extract of the aerial part of this plant (Krisanapun et al. 2012). The extract, when tested using the inverted intestinal sac method, at a concentration of 5 mg/mL, lowered glucose absorption from the intestine. Further, the extract (2.5–5 mg/mL) enhanced glucose consumption by the isolated rat diaphragm. Thus, the hypoglycemic activity of the plant extract may result from inhibition of glucose absorption and enhancement of glucose consumption (Krisanapun et al. 2012).

Further studies are warranted on this plant to determine its possible use as an anti-DM medicine. These include isolation of active principles, detailed safety evaluation, and follow-up clinical studies.

Pisonia alba Span., Nyctaginaceae
Common Name: Lettuce tree
The ethanol extract of *Pisonia alba* showed α-glucosidase inhibitory activity in a concentration-dependent manner *in vitro*. Administration of the extract (250 and 500 mg/kg) for 15 days to alloxan-induced diabetic rats resulted in highly significant decrease in blood glucose, serum GPT, GOT, ALP, cholesterol, and TG levels and increase in HDL levels as compared to diabetic control rats (Christudas et al. 2009).

Pistacia atlantica Dest., Anacardiaceae
Pistacia atlantica is used in Jordan as an ethnomedicine to control diabetes. *In vitro* studies using MIN6 β-cells in culture have shown that the extract (10–500 μg/mL) potentiated calcium-stimulated insulin secretion from the β-cells. Further, the plant extract inhibited dose-dependent glucose movement *in vitro* as effectively as gaur gum diffusional hindrance in a simple glucose dialysis model (Kasabri et al. 2012a).

Pistacia lentiscus L. (var. *chia*), Anacardiaceae
Pistacia lentiscus is used in Indian traditional medicine as carminative, diuretic, stimulant, astringent, etc. The antidiabetic compound oleanolic acid has been reported in this plant. Among other things, oleanolic acid is an agonist of PPAR-γ (review: Wang et al. 2014).

Pistacia vera L., Anacardiaceae
The extract of *Pistacia vera* exhibited marked inhibition on α-glucosidase activity (Gholamhoseinian et al. 2008).

Plantago ovata Forssk., Plantaginaceae
Common Name: Desert Indian wheat
Plantago ovata is used in traditional medicine to treat DM. Psyllium, a bulk-forming laxative derived from *P. ovate*, is high in both fiber and mucilage. In a human trial, patients were randomly selected from an outpatient clinic of primary care to participate in a double-blind placebo-controlled study in which psyllium or placebo was given in combination with their antidiabetic drugs. Forty-nine subjects were included in the study and were given diet counseling before the study and throughout the 8-week treatment period. The test products (psyllium or placebo) were supplied to subjects in identically labeled foil packets containing a 5.1 g dose of the product, to consume two doses per day, half an hour before breakfast and dinner. Fasting blood glucose, and HbA1c, showed a significant reduction, whereas HDL-C increased significantly following psyllium treatment. The LDL/HDL ratio was significantly decreased. The study showed that psyllium treatment to type 2 diabetes was safe and well tolerated and improved glycemic control (Ziai et al. 2005). In another study, the ability of psyllium fiber to reduce postprandial serum glucose and insulin concentrations was evaluated in 18 NIDDM patients in a crossover design. Psyllium fiber or placebo was administered twice during each 15 h crossover phase, immediately before breakfast and dinner. No psyllium fiber or placebo was given at lunch, which allowed measurement of residual or second-meal effects. For meals eaten immediately after psyllium ingestion, maximum postprandial glucose elevation was reduced by 14% at breakfast and 20% at dinner relative to placebo. Postprandial serum insulin concentrations measured after breakfast was reduced by 12% relative to placebo. Second-meal effects after lunch showed a 31% reduction in postprandial glucose elevation relative to placebo. No significant differences in effects were noted between patients whose diabetes was controlled by diet alone and those whose diabetes was controlled by oral hypoglycemic drugs. Thus, the study indicated that psyllium as a meal supplement reduced proximate and second-meal postprandial glucose and insulin concentrations in non-insulin-dependent diabetics (Pastors et al. 1991).

Plectranthus amboinicus (Lour.) Spreng, Lamiaceae

Synonyms: *Coleus amboinicus, Coleus aromaticus, Plectranthus aromaticus*

Common Names: Indian mint, Spanish thyme, Cuban oregano, Indian borage, etc.

Plectranthus amboinicus is a tender fleshy perennial medicinal herb. The ethanol extract of *P. amboinicus* leaf (200 and 400 mg/kg, p.o., daily for 15 days) possessed an insulinotropic effect and exhibited hypoglycemic and antihyperlipidemic activities in normal and alloxan-induced diabetic rats, respectively. Furthermore, the effect of the ethanol extract of *P. amboinicus* showed profound elevation of serum amylase and reduction of serum lipase. The extract treatment resulted in normalization of damaged pancreatic architecture in rats with DM. Thus, the hypoglycemic and antihyperlipidemic effects of the extract were mediated through the restoration of the functions of pancreatic tissues and the insulinotropic effect (Viswanathaswamy et al. 2011a).

Clinical Trial: In a single-center, randomized, controlled, open-label clinical study, the effects of a topical cream containing *P. amboinicus* and *C. asiatica* were evaluated and compared to the effects of hydrocolloid fiber wound dressing for diabetic foot ulcers. Twenty-four type 1 or type 2 diabetes patients aged 20 years or older with Wagner grade 3 foot ulcers were selected for the study. Twelve randomly assigned patients were treated with WH-1 cream containing *P. amboinicus* and *C. asiatica* twice daily for 2 weeks. Another 12 patients were treated with hydrocolloid fiber dressings changed at 7 days or when clinically indicated. Wound condition and safety were assessed at days 7 and 14 and results were compared between groups. No statistically significant differences were seen in percent changes in wound size at 7- and 14-day assessments of WH-1 cream and hydrocolloid dressing groups. A slightly higher proportion of patients in the WH-1 cream group (10 of 12; 90.9%) showed improvement compared to the hydrocolloid fiber dressing group but without statistical significance. For treating diabetic foot ulcers, *P. amboinicus* and *C. asiatica* cream is a safe alternative to hydrocolloid fiber dressing without significant difference in effectiveness (Yuan-Sung et al. 2012).

Plumeria rubra L., Apocynaceae

Extracts from *Plumeria rubra* (whole plant) lowered blood glucose in experimental animals (Marles and Farnsworth 1995).

Polyalthia longifolia Sonn. var. angustifolia, Annonaceae

The hypoglycemic and antihyperglycemic activities of various solvent extracts of *Polyalthia longifolia* leaf were evaluated in alloxan-induced experimental diabetes in rats. *P. longifolia* extracts and powder (daily administration for 7 days) produced glucose-lowering activity. However, the extracts did not significantly modify any of the biochemical parameters. Hence the extracts and crude powder are devoid of antidiabetic properties, but have gross glucose-lowering properties. The extracts also showed antihyperglycemic effects against sucrose loading-induced hyperglycemia in rats (Nair et al. 2007). In another study, *P. longifolia* leaf (500 mg/kg, ethanol extract) exhibited antihyperglycemic effect in oral glucose tolerance test in rats. The extract also showed analgesic activity in rats (Khan et al. 2013). The ethanol and chloroform extracts of *P. longifolia* leaves showed α-amylase and glucosidase enzyme inhibitory activity. The extracts protected streptozotocin-induced type 1 diabetic rats (Sivashanmugan and Chatterjee 2013).

Polygala senega L. var. *latifolia* Torry et Gray, Polygalaceae

Synonyms: *Polygala senega* L., *Polygala rosea* Steud., *Senega officinalis* Spach (monograph: WHO 2002)

Vernacular Names: Senega snake root, mountain flax, bambara, bulughâ lon, etc.

Polygala senega is a perennial herbaceous plant with numerous stems sprouting from a single, thick gnarled crown arising from a conical, twisted, branched yellow root. The aerial portion consists of several erect or ascending, smooth stems up to 15–40 cm high, bearing alternate, lance-shaped, or oblong-lanceolate leaves with serrulate margins. Inflorescence is a spike of small, white flowers. The plant is indigenous to eastern Canada and northeastern United States (monograph: WHO 2002). It is still used in modern herbalism where it is valued mainly as an expectorant and stimulant to treat bronchial asthma, chronic bronchitis, and whooping cough. It was used by the North American Indians in the treatment of snakebites. A tea made from the bark has been drunk to bring about a miscarriage. The dried root is used as a stimulant and expectorant (Daniel 1998).

Four triterpenoid glycosides isolated from the rhizomes of *P. senega* var. *latifolia* were tested for hypoglycemic activity in normal and KK-Ay diabetic mice. Two of the compounds, senegins II and desmethoxysenegin II, reduced the blood glucose of normal mice 4 h after intraperitoneal administration and also significantly lowered the glucose level of KK-Ay mice under similar conditions. The other compounds were inactive when tested against normal mice (Kako et al. 1997). Senegin II (the main component of *P. senega*) (2.5 mg/kg) significantly reduced the level of blood glucose in glucose-loaded normal mice from 12.2 to 7.3 mmol/L 4 h after intraperitoneal administration and also lowered the blood glucose of KK-Ay mice from 24.2 to 7.9 mmol/L under similar conditions (Kako et al. 1995). In another study, intraperitoneal administrations of the n-butanol extract of the roots (10 mg/kg) reduced blood glucose levels in healthy mice and in mice with streptozotocin-induced hyperglycemia. One of the active components of the hypoglycemic effect was identified as a triterpenoid glycoside, senegin II (Kako and Miura 1994). Intragastric administration of a saponin fraction of a root extract reduced glucose-induced hyperglycemia in rats at a dose of 200 mg/kg. E and Z-senegasaponins and E and Z-senegins II, III, and IV were found to exhibit hypoglycemic activity in the oral glucose tolerance test (Yoshikawa et al. 1996b). The mode of action of escins Ia and IIa and E and Z-senegin II from the plant for the inhibitory effect on the increase in serum glucose levels in oral glucose-loaded rats was studied. Although these compounds inhibited the increase in serum glucose levels in oral glucose-loaded rats, these compounds did not lower serum glucose levels in normal or intraperitoneal glucose-loaded rats. Furthermore, the compounds suppressed gastric emptying in rats and also inhibited glucose uptake in the small intestine of rats *in vitro*. These results indicated that the compounds given orally have neither insulin-like activity nor insulin-releasing activity. The compounds inhibited glucose absorption by suppressing the transfer of glucose from the stomach to the small intestine and by inhibiting the glucose transport system at the small intestinal brush border (Matsuda et al. 1998). But in an earlier study, administration (i.p.) of a saponin fraction of the root extract (25 mg/kg) significantly increased the plasma levels of adrenocorticotropic hormone, cortisone, and glucose in rats (Yokoyama et al. 1982).

The major pharmacological activities reported for this plant is as an expectorant. Other reported activities include lowering of blood cholesterol and TG, inhibition of stress-induced gastric ulcers, inhibition of edema, diuretic activity, and anticancer activity (Paul et al. 2011). As extracts of the root have been shown to stimulate uterine contractions in animal models, Radix Senega should not be taken during pregnancy. The LD_{50} of the root was 17 g/kg body weight after intragastric administration to mice. The LD_{50} of the root bark was 10 g/kg body weight and that of the root core (which had the lowest saponin concentration of the three root samples) was 75 g/kg body weight (monograph: WHO 2002).

Phytochemicals reported include senegins and senegasaponins isolated from the root, (Yoshikawa et al. 1996), methyl salicylate, and triterpene saponins. The saponins are 3-glucosides of presenegenin (monograph: WHO 2002).

Polygonatum odoratum (Mill.) Druce, Asparagaceae/Liliaceae

Synonym: *P. officinale* All.

Common Name: Solomon's seal

Polygonatum odoratum (angular Solomon's seal) is a perennial herb. It is native to Europe and Asia. The methanol extracts of *P. officinale* rhizomes (800 mg/kg) reduced blood glucose levels of normal mice as well as streptozotocin-induced diabetic mice 4 h after i.p. administration. The extract also suppressed epinephrine-induced hyperglycemia in mice (Kato and Miura 1994). The total flavonoid fraction from the plant (50–200 mg/kg) exhibited dose-dependent antihyperglycemic effects in both streptozotocin-induced and alloxan-induced diabetic rats. The flavonoids could also increase insulin levels in diabetic rats (Shu et al. 2009). Further, the flavonoid fraction inhibited α-amylase activity under *in vitro* conditions (Shu et al. 2009). Interestingly, saponin-rich fractions from *P. odoratum* (*P. officinale*) showed potential hypoglycemic effects. This study suggests that the saponin in this herb is also important in exhibiting antidiabetic activity and the flavonoid contributed more to the antioxidant activity. However, both saponin and flavonoid fractions inhibited *in vitro* α-amylase activity (Deng et al. 2012). In another study, the effect of the extract of *P. odoratum* on metabolic disorders in diet-induced C57BL/6 obese mice was examined. In the preventive experiment, the extract blocked body weight gain and lowered serum total cholesterol, TG, and fasting blood glucose, improved glucose tolerance and reduced the levels of serum insulin and leptin, and increased serum adiponectin levels in mice fed with an HFD. In the therapeutic study, the extract treatments for 2 weeks to obese mice reduced serum TG and fasting blood glucose and improved glucose tolerance. Gene expression analysis showed that the mRNA levels of PPAR-γ and PPAR-α and their downstream target genes in mice livers, adipose tissues, and HepG2 cells were increased (Gu et al. 2013). It appears that this is an important antidiabetic plant with more than one active principles and mechanisms of action. Detailed studies and follow-up clinical trials are required to determine its utility as an anti-DM medicine.

Polygonum hyrcanicum Rech. f., Polygonaceae
A methanol extract from the flowering aerial parts of the plant and its phenolic compounds showed notable α-glucosidase inhibitory activity (IC_{50}, 15 µg/mL). The methanol extract also showed *in vitro* antioxidant activities (Moradi-Afrapoli et al. 2012).

Polygonum multiflorum Thunb., Polygonaceae
Polygonum multiflorum is an important traditional medicinal plant. Slimax (a Chinese anti-DM preparation containing water extracts of the aerial parts of this plant) favorably regulated glucose utilization and lipid metabolism in obese animals (Vijaya et al. 1995).

Polygonum senegalensis Hausa, Polygonaceae
The hydroalcoholic extract of *Polygonum senegalensis* leaves exhibited potent α-glucosidase inhibition and antioxidant activities *in vitro* (Bothon et al. 2013).

Pomaderris kumeraho A.Cunn. ex Fenzl, Rhamnaceae
Pomaderris kumeraho (common name: kumarahou) is a small shrub occurring in New Zealand. The leaves contain antidiabetic compounds quercetin, kaempferol, and ellagic acid (review: Ghisalberti 2005).

Pongamia pinnata (L.) Pierre, Polygalaceae
Synonyms: *Pongamia glabra* Vent., *Pongamia pinnata* (L.) Merr., *Galedupa indica* Lamk, *Derris indica* (Lam.) Bennett, *Millettia novo-guineensis* Kane. & Hat. (reviews: SatishKumar 2011; Bhalerao and Sharma 2014) (Figure 2.77)

(a)　　　　　　　　　　　　　　　(b)

FIGURE 2.77 *Pongamia pinnata.* (a) Twig with flowers. (b) Twig with fruits.

Common Names: Indian beech, karanja, pongam, etc.

Description: *Pongamia pinnata* is a fast-growing medium-sized, evergreen, perennial, and deciduous tree, which spreads, forming a broad, spreading canopy casting moderate shade. The leaves are alternate, odd pinnately compound, 5–10 cm long, evergreen, and hairless; the flower is lavender, pink, and white, 2–4 together, short-stalked, pea shaped, and 15–18 mm long; the pods are 3–6 cm long and 2–3 cm wide, smooth, brown, thick-walled, hard, indehiscent, and 1–2 seeded; the seed is ovoid or elliptical, bean-like, 10–15 cm long, and dark brown; in the roots, the taproot is thick and long, with numerous lateral roots; the bark is thin gray to grayish-brown and yellow on the inside (review: Bhalerao and Sharma 2014).

Distribution: *P. pinnata* is native to Bangladesh, India, Myanmar, Nepal, and Thailand and exotic to Australia, China, Egypt, Fiji, Indonesia, Japan, Malaysia, Mauritius, New Zealand, Pakistan, the Philippines, Seychelles, Solomon Islands, Sri Lanka, Sudan, and the United States (Orwa et al. 2009).

Traditional Medicinal Uses: All parts of the plant are used in traditional medicine. The juice of the root is used for treating gonorrhea and cleansing foul ulcers and closing fistulous sores; the roots are bitter anthelmintic and used in vaginal and skin diseases; the juice of the leaves is used for cold, cough, diarrhea, dyspepsia, flatulence, gonorrhea, and leprosy; the leaves are used as an anthelmintic, digestive, and laxative and for inflammations, piles, wounds, rheumatism, and herpes; the fruits are used for abdominal tumors, ailments of the female genital tract, leprosy, piles, ulcers, etc.; the seeds are used in hypertension, skin ailments, rheumatic arthritis, bronchitis, whooping cough, chronic fever, and inflammation and as a febrifuge, tonic, etc.; the oil is styptic, anthelmintic, and good in leprosy, piles, ulcers, chronic fever, and liver pain; the bark is used for piles, ulcers, chronic fever, liver pain, and mental disorder; the flower is used to quench dipsia in diabetes and for bleeding piles (review: Bhalerao and Sharma 2014).

Pharmacology

Antidiabetes: The antidiabetic activity of petroleum ether, chloroform, alcohol, and water extracts of *P. pinnata* leaf was investigated in alloxan-induced diabetic albino rats. The ethanol and water extracts decreased blood glucose levels in alloxan-induced diabetic rats (Sikarwar and Patil 2010). The methanol extract of the leaf showed anti-DM activity in streptozotocin-induced diabetic rats as well (Kavipriya et al. 2013).

In another study, oral administration of the ethanol extract of flowers (300 mg/kg) resulted in antihyperglycemic and anti-LPO effects along with enhancement in the antioxidant defense system in alloxan-induced diabetic rats. The efficacy of the extract was comparable to that of glibenclamide. The extract did not exhibit a hypoglycemic effect in normal rats. The authors suggest the use of *P. pinnata* as a safe alternative antihyperglycemic drug for diabetic patients (Punitha and Manohar 2006).

Cycloart-23-ene-3β, 25-diol (also called B2) isolated from the stem bark of *P. pinnata* was evaluated and showed significant antidiabetic activity against streptozotocin–nicotinamide-induced diabetic mice (Badole and Bodhankar 2009a). B2 treatment reduced serum glucose in an acute study. In a chronic study (treatment for 28 days), increase in body weight and decrease in food and water intake were observed; increased glucose utilization was observed in oral glucose tolerance test. B2 increased serum and pancreatic insulin levels and decreased serum levels of glycosylated hemoglobin, cholesterol, TGs, AST, ALT, ALP, globulin, bilirubin, lactate dehydrogenase, urea, and uric acid. Besides, B2 treatment decreased liver MDA and increased superoxidase dismutase and reduced GSH levels. Histologically, focal necrosis was observed in the diabetic mouse pancreas but was less obvious in treated groups. The mechanism of B2 appears to be due to increased pancreatic insulin secretion and antioxidant activity (Badole and Bodhankar 2010). B2 isolated from the flowers at a dose of 1 mg/kg protected vital organs from diabetic complications in streptozotocin–nicotinamide-induced diabetic rats (Badole et al. 2011).

In another study, the aforementioned authors reported that concomitant administration of the petroleum ether extract of the stem bark of *P. pinnata* with glyburide, pioglitazone, or metformin, but not repaglinide, showed a synergistic antihyperglycemic effect in alloxan-induced diabetic mice (Badole and Bodhankar 2009b).

The flavonoids pongamol and karanjin were isolated from the *P. pinnata* fruit. In streptozotocin-induced diabetic rats, treatment with pongamol and karanjin (single dose 50 and 100 mg/kg) lowered the blood glucose levels 6 h postoral administration. Additionally, these compounds (100 mg/kg daily for 10 days) lowered the blood glucose level in type 2 diabetes *db/db* mice (Tamrakar et al. 2008). Both pongamol and karanjin exhibited PTP1B inhibitory activity (Jiany et al. 2012).

Other Activities: Other pharmacological activities reported include antiplasmodial activity against *Plasmodium falciparum*, anti-inflammatory activity, antimicrobial effect of the crude leaf extract, anti-ulcer activity of the plant root, antioxidant activity, anti-LPO activity, antidiarrheal activity, antibacterial activity, antifungal activity, antiviral activity, protective effect against nephrotoxicity, nootropic activity, antinociceptive activity, and antilice activity (Sangwan et al. 2010; review: Bhalerao and Sharma 2014; Yadav et al. 2004b).

Toxicity: The results of several toxicity studies indicated that the extracts and single compounds isolated from this species did not show any significant toxicity. However, further studies on the toxicity of the other compounds isolated from this plant are required to ensure their eligibility to be used as sources of drugs (review: AlMugarrabun et al. 2013).

Phytochemicals: This plant contains large amounts of prenylated flavonoids such as furanoflavones, furanoflavonols, chromenoflavones, furanochalcones, pyranochalcones, pongaflavanol and tunicatachalcone, furanoflavonoids (pongapinnol A to D), coumestan, pongacoumestan, and structurally unusual flavonoids, pongamones A to E. Sterols, sterol derivatives, fatty acid, karangin, pongamol, pongagalabrone and pongapin, pinnatin, and kanjone have been isolated from the seeds. The metabolites β-sitosteryl acetate, galactoside, and stigmasterol and its galactoside are also reported from this plant. The leaves and stem of the plant contain flavone and chalcone derivatives such as pongone, galbone, pongalabol, and pongagallone A and B (review: Bhalerao and Sharma 2014).

Populus balsamifera L., Salicaceae

Common Name: White spruce

Resin from the buds of *Populus balsamifera* has long been used in traditional medicine of native North Americans for its anti-inflammatory properties, to treat sore throats, coughs, chest pain, and rheumatism.

The antidiabetic activity of this plant has been reviewed recently (Eid and Haddad 2014). The ethanol extract of *P. balsamifera* exhibited complete inhibition of adipogenesis in the 3T3-L1 cell bioassay. Detailed mechanistic studies demonstrated that *P. balsamifera* exerted this action by intervening in the earliest events of the adipogenetic program and kept cells in a fibroblastic undifferentiated state. The extract also influenced PPAR-γ, a key transcription factor controlling the adipogenetic program. Salicortin, an ester of salicylic alcohol glucoside and an abundant component of most species of *Salix* and *Populus* genera, was identified as the inhibitor of adipogenesis. The diet-induced obese (DIO) mouse model was used to examine the effects of the plant and its active component *in vivo*. When given at the onset of the HFD (125 and 250 mg/kg), the *P. balsamifera* extract reduced the weight gain observed in DIO controls by nearly 50% as compared to chow-fed normal mice. This effect was dose dependent and accompanied by a decrease of glycemia, insulinemia, hepatic steatosis, and leptin/adiponectin ratio, all indicative of improved insulin sensitivity. The *P. balsamifera* crude extract exerted a partial yet significant preventative action against diet-induced obesity. A treatment study was then carried out in the same DIO model where mice were allowed to become obese and insulin resistant before the onset of treatment with the plant's crude extract and the corresponding dose of salicortin (12.5 mg/kg). Both the crude extract and salicortin reduced body weight, glycemia, insulinemia, liver TGs, and leptin/adiponectin ratio. However, the profile of activities of the crude extract differed somewhat from that of salicortin. Salicortin's action on body weight gain was immediate and maintained throughout the treatment period, whereas the crude extract caused a small initial drop and only began decreasing weight gain significantly after several weeks. In contrast, the crude extract of *P. balsamifera* had a sustained effect on blood glucose, whereas salicortin's effects were initially equivalent but waned with time. *P. balsamifera* extract increased the components involved in insulin-responsive fatty acid oxidation in the skeletal muscle, liver, and adipose tissue. It also improved hepatic and adipose tissue insulin signaling proteins. The effect of salicortin was similar but generally weaker. Thus, the bark extract of this plant exerted a promising activity as an antiobesity agent. It appears to act by suppressing adipogenesis, by improving insulin resistance, and by favoring energy wastage mechanisms such as fat oxidation. Salicortin can explain part of the plant's effects, notably on body weight loss, but its effect on glucose and lipid homeostasis appears to be weaker and more transient (Eid and Haddad 2014). This suggests that there are better anti-DM compounds in this extract or compounds act synergistically to give better results.

Portulaca oleracea L., Portulacaceae
The polysaccharide fraction from *Portulaca oleracea* (whole plant) exhibited hypoglycemic and hypo-lipidemic activities in alloxan-induced diabetic mice (Fenglin et al. 2009b).

Posidonia oceanic (L.) Delile, Posidoniaceae
Common name: Neptune grass
Decoction of the leaves of *Posidonia oceanic* has been used as a remedy for diabetes mellitus and hyper-tension by villagers living by the sea coast of Western Anatolia. Oral administration of extract for 15 days (50, 150, and 250 mg/kg) resulted in a dose-dependent decrease in blood glucose in alloxan diabetic rats. The extract also showed vasoprotective and antioxidant effects (Gokce and Haznedaroglu 2008).

Potentilla chinensis Ser., Rosaceae
Synonym: *Potentilla chinensis* ssp. *trigonodonta* Handel-Mazz.
Potentilla chinensis is a medicinal herb distributed in temperate, arctic, and Alpine zones of the Northern Hemisphere. It contains hydrolyzable tannins, α-hydroxyoleanolic acid, α-hydroxyursolic acid, β-sitosterol, ursolic acid, oleanolic acid, etc.; β-sitosterol, ursolic acid, and oleanolic acid are known anti-DM compounds (Tomczyk and Latte 2009). Trans-tiliroside, an active principle of *P. chinensis*, decreased blood glucose and lipids (cholesterol, LDL, and TG) in alloxan-induced diabetic mice and streptozotocin-induced diabetic rats (Qiao et al. 2011).

Potentilla discolor Bunge, Rosaceae
Potentilla discolor, a Chinese folk medicine, has been used for the treatment of diabetes for many years.

A study reported the antidiabetic effect of *P. discolor* on normal and alloxan-induced diabetic mice, as well as the possible constituents and mechanism responsible for this activity. The standardized extract of *P. discolor* had little effect on the glucose levels of normal mice, while a dose-dependent hypoglycemic effect was observed in diabetic mice. Glucose tolerance test data indicated that there was a significantly higher rate of glucose disposal after the treatment. Phytochemical characterization indicated that the major components of the extract were triterpenes and flavonoids that could inhibit the activity of glycogen phosphorylase *in vitro*. Tormentic acid, asiatic acid, and potengriffioside A exhibited inhibitory effects on this enzyme. The inhibition on glycogen phosphorylase may be one molecular mechanism through which the extract ameliorated hyperglycemia (Yang et al. 2010b). The effects of the total flavonoid extract and total triterpenoid extract of *P. discolor* on blood glucose, lipid profiles, and antioxidant parameters on HFD-fed and streptozotocin-induced diabetic rats were evaluated (Li et al. 2010c). The diabetic rats treated with triterpenoid extract or flavonoid extract for 15 days showed a decrease in blood glucose and glycosylated serum protein compared to diabetic control rats. Further, the levels of serum total cholesterol, TGs, and LDL in the triterpenoid- or flavonoid-treated diabetic rats were lower and the HDL level was higher than in diabetic control rats. Besides, the treatments increased antioxidant enzymes and reduced oxidative stress in diabetic rats. Histopathologic examination also showed that the extracts have protective effects on β-cells in diabetic rats; the treatments also increased serum insulin levels (Zhang et al. 2010a; Li et al. 2010c). A decoction from *P. discolor* caused a decrease in the mRNA expression of insulin degrading enzymes in type 2 diabetic rats (Guo et al. 2005).

In another study, the antihyperglycemic effects of an active extract (decoction) from *P. discolor* were investigated in *ob/db* mice. Four week treatment with the decoction ameliorated the development of hyperlipidemia, LPO, and hyperglycemia associated with hyperphagia and polydipsia. These findings clearly provided evidence regarding the anti-DM potentials of *P. discolor* decoction (Song et al. 2012a).

In a comprehensive analysis of phytochemicals in *P. discolor*, 35 components were identified or characterized in *P. discolor* decoction. One of the major components in the decoction was a flavonoid sulfate. Sulfated flavonoids have been reported to improve the complications of DM by inhibition of the aldose reductase in both experimental animals and clinical trials. Therefore, the sulfated flavonoid in the decoction may in part contribute to the antihyperglycemic effect of *P. discolor* (Song et al. 2012a).

Potentilla fulgens Wall. ex Hook., Rosaceae

Potentilla fulgens is known as bauitara in India. It is an erect, perennial herb with thick root stock. The leaves are pinnately compound, with adnate stipules to the root stock; the flowers are yellow or orange-yellow in panicle or corymbose inflorescence. The plant is distributed in the temperate Himalayas from Kunwar to Sikkim and in the Khasia Mountains. It is dentifrice and used for toothache, diabetes, and diarrhea in traditional medicine (*The Wealth of India* 1969).

Taproots of *P. fulgens* are traditionally used by local practitioners for various types of ailments including diabetes. The crude methanol extract of the roots was tested for its effects in normoglycemic and alloxan-induced diabetic mice. The hypoglycemic activity was observed to be dose and time dependent. The extract reduced blood glucose level 2 h following administration in both normal and alloxan-induced diabetic mice. In alloxan-induced diabetic mice, blood glucose was reduced by 63%, while in normal mice, a 31% reduction was observed 24 h after the extract administration. Further, in diabetic mice, a prolonged antihyperglycemic action was observed where glucose levels were found to be significantly low (79%) when compared with control even on the third day. Glucose tolerance was also improved in both normal and diabetic mice (Syiem et al. 2002). Administration of different solvent fractions of *P. fulgens* to normoglycemic and diabetic mice resulted in varying degrees of inhibition of aldose reductase and sorbitol dehydrogenase activities in the liver, kidney, and eye. Several biochemical features implicate the polyol pathway as a plausible and important contributor to complications of diabetes (Syiem and Majaw 2011). Besides, the methanol extract of *P. fulgens* exerted an antihyperlipidemic effect and improved hexokinase activity in alloxan-induced diabetic mice in a tissue-specific manner (Syiem et al. 2009).

The antitumor activity of the water extract of the root of *P. fulgens* against murine ascites Dalton's lymphoma in mice has been reported (review: Ajikumaran and Subramoniam 2005).

Poterium ancisroides Desf., Rosaceae

Tormentic acid from *Poterium ancisroides* (aerial parts) stimulated insulin secretion in isolated rat islets of Langerhans (Ivorra et al. 1989).

Poterium spinosum L., Rosaceae, thorny burnet

Poterium spinosum is a shrub commonly found growing in the Mediterranean region. The root bark of this plant is used in folklore medicine to treat diabetes. The water extract of the root bark showed hypoglycemic activity in alloxan-induced diabetic rats; it improved oral glucose tolerance also. Seasonal variation in efficacy was also reported (review: Haddad et al. 2005).

Poupartia birrea (Hochst) Aubr., Anacardiaceae

Poupartia birrea leaves are used in traditional medicine to treat diabetes in Africa. The plant is reported to have hypoglycemic activity. Further, the leaf extract of this plant inhibited aldose reductase activity (review: Haddad et al. 2005).

Pouteria ramiflora (Mart.) Radlk, Sapotaceae

Pouteria ramiflora extract inhibited the activity of human salivary α-amylase activity. Further, it exhibited blood glucose-lowering activity in mice (De Gouveia et al. 2013). In another study, administration of the alcohol extract of *P. ramiflora* to streptozotocin-induced diabetic rats resulted in improved glycemic level, increased GSH peroxidase activity, decreased superoxide dismutase activity, and reduced LPO and antioxidant status. The extract also restored myosin-Va expression and the nuclear diameters of pyramidal neurons of the CA3 subregion and that of the polymorphic cells of the hilus. *P. ramiflora* extract exerted a neuroprotective effect against oxidative damage and myosin-Va expression and was able to prevent hippocampal neuronal loss in the CA3 and hilus subfields of diabetic rats (da Costa et al. 2013).

Premna corymbosa (Burm. f.) Rottl. & Willd., Verbenaceae
The root extract of *Premna corymbosa* (200–400 mg/kg) produced marked reduction in blood glucose concentrations in a dose-dependent manner in both normoglycemic and hyperglycemic rats (Thiruvenkata and Jayakar 2010).

Premna latifolia Thwartes, Verbenaceae
The isolation of known anti-DM compounds such as β-sitosterol and its glucoside and stigmasterol was reported. Further, flavone glycosides with antioxidant property are present in the leaves of this plant (Ghosh et al. 2014a)

Premna serratifolia L., Verbenaceae
Synonyms: *Premna obtustifolia* L., *Premna integrifolia* L.
Premna integrifolia is used in indigenous traditional medicine in India. In a recent study, the methanol extract of *P. integrifolia* (*P. serratifolia*) bark showed potential glucose-lowering effect in glucose tolerance test in normal rats compared to untreated control normal rats. At a dose of 300 mg/kg (daily for 7 days), the methanol extract caused substantial fall in blood glucose in alloxan-induced diabetic rats. The extract also showed *in vitro* antioxidant activity (Majumder and Akter 2014). Further, the chloroform–methanol extract of *P. integrifolia* (*P. serratifolia*) possessed antiobesity activity in mice fed with cafeteria diet (Mali et al. 2013). The blood glucose-lowering effect of *P. integrifolia* in streptozotocin-induced type 1 and type 2 diabetic rats has been reported (Alamgir et al. 2001). Ethanol extracts of this plant (250 mg/kg) also exhibited a blood glucose-lowering effect in alloxan-induced diabetic rats (Kar et al. 2003). The mechanism of action and active principles involved are not known.

Prosopis farcta (Sol. ex Russell) Macbride, Mimosaceae
In Jordan, *Prosopis farcta* root is used in ethnomedicine to control DM. The hypoglycemic effect of this plant was reported in rats (Afifi 1993; Al-Aboudi and Afifi 2011). Oral administration of *P. farcta* fruit dusk powder and root water extract accelerated cutaneous wound healing in streptozotocin-induced diabetic rats (Ranjbar-Heidari et al. 2012).

Prosopis glandulosa Torr., Fabaceae
Common Name: Honey mesquite
Prosopis glandulosa is used in traditional medicine to treat diabetes and certain other diseases. Diavite™ (a product consisting solely of the dried and ground pods of *P. glandulosa*) is currently marketed as a food supplement with glucose-stabilizing properties. However, these are anecdotal claims lacking scientific evidence. Experiments on animals support its anti-DM property. *P. glandulosa* treatment (100 mg/kg for 8 days) moderately lowered blood glucose levels in both alloxan-induced diabetic rats and high-caloric-diet-fed insulin-resistant type 2 diabetic Zucker fa/fa rats and also stimulated insulin secretion. Further, the treatment led to the formation of small β-cells and improved insulin sensitivity of isolated cardiomyocytes (George et al. 2011).

Protium heptaphyllum (Aubl.) L. Marchand, Burseraceae
Administration of pentacyclic triterpenes, α,β-amyrins (10, 30, and 100 mg/kg), from the resin of *Protium heptaphyllum* to streptozotocin-induced diabetic mice resulted in significant reduction in blood glucose, total cholesterol, and serum TG compared to untreated diabetic control mice. Unlike glibenclamide, α,β-amyrin did not lower blood sugar levels in normoglycemic mice but showed improved glucose tolerance in glucose-loaded normal mice. Further, the plasma insulin level and histopathological analysis of the pancreas revealed the beneficial effects of these triterpenes in the preservation of β-cell integrity. Treatment with α,β-amyrin (100 mg/kg) to HFD-fed mice resulted in a marked reduction in

serum total cholesterol and TG, LDL, and VLDL along with a concomitant increase in HDL compared to untreated HFD-fed mice. The others suggest that α,β-amyrin could be lead compounds for improved drug development to treat DM and atherosclerosis (Santos et al. 2012). Triterpene compounds oleanolic acid, maslinic acid, asiatic acid, ursolic acid, and astragaloside IV isolated from *P. heptaphyllum* showed anti-DM activities in rodent models of DM (review: Nazaruk and Borzym-Kluczyk 2014). Thus, this plant appears to be very promising for further studies.

Prunella vulgaris L., Labiatae

Synonyms: *Prunella vulgaris* L. var. *elongata* Benth., *Prunella vulgaris* L. var. *lanceolata* (W. Bartram) Fernald, *Prunella pennsylvanica* Willd. var. *lanceolata* W.P.C. Barton, *Prunella vulgaris* var. *calvescens* (Figure 2.78)

Local Names: Lance self-heal, mountain self-heal, narrow leaf self-heal, lanceleaf self-heal, American self-heal, self-heal, heal-all, etc.

Distribution: The exact origin of this plant species is not clear. *Prunella. vulgaris* is widely distributed in places such as the warm regions of Europe and Asia, northwest region of Africa, and northern region of America (review: Meng et al. 2014).

Description: *P. vulgaris* is a perennial herb with branches decumbent to 50 cm long, often with adventitious roots and glabrescent except for a few simple hairs mainly near the nodes. The leaves have ovate, rarely narrow-ovate, 1.5–6 cm long, 0.7–2.5 cm wide lamina and 0.2–2 cm long petiole. The calyx is 6–7 mm long; and the lower lip has narrow-ovate teeth. Inflorescence is 1–6 cm long; and corolla is 8–12 mm long in deep purple-blue. *P. vulgaris* tastes bitter and has a pungent aroma.

Traditional Medicinal Uses: *P. vulgaris* is an important traditional medicinal plant in oriental medicine. It is often used in the treatment of headache, dizziness, swollen eyes, scrofula, cecidium, swollen and painful mastitis, breast cancer, goiter, and tuberculosis. It has long been used as a folk medicine for alleviating sore throat and reducing fever by Europeans and Chinese (review: Meng et al. 2014).

Pharmacology

Antidiabetes: The water–ethanol extract of *P. vulgaris* spikes (100 mg/kg) significantly suppressed the rise in blood glucose in acute glucose tolerance test in mice. Further, this dose of the extract significantly decreased blood glucose levels in streptozotocin-induced diabetic mice. The extract enhanced

(a) (b)

FIGURE 2.78 *Prunella vulgaris*. (a) Twig with flowers. (b) Flowers. (From Wikimedia Foundation, Inc.)

and prolonged the hypoglycemic effects of exogenous insulin in streptozotocin-induced diabetic mice. The extract treatment did not increase blood insulin levels in diabetic mice. Thus, the extract increased insulin sensitivity (Zheng et al. 2007).

Caffeic acid ethylene ester, isolated from *P. vulgaris*, showed a potent noncompetitive inhibition of rat lens aldose reductase and human recombinant aldose reductase. Furthermore, this compound inhibited galactitol formation in a rat lens incubated with a high concentration of galactose. Also it has antioxidative as well as AGE inhibitory effects. As a result, this compound could be offered as a leading compound for further study as a new natural product drug for diabetic complications (Li et al. 2012).

In another study, the effects of a triterpenic acid from *P. vulgaris* on streptozotocin-induced diabetic rats were evaluated. Administration of the triterpenic acid (50–200 mg/kg, p.o., daily for 6 weeks) to streptozotocin-induced diabetic rats resulted in decreased levels of blood glucose, fructosamine, MDA, NO, and the activity of nitric oxide synthase in serum compared to diabetic control. The activity of SOD in serum and body weight increased significantly compared with the streptozotocin-induced diabetic control group. Histopathological examination also showed the protective effect of triterpenic acid on pancreatic β-cells (Zhou et al. 2013).

It is of interest to note that this plant contains several antidiabetic compounds with different mechanisms of actions. Further studies are warranted to study all fully active compounds and the efficacy of their various combinations.

Other Activities: Pharmacological studies revealed several biological activities of this plant including systemic anaphylaxis inhibition, antihypertensive activity, UV radiation photoprotection, immune modulation, antioxidative action, antihyperlipidemic activity, anti-inflammatory activity, hepatoprotective activity, antineoplastic activities, antiviral activity, and antibacterial effects (review: Meng et al. 2014).

Phytochemicals: The major classes of phytochemicals isolated from this plant include triterpenoids, sterols, phenylpropanoids, flavonoids, coumarins, fatty acids, and volatile oils (review: Meng et al. 2014).

Prunus amygdalus Batsch., Rosaceae (Subfamily Spiraeoideae)

Synonyms: *Prunus communis* Fritsch., *Prunus dulcis* (Muller) D.A. Webb.

Common Names: Almond, suphala, badam, vadamkottai, vadam-kotta, etc.

Description: The almond tree is a small deciduous tree that grows to between 4 and 10 m in height, with a trunk of up to 30 cm in diameter. The young twigs are green at first and become purplish when they are exposed to sunlight and then gray in their second year. The leaves are 3–5 in. long with serrated margins and 2.5 cm petioles. The flowers are pale pink and 3–5 cm in diameter with five petals; they are produced singly or in pairs before the leaves in early spring. Almonds begin to bear an economic crop in the third year after planting of the trees. The trees reach the full bearing status 5–6 years after their planting. In botanical terms, the almond is not a nut, but a drupe that is 3.5–6 cm long. The fruit consists of an outer hull and a hard shell with the seed inside. Almonds are commonly sold shelled or unshelled.

There are three varieties of almonds, all of which produce nuts, but some are edible and some are not. One almond variety produces the sweet edible nut; bitter or semibitter nut producing varieties are not edible. Two major types of sweet almonds (*Prunus amygdalus* dulcis and *Prunus amygdalus* amara) are grown commercially (review: Rao and Lakshmi 2012).

Distribution: Almond trees are native to the region that extends from India to Persia. The tree had spread to east and west of its native region thousands of years before Christ. It is cultivated in the cooler parts of India.

Traditional Medicinal Uses: *P. amygdalus* is a sedative, demulcent, emollient, aphrodisiac, stimulant, bitter, laxative, sweet, stimulant nervine tonic, lithontriptic, and diuretic and used in the treatment of irritable sores, skin eruption, peptic ulcers, leprosy, headache, biliousness, intestinal ailments, dry cough, colic, inflammation, eye diseases, bronchitis, earache, sore throat, crack skin, neuralgia, and menstrual pain (*The Wealth of India* 1969; Kirtikar and Basu 1975). It is used in the traditional system of medicine to treat diabetes in India (Shah et al. 2011).

Pharmacology

Antidiabetes: The seeds exhibit hypoglycemic activity in albino rabbits (Teotia and Singh 1997). The ethanol extract of this plant leaf, flower, and seed showed anti-DM activity in streptozotocin-induced diabetic mice. Administration of the extracts (250 or 500 mg/kg, p.o., for 20 days) resulted in reduction in blood glucose, serum cholesterol, TG, creatine, urea, and ALP, whereas an increase in serum HDL, total protein, and body weight was observed in the treated groups. The extracts from all the three plant parts (leaf, flower, and seed) were found to be active (Shah et al. 2011). Further studies are required to identify the active principles and their mechanism of action. The plant contains known antidiabetes compound such as β-sitosterol.

Clinical Trials: Almonds lowered postprandial glycemia, insulinemia, and oxidative stress in healthy individuals (Jenkins et al. 2006). Fifteen healthy individuals, seven men and eight women, about 26 years of age were studied. All the subjects completed five study sessions, each lasting 4 h, with a minimum 1 week washout between the tests. The subjects consumed a control meal on two occasions, and almond, parboiled rice, and instant mashed potato meals only once. The blood glucose concentration over the 4 h testing for each meal revealed that the almond and rice meals showed lower values than that of the instant mashed potato meal. Glycemic indices for the rice (38 ± 6) and almond meals (55 ± 7) were less than for the potato meal (94 ± 11) as were the postprandial areas under the insulin concentration time curve. No postmeal treatment differences were seen in total antioxidant capacity. However, the serum protein thiol concentration increased following the almond meal, indicating less oxidative protein damage. The authors conclude that almonds are likely to lower the risk of oxidative damage to proteins by decreasing the glycemic excursion and by providing antioxidants. These actions may relate to mechanisms by which nuts are associated with a decreased risk of CHD (Jenkins et al. 2006).

Studies on diabetic individuals with different doses of almond are urgently required to determine its utility as a nutraceutical for the benefit of diabetic patients.

Other Activities: Almonds have beneficial effects on serum lipids in healthy adults (Lovejoy et al. 2002) and have antiobesity activity (Wien et al. 2003). A decrease in serum total cholesterol, TG, LDL-C, and VLDL-C and an increase in phospholipid, fecal sterol, and HDL-C are reported in seed-fed rabbits (Teotia et al. 1997). Other reported activities include immunostimulation, antioxidant activity, hepatoprotection, aphrodisiac action, and memory enhancing property. Almonds significantly increased the sperm motility and sperm contents in the epididymides and vas deferens in rats (review: Rao and Lakshmi 2012).

Phytochemicals: Amygdalin, almond oil, glucoside amygdalin, hydrocyanic acid, cresol, eugenol, caproic acid, phenyl acetic acid, I-decanol, geraniol, phenyl ethyl alcohol, quercetin, cyaniding, kaempferol, caffeic acid (*The Wealth of India* 1969), phenolic compounds from almond skins (Sang et al. 2002), 4‴-methoxy-2″,3″-dihydroflavonyl-4″,4″-oxymethyl-5,7dimethoxy flavone, coumaric acid, 2-methylnonacosan-3-one, n-octacosanol, n-triacontane, β-sitosterol, procyanidin dimmer, (+) catechin, (−)epicatechin, prunasin, and daucosterin were isolated (review: Ajikumaran and Subramoniam 2005).

Prunus davidiana Fr., Rosaceae

The methanol extract of *Prunus davidiana* stem and its main component prunin (naringenin 7-*O*-β-D-glucoside) improved hyperglycemia and hyperlipidemia in streptozotocin-induced diabetic rats (Choi et al. 1991).

Prunus divaricata Ledeb., Rosaceae

Prunus divaricata is a small tree cultivated in Iran, Middle East, and Central Asia. The antidiabetic and antihyperglycemic effects of the plant fruits were evaluated in normal and streptozotocin-induced diabetic rats. In diabetic rats, the fruit extract (400 mg/kg) and juice (800 mg/kg) prevented weight loss but had no effect on the body weight of normal rats. A single dose had no effect on glucose levels in normal rats, but repeated high doses of the extract (400 mg/kg) for 27 days induced hypoglycemia. In diabetic rats, repeated administration of the extract resulted in marked decrease in blood glucose levels at lower doses (100 and 200 mg/kg extract) also. Besides, repeated treatment reduced TG and cholesterol levels in diabetic rats (Minaiyan et al. 2013).

Prunus mume Siebold, Rosaceae

Common Name: Asian plum

Concentrated plum juice (*Prunus mume*) decreased blood glucose and plasma TG concentrations in rats. Plum treatment for 2 weeks reduced AUCs for glucose and insulin during a glucose tolerance test. In *db/db* mice, plum decreased these AUCs and also blood glucose during an insulin tolerance test. Plum treatment significantly increased plasma adiponectin concentrations and PPAR-γ mRNA expression in adipose tissue from obese type 2 diabetic rats. Plum may increase insulin sensitivity in these rats via adiponectin-related mechanisms (Utsunomiya et al. 2005; review: El-Abhar and Schaalan 2014).

Prunus persica (L) Batsch, Rosaceae

The methanol extract of the plant leaves (500 mg/kg) reduced blood glucose levels remarkably in alloxan-induced hyperglycemic rats, and the water extract was less effective compared to the methanol extract. In normal glucose-loaded rats, there was suppression of postprandial spike initially by the extract and later maintenance of blood levels was noticed (Chatragadda et al. 2014). This appears to be a preliminary study and follow-up studies are required.

Psacalium decompositum A. Gray, Compositae

Synonym: *Cacalia decomposita* A. Gray

Psacalium decompositum (*Cacalia decomposita*) is a medicinal plant used in Mexico for the treatment of DM. A water fraction *P. decompositum* roots containing carbohydrates exhibited hypoglycemic effects in experimental animals. The chemical structures of the major compounds in the fraction correspond with fructan-type oligosaccharides. These compounds showed a hypoglycemic effect in healthy and diabetic mice (Jimenez-Estrada et al. 2011). Intraperitoneal administration of the root decoction of *P. decompositum* reduced the glycemia in normal mice. It lowered the hyperglycemic peak by 17% in rabbits with temporal glycemia (review: Mohammed et al. 2013). *In vivo* bioassay-guided fractionation of a water extract of the roots of *P. decompositum* led to the isolation of two new eremophilanolides, 3-hydroxycacalolide and epi-3-hydroxycacalolide. A 1:1 mixture of these compounds exhibited antihyperglycemic activity when tested at a dose of 1.09 mmol/kg in *ob/ob* type 2 DM mice. The known furanoeremophilane cacalol isolated from the CH_2Cl_2 extract of the plant roots also possessed antihyperglycemic activity (Inman et al. 1999). Novel hypo-glycemically active eremophilanolide sesquiterpene, obtained from the roots of *P. decompositum*, have been patented for their use as hypoglycemic agents in the treatment of diabetes (Inman et al. 1998). Mechanisms of action, detailed toxicity study, and clinical trials are needed to establish the therapeutic value, if any.

Psacalium peltatum (H.B.K.) Cass, Compositae

Psacalium peltatum is a folk medicinal plant used in the treatment of DM, rheumatic pains, gastrointestinal problems, kidney ailments, etc.

The water decoction, methanol extract, and water extract exhibited hypoglycemic activity in fasting healthy mice. Two active fractions were isolated from the methanol extract that reduced glucose levels in healthy mice (Contreras-Weber et al. 2002). A hypoglycemic fructan (a carbohydrate) fraction from *P. peltatum* roots exhibited anti-inflammatory and antioxidant activities in streptozotocin-induced diabetic mice (Alarcon-Aguilar et al. 2010). The fraction (when administered orally daily for 15 days before streptozotocin challenge and for 33 days after the challenge) significantly reduced glycemia, glycated hemoglobulin levels, and the proinflammatory cytokine TNF-α and increased the production of anti-inflammatory cytokine IL-10 and the immune modulator IFN-γ. This plant may be useful in preventing insulin resistance and complication of diabetes to some extent (Alarcon-Aguilar et al. 2010).

Pseudarthria viscida (L.) Wight & Arn., Fabaceae

The ethanol extract of *Pseudarthria viscida* root showed antidiabetic activity against alloxan-induced diabetes in rats (Masirkar et al. 2008).

Pseuderanthemum palatiferum (Nees) Radlk, Acanthaceae

The leaves of *Pseuderanthemum palatiferum* are recommended in the folk medicine of Vietnam and Thailand for promoting and treating various diseases including hypertension, diarrhea, arthritis, hemorrhoids, stomachache, tumors, colitis, bleeding, wounds, constipation, flu, colon cancer, nephritis, and diabetes. The ethanol (80%) extract of *P. palatiferum* leaves (250, 500, and 1000 mg/kg, p.o., daily for 14 days) did not significantly change the levels of blood glucose, insulin, and biochemical parameters in normal rats. However, in diabetic rats, the treatment decreased blood glucose levels and increased serum insulin levels at the end of the experiment. The hypoglycemic effect of the extract was more than that of glibenclamide. Further, the extract treatment increased HDL and decreased total cholesterol, TG, LDL, blood urea nitrogen, and ALP in diabetic rats (Padee et al. 2010). In another study, single oral administrations of the water extract (250 and 500 mg/kg) of *P. palatiferum* leaf significantly decreased blood glucose levels in streptozotocin-induced diabetic rats after 30 min, but not in normal rats. The extract also showed *in vitro* antioxidant activity (Chayarop et al. 2011).

Pseudocedrela kotschyi (Schweinf.) Harms, Meliaceae

The water extract (200 mg/kg) of leaf of *Pseudocedrela kotschyi* reduced blood glucose levels in alloxan-induced diabetic rats (Georgewill and Georgewill 2009). In a recent study, alloxan-induced diabetic rats treated with the water extract of the root (250 and 500 mg/kg, p.o., for 15 days) showed weight gain, dose-dependent decrease in glycemia, and regeneration of β-cells. Further, the treatment ameliorated dyslipidemia in diabetic rats (Mbaka et al. 2014). The hydroalcoholic extract of *P. kotschyi* root exhibited α-glucosidase inhibition and antioxidant activities *in vitro* (Bothon et al. 2013). Thus, the plant root exhibits properties such as lowering blood glucose, α-glucosidase inhibition, regeneration of β-cells, and antioxidant activity. This plant is promising for further studies.

Pseudolarix amabilis (J. Nelson) Rehder, Pinaceae

Pseudolarix amabilis is a traditional medicinal plant primarily used to treat skin infections. Pseudolaric acid B is present in the extracts of the root and trunk barks of this plant. Pseudolaric acid and its analogues are shown to have PPAR-γ agonist activity (Wang et al. 2014).

Psidium guajava L., Myrtaceae

Synonyms: *P. aromaticum* L., *P. cujavillus* Burm.f., *P. pomiferum* L., *P. pyriferum* L., *P. pumilum* Vahl (monograph: WHO 2009) (Figure 2.79)

Common Names: Common guava, guave, madhur amla, amrud, koyya, perakayya, etc.

Description: *Psidium guajava* is an arborescent shrub or small tree. The leaves are opposite, entire, light green, glabrous or nearly so above and softly pubescent beneath; the flowers are large, white, and fragrant. The fruit is a globose or pear-shaped berry with many seeds with hard testa. Horticulture varieties with large fruits with an edible pink mesocarp are common. They contain many seeds and the calyx persists in the fruit.

(a)   (b)   (c)

FIGURE 2.79 *Psidium guajava.* (a) Young plant. (b) Flower. (c) Fruits.

Distribution: *Psidium guajava* is indigenous to Mexico and other parts of tropical America and cultivated and naturalized in most tropical countries including Asia.

Ethnomedical Uses: *P. guajava* is acrid, sour, cooling, aphrodisiac, astringent, antispasmodic, antiseptic, antibacterial, emmenagogue, hemostat, laxative, and vermifuge and is used in headache, ringworm, wounds, diabetes, colic, bleeding of gums, abdominal pain, diarrhea, dysentery, ulcers, carbuncles, catarrh, common cold, cough, deafness, vaginitis, dermatosis, dropsy, dyspepsia, hysteria, indigestion, digestive ailments, cholera, diarrhea, leukorrhea, piles, scabies, sores, sprains, cerebral infections, nephritis, cachexia, rheumatism, epilepsy, convulsions, toothache, gum boil, bronchitis, eye sores, colic, and gum bleeding (Kirtikar and Basu 1975; monograph: WHO 2009).

Pharmacology

Antidiabetes: Intragastric administration of a 50% ethanol extract of the leaves to rats, at a dose of 200 mg/kg, prevented alloxan-induced hyperglycemia. The extract did not stimulate insulin synthesis. A butanol fraction of the 50% ethanol extract reduced alloxan-induced hyperglycemia in rats when administered at a dose of 25.0 mg/kg by gastric lavage. The water extract of leaves showed antidiabetic activity in alloxan-induced diabetic rats (Mukhtar et al. 2004b). The leaf extract of *P. guajava* increased the plasma insulin levels and glucose utilization in type 2 diabetic rats (Shen et al. 2008). Total triterpenoids from of *P. guajava* leaves exhibited nephroprotective activity in streptozotocin-induced type 2 diabetic rats (Kuang et al. 2012). The methanol extract the of leaves of this plant showed beneficial effects by preventing diabetic cardiovascular complications in streptozotocin-induced diabetic female rats (Soman et al. 2010; Soman et al. 2013). Sesquiterpenoid-based meroterpenoids, psidials B and C, isolated from the leaves of *P. guajava*, showed PTP1B inhibitory activity at 10 µmol/L (Jiang et al. 2012). Further, studies established that the extracts of the plant leaves and fruits have antidiabetic and hypoglycemic effects in rodents (Mukhtar et al. 2004; Oh et al. 2005). A marked glucose-lowering effect of guava fruit juice (1 g/kg, i.p.) was shown in alloxan-induced diabetic mice (Cheng and Yang 1983). Guava fruit juice reduced blood glucose levels in diabetic mice (Yusof and Said 2004).

In view of the likely food medicine value of ripe and unripe fruits of *P. guajava*, there is a need to investigate in details the anti-DM activity of the fruits of this plant.

Clinical Trials: In a clinical trial, the blood glucose-lowering effect of guava fruit juice (p.o.) was observed in maturity-onset diabetic patients and healthy volunteers. The effective duration of guava is more transient and is less potent than chlorpropamide and metformin (Cheng and Yang 1983).

Other Activities: The extracts of *P. guajava* leaves have several pharmacological activities such as antioxidant, antidiarrhea, antibacterial, anticough, neuroactive, cardioprotective, and anti-acute diarrheal activities, and the inhibition of gastrointestinal release of acetylcholine by quercetin from the extract has been suggested as at least one of the mechanisms of the antidiarrheal activity (Lutterodt 1992; Jaiarj et al. 1999; Shaheen et al. 2000). Further, the leaf extract showed inotropic effects on guinea pig atrium (Conde et al. 2003). The essential oil of the leaves showed analgesic and antipyretic activities (monograph: WHO 2009).

Intestinal antispasmodic effect of the extract is also known (Lozoya et al. 2002). The fruit is a good new source of antioxidant dietary fiber and the leaf has a very potent antioxidant property (Jimenez-Escrig et al. 2001; Qian and Nihorimbere 2004). Clinical study has shown anti-infantile rotaviral enteritis activity (Wei et al. 2000). The antimutagenic effects of guava have been reported (Grover and Bala 1993). Cardioprotective effects of the extract from this plant against ischemia–reperfusion injury in perfused rat hearts have been reported (Yamashiro et al. 2003).

Toxicity: The leaf extract has *in vitro* cytotoxicity; a narcotic-like principle from the plant leaf affects locomotor activity in mice (Lutterodt and Maleque 1988). Cytotoxic phenylethanol glycosides are reported from the seeds. Intragastric administration of the water extract of the leaves to rats exhibited a median lethal dose of 50.0 g/kg. In chronic toxicity tests, the water extract of the leaves was administered by gavage to 128 rats of both sexes at doses of 0.2, 2.0, and 20.0 g/day (1, 10, and 100 times of the normal therapeutic dose for the treatment of diarrhea) for 6 months. The results showed that the body weight gains in male rats were lower in all treated animals. Significant increases in white blood cell count, ALP, SGPT, and serum blood urea nitrogen levels were observed. Serum sodium and cholesterol levels were significantly reduced indicating signs of hepatotoxicity. In female rats, serum sodium, potassium, and albumin levels increased, while levels of platelets and serum globulin were significantly decreased. Histopathological assessment showed a mild degree of fatty change and hydronephrosis in male rats and nephrocalcinosis and pyelonephritis in female rats (monograph: WHO 2009).

Phytochemicals: Quercetin glycosides and triterpenoids from leaves, carotenoids in fruits (Mercadante et al. 1999; Begum et al. 2002), volatile and aroma active compounds from the leaf including pentacyclic triterpenoid guajanoic acid, β-sitosterol, uvaol , oleanolic acid and ursolic acid (Begum et al. 2004), flavonol glycosides (Liang et al. 2005), avicularin, arabinan, leucocyanidin, luteic acid, ellagic acid, rhamnose, amritoside, arjunolate, sesquiguavaene, ester of hexahydroxydiphenic acid and arabinose, myrcene, 1,8-cineole, α-pinene, caryophyllene, caryophyllene oxide, eugenol, β-pinene, limonene, menthol, α-terpinyl acetate, isopropyl alcohol, longicyclene, β-bisabolene, β-copaene, β-caryophyllene, farnesene, β-humulene, α- and β-selinene, cardinene, curcumene, and guajaverin were isolated (monograph: WHO 2009).

Psittacanthus calyculatus (DC.) Don., Loranthaceae

The aerial parts of *Psittacanthus calyculatus* are used in the treatment of diabetes and hypertension in the traditional Mexican medicine. The methanol extracts (200 mg/kg) of the aerial parts of this plant produced significant hypoglycemic activity in streptozotocin-induced diabetic rats when compared with diabetic control (Guillermo et al. 2012).

Psoralea corylifolia L., Leguminosae

This is a medicinally important plant in India and China. The seeds of this plant are used in traditional medicine as a laxative, aphrodisiac, anthelmintic, diuretic, and diaphoretic and are also used to treat leprosy and inflammation of the skin. The whole plant is also used traditionally in the treatment of DM. The seed of *P. corylifolia* is commonly used in traditional Chinese medicine against gynecological bleeding, vitiligo, and psoriasis.

The water extract of the seed of *P. corylifolia* showed antidiabetes activity against streptozotocin-induced diabetic rats (Kamboj et al. 2011). The extract at a dose of 20 mg/kg (daily administration for 21 days) resulted in almost normalization of blood glucose levels, liver carbohydrate metabolizing enzymes (hexokinase, glucose-6-phosphatase, and glucose-6-phosphate dehydrogenase), and antioxidant enzymes (catalase, peroxidase, and superoxide dismutase) (Ghosh et al. 2009). A recent study confirmed the anti-DM activity of the seed extract of this plant in streptozotocin-induced diabetic rats. The oral administration of the extract improved hyperglycemia and glucose tolerance in diabetic rats. Further, the treatment increased serum insulin levels (Seo et al. 2014). Treatment with the extract under *in vitro* conditions inhibited hydrogen peroxide–induced apoptosis in INS-1 cells. The extract decreased ROS levels. Psoralen and isopsoralen (coumarins present in the extract) showed preventive effects against hydrogen peroxide–induced cell death (Seo et al. 2014). *In vitro* studies (glucose uptake by yeast cells, α-amylase inhibition assay, glycosylation of hemoglobin assay, and glucose diffusion assay) also showed that the methanol extract of the plant possesses antidiabetes activity (Suhashini et al. 2014). When mice were treated with *P. corylifolia* (100–300 mg/kg), the treatment reduced dexamethasone-induced insulin resistance as judged from a significant reduction in the levels of plasma glucose, serum TG, and lipid peroxide; the treatment also stimulated glucose uptake in skeletal muscles (Tayade et al. 2012). Bioassay-guided fractionation of the ethyl acetate–soluble extract of *P. corylifolia* produced psoralidin, which is a noncompetitive PTP1B inhibitor with an IC_{50} value of 9.4 μmol/L. Another inhibitor of PTP1B, bakuchiol, was also isolated from the plant seed (Jiany et al. 2012).

Follow-up studies are warranted to determine this plant's utility to develop a drug for DM. Flavonoids and tannins were reported from this plant.

Pterocarpus marsupium Roxb., Fabaceae (Figure 2.80)

Common Names: Indian kino tree, bijasal, vengai, venga, etc.

Description: *Pterocarpus marsupium* is an erect, moderate- to large-sized deciduous tree. The leaves are imparipinnately compound, exstipulate, and alternate, and the leaflets are usually 5–7, oblong, and coriaceous. Yellowish flowers are borne in copious panicle racemes, with pedicels shorter than the calyx. The pods are orbicular, flat, winged with 1–2 seeds, convex, and bony.

Distribution: *Pterocarpus marsupium* is distributed in the Western Peninsula of India and Ceylon.

Traditional Medicinal Uses: *P. marsupium* is used as an astringent, hemostat, tonic, styptic, vulnerant, antipyretic, and anthelmintic and used in the treatment of leukorrhea, passive hemorrhages, toothache, diarrhea, dysentery, fever, boils, sores, skin diseases, diabetes, biliousness, ophthalmia, gleet, and urinary discharge (Kirtikar and Basu 1975).

Pharmacology

Antidiabetes: The water extract of the bark showed hypoglycemic activity in normal and alloxan-induced diabetic rats (Vats et al. 2002). The bark extract of *P. marsupium* exhibited antidiabetic activity and corrected the metabolic alterations in diabetic rats, which activity resembled that of insulin (Dhanabal et al. 2006). (–) epicatechin from the water extract of the stem bark showed antidiabetic activity in alloxan-induced diabetic rats and stimulated isolated rat islets (Sheehan et al. 1983; Ahmed et al. 1991). The hypoglycemic effect of phenolic compounds from heartwood has been reported (Manickam et al. 1997). Pterostilbene improved the activities of key enzymes of glucose metabolism in the liver of diabetic rats (Pari and Satheesh 2006). An isoflavone from the methanol extract of the plant upregulated Glut-4 and PPAR-γ on L6 myotubes (Anandharajan et al. 2005; Tripathi and Joshi 1988).

Marsupin and pterostilbene, phenolic constituents of the heartwood of *P. marsupium*, lowered blood glucose levels of streptozotocin-induced hyperglycemic rats, which was comparable to that of metformin (Manickam et al. 1997).

The aforementioned studies suggest that the plant possesses several active principles and different mechanisms of action. Further studies are needed to determine its therapeutic utility.

FIGURE 2.80   *Pterocarpus marsupium.*

Other Activities: The ethyl acetate extract of the plant heartwood and its flavonoid constituents produced a significant reduction of serum lipids (Jahromi and Ray 1993).

Phytochemicals: L-Epicatechin, liquiritigenin, isoliquiritigenin, alkaloid, resin, essential oil, kinotannic acid, kinonin, kino red, pyrocatechin, protocatechuic acid, gallic acid, flavanone glycosides, pterostilbene, trihydroxychalcone, marsupol isoflavanoid, benzofuranone marsupin, pterocarpol, B-endesmol, propterol-B, pterostilbene, 4,2′,4′-trihydroxychalcone, 8-c-β-D-glucopyranosyl-3,7,4′-trihydroxyflavone, 3,7,3′-tetrahydroxyflavone, 3′-C-β-D-glucopyranosyl-α-hydroxydihydrochalcon, 7-hydroxy-6,8-dimethylflavone-7-*O*-β-arabinopyranoside, 7,8,4′-trihydroxy-3′,5′-dimethoxyflavone-4′-*O*-β-D-glucopyranoside, retusin-8-*O*-α-L-arabinopyranoside, lupeol, and naringenin were isolated (review: Ajikumaran and Subramoniam 2005).

Pterocarpus santalinus Linn.f., Fabaceae

Pterocarpus santalinus is commonly known as red sandalwood, agarugandha, raktachandan, sivappu chandanam, raktachandanam, etc. *P. santalinus* is a medium-sized deciduous tree with deeply clefted blackish-brown bark and dark purple heartwood. The leaves are compound with ovate, obtuse leaflets. Yellow flowers are produced in simple or sparingly branched racemes. Pods have narrow wings, a hard central region, and a beak at the basal corner. *P. santalinus* is distributed in the Western Peninsula of India. In traditional medicine, it is used to treat many conditions such as inflammation, headache, bilious affections, skin diseases, chronic dysentery, and vomiting (review: Ajikumaran and Subramoniam 2005).

Oral administration of the bark (alcohol extract) of the plant showed antihyperglycemic activity in diabetic rats (Kameswara Rao et al. 2001). The plant bark exhibited promising antidiabetes activity in streptozotocin-induced diabetic rats and reduced the levels of blood glucose and glycosylated hemoglobin and improved insulin and lipid profiles (Kondeti et al. 2010). Antidiabetic compounds such as β-sitosterol and β-amyrin were reported from this plant.

The plant has wound healing property. A lignan from the heartwood inhibited TNF-α production and T-cell proliferation.

Phytochemicals reported include desoxysantalin, isoflavonesantal, pterostilbene, pterocarpin, homopterocarpin, isoflavone glucoside, lignan, aurone glycosides, santalin A and B, pterocarptriol, isopterocarpolone, pterocarpodiolone, β-eudesmol, pterocarpol, cryptomeridiol, acetyl oleanolic acid, erythrodiol, lupeol, epilupeol, lupenone, lup-(20)29-ene, β-amyrone, β-sitosterol, β-amyrin, stigmasterol, betulin, and eudesmol (review: Ajikumaran and Subramoniam 2005).

Pteronia divaricata (P.J.Bergius) Less., Asteraceae

Pteronia divaricata is traditionally used in South Africa to treat diabetes. The extract of this plant inhibited α-glucosidase and α-amylase enzymes *in vitro* (Deutschlander et al. 2009).

Pueraria lobata (Willd.) Ohwi, Leguminosae

Common Name: Kudzu vine

The biologically active compound puerarin was isolated from *Pueraria lobata* root. This compound showed antihyperglycemic activity in streptozotocin-induced diabetic mice. Serum insulin level was increased by the treatment. Histopathological examination of the pancreas revealed that the compound alleviated streptozotocin-induced lesions in the pancreas. Further, the levels of IRS-1 and IGF-1 in the pancreas were increased. Endogenous mRNA levels of skeletal muscle IR and PPAR-γ were increased in the treated animals. Besides, the compound improved the dyslipidemia in diabetic mice (Wu et al. 2013). Puerarin (80 mg/kg) provided protection from diabetic nephropathy in streptozotocin-induced diabetic rats (Li et al. 2009). Further, the compound ameliorated retinal microvascular dysfunction in retinal pigment epithelial cells in both C57BL/6 mice and streptozotocin-induced diabetic rats (Hao et al. 2010). The compound appears to be promising for further studies. Kakonin (a flavonoid) isolated from the root of this plant also exhibited a blood glucose-lowering effect in alloxan- or adrenaline-induced diabetic mice (Shen and Xie 1985).

Pueraria thomsonii Benth., Leguminosae

Extracts of *Pueraria thomsonii* significantly activated PPAR-α and PPAR-γ. Bioassay-guided fractionation resulted in the isolation of daidzein from *P. thomsonii* as the PPAR-activating compound (Shen et al. 2006). A flavonoid (tectorigenin) isolated from the flowers of this plant showed hypoglycemic and hypolipidemic activity (Lee et al. 2000). *Pueraria* flavonoid showed hypoglycemic and hypolipidemic effects in hyperlipidemic diabetic mice (Sun et al. 2002).

Pueraria thunbergiana Siebold & Zuccarini, Fabaceae

Kaikasaponin III isolated from the flowers of *Pueraria thunbergiana* (10 mg/kg, i.p., for 7 days) showed potent hypoglycemic and hypolipidemic effects in the streptozotocin-induced diabetic rat. The compound protected Vero cell from injury by hydrogen peroxide (Lee et al. 2000). Other reported activities of this compound include antimutagenic, anti-lipid peroxidative and hepatoprotective effects (Park et al. 2002).

Pulicaria incisa (Lam.) DC., Asteraceae

Pulicaria incisa (aerial part) is used in Egyptian traditional medicine to treat DM (Haddad et al. 2005). The plant extract exhibited transient hypoglycemic effects in normal rats; the effect appeared 1 h after administration (Shabana et al. 1990).

Punica granatum L., Lythraceae (Formerly Punicaceae)

Synonym: *Punica nana* L. (Figure 2.81)

Common Names: Pomegranate, wild pomegranate, bijapura, anar, anara, madulai, matalam, etc.

Description: *Punica granatum* is a shrub or small tree with a smooth, dark gray bark attaining a height of 10 m. The branchlets are sometimes spinescent with oblong or obovate leaves, shining above. The scarlet red or yellow flowers are produced solitarily or with 1–4 flowers together. The fruit is a globose balusta, crowned with persistent calyx, rind woody, and coriaceous and an interior septa containing variously colored seeds with fleshy testa. The seeds are numerous, red, pink, or yellowish-white (monograph: WHO 2009).

Distribution: *Punica granatum* is considered to be native to Iran, Afghanistan, and Baluchistan, cultivated throughout India, and found in the wild in the warmer valleys and outer hills of the Himalayas between altitudes of 900 and 1800 m.

Traditional Medicinal Uses: All parts of this plant such as the leaves, fruit, fruit peel, and seeds are used in traditional medicine. It is an astringent, refrigerant, cardiotonic, stomachic, stimulant, taeniafuge, vermifuge, anthelmintic, and purgative and used in diarrhea, dysentery, leukorrhea, scorpion

(a)   (b)   (c)

FIGURE 2.81 *Punica granatum.* (a) Twig. (b) Flower. (c) Fruits.

venom, worms, dyspepsia, hydrocele, inflammation, sore throat, sore eye, brain disease, spleen complaints, chest troubles, scabies, bronchitis, earaches, biliousness, liver and kidney disorders, asthma, dermatosis, metrorrhagia, piles, tapeworm attack, and tumor. The dried root or trunk bark is used orally to treat dyspepsia, sore throat, menorrhagia, leukorrhea, and ulcers (monograph: WHO 2009).

Pharmacology

Antidiabetes: The plant flower, fruit, and seed possess antidiabetic and antihyperglycemic properties. The flower extract is a potent α-glucosidase inhibitor that improves postprandial hyperglycemia in Zucker diabetic fatty rats (Li et al. 2005b). The flower extract lowered blood glucose level in normal and alloxan-induced diabetic rats (Jafri et al. 2000). The seed has glucose-lowering effects in streptozotocin-induced diabetic rats (Das et al. 2001). The ethanol extract of the leaves of this plant showed anti-DM and antihyperlipidemic activities in alloxan-induced NIDDM rats; the extract increased glycogen content in the liver and muscles (Das and Barman 2012).

Several studies suggest that the peel extract can also ameliorate the complications of DM. The methanol extract of the fruit peel (200 mg/kg) increased the activities of antioxidant enzymes and decreased oxidative stress in the kidney and liver of alloxan-induced diabetic rats (Parmar and Kar 2007). Further, the methanol extract reduced the levels of lipid peroxides in the heart, kidney, and liver and improved the levels of serum insulin and thyroid hormones in diabetic rats (Parmar and Kar 2008a). Besides, the peel extract favorably improved the lipid profiles and oxidative potential in experimental diabetic rats (Althunibat et al. 2010).

An antidiabetes principle has been isolated from the methanol extract of fruit rinds of *P. granatum*; the compound, valoneic acid dilactone (25 or 50 mg/kg, p.o.), showed promising dose-dependent antidiabetic activity (Jain et al. 2012). The methanol extract (400 mg/kg) or valoneic acid dilactone treatment resulted in very minimal acinar damage and an adequate number of pancreatic islets in diabetic rats. Besides, the extract and the active compound inhibited α-amylase activity. In addition to this, the extract and valoneic acid dilactone were found to inhibit PTP1B activity that mimicked insulin action to some extent. Thus, valoneic acid dilactone was found to act through several mechanisms that include inhibition of aldose reductase, α-amylase, and PTP1B (Jain et al. 2012). There could be more anti-DM compounds in the peel. In fact, the presence of antidiabetic compounds such as ursolic acid, gallic acid, ellagic acid, kaempferol, and epigallocatechin-3-gallate in this plant has been reported. Detailed studies including toxicity evaluation and subsequent clinical studies appear to be rewarding.

Clinical Studies: Concentrated pomegranate juice improved the lipid profiles in diabetic patients with hyperlipidemia (Esmaillzadeh et al. 2004).

Other Activities: Various parts of the plants, including the fruit juice and peel, have many pharmacological properties such as antioxidant, hypolipidemic, anti-inflammatory, antiviral, antimicrobial, immunomodulatory, and cardiovascular protective effects (review: Middha et al. 2013). Pomegranate flower improved cardiac lipid metabolism in a diabetic rat model (Huang et al. 2005). The juice consumption for 3 years by patients with carotid artery stenosis reduced common carotid intima-media thickness, blood pressure, and LDL oxidation (Aviram et al. 2004). In atherosclerotic mice and in humans, pomegranate juice flavonoids inhibit LDL oxidation and cardiovascular diseases (Aviram et al. 2004). Several studies strongly suggest that the fruit extracts and seed have anticancer activity (Toi et al. 2003; Albrecht et al. 2004; Kohno et al. 2004; Mehta and Lansky et al. 2004; Afaq et al. 2005). Pomegranate extract is active against bacteria including drug-resistant bacteria (Voravuthikunchai and Kitpipit 2005). The fruit juice has anti-HIV 1 activity (Neurath et al. 2004). Seed extracts have antidiarrheal activity in rats (Das et al. 1999). It also has a gastric protective effect (Ajaikumar et al. 2005). Pomegranate extract improved a depressive state and bone properties in menopausal syndrome model ovariectomized mice (Mori-Okamoto et al. 2004).

Toxicity: Pomegranate ellagitannin punicalagin is not toxic to rats contrary to earlier reports about its toxicity to cattle. Repeated oral administration of a 6% punicalagin-containing diet for 37 days did not result in any significant adverse effects except a decrease in blood urea and TG and initial reduction in growth rate. In animal experiments, intragastric administration of very large doses of the alkaloids from the bark caused respiratory arrest and death (monograph: WHO 2009).

Phytochemicals: Compounds from the seed include coniferyl 9-*O*-[β-D-apiofuranosyl(1 → 6)]-*O*-β-D-glucopyranoside, sinapyl 9-*O*-[β-D-apiofuranosyl(1 → 6)]-*O*-β-D-glucopyranoside, 3,3′-di-*O*-methylellagic acid, 3,3′,4′-tri-*O*-methylellagic acid, phenethyl rutinoside, icariside D1, and daucosterol. The fruit contains anthocyanidins: delphinidin, cyanidin, and pelargonidin, punic acid, estrone, isomer of eleostearic acid, 3,5-diglucoside, delphinidin diglucoside, isoquercitrin, petunidin derivatives, pelletierine, pseudopelletierine, methyl pelletierine, methyl isopelletierine, friedelin, betulinic acid, ursolic acid, 2–92-(propenyl)-Δ-piperidine, cyanidin-3-glucoside, cyanidin-3,5-diglucoside, granatin A and B, corilagin, strictinin, punicalin, punicalagin, nonadecanoic acid, heneicosanoic acid, tricosanoic acid, 13-methyl stearic acid, 4-methyl lauric acid (Rastogi and Mehrotra 1999). The plant contains betainic alkaloid, punicalagin, ellagic acid, tannins, flavonoids, etc. Interestingly, the fruit peel contains bioactive compounds such as casuarine, corilagin, ellagic acid, gallic acid, methyl gallate, granatin A and B, peduncularin, punicalagin, and punicalin. The flavonoids of pomegranate peel include catechin, cyanidin, epicatechin, epigallocatechin 3-gallate, flavan-3-ol, kaempferol, kaempferol-3-*O*-glucoside, kaempferol-3-*O*-rhamnoglycoside, luteolin, luteolin-3–0-glucoside, naringin, pelargonidin, quercetin, and rutin (monograph: WHO 2009; review: Middha et al. 2013).

Quassia amara L., Simaroubaceae

Quassia amara is a shrub occurring in many parts of the world including Central America and South America. Standardized methanol extract of *Q. amara* (100 and 200 mg/kg, p.o., daily for 14 days) reduced elevated fasting blood glucose levels in nicotinamide–streptozotocin-induced diabetic rats. In the oral glucose tolerance test, the extract treatment significantly increased the glucose tolerance compared with the vehicle control group. Further, the extract effectively normalized dyslipidemia associated with streptozotocin-induced diabetes (Husain et al. 2011). In another study, the antihyperglycemic effect of wood powder of *Q. amara* was evaluated in normal and in alloxan-induced diabetic rats. The powder of wood (200 mg/kg) showed antihyperglycemic effects, similar to metformin, only in diabetic animals, when compared to the diabetic control group. Further, the supplementation decreased plasma cholesterol and TG levels and increased the HDL level in diabetic animals; it also decreased LPO (Ferreira et al. 2013). The plant appears to be promising for further study.

Quercus infectoria Olivier, Fagaceae

Hexagalloylglucose isolated from the galls of *Quercus infectoria* inhibited α-glucosidase activity *in vitro* (Hwang et al. 2000).

Quillaja saponaria Molina, Rosaceae

Quillaja saponaria (plant powder) supplementation with the diet (100 ppm in the standard feed) for 3 weeks resulted in decrease in blood glucose and increase in insulin levels in streptozotocin-induced diabetic rats (Fidan and Dundar 2008).

Raphanus sativus L., Brassicaceae

Raphanus sativus is an edible plant. Oral administration of alcohol extract of *R. sativus* (whole plant) for 21 days to streptozotocin induced diabetic rats resulted in a decrease in the levels of blood glucose and lipids (Taniguchi et al. 2006). Furthermore, water extract of the leaves of this plant showed significant protection and reduction in blood glucose levels in alloxan-induced diabetic rats (Macharala et al. 2012). The root juice of this plant (100–400 mg/kg, p.o.) also lowered blood glucose levels in normal and streptozotocin diabetic rats 3 h after administration (Shukla et al. 2011).

Raphia gentiliana De Wild., Arecaceae

Raphia gentiliana fruits have been used in Congolese traditional medicine. The fruit extract (250 mg/kg, p.o.) showed antihyperglycemic activity in NMRI (Naval Medical Research Institute) mice. In humans the fruit also showed decrease in blood glucose levels in 45 participating individuals (Mpiana et al. 2013).

Raphia hookeri G.Mann. & H.Wendl., Arecaceae
In alloxan-induced diabetic rats, *Raphia hookeri* root extract (50, 100, and 200 mg/kg, p.o.) exhibited significant dose-dependent decrease in glycemia from day 3 to 15 with the highest dose exerting 87% decrease on day 15. Total cholesterol, TG, and LDL-C levels decreased with the dose, while HDL-C showed increase in diabetic rats. The extract stimulated insulin secretion and β-cell survival in diabetic rats. In normal rats, the root extract (500 mg/kg) caused hypoglycemia after 4 h with maximum decrease (54.7%) observed after 8 h (Mbaka et al. 2011).

Rosmarinus officinalis L., Lamiaceae
The ethanol extracts of *Rosmarinus officinalis* (100 and 200 mg/kg) showed an antihyperglycemic effect in alloxan-induced diabetic rabbits. This effect was accompanied by a significant increase in serum insulin levels in diabetic rabbits. The extract could also inhibit LPO and activate the antioxidant enzymes (Bakirel et al. 2008).

Rauvolfia serpentina Benth. ex Kurz, Apocynaceae
Extracts of *Rauvolfia serpentina* (Indian snakeroot or sarpagandha), a medicinal plant endemic to the Himalayan mountain range, have been known to be effective in alleviating diabetes and its complications. The methanol extract of *R. serpentina* root (10, 30, and 60 mg/kg, p.o., daily for 14 days) improved the glycemic, antiatherogenic, coronary risk, and cardioprotective indices in alloxan-induced diabetic mice. The treatment improved body weight and blood insulin levels (Azmi and Qureshi 2012). In another study, an extensive library of *R. serpentina* molecules was compiled and computationally screened for inhibitory action against aldose reductase. Plant-derived indole alkaloids and their structural analogues were obtained as potential aldose reductase inhibitors from a data set of *R. serpentina* molecules (Pathania et al. 2013).

Rehmannia glutinosa Libosch, Scrophulariaceae

Latin Name: Radix rehmanniae
An oligosaccharide of *Rehmannia glutinosa* (root) exerted a significant hypoglycemic effect in normal and alloxan-induced diabetic rats (Zhang et al. 2004). The water extract of *R. glutinosa* stimulated expression of mRNA and protein of proinsulin in the pancreas of type 2 diabetic rats. This could be one of the plant's mechanisms of action (Meng et al. 2008). The extracts of *R. glutinosa* improved glucose-stimulated insulin secretion and β-cell proliferation through IRS2 induction (Park et al. 2008).

Retama raetam (Forssk.) Webb, Fabaceae

Common Name: Weeping broom
Retama raetam is a desert shrub occurring in Egypt and the Mediterranean region. *R. raetam* fruits are used in Saudi and Egypt traditional medicine for the treatment of diabetes. The effect of the water extract of *R. raetam* on the blood glucose levels was investigated in fasting normal and streptozotocin-induced diabetic rats after single and repeated oral administration. The water extract (20 mg/kg) significantly reduced the blood glucose level in normal rats 6 h after a single oral administration and 2 weeks after repeated oral administration. This hypoglycemic effect is more pronounced in streptozotocin-induced diabetic rats. The extract had no effect on basal plasma insulin levels indicating that the underlying mechanism of activity is extrapancreatic (Maghrani et al. 2003). In another study, administration of the methanol extract of *R. raetam* fruit (100, 250, or 500 mg/kg/day for 4 consecutive weeks) to streptozotocin-induced diabetic rats resulted in a dose-dependent reduction in the levels of blood glucose. In oral glucose tolerance test, the extract (250 or 500 mg/kg) caused significant reduction in blood glucose levels at 30 and 60 min after glucose challenge. Further, the extract (500 mg/kg/day for 4 consecutive weeks) significantly increased serum insulin level. The extract significantly inhibited glucose absorption by rat isolated intestine. The extract altered neither the glucose uptake by rat isolated psoas muscle nor the activity of hepatic microsomal glucose-6-phosphatase. In conclusion, the methanol extract of the fruit ameliorated streptozotocin-induced diabetes in rats. This can be attributed, at least partly,

to stimulating pancreatic insulin release and reducing intestinal glucose absorption (Algandaby et al. 2010). The flower of *R. raetam* also showed hypoglycemic and lipid-lowering properties in normal and diabetic rats (Maghrani et al. 2005). In normal rats, the water extract (20 mg/kg, p.o.) of the flower induced a significant decrease in plasma TG and cholesterol levels 1 week after repeated administration. This reduction was maintained for the second week of treatment. In diabetic rats, the extract caused a significant decrease of plasma TGs and cholesterol levels after a single and repeated oral administration. Further, the repeated administration of the extract caused a significant decrease of body weight 1 week after treatment in diabetic rats (Maghrani et al. 2004a). The mechanism of action of different extracts and plant parts is not clear. Further, the active principles involved remain to be studied.

Rhazya stricta Decne, Apocynaceae

The water extract of *Rhazya stricta* leaf (0.5–2 g/kg, p.o., daily for 28 days) did not significantly influence plasma glucose or insulin concentrations in normal and streptozotocin-induced diabetic rats (Wasfi et al. 1994; Tanira et al. 1996). But treatment with a high dose of the extract (4 g/kg) produced a significant and short-lived increase in plasma insulin concentration, accompanied by a significant reduction in plasma glucose concentration. In diabetic rats loaded orally with glucose, *R. stricta* extract (8 g/kg, p.o.) produced significant decreases (16%–32%) in plasma glucose concentration, at 0.5 and 1 h following the treatment. The plant extract did not significantly affect plasma glucose concentrations in nondiabetic control rats (Tanira et al. 1996; Ali 1997). The anti-DM effect is also confirmed by another study. Oral doses of the leaf extract (0.5, 2, and 4 g/kg) reduced the plasma glucose and increased insulin levels at 1 and 2 h following administration to streptozotocin-treated rats. Concomitant oral treatment with *R. stricta* extract (0.5, 2, and 4 g/kg) and glibenclamide (2.5, 5, and 10 mg/kg) resulted in exacerbation of the effect on glucose and insulin concentrations caused by either the extract or glibenclamide when given separately. These results seem to imply that coadministration of the extract with sulfonylurea drugs might adversely interfere with glycemic control in diabetic subjects (Ali 1997). The plant's extract decreased insulin resistance in rats (Baeshen et al. 2010). The active ingredients in the extract responsible for this interaction are not known.

Rheum emodi Wall. ex Meisn., Polygonaceae

Synonym: *Rheum australe* D. Don

Common Name: Himalayan rhubarb

Oral administration of 75% ethanol extract (250 mg/kg) of *Rheum emodi* rhizome for 30 days resulted in decrease in the activities of glucose-6-phosphatase, fructose-1,6-disphosphatase, and aldolase reductase and an increase in the activity of phosphoglucoisomerase and hexokinase in the liver and kidneys of alloxan-induced diabetic rats. The treatment also reduced blood glucose levels (Radhika et al. 2010). Further, the rhizome extract exhibited α-glucosidase inhibitory activity. The methanol extract of the rhizome of *R. emodi* displayed mild intestinal α-glucosidase inhibitory activity. However, fractionation of the active extract led to the isolation of several potent molecules in excellent yields, displaying varying degrees of inhibition of α-glucosidase (Suresh Babu et al. 2004).

Rheum officinale Baill., Polygonaceae

Rhein from the roots of *Rheum officinale* (150 mg/kg) provided protection against diabetic nephropathy progression and ameliorated dyslipidemia in diabetic *db/db* mice (Gao et al. 2010).

Rheum palmatum Linn., Polygonaceae

Rheum palmatum has been widely applied in the clinical treatment of DM in traditional medicine. Emodin, the major bioactive component of *R. palmatum*, exhibited the competency to activate PPAR-γ *in vitro*. Intraperitoneal administration of emodin for 3 weeks to HFD-fed and low-dose streptozotocin-induced diabetic rats resulted in significant reduction in the levels of blood glucose, TG, and total cholesterol. Serum HDL-C concentration was elevated. The glucose tolerance and insulin sensitivity in the emodin-treated group were significantly improved. Furthermore, the results of quantitative RT-PCR analysis showed that emodin significantly elevated mRNA expression level of PPAR-γ and regulated the mRNA expressions of LPL, FAT/CD36, resistin, and FABPs (ap2) in liver and adipocyte tissues (Xue et al. 2010).

Rheum ribes L., Polygonaceae

Rheum ribes is traditionally utilized in the treatment of diabetic patients in Jordan, Syria, and elsewhere. The stem of this plant is edible. In an *in vitro* study, the water extract of this plant augmented pancreatic MIN6 β-cell expansion and stimulated insulin secretion (Kasabri et al. 2012b). Further studies are required to determine its *in vivo* efficacy.

Rhinacanthus nasutus (L.) Kurz, Acanthaceae

The methanol extract of *Rhinacanthus nasutus* leaf (200 mg/kg, p.o.) reduced blood glucose levels 8 h after administration in both fasted normal and streptozotocin-induced diabetic rats. Thus, the plant extract was capable of ameliorating hyperglycemia (200 mg/kg) in streptozotocin-induced diabetic rats (Rao and Naidu 2010). Administration of the methanol extract of *R. nasutus* (200 mg/kg, p.o., daily for 30 days) ameliorated the altered levels of carbohydrate, glycogen, proteins, and AST and ALT observed in streptozotocin-induced diabetic rats. The study indicated that *R. nasutus* restored the overall metabolism and liver function in experimental diabetic rats (Rao et al. 2013a). In another study, administration of *R. nasutus* methanol extract altered the activities of oxidative enzymes in a positive manner in streptozotocin-induced diabetic rats, indicating that *R. nasutus* improved mitochondrial energy production. The authors suggest that *R. nasutus* should be further explored for its role in the treatment of DM (Rao et al. 2013b).

Rhizophora apiculata Blume, Rhizophoraceae

The alcohol extract of *Rhizophora apiculata* leaves showed potential hypoglycemic action in fed rats, glucose-loaded rats, and streptozotocin-induced diabetic rats (Sur et al. 2004). *In silico* analysis shows that three compounds from *R. apiculata* could serve as potential agonists of PPAR-γ. *In silico* analysis of major compounds, isolated from the leaves, on human PPAR-γ protein was carried out by AutoDock 4.0. Compared to thiazolidinediones, *R. apiculata*-derived ligands act as potential agonists. The molecular interaction of ligands was at residues of TYR473, ILE326, ARG288, HIS323, and ARG 288 to activate the action of the PPAR-γ protein (Selvaraj et al. 2014).

Rhizophora mangle L., Rhizophoraceae

Common Name: Red mangrove

Rhizophora mangle is traditionally used in Mexico to treat type 2 diabetes. The water and ethanol/water extracts (5.9 and 59 mg/kg) of the *R. mangle* cortex produced a significant hypoglycemic effect in nicotinamide–streptozotocin-induced diabetic rats, similar to that achieved with glibenclamide (Andrade-Cetto and Mares 2012).

Rhizophora mucronata Lam., Rhizophoraceae

Synonym: *Rhizophora mangle* Roxb.

Rhizophora mucronata (true mangrove) is a large, evergreen shrub with leaf scars on branches. The leaves are ovate or elliptic with rather large flowers. It is distributed in the tidal shores from the mouth of the Indus to Malacca and Ceylon. The plant has numerous traditional uses (Kirtikar and Basu 1975). *R. mucronata* is used as a remedy to control diabetes in the Philippines.

The plant, when administered in the traditional usage form, showed antihyperglycemic activity in glucose tolerance test in rabbits (Alarcon-Aguilara et al. 1998). *B. sensitivum* (80% aqueous ethanol extract of the stem bark) extract manifested very promising α-glucosidase inhibitory activity with an IC$_{50}$ value of 2.24 ± 1.58 µg/mL. This could be one of the mechanisms of its antidiabetes activity (Lawag et al. 2012).

In a clinical study, the water extract of the stem bark has been reported to have surgical open wound healing property, and in rats, it promoted healing of gastric ulceration induced by ethanol–hydrochloric acid. The bark extract has antimicrobial activities. Phytochemicals reported include triterpenoids, tannins (catechol type), and gum 25(S) spirost-5-ene-23β,14α-diol (review: Ajikumaran and Subramoniam 2005).

Rhododendron groenlandicum (Oeder) Kron & Judd, Ericaceae

Common Name: Labrador tea

Indigenous populations of North America used the leaves of *Rhododendron groenlandicum* to prepare herbal tea to treat asthma, colds, scurvy, fever, and stomach and kidney problems. Recent studies reported activities such as antiparasitic, antitumor, decongestant, diuretic, sedative, and stomachic. Ethnomedical exploration revealed the use of this plant as an anti-DM agent. The antidiabetic activity of this plant has been reviewed recently (Eid and Haddad 2014). Following the screening assays for antidiabetic activities *in vitro*, the plant emerged as a promising species with potential antidiabetic activity in adipocytes, skeletal muscle cells, and hepatocytes. It also exerted moderate to strong inhibition of glucose transport in Caco-2 intestinal cells and the effect waned with longer incubation times. Of note, the adipogenic activity of the plant was comparable to that of PPAR-γ activator rosiglitazone.

The plant extract triggered a significant suppression of glucose-6-phosphatase activity and concomitant stimulation of AMPK in cultured hepatocytes *in vitro*.

R. groenlandicum extract was found to be a potent inhibitor of repaglinide metabolism, consistent with interference with CYP3A4 activity. When brewed as a hot-water decoction, Labrador tea showed an increasing capacity to inhibit CYP3A4 and part of this effect could be related to the presence of quercetin aglycone. Hence, Labrador tea preparations should be used with caution when consumed with prescription drugs (Eid and Haddad 2014).

Rhododendron tomentosum Harmaja, Ericaceae

Synonym: *Ledum palustre* L. ssp. *decumbens* (Aiton) Hultén.

Common Name: Marsh Labrador tea or northern Labrador tea

Rhododendron tomentosum is one of the important ethnomedical plants for treating DM in Canada. It showed antidiabetic activities in skeletal muscle cells and adipocytes in culture. Besides, it protected against glucose toxicity and glucose deprivation in neuronal precursor cells *in vitro*. The plant showed potent antioxidant property and is rich in phenolic compounds such as catechin, procyanidins, and caffeic acid derivatives. The plant extract inhibited AGE formation. In a dose-dependent manner, *R. tomentosum* inhibited glucose uptake by intestinal cells *in vitro*. This effect was coupled to a decreased expression of sodium-dependent glucose transporter-1. This transporter is located at the intestinal brush border and mediates the transport of glucose into the intestinal cells. The plant extracts improved oral glucose tolerance compared to untreated control rats. Inhibition of glucose absorption may explain this effect. Thus, this plant has promising antidiabetic properties with more than one mechanisms of action (review: Eid and Haddad 2014). However, more in-depth *in vivo* studies are required to establish its therapeutic benefits.

Rhododendron brachycarpum G. Don. (*Rhododendron fauriei* Franch.), Ericaceae

Rhododendron brachycarpum is an evergreen shrub. Bioassay-guided fractionation on the leaves of *R. brachycarpum* yielded seven PTP1B inhibitory triterpenoids, including a new triterpene, rhododendric acid A. Their PTP1B inhibitory potency and their lipophilicity were investigated to provide a feasible scaffold that may overcome the innate limitation of the previously reported PTP1B inhibitors (Choi et al. 2012). Corosolic acid is also reported from this plant, which inhibited PTP1B activity.

Rhus chinensis Mill., Anacardiaceae

Common name: Chinese sumac

The root of *Rhus chinensis* showed hypoglycaemic activity in experimental animals (Jain and Sharma 1967b).

Rhus chirindensis Baker. F., Anacardiaceae
Rhus chirindensis is used in ethnomedicine in the management and/or control of type 2 DM in some rural communities of South Africa. The water extract of *R. chirindensis* stem bark (50–800 mg/kg, p.o.) caused dose-dependent hypoglycemia in normal (normoglycemic) and diabetic rats. Further, the water extract of *R. chirindensis* exhibited analgesic and anti-inflammatory and hypoglycemic properties (Ojewole 2007).

Rhus coriaria L., Anacardiaceae
Common Name: Sumac
The fruits of *Rhus coriaria* are traditionally used as a dining table spice in Iran and are recommended for DM patients. A single administration of the ethanol extract of the plant fruit to rats resulted in a 24% decrease in PPG levels. Repeated administration for 21 days also lowered PPG (by 25%). The extract markedly increased the levels of HDL and decreased LDL. It showed antioxidant effects by elevating the activities of SOD and catalase. Besides, the extract inhibited maltase and sucrose activities under *in vitro* conditions (Mohammadi et al. 2010). Thus, the fruit extract may be beneficial to DM patients. The ethyl acetate extract of this plant inhibited α-amylase activity with an IC50 value of 28 μg/mL (Giancarlo et al. 2006).

A recent clinical study has shown that daily intake of 3 g sumac powder for 3 months reduced insulin resistance and oxidative stress, which may be beneficial for diabetic patients to make them less susceptible to cardiovascular diseases (Rahideh et al. 2014).

Rhus verniciflua Stokes, Anacardiaceae
Rhus verniciflua has shown the potential against diabetes in many animal models. In streptozotocin-induced rat model, *R. verniciflua* exhibited a decrease in blood glucose levels and blood thiobarbituric acid-reactive substance concentrations. Another study examined the modulatory effects of *R. verniciflua* against hyperlipidemia in WR-1339-induced hyperlipidemic mice model. This remedy decreased plasma lipid levels (total cholesterol, TG, and LDL) and inhibited the activity of 3-hydroxy-3-methylglutaryl CoA reductase and the levels of TBARS (Jung et al. 2006a). The crude extract of *R. verniciflua* stems showed strong α-glucosidase inhibitory effects (Kim et al. 2011a).

Rhynchelytrum repens (Willd.) C.E. Hubb., Poaceae
Common Name: Ruby grass
Rhynchelytrum repens is used as a hypoglycemic drug in folk medicine. The presence of β-glucan in leaf blades, sheaths, stems, and young leaves of this plant was confirmed. These fractions reduced blood sugar to normal levels for approximately 24 h in streptozotocin-induced diabetic rats. This performance was better than that obtained with pure β-glucan from barley, which decreased blood sugar levels for about 4 h. Thus, β-glucan is one of the blood glucose-lowering compounds in this plant (DePaula et al. 2005).

Ricinus communis L., Euphorbiaceae
Synonyms: *R. speciosus* Burm., *R. viridis* Willd. (monograph: WHO 2009)
Ricinus communis is commonly known as castor oil plant, amangula, erandi, amanakku, etc. It is an evergreen bush or small tree with glaucous shoots. The leaves are green or reddish, alternate, broad, palmately lobed, and serrate. Large, monoecious flowers are produced in terminal subpanicled racemes. It is considered as native to Africa but has been adapted and cultivated in subtropical to temperate areas of the world.

In traditional medicine, *R. communis* is a cathartic, contraceptive, cyanogenetic, emollient, laxative, vermifuge, heating, purgative, aphrodisiac, alternative, and galactagogue and used in inflammation, pains, and many other diseases (Kirtikar and Basu 1975; review: Ajikumaran and Subramoniam 2005).

The ethanol extract (50% v/v with water) of *R. communis* roots at an effective dose of 500 mg/kg showed maximum hypoglycemic effect at the eighth hour up to which the study has been conducted. Administration of the effective dose of the extract to diabetic rats for 20 days showed favorable effects not only on fasting blood glucose but also on the total lipid profile and liver and kidney functions on days 10 and 20. The extract seemed to have a high margin of safety as no mortality and no statistically significant difference in ALP, serum bilirubin, creatinine, SGOT, SGPT, and total protein was observed even after the administration of the extract at a dose of 10 g/kg. *R. communis* seems to have a promising value for the development of a potent phytomedicine for diabetes (Shokeen et al. 2008). Oral administration of the water extract of *R. communis* root (500 mg/kg) to alloxan-induced diabetic rats resulted in 62% reduction in the levels of fasting blood glucose (glibenclamide exhibited 69% reduction). Further, the root extract caused a significant reduction in the coronary risk index of rats (as judged from the lipid profiles). Tannin, saponin, cardenolide, and alkaloid were detected in the plant root (Mathew et al. 2012; Obaineh et al. 2012). The antidiabetic property of the *R. communis* leaf was also reported earlier (Jain and Sharma 1967b).

Lectins with immunomodulatory and antitumor activities have been isolated from the seeds.

Ricin is a highly toxic protein produced by the castor bean. The plant is nephrotoxic to cattle.

Phytochemicals reported include hydrocyanic acid, ricinine, albumin, ricin, ricinoleic acid, tristearin, triricinolein, β-sitosterol, squalene, tocopherol, diricinoleins, linoleo-diricinolein, oleo-diricinolein, monoricinoglycerides, triricinolein, phenolics, lectins, β-amyrin, quercetin, rutin, hyperoside, *N*-demethylricinine, 3-*O*-β-ᴅ-xylopyranoside, 3-*O*-β-ᴅ-glucopyranoside, and 3-*O*-β-ᴅ-rutinoside of kaempferol (review: Ajikumaran and Subramoniam 2005).

Robinia pseudoacacia var. umbraculifera DC., Fabaceae

In the traditional medicine of India, *Robinia pseudoacacia* is used as a laxative, antispasmodic, diuretic, etc. Amorphastilbol isolated from the seed of *R. pseudoacacia* exhibited dual PPAR-γ/PPAR-α agonist effects (Wang et al. 2014). Synthetic derivatives of this compound are being tested for the development of improved activity.

Rosa damascena Mill., Rosaceae

Rosa damascena is an ornamental plant, and besides its perfuming effect, several pharmacological properties including anti-HIV, antibacterial, antioxidant, antitussive, hypnotic and antidiabetic, and relaxant effect on tracheal chains have been reported for this plant (review: Boskabady et al. 2011). The methanol extract of the *R. damascena* flower exhibited an intensive inhibitory effect on α-glucosidase activity. Its inhibition was found to be noncompetitive. Oral administration of the extract (100–1000 mg/kg) significantly decreased blood glucose levels after maltose loading in normal and diabetic rats in a dose-dependent manner. These results suggest that *R. damascena* might exert an antidiabetic effect by suppressing carbohydrate absorption from the intestine thereby reducing the postprandial glucose level (Gholamhoseinian et al. 2009).

Rosmarinus officinalis L., Lamiaceae

Local/Vernacular Names: Common rosemary, romani, alecrim, etc.

Rosmarinus officinalis is a busy, much branched, and perennial subshrub. It is a thorny plant widely distributed in Europe and Southeastern Asia. The aerial parts are mostly used in traditional medicine. The ethanol extract of the leaves of this plant (200 mg/kg) exhibited antihyperglycemic activity and increased the levels of insulin in the blood of alloxan-induced diabetic rabbits. Further, the extract inhibited LPO and activated antioxidant enzymes in diabetic rabbits (Bakirel et al. 2008). The water extract of *R. officinalis* (200 mg/kg for 21 days) reduced the blood sugar levels and oxidative stress in streptozotocin-induced diabetic rats compared to untreated diabetic rats (Khalil et al. 2012). Carnosic acid and carnosol, phenolic diterpene compounds of *R. officinalis*, activated the human PPAR-γ (review: Wang et al. 2014). This could be one of the mechanisms of the antidiabetes actions of this plant.

Clinical Trial: The effects of *R. officinalis* leaf powder on the glucose level and lipid profile in normal adult humans were investigated. The leaf powder (5 or 10 g/day daily for 4 weeks) exhibited a dose-dependent decrease in blood glucose. Values for total cholesterol, TGs, and lipid peroxides were very significantly lower in the treated groups (Labban et al. 2014a).

Rubia cordifolia L., Rubiaceae

The alcohol extract of *Rubia cordifolia* root reduced blood glucose levels in alloxan-induced diabetic rats. The extract also exhibited antistress and nootropic activities (Patil et al. 2006). In another study, administration of the water extract of *R. cordifolia* root (1 g/kg, oral, daily for 8 weeks) to streptozotocin-induced diabetic rats led to normalization of observed hyperglycemia and hypertriglyceridemia, enhanced transaminases of the liver and kidney, and loss of body weight (Baskar et al. 2008). In another study investigating the roots of *R. cordifolia*, the results showed significant antiglycation, antioxidant, and antidiabetic activities (Rani et al. 2013).

The glucose- and lipid-lowering effect of the plant leaf was also demonstrated. The alcohol extract of leaves of *R. cordifolia* (200 and 400 mg/kg daily for 15 days) decreased blood glucose levels in normal rats. In alloxan-induced diabetic rats, the treatment decreased the blood glucose level, which was comparable to the effect of glibenclamide. In addition, the extract also showed a favorable effect on glucose disposition in glucose-fed hyperglycemic rats after 90 min of glucose administration. Serum cholesterol and TG levels were decreased, whereas serum HDL and protein levels were increased in treated diabetic rats. The histopathological study showed the regeneration of β-cells in both extract- and glibenclamide-treated groups (Viswanathaswamy et al. 2011b).

Rubus fructicosis L., Rosaceae

Common Name: Blackberry

Rubus fructicosis is an important traditional medicinal plant used to treat various diseases including diabetes. It is a deciduous shrub growing up to 3 m. The effect of oral administration of the water extract of *R. fructicosis* leaves on blood glucose levels in normal and streptozotocin-induced diabetic rats was studied. In normal rats as well as streptozotocin-induced diabetic rats, single and repeated oral administration of the extract significantly lowered blood glucose levels. The extract treatments did not affect insulin secretion in both normal and streptozotocin-induced diabetic rats (Jouad et al. 2002a). Tea made from *R. fructicosis* leaves decreased diabetic symptoms associated with chromium- and zinc-dependent diabetes. Daily administration of leaves (5 g/kg) decreased 50% glucose-induced hyperglycemia in alloxan-induced diabetic rabbits. Administration of the fruit (blackberry) extracts to diabetic rats in drinking water for 5 weeks significantly decreased glucose level from 20 to 15 mmol/L. Extracts (*n*-hexane and chloroform) of blackberry inhibited α-glucosidase activity. Pressed residue of two blackberry cultivars exhibited stronger α-glucosidase inhibitory activity. Collectively, the inhibition of intestinal α-glucosidase and pancreatic α-amylase activities as well as a rich profile of antioxidant bioactive constituents indicates that berry fruit is a promising dietetic therapy for DM (review: Zia-ul-haq et al. 2014). However, controlled clinical studies on this functional food are lacking.

Rubus ulmifolius Schott, Rosaceae

Rubus ulmifolius is a wild blackberry known by the English common name elm leaf blackberry or thornless blackberry. It is a brambly shrub with compound leaves and small white or pink flowers native to Europe and North Africa. The dried leaf infusion (20%) of this plant did not modify the glycemia in normal rats, but in streptozotocin-induced as well as alloxan-induced diabetic rats, the infusion elicited remarkable blood glucose-lowering effects (Lemus et al. 1999).

Ruellia tuberosa L., Acanthaceae

Administration of the optimum dose (500 mg/kg) of the methanol extract of *Ruellia tuberosa* to normal and diabetic rabbits showed a significant blood glucose-lowering effect. The ethyl acetate fraction of the methanol extract showed antidiabetic activity (28.6% decrease in blood glucose) at a dose of 100 mg/kg. The results were compared with the standard drug tolbutamide. The fraction also showed antioxidant activity. Phytochemical investigation of the fraction indicated the presence of flavonoids (Shahwar et al. 2011).

Rumex patientia L., Polygonaceae

The effect of subchronic feeding of *Rumex patientia* seeds (seed powder mixed with standard pellet food at a weight ratio of 6% for 4 weeks) was evaluated on serum glucose and lipid profile in streptozotocin-induced diabetic rats. Serum glucose was significantly lower in seed powder-treated diabetic rats as compared to untreated diabetic rats. Serum total cholesterol and TG levels did not show significant reductions in treated diabetic rats as compared to untreated diabetic rats. Serum HDL-C increased and LDL-C showed a reduction in treated diabetic rats. The seed also attenuated the increased MDA content and reduced activity of superoxide dismutase in hepatic tissue (Sedaghat et al. 2011).

Ruta graveolens L., Rutaceae

Administration of *Ruta graveolens* infusion (125 mg/kg, p.o., daily for 30 days) or rutin (50 mg/kg) led to significant amelioration of hyperglycemia, hyperlipidemia, serum insulin and C-peptide concentrations, liver glycogen content, and the activities of hexokinase, glucose-6-phosphatase, and glycogen phosphorylase as well as oxidative stress in nicotinamide–streptozotocin-induced diabetic rats. The *in vitro* and *in situ* studies indicated that the infusion and rutin enhanced insulin release from isolated islets and insulin binding to its receptors in rat diaphragm; the infusion or rutin decreased intestinal glucose and cholesterol absorption. Further, the treatments decreased resistin expression in adipose tissue; rutin, but not the infusion, increased adipose tissue PPAR-γ expression. Thus, *R. graveolens* and rutin exhibit anti-DM activity through multiple mechanisms (Ahmed et al. 2010). Rutin is present in many plants.

Saccharum officinarum L., Poaceae

Common Name: Sugarcane

The origin of *Saccharum officinarum* is unknown; it is cultivated in the hotter parts of India and elsewhere. In traditional medicine, it is used as a stimulant and in burns, asthma, fatigue, leprosy, inflammations, ulcers of the skin, and diarrhea (Kirtikar and Basu 1975). Hypoglycemic activity of glycon A, B, C, D, and E from the *S. officinarum* stem has been reported (Chohachi et al. 1985).

Other activities of this plant's stem include hepatoprotection, antioxidant activity, anti-inflammatory activity, immunomodulatory action, antithrombosis, and other beneficial cardiovascular actions. Phytochemicals reported include phenolic compounds, 5-*O*-methylapigenin, 3′,4′,5,7-tetrahydroxy-3–6-dimethoxyflavone, 5,7-*O*-dimethylapigenin-4′-*O*-β-D-glucopyranoside, saccharans (A, B, C, D, E, and F), neocarlinoside, neoisoschaftoside, orientin, swertiajaponin, vicemin, *p*-coumaric acid, ferulic acid, caffeic acid, and 3,4-di-*O*-methyl caffeic acid (review: Ajikumaran and Subramoniam 2005).

Salacia chinensis L., Celastraceae or Hippocratea

Synonym: *Salacia prinoides* DC.

Common Names: Akan pelanduk, cherukoranti, saptrangi, anok, daun puyu, ki telor

Salacia chinensis is a small struggling tree or large climbing shrub with simple, petiolate, opposite, exstipulate, and very coriaceous leaves and 3–6 pediccllate and mostly axillary flowers. The fruit is globose and one-celled bearing one seed. It is distributed in Eastern and Western Indian Peninsula and also found in Ceylon, Java, and the Philippine islands. In traditional medicine, it is an abortifacient and used in amenorrhea, dysmenorrhea, etc. The plant is used in the treatment of diabetes in Southeast Asia.

Hypoglycemic activity of the root bark of *S. prinoides* (*S. chinensis*) in rats has been reported (Pillai et al. 1979). The methanol extract from the stems of *S. chinensis* showed potent antihyperglycemic effects in oral sucrose- or maltose-loaded rats; the extract inhibited intestinal α-glucosidase and rat lens aldose reductase. Further, it inhibited formation of AGEs (Yoshikawa et al. 2003). A recent report indicated that mangiferin present in *S. chinensis* possessed antidiabetic properties and is nontoxic in chemically induced diabetic rats (Sellamuthu et al. 2014). Mangiferin is an important antidiabetes molecule (see under *M. indica*). Besides, 3-β,22-β-dihydroxyoleon-12-en-29-oic acid, tingenone, tingenine B, regeol A, triptocalline A, and mangiferin showed an inhibitory effect on rat lens aldose reductase (Morikawa et al. 2003; Kishi et al. 2003).

The stem extract inhibited nitric oxide production from lipopolysaccharide-activated mouse peritoneal macrophage (Yoshikawa et al. 2003).

Phytochemicals reported include triterpenes, leucopelargonidin, dulcitol, gutta, friedel-1-en-3-one, friedelan-1,3-dione-7∞-1ol, friedel-1-en-3-one, friedelan-1,3-dione-24al, other friedalan derivatives, mangiferin, 3-β, 22-β-dihydroxyoleon-12-en-29-oic acid, tingenone, tingenine B, regeol A, and triptocalline A; friedelane-type triterpenes including salasones A, B, and C, eudesmane-type sesquiterpene including salasol A, and norfriedelane-type triterpene (salaquinone A) were isolated from the stems (Morikawa et al. 2003).

Salacia macrosperma Wight., Celastraceae

Common Name: Anakoranti

Salacia macrosperma is a diffuse rambling shrub found in the Western Peninsula of India and elsewhere. In traditional medicine, the root is used in the treatment of diabetes. The root extract showed antidiabetic activity in rats (Venkateswarlu et al. 1993). The alcohol extract of the roots of *S. macrosperma* and fractions of the extract showed anti-DM activity in alloxan-induced diabetic rats as evidenced from a decrease in the levels of glucose, cholesterol, lipids, and free fatty acids (Venkateshwarulu et al. 1993). Saptarangi quinone A, B and C, salacia quinonemethide, pristimerin, tingenone, hydroxyl tingenone, and salaspermic acid were reported from this plant.

Salacia oblonga Wall. ex Wight. & Arn., Celastraceae (Figure 2.82)

Common Names: Chundan, ponkoranti, etc.

Description: *Salacia oblonga* is a small glabrous shrub or tree with simple, opposite, petiolate, elliptic oblong, slightly serrate, subcoriaceous, and exstipulate leaves and short peduncles bearing congested cymes; the flowers are sometimes sessile in the axils of the leaves owing to the extreme shortness of the common peduncle. The fruit is globose, tuberculate, and light brown or orange when ripe, bearing about eight seeds.

Distribution: *S. oblonga* is distributed in the Western Peninsula from Konkan southward and also found in the hotter parts of Ceylon.

Traditional Medicinal Uses: It is used in the treatment of rheumatism, gonorrhea, skin disease, asthma, ear disease, and excess thirst (*The Wealth of India* 1972).

FIGURE 2.82 *Salacia oblonga.*

Pharmacology

Antidiabetes: *S. oblonga* (hot-water extract of roots) improved cardiac fibrosis and inhibited post-prandial hyperglycemia in obese Zucker rats (Li et al. 2004). Tea prepared from *S. oblonga* suppressed glucose absorption from the intestine (Matsuura et al. 2004). The water extract of the plant showed hypoglycemic and antioxidant activity in streptozotocin-induced diabetic rats and normal rats (Augusti et al. 1995; Krishnakumar et al. 1999). The aqueous methanol extract of the plant inhibited α-glucosidase and aldose reductase, which is caused by salacinol and kotalanol, respectively (Matsuda et al. 1999).

The water extract of the root showed antihyperlipidemic and antiobesity activities in Zucker diabetic fatty rats (Huang et al. 2006). Besides, *S. oblonga* root extract showed hypoglycemic and antilipid peroxidative activities in diabetic rats (Krishnakumar et al. 1999, 2000). The root extract inhibited cardiac fibrosis and cardiac hypertrophy in Zucker diabetic fatty rats, inhibiting the expression of angiotensin II type 1 receptor in the heart of fatty rats (Huang et al. 2008). *S. oblonga* (root extract) exhibited antidiabetes and hypolipidemic activities in streptozotocin-induced diabetic rats. The serum insulin level increased and the plasma HbA1c level decreased in the extract-treated diabetic animals compared to untreated diabetic control animals (Bhat et al. 2012). The plant is reported to contain the anti-DM compound mangiferin (Matsuda et al. 1999).

Clinical Studies: The hot-water extract of the root is an α-glucosidase inhibitor and the extract at a higher dose reduced the plasma glucose and postprandial serum insulin incremental in humans (Heacock et al. 2005).

Other Activities: In the cotton pellet granuloma assay, the root bark showed anti-inflammatory property in rats (Ismail et al. 1997).

Toxicity: The salacinol extract, even at high doses, did not result in clinical chemistry or histopathologic indications of toxic effect in rats (Wolf and Weisbrode 2003).

Phytochemicals: Polyphenols mangiferin, catechins, and catechin dimers, friedelane-type triterpene, salacinol, and kotalanol were isolated (Matsuda et al. 1999).

Salacia reticulata Wight, Celastraceae/Hippocrateaceae
Salacia reticulata (commonly known as ekanayakam, saptarangi, and ponkoranti) is a shrub used for the treatment of diabetes in Ayurveda. It occurs in India, Sri Lanka, etc. Salacinol and kotanol, potent α-glucosidase inhibitors, have been isolated from *S. reticulata*. A polyhydroxylated cyclic 13-membered sulfoxide was also isolated from the water extract of *S. reticulata*. This compound exhibited greater α-glucosidase inhibitory activity than the inhibitory activity of salacinol and kotanol (Ozaki et al. 2008). Salacinol bound to the enzyme with a binding mode that is similar to that of casuarine (Nakamura et al. 2010).

Clinical Trials: The water extract of *S. reticulata* stem (240 mg/day for 6 weeks) decreased fasting levels of blood glucose and HbA1c levels in type 2 diabetic patients. In another study, a reduction in HbA1c levels has been reported in patients receiving a preparation of *S. reticulata* tea for 3 months (review: Ghorbani 2013). The usefulness of *S. reticulata* consumption (2 g/day for 3 months) in the management of type 2 DM has been observed in a preliminary clinical study of 30 patients (Radha and Amrithaveni 2009). In view of these reported effects on type 2 DM patients, follow-up studies are warranted; these include the mechanism of action (other than α-glucosidase inhibition), active principles involved, and toxicity, if any.

Salmalia malabarica (D.C.) Schott & Endl., Malvaceae
The sepal (hexane fraction from the hydromethanol extract) of this plant caused a marked reduction in fasting blood glucose levels and lipid profiles in streptozotocin-induced diabetic rats (De et al. 2012).

Salpianthus macrodonthus Stand., Nyctaginaceae
Salpianthus macrodonthus is a Mexican traditional medicinal plant. The plant extracts improved glucose tolerance in rabbits (Roman-Ramos et al. 1991).

Salvadora oleoides Decne, Salvadoraceae

Local Names: Jall, mitha jall, etc.

The ethanol extract of the aerial part of *Salvadora oleoides* (a small tree) produced significant reduction in blood glucose level and also had beneficial effects on the lipid profile in euglycemic as well as alloxan-induced diabetic rats at the end of the treatment period (21st day). However, the reduction in the blood glucose and improvement in the lipid profile was less than that achieved with the standard drug tolbutamide (Yadav et al. 2008).

Salvadora persica L., Salvadoraceae

Salvadora persica hydroalcoholic root extract was not toxic at doses up to 1200 mg/kg. Significant reduction of blood glucose and lipid profile in streptozotocin-induced diabetic rats treated with the hydroalcoholic root extract (400 mg/kg for 21 days) was observed compared to untreated diabetic control rats. The glibenclamide- and root extract-treated groups' peak values of blood glucose significantly decreased from 15.6–5.9 mmol/L to 15.8–8.3 mmol/L, respectively (Hooda et al. 2014). In another study, the water extract of the plant root was evaluated in streptozotocin-induced diabetic rats for its antidiabetic activity. The extract (500 mg/kg) of *S. persica* used in the treatment for 28 days resulted in marked reduction in the levels of blood glucose, total cholesterol, TGs, and LDL and elevation of HDL in comparison with the diabetic control. It also accelerated the regeneration of β-cells in treated diabetic rats compared to diabetic control rats (Khan et al. 2014).

Salvia acetabulosa L., Lamiaceae

Salvia acetabulosa is used to treat, among other things, diabetes in traditional medicine. The plant extract showed potent α-amylase inhibitory activity *in vitro* (review: El-Abhar and Schaalan 2014).

Salvia aegyptiaca L., Labiatae

The aerial parts of *Salvia aegyptiaca* are used in traditional medicine to treat diabetes in Egypt and elsewhere. The plant extract exhibited persistent hypoglycemic effects in normal rats. In alloxan-induced diabetic rats, *S. aegyptiaca* extract showed hypoglycemic effects more potent than those of Daonil (Shabana et al. 1990).

Salvia coccinea Buchoz ex Etl., Labiatae/Lamiaceae

Under *in vitro* conditions, the ethanol extract of this plant showed insulin secretagogue activity in INS-1 cells at the 1 μg/mL level (Hussain et al. 2004). The plant extract showed *in vitro* antioxidant properties (Yadav and Mukundan 2011).

Salvia fruticosa Mill, Lamiaceae

Salvia fruticosa has been used as a hypoglycemic agent in folk medicine. A 10% infusion of *S. fruticosa* leaves was tested (250 mg/kg) on normoglycemic rabbits and in rabbits made hyperglycemic by alloxan administration. The infusion caused a significant reduction in blood glucose levels in alloxan hyperglycemic rabbits, but not in normoglycemic animals. The antihyperglycemic effect was evoked by single oral doses of infusion in both normoglycemic and alloxan-induced diabetic rabbits orally loaded with glucose. However, in these animals, *S. fruticosa* infusion did not modify plasma insulin levels. Moreover, the hypoglycemic effect of the drug was not evoked in rabbits that received the glucose load intravenously. These data strongly suggest that *S. fruticosa* treatment produces a reduction in blood glucose mainly by reducing intestinal absorption of glucose (Perfumi et al. 1991).

Salvia lavandulifolia Vahl., Lamiaceae

Common Name: Spanish sage

Salvia lavandulifolia extract exerted hypoglycemic activity. Studies have suggested that the hypoglycemic action of *S. lavandulifolia* may be the result of several synchronous mechanisms: (1) potentiation of insulin release induced by glucose, (2) increased peripheral uptake of glucose, (3) decreased intestinal absorption of glucose, and (4) hyperplasia of the pancreatic islet β-cells (seen after chronic treatment) (Zarzuelo et al. 1990). *S. lavandulifolia* ssp. *oxyodon* extract (10 mg/kg) produced an increase in Langerhans islet number and size and pancreatic insulin content in streptozotocin-induced diabetic rats. When both *Salvia* extract (10 mg/kg) and glibenclamide (1 mg/kg) were administered concurrently to streptozotocin-induced diabetic rats, a significant decrease in glycemic level and mortality index was achieved (Jimenez et al. 1995).

Salvia leriifolia Benth., Lamiaceae
Salvia leriifolia is a traditional medicinal plant. The leaf and seed extracts of this plant exhibited antihyperglycemic effect in mice (Hosseinzadeh et al. 1998).

Salvia miltiorrhiza Bunge, Lamiaceae
Total polyphenolic acid fraction from *Salvia miltiorrhiza* (187 mg/kg, p.o., daily for 28 days) induced a significant decrease in fasting blood glucose, fasting blood insulin, total cholesterol, TG, and blood urea nitrogen and an obvious increase in the insulin sensitivity index in diabetic rats induced by an HFD and a low dose of streptozotocin. The study suggested that polyphenolic acid fraction has an antidiabetic potential *in vivo* (Huang et al. 2012). A recent study showed that intraperitoneal administration of *S. miltiorrhiza* tea improved the learning and memory decline of streptozotocin-induced diabetic rats. The changes in expression of mitogen-activated protein kinase phosphatase-1 under hyperglycemia may play a role in the protective effects of *S. miltiorrhiza* against dementia in diabetic rats (Cai et al. 2014). Bioassay-guided fractionation and purification of the methanol extract of the dried root of *S. miltiorrhiza* led to the isolation of three abietane diterpene PTP1B inhibitors, namely, isotanshinone IIA, dihydroisotanshinone I, and isocryptotanshinone (Jiany et al. 2012). Danshenol A (triterpene) was isolated from the root and rhizome of this plant, which inhibited lens aldose reductase (Angel de la Fuente and Manzanaro 2003).

Salvia officinalis L., Lamiaceae (Figure 2.83)
Common Names: Common sage, garden sage, true sage, kitchen sage, etc. (it is an important culinary herb in Mediterranean cuisines)

Description: The leaves are simple (lobed or unlobed but not separated into leaflets) and opposite; there are two leaves per node along the stem; the edge of the leaf blade has teeth; the flower is bilaterally

(a)   (b)

FIGURE 2.83   *Salvia officinalis*. (a) Whole plant. (b) Flower (close-up).

symmetrical; there are five petals, sepals, or tepals in the flower; the petals or the sepals are fused into a cup or tube; there is one or two stamen; the fruit is dry but does not split open when ripe.

Distribution: *Salvia officinalis* is native to the Mediterranean region. Cultivated forms include purple sage and red sage.

Traditional Medicinal Uses: The plant is traditionally used for the treatment of digestive and circulation disturbances, bronchitis, cough, asthma, angina, inflammation, depression, excessive sweating, and skin diseases. Essential oils of the plant have been used in the treatment of a wide range of diseases. The plant is also a traditional remedy against diabetes in many countries (review: Hamidpour et al. 2014).

Pharmacology

Antidiabetes: The blood glucose-lowering effect of this plant's extracts has been demonstrated in animal studies. The effects of essential oil and the methanol extract of the plant leaf on healthy and streptozotocin-induced diabetic rats were investigated. The essential oil of sage (0.042, 0.125, 0.2, and 0.4 mL/kg, i.p.) did not change serum glucose, while the plant extract (100, 250, 400, and 500 mg/kg, i.p.) significantly decreased serum glucose in diabetic rats in 3 h without affecting insulin release from the pancreas, but not in healthy rats. Thus, sage extract has hypoglycemic effect on diabetic animals (Eidi et al. 2005).

In another study, the antidiabetic effect of sage ethanol extract was examined in normal and streptozotocin-induced diabetic rats. Oral administration of sage extract (0.1, 0.2, and 0.4 g/kg for 14 days) to streptozotocin-induced diabetic rats resulted in a significant reduction in serum glucose, TGs, total cholesterol, urea, uric acid, creatinine, AST, and ALT and increased plasma insulin in streptozotocin-induced diabetic rats. These effects did not occur in normal rats. The antidiabetic activity of the extract was comparable to that of glibenclamide (Eidi and Eidi 2009).

Another study suggests metformin-like action for *S. officinalis* tea. Replacing water with an infusion of common sage tea for 14 days lowered the fasting plasma glucose level in normal mice, but had no effect on glucose clearance in response to an intraperitoneal glucose loading. This indicated effects on gluconeogenesis at the level of the liver. Primary cultures of hepatocytes from healthy, sage-tea-drinking rats showed, after stimulation, a high glucose uptake capacity and decreased gluconeogenesis in response to glucagon. Essential oil from sage further increased hepatocyte sensitivity to insulin and inhibited gluconeogenesis. Overall, these effects resemble those of metformin, a known inhibitor of gluconeogenesis. In primary cultures of rat hepatocytes, isolated from streptozotocin-induced diabetic rats, none of these activities was observed. These results seem to indicate that sage tea does not possess antidiabetic effects at this level. However, its effects on fasting glucose levels in normal animals and its metformin-like effects on rat hepatocytes suggest that sage may be useful as a food supplement in the prevention of type 2 DM by lowering the plasma glucose of individuals at risk (Lima et al. 2006). Another study evaluated the antidiabetic effects of subchronic administration of water and ethanol extracts of *S. officinalis* leaves in streptozotocin-induced diabetic rats. The water and alcohol extracts (430 mg/kg, i.p., daily for 6 days) did not show hypoglycemic activity in streptozotocin-induced diabetic rats. Further studies are necessary to prove the antidiabetic effect of *S. officinalis* and its active components (Hajzadeh et al. 2011). However, extract from *S. officinalis* was found to activate PPAR-γ, which is accompanied by improvement in the HDL/LDL ratio and lower level of TGs in the serum; the extract also reduced insulin resistance and the size of the adipose tissue (Christensen et al. 2010). There are several compounds in this plant that activate PPAR-γ. These active molecules include carnosic acid and carnosol present in the ethanol extract, as well as 12-*O*-methyl carnosic acid and linolenic acid (present in the DCM extract) (review: Wang et al. 2014). It should be noted that the plant contains ursolic acid, rosmarinic acid, etc. These compounds have beneficial effects in DM. Studies on other active principles, if any, are not available.

Human Clinical Trial: Sage tea drinking (300 mL twice a day) improved lipid profiles and antioxidant status in humans without causing any hepatotoxicity or causing any adverse effects such as change in blood pressure and heart rate (Sa et al. 2009). In a double-blind clinical trial on type 2 diabetic patients who had not reached the ideal control of the disease, the effect of *S. officinalis* on blood glucose, glycosylated hemoglobin (HbA1c), lipid profile, and liver and kidney function tests was evaluated. Patients were randomly divided into two equal groups of case and control. The case group received *S. officinalis* and the control group received placebo tablets three times a day for 3 months. The 2 h postprandial blood sugar and cholesterol levels were significantly decreased in *S. officinalis*-treated patients compared to the control group. There were no significant changes in glycosylated hemoglobin and fasting blood sugar between the two groups. It is concluded that *S. officinalis* might be beneficial in diabetic patients to reduce the levels of 2 h postprandial glucose and cholesterol. However, higher doses might be needed to decrease fasting blood glucose and glycosylated hemoglobin (Behradmanesh and Rafieian-kopaei 2013).

Other Activities: Other established pharmacological properties include antioxidant activity of rosmarinic acid and carnosic acid, anti-inflammatory activity, memory enhancing property, anticancer property, antiobesity, antibacterial property, antihyperlipidemic activity, and antidiarrheal activity (review: Hamidpour et al. 2014).

Toxicity: The normal usage of sage is safe. Using *S. officinalis* in excess amount may be harmful due to the toxic α- and β-thujones. The related species, *S. lavandulaefolia* (Spanish sage), has almost similar phytochemical composition, but it is devoid of thujones (review: Hamidpour et al. 2014).

Phytochemicals: The major constituents of *S. officinalis* are 1,8-cineole, camphor, borneol, bornyl acetate, camphene, thujones, linalool, α- and β-caryophyllene, α-humulene, α- and β-pinene, viridiflorol, pimaradiene, salvianolic acid, rosmarinic acid , carnosolic acid, ursolic acid, carnosic acid, caffeic acid, etc. (review: Hamidpour et al. 2014).

Salvia splendens Sellow ex J.A. Schultes, Labiatae

Common Name: Scarlet sage

The water and methanol extracts of the aerial parts of the plant showed antihyperglycemic activity in streptozotocin-induced diabetic rats. No significant effect was found in normal rats (Mahesh kumar et al. 2010).

Sambucus adnata Wall. ex DC., Caprifoliaceae

The methanol extracts from the whole plants of *Sambucus adnata* showed significant PTP1B inhibitory activity. A chemical study on the extract resulted in the isolation of 13 compounds, including a novel triterpene. The structure of the novel compound was determined as 1α,3β-dihydroxy-urs-12-en-11-one-3-yl palmitate. Among the isolated compounds, ursolic acid, oleanolic acid, and (±)-boehmenan showed the most potent PTP1B inhibitory activity *in vitro* with IC_{50} values of 4.1, 14.4, and 43.5 μM, respectively. The kinetic analysis indicated that one of the isolates, (±)-boehmenan, inhibited PTP1B activity in a competitive manner (Sasaki et al. 2011).

Sambucus nigra L., Caprifoliaceae

Common Name: Black elderberry

Sambucus nigra is a traditional medicinal plant used, among other things, to treat DM. The plant extract exhibited insulin-like and insulin-releasing actions *in vitro* (Gray et al. 2000). α-Linolenic acid, linoleic acid, and naringenin separated from the methanol extract of *S. nigra* flowers activated human PPAR-γ (review: Wang et al. 2014).

Sanguisorba minor Scop., Rosaceae

The extract of *Sanguisorba minor* showed promising *in vitro* inhibition of α-amylase activity (Gholamhoseinian et al. 2008).

Sansevieria senegambica Baker., Agavaceae

The water extract of *Sansevieria senegambica* rhizome (150 and 200 mg/kg, p.o.) exerted a dose-dependent hypoglycemic and cardioprotective (against dyslipidemic conditions) potential, as well as protection against hypocalcemia in alloxan-induced diabetic rats (Ikewuchi 2010). In another study, the effects of the water extract of the rhizomes of *S. senegambica* on the hematology, plasma biochemistry, and ocular indices of oxidative stress were investigated in alloxan-induced diabetic rats. Oral administration of the extract (100, 200, and 300 mg/kg) resulted in significant reduction in the levels of ocular MDA, atherogenic indices, and total white cell and monocyte counts. Further, the treatment reduced the plasma levels of glucose, TG, total cholesterol, VLDL, LDL and total and conjugated bilirubins, total protein, sodium, urea, blood urea nitrogen, and plasma activities of ALP, ALT, and AST compared to untreated diabetic rats. The treatment significantly increased hematocrit; hemoglobin concentration; red cell, lymphocyte, and platelet counts; mean cell volume; ocular vitamin C content and plasma levels of HDL-C, potassium, chloride, and bicarbonate; and ocular activities of catalase and superoxide dismutase. Analysis of the flavonoid fraction showed the presence of apigenin, quercetin, kaempferol, (–)-epicatechin, naringenin, biochanin, (+)-catechin, and other minor compounds. Thus, the extract containing bioactive compounds showed hypoglycemic, hypolipidemic, antianemic, immune-modulating, hepatorenal protective, and cardioprotective potentials (Chigozie and Chidinma 2012).

Santalum album L., Santalaceae
The heart wood extract of *Santalum album* showed potent antihyperglycemic and antihyperlipidemic activities in streptozotocin-induced diabetic rats (Kulkarni et al. 2012).

Sarcocephalus latifolius (Sm.) Bruce, Rubiaceae
The water extracts of *Sarcocephalus latifolius* roots (250 mg/kg) lowered blood sugar levels in alloxan-induced diabetic rats (Iwueke and Nwodo 2008). The extract (250 mg/kg) caused a marked lowering of blood sugar level in alloxan-induced diabetic rats within 6 h from (mean mmol/L) 14.5 to 3.6 after 3 weeks. The same dose did not show blood sugar-lowering effect in normoglycemic rats. The extract significantly restored both hepatic glycogen content and the activities of hexokinase, glucokinase, and phosphofructokinase in diabetic rats (Iwueke and Nwodo 2008). Urgent follow-up studies are required in view of the promising results reported.

Sarcopoterium spinosum (L.) Spach, Rosaceae
Common Name: Thorny burnet
Sarcopoterium spinosum, a common plant in the Mediterranean region, is widely used as an antidiabetic drug by Bedouin healers. *S. spinosum* is an abundant plant in Israel and was suggested to have a beneficial effect on diabetes symptoms. In pancreatic β-cells, *S. spinosum* extract increased basal insulin secretion but reduced glucose-/forskolin-induced insulin secretion. The extract had insulin-like effects on metabolic pathways in classic insulin-responsive tissues (Rosenzweig et al. 2007). In another study, the water extract of *S. spinosum* increased basal and glucose-/forskolin-induced insulin secretion in RINm cells (rat insulinoma β-cell line) and increased cell viability. Further, the extract inhibited lipolysis in 3T3-L1 adipocytes and induced glucose uptake in these cells as well as in AML-12 hepatocytes and L6 myotubes. GSK3β phosphorylation was also induced in L6 myotubes, suggesting increased glycogen synthesis. *S. spinosum* extract had a preventive effect on the progression of diabetes in KK-A(y) mice. Catechin and epicatechin were detected in the extract (Smirin et al. 2010).

Sarcostemma secamone (L.) Bennett, Asclepiadaceae
The ethanol extract of *Sarcostemma secamone* whole plant (150 and 300 mg/kg daily for 14 days) showed hypoglycemic and hypolipidemic effects in alloxan-induced diabetic rats. Treatment with the extract increased body weight, decreased blood glucose and glycosylated hemoglobin, and remarkably improved lipid profiles and other serum biochemical parameters compared to untreated diabetic control animals (Mohan et al. 2013b). In an acute toxicity study, the ethanol extract of *S. secamone* whole plant was nontoxic at 2000 mg/kg in rats (Mohan et al. 2013b).

Sarracenia purpurea L., Sarraceniaceae
Common Name: Purple pitcher plant
Sarracenia purpurea has a long history of medicinal use by Native Americans to alleviate several ailments. The antidiabetic activity of this plant has been reviewed recently (Eid and Haddad 2014). The alcohol extract of this plant was more potent than metformin in stimulating insulin-dependent glucose uptake in skeletal muscle cells *in vitro*. This effect was attributed to a metformin-like action on the AMPK pathway. The extract also demonstrated cytoprotective activity in a neuronal precursor cell

model for diabetic neuropathy. The bioassay-guided fractionation study resulted in the isolation of 10 known compounds and one new compound.

The following compounds showed significant stimulation of glucose uptake by cultured skeletal muscle cells: 7-β-*O*-methylmorroniside, rutin, kaempferol-3-*O*-rutinoside, kaempferol-3-*O*-(6″-caffeoyl glucoside), morroniside, goodyeroside, and quercetin-3-*O*-galactoside. Quercetin-3-*O*-galactoside and morroniside were responsible for the cytoprotective effect of *S. purpurea* in neuronal precursor cells. *S. purpurea* only moderately affected repaglinide metabolism and left that of gliclazide unaffected (Eid and Haddad 2014).

Satureja khuzestanica Jamzad, Lamiaceae
Satureja khuzestanica is widely distributed in the southern part of Iran. It is known to have high levels of antioxidant property. *S. khuzestanica* essential oil decreased the normal blood lipid peroxide level and increased total antioxidant power in rats. Significant decreases in fasting blood glucose and TG levels were observed with the essential oil in streptozotocin-induced diabetic and hyperlipidemic rats (Abdollahi et al. 2003).

Clinical Trial: The effect of *S. khuzestanica* (tablets containing 250 mg dried leaves) supplementation on hyperlipidemic type 2 DM patients was studied in a double-blind, placebo-controlled clinical trial. The herbal drug was administered daily for 60 days. Blood samples were collected at baseline and at the end of the study. The treatment resulted in significant decrease in total cholesterol and LDL-C, while it increased HDL-C and antioxidant power as compared to the baseline values. The treatment did not influence the levels of serum glucose. Thus, the herbal drug showed antihyperlipidemic activity (Vosough-Ghanbari et al. 2010).

Saururus chinensis (Lour.) Baill., Saururaceae
Saururus chinensis is a traditional medicinal plant used, among other things, for inflammatory diseases. Saurufuran A, a new furano-diterpene, isolated from the roots of *S. chinensis* exhibited an agonist effect on PPAR-γ (review: Wang et al. 2014).

Saussurea lappa C.B. Clarke, Asteraceae
Synonyms: *Saussurea costus* (Falc.) Lipsch., *Aucklandia costus* Falc., *Aucklandia lappa* Decne (review: Wei et al. 2014).

Local Name: Kuth

Description: *Saussurea costus* is an erect, robust, pubescent, perennial herb, with a simple stout stem 1–2 m high. The leaves are membranous, scaberulous above, glabrate beneath, auricled at the base, and irregularly toothed; basal ones are very large, 0.50–1.25 m long, with a long winged petiole; the upper leaves are smaller, subsessile, or shortly petioled; two small lobes at the base of these leaves are almost clasping the stem. Flower heads are stalkless, bluish-purple to almost black, hard, rounded, 2.4–3.9 cm across, and often 2–5 clustered together in the axils of leaves or terminal. Receptacle bristles are very long; the corolla is about 2 cm long, tubular, blue-purple or almost black. Achenes have a curved compressed tip narrowed, with one rib on each face. The roots are stout and dark brown or gray (review: Pandey et al. 2007).

Distribution: *Saussurea costus* is distributed in the Himalayas and native to the Himalayan region. It is considered as an endangered medicinal plant.

Traditional Medicinal Uses: The medicinal properties of *S. costus* are well documented in different systems of medicine. Its roots have a strong and sweet aromatic odor with bitter tastes and used as an antiseptic and in controlling bronchial asthma (Anonymous 1972). It is also used for curing various diseases like dysentery, rheumatism, cholera, jaundice, cold, fever, and stomachache. *S. costus* was found as most effective for obese diabetics when a detailed survey and clinical study on potent hypoglycemic plants of different regions from India was undertaken to find antidiabetic plants used in Indian folklore and by different tribes (Upadhyay et al. 1996).

Antidiabetes: The alcohol extract of the root of this plant, when administered orally to rats for 7 days, showed significant hypoglycemic activity without an increase in plasma insulin level (Chaturvedi et al. 2006). The hypoglycemic activity of the plant in rats has been reported first in 1995. The plant extract showed maximum response in reducing blood glucose levels from 4 to 8 h after administration (Chaturvedi et al. 1995).

PTP1B is a key enzyme in insulin signal transduction and the inhibition could improve insulin action. Three anthraquinones isolated from the ethanol extract of the roots of *S. costus* showed moderate bio-activity against hPTP1B *in vitro* (Li et al. 2006b). In another study, activity-guided fractionation of the methanol extract of *S. lappa* root led to the isolation of four compounds (betulinic acid, methyl ester of betulinic acid, mokkolactone, and dehydrocostus lactone) that inhibit PTP1B. The authors suggest that betulinic acid and its methyl ester, as well as the two guaiane sesquiterpenoids, are potential lead moieties for the development of new PTP1B inhibitors to treat diabetes (Choi et al. 2009). Further, chrysophanol and its glucopyranoside derivatives with PTP1B inhibitory activity were obtained from the ethanol extract of the root (Jiany et al. 2012).

Since this plant has been used in traditional and folk medicine successfully (Upadhyay et al. 1996), further studies are warranted to determine its efficacy in various diabetes animal models.

Other Activities: *In vitro* and *in vivo* models have convincingly demonstrated the ability of *S. lappa* to exhibit anti-inflammatory, antiulcer, anticancer, hepatoprotective, and spasmolysis activities. It promoted gastric dynamics, lowered blood pressure, prevented blood coagulation, and provided resistance, to some extent, to microorganisms including *Helicobacter pylori*. Other properties reported include analgesic activity, antioxidant activity, immunomodulation, antiangiogenesis, and antimigraine (review: Wei et al. 2014).

Phytochemicals: Sesquiterpene lactone and triterpenes are the major active ingredients of *S. lappa*. It also contains alkaloids, anthraquinones, and flavonoids (review: Wei et al. 2014).

Schisandra arisanensis Hayata, Schisandraceae

Schisandra arisanensis is a traditional Chinese medicinal plant. The plant extract dose dependently attenuated cytokine-mediated cell death and apoptosis in insulin-secreting cells (BRIN-BD11 cells). In addition, the extract provided some insulinotropic effects that reactivated the abolished insulin exocytosis in cytokine-treated BRIN-BD11 cells. Schiarisanrin A and B isolated from the extract showed a dose-dependent protective effect against cytokine-mediated β-cell death (Hsu et al. 2012).

Schisandra chinensis (Turcz.) Baillon, Schisandraceae

Fructus Schisandrae (the fruit of *Schisandra chinensis*) has been traditionally used as a hypoglycemic agent in Asia. The fruit extracts improved insulin-stimulated glucose uptake in cell-based assays. A lignan-rich fraction obtained from 70% ethanol extract improved glucose homeostasis by increasing glucose disposal rates and enhancing hepatic insulin sensitivity by working as a PPAR-γ agonist in type 2 diabetic rats. The fraction increased the glucose disposal rates of pancreatectomized diabetic rats as much as rosiglitazone during euglycemic hyperinsulinemic clamp. This fraction contained schizandrin, gomisin A, and angeloylgomisin H (Kwon et al. 2011).

Schkuhria pinnata (Lam.) Kuntze ex Thell./*Schkuhria advena* Thell., Asteraceae

The acetone extract of *Schkuhria pinnata* inhibited the activities of α-glucosidase and α-amylase. However, cytotoxicity study revealed that acetone extracts of *S. pinnata* are toxic and raise concern for chronic use (Deutschlander et al. 2009).

Scilla peruviana L., Asparagaceae/Hyacinthaceae

Common Name: Caribbean Lily

Scilla peruviana contains glycosidase-inhibiting pyrrolidines and pyrrolizidines with a long side chain. Some of these compounds are potent inhibitors of yeast α-glucosidase, with IC_{50} values about 6.3 μM (review: El-Abhar and Schaalan 2014).

Sclerocarya birrea (A. Rich.) Hochst., Anacardiaceae

The stem bark (water extract) of *Sclerocarya birrea* showed hypoglycemic activity in normal and alloxan-induced diabetic rats (Ojewole 2003). The water and methanol extracts of the bark of this plant were tested for their antidiabetic effect *in vitro*. The extracts inhibited the activities of α-amylase and α-glucosidase in a concentration-dependent manner. This effect was comparable to that of acarbose. The methanol extract of the bark also displayed strong free radical scavenging activity and the water extract also showed some level of antioxidant activity. Both extracts increased glucose uptake in C2C12, 3T3-L1, and HepG2 cells. However, insulin secretion from RIN-m5F cells was not influenced (Mousinho et al. 2013).

Scoparia dulcis L., Scrophulariaceae (Figure 2.84)

Common/Local Names: Sweet broomweed, mithipatti, kallurukki, etc.

Description: *Scoparia dulcis* is a branched, erect, or ascending leafy herb, 25–90 cm in height; the leaves are opposite, toothed, lanceolate or elliptic-lanceolate, and narrowed into a short petiole. The flowers are white, small, numerous, and pedicellate. The fruit is a globose capsule bearing many seeds.

Distribution: *Scoparia dulcis* is widely spread in India and also in tropical America and sporadically in Africa, Asia, and Australia. It is a perennial herb widely distributed in many tropical countries.

Traditional Medicinal Uses: *S. dulcis* is an emetic, astringent, emollient, depurative, antiseptic, diuretic, expectorant, and purgative and used in the treatment of kidney stones, urinary bladder stones, toothache, blennorrhagia, excessive menstruation, stomach trouble, sore throat, amygdalosis, bruises, conjunctivitis, cough, earache, diabetes, dysmenorrhea, eczema, fever, gonorrhea, gravel, headache, inflammation, jaundice, labor, marasmus, menorrhea, metroxenia, ophthalmia, piles, rash, sores, spasms, syphilis, and tumor. In India, the plant has been used as a traditional remedy for DM (review: Saikia et al. 2011).

Pharmacology

Antidiabetes: In alloxan-induced diabetic rats, the plant extract exhibited hypoglycemic activity (Pari and Venkateswaran 2002). Further, diabetic complications such as hyperlipidemia and oxidative stress are ameliorated by the extract in toxic chemical-induced diabetic rats (Latha and Pari 2003). The water extract of *S. dulcis* reduced hyperglycemia and increased plasma insulin and some polyol pathway enzymes in experimental rat diabetes (Latha and Pari 2004). The administration of the plant extract (200 mg/kg) significantly decreased the blood glucose with a significant increase in the plasma insulin level in streptozotocin-induced diabetic rats at the end of the 15-day treatment. The extract induced stimulation of insulin secretion from isolated pancreatic islet cells indicating its insulin secretagogue activity. It protected against streptozotocin-induced cytotoxicity and nitric oxide production in the rat insulinoma cell line RINm5F (Latha et al. 2004). Flow cytometric assessment demonstrated that the extract

FIGURE 2.84   *Scoparia dulcis.*

suppressed the level of streptozotocin-induced intracellular oxidative stress in RINm5F cells (Latha et al. 2004). Oral administration of the water extract of *S. dulcis* for 6 weeks to streptozotocin-induced diabetic rats led to decreased levels of blood glucose and plasma glycoproteins. The levels of plasma insulin and tissue sialic acid were increased, whereas the levels of tissue hexose, hexosamine, and fructose were near normal. This study indicates that the extract possessed a significant beneficial effect on glycoproteins in addition to its antidiabetic effect (Latha and Pari 2005). The water extract of this plant successfully stimulated glucose uptake activity as potent as insulin at 480 min on L6 myotubes. Furthermore, it stimulated GLUT4 expression and translocation to plasma membranes. Even in the insulin-resistant stage, the extract-treated L6 myotubes were found to be more capable at glucose transport than insulin treatment.

Oral administration of the water extract of this herb for 6 weeks to diabetic rats significantly increased the plasma insulin and antioxidants and decreased LPO (Pari and Latha 2004). In diabetic rats, the levels of blood glucose, glucose-6-phosphatase, and fructose-1, 6-bisphosphatase were significantly increased and plasma insulin, hexokinase, glucose-6-phosphate dehydrogenase, and glycogen were significantly decreased. Diabetic rats treated with the extract significantly reversed all these changes to near normal. The water extract of the plant demonstrated better results compared to the ethanol and chloroform extracts (Pari and Latha 2004). Besides, the water extract treatment for 6 weeks to streptozotocin-induced diabetic rats significantly reduced blood glucose, sorbitol dehydrogenase, glycosylated hemoglobin, TBARS, and hydroperoxides and significantly increased plasma insulin, GPx, GST, and GSH activities in the liver. The effect was compared with that of glibenclamide. The effect of the extract may have been due to the decreased influx of glucose into the polyol pathway leading to increased activities of antioxidant enzymes and plasma insulin and decreased activity of sorbitol dehydrogenase (Latha and Pari 2004). Oral administration of the water extract of *S. dulcis* plant to streptozotocin-induced diabetic rats for 6 weeks resulted in a significant reduction in blood glucose, serum and tissue cholesterol, TGs, free fatty acids, phospholipids, HMG-CoA reductase activity, and VLDL and LDL-C levels (Pari and Latha 2006). Oral administration of the water extract (450 mg/kg) for 45 days to alloxan-induced diabetic rats resulted in a significant reduction in blood glucose and glycosylated hemoglobin and an increase in total hemoglobin; there was a significant improvement in glucose tolerance in the treated diabetic animals (Pari and Venkateswaram 2002).

The antihyperglycemic effect of scoparic acid D, a diterpenoid isolated from the ethanol extract of *S. dulcis*, in streptozotocin-induced diabetic rats was evaluated. Scoparic acid D treatment resulted in decreased levels of glucose as compared with diabetic control rats. The improvement in blood glucose levels of the compound-treated rats was associated with a significant increase in plasma insulin levels. Further, the effect of the scoparic acid was tested on streptozotocin-treated rat insulinoma cell lines (RINm5F cells) and isolated islets *in vitro*. Scoparic acid D evoked twofold stimulation of insulin secretion from isolated islets, indicating its insulin secretagogue activity. A glycoside amellin was reported to ameliorate complications of DM (Latha et al. 2009). The flavonoids from the methanol extract of the aerial parts of this plant extract exhibited significant glucose-lowering activity in normal, glucose-loaded, and streptozotocin-induced diabetic rats when compared with glibenclamide (Sharma and Shah 2010). These studies suggest the likely utility of this plant in the management of DM. There could be more than one active principle. Identification of all active principles other than scoparic acid D and amellin detailed toxicity studies and clinical trials are warranted.

Other Activities: The water extract of the plant showed antioxidant activity. Acetylated flavonoid glycosides from this plant potentiated nerve growth factor (NGF) action. Cytotoxic diterpenes are reported. Scopadulcic acid B, an isolate from the plant, showed *in vivo* and *in vitro* activity against the herpes simplex virus. Scopadulcic acid A (an isolate from this plant) has antiviral activity against herpes simplex virus type 1, antitumor activity in various human cell lines, and *in vitro* antimalarial activity. Scoparinol, a diterpene isolated from the plant, showed significant analgesic, anti-inflammatory, and diuretic activities in animals. The antitumor-promoting activity of scopadulcic acid B from the plant has been reported. Scopadulciol from this plant is an inhibitor of gastric H+ and K(+)-ATPase (review: Saikia et al. 2011).

Phytochemicals: More than 60 compounds have been isolated and identified from *S. dulcis*. These include terpenoids, flavonoids, and steroids. A number of bioactive compounds isolated from this plant include scoparic acid A, scoparic acid B, scopadulcic acid A and B, scoparic acid D, scopadulciol, scopadulin, glutinol (a major triterpene), scoparinol (a diterpene), and acetylated flavonoid glycosides.

Besides, flavone cirsitakaoside, catecholamines, scopadulcinol, ifflaionic acid, flavanone glycoside-8-hydroxytricetin 7-glucoronide, ketone, dulcitone, amellin, triterpene, dulcinol, hexacosanol, β-sitosterol, D-mannitol, 6-methoxybenzoxazolinone, scutellarein, and 7-*O*-methylether were reported from this plant (review: Saikia et al. 2011).

Scrophularia deserti Del., Scrophulariaceae

Five iridoid glycosides, including the two new compounds scropolioside D and harpagoside B, were isolated from the aerial parts of *Scrophularia deserti*. Scropolioside D (10 mg/kg) reduced blood glucose levels in alloxan-induced diabetic rats and harpagoside B was found to possess anti-inflammatory activity (Ahmed et al. 2003).

Scrophularia ningpoensis Hemsl., Scrophulariaceae

A new phenylpropanoid glycoside, designated scrophuside, and two new iridoid glycosides (ningposide I and ningposide II), along with 12 known iridoid and phenylpropanoid glycosides, were obtained from the roots of *Scrophularia ningpoensis*. Most of the obtained compounds showed α-glucosidase inhibitory activity (Hua et al. 2013). Scrophuside and iridoid glycosides ningposide 1 and 2 from the roots of *S. ningpoensis* showed marked α-glucosidase inhibitory activity (review: Firdous 2014).

Scutellaria baicalens Georgia, Lamiaceae

The methanol extract of *Scutellaria baicalens* showed potent inhibitory activity against rat intestinal sucrase. The active principle was identified as bailalein (5,6,7-trihydroxyflavone). It inhibited human intestinal sucrase expressed in Caco-2 cells (Nishioka et al. 1988). The extract of *S. baicalens* improved glucose-stimulated insulin secretion and β-cell proliferation through IRS2 induction (Park et al. 2008).

Sechium edule (Jacq.) Sw., Cucurbitaceae

Common Name: Sayote

Administration of the ethanol extract of *Sechium edule* fruits (100 and 200 mg/kg, p.o.) to alloxan-induced diabetic rats prevented body weight loss and decreased the blood glucose level on the 0th, 7th, 14th, and 21st day. The lipid profile of diabetic rats was also improved upon administration of the ethanol extract of the fruits (Maity et al. 2013).

Securigera securidaca (L.) Degen & Dörfl., Fabaceae

The seeds of this plant are used as an anti-DM agent in Indian traditional medicine. The plant seed water and ethanol extracts (i.p. as well as oral administration) showed antihyperglycemic activity in alloxan-induced diabetic mice. However, in normoglycemic and glucose-induced hyperglycemic mice, the extracts did not reduce blood glucose levels significantly. Glibenclamide was not able to lower blood glucose in alloxan-induced diabetic mice, whereas it significantly lowered the blood sugar in normoglycemic mice. This indicates that the mechanism of action of the seed extracts is different from that of sulfonylurea drugs (Hosseinzadeh et al. 2002). These results were confirmed by another study wherein administration of the water extract of the seed resulted in a reduction in the levels of blood glucose in alloxan-induced diabetic rats, but the extract was devoid of hypoglycemic activity in normal rats (Porchezhian and Ansari 2001). In contrast, the hydroalcoholic extract of the seed (200–800 mg/kg, oral, or 400 mg/kg, i.p.) of this plant did not influence blood glucose levels in both normal and streptozotocin-induced diabetic rats (Minaiyan et al. 2006). Further studies are required to understand the mechanism of action of this plant extract and the active principles involved.

Selaginella tamariscina (Beauv.) Spring, Selaginellaceae

Selaginella tamariscina is a medicinal fern used in traditional medicine to treat infectious diseases, malignant tumors, etc. In HepG2 cells, the ethanol extract of *S. tamariscina* showed a better hypoglycemic effect than the water extract. The 80% ethanol fraction from the ethanol extract had a strong hypoglycemic activity in HepG2 cells. In low-dose streptozotocin and HFD-induced type 2 diabetic rats, both ethanol extract and water extract were able to ameliorate the fasting blood glucose level and improve oral glucose tolerance. Total cholesterol, TG, LDL, free fatty acids, TNF-α, ALT, AST, blood urea nitrogen, and MDA levels in serum were lowered in treated diabetic rats. HDL, insulin, and superoxide dismutase levels in serum as well as the hepatic glycogen content in diabetic rats were elevated by the treatment (Zheng et al. 2011a). In another study, the total flavonoids of *S. tamariscina* possessed antidiabetic activities. The flavonoid fraction also improved the oral glucose tolerance. Furthermore, the flavonoids increased the protein expression of PPAR-γ in adipose tissue and increased the protein expressions of IRS-1 in hepatic and skeletal muscle tissues. These beneficial effects were associated with increased superoxide dismutase and decreased MDA in serum (Zheng et al. 2011b). In another study, treatment with the total flavonoid fraction from *S. tamariscina* for 6 weeks to streptozotocin-induced diabetic mice decreased the concentration of fasting blood glucose, total cholesterol, TGs, and LDL-C, while it increased the levels of insulin and HDL-C compared to untreated diabetic mice. The treatment also improved oral glucose tolerance. Furthermore, both the free fatty acid levels in the liver and hepatic steatosis were ameliorated by the treatment. These changes may be associated with decreased production of visfatin. Administration of the fraction also significantly decreased the levels of MDA, nitric oxide, and inducible nitric oxide synthase and increased the content of GSH and the activity of superoxide dismutase, catalase, GSH peroxidase, and GSH-S-transferase in the liver. Thus, the flavonoid fraction showed an excellent effect in reducing the high blood glucose level but had no effect on normal mice blood glucose level (Zheng et al. 2013).

From the methanol extract of *S. tamariscina*, amentoflavone was obtained and characterized as a noncompetitive PTP1B inhibitor (IC_{50} value = 7.3 μmol/L). Treatment with the compound of 32D cells overexpressing the IR resulted in a dose-dependent increase in tyrosine phosphorylation of IR, possibly through inhibition of PTP1B to enhance insulin-induced intracellular signaling (review: Jiany et al. 2012). Follow-up detailed studies including detailed long-term toxicity evaluations are warranted.

Semecarpus anacardium L.F., Anacardiaceae (Figure 2.85)

Common/Vernacular Names: Marking nut tree, marany nut, oriental cashew, ballatak, bhallatakah, bhela, bhilva, bhilwa, senkottai, erimugi, ceru, allakkuceru, etc.

Description: *Semecarpus anacardium* is a medium-sized to large tree, with a gray bark exfoliating in small irregular flakes. The leaves are simple, alternate, obviate oblong, rounded at the apex, coriaceous,

FIGURE 2.85 *Semecarpus anacardium.*

glabrous above and more or less pubescent beneath, with main nerves in 15–25 pairs. The flowers are greenish-white and fascicled in pubescent pedicles. The fruits are obliquely ovoid or oblong drupes and 2–5 cm long. The upper portion of the fruit is cup shaped, smooth, fleshy, orange-red in color, and sweet and edible when ripe. It is formed of the thickened disc and calyx base. The lower base consists of smooth, black shining pericarp that is thick, containing between its outer and inner laminae oblong cells full of a corrosive resinous juice. This juice is white when the fruit is immature, but brownish or quite black when the fruit is ripe. The nut weighs an average of 3.5 g. Marking nut tree is similar to the cashew nut tree, having an edible false fruit that is orange and fleshy, like the cashew nut; the true fruit is black oily and bitter, the kernels of the nut are edible, but the juice of the nut is highly vesicant and has been traditionally used to mark cloth by washer men (review: Poornima et al. 2013).

Distribution: *S. anacardium* is distributed in the sub-Himalayan region and tropical and central parts of India (review: Jain and Sharma 2013).

Traditional Medicinal Uses: It is a well-known medicinal plant in Ayurveda. The parts used are nut shells, nuts, oil, leaves, and twigs. It is used extensively in piles, skin diseases, etc. The tree bark exudes a gum resin used in leprosy, venereal infections, and nervous debility; the juice from the nut is used in ascites, rheumatism, asthma, neuralgia, epilepsy, and psoriasis, as well as for warts and tumors and is used against epidermal carcinoma; when the nut is bruised, the exudate is used as an abortifacient and a vermifuge. The resinous juice is applied to heel cracks and to treat tumors and malignant growths. The fruit is an astringent, alternative, antirheumatic, carminative, and rubefacient. It is also used in anorexia, cough, asthma, indigestion, leprosy, leukoderma, piles, and various nervous disorders (Warrier 2002; review: Poornima et al. 2013).

Pharmacology

Antidiabetes: The ethanol extract of dried nuts (100 mg/kg) of *S. anacardium* reduced blood glucose levels in both normal and streptozotocin-induced diabetic rats. In normal rats it showed hypoglycemic action, while in the diabetic rats it exhibited antihyperglycemic activity. The antihyperglycemic activity of *S. anacardium* was comparable to that of tolbutamide, a sulfonylurea derivative used in DM (Arul et al. 2004). *S. anacardium* nut milk (300 mg/kg) treatment to streptozotocin-induced diabetic rats resulted in an increase in body weight and insulin and C-peptide levels and a decrease in blood glucose levels. The mRNA levels of O-GlcNAse and the protein levels of PPAR-γ were also increased in the treated group (Jaya et al. 2010). In a follow-up study, the effect of the nut milk on carbohydrate metabolism and energy production in diabetic rats was evaluated. In nut milk-treated (300 mg/kg, for 21 days) streptozotocin-induced diabetic rats, the levels of the enzymes involved in glycolysis and TCA cycle increased, while those of gluconeogenesis decreased. The activities of the mitochondrial complexes were also favorably modulated. The expressions of PI3K and Akt also increased in the skeletal muscle. These effects may be attributed to the hypoglycemic and the antioxidative activity of the nut milk (Jaya et al. 2011). A few more studies confirmed the anti-DM activities of the nut milk. The antidiabetic and antioxidant potential of the nut milk was evaluated in type 2 diabetic rats induced by feeding an HFD for 2 weeks followed by a single injection (i.p.) of streptozotocin. The nut milk (200 mg/kg) treatment for 30 days reduced and normalized blood glucose levels and also decreased the levels of HbA1c as compared with that of diabetic control rats. The treatment also increased the levels of antioxidant enzymes while decreasing the levels of LPO (Khan et al. 2012a). In another study, HFD-fed streptozotocin-induced diabetic rats were treated with the nut milk at four different concentrations (100, 200, 300, and 400 mg/kg for 30 days). In the treated diabetic animals, the blood glucose level and glycosylated hemoglobin levels were lowered, and also an improvement in glucose tolerance was observed. The effect of the drug was found to be effective at the 200 mg/kg dosage (Khan et al. 2012b).

In another study by the same investigators, the nut milk (200 mg/kg; 30 days) reduced the blood glucose levels and decreased the levels of HbA1c and glucose intolerance in streptozotocin-induced diabetic rats. The treatment also decreased the increase in the lipid profile. The levels of urea, uric acid, and creatinine were restored to near normal levels in the treated animals when compared with diabetic control rats. The histopathological abnormalities were also found to be normalized after treatment with the nut milk extract (Khan et al. 2013).

The ethanol extract of the stem bark of *S. anacardium* also showed antidiabetic activity.

The extract (200 and 400 mg/kg) reduced blood sugar level in alloxan-induced diabetic and glucose-induced hyperglycemic (normohyperglycemic) rats. The phytochemical screening of the stem bark extract showed the presence of steroids, triterpenoids, flavonoids, glycosides, saponins, and tannins (Ali et al. 2012). In spite of the promising anti-DM activities, nothing is known regarding the active principles and long-term toxicity, if any.

Other Activities: *S. anacardium* nut extract oil fraction at a dose of 10 mg/kg body weight significantly reduced serum cholesterol levels and increased HDL-C levels in the rat fed with an atherogenic diet (Tripathy et al. 2004). Other pharmacological properties reported include antineoplastic activity of the nut milk extract, antibacterial activity of the dry nut (alcohol extract), neuroprotective activity, anti-inflammatory activity of the nut extract, antiarthritic activity of the nut milk, antioxidant activity, anticarcinogenic activity of the nut, and a contraceptive agent-like action (review: Poornima et al. 2013).

Phytochemicals: The most significant components of *S. anacardium* are bhilawanols, phenolic compounds, and biflavonoids. Nut shells contain biflavonoids: biflavones A, C, A1, and A2, tetrahydrorobustaflavone B (tetrahydro-amentoflavone), jeediflavone, semecarpuflavone, and gulluflavone. Oil from nuts contains a mixture of phenolic compounds mainly of 1, 2-dihydroxy-3 (pentadecadienyl-8, 11)benzene and 1, 2-dihydroxy-3(pentadecadienyl-8′, 11′)-benzene (Majumdar et al. 2008). (7;Z,10;Z)-3-pentadeca-7,10-dienyl-benzene-1,2-diol, (8;Z)-3-pentadec-10-enyl-benzene-1,2-diol, and 3-pentadecyl-benzene-1,2-diol were isolated from the hexane extract (Zhao et. al. 2009). The kernel of the nut contains a small quantity of sweet oil; the pericarp of the fruit contains a bitter and powerful astringent principle. The black corrosive juice of the pericarp contains a tarry oil consisting of 90% anacardic acid and 10% of a higher, nonvolatile alcohol called cardol (review: Poornima et al. 2013).

Senna siamea Lam., Caesalpiniaceae

Senna siamea is traditionally used for the treatment of many diseases including diabetes. The methanol extracts of its leaves and stem bark exhibited antidiabetic activities (antihyperglycemic and antihyperlipidemic effects) in alloxan-induced diabetic rats (Mohammed and Atiku 2012).

Senna tora (L.) Roxb., Fabaceae

Synonyms: *Cassia tora* L., *Cassia obtusifolia* L.

Cassia tora or *Senna tora* is a traditional medicinal plant distributed in tropical and Asian countries. The seeds are reputed in Chinese medicine. Several pharmacological properties have been reported for this plant. There are two reports on the antidiabetic properties of this plant. In one study, the methanol extract from the plant seeds and its fractions were evaluated for glucose-lowering activity (single dose or repeated doses) in normal and alloxan-induced diabetic rats. The n-butanol fraction of the methanol extract was found to be the active fraction. The study showed that the methanol extract (200 mg/kg) or the butanol fraction (100 mg/kg) had blood glucose-lowering activity in normal and diabetic rats in acute and prolonged treatment. The napthopyrone glycoside was isolated from the *n*-butanol fraction of the seed as an active constituent (Chaurasia et al. 2011). In another report, one of the fractions from the n-butanol-soluble portion of the ethanol extract of *C. tora* seeds (100, 250, and 500 mg/kg, single dose or multiple dose, for 2 weeks) decreased elevated blood glucose levels and markers of hepatic and renal dysfunctions in alloxan-induced diabetic rats. Further, the treatment improved lipid profiles. The fraction also decreased glucose levels in normal rats. The authors suggest that these effects could be due to the regeneration of existing β-cells and possibly some more unknown mechanisms of action (Jain et al. 2011).

Sesamum indicum L., Pedaliaceae

Common names of *Sesamum indicum* include gingelly, homadhanya, bariktel, and ellu. It is an erect herb of 30–60 cm height cultivated throughout the warmer parts of India and also in all hot countries. The seeds yield a very valuable edible oil. In traditional medicine it is used to treat various ailments (Kirtikar and Basu 1975).

The hot-water extract of defatted seed from the plant possessed hypoglycemic effect in genetically diabetic mice (Takeuchi et al. 2001). Administration of *S. indicum* seed to streptozotocin-induced diabetic rats decreased the levels of blood glucose and glycosylated hemoglobin and increased serum insulin and hemoglobin levels in treated rats compared to diabetic control animals. Further, the treatment improved the liver glycogen levels and body weight of diabetic rats (Bhuvaneswari and Krishnakumari 2012).

The plant seed contains several antioxidant compounds. Sesamol isolated from the plant has potent antioxidant and free radical scavenging activities.

Phytochemicals isolated include anthraquinones, hydroxysesamone, 2,3-epoxysesamone, sesamol, sesamol dimer, sesamin, sesamolin, sesaminol triglucoside, sesaminol diglucoside, furanofuran lignans, flavonoids, naphthoquinone, and phenylethanoid glycosides (review: Ajikumaran and Subramoniam 2005).

Sesbania sesban (L.) Merr., Fabaceae

Synonyms: *Sesbania sesban* (L.) Merr. ssp. *sesban, Sesbania aegyptiaca* Poiret, *Sesbania aegyptiaca* (Pers) Sted., *Emerus sesban* (L.) Kuntze, *Aeschynomene sesban* L., *Sesbania confaloniana* (Chiov.) Chiov, etc.

Sesbania sesban is widely distributed in India and other tropical countries. Different parts of this plant are used in traditional medicine. Administration of 80% alcohol extract of *S. aegyptiaca* (200 mg/kg daily for 21 days) to alloxan-induced diabetic rats resulted in a decrease in blood glucose levels compared to diabetic control rats (Prakash et al. 2009). The water extract of *S. sesban* leaves also showed anti-DM activity in streptozotocin-induced diabetic rats (Pandhare et al. 2011). In the chronic model, the water extract was administered to streptozotocin-induced diabetic rats at a dose of 500 mg/kg daily, p.o., for 30 days. The treatment resulted in an increase in body weight, liver glycogen content, and serum insulin and HDL levels and a decrease in blood glucose, glycosylated hemoglobulin, total cholesterol, and serum TGs compared to untreated diabetic rats (Pandhare et al. 2011). The following study is a similar one reported in the same year. The water extract of *S. sesban* leaf (250 and 500 mg/kg, p.o., daily for 30 days) exhibited antidiabetic activity in streptozotocin-induced diabetic rats as evidenced by increase in body weight, liver glycogen, serum insulin levels, and HDL levels and a decrease in blood glucose, glycosylated hemoglobin, total cholesterol, and TGs (Ramdas et al. 2011). The bioactive oleanolic acid has been isolated from this plant (review: Gomase 2012). Both water and alcohol extracts (500 mg/kg in each case) of the plant root also showed antidiabetic activity in streptozotocin-induced diabetic rats (Choudhary et al. 2014).

Setaria megaphylla (Steud.) Dur. & Schinz., Poaceae

Treatment of alloxan-induced diabetic rats with the ethanol extract of *Setaria megaphylla* root caused a significant reduction in the fasting blood glucose levels of diabetic rats in both acute study and prolonged treatment (15 days). The activity of the extract was comparable to that of glibenclamide. Further, *S. megaphylla* treatment showed a considerable lowering of serum total cholesterol, TGs, LDL-C, and VLDL-C and an increase in HDL-C in the treated diabetic group. However, the ethanol extract of *S. megaphylla* root was found to be slightly toxic with an LD_{50} value of about 2 g/kg (Okokon et al. 2007c). The ethanol extract of the leaves of this plant also showed antidiabetic activities. The extract produced reduction in blood glucose levels of both normal and alloxan-induced diabetic rats after a single dose of the extract (100, 200, and 300 mg/kg) and after daily administration for 1 week (Okokon and Antia 2007).

Sida acuta Burm.f., Malvaceae

Sida acuta is claimed in folk medicine as an effective oral hypoglycemic agent. Both the water extract and methanol extracts of its leaf (400 mg/kg) significantly increased the tolerance for glucose in glucose-fed normal rabbits. Both extracts (water and methanol) reduced blood glucose level in alloxan-induced rabbits. The water and methanol extracts (400 mg/kg, p.o.) produced a significant decrease in blood sugar at 4 h with a percentage glycemic change of 30% and 20%, respectively. The antihyperglycemic actions of the water and methanol extracts were sustained up to 8 h with a percentage glycemic change of 46% and 45%, respectively. Glibenclamide produced significant glucose reduction in alloxanized rabbits at 2 h with a percentage glycemic change of 24%. Thus, the crude leaf extracts of *S. acuta* possess antihyperglycemic activity in diabetic and normal glucose-fed rabbits (Okwuosa et al. 2011).

Sida rhombifolia L. ssp. retusa, Malvaceae

The water extract of *Sida rhombifolia* ssp. *retusa* leaves (200 mg/kg, p.o.) showed significant hypoglycemic effects (90 min after the extract administration) in normal and streptozotocin-induced diabetic rats. In a hypolipidemic study also, the water extract (200 mg/kg) showed reduction in TGs (Dhalwal et al. 2010).

Sida tiagii Bhandari, Malvaceae

Local Name: Kharenti

S. tiagii is a native species of the Indian and Pakistan desert area. The fruit extract of *S. tiagii* showed substantial improvement in blood glucose levels, glycated hemoglobin level, and liver glycogen content in N5-streptozotocin type 2 diabetic rats (Datusalia et al. 2012).

Siegesbeckia glabrescens Makino

Kaurane-type diterpenes, isolated from the methanol extract of *Siegesbeckia glabrescens* (aerial part), exhibited a PTP1B inhibitory activity (IC_{50} = 8.7 μmol/L) (Jiany et al. 2012).

Silene succulenta Forssk., Aizoaceae

Silene succulenta is a wild Egyptian medicinal plant. The extract of this plant caused transient hypoglycemic effects that appeared only 1 h after administration in rats (Shabana et al. 1990).

Silybum marianum (Linn.) Gaertn., Asteraceae (Figure 2.86)

Common Names: Holy thistle, milk thistle, lady's thistle, etc.

Description: *Silybum marianum* is an erect, shining annual or biennial with 30–120 cm height and a grooved but not winged stem. Large leaves are alternate with strong spines. The head is large, solitary, and terminal with an intruding base. Involucre of bracts is coriaceous with spine. The receptacle is fleshy and the flowers are rose-purple in color. The fruit is an achene, transversely wrinkled, black or gray with white pappus hairs.

(a)

(b)

FIGURE 2.86 *Silybum marianum.* (a) Leaves. (b) Flower.

Distribution: *S. marianum* is found in Europe, North Africa, and North India.

Ethnomedical Uses: *S. marianum* is a well-known hepatoprotective plant. It is sudorific, depurative, aperient, demulcent, alternative, cholagogue, and emmenagogue and used in the treatment of hemorrhage, asthma, dropsy, cancer, jaundice, fever, and leukorrhea (Kirtikar and Basu 1975).

Pharmacology

Antidiabetes: The hypoglycemic effect of the water extract of *Silybum marianum* (aerial part) was investigated in normal and streptozotocin-induced diabetic rats. Oral administration of the water extract (20 mg/kg, a single dose or 15 daily doses) produced a significant decrease of blood glucose levels in both normal and streptozotocin-induced diabetic rats. In addition, no changes were observed in basal plasma insulin concentrations after the extract treatments in both normal and streptozotocin-induced diabetic rats indicating that the extract exerted its pharmacological activity without affecting insulin secretion (Maghrani et al. 2004c). It can prevent the complications of diabetes to a large extent. The aerial part of this plant showed hypoglycemic activity in normal rats and an antihyperglycemic activity in type 1 diabetic rats without affecting insulin secretion (Eddouks et al. 2004). It has been reported that silymarin (flavonoid from the plant) has a favorable impact on glycemic and lipidemic control in type 2 DM with cirrhosis and may reduce fat-induced insulin resistance (McCarty 2005). Silymarin, from the seeds of *S. marianum*, prevented the progression of diabetic nephropathy in streptozotocin-induced diabetic rats (Vessal et al. 2010). Isosilybin A, a phenolic mixture from the seeds of *S. marianum*, activated PPAR-γ (Wang et al. 2014). Thus, it appears that in addition to the hepatoprotective silymarin, specific antidiabetic principles are present in this plant; the plant is very promising for its use as an adjuvant therapy to DM patients with impaired liver function.

Clinical Studies: A few clinical studies on type 2 DM patients have shown the beneficial effects of the hepatoprotective principle silymarin (flavonolignans) isolated from *S. marianum* in lowering blood glucose and lipid levels. In a randomized, double-blind clinical study, silymarin (200 mg, thrice a day, for 2 months) reduced fasting blood glucose, HbA1c, total cholesterol, and LDL in type 2 DM patients (30 patients) receiving conventional therapy. Similar results were obtained in another clinical trial on 25 type 2 DM patients wherein silymarin (200 mg, thrice a day, for 4 months) exhibited reduction in the levels of blood glucose, lipids, and hepatic marker enzymes for liver function. Beneficial effects of silymarin (200 mg/day) on fasting blood glucose, postprandial glucose, and HbA1c levels have been observed in type 2 DM patients maintained on glibenclamide (review: Ghorbani 2013). Silymarin administration to diabetic patients with liver disease could reduce insulin resistance and the need for exogenous insulin administration (Jose et al. 2011; Ghorbani 2013). In another study, the oral intake of silymarin (600 mg/day for 4 months) decreased blood glucose, HbA1c, and glucosuria in insulin-treated diabetic patients with alcoholic liver cirrhosis (Ghorbani 2013). It should be noted that all these clinical studies were done using silymarin; *S. marianum* is reported to contain other antidiabetic molecules such as isosilybin A, a phenolic mixture that activated PPAR-γ (Wang et al. 2014). Further studies could possibly lead to the development of antidiabetic combinations/fractions from this plant.

Other Activities: It is a well-known and widely used hepatoprotective plant. The hepatoprotective property of silymarin has been established in various animal models (review: Subramoniam and Pushpangadan 1999). Silymarin and its most active constituent silybin have been reported to work as antioxidants, a scavenger of free radicals, and an inhibitor of LPO. They protect against genomic injury, increase hepatocyte protein synthesis, decrease the activity of tumor promoters, stabilize mast cells, chelate iron, and slow calcium metabolism. The plant extract has immunomodulatory properties also. Silybin inhibited human T-lymphocyte activation and neutrophil function. However, silibinin (Legalon-70) enhanced the motility of human neutrophils. Silymarin is also a potent hypocholesterolemic drug and inhibits carcinogenesis. Patients infected with hepatitis B and C might benefit from silymarin (Subramoniam and Pushpangadan 1999). Several studies showed that silymarin has hepatoprotective, anti-inflammatory, and regenerative properties. It also has an anti-Gram-positive bacterial activity (review: Ajikumaran and Subramoniam 2005).

Toxicity: Silymarin is considered as safe at reasonable doses.

Phytochemicals: Silymarin, the principal active constituent of milk thistle, is an isomeric mixture of flavonolignans; the main components of silymarin are silybin (also known as silibinin or silybinin), isosilybin, silychristin, and silydianin. Silymarin is localized mainly in the external cover of the seeds (1.5%–3%) (Noor et al. 2014). Other constituents are essential fatty acids, flavonoids (taxifolin, kaempferol, and quercetin), 12 different polyacetylenes, one polyene, myristic acid, taxifolin, silybonol, 2–3-dehydrosilimarin, 2–3-dehydrosilychristin, silymonin, silandrin, D-catechin, tyramine, γ-linoleic acid, and betamine (review: Ajikumaran and Subramoniam 2005).

Siraitia grosvenorii (Swingle) C. Jeffrey ex. A.M. Lu. and Zhi Y., Cucurbitaceae (*Momordica grosvenorii* Swingle)

Siraitia grosvenorii (= *Momordica grosvenorii*) is a traditional medicinal herb used as a substitute sugar for obese and diabetes patients. Recently, improved glucose tolerance, lipid utility, and increased insulin sensitivity were observed on several diabetic rodent models treated with crude mogrosides isolated from the fruit of *S. grosvenorii*. Mogrosides lowered oxidative stress and serum glucose and lipid levels in alloxan-induced diabetic mice (Xiang-Yang et al. 2008). Crude mogrosides provided five new cucurbitane triterpenoids. The main aglycone mogrol and two cucurbitane triterpenoids were found to be potent AMPK activators in the HepG2 cell line. This result suggests that AMPK activation by the mogroside aglycones contributes at least partially to the antihyperglycemic and antilipidemic properties of *S. grosvenorii* (Chen et al. 2011a). This plant has more than one group of active principles. Polysaccharides from *S. grosvenorii* showed a glucose-lowering effect on hyperglycemic rabbits induced by feeding high-fat/high-sucrose chow; its mechanism may be related to the amelioration of lipid metabolism and restoring the blood lipid levels of hyperglycemic rabbits (Lin et al. 2007).

Smallanthus sonchifolius (Poepp & Endl.) H. Robinson, Asteraceae

Synonym: *Polymnia sonchifolia* Poeppig and Endlicher (Figure 2.87)

Common Name: Yacon

Description: *Smallanthus sonchifolius* is a perennial herb 1.5–3.0 m in height with a root system composed of 4–20 oval or fusiform tubers; it produces big underground edible tubers. The aerial part is formed by large leaves and flowers with yellow radially arranged petals.

(a) (b)

FIGURE 2.87 *Smallanthus sonchifolius.* (a) Leaves. (b) Roots.

Distribution: Yacon is native to the Andes.

Traditional Medicinal Uses: Yacon is a traditional crop of the original populations of Peru where they still use this in folk medicine.

Pharmacology

Antidiabetes: Treatment with the hydroethanolic extract (400 mg/kg) of yacon leaves for 14 days reduced glycemia in streptozotocin-induced diabetic rats and nondiabetic normal rats. Besides, the extract restored the activities of plasma enzymes (AST, ALT, ALP) that were altered and improved weight gain in diabetic rats. The effect of the extract is not related to interference of the extract with the intestinal absorption of carbohydrates. The hot- and cold-water extracts were inactive (Baroni et al. 2008). The potential of yacon tubers or its leaves to treat hyperglycemia and kidney problems seems to be related mostly to oligofructan (Valentova and Ulrichova 2003). In addition, studies have reported that tea made from an infusion of yacon leaves reduced glycemia and increased concentration of insulin in the plasma of diabetic rats (Aybar et al. 2001). Yacon extract reduced the proportion of glucose in cultures of hepatocytes acting in a manner similar to insulin (Valentova et al. 2004). When different extracts were tested, the methanol, butanol, and chloroform extracts of yacon leaves showed hypoglycemic activity in rats at minimum doses of 50, 10, and 20 mg/kg, respectively. Oral administration of a single dose of each extract lowered fasting blood glucose levels of normal healthy rats, whereas each extract reduced significantly the hyperglycemic peak after food ingestion. Daily administration of each extract for 8 weeks produced an effective glycemic control in streptozotocin-induced diabetic rats with an increase in the plasma insulin level. In another study, water decoction of the leaves (70 mg/kg) showed antidiabetic and hypolipidemic activities in streptozotocin-induced diabetic rats. Further, the decoction prevented early diabetic nephropathy in diabetic rats (Honore et al. 2012).

Phytochemical analysis of the most active extract (butanol extract) showed the presence of caffeic, chlorogenic, and three dicaffeoylquinic acids as major compounds. Additionally, enhydrin, the major sesquiterpene lactone of yacon leaves, was also effective in reducing postprandial glucose and useful in the treatment of diabetic animals (minimum dose 0.8 mg/kg) (Genta et al. 2010). It is of interest to note that antidiabetic compounds such as chlorogenic, quercetin, and ferulic acid are reported in this plant (Pedreschi et al. 2003) in addition to enhydrin and oligofructan.

Other Activities: Other important properties include skin rejuvenation and cytoprotective activity. Several studies have shown that the leaves of this plant possess activities such as inhibition of migration of polymorphonuclear luekocytes, immunomodulation, antioxidant, etc. (Valentova et al. 2003).

Phytochemicals: Phytochemical studies have demonstrated that yacon leaves and stems are rich in proteins, phenolic compounds such as caffeine, chlorogenic acid, ferulic acid, and flavonoids such as quercetin (Vaalentova et al. 2003). The roots of the plant contain fructose, glucose, and fructooligosaccharides that after fermentation favor the development of bacterial flora (Pedreschi et al. 2003).

Smilax glabra Roxb., Liliaceae

Smilax glabra is a traditional Chinese herb that has been used in folk medicine for the treatment of diabetes. The hypoglycemic effect of the methanol extract of *S. glabra* rhizomes has been shown in normal and KK-Ay mice (one of the animal models of NIDDM). Further, the extract of the rhizome of this plant protected human umbilical vein endothelial cells from AGE-induced endothelial dysfunction *in vitro* (review: Ghisalberti 2005; Sang et al. 2014).

Solanum lycocarpum St. Hill, Solanaceae

Common Name: Wolf fruit

Complex polysaccharides from *Solanum lycocarpum* green fruits slowed gastric emptying and lowered blood glucose and cholesterol levels (Dall'Agnol and von Poser 2000).

Solanum nigrum L., Solanaceae

Common Name: Black nightshade or houndsberry

Solanum nigrum is a weed found in most parts of Europe and the African continent. It contains a toxic glycol alkaloid with a highest concentration in unripe berries (Atanu et al. 2011). However, the plant is known to have several phytochemicals with beneficial pharmacological properties. The leaves of this plant are used in ethnomedicine to treat DM. Validation of the ethnomedical use of the leaves of *S. nigrum* using the oral glucose tolerance test showed no significant glucose-lowering effect (Villasenor and Lamadrid 2006; Atanu et al. 2011). Further studies, especially to detect delayed anti-DM effects, if any, are required to educate the traditional users.

Solanum torvum Sw., Solanaceae

The methanol extract of *Solanum torvum* fruit (100 and 200 mg/kg) containing phenolic compounds exhibited anti-DM and antioxidant properties in streptozotocin-induced diabetic rats (Gandhi et al. 2011).

Solanum tuberosum L., Solanaceae

The polyphenol-rich Irish potato extract (*Solanum tuberosum*) decreased body weight gain and adiposity and improved glucose control in the mouse model of diet-induced obesity. The authors suggest that the potato extract may serve as part of a preventative dietary strategy against the development of obesity and type 2 diabetes (Kubow et al. 2014).

Solanum virginianum L., Solanaceae

Synonyms: *S. surattense* Burm.f., *S. xanthocarpum* Schrad & Wendl.

Common Names: Spiny nightshade, yellow fruited nightshade, kantkari, kateri, kateli, kantankattiri, kantkariccunta, etc.

Description: *Solanum virginianum* is a very prickly, diffuse, bright green perennial herb, somewhat woody at the base; the leaves are usually ovate or elliptic, armed on the midrib and often on the nerves with long yellow sharp prickles; the base is usually rounded with unequal sides. The berries have green and white strips when young but yellow when mature. They are 1.3–2 cm in diameter, yellow or white with green veins, surrounded by the enlarged calyx. The seeds are 2.5 mm in diameter and glabrous. The ovary is ovoid and glabrous; and the style is glabrous (review: Parmar et al. 2010).

Distribution: The plant is distributed in Ceylon, Asia, Malaya, Tropical Australia, and Polynesia. It occurs throughout India, in dry situations as a weed along the roadsides and wastelands (review: Parmar et al. 2010).

Traditional Medicinal Uses: In Ayurveda, this plant is described as pungent, bitter, digestive, alternative, and astringent. The root decoction is used as a febrifuge, diuretic, and expectorant; the extract of the entire plant and fruits has been used for bronchial asthma, tympanitis, misperistalsis, piles, dysuria, and rejuvenation. The whole plant is used traditionally for cough and asthma; decoction of the plant is used in gonorrhea; the paste of leaves is applied to relieve pains; the seeds act as an expectorant in cough and asthma; the roots are used as an expectorant and diuretic; it is useful in the treatment of catarrhal fever, coughs, asthma, and chest pain (review: Reddy and Reddy 2014).

Antidiabetes: The water extract of the fruits (100 and 200 mg/kg) of *S. surattense* (*S. xanthocarpum*) showed significant hypoglycemic activity in both normal and streptozotocin-induced diabetic rats. The activity showed by the extract was comparable to that of the standard oral hypoglycemic agent glibenclamide. The experimental results indicated that it exhibited a potent blood glucose-lowering property in both normal and streptozotocin-induced diabetic rats (Gupta et al. 2005). The antihyperglycemic activity of the alcohol extract in streptozotocin-induced diabetic rats was associated with an increase in plasma insulin. Though the exact mechanism of action is not known, it could be due to increased pancreatic secretion of insulin from existing β-cells (Sridevi et al. 2011). Treatment with β-sitosterol (phytosterol) isolated from *S. surattense* (*S. xanthocarpum*) (10, 15, and 20 mg/kg, p. o.) for 21 days resulted in a dose-dependent decrease in glycated hemoglobin, serum glucose, and nitric oxide with concomitant increase in serum insulin levels in streptozotocin-induced diabetic rats. Further, the treatment increased pancreatic antioxidant levels with a concomitant decrease in thiobarbituric acid-reactive substances (Gupta et al. 2011). This plant may contain more than one active principle. Further studies are warranted to identify all the anti-DM principles present in this plant and to determine their therapeutic potential as an individual compound and in combination.

Other Activities: Other pharmacological properties reported for this plant include hepatoprotective activity, apoptosis induction in cancer cells, antiasthmatic activity, antifilarial activity, anti-inflammatory activity, antifertility activity, and hypolipidemic activity (review: Reddy and Reddy 2014).

Phytochemicals: *S. xanthocarpum* species contains flavones and phenolic acids (apigenin, scopoletin, esculetin, methyl caffeate, caffeic acid, etc.), coumarin, steroids and triterpenoids (carpesterol, campesterol, β-sitosterol, lupeol, cycloartenol, stigmasterol), steroidal alkaloids and glycoalkaloids (solasodine, diosgenis, tomatidenol, α-solamargine), and fatty acids (Reddy and Reddy 2014).

Sonchus oleraceus L., Asteraceae

Common Name: Sowthistle

The effect of the oral administration of an infusion of *Sonchus oleraceus* (100 mg/kg) and its active principle esculetin (6 mg/kg) for 4 weeks to streptozotocin-induced diabetic rats was evaluated for their antidiabetic activity. The treatment to diabetic rats with *S. oleraceus* infusion and esculetin resulted in a marked amelioration of the impaired glucose tolerance and the lowered insulin and C-peptide levels. The impoverished liver glycogen content and elevated liver glucose-6-phosphatase and serum AST and ALT activities of diabetic rats were profoundly corrected as a result of treatments with the plant infusion and esculetin. Also, these treatments led to decrease in serum total lipid, total cholesterol, TG, LDL, and VLDL levels and increase in the HDL level. The antioxidant defense system was potentially improved in diabetic rats as a result of the treatments. Esculetin seemed to be the most effective. However, further clinical studies are required to assess the efficacy and safety of these treatments in diabetic human beings (Hozayen et al. 2011). In another study, the antioxidant and antidiabetic potential of the hydroethanolic extracts of *S. oleraceus* (whole plant) was evaluated in streptozotocin-induced diabetic rats. The hydroethanolic extract of the whole plant (150 mg/kg) exhibited significant antidiabetic activities. Further, the extract showed significant reduction in the levels of MDA and hydrogen peroxides and a substantial increase in catalase activity in diabetic rats. The hydroethanolic extract also showed *in vitro* antioxidant activity (Teugwa et al. 2013).

Sorbus commixta Hedl., Rosaceae

A bioassay-guided fractionation of the methanol extract of the stem barks of *Sorbus commixta* resulted in the isolation of two lupine-type triterpenes, lupenone and lupeol. Both compounds inhibited PTP1B in a noncompetitive manner with IC_{50} values of 13.7 and 5.6 μmol/L, respectively (Jiany et al. 2012).

Sorbus decora C.K. Schneid, Rosaceae

Sorbus decora is used traditionally by the Eeyou Istchee Cree First Nations to control DM. Bioassay-guided fractionation of 80% ethanol extract in water of the stem bark of this plant led to the isolation of three new pentacyclic triterpenes. One of them, 23,28,-dihydroxylupan-20(29)-ene-3-β-caffeate, significantly enhanced glucose uptake in C2C12 cells. Catechin and epicatechin were also isolated from the extract (Guerrero-Analco et al. 2010).

Sorghum bicolor (L.) Moench, Gramineae

Synonyms: *S. vulgare* Pers., *Andropogon sorghum* (L.) Brot.

Common Names: Sorghum, durra, jowari, milo, etc.

Sorghum is an important cereal crop. It also has a nutraceutical value. In a study, the antidiabetic effects of the phenolic extracts of three varieties (*Hwanggeumchal sorghum*, *Chal sorghum*, and *Heuin sorghum*) of sorghum from Korea were studied in normal and streptozotocin-induced diabetic rats. The phenolic extracts from *H. sorghum* exhibited higher antidiabetic activity compared to the other two varieties. Administration of the phenolic extracts (100 and 250 mg/kg for 14 days) to streptozotocin-induced diabetic rats showed significant hypoglycemic activity; the treatment significantly decreased the serum total cholesterol, TGs, urea, uric acid, creatinine, AST, and ALT, while it increased the serum insulin in diabetic rats, but not in normal rats. A total of 29 phenolic components were detected in the sorghum studied (Chung et al. 2011). The effects of the oral administration of the sorghum extract on hepatic gluconeogenesis and the glucose uptake of muscles in streptozotocin-induced diabetic were investigated. Administration of the sorghum extract (0.4 and 0.6 g/kg) reduced the concentration of TGs, total cholesterol and LDL-C, and glucose and improved (i.p.) glucose tolerance. In addition, administration of the extract significantly reduced the expression of phosphoenolpyruvate carboxykinase and the phosphor p38/p38 ratio, while it increased the phosphor AMPK/AMPK ratio, but GLUT4 translocation and the phosphor Akt/Akt ratio were not significantly increased. The authors suggest that the hypoglycemic effect of sorghum was related to hepatic gluconeogenesis but not the glucose uptake of skeletal muscle (Kim and Park 2012). Sorghum extracts exerted antidiabetic effects through a mechanism that improved insulin sensitivity via PPAR-γ. Serum insulin level was significantly lower in HFD-fed mice administered 1% sorghum extract than that in control HFD-fed mice. PPAR-γ expression was significantly higher, whereas the expression of TNF-α was significantly lower in mice given 1% extract compared to control HFD-fed mice. Adiponectin expression was also significantly higher in mice given the extract (Park et al. 2012). Thus, sorghum is very promising for the development of dietary supplements for diabetic patients.

Spathodea campanulata Buch.-Ham ex DC., Bignoniaceae

Spathodea campanulata is a common medicinal plant in Uganda. It is popular for its use in the treatment of diabetes and some other ailments. *S. campanulata* bark decoction showed hypoglycemic activity in streptozotocin-induced diabetic mice (Niyonzima et al. 1990). In another study also, decoction of the stem bark of *S. campanulata* exerted a hypoglycemic activity in streptozotocin-induced diabetic mice. Later, the decoction, washed with chloroform, was divided in water and butanol fractions. Both exhibited hypoglycemic activity but were devoid of any influence on insulin levels in diabetic mice. The decoction also decreased blood glucose levels in a glucose tolerance test in normal mice (Niyonzima et al. 2006). A recent study also showed moderate level of blood glucose-lowering effect of this plant. In normal rats the aqueous methanol extract of the stem bark of *S. campanulata* (800 mg/kg) reduced blood glucose levels 2 h after administration. In an oral glucose tolerance test, the extract reduced glucose-induced glycemia in a moderate manner. In alloxan-induced diabetic rats, the extract (200, 400, and 800 mg/kg) caused reduction in hyperglycemia by 10%, 29%, and 4%, respectively. Prolonged treatment (daily dose for 17 days) with the extract to alloxan-induced diabetic rats reduced blood glucose levels by 33% (200 mg/kg), 66% (400 mg/kg), and 42.9% (800 mg/kg) (Tanayen et al. 2014).

Spergularia purpurea Pers. G. Don., Caryophyllaceae

Spergularia purpurea is a medicinal herb. In normoglycemic rats, the water extract of *S. purpurea* (whole plant) produced a significant lowering of glycemia 4 h after single oral administration and 1 week after repeated oral administration. In streptozotocin-induced diabetic rats, the water extract caused a marked decrease in glycemia.

Moreover, 2 weeks after repeated oral administration of the extract, glycemia was normalized in diabetic rats (Jouad et al. 2000). Another study suggests that the hypoglycemic effect of *S. purpurea* in streptozotocin mice could be due to the inhibition of endogenous glucose production (Eddouks et al. 2003).

Sphagneticola trilobata (L.) Pruski, =*Wedelia trilobata* (L.) Hitchcock, Asteraceae

Common Name: Singapore daisy

This is a creeping herb native to the tropics of Central America and West Indies. Oral administration of the water extract of *Wedelia trilobata* (50 mg/kg) to streptozotocin-induced diabetic rats reduced blood glucose levels and body weight gain. Further, the treatment resulted in a marked restoration of decreased vitamin C and reduced GSH in liver and kidney tissues of treated diabetic rats. *In vitro*, the extract caused an inhibition of LPO. Besides, the extract of *W. trilobata* caused a reduction in the high levels of thiobarbituric acid-reactive substances observed in the liver, kidney, and testes as well as high serum TG, ALT, and AST of diabetic rats (Kade et al. 2010). In another study, the extract of *W. trilobata* inhibited α-glucosidase activity; this effect was comparable to that of acarbose (Rungprom et al. 2010).

Sphenocentrum jollyanum Pierre, Menispermaceae

Sphenocentrum jollyanum is used in traditional medicine to control diabetes. Oral administration of the methanol extract of *S. jollyanum* root (50, 100, and 200 mg/kg) to alloxan-induced diabetic rabbits produced a significant dose-dependent decrease in glycemia from day 3 of treatment with comparatively higher activity than glibenclamide. The treatment also improved oral glucose tolerance in diabetic rabbits. The extract also exhibited a marked hypoglycemic effect in normal rats (500 mg/kg) with the lowest glycemic reduction (30.1%), 3.0 mmol/L (55 mg/dL). The treatment also improved the pancreatic histology in treated diabetic rabbits (Mbaka et al. 2009). In a recent study also, the methanol extract of *S. jollyanum* root (200 mg/kg daily for 2 weeks) decreased the levels of blood glucose and serum lipids in streptozotocin-induced diabetic rats (Alese et al. 2014). The petroleum ether extract of *S. jollyanum* seed (300, 600, and 1200 mg/kg) also attenuated hyperglycemia in a dose-dependent manner in alloxan-induced diabetic rabbits. The tissue morphology of the extract-treated diabetic rabbits showed necrotic β-cells with some viable cells at the periphery (Mbaka et al. 2010). In another study, oral administration of the ethanol extract of *S. jollyanum* (200 mg/kg daily for 2 weeks) to streptozotocin-induced diabetic rats resulted in the recovery of β-cells and significant fall in blood glucose levels compared to diabetic control rats. These effects of the extract were comparable to those of glibenclamide (Alese et al. 2013). The active principles involved remain to be studied.

Sphenostylis stenocarpa (Hochst. ex. A. Rich.) Harms, Leguminosae

Common Name: African yam bean

The methanol extract of the seeds of *Sphenostylis stenocarpa* decreased blood glucose levels in alloxan-induced diabetic rats (Ubaka and Ukwe 2010).

Spinacia oleracea L., Chenopodiaceae

Common Names: Garden spinach, chhurika, isfinaji, vasaiyilaikkirai, etc.

Description: *S. oleracea* is an erect annual herb with leaves that are alternate, deltoid, ovate, acuminate, acutely and broadly pinnatifidly lobed, and smooth on both surfaces. The flowers are dioecious and ebracteate; male flowers are in terminal leafless spikes while females are in axillary clusters. The fruit is an utricle, fruiting perianth free.

Distribution: Its origin is unknown, but *Spinacia oleracea* is cultivated throughout India and elsewhere.

Traditional Medicinal Uses: *S. oleracea* acts as an astringent, demulcent, sweet, cooling, laxative, alexipharmic, emollient, antipyretic, diuretic, maturant, laxative, and anthelmintic and used in the treatment of asthma, leprosy, biliousness, inflammation, sore throat, joint pain, thirst, lumbago, vomiting, flatulence, urinary disease, jaundice, anemia, fatigue, rachitis, and tuberculosis (Kirtikar and Basu 1975).

Antidiabetes: The fruit and leaf of this plant have antihyperglycemic activity (Roman-Ramos et al. 1995). The ethanol and water extracts of its leaves (200 and 400 mg/kg, p.o.) produced a significant reduction in fasting blood glucose levels in normal as well as in alloxan-induced diabetic rats. The treatment also improved the body weight and lipid profiles. Further, histopathological studies of the pancreas of these animals showed regeneration of β-cells in extract-treated alloxan-induced diabetic rats (Gomathy et al. 2010). Another study also reported the antidiabetic activity of *S. oleracea* leaf, whose extract (70% ethanol) lowered blood glucose levels in alloxan-induced diabetic rats. The body weight of treated diabetic rats was restored to near normal levels in 9 days. The treatment also lowered the levels of serum GPT and GOT suggesting improvement in liver function in treated alloxan-induced diabetic rats (Jaikumar and Loganathan 2010). Thus, this nutraceutical has beneficial effects in diabetic animals. However, the active principles involved and the details of the mechanism of action remain to be studied.

Other Activities: Spinach leaves exhibit antioxidative, antiproliferative, and anti-inflammatory properties and exert beneficial effects, such as central nervous system protection, anticancer activity, and antiaging functions. The antioxidant potential of spinach can be affected by processing. Glycolipids from spinach inhibited *in vitro* and *ex vivo* angiogenesis. *S. oleracea* protects mouse against gamma radiations by virtue of its antioxidant actions. Dietary supplementation with spinach reduced ischemic brain damage. An active molecular complex from the mesophyll tissue of *S. oleracea* modulated the release of TNF and IL-10 and pro- and anti-inflammatory cytokines. No target organ toxicity or side effects have been reported; it is a leafy vegetable.

Phytochemicals reported include bioactive glycolipids and flavonoids including glucuronic acid derivatives of flavonoids, *N-N*-dimethyl histamine, trimethyl histamine, *N*-methyl histamine, acetylcholine, α-sitosterol, 7-stigmasterol, patuletin, quercetin, 3-hydroxytyramine, saponin, catechol, lecithin, colamine, cephalin, and phosphatidylglycerol (review: Ajikumaran and Subramoniam 2005).

Spondias mombin L., Anacardiaceae

Spondias mombin is a traditional medicinal plant used in the treatment of diabetes and other diseases. The methanol extract of the leaves of *S. mombin* and its chloroform fraction exhibited significant blood glucose-lowering effect in the oral glucose tolerance test in alloxan-induced diabetic rats. The chloroform fraction of the methanol extract showed a peak hypoglycemic effect of about 61% at 4 h (Adediwura and Kio 2009). In another study, the ethanol extract of *S. mombin* seeds significantly reduced the levels of glucose and protein in the blood of treated alloxan-induced diabetic rats. However, the treatment did not substantially influence the activities of serum transaminase and levels of lipid peroxides (Iweala and Oludare 2011).

Stachytarpheta angustifolia Mill. Vahl., Verbenaceae

Oral administration of the methanol extract of *Stachytarpheta angustifolia* plant (1000 and 1300 mg/kg daily for 28 days) to streptozotocin-induced diabetic rats resulted in a significant reduction in the levels of plasma glucose, glycosylated hemoglobin, glucose-6-phosphatase, lipid peroxides, triacylglycerols, cholesterol, and LDL-C with a significant increase in body weight, plasma insulin, and HDL-C level. Thus, the plant extract treatment prevented the progression of diabetes-associated symptoms in streptozotocin-induced diabetic rats (Garba et al. 2013).

Stephania japonica (Thunb.) Miers., Minispermaceae

Synonym: *Stephania hernandifolia* (Willd.) Walp.

Stephania japonica is commonly known as tape vine, ambashtha, pataklannu, etc.

It is a slender twining shrub with glabrous branchlets; leaves simple, petiolate, ovate or subdeltoid, acute, obtuse or acuminate, glabrous on both sides. Flowers are unisexual, born in axillary stalked umbellate heads; drupes red, subglobose, bearing almost annular seeds.

It is distributed in the East and West coast of India, Sikkim, Nepal, Singapore, Ceylon, Tropical Australia, Africa and Malay Islands. In traditional medicine, it is used in the treatment of fever, enterosis, mastitis, diarrhoea, urinary disease and dyspepsia (Kirtikar and Basu 1975). In Darjeeling, India, herbalists use this plant for treating various diseases, one of which is diabetes mellitus (Sharma et al. 2010).

The ethanol and water extracts of *S. hernandifolia* corm (400 mg/kg) lowered blood glucose levels in streptozotocin-induced diabetic rats. Both extracts showed significant antihyperglycemic and antioxidant activities in streptozotocin-induced diabetic rats compared to the standard drug. The ethanol extract exhibited more significant antidiabetic and antioxidant activities than the water extract (Sharma et al. 2010c). The bulb of *S. hernandifolia* showed significant hypoglycemic effect in IDDM model rats whereas the bulb has no glucose lowering effect in nondiabetic and NIDDM model rats (Mosihuzzaman et al. 1994). The methanol extract of *S. japonica* tendril (250 and 500 mg/kg) reduced blood glucose levels in alloxan-induced hyperglycemic rats. The treatment reduced serum lipids also; this effect was comparable to that of metformin (Sultana et al. 2012).

Other activities reported for methanol extract of the plant leaf include testicular function inhibition in rats.

Reported phytochemicals include, epistephanine, 4 demethylhasubanonine, an alkaloid, hypoepistephanine, metastephanine, stephanine, stephanoline, homostephanoline, protostephanine, hasubanonine, stephanne chloride, obamegine, insucarine, cylantholine, isoterilobine, stebisamine, hernandoline, protometastephanine, aknadine, aknainine, aknadicine, cycleanine, isochondrodendrine, D-fagchinoline, D-tetrandrine, L-quercitol, magnoflorine (review: Ajikumaran and Subramoniam 2005).

Stephania tetrandra Moore, Menispermaceae
In streptozotocin-induced diabetic mice, the water extract of *Stephania tetrandra* reduced blood glucose in a dose-dependent manner in the range of 0.48–16 mg/kg, which is due to fangchinoline (Tsutsumi et al. 2003). The main bisbenzylisoquinoline alkaloid fangchinoline (0.3–3 mg/kg) significantly brought down the blood glucose level of diabetic mice in a dose-dependent manner. The effect of fangchinoline was 3.9-fold greater than that of the water extract of *S. tetrandra*. However, another main compound, tetrandrine (1–100 mg/kg), did not show any effect (Tsutsumi et al. 2003). The water extract of *A. membranaceus* did not affect singly but potentiated the antihyperglycemic action of fangchinoline (0.3 mg/kg) in streptozotocin-induced diabetic ddY mice. Fangchinoline (0.3–3 mg/kg), a main constituent of *S. tetrandra*, decreased the high level of blood glucose and increased the low level of blood insulin in streptozotocin-induced diabetic mice. Fangchinoline appears to be an effective insulin secretagogue in diabetic rats at very low oral doses. Formononetin and calycosin (0.03–0.1 mg/kg) isoflavones from *A. membranaceus* alone did not affect the blood glucose or blood insulin level of diabetic mice. These compounds (0.03–0.1 mg/kg) potentiated the antihyperglycemic action of fangchinoline (0.3 mg/kg). Further, formononetin (0.1 mg/kg) facilitated the fangchinoline-induced insulin release (Ma et al. 2007). Treatment of BB rats from 35 to 120 days of age with a *S. tetrandra* alkaloid, tetrandrine (20 mg/kg/day), a novel anti-inflammatory compound, reduced the cumulative incidence of spontaneous diabetes from 75.5% to 10.9%. Histological examination of the pancreases from tetrandrine-treated rats showed a lesser degree of insulitis than controls. These results provide some hope that tetrandrine may be of value in preventing diabetes and treating newly diagnosed diabetic subjects, either by itself or in combination with a more potent immunosuppressive agent (Lieberman et al. 1992).

Sterculia guttata Roxb., Sterculiaceae
The extract of *Sterculia guttata* showed hypoglycemic activity in rats (Marles and Farnsworth 1995).

Sterculia villosa Roxb., Sterculiaceae
The ethanol extract of *Sterculia villosa* barks (100, 200, and 400 mg/kg) showed significant antidiabetic activity in a dose-dependent manner in diabetic control rats. Further, the extract showed an anti-inflammatory activity against carrageenan-induced paw edema. The acute oral toxicity evaluation showed that the ethanol extract was safe up to 4 g/kg in rats (Hossain et al. 2012a).

Stereospermum suaveolens DC., Bignoniaceae
The bark extract (ethyl acetate fraction) of *Stereospermum suaveolens* reduced fasting blood glucose levels and pancreatic oxidative stress and increased liver glycogen content in streptozotocin-induced diabetic animals (Balasubramanian et al. 2012).

Stereospermum tetragonum DC., Bignoniaceae

Synonyms: *Stereospermum personatum* (Hassk.) Chatterj., *Stereospermum colais* Mabb., *Bignonia caudata DC.*, *Bignonia colais* Buch.-Ham ex Dillw. (Figure 2.88)

Common Names: Yellow snake tree, trumpet flower tree, ambu, etc.

Description: *Stereospermum tetragonum* is a large deciduous tree up to 18 m high and about 1.8 m in girth with a clear bole of about 9 m. The leaves are pinnately compound, with elliptic-acuminate leaflets. The flowers are borne in large, lax, and terminal panicles. The corolla is tubular, thinly villous within and without tuberose; the lobes are yellow or pale rose and crisped. The capsules are loculicidally two valved. The seeds are compressed or subtigonous with a membranous wing on each side (Binokinsley 2014).

Distribution: *S. tetragonum* is globally distributed in Indo-Malaysia and found in most regions of India. This medicinal tree grows throughout tropical parts of the Indian subcontinent, particularly in sandy soils of river beds in Northern India and parts of Tamil Nadu (Binokinsley 2014).

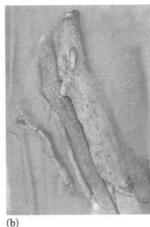

(a)   (b)

FIGURE 2.88   *Stereospermum tetragonum.* (a) Twig. (b) Root.

Traditional Medicinal Uses: *S. tetragonum* root is used in folk medicine to treat DM in certain remote villages in the state of Tamil Nadu. In ethnomedical practices, the plant is also used as a diuretic, antiulcer, antipyretic, etc. The stem bark is mostly used in traditional medicine such as Ayurveda; the root is also used in traditional medicine.

Pharmacology

Antidiabetes: The water extract of the root, but not hexane and ethanol extracts, showed antihyperglycemic activity at the 50 mg/kg level. The alcohol-soluble fraction of the water extract (active fraction) showed marked concentration-dependent antihyperglycemic activity. In alloxan-induced diabetic rats, the active fraction (25 mg/kg) was found to be as good as 5 IU/kg (i.p.) of insulin. The active fraction (25 mg/kg) showed promising activity against streptozotocin-induced type 2 DM also (Bino Kingsley et al. 2012, 2014a). The fraction did not influence insulin release from the isolated and cultured islet β-cells. However, it (40 μg/mL) stimulated glucose uptake by isolated rat hemidiaphragm. At 40 μg/mL, the active fraction markedly increased the levels of GLUT4 and PPAR-γ in cultured A432 human skin carcinoma cells as measured by ELISA. However, the *in vivo* relevance of these observations in increasing glucose utilization remains to be investigated. In contrast to oral administration of glucose, the active fraction did not substantially influence serum glucose levels, when glucose was loaded through the intraperitoneal route. Further, in the fasted rats, the active fraction did not significantly influence serum glucose levels in orally glucose-loaded rats. Therefore, the drug may be stimulating the release of one or more factors from the gastrointestinal tract of fed rats that may block the absorption/transport of glucose directly or indirectly from the intestinal tract (Bino Kingsley et al. 2013a).

Two novel antihyperglycemic principles, active at 2 mg/kg level, were isolated from the active fraction by activity-guided column chromatography. The structures of these compounds have been elucidated. One of them was identified with spectral data as a novel iridoid glycoside (1,4a,5,7a-tetrahydro-5-hydroxy-7-(hydroxymethyl)-1-(tetrahydro-6-(hydroxymethyl)-3,4,5-trimethoxy-2H-pyran-2yloxy) cyclopenta[c]pyran-4-carboxylic acid), and the other was identified as a lapachol-like compound, a derivative of naphthoquinone (5,8-dihydro-7-isopentyl-2,3,5,8-tetramethoxynaphthalene-1,4,6-triol). *In silico* docking studies showed binding sites in PPARs for these two compounds (Bino Kingsley 2014).

The mechanism of action of the antidiabetic molecules appears to be the inhibition of glucose absorption from the gut as well as activating PPAR-γ and GLUT4 (Bino Kingsley et al. 2014b). The plant is promising for the development of valuable antihyperglycemic agents to treat DM as a monotherapy or as combination therapy (Bino Kingsley et al. 2013a).

Other Activities: Pharmacological evaluation on this plant is lacking other than the antidiabetes studies. The active fraction showed remarkable antioxidant activity.

Phytochemicals: Phytochemical analysis showed the presence of tannins, phenol, glycosides, terpenoids, and coumarins, in the active fraction (Bino Kingsley et al. 2013b).

Toxicity: The active fraction from the water extract (500 mg/kg) did not exhibit any adverse toxicity in the limited subacute toxicity evaluation in mice (Bino Kingsley et al. 2013b).

Stevia rebaudiana (Bert.) Bertoni, Asteraceae (Figure 2.89)

Common Names: Stevia, sweet leaf, sweet herb of Paraguay, honey leaf, candy leaf, caa-he-éé, etc.

Description: *Stevia rebaudiana* is a perennial semishrub up to 30 cm in height. The leaves are sessile, 3–4 cm long, elongate with blunt-tipped lamina, serrate margin from the middle to the tip, and entire below. The upper surface of the leaf is slightly glandular pubescent. The stem is weak-pubescent at the bottom and woody. The rhizome has slightly branching roots. The flowers are light purple, pentamerous, and composite surrounded by involucres of epicalyx with sympodial cymes. The fruit is a five-ribbed spindle-shaped achene. The leaf is edible (review: Madan et al. 2010).

Distribution: *S. rebaudiana* is native to the valley of Rio Monday in Northeastern Paraguay in South America where it grows in sandy soils near streams on the edge of marshland. Now it is commercially cultivated in many countries including Brazil, China, Korea, Mexico, the United States, Indonesia, Japan, Tanzania, and Canada (review: Madan et al. 2010).

Traditional Medicinal Uses: *Stevia rebaudiana* has been used to sweeten tea for centuries dating back to the Guarani Indians of South America.

Pharmacology

Antidiabetes: The plant possesses anti-DM phytochemicals. Stevioside isolated from *S. rebaudiana* showed anti-DM activity in rats. The sweet compound stevioside has promising antidiabetes activity. Under *in vitro* conditions, stevioside and steviol enhanced insulin secretion from mouse islets in the presence of glucose (Jeppesen et al. 2000). In an intravenous (IV) glucose tolerance test,

FIGURE 2.89   *Stevia rebaudiana.*

stevioside significantly suppressed the increase in glucose in type 2 Goto-Kakizaki rats. Further, in the glucose tolerance test, the compound also suppressed the levels of glucagon. In normal rats, stevioside enhanced insulin levels above baseline during the glucose tolerance test, without altering the blood glucose response and the glucagon levels. Thus, stevioside showed antihyperglycemic and insulinotropic actions (Jeppesen et al. 2002). The effect of this compound on skeletal muscle glucose transport was studied in both insulin-sensitive lean and insulin-resistant obese. Acute oral stevioside administration increased whole body insulin sensitivity. Further, low concentrations of stevioside improved *in vitro* insulin action in skeletal muscle glucose transport in both lean and obese rat skeletal muscle (Lailed et al. 2004). Stevioside (0.5 mg/kg) lowered blood glucose levels in streptozotocin-induced diabetic rats, peak at 90 min. Stevioside administration twice daily also demonstrated dose-dependent effects in both streptozotocin and alloxan-induced diabetic rats. Further, stevioside reduced the rise in glucose during glucose tolerance testing in normal rats. Stevioside dose dependently decreased protein as well as mRNA levels of phosphoenol pyruvate carboxylase after 15 days of treatment (Chen et al. 2005). The authors conclude that stevioside is able to decrease blood glucose levels by enhancing not only insulin secretion but also insulin utilization in insulin deficient rats.

Repeated oral administration of stevioside delayed the development of insulin resistance in rats on a high-fructose diet. Further, oral administration of stevioside (0.2 mg/kg), three times daily, to streptozotocin-induced diabetic rats for 10 days resulted in an increase in the response to exogenous insulin (Chung et al. 2005). The combination of stevioside with a dietary supplement of soy protein exhibited beneficial effects in the development of metabolic syndrome and type 2 DM in Zucker rats as evidenced by amelioration of hyperglycemia, hyperlipidemia, and hypertension (Dyrskoy et al. 2005). Besides, both stevioside and GLP-1 counteracted glyburide-induced desensitization of glucose-stimulated insulin secretion (Chen et al. 2006).

Under *in vitro* conditions, in the presence of glucose, rebaudioside stimulated insulin secretion from isolated mouse islets in a dose-dependent manner. This effect of rebaudioside was dependent on the presence of extracellular calcium. Thus, this compound also possesses insulinotropic property (Abudula et al. 2004). The reduction of glycemia caused by the treatment with the plant leaves in 15 h fasted rats was mediated, at least in part, by an inhibition of hepatic gluconeogenesis. This effect did not involve stevioside and the activation of PPAR-γ receptors (review: Madan et al. 2010).

Medium polar extract of the leaves of this plant also showed antidiabetes activity in alloxan-induced diabetic rats (Misra et al. 2011).

Clinical Trial: Stevioside showed positive effects in type 2 diabetic patients. When a standard test meal was supplemented with 1 g of stevioside, and blood samples were drawn 30 min before and at 240 min after ingestion of a test meal, a reduction in PPG levels was observed (Gregersen et al. 2004). Many more clinical studies as per modern standards using isolated active principles and crude extracts are required to establish the utility of this plant for developing medicine and/or food supplements to type 2 DM patients.

Other Activities: Several other pharmacological properties, in addition to the use as a sweetener, were observed. The reported activities include antioxidant, anti-inflammatory, immunomodulatory, antibacterial, antiviral, diuretic and natriuretic, amelioration of hypertension, and gastroprotective activities (review: Madan et al. 2010).

Toxicity: Studies on animals did not show any adverse effects at therapeutically relevant doses. However, stevia treatment tended to decrease the plasma testosterone level in rats. Further studies on humans regarding the safety of stevia are recommended (review: Madan et al. 2010).

Phytochemicals: Stevia glycosides such as dulcoside A, rebaudiosides A to E, steviobioside, and stevioside were isolated from this plant. Amyrin acetate, three esters of lupeol, and steroids like sitosterol, stigmasterol, and campesterol were isolated from the leaves. In addition to the sweet diterpenoid glycosides, labdane-type diterpenes and other diterpenes have been isolated. Besides, flavonoid glycosides have been isolated from the aqueous methanol extract of leaves (apigenin-4-*O*-glucoside, luteolin-7-*O*-glucoside, kaempferol-3-*O*-rhamnoside, quercetin, quercetin-3-*O*-glucoside and quercetin-3-*O*-arabinoside, centauredin, etc.).

The essential oil contained sesquiterpenes and monoterpenes. Stevioside is the major constituent (3%–8%) of dried leaves. This glycoside is the sweetest glycoside, about 300 times sweeter than sucrose (review: Madan et al. 2010).

Streblus asper Lour, Moraceae
Streblus asper is used in Indian traditional medicine to treat DM. α-Amyrin acetate isolated from this plant's stem bark and the petroleum ether extract of the stem bark exhibited an antidiabetic effect against streptozotocin-induced diabetic rats (Karan et al. 2013). The petroleum ether extract of the stem bark (500 mg/kg) or α-amyrin acetate treatment (75 mg/kg) resulted in normalization of blood glucose levels and lipid parameters in diabetic rats (Karan et al. 2013). From the bark, α-amyrin acetate, lupeol acetate, β-sitosterol, α-amyrin, and diol have been isolated (Rastogi et al. 2006). α-Amyrin, α-amyrin acetate, and β-sitosterol are reported to have hypoglycemic activity. Thus, the plant appears to have several compounds active against DM.

Strychnos henningsii Gilg., Loganiaceae
Strychnos henningsii is recommended among other remedies for the treatment of DM in traditional medicine in Southern Africa. The antidiabetic activity of *S. henningsii* was assessed by intraperitoneally injecting varying doses of the water extract of the plant into alloxan-induced diabetic mice. The extract showed hypoglycemic activity (Piero et al. 2011). In another study, the water extract of the stem bark of this plant (250 mg/kg) decreased the blood glucose level, food and water intake, and TG levels in streptozotocin–nicotinamide-induced type 2 diabetic rats (Sunday et al. 2012).

Strychnos nux-vomica L., Loganiaceae
Strychnos nux-vomica is commonly known as kuchla. The seeds are used traditionally to treat diabetes and asthma, as an aphrodisiac, and to improve appetite. The hydroalcoholic and water extracts of the seeds of this plant showed antidiabetes activity in alloxan-induced diabetic rats (Bhati et al. 2012). The seed extract also showed free radical scavenging activity (Chitra et al. 2010).

Strychnos potatorum L.f., Loganiaceae
Strychnos potatorum has been used in folk medicine to treat diabetes. *S. potatorum* seed (100 mg/kg, p.o.) significantly reduced fasting blood sugar in streptozotocin-induced diabetic rats; this effect was comparable with that of glipizide (40 mg/kg, p.o.), an established hypoglycemic drug. The treatment also increased body weight along with decreased food and water intake in streptozotocin-induced diabetic rats (Biswas et al. 2012). Another report showed essentially the same results. The fried seed powder suspension (100 mg/kg) of *S. potatorum* reduced blood glucose levels significantly in streptozotocin-induced diabetic rats, which was comparable to the effect of glipizide (Biswas et al. 2014). In another study, a significant reduction in the blood glucose level was observed in streptozotocin–nicotinamide-induced diabetic animals treated with different doses of the ethanol (50%) extract of *S. potatorum* seeds (200 and 400 mg/kg, p.o., daily for 21 days). Treatment with the extract significantly increased the levels of reduced GSH and serum antioxidant enzymes in diabetic animals. Thus, the extract showed antioxidant and glucose-lowering effects in diabetic rats (Mishra et al. 2013). The plant seed is promising for follow-up studies.

Styrax japonica Siebold & Zucc., Styracaceae
Triterpenoids and a sterol from the stem bark of *Styrax japonica* inhibited PTP1B activities (Jiany et al. 2012).

Suaeda fruticosa L. Forsk., Amaranthaceae
The effect of *Suaeda fruticosa* (water extract prepared in an infusion at 10%) on lipid and carbohydrate metabolism in hypercholesterolemic and insulin-resistant sand rats was studied. Oral administration of the extract (1.5 mL/100 g) to hypercholesterolemic and insulin-resistant rats resulted in marked reduction in the levels of blood glucose. Furthermore, the treatment led to a decrease in plasma levels

of insulin, total cholesterol (50%), LDL-C (55%), VLDL-C (49%), oxidized LDL (40%), TGs (57%), and free fatty acids (36%). Thus, the plant extract showed hypoglycemic and hypolipidemic activities in type 2 diabetic sand rats (Bennani-Kabachi et al. 1999). The hypoglycemic effect was shown in streptozotocin-induced diabetic rats also (Benwahhoud et al. 2001). The active principles involved and details of the mechanism of action remain to be studied, among other things, to determine its utility as a therapeutic agent.

Sutherlandia frutescens R.Br. var. *incana* E. Mey., Fabaceae

Synonyms: *Lessertia frutescens* (L.) Goldblatt and J. Comanning, *Colutea frutescens* L.
Sutherlandia frutescens has been used as a potential hypoglycemic agent for the treatment of type 2 DM. When the crushed leaf of this plant was administered for 8 weeks in drinking water to rats fed an HFD, there was a significant increase in glucose uptake into muscle and adipose tissue and a decrease in intestinal glucose uptake compared to control animals. Further, HFD-fed rats receiving the leaf displayed normal levels of insulin and body weight compared to the fatty controls (Chadwick et al. 2007). Further studies are warranted to elucidate the exact mechanisms of action and efficacy in humans. The anti-DM compound pinitol has been reported from this plant also (see under *Bougainvillea spectabilis*).

In addition to the antidiabetic properties *in vitro* and *in vivo*, studies have shown antiproliferative, anti-HIV, anti-inflammatory, analgesic, antibacterial, antistress, anticonvulsant, and antithrombotic activities of this plant (review: van Wyk and Albrecht 2008).

Swertia bimaculata (Siebold & Zucc.) C. B. Clarke., Gentianaceae

The ethanol and dichloromethane extracts of *Swertia bimaculata* had the most effective potency in glucose consumption in 3T3-L1 adipocyte. The dichloromethane extract and corymbiferin (the most abundant component of the dichloromethane extract) displayed remarkable antidiabetic effects in high-fat- and high-sucrose-fed low-dose streptozotocin-induced diabetic rats (Liu et al. 2013b). The treatments decreased fasting blood glucose levels, increased serum insulin levels, and improved oral glucose tolerance. Further, the treatments improved the lipid profile, improved insulin sensitivity, and decreased oxidative stress in diabetic rats (Liu et al. 2013).

Swertia chirayita (Roxb. ex Flm.) Krast., Gentianaceae

Swertia chirayita is commonly known as bitter stick, anaryatika, charayatah, chiretta, chirata, nilavembu, nilveppu, etc. It is an annual, robust, stemmed, erect herb present in temperate Himalayas, Kashmir, Bhutan, and Khasi Hills. In ethnomedicine, this plant is used to treat various ailments (Kirtikar and Basu 1975).

The alcohol extract (250 mg/kg once daily for 2 weeks) of *S. chirayita* exhibited a hypoglycemic effect in alloxan-induced diabetic rats (Kar et al. 2003). The hexane fraction (250 mg/kg orally for 28 days) of *S. chirayita* lowered the blood sugar of albino rats with increased glycogen content of the liver and insulin release from pancreatic β-cells (Chandrasekar et al. 1990). Swerchirin (1,8-dihydroxy-3,5-dimethoxy xanthone) isolated from the hexane fraction of *S. chirayita* at a dose of 50 mg/kg (p.o.) lowered the blood glucose levels of streptozotocin-induced diabetic rats (Saxena et al. 1991). It is reported that the fraction lowered the blood glucose level by stimulating insulin release from islets of Langerhans and did not influence intestinal glucose absorption (Saxena et al. 1993). Xanthone was isolated from the hexane fraction of *S. chirayita* and identified as 1,8-dihydroxy-3,5-dimethoxy xanthone (swerchirin). Swerchirin exhibited significant blood sugar-lowering effect in fasted, fed, glucose-loaded, and tolbutamide-pretreated albino rat models (Bajpai et al. 1991). Swerchirin isolated from the plant was found to be a superior blood sugar-lowering agent over tolbutamide (Saxena et al. 1996). Thus, swerchirin appears to be promising for further studies. Further, the alcohol extract of this plant is reported to contain the anti-DM compound mangiferin. However, swertiamarin present in the alcohol extract of this plant is a CNS depressing compound (Bhattacharya et al. 1976).

Other Activities: The aqueous extract of *S. chirayita* stem modulated the levels of IL-1β, IL-6, IL-10, IFN-γ, and TNF-α in arthritic mice. *S. chirayita* inhibited LPO and modulated detoxifying enzymes in the liver; the chemopreventive potential of the plant has been suggested. The anticarcinogenic activity of *S. chirayita* on the DMBA-induced mouse skin carcinogenesis model has been reported. Amarogentin, a secoiridoid glycoside from this plant, inhibited topoisomerase I from *Leishmania donovani*. An antihepatotoxic activity of this plant has also been reported (review: Ajikumaran and Subramoniam 2005).

Toxicity: The CNS depressing effect was detected in swertiamarin, whereas the same effect was reversed by another CNS stimulating compound mangiferin; these two compounds are present in the alcohol extract of the plant (Bhattacharya et al. 1976).

Phytochemicals: Reported phytochemicals include secoiridoid glycoside, swertiamarin, mangiferin, amarogentin, ophelic acid, chiratin, amargogentin, gentiopicrin, phenols, xanthone, swerchirin, swertinin, swertianin, decussating, isobellidifolin, friedelin, β-sitosterol, nine different derivatives of tetraoxygenated xanthone, chiratanin, and swertanone (review: Ajikumaran and Subramoniam 2005).

Swertia corymbosa (Griseb.) Wight ex C.B. Clarke, Gentianaceae
Swertia corymbosa (local name: shirattakuchi) has a long history of use in Ayurveda herbal preparations in Indian. *S. corymbosa* has been used in folklore medicine for the treatment of diabetes. In an *in vitro* antidiabetic study, the methanol extract of the plant was found to be a potent inhibitor of α-glucosidase and α-amylase activity. Oral administration of the methanol extract (aerial parts) of *S. corymbosa* (125, 250, and 500 mg/kg for 28 days) to streptozotocin-induced diabetic rats caused a significant decrease in the levels of blood glucose, total cholesterol (TC), serum TGs, LDL-C, and MDA and a significant increase in the levels of HDL-C, serum insulin, and body weight. Furthermore, the activities of antioxidative enzymes, including SOD, GPx, GSH, and CAT, were enhanced by the methanol extract. Histopathological studies of the pancreas showed the regeneration of β-cells by the extract in streptozotocin-induced diabetic rats (Mahendran et al. 2014).

Swertia japonica (Schult.) Makino., Gentianaceae
Bellidifolin (a xanthone) isolated from the ethyl acetate fraction of *S. japonica* showed a potent and dose-dependent hypoglycemic activity in streptozotocin-induced diabetic rats, both in i.p. and p.o. administration (Basnet et al. 1994).

Swertia mussotii Franch, Gentianaceae
Swertia mussotii is an herb used in Tibetan folk medicine. The most abundant active compound in this herb is mangiferin. Mangiferin is an antidiabetic compound (see under *Mangifera indica*). Other reported phytochemicals in this plant include swertiamarin, swertisin, oleanolic acid, and 1,5,8-trihydroxy-3-methoxyxanthone (Yang et al. 2005). Oleanolic acid, present in several plants, has several pharmacological activities including a hypoglycemic effect.

Swertia punicea Hemsl., Gentianaceae
Swertia punicea is a Chinese traditional medicinal herb. Two antidiabetic compounds, methylswertianin and bellidifolin, were isolated from the active ethyl acetate fraction of *S. punicea*. These compounds reduced fasting blood glucose in streptozotocin-induced type 2 diabetic male mice. The compounds also improved oral glucose tolerance and lowered fasting serum insulin levels. Further, treatment with these compounds resulted in reduction in the serum levels of total cholesterol, LDL, and TG and increase in the levels of HDL. The compounds improved insulin resistance by enhancing insulin signaling. The expression levels of the IR-α subunit, IR substrate 1, and phosphatidylinositol 3-kinase were also increased by the compounds. Besides, increased glycogen content, decreased glucokinase activities, and increased glucose-6-phosphatase activities were observed in the compound-treated animals (Tian et al. 2010).

Swietenia macrophylla King, Meliaceae
Swietenia macrophylla is a tree occurring in Indonesia and many other countries. The methanol extract (150 and 300 mg/kg daily for 12 days) of *S. macrophylla* seeds showed antidiabetic activities in streptozotocin- and nicotinamide-induced type 2 diabetic rats. Treatment with the

extract decreased fasting blood glucose levels, increased liver glycogen levels, and improved the lipid profile in diabetic rats compared to diabetic control rats (Maiti et al. 2008). Besides, the alcohol extract of seeds markedly reduced blood glucose levels, lipids, and oxidative stress in streptozotocin-induced diabetic rats (Kalpana and Pugalendi 2011). In another study, petroleum ether extracts of *S. macrophylla* seeds showed antihyperglycemic activity in glucose-loaded hyperglycemic rats and the extract contained several compounds, including fucosterol and β-sitosterol (Hashim et al. 2013). β-Sitosterol is a known antidiabetic compound. The plant may contain several antidiabetic compounds.

Swietenia mahagoni Jacq., Meliaceae

Swietenia mahagoni is a tree native to Southern Florida, Cuba, the Bahamas, Hispaniola, and Jamaica. The methanol extract of *S. mahagoni* bark (25 and 50 mg/kg daily for 15 days) showed antidiabetic activity in streptozotocin-induced diabetic rats. Oral administration of the extract to diabetic rats resulted in a highly significant reduction in blood glucose levels. Body weights were significantly reduced in streptozotocin-induced diabetic rats when compared to normal rats; the extract treatment restored the body weight. Besides, the extract exhibited an antioxidant activity in treated diabetic rats (Panda et al. 2010).

The aqueous methanol (2:3) extract of *S. mahagoni* seed showed antidiabetic, antioxidant, and antihyperlipidemic activities in streptozotocin-induced diabetic rats. Feeding the diabetic rats with the seed extract (250 mg/kg) for 21 days lowered the blood glucose level. Moreover, activities of antioxidant enzymes like catalase, peroxidase, and levels of the products of free radicals like conjugated dienes and thiobarbituric acid-reactive substances in liver, kidney, and skeletal muscles were corrected toward the normal control values after this extract treatment in diabetic rats. Furthermore, the seed extract corrected the levels of serum urea, uric acid, creatinine, cholesterol, TG, and lipoproteins toward the control levels in diabetic rats (De et al. 2011). In a recent study also, the n-hexane fraction of the hydromethanol (2:3) extract of *S. mahagoni* seed exhibited an antidiabetic activity in streptozotocin-induced diabetic rats. Treatment with the extract resulted in a decrease in the levels of blood glucose and glycated hemoglobin and increase in the activities of hepatic hexokinase, glucose-6-phosphate dehydrogenase, catalase and peroxidase, and superoxide dismutase in diabetic rats compared to diabetic control animals. Further, the treatment decreased the levels of serum urea, uric acid, and creatinine (Bera et al. 2015). The active principles involved are not known.

Symplocos cochinchinensis (Lour.) S. Moore, Symplocaceae

The leaf extract of this plant showed promising reduction in plasma insulin and cholesterol levels and an increase in the liver glycogen content in type 2 diabetic rats (Sunil et al. 2011). A recent study gave some insights regarding the mechanism of action of this plant. The ethanol extract of *Symplocos cochinchinensis*, under *in vitro* condition, exhibited activities like α-glucosidase inhibition, insulin-dependent glucose uptake (threefold increase) in L6 myotubes, regeneration of RIN-m5F cells (3.5-fold increase), and reduced TG accumulation (22% decrease) in 3T3L1 cells. Further, the extract decreased hyperglycemia-induced generation of ROS in HepG2 cells (59% decrease) with moderate antiglycation and PTP1B inhibition (Antu et al. 2014). Chemical characterization by HPLC revealed the presence of β-sitosterol, phloretin 2′-glucoside, oleanolic acid, etc., in the ethanol extract. The extract showed significant antihyperglycemic activity *in vivo* in diabetic models. The authors concluded that the extract mediates the antidiabetic activity mainly via α-glucosidase inhibition and improved insulin sensitivity, with moderate antiglycation and antioxidant activity (Antu et al. 2014). Oleanolic acid and β-sitosterol reported in the extract are known anti-DM compounds.

Symplocos paniculata (Thunb.) Miq., Symplocaceae

Three triterpenes, ursolic acid, corosolic acid, and 2α, 3α, 19α, 23-tetrahydroxyurs-12-en-28-oic acid, were isolated from methanol extract of the leaves and stems of *Symplocos paniculata* using an *in vitro* PTP1B inhibitory assay. Furthermore, ursolic acid was found to stimulate glucose uptake in L6 myotubes and facilitate glucose transporter isoform 4 translocation in CHO/hIR cells via enhancing IR phosphorylation (Jiany et al. 2012).

Symplocos theaefolia D. Don, Symplocaceae

Oral administration of *Symplocos theaefolia* leaf extract showed a glucose-lowering effect in normal rats (review: Marles and Farnsworth 1995).

Syzygium alternifolium (Wt.) Walp, Myrtaceae

The water, ethanol, and hexane extracts of *Syzygium alternifolium* seeds were tested for their blood glucose-lowering effects in both normal and alloxan-induced diabetic rats. The blood glucose levels were measured at 0, 1, 3, 5, and 7 h after the treatment. The water extract of *S. alternifolium* (750 mg/kg) showed maximum blood glucose-lowering effect in both normal and alloxan-induced diabetic rats. The ethanol and hexane extracts also showed hypoglycemic and antihyperglycemic activity, but the effect was less than that of the water extract (Rao and Rao 2001).

Syzygium cordatum Hochst. ex C. Krauss, Myrtaceae

The water extract of the leaf of *Syzygium cordatum* exhibited blood glucose-lowering activity in both normal and streptozotocin-induced diabetic rats. The same extract has antidiarrheal activity also (Deliwe and Amabeoku 2013). A previous study has concluded that *S. cordatum* leaf extract (60 mg/kg) is effective in mild DM or in cases of glucose tolerance impairment (Musabayane et al. 2005). The antidiabetes activity of *S. cordatum* leaf extract against streptozotocin-induced DM in rats is attributed at least in part to oleanolic acid and ursolic acid present in the leaf. Oleanolic acid (60 mg/kg) is shown to improve kidney function in normal as well as streptozotocin-induced diabetic rats (Mapanga et al. 2009).

Syzygium cumini (L.) Skeels., Myrtaceae

Synonym: *Eugenia jambolana* (Lam.) DC. (Figure 2.90)

Common Names: Blackberry, black plum (Eng.), Jambah, jamun, arugadam, njaval (Indian languages)

Description: *Syzygium cumini* is a tree of 25–35 m height and glabrous branchlets; the leaves are decussate, elliptic or ovate-lanceolate, coriaceous, and glossy above with entire margin; the flowers are in axillary or terminal cymose panicle. The fruit is a globose berry with a solitary seed.

Distribution: *S. cumini* occurs in India, subtropical Himalaya, Sri Lanka, Malaysia, and Australia.

Traditional Medicinal Uses: The bark, fruit, flower, seeds, etc., are used in traditional medicine. It is used in Ayurveda and Unani medicines against diabetes, dysentery, asthma, sore throat, and bronchitis. It has acrid, sweet, digestive, astringent, anthelmintic, tonic, diuretic, refrigerant, and carminative properties and is used to treat thirst, biliousness, blood diseases, ulcers, stopping urinary discharge, diarrhea, anemia, cancer, and colic (Kirtikar and Basu 1975; Gupta and Saxena 2011).

Pharmacology

Antidiabetes: Several studies have established the antidiabetic properties of *S. cumini*. The seed kernel (water suspension, 4 g/kg) showed blood glucose-lowering effects in alloxan-induced diabetic

(a)   (b)

FIGURE 2.90 *Syzygium cumini*. (a) Twig with flowers. (b) Twig with fruits.

rats (Nair and Santhakumari 1986; Sharma et al. 2003). In another study, defatted seeds and water-soluble fiber from the seeds exhibited hypoglycemic effects in alloxan-induced diabetic rats (Pandey and Khan 2002). Further, the seed extracts reduced tissue damage in diabetic rat brain (Stanely et al. 2003). The water extract of the *S. cumini* seed at a dose of 250 mg/kg (p.o.) was found to exert more prominent blood glucose-lowering activity than that of glibenclamide in experimental diabetic conditions (Prince et al. 1998). The extract also showed hypolipidemic effects in alloxan-diabetic rats (Prince and Menon 1997). However, the water-insoluble portion of the seed did not exhibit anti-diabetic activity. In another study, the seed extract showed antidiabetes and antiulcer effects in MD rats (Chaturvedi et al. 2009). The antidiabetic and antihyperlipidemic effects of the alcohol extract of *S. cumini* seeds in alloxan-induced diabetic albino rats were shown in another study (Prince et al. 2004). Two active principles, ferulic acid and cuminoside, were isolated from *S. cumini* seeds. The ferulic acid fraction showed anti-DM activity against streptozotocin-induced diabetes in rats (Mandal et al. 2008). Cuminoside isolated from the seed also exhibited anti-DM activity in rats (Farswan et al. 2009). The water extract of the seeds of *S. cumini* was reported to inhibit pancreatic α-amylase *in vitro* (review: Sharma et al. 2012). In an *in vitro* study, the methanol extract of *S. cumini* stimulated glucose uptake by activating GLUT4, PI3K, and PPAR-γ in L6 myocytes (Anandharajan et al. 2006).

The hypoglycemic and antihyperglycemic activities of the 90% ethanol extract of the whole *S. cumini* fruit in normal and streptozotocin-induced diabetic rats have been reported (Gupta and Saxena 2010). The fruit extract exhibited blood glucose-lowering activity in fasted, fed, and streptozotocin-induced diabetic rats at a single oral dose of 250 mg/kg. Further, it increased glycogen content in muscles along with marked degranulation in pancreatic β-cells. The authors suggest that the whole fruit may stimulate insulin secretion (Gupta and Saxena 2011). The antihyperglycemic effect of the water extract of *E. jambolana* has been shown in the fruit pulp in normal and alloxan-induced diabetic rabbits (Sharma et al. 2006). An antihyperglycemic compound, α-hydroxy succinamic acid, was isolated from the fruits of *S. cumini*; this compound attenuated renal dysfunction in streptozotocin-induced diabetic rats (Tanwar et al. 2010). The fruit appears to be an attractive antidiabetic food.

Oral administration of the water extract of the plant bark (300 mg/kg) for 45 days exhibited anti-DM activity in streptozotocin-induced diabetic rats. The treatment resulted in significant reduction in blood glucose and urine sugar levels in diabetic rats. Besides, the treatment increased the levels of plasma insulin and C-peptide in diabetic rats compared to untreated diabetic control animals (Saravanan and Leelavinothan 2006). Extension of these studies revealed the dose-dependent effect of the bark extract. Further, the treatment, as expected, improved glucose tolerance and restored to almost normal level the decrease in body weight and increase in food and fluid intake observed in diabetic rats (Saravanan and Pari 2008). The bark decreased blood glucose and elevated plasma insulin and C-peptide levels in streptozotocin-induced diabetic rats (Saravanan et al. 2006). The methanol extract of the root of *S. cumini* also showed both hypoglycemic and hypolipidemic properties in alloxan-induced diabetic rats (Sharma et al. 2008). Anti-DM compounds such as stigmasterol, β-sitosterol, and lupeol have been reported in the leaves (Alam et al. 2012).

Identification of all active principles and their distributions in the fruit, seed, bark, and root will shed more light on the antidiabetic properties of this plant. Although this plant has been subjected to many studies, further studies are required to confirm the actions of the fruit components and determine the mechanisms of actions and multiple compounds involved in mediating the antidiabetes activities of *S. cumini*. It should be noted that anti-DM compounds, ellagic acid, gallic acid, quercetin, β-sitosterol, ferulic acid, kaempferol glycosides, etc., were isolated from this plant.

Clinical Studies: Most of the available clinical studies were not carried out following currently acceptable procedures. The seeds appear to be beneficial to type 2 diabetic patients (Teixeira et al. 2004). In a clinical study, jamun seed powder produced good symptomatic relief along with reduction in blood glucose levels (Kohli and Singh 1993).

Other Activities: The antibacterial activity of *S. cumini* leaf essential oils has been reported (Shafi et al. 2002). The seed inhibited autacoid-induced inflammation in rats (Muruganandan et al. 2002). The leaf extract has radioprotective effects (Jagetia and Baliga 2003; Jagetia et al. 2005). Tannins extracted from the plants have gastroprotective activities (Ramirez and Rao 2003). Other reported activities include antiallergic, anti-inflammatory, antinociceptive, chemopreventive, and antifungal (review: Sharma et al. 2012).

Phytochemicals: The plant is rich in anthocyanin-containing compounds, glucoside, ellagic acid, isoquercetin, kaempferol, gallic acid, and myricetin. Other reported compounds include myricetin-3-L-arabinoside, dihydromyricetin, quercetin-3-D-galactoside, betulinic acid, friedelin, friedelinol, kaempferol-3-*O*-glucoside, quercetin, β-sitosterol, stigmasterol, lupeol, heptacosane, triacontane, hentriacontane, octacosanol, triacosanol, dotriacosanol, crotegolic acid, delphinidine-3-gentiobioside, malvidin-3-laminaribioside, petunidin-3-gentiobioside, corilagin, ellagitannins, 3,6-hexahydroxydiphenoyl glucose, caffeic acid, ferulic acid, guaiacol, resorcinol dimethyl ether, veratrole, vernolic acid, and malvalic acid (Rastogi and Mehrotra 1999; Gupta and Saxena 2011).

Syzygium malaccense (L) Merr. & Perry, Myrtaceae

Synonyms: *Eugenia malaccensis* L., *Jambosa malaccensis* (Figure 2.91)

Common Names: Malay apple, rose apple, mountain apple, was jumbo, etc.

Description: Malay apple is an evergreen tree with a spreading but cone-shaped crown; it usually grows 5–20 m tall. The straight, cylindrical bole can usually be 20–45 cm in diameter often branching from near the ground though sometimes free of branches for 10–15 m, with buttresses at the base. The fruit is edible. When in flower this is one of the most delightful of trees. The fruit is a berry, ellipsoid, 5–8 cm in diameter, crowned by the incurved nonfleshy calyx segments, and dark red or purplish-yellow or yellow-white in color; the flesh is 0.5–2.5 cm thick, juicy, white, and fragrant; it has one seed per fruit that is globose, 2.5–3.5 cm in diameter, and brown in color.

Distribution: Originally found in Malaysia and India, the tree is sometimes cultivated, usually in home gardens, both as an ornamental and in the tropics for its edible fruit.

Traditional Medicinal Uses: The leaves are used as a tonic, and the old fruit is used as a remedy for sore throats. A preparation of the root is a remedy for itching. The root is a diuretic and is given to alleviate edema. The root bark is useful against dysentery and also serves as an emmenagogue and abortifacient. Cambodians take a decoction of the fruit, leaves, or seeds as a febrifuge. In Brazil, various parts of the plant are used as remedies for constipation, diabetes, coughs, pulmonary catarrh, headache, and other ailments.

Pharmacology

Antidiabetes: The bark extract of *Syzygium malaccense* has been shown to effectively serve as a hypoglycemic agent that decreased the fasting blood sugar level, increased liver glycogen content, and reduced

(a)   (b)

FIGURE 2.91   *Syzygium malaccense*. (a) Twig with flowers. (b) Twig with fruits.

diabetes-induced hyperlipidemia in streptozotocin-induced diabetic rats (Bairy et al. 2005). The extract is believed to be able to prevent the development of diabetes-induced cataractogenesis based on its strong inhibitory effect toward aldose reductase (Guzman and Guerrero 2005). *S. malaccense* leaf extract was found to be an excellent scavenger of DPPH and 2,2-Azinobis(3-ethylbenzthiazoline)-6-sulfonic acid (ABTS). It also inhibited α-glucosidase more significantly than the positive control, acarbose, but was a poor α-amylase inhibitor. Myricitrin was identified as the major bioactive compound present in the extract. The presence of the potent antioxidant and antihyperglycemic agent myricitrin in the *S. malaccense* leaf extract indicates the potential use of the extract in the management of DM and its related complications (Arumugam et al. 2014). It should be noted that the leaf contains anti-DM compounds ursolic acid and β-sitosterol also. However, all the active principles involved and details of mechanism of action remain to be studied in detail to determine its possible therapeutic utility. Further, the anti-DM activity of the edible fruit, if any, remains to be studied.

Other Activities: The plant leaves possess antioxidant activity (Savitha et al. 2011). The plant is reported to have anti-inflammatory activity (Noreen et al. 1998).

Phytochemicals: The plant leaves contain phenolics and flavonoids (Savitha et al. 2011). The unripe and ripe fruits contain flavonoids, tannins, phenolics alkaloids, and saponins (Oyinlade 2014). Leaf oil is largely composed of monoterpenes (30% sesquiterpenes, 9% caryophyllene). Ursolic acid, B-sitosterol, and sitos-4-en-3-one were isolated from the leaf (Ismail et al. 2010).

Syzygium samarangense (Blume) Merrill and Perry cv. Pink, Myrtaceae

Common Name: Pink wax apple
An active fraction from wax apple (*Syzygium samarangense*) fruit extract ameliorated insulin resistance via modulating the insulin signaling and inflammation pathway in TNF-α-treated FL83B mouse hepatocytes (Shen et al. 2012). Vescalagin isolated from the *S. samarangense* fruit enhanced glucose uptake in insulin-resistant FL83B cells. In a follow-up study, vescalin showed hypotriglyceridemic and hypoglycemic effects in high-fructose diet-induced diabetic rats. Fasting blood glucose, C-peptide, fructose amine, TG, and free fatty acid contents decreased in vescalin-treated (4-week treatment, daily, 30 mg/kg) high-fructose diet-fed rats (Shen and Chang 2013). 2′,4′-Dihydroxy-3′,5′-dimethyl-6′-methoxychalcone (compound 1), its isomeric flavanone 5-*O*-methyl-4′-desmethoxymatteucinol (compound 2), and 2′4′-dihydroxy-6′-methoxy-3′-methylchalcone were isolated from the leaves of *S. samarangense*. In an oral glucose tolerance test, compounds 1 and 2 (1 mg/20 g/mouse) significantly lowered the blood glucose levels in mice when administered 15 min after a glucose load. When coadministered with glucose, only compound 1 showed a significant lowering of glucose levels 45 min after its oral administration. When administered 15 min before glucose, none of the flavonoids showed a positive effect (Resurreccion-Magno et al. 2005).

Tabernaemontana divaricata (L.) Roemer & Schultes, Apocynaceae

Tabernaemontana divaricata is a shrub native to Asia. The methanol extract of *T. divaricata* (300 and 400 mg/kg, i.p.) leaves showed antidiabetic activity in alloxan-induced diabetic mice. The extract (400 mg/kg) reduced maximum blood glucose level 12 h after administration from about 14 to 7 mmol/L. The extract also showed low *in vitro* cytotoxicity (Rahman et al. 2012). In a recent study, the methanol extract of *T. divaricata* (100 and 200 mg/kg daily for 21 days) reduced blood glucose levels, improved lipid profiles, and ameliorated oxidative stress in alloxan-induced diabetic rats (Kanthlal et al. 2014). Conophylline, a vinca alkaloid, was separated from the leaves of *T. divaricata*. Conophylline, from *Ervatamia microphylla*, is known to induce the differentiation of pancreatic precursor cells to insulin-producing cells *in vitro*. In a study, this alkaloid decreased the blood glucose level and increased the plasma insulin level in streptozotocin-induced diabetic rats after repetitive administration (0.11 and 0.46 mg/kg/day for 15 days). Conophylline also decreased the fasting blood glucose level in Goto-Kakizaki rats in a dose-dependent manner after daily administration for 42 days. These results suggest that the extract from conophylline-containing leaves may be useful as a functional food for the treatment of type 2 DM (Fujij et al. 2009). The methanol extract of *T. divaricata* aerial parts exerted antiobese and hypolipidemic effects in rats fed on atherogenic diet (Kanthalal et al. 2012).

Tabernanthe iboga Vault, Apocynaceae

Tabernanthe iboga is used in traditional medicine to treat diabetes. The water extract of *T. iboga* stimulated insulin secretion in a dose-dependent manner in the presence of 11.1 mM of glucose from the rat islets *in vitro* (Souza et al. 2011). Furthermore, the insulinotropic effect of the extract (1 µg/mL) was potentiated in K(+)-depolarized media as well as in the presence of 2.8 and 16.8 mM glucose. In contrast, in the same conditions, the extract failed to stimulate the high K(+) medium-induced insulin release. The extract significantly amplified the insulin secretion induced by either IBMX or tolbutamide. Calcium removal inhibited the insulinotropic effect of the extract. The extract increased glucose-induced $^{45}Ca^{2+}$ uptake in rat islets. Thus, stimulation of insulin secretion by the extract might involve the closure of K^+-ATP and the intensification of calcium influx through voltage-sensitive Ca^{2+} channels (Souza et al. 2011).

Talinum cuneifolium L., Portulacaceae

The glucose-lowering effect of the water extract of *Talinum cuneifolium* (whole plant) was evaluated in normal, glucose-fed, and alloxan-induced diabetic rats. The extract (200 and 400 mg/kg, p.o., daily for 7 days) caused a marked reduction in blood glucose levels (Janapati et al. 2008a). In another study, the alcohol extract (200 and 400 mg/kg, p.o., daily for 7 days) significantly lowered the blood glucose levels of alloxan-induced diabetic rats. The extract showed the presence of steroid, flavonoids, tannins, and alkaloids. Based on preliminary toxicity evaluation, the extract was nontoxic (Janapati et al. 2008b). In another study, administration of the ethanol extracts (200 and 400 mg/kg, p.o.) for 30 days to streptozotocin-induced diabetic rats resulted in significant dose-dependent reduction in blood glucose levels. Further, the serum levels of TGs, total cholesterol, LDL-C, VLDL-C, and uric acid were lowered in treated diabetic rats, while the levels of HDL-C, total protein, albumin, and globulin were increased (Manohar et al. 2011). Follow-up studies should include identification of active principles and mechanism of action.

Tamarindus indica L., Fabaceae (Figure 2.92)

Common Names: Tamarind tree, abdika, amli, ambilam, valan puli, puli, etc.

Description: *Tamarindus indica* is a large unarmed tree with longitudinally and horizontally fissured brownish or gray bark. The leaves are abruptly paripinnate and the leaflets are 20–40, glabrescent, close, obtuse, opposite, and oblong; the flowers are few together in copious lax racemes at the end of the branchlets. The pedicels are articulated at the base of the calyx. The bracts are boat shaped, enclosing the buds, and are caducous. The petals are yellow with red stripes. The fruit is a pod bearing 3–10 seeds. The fruit and seeds are edible.

(a)

(b)

FIGURE 2.92 *Tamarindus indica.* (a) Twig with fruits. (b) Flowers.

Distribution: Distributed throughout India and also in the tropics, *T. indica* is probably indigenous to Africa.

Traditional Medicinal Uses: *Tamarindus indica* has sour, laxative, tonic, anthelmintic, astringent, aphrodisiac, refrigerant, carminative, abortifacient, diuretic, emetic, and purgative properties and is used to treat paralysis, urinary discharge, gonorrhea, tumor, ringworm, small pox, disease of the blood, ophthalmia, earache, snakebite, biliousness, cough, fractures, ulcers, vaginal discharge, vomiting, scabies, sore throat, stomatitis, giddiness, vertigo, boils, conjunctivitis bleeding piles, diarrhea, asthma, amenorrhea, sprains, fever, dyspepsia, gastritis, dysentery, carbuncles, chills, collyrium, common cold, dermatosis, dropsy, diabetes, headache, hypertension, itch, inflammation, mucositis, rheumatism, sores, thrush, and wounds (Kirtikar and Basu 1975). The fruit and stem bark extracts are used in folklore medicine in India for the management of DM (Ghosh et al. 2004; Maiti et al. 2004).

Pharmacology

Antidiabetes: The water extract of the seed of *T. indica* is reported to have potent antidiabetogenic activity in streptozotocin-induced diabetic rats (Ghosh et al. 2004; Maiti et al. 2004). Methanol extracts (200 mg/kg) of the seeds and fruits of *T. indica* showed antihyperglycemic effects in oral glucose tolerance test in mice. The methanol extract of seeds showed greater antihyperglycemic effects compared to the methanol extracts of fruits (Roy et al. 2010). The ethanol extract of the fruit pulp (300 and 500 mg/kg, p.o., daily for 14 days) showed antidiabetic effects as judged from marked decrease in blood glucose levels and improvement in lipid profiles in alloxan-induced diabetic rats. The treatment also improved liver function (Koyagura et al. 2013). In antihyperglycemic activity tests conducted with glucose-loaded mice, the methanol extract of the leaves of *T. indica* (50, 100, and 400 mg/kg) dose dependently reduced blood glucose concentrations. The extract also exhibited antinociceptive effects (Mukti et al. 2013). A study suggests that *T. indica* seed powder possesses antihyperglycemic properties in streptozotocin type 2 diabetic rats that are at least partly due to its inhibitory effect on intestinal glucose absorption. This effect cannot be attributed to the acceleration of intestinal transit (Parvin et al. 2013). In a recent study, the stem bark extract of *T. indica* (250–1000 mg/kg) significantly lowered elevated blood glucose concentrations in experimental diabetes animal models. The extract also improved oral glucose tolerance in normoglycemic animals. The extract is practically nontoxic when administered orally (Yerima et al. 2014). Thus, the investigations show that the bark, seed, fruit, and leaves of this plant ameliorate hyperglycemia. However, the active principles involved and details of the mechanism of action remain to be studied to determine, among other things, its utility as a therapeutic agent. It should be noted that the fruit and seed are edible and nontoxic.

Other Activities: Proteinaceous inhibitors with high inhibitory activities against human neutrophil elastase are detected in the seeds. Dietary tamarind decreased blood cholesterol levels in hens. Tamarind costimulated MNU-induced colonic cell proliferation and may serve as a risk factor in colon cancer. The extract of the seed coat inhibited nitric oxide production by murine macrophages *in vitro* and *in vivo*. The methanol extract of fruits contained L-(–)-di-n-butyl malate that exhibited a pronounced cytotoxic activity against sea urchin embryo cells. Polysaccharide from *T. indica* has an immunomodulatory property (review: Ajikumaran and Subramoniam 2005).

Phytochemicals: Phytochemicals isolated include tartaric acid, malic acid, lactic acid, oxalic acid, quinic acid, citric acid, pipecolinic acid, cardenolide from the seed, L-(–)-di-n-butyl malate from the fruit, and hordenine (Rastogi and Mehrotra 1999).

Taraxacum officinale F.H. Wigg, Asteraceae

The effects of methanol and water extracts of the leaves and roots of *Taraxacum officinale* on blood glucose levels of normal and streptozotocin-induced diabetic rats were evaluated. Oral administration of the extracts (300 and 500 mg/kg twice daily for 21 days) decreased fasting blood glucose levels in normal and diabetic rats. Although both extracts of the leaf and root exerted hypoglycemic properties, the ethanol extract of the root was relatively more potent (Chinaka et al. 2012). Besides, under *in vitro* conditions, the ethanol extract of *T. officinale* leaf showed insulin secretagogue activity in INS-1 cells at the 40 μg/mL level (Hussain et al. 2004).

Tarchonanthus camphoratus L., Asteraceae

Tarchonanthus camphoratus is used in South Africa in traditional medicine to treat DM. The water extracts of this plant (131.5%) showed significant increase in glucose utilization in Chang liver cells. Further, the ethanol extracts of *T. camphorates* showed marked increase in the utilization of glucose in C2C12 muscle cells in culture. These observations suggest the anti-DM potential of this plant (Huyssteen et al. 2011).

Taxus yunnanensis W.C. Cheng and L.K. Fu, Taxaceae

This plant is a synonym of *Taxus wallichiana* Zucc. Acute intraperitoneal administration of alcohol extract, methanol extract and isolated compounds from the wood of this plant decreased blood glucose levels in streptozotocin-induced diabetic rats (Banskota et al. 2006).

Tecoma stans (L.) Kunth., Bignoniaceae

Common Names: Yellow bell, ginger thomas, manjarali

Tecoma stans is a shrub of 4–6 m height. It is a common garden shrub in India and indigenous from South Florida to the West Indies and South America. In traditional medicine, it is a diuretic, tonic, and vermifuge and used in rat bite, snakebite, scorpion sting, alcoholism, biliousness, diabetes, dysentery, gastritis, indigestion, stomachache, and syphilis (Kirtikar and Basu 1975).

Infusion of *T. stans* leaves in normal dogs produced an early hyperglycemic response with arterial hypotension followed by a slow decrease of the serum glucose level (Lozoya-Meckes and Mellado-Campos 1985). The oxalate salt of tecostanine (an alkaloid from leaves) in normoglycemic and hyperglycemic rats did not alter the serum glucose levels (Costantino et al. 2003). Tecomine (alkaloid) isolated from this plant increased glucose uptake by normal rat white adipocytes. *In vivo*, it decreased blood cholesterol levels without much effect on glycemia in *db/db* mice (Costantino et al. 2003).

The water extract of *T. stans* decreased hyperglycemic peak values in both normal and streptozotocin-induced diabetic rats in a magnitude similar to that of acarbose. The acute and subchronic administration of the extract (500 mg/kg) did not influence fasting glucose levels; subchronic administration reduced blood cholesterol and TG levels (Aguilar-Santamaria et al. 2009). An *in vitro* intestinal brush border preparation showed a dose-dependent inhibition of glucose release from starch (Aguilar-Santamaria et al. 2009). 5-β-Hydroxyskitanthine and boschniakine from this plant exhibited an insulin mimic action (Pandey et al. 2013).

Methanol extracts of the stem bark showed antibacterial and antifungal activity, but the leaf extract was found to be effective against *Candida albicans* only. Phytochemicals identified from this plant include indole, tryptamine, skatole, chrysoeriol, hyperoside, luteolin, 5-dehydroskytanthine, Δ-skytanthine, 5-hydroskytanthine, *N*-normethylskytanthine, tecostanine, boschniakine, tecomanine, 4-noractinidine, boschniakine, and tecomine (Rastogi and Mehrotra 1999; Costantino et al. 2003).

Tectona grandis L., Verbenaceae (Figure 2.93)

Local Names: Indian teak, teak, sagwan, sagon, etc.

Distribution: *Tectona grandis* is native to India and Myanmar and Southeast Asian countries. It is now one of the most important species of tropical plantation forestry. Teak has been introduced as a plantation species in as many as 36 tropical countries across tropical Asia, Africa, and South and Central America. It is one of the most widely cultivated high-value hardwoods (review: Nidavani and Mahalakshmi 2014).

Description: *T. grandis* is a large, deciduous tree reaching over 30 m in height. The crown is open with many small branches; the stem is usually cylindrical but becomes fluted and slightly supported at the base when mature and the bark of the stem is light brown or gray and distinctly fibrous with shallow, longitudinal fissures. The leaves are about 30 × 20 cm, shiny above, hairy below, vein network clear, broadly ovate or oval with shortly pointed or blunt tip and taping base, with minute glandular dots; the leaves shed for 3–4 months during the latter half of the dry season, leaving the branchlets bare. The root system is superficial, often no deeper than 50 cm, but the roots may extend laterally up to 15 m from the stem. The flowers are small, about 8 mm across, mauve to white, and arranged in large, flowering heads,

(a)   (b)

FIGURE 2.93 *Tectona grandis.* (a) Twig. (b) Tree.

about 45 cm long, and are found on the topmost branches in the unshaded part of the crown. The fruit is a drupe with four chambers and is round, hard, and woody, enclosed in an inflated, bladder-like covering; it is pale green at first and is then brown when mature.

Each fruit may contain 0–4 seeds (review: Nidavani and Mahalakshmi 2014).

Traditional Medicinal Use: *T. grandis* has a folk reputation as a hypoglycemic agent. Traditionally, it is used against bronchitis, biliousness, hyperacidity, diabetes, leprosy, and helminthiasis and is used as an astringent. A wood powder paste has been used against bilious headache and swellings. According to Ayurveda, the teak wood is acrid, cooling, laxative, sedative to gravid uterus, and useful in the treatment of piles, leukoderma, and dysentery. It is believed to possess anthelmintic and expectorant properties. *T. grandis* leaf extracts are widely used in folklore medicine for the treatment of various kinds of wounds, especially burn wounds. The plant is also used in the treatment of urinary discharge, bronchitis, and scabies and used as a diuretic, antidiabetic, analgesic, and anti-inflammatory (review: Khera and Bhargava 2013; Nidavani and Mahalakshmi 2014).

Pharmacology

Antidiabetes: The antihyperglycemic effect of *T. grandis* bark extract was evaluated in control and alloxan-induced diabetic rats. Oral administration of the bark suspension of *T. grandis* (2.5 and 5 g/kg for 30 days) resulted in a marked reduction in blood glucose (from 13.88 to 2.77 mmol/L, mean) (Varma and Jaybhaye 2010). Administration of the methanol extract of *T. grandis* bark (300 mg/kg daily for 20 days) to alloxan-induced diabetic rats resulted in marked reduction in the levels of blood glucose, glycosylated hemoglobin, total cholesterol, urea, serum creatinine, aspartate transaminase, alanine transaminase, lactate dehydrogenase, and thiobarbituric acid-reactive substances; the activities of superoxide dismutase and catalase were significantly increased compared to untreated diabetic rats. Thus, the methanol extract of *T. grandis* bark showed potent antioxidant and antidiabetic effects in alloxan-induced diabetic rats (Rajaram 2013). In another study, the hydromethanol extract of the bark of *T. grandis* exerted remedial effects on hyperglycemia and oxidative stress in streptozotocin-induced diabetic rats. A significant recovery was observed in the activities of antioxidant enzymes, as well as the levels of conjugated dienes and thiobarbituric acid-reactive substances in the liver and kidney of extract-treated diabetic rats compared with untreated diabetic rats. Extract-treated diabetic animals also showed a significant correction in the activities of GOT and GPT in the serum compared to untreated diabetic animals (Bera et al. 2011). The methanol extract of *T. grandis* root (500 mg/kg) also exhibited significant hypoglycemic activity in alloxan-induced diabetic rats (Sharma and Samanta 2011). The *T. grandis* flower (methanol extract) showed anti-DM and antioxidant activities in streptozotocin-induced diabetic rats (Ramachandran et al. 2011).

There is a need to test the anti-DM properties of the leaves that could be collected in a nondestructive manner. Since the activity is found in the bark, root, and flower, the leaf is likely to have an anti-DM property. Further studies should include identification of active principles and a detailed toxicity evaluation.

Other Activities: Other pharmacological activities of this plant include antioxidant, antibacterial, antifungal, antiasthmatic, anti-inflammatory, diuretic, antitumor, wound healing, antipyretic, antimetastatic, hair growth activity, antinociceptive, and antiulcer properties (review: Khera and Bhargava 2013).

Phytochemicals: Several classes of phytochemicals, such as alkaloids, glycosides, saponins, steroids, and flavonoids, have been reported in this plant. The various phytoconstituents isolated from *T. grandis* include juglone (antimicrobial molecule), betulinic aldehyde (antitumor compound), lapachol (antiulcerogenic compound), tectoquinone, 5-hydroxylapachol, tectol, betulinic acid, betulinic aldehyde, squalene, lapachol, acetovanillone, E-isofuraldehyde, evofolin, syringaresinol, medioresinol, balaphonin, lariciresinol, zhebeiresinol, 1-hydroxypinoresinol, tectonoelin A and tectonoelin B, 9,10 dimethoxy-2 methylanthra-1,4-quinone, 5-hydroxylapachol, tecomaquinone, methylquinizarin, and dehydro-α-lapachone. Teak wood contains naphthoquinone (lapachol, deoxylapachol, 5-hydroxylapachol), naphthoquinone derivatives (α-dehydro-α-lapachone, β-dehydrolapachone, tectol, dehydrotectol), anthraquinones (tectoquinone, 1-hydroxy-2-methylanthraquinone, 2-methyl quinizarin, pachybasin), and also obtusifolin, betulinic acid, trichione, β-sitosterol, and squalene. The roots are rich in lapachol, tectol, tectoquinone, β-sitosterol, diterpenes, and tectograndinol (review: Khera and Bhargava 2013; Nidavani and Mahalakshmi 2014).

Telfairia occidentalis Hook. f., Cucurbitaceae

Telfairia occidentalis is a popular vegetable in West Africa. The water extract of the leaf of *T. occidentalis* reduced glucose level significantly in streptozotocin-induced diabetic and glucose-induced hyperglycemic rats (Aderibigbe et al. 1999). The seed of the plant has also been reported to possess antidiabetic activity (Eseyin et al. 2007). In a study on rabbits, the effects of the methanol extract of *T. occidentalis* on plasma cholesterol, glucose, protein, TG, and creatinine levels were studied in alloxan-induced diabetic rabbits. The extract treatment ameliorated polydipsia; reduced the plasma glucose, cholesterol, creatinine, and TG levels of diabetic rabbits; but had no effect on the plasma protein level (Nwozo et al. 2004). The fruit extract exhibited a glucose-lowering effect when administered 45 min before glucose loading, and the extract was not effective when administered simultaneously (Eseyin et al. 2010a).

In another study, it was reported that water extracts of the leaf of *T. occidentalis* (500 mg/kg) showed promising antihyperglycemic activity, whereas the ethanol extract of the seed did not lower blood glucose concentrations, but the ethanol extract of the leaf reduced blood glucose concentrations significantly in glucose tolerance test in rats (Eseyin et al. 2010b). The antidiabetic activity of the ethanol leaf extract of the plant was traced to the ethyl acetate fraction of the leaf (Eseyin et al. 2010c). An *in vitro* study has shown that unprocessed *T. occidentalis* leaf reduced $Fe(3+)$ to $Fe(2+)$ and also inhibited α-amylase and α-glucosidase activities in a dose-dependent manner. However, blanching of the leafy vegetables caused a significant increase in the antioxidant properties but decreased their ability to inhibit α-amylase and α-glucosidase activities (Oboh et al. 2012). In a recent study, it was shown that the polysaccharide fraction of the leaf extract exhibited blood glucose-lowering effects in normal and streptozotocin-induced diabetic rats (Eseyin et al. 2014).

Tephrosia purpuria (L.) Pers, Fabaceae (Figure 2.94)

Common Names: Sarboka (Arab), sharapunkha, sarphank, kolingi, kottamari (Indian languages)

Description: This plant species is a highly branched, suberect, and herbaceous perennial herb. The leaves are up to 7 cm with 4–9 leaflet pairs that obovate, pubescent, base cuneate, and mucronate with an entire margin and an obtuse apex. Inflorescence is an axillary or leaf-opposed pseudo-raceme;

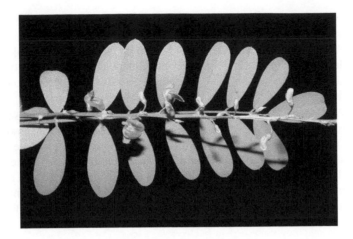

FIGURE 2.94 *Tephrosia purpuria.*

flowers are red or purple and 4–8 mm long. The corolla is bluish-pink or purple; stamens 10 alternatively longer and shorter. The ovary is 5 mm and apprised pubescent; the style is 3 mm and glabrous. The pod is 4 × 0.4 cm, with ca. 7, ovoid, 3.5 mm seeds and a strophiole in the middle of the seed (review: Sharma et al. 2013).

Distribution: This herb is found in tropical regions of the world.

Traditional Medicinal Uses: In traditional medicine the plant is used in wound healing and as tonic and diuretic; it is also used to treat several ailments such as boils, pimples, bleeding piles, and inflammation. Decoction of the roots is used in dyspepsia, diarrhea, rheumatism, asthma, and urinary disorders and the roots are given with black pepper in colic. A liniment prepared from the roots is used in elephantiasis and the pulverized roots are smoked for relief from asthma and cough; decoction of the pods is used as a vermifuge and to stop vomiting. The seed oil is used against scabies, itch, eczema, and other skin eruptions. Fresh root bark, ground and made into a pill, with a little black pepper, is frequently given in cases of obstinate colic. It is considered as a cardiac tonic. In Ceylon, it is employed as an anthelmintic for children (review: Sharma et al. 2013).

Pharmacology

Antidiabetes: Several studies on experimental animals have shown the anti-DM and hypoglycemic property of *Tephrosia purpuria* extracts. Oral administration of the ethanol extract of the seed of *T. purpuria* (300 mg/kg) to streptozotocin-induced diabetic rats showed antihyperglycemic and anti-LPO activities. Besides, the ethanol extract showed marked increase in the levels of antioxidants including the activities of antioxidant enzymes. These effects were comparable to those of glibenclamide (Pavana et al. 2007a).

In another study, the extracts of the root were evaluated for hypoglycemic and antidiabetic activity against normal as well as alloxan-induced diabetic rats. Single dose administration of the extracts did not show any hypoglycemic or antidiabetic activity, whereas repeated dose administration for 7 days of the alcohol and hydroalcoholic extracts (but not petroleum ether extract) showed significant antidiabetic activity in diabetic rats and hypoglycemic activity in normal rats (Joshi et al. 2008).

The water extract of the leaf also showed promising antihyperglycemic and antilipid peroxidative activities in streptozotocin-induced diabetic rats as judged from the levels of glucose, glycosylated hemoglobin, and thiobarbituric acid-reactive substances, enzymatic and nonenzymatic antioxidants in the blood and liver, and hexokinase and glucose-6-phosphatase activities (Pavana et al. 2007b). In another study also, the hydroalcoholic extract of the leaf powder showed activity against streptozotocin-induced diabetes in rats at a dose of 400 mg/kg (Vijayakumar et al. 2014).

Thus, the plant extracts from the root, seed, and leaf possess antidiabetic and antioxidant properties. There is an urgent need to identify the active principles involved and their mechanism of action.

The plant is reported to contain compounds such as rutin, β-sitosterol, and lupeol with anti-DM properties. This plant appears to be very promising for further studies.

Other Activities: Antimicrobial activity and wound healing property are two major pharmacological properties of this plant. Other reported activities include immune-modulation activity, antioxidant activity, antiulcer, anti-inflammatory action, antileishmanial activity, hepatoprotection, anticarcinogenic action, and analgesic activity (review: Sharma et al. 2013b).

Phytochemicals: Phytochemical investigation of this plant has revealed the presence of glycosides, carotenoids, isoflavone, flavanones, chalcones, flavonol, and sterols. The plant showed the presence of glycosides such as rutin and quercetin, retinoids like deguelin, elliptone, rotenone and tephrosin, flavonoids (like purpurin, purpurenone, and purpuritenin), and sterols such as sitosterol. An isoflavone, 7,4-dihydroxy-3,5-dimethoxyisoflavone, and chalcone, (+)-tephropurpurin, are also reported to be present. The major constituents are rutin, quercetin, retinoids deguelin, elliptone, rotenone, tephrosin, and lupeol, and the minor compounds are flavanones, lanceolatin (A, B, and C), isolonchocarpin, and purpurin from the root and from entire plant is pongamol (review: Sharma et al. 2013).

Tephrosia villosa Pers, Leguminosae/Fabaceae
The ethanol extract of this plant's leaves lowered blood glucose levels in alloxan-induced diabetic rats (Ahmad et al. 2009).

Terminalia arjuna Roxb., Combretaceae
Terminalia arjuna is an important medicinal treatment used in Ayurveda and local health traditions. Oral administration of the ethanol extract of the bark of *T. arjuna* (250 and 500 mg/kg daily for 30 days) to alloxan-induced diabetic rats resulted in significant decrease of blood glucose from 16.77 to 4.55 mmol/L and a decrease in the activities of glucose-6-phosphatase, fructose-1,6-disphosphatase, and aldolase and an increase in the activity of phosphoglucoisomerase and hexokinase in tissues. Thus, the study showed that the bark extract of *T. arjuna* possessed potent anti-DM activity (Ragavan and Krishnakumari 2006). In another study, the ethanol extract of *T. arjuna* bark (500 mg/kg) showed significant decrease in blood glucose levels in alloxan-induced diabetic rats. Further, the extract treatment resulted in a significant increase in glycogen content in the liver, cardiac muscle, and skeletal muscle in diabetic rats. Besides, the bark extract reduced adrenaline-induced hyperglycemia and intestinal glucose absorption. In normal rats also the extract exhibited blood glucose-reducing effects (Barman and Das 2012). The hypoglycemic effect of the acetone extract of *T. arjuna* bark in high-fructose-fed streptozotocin-induced type 2 diabetic rats was also studied. Oral administration of 500 mg/kg of the acetone extract to type 2 diabetic rats lowered blood glucose, improved oral glucose tolerance, and decreased the levels of urine glucose and ketone bodies compared to diabetic control rats. These effects of the extract were comparable to those of glimepiride (Kumar et al. 2013a). The leaf extract of *T. arjuna* also demonstrated remarkable antihyperglycemic and antioxidant activities in streptozotocin-induced diabetic rats (Biswas et al. 2011). Although studies have established the anti-DM activity of *T. arjuna*, the active principles involved and details of the mechanism of action are not known.

Terminalia bellirica (Gaertn.) Roxb., Combretaceae
Terminalia bellirica is extensively used in Indian traditional systems of medicine to treat various diseases including DM. The fruit of *T. bellirica* lowered the serum glucose levels in alloxan-induced diabetic rats; further, it showed an antioxidant activity (Sabu and Kuttan 2009). In another study, administration of *T. bellirica* fruit (hexane extract, 200 mg/kg; ethyl acetate extract, 300 mg/kg; and methanol extract, 300 mg/kg) for 60 days to streptozotocin-induced diabetic rats resulted in an increase in the plasma insulin, C-peptide, and glucose tolerance levels compared to the diabetic control; the effect was more pronounced in the methanol extract-treated rats. In addition, the plant extracts significantly increased body weight and serum total protein and significantly decreased the serum levels of total cholesterol, TGs, LDL-C, urea, uric acid, and creatinine in diabetic rats. The authors attribute these beneficial therapeutic effects of the fruit extracts to the synergistic action of more than one bioactive compound (Latha and Daisy 2010). *T. bellirica* (decoction of dried fruits) stimulated the secretion and action of insulin and inhibited starch digestion and protein glycation *in vitro* (Kasabri et al. 2010). Gallic acid (5–20 mg/kg)

isolated from the fruits of *T. bellirica* showed insulin secretagogue and antihyperlipidemic effects in streptozotocin-induced diabetic rats (Latha and Daisy 2011). Gallotannins present in the fruits of this plant increased PPAR-α and PPAR-γ levels and stimulated glucose uptake without enhancing adipocyte differentiation (Yang et al. 2013).

Terminalia catappa L., Combretaceae

The water extract of *Terminalia catappa* leaves produced a significant antidiabetic activity in alloxan-induced diabetic rats as judged from the blood glucose levels and serum biochemical parameters. The extract also caused regeneration of β-cells of the pancreas in diabetic rats (Syed et al. 2005). In another study, the water extract of *T. catappa* leaves exhibited a concentration-dependent blood glucose-lowering effect in glucose-fed hyperglycemic rabbits (Nguessan et al. 2011). In a recent study, the antidiabetic activity of *T. catappa* bark extracts in alloxan-induced diabetic rats was evaluated. The bark extract (400 mg/kg daily for 21 days) exhibited significant antihyperglycemic activity and improvement in parameters like body weight and lipid profile (Abdul and Jyoti 2014). Follow-up studies are needed, among other things, to establish the safety and therapeutic value.

Terminalia chebula Retz. or *Terminalia chebula* Retz var. *tomentella* Kurt, Combretaceae (Figure 2.95)

Common Names: Ink-nut tree (Eng.), harad, hadukkaya, kadukkaai, haritaki, etc. (Indian languages), myrobalan (French), sa mao tchet (Cambodia), etc. (its edible fruits are locally known as *haritaki*)

Description: *Terminalia chebula* is a flowering evergreen tree; it is a medium to large, highly branched deciduous tree with a height up to 30 m and girth 1–1.5 m. The leaves are 10–30 cm long and elliptical with an acute tip and cordate base. The vasculature of the leaves consists of 6–8 pairs of veins. The flowers are short stalked, monoecious, and dull white to yellow with a strong unpleasant odor and are found in simple terminal spikes or short panicles. The fruits are 3–6 cm long and 1.3–1.5 cm broad with yellowish-green ovoid drupes containing one oval seed (review: Gupta 2012).

Distribution: *T. chebula* is a native of Asia and is still found in many Asian countries. It is capable of growing in a variety of soils, clayey as well as shady. The trees may grow at places up to a height of about 2000 m from the sea level and in areas with an annual rainfall of 100–150 cm and temperature of 0°C–17°C. In India, it grows in deciduous forests of many states (review: Gupta 2012).

FIGURE 2.95 *Terminalia chebula.*

Traditional Medicinal Uses: The fruits of this plant are edible. The fruit and, to some extent, the leaf and bark are used in traditional medicine. It is used in Thai traditional medicine as a carminative, astringent, and expectorant. In Ayurveda, the "triphala," an herbal preparation of "three fruits" from *T. chebula*, *T. bellirica*, and *E. officinalis*, is used as a laxative in chronic constipation, a detoxifying agent of the colon, in digestive problems (poor digestion and assimilation), and a rejuvenator of the body. It is extensively used in Ayurveda, Siddha, Unani, and as homeopathic medicines in India. It is a top-listed plant in Ayurvedic *Materia Medica* for the treatment of asthma, bleeding piles, sore throat, vomiting, and gout (review: Gupta 2012). It is also used as an anthelmintic, cardiotonic, and diuretic and as a treatment for diabetes, ulcers, and urinary disorders (monograph: WHO 2009).

Pharmacology

Antidiabetes: Oral administration of 75% methanol extract of *T. chebula* (100 mg/kg) reduced the blood sugar level in normal and alloxan-induced diabetic rats significantly within 4 h. Continued daily administration of the drug produced a sustained effect (Sabu and Kuttan 2002). The chloroform extract of *T. chebula* seeds (100, 200, and 300 mg/kg body weight) produced a dose-dependent reduction in blood glucose of streptozotocin-induced diabetic rats; this effect was comparable with that of the standard drug glibenclamide in a short-term study. It also produced significant reduction in blood glucose when administered daily for 8 weeks at a dose of 300 mg/kg. Besides, it exhibited remarkable renoprotective effects in these diabetic animals (Rao and Nammi 2006). The water extract of the dry fruits of *T. chebula* improved glucose tolerance and brought down fasting blood glucose levels in diabetic rats (Murali et al. 2004). Oral administration of the ethanol extract of the fruits (200 mg/kg for 30 days) reduced the levels of blood glucose and glycosylated hemoglobin in streptozotocin-induced experimental diabetic rats (Senthilkumar et al. 2006). In a similar study, the water extract of *T. chebula* (200 mg/kg for 2 months) reduced the elevated blood glucose and increased glycosylated hemoglobin. The same dose also markedly decreased blood lipids and increased serum insulin levels. *In vitro* studies with pancreatic islets showed that the insulin release was nearly two times more than that in untreated diabetic animals. The treatment did not have any unfavorable effect on liver and kidney function tests (Murali et al. 2007). The ethanol extract of the fruit showed antidiabetic and antioxidant activities in streptozotocin-induced diabetic rats (Senthilkumar 2007; Senthilkumar and Subramanian 2008). Studies on the effect of *T. chebula* on the levels of glycoproteins in streptozotocin-induced diabetes in rats have shown amelioration of the changes induced by streptozotocin on the levels of the glycoproteins by the treatment (Senthilkumar and Subramanian 2008). The plant extract possessed antioxidant and hypolipidemic and hypocholesterolemic activities (review: Gupta 2012).

The aforementioned studies indicate that active principles soluble in water, methanol, and chloroform are present in this plant fruit, seed, etc. The plant is reported to contain anti-DM compounds, gallic acid, ellagic acid, etc. Further, studies are needed to identify these active principles and determine their therapeutic utility. The edible fruit appears to be suitable for the development of dietary supplements.

Toxicity: Dietary administration of the fruit to rats, as 25% of the diet, produced hepatic lesions, which included centrilobular vein abnormalities and centrilobular sinusoidal congestion. Marked renal lesions were also observed and included marked tubular degeneration, tubular casts, and intertubular congestion. A brown pigmentation of the tail and limbs was also observed after 10 days. The median lethal dose of a 50% ethanol extract of the fruit was 175.0 mg/kg after intraperitoneal administration (monograph: WHO 2009).

Other Activities: Other reported pharmacological properties of this plant include antioxidant, antibacterial, antifungal, antiviral, antiamoebic, antiplasmodial, anthelmintic, immunomodulatory, antiarthritis, antimutagenic and anticarcinogenic, cancer preventive, antiaging, antianaphylactic and adaptogenic, antiulcer, wound healing, antinociceptive, radioprotective, hepatoprotective, cardioprotective, antihyperlipidemic, antiatherosclerosis, and spermatogenic activities (monograph: WHO 2009; review: Gupta 2012).

Phytochemicals: In *T. chebula*, the major class of phytoconstituents is hydrolyzable tannins (which may vary from 20% to 50%). These tannins contain phenolic carboxylic acid like gallic acid, ellagic acid, chebulic acid, and gallotannins such as 1,6-di-*O*-galloyl-β-D-glucose, 3,4,6-tri-*O*-galloyl-β-D-glucose, 2,3,4,6-tetra-*O*-galloyl-β-D-glucose, and 1,2,3,4,6-penta-*O*-galloyl-β-D-glucose. Ellagitannin such as punicalagin, casurarinin, corilagin, and terchebulin and others such as chebulanin, neochebulinic acid, chebulagic acid, and chebulinic acid were reported. The tannin content varies with the geological variation. Flavonol glycosides, triterpenoids, coumarin conjugated with gallic acid called chebulin, as well as phenolic compounds were also isolated. There are seven varieties of *T. chebula*, all of which are more or less used in a similar fashion but vary in specific usages and quality (review: Rathinamoorthy and Thilagavathi 2014).

Terminalia pallida Brandis, Combretaceae
In ethnomedicine, the powder of *Terminalia pallida* fruit is given by tribal people in Andhra Pradesh, India (where locally it is known as tellakaraka). In a follow-up pharmacological study, oral administration of the ethanol extract (0.5 g/kg) of *T. pallida* fruit exhibited a significant antihyperglycemic activity in alloxan-induced diabetic rats, whereas in normal rats no hypoglycemic activity was observed (Rao et al. 2003).

Terminalia paniculata Roth, Combretaceae
The water extract of *Terminalia paniculata* bark (oral daily administration for 28 days) exhibited significant hypoglycemic, hypolipidemic, and antioxidant activities in streptozotocin-induced diabetic rats. Oral administration of the extract did not exhibit toxicity and death at a dose of 2 g/kg (Ramachandran et al. 2012). In another study, oral treatment of the water extract of *T. paniculata* bark (100 and 200 mg/kg) to streptozotocin–nicotinamide-induced type 2 diabetic rats decreased blood glucose and glycosylated hemoglobin levels in treated diabetic rats compared to diabetic control rats. Further, body weight, total protein, insulin, and hemoglobin levels were increased in the extract-treated diabetic rats compared to diabetic control rats. A significant reduction of total cholesterol and TGs and increase in HDL levels were observed in type 2 diabetic rats after extract administration. The presence of biomarkers gallic acid, ellagic acid, catechin, and epicatechin in the extract was confirmed in HPLC analysis. The extract and gallic acid showed significant enhancement of glucose uptake action in the presence of insulin in muscle cells than the vehicle control *in vitro*. Also, the extract inhibited pancreatic α-amylase and α-glucosidase enzymes (Ramachandran et al. 2013).

Terminalia superba Engl. & Diels, Combretaceae
Stem bark extracts of *Terminalia superba* is used in Africa for the treatment of various ailments, including DM. In a study, the antidiabetic effects of the methanol/methylene chloride extract of the stem bark were evaluated in streptozotocin-induced diabetic rats. Administration of the extract (300 mg/kg daily for 14 days) resulted in marked reduction in blood glucose levels in diabetic rats compared to diabetic control rats. The treatment improved body weight also (Kamtchouing et al. 2005). In another study, the water extract of the roots of *T. superba* was investigated for its effect on glucose homeostasis and oxidative stress in alloxan-induced diabetic extracts. The extract (300 mg/kg) demonstrated significant hypoglycemic and *in vivo* antioxidant activities in alloxan-induced diabetes (Momo and Oben 2009). The antidiabetic effect of the methanol/methylene chloride extract of *T. superba* leaves was studied using streptozotocin-induced diabetic rats. Administration of the extract (200 and 400 mg/kg daily for 2 weeks) resulted in normalization of fasting blood glucose levels, reduction in polyphagia and polydipsia, and increase in body weight gain in treated diabetic rats (Padmashree et al. 2010).

Thus, the stem bark, root, and leaves of this plant showed anti-DM activity in experimental animals. Further studies are warranted to determine its utility as a therapeutic agent.

Tetracera indica Merr., Dilleniaceae

Common Name: Sand piper vine
In folk medicine the leaves of *Tetracera indica* are used to treat diabetes in Malaysia. Water and methanol extracts of *T. indica* leaves (250 and 500 mg/kg) showed antihyperglycemic activity in alloxan-induced diabetic rats after a single dose administration. In normal rats the extracts did not show hypoglycemic activity. *In vitro*, the water extract was found to reduce TG accumulation in 3T3-L1 cells, whereas the methanol extract induced lipid accumulation (Ahmed et al. 2012).

Tetracera scandens (L.) Merr., Dilleniaceae

Tetracera scandens is a traditional Vietnamese medicinal plant. Water and ethanol extracts of the leaves of *T. scandens* reduced blood glucose levels in alloxan-induced diabetic rats without causing any hypoglycemic effect in normal rats. The highest antihyperglycemic effect (62.5%) was observed for the water extract at 250 mg/kg and methanol extract (36.5%) at 500 mg/kg after 8 h of administration (Umar et al. 2010). Several flavonoids isolated from the methanol extract of *T. scandens* exhibited PTP1B inhibitory activity. Besides, two of the flavonoids exhibited glucose uptake activity in basal and insulin-stimulated L6 myotubes; these compounds stimulated AMPK phosphorylation and inhibited GLUT4 and GLUT1 mRNA expressions. Further, genistein derivatives from *T. scandens* stimulated glucose uptake in L6 myotubes (Jiany et al. 2012).

Tetrapleura tetraptera (Schum and Thonn) Taub, Fabaceae

This plant is used in African traditional medicine for the management of an array of human ailments, including arthritis and other inflammatory conditions, diabetes, asthma, hypertension, and epilepsy. The water extract of the fruit showed anti-DM activity in streptozotocin-induced diabetic rats. Further, the extract exhibited anti-inflammatory activity against egg albumin-induced pedal edema in rats (Ojewole and Adewunmi 2004). The methanol extract of *Tetrapleura tetraptera* leaves also showed anti-DM activity in alloxan-induced diabetic rats (Atawodi et al. 2014).

Teucrium capitatum L., Lamiaceae

Synonym: *Teucrium polium* L.

The alcohol extract of the aerial parts of the plant showed an insulinotropic effect in INS-1E cells *in vitro*. Intragastric administration of the extract (125 mg/kg) in both normoglycemic and streptozotocin-induced hyperglycemic rats was more efficient in lowering the blood glucose than intraperitoneal injection (35% vs. 24% reduction) with the highest effect (50% reduction) 8 h after administration. After 10 days of treatment, the magnitude of the effect was comparable to that of 2.5 mg/kg of glibenclamide (38% reduction). No effect was seen on blood lipid profiles. In oral glucose tolerance test, the extract lowered blood glucose levels. The treatment reduced hepatic glycogen and tended to normalize the activity of gluconeogenic enzymes (Stefkov et al. 2011).

Teucrium cubense Jacq., Labiatae

Teucrium cubense is native to a section of the Americas and is used in traditional medicine. The water extract of *T. cubense* induced glucose uptake in insulin-sensitive and insulin-resistant murine and human adipocytes without affecting TG accumulation (Zapata-Bustos et al. 2009; Alonso-Castro et al. 2010).

Teucrium oliverianum R. Br., Lamiaceae

Teucrium oliverianum is used in traditional medicine to treat DM in Middle East. Administration of the decoction and ethanol extract of *T. oliverianum* to alloxan-treated mice produced a reduction of blood sugar levels. Phytochemical analysis revealed the presence of alkaloids, cardiac glycosides, flavonoids, sterols/triterpenes, volatile oil, coumarins, tannins, and saponins (Ajabnoor et al. 1984).

Teucrium polium (L.) Tasuch, Lamiaceae

Teucrium polium is a flowering plant found abundantly in the wild in Southwestern Asia, Europe, and North Africa. In traditional Iranian medicine, *T. polium* leaf is used for treating abdominal pain, indigestion, common cold, type 2 DM, etc. The main groups of chemicals isolated from these plants are terpenoids and flavonoids. These compounds possess a broad spectrum of pharmacological activity including hypoglycemic, antioxidant, and hypolipidemic properties. However, further studies are required to identify the active antidiabetic compounds and determine their utility as an anti-DM therapeutic agent (review: Bahramikia and Yazdanparast 2012).

Theobroma cacao L., Sterculiaceae

The cocoa extract was reported to possess various medicinal properties including the hypoglycemic effect. In hyperglycemic rats, 3.0% cocoa extract had reduced the glucose level significantly at 60 min in oral glucose tolerance test as compared to control. However, in a 2-week study with 3.0% cocoa powder, there was no significant difference in plasma glucose levels and lipid profiles in hyperglycemic and normal rats, which were given basal diet enriched with 3.0% cocoa powder extract as compared to the control diet (basal diet) (Amin and Azli 2004). In another study, administration of cocoa extract reduced blood glucose level in normal, obese, obese diabetic, and diabetic rats. It is believed that the rich polyphenol contents could have contributed these effects (Jin et al. 2012). In a recent *in vitro* study, polyphenol-rich extracts from cocoa bean inhibited α-amylase, α-glucosidase, and angiotensin-1-converting enzyme activities in a dose-dependent manner and also showed dose-dependent *in vitro* free radicals scavenging ability (Oboh et al. 2014). Detailed studies with cocoa bean powder, standardized extract, and active molecules (to be identified) are required to determine the therapeutic value of cocoa in diabetes.

Thespesia populnea Soland ex Correa, Malvaceae

All the parts of *Thespesia populnea* are used in the Indian system of traditional medicine. In one study, the ethanol extract of the stem bark and leaf showed a glucose-lowering effect in streptozotocin-induced diabetic rats (Parthasarathy et al. 2009). In another study, the ethanol extract of *T. populnea* flower (250 and 500 mg/kg daily for 10 days) showed a dose-dependent decrease in blood glucose levels and increase in body weight in treated alloxan-induced diabetic rats. Further, the treatment decreased the levels of serum urea, creatinine, and cholesterol in diabetic rats. This effect of the ethanol extract at 500 mg/kg was comparable with that of the reference standard glibenclamide (Shivkumar et al. 2008).

Thunbergia laurifolia L., Acanthaceae

Common Name: Laurel clock vine or blue trumpet vine

The antidiabetic effects of the leaf extract of *Thunbergia laurifolia* (purple flower strain) were studied. The study showed that a 15-day treatment with the extract (60 mg/mL/day) decreased the levels of blood glucose in alloxan-induced diabetic rats. The recovery of some β-cells to a considerable extent was found in treated diabetic rats (Aritajat et al. 2004).

Thymelaea hirsuta L. Endl., Thymelaeaceae

Thymelaea hirsuta is a medicinal plant exhibiting an antihyperglycemic effect and is used in Morocco to treat diabetes. The polyphenol-rich fraction from *T. hirsuta* (80 mg/kg/day for 21 days) showed antidiabetic and antihypertensive activities in streptozotocin-induced diabetic NO-deficient hypertensive rats. The treatment reduced blood glucose levels and increased liver glycogen content in diabetic rats. Moreover, the fraction significantly reduced the amount of glucose absorbed in *in situ* perfused jejunum segments compared with control. The authors conclude that *T. hirsuta* may be useful as a food supplement for the prevention of type 2 diabetes and hypertension (Bnouham et al. 2007, 2012). The fractions of *T. hirsuta* induced, *in vitro*, a significant inhibition of α-glucosidase. The ethyl acetate fraction had high activity and its inhibition mode was noncompetitive. The fraction (50 and 100 mg/kg) decreased significantly, *in vivo*, the postprandial hyperglycemia after sucrose loading in normal and diabetic mice. Moreover, 50 mg/kg of this fraction significantly decreased intestinal glucose uptake, *in situ*, in rats. Thus, the antihyperglycemic effect of *T. hirsuta* can be explained, in part, by the inhibition of intestinal α-glucosidase and intestinal glucose absorption (Abid et al. 2013).

Thymus capitatus Hoff et Link., Labiatae

The aerial parts of *Thymus capitatus* are used in local health traditions to treat diabetes in Egypt. In alloxan-induced diabetic rats, thyme oil (0.1 mL/kg, p.o., for 28 days) exhibited antihyperglycemic and antilipidemic effects (Benkhayal et al. 2010). In another study, *T. capitatus* extract showed hypoglycemic effects in rats (Shabana et al. 1990).

Thymus serpyllum L., Lamiaceae

The water extract of *Thymus serpyllum* reduced the levels of blood glucose in alloxan-induced diabetic rabbits. This decrease in glucose level was comparable to that of glibenclamide and acarbose. The treatment maintained the weight of diabetic rabbits and ameliorated diabetes-induced hematological disturbances in rabbits (Alamgeer et al. 2012). In another study, the water extract of *T. serpyllum* was evaluated for its hypoglycemic activity in 10% D-glucose-treated mice. The blood glucose level of glucose-fed mice treated with the water extract for 1 month remained within the normal range, while the blood glucose level of glucose-fed untreated mice was increased as compared to control. The water extract also exhibited a significant increase in the weight of the pancreas, heart, and kidneys of glucose-treated mice (Alamgeer et al. 2013).

Thymus vulgaris L., Lamiaceae

Thymus vulgaris is a medicinal herb. Alloxan-induced diabetic rats fed 10% thyme (*T. vulgaris*) showed decrease in blood glucose, lipids, uric acid, and creatinine and increase in HDL-C levels compared to diabetic control rats. The treatment also improved liver function (Hanna et al. 2014). The plant may also ameliorate diabetic complications. *T. vulgaris* (water extract) treatment to streptozotocin-induced experimental diabetic rats resulted in improvements in biochemical, physiological, and histological changes in DM-related disorders, such as neuropathy and nephropathy (Akan 2014). In another study, oral administration of the crude water extract of *T. vulgaris* (500 mg/kg) to alloxan-induced diabetic rats resulted in marked reduction in the levels of fasting blood glucose, total cholesterol, LDL, VLDL, and TGs and increase in the levels of HDL compared to diabetic control rats (Ekoh et al. 2014). Carvacrol, a component of thyme oil, activated PPAR-α and PPAR-γ and suppressed COX-2 expression (Wang et al. 2014). Thus, recent studies have established the antidiabetes and hypolipidemic properties of this spice.

Tinospora cordifolia (Willd.) Miers. ex Hook. F. & Thoms., Menispermaceae (Figure 2.96)

Common Names: Gulancha tinospora, guduchi, amrita, giloe, amridavalli, amrytu, etc.

Description: *Tinospora cordifolia* is a large deciduous climbing shrub with glabrous shoot and corky bark. Stems are succulent having long filiform fleshy aerial roots. The leaves are pear shaped, membranous, juicy, petiolate, cordate, and glabrous. It has greenish flowers borne on the axillary, terminal, or the old stem as racemes. Flowers are unisexual. Male flowers are fascicled, while female flowers are usually solitary and glabrous. The fruit is a reddish drupe with curved seeds (review: Sankhala et al. 2012).

Distribution: The shrub is native to India and is distributed throughout tropical India, from Kumaon to Assam and from Bihar to Konkan. It is also widely distributed in Ceylon, Burma, and China.

(a)

(b)

FIGURE 2.96 *Tinospora cordifolia.* (a) Plant (Climbing shrub). (b) Fruits and leaves.

Traditional Medicinal Uses: In Ayurvedic medicine, *T. cordifolia* is used for treating DM. In traditional medicine, it is an emetic, aphrodisiac, sedative, stomachic, tonic, diuretic, appetizer, antipyretic, expectorant, and alternative and is used in the treatment of asthma, malaria, jaundice, giddiness, skin diseases, diabetes, vomiting, vaginal and urethral discharge, enlarged spleen, cough, piles, chronic fever, anemia, snakebite, fractures, syphilis, diarrhea, dysentery, rheumatism, and gonorrhea (Kirtikar and Basu 1975; review: Sankhala et al. 2012).

Pharmacology

Antidiabetes: The alcohol extract of *T. cordifolia* roots exhibited hypoglycemic and hypolipidemic actions in alloxan-induced diabetes in rats (Grover et al. 2002; Stanely et al. 2003). Oral administration of the plant root (water extract) to alloxan-induced diabetic rats caused a significant reduction in blood glucose and brain lipids. Further, the water extract caused an increase in body weight, total hemoglobin, and hepatic hexokinase. The root water extract also lowered hepatic glucose-6-phosphatase and serum ALP, acid phosphatase, and lactate dehydrogenase in diabetic rats. The root extract showed antioxidant activity also (Stanely and Menon 2001). In a follow-up study, oral administration of the root extract for 6 weeks resulted in a marked reduction in blood and urine glucose levels in alloxan-induced diabetic rats. The extract also prevented a decrease in body weight (Standly and Menon 2003). *T. cordifolia* (stem) extract prevented high-fructose diet-induced insulin resistance and oxidative stress in male rats (Reddy et al. 2009). Blood glucose levels are reduced without a significant effect on total lipid by the leaf extracts in normal and alloxan-induced diabetic rabbits (Wadood et al. 1992). Tinosporaside, a triterpene isolated from stem of this plant, exhibited antihyperglycemic activity in streptozotocin-induced diabetic rats, and in comparison it was better than metformin (review: Arif et al. 2014).

The plant possesses hypoglycemic, antihyperlipidemic, and antioxidant properties. The antidiabetic compounds berberine and β-sitosterol are present in this plant also (review: Sankha et al. 2012). It is worthwhile to carry on detailed studies on its usefulness as an anti-DM medicine.

Other Activities: The alcohol extract of *T. cordifolia* roots almost restored the antioxidant defense in alloxan-induced diabetes in rats. The herb exhibited strong free radical scavenging properties against reactive oxygen and nitrogen species. The plant extract has radioprotective potential. The plant (water extract) treatment altered lethal effects of gamma rays in mice. Hepatoprotective and immunomodulatory properties of the plant in CCl_4-intoxicated mature albino rats have been reported. *T. cordifolia* extract showed a wide spectrum of immunoaugmentary effects. It upregulated antitumor activity of tumor-associated macrophages and several studies suggest the anticancer property of this plant (Subramanian et al. 2002; Leyon and Kuttan 2004a,b; Nair et al. 2004; Prince et al. 2004; Singh et al. 2004; Badar et al. 2005; review: Sankhala et al. 2012). Other reported activities include anti-inflammatory activity, antipyretic action, relaxation of intestinal and uterine smooth muscles, temporary decrease in blood pressure, antiulcer activity, and antiamoebic activity (review: Sankhala et al. 2012).

Toxicity: The stem extract possesses an antifertility effect in male rats (Gupta and Sharma 2003).

Phytochemicals: The plant contains different classes of phytochemicals such as alkaloids, glycosides, diterpenoid lactones, sesquiterpenoids, and steroids. Important alkaloids isolated from the stem are berberine, palmitine, choline, and tinosporin; palmatine (alkaloid) was isolated from the root. Glycosides present in the stem are 18-norclerodane glycoside, furanoid diterpene, glucoside, tinocordiside, and syringing. Furanolactone, tinosporin, and columbin are diterpenoid lactones isolated from the whole plant. Steroids isolated from the aerial parts of the plant include ecdysterone, makisterone A, giloinsterol, β-sitosterol, and γ-sitosterol (review: Sankha et al. 2012).

Tinospora crispa (L.) Hook. f. & Thomson
Borapetol B, a compound isolated from *Tinospora crispa*, stimulated insulin release from pancreatic β-cells (Lokman et al. 2013). Acute oral administration of the compound improved glucose levels in spontaneously type 2 DM Goto-Kakizaki rats. Plasma insulin levels were increased by twofold in treated diabetic rats compared to the placebo group at 30 min. Further, the compound dose dependently increased insulin secretion from islets isolated from diabetic rats at 3.3 and 16.7 nM glucose in the medium. Thus, the study provides evidence that borapetol B stimulated insulin release, which could be its mechanism of anti-DM activity (Lokman et al. 2013).

Triumfetta rhomboidea Jacq., Poaceae
The ethanol extract of *Triumfetta rhomboidea* reduced blood glucose levels in alloxan-induced diabetic rats (Duganath et al. 2011).

Tithonia diversifolia (Hemsl.) A. Gray, Asteraceae
Nitobegiku (*Tithonia diversifolia*) has been used as a traditional medicinal plant for diabetes. *T. diversifolia* (80% alcohol extract of the whole plant; 500 mg/kg) reduced the blood glucose of KK-Ay mice (an animal model of type 2 DM) 7 h after a single oral administration. Besides, treatment with the extract resulted in improvement in insulin resistance and a decrease in plasma insulin in KK-Ay mice. The extract did not influence blood glucose levels in normal ddY mice (Miura et al. 2005a). Three new germacrene sesquiterpenes obtained from *T. diversifolia* significantly increased glucose uptake in 3T3-L1 adipocytes (Zhao et al. 2012b).

Toddalia asiatica (L.) Lam., Rutaceae
The extract of this plant decreased blood glucose levels and increased insulin levels in streptozotocin-induced diabetic rats. Further, the treatment increased body weight and liver glycogen content in the treated animals compared to untreated diabetic control animals (Stephen et al. 2012).

Toona ciliata var. *pubescens* (Franchet) Handel-Mazzetti, Meliaceae
The stem bark of *Toona ciliata* var. *pubescens* yielded (Z)-aglawone, which inhibited PTP1B activity in a competitive manner with an IC_{50} value of 1.12 µg/mL. It is interesting to note that the (E)-isomer of this compound showed no inhibition of PTP1B (Jiany et al. 2012).

Toona sinensis Roem., Meliaceae
Toona sinensis leaf extract (50% alcohol/water) (1–100 µg/mL) increased glucose uptake in basal and insulin-stimulated 3T3-L1 adipocytes. The enhancement of cellular glucose uptake by the extract in insulin-stimulated 3T3 L1 (but not in the absence of insulin) was inhibited by protein synthesis inhibitors (Yang et al. 2003). In another study, administration of *T. sinensis* leaf to alloxan-induced diabetic rats resulted in a reduction in plasma glucose and improvement in plasma insulin levels compared to diabetic control rats. *T. sinensis* leaf exerted the hypoglycemic effect via an increment of insulin to mediate the adipose GLT4 mechanisms in adipose tissue (Wang et al. 2008b).

Tournefortia hartwegiana Steud, Boraginaceae
Tournefortia hartwegiana is used in traditional medicine for the treatment of diabetes, diarrhea, and kidney pain in Morelos, Mexico. Administration of the methanol extract from the aerial parts of *T. hartwegiana* (310 mg/kg/day) for 10 days to normoglycemic and alloxan-induced diabetic rats significantly lowered blood glucose levels (37% and 36%, respectively). The antidiabetic and hypoglycemic activities of the methanol extract were similar to those produced by metformin at 120 mg/kg (Ortiz-Andrade et al. 2005). In another study, the effect of the methanol extract of *T. hartwegiana* (310 mg/kg) on α-glucosidase activity was investigated. In oral glucose tolerance test, the extract (intragastric administration) significantly suppressed the increase in plasma glucose level after glucose administration. The extract inhibited α-glucosidase activity *in vitro* in a concentration-dependent manner (review: El-Abhar and Schaalan 2014). These results suggest that the extract might exert its antidiabetic effect partly by suppressing carbohydrate absorption from the intestine, thereby reducing the postprandial increase of blood glucose. However, the bioactivity-guided fractionation of this extract led to the isolation of bioactive compounds (β-sitosterol, stigmasterol, lupeol, ursolic acid, oleanolic acid, saccharose, and myo-inositol), which include known antidiabetic compounds (Ortiz-Andrade et al. 2007).

Tournefortia hirsutissima L., Boraginaceae
The plant is used in traditional medicine for the treatment of DM, diarrhea, and kidney pain in Mexico. Oral administration of the water (20 and 80 mg/kg) and butanol extracts (8 and 80 mg/kg) of *Tournefortia hirsutissima* significantly lowered plasma glucose levels in streptozotocin-induced diabetic rats within

3 h (Andrade-Cetto et al. 2007). The methanol extract from the aerial parts of this plant at a dose of 310 mg/kg (daily for 10 days, p.o.) also showed a glucose-lowering effect in normoglycemic and alloxan-induced diabetic rats (Ortiz-Andrade et al. 2005).

Tragia involucrata L., Euphorbiaceae, Indian stinging nettle

Tragia involucrata is used in Indian traditional medicine for the treatment of bronchitis, asthma, skin diseases, diabetes, etc. The water–ethanol extract of *T. involucrata* (whole plant), when administered to streptozotocin–nicotinamide-induced type 2 diabetic rats (250 and 500 mg/kg daily for 28 days), markedly decreased blood glucose and HbA1c levels and increased body weight in diabetic rats. The treatment also improved the serum lipid profile (Farook and Atlee 2011ab). In another study, the alcohol extract (80%) of *T. involucrata* at a dose of 200 mg/kg (oral, daily for 21 days) showed hypoglycemic and antioxidant activities in alloxan-induced diabetic rats (Prakash et al. 2009).

Treculia africana Decne., Moraceae, African breadfruit

Treculia africana, the African breadfruit tree, is native to East Africa. The water extract of the *T. africana* root (200 mg/kg daily for 5 weeks) markedly decreased blood glucose, hemoglobin glycosylation, and plasma LPO in streptozotocin-induced diabetic rats (Ojieh et al. 2009). The diethyl ether-soluble fraction (10 mg/kg daily for 10 days) of the hydroacetone root bark extract of *T. africana* was found to exhibit the highest activity, giving a 69.4% reduction in the blood sugar level that was in comparable range with the reference standard glibenclamide (0.5 mg/kg) that reduced blood sugar levels by 65.8% below the initial baseline values (Oyelola et al. 2007). In a recent study, the hypoglycemic effect of the ethyl acetate-soluble portion of the water–acetone extract of the root bark and stem bark of *T. africana* was compared in alloxan-induced diabetic rats. Both the stem and root extracts showed a sustainable and better reduction of blood glucose level when compared with glibenclamide. However, the activity appeared more in the root bark of the plant (Olatunji et al. 2014). There is a need for further studies on this plant including the details of the mechanism of action and identification of active principles.

Trema orientalis (L.) Blume, Ulmaceae

Alcohol extract of the stem of this plant (acute treatment, p.o.) decreased blood glucose levels in streptozotocin-induced diabetic rats (Dimo et al. 2006).

Tribulus terrestris L., Zygophyllaceae

Synonym: *Tribulus lanuginosus* L.

Tribulus terrestris is a perennial, decumbent herb native to the Mediterranean region but also found throughout the world. The fruit is mostly used in traditional medicine. The protective effects of *T. terrestris* in streptozotocin-induced diabetic rats were studied. The *T. terrestris* extract (2 g/kg) significantly decreased the levels of ALT and creatinine in the serum in diabetic groups and lowered the MDA level and increased reduced GSH level in the liver in diabetic and nondiabetic groups (Amin et al. 2006). Saponins from *T. terrestris* could significantly reduce the levels of serum glucose in normal and alloxan-induced diabetic mice. The saponins could also decrease the content of serum cholesterol and TG in diabetic rats (Li et al. 2002). Other reported activities of this plant include CNS stimulation, aphrodisiac activity, and diuretic action (monograph: WHO 2009). The plant may also have insulin mimetic or insulin secretory action (review: El-Abhar and Schaalan 2014).

Trichosanthes cucumerina L., Cucurbitaceae

Synonym: *Trichosanthes anguina* L.

Trichosanthes cucumerina is an annual, dioecious climber and its common names include snake gourd, Chinese gourd, and melon-melonan. Its unripe fruits are used as vegetables. It is widely distributed in Asian countries and is cultivated for food and medicine. The whole plant of *T. cucumerina* has medicinal properties. Aerial parts of this plant (along with or without other plant materials) are used in traditional medicine for the treatment of diabetes and other ailments such as indigestion, bilious fevers, boils, sores, skin eruptions, and ulcers. The root is used as a cure for bronchitis, headache, and boils. Both the fruit and root are considered to be cathartic. The fruits are used as anthelmintic. The seeds are used for stomach disorders and considered as antifebrile (review: Arawwawala et al. 2013).

The hot-water as well as ethanol extract of the aerial parts of *T. cucumerina* reduced serum glucose levels in normoglycemic rats. In streptozotocin-induced type 1 and type 2 diabetic rats, no immediate hypoglycemic effect was observed with the hot-water extract administration. However, continuous daily oral administration resulted in gradual reduction in serum glucose levels. Repeated administration of the extract for 28 days to streptozotocin-induced diabetic rats resulted in marked decrease in blood glucose levels and increase in body weight, liver glycogen content, and adipose tissue TG levels as compared to untreated diabetic control rats. In normal rats also, administration of the extract for 28 days increased the levels of liver glycogen content and adipose tissue TG. The extract failed to inhibit intestinal glucose uptake. The authors think that it may stimulate insulin secretion (Arawwawala et al. 2009). In streptozotocin-induced type 1 and 2 diabetic rats, the hot-water extract of the aerial parts (when administered orally daily for 28 days) lowered serum TGs, total cholesterol, and LDL-C, while it increased HDL-C (Arawwawala et al. 2012). Another study also reported improvement in glucose tolerance and increase in tissue glycogen content in *T. cucumerina*-treated streptozotocin-induced type 2 diabetic rats (Kirana and Srinivasan 2008). The glucose-lowering activity of the alcohol extract of the seed of this plant was also shown in alloxan-induced diabetic rats (Kar et al. 2003). Further, the extract exhibited beneficial effects on the lipid profile in diabetic rats. Although these studies on animals have established the anti-DM properties of the extract of the aerial parts of this plant, the anti-DM property of its edible fruit (vegetable) remains to be studied for its possible use as a food supplement to diabetes patients. Besides, the active principles involved in the hypoglycemic and antidiabetic activities are not known. There is an urgent need for follow-up studies on the anti-DM properties of this plant.

Other confirmed pharmacological properties include anti-inflammatory activity, gastroprotection from indomethacin- and alcohol-induced gastric ulcers, antimicrobial activity, and antioxidant activities. *T. cucumerina* contains alkaloids, glycosides, tannins, flavonoids, phenols, and sterols (review: Arawwawala et al. 2013).

Trichosanthes dioica Roxb., Cucurbitaceae

Synonyms: *Trichosanthes cucumeroides* (Ser.) Maxim., *Trichosanthes ovigera*

Common Name: Sespadula

Trichosanthes dioica is used to treat DM, epilepsy, alopecia, and skin disease in folklore medicine. The leaf extract of the plant is used in DM. The water extract of the leaves of *T. dioica* (800 and 1600 mg/kg, p.o.) reduced blood glucose significantly when compared to control in glucose-loaded, normal, and streptozotocin-induced diabetic rats; but it was not as effective as glibenclamide (Adiga et al. 2010). In another study, the plant extract exhibited antidiabetes and antihyperlipidemic activity in normal, mild diabetic, and severe diabetic animal models (Rai et al. 2013).

Trichosanthes kirilowii Maxim., Cucurbitaceae

Synonym: *Trichosanthes japonica* E As

Common Name: Chinese cucumber

Trichosanthes kirilowii is an important Chinese medicinal plant with many pharmacological properties. The roots of *T. kirilowii* exhibited hypoglycemic effects in mice. Bioactivity-guided fractionation identified five glycans (trichosans A, B, C, D, and E). These fractions showed hypoglycemic activity in normal mice, but only one glycan (trichosan A) exhibited a blood glucose-lowering effect in alloxan-induced diabetic mice (Hikino et al. 1989).

Tridax procumbens L., Asteraceae

Tridax procumbens is employed as an indigenous medicine for a variety of ailments.

The tribal inhabitants of Rajasthan (India) use the leaf powder (along with other herbs), given orally, to treat diabetes. Oral (acute and subchronic) administration of *T. procumbens* whole plant, 50%

methanol extract (250 and 500 mg/kg), showed a significant reduction in fasting blood glucose levels in alloxan-induced diabetic rats; however, the extract did not lower blood glucose levels in normal rats. The subchronic treatment improved body weight and oral glucose tolerance in diabetic rats (Pareek et al. 2009). However, in another study, the leaf extract of *T. procumbens* (200 mg/kg) showed a blood glucose-lowering effect in normal rats (Bhagwat et al. 2008). The hypoglycemic, hypolipidemic, and antioxidant properties of *T. procumbens* leaf extract in alloxan-induced diabetic rats were evaluated in another study. Oral administration of the ethanol extract of the leaves (400 mg/kg) to alloxan-induced diabetic rats for 30 days resulted in reversal to near normal levels of the altered levels of blood glucose, glycosylated hemoglobin, plasma insulin, urea, uric acid, creatinine, and protein found in untreated diabetic control rats. The treatment also normalized the serum lipid profile, the levels of lipid peroxides in the plasma, and pancreatic tissues of diabetic rats. Thus, *T. procumbens* leaves exhibited antilipidemic and antioxidant effects in addition to its antidiabetic activity (Subramanian et al. 2011).

Trifolium alexandrinum L., Papilionaceae

Trifolium alexandrinum is a traditional medicinal plant, and the seeds of this plant are used in folk medicine in Iran to control DM (review: Haddad et al. 2005). Daily intake of *T. alexandrinum* flower head extracts (water, hexane, and ethanol) in drinking water for 4 weeks (immediately after diabetes induction with streptozotocin) caused significant decrease in glucose and glycated hemoglobin levels and increase in insulin level. The treatment also greatly improved the levels of serum lipid parameters and significantly decreased LPO in addition to the increase of hepatic GSH content significantly (Amer et al. 2004).

Trifolium pratense L., Fabaceae

Trifolium pratense (the red clover) is used in traditional medicine, among other things, as an antispasmodic, sedative, and anti-inflammatory agent. The red clover extracts and the compounds genistein and biochanin A were potent activators of PPAR-γ ligands. Several metabolites exerted higher binding affinities or transactivational activities than their precursor molecules. 6-Hydroxydaidzein exerted a more than 100-fold higher binding affinity than its precursor daidzein. The maximal transactivational activity of 6-hydroxydaidzein and 3′-hydroxygenistein exceeded even that of rosiglitazone, a known PPAR-γ agonist (review: Wang et al. 2014). In view of these observations, follow-up antidiabetic studies in *in vivo* models are warranted.

Trigonella berythea Boiss. & Blanche, Fabaceae

A study on complementary and alternative medicine use among Palestinian diabetic patients showed that out of 1883 patients with diabetes interviewed, 191 used *Trigonella berythea* decoction as a treatment. More than 70% of those using complementary and alternative medicine reported positive benefits (Ali-Shtayeh et al. 2011).

Trigonella foenum-graecum L., Leguminosae (Figure 2.97)

Common Names: Fenugreek (Eng.), methi, uluva, methika, etc. (Indian languages)

Description: Fenugreek is an erect annual, aromatic herb, growing about 2 ft high; flowers are axillary in position, sessile with much exerted corolla. The pod is 5–7.5 cm long with a long persistent beak. The seeds are brownish, oblong, and rhomboidal, with a deep furrow dividing them into two unequal lobes. They are contained, 10–20 together, in long, narrow, sicklelike pods.

Distribution: The plant is native to North Africa and countries bordering the eastern Mediterranean; fenugreek grows in open areas and is widely cultivated, notably, in India. The seeds are collected during autumn.

Traditional Medicinal Uses: Fenugreek seeds and leaves are used as food and medicine. Fenugreek is known in Ayurveda as methika and has been used in a number of Ayurvedic preparations for its carminative, galactagogue, and antidiabetic properties. The leaves are given internally for vitiated conditions of pitta. The leaves can be applied externally in the form of poultice for swelling, burns, boils, abscesses, and ulcers. The seeds are good for fever, vomiting, anorexia, cough, bronchitis, and colonitis. An infusion of the seed is a good cool drink for small pox patients. An aqueous extract of the seeds possesses

(a)

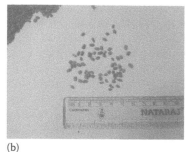
(b)

FIGURE 2.97   *Trigonella foenum-graecum.* (a) Herb. (b) Seeds.

an antibacterial property (Warrier 2002). Fenugreek has a long history of dubious indications, including fevers, colic, flatulence, dyspepsia, dysentery, cough, tuberculosis, edema, rickets, leg ulcers, gout, diabetes, and baldness.

Pharmacology

Antidiabetes: The seeds of fenugreek have been studied for their antidiabetic activity in animals as well as in humans. The activity is reported to be mainly due to saponins and high fiber content. Its regular consumption has been suggested to be beneficial in the management of diabetes and the prevention of atherosclerosis and coronary heart disease (review: Patil and Jain 2014).

The ethanol extract of *Trigonella foenum-graecum* seeds exhibited hypoglycemic effects in alloxan-induced diabetic rats. The most effective dose was 1 g/kg (Mowla et al. 2009). In another study, the water and ethanol extracts of fenugreek exhibited good antidiabetic activity in alloxan-induced diabetic rats (Rajarajeswari et al. 2012).

Fenugreek's major free amino acid, 4-hydroxyisoleucine, has been shown to stimulate insulin secretion from perfused pancreas *in vitro* (review: Patel et al. 2012a). Fenugreek seeds contain 4-hydroxy isoleucine in 2 diastero isomers: the major isomer being the (2S, 3R, and 4S) configuration and the minor isomer being the (2R, 3R, and 4S) configuration. The ability of the major isomer to stimulate glucose-induced insulin secretion in micromolar concentrations was shown (Sauvaire et al. 1998). The semisynthetic derivatives of this compound were reported to have more antidiabetic activity than the parental compound (Sridevi et al. 2014).

Trigonelline is a major alkaloid present in fenugreek. The isolated pure trigonelline (10 mg/kg twice a day for 4 weeks) exhibited a significant hypoglycemic effect in normal and alloxan-induced diabetic rabbits, but its effect was greater in diabetic animals (Al-Khateeb et al. 2012). In another study, administration of trigonelline to alloxan-induced diabetic rats helped protect β-cells from death and damage. Further, trigonelline treatment decreased intestinal α-amylase, maltase, and lipase; the treatment resulted in decrease in blood glucose, cholesterol, and TGs in diabetic rats. Trigonelline was also found to efficiently protect the liver and kidney functions of diabetic rats (Hamden et al. 2013).

An antihyperglycemic compound named GII was purified from the water extract of the seeds of *T. foenum-graecum* and shown to be different from trigonelline isolated earlier from the same plant. GII (50 mg/kg, p.o.) reduced blood glucose in glucose tolerance tests in subdiabetic and moderately diabetic rabbits. Treatment for 7 days of subdiabetic rabbits with GII (50 mg/kg, p.o.) improved glucose tolerance without reducing fasting blood glucose, which was nearly normal (Moorthy et al. 2010a). The mechanism of action of GII (100 mg/kg, p.o., for 15 days) seeds was studied in subdiabetic and moderately diabetic rabbits. GII seems to decrease the lipid content of the liver and stimulate the enzymes of glycolysis (except glucokinase) and inhibit the enzymes of gluconeogenesis in the liver of diabetic, especially moderately diabetic, rabbits (Moorthy et al. 2010b). In another study, administration of GII (50 mg/kg for 15 days) to subdiabetic and moderately diabetic rabbits or (50 mg/kg for 30 days) to SD rabbits corrected or almost normalized the altered serum lipids, tissue lipids, liver glycogen, enzymes of glycolysis, gly-coneogenesis, glycogen metabolism, polyol pathway, and antioxidant enzymes. Histopathological abnor-malities seen in the pancreas, liver, heart, and kidneys were normalized by the treatment. The compound increased serum insulin levels, increased sensitivity of tissues to insulin action, and stimulated the activ-ity of enzymes in glucose utilization (Puri et al. 2011). GII (50 mg/kg) brought down the elevated fasting blood glucose levels to normal in 12 (subdiabetic), 15 (moderately diabetic), and 28 days (severs diabetic) of treatment. It improved serum HbA1C and insulin levels also in these rabbits. Intermittent therapy once a week for 6 weeks with GII at the same dose brought down the fasting blood glucose levels to normal in the subdiabetic and in the moderately diabetic rabbits. After stopping therapy of the subdiabetic and moderately diabetic rabbits whose fasting blood glucose levels came to normal after treatment with GII 50 mg/kg, the values remained normal for 1 week and showed a tendency to increase only after 15 days (Puri et al. 2012).

GII, 4-hydroxyisoleucine, and trigonelline are known active compounds; there could be more bioac-tive molecules involved in the anti-DM activity of fenugreek. Details of the efficacy and long-term toxic-ity, if any, of high doses, in particular, including toxicity and efficacy of active principles individually and in combination should form priority area of future study.

Clinical Trials: Several human clinical studies have shown the usefulness of fenugreek seeds in the management of both type 1 and type 2 DM. Fenugreek has been shown to reduce fasting and PPG levels in diabetic patients. A metabolic study was carried out, and diets with or without 25 g fenugreek were given randomly to 10 non-insulin-dependent diabetics, each for 15 days, in a crossover design. An intravenous glucose tolerance test at the end of each study period indicated that fenugreek in the diet significantly reduced the area under the plasma glucose curve and half-life and increased the metabolic clearance rate. In addition, it increased erythrocyte IRs. Thus, fenugreek may exert its hypoglycemic effect by acting at the IR as well as at the gastrointestinal level (Raghuram et al. 1994).

In a study, defatted seeds (100 g/day for 10 days) significantly reduced fasting blood glucose levels, TG, total cholesterol, LDL, VLDL, and glucosuria in type 1 DM patients. In another study, patients with type 2 DM were placed on seeds (10 g/day) soaked in hot water (11 subjects) or mixed with yogurt (7 sub-jects). After 8 weeks, fasting blood glucose levels, TG, total cholesterol, and VLDL decreased in patients who received the seeds in soaked form (Kassaian et al. 2009). In another study also, consumption of the seeds (100 g/day for 10 or 20 days) led to improvement in glucose tolerance and decrease in the levels of blood glucose, TG, total cholesterol, and VLDL in type 2 diabetic patients. Administration of 20 g/day of the seeds to 20 type 2 DM patients for 14 weeks resulted in marked decrease in PPG levels. In a double-blind placebo trial, 46 type 2 DM patients were given sulfonylurea drug plus the seeds (in the form of a pill; 6 pills 3 times/day) or sulfonylurea drug plus placebo (23 patients). After 12 weeks, the combined therapy was found to be more effective. With a similar trial design, in another study, administration of 1 g/day hydroalcoholic extract of the seeds for 2 months to 12 newly diagnosed type 2 diabetic patients did not result in improvement in glucose tolerance and fasting blood glucose levels. The active principles in the soaked seed and the extract could differ (review: Ghorbani et al. 2013). In one study, a single dose of whole seeds, defatted seeds, gum isolate, and cooked seeds (but not degummed seeds) was also able to prevent the rise of plasma glucose after meal or glucose ingestion in nondiabetic subjects.

Unlike the seeds, the effect of *T. foenum-graecum* leaves on reducing blood glucose level was not con-sistent. They differed in two different studies. This discrepancy may be due to the methodological issues such as a difference in the methods of extract preparation (review: Ghorbani et al. 2013).

Other Activities: Other reported activities include antioxidant activity, anti-inflammatory activity, and antineoplastic effect. An aqueous fraction of fenugreek has been reported to have the highest antioxidant activity. The seed extract at high concentrations has been shown to act as a scavenger of free radicals. Fenugreek leaf extract possessed anti-inflammatory as well as antipyretic properties (Ahmadiani et al. 2001). The alcohol extract of the seed inhibited the growth of Ehrlich ascites carcinoma cells in mice. Fenugreek seed extract also significantly inhibited the 7,12-dimethylbenz(α)anthracene-induced mam-mary hyperplasia in rats (review: Patil and Jain 2014).

Toxicity: There is little evidence to suggest that the spice is toxic or that it has significant anticoag-ulant or hormonal effects. No acute toxicity was observed for the ethanol extract of the seeds when administered orally at a high dose of 3 g/kg body weight (Mowla et al. 2009). Its effect on other diseases such as arthritis and safe dose levels is to be studied. Prolonged use of this plant seed may result in joint pain or arthritis (author's observation on two cases).

Phytochemicals: The ethanol extract of the seeds showed the presence of alkaloids, steroids, and carbohydrates (Mowla et al. 2009). The seeds of *T. foenum-graecum* contain sapogenins such as diosgenin, hederagin, tigogenin, neotigogenin, yuccagenin, gitogenin, smilagenin, sarsasapogenin, and yamogenin together with glycosides like foenugracein; trigoneosides 1a, 1b, 11a, 111a, IVa, Vb, VI, VIIb, and VIIIb; trigofoenosides A, B, C, and D; and fenugrin B. Several coumarin compounds have been identified in the seeds, namely, 3,4,7-trimethyl coumarin, 4-methyl coumarin, and trigocoumarin. The seeds also contain several alkaloids such as trigonelline, gentianine, carpaine, and C-type glycoside flavones such as vitexin, isoorientin, isovitexin, saponaretin, and tricin. Minor steroidal sapogenin such as smilagenin, sarsasapogenin, and yuccagenin were also identified in the seeds (review: Patil and Jain 2014).

Triplochiton scleroxylon K. Schum, Malvaceae

Triplochiton scleroxylon is used in the rural areas of Nigeria to treat diabetes. Exposure of normal and streptozotocin-induced diabetic rats to the water or 50% ethanol extracts of *T. scleroxylon* bark (200 mg/kg, p.o.) for 28 days resulted in marked decrease in the levels of plasma glucose and MDA concentrations in both normal and streptozotocin-induced diabetic rats. However, the water extract demonstrated a greater reduction in plasma glucose and MDA concentrations than the 50% ethanol extract (Proph and Onoagbe 2012). The extracts exhibited an antihyperlipidemic effect also. In another study, almost the same results were reported wherein the water and 50% ethanol stem bark extracts of *T. scleroxylon* (200 mg/kg daily for 28 days) attenuated hyperglycemia and hyperlipidemia in streptozotocin-induced diabetic rats (Prohp et al. 2012).

In a recent report, oral administration of the water and ethanol extracts of *T. scleroxylon* bark (200 mg/kg twice daily for 28 days) to normal and streptozotocin-induced diabetic rats resulted in significant lowering of serum cholesterol, TGs, and phospholipid concentrations in normal and diabetic rats compared to control and diabetic control animals. The extracts did not cause adverse histological changes in rat livers. The water extract was more effective in reducing the sternness of necrosis in the kidneys than the ethanol extract (Proph et al. 2014).

Triticum aestivum L. (*Trt. hybernum* L.), Gramineae

Synonyms: *Triticum vulgare* Vill., *Triticum sativum* Lam., *Triticum sphaerococcum* Percival

Triticum aestivum is the common bread wheat. The ethanol extract of the young plant (wheat grass), when administered daily for 30 days at a dose of 100 mg/kg, p.o., exhibited marked antihyperglycemic, antihypolipidemic, and antioxidant activities in streptozotocin-induced diabetic rats (Mohan et al. 2013c). The levels of total cholesterol, TGs, LDL, and VLDL were reduced and the HDL level was increased by the treatment with the ethanol extract (Mohan et al. 2013c). However, a report suggests the possible involvement of a protein from wheat in the immune-system mediated development of type 1 DM (MacFarlane et al. 2003).

Triticum repens L., Gramineae

The hypoglycemic effect of the water extract of *Triticum repens* rhizomes was investigated in normal and streptozotocin-induced diabetic rats. After a single oral administration of the water extract (20 mg/kg), a highly significant decrease on blood glucose levels in diabetic rats was observed; the blood glucose levels were normalized after 2 weeks of daily oral administration of the extract. Significant reduction on the blood glucose levels were noticed in normal rats after both acute and chronic treatment. However, no changes were observed in basal plasma insulin concentrations after treatment in either normal or diabetic rats indicating that the underlying mechanism of this pharmacological activity seems to be independent of insulin secretion (Eddouks et al. 2005d). Besides, the water extract of the *T. repens* rhizome exhibited lipid- and body weight-lowering activities in severe streptozotocin-induced hyperglycemic rats after repeated oral administration of the extract at a dose of 20 mg/kg. A strong decrease in cholesterol level was observed 6 h after a single oral administration of the water extract also (Maghrani et al. 2004b).

Tulbaghia violacea Harv., Amaryllidaceae

Common Names: Society garlic or pink agapanthus
Tulbaghia violacea is used in South Africa in traditional medicine to treat DM. The ethanol extract of *T. violacea* produced a marked increase in glucose utilization in C2C12 muscle cells. However, further *in vivo* studies are required to confirm its antidiabetic property (Huyssteen et al. 2011). In an MSc thesis, an organic extract of *T. violacea* was shown to improve glucose-stimulated insulin secretion in INS-1 pancreatic β-cells and glucose uptake in Chang liver cells. The extract had no effect on the glucose uptake in 3T3-L1, an adipose cell line, and reduced glucose utilization in C2C12, a skeletal muscle cell line. It was observed that the extract increased the membrane potential and GLUT-2 expression in INS-1 cells cultured at hyperglycemic levels; however, at normoglycemic levels, a reduction was observed. The oxygen consumption increased at both glycemic levels due to treatment with the extract (Davison 2009).

Turnera diffusa Willd., Turneraceae

Synonyms: *Turnera aphrodisiaca* Ward, *Turnera microphylla*
Turnera diffusa is a Mexican medicinal plant used to control diabetes. The aerial part of the *T. diffusa* water–ethanol extract was evaluated for its blood glucose-lowering activity in normal and alloxan-induced diabetic mice. The extracts of *T. diffusa* did not show any hypoglycemic activity (Alarcon-Aguilar et al. 2002b). Although the tested extracts showed negative results, the plant powder and other extracts are to be tested. Further, a delayed effect of the plant cannot be ruled out.

Tussilago farfara L., Asteraceae

Tussilago farfara is commonly known as coltsfoot. It traditionally had medicinal uses. The flower buds of *T. farfara* inhibited α-glucosidase (review: El-Abhar and Schaalan 2014). However, this plant is reported to have hepatotoxic pyrrolizidine alkaloids.

Uncaria tomentosa (Willd.) DC, Rubiaceae

In a study, the immunomodulatory potential of *Uncaria tomentosa* aqueous ethanol extract on the progression of immune-mediated diabetes was evaluated. The extract was effective to prevent the progression of immune-mediated diabetes in mice by distinct pathways (Domingues et al. 2011).

Urtica dioica L., Urticaceae

Common Name: Nettle
Urtica dioica has been used in traditional medicine as an antihypertensive, antihyperlipidemic, and antidiabetic herbal medicine. The effect of the hydroalcoholic extract of *U. dioica leaf* on fructose-induced insulin-resistant rats was studied. The extract (50, 100, and 200 mg/kg) significantly decreased serum glucose, insulin, LDL and leptin, LDL/HDL ratio, and insulin resistance index in fructose-fed insulin-resistant rats (Ahangarpour et al. 2012). The authors suggest that *U. dioica* extract may be useful in improving type 2 DM.

In another study, the effect of dried *U. dioica* leaf alcohol and water extracts on the number and diameter of the islets and histological parameters in streptozotocin-induced diabetic rats were evaluated. The pancreas from diabetic rats showed injury of the pancreatic tissue, while the pancreas from diabetic rats treated with dried *U. dioica* leaf extracts for 8 weeks showed slight to moderate rearrangement of islets. Thus, *U. dioica* leaf alcohol and water extracts could cause repair of pancreatic tissue in streptozotocin-induced diabetic experimental model (Qujeq et al. 2013).

The effect of the water extract of *U. dioica* on the glycemic status, body weight, and lipidemic status was evaluated in type 2 diabetic rats. The water extract of *U. dioica* leaves (1.25 g/kg daily for 14 days) markedly lowered fasting serum glucose levels in streptozotocin-induced type 2 diabetic rats. There was a significant increase in the body weight of the treated group in comparison to the diabetic control group. The favorable changes observed on the lipid profile fell short of statistical significance (Das et al. 2009). In another study, treatment with the water extract was continued for 28 days. In this case, cholesterol levels were significantly lowered in the extract-treated diabetic rats (Das et al. 2011b).

Urtica parviflora Roxb., Urticaceae

The water extract of the leaves of *Urtica parviflora* exhibited hypoglycemic effects in fasted normal rats. Further, the extract improved oral glucose tolerance in glucose-loaded rats (Sah et al. 2010). Follow-up studies on diabetic animals are warranted.

Urtica pilulifera L., Urticaceae

Lectin isolated from *Urtica pilulifera* seeds (100 mg/kg, i.p., for 30 days) exhibited significant hypoglycemic activity in streptozotocin-induced diabetic rats (Kavalal et al. 2003).

Uvaria rufa Blume, Annonaceae

Preliminary screening of the ethyl acetate extract of the leaves of *Uvaria rufa* indicated its AGE inhibitory activity. Fractionation of the extract yielded five flavonol glycosides including isoquercitrin and its 6-acetate derivative. Isoquercitrin and its 6-acetate derivative inhibited the formation of AGEs in the bovine serum albumin–glucose assay with 50% inhibitory concentrations of 8.4 and 6.9 µM, respectively (Deepralard et al. 2009).

Vaccinium angustifolium Ait., Ericaceae

Synonyms: *V. brittonii* Porter ex. C. Bicknell, *V. angustifolium* Aiton var. *hypolasium* Fernald, *V. angustifolium* Aiton var. *laevifolium* House, *V. angustifolium* Aiton var. *nigrum* (A.W.Wood) Dole, *Vaccinium brittonii* Porter ex E.P. Bicknell, *Vaccinium lamarckii* Camp, *Vaccinium nigrum* (A.W.Wood) Britton, *Vaccinium pensilvanicum* Lam., *Vaccinium pensilvanicum* Lam. var. *angustifolium* (Aiton) A. Gray, *Vaccinium pensilvanicum* Lam. var. *nigrum* A.W. Wood (Figure 2.98)

Common Names: Lowbush blueberry, early lowbush blueberry, low sweet blueberry, etc.

Description: Early lowbush blueberry is an erect, low-growing, variable shrub that reaches 50–60 cm in height. It usually forms dense, extensive colonies. The roots are shallow and fibrous but may possess a taproot, which can extend to 1 m in depth. Woody rhizomes average 4.5 mm in diameter and 6 cm in depth. The leaves are deciduous, smooth, and narrowly elliptical with tiny, sharp teeth. The flowers are borne in short, few-flowered terminals or axillary racemes. The fruit is a globular berry averaging 4–11 mm in diameter.

Distribution: *V. angustifolium* is extensively harvested from cultivated and wild plants in New England (especially Maine) and in Quebec and the Canadian Maritime Provinces. It can be grown as a potted patio plant or as a foreground plant in the edible garden or landscape. Very few historical records exist

FIGURE 2.98   *Vaccinium angustifolium.*

on ancient blueberry culture in Greek and Roman empires. Those cultures did use parts of the blueberry plants and fruit to eat or to treat ailments.

Traditional Medicinal Uses: Various members of the *Vaccinium* genus have been used in traditional medicine to treat diabetes (Martineau et al. 2006).

Pharmacology

Antidiabetes: Various members of *Vaccinium* genus are reported to possess anti-DM activity. A survey identified *V. angustifolium* (Canadian lowbush blueberry) as one of the most highly recommended traditional anti-DM plants (Martineau et al. 2006). The root, stem, and leaf extracts significantly enhanced glucose transport in cultured C2C12 cells by 15%–25% in the presence as well as absence of insulin after 20 h of incubation; no enhancement resulted from 1 h exposure. In 3T3 cells, only the root and stem extracts enhanced uptake and this effect was grater after 1 h than after 20 h; the uptake was increased up to 75% in the absence of insulin. Glucose-stimulated insulin secretion was potentiated, to a small extent, in growth-arrested β-TC-tet cells incubated overnight with the leaf or stem extract. Interestingly, the fruit extracts were found to increase ^{3}H-thymidine incorporation in replicating β-TC-tet cells by 2.8-fold. The stem leaf and fruit extracts reduced apoptosis by 20%–33% in PC12 cells exposed to elevated levels of glucose for 96 h. Further, lipid accumulation in differentiating 3T3-L1 cells was increased by root stem and leaf extracts by as much as 6.5-fold by a 6-day period. The authors conclude that *V. angustifolium* contains several active principles with insulin-like and glitazone-like properties (Martineau et al. 2006).

Anthocyanins from blueberry have alleviated symptoms of hyperglycemia in diabetic C57b1/6J mice. The antidiabetic activity of different anthocyanin-related extracts was evaluated using the pharmaceutically acceptable self-microemulsifying drug delivery system Labrasol. Treatment with a phenolic-rich extract and an anthocyanin-enriched fraction (500 mg/kg, p.o.) formulated with Labrasol lowered elevated blood glucose levels in diabetic mice by 33% and 51%, respectively. The extracts were not significantly hypoglycemic when administered without Labrasol, most likely due to an increase in the bioavailability of the administered preparations. The phenolic-rich extract contained 287 mg/g anthocyanins, while the anthocyanin-enriched fraction contained 595 mg/g (cyanidin-3-glucoside equivalents). The greater hypoglycemic activity of the anthocyanin-enriched fraction compared to the initial phenolic-rich extract suggested that the activity was due to the anthocyanin components. Treatment with the pure anthocyanins (300 mg/kg), delphinidin-3-*O*-glucoside and malvidin-3-*O*-glucoside, formulated with Labrasol, showed that malvidin-3-*O*-glucoside was very significantly hypoglycemic while delphinidin-3-*O*-glucoside was not (Grace et al. 2009b).

Chlorogenic acid is reported from the blueberry fruit (Rodriguez-Mateos et al. 2012). Chlorogenic acid is a known antidiabetes molecule.

Clinical Trial: A study on diabetic patients reported in 1928 showed the beneficial effects of blueberry leaf extract in certain cases of diabetes and its action was not consistent. The utility of the extract was more pronounced in the case of mild diabetes in elderly patients (Wilson 1928). Blueberries decreased insulin resistance in a human clinical trial. The study showed that consuming blueberry for 6 weeks (twice a day) improved insulin sensitivity in obese insulin-resistant type 2 DM patients (Stull et al. 2010).

Other Activities: The plant is reported to have a promising anticancer activity (review: Johnson and Arjmandi 2013). Other reported activities include improvement in memory in old rats, improvement in age-related declines in neuronal signal transduction, antihypertension (Rodriguez-Mateos et al. 2013), and anti-inflammation (Esposito et al. 2014).

Phytochemicals: Proanthocyanidin oligomers (delphinidin-3-*O*-glucoside and malvidin-3-*O*-glucoside), flavanol oligomers, phenolics, chlorogenic acid, and pterostilbene were reported from blueberry (Rodriguez-Mateos et al. 2012).

Vaccinium arctostaphylos L., Ericaceae

Common Names: Broussa tea, Caucasian whortleberry, cranberry (Eng.), qaraqat (in Iran)

Description: *Vaccinium arctostaphylos* is a deciduous tree. The flowers are hermaphroditic (having both male and female organs) and are pollinated by insects. The fruit is juicy with a slightly acid flavor; it makes an acceptable fruit to nibble on. The pear-shaped fruit is small, about 8–10 mm in size.

Distribution: It is a native of Asia (temperate region). The plant occurs in Iran, Turkey, Armenia, Azerbaijan, Georgia, Russian Federation, Europe, etc. It is also cultivated.

Traditional Medicinal Uses: *V. arctostaphylos* is used in traditional medicine for the treatment of DM in Iran and elsewhere. In traditional Iranian medicine, the decoction from the berries has been used as an antidiabetic and antihypertensive agent for a long time (Soltani et al. 2014).

Antidiabetes: The extract obtained from *V. arctostaphylos* berries showed an inhibitory effect on pancreatic α-amylase *in vitro*. Activity-guided purification of the extract led to the isolation of malvidin-3-*O*-β-glucoside as the amylase inhibitor (Nickavar and Amin 2010). The ethanol extract of the fruits showed antihyperglycemic, antioxidant, and TG-lowering activities in alloxan-induced diabetic rats (Feshani et al. 2011). In another study, oral administration of *V. arctostaphylos* fruit and leaf extracts (250, 500, and 1000 mg/kg daily for 2 months) significantly reduced the fasting glucose and HbA1c levels but significantly increased the insulin levels without any significant effects on serum GOT, GPT, and creatinine levels in alloxan-induced diabetic rats compared with diabetic control rats (Kianbakht and Hajiaghaee 2013). Anthocyanin glycoside isolated from this plant inhibited pancreatic α-amylase (Nickavar and Amin 2010). The anthocyanin protected β-cells from oxidative stress and increased secretion of insulin; further, it improved insulin resistance (Sancho and Pastore 2012).

Clinical Trial: In a randomized, double-blind, placebo-controlled clinical trial consisting of 37 patients aged 40–60 years with type 2 diabetes who were resistant to conventional oral antihyperglycemic drugs, the hydroalcoholic extract of *V. arctostaphylos* fruit (1 capsule = 350 mg every 8 h for 2 months) in combination with antihyperglycemic drugs lowered the blood levels of fasting glucose, 2 h postprandial glucose, and HbA1c without any significant effects on the liver/kidney function compared with the placebo group. No adverse effects were reported (Kianbakht et al. 2013).

Other Activities: In addition to its antidiabetic property, the fruit extract of *V. arctostaphylos* exhibited antilipidemic and antioxidative properties in a randomized, double-blind, placebo-controlled clinical trial in hyperlipidemic adult patients. The fruit extract of *V. arctostaphylos* has beneficial effects on the serum lipid profile and oxidative stress in hyperlipidemic adult patients (Soltani et al. 2014). Other reported activities include antimicrobial effects, antihyperlipidemic property, hypotensive activity without changing the heart rate, and antioxidant activity (Saral et al. 2014).

Phytochemicals: The presence of flavonol glycosides, coumarins, phenolic acids, and their derivatives in the leaves was reported. Phenolic acids and their derivatives in unripe fruits and essential oil in the flowering aerial parts were also reported. Anthocyanins, delphinidin 3-*O*-β-glucoside, petunidin-3-*O*-β-glucoside, and malvidin-3-*O*-β-glucoside were reported from the berry (Nickavar and Amin 2004, 2010).

Vaccinium bracteatum Thunb., Ericaceae
Vaccinium bracteatum is a traditional Chinese herbal medicine. Oral administration of *V. bracteatum* leaf (water extract) for 4 weeks to streptozotocin-induced diabetic mice resulted in partial recovery of body weight and marked reduction in blood glucose levels compared to diabetic control mice. The treatment also increased blood insulin levels and improved the lipid profile in diabetic mice (Wang et al. 2010c). The polysaccharide isolated from the leaves of this plant exhibited a dose-dependent decrease in the blood glucose level in streptozotocin-induced diabetic mice. Further, the treatment marginally increased the levels of insulin in the blood. The treatment improved the lipid profile also in diabetic rats (Wang et al. 2013).

Vaccinium myrtillus L., Ericaceae
Synonyms: *V. angelosums* Dulac, *V. montanum* Salisb., *Myrtillus niger* Gilib. (monograph: WHO 2009)

Common Names: Bilberry, blueberry

Vaccinium myrtillus is a trailing shrub forming large colonies from creeping rhizomes; the fruits and other parts are used in traditional medicine (monograph: WHO 2009). *V. myrtillus* leaf is one of the most frequently used antidiabetic remedies of plant origin before the discovery of insulin. During the last century, many animal, clinical, and phytochemical studies have been undertaken with this plant and its extracts. Overall, it must be concluded that the results were more or less disappointing and could not support the traditional use of bilberry leaves against DM, which is sometimes recommended even up to the present day (Helmstadter and Schuster 2010). A recent study has shown that a combined extract obtained from *V. myrtillus* leaves and *Humulus lupulus* cone exhibited hypoglycemic and cholesterol-lowering properties in streptozotocin-induced diabetic rats (Bubueanu and Campeanu 2014).

Vaccinium vitis-idaea L., Ericaceae
Common Names: Lingonberry, mountain cranberry
In Scandinavian countries, *Vaccinium vitis-idaea* berries are used to prepare traditional food and herbal remedies. The antidiabetic activity of this plant has been reviewed recently (Eid and Haddad 2014). The berry ethanol extract inhibited hepatic glucose-6-phosphatase, an effect coupled to increased AMPK phosphorylation. *V. vitis-idaea* was also as potent as metformin at enhancing basal or insulin-stimulated glucose uptake in cultured myocytes. Recently, this effect was confirmed to implicate the enhanced translocation of GLUT4 transporters to the myocyte cell surface and was found to involve the activation of AMPK. An activity-guided fractionation procedure was guided by muscle cell glucose uptake resulting in the isolation and identification of quercetin-3-*O*-galactoside, quercetin, and quercetin-3-*O*-glucoside as the active constituents. Similar to the crude berry extract, the active quercetin glycosides and the aglycone also stimulated the AMPK pathway. To validate the effects of *V. vitis-idaea*, the berry extract was administered to KK-Ay as well as DIO mice. In both models, the plant extract reduced glycemia and increased skeletal muscle GLUT4 protein expression. In the DIO mouse model, this was accompanied by enhanced Akt and AMPK phosphorylation in both skeletal muscle and liver. In the liver, the berry extract significantly reduced TG accumulation and hence hepatic steatosis. Interestingly, in KK-Ay animals, the body weight was significantly reduced by the treatment and was associated with a decrease in food intake. Unlike DIO animals, KK-Ay mice have a defect in the agouti-related peptide (ARP) pathway involved in the feeding behavior and that causes the hyperphagia forming the basis of the diabetic status of this model. Thus, *V. vitis-idaea* could reduce hepatic glucose production and increase muscle glucose disposition, both of which forecast a positive action to enhance insulin sensitivity in diabetic patients (Eid and Haddad 2014).

Valeriana edulis Nutt, Caprifoliaceae
Extracts from *Valeriana edulis* root lowered blood glucose in experimental animals (Marles and Farnsworth 1995).

Valeriana mexicana DC, Caprifoliaceae
Extracts from *Valeriana mexicana* root exhibited hypoglycemic activity in experimental animals (Marles and Farnsworth 1995).

Valeriana officinalis L., Caprifoliaceae
The root extract of *Valeriana officinalis* showed hypoglycemic activity in experimental animals (Marles and Farnsworth 1995).

Verbesina crocata Less, Asteraceae/Compositae
Extracts of the leaf and flower of *Verbesina crocata* contain daucosterol, galegine, and lupeol and lupeol acetate. These compounds decrease blood glucose levels (Marles and Farnsworth 1995).

Verbesina persicifolia DC, Compositae
Oral administration of 100 and 150 mg/kg of the chloroform extract of *Verbesina persicifolia* produced a significant hypoglycemic effect in normal as well as in diabetic mice and rats. In addition, the extracts altered glucose tolerance in alloxan-induced diabetic rats, enhanced glucose uptake in skeletal muscles, and significantly inhibited glycogenolysis in the liver. These results indicate that the hypoglycemic effect may be similar to tolbutamide (Perez et al. 1996).

Vernonia amygdalina Del., Asteraceae

Vernonia amygdalina is a small shrub with medicinal properties. In an oral glucose tolerance test, 400 mg/kg of the ethanol extract of *V. amygdalina* exhibited a significant improvement in the glucose tolerance of streptozotocin-induced diabetic rats. A 28-day treatment with 400 mg/kg of the extract resulted in a 32.1% decrease in the fasting blood glucose compared to diabetic control. The treatment also caused significant decrease in TG and total cholesterol levels. Besides, it showed a protective effect over pancreatic β-cells against streptozotocin-induced damage, causing a slight increase in the insulin level compared to the diabetic control. The extract administration also showed positive regulation of the antioxidant system, both enzymatic and nonenzymatic. Furthermore, the extract was found to increase the expression of GLUT4 (24%) in rat skeletal muscle. Further, tissue fractionation revealed that it can increase the GLUT4 translocation (35.7%) to plasma membrane. This observation is in line with the restoration in skeletal muscle glycogenesis of the extract-treated group. However, no alteration was observed in GLUT1 expression. In addition, the extract treatment suppressed (40% inhibition) one of the key hepatic gluconeogenic enzymes, glucose-6-phosphatase (Ong et al. 2011). In another study, oral administration of the ethanol extract of *V. amygdalina* at a dose of 400 mg/kg daily for 42 days resulted in normalization of blood glucose levels of streptozotocin-induced diabetic rats. In the plant extract-treated groups, hepatic microanatomy was comparable to that of normal control rats (Akinola et al. 2009). In another report, the combinations of the water extract of the leaves of *V. amygdalina* and metformin caused more reduction in glycemia compared to any of the agents acting alone in both normal and alloxan-induced diabetic rats (Michael et al. 2010). The ethanol extract of *V. amygdalina* contained a high level of polyphenols, mainly 1,5-dicaffeoylquinic acid, dicaffeoylquinic acid, chlorogenic acid, and luteolin-7-*O*-glucoside. These compounds include known antidiabetic compounds.

Vernonia anthelmintica (L.) Willd., Asteraceae

Common Names: Ironweed, pinyin (in China), purple fleabane (Eng.), karinjeeragum (Malayalam)
In traditional medicine, *Vernonia anthelmintica* is used as an anthelmintic medicine and is also used for the treatment of asthma, hiccup, inflammatory swellings, sores, and itching of the eye (review: Srivastava et al. 2014).

Ethanolic extracts of the seeds of this plant (500 mg/kg) produced the maximum fall of the blood glucose levels in streptozotocin-induced diabetic rats after 6 h of treatment. Bioassay-guided fractionation of the ethanol extract resulted in five fractions. One of them (A2) at a dose of 100 mg/kg showed maximum antihyperglycemic activity that is significantly higher than that of glibenclamide (20 mg/kg). Administration of this active fraction for 45 days (100 mg/kg) resulted in marked reduction in plasma glucose, HbA1c, cholesterol, TGs, LDL, VLDL, free fatty acids, phospholipids, and HMG-CoA reductase in streptozotocin-induced diabetic rats. Besides, the levels of plasma insulin, protein, HDL, and liver glycogen content were restored to normal levels in treated diabetic rats (Kumar et al. 2009a). Essentially the same results have been reported again with the same title (Fatima et al. 2010). This plant appears to be very promising for detailed studies.

Other reported pharmacological properties include anti-inflammatory, antiarthritic, antimicrobial, diuretic, analgesic, and antipyretic activities (Srivastava et al. 2014). Nine highly oxygenated stigmastane-type steroids and two stigmastane-type steroidal glycosides were isolated from the aerial parts of this plant. Further, vernoanthelsterone A (a steroid) and five other steroids were isolated from this plant (Srivastava et al. 2014).

Vernonia colorata Schreb., Compositae

The leaf (acetone extract) of *Vernonia colorata* showed hypoglycemic activity in normal rats and antidiabetic activity in alloxan-induced diabetic rats (Sy et al. 2005).

Viburnum foetens Decne., Caprifoliaceae

The ethanol extract of *Viburnum foetens* stimulated insulin release from INS-1 cells in the presence of 5.5 mM glucose. Insulin secretagogue activity of the extract was found at 40 μg/mL (Hussain et al. 2004).

***Viburnum lantana* L.**, Caprifoliaceae

Common name: Wayfaring tree
The extracts of *Viburnum lantana* exerted high antioxidant activity. Furthermore, the plant is rich in the anti-diabetic compound chlorogenic acid (Erdogan-Orhan et al. 2011).

***Viburnum opulus* L.**, Adoxaceae (Formerly Caprifoliaceae)
Viburnum opulus is an ornamental as well as medicinal plant. *V. opulus* fruit (water extract, 100 mg/kg) showed anti-DM activity in alloxan-induced diabetic rats (Sharma and Arya 2011). Further, this plant is rich in chlorogenic acid, an antidiabetic compound (Erdogan-Orhan et al. 2011).

***Vigna trilobata* (L.) Verdc.**, Fabaceae

Synonyms: *Phaseolus trilobatus* (L.) Schreb, *Phaseolus trilobus* Ait., *Dolichos trilobatus* L.
Repeated oral administration of the methanol extract of the root of *P. trilobus* (400 mg/kg) to streptozotocin-induced diabetic rats resulted in a significant increase in body weight and decrease in the blood glucose level of diabetic rats, which is comparable to that of the standard drug glibenclamide (Kaur et al. 2012). Acute toxicity studies of the dried extract of roots at an orally administered dose of 2000 mg/kg showed neither lethality nor any adverse reaction and behavioral change, indicating the dose to be safe (Kaur et al. 2012).

***Vigna unguiculata* (L.) Walp.**, Fabaceae

Common Name: Cowpea
Vigna unguiculata seed is used in traditional medicine. The anti-DM effect of *V. unguiculata* seed oil was evaluated in alloxan-induced diabetic rats. The seed oil (200 mg/kg for 21 days) reduced the levels of blood glucose, cholesterol, LDL, and TG and increased the level of HDL in alloxan-induced diabetic rats. Further, the treatment decreased the levels of liver function marker enzymes in the blood suggesting improvement in liver function (Ashraduzzaman et al. 2011).

***Vinca erecta* Regel & Schmalh**, Apocynaceae
Vinca alkaloids have been reported to have anti-DM activity. Extracts from *Vinca erecta* (whole plant) have anti-DM activity (Marles and Farnsworth 1995).

***Vinca minor* L.**, Apocynaceae

Common Name: Periwinkle
Vincamine, an alkaloid found in the leaves of this plant, is reported to have hypoglycemic and other properties such as pronounced cerebrovasodilatory and neuroprotective activity (De and Saha 1975; Farahanikia et al. 2011).

***Viscum album* L.**, Viscaceae

Common Name: African mistletoe
The leaf and stem of this plant showed anti-DM activity including antihyperlipidemic activity in streptozotocin-induced diabetic rats (Adaramoye et al. 2012).

***Vitex doniana* Sweet.**, Lamiaceae/Labiatae
Vitex doniana is a tropical medicinal plant endowed with important pharmacological properties. The effect of the water extract of *V. doniana* leaves on alloxan-induced DM in rats was evaluated. The leaf extract (50 and 100 mg/kg, i.p.) caused remarkable decreases in blood sugar levels; at 50 mg/kg dose, the leaf extract reduced blood glucose levels from 27.38 to 4.72 mmol/L after 4 days. The leaf extract, at both doses tested, was more potent than the reference drug glibenclamide (0.3 mg). The extract exhibited *in vivo* antioxidant activity also. Quantitative phytochemical analyses of the extract showed the presence of saponins, flavonoids, alkaloids, and cardiac glycosides (Ezekwesili et al. 2012). Further studies are warranted on this promising plant.

Vitex lucens Kirk, Lamiaceae

Synonym: *Vitex littoralis* A. Cunn.

Vitex lucens is an evergreen tree endemic to New Zealand and contains vitexin and p-hydroxy benzoic acid, antidiabetic compounds (Ghisalberti 2005).

Vitex negundo L., Lamiaceae/Verbenaceae

In traditional medicine, *Vitex negundo* alone or in combination with other medicinal plants is used for the treatment of various diseases. The water and ethanol extracts of *V. negundo* leaf showed an antidiabetic activity in alloxan-induced diabetic rats; the water extract showed better activity than the ethanol extract (Prasannaraja et al. 2012).

Idopyranose isolated from *V. negundo* exhibited antidiabetic activity against streptozotocin-induced diabetic rats. Interestingly, liver, kidney, and pancreatic sections of diabetic mice fed with the isolated 1, 2 disubstituted idopyranose showed regeneration of hepatocytes, nephrocytes, as well as β-cells, and the acinar region appeared normal with increased numbers of β-cells. The expression of the nuclear factor-kappa B (NF-κB) and inducible nitric oxide synthase was increased in diabetic rats, and treatment with idopyranose inhibited its expression (Manikandan et al. 2011). Iridoid glucoside, isolated from the leaves of *V. negundo*, exhibited antihyperglycemic effects in streptozotocin-induced diabetic rats (Sundaram et al. 2012). The levels of blood glucose, plasma, and tissue glycoproteins such as hexose, hexosamine, fucose, and sialic acid were significantly increased, whereas plasma insulin levels were significantly decreased in streptozotocin-induced diabetic rats. Oral administration of iridoid glucoside (50 mg/kg, once daily) to diabetic rats for 30 days reversed the aforementioned hyperglycemia-induced biochemical changes to near normal levels (Sundaram et al. 2012). Chloroform, ethyl acetate, and n-butanol extracts of *V. negundo* leaf showed significant α-amylase inhibition *in vitro*; out of these three extracts, chloroform extracts showed a better inhibitory action (Devani et al. 2013). The plant possesses more than one active principle and mechanisms of action. Further studies on this plant may be rewarding.

Vitis vinifera L., Vitaceae

The grape-skin extract inhibited α-glucosidase activity and suppressed postprandial glycemic response in streptozotocin-induced diabetic mice (Zhang et al. 2011). The grape skin extract activated the insulin-signaling cascade and reduced hyperglycemia in alloxan-induced diabetic mice. The IR content and Akt phosphorylation were significantly greater in the extract-treated diabetic mice (gastrocnemius muscles) compared with the untreated alloxan-induced diabetic mice. Further, the treatment improved the glucose transporter (GLUT-4) content. The extract treatment did not change glucose-induced insulin secretion in isolated pancreatic islets (Soares de Moura et al. 2012). Ellagic acid, epicatechin gallate, and flavonoids present in grapes activated PPAR-γ (Wang et al. 2014). The grape seed extract could play a role in the management of peripheral neuropathy, similar to other antioxidants known to be beneficial for diabetic peripheral neuropathy (Jin et al. 2013).

The antidiabetes and antioxidant activities of this plant leaf (ethanol extract; 250 mg/kg) in streptozotocin-induced diabetic rats have been reported (Sendogdu et al. 2006). In another study, the anti-DM activities of the chloroform and ethanol extracts of the *Vitis vinifera* stem bark were investigated. Fasting blood glucose levels were reduced significantly in diabetic rats upon treatment with the stem bark extracts (200 mg/kg). Pretreatment with plant extracts improved the SOD, catalase, and peroxidase levels and reduced LPO in diabetic animals comparable to the standard drug-treated group of animals (Yarapa et al. 2012).

Vittadinia australis A. Rich, Asteraceae

The root extract of this plant has been reported to decrease blood glucose in rodents (Marles and Farnsworth 1995).

Wattakaka volubilis (L.f.) Stapf., Asclepiadaceae

Wattakaka volubilis is a traditional medicinal plant in India used to treat various diseases such as diabetes and inflammation and is utilized as an analgesic. The methanol extract of *W. volubilis* (50, 100, and 200 mg/kg daily for 28 days) showed hypoglycemic and antihyperlipidemic effects in alloxan-induced diabetic rats. An acute oral toxicity study in rats with a high dose (1g/kg) did not show any toxicity (Arunkumar et al. 2010). A recent study confirmed the antidiabetic and antioxidant property of the plant leaf. The ethanol extract of *W. volubilis* leaves (150 mg/kg daily for 14 days) lowered blood glucose levels and decreased the activities of serum transaminases and ALP in alloxan-induced diabetic rats compared to diabetic control rats. Further, the treatment increased serum insulin levels and antioxidant enzymes (Vishnusithan et al. 2014).

Wedelia paludosa DC, Asteraceae (Present Name: *Acmella brasiliensis* Spreng.)
The extract of the Brazilian medicinal plant *Wedelia paludosa* (*Acmella brasiliensis*) lowered blood glucose levels in alloxan-induced diabetic rats (Novaes et al. 2001). Kaurenoic acid, a bioactive diterpene, was isolated from the leaves, flowers, stems, and roots of *W. paludosa*. The concentration of this compound was higher in the roots and stems during the autumn season. The pharmacological evaluation showed that kaurenoic acid is responsible, at least in part, for the hypoglycemic potential detected in this plant (Bresciani et al. 2004).

Weigela subsessilis L.H.Bailey, Caprifoliaceae
Weigela subsessilis is a Korean endemic plant. The water extract of this plant is used as an *anti-DM* in traditional medicine. 24-Norursane triterpenes, ilekudinols A and B, isolated from the methanol extract of the leaves and stems of *W. subsessilis*, inhibited PTP1B activity (Jiany et al. 2012).

Withania coagulans Dunal, Solanaceae
The water extract of *Withania coagulans* (fruit) partially reversed nicotinamide–streptozotocin-induced diabetes in rats (Shukla et al. 2012b).

Withania somnifera (L.) Dunal, Solanaceae
Synonyms: *Physalis somnifera* L. (monograph: WHO 2009)

Common Names: Winter cherry (Eng.), ashwagandha (Sanskrit)
Withania somnifera is a woody herb or shrub growing from a long tuberous taproot; it is widespread from the Mediterranean coast to India in semiarid habitats; it is one of the important plants in Indian traditional medicine with several validated pharmacological properties. The root and leaf extracts of *W. somnifera* showed hypoglycemic and hypolipidemic effects in alloxan-induced diabetic rats (Udayakumar et al. 2009).

Woodfordia fruticosa (L.) Kurz., Lythraceae
The ethanol extract of *Woodfordia fruticosa* flowers (250 and 500 mg/kg) significantly reduced fasting blood glucose level and increased insulin level after 21 days' treatment in streptozotocin-induced diabetic rats. The extract also increased catalase, superoxide dismutase, GSH reductase, and GSH peroxidase activities and reduced LPO. Glycolytic enzymes showed an increase in their levels while a decrease was observed in the levels of the gluconeogenic enzymes in the ethanol extract–treated diabetic rats. The extract has a favorable effect on the histopathological changes of pancreatic β-cells in streptozotocin-induced diabetic rats. The results suggest that *W. fruticosa* possesses a potential antihyperglycemic effect by regulating glucose homeostasis and antioxidant status in streptozotocin-induced diabetic rats (Verma et al. 2012a). Further, in a screening study, the leaf extract of *W. fruticosa* (250 or 500 mg/kg, p.o.) significantly reduced PPG levels in normal glycemic rats (Arya et al. 2012). The presence of known anti-DM compounds such a gallic acid, oleanolic acid, and β-sitosterol in this plant has been reported.

Wrightia tinctoria L., Apocynaceae
Wrightia tinctoria is used in folk medicine to treat piles, fever, colic, inflammation, diabetes, etc. The chloroform extract of *W. tinctoria* leaves showed a significant antidiabetic activity in alloxan-induced diabetic rats (Shruthi et al. 2012). Oral administration of the petroleum ether extract of *W. tinctoria* leaves (200 and 400 mg/kg daily for 14 days) exhibited hypoglycemic and hypolipidemic activity in alloxan-induced

diabetes in rats (Raj et al. 2009). In another study, the petroleum ether extract (200 and 400 mg/kg/day for 14 days) exhibited hypoglycemic and hypolipidemic activities in streptozotocin-induced diabetic rats (Raj et al. 2010). The fruit of this plant also exhibited antidiabetic activity. Oral administration of the methanol extract (300 mg/kg) and ethyl acetate extract (200 mg/kg) of *W. tinctoria* exhibited highly significant hypoglycemic activity in normal rats and antihyperglycemic activity in alloxan-induced diabetic rats. The maximum reduction in blood glucose level was observed after 4 h in the case of methanol and ethyl acetate extracts with a percentage protection of 37% and 42%, respectively. In the long-term treatment of alloxan-induced diabetic rats, both of the extracts showed a significant antidiabetic activity comparable to that of glibenclamide (Sandhyarani et al. 2012).

Xanthium strumarium L., Compositae

Synonyms: *Xanthium macrocarpum* DC, *Xanthium natalense* Widder, *Xanthium pungens* Wall.

Common Name: Noogoora burr

Xanthium strumarium is a traditional medicinal plant used to treat many diseases in many parts of the world. The plant has exhibited potent hypoglycemic activity in the rat. Recently, oral administration of the alcohol and water extracts of *X. strumarium* (250 and 500 mg/kg) daily for 21 days to streptozotocin-induced diabetic rats resulted in a decrease in the levels of blood glucose, cholesterol, and TG. Further, the treatment improved the histological appearance of pancreatic β-cells in treated diabetic rats compared to diabetic control rats. The efficacy of the alcohol extract was found to be better than that of the water extract (Suresh et al. 2014). A compound isolated from the seeds of *X. strumarium* exhibited potent hypoglycemic activity in the rat. Partial characterization of the crystalline compound showed that it differed from all compounds previously isolated from *X. strumarium* (Kupiecki et al. 1974). The antihyperglycemic effect of caffeic acid and phenolic compounds present in the fruit of *X. strumarium* was investigated. After an intravenous injection of caffeic acid into diabetic rats of both streptozotocin-induced and insulin-resistant models, a dose-dependent decrease of plasma glucose was observed. However, a similar effect was not produced in normal rats (Hsu et al. 2000). Further, this compound reduced the elevation of plasma glucose level in insulin-resistant rats receiving a glucose tolerance test. Also, the glucose uptake into isolated adipocytes was raised by caffeic acid in a concentration-dependent manner. Increase of glucose utilization by caffeic acid seems to be responsible for the lowering of plasma glucose (Hsu et al. 2000). It seems that several antidiabetic compounds are present in this plant. The details of the mechanism of action of these compounds (individually and in combination) are not clear. Further studies are warranted.

Xanthocercis zambesiaca (Baker) Dumaz-le-Grand, Leguminosae

Common Name: Nyala tree

The aqueous methanol extract of the leaves and root of *Xanthocercis zambesiaca* and eight structurally related nitrogen-containing sugars, fagomine (1), 4-*O*-β-D-glucopyranosylfagomine (2), 3-*O*-β-D-glucopyranosylfagomine (3), 3-epifagomine (4), 2,5-dideoxy-2,5-imino-D-mannitol (5), castanospermine (6), α-homonojirimycin (7), and 1-deoxynojirimycin (8), has been evaluated for their antihyperglycemic effects in streptozotocin-induced diabetic mice. The insulin-releasing effects of fagomine were also investigated. The blood glucose level fell after i.p. injection of the extract (50 mg/kg). Compounds 1, 2, 5, and 6 reduced the blood glucose level after i.p. injection of 150 μmol/kg. Compound 1 increased plasma insulin level in streptozotocin-induced diabetic mice and potentiated the 8.3 mM glucose-induced insulin release from rat isolated perfused pancreas. The fagomine-induced potentiation of insulin release may partly contribute to the antihyperglycemic action (Nojima et al. 1998).

Xanthosoma sagittifolium (L.) Schott, Araceae

Synonym: *Xanthosoma violaceum* Schott

Xanthosoma sagittifolium (*X. violaceum*) is used in traditional medicine to treat diabetes and other diseases in Bangladesh. The methanol extract (50, 100, 200, and 400 mg/kg body) of the aerial parts of the plant reduced blood sugar levels compared to control animals in oral glucose tolerance test; the effect was concentration dependent (Faisal et al. 2014).

Yucca schidigera Roezl ex Ortgies, Agavaceae
Yucca schidigera plant powder supplementation with the diet (100 ppm in the standard feed) for 3 weeks resulted in decrease in blood glucose and increase in insulin levels in streptozotocin-induced diabetic rats. The treatment improved lipid profiles and reduced oxidative stress (Fidan and Dundar 2008).

Zaleya decandra (L.) Burm. f., Aizoaceae
The ethanol extract of *Zaleya decandra* roots reduced oxidative stress and improved the lipid profiles in alloxan-induced diabetic rats (Meenakshi et al. 2010).

Zataria multiflora Boiss, Lamiaceae
Zataria multiflora is a thyme-like plant that grows wild in Iran, Pakistan, and Afghanistan. Oral administration of Z. *multiflora* essential oil to streptozotocin-induced diabetic rats resulted in a marked reduction in plasma glucose, ALP, AST, and ALT and a significant increase in total protein and insulin (Kavoosi 2011). Z. *multiflora* essential oil reduced diabetic damages and prevented the progression of streptozotocin-induced diabetes in rats and reduced hepatic and pancreas injury. Besides, Z. *multiflora* extract increased insulin sensitivity and PPAR-γ gene expression in high-fructose-fed insulin-resistant rats (Mohammadi et al. 2014). The water extract of Z. *multiflora* reduced microvascular permeability in streptozotocin-induced diabetic rats (Sepehri et al. 2014).

Zea mays L., Poaceae
Common Names: Maize plants, corn
Corn silk is well known and frequently used in traditional Chinese herbal medicines. Corn silk extract markedly reduced hyperglycemia in alloxan-induced diabetic mice. The action of corn silk extract on glycemic metabolism is not via increasing glycogen and inhibiting gluconeogenesis but through increasing the insulin level as well as recovering the injured β-cells. The results suggest that corn silk extract may be used as a hypoglycemic food or medicine for hyperglycemic people (Guo et al. 2009). In another study, treatment with 100–500 mg/kg of polysaccharides of corn silk to streptozotocin-induced diabetic rats showed the possibility of significant reduction in blood glucose level and serum lipids. The polysaccharide also showed good antidepressant activity (Zhao et al. 2012a). Corn also has aldose reductase inhibitory activity. Seven nonanthocyanin phenolic compounds and 5 anthocyanins were isolated through bioassay-guided fractionation of the ethanol extract of the kernel from purple corn. These compounds were investigated by rat lens aldose reductase inhibitory assays. Hirsutrin, one of 12 isolated compounds, showed the most potent competitive inhibition of aldose reductase (IC_{50} 4.78 µM). Furthermore, hirsutrin inhibited galactitol formation in rat lens and erythrocytes sample incubated with a high concentration of galactose; this finding indicates that hirsutrin may effectively prevent osmotic stress in hyperglycemia. Therefore, hirsutrin derived from *Zea mays* may be a therapeutic agent against diabetes complications (Kim et al. 2013).

Zhumeria majdae Rech. f., Lamiaceae
The extract of *Zhumeria majdae* showed promising *in vitro* inhibition of α-amylase activity (Gholamhoseinian et al. 2008).

Zingiber officinale Rosc., Zingiberaceae (Figure 2.99)
Common Names: Ginger (Eng.), inchi, ardraka, adark, allam, inji, etc. (in India)
Distribution: *Zingiber officinale* is a rhizomatous biennial plant with lanceolate leaves, glabrous beneath, clasping the stem by their long sheath. Spikes produced from the root stock are more or less elongated peduncles with sheathing scarious bract leaves. The fruit is a dehiscent capsule.
Distribution: Z. *officinale* is widely cultivated in India and other parts of tropical Asia.
Traditional Medicinal Uses: It is an antiseptic, anodyne, aperitif, aphrodisiac, astringent, carminative, digestive, expectorant, pediculicide, rubefacient, sialagogue, sternutatory, stimulant, stomachic, and sudorific and used in asthma, amenorrhea, backache, bronchitis, cancer, colic, chills, cataplasm, cholera, dog bite, congestion, cough, diarrhea, dysentery, fever, fistula, dyspepsia, fatigue, flatulence, gingivitis,

(a)   (b)

FIGURE 2.99   *Zingiber officinale.* (a) Herb. (b) Rhizome.

flu, headache, hepatosis, gout, indigestion, laryngitis, malaria, nausea, paralysis, phthisis, rabies, puerperium, rheumatism, snakebite, rhinosis, sores, menstruation ailments, stomachache, swelling, toothache, and tetanus (Kirtikar and Basu 1975).

Pharmacology

Antidiabetes: In alloxan-induced diabetic rats, ginger lowered blood glucose levels (Kar et al. 2003). It also showed antidiabetic activity in streptozotocin-induced type 1 diabetic rats (Akhani et al. 2004). Further, ginger enhanced insulin sensitivity in adipocytes (Sekiya et al. 2004). The ethanol extract of ginger exhibited significant lipid-lowering and anti-LPO activities in diabetic rats (Bhandari et al. 2005).

The rhizome extract of this plant exhibited antidiabetes and antioxidant effects in alloxan-induced and insulin-resistant diabetic male rats. The treatment reduced fasting blood glucose levels and increased the insulin level and also enhanced insulin sensitivity in these diabetic rats (Iranloye et al. 2011). The alcohol extract of the rhizome attenuated progression of diabetic structural nephropathy in streptozotocin-induced diabetic rats (Ramudu et al. 2011). 6-Shogaol and 6-gingerol, the pungent of ginger, inhibited TNF-α-mediated downregulation of adiponectin expression via different mechanisms in 3T3-L1 adipocytes (Wang et al. 2014). These studies suggest the potential of ginger for the preparation of a food supplement to diabetic patients. The active principles involved and their dose response remain to be studied.

Other Activities: Ginger extract inhibited the production of prostaglandins and leukotrienes (Kiuchi et al. 1992). Ginger also has antiplatelet activity (Subramoniam and Satyanarayana 1989; Guh et al. 1995) and its extracts have beneficial effects on osteoarthritis and inflammation (Bliddal et al. 2000; Altman and Marcussen 2001). The extract has hypocholesterolemic, antioxidant, immunomodulatory, and antiatherosclerotic activities in mouse model. Ginger extract components suppressed induction of chemokine expression in human synoviocytes (Phan et al. 2005). It is a potential anti-inflammatory and antithrombotic agent. The rhizome juice increased sperm count and motility. In a clinical study, ginger has been reported to be useful for nausea and vomiting in pregnancy. Ginger has antiviral properties and antirhinoviral sesquiterpenes have been isolated. The essential oils of ginger have antimicrobial activity. Ginger juice inhibited rat ileal motility *in vitro* (monograph: WHO 1999).

Toxicity: The extract of ginger appears to be safe in rats (monograph: WHO 1999). Further, ginger is used as an ingredient of diets from ancient time onward without any recorded toxicity at reasonable doses.

Phytochemicals: Phytochemicals reported include sesquiterpene alcohol, chavicol (phenol), esters of acetic and caprylic acid, α- and β-zingiberene, ar-curcumine, farnesene, β-bisabolene, r-selinene, β-elemene, β-sesquiphellandrene, β-eudesmol, zingiberol, camphene, α and β-pinene, cumene, myrcene, limonene, p-cymene, β-phellandrene, bornyl acetate, linalool, n-nonanal, D-decanyl, methyl heptenone, 1–8-cineole, borneol, gingerol, shogaol, zingerone, paradol (*The Wealth of India* 1976), oleoresin, gingerol, gingesulfonic acid (antiulcer principle), gingerglycolipids (A, B, and C), monoacyldigalactosylglycerols, and geranyl disaccharide (monograph: WHO 1999).

Zingiber zerumbet (L.) Roscoe ex Sm., Zingiberaceae

Common Name: Bitter ginger

The ethanol extract of the rhizome of the plant attenuated streptozotocin-induced nephropathy in rats (Tzeng et al. 2014). The plasma glucose, creatine, blood urea nitrogen, urine protein, and the ratio of kidney weight to body weight were elevated in diabetic rats. These parameters were reduced with the extract as well as metformin in a similar manner. The AMPK protein phosphorylation and expression levels were reduced in diabetic renal tissue. The extract treatment reversed this reduction (Tzeng et al. 2014). The authors suggest that the renoprotective effects of the extract may be similar to that of metformin. Prevention of AMPK dephosphorylation and upregulation of the expression of renal nephrin and podocin could be the mechanism. One study reported insignificant antidiabetic activity of the water extract of *Zingiber zerumbet* in streptozotocin-induced diabetic rats (Dal et al. 2010). The difference could be due to the difference in the extract used in these studies.

Ziziphus jujuba Mill., Rhamnaceae

Synonyms: *Zizyphus sativa* Gaertn; *Z. vulgaris* Lam.; *Rhamnus ziziphus* L., [Ziziphus = Zizyphus] (Figure 2.100)

Common Names: Jujube plum, bar, Chinese jujube (Eng.), jujube (French), manzanas (Spain), ilanthai pazhalam (Tamil), etc.

Description: Mansanitas is a small tree, 5–10 m high. Its branches are armed with short, sharp spines. Its leaves are elliptic ovate, 5–8 cm long, 3–5 cm wide, rounded at the base, green and smooth on the upper surface, and densely covered with woolly, pale hairs beneath. Its flowers are greenish white, about 7 mm in diameter, borne on axillary cymes 3 cm in diameter or less. Its fruit is fleshy and mealy, smooth, orange or red, ovoid or somewhat rounded, with a bony and irregularly furrowed stone within.

Distribution: Introduced from tropical Asia, *Z. jujuba* is widely scattered in the Philippines as a semicultivated tree. It grows in Europe, Southern and Eastern Asia, China, Australia, etc.

FIGURE 2.100 *Ziziphus jujuba.*

Traditional Medicinal Uses: It has a long history of use as a fruit and a remedy. Its root is considered as purgative. Its fruit of wild variety is acidic and astringent; fruit is considered to have nourishing, mucilaginous, pectoral, styptic, digestive and blood purifying, tonic, aphrodisiac, anxiolytic, hypnotic sedative, anticancer, antifungal, and antiulcer properties. Based on a recent ethnobotanical survey, *Z. jujuba* is one of the most important antidiabetic plants used for the treatment of diabetes in Manisa, Turkey (Durmuskahya and Ozturk 2013).

Pharmacology

Antidiabetes: The alcohol extract of *Z. jujuba* leaves (100–400 mg/kg) exhibited a dose-dependent blood glucose-lowering effect in normal rats. The effect was most pronounced at 6 h with blood glucose returning to normal values at 24 h. However, the extract was devoid of significant acute glucose-lowering activity in alloxan-induced diabetic rats (Anand et al. 1989). The methanol extract of this plant root (100 mg/kg; p.o., daily for 2 weeks) showed antihyperglycemic and antioxidant activities in alloxan-induced diabetic rats. The treatment reduced serum glucose levels and increased serum insulin levels; further, the treatment lowered lipid profiles (Hussen et al. 2006).

The hydroalcoholic extract of the plant leaves reduced the levels of blood glucose, TGs, cholesterol, and VLDL and increased the levels of HDL in alloxan-induced diabetic rats compared to the untreated diabetic rats (Shirdel et al. 2009). In another study, ethanol extract of *Z. jujuba* leaves showed anti-DM activity in diabetic rats; this activity was comparable with that of glibenclamide (Shirdel and Mirbadalzadeh 2011).

Other Activities: Other pharmacological properties of this plant include sedative effect, anxiolytic effect, anticancer activity, antioxidant activity, immunostimulant effects, wound healing activity, cardiovascular activity, anti-inflammatory activity, antiulcer activity, antifungal activity, antidiarrheal activity, and antiobese activity (review: Preeti and Tripathi 2014). *Z. jujuba* powder exhibited hypolipidemic and antiobesity properties, and did not show any negative impact on liver function as measured by ALT and AST (Mostafa and Labban 2013).

Phytochemicals: The plant contains alkaloids, glycosides, saponins, terpenoids, and phenolic compounds (Preeti and Tripathi 2014). Its bark contains much tannin and a crystalline principle, ziziphic acid. Its fruit contains mucilage, fruit acids, etc. Studies have yielded various chemical substances like mauritine-A, amphibine-H, jubanine-A, jubanine-B, mucronine-D, nummularine-B, sativanine-E, frangufoline, ziziphine-A to Q, betulinic acid, colubrinic acid, alphitolic acid, 3-*O*-cis-p-coumaroylalphitolic acid, 3-*O*-trans-p-Â-coumaroylalphitolic acid, 3-*O*-cis-p-coumaroylmaslinic acid, 3-*O*-trans-p-coumaroylmaslinic acid, oleanolic acid, betulonic acid, oleanonic acid, zizyberenalic acid, betulinic acid, jujubosides A, B, A1, B1 and C, acetyljujuboside B, the protojujubosides A, B and B1, and ziziphin from the dried leaves of *Z. jujuba* (reviews: Gao et al. 2013; Preeti and Tripathi 2014).

Zizyphus lotus L., Rhamnaceae

Synonym: *Rhamnus lotus* L., lotus jujube

Zizyphus lotus is a traditional medicinal plant used for its multiple therapeutic properties. Recently, the antidiabetic and antioxidant effects of water extracts of different parts of this plant were evaluated in streptozotocin-induced diabetic rats. Oral administration of the extracts (300 mg/kg, daily) of leaf and root, but not seed, for 21 days to the diabetic rats resulted in lowering of blood glucose levels; the treatment also improved antioxidant status in the pancreas, liver, and erythrocytes. Thus, the leaf and root extracts showed antihyperglycemic and antioxidant activities (Benammar et al. 2014).

Ziziphus mauritiana Lam., Rhamnaceae, Indian jujube

Ziziphus mauritiana [syn: *Z. jujuba* (L.) Lam., *Rhamnus mauritiana* Soyer-Willemet, etc.] is a fruit tree that has been used in folkloric medicine for many ailments and diseases. The fruit, seed, and leaves of this plant showed antidiabetic activity in animals. The water extract and the nonpolysaccharide fraction of the water extract of the fruit were found to exhibit significant antihyperglycemic and hypoglycemic activities. The petroleum ether extract was found to exhibit only antihyperglycemic effect.

Treatment of alloxan-induced diabetic rats with petroleum ether extract, water extract (400 mg/kg), and nonpolysaccharide fraction of water extract (200 mg/kg) of this plant restored the elevated biochemical parameters, glucose, urea, creatinine, serum cholesterol, serum TG, HDL, LDL, hemoglobin, and glycosylated hemoglobin significantly to near normal levels. Comparatively, the nonpolysaccharide fraction of the water extract was found to be more effective, followed by the water extract, and the petroleum ether extract. The activity of the nonpolysaccharide fraction was comparable to that of the standard drug glibenclamide (Jarald et al. 2009a). Oral administration of water–ethanol seed extract of Z. *mauritiana* alone (100, 400, and 800 mg/kg) or in combination with glyburide (800 mg/kg seed extract and 10 mg/kg glyburide) reduced the blood glucose level in all the diabetic mice after acute and subacute (28 days) treatment. Further, the extract reduced the weight loss and mortality rate during the subacute study. The combination treatment exerted more effect than any of the single agent. The extract also augmented the glucose tolerance in both normal and diabetic mice (Bhatia and Mishra 2010). In another study, oral administration of ethanol extract of Z. *mauritiana* leaves to alloxan-induced diabetic rats for 4 weeks reduced dose dependently the blood glucose level and this effect was comparable to that of metformin. Further, the treatment increased oral glucose tolerance and reduced elevated serum cholesterol, urea, creatinine, and GPT levels in the diabetic rats (Une et al. 2013).

Ziziphus mucronata Willd. ssp. *mucronata*, Rhamnaceae

Water and methanol extracts of the bark of *Ziziphus mucronata* were tested for their antidiabetic effect *in vitro*. The extracts inhibited the activities of α-amylase and α-glucosidase in a concentration-dependent manner. This effect was comparable to that of acarbose. The extracts showed antioxidant activity also. Both the extracts increased glucose uptake in C2C12, 3T3-L1, and HepG2 cells. However, insulin secretion from RIN-m5F cells was not influenced by the extracts (Mousinho et al. 2013).

Ziziphus rugosa Lamk., Rhamnaceae

Local names: churna, kattilandai, malantudali, etc.

Ziziphus rugosa is a large straggling evergreen shrub or small tree, often a climber; armed with curved, short, solitary prickles. Its fruit is obovoid or globose drupe, white when ripe, fleshy, one seeded, and has a very thin crustaceous stone. It is found in India, Burma, Ceylon, etc. In traditional medicine, it is used in the treatment of diarrhea, carbuncles, glossitis, inflammation, gingivitis, inflammation, glossitis, infection of teeth, menorrhagia, etc.

The plant bark extract is reported to have antidiabetes activity (Khosa et al. 1983). Further, this plant contains phytochemicals with antidiabetic activities such as β-sitosterol, oleanolic acid, kaempferol, and quercetin. Reported phytochemicals include ziziphoside, betalic acid, oleanolic acid, alphitolic acid, 2-α-hydroursolic acid, kaempferol, kaempferol-3-rhamnoside, quercetin, quercetin-3-rhamnoside, myricetin, myricetin-3-rhamnoside, β-sitosterol, n-nonacosane, octacosanol, apigenin, betulin, betulinic acid, vanillic acid, rugosamine-A, rugosamine-B, nummularine-p, and sativanine (review: Ajikumaran and Subramoniam 2005).

Ziziphus sativa Gaertn, Rhamnaceae

In normal rats, alcohol extract of Z. *sativa* leaves (single oral dose; 100–400 mg/kg) exhibited a dose-dependent lowering of blood glucose at 2, 4, and 6 h after administration. The effect was most pronounced at 6 h with blood glucose returning to control values at 24 h. However, in the alloxan-induced diabetic rats, the extract as well as the reference drug tolbutamide did not influence glucose levels significantly (Anand et al. 1989).

Ziziphus spina-christi (L.) Willd., Rhamnaceae (Figure 2.101)

Common Names: Christ's thorn jujube; known as Sedr in Iran

Description: *Ziziphus spina-christi* is a shrub, sometimes a tall tree, reaching a height of 20 m and a diameter of 60 cm; its bark is light gray, very cracked, scaly; trunk twisted; very branched, crown thick; shoots whitish, flexible, drooping; thorns in pairs, one straight and the other curved. Its leaves are

FIGURE 2.101   *Ziziphus spina-christi* (twig with flowers).

glabrous on the upper surface, finely pubescent below, ovate lanceolate or ellipsoid, apex acute or obtuse, margins almost entire; lateral veins conspicuous. Its flowers are in cymes, subsessile, peduncle 1–3 mm. Its fruit is about 1 cm in diameter. [*Z. spina-christi* has very nutritious fruits that are usually eaten fresh] (review: Asgarpanah and Haghighat 2012).

Distribution: *Z. spina-christi* is a deciduous tree and native to the warm-temperate and subtropical regions, including North Africa, South Europe, Mediterranean, Australia, tropical America, South and East Asia, and Middle East (Yossef et al. 2011).

Traditional Medicinal Uses: The leaves of this plant are applied locally to sores, as poultices, and are helpful in liver troubles, asthma, and fever; the roots are used to cure and prevent skin diseases. The fruits are applied on cuts and ulcers. They are also used to treat pulmonary ailments, dysentery, and fevers. The seeds are sedative and are sometimes taken with buttermilk to halt nausea, vomiting, and abdominal pains associated with pregnancy (review: Asgarpanah and Haghighat 2012).

Pharmacology

Antidiabetes: Pretreatment with *Z. spina-christi* leaf butanol extract (100 mg/kg) or christinin-A, the major saponin glycoside of the leaves, potentiated glucose-induced insulin release in normal rats. In type 2, but not in type 1 diabetic rat, pretreatment with the butanol extract or christinin-A improved the oral glucose tolerance and potentiated glucose-induced insulin release. Treatment either with butanol extract (100 mg/kg) or christinin-A reduced the serum glucose level and increased the serum insulin level of nondiabetic and type 2 diabetic rats but not of type 1 diabetic rats. Pretreatment of nondiabetic normal and type 2 diabetic rats either with 100 mg/kg butanol extract or christinin-A also enhanced the glucose lowering and insulinotropic effect of gliben-clamide. The hyperglycemic and hypoinsulinemic effects of 30 mg/kg diazoxide in nondiabetic control and type 2 diabetic rats were inhibited and antagonized, respectively, by pretreatment with the butanol extract or christinin-A. Treatment of rats with butanol extract (100 mg/kg) for 3 months produced no functional or structural disturbances in the liver and kidney. In addition, the oral LD_{50} of the butanol extract in mice was 3.8 g/kg, while that of glibenclamide was 3.2 g/kg. Thus, *Z. spina-christi* leaves appear to be a safe alternative to lower blood glucose. The safe insulinotropic and subsequent hypoglycemic effects of *Z. spina-christi* leaves may be due to a sulfonylurea-like activity (Abdel-Zaher et al. 2005). The leaf extract of this plant showed antihyperglycemic and anti-oxidant activities in alloxan-induced diabetic rats. The treatment reduced serum glucose levels and increased serum insulin levels (Hussen et al. 2006).

  Z. spina-christi leaf extract (plain and formulated), when administered (p.o.) for 28 days, reduced blood glucose level with increase in serum insulin and C-peptide levels. Marked elevation in total antioxidant capacity with normalization of percentage of glycated hemoglobin (HbA1C%) was observed. Moreover, the treatment reduced the elevated blood lactate level and increased the reduced blood pyruvate content of diabetic rats. In line with amelioration of the DM state, the extract restored liver and muscle glycogen content together with significant decrease of hepatic glucose-6-phosphatase and increase in glucose-6-phosphate dehydrogenase activities. *In vitro* experiments showed a dose-dependent inhibitory activity of the extract against α-amylase enzyme with IC_{50} at 0.3 mg/mL. Such finding has been supported by the *in vivo* suppression of starch digestion and absorption by the extract in normal rats. The results revealed that *Z. spina-christi* leaf extract improved glucose utilization in diabetic rats by increasing insulin secretion, (which may be due to both saponin and polyphenol content), and controlling hyperglycemia through attenuation of meal-derived glucose absorption that might be attributed to the total polyphenols (Michel et al. 2011). The anti-DM effects of the fruit remain to be studied. The fruit is likely to have an anti-DM activity because it contains rutin and quercetin.

Other Activities: Important published pharmacological properties of this plant include antibacterial and antifungal properties, antinociceptive activity, antioxidant activity, antiplasmodial, antischistosomiasis, analgesic, and anticonvulsant activities (review: Asgarpanah and Haghighat 2012).

Phytochemicals: The extract of *Z. spina-christi* was shown to contain butic acid, ceanothic acid (a ring-A homologue of betulinic acid), cyclopeptides, saponin glycoside, flavonoids, and mucilage. Cardiac glycosides, polyphenols (such as tannins), christinin-A (saponins), dodecaacetylprodelphinidin B3 and C-glycoside, 3′,5′-di-C-β-D-glucosylphloretin (flavonoids) are reported from the leaves (Patel et al. 2012). Geranyl acetate, methyl hexadecanoate methyl octadecanoate, farnesyl acetone C, hexadecanol and ethyl octadecanoate were characterized as the main components of essential oil from the leaves. Zizyphine-F, jubanine-A, and amphibine-H and a new peptide alkaloid spinanine-A have been isolated from the stem bark. New flavonoid, quercetin 3-xylosy (1 → 2) rhamnoside-4′-rhamnoside accompanying with rutin, hyperin, quercetin, apigenin-7-*O*-glucoide, isovitexin, and quercetin-3-*O* lucoside-7-*O*-rhamnoside were characterized from *Z. spina-christi* fruits. In addition, 4-hydroxymethyl-1-methyl pyrrolidine-2-carboxylic acid and 4-hydroxy-4-hydroxymethyl-1-methyl pyrrolidine-2-carboxylic acid were characterized as two new cyclic amino acids from *Z. spina-christi* seeds (review: Asgarpanah and Haghighat 2012).

Zygophyllum album L., Zygophyllaceae

Zygophyllum album is an aromatic medicinal shrub occurring in Algeria and elsewhere. Water extract of *Z. album* showed antihyperglycemic, antihyperlipidemic, and antioxidant activities in streptozotocin-induced diabetic mice (Ghoul et al. 2012). The ethanol extract of *Z. album* also exhibited antihyperglycemic and antihyperlipidemic activities in streptozotocin-induced diabetic mice. The treatment did not increase insulin levels significantly (Ghoul et al. 2013). Extracts of *Z. album* (rich in flavonoids and phenolic compounds) exhibited an inhibitory activity on pancreatic lipase *in vitro* (IC_{50} 91 μg/mL). *In vivo* administration of this extract to HFD-rats lowered body weight and serum leptin level; the treatment inhibited lipase activity of obese HDF-rats leading to notable decrease of total cholesterol, TGs, and LDL levels accompanied with an increase in HDL in serum and liver. Moreover, the treatment helped to protect liver tissue from the appearance of fatty cysts. Besides, the extract modulated key enzyme related to hypertension such as ACE in serum of HFD animals and improved some of the serum electrolytes such as Na^+, K^+, Cl^-, Ca^{2+}, and Mg^{2+}. Moreover, the treatment reverted back near to normal values of liver–kidney dysfunction indices (Mnafgui et al. 2012).

Data from *in vitro* indicated that each extract from the medicinal plant showed moderate inhibition of α-amylase enzyme except the ethyl acetate extract that was ineffective. The powerful inhibition was achieved by ethanol extract of *Z. album*. *In vivo* studies showed that the extract decreased α-amylase levels in the serum, pancreas, and intestine of diabetic rats, associated with considerable reduction in blood glucose. Moreover, it helped to protect the structure and function of the β-cells. Interestingly, *Z. album* extract had a potent anti-inflammatory effect that is manifested by decreases in CRP and TNF-α levels (Mnafgui et al. 2014). Besides, vasorelaxant effects of water extract of *Z. album* in streptozotocin-induced diabetic rats have been reported (Ghoul and Ben-Attia 2014). The aforementioned studies suggest that it is a potential plant for further studies.

Zygophyllum gaetulum Emb. & Maire, Zygophyllaceae

Zygophyllum gaetulum is a well known traditional anti-DM plant in the south of Morocco. It is used as a condiment and food in Morocco and Algeria. The water infusion of the aerial parts of *Z. gaetulum* (700 mg/kg, p. o.) showed hypoglycemic activity in normal and alloxan-induced diabetic rats. The butanol-soluble fraction of water extract and another fraction that was obtained as precipitate upon reduction in volume of the water fraction caused a significant reduction in blood glucose concentrations. It produced a significant increase in insulin levels in normal healthy rats (Jaouhari et al. 2000).

In a human clinical study on newly diagnosed type 2 diabetics, *Z. gaetulum* decoction was compared with the sulfonylurea glipizide and a water placebo. A single oral intake of the plant had a 30% hypoglycemic action that developed more slowly but became comparable to glipizide, as it did during a 3-week treatment, maintaining normal fasting and postprandial glycemia (review: Huddad et al. 2005). Thus, *Z. gaetulum* is a promising nontoxic anti-DM plant for further studies.

2.3 Anti-DM Plants Not Exposed to Scientific Studies

Although substantial scientific studies were carried out on the traditional medicinal plants, hundreds of plants used in traditional medicine have not yet been subjected to scientific/pharmacological validation. An incomplete list of such plants is given in Table 2.1, among other things, to facilitate research on these plants.

2.4 Conclusion

More than 1085 plants were subjected to at least one scientific study to show their hypoglycemic and/or other beneficial effects in DM based on animal experiments, *in vitro* assays, and/or human clinical studies. In most of the cases, the studies have not reached logical conclusions. Comparative studies are not available to select the best antidiabetic plants for drug development. The available comparative studies are inadequate to make meaningful comparisons. For example, a single type of extract (water or alcohol–water or alcohol extract) was used to compare different groups of antidiabetic plants. The anti-DM compounds known are of diverse chemical types and their solubilities differ; some of them are soluble in highly nonpolar solvents like hexane and petroleum ether. Therefore, the appropriate parts of the plant, extract, and optimum dose have to be determined in each case for meaningful comparison of their hypoglycemic activity in normal animals and antidiabetic activity in type 1 and type 2 diabetic models. In certain cases, very high doses may have adverse effects or no effect, whereas low doses will be beneficial (Calabrese et al. 2010). Further, some phytochemicals/extracts may show delayed effects, not acute effects. Depending on the mechanism of actions, the efficacy of herbal drugs will differ in type 1 and type 2 models. Thus, meaningful comparative studies remain to be done. Further, target-oriented studies will determine the therapeutic value of many of these plants. Numerous leads were obtained and there is an urgent need for the development of these existing leads and to improve the use of proven anti-DM species. Many of the ethnomedical plants are poorly studied. Further, ethnomedical exploration remains to be done to find out unrecorded ethno-DM plants, particularly in African and South American countries. Since, even today, majority of the world population are using herbal drugs to control diabetes, research should be focused on the safety and efficacy aspects of these plant drugs on one hand and, on the other hand, development of internationally acceptable drugs including new chemical entity drugs and combination drugs from the traditional anti-DM plants. The information provided in this chapter is aimed at facilitating such research and development.

3

Nutraceuticals for Diabetes Mellitus (Functional Food for Diabetes Mellitus)

3.1 What Are Nutraceuticals?

Certain ingredients of diet or edible items have medicinal values. Such items are known as medicinal food or functional food or bioactive food. Bioactive molecules present in functional food are nutraceuticals; functional foods are also considered as nutraceuticals or food medicines. Plant-based nutraceuticals are plant products with nutritional and medicinal values. In other words, these are food (ingredients of diet) with pharmaceutical properties (bioactivities). Nutraceuticals have health benefits when consumed to the optimum levels. The term "nutraceuticals" emerged in 1979 (Brower 1998). Vegetables (like bitter gourd), fruits (like grapes and papaya), rice bran, gooseberry, and spices (such as turmeric and garlic) are examples of nutraceuticals. These ingredients of food have biologically active molecules. For example, curcumin from turmeric has antioxidant, antidiabetic, anti-inflammatory, and cancer-preventive properties. Similarly, sulfur compounds in garlic extract have hypolipidemic property. Grapes and peanuts contain, among other things, pharmacologically active resveratrol. Herbal drugs that are not edible (not ingredients of diet) are not nutraceuticals. Plant foods are edible parts of plants (as is or after cooking) accepted by any community through custom, habit, and tradition, as appropriate, desirable food or ingredients of diet. Generally, food items provide nutrients to the body without any short- or long-term adverse effects to health and well-being.

Certain pharmacy and biotech companies erroneously extended the term nutraceutical even to isolated compounds from wild plants that are not edible (Kamboj 2000). For example, docosahexaenoic acid, a cardiovascular stimulant from algae, was marketed as a nutraceutical. Many herbal preparations are being marked as nutraceuticals without following the minimum standards for herbal drugs established by the WHO (WHO 1991). This is a dangerous trend considering human health. The Dietary Supplement Health and Education Act passed by the United States in 1994 permits making unprecedented claims about the health benefits of food or dietary supplements. In view of this, unfortunately, many herbal remedies and isolated compounds are marketed as nutraceuticals by some of the pharmacy companies. Regulatory agents should stop this dangerous trend as was done by U.S. FDA by banning the so-called dietary supplement Cholestin (lovastatin).

3.2 Nutraceuticals for Diabetes Mellitus

Edible plant parts with antidiabetic properties are considered as plant-based anti-diabetes mellitus (DM) nutraceuticals. Further, anti-DM plants with their anti-DM activities only in the nonedible parts are not considered as plants with nutraceuticals. Food items with anti-DM properties and consumed for a long time without any known adverse effects (when taken in reasonable quantities) are real nutraceuticals for DM. Only those parts of plants that are consumed normally as human food in any parts of the world are to be used in the preparation of nutritional medicine (dietary formulation) for DM patients. There are many plants with anti-DM compounds in edible parts. Most of these plants are listed in Table 3.1. There are vegetables, fruits, tubers, spices, pulses, etc., with considerable anti-DM properties. Some examples are shown in Figure 3.1. Vegetables include bitter melon (unripe fruit), cucumber, cluster beans, ivy guard,

TABLE 3.1

Plants with Anti-Diabetes Mellitus Properties in Edible Parts

Number	Botanical Name	Common Name	Family	Edible Part with Anti-diabetes Activity
1.	*Abelmoschus esculentus* (L.) Moench	Lady's finger, okra	Malvaceae	Fruit
2.	*Allium cepa* L.	Onion	Liliaceae	Bulb
3.	*Allium sativum* L.	Garlic	Alliaceae or Liliaceae	Bulb
4.	*Amaranthus caudatus* L.	Pendant amaranth	Amaranthaceae	Leaves
5.	*Amaranthus spinosus* L.	Spiny amaranth	Amaranthaceae	Leaves and stem
6.	*Amaranthus viridis* L.	Amaranth (shak in Bengali)	Amaranthaceae	Leaves
7.	*Anacardium occidentale* L.	Cashew nut tree	Anacardiaceae	Nut (kernel)
8.	*Apium graveolens* L.	Wild celery	Apiaceae	Leaf stalks and leaves
9.	*Arachis hypogaea* L.	Earth nut, ground nut	Fabaceae	Nut, seed oil
10.	*Arctium lappa* L.	Burdock	Asteraceae	Roots
11.	*Artemisia dracunculus* L.	Tarragon, dragon herb	Asteraceae	Leaves
12.	*Aspalathus linearis* (Burm.f.) R. Dahlgren	Rooibos tea	Fabaceae	Leaf tea
13.	*Bidens pilosa* L.	Spanish needles, beggar's ticks, etc.	Asteraceae	Leaves
14.	*Brassica juncea* Czern. & Coss.	Indian mustard	Brassicaceae	Seeds, leaves
15.	*Brassica nigra* (L.) Koch	Black mustard	Brassicaceae	Leaf, seed
16.	*Brassica oleracea* L.	Cauliflower	Brassicaceae	Cauliflower (bulb-like stem), vegetable
17.	*Cajanus cajan* (L.) Millsp.	Pigeon pea, red gram, etc.	Fabaceae	Seeds, pods, young leaves
18.	*Camellia sinensis* (Linn.) Kuntze	Tea plant	Theaceae	Tea leaf
19.	*Canavalia ensiformis* (L.) DC	Jack bean	Fabaceae	Seeds, pods
20.	*Capsicum annum* L. and *Capsicum frutescens* L.[a]	Capsicum, chili, red chili	Solanaceae	Fruits (green and dried red chili)
21.	*Carica papaya* L.	Papaya	Caricaceae	Unripe fruit
22.	*Carum carvi* L.	Caraway	Apiaceae	Seed, leaves, etc.
23.	*Caylusea abyssinica* Fresen. Fisch. & Mey	—	Resedaceae	Leaves
24.	*Cinnamomum cassia* (Nees & T. Nees) J. Presl.	Cinnamon, Chinese cinnamon	Lauraceae	Inner bark (common spice)
25.	*Cinnamomum verum* J.S. Presl.	Ceylon cinnamon, true cinnamon	Lauraceae	Inner bark (common spice)
26.	*Citrullus lanatus* (Thunb.) Matsumara & Makai	Watermelon	Cucurbitaceae	Fruit (rind of the fruit is anti-DM)
27.	*Citrus lemon* (L.) Burm.f.	Lemon	Rutaceae	Fruit juice, lemon peel
28.	*Citrus limetta* Risso	Pomona sweet lemon	Rutaceae	Fruit, fruit peel
29.	*Citrus maxima* (Burm.) Merr.	Pomelo	Rutaceae	Fruits, fruit peel
30.	*Coccinia indica* Wight & Arn.	Scarlet vine, ivy gourd	Cucurbitaceae	Unripe fruit (vegetable)
31.	*Cocos nucifera* L.	Coconut tree	Arecaceae	Mature coconut water, tender inflorescence
32.	*Coffea arabica* L.	Coffee, Arabica coffee	Rubiaceae	Coffee bean
33.	*Colocasia esculenta* (L.) Schott	Cocoyam	Araceae	Corm and leaves

(Continued)

TABLE 3.1 (*Continued*)

Plants with Anti-Diabetes Mellitus Properties in Edible Parts

Number	Botanical Name	Common Name	Family	Edible Part with Anti-diabetes Activity
34.	*Coriandrum sativum* L.	Coriander	Apiaceae	Seed, aerial parts of young plant, etc.
35.	*Cucumis melo* L.	Musk melon, oriental melon	Cucurbitaceae	Fruits
36.	*Cucumis sativus* L.	Cucumber	Cucurbitaceae	Fruits, seeds (vegetable)
37.	*Cucumis trigonus* Roxb.	Wild cucumber	Cucurbitaceae	Fruits
38.	*Cucurbita ficifolia* Bouche	Malabar gourd, fig leaf gourd	Cucurbitaceae	Fruits
39.	*Cucurbita maxima* Duchesne	Pumpkin, autumn squash	Cucurbitaceae	Fruit (vegetable), seed
40.	*Cucurbita moschata* Duchesne ex Poiret	Crookneck squash	Cucurbitaceae	Fruits, seeds
41.	*Cucurbita pepo* L.	Pumpkin, common pumpkin	Cucurbitaceae	Fruits
42.	*Cuminum cyminum* L.	Cumin, huminum	Apiaceae/ Umbelliferae	Seeds (condiment)
43.	*Curcuma longa* L.	Turmeric	Zingiberaceae	Rhizome
44.	*Cyamopsis tetragonoloba* (L.) Taub.	Cluster beans	Fabaceae	Beans (vegetable)
45.	*Cydonia oblonga* Miller	Quince	Rosaceae	Leaves
46.	*Cynara scolymus* L.	Artichoke	Asteraceae	Leaves
47.	*Daucus carota* L.	Carrot	Apiaceae	Storage root (carrot)
48.	*Dioscorea alata* L.	Winged yam, water yam	Dioscoreaceae	Tuber
49.	*Dioscorea bulbifera* L. and related spp.	Air potato, potato yam	Dioscoreaceae	Tuber
50.	*Eleusine coracana* (L.) Gaertner	Finger millet	Poaceae	Seed coat
51.	*Eruca sativa* Mill.	Rocket (roquette)	Brassicaceae	Leaf (vegetable)
52.	*Ficus carica* L.	Edible fig, common fig	Moraceae	Fruit
53.	*Filipendula ulmaria* (L.) Maxim	Meadowsweet	Rosaceae	Tea from leaf, root and flower
54.	*Fraxinus excelsior* L.	Common ash, English ash	Oleaceae	Seed
55.	*Girardinia heterophylla* Decne	Himalayan nettle	Urticaceae	Leaves
56.	*Glycine max* (L.) Merr.	Soya bean	Fabaceae	Seed
57.	*Glycyrrhiza glabra* L.	Liquorice or licorice	Fabaceae	Root/rhizome
58.	*Glycyrrhiza uralensis* Fisch.	Chinese licorice, licorice	Fabaceae	Root
59.	*Gongronema latifolium* Benth.	Utalis, arokeke (in Nigeria)	Asclepiadaceae	Leaf (vegetable)
60.	*Grewia asiatica* L.	Phalsa or fasla	Tiliaceae	Fruit
61.	*Helianthus tuberosus* L.	Jerusalem artichoke, sunchoke	Asteraceae	Tubers
62.	*Hordeum vulgare* L.	Barley	Poaceae	Barley seeds
63.	*Humulus lupulus* L.	Hop, hops	Cannabaceae	Yong shoots, inflorescence
64.	*Ipomoea batatas* L.	Sweet potato, patate douce	Convolvulaceae	Tuberous root
65.	*Ipomoea digitata* L.	Wild yam	Convolvulaceae	Tubers
66.	*Juglans regia* L.	Walnut	Juglandaceae	Nut, nut oil[a]
67.	*Kalanchoe crenata* Andr. Haw.	Air plant, miracle leaf	Crassulaceae	Leaf
68.	*Kochia scoparia* (L.) Schrad.	Burning bush, ragweed	Chenopodiaceae	Seeds, leaves
69.	*Lagenaria siceraria* L.	Bottle gourd	Cucurbitaceae	Fruit (vegetable)

(*Continued*)

TABLE 3.1 (*Continued*)

Plants with Anti-Diabetes Mellitus Properties in Edible Parts

Number	Botanical Name	Common Name	Family	Edible Part with Anti-diabetes Activity
70.	*Lepidium sativum* L.	Garden cress	Brassicaceae	Fresh or dried seeds, pods[d]
71.	*Leptadenia hastata* Pers.	Hagalhadjar (Arabic)	Asclepiadaceae	Leaf?
72.	*Lupinus mutabilis* Sweet	Hanchcoly, white lupin	Fabaceae	Edible seeds
73.	*Mangifera indica* L.	Mango tree	Anacardiaceae	Seed kernel, mango peel
74.	*Medicago sativa* L.	Forage crop, alfalfa	Leguminosae	Seed
75.	*Melothria maderaspatana* (L.) Cogn.	Madras pea pumpkin	Cucurbitaceae	Leaf (vegetable)
76.	*Mentha piperita* (Ehrhart) Briquet	Mint	Labiatae	Leaf
77.	*Momordica charantia* L.	Bitter melon, bitter gourd	Cucurbitaceae	Unripe fruits (vegetable)
78.	*Murraya koenigii* (Linn.) Spreng.	Curry leaf, kari patta	Rutaceae	Leaf
79.	*Musa paradisiaca* L.	Banana	Musaceae	Unripe fruit, flowers
80.	*Musa sapientum* L.	Banana	Musaceae	Unripe fruit, flowers
81.	*Nelumbo nucifera* Gaertn.	Sacred lotus, Indian lotus	Nymphaeaceae	Rhizome, young leaves, flower
82.	*Opuntia dillenii* Haw.[b]	Brittle prickly pear	Cactaceae	Fruit
83.	*Opuntia robusta* J.C. Wendl	Wheel cactus, Camuesa	Cactaceae	Pulp
84.	*Opuntia streptacantha* Lemaire	Prickly pear cactus, nopal[c]	Cactaceae	Cladodes (stem)
85.	*Oryza sativa* L. (brown and related variety)	Brown rice	Poaceae	Rice bran
86.	*Phaseolus mungo* L. or *Vigna mungo* (L.) Hepper	Black gram	Fabaceae	Seed (pulse)
87.	*Phaseolus vulgaris* L.	Kidney beans, common bean	Fabaceae	Pods, seeds
88.	*Phyllanthus emblica* L.	Indian gooseberry, amala	Euphorbiaceae	Fruit
89.	*Picea glauca* (Moench) Voss	White spruce	Pinaceae	Needle, cones?
90.	*Piper longum* L.	Long pepper, thippali	Piperaceae	Fruit, dried fruit
91.	*Piper sarmentosum* Roxb.	Chaplu, wild betel	Piperaceae	Fruit, leaves
92.	*Prunus amygdalus* Batsch	Almond, suphala	Rosaceae	Nut
93.	*Psidium guajava* L.	Guava	Myrtaceae	Fruit juice
94.	*Punica granatum* L.	Pomegranate	Lythraceae	Seed, fruit
95.	*Salvia officinalis* L.	Garden sage, true sage, etc.	Lamiaceae	Leaves
96.	*Smallanthus sonchifolius* (Poepp & Endl.) H. Robinson	Yacon	Asteraceae	Tubers
97.	*Solanum tuberosum* L.	Irish potato	Solanaceae	Potato
98.	*Sorghum bicolor* (L.) Moench	Sorghum, durra	Gramineae	Seed (grain)
99.	*Spinacia oleracea* L.	Garden spinach	Chenopodiaceae	Leaves
100.	*Stevia rebaudiana* (Bert.) Bertoni	Stevia, sweet leaf, sweet herb of Paraguay	Asteraceae	Leaves
101.	*Syzygium cumini* (L.) Skeels.	Blackberry, black plum	Myrtaceae	Fruit
102.	*Syzygium samarangense* (Blume) Merr. and Perry	Wax apple	Myrtaceae	Fruit
103.	*Tamarindus indica* L.	Tamarind tree	Fabaceae	Seed, fruit pulp

(Continued)

TABLE 3.1 (*Continued*)

Plants with Anti-Diabetes Mellitus Properties in Edible Parts

Number	Botanical Name	Common Name	Family	Edible Part with Anti-diabetes Activity
104.	*Telfairia occidentalis* Hook. f.	Fluted gourd	Cucurbitaceae	Seed, leaf
105.	*Terminalia chebula* Retz	Inknut tree	Combretaceae	Fruit and nut
106.	*Theobroma cacao* L.	Cocoa	Sterculiaceae	Bean
107.	*Trichosanthes cucumerina* L.	Snake gourd	Cucurbitaceae	Seed, fruit?
108.	*Trigonella foenum-graecum* L.	Fenugreek	Leguminosae	Seed, leaf
109.	*Vaccinium angustifolium* Ait.	Lowbush blueberry	Ericaceae	Fruit
110.	*Vaccinium arctostaphylos* L.	Broussa tea, whortleberry, etc.	Ericaceae	Fruit
111.	*Vaccinium vitis-idaea* L.	Lingonberry, mountain cranberry	Ericaceae	Berries
112.	*Vitis vinifera* L.	Grape	Vitaceae	Fruit skin
113.	*Zea mays* L.	Maize plant, corn	Poaceae	Corn silk tea
114.	*Zingiber officinale* Rosc.	Ginger	Zingiberaceae	Rhizome
115.	*Ziziphus mauritiana* Lam.	Indian jujube	Rhamnaceae	Fruit

[a] Both species have numerous varieties and hybrids and are difficult to distinguish (Zhigila et al. 2014).
[b] Very related species are also edible with antidiabetic properties.
[c] Used generally to refer *Opuntia* sp.
[d] Anti-diabetes mellitus activity of edible leaf remains to be studied.

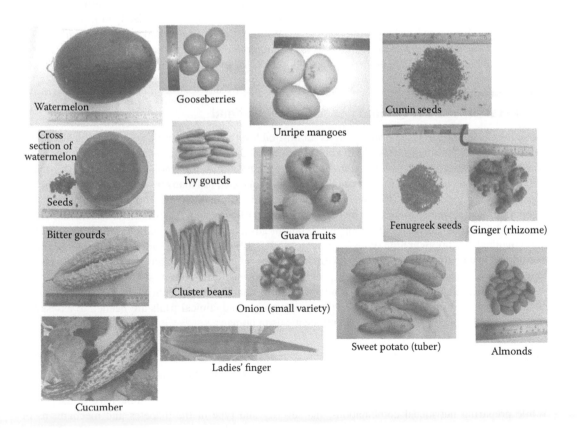

FIGURE 3.1 Examples of some commonly consumed plant food items with anti-diabetes mellitus properties.

banana (unripe fruit), carrot, soya been, prickly pear stem (cladodes), bottle gourd, ladies' finger (okra), onion, Malabar gourd, cauliflower, jack bean, pigeon pea (seeds and pods), green chili, and unripe papaya fruit. Leafy vegetables include garden spinach leaves, tarragon (*Artemisia dracunculus* leaves), stevia leaves, *Bidens pilosa* leaves, nettle leaves, garden sage leaves, wild celery leaves, and Madras pea pumpkin leaves. Examples of ripe fruits (edible as such) with antidiabetic activities are watermelon (fruit rim), sweet lemon, *Citrus maxima* (pummel) fruit, musk melon, *Ficus carica* fruit, guava fruit, blackberry, lowbush blueberry, brown grapefruit skin, and mango fruit peel. Tubers with antidiabetic properties include sweet potato, yacon tubers, air potato tubers, cocoyam, and water yam tubers. Nuts such as almond, cashew nut kernel, walnut, and ground nut have antidiabetic agents. Spices beneficial to diabetic patients include fenugreek seeds, cinnamon bark, coriander seed, cumin seeds, turmeric, curry leaf, tamarind pulp, mint leaf, dried red chili, garlic, mustard, and asafetida. Spices are normally added in small amounts in various dishes and food preparations for taste, flavor, etc. Other important edible items with antidiabetic properties include gooseberry, ginger, coconut water (from mature coconut), coffee, white lupin seeds, and black gram (pulse). Nutraceuticals are in great demand, considering their safety and health benefits, particularly, in the developed world. The nutraceutical market in the United States alone was about \$80–\$250 billion (Brower 1998). Recent reports on sales of nutraceuticals are not available. It may be about 40 times higher. Nutraceuticals do not involve regulatory clearances and offer large market and health promotion role. This is even truer in the case of nutraceuticals for DM.

3.3 Dietary Supplements for Diabetes Mellitus

The edible plant parts with anti-DM activities can be used in the preparation of many traditional as well as modern recipes and value-added dishes. Anti-DM teas, fruit juices, drinks, soups, salads, vegetable preparations, relishes, side dishes, etc., can be made available judiciously to diabetic patients as nutritional supplements to control diabetes. Antidiabetic nutraceuticals can be incorporated appropriately in food items such as biscuits, cookies, bread, pickles, jams, and sauces. When cooked food items are prepared from anti-DM plant parts, care should be taken to see the effect of cooking on the antidiabetes activity. Heating can destroy/decrease or increase the anti-DM properties depending on the heat sensitivity of the active principles involved.

3.4 Fixation of Dose/Quantity of Antidiabetic Food

Medicines have to be used at optimum levels for desired health benefits; similarly, nutraceuticals should be consumed at optimum levels. Animal experiments and clinical trials are required, among other things, to determine the optimum levels. For example, resveratrol, a phenolic compound (3,4,5-trihydroxystilbene) and a nutraceutical present in grapes, peanuts, etc., has many beneficial biological effects, dependent on the dose used. Too much could be harmful. Animal experiments have shown that this compound has a protective effect at low doses against cardiovascular injury, gastric lesions, ischemic stroke, Alzheimer's disease, and osteoporosis, but an adverse or no beneficial effect was observed in these medical conditions at high doses (Calabrese et al. 2010). In cell proliferation assays, under *in vitro* conditions, resveratrol stimulated the growth of a variety of cell types including cancer cells, whereas high concentrations inhibited cancer cell proliferation (review: Calabrese et al. 2010). Although many spices are potent antidiabetic agents, it is uncertain if the quantity of spice normally added in the diet is enough to have substantial anti-DM effects. Epidemiologic and clinical trials are needed to assess the nature of the dose–response of anti-DM nutraceuticals in humans.

3.5 Combination of Nutraceuticals (Combination of Functional Food)

Tailor-made nutritional supplements can be prepared with selected items of the edible parts of the plants. While preparing nutritional combinations, the severity and type of the diabetes and mechanisms of actions of different ingredients of the combination should be taken into account.

Small amounts of different food medicines could be better than larger amounts of a single food item with anti-DM property. Intense stimulation through one target or mechanism could be harmful in the long run. Functional foods acting through different target molecules involved in the complex DM have to be selected to prepare combination-nutritional supplements. For example, a combination of three or four different plant parts that act through different mechanisms (such as stimulation of β-cell proliferation and function, decrease in insulin resistance, inhibition of sugar absorption from intestine, and decrease in glucagon levels) could prove to be very effective. There are many food-stuffs that act through different mechanisms. For example, *Artemisia herba-alba* leaf reduces insulin resistance; coriander seed stimulates insulin release from β-cells and improves β-cell integrity and function; *Phyllanthus emblica* fruit (gooseberry) reduces intestinal absorption of glucose and inhibits lens aldose reductase activity (Chapter 2). An edible plant part that acts through multiple pathways involved in DM may not be ideal for combination-food supplement development. For example, bitter melon fruit has multiple mechanisms of action, and this vegetable could alone prove to be an effective food medicine. Value-added food supplements or side dishes could be prepared rationally with this vegetable. In most cases, a combination can give added effects. However, in some cases, it can lead to synergistic effects. Therefore, animal experiments and clinical trials are required to establish combined effects of nutraceuticals. It should be noted that, in certain cases, a combination of nutraceuticals or even a very high intake of a particular nutraceutical can lead to extreme hypoglycemia. Care should be taken for possible adverse intranutraceutical interactions as well as interaction of nutraceuticals with conventional anti-DM drugs.

3.6 Research and Development on Anti-Diabetes Mellitus Plant Foods

Studies on anti-DM nutraceuticals are rewarding considering the safety of nutraceuticals (food medicines) in the long run. Nutritional supplements can be formulated for the prevention of diabetes and for controlling/managing DM and, possibly, even to cure certain types of type 2 DM. There are a good number of anti-DM plants, but their edible parts are not evaluated for anti-DM activities. These include *Annona squamosa* (fruit), *Annona muricata* (fruit), *Cocos nucifera* (tender coconut, coconut kernel), *Mangifera indica* (unripe fruit at various stages of maturity), *Trichosanthes cucumerina* (fruit cortex, etc.), *Anacardium occidentale* (the tender seed [nut] without the shells and the cashew apples [fresh or cooked]), *Artocarpus heterophyllus* (unripe jack fruit at various stages of maturity), *Juglans regia* (walnut kernel), *Syzygium malaccense* (fruit), and *Ziziphus jujuba* (fruit). Plants reported to have anti-DM properties in nonedible parts may, possibly, have anti-DM activity in the edible parts as well. The presence of anti-DM compounds in sweet fruits cannot be ruled out. Further, the anti-DM properties of fresh and cooked edible parts are to be compared for efficacy. Among other things, for the preparation of nutraceutical combination and for better management of diabetes, knowledge of the mechanisms of action of nutraceuticals is required. In the case of many important plant food items with anti-DM activity, the mechanisms of action are not known. For example, cucumber is an important vegetable with hypoglycemic and hypolipidemic activity, but its mechanism of action is not known. The detailed mechanism of action of ivy gourd (fruit of *Coccinia indica*) is not clear. It should be noted that anti-DM nutraceuticals, in most cases, possess pharmacological properties other than anti-DM activities. Their beneficial and possible adverse effects on human health and well-being should be accounted for while consuming nutraceuticals. Future research should be focused toward these areas.

In many fruits and vegetables and other edible anti-DM plant parts, common phytochemicals with anti-DM and other beneficial pharmacological properties such as rutin, quercetin, β-sitosterol, oleanolic acids, gallic acids, ellagic acids, and resveratrol are present. Consumption of such nutraceuticals as part of one's diet on a regular basis in required amounts is beneficial to DM patients and DM-prone individuals. It is also possible that consumption of a diet containing phytochemicals, which improve the health of the β-cells and protect them from oxidative stress and mediated damage including apoptosis, could prevent and cure certain types of DM. This is an attractive area for research.

3.7 Conclusion

This chapter provides valuable information to translate new research findings and knowledge on edible anti-DM plant parts into practical and personalized recommendations for preventing diabetes and treating pathologies in diabetes. Edible anti-DM medicines are time tested for safety for longtime use. They could prove to be safer than other anti-DM agents. For the best health benefits, optimum dose should be fixed based on human studies. Food supplements should be prepared, and the dose should be fixed considering the mechanisms of action of nutraceuticals and the type and severity of the DM in question. Further, urgent research is warranted regarding the optimum amount of food medicines to be consumed and mechanisms of actions of important antidiabetic plants. Pharmacological properties, other than anti-DM, of the food medicines should also be taken into consideration. Proper use of nutritional medicines could prove to be an excellent way of controlling diabetes to a large extent.

References

Aakruti, A.K., Swati, R.D., and Vilasrao, J.K. 2013. Overview of Indian medicinal tree: *Bambusa bambos* (Druce). *Int. Res. J. Pharm.* 4: 52–56.

Abass, O.A. 2012. Therapeutic effect of *Artemisia herba-alba* aqueous extract added to classical therapy of acquired hyperlipidemia. *Iraqi J. Comm. Med.* 2012: 320–323.

Abate, N. and Chandalia, M. 2007. Ethnicity type 2 diabetes and migrant Asian Indians. *Indian J. Med. Res.* 125: 251–258.

Abdel, W., Wassel, S.M., Ammer, G.M., Ammern, N.M., and Hanna, T. 1987. Isolation of active anti-diabetic principles from aerial parts of *Bauhinia purpuria* and *Bauhinia veriegata*. *Herba Ung.* 26: 27–31.

Abdel-Hassan, I.A., Abdel-Barry, J.A., and Tariq, M.S. 2000. The hypoglycaemic and antihyperglycaemic effect of *Citrullus colocynthis* fruit aqueous extract in normal and alloxan diabetic rabbits. *J. Ethnopharmacol.* 7: 325–330.

Abdel-Rahim, E.A., El-Saadany, S.S., and Wasif, M.M. 1986. Biochemical dynamics of hypocholesterolemic actions of *Balanites aegyptiaca* fruit. *Food Chem.* 20: 69–78.

Abdel-Sattar, E.A., Abdallah, H.M., Khedr, A., Abdel-Naim, A.B., and Shehata, I.A. 2013. Antihyperglycemic activity of *Caralluma tuberculata* in streptozotocin-induced diabetic rats. *Food Chem. Toxicol.* 59: 111–117.

Abdel-Sattar, E.A., Harraz, F.M., Ghareib, S.A., Elberry, A.A., Gabr, S., and Suliaman, M.I. 2011. Antihyperglycaemic and hypolipidaemic effects of the methanolic extract of *Caralluma tuberculata* in streptozotocin-induced diabetic rats. *Nat. Prod. Res.* 25: 1171–1179.

Abdillah, S., Budiady, I., and Winarno, H. 2008. Hypoglycemic and anti-hyperlipidemic effects of henna leaves extract (*Lawsonia inermis* Linn.) on alloxan induced diabetic mice. *Jordan J. Pharm. Sci.* 1: 126–131.

Abdollahi, M., Salehnia, A., Mortazavi, S.H., Ebrahimi, M., Shafiee, A., Fouladian, F. et al. 2003. Antioxidant, antidiabetic, antihyperlipidemic, reproduction stimulatory properties and safety of essential oil of *Satureja khuzestanica* in rat *in vivo*: A toxicopharmacological study. *Med. Sci. Monit.* 9: 331–335.

Abdul, V.A. and Jyoti, H. 2014. Anti-diabetic activity of *Terminalia catappa* bark extracts in alloxan induced diabetic rats. *Res. Rev. J. Pharmacol. Toxicol. Stud.* 2: 37–42.

Abdulla, I.Z.A. 2008. Evaluation of hypoglycemic activity of *Opuntia dillenii* Haw fruit juice in streptozotocin-induced diabetic rats. *Egypt. J. Hosp. Med.* 33: 544–558.

Abdul-Muneer, M.T., Shenoy, A., Hegde, K., Aamer, S., and Shabaraya, A.R. 2014. Evaluation of the anti-diabetic activity of ethanolic extract of *Citrus maxima* stem bark. *Int. J. Pharm. Chem. Sci.* 3: 642–650.

Abeer, A.E., Doaa, A.G., Eman, E.M.S., Marwa, M.A., and Maha, A.E.D. 2013. *In vitro* biological assessment of *Berberis vulgaris* and its active constituent, berberine: Antioxidants, anti-acetylcholinesterase, anti-diabetic and anticancer effects. *BMC Complement. Altern. Med.* 13: 218–225.

Abesundara, K.J., Matsui, T., and Matsumoto, K. 2004. Alpha-glucosidase inhibitory activity of some Sri Lanka plant extracts, one of which, *Cassia auriculata*, exerts a strong antihyperglycemic effect in rats comparable to the therapeutic drug acarbose. *J. Agric. Food. Chem.* 52: 2541–2545.

Abid, S., Lekchiri, A., Mekhfi, H., Ziyyat, A., Legssyer, A., and Aziz, M. 2013. Inhibition of alpha-glucosidase and glucose intestinal absorption by *Thymelaea hirsuta* fractions. *J. Diabetes.* 6: 351–359.

Ablat, A., Mohamad, J., Awang, K., Shilpi, J.A., and Arya, A. 2014. Evaluation of antidiabetic and antioxidant properties of *Brucea javanica* seed. *Sci. World J.* 2014: Article ID 786130, 8 pages.

Abozid, M.M. 2010. Hypoglycemic and hypolipidemic effect of chicory (*Cichorium intybus* L.) herb in diabetic rats. *Menofia J. Agric. Res.* 35: 1201–1208.

Abraham, Z., Bhakuri, D.S., Garg, H.S., Mehrolra, B.N., and Patnaik, G.K. 1986. Screening of Indian plants for biological activity. Part XII. *Indian J. Exp. Biol.* 24: 48–68.

Abu-Amara, T., Helal, E., Abou-Aouf, N., Khattab, S.M., and Meselhy, A.R.A. 2014. Effects of *Ambrosia maritime*, L. (Damsissa) on some biochemical and histological parameters of diabetic albino rats. *Egypt. J. Hosp. Med.* 57: 612–629.

Abuh, F.Y., Wambebe, C., Rai, P.P., and Sokomba, E.N. 2006. Hypoglycaemic activity of *Anthocleista vogelii* (Planch) aqueous extract in rodents. *Phytother. Res.* 4: 20–24.

Abu-Zaiton, A.S. 2010. Anti-diabetic activity of *Ferula assafoetida* extract in normal and alloxan-induced diabetic rats. *Pak. J. Biol. Sci.* 13: 97–100.

Abyar, M.J., Riera, A.S., Grau, A., and Sanches, S.S. 2001. Hypoglycemic effect of the water extract of *Smallanthus sonchifolius* (Yacon) leaves in normal and diabetic rats. *J. Ethnopharmacol.* 74: 125–132.

Achrayya, S., Dash, G.K., Mondal, S., and Dash, S.K. 2010. Studies on glucose lowering efficacy of the *Anthocephalus cadamba* (Roxb.) Miq. Roots. *Int. J. Pharm. BioSci.* 6: 1–9.

Acosta-Patino, J.L., Jimenez-Balderas, E., Juarez-Oropeza, M.A., and Diaz-Zagoya, J.C. 2001. Hypoglycemic action of *Cucurbita ficifolia* on type 2 diabetic patients with moderately high blood glucose levels. *J. Ethnopharmacol.* 77: 99–101.

Adaramoye, O., Amanlou, M., Habibi-Rezaei, M., Pasalar, P., and Ali, M.M. 2012. Methanolic extract of African mistletoe (*Viscum album*) improves carbohydrate metabolism and hyperlipidemia in strepto-zotocin-induced diabetic rats. *Asian Pac. J. Trop. Med.* 5: 427–433.

Adebajo, A.C., Iwalewa, E.O., Obuotor, E.M., Ibikunle, G.F., Omisore, N.O., Adewunmi, C.O. et al. 2008. Pharmacological properties of the extract and some isolated compounds of *Clausena lansium* stem bark: Anti-trichomonal, antidiabetic, anti-inflammatory, hepatoprotective and antioxidant effects. *J. Ethnopharmacol.* 122: 100–109.

Adebayo, A.G. 2009. Inventory of antidiabetic plants in selected districts of Lagos State, Nigeria. *J. Ethnopharmacol.* 121: 135–139.

Adedapo, A., Ofuegbe, S., and Oguntibeju, O. 2014. The antidiabetic activities of the aqueous leaf extract of *Phyllanthus amarus* in some laboratory animals, Chapter 5. In: *Antioxidant-Antidiabetic Agents and Human Health*, Oguntibeju, O. (ed.). Croatia: InTech. Published under CC BY 3.0 license, http://dx.doi.org/10.5772/57030.

Adedapo, A.A., Ofuegbe, S.O., and Adeyemi, A.A. 2013. The anti-diabetic activities of the methanol leaf extract of *Phyllanthus amarus* in some laboratory animals. *Asian J. Med. Sci.* 14: 23–34.

Adediwura, F.-J. and Kio, A. 2009. Antidiabetic activity of *Spondias mombin* extract in NIDDM rats. *Pharm. Biol.* 47: 215–218.

Adejuwon, A.A., Crooks, P.A., Fadhel-Albayati, Z., Miller, A.F., Zito, S.W., and Adeyemi, O.O. 2013. Antihyperglycemic profile of erinidine isolated from *Hunteria umbellata* seed. *Afr. J. Tradit. Complement. Altern. Med.* 10: 189–202.

Adeneye, A.A. 2008. Hypoglycemic and hypolipidemic effects of methanol seed extract of *Citrus paradisi* Macfad (Rutaceae) in alloxan-induced diabetic Wistar rats. *Nig. Q. J. Hosp. Med.* 18: 211–215.

Adeneye, A.A. and Adeyemi, O.O. 2009a. Further evaluation of antihyperglycaemic activity of *Hunteria umbellata* (K. Schum) Hallier f. seed extract in experimental diabetes. *J. Ethnopharmacol.* 126: 238–243.

Adeneye, A.A. and Adeyemi, O.O. 2009b. Hypoglycmic effects of the aqueous seed extract of *Hunteria umbellata* in normoglycemic and glucose and nicotine-induced hypoglycemic rats. *Int. J. Appl. Res. Nat. Prod.* 2: 9–18.

Adeneye, A.A., Adeyemi, O.O., Agbaie, E.O., and Banjo, A.A.F. 2010. Evaluation of the toxicity and revers-ibility profile of the aqueous seed extract of *Hunteria umbellata* (K. Schum.) Hallier f. in rodents. *Afr. J. Tradit. Complement. Altern. Med.* 7: 350–369.

Adeneye, A.A., Adenemi, O.O., Agbaje, E.O., and Sofidiya, M.O. 2011. The novel antihyperglycaemic action of *Hunteria umbellata* seed fractions mediated via intestinal glucose uptake inhibition. *Afr. J. Tradit. Complement. Altern. Med.* 9: 17–24.

Adeneye, A.A. and Agbaje, E.O. 2007. Hypoglycemic and hypolipidemic effects of fresh leaf aqueous extract of *Cymbopogon citrates* Stapf in rats. *J. Ethnopharmacol.* 112: 440–444.

Aderibigbe, A., Lawal, B., and Oluwagbemi, J. 1999. The antihyperglycaemic effect of *Telfairia occidentalis* in mice. *Afr. J. Med. Med. Sci.* 28: 115–121.

Aderibigbe, A.O., Emudianughe, T.S., and Lawal, B.A. 2001. Evaluation of the antidiabetic action of *Mangifera indica* in mice. *Phytother. Res.* 15: 456–458.

Adewale, O.B. and Oloyede, O.I. 2012. Hypoglycemic activity of aqueous extract of the bark of *Bridelia ferruginea* in normal and alloxan-induced diabetic rats. *Prime Res. Biotechnol.* 2: 53–56.

Adewole, S.O., Adenowo, T., Naicker, T., and Ojewole, J.A. 2011. Hypoglycaemic and hypotensive effects of *Ficus exasperata* Vahl. (Moraceae) leaf aqueous extract in rats. *Afr. J. Tradit. Complement. Altern. Med.* 8: 275–283.

Adewole, S.O. and Caxton-Martins, E.A. 2006. Morphological changes and hypoglycemic effects of *Annona muricata* Linn. (Annonaceae) leaf aqueous extract on pancreatic β-cells of streptozotocin-treated diabetic rats. *Afr. J. Biomed. Res.* 9: 173–180.

Adewole, S.O. and Ojewole, J.A.O. 2009. Protective effects of *Annona muricata* Linn. (Annonaceae) leaf aqueous extract on serum lipid profiles and oxidative stress in hepatocytes of streptozotocin-treated diabetic rats. *Afr. J. Tradit. Complement. Altern. Med.* 6: 30–41.

Adewole, S.O., Ojo, S.K., Adenowo, T.K., Salako, A.A., Naicker, T., and Ojewole, J.A. 2012. Effects of *Ficus exasperata* Vahl. (Moraceae) leaf aqueous extract on the renal function of streptozotocin-treated rats. *Folia Morphol. (Warsz)* 71: 1–9.

Adewoye, E.O. and Olorunsogo, O.O. 2010. Hypoglycaemic and biochemical properties of *Cnestis ferruginea*. *Afr. J. Tradit. Complement. Altern. Med.* 7: 185–194.

Adewunmi, C.O. and Ojewole, J.A. 2004. Anti-inflammatory and hypoglycemic effects of *Tetrapleura tetraptera* (Taub.) (Fabaceae) fruit aqueous extract in rats. *J. Ethnopharmacol.* 95: 177–182.

Adeyemi, D., Komolafe, O., Adewole, S., Obuotor, E., and Adenowo, T. 2007. Effects of *Annona muricata* (Linn.) on the morphology of pancreatic islet cells of experimentally-induced diabetic Wistar rats. *Int. J. Altern. Med.* 5: 7–16.

Adeyemi, D.O., Komolafe, O.A., Adewole, O.S. Obuotor, E.M., and Adenowo, T.K. 2009. Anti hyperglycemic activities of *Annona muricata* (Linn.). *Afr. J. Tradit. Complement. Altern. Med.* 6: 62–69.

Adeyemi, D.O., Ukwenya, V.O., Obuotor, E.M., and Adewole, S.O. 2014. Anti-hepatotoxic activities of *Hibiscus sabdariffa* L. in animal model of streptozotocin diabetes-induced liver damage. *BMC Complement. Altern. Med.* 14: 277–282.

Adeyemi, O.O., Adeneye, A.A., and Alabi, T.E. 2011. Analgesic activity of aqueous seed extract of *Hunteria umbellata* K. Schum. Hallier f. in rodents. *Indian J. Exp. Biol.* 49: 698–703.

Adiga, S., Bairy, K.L., Meharban, A., and Punita, I.S.R. 2010. Hypoglycemic effect of aqueous extract of *Trichosanthes dioica* in normal and diabetic rats. *Int. J. Diab. Dev. Countries* 30: 38–44.

Adnyana, I.K., Yulinah, E., Yuliet, K., and Kurniati, N.F. 2013. Antidiabetic activity of aqueous leaf extracts of *Guazuma ulmifolia* Lamk., ethanolic extracts of *Curcuma xanthorrhiza* and their combinations in alloxan-induced diabetic mice. *Res. J. Med. Plant* 7: 158–164.

Afaq, F., Saleem, M., Krueger, C.G., Reed, J.D., and Mukhtar, H. 2005. Anthocyanin and hydralyzable tannin rich pomegranate fruit extract modulates MAPK and NF-kappa B pathways and inhibits skin tumerogenesis in CD-1 mice. *Int. J. Cancer* 113: 423–433.

Afifi, F. 1993. Hypoglycemic effects of *Prosopis farcta*. *Int. J. Pharmacogn.* 31: 161–164.

Afifi, F., Saket, M., Jaghabir, M., and Al-Eisawi, D.1997. Effect of *Varthemia iphionoides* on blood glucose level of normal rats and rats with streptozocin-induced diabetes mellitus. *Curr. Ther. Res.* 58: 888–892.

Afify, A.E.M.R., Romeilah, R.R.M., Osfor, M.M.H., and Elbahnasawy, A.S.M. 2013. Evaluation of carrot pomace (*Daucus carota* L.) as hypocholesterolemic and hypolipidemic agent on albino rats. *Not. Sci. Biol.* 5: 7–14.

Afolayan, A.J. and Sunmonu, T.O. 2011. *Artemisia afra* Jacq ameliorates oxidative stress in the pancreas of streptozotocin-induced diabetic Wistar rats. *Biosci. Biotechnol. Biochem.* 75: 2083–2086.

Agrawal, N.K., Gupta, U., Misra, P., Singh, S.P., and Verma, R.C. 2014. Antidiabetic effects of *Acacia tortilis* seed extract in normal and alloxan-induced diabetic rats. *Int. J. Pharm. Sci. Res.* 4: 1392–1397.

Agarwal, P., Rai, V., and Singh, R.B. 1996. Randomized placebo-controlled, single blind trail of holy basil leaves in patients with non insulin dependent diabetes mellitus. *Int. J. Clin. Pharmacol. Ther.* 34: 406–409.

Agarwal, P., Sharma, B., Jain, S.K., Fatima, A., and Alok, S. 2014. Effect of *Convolvulus pluricaulis* Chois. on blood glucose and lipid profile in streptozotocin induced diabetic rats. *Int. J. Pharm. Sci. Res.* 5: 213–219.

Agarwal, V., Sharma, A.K., Upadhyay, A., Singh, G., and Gupta, R. 2012. Hypoglycemic effects of *Citrullus colocynthis* roots. *Acta Pol. Pharm.* 69: 75–79.

Agarwala, I.P., Achar, M.V., Boradkar, R.V., and Roy, N. 1968. Galactagogue action of *Cuminum cyminum* and *Nigella sativa*. *Indian J. Med. Res.* 56: 841–844.

Agbai, E.O. and Nwanegwo, C.O. 2013. Effect of methanolic extract of *Annona muricata* seed on liver function enzymes in alloxan-induced diabetic male mice. *J. Sci. Multidiscip. Res.* 5: 74–83.

Aguh, B.I., Nock, I.H., Ndams, I.S., Agunu, A., and Ukwubile, C.A. 2013. Hypoglycemic activity of *Bauhinia rufescens* in alloxan-induced diabetic rats. *Int. J. Adv. Pharm. Biol. Chem.* 2: 249–255.

Aguilar, F.J.A., Ramos, R.R., Estrada, M.J., Chilpa, R.R., Peredes, B.G., and Flores, S.J.L. 1997. Effects of three Mexican medicinal plants (Asteraceae) on blood glucose levels on healthy mice and rabbits. *J. Ethnopharmacol.* 55: 171–177.

Aguilar-Santamaría, L., Ramírez, G., Nicasio, P., Alegría-Reyes, C., and Herrera-Arellano, A. 2009. Antidiabetic activities of *Tecoma stans* (L.) Juss. ex Kunth. *J. Ethnopharmacol.* 124: 284–288.

Agunbiade, O.S., Ojezele, O.M., Ojezele, J.O., and Ajaye, A.Y. 2012. Hypoglycaemic activity of *Commelina africana* and *Ageratum conyzoides* in relation to their mineral composition. *Afr. Health Sci.* 12: 198–203.

Agunu, A., Abubaker, M.S., Ibrahim, G., and Salisu, Y. 2011. Hypoglycaemic effects of *Acacia albida* Del. (Mimosaceae) methanol root bark extract. *Nig. J. Pharm. Sci.* 8: 66–72.

Ahalya, B., Shankar, K.R., and Kiranmayi, G.V.N. 2014. Exploration of anti-hyperglycemic and hypolipidemic activities of ethanolic extract of *Annona muricata* bark in alloxan induced diabetic rats. *Int. J. Pharm. Sci. Rev. Re*s. 25: 21–27.

Ahangarpour, A., Haidari, H., Fatemeh, R.A.K., Pakmehr, M., Shahbazian, H., Ahmadi, I. et al. 2014. Effect of *Boswellia serrata* supplementation on blood lipid, hepatic enzymes and fructosamine levels in type 2 diabetic patients. *J. Diabetes Metab. Disord.* 13: 29.

Ahangarpour, A., Mohammadian, M., and Dianat, M. 2012. Antidiabetic effect of hydroalcholic *Urtica dioica* leaf extract in male rats with fructose-induced insulin resistance. *Iran J. Med. Sci.* 37: 181–186.

Ahlema, S., Khaleda, H., Wafaa, M., Sofianeb, B., Mohameda, D., and Claude, M.J. 2009. Oral administration of *Eucalyptus globulus* extract reduces the alloxan-induced oxidative stress in rat. *Chem. Biol. Interact.* 181: 71–76.

Ahmad, A., Balakrishnan, B.R., Akhtar, R., and Pimprikar, R. 2009. Anti-diabetic activity of leaves of *Tephrosia villosa* Pers in alloxan-induced diabetic rats. *J. Pharm. Res.* 2: 528–531.

Ahmad, F., Khan, M.M., Rastogi, A.K., Chaubey, M., and Kidwai, J.R. 1991. Effects of epicatechin on cAMP content, insulin release and conservation of proinsulin to insulin in mature and immature rat islets *in vitro*. *Indian J. Exp. Biol.* 29: 516–520.

Ahmad, I., Ibrar, M., Barkatullah, Muhammad, N., Muhammad, Z., and Ali, N. 2013. Pharmacognostic and hypoglycemic studies of *Achyranthus aspera* L. *J. Pharmacogn. Phytother.* 5: 127–131.

Ahmad, M., Akhtar, M.S., Malik, T., and Gilani, A.H. 2000a. Hypoglycaemic action of the flavonoid fraction of *Cuminum nigrum* seeds. *Phytother. Res.* 14: 103–106.

Ahmad, N., Hassan, M.R., Halder, H., and Bennoor, K.S. 1999. Effects of *Momordica charantia* (Karolla) extracts on fasting and postprandial serum glucose levels in NIDDM patients. *Bangladesh Med. Res. Counc. Bull.* 25: 11–13.

Ahmad, R., Srivastava, S.P., Maurya, R., Rajendran, S.M., Arya, K.R., and Srivastava, A.K. 2008. Mild anti-hyperglycaemic activity in *Eclipta alba*, *Berberis aristata*, *Betula utilis*, *Cedrus deodara*, *Myristica fragrans* and *Terminalia chebula*. *Indian J. Sci. Technol.* 1: 1–6.

Ahmad, W., Khan, I., Khan, M.A., Ahmad, M., Subhan, F., and Karim, N. 2014. Evaluation of antidiabetic and antihyperlipidemic activity of *Artemisia indica* Linn. (aerial parts) in streptozotocin induced diabetic rats. *J. Ethnopharmacol.* 151: 618–623.

Ahmed, B., Al-Rehaily, A.J., Al-Howiriny, T.A., El-Sayed, K.A., and Ahmad, M.S. 2003. Scropolioside-D2 and harpagoside-B: Two new iridoid glycosides from *Scrophularia deserti* and their antidiabetic and antiinflammatory activity. *Biol. Pharm. Bull.* 26: 462–467.

Ahmed, I., Adeghate, E., Cummings, E., Sharma, A.K., and Singh, J. 2004. Beneficial effects and mechanism of action of *Momordica charantia* juice in the treatment of streptozotocin induced diabetic mellitus in rats. *Mol. Cell. Biochem.* 261: 63–70.

Ahmed, I., Adeghate, E., Sharma, A.K., Pallot, D.J., and Singh, J. 1998. Effects of *Momordica charantia* fruit juice on islet morphology in the pancreas of streptozotocin diabetic rats. *Diabetes Res. Clin. Pract.* 40: 145–151.

Ahmed, I., Lakhani, M.S., Gillett, M., John, A., and Raza, H. 2001. Hypotriglyceridemic and hypocholesterolic effects of anti diabetic (karela) fruit extract in streptozotocin induced diabetic rats. *Diabetic Res. Clin. Pract.* 51: 155–161.

Ahmed, M., Ahamed, R.N., Aladakatti, R.H., and Ghosesawar, M.G. 2002. Reversible anti-fertility effect of benzene extract of *Ocimum sanctum* leaves on sperm parameters and fructose content in rats. *J. Basic Clin. Physiol. Pharmacol.* 13: 51–59.

Ahmed, M.F., Kazim, S.M., Ghori, S.S., Mehjabeen, S.S., Ahmed, S.R., Ali, S.M. et al. 2010a. Antidiabetic activity of *Vinca rosea* extracts in alloxan-induced diabetic rats. *Int. J. Endocrinol.* 2010: 841090, 6 pages.

Ahmed, O.M., Moneim, A.A., Yazid, I.A., and Mahmoud, A.M. 2010b. Anti-hyperglycemic, anti-hyperlip-idemic and antioxidant effects and the probable mechanisms of action of *Ruta graveolens* infusion and rutin in nicotinamide-streptozotocin-induced diabetic rats. *Diabetol. Croat.* 39: 15–35.

Ahmed, Q.U., Dogarai, B.B.S., Amiroudine, M.Z.A.M., Taher, M., Latip, J., Umar, J. et al. 2012. Anti-diabetic activity of the leaves of *Tetracera indica* Merr. (Dilleniaceae) *in vivo* and *in vitro*. *J. Med. Plants Res.* 6: 5912–5922.

Ahmed, R.S., Seth, V., Pasha, S.T., and Banerjee, B.D. 2000b. Influence of dietary ginger (*Zingiber officinale* Rosc.) on oxidative stress induced by malathion in rats. *Food Chem. Toxicol.* 38: 443–450.

Ahmed, S., Lin, H.-C., Nizam, I., Khan, N.A., Lee, S.-H., and Khan, N.U. 2014. A trimeric proanthocyanidin from the bark of *Acacia leucophloea* Willd. *Rec. Nat. Prod.* 8: 294–298.

Ahmed, Z., Chishti, A.M., Johri, R.K., Bhagat, A., Gupta K.K., and Ram, G. 2009. Antihyperglycemic and antidyslipidemic activity of aqueous extract of *Dioscorea bulbifera* tubers. *Diabetol. Croat.* 38: 63–72.

Ahn, J., Choi, W., Kim, S., and Ha, T. 2011. Anti-diabetic effect of watermelon (*Citrullus vulgaris* Schrad) on streptozotocin-induced diabetic mice. *Food Sci. Biotechnol.* 20: 251–254.

Ahn, J., Um, M.Y., Lee, H., Jung, C.H., Heo, S.H., and Ha, T.Y. 2013. Eleutheroside E, an active component of *Eleutherococcus senticosus*, ameliorates insulin resistance in type 2 diabetic db/db mice. *Evid. Based Complement. Altern. Med.* 2013: 934183, 9 pages.

Ahrén, B. 2007. DPP-4 inhibitors. *Best Pract. Res. Clin. Endocrinol. Metab.* 21: 517–533.

Ahuja, K.D., Robertson, I.K., Geraghty, D.P., and Ball, M.J. 2006. Effects of chili consumption on postpran-dial glucose, insulin, and energy metabolism. *Am. J. Clin. Nutr.* 84: 63–69.

Aissaoui, A., Zizi S., Israili, Z.H., and Lyoussi, B. 2011. Hypoglycemic and hypolipidemic effects of *Coriandrum sativum* L. in Meriones shawi rats. *J. Ethnopharmacol.* 137: 652–661.

Aja, P.M., Nawafor, E.J., Ibiam, A.U., Orji, O.U., Ezeani, N., and Nwali, B.U. 2013. Evaluation of anti-diabetic and liver enzymes activity of aqueous extracts of *Moringa oleifera* and *Bridelia ferruginea* leaves in alloxan induced diabetic albino rats. *Int. J. Biochem. Res. Rev.* 3: 248–258.

Ajabnoor, M.A., Al-Yahya, M.A., Tariq, M., and Jayyab, A.A. 1984. Anti-diabetic activity of *Teucrium oliveri-anum*. *Fitoterapia* 55: 227–230.

Ajaikumar, K.B., Asheef, M., Babu, B.H., and Padikkala, J. 2005. The inhibition of gastric mucosal injury by *Punica granatum* L. methanolic extract. *J. Ethnopharmacol.* 96: 171–176.

Ajani, H., Patel, H.P., Shah, G.B., Acharya, S.R., and Shah, S.K. 2009. Evaluation of antidiabetic effect of methanolic extract of *Inula racemosa* root in rats. *Pharmacologyonline* 3: 118–129.

Ajayi, A.F., Akhigbe, R.E., Adewumi, O.M., Okeleji, L.O. Majaidu, K.B., and Olaleye, S.B. 2012. Effect of ethanolic extract of *Cryptolepis sanguinolenta* stem on *in vivo* and *in vitro* glucose absorption and trans-port: Mechanism of its antidiabetic activity. *Indian J. Endocrinol. Metab.* 16: S91–S96.

Ajikumaran, N.S. 2008. Studies on the anti-diabetic properties of a folklore medicinal fern, *Hemionitis arifo-lia* (Burm.) Moore. PhD thesis, University of Kerala, Trivandrum, India.

Ajikumaran, N.S., Sabulal, B., Radhika, S., Arunkumar, R., and Subramoniam, A. 2014. Promising anti-diabetes mellitus activity of β-amyrin palmitate isolated from *Hemidesmus indicus* roots. *Eur. J. Pharmacol.* 734: 77–82.

Ajikumaran, N.S., Shylesh, B.S., Gopakumar, B., and Subramoniam, A. 2006. Anti-diabetic and hypoglycae-mic properties of *Hemionitis arifolia* (Burm.) Moore in rats. *J. Ethnopharmacol.* 106: 192–197.

Ajikumaran, N.S. and Subramoniam, A. 2005. Indian medicinal plants with anti-diabetes properties. In: *Modern and Alternative Medicine for Diabetes*, Khan, I.A., Khanum, A., and Khan, A.H. (eds.). Hydrabad, India: Ukaaz Publications, pp. 43–193.

Ajila, C.M., Rao, L.J., and Prasada Rao, U.J.S. 2010. Characterization of bioactive compounds from raw and ripe *Mangifera indica* L. fruit peel extracts. *Food Chem. Toxicol.* 48: 3406–3411.

Ajithadas, A., Vijayalakshmi, K., Karthikeyan, V., Nandhini, S.R., and Jegadeesh, S. 2014. Water melon plant (*Citrullus lanatus*): Pharmacognostical standardization and phytochemical screening of its leaves. *Pharmanest* 5: 2184–2191.

Akah, P.A., Uzodinma, S.U., and Okolo, C.E. 2011. Anti-diabetic activity of aqueous and methanol extract and fractions of *Gongronema latifolium* (Asclepidaceae) leaves in alloxan diabetic rats. *J. Appl. Pharm. Sci.* 1: 99–102.

Akan, Z. 2014. Improvements of diabetes mellitus related nephropathy and blood bio-chemicals by the *Thymus vulgaris* L. and *Thymbra spicata* L. *Med. Sci. Discov.* 1: 51–60.

Akanksha, Srivastava, A.K., and Maurya, R. 2010. Antihyperglycemic activity of compounds isolated from Indian medicinal plants. *Indian J. Exp. Biol.* 48: 294–298.

Akase, T., Shimada, T., Terabayashi, S., Ikeya, Y., Sanada, H., and Aburada, M. 2011. Antiobesity effects of *Kaempferia parviflora* in spontaneously obese type 2 diabetic mice. *J. Nat. Med.* 65: 73–80.

Akhani, S.P., Vishwakarma, S.L., and Goyal, R.K. 2004. Antidiabetic activity of *Zingiber officinale* in streptozotocin induced type I diabetic rats. *J. Pharm. Pharmacol.* 56: 101–105.

Akhlaghi, F., Rajaei, Z., Hdjzadeh, M., Irasshahi, M., and Alizadeh, M. 2012. Anti-hyperglycemic effect of Asafoetida (*Ferula assafoetida* oleo-gum-resin) in streptozotocin-induced diabetic rats. *World Appl. Sci. J.* 17: 157–162.

Akhtar, M.S. and Ali, M.R. 1985. Study of hypoglycaemic activity of *Cuminum nigrum* seeds in normal and alloxan diabetic rabbits. *Planta Med.* 51: 81–85.

Akhtar, M.S. and Bano, H. 2002. Hypoglycemic effect of powdered *Alstonia scholaris* (Satona). *Prof. Med. J.* 9: 268–271.

Akhtar, M.S. and Iqbal, J. 1991. Evaluation of hypoglycemic effect of *Achyranthes aspera* in normal and alloxan diabetic rats. *J. Ethnopharmacol.* 31: 49–57.

Akhtar, M.S., Khan, Q.M., and Khaliq, T. 1984. Effects of *Euphorbia prostrata* and *Fumaria parviflora* in normoglycaemic and alloxan-treated hyperglycaemic rabbits. *Planta Med.* 50: 138–142.

Akhtar, M.S. and Qureshi, A.O. 1988. Phytopharmacological evaluation of *Ficus glomerata*, Roxb. fruit for hypoglycaemic activity in normal and diabetic rabbits. *Pak. J. Pharm. Sci.* 1: 87–96.

Akhtar, M.S. and Qureshi, A.Q. 1998. Phytopharmacological evaluation of *Ficus glomerata* Roxb. fruit for hypoglycemic activity in normal and diabetic rabbits. *Pak. J. Pharm. Sci.* 1: 877–889.

Akhtar, M.S., Qureshi, A.Q., and Iqbal, J. 1990. Anti-diabetic evaluation of *Mucuna pruriens* L. seed. *J. Pak. Med. Assoc.* 40: 147–150.

Akhtar, M.S., Sajid, S.M., Akhtar, M.S., and Ahmad, M. 2008. Hypoglycemic effect of *Berberis aristata* root, its aqueous and methanolic extracts in normal and alloxan induced diabetic rabbits. *Pharmacologyonline* 2: 845–856.

Akinmoladun, A.C., Farombi, E.O., and Oguntibeju, O.M. 2014. Antidiabetic botanicals and their potential benefits in the management of diabetes mellitus. In: *Anti-Oxidants-Antidiabetic Agents and Human Health*, Oguntibeju, O. (ed.). Croatia: InTech. http://dx.doi.org/10.5772/57339.

Akinola, O.S., Akinola, O.B., and Caxton-Martins, E.A. 2009. *Vernonia amygdalina* upregulates hepatic enzymes and improves liver microanatomy in experimental diabetes mellitus. *Pharmacologyonline* 2: 1231–1242.

Akinpelu, D.A. 2001. Antimicrobial activity of *Anacardium occidentale* bark. *Fitoterapia* 72: 286–287.

Akpan, E.J., Okokon, J.E., and Offong, E. 2012. Antidiabetic and hypolipidemic activities of ethanolic leaf extract and fractions of *Melanthera scandens*. *Asian Pac. J. Trop. Biomed.* 2: 523–527.

Akter, A., Rahman, S., Morshed, M.T., Hossain, S., Jahan, S., Swarna, A. et al. 2013. Evaluation of antihyperglycemic and antinociceptive potential of *Colocasia esculenta* (L.) Schott (Araceae) leaves. *Adv. Nat. Appl. Sci.* 7: 143–148.

Al-Aboudi, A. and Afifi, F.U. 2011. Plants used for the treatment of diabetes in Jordan: A review of scientific evidence. *J. Pharmacol. Biol.* 49: 221–239.

Alam, M.R., Rahman, A.B., Moniruzzaman, M., Kadir, M.F., Haque, M.A., Alvi, M.R. et al. 2012. Evaluation of anti-diabetic phytochemicals in *Syzygium cumini* (L.) Skeels (Family: Myrtaceae). *J. Appl. Pharm. Sci.* 2: 94–98.

Al-Mugarrabun, L.M., Ahmat, N., Ruzaina, S.A., Ismail, N.H., and Sahidin, I. 2013. Medicinal uses, phytochemistry and pharmacology of *Pongamia pinnata* (L.) Pierre: A review. *J. Ethnopharmacol.* 150: 395–420.

Alamgeer, Mushtaq, M.N., Bashir, S., Rashid, M., Malik, M.N.H., Ahmed, T. et al. 2013. Evaluation of hypoglycemic activity of *Thymus serpyllum* Linn. in glucose treated mice. *Int. J. Basic Med. Sci. Pharm.* 3: 33–36.

Alamgeer, Mushtaq, M.N., Bashir, S., Rashid, M., Malik, M.N.H., Ghumman, S.K. et al. 2012. Hypoglycemic and hematological effects of aqueous extract of *Thymus serpyllum* Linn. in alloxan-induced diabetic rabbits. *Afr. J. Pharm. Pharmacol.* 6: 2845–2850.

Alamgir, M., Rokeya, B., Hannan, J.M., and Choudhuri, M.S. 2001. The effect of *Premna integrifolia* Linn. (Verbenaceae) on blood glucose in streptozotocin induced type 1 and type 2 diabetic rats. *Pharmazie* 56: 903–904.

Alarcon-Aguilar, F.J., Campos-Sepulveda, A.E., Xolalpa-Molina, S., Hernandez-Galicia, E., and Roman-Ramos, R. 2002a. Hypoglycaemic activity of *Ibervillea sonorae* roots in healthy and diabetic mice and rats. *Pharm. Biol.* 40: 570–575.

Alarcon-Aguilar, F.J., Fortis-Barrera, A., Angeles-Mejia, S., Banderas-Dorantes, T.R., Jasso-Villagomez, E.I., Almanza-Perez, J.C. et al. 2010. Anti-inflammatory and antioxidant effects of a hypoglycemic fructan fraction from *Psacalium peltatum* (H.B.K.) Cass. in streptozotocin-induced diabetes mice. *J. Ethnopharmacol.* 132: 400–407.

Alarcon-Aguilar, F.J. and Roman-Ramos, R. 2005. Antidiabetic plants in Mexico and Central America. In: *Traditional Medicines for Modern Times: Antidiabetic Plants,* Soumyanath, A. (ed.). Boca Raton, FL: CRC Press, pp. 221–224.

Alarcon-Aguilar, F.J., Roman-Ramos, R., Flores-Saenz, J.L., and Aguirre-Garcia, F. 2002b. Investigation on the hypoglycaemic effects of extracts of four Mexican medicinal plants in normal and alloxan-diabetic mice. *Phytother. Res.* 16: 383–386.

Alarcon-Aguilar, F.J., Roman-Ramos, R., Flores-Saenz, J.L., Aguirre-Garcia, F., Contreras-Weber, C.C., and Flores-Saenz, J.L. 1998. Study of the anti-hyperglycemic effects of plants used as anti-diabetics. *J. Ethnopharmacol.* 61: 101–110.

Alarcon-Aguilar, F.J., Valdes-Arzata, A., Xolalpa-Molina, S., Banderas-Dorantes, T., Jimenez-Estrada, M., Hernandez-Galicia, E. et al. 2003. Hypoglycemic activity of two polysaccharides isolated from *Opuntia ficus-indica* and *O. streptacantha. Proc. West. Pharmacol. Soc.* 46: 139–142.

Alariri, K., Meng, K.Y., Atangwho, I.J., Asmawi, M.Z., Sadikun, A., Murugaiyah, V. et al. 2013. Hypoglycemic and anti-hyperglycemic study of *Gynura procumbens* leaf extracts. *Asian Pac. J. Trop. Biomed.* 3: 358–366.

Albrecht, M., Jiang, W., Kumi-Diaka, J., Lansky, E.P., Gommersall, L.M., Patel, A. et al. 2004. Pomegranate extracts potently suppress proliferation, xenograft growth, and invasion of human prostate cell cancer. *J. Med. Food* 7: 274–283.

Alese, M.O., Adewole, O.S., Ijomone, O.M., Ajayi, S.A., and Alese, O.O. 2014. Hypoglycemic and hypolipidemic activities of methanolic extract of *Sphenocentrum jollyanum* on streptozotocin-induced diabetic Wistar rats. *Eur. J. Med. Plants* 4: 353–364.

Alese, M.O., Adewole, S.T., Ijomone, M.O., Ajayi, S.A., and Omonisi, A. 2013. Histological studies of β-cells of streptozotocin-induced diabetic Wistar rats treated with ethanolic extract of *Sphenocentrum jollyanum. J. Pharm. Sci. Innov.* 2: 8–12.

Alexander-Lindo, R.L., Morrison, E.Y., and Nair, M.G. 2004. Hypoglycemic effect of stigmast-4-en-3-one and its corresponding alcohol from the bark of *Annarcardium occidentale* (chasew). *Phytother. Res.* 18: 403–407.

Algandaby, M.M., Alghamdi, H.A., Ashour, Q.M., Abdel-Naim, A.B., Ghareib, S., Abdel-Sattar, E.A. et al. 2010. Mechanisms of the antihyperglycemic activity of *Retama raetam* in streptozotocin-induced diabetic rats. *Food Chem. Toxicol.* 48: 2448–2453.

Al-Ghaithi, F., El-Ridi, M.R., Adeghate, E.E., and Amiri, M.H. 2004. Biochemical effects of *Citrullus colocynthis* in normal and diabetic rats. *Mol. Cell. Biochem.* 261: 143–149.

Ali, B.H. 1997. The effect on plasma glucose, insulin, and glucagon levels of treatment of diabetic rats with the medicinal plant *Rhazya stricta,* and with glibenclamide, alone and in combination. *J. Pharm. Pharmacol.* 49: 1003–1007.

Ali, B.H., Al-Qarawi, A.A., Bashir, A.K., and Tanira, M.O. 2000. Phytochemistry, pharmacology, and toxicity of *Rhazya stricta* Decne: A review. *Phytother. Res.* 14: 229–234.

Ali, K.M., Bera, T.K., Mandal, S., Barik, B.R., and Ghosh, D. 2010. Attenuation of diabetic disorders in experimentally induced diabetic rat by methanol extract of seed of *Holarrhena antidysenterica. Int. J. PharmTech Res.* 1: 1205–1211.

Ali, K.M., Chatterjee, K., De, D., Bera, T.K., and Ghosh, D. 2009. Efficacy of aqueous extract of seed of *Holarrhena antidysenterica* for the management of diabetes in experimental model rat: A correlation study with antihyperlipidemic activity. *Int. J. Appl. Res. Nat. Prod.* 2: 13–21.

Ali, L., Khan, AK., Mamun, M.I., Mosihuzzaman, N., Nur-e-Alam, M., and Rokeya, B. 1993. Studies on hypoglycemic effects of fruit pulp, seed, and whole plant of *Momordica charantia* on normal and diabetic model rats. *Planta Med.* 59: 408–412.

Ali, M.A., Sultana, M.C., Rahman, B.M., Khatune, N.A., and Waheed, M.I.I. 2012a. Antidiabetic activity of ethanol extract of *Semecarpus anacardium* (Linn.) stem barks in normal and alloxan induced diabetic rats. *Int. J. Pharm. Sci. Res.* 3: 2680–2685.

Ali, M.S., Sagar, H.A., Khatun, M.C.S., Azad, A.K., Begum, K., and Wahed, M.I.I. 2012b. Anti-hyperglycemic and analgesic activities of ethanol extract of *Cassia fistula* (L.) stem bark. *Int. J. Pharm. Sci. Res.* 3: 416–423.

Ali, S., Igoli, J., Clements, C., Semaan, D., Alamzeb, M., Rashid, M.U. et al. 2013. Antidiabetic and anti-microbial activities of fractions and compounds isolated from *Berberis brevissima* Jafri and *Berberis parkeriana* Schneid. *Bangladesh J. Pharmacol.* 8: 336–342.

Ali-Shtayeh, M.S., Jamous, R.M., and Jamous, R.M. 2011. Complementary and alternative medicine use amongst Palestinian diabetic patients. *Complement. Ther. Clin. Pract.* 18: 16–21.

Al-Khateeb, E., Hamadi, S.A., Al-Hakeemi, A.A.N., Abu-Taha, M., and Al-Rawi, N. 2012. Hypoglycemic effect of trigonelline isolated from Iraqi fenugreek seeds in normal and alloxan-diabetic rabbits. *Eur. Sci. J.* 8: 16–24.

Almasri, M., Mohammad, M.K., and Tahaa, M.O. 2009. Inhibition of dipeptidyl peptidase IV (DPP IV) is one of the mechanisms explaining the hypoglycemic effect of berberine. *J. Enzyme Inhib. Med. Chem.* 24: 1061–1066.

Almeida, R.N., Filho, J.M.B., and Naik, S.R. 1985. Chemistry and pharmacology of an ethanol extract of *Bumelia sartorum*. *J. Ethnopharmacol.* 14: 173–185.

Al-Nozha, M.M., Al-Maatoug, M.A., Al-Mazrou, Y.Y., Al-Harthi, S.S., Arafah, M.R., Khalil, M.Z. et al. 2004. Diabetes mellitus in Saudi Arabia. *Saudi Med. J.* 25: 1603–1610.

Alonso-Castro, A.J., Miranda-Torres, A.C., Gonzalez-Chavez, M.M., and Salazar-Olivo, L.A. 2008. Chlorogenic acid stimulate 2-NBD glucose uptake in both insulin-sensitive and insulin-resistant 3T3 adipocytes. *J. Ethnopharmacol.* 120: 458–464.

Alonso-Castro, A.J. and Salazar-Olivo, L.A. 2008. The anti-diabetic properties of *Guazuma ulmifolia* Lam are mediated by the stimulation of glucose uptake in normal and diabetic adipocytes without inducing adipogenesis. *J. Ethnopharmacol.* 118: 252–256.

Alonoso-Castro, A.J., Zapata-Bustos, R., Gomez-Espinoza, G., and Salazar-Olivo, L.A. 2012. Isoorientin reverts TVF-α-induced insulin resistance in adipocytes activating the insulin signaling pathway. *Endocrinology* 153: 5222–5230.

Alonso-Castro, A.J., Zapata-Bustos, R., Romo-Yañez, J., Camarillo-Ledesma, P., Gomez-Sanchez, M., and Salazar-Olivo, L.A. 2010. The antidiabetic plants *Tecoma stans* (L.) Juss. ex Kunth (Bignoniaceae) and *Teucrium cubense* Jacq (Lamiaceae) induce the incorporation of glucose in insulin-sensitive and insulin-resistant murine and human adipocytes. *J. Ethnopharmacol.* 127: 1–6.

Al-Romaiyan, A., Jayasri, M.A., Mathew, T.L., Huang, G.C., Amiel, S., Jones, P.M. et al. 2010. *Costus pictus* extracts stimulate insulin secretion from mouse and human islets of Langerhans *in vitro*. *Cell. Physiol. Biochem.* 26: 1051–1058.

Al-Snafi, A.E. 2013. The pharmaceutical importance of *Althaea officinalis* and *Althaea rosea*: A review. *Int. J. PharmTech Res.* 5: 1378–1385.

Althunibat, O.Y., Al-Mustafa, A.H., Tarawneh, K., Khleifat, K.M., Ridzwan, B.H., and Qaralleh, H.N. 2010. Protective role of *Punica granatum* L. peel extract against oxidative damage in experimental diabetic rats. *Process Biochem.* 45: 581–585.

Altman, R.D. and Marcussen, K.C. 2001. Effect of ginger extract on knee pain in patients with osteoarthritis. *Arthritis Rheum.* 44: 2531–2538.

Al-Waili, N.S. 1986. Treatment of diabetes mellitus by *Artemisia herba-alba* extract: Preliminary study. *Clin. Exp. Pharmacol. Physiol.* 13: 569–573.

Aly, H.F., Ebrahim, M.E., Metawaa, H.M., Hosni, E.A.A., and Ebrahim, F.M. 2010. *In vitro* and *in vivo* evaluation of the anti-diabetic effect of different extracts of *Nepeta cataria* in streptozotocin-induced diabetic rats. *J. Am. Sci.* 6: 364–386.

Al-Youssef, H.M. and Hassan, W.H.B. 2014. Phytochemical and pharmacological aspects of *Carissa edulis* Vahl: A review. *Int. J. Curr. Res. Chem. Pharm. Sci.* 1: 12–24.

Amalraj, T. and Ignacimuthu, S. 1998a. Hypoglycemic activity of *Cajana cajan* (seeds) in mice. *Indian J. Exp. Biol.* 36: 1032–1033.

Amalraj, T. and Ignacimuthu, S. 1998b. Evaluation of the hypoglycemic effect of *Memecylon umbellatum* in normal and alloxan diabetic mice. *J. Ethnopharmacol.* 62: 247–250.

Amer, M., El-Habibi, S., and El-Gendy, A. 2004. Effects of *Trifolium alexandrinum* extracts on streptozotocin-induced diabetes in male rats. *Ann. Nutr. Metab.* 48: 343–347.

Amer, N. 2012. Effects of soybean seed on glucose levels, lipid profiles and histological structures of the liver in alloxan-induced diabetic albino rats. *Tikrit J. Pure Sci.* 17: 1–5.

Amin, A., Lofty, M., Shafiullah, M., and Adeghate, E. 2006. The protective effect of *Tribulua terrestris* in diabetes. *Ann. N.Y. Acad. Sci.* 1084: 391–401.

Amin, H.A.F. and Azli, R. 2004. Effect of cocoa powder extract on plasma glucose levels in hyperglycaemic rats. *Nutr. Food Sci.* 34: 116–121.

Amorim, C.Z., Marques, A.D., and Cordeiro, R.S. 1991. Screening of the antimalarial activity of plants of Cucurbitacea family. *Mem. Inst. Oswaldo Cruz.* 86: 177–180.

Amudhan, M.S., Begum, H.V., and Hebbar, K.B. 2012. A review on phytochemical and pharmacological potential of *Areca catechu* L. seed. *Int. J. Pharm. Sci. Res.* 3: 4151–4157.

Analco, J.A.G., Martineau, L., Saleem, A., Madiraju, P., Muhammad, A., Durst, T., Haddad, P., and Arnason, J.T. 2010. Bioassay-guided isolation of the antidiabetic principle from *Sorbusdecora* (Rosaceae) used traditionally by the Eeyou Istchee Cree First Nations. *J. Nat. Prod.* 73: 1519–1523.

Anand, K.K., Singh, B., Grand, D., Chandan, B.K., and Gupta, V.N. 1989. Effect of *Zizyphus sativa* leaves on blood glucose levels in normal and alloxan-diabetic rats. *J. Ethnopharmacol.* 27: 121–127.

Anand, P., Murali, K.Y., Tandon, V., Chandra, R., and Murthy, P.S. 2007. Preliminary studies on antihyperglycemic effect of aqueous extract of *Brassica nigra* (L.) Koch in streptozotocin induced diabetic rats. *Indian J. Exp. Biol.* 45: 696–701.

Anand, P., Murali, K.Y., Tandon, V., Murthy, P.S., and Chandra, R. 2009. Insulinotropic effect of aqueous extract of *Brassica nigra* improves glucose homeostasis in streptozotocin induced diabetic rats. *Exp. Clin. Endocrinol. Diabetes* 117: 151–156.

Ananda Prabu, K., Kumarappan, C.T., Christudas, S., and Kalaichelvan, V.K. 2012. Effect of *Biophytum sensitivum* on streptozotocin and nicotinamide induced diabetic rats. *Asian Pac. J. Trop. Biomed.* 2: 31–35.

Anandharajan, R., Jaiganesh, S., Shankernarayanan, N.P., Viswakarma, R.A., and Balakrishnan, A. 2006. *In vitro* glucose uptake activity of *Aegles marmelos* and *Syzygium cumini* by activation of Glut4, PIP3 kinase and PPARgamma in L6 myocytes. *Phytomedicine* 13: 434–441.

Anandharajan, R., Pathmanathan, K., Shankernarayanan, N.P., Vishwakarma, R.A., and Balakrishnan, A. 2005. Upregulation of Glut-4 and PPAR gamma by an isoflavone from *Pterocarpus marsupium* on L6 myotubes: A possible mechanism of action. *J. Ethnopharmacol.* 97: 253–260.

Ananthi, J., Prakasam, A., and Pugalendi, K.V. 2003. Antihyperglycemic activity of *Eclipta alba* leaf on alloxan-induced diabetic rats. *Yale J. Biol. Med.* 76: 97–102.

Andallu, B., Suryakantham, V., Lakshmi Srikanthi, B., and Reddy, G.K. 2001. Effect of mulberry (*Morus indica* L.) therapy on plasma and erythrocyte membrane lipids in patients with type 2 diabetes. *Clin. Chem. Acta* 314: 47–53.

Andallu, B. and Varadacharyulu, N.C. 2003. Anti-oxidant role of (*Morus indica* L.) leaves instreptozotocin-diabetic rats. *Clin. Chem. Acta* 338: 3–10.

Anderson, R.A., Broadhurst, C.L., Polansky, M.M., Schmidt, W.E., Khan, A., Flanagan, V.P. et al. 2004. Isolation and characterization of polyphenol type-A polymers from cinnamon with insulin-like biological activity. *J. Agric. Food Chem.* 52: 65–70.

Anderson, R.A. and Polansky, M.M. 2002. Tea enhances insulin activity. *J. Agric. Food Chem.* 50: 7182–7186.

Andrade-Cetto, A. 2011. Inhibition of gluconeogenesis by *Malmea depressa* root. *J. Ethnopharmacol.* 137: 930–933.

Andrade-Cetto, A. 2013. Hypoglycemic effect of *Malmea depressa*, a plant used in the treatment of type 2 diabetes in Yucatan, Mexico. *Planta Med.* 79: PB4.

Andrade-Cetto, A., Becerra-Jimenez, J., and Cardenas-Vazquez, R. 2008b. Alfa-glucosidase-inhibiting activity of some Mexican plants used in the treatment of type 2 diabetes. *J. Ethnopharmacol.* 116: 27–32.

Andrade-Cetto, A. and Mares, M.L.R. 2012. Hypoglycemic effect of the *Rhizophora mangle* cortex on streptozotocin-nicotinamide-induced diabetic rats. *Pharmacologyonline* 3: 1–5.

Andrade-Cetto, A., Martinez-Zurita, E., Soto-Constantino, A., Revilla-Monsalve, C., and Wiedenfeld, H. 2008a. Chronic hypoglycemic effect of *Malmea depressa* root on n5-streptozotocin diabetic rats. *J. Ethnopharmacol.* 116: 358–362.

Andrade-Cetto, A., Revilla-Monsalve, C., and Wiedenfeld, H. 2007. Hypoglycemic effect of *Tournefortia hirsutissima* L., on streptozotocin diabetic rats. *J. Ethnopharmacol.* 112: 96–100.

Andrade-Cetto, A. and Vazquez, R.C. 2010. Gluconeogenesis inhibition and phytochemical composition of two *Cecropia* species. *J. Ethnopharmacol.* 130: 93–97.

Andrade-Cetto, A. and Wiedenfeld, H. 2001. Hypoglycemic effect *Cecropia obtusifolia* on streptozotocin diabetic rats. *J. Ethnopharmacol.* 78: 145–149.

Andrade-Cetto, A. and Wiedenfeld, H. 2004. Hypoglycemic effect of *Acosmium panamense* bark on streptozotocin diabetic rats. *J. Ethnopharmacol.* 90: 217–220.

Andrade-Cetto, A. and Wiedenfeld, H. 2011. Anti-hyperglycemic effect of *Opuntia streptacantha* Lem. *J. Ethnopharmacol.* 133: 940–943.

Andrade-Cetto, A., Zurita, E.M., and Wiedenfeld, H. 2005. Hypoglycemic effect of *Malmea depressa* root on streptozotocin diabetic rats. *J. Ethnopharmacol.* 100: 319–322.

Angel, J., Sailesh, K.S., and Mukkadan, J.K. 2013. A study on anti-diabetic effect of peppermint in alloxan induced diabetic model of Wistar rats. *J. Clin. Biomed. Sci.* 3: 177–181.

Angel de la Fuente, J. and Manzanaro, S. 2003. Aldose reductase inhibitors from natural sources. *Nat. Prod. Rep.* 20: 243–251.

Anila, L. and Vijayalakshmi, N.R. 2002. Flavonoids from *Emblica officinalis* and *Mangifera indica*: Effectiveness for dyslipidemia. *J. Ethnopharmacol.* 79: 81–87.

Anis, E., Anis, I., Ahmed, S., Mustafa, G., Malik, A., Afza, N., Hal, S.M., Shahzad-ul-hussan, S., and Choudhary, M.I. 2002. Alpha glucosidase inhibitory constituents from *Cuscuta reflexa*. *Chem. Pharm. Bull.* 50: 112–114.

Anthony, O.E., Ese, A.C., and Lawrence, E.O. 2013. Regulated effects of *Capsicum frutescens* supplemented diet on fasting blood glucose level, biochemical parameters and body weight in alloxan induced diabetic Wistar rats. *Br. J. Pharm. Res.* 3: 496–507.

Antia, B.S, Okokon, J.E., and Okon, P.A. 2005. Hypoglycemic activity of aqueous leaf extract of *Persea americana* Mill. *Indian J. Pharmacol.* 37: 325–326.

Antu, K.A., Riya, M.P., Mishra, A., Anilkumar, K.S., Chandrakanth, C.K., Tamrakar, A.K. et al. 2014. Antidiabetic property of *Symplocos cochinchinensis* is mediated by inhibition of alpha glucosidase and enhanced insulin sensitivity. *PLOS One.* 9: e105829.

Anu, S.J. and Rao, J.M. 2002. Oxanthrone esters from the roots of *Cassia kleinii*. *Phytochemistry* 59: 425–427.

Anupama, V., Narmadha, R., Gopalakrishnan, V.K., and Devaki, K. 2012. Enzymatic alteration in the vital organs of streptozotocin diabetic rats treated with aqueous extract of *Erythrina variegata* Bark. *Int. J. Pharm. Pharm. Sci.* 4: 134–147.

Anurakkun, N.J., Bhandari, M.R., and Kawabata, J. 2007. α-Glucosidase inhibitors from devil tree (*Alstonia scholaris*). *Food Chem.* 103: 319–323.

Aragao, D.M., Guarize, L., Lanini J., da Costa, J.C., Garcia, R.M., and Scio, E. 2010. Hypoglycemic effects of *Cecropia pachystachya* in normal and alloxan-induced diabetic rats. *J. Ethnopharmacol.* 128: 629–633.

Aragno, M., Mastrocola, R., Catalano, M.G., Brignardello, E., Danni, O., and Boccuzzi, G. 2004. Oxidative stress impairs skeletal muscle repair in diabetic rats. *Diabetes* 53: 1082–1088.

Arai, H., Hirasawa, Y., Rahman, A., Kusumawati, I., Zaini, N.C., Sato, S. et al. 2010. Alstiphyllanines E-H, picraline and ajmaline-type alkaloids from *Alstonia macrophylla* inhibiting sodium glucose cotransporter. *Bioorg. Med. Chem.* 18: 2152–2158.

Arambewela, L.S., Arawwawala, L.D., and Ratnasooriya, W.D. 2005. Anti-diabetic activities of aqueous and ethanolic extracts of *Piper betle* leaves in rats. *J. Ethnopharmacol.* 102: 239–245.

Aransiola, E.F., Daramola, M.O., Iwalewa, E.O., Seluwa, A.M., and Olufowobi, O.O. 2014. Anti-diabetic effect of *Bryophyllum pinnatum* leaves. *Int. J. Biol. Vet. Agric. Food Eng.* 8: 89–93.

Arawwawala, M., Thabrew, I., and Arambewela, L. 2009. Anti-diabetic activity of *Trichosanthes cucumerina* in normal and streptozotocin-induced diabetic rats. *Int. J. Biol. Chem. Sci.* 3: 287–296.

Arawwawala, M., Thabrew, I., and Arambewela, L. 2012. Lipid lowering effect of hot water extract of *Trichosanthes cucumerina* Linn. and anti-hyperglycemic activity of its fractions on streptozotocin-induced diabetic rats. *Isr. J. Plant Sci.* 60: 447–455.

Arawwawala, M., Thabrew, I., and Arambewela, L. 2013. A review of the pharmacological properties of *Trichosanthes cucumerina* Linn. of Sri Lankan origin. *Unique J. Pharm. Biol. Sci.* 1: 3–6.

Ardestani, A. and Yazdanparast, R. 2007. *Cyperus rotundus* suppresses AGE formation and protein oxidation in a model of fructose-mediated protein glycoxidation. *Int. J. Biol. Macromol.* 41: 572–578.

Arif, T., Sharma, B., Gahlauit, A., Kumar, V., and Dabur, R. 2014. Antidiabetic agents from medicinal plants: A review. *Chem. Biol. Lett.* 1: 1–13.

Aritajat, S., Wutteerapol, S., and Saenphet, K. 2004. Anti-diabetic effect of *Thunbergia laurifolia* Linn. aqueous extract.. *S.E. Asian J. Trop. Med. Public Health* 35: 53–58.

Arokiyaraj, S., Balamurugan, R., and Augustin, P. 2011. Antihyperglycemic effect of *Hypericum perforatum* ethyl acetate extract on streptozotocin-induced diabetic rats. *Asian Pac. J. Trop. Biomed.* 1: 386–390.

Arseculeratne, S.N., Gunatilaka, A.A., and Panabokke, R.G. 1981. Studies on medicinal plants of Sri Lanka: Occurrence of pyrrolizidine alkaloids and hepatotoxic properties in some traditional medicinal herbs. *J. Ethnopharmacol.* 4: 159–177.

Arul, B., Kothai, R., and Chritina, A.J. 2004. Hypoglycemic and antihyperglycemic effect of *Semecarpus anacardium* L. in normal and streptozotocin induced diabetic rats. *Methods Find. Exp. Clin. Pharmacol.* 26: 759–762.

Arul, B., Kothai, R., and Christina, A.M.J. 2006. Hypoglycemic activity of *Casearia esculenta* Roxb. in normal and diabetic albino rats. *Indian J. Pharm. Res.* 5: 47–51.

Arulmozhi, S., Mazumder, P.M., Lohidasan, S., and Thakurdesai, P. 2010. Antidiabetic and antihyperlipidemic activity of leaves of *Alstonia scholaris* Linn. R. Br. *Eur. J. Integr. Med.* 2: 23–32.

Arulselvan, P. and Subramanian, S.P. 2007. Beneficial effects of *Murraya koenigii* leaves on antioxidant defense system and ultra structural changes of pancreatic β-cells in experimental diabetes in rats. *Chem. Biol. Interact.* 165: 155–164.

Arumugam, B., Manaharan, T., Heng, C.K., Kuppusamy, U.R., and Palanisamy, U.D. 2014. Antioxidant and antiglycemic potentials of a standardized extract of *Syzygium malaccense*. *Food Sci. Technol.* 59: 707–712.

Arumugam, S. and Siddhuraju, P. 2013. Protective effect of bark and empty pod extracts from *Acacia auriculiformis* against paracetamol intoxicated liver injury and alloxan-induced type-2 diabetes. *Food Toxicol.* 56: 162–170.

Arun, N. and Nalini, N. 2002. Efficacy of turmeric on blood sugar and polyol pathway in diabetic albino rats. *Plant Food Hum. Nutr.* 57: 41–52.

Aruna, A., Vijayalakshmi, K., and Karthikeyan, V. 2014. Anti-diabetic screening of methanolic extract of *Citrullus lanatus* leaves. *Am. J. PharmTech Res.* 4: 295–323.

Arunachalam, K. and Parimelazhagan, T. 2012. Antidiabetic activity of aqueous root extract of *Merremia tridentata* (L.) Hall. f. in streptozotocin-induced-diabetic rats. *Asian Pac. J. Trop. Med.* 5: 175–179.

Arunachalam, K. and Parimelazhagan, T. 2013. Antidiabetic activity of *Ficus amplissima* Smith. bark extract in streptozotocin induced diabetic rats. *J. Ethnopharmacol.* 147: 302–310.

Arunkumar, R., Ahmed, A.B.A., Mani, V.P., and Bastin, T.M.M.J. 2010. Anti-hyperlipidemic and hypoglycemic activity of *Wattakaka volubilis* methanol extract in alloxan-induced diabetic rats. *J. Pharm. Res.* 3: 1913–1915.

Arva, H.R., Bhaskar, J.J., Salimath, P.V., and Mallikarjuna, A.S. 2013. Anti-diabetic effects of elephant-food yam (*Amorphophallus paeoniifolious* (Dennst.) Nicolson) in streptozotocin-induced diabetic rats. *Int. J. Biomed. Pharm. Sci.* 7: 6.

Arya, A., Abdullah, M.A., Haerian, B.S., and Mohd, M.A. 2012. Screening for hypoglycemic activity on the leaf extracts of nine medicinal plants: *In vivo* evaluation. *E-J. Chem.* 9: 1196–1205.

Arya, S., Saini, J., and Singh, S. 2013. Antidiabetic activities of *Cassia occidentalis*. *Recent Res. Sci. Technol.* 5: 51–53.

Asgarpanah, J. and Haghighat, E. 2012. Phytochemistry and pharmacological properties of *Ziziphus spina christi* (L.) Willd. *Afr. J. Pharm. Pharmacol.* 6: 2332–2339.

Asgary, S., Rahimi, P., Mahzouni, P., and Madani, H. 2012. Antidiabetic effect of hydroalcoholic extract of *Carthamus tinctorius* L. in alloxan-induced diabetic rats. *J. Res. Med. Sci.* 17: 386–392.

Asha, M.K., Prashanth, D., Murali, B., Padmaja, R., and Amit, A. 2001. Anthelminthic activity of essential oil of *Ocimum sanctum* and eugenol. *Fitoterapia* 72: 669–670.

Ashcroft, F.M. and Rorsman, P. 2012. Diabetes mellitus and the β cell: The last ten years. *Cell* 148: 1160–1171.

Ashok, P. and Meenakshi, B. 2004. Contraceptive effect of *Curcuma longa* L. in male albino rats. *Asian J. Androl.* 6: 71–74.

Ashok Kumar, B.S., Lakshman, K., Jayaveea, K.N., Sheshadri S.D., Saleemulla, K., Thippeswamy, B.S., and Veerapur, V.P. 2012. Antidiabetic, antihyperlipidemic and antioxidant activities of methanolic extract of *Amaranthus viridis* Linn. in alloxan-induced diabetic rats. *Exp. Toxicol. Pathol.* 64: 75–79.

Ashok Kumar, B.S., Lakshman, L., Nandeesh, R., Arunkumar, P.A., Manoj, B., Kumar, V. et al. 2011. *In vitro* alpha-amylase inhibition and *in vivo* antioxidant potential of *Amaranthus spinosus* in alloxan-induced oxidative stress in diabetic rats. *Saudi J. Biol. Sci.* 18: 1–5.

Ashraduzzaman, M.D., Alam, M.D.A., Khatun, S., Banu, S., and Absar, N. 2011. *Vigna unguiculata* (L.) Walp. seed oil exhibiting anti-diabetic effects in alloxan-induced diabetic rats. *Malaysian J. Pharm. Sci.* 9: 13–23.

Aslan, M., Orhan, N., Orhan, D.D., and Ergun, F. 2010. Hypoglycemic activity and antioxidant potential of some medicinal plants traditionally used in Turkey for diabetes. *J. Ethnopharmacol.* 128: 384–389.

Aslan, M., Sezik, E., and Yeilada, E. 2003. Effect of *Hibiscus esculentus* L. seeds on blood glucose levels in normoglycaemic, glucose-hyperglycaemic and streptozotocin-induced diabetic rats. *Gazi Universitesi. Eczacilik Fakultesi Dergisi.* 20: 1–7.

Asok Kumar, K., Maheswari, M.U., and Sivashanmugam, A.T. 2007. Hypoglycemic effect of *Ficus microcarpa* leaves (Chinese banyan) on alloxan-induced diabetic rats. *J. Biol. Sci.* 7: 321–326.

Atanu, F.O., Ebiloma, U.G., and Ajayi, E.I. 2011. A review on the pharmacological aspects of *Solanum nigrum* L. *Biotechnol. Mol. Biol. Rev.* 6: 1–7.

Atawodi, S.E.-O., Yakubu, O.E., Liman, M.L., and Iliemene, D.U. 2014. Effect of methanolic extract of *Tetrapleura tetraptera* (Schum and Thom) Taub leaves on hyperglycemia and indices of diabetic complications in alloxan-induced diabetic rats. *Asian Pac. J. Trop. Biomed.* 3: 272–278.

Atolani, O., Olatunji, G.A., and Fabyk, O.A. 2009. *Blighia sapida*: The plant and its hypoglycins, an overview. *J. Sci. Res.* 39: 15–25.

Attele, A.S., Zhou, Y.P., Xie, J.T., Wu, J.A., Zhang, L., Dey, L. et al. 2002. Antidiabetic effects of *Panax ginseng* berry extract and the identification of an effective component. *Diabetes* 51: 1851–1858.

Augusti, K.T., Daniel, R.S., Cherian, S., Sheela, C.G., and Nair, C.R. 1994. Effect of leucopelargonin derivative from *Ficus bengalensis* L. on diabetic dogs. *Indian J. Med. Res.* 99: 82–86.

Augusti, K.T., Joseph, P., and Babu, T.D. 1995. Biologically active principles isolated from *Salacia oblonga* Wall. *Indian J. Physiol. Pharmacol.* 39: 415–417.

Augusti, K.T. and Sheela, C.G. 1996. Antiperoxide effect of S-allyl cysteine sulfoxide, an insulin secretagogue in diabetic rats. *Experientia* 52: 115–120.

Austin, A. 2008. A review on Indian Sarsaparilla, *Hemidesmus indicus* (L.) R. Br. *J. BioSci.* 8: 1–12.

Auwal, M.S., Tijjani, A.N., Lawan, F.A., Mariga, I.A., Ibrahim, A., Njobdi, A.B. et al. 2012. The quantitative phytochemistry and hypoglycemic properties of crude mesocarp extract of *Hyphaene thebaica* (doump-alm) on normoglycemic Wistar albino rats. *J. Med. Sci.* 12: 280–285.

Auwerx, J. 1999. PPARγ, the ultimate thrifty gene. *Diabetologia* 42: 1033–1049.

Aviram, M., Rosenblat, M., Gaitini, D., Nitecki, S., Hoffman, A., Dornfeld, L. et al. 2004. Pomegranate juice consumption for 3 years by patients with carotid artery stenosis reduces common carotid intima media thickness, blood pressure and LDL oxidation. *Clin. Nutr.* 23: 423–433.

Awad, N.E., Seida, A.A., Shaffie, Z.E.N., and El-Aziz, A.M.A. 2012. Hypoglycemic activity of *Artemisia herba-alba* (Asso.) used in Egyptian traditional medicine as hypoglycemic remedy. *J. Appl. Pharm. Sci.* 2: 30–39.

Azemi, M.E., Namjoyan, F., Khodayar, M.J., Ahmadpour, F., Padok, A.D., and Panahi, M. 2012. The antioxidant capacity and anti-diabetic effect of *Boswellia serrata* Triana and Planch aqueous extract in fertile female diabetic rats and the possible effects on reproduction and histological changes in the liver and kidneys. *Jundishapur J. Nat. Pharm. Prod.* 7: 168–175.

Azevedo, C.R., Maciel, F.M., Silva, L.B., Ferreira, A.T., da Cunha, M., Machado, O.L. et al. 2006. Isolation and intracellular localization of insulin-like proteins from leaves of *Bauhinia variegata*. *Braz. J. Med. Biol. Res.* 39: 1435–1444.

Azmi, M.B. and Qureshi, S.A. 2012. Methanolic extract of *Rauwolfia serpentina* Benth improves the glycemic, antiatherogenic and cardioprotective indices in alloxan-induced diabetic mice. *Adv. Pharmacol. Sci.* 2012: Article ID 376429, 11 pages.

Babu, V., Gangadevi, T., and Subramoniam, A. 2002. Anti-hyperglycaemic activity of *Cassia kleinii* leaf extract in glucose fed normal rats and alloxan-induced diabetic rats. *Indian J. Pharmacol.* 34: 409–415.

Babu, V., Gangadevi, T., and Subramoniam, A. 2003. Antidiabetic activity of ethanol extract of *Cassia kleinii* leaf in streptozotocin-induced diabetic rats and isolation of an active fraction and toxicity evaluation of the extract. *Indian J. Pharmacol.* 35: 290–296.

Bachri, M.S., Jang, H.W., Choi, J., and Park, J. 2010. Protective effect of white-skinned sweet potato (*Ipomoea batatas* L.) from Indonesia on streptozotocin-induced oxidative stress in rats. *J. Life Sci.* 20: 1569–1576.

Badal, R.M., Badal, D., Badal, P., Khare, A., Srivastava, J., and Kumar, V. 2011. Pharmacological action of *Mentha piperita* on lipid profile in fructose-fed rats. *Iran J. Pharm. Res.* 10: 843–848.

Badar, V.A., Thawani, V.R., Wakode, P.T., Shrivastava, M.P., Gharpure, K.J., Hingorani, L.L. et al. 2005. Efficacy of *Tinospora cordifolia* in allergic rhinitis. *J. Ethnopharmacol.* 96: 445–449.

Badole, S., Patel, N., Bodhankar, S., Jain, B., and Bhardwaj, S. 2006. Antihyperglycemic activity of aqueous extract of leaves of *Cocculus hirsutus* (L.) Diels in alloxan-induced diabetic mice. *Indian J. Pharmacol.* 38: 49–53.

Badole, S.L. and Bodhankar, S.L. 2009a. Investigation of antihyperglycaemic activity of aqueous and petroleum ether extract of stem bark of *Pongamia pinnata* on serum glucose level in diabetic mice. *J. Ethnopharmacol.* 123: 115–120.

Badole, S.L. and Bodhankar, S.L. 2009b. Concomitant administration of petroleum ether extract of the stem bark of *Pongamia pinnata* (L.) Pierre with synthetic oral hypoglycaemic drugs in alloxan-induced diabetic mice. *Eur. J. Integr. Med.* 1: 73–79.

Badole, S.L. and Bodhankar, S.L. 2009c. Investigation of antihyperglycemic activity of *Glycine max* (L.) Merr. on serum glucose level in diabetic mice. *J. Complement. Integr. Med.* 6: article 4.

Badole, S.L. and Bodhankar, S.L. 2010. Antidiabetic activity of cycloart-23-ene-3beta, 25-diol (B2) isolated from *Pongamia pinnata* (L. Pierre) in streptozotocin-nicotinamide induced diabetic mice. *Eur. J. Pharmacol.* 632: 103–109.

Badole, S.L., Bodhankar, S.L., and Raut, C.G. 2011. Protective effect of cycloart-23-ene-3 β, 25-diol(B2) isolated from *Pongamia pinnata* L. Pierre on vital organs in streptozotocin-nicotinamide-induced diabetic mice. *Asian Pac. J. Trop. Biomed.* 1: S186–S190.

Baeshen, N.A., Lari, S.A., Al Doghaither, H.A.R., and Ramadan, H.A.I. 2010. Effect of *Rhazya stricta* extract on rat adiponectin gene and insulin resistance. *J. Am. Sci.* 6: 1237–1245.

Bahramikia, S. and Yazdanparast, R. 2012. Phytochemistry and medicinal properties of *Teucrium polium* L. (Lamiaceae). *Phytother. Res.* 26: 1581–1593.

Bai, N., He, K., Roller, M., Zheng, B., Chen, X., Shao, Z. et al. 2008. Active compounds from *Lagerstroemia speciosa*, insulin-like glucose uptake stimulatory/inhibitory and adipocyte differentiation inhibitory activities in 3T3-L1 cells. *J. Agric. Food Chem.* 56: 11668–11674.

Bairy, I., Reeja, S., Siddharth, Rao, P.S., Bhat, M., and Shivananda, P.G. 2002. Evaluation of antibacterial activity of *Mangifera indica* on aerobic dental microglora based on *in vitro* studies. *Indian J. Pathol. Microbiol.* 45: 307–310.

Bairy, K.L., Sharma, A., and Shalini, A. 2005. Evaluation of the hypoglycemic, hypolipidemic and hepatic glycogen raising effects of *Syzygium malaccense* on streptozotocin induced diabetic rats. *J. Nat. Remedies* 5: 46–51.

Bajpai, M.B., Asthana, R.K., Sharma, N.K., Chatterjee, S.K., and Mukherjee, S.K. 1991. Hypoglycemic effect of swerchirin from the hexane fraction of *Swertia chirayita*. *Planta Med.* 57: 102–104.

Bakirel, T., Bakirel, U., Keles, Oya, O.U., Ulgen, S.G., and Yardibi, H. 2008. *In vivo* assessment of antidiabetic and anti-oxidant activities of rosemary (*Rosamarinus officinalis*) in alloxan diabetic rabbits. *J. Ethnopharmacol.* 116: 64–73.

Bako, H.Y., Mohammad, J.S., Waziri, P.M., Bulus, T., Gwarzo, M.Y., and Zubairu, M.M. 2014. Lipid profile of alloxan-induced diabetic Wistar rats treated with methanolic extract of *Adansonia digitata* fruit pulp. *Sci. World J.* 9: 19–24.

Balakrishna, A., Kokilavani, R., Gurusamy, K., Teepa, K.S., and Sathya, M. 2011. Effect of ethanolic fruit extract of *Cucumis trigonus* Roxb. on anti-oxidants and lipid peroxidation in urolithiasis induced Wistar albino rats. *Anc. Sci. Life* 31: 10–16.

Balakrishnan, S. and Pandhare, R. 2010. Anti-hyperglycemic and anti-hyperlipidemic activities of *Amaranthus spinosa* Linn. extract on alloxan-induced diabetic rats. *Malays. J. Pharm. Sci.* 8: 13–22.

Balamurugan, R., Duraipandiyan, V., and Ignacimuthu, S. 2011. Antidiabetic activity of γ-sitosterol isolated from *Lippianodi flora* L. in streptozotocin induced diabetic rats. *Eur. J. Pharmacol.* 667: 410–418.

Balamurugan, R. and Ignachimuthu, S. 2011. Antidiabetic and hypolipidemic effect of methanol extract of *Lippia nodiflora* L. in streptozotocin induced diabetic rats. *Asian Pac. J. Trop. Biomed.* 1: 530–536.

Balanehru, S. and Nagarajan, B. 1991. Protective effect of oleanolic acid and ursolic acid against lipid peroxidation. *Biochem. Int.* 24: 981–990.

Balaraman, A.K., Singh, J., Dash, S., and Maity, T.K. 2010. Antihyperglycemic and hypolipidemic effects of *Melothria maderaspatana* and *Coccinia indica* in Streptozotocin induced diabetes in rats. *Saudi Pharm. J.* 18: 173–178.

Balaraman, A.K., Singh, J., Maity, T.K., Selvan, V.T., and Palanisamy, S.B. 2011. Ethanol extract of *Melothria maderaspatana* inhibits glucose absorption and stimulates insulin secretion in male C57BL/6 mice. *Int. Proc. Chem. Biol. Environ. Eng.* 5: 429–433.

Balasubramanian, T., Chatterjee, T.K., Senthilkumar, G.P., and Mani, T. 2012. Effect of potent ethyl acetate fraction of *Stereospermum suaveolens* extract in streptozotocin-induced diabetic rats. *Sci. World J.* Article ID 413196.

Balasubramanyam, M., Miranda, P., and Mohan, V. 2001. Carbohydrate metabolism and diabetes: Links to etiology of diabetic complications and therapy. *Trends Carbohydr. Chem.* 6: 15–29.

Balasubramanyam, M. and Mohan, V. 2000. Current concept of PPAR-signaling in diabetes mellitus. *Curr. Sci.* 79: 1440–1445.

Baldeon, M.E., Castro, J., Villacres, E., Narvaez, L., and Fornasini, M. 2012. Hypoglycemic effect of cooked *Lupinus mutabilis* and its purified alkaloids in subjects with type-2 diabetes. *Nutr. Hosp.* 27: 1261–1266.

Bandara, T., Uluwaduge, I., and Jansz, E.R. 2012. Bioactivity of cinnamon with special emphasis on diabetes mellitus: A review. *Int. J. Food Sci. Nutr.* 63: 380–386.

Bandawane, D., Juvekar, A., and Juvekar, M. 2011. Antidiabetic and antihyperlipidemic effect of *Alstonia scholaris* Linn. bark in streptozotocin induced diabetic rats. *Indian J. Pharm. Edu. Res.* 45: 114–120.

Bandyopadhyay, U., Biswas, K., Sengupta, A., Moitra, P., Dutta, P., Sarkar, D. et al. 2004. Clinical studies on the effect of neem (*Azadirachta indica*) bark extract on gastric secretion and gastroduodenal ulcer. *Life Sci.* 75: 2867–2878.

Banerjee, S., Singh, H., and Chatterjee, T.K. 2013. Evaluation of anti-diabetic and anti-hyperlipidemic potential of methanolic extract of *Juniperus communis* (L.) in streptozotocin-nicotinamide induced diabetic rats. *Int. J. Pharm. BioSci.* 4: 10–17.

Bangar, A.B. and Saralaya, M.G. 2011. Anti-hyperglycemic activity of ethanol extract and chloroform extract of *Indigofera tinctoria* leaves in streptozotocin induced diabetic mice (Family-Papilionaceae). *Res. J. Pharm. Biol. Chem. Sci.* 2: 445–455.

Bansal, P., Paul, P., Mudgal, J., Nayak, P.G., Pannakal, K.I., and Unnikrishnan, M.K. 2012. Antidiabetic, antihyperlipidemic and antioxidant effects of the flavonoid rich fraction of *Pilea microphylla* (L.) in high fat diet/streptozotocin-induced diabetes in mice. *Exp. Toxicol. Pathol.* 64: 651–658.

Banskota, N.T., Nauyen, Y., Tezuka, T., Nobukawa, S., and Kadota, S. 2006. Hypoglycemic effect of the wood of *Taxus yunnanensis* on streptozotocin-induced diabetic rats and its active components. *Phytomedicine* 13: 109–914.

Banu, S., Jabir, N.R., Manjunath, N.C., Khan, M.S., Ashraf, G.M., Kamal, M.A. et al. 2014. Reduction of post-prandial hyperglycemia by mulberry tea in type-2 diabetes patients. *Saudi J. Biol. Sci.* 22: 32–36.

Banu, S.M., Selvendiran, K., Singh, J.P., and Sakthisekaran, D. 2004. Protective effect of *Emblica officinalis* ethnolic extract against 7,12-dimethyl benz(a) anthrazene (DMBA) induced genotoxicity in Swiss albino mice. *Hum. Exp. Toxicol.* 23: 527–531.

Bao, H. and Chen, L. 2011. Icariin reduces mitochondrial oxidative stress injury in diabetic rat hearts. *Zhongguo Zhong Yao Za Zhi* 36: 1503–1507.

Baral, R. and Chattopadhyay, U. 2004. Neem (*Azadirachta indica*) leaf mediated immune activation causes prophylactic growth inhibition of murine Ehrlich carcinoma and B16 melanoma. *Int. Immunopharmacol.* 4: 355–366.

Baratil, S., Homtaze, H., and Azhdarymamoreh, M. 2012. The effect of hydro-alcoholic extract of *Moris nigra* leaf on lipids and sugar in serum of diabetic rats. *Asian J. Biomed. Pharm. Sci.* 2: 38–40.

Barbalho, S.M., Damasceno, D.C., Spada, A.P.M., da Silva, V.S., Martuchi, K.A., Oshiiwa, M. et al. 2011. Metabolic profile of offspring from diabetic Wistar rats treated with *Mentha piperita* (peppermint). *Evid. Based Complement. Altern. Med.* 2011: 6. http://dx.doi.org/10.1155/2011/430237.

Barik, R., Jain, S., Qwatra, D., Joshi, A., Tripathi, G.S., and Goyal R. 2008. Antidiabetic activity of aqueous root extract of *Ichnocarpus frutescens* in streptozotocin-nicotinamide induced type-2 diabetes in rats. *Indian J. Pharmacol.* 40: 19–22.

Barman, S. and Das, S. 2012. Hypoglycemic effect of ethanolic extract of bark of *Terminalia arjuna* Linn. in normal and alloxan-induced noninsulin-dependent diabetes mellitus albino rats. *Int. J. Green Pharm.* 6: 279–284.

Baroni, S., Suzuki-Kemmelmeir, F., Caparroz-Assef, S.M., Cuman, R.K.N., and Bersani-Amado, C.A. 2008. Effects of crude extracts of leaves of *Smallanthus sonchifolius* (yocon) on glycemia in diabetic rats. *Braz. J. Pharm. Sci.* 44: 521–530.

Bartolome, A.P., Villaseñor, I.M., and Yang, W.-C. 2013. *Bidens pilosa* L. (Asteraceae): Botanical properties, traditional uses, phytochemistry, and pharmacology. *Evid. Based Complement. Altern. Med.* 2013: Article ID 340215, 51 pages.

Basak, S.S. and Canadan, F. 2010. Chemical composition and *in vitro* anti-oxidant and anti-diabetic activities of *Eucalyptus camaldulensis* Dehnx essential oil. *J. Iran Chem. Soc.* 7: 216–226.

Basch, W.E., Gabardi, S., and Ulbricht, C. 2003. Bitter melon (*Momordica charantia*): A review of efficacy and safety. *Am. J. Health Syst. Pharm.* 60: 356–359.

Bashir, R., Alam, B., Javed, I., Jhammad, F., Sindhu, Z.U.D., Sarfraz, M., and Fayyaz, A. 2013. Anti-diabetic efficacy of *Mimosa pudica* (lajwanti) root in albino rabbits. *Int. J. Agric. Biol.* 15: 782–786.

Baskar, R., Bhakshu, L.M., Bharathi, G.V., Reddy, S.R., Karuna, R., Reddy, G.K. et al. 2008. Antihyperglycemic activity of aqueous root extract of *Rubia cordifolia* in streptozotocin-induced diabetic rats. *Pharm. Biol.* 44: 475–479.

Baskaran, K., Ahamath, B.K., Shanmugasundaram, K.R., and Shanmugasundaram, E.R.B. 1990. Antidiabetic effect of a leaf extract from *Gymnema sylvestre* in non-insulin-dependent diabetes mellitus patients. *J. Ethnopharmacol.* 30: 295–305.

Basnet, P., Kadota, S., Shimizu, M., and Namba, T. 1994. Bellidifolin a potent hypogycemic agent in streptozotocin (STZ)-induced diabetic rats from *Swertia japonica*. *Planta Med.* 60: 507–511.

Basnet, P., Kadota, S., Terashima, S., Shimizu, M., and Namba, T. 1993. Two new 2-arylbenzofuran derivatives from hypoglycemic activity-bearing fractions of *Morus insignis*. *Chem. Pharm. Bull.* 41: 1238–1243.

Bates, S.H., Jones, R.B., and Bailey, C.J. 2000. Insulin-like effect of pinitol. *Br. J. Pharmacol.* 130: 1944–1948.

Battram, D.S., Arthur, R., Weekes, A., and Graham, T.E. 2006. The glucose intolerance induced by caffeinated coffee ingestion is less pronounced than that due to alkaloid caffeine in men. *J. Nutr.* 136: 1276–1280.

Bavarva, J.H. and Narasimhacharya, A.V. 2008. Antihyperglycemic and hypolipidemic effects of *Costus speciosus* in alloxan induced diabetic rats. *Phytother. Res.* 22: 620–626.

Bavarva, J.H. and Narasimhacharya, A.V. 2013. Systematic study to evaluate anti-diabetic potential of *Amaranthus spinosus* on type-1 and type-2 diabetes. *Cell. Mol. Biol.* 2: 1818–1825.

Bavarva, J.H. and Narasimhacharya, A.V. 2005. Antidiabetic and antihyperlipidemic effects of ethanolic extract of *Phyllanthus niruri* L. leaves. *J. Cell Tissue Res.* 5: 461–464.

Baxi, B.B., Singh, P.K., Doshi, A.A., Arya, S., Mukherjee, R., and Ramachandran, A.V. 2010. *Medicago sativa* leaf extract supplementation corrects diabetes induced dyslipidemia, oxidative stress and hepatic renal functions and exerts anti-hyperglycemic action as effective as metformin. *Ann. Biol. Res.* 1: 107–119.

Bayat, A., Jamali, Z., Hajianfar, H., and Heidari-Beni, M. 2014. Effects of *Cucurbita ficifolia* intake on type 2 diabetes: Review of current evidences. *Shiraz E. Med. J.* 15: 1–3.

Bean, M.F., Anthoun, M., Abramson, D., Chang, C.J., McLaughlin, J.L., and Cassady, J.M. 1985. Cucurbitacin B and isocucurbitacin B: Cytotoxin components of *Halicteres isora*. *J. Nat. Prod.* 48: 500.

Becerra-Jimenez, J. and Andrade-Cetto, A. 2012. Effect of *Opuntia streptacantha* Lem. on alpha-glucosidase activity. *J. Ethnopharmacol.* 139: 493–496.

Beghalia, M., Ghalem, S., Allali, H., Belouatek, A., and Marouf, A. 2008. Inhibition of calcium oxalate monohydrate crystal growth using Algerian medicinal plants. *J. Med. Plants Res.* 2: 66–70.

Behera, B. and Yadav, D. 2013. Current researches on plants having antidiabetic potential: An overview. *Res. Rev. J. Bot. Sci.* 2: 4–17.

Behradmanesh, S., Derees, F., and Rafieian-kopaei, M. 2013. Effect of *Salvia officinalis* on diabetic patients. *J. Ren. Inj. Prev.* 2: 51–54.

Belemkar, S., Dhameliya, K., and Pata, M.K. 2013. Comparative study of garlic species (*Allium sativum* and *Allium porrum*) on glucose uptake in diabetic rats. *J. Talibah Univ. Med. Sci.* 8: 80–85.

Belfiore, A. and Malaguarnera, R. 2011. Insulin receptor and cancer. *Endocr. Relat. Cancer* 18: R125–R147.

Bellamkonda, R., Rasineni, K., Reddy, S., Kasetti, R.B., Pasurla, R., Chippada, A.P., and Desireddy, S. 2010. Antihyperglycemic and antioxidant activities of alcoholic extract of *Commiphora mukul* gum resin streptozotocin-induced diabetic rats. *Planta Med.* 76: 412–417.

Bello, A., Aliero, A.A., Saidu, Y., and Muhammad, S. 2011a. Hypoglycemic and hypolipidemic effects of *Leptadenia hastata* (Pers.) Decne in alloxan-induced diabetic rats. *Nig. J. Basic Appl. Sci.* 19: 187–192.

Bello, A., Aliero, A.A., Saidu, Y., and Muhammad, S. 2011b. Phytochemical screening, polyphenolic content and alpha-glycosidase inhibitory potential of *Leptadenia hastata* (Pers.) Decne. *Nig. J. Basic Appl. Sci.* 19: 181–186.

Beltran, A.E., Alvarez, Y., Xavier, F.E., Hernanz, R., Rodriguez, J., Nunez, A.J. et al. 2004. Vascular effects of the *Mangifera indica* L. extract (Vimang). *Eur. J. Pharmacol.* 499: 297–305.

Benammar, C., Baghdad, C., Belarbi, M., Subramaniam, S., Hichami, A., and Khan, N.A. 2014. Antidiabetic and antioxidant activities of *Zizyphus lotus* L aqueous extracts in Wistar rats. *J. Nutr. Food Sci.* S8: 1–6.

Bendjeddou, D., Lalaoui, K., and Satta, D. 2003. Immunostimulating activity of the hot water soluble polysaccharide extracts of *Anacyclus pyrethrum*, *Alpinia galanga* and *Citrullus colocynthis*. *J. Ethnopharmacol.* 88: 155–160.

Benkhayal, F.A., Al-Gazwi, S.M., Ramesh, S., and Kumar, S. 2010. Biochemical studies on the effect of volatile oil of *Thymus capitatus* in alloxan-induced diabetic rats. *Curr. Trends Biotechnol. Pharm.* 4: 519–524.

Benkhayal, F.A., Musbah, E., Ramesh, S., and Dhayabaran, D. 2009. Biochemical studies on the effect of phenolic compounds extracted from *Myrtus communis* in diabetic rats. *TamilNadu J. Vet. Anim. Sci.* 5: 87–93.

Bennani-Kabachi, N., el-Bouayadi, F., Kehel, L., Fdhil, H., and Marquié, G. 1999. Effect of *Suaeda fruticosa* aqueous extract in the hypercholesterolaemic and insulin-resistant sand rat. *Therapie* 54: 725–730.

Benwahhoud, M., Jouad, H., Eddouks, M., and Lyoussi, B. 2001. Hypoglycemic effect of *Suaeda fruticosa* in streptozotocin-induced diabetic rats. *J. Ethnopharmacol.* 76: 35–38.

Bera, S., Chatterjee, K., De, D., Ali, K.M., and Ghosh, D. 2011. Effect of the hydro-methanolic (2:3) extract of the bark of *Tectona grandis* L. on the management of hyperglycemia and oxidative stress in streptozotocin-induced diabetes in rats. *J. Nat. Pharm.* 2: 196–199.

Bera, T.K., Chatterjee, K., and Ghosh, D. 2015. Remedial hypoglycemic activity of n-hexane fraction of hydro-methanol extract of seed of *Swietenia mahagoni* (L.) Jacq. in streptozotocin-induced diabetic rat: A comparative evaluation. *J. Herbs Spices Med. Plants* 2015: 21. DOI: 10.1080/10496475.2014.895920.

Berger, J. and Moller, D.E. 2002. The mechanisms of action of PPARs. *Annu. Rev. Med.* 53: 409–435.

Bhagwat, D.A., Killedar, S.G., and Adnaik, R.S. 2008. Anti-diabetic activity of leaf extract of *Tridax procumbens*. *Int. J. Green Pharm.* 2: 126–128.

Bhalerao, S.A. and Sharma, A.S. 2014. Ethnopharmacology, phytochemistry and pharmacological evaluation of *Pongamia pinnata* (L.) Pierre. *Int. J. Curr. Res. Biosci. Plant Biol.* 3: 50–60.

Bhandari, U. and Ansari, M.N. 2009. Ameliorative effect of ethanol extract of *Embelia ribes* fruits on isoproterenol-induced cardiotoxicity in diabetic rats. *Pharm. Biol.* 47: 669–674.

Bhandari, U., Kanojia, R., and Pillai, K.K. 2005. Effect of ethanolic extract of *Zingiber officinale* on dyslipideamia in diabetic rats. *J. Ethnopharmacol.* 97: 227–230.

Bhandarkar, M. and Khan, A. 2004. Antihepatotoxic effect of *Nymphaea stellata* Willd. against carbon tetrachloride-induced hepatic damage in albino rats. *J. Ethnopharmacol.* 91: 61–64.

Bharti, S.K., Kumar, A., Prakash, O., Krishnan, S., and Gupta, A.K. 2013. Essential oil of *Cymbopogon citratus* against diabetes: Validation by *in vivo* experiments and computational studies. *J. Bioanal. Biomed.* 5: 194–203.

Bhaskar, A. and Vidhya, V.G. 2012. Hypoglycemic and hypolipidemic activity of *Hibiscus rosa sinensis* Linn. on streptozotocin-induced diabetic rats. *Int. J. Diab. Dev. Countries* 32: 214–219.

Bhaskar, A., Vidhya, V.G., and Ramya, M. 2008. Hypoglycemic effect of *Mucuna pruriens* seed extract on normal and streptozotocin-diabetic rats. *Fitoterapia* 79: 539–543.

Bhaskar, V.H. and Ajay, S.S. 2009. Evaluation of antihperglycemic activity of extracts of *Calotropis procera* (Ait.) R.Br on streptozotocin induced diabetic rats. *Global J. Pharmacol.* 3: 95–98.

Bhaskara, R.R., Murugesn, T., Sinha, S., Saha, B.P., Pal, M., and Mandal, S.S. 2002. Glucose lowering efficacy of *Ficus racemosa* bark extract in normal and alloxan diabetic rats. *Phytother. Res.* 16: 590–592.

Bhat, B.M., Raghuveer, C.V., D'Souza, V., and Manj, P.A. 2012. Antidiabetic and hypolipidemic effect of *Salacia oblonga* in streptozotocin-induced diabetic rats. *Clin. Diagn. Res.* 6: 1685–1687.

Bhat, M., Kothiwale, S.K., Tirmale, A.R., Bhargava, S.Y., and Joshi, B.N. 2011a. Antidiabetic properties of *Azardiracta indica* and *Bougainvillea spectabilis in vivo* studies in murine diabetes model. *Evid. Based Complement. Altern. Med.* 2011: Article ID 561625.

Bhat, M., Zinjarde, S.S., Bhargava, S.Y., Kumar, A.R., and Joshi, B.N. 2011b. Anti-diabetic Indian plants: A good source of potent amylase inhibitors. *Evid. Based Complement. Altern. Med.* 2011: 810207.

Bhat, M.Z.A., Ali, M., and Mir, S.R. 2013. Anti-diabetic activity of *Ficus carica* L. stem barks and isolation of two new flavonol esters from the plant by using spectroscopical techniques. *Asian J. Biomed. Pharm. Sci.* 3: 22–28.

Bhatti, R., Singh, A., Saharan, V.A., Ram, V., and Bhandari, A. 2012. *Strychnos nux-vo*mica seeds: Pharmacognostical standardization, extraction and anti-diabetic activity. *J. Ayurveda Integr. Med.* 3: 80–84.

Bhatia, A. and Mishra, T. 2010. Hypoglycemic activity of *Ziziphus mauritiana* aqueous ethanol seed extract in alloxan-induced diabetic mice. *Pharm. Biol.* 48: 604–610.

Bhatia, B., Kinja, K., Bishnoi, H., Savita, S., and Gnaneshwari, D. 2011. Anti-diabetic activity of the alcoholic extract of aerial parts of *Boerhaavia diffusa* in rats. *Recent Res. Sci. Technol.* 3: 4–7.

Bhatia, D., Gupta, M.K., Bharadwaj, A., Pathak, M., Kathiwas, G., and Singh, M. 2008. Anti-diabetic activity of *Centratherum anthelminticum* Kuntze on alloxan induced diabetic rats. *Pharmacologyonline* 3: 1–5.

Bhatt, M., Gahlot, M., Juyal, V., and singh, A. 2011. Phytochemical investigation and antidiabetic activity of *Adhatoda zeylanica*. *Asian J. Pharm. Clin. Res.* 4: 27–30.

Bhatt, N.M., Barua, S., and Gupta, S. 2009. Protective effect of *Enicostemma littorale* Blume on rat model of diabetic neuropathy. *Am. J. Infect. Dis.* 5: 99–105.

Bhattacharya, S., Rasmussen, M.K., Christensen, L.P., Young, J.F., Kristiansen, K., and Oksbjerg, N. 2014. Naringenin and falcarinol stimulate glucose uptake and TBC1D1 phosphorylation in porcine myotube cultures. *J. Biochem. Pharmacol. Res.* 2: 91–98.

Bhattacharya, S.K., Reddy, P.K., Ghosal, S., Singh, A.K., and Sharma, P.V. 1976. Chemical constituents of Gentianaceae XIX: CNS-depressant effects of swertiamarin. *J. Pharm. Sci.* 65: 1547–1549.

Bhatti, R., Sharma, S., Singh, J., and Ishar, M.P.S. 2011. Ameliorative effect of *Aegle marmelos* leaf extract on early stage alloxan-induced diabetic cardiomyopathy in rats. *Pharm. Biol.* 49: 1137–1143.

Bhoumik, A., Ali Khan, L., Akhter, M., and Rokeya, B. 2009. Studies on the anti-diabetes effects of *Mangifera indica* stem barks and leaves on type 1 and type 2 diabetic models of rats. *Bangladesh J. Pharmacol.* 4: 110–114.

Bhusal, A.B., Jammarkattel, N., Shrestha, A., Lamsal, N.K., Shakya, S., and Rajbhandari, S. 2014. Evaluation of antioxidative and antidiabetic activity of bark of *Holarrhena pubescens* Wall. *J. Clin. Diagn. Res.* 8: 5–8.

Bhuvaneswari, P. and Krishnakumari, S. 2012. Antihyperglycemic potential of *Sesamum indicum* (Linn.) seeds in streptozotocin induced diabetic rats. *Int. J. Pharm. Pharm. Sci.* 4: 527–531.

Bicalho, B. and Rezende, C.M. 2001. Volatile compounds of cashew apple (*Anacardium occidentale* L.). *Z. Naturforsch. C.* 56: 35–39.

Bidkar, J.S., Ghanwat, D.D., Bhujbal, M.D., and Dama, G.Y. 2012. Anti-hyperlipidemic activity of *Cucumis melo* fruit peel extracts in high cholesterol diet induced hyperlipidemia in rats. *J. Complement. Interg. Med.* 9. DOI: 10.1515/1553-3840.1580.

Bilbis, L.S., Shehu, R.A., and Abubarkar, M.G. 2002. Hypoglycemic and hypolipidemic effects of aqueous extract of *Arachis hypogaea* in normal and alloxan induced diabetic rats. *Phytomedicine* 9: 553–555.

Binokingsley, R. 2014. Studies on the anti-diabetic potentials of selected ethnomedicinal plants. PhD thesis submitted to SASTRA Deemed University, Thanjavur, India.

Bino Kingsley, R., Ajikumaran Nair, S., Anil John, J., Manisha, M., Brindha, P., and Subramoniam, A. 2012. Anti-diabetes mellitus activity of *Stereospermum tetragonum* DC. in alloxan diabetic rats. *J. Pharmacol. Pharmacother.* 3: 191–193.

Bino Kingsley, R., Arun, K.P., Stalin, S., and Brindha, P. 2014b. *Stereospermum tetragonam* as an antidiabetic agent by activating PPARγ and GLUT4. *Bangladesh J. Pharmacol.* 9: 250–256.

Bino Kingsley, R., Brindha, P., and Subramoniam, A. 2014a. Anti-hyperglycemic screening of ethno-medicinal plants. *Int. J. PharmTech Res.* 6: 494–499.

Bino Kingsley, R., Brindha, P., and Subramoniam, A. 2013b. Toxicity studies of the active fraction of *Streospermum tetragonum* DC. *Int. J. Pharm. Pharm. Sci.* 5(Suppl. 3): 648–651.

Bino Kingsley, R., Manisha, M., Brindha, P., and Subramoniam, A. 2013a. Anti-diabetes mellitus activity of active fraction of *Stereospermum tetragonum* DC. and isolation of active principles. *J. Young Pharm.* 5: 7–12.

Binutu, O.A. and Lajubutu, B.A. 1994. Antimicrobial potentials of some plant species of the Bignoniaceae family. *Afr. J. Med. Med. Sci.* 23: 269–273.

Biradar, S.M., Rangani, A.T., Kulkarni, V.H., Joshi, H., Habbu, P.V., and Smita, D.M. 2010. Prevention of onset of hyperglycemia by extracts of *Argyriea cuneata* on alloxan-induced diabetic rats. *J. Pharm. Res.* 3: 2186–2187.

Bisby, F. 1994. *Phytochemical Dictionary of the Leguminosae.* London, U.K.: Chapman & Hall.

Bisht, S. and Sisodia, S.S. 2010. *Coffea arabica*: A wonder gift to medical science. *J. Nat. Pharm.* 1: 58–65.

Biswas, A., Chatterjee, S., Chowdhury, R., Sarkar, D., Chatterjee, M., and Das, J. 2012. Antidiabetic effect of seeds of *Strychnos potatorum* Linn. in a streptozotocin-induced model of diabetes. *Acta Pol. Pharm.* 69: 939–943.

Biswas, A., Goswami, T.K., Ghosh, A., Paul, J., Banerjee, K., and Halder, D. 2014. Hypoglycemic effects of *Strychnos potatorum* Linn were compared with glipizide on male diabetic rats. *Indian Med. Gaz.* 2014: 297–304.

Biswas, M., Kar, B., Bhattacharya, S., Kumar, R.B., Ghosh, A.K., and Haldar, P.K. 2011. Antihyperglycemic activity and antioxidant role of *Terminalia arjuna* leaf in streptozotocin-induced diabetic rats. *Pharm. Biol.* 49: 335–340.

Bliddal, H., Rosetzsky, A., Schlichting, P., Weidner, M.S., Anderson, L.A., Ibfelt, H.H. et al. 2000. A randomized, placebo-controlled, cross-over study of ginger extracts and ibuprofen in osteoarthritis. *Osteoarthritis Cartilage* 8: 9–12.

Bnouham, M., Benalla, W., Bellahcen, S., Hakkou, Z., Zivyat, A., Mekhfi, H. et al. 2012. Anti-diabetic and anti-hypertensive effect of polyphenol rich fraction of *Thymelaea hirsute* L. in a model of neonatal streptozotocin-diabetic and N(G)-nitro-L-arginine methyl ester-hypertensive rats. *J. Diabetes* 4: 307–313.

Bnouham, M., Merhfour, F.Z., Legssyer, A., Mekhfi, H., Maallem, S., and Ziyyat, A. 2007. Antihyperglycemic activity of *Arbutus unedo*, *Ammoides pusilla* and *Thymelaea hirsuta*. *Pharmazie* 62: 630–632.

Bnouham, M., Ziyyat, A., Mekhfi, H., Tahri, A., and Legssyer, A. 2006. Medicinal plants with potential antidiabetic activity—a review of ten years of herbal medicine research (1990–2000). *Int. J. Diabetes Metab.* 14: 1–25.

Boby, R.G. and Leelamma, S. 2003. Black gram fiber (*Phaseolus mungo*): Mechanism of hypoglycemic action. *Plant Foods Hum. Nutr.* 58: 7–13.

Bohm, H. 2008. *Opuntia dillenii*—An interesting and promising Cactaceae taxon. *J. PACD* 10: 148–170.

Bolkent, S., Yanardag, R., Tabakoglu-Oguz, A., and Ozsov-Sacan, O. 2000. Effects of chard (*Beta vulgaris* L. var. cicla) extract on pancreatic β-cells in streptozotocin-diabetic rats: A morphological and biochemical study. *J. Ethnopharmacol.* 73: 251–259.

Boopathy, R.A., Elanchezhiyan, C., Manoharan, S., Suresh, K., Balakrishnan, S., and Sethupathy, S. 2009. Evaluation of anti-hyperglycemic potential of *Helicterees isora* in streptozotocin-induced diabetic Wistar rats. *J. Cell Tissue Res.* 9: 1925–1928.

Boopathy, R.A., Elanchezhiyan, C., and Sethupathy, S. 2010. Anti-hyperlipidemic activity of *Helicterees isora* fruit extract on streptozotocin-induced diabetic male Wistar rats. *Eur. Rev. Med. Pharmacol. Sci.* 14: 191–196.

Bopanna, K.N., Kannan, J., Gadgil, S., Balaraman, R., and Rathod, S.P. 1997. Anti-diabetic and antihyperlipidemic effects of neem seed kernel powder on alloxan diabetic rabbits. *Indian J. Pharmacol.* 29: 162–167.

Borniquel, S., Jansson, E.A., Cole, M.P., Freeman, B.A., and Lundberg, J.O. 2010. Nitrated oleic acid up-regulates PPAR-gamma and attenuates experimental inflammatory bowel disease. *Free Radic. Biol. Med.* 48: 499–505.

Boskabady, M.H., Shafei, M.N., Saberi, Z., and Amini, S. 2011. Pharmacological effects of *Rosa damascene*. *Iran J. Basic Med. Sci.* 14: 295–307.

Bothon, F.T.D., Debiton, E., Avlessi, F., Forestier, C., Teulade, J.-C., and Sohounhloue, D.K.C. 2013. *In vitro* biological effects of two anti-diabetic medicinal plants used in Benin as folk medicine. *BMC Complement. Altern. Med.* 13: 51.

Boudjelal, A., Henchiri, C., Siracusa, L., Sari, M., and Ruberto, G. 2012. Composition analysis and *in vivo* anti-diabetic activity of wild Algerian *Marrubium vulgare* L. infusion. *Fitoterapia* 83: 286–292.

Brahmachari, H.D. and Augusti, K.T. 1961. Anti-diabetic activity of bark of *Ficus religiosa*. *Indian J. Pharmacol.* 13: 128–132.

Bräunlich, M., Slimestad, R., Wangensteen, H., Brede, C., Malterud, K.E., and Barsett, H. 2013. Extracts, anthocyanins and procyanidins from *Aronia melanocarpa* as radical scavengers and enzyme inhibitors. *Nutrients* 5: 663–678.

Breen, D.M., Rasmussen, B.A., Kokorovic, A., Wang, R., Cheung, G.W., and Lam, T.K. 2012. Jejunal nutrient sensing is required for duodenal-jejunal bypass surgery to rapidly lower glucose concentrations in uncontrolled diabetes. *Nat. Med.* 18: 950–955.

Bresciani, L.F.V., Yunes, R.A., Burger, C., Oliveira, L.E.D., Bof, K.L., and Cechinel-Filho, V. 2004. Seasonal variation of kaurenoic acid, a hypoglycemic diterpene in *Wedelia paludosa* (*Acmela brasiliensis*) (Asteraceae). *Z. Naturforsch C.* 59: 229–232.

Broadhurst, C.L., Polansky, M.M., and Anderson, R.A. 2000. Insulin-like biological activity of culinary and medicinal plant aqueous extracts *in vitro*. *J. Agric. Food Chem.* 48: 849–852.

Brower, V. 1998. Nutraceuticals: Poised for a healthy slice of the healthcare market. *Nat. Biotechnol.* 16: 721–731.

Brown, P.N. and Roman, M.C. 2008. Determination of hydrastine and berberine in goldenseal raw materials, extracts, and dietary supplements by high-performance liquid chromatography with UV: Collaborative Study. *J. AOAC Int.* 91: 694–701.

Budhwani, A.B.K., Shrivastava, B., Singhai, A.K., and Gupta, P. 2012. Antihyperglycemic activity of ethanolic extract of leaves of *Dioscorea japonica* in streptozotocin-induced diabetic rats. *Res. J. Pharm. Technol.* 5: 553–557.

Bustanji, Y., Taha, M.O., Al-Masri, I.M., and Mohammad, M.K. 2009. Docking simulations and *in vitro* assay unveil potent inhibitory action of papaverine against protein tyrosine phosphatase 1B. *Biol. Pharm. Bull.* 32: 640–645.

Bwititi, P., Musabayane, C.T., and Nhachi, C.F. 2000. Effects of *Opuntia megacantha* on blood glucose and kidney function in streptozotocin diabetic rats. *J. Ethnopharmacol.* 69: 247–252.

Bwititi, P.T., Machakaire, T., Nhachi, C.B., and Musabayane, C.T. 2001. Effects of *Opuntia megacantha* leaves extract on renal electrolyte and fluid handling in streptozotocin (STZ)-diabetic rats. *Ren. Fail.* 23: 149–158.

Cai, H., Lian, L., Wang, Y., Yu, Y., and Liu, W. 2014. Protective effects of *Salvia miltiorrhiza* injection against learning and memory impairments in streptozotocin-induced diabetic rats. *Exp. Therapeut. Med.* 8: 1127–1130.

Calabrese, F.J., Matton, M.P., and Calabrese, V. 2010. Resveratrol commonly displays hormesis: Occurrence and biomedical significance. *Hum. Exp. Toxicol.* 29: 980–1015.

Campbell-Tofte, J.I.A., Molgaard, P., and Winther, K. 2013. Harnessing the potential clinical use of medicinal plants as anti-diabetic agents. *Botanics: Targets Ther.* 2: 7–19.

Cao, H., Graves, D.J., and Anderson, R.A. 2010. Cinnamon extract regulates glucose transporter and inslin-signalling gene expression in mouse adipocytes. *Phytomedicine* 17: 1027–1032.

Cao, J., Li, C., Zhang, P., Cao, X., Huang, T., Bai, Y. et al. 2012. Antidiabetic effect of burdock (*Arctium lappa* L.) root ethanolic extract on streptozotocin-induced diabetic rats. *Afr. J. Biotechnol.* 11: 9079–9085.

Carney, J.R., Krenisky, J.M., Williamson, R.T., and Luo, J. 2002. Achyrofuran, a new antihyperglycemic dibenzofuran from the South American medicinal plant *Achyrocline satureioides*. *J. Nat. Prod.* 65: 203–205.

Casanova, L.M., da Silva, D., Sola-Penna, M., Camargo, L.M., Celestrini, D.M., Tinoco, L.W. et al. 2014. Identification of chicoric acid as a hypoglycemic agent from *Ocimum gratissmum* leaf extract in a biomonitoring in vivo study. *Fitoterapia* 93: 132–141.

Cazarolli, L., Zanatta, L., Jorge, A.P., de Sousa, E., Woehl, V.M., Pizzolatti, M.G. et al. 2006. Follow-up studies on glycosylated flavonoids and their complexes with vanadium: Their anti-hyperglycemic potential role in diabetes. *Chem. Biol. Interact.* 163: 177–191.

Cemek, K., Kaga, S., Simsek, N., Buyukokuroglu, M.E., and Konuk, M. 2008. Antihyperglycemic and anti-oxidative potential of *Matricaria chamomilla* L. in sterptozotocin-induced diabetic rats. *J. Nat. Med.* 62: 284–293.

Cetto, A.A. and Heinrich, M. 2005. Mexican plants with hypoglycaemic effect used in the treatment of diabetes. *J. Ethnopharmacol.* 99: 325–348.

Cetto, A.A., Wiedenfeld, H., Revilla, M.C., and Sergio, I.A. 2000. Hypoglycemic effect of *Equisetum myriochaetum* aerial parts on streptozotocin diabetic rats. *J. Ethnopharmacol.* 72: 129–133.

Chabane, D., Saidi, F., Rouibi, A., and Azine, K. 2013. Hypoglycemic activity of the aqueous extract of *Ajuga iva* L. in diabetic rats induced by alloxan. *Afr. Sci.* 9: 120–127.

Chackrewarthy, S., Thabrew, M.I., Weerasuriya, M.K., and Jayasekera, S. 2010. Evaluation of the hypoglycemic and hypolipidemic effects of an ethylacetate fraction of *Artocarpus heterophyllus* (Jak) leaves in streptozotocin-induced diabetic rats. *Pharmacogn. Mag.* 6: 186–190.

Chadwick, W.A., Roux, S., van de Venter, M., Louw, J., and Oelofsen, W. 2007. Anti-diabetic effects of *Sutherlandia frutescens* in Winstar rats fed a diabetogenic diet. *J. Ethnopharmacol.* 109: 121–127.

Chakrabarti, R., Vikramadithyan, R.K., Mullangi, R., Sharma, V.M., Jagadheshan, H., Roa, Y.N. et al. 2002. Antidiabetic and hypolipidemic activity of *Helicteres isora* in animal models. *J. Ethnopharmacol.* 81: 343–349.

Chakrabarti, S., Biswas, T.K., Rokeya, B., Ali, L., Mosihuzzaman, M. Nahar, N. et al. 2003. Advanced studies on the hypoglycemic effects of *Caesalpinia bonducella* F. in type1 and 2 diabetes in Long Evans rats. *J. Ethnopharmacol.* 84: 41–46.

Chakrabarthi, S., Biswas, T.K., Seal, T., Rokeya, B., Ali, L., AzadKhan, A.K. et al. 2005. Antidiabetic activity of *Caesalpinia bonducella* F. in chronic type 2 diabetic model in Long Evans rats and evaluation of insulin secretagogue property of its fraction on isolated islets. *J. Ethnopharmacol.* 97: 117–122.

Chakraborty, U. and Das, H. 2010. Anti-diabetic and anti-oxidant activities of *Cinnamomum tamala* leaf extracts in streptozotocin-treated diabetic rats. *Global J. Biotechnol. Biochem.* 5: 12–18.

Chakravarty, S. and Kalita, J.C. 2012. Antihyperglycaemic effect of flower of *Phlogacanthus thyrsiflorus* Nees on streptozotocin induced diabetic mice. *Asian Pac. J. Trop. Biomed.* S1357–S1361.

Chakravarty, S. and Kalita, J.C. 2014. Evaluation of antidiabetic, hypolipidemic and hepatoprotective activity of *Phlogacanthus thyrsiflorus* Nees in streptozotocin induced diabetic mice: A 7-days intensive study. *Int. J. PharmTech Res.* 6: 345–350.

Chan, H.-H., Sun, H.-D., Reddy, M.V.B., and Wu, T.-S. 2010. Potent α-glucosidase inhibitors from the roots of *Panax japonicas* C. A. Meyer var. major. *Phytochemistry* 71: 1360–1364.

Chan, J.C., Lau, C.B., Chan, J.Y., Fung, K.P., Leung, P.C., Liu, J.Q. et al. 2015. The anti-guconeogenic activity of cucurbitacins from *Momordica charantia*. *Planta Med.* 81: 327–332.

Chandira, M. and Jayakar, B. 2010. Formulation and evaluation of herbal tablets containing *Ipomoea digitata* Linn. extract. *Int. J. Pharm. Sci. Rev. Res.* 3: 101–110.

Chandramohan, G., Al-Numair, K.S., Sridevi, M., and Pugalendi, K.V. 2010. Antihyperlipidemic activity of 3-hydroxymethyl xylitol, a novel antidiabetic compound isolated from *Casearia esculenta* (Roxb.) root, in streptozotocin-diabetic rats. *J. Biochem. Mol. Toxicol.* 24: 95–101.

Chandramohan, G., Ignacimuthu, S., and Pugalendi, K.V. 2008. A novel compound from *Casearia esculenta* (Roxb.) root and its effect on carbohydrate metabolism in streptozotocin-diabetic rats. *Eur. J. Pharmacol.* 590: 437–443.

Chandrasekar, B., Bajpai, M.B., and Mukherjee, S.K. 1990. Hypoglycemic activity of *Swertia chirayita* (Roxb ex Flem) Karst. *Indian J. Exp. Biol.* 28: 616–618.

Chandrasekar, B., Mukherjee, B., and Mukherjee, S.K. 1989. Blood sugar lowering potentiality of selected Cucurbitaceae plants of Indian origin. *Indian J. Med. Res.* 90: 300–305.

Chandrashekar, K.S. and Prasanna, K.S. 2009. Hypoglycemic effect of *Leucas lavandulaefolia* Willd in alloxan-induced diabetic rats. *J. Young Pharm.* 1: 326–329.

Chang, C.L., Chang, S.L., Lee, Y.M., Chiang, Y.M., Chuang, D.Y., Kuo, H.K. et al. 2007. Cytopiloyne, a polyacetylenic glucoside, prevents type 1 diabetes in nonobese diabetic mice. *J. Immunol.* 178: 6984–6993.

Chang, C.L., Kuo, H., Chang, S.L., Chiang, Y.M., Lee, T.H., Wu, W.M. et al. 2005. The distinct effects of a butanol fraction of *Bidens pilosa* plant extract on the development of Th1-mediated diabetes and Th2-mediated airway inflammation in mice. *J. Biomed. Sci.* 12: 79–89.

Chang, C.L., Shen, C.C., Ni, C.L., and Chen, C.C. 2012. A new sesquiterpene from *E. scaber. Hiromitsu J.* 65: 49–56.

Chang, C.L.T., Lin, Y., Bartolome, A.P., Chen, Y.-C., Chiu, S.-C., and Yang, W.-C. 2013a. Herbal therapies for type-2 diabetes mellitus: Chemistry, biology, and potential application of selected plants and compounds. *Evid. Based Complement. Altern. Med.* 2013: 378657.

Chang, C.L.T., Liu, H.Y., and Kuo, T.F. 2013b. Anti-diabetic effect and mode of action of cytopiloyne. *Evid. Based Complement. Altern. Med.* 2013: Article ID 685642, 13 pages.

Chang, S.H., Liu, C.J., Kuo, C.H. Chen, H., Lin, W.Y., Teng, K.Y. et al. 2011. Garlic oil alleviates MAPKs- and IL-6-mediated diabetes-related cardiac hypertrophy in STZ-induced DM rats. *Evid. Based Complement. Altern. Med.* 950150: 1–11.

Chang, S.L., Chang, C.L., Chiang, Y.M., Hsieh, R.H., Czeng, C.R., Wu, T.K. et al. 2004. Polyacetylenic compounds and butanol fraction from *Bidens pilosa* can modulate the differentiation of helper T cells and prevent autoimmune diabetes in non-obese diabetic mice. *Planta Med.* 70: 1045–1051.

Chang, Y., Chang, T.-C., Lee, J.-J., Chang, N.-C., Huang, Y.-K. Choy, C.-S. et al. 2014. Sanguis draconis, a dragon's blood resin, attenuates high glucose-induced oxidative stress and endothelial dysfunction in human umbilical vein endothelial cells. *Sci. World J.* 2014: Article ID 423259, 10 pages.

Chang-Hwa, J., Ho-Moon, S., In-Wook, C., Hee-Don, C., and Hong-Yon, C. 2005. Effects of wild ginseng (*Panax ginseng* C.A.Meyer) leaves on lipid peroxidation levels and antioxidant enzyme activities in streptozotocin-diabetic rats. *J. Ethnopharmacol.* 98: 245–250.

Chao, L.K., Chang, W.T., Shih, Y.W., and Huang, J.S. 2010. Cinnamaldehyde impairs high glucose-induced hypertrophy in renal interstitial fibroblasts. *Toxicol. Appl. Pharmacol.* 244: 174–180.

Chatragadda, U., Kodandaram, N., and Bowjanku, V. 2014. Pharmacological evaluation on glucose lowering efficacy of leaves of *Prunus persica*. *Int. J. Innovative Pharm. Sci. Res.* 2: 1321–1336.

Chattopadhyay, R.R. 1993. Hypoglycemic effect of *Ocimum sanctum* leaf extract in normal and streptozotocin diabetic rats. *Indian J. Exp. Biol.* 31: 891–893.

Chattopadhyay, R.R. 1999a. A comparative evaluation of some blood sugar lowering agents of plant origin. *J. Ethnopharmacol.* 67: 367–372.

Chattopadhyay, R.R. 1999b. Possible mechanism of antihyperglycemic effect of *Azadirachta indicia* leaf extract: Part V. *J. Ethnopharmacol.* 67: 373–376.

Chattopadhyay, R.R., Sarkar, S.K., Ganguly, S., Medda, C., and Basu, T.K. 1992. Hepatoprotective activity of *Ocimum sanctum* leaf extract against paracetamol induced hepatic damage in rats. *Indian J. Pharmacol.* 24: 163–165.

Chaturvedi, A., Bhawani, G., Agarwal, P.K., Goel, S., Singh, A., and Goel, R.K. 2009. Antidiabetic and antiulcer effects of extract of *Eugenia jambolana* seed in mild diabetic rats: Study on gastric mucosal offensive acid-pepsin secretion. *Indian J. Physiol. Pharmacol.* 53: 137–146.

Chaturvedi, G.N., Subramaniyam, P.N., Tiwari, S.K., and Singh, K.P. 1984. Experimental and clinical studies of diabetes mellitus: Evaluating the efficacy of an indigenous oral hypoglycemic drug—Arani. *Anc. Sci. Life* 3: 216–224.

Chaturvedi, M., Mali, P.C., and Ansari, A.S. 2003. Induction of reversible antifertility with a crude ethanol extract of *Citrullus colocynthis* Schrad fruit in male rats. *Pharmacology* 68: 38–48.

Chaturvedi, P., George, S., Milinganyo, M., and Tripathy, Y.B. 2005. Microsomal triglyceride transfer protein gene expression and ApoB secretion are inhibited by bitter melon in HepG2 cells. *J. Nutr.* 135: 702–706.

Chaturvedi, P., Tripathy, P., Pandey, S., Singh, U., and Tripathy, Y.B. 2006. Effect of *Saussurea lappa* alcohol extract on different endocrine glands in relation to glucose metabolism in the rat. *Phytother. Res.* 7: 205–207.

Chauhan, A., Sharma, P.K., Srivastava, P., Kumar, N., and Dudhe, R. 2010a. Plants having potential antidiabetic activity: A review. *Der Pharmacia Lett.* 2: 369–387.

Chauhan, S.P., Sheth, N.R., Jivani, N.P., Rathod, I.S., and Shah, P.I. 2010b. Biological actions of *Opuntia* species. *Syst. Rev. Pharm.* 1: 146–151.

Chaurasia, B., Dhakad, R.S., Dhakar, V.K., and Jain, P.K. 2011. Preliminary phytochemical and pharmacological (antidiabetic) screening of *Cassia tora* Linn. *Int. J. Pharm. Life Sci.* 2: 759–766.

Chaurasia, J.K. 2013. A review on *Cinnamomum tamala* (Tejpat) with reference to their chemical composition and biological activity. *Int. J. Adv. Res. Technol.* 1: 7–9.

Chawla, A., Chawla, P., and Roy, M.R.C. 2011. *Asparagus racemosus* (Willd): Biological activities and its active principles. *Indo-Global J. Pharm. Sci.* 1: 113–120.

Chawla, R., Thakur, P., Chowdhry, A., Jaiswal, S., Sharma, A., Goel, R. et al. 2013. Evidence based herbal drug standardization approach in coping with challenges of holistic management of diabetes: A dreadful lifestyle disorder of 21st century. *J. Diabetes Metab. Disord.* 12: 35.

Chayarop, K., Temsiririrkkul, R., Peungvicha, P., Wongkrajang, Y., Chuakul, S., and Ruangwises, N. 2011. Anti-diabetic effects and *in vitro* anti-oxidant activity of *Pseuderanthemum palatiferum* (Nees) Radlk. ex Lindau leaf aqueous extract. *Mahidol Univ. J. Pharm. Sci.* 38: 13–22.

Chempakam, B. 1993. Hypoglycaemic activity of arecoline in betel nut, *Areca catechu* L. *Indian J. Exp. Biol.* 31: 474–475.

Chen, C., Zhang, Y., and Huang, C. 2010. Berberine inhibits PTP 1B activity and mimics insulin action. *Biochem. Biophys. Res. Commun.* 397: 543–547.

Chen, F., Xiong, H., Wang, J., Ding, X., Shu, G., and Mei, Z. 2013. Antidiabetic effect of total flavonoids from Sanguish draxonis in type 2 diabetic rats. *J. Ethnopharmacol.* 149: 723–736.

Chen, J., Li, W.L., Wu, J.L., Ren, B.R., and Zhang, H.Q. 2008. Hypoglycemic effects of a sesquiterpene glycoside isolated from leaves of loquat (*Eriobotrya japonica* (Thunb.) Lindl.). *Phytomedicine* 15: 98–102.

Chen, L. and Kang, Y. 2013. *In vitro* inhibitory effect of oriental melon (*Cucumis melo* L. var. *makuwa* Makino) seed on key enzyme linked to type 2 diabetes: Assessment of anti-diabetic potential of functional food. *J. Funct. Foods* 5: 981–986.

Chen, L., Maghano, D.J., and Zimmet, P.Z. 2012. The worldwide epidemiology of type 2 diabetes mellitus—Present and future perspectives. *Nat. Rev. Endocrinol.* 8: 228–236.

Chen, Q.M. and Xie, M.Z. 1986. Studies on the hypoglycaemic effect of *Coptis chinensis* and berberine. *Acta Pharm. Sin.* 21: 401–406.

Chen, T.H., Chen, S.C., Chan, P., Chu, Y.L., Yang, H.Y., and Cheng, J.T. 2005. Mechanism of hypoglycemic effect of Stevioside, a glycoside of *Stevia rebaudiana*. *Planta Med.* 71: 108–113.

Chen, X., Jin, J., Tang, J., Wang, Z., Wang, J., Jin, L., and Lu, J. 2011c. Extraction, purification, characterization and hypoglycemic activity of a polysaccharide isolated from the root of *Ophiopogon japonicas*. *Carbohydr. Polymer.* 83: 749–754.

Chen, X., Liu, Y., Bai, X., Wen, L., Fang, J., Ye, M., and Chen, J. 2009. Hypoglycemic polysaccharides from the tuberous root of *Liriope spicata*. *J. Nat. Prod.* 72: 1988–1992.

Chen, X.B., Zhuang, J.J., Liu, J.H., Lei, M., Ma, L., Chen, J. et al. 2011a. Potential AMPK activators of cucurbitane triterpenoids from *Siraitia grosvenorii* Swingle. *Bioorg. Med. Chem.* 19: 5776–5781.

Chen, Y., Huang, P., Lin, F., Chen, W., Chen, Y., Yin, W. et al. 2011b. Magnolol: A multifunctional compound isolated from the Chinese medicinal plant *Magnolia officinalis*. *Eur. J. Integr. Med.* 3: e317–e324.

Chen, Z., Wang, X., Yie, Y., Huang, C., and Zhang, S. 1994. Study of hypoglycemia and hypotension function of pumpkin powder on human. *Jiangxi. Chin. Med.* 25: 55–60.

Cheng, D., Liang, B., and Li, Y. 2013b. Antihyperglycemic effect of *Ginkgo biloba* extract in streptozotocin-induced diabetes in rats. *Biomed. Res. Int.* 2013: 162724.

Cheng, J.L. and Yang, R.S. 1983. Hypoglycemic effect of guava juice in mice and human subjects. *Am. J. Chin. Med.* 11: 74–76.

Cherian, S. and Augusti, K.T. 1993. Antidiabetic effects of a glycoside of leucopelargonidin isolated from *Ficus bengalensis* Linn. *Indian J. Exp. Biol.* 31: 26–29.

Cherian, S., Kumar, R.V., Augusti, K.T., and Kidwai, J.R. 1992. Antidiabetic effect of a glycoside of pelargonidin isolated from the bark of *Ficus Bengalensis* L. *Indian J. Biochem. Biophys.* 29: 380–382.

Chi, J.F. and Hu, S.H. 1997. Effects of 8-oxoberberine on sodium current in rat ventricular and human atrial myocytes. *Can. J. Cardiol.* 13: 1103–1110.

Chiang, Y.-M., Chang, C.L.-T., Chang, S.-L., Yang, W.-C., and Shyur, L.-F. 2007. Cytopiloyne, a novel polyacetylenic glucoside from *Bidens pilosa*, functions as a T helper cell modulator. *J. Ethnopharmacol.* 110: 532–538.

Chien, S.-C., Young, P.H., Hsu, Y.-J., Chen, C.-H., Tien, Y.-J., Shiu, S.-Y. et al. 2009. Anti-diabetic properties of three common *Bidens pilosa* variants in Taiwan. *Phytochemistry* 70: 1246–1254.

Chigozie, I.J. and Chidinma, I.C. 2012. Hypoglycemic, hypocholesterolemic and ocular-protective effects of an aqueous extract of the rhizomes of *Sansevieria senegambica* Baker (Agavaceae) on alloxan-induced diabetic Wistar rats. *Am. J. Biochem. Mol. Biol.* 2: 48–66.

Chika, A. and Bello, S.O. 2010. Antihyperglycaemic activity of aqueous leaf extract of *Combretum micranthum* (Combretaceae) in normal and alloxan-induced diabetic rats. *J. Ethnopharmacol.* 129: 34–37.

Chikhi, I., Allali, H., Dib, M.E.A., Medjdoub, H., and Tabti, B. 2014. Antidiabetic activity of aqueous leaf extract of *Atriplex halimus* L. (Chenopodiaceae) in streptozotocin-induced diabetic rats. *Asian Pac. J. Trop. Dis.* 4: 181–184.

Chime, S.A., Onyishi, I.V., Ugwoke, P.U., and Attama, A.A. 2014. Evaluation of the properties of *Gongronema latifolium* in phospholipon 90H based solid lipid microparticles (SLMs): An antidiabetic study. *J. Diet. Suppl.* 11: 7–18.

Chinaka, N.C., Uwakwe, A.A., and Chuku, L.C. 2012. Hypoglycemic effects of aqueous and ethanolic extracts of dandelion (*Taraxacum officinale* F.H. Wigg.) leaves and roots on streptozotocin-induced albino rats. *Global J. Res. Med. Plants Indigenous Med.* 1: 211–217.

Chinese Pharmacopoeia Committee. 2005. *Chinese Pharmacopoeia* (Part 3). Beijing, China: Chemical Industry Press.

Chitra, V., Venkada, K.R., Ch, H., Verma, P., Krishna Raju, M.V.R., and Jeya Prakash, K. 2010. Study of anti-diabetic and free radical scavenging activity of the seed extract of *Strychnos nuxvomica*. *Int. J. Pharm. Pharmaceut. Sci.* 2: 106–110.

Cho, H., Mu, J., Kim, J.K., Thorvaldsen, J.L., Chu, Q., Crenshaw, E.B. et al. 2001. Insulin resistance and a diabetes mellitus-like syndrome in mice lacking the protein kinase Akt2 (PKB beta). *Science* 292: 1728–1731.

Cho, W.C., Chung, W.S., Lee, S.K., Leung, A.W., Cheng, C.H., and Yue, K.K. 2006. Gingenoside Re of *Panax ginseng* possesses significant anti-oxidant and anti-hyperlipidemic effects in streptozotocin-induced diabetic rats. *Eur. J. Pharmacol.* 550: 173–179.

Chohachi, K., Yusuka, S., Kajna, S., Eiko, M., and Hiroshi, S. 1985. Hypoglycemic activity of Glycan A, B, C, D and E from *Saccharum officinarum*. *Planta Med.* 51: 113–115.

Choi, E.-K., Kim, K.-S., Yang, H.J., Shin, M.H., Suh, H.-W., Lee, K.-B. et al. 2012a. Hexane fraction of *Citrus aurantium* L. stimulates glucagon-like peptide-1 (GLP-1) secretion *via* membrane depolarization in NCI-H716 cells. *BioChip J.* 6: 41–47.

Choi, H.S., Kim, Y.H., Han, J.H., and Park, S.H. 2008. Effects of *Eleutherococcus senticosus* and several oriental medicinal herbs extracts on serum lipid concentrations. *Korean J. Food Nutr.* 21: 210–217.

Choi, J.S., Yokozawa, T., and Oura, H. 1991. Improvement of hyperglycemia and hyperlipidemia in strep-tozotocin-diabetic rats by a methanolic extract of *Prunus davidiana* stems and its main component, prunin. *Planta Med.* 57: 208–211.

Choi, J.Y., Na, M., Hwang, I.H., Lee, S.H., Bae, E.Y., Kim, B.Y., and Ahn, J.S. 2009. Isolation of betulinic acid, its methyl ester and guaiane sesquiterpenoids with protein tyrosine phosphatase 1B inhibitory activity from the roots of *Saussurea lappa* C.B.Clarke. *Molecules* 14: 266–272.

Choi, S.E., Shin, H.C., Kim, H.E., Lee, S.J., Jang, H.J., Lee, K.W. et al. 2007. Involvement of Ca^{2+}, CaMK II and PKA in EGb761-induced insulin secretion in INS-1 cells. *J. Ethnopharmacol.* 110: 49–55.

Choi, Y.H., Zhou, W., Oh, J., Choe, S., Lim, D.W., and Lee, S.H. 2012b. Rhododendric acid A, a new ursane-type PTP1B inhibitor from the endangered plant *Rhododendron brachycarpum* G. Don. *Bioorg. Med. Chem. Lett.* 22: 6116–6119.

Chopade, B.A., Ghosh, S., Ahire, M., Patil, S., Jabgunde, A., Dusane, M.B. et al. 2012. Antidiabetic activity of *Gnidia glauca* and *Dioscorea bulbifera*: Potent amylase and glucosidase inhibitors. *Evid. Based Complement. Altern. Med.* 2012: 1–10, Article ID 929051.

Chothani, D.L. and Vaghasiya, H.U. 2011. A review on *Balanites aegyptiaca* Del (desert date): Phytochemical constituents, traditional uses and pharmacological activity. *Pharmacogn. Rev.* 5: 55–62.

Choudhary, M., Aggarwal, N., Choudhary, N., Gupta, P., and Budhwaar, V. 2014. Effect of aqueous and alcoholic extract of *Sebania sesban* (Linn.) Merr. root on glycemic control in streptozotocin-induced diabetic mice. *Drug Dev. Ther.* 5: 115–122.

Choudhary, N.K., Jha, A.K., Sharma, S., Goyal, S., and Dwivedi, J. 2011. Anti-diabetic potential of chlo-roform extract of flowers of *Calotropis gigantea*: An *in vitro* and *in vivo* study. *Int. J. Green Pharm.* 5: 296–301.

Choudhury, K.D. and Basu, N.K. 1967. Phytochemicals and hypoglycemic investigation of *Casearia escu-lenta*. *J. Pharm. Sci.* 56: 1405–1409.

Christudas, S., Gopalakrishnan, L., Mohanraj, P., Kaliyamoorthy, K., and Agastian, P. 2009. α-glucosidase inhibitory and antidiabetic activities of ethanol extract of *Pisinia alba* Span. leaves. *Int. J. Interact. Biol.* 6: 41–45.

Chucla, M.T., Lamela, M., Gata, A., and Cadavid, I. 1998. *Centaurea corcubionensis*: A study of its hypogly-cemic activity in rats. *Planta Med.* 54: 107–109.

Chung, I.M., Kim, E.H., Yeo, M.-A., Kim, S.-J., Seo, M.-C., and Moon, H.-I. 2011. Anti-diabetic effects of three Korean sorghum phenolic extracts in normal and streptozotocin-induced diabetic rats. *Food Res. Int.* 44: 127–132.

Chung, M.J., Cho, S.Y., Bhuiyan, M.J., Kim, K.H., and Lee, S.J. 2010. Anti-diabetic effects of lemon balm (*Melissa officinalis*) essential oil on glucose- and lipid-regulating enzymes in type 2 diabetic mice. *Br. J. Nutr.* 104: 180–188.

Chung, S.H, Choi, C.G., and Park, S.H. 2001. Comparison between white ginseng radix and rootlet for anti-diabetic mechanism in KKAy mice. *Arch. Pharm. Res.* 24: 214–218.

Claudia, M.N.E., Ines, F.D.G., Dagobert, T., and Julius, O.E. 2007. Effects of aqueous and methanol/methylene-chloride extracts of *Laportea ovalifolia* (Urticaceae) on blood glucose level in rats. *Pharmacologyonline* 3: 105–118.

Cleasby, M.E., Dzamko, N., Hegarty, B.D., Cooney, G.J., Kraegen, E.W., and Ye, J.M. 2004. Metformin prevents the development of acute lipid-induced insulin resistance in the rat through altered hepatic signaling mechanisms. *Diabetes* 53: 3258–3266.

Coman, C., Rugina, O.D., and Socaciu, C. 2012. Plants and natural compounds with anti-diabetic action. *Not. Bot. Hortic. Agrobo.* 40: 314–325.

Comelli, F., Bettoni, I., Colleoni, M., Giagnoni, G., and Costa, B. 2009. Beneficial effects of a *Cannabis sativa* extract treatment on diabetes-induced neuropathy and oxidative stress. *Phytother. Res.* 23: 1678–1684.

Campbell, W.W., Haub, M.D., Fluckery, J.D., Ostlund, R.E. Jr., Thyfault, J.P., Morse-Carrithers, H. et al. 2004. Pinitol supplementation does not affect insulin-mediated glucose metabolism and muscle insulin receptor content and phosphorylation in older humans. *J. Nutr.* 134: 2998–3003.

Comstock, S.S., Gershwin, L.J., and Teuber, S.S. 2010. Effect of wanut (*Juglans regia*) polyphenolic compounds on ovalbumin-specific IgE induction in female BALB/c mice. *Ann. N.Y. Acad. Sci.* 1190: 58–69.

Conde, G. E.A., Nascimento, V.T., and Santiago, S.A.B. 2003. Ionotropic effect of extracts of *Psidium guajava* L. (guava) leaves on the guinea pig atrium. *Braz. J. Med. Biol. Res.* 36: 661–668.

Cong, W., Tao, R., Tian, J., Zhao, J., Liu, Q., and Ye, F. 2011. EGb761, an extract of *Ginkgo biloba* leaves reduces insulin resistance in a high-fat-fed mouse model. *Acta Pharm. Sin. B* 1: 14–20.

Contreras-Weber, C., Perez-Gutierrez, S., Alarcon-Aguilar, F., and Roman-Ramos, R. 2002. Anti-hyperglycemic effect of *Psacalium peltatum*. *Proc. West Pharmacol. Soc.* 45: 134–136.

Contu, S. 2012. *Bauhinia forficata*. In: *IUCN 2013*. IUCN Red List of Threatened Species, Version 2013. 2.

Cooper, M.E., White, M.F., Zick, Y., and Zimmet, P. 2012. *Type 2 Diabetes Mellitus*. Novo Nordisk, Denmark: Nature Publishing Group. http://www.Nature.com/nrendo/posters/type2diabetesmellitus/.

Coskun, O., Kanter, M., Korkmaz, A., and Oter, S. 2005. Quercetin, a flavonoid antioxidant, prevents and protects streptozotocin-induced oxidative stress and β-cell damage in rat pancreas. *Pharmacol. Res.* 51: 117–123.

Costa, M.A.C., Paul, M.I., Alves, A.A.C., and van der Vyver, L.M. 1978. Aliphatic and triterpenoid compounds of Ebenaceae species. *Rev. Port. Farm.* 28: 171–174.

Costantino, L., Raimondi, L., Pirisino, R., Brunetti, T., Pessotto, P., Giannessi, F., Lins, A.P., Barlocco, D., Antolini, L., and Abady, S.A.E. 2003. Isolation and pharmacological activities of the *Tecoma stans* alkaloids. *Farmaco* 58: 781–785.

Crawford, P. 2009. Effectiveness of cinnamon for lowering hemoglobin A1C in patients with type 2 diabetes: A randomized, controlled trial. *J. Am. Board Fam. Med.* 22: 507–512.

Cristians, S., Guerrero-Analco, J.A., Perez-Vasquez, A., Palacios-Espinosa, F., Ciangherotti, C., Bye, R. et al. 2009. Hypoglycemic activity of extracts and compounds from the leaves of *Hintonia standleyana* and *H. latiflora*: Potential alternatives to the use of the stem bark of these species. *J. Nat. Prod.* 72: 408–413.

Cummings, D.E. 2006. Ghrelin and the short and long-term regulation of appetite and body weight. *Physiol. Behav.* 89: 71–84.

Cunha, A.M., Menon, S., Couto, A.G., Burger, C., and Biavatti, M.W. 2010. Hypoglycemic activity of dried extracts of *Bauhinia forficata*. *Phytomedicine* 17: 37–41.

Cunha, W.R., Arantes, G.M., Ferreira, D.S., Lucarini, R., Silva, M.L.A., Furtado, N.A.J.C. et al. 2008. Hypoglycemic effect of *Leandra lacunosa* in normal and alloxan-induced diabetic rats. *Fitoterapia* 79: 356–360.

Curcio, S.A., Stefan, L.F., Randi, B.A., Dias, M.A., da Silva, R.E., and Caldeira, E.J. 2012. Hypoglycemic effects of an aqueous extract of *Bauhinia forficata* on the salivary glands of diabetic mice. *Pak. J. Pharm. Sci.* 25: 493–499.

daCosta, A.V., Calabria, L.K., Furtado, F.B., de Gouveia, N.M., Oliveira, R.J., de Oliveira, V.N. et al. 2013. Neuroprotective effects of *Pouteria ramiflora* (Mart.) Radlk (Sapotaceae) extract on the brains of rats with streptozotocin-induced diabetes. *Metab. Brain Dis.* 28: 411–419.

daCunha, A.M., Menon, S., Menon, R., Couto, A.G., Burger, C., and Biavatti, M.W. 2010. Hypoglycemic activity of dried extracts of *Bauhinia forficata* Link. *Phytomedicine* 17: 37–41.

Dahake, A.P., Chakma, C.S., Chakma, R.C., and Bagherwal, P. 2010. Antihyperglycemic activity of methanolic extract of *Madhuca longifolia* bark. *Diabetol. Croat.* 39: 3–8.

Daisy, P., Eliza, J., and Farook, K.A.M. 2009b. A novel dihydroxygymnemic triacetate isolated from *Gymnema sylvestre* possessing normoglycemic and hypolipidemic activity on streptozotocin-induced diabetic rats. *J. Ethnopharmacol.* 126: 339–344.

Daisy, P., Jasmine, R., Ignacimuthu, S., and Murugan, E. 2009a. A novel steroid from *Elephantopus scaber* L., an ethnomedical plant with antidiabetic activity. *Phytomedicine* 16: 252–257.

Daisy, P. and Priya, C.E. 2010. Hypolipidemic and renal functionality potentials of the hexane extract fractions of *Elephantopus scaber* L. *Int. J. Biomed. Sci.* 6: 241–245.

Daisy, P., Priya, C.E., and Vargese, L. 2011. A study on the regenerative potential of the root and leaf extracts of *Elephantopus scaber* L.: An anti-diabetic approach. *Afr. J. Pharm. Pharmacol.* 5: 1832–1837.

Daisy, P. and Saipriya, K. 2012. Biochemical analysis of *Cassia fistula* aqueous extract and phytochemically synthesized gold nanoparticles as hypoglycemic treatment for diabetes mellitus. *Int. J. Nanomed.* 7: 1189–1202.

Dall'Agnol, R. and von Poser, G.L. 2000. The use of complex polysaccharides in the management of metabolic diseases: The case of *Solanum lycocarpum* fruits. *J. Ethnopharmacol.* 71: 337–341.

Dallak, M.A., Bin-Jaliah, I., Al-Khateeb, M.A., Nwoye, L.O., Shatoor, A.S., Soliman, H.S. et al. 2010. *In vivo* acute effects of orally administered hydro-ethanol extract of *Catha edulis* on blood glucose levels in normal, glucose-fed hyperglycemic, and alloxan-induced diabetic rats. *Saudi Med. J.* 31: 627–633.

Daniel, R.S., Devi, K.S., Augusti, K.T., and Sudhakaran, N.C.R. 2003. Mechanism of action of antiatherogenic and related effects of *Ficus begalensis* L. flavanoids in experimental animals. *Indian J. Exp. Biol.* 41: 296–303.

Daniel, R.S., Mathew, B.C., Devi, K.S., and August, K.T. 1998. Antioxidant effect of two flavonoids from the bark of *Ficus bengalensis* L. in hyperlipidemic rats. *Indian J. Exp. Biol.* 36: 902–906.

Danish, M., Singh, P., Mishra, G., Srivastava, S., Jha, K.K., and Khosa, R.L. 2011. *Cassia fistula* Linn. (amulthus)—An important medicinal plant: A review of its traditional uses, phytochemistry and pharmacological properties. *J. Nat. Prod. Plant Resour.* 1: 101–118.

Danmalam, U.H., Abdullahi, L.M., Agunu, A., and Musa, K.Y. 2009. Acute toxicity studies and hypoglycemic activity of the methanol extract of the leaves of *Hyptis suaveolens* Poit. (Lamiaceae). *Nig. J. Pharm. Sci.* 8: 87–92.

Darvhekar, V.M., Tripathy, A.S., Jyotishi, S.G., and Mazumder, P.M. 2013. Evaluation of diabetic gastroparesis effect of *Musa sapientum* L. bark juice by reducing the level of oxidative stress. *Orient. Pharm. Exp. Med.* 13: 29–34.

Das, A.K., Manal, S.C., Banerjee, S.K., Sinha, S., Saha, B.P., and Pal, M. 2001. Studies on hypoglycemic activity of *Punica granatum* seed in streptozotocin induced diabetic rats. *Phytother. Res.* 15: 628–629.

Das, A.K., Mandal, S.C., Banerjee, S.K., Sinha, S., Das, J., Saha, B.P., and Pal, M. 1999. Studies on antidiarrhoeal activity of *Punica granatum* seed extract in rats. *J. Ethnopharmacol.* 68: 205–208.

Das, L., Gunindro, N., Ghosh, R., Roy, M., and Debbarma, A. 2014. Mechanism of action of *Azadirachta indica* Linn. (Neem) aqueous leaf extract as hypoglycemic agent. *Indian Med. Gaz.* 2014: 29–32.

Das, M., Sarma, B.P., Khan, A.K.A., Mosihuzzaman, M., Nahar, N., Ali, L. et al. 2009. The antidiabetic and antilipidemic activity of aqueous extract of *Urtica dioica* L. on type-2 diabetic model rats. *J. Bio-Sci.* 17: 1–6.

Das, M., Sarma, B.P., Rokeya, B., Parial, R., Nahar, N., Mosihuzzaman, M. et al. 2011b. Anti-hyperglycemic and anti-hyperlipidemic activity of *Urtica dioica* on type-2 diabetic model rats. *J. Diabetol.* 2: 1–6.

Das, S. and Barman, S. 2012. Antidiabetic and antihyperlipidemic effects of ethanolic extract of leaves of *Punica granatum* in alloxan-induced non-insulin-dependent diabetes mellitus albino rats. *Indian J. Pharmacol.* 44: 219–224.

Das, S., Bhattacharya, S., Prasanna, A., Suresh Kumar, R.B., Pramanik, G., and Haldar, P.K. 2011a. Preclinical evaluation of antihyperglycemic activity of *Clerodendron infortunatum* leaf against streptozotocin-induced diabetic rats. *Diabetes Ther.* 2: 92–100.

Datusalia, A.K., Dora, C.P., and Sharma, S. 2012. Acute and chronic hypoglycemic activity of *Sida tiagii* fruits in N5-streptozotocin diabetic rats. *Acta Pol. Pharm.* 69: 699–706.

Davison, C. 2009. A biochemical study of the antidiabetic and anticoagulant effects of *Tulbaghia violacea*. MSc thesis, Nelson Mandela Metropolitan University, Western Cape, South Africa.

Dayuan, W., Hong, Y., Shengshan, H., Ping, Q., and Hongwen, L. 1995. Hypoglycemic effect of total saponins of *Aralia decaisneana* Hance. *Chinese Pharm. J.* 7. (Article in Chinese.)

De, A.U. and Saha, B.P. 1975. Indolizines II: Search for potential oral hypoglycemic agents. *J. Pharm. Sci.* 64: 49–55.

De, D., Ali, K.M., Chatterjee, K., Bera, T.K., and Ghosh, D. 2012. Antihyperglycemic and antihyperlipidemic effects of n-hexane fraction from the hydro-methanolic extract of sepals of *Salmalia malabarica* in streptozotocin-induced diabetic rats. *J. Complement. Integr. Med.* 9: Article 12.

De, D., Chatterjee, K., Ali, K.M., Bera, T.K., and Ghosh, D. 2011. Antidiabetic potentiality of the aqueous-methanolic extract of seed of *Swietenia mahagoni* (L.) Jacq. in streptozotocin-induced diabetic male albino rat: A correlative and evidence-based approach with antioxidative and antihyperlipidemic activities. *Evid. Based Complement. Altern. Med.* 2011: Article ID 892807, 11 pages.

de Bock, M., Derraik, J.G.B., Brennan, C.M., Biggs, J.B., Morgan, P.E., Hodgkinson, S.C. et al. 2013. Olive (*Olea europaea* L.) leaf polyphenols improve insulin sensitivity in middle-aged overweight men: A randomized, placebo-controlled, crossover trial. *PLoS ONE* 8: e57622. DOI: 10.1371/journal.pone.0057622.

Dechandt, C.R.P., Siqueire, J.T., de Souze, D.L.P., Araujo, L.C.J., da Silva, V.C., de Sousa Junior, P.T. et al. 2013. *Combretum lanceolatum* flowers extract shows antidiabetic activity through activation of AMPK by quercetin. *Revista Brasileira de Farmacognosia* 23: 291–300.

Deepralard, K., Kawanishi, K., Moriyasu, M., Pengsuparp, T., and Suttisri, R. 2009. Flavonoid glucosides from the leaves of *Uvaria rufa* with advanced glycation end-products inhibitory activity. *Thai. J. Pharm. Sci.* 33: 84–90.

DeFronzo, R.A. 2004. Pathogenesis of type 2 diabetes mellitus. *Med. Clin. North Am.* 88: 787–835.

De Gouveia, N.M., Albuquerque, C.L., Espindola, L.S., and Espindola, F.S. 2013. *Pouteria ramiflora* extract inhibits activity of human salivary alpha-amylase and decreases glycemic level in mice. *An. Acad. Bras. Cienc.* 85: 1141–1148.

Dehkordi, F.R. and Enteshari, A. 2013. The in vitro effect of *Melissa officinalis* aqueous extract on aortic reactivity in rats with subchronic diabetes. *J. Basic Clin. Pathophysiol.* 2: 44–49.

Dejkhamron, P., Menon, R.K., and Sperling, M.A. 2007. Childhood diabetes mellitus: Recent advances and future prospects. *Indian J. Med. Res.* 125: 231–250.

Dela Fuente, J.A. and Manzanaro, S. 2003. Aldose reductase inhibitors from natural sources. *Nat. Prod. Rep.* 20: 243–251.

Deliwe, M. and Amabeoku, G.J. 2013. Evaluation of anti-diarrhoeal and anti-diabetic activities of the leaf aquous extract of *Syzygium cordatum* Hoscht. Ex. C.Krauss (Mytraceae) in rodents. *Int. J. Pharmacol.* 9: 125–131.

Deng, X.Y., Chen, Y.S., Zhang, W.R., Chen, B., Qiu, X.M., He, L.H. et al. 2011. Polysaccharide from *Gynura divaricata* modulates the activities of intestinal disaccharidases in streptozotocin-induced diabetic rats. *Br. J. Nutr.* 106: 1323–1329.

Deng, Y., He, K., Ye, X., Chen, X., Huang, J., Li, X. et al. 2012. Saponin rich fractions from *Polygonatum odoratum* (Mill.) Druce with more potential hypoglycemic effects. *J. Ethnopharmacol.* 141: 228–233.

DePaula, A.C.C.F.F., Sousa, R.V., Figueiredo-Ribeiro, R.C.L., and Buckeridge, M.S. 2005. Hypoglycemic activity of polysaccharide fractions containing ß-glucans from extracts of *Rhynchelytrum repens* (Willd.) C.E. Hubb., Poaceae. *Braz. J. Med. Biol. Res.* 38: 885–893.

DeSales, P.M., de Souza, P.M., Simeoni, l.A., Magalhaes, P.O., and Silvera, D. 2012. α-amylase inhibitors: A review of raw material and isolated compounds from plant source. *J. Pharm. Pharm. Sci.* 15: 141–183.

Deshmukh, S.A. and Gaikwad, D.K. 2014. A review of the taxonomy, ethnobotany phytochemistry and pharmacology of *Basella alva* (Basellaceae). *J. Appl. Pharm. Sci.* 4: 153–165.

Deshmukh, T.A., Yadav, B.V., Badole, S.L., Bodhankar, S.L., and Dhaneshwa, S.R. 2008. Antihyperglycaemic activity of alcoholic extract of *Aerva lanata* (L.) A. L. Juss. Ex J. A. Schultes leaves in alloxan induced diabetic mice. *J. Appl. Biomed.* 6: 81–87.

Desire, D.D.P., Florine, N.D., Theophile, D., Leonard, T., Florence, N.T., Marie-Claire, T. et al. 2009. Hypoglycemic and hypolipidemic effects of *Irvingia gabonensis* (Irvingiaceae) in diabetic rats. *Pharmacologyonline* 2: 957–962.

Deutschlander, M.S. 2010. Isolation and identification of a novel anti-diabetic compound from *Euclea undulate* Thunb. PhD thesis, University of Pretoria, Pretoria, South Africa. http://upetd.up.ac.za/thesis/available/etd-10232010-172051.

Deutschlander, M.S., Lall, N., Van de Venter, M., and Dewanjee, S. 2012. The hypoglycemic activity of *Euclea undulate* Thunb. var. *myrtina* (Ebenaceae) root bark evaluated in streptozotocin-nicotinamide induced type 2 diabetes rat model. *S. Afr. J. Bot.* 80: 9–12.

Deutschlander, M.S., Lall, N., Van de Venter, M., and Hussein, A.A. 2011. Hypoglycaemic evaluation of a new triterpene and other compounds isolated from *Euclea undulata* Thunb. var. *myrtina* (Ebenaceae) root bark. *J. Ethnopharmacol.* 16: 1091–1095.

Deutschlander, M.S., van de Venter, M., Roux, S., Louw, J., and Lall, N. 2009. Hypoglycaemic activity of four plant extracts traditionally used in South Africa for diabetes. *J. Ethnopharmacol.* 124: 619–624.

Devaki, K., Beulah, U., Akila, G., Narmadha, R., and Gopalakrishnan, V.K. 2011a. Effect of aqueous leaf extract of *Bauhinia tomentosa* on GTT of normal and diabetic rats. *Pharmacologyonline* 3: 195–202.

Devaki, K., Beulah, U., Akila, G., Narmadha, R., and Gopalakrishnan, V.K. 2011b. Glucose lowering effect of aqueous extract of *Bauhinia tomentosa* L. on alloxan induced type 2 diabetes mellitus in Wistar albino rats. *J. Basic Clin. Pharm.* 2: 167–174.

Devani, U., Pandita, N., and Kachwala, Y. 2013. Evaluation of inhibitory activity of *Vitex negundo* and *Terminalia chebula* by alpha amylase inhibition assay in management of diabetes. *Asian J. Plant Sci. Res.* 3: 6–14.

Devi, B., Sharma, N., Kumar, D., and Jeet, K. 2013. *Morus alba* L.: A phytopharmacological review. *Int. J. Pharm. Pharm. Sci.* 5: 14–18.

Devi, V.D. and Urooj, A. 2008. Hypoglycemic potential of *Morus indica* L. and *Costus igneus* Nak—A preliminary study. *Indian J. Exp. Biol.* 46: 614–616.

Devi, V.D. and Urooj, A. 2011. Evaluation of anti hyperglycemic and anti lipid peroxidative effect of *Costus igneus* Nak in streptozotocin induced diabetic rats. *Int. J. Curr. Res.* 3: 4–8.

Dewanjee, S., Das, A.K., Sahu, R., and Gangopadhyay, M. 2009a. Antidiabetic activity of *Diospyros peregrine* fruit: Effect on hyperglycemia, hyperlipidemia and augmented oxidative stress in experimental type 2 diabetes. *Food Chem. Toxicol.* 47: 2679–2685.

Dewanjee, S., Sahu, R., Mandal, V., Maiti, A., and Mandal, S.C. 2009b. Antidiabetic and antioxidant activity of the methanol extract of *Diospyros peregenia* fruit on Type I diabetic rats. *Pharm. Biol.* 47: 1149–1153.

Dewiyanti, I.D., Filailla, E., and Yuliani, T. 2012. The anti-diabetic activity of Cocor Bebek leaves (*Kalanchoe pinnata* Lam. Pers.) ethanol extract from various areas. *J. Trop. Life Sci.* 2: 37–39.

Dey, B. and Mitra, A. 2013. Chemo-profiling of eucalyptus and study of its hypoglycemic potential. *World J. Diabetes* 4: 170–176.

Dey, L., Xie, J.T., Wang, A., Wu, J., Maleckar, S.S., and Yuan, C.S. 2003. Antihyperglycemic effects for ginseng: Comparison between root and berry. *Phytomedicine* 10: 600–605.

Dhalwal, K., Shinde, V.M., Singh, B., and Mahadik, K.R. 2010. Hypoglycemic and hypolipidemic effect of *Sida rhombifolia* ssp. *retusa* in diabetic induced animals. *Int. J. Phytomed.* 2: 160–165.

Dhanabal, S.P., Kokate, C.K., Ramanathan, M., Kumar, E.P., and Suresh B. 2006. Hypoglycaemic activity of *Pterocarpus marsupium* Roxb. *Phytother Res.* 20: 4–8.

Dhanabal, S.P., Mohan Marugaraja, M.K., Ramanathan, M., and Suresh, B. 2007. Hypoglycemic activity of *Nymphaea stellata* leaves ethanolic extract in alloxan induced diabetic rats. *Fitoterapia* 78: 288–291.

Dhanabal, S.P., Mohan Marugaraja, M.K., and Suresh, B. 2008. Antidiabetic activity of *Clerodendron phlomoidis* leaf extract in alloxan-induced diabetic rats. *Indian J. Pharm. Sci.* 70: 841–844.

Dhanabal, S.P., Sureshkumar, M., Ramanathan, M., and Suresh, B. 2005. Hypoglycemic effect of ethanolic extract of *Musa sapientum* on alloxan induced diabetes mellitus in rats and its relation with antioxidant potential. *J. Herbal Pharmacother.* 5: 7–19.

Dhandapani S., Subramanaian, V.R., Rajagopal, S., and Namashivayam, N. 2002. Hypolipidemic effect of *Cuminum cyminum* L. on alloxan induced diabetic rats. *Pharmacol. Res.* 46: 251–255.

Dharmani, P., Kuchibhotla, V.K., Maurya, R., Srivastava, S., Sharma, S., and Palit, G. 2004. Evaluation of anti-ulcerogenic and ulcer healing properties of *Ocimum sanctum* L. *J. Ethnopharmacol.* 93: 197–206.

Dharmappa, K.K., Kumar, R.V., Nadaraju, A., Mohamed, R., Shivaprasad, H.V., and Vishwanath, B.S. 2009. Anti-inflammatory activity of oleanolic acid by inhibition of secretory phospholipase A_2. *Planta Med.* 75: 211–215.

Dheer, R. and Bhatnagar, P. 2010. A study of the antidiabetic activity of *Barleria prionitis* Linn. *Indian J. Pharmacol.* 42: 70–73.

Dias, A.S., Porwski, M., Alonso, M., Marroni, N., Collado, P.S., and Gonzalez-Gallego, J. 2005. Quercetin decreases oxidative stress, NF-κB activation and iNOS overexpression in liver of streptozotocin-induced diabetic rats. *J. Nutr.* 135: 2299–2304.

Diatewa, M., Samba, C.B., Assah, T.C.H., and Abena, A.A. 2004. Hypoglycemic and antihyperglycemic effects of diethyl ether fraction isolated from the aqueous extract of the leaves of *Cogniauxia podolaena* Baillon in normal and alloxan-induced diabetic rats. *J. Ethnopharmacol.* 92: 229–232.

Diaz-Flores, M., Angele-Mejia, S., Baiza-Gutman, L.A., Medina-Navarro, R., Hernandez-Saavedra, D., Ortega-Camarillo, C. et al. 2012. Effect of an aqueous extract of *Cucurbita ficifolia* Bouche on the glutathione redox cycle in mice with streptozotocin-induced diabetes. *J. Ethnopharmacol.* 144: 101–108.

Dimo, T., Nagueguim, F.T., Kamtchouing, P., Dongo, E., and Tan, P.V. 2006. Glucose lowering effect of the aqueous stem bark extract of *Trema orientalis* (L.) Blume in normal and streptozotocin-induced diabetic rats. *Die Pharmazie* 61: 233–236.

Dineshkumar, B., Mithra, A., and Mahadevappa, M. 2010. Antidiabetic and hypolipidemic effects of mahanimbine (carbazole alkaloid) from *Murraya koenigii* (Rutaceae) leaves. *Int. J. Phytomed.* 2: 22–30.

Dinka, T., Tadesse, S., and Asres, K. 2010. Antidiabetic activity of the leaf extracts of *Pentas schimperiana* subsp. *schimperiana* (A. Rich) Vatke on alloxan-induced diabetic mice. *Ethiop. Pharm. J.* 28: 12–20.

Dirsch, V.M., Kiemer, A.K., Wagner, H., and Vollmar, A.M. 1998. Effect of allicin and ajoene, two compounds of garlic, on inducible nitric oxide synthase. *Atherosclerosis* 139: 333–339.

Diwan, F.H., Abdel-Hassan, I.A., and Mohammed, S.T. 2000. Effect of saponin on mortality and histopathological changes in mice. *East. Mediterr. Health J.* 6: 345–351.

Dixit, R.D. 1984. *A Census of the Indian Pteridophytes.* Howrah (Calcutta), India: Flora of India, Series-4, Botanical Survey of India, pp. 1–177.

Djilani, A., Toudert, N., and Djilani, S. 2011. Evaluation of the hypoglycemic effect and anti-oxidant activity of methanol extract of *Ampelodesma mauritanica* roots. *Life Sci. Med. Res.* 2011: 1–6.

Dogrul, A., Gul, H., Yildiz, O., Bilgin, F., and Guzeldemir, M.E. 2004. Cannabinoids blocks tactile allodynia in diabetic mice without attenuation of its antinociceptive effect. *Neurosci. Lett.* 368: 82–86.

Domingues, A., Sartori, A., Golim, M.A., Valente, L.M., da Rosa, L.C., Ishikawa, L.L. et al. 2011. Prevention of experimental diabetes by *Uncaria tomentosa* extract: Th2 polarization, regulatory T cell preservation or both? *J. Ethnopharmacol.* 137: 635–642.

Donato, A.M. and Morretes, B.L. 2011. Leaf morphoanatomy of *Myrcia multiflora* (Lam.) DC. -Myrtaceae. *Rev. Bras. Plantas Med.* 13: 43–51.

Dong, H.-Q., Li, M., Zhu, F., Liu, F.-L., and Huang, J.-B. 2012. Inhibitory potential of trilobatin from *Lithocarpus polystachyus* Rehd against α-glucosidase and α-amylase linked to type 2 diabetes. *Food Chem.* 130: 261–266.

Dong, L., Liu, M., Song, G.Y., and Wang, J. 2009. Protective effects of *Ginkgo biloba* extract on apoptosis of pancreatic β-cells in insulin resistance rats induced by high fat diet. *J. Pract. Med.* 25: 3761–3763.

Donnapee, S., Li, J., Yang, X., Ge, A., Donkor, P.W., Gao, X. et al. 2014. *Cuscuta chinensis* Lam.: A systematic review on ethnopharmacology, phytochemistry and pharmacology of important traditional herbal medicine. *J. Ethnopharmacol.* 157: 292–308.

Dos Anjos, P.J.C., Pereire, P.R., Moreira, I.J.A., Serafini, M.R., Araujo, A.A.S., da Silva, F.A. et al. 2013. Antihypertensive effect of *Bouhinia forficata* aqueous extract in rats. *J. Pharmacol. Toxicol.* 8: 82–89.

Doss, A. and Anand, S.P. 2014. Evaluation of anti-diabetic activity of methanol and aqueous extracts of *Asteracantha longifolia* (Linn.) Nees. *Res. J. Pharmacol.* 8: 1–5.

Doss, A. and Dhanabalan, R. 2008. Anti-hyperglycemic and insulin release effects of *Coccinia grandis* (L.) Voigt leaves in normal and alloxan diabetic rats. *Ethnobot. Leaflets* 12: 1172–1175.

Drucker, D.J. 2006. The biology of incretin hormones. *Cell Metab.* 3: 153–165.

Duan, S.Z., Ivashchenko, C.Y., Usher, M.G., and Mortensen, R.M. 2008. PPAR-γ in the cardiovascular system. *PPAR Res.* 2008: 745804.

Duganath, N., Krishna, D.R., Reddy, G.D., Sudheera, B., Mallikarjun, M., and Beesetty, P. 2011. Evaluation of anti-diabetic activity of *Triumfetta rhomboidea* in alloxan induced Wistar rats. *Res. J. Pharm. Biol. Chem. Sci.* 2: 721–726.

Duke, J.A. 1983. *Handbook of Energy Crops*. West Lafayette, IN: Purdue University, New CROPS web site, http://www.hort.purdue.edu/newcrop/duke_energy/dukeindex.html.

Duke, J.A. 1985. *Handbook of Medicinal Herbs*. Boca Raton, FL: CRC Press.

Duke, J.A. 1992. *Handbook of Phytochemical Constituent of GRAS Herbs and Other Economic Plants*. Boca Raton, FL: CRC Press.

Duke, J.A. and Judith, L.D. 1993. *Handbook of Alternative Cash Crops*. Boca Raton, FL: CRC Press.

Duke, J.A. and Wain, K.K. 1981. Medicinal plants of the world. Computer index with more than 85,000 entries, 3 vols. Beltsville, MD: Agricultural Research Service, Plants Genetics and Germplasm Institute.

Durmuskahya, C. and Ozturk, M. 2013. Ethnobotanical survey of medicinal plants used for the treatment of diabetes in Manisa, Turkey. *Sains Malaysiana* 42: 1431–1438.

Dzeufiet, P.D., Ohandja, D.Y., Tedong, L., Asongalem, E.A., Dimo, T., Sokeng, S. et al. 2006. Antidiabetic effect of *Ceiba pentandra* extract on streptozotocin-induced non-insulin-dependent diabetic (NIDDM) rats. *Afr. J. Tradit. Complement. Altern. Med.* 4: 47–54.

Ebrahimpur, M.R., Khaksar, Z., and Noorafsan, A. 2009. Antidiabetic effect of *Otostegia persica* oral extract on streptozotocin-diabetic rats. *Res. J. Biol. Sci.* 4: 1227–1229.

Eddouks, M., Jouad, H., Maghrani, M., Lemhadri, A., and Burcelin, R. 2003. Inhibition of endogenous glucose production accounts for hypoglycemic effect of *Spergularia purpurea* in streptozotocin mice. *Phytomedicine* 10: 594–599.

Eddouks, M., Lemhadri, A., and Michel, J.B. 2004a. Caraway and caper: Potential anti-hyperglycaemic plants in diabetic rats. *J. Ethnopharmacol.* 94: 143–148.

Eddouks, M., Jouad, H., Maghrani, M., Zeggwagh, N.A., and Lemhadri, A. 2004b. Antihyperglycemic activity of the aqueous extract of *Origanum vulgare* growing wild in Tafilalet region. *J. Ethnopharmacol.* 92: 251–256.

Eddouks, M., Lemhadri, A., and Michel, J.B. 2005a. Hypolipidemic activity of aqueous extract of *Capparis spinosa* L. in normal and diabetic rats. *J. Ethnopharmacol.* 98: 345–350.

Eddouks, M., Lemhadri, A., Zeggwagh, N.A., and Michel, J.B. 2005b. Potent hypoglycaemic activity of the aqueous extract of *Chamaemelum nobile* in normal and streptozotocin-induced diabetic rats. *Diabetes Res. Clin. Pract.* 67: 189–195.

Eddouks, M. and Maghrani, M. 2004. Phlorizin-like effect of *Fraxinus excelsior* in normal and diabetic rats. *J. Ethnopharmacol.* 94: 149–154.

Eddouks, M. and Maghrani, M. 2008. Effect of *Lipidium sativam* L. on renal glucose reabsorption and urinary TGF-beta 1 levels in diabetic rats. *Phytother. Res.* 22: 1–5.

Eddouks, M., Maghrani, M., and Michel, J.B. 2005d. Hypoglycemic effect of *Triticum repens* P. Beauv in normal and diabetic rats. *J. Ethnopharmacol.* 102: 228–232.

Eddouks, M., Maghrani, M., Zeggwagh, N.A., Lemhadri, A., El Amraoui, M., and Michel, J.B. 2004. Study of the hypoglycemic activity of *Fraxinus excelsior* and *Silybum marianum* in an animal model of type 1 diabetes mellitus. *J. Ethnopharmacol.* 91: 309–316.

Eddouks, M., Maghrani, M., Zeggwagh, N.A., and Michel, J.B. 2005c. Study of the hypoglycemic activity of *Lepidium sativum* L. aqueous extract in normal and diabetic rats. *J. Ethnopharmacol.* 97: 391–395.

Edem, D.O. 2010. Anti-hyperlipidemic effects of ethanol extract of alligator pear seed (*Persea americana* Mill.) in alloxan-induced diabetic rats. *Pharmacologyonline* 1: 901–908.

Edet, E.E., Edet, T.E., Akpanabiatu, M.I., David-Oku, E., Atangwho, I.J., Igile, G. et al. 2013. Toxicopathological changes and phytochemically-induced alleviation in diabetic rats treated with *Gongronema latifolium* leaf extracts. *J. Med. Med. Sci.* 4: 204–213.

Edim, E.H., Egomi, U.G., Ekpo, U.F., and Archibong, O.E. 2012. A review on *Gongronema latifolium* Benth: A novel antibiotic against Staphylococcus aureus related infections. *Int. J. Biochem. Biotechnol.* 1: 204–208.

Egua, M.O., Etuk, E.U., Bello, S.O., and Hassan, S.W. 2013. Antidiabetic activity of ethanolic seed extract of *Corchorus olitorius*. *Int. J. Sci.: Basic Appl. Res.* 12: 8–21.

Egua, M.O., Etuk, E.U., Bello, S.O., and Hassan, S.W. 2014. Antidiabetic potential of liquid-liquid partition fractions of ethanolic seed extract of *Corchorus olitorius*. *J. Pharmacog. Phytother.* 6: 4–9.

Eid, H.M. and Haddad, P.S. 2014. Mechanisms of action of indigenous anti-diabetic plants from the Boreal forest of Northeastern Canada. *Adv. Endocrinol.* 2014: Article ID 272968, 11 pages.

Eidi, A. and Eidi, M. 2009. Antidiabetic effects of sage (*Salvia officinalis* L.) leaves in normal and streptozotocin-induced diabetic rats. *Diabetes Metabol. Syndr. Clin. Res. Rev.* 3: 40–44.

Eidi, A., Eidi, M., Givian, M., and Abaspour, N. 2009a. Hypolipidemic effects of alcoholic extract of euca-lyptus (*Eucalyptus globulus* Labill.) leaves on diabetic and non-diabetic rats. *Iranian J. Diabetes Lipid Disord.* 8: 105–112.

Eidi, A., Eidi, M., and Darzi, R. 2009b. Antidiabetic effect of *Olea europaea* L. in normal and diabetic rats. *Phytother. Res.* 23: 347–350.

Eidi, M., Eidi, A., Saeidi, A., Monanaei, S., Sadeghipour, A., Bahar, M. et al. 2008. Effect of coriander seed (*Coriandrum sativum* L.) ethanol extract on insulin release from pancreatic beta cells in streptozotocin-induced diabetic rats. *Phytother. Res.* 23: 404–406.

Eidi, M., Eidi, A., and Zamanizadeh, H. 2005. Effect of *Salvia officinalis* L. leaves on serum glucose and insulin in healthy and streptozotocin-induced diabetic rats. *J. Ethnopharmacol.* 100: 310–313.

Einstein, J.W., Rais, M.M., and Mohd, M.A. 2013. Comparative evaluation of the antidiabetic effects of differ-ent parts of *Cassia fistula* Linn., a Southeast Asian plant. *J. Chem.* 2013: Article ID 714063, 10 pages.

Ekeanyanwu, R.C., Udeme, A.A., Onuigbo, A.O., and Etienajirhvwe, O.F. 2012. Anti-diabetic effect of etha-nol leaf extract of *Cissempelos owariensis* on allonxan induced diabetic rats. *Afr. J. Biotechnol.* 11: 6758–6762.

Ekoh, S.N., Akubugwo, E.I., Ude, V.C., and Edwin, N. 2014. Anti-hyperglycemic and anti-hyperlipidemic effect of spices (*Thmus vulgaris, Murraya koenigii, Ocimum gratissimum* and *Piper guineense*) in alloxan-induced diabetic rats. *Int. J. Biosci.* 4: 178–187.

El-abhar, H.S. and Schaalan, M.F. 2014. Phytotherapy in diabetes: Review on potential mechanistic perspec-tives. *World J. Diabetes* 5: 176–197.

El-Amin, M., Virk, P., Elobeid, M.A., Almarhoon, Z.M., Hassan, Z.K., Omer, S.A. et al. 2013. Anti-diabetic effect of *Murraya koenigii* (L.) and *Olea europaea* (L.) leaf extracts on streptozotocin induced diabetic rats. *Pak. J. Pharm. Sci.* 26: 359–365.

Elberry, A.A., Harraz, F.M., Ghareib, S.A., Gabr, S.A., Nagy, A.A., and Abdel-Sattar, E. 2011. Methanolic extract of *Marrubium vulgare* ameliorates hyperglycemia and dyslipidemia in streptozotocin-induced diabetic rats. *Int. J. Diab. Mellitus.* 3: 37–44.

El-Demerdash, F., Yousef, M.I., and El-Naga, N.L. 2005. Biochemical study on the hypoglycemic effects of onion and garlic in alloxan-induced diabetic rats. *Food Chem. Toxicol.* 43: 57–63.

Eleazu, C.O., Iroaganachi, M., and Eleazu, K.C. 2013. Ameliorative potentials of cocoyam (*Colocasia escu-lenta* L.) and unripe plantain (*Musa paradisiaca* L.) on the relative tissue weights of streptozotocin-induced diabetic rats. *J. Diabetes Res.* 2013: 160964.

El-Fiky, F.K. 1996. Effect of *Luffa aegyptiaca* (seeds) and *Carissa edulis* (leaves) extracts on blood glucose level of normal and streptozotocin diabetic rats. *J. Ethnopharmacol.* 50: 43–47.

El-Hilaly, J., Israili, Z.H., and Lyoussi, B. 2004. Acute and chronic toxicological studies of *Ajuga iva* in experimental animals. *J. Ethnopharmacol.* 91: 43–50.

El-Hilaly, J. and Lyouss, B. 2002. Hypoglycaemic effect of the lyophilised aqueous extract of *Ajuga iva* in normal and streptozotocin diabetic rats. *J. Ethnopharmacol.* 80: 109–113.

El-Hilaly, J., Tahraoui, A., Israili, Z.H., and Lyoussi, B. 2007. Acute hypoglycemic, hypocholesterolemic and hypotriglyceridemic effects of continuous intravenous infusion of a lyophilised aqueous extract of *Ajuga iva* L. Schreber whole plant in streptozotocin-induced diabetic rats. *Pak. J. Pharm. Sci.* 20: 261–268.

Eliza, J., Daisy, P., and Ignacimuthu, S. 2008. Influence of *Costus speciosus* (Voen) Sm. rhizome extracts on biochemicals parameters in streptozotocin induced diabetic rats. *J. Health Sci.* 54: 675–681.

Eliza, J., Daisy, P., and Ignacimuthu, S. 2009a. Antidiabetic and antilipidemic effect of eramanthin from *Costus speciosus* (Koen) Sm., in STZ-induced diabetic rats. *Chem. Biol. Interact.* 182: 67–72.

Eliza, J., Daisy, P., Ignacimuthu, S., and Duraipandiyan, V. 2009b. Normo-glycemic and hypolipidemic effect of costunolide isolated from *Costus speciosus* (Koen) Sm., in streptozotocin-induced diabetic rats. *Chem. Biol. Interact.* 179: 329–334.

Eliza, J., Daisy, P., and Ignacimuthu, S. 2010. Antioxidant activity of costunolide and eremanthin isolated from *Costus speciosus* (Koenex.Retz) Sm. *Chem. Biol. Interact.* 188: 467–472.

Ellis, L., Clauser, E., Morgan, D.O., Edery, M., Roth, R.A., and Rutter, W.J. 1986. Replacement of insulin receptor tyrosine residues 1162 and 1163 compromises insulin stimulated kinase activity and uptake of 2-deoxy glucose. *Cell* 45: 721–732.

El-Abhar, H.S. and Schaalan, M.F. 2014. Phytothraphy in diabetes: Review on potential mechanistic perspectives. *World J. Diabetes* 5: 176–197.

El-Mehiry, H.F., Helmy, H.M., and El-Ghany, M.A.A. 2012. Anti-diabetic and antioxidative activity of physalis powder or extract with chromium in rats. *World J. Med. Sci.* 7: 27–33.

El-Missiry, M.A. and El-Gindy, A.M. 2000. Amelioration of alloxan induced diabetes mellitus and oxidative stress in rats by oil of *Eruca sativa* seeds. *Ann. Nutr. Metab.* 44: 97–100.

El-Mostafa, K., Kharrassi, Y.E., Badreddline, A., Andreoletti, P., Vamecq, J., Kebbaj, M.S.E. et al. 2014. Nopal cactus (*Opuntia ficus-indica*) as a source of bioactive compounds for nutrition, health and disease. *Molecules* 19: 14879–14901.

El-Remessy, A.B., Al-Shabrawey, M., Khalifa, Y., Tsai, N.T., Caldwell, R.B., and Liou, G.I. 2006. Neuroprotective and blood-retinol barrier preserving effects of cannabidiol in experimental diabetes. *Am. J. Pathol.* 168: 235–244.

El-Shenawy, N.S. and Abdel-Nabi, I.M. 2006. Hypoglycemic effect of *Cleome droserifolia* ethanolic leaf extract in experimental diabetes, and on non-enzymatic anti-oxidant, glycogen, thyroid hormone and insulin levels. *Diabetol. Croat.* 35: 15–22.

El-Shobaki, F.A., El-Bahay, A.M., Esmail, R.S.A., Abd-El-Megeid, A.A., and Esmail, N.S. 2010. Effects of figs fruit (*Ficus carica* L.) and its leaves on hyperglycemia in alloxan-diabetic rats. *World J. Dairy Food Sci.* 5: 47–57.

El-Wahab, A., Abeer, E., Ghareeb, D.A., Sarhan, E.E., Abu-Serie, M.M., and El-Demellawy, M.A. 2013. *In vitro* biological assessment of *Berberis vulgaris* and its active constituent, berberine: Antioxidants, anti-acetylcholinesterase, anti-diabetic and anticancer effects. *BMC Complement. Altern. Med.* 13: 218.

El-Zalabani, S.M., S.M.E., Hetta, M.H., Ross, S.A., Youssef, A.M.A., Zaki, M.A., and Isma, A.S. 2012. Antihyperglycemic and antioxidant activities and chemical composition of *Conyza dioscoridis* (L.) Desf. DC. growing in Egypt. *Aust. J. Basic Appl. Sci.* 6: 257–265.

Emekli-Alturfan, E., Kasikci, E., and Yarat, A. 2008. Peanut (*Arachis hypogaea*) consumption improves glutathione and HDL-cholesterol levels in experimental diabetes. *Phytother. Res.* 22: 180–184.

Emmanuel, S., Rani, M.S., and Sreekanth, M.R. 2010. Anti-diabetic activity of *Cassia occidentalis* Linn. in streptozotocin-induced diabetic rats: A dose dependant study. *Int. J. Pharm. Biosci.* 1: B14–B25.

Envikwola, O., Addy, E.O., and Adoga, G.I. 1991. Hypoglycemic effect of *Canavalia ensiformis* (Leguminosae) in albino rats. *Discov. Innovat.* 3: 61–63.

Erdogan-Orhan, I., Altun, M.L., Sever-Yilmaz, B., and Saltan G. 2011. Anti-acetylcholinesterase and antioxidant assets of the major components (salicin, amentoflavone, and chlorogenic acid) and the extracts of *Viburnum opulus* and *Viburnum lantana* and their total phenol and flavonoid contents. *J. Med. Food* 14: 434–440.

Erhirhie, E.O. and Ekene, N.E. 2013. Medicinal values of *Citrullus lanatus* (watermelon). *Int. J. Res. Pharm. Biomed. Sci.* 4: 1305–1312.

Eseyin, O.A., Ebong, P., Ekpo, A., Igboasoiyi, A., and Oforah E. 2007. Hypoglycemic effect of the seed extract of *Telfairia occidentalis* in rat. *Pak. J. Biol. Sci.* 10: 498–501.

Eseyin, O.A., Ebong, P., Eyong, E., Awofisayo, O., and Agboke, A. 2010a. Effect of *Telfairia occidentalis* on oral glucose tolerance in rats. *Afr. J. Pharm. Pharmacol.* 4: 368–372.

Eseyin, O.A., Ebong, P., Eyong, E.U., Umoh, E., and Agboke, H. 2010c. Hypoglycaemic activity of ethyl acetate fraction of the leaf extract of *Telfairia occidentalis*. *Pak. J. Pharm. Sci.* 23: 341–343.

Eseyin, O.A., Ebong, P., Eyong, U.E., Umoh, E., and Attih, E. 2010b. Comparative hypoglycaemic effects of ethanolic and aqueous extracts of the leaf and seed of *Telfairia occidentalis*. *Turk. J. Pharm. Sci.* 7: 29–34.

Eseyin, O.A., Sattar, M.A., Rathore, H.A., Ahmad, A., Afzal, S., Lazhari, M. et al. 2014. Hypoglycemic potential of polysaccharides of the leaf extract of *Telfairia occidentalis*. *Ann. Res. Rev. Biol.* 4: 1813–1826.

Esmaillzadeh, A., Tahbaz, F., Gaieni, I., Alavi-Majd, H., and Azadbakht, L. 2004. Concentrated pomegranate juice improves lipid profiles in diabetic patients with hyperlipidemia. *J. Med. Food* 7: 305–308.

Esposito, A.M., Diaz, A., de-Dracia, I., de-Tello, R., and Gipta, M.P. 1991. Evaluation of traditional medicine: Effects of *Cajanus cajan* L. and of *Casia fistula* L. on carbohydrate metabolism in mice. *Rev. Med. Panama* 16: 39–45.

Esposito, D., Chen, A., Grace, M.H., Komarnytsky, S., and Lila, M.A. 2014. Inhibitory effects of wild blueberry anthocyanins and other flavonoids on biomarkers of acute and chronic inflammation *in vitro*. *J. Agric. Food Chem.* 62: 7022–7028.

Estrada, O., Hasegawa, M., Gonzalez-Mujica, F., Motta, N., Ferdomo, E., Solorzano, A. et al. 2005. Evaluation of flavonoids from *Bauhinia megalandra* leaves as inhibitors of glucose-6-phosphatase system. *Phytother. Res.* 19: 859–863.

Eunjin, S., Chan-Sik, K., Young, S.K., Dong, Ho, J., Dae, S.J., Yun, M.L., and Jin, S.K. 2007. Effects of magnolol (5,5′-diallyl-2,2′-dihydroxybiphenyl) on diabetic nephropathy in type 2 diabetic Goto–Kakizaki rats. *Life Sci.* 80: 468–475.

Eunjin, S., Junghyun, K., Chan-Sik, K., Young, S.K., Dae, S.J., and Jin, S.K. 2010. Extract of the aerial parts of *Aster koraiensis* reduced development of diabetic nephropathy *via* anti-apoptosis of podocytes in streptozotocin-induced diabetic rats. *Biochem. Biophys Res. Commun.* 391: 733–738.

Evans, J.L. and Rushakoff, R.J. 2007. Oral pharmacological agents for type 2 diabetes: Sulfonylureas, meglitinides, metformin, thiazolidinediones, α-glucosidase inhibitor and emerging approaches. http://www.endotext.org/diabetes/diabetes16/diabetes16.htm.

Eydi, M., Fatemeh, S., and Ebrahimi, S. 2007. Hypolipidemic effects of *Allium porrum* L. leaves in healthy and streptozotocin-induced diabetic mice. *J. Med. Plants* 6: 85–91.

Ezeigbo, I.I. and Asuzu, I.U. 2011. Anti-diabetic activities of the methanol leaf extracts of *Hymenocardia acida* (Tul.) in alloxan-induced diabetic rats. *Afr. J. Tradit. Complement. Altern. Med.* 9: 204–209.

Ezejiofor, A.N., Okorie, A., and Orisakwe, O.E. 2013. Hypoglycaemic and tissue-protective effects of the aqueous extract of *Persea americana* seeds on alloxan-induced albino rats. *Malays. J. Med. Sci.* 20: 31–39.

Ezekwesili, C.N., Ogbunugafor, H.A., and Ezekwesili-Ofili, J.O. 2012. Anti-diabetic activity of aqueous extracts of *Vitex doniana* leaves and *Cinchona calisaya* bark in alloxan-induced diabetic rats. *Int. J. Trop. Dis. Health* 2: 290–300.

Ezike, A.C., Akah, P.A., Okoli, C.C., and Okpala, C.B. 2010. Experimental evidence for the antidiabetic activity of *Cajanus cajan* leaves in rats. *J. Basic Clin. Pharm.* 1: 25–30.

Fai, Y.M. and Tao, C.C. 2009. A review of presence of oleanoic acid in natural products. *Nat. Proda Med.* 28: 277–290.

Faisal, M., Hossain, A.I., Rahman, S., Jahan, R., and Rahmatullah, M. 2014. A preliminary report on oral glucose tolerance and anti-nociceptive activity tests conducted with methanol extract of *Xanthosoma violaceum* aerial parts. *BMC Complement. Altern. Med.* 14: 335–341.

Falla, H.H., Kianbakht, S., and Heshmat, R. 2012. *Cynara scolymus* L. in treatment of hypercholesterolemic type 2 diabetic patients: A randomized double-blind placebo-controlled clinical trial. *J. Med. Plants* 11: 58–63.

Falodun, A., Nworgu, Z.A.M., and Ikponmwonsa, M.O. 2006. Phytochemical components of *Hunteria umbellata* (K. Schum) and its effect on isolated non-pregnant rat uterus in oestrus. *Pak. J. Pharm. Sci.* 19: 256–258.

Fang, X.K., Gao, J., and Zhu, D.N. 2008a. Kaempferol and quercetin isolated from *Euonymus alatus* improve glucose uptake of 3T3-L1 cells without adipogenesis activity. *Life Sci.* 82: 615–622.

Fang, X.K., Gao, Y., Yang, H.Y., Lang, S.M., Wang, O.J., Yu, B.Y. et al. 2008b. Alleviating effects of active fraction of *Euonymus alatus* abundant in flavonoids on diabetic mice. *Am. J. Chin. Med.* 36: 125–140.

Farahanikia, B., Akbarzadeh, T., Jahangirzadeh, A., Yassa, N., Shams, A.M.R., Mirnezami, T. et al. 2011. Phytochemical investigation of *Vinca minor* cultivated in Iran. *Iranian J. Pharm. Res.* 10: 777–785.

Farias, R.A.F., Rao, V.S.N., Viana, G.S.B., Silveira, E.R., Maciel, M.A.M., and Pinto, A.C. 1997. Hypoglycaemic effect of trans-Dehydrocrotonin, a Nor-clerodane diterpene from *Croton cajucara*. *Planta Med.* 63: 558–560.

Farook, S.M. and Atlee, W.C. 2011a. Antidiabetic and hypolipidemic potential of *Tragia involucrata* Linn. in streptozotocin-nicotinamide induced type-2 diabetic rats. *Int. J. Pharm. Pharm. Sci.* 3: 103–109.

Farook, S.M. and Atlee, W.C. 2011b. Anti-oxidant potential of *Tragia involucrata* Linn. on streptozotocin-induced oxidative stress in rats. *Int. J. Pharm. Sci. Res.* 2: 1530–1536.

Farswan, M., Mazumder, P.M., Parcha, V., and Upaganlawar, V. 2009. Modulatory effect of *Syzygium cumini* seeds and its isolated compound on biochemical parameters in diabetic rats. *Int. J. Green Pharm.* 5: 127–133.

Fathiazad, F., Hamedeyazdan, S., Khosropanah, M.K., and Khaki, A. 2013. Hypoglycemic activity of *Fumaria parviflora* in streptozotocin-induced diabetic rats. *Adv. Pharm. Bull.* 3: 207–210.

Fatima, A., Agarwal, P., and Singh, P.P. 2012. Herbal option for diabetes: An overview. *Asian Pac. J. Trop. Dis.* S536–S544.

Febrinda, A.E., Yuliana, N.D., Ridwan, E., Wresdiyati, T., and Astawan, M. 2014. Hyperglycemic control and diabetes complication preventive activities of Bawang Dayak (*Eleutherine palmifolia* L. Merr.) bulbs extracts in alloxan-diabetic rats. *Int. Food Res. J.* 21: 1405–1411.

Fenglin, L., Qingwang, L., Dawei, G., and Yong, P. 2009a. The optimal extraction parameters and anti-diabetic activity of flavonoids from *Ipomoea batatas* leaf. *Afr. J. Tradit. Complement. Altern. Med.* 6: 195–202.

Fenglin, L., Qingwang, L., Dawei, G., Yong, P., and Caining, F. 2009b. Preparation and antidiabetic activity of polysaccharide from *Portulaca oleracea* L. *Afr. J. Biotechnol.* 8: 569–573.

Fernandes, A.C., Almeida, C.A., Albano, F., Laranja, G.A., Felzenszwalb, I., Lage, C.L. et al. 2002. Norbixin ingestion did not induce any detectable DNA breakage in liver and kidney but caused a considerable impairment in plasma glucose levels of rats and mice. *J. Nutr. Biochem.* 13: 411–420.

Fernando, M.R., Thabrew, M.I., and Karunanayake, E.H. 1990. Hypoglycemic activity of some medicinal plants in Sri Lanka. *Gen. Pharmacol.* 2: 779–782.

Fernando, M.R., Wickramasinghe, N., Thabrew, M.I., Ariyananda, P.L., and Karunanayayake, E.H. 1991. Effect of *Artocarpus heterophyllus* and *Asteracanthus longifolia* on glucose tolerance in normal human subjects and in maturity-onset diabetic patients. *J. Ethnopharmacol.* 31: 277–282.

Fernandopulle, B.M.R., Karunanayake, E.H., and Ratnasooriya, W.D. 1994. Oral hypoglycaemic effects of *Momordica dioica* in the rat. *Med. Sci. Res.* 22: 137–139.

Ferreira, E.A., Gris, E.F., Felipe, K.B., Correia, J.F.G., Cargnin-Ferreira, E., Filho, D.W. et al. 2010. Potent hepatoprotective effect in CCl$_4$-induced hepatic injury in mice of phloroacetophenone from *Myrcia multiflora*. *Libyan J. Med.* 5: 1–27.

Ferreira, E.A., Gris, E.F., Rebello, J.M., Correia, J.F., de Oliveira, L.F., Filho, D.W. et al. 2011. The 2′,4′,6′-tri-hydroxyacetophenone isolated from *Myrcia multiflora* has antiobesity and mixed hypolipidemic effects with the reduction of lipid intestinal absorption. *Planta Med.* 77: 1569–1574.

Ferreira, F.M., Peixoto, F., Nunes, E., Sena, C., Seica, R., and Santos, M.S. 2010. MitoTea: *Geranum robetia- num* L. decoctions decrease blood glucose levels and improve liver mitochondrial oxidative phosphory- lation in diabetic Goto-Kakizaki rats. *Acta Biochica Polonica* 57: 399–402.

Ferreira, S.F., Azevedo, S.C.S.F., Vardanega-Peicher, M., Pagadigorria, C.L.S., and Garcia, R.F. 2013. Anti- hyperglycemic effect of *Quassia amara* (Simaroubaceae) in normal and diabetic rats. *Rev. Bras. Pl. Med.* 15: 368–372.

Ferreres, F., Gil-Izquierdo, A., Vinholes, J., Silva, S.T., Valentao, P., and Andrade, P.B. 2012. *Bauhinia for- ficata* link authenticity using flavonoids profile: Relation with their biological properties. *Food Chem.* 134: 894–904.

Feshani, A.M., Kouhsari, S.M., and Mohammadi, S. 2011. *Vaccinium arctostaphylos*, a common herbal medi- cine in Iran: Molecular and biochemical study of its antidiabetic effects on alloxan-diabetic Wistar rats. *J. Ethnopharmacol.* 133: 67–74.

Fidan, A.F. and Dundar, Y. 2008. The effects of *Yucca schidigera* and *Quillaja saponaria* on DNA damage, protein oxidation, lipid peroxidation, and some biochemical parameters in streptozotocin-induced dia- betic rats. *J. Diabetes Complications* 22: 348–356.

Figueroa-Valverde, L., Diaz-Cedillo, F., Camacho-Luis, A., and Mamos, M.L. 2009. Induced effects by *Ruta graveolens* L., Rutaceae, *Cnidoscolus chayamansa* McVaugh, Euphorbiaceae, and *Citrus aurantium* L., Rutaceae, on glucose, cholesterol and triacylglycerides levels in a diabetic rat model. *Rev. Bras. Parmacogn.* 19: 898–907.

Firdous, S.M. 2014. Phyto-chemicals for the treatment of diabetes. *EXCLI J.* 13: 451–453.

Fisman, E.Z., Motro, M., and Tenenbaum, A. 2008. Non-insulin antidiabetic therapy in cardiac patients: Current problems and future prospects. *Adv. Cardiol.* 45: 154–170.

Fleischer, T.C., Sarkodie, J.A., Komlaga, G., Kuffour, G., Dickson, R.A., and Mensah, M.L.K. 2011. Hypoglycaemic and antioxidant activities of the stem bark of *Morinda lucida* Benth in streptozotocin induced diabetic rats. *Pharmacogn. J.* 1: 23–29.

Florence, N.T., Benoit, M.Z., Jonas, K., Alexandra, T., Desire, D.D., Pierre, K. et al. 2014. Antidiabetic and antioxidant effects of *Annona muricata* (Annonaceae), aqueous extract on streptozotocin-induced dia- betic rats. *J. Ethnopharmacol.* 151: 784–790.

Florian, J.C., Valdivia, J.B., Guevara, L.C., and Llanos, D.C. 2013. Anti-diabetic effect of *Coffea arabica* in alloxan-induced diabetic rats. *Emir. J. Food Agric.* 25: 772–777.

Fondjo, F.A., Kamgang, R., Oyono, J.-L.E., and Yonkeu, J.N. 2012. Anti-dyslipidemic and anti-oxidant potentials of methanol extract of *Kalanchoe crenata* whole plant in streptozotocin-induced diabetic nephropathy in rats. *Trop. J. Pharm. Res.* 11: 767–775.

Fornasini, M., Castro, J., Villacres, E., Narvaez, L., Villamar, M.P., and Baldon, M.E. 2012. Hypoglycemic effect of *Lupinus muta*bilis in healthy volunteers and subjects with dysglycemia. *Nutr. Hosp.* 27: 425–433.

Franca, F., Cuba, C.A., Moreira, E.A., Miguel, O., Almeida, M., das Virgens, M.L. et al. 1993. An evaluation of the effect of a bark extract from the cashew (*Anacardium occidentale* L.) on infection by *Leishmania* (Viannia) *braziliensis*. *Rev. Soc. Bras. Med. Trop.* 26: 151–155.

Frankish, N., de Sousa, F., Mills, C., and Sheridan, H. 2010. Enhancement of insulin release from the beta-cell line INS-1 by an ethanolic extract of *Bauhinia variegata* and its major constituent roseoside. *Planta Med.* 76: 995–997.

Frati-Munari, A.C., Del Valle-Martinez, L.M., Ariza-Andraca, C.R., Islas-Andrade, S., and Chavez-Negrete, A. 1989. Hypoglycemic action of different doses of nopal (*Opuntia streptacantha* Lemaire) in patients with type 2 diabetes mellitus. *Arch. Invest. Med.* 20: 197–201.

Fred-Jaiyesimi, A.A. and Abo, K.A. 2009. Hypoglycaemic effects of *Parkia biglobosa* (Jacq) Benth seed extract in glucose-loaded and NIDDM rats. *Int. J. Biol. Chem. Sci.* 3: 545–550.

Frhirhie, E.O., Ewhre, L.O., Ilevbare, F.R., and Okparume, D.E. 2013. The roles of capsicum in diabetes mellitus. *Continental J. Pharmacol. Toxicol. Res.* 6: 22–27.

Frias, A.C. and Sgarbieri, V.C. 1998a. Guar gum effects on food intake, blood serum lipids and glucose levels of Wistar rats. *Plant Food Hum. Nutr.* 53: 15–28.

Frias, A.D. and Sgarbieri, V.C. 1998b. Guar gum effect on blood serum lipids and glucose concentration of Wistar diabetic rats. *Ciencia Techno De Alimentos* 18: 1–10.

Fuangchana, A., Sonthisombata, P., Seubnukarnb, T., Chanouanc, R., Chotchaisuwatd, P., Sirigulsatiene, V. et al. 2011. Hypoglycemic effect of bitter melon compared with metformin in newly diagnosed type 2 diabetic patients. *J. Ethinopharmacol.* 134: 422–428.

Fuentes, N.L., Sagua, H., Morales, G., Borquez, J., San Martin, A., Soto, J., and Loyola, L.A. 2005. Experimental anti- hyperglycemic effect of diterpenoids of *Illareta azorella* compacta (Umbelliferae) Phil in rats. *Phytother. Res.* 19: 713–716.

Fuentes, O., Arancibia-Avila, P., and Alarcon, J. 2004. Hypoglycemic activity of *Bauhinia candicans* in diabetic induced rabbits. *Fitoterapia* 75: 527–532.

Fujij, M., Takei, I., and Umezawa, K. 2009. Antidiabetic effect of orally administered conophylline-containing plant extract on streptozotocin-treated and Goto-Kakizaki rats. *Biomed. Pharmacother.* 63: 710–716.

Fukuda, T., Ito, H., and Yoshida, T. 2004. Effect of the walnut polyphenol fraction on oxidative stress in type 2 diabetic mice. *Biofactors* 2: 251–253.

Fukushima, M., Matsuyama, F., Ueda, N., Egawa, K., Takemoto, J., and Kajimoto, Y. 2006. Effects of corosolic acid on post-challenge plasma glucose levels. *Diabetes Res. Clin. Pract.* 73: 174–177.

Furukawa, F., Nishikawa, A., Chihara, T., Shimpo, K., Beppu, H., Kuzuya, H. et al. 2002. Chemopreventive effects of *Aloe arborescens* on N-nitrobis (2-oxypropyl) aminc induced pancreatic carcinogenis in hamsters. *Cancer Lett.* 178: 117–122.

Gacche, R.N. and Dhole, N.A. 2011. Aldolase reductase inhibitory, anti-cataract and antioxidant potential of selected medicinal plants from the Marathwada region, India. *Nat. Prod. Res.* 25: 760–763.

Gagandeep, Dhanalakshmi, S., Mendiz, E., Rao, A.R., and Kale, R.K. 2003. Chemopreventive effects of *Cuminum cyminum* in chemically induced forestomach and uterine cervix tumors in murine model systems. *Nutr. Cancer* 47: 171–180.

Gagtap, A.G. and Patil, P.B. 2010. Anti-hyperglycemic activity and inhibition of advanced glycation end producr formation by *Cuminum cyminum* in streptozotocin-induced diabetic rats. *Food Chem. Toxicol.* 48: 2030–2036.

Gajalakshmi, S., Vijayalakshmi, S., and Rajeswari, D.V. 2012. Phytochemical and pharmacological properties of *Annona muricata*: A review. *Int. J. Pharm. Pharm. Sci.* 4: 3–6.

Gajalakshmi, S., Vijayalakshmi, S., and Rajeswari, D.V. 2013. Pharmacological activities of *Balanites aegyptiaca* (I.)—A perspective review. *Int. J. Pharm. Sci. Rev. Res.* 22: 117–120.

Gallagher, A.M., Flatt, P.R., Duffy, G., and Abdel-Wahab, Y.H. 2003. The effects of traditional anti-diabetic plants on *in vitro* glucose diffusion. *Nutr. Res.* 23: 413–424.

Gandhi, G.R., Ignacimuthu, S., and Paulraj, M.G. 2011. *Solanum torvum* Swartz. fruit containing phenolic compounds shows antidiabetic and antioxidant effects in streptozotocin induced diabetic rats. *Food Chem. Toxicol.* 49: 2725–2733.

Gandhi, G.R. and Sasikumar, P. 2012. Antidiabetic effect of *Merremia emarginata* Burm. f. in streptozotocin induced diabetic rats. *Asian Pac. J. Trop. Biomed.* 2: 281–286.

Gandhi, G.R., Vanlalhruaia, P., Stalin, A., Irudayaraj, S., Ignacimuthu, S., and Paulraj, M.G. 2014. Polyphenols rich *Cyamopsios tetragonoloba* (L.) Taub. beans show hypoglycemic and β-cell protective effects in type 2 diabetic rats. *Food Chem. Toxicol.* 66: 358–365.

Ganesh, T., Sen, S., Thilagam, E., Thamotharan, G., loganathan, T., and Chakraborty, R. 2010. Pharmacognostic and anti-hyperglycemic evaluation of *Lantana camara* (L.) var. aculeate leaves in alloxan-induced hyperglycemic rats. *Int. J. Res. Pharm. Sci.* 1: 247–252.

Ganju, L., Karan, D., Chanda, S., Srivastava, K.K., Sawhney, R.C., and Selvamurthy, W. 2003. Immuno-modulatory effects of agents of plant origin. *Biomed. Pharmacother.* 57: 296–300.

Ganu, G. and Jadhav, S. 2010. *In vitro* antioxidant and *in vivo* antihyperglycemic potential of *Mimusops elengi* L. in alloxan-induced diabetes in mice. *J. Complement. Integr. Med.* 7: ISSN (Online) 1553-3840. DOI: 10.2202/1553-3840.1358.

Ganu, G.P., Jadhav, S.S., and Deshpande, A.D. 2010. Antioxidant and antihyperglycemic potential of metha-nolic extract of bark of *Mimusops elengi* L. in mice. *Int. J. Phytomed.* 2: 116–123.

Gao, D., Li, Q., Li, Y., Liu, Z., Fan, Y., Liu, Z. et al. 2009. Antidiabetic and antioxidant effects of oleanolic acid from *Ligustrum lucidum* Ait in alloxan-induced diabetic rats. *Phytother. Res.* 23: 1257–1262.

Gao, Q., Qin, W.S., Jia, Z.H., Zheng, J.M., Zeng, C.H., Li, L.S. et al. 2010. Rhein improves renal lesion and ameliorates dyslipidemia in db/db mice with diabetic nephropathy. *Planta Med.* 76: 27–33.

Gao, Q.H., Wu, C.S., and Wang, M. 2013a. The jujube (*Ziziphus jujuba* Mill.) fruit: A review of current knowl-edge of fruit composition and health benefits. *J. Agric. Food Chem.* 61: 3351–3363.

Gao, X., Li, B., Jiang, H., Liu, F., Xu, D., and Liu, Z. 2007. *Dioscorea opposita* reverses dexamethasone induced insulin resistance. *Fitoterapia* 78: 12–15.

Gao, Y., Yong, M.F., Su, Y.P., Jiang, H.M., You, X.J., Yang, Y.L. et al. 2013b. Gingenoside reduces insulin resis-tance through activation of PPARγ pathway and inhibition of TNF-α production. *J. Ethnopharmacol.* 147: 509–516.

Garba, A., Mada, S.B., Ibrahim, G., Dauran, I.A., and Hamza, A.B. 2013. Studies on hypoglycemic and hypolipidemic effect of methanolic extract of *Stachytarpheta angustifolia* (Mill) plant in streptozotocin induced diabetic rats. *Asian J. Biol. Sci.* 6: 161–167.

García López, P.M., de la Mora, P.G., Wysocka, W., Maiztegui, B., Alzugaray, M.E., Del Zotto, H., and Borelli, M.I. 2004. Quinolizidine alkaloids isolated from *Lupinus* species enhance insulin secretion. *Eur. J. Pharmacol.* 504: 139–142.

Garcia, D., Escalante, M., Delgado, R., Ubeira, F.M., and Leiro, J. 2003. Anthelmintic and antiallergic activities of *Mangifera indica* L. stem bark components Vimang and mangiferin. *Phytother. Res.* 17: 1203–1208.

Garg, M., Garg, C., Dhar, V.J., and Kalia, A.N. 2010. Standardized alcoholic extract of *Phyllanthus fraternus* exerts potential action against disturbed biochemical parameters in diabetic animals. *Afr. J. Biochem. Res.* 4: 186–190.

Garrido, G., Blanco-Molina, M., Sancho, R., Macho, A., Delgado, R., and Munoz, E. 2005. An aqueous stem bark extract of *Mangifera indica* (Vimang(®)) inhibits T cell proliferation and TNF-induced activation of nuclear transcription factor NF-kappa B. *Phytother. Res.* 19: 211–215.

Garrido, G., Gonzalez, D., Lemus, Y., Garcia, D., Lodeiro, L., Quintero, G. et al. 2004. *In vivo* and *in vitro* anti-inflammatory activity of *Mangifera indica* L. extract (Vimang). *Parmacol. Res.* 50: 143–149.

Gavimath, C.C., Kangralkar, V.A., Jadhav, N.A., and Burli, S.C. 2010. Effect of essential oil of *Citrus reticu-lata* on blood glucose and lipid profile in alloxan induced diabetic rats. *Int. J. Pharm. Appl.* 1: 1–5.

Gayathri, M. and Kannabiran, K. 2008. Hypoglycemic activity of *Hemidesmus indicus* R. Br. on streptozoto-cin-induced diabetic rats. *Int. J. Diabetes Dev. Countries* 28: 6–10.

Gayathri, M. and Kannabiran, K. 2009. Antidiabetic activity of 2-hydroxy 4-methoxy benzoic acid isolated from the roots of *Hemidesmus indicus* on streptozotocin-induced diabetic rats. *Int. J. Diabetes Metabol.* 17: 53–57.

Gayathri, V., Lekshmi, P., and Padmanabhan, R.N. 2011. Anti-diabetes activity of ethanol extract of *Centella asiatica* (L.) Urban (whole plant) in streptozotocin-induced diabetic rats, isolation of an active fraction and toxicity evaluation of the extract. *Int. J. Med. Arom. Plants* 1: 278–286.

Gebhardt, R. 1993. Multiple inhibitory effect of garlic extract on cholesterol biosynthesis in hepatocytes. *Lipids* 2: 613–619.

Gebreyohannis, T., Shibeshi, W., and Asres, K. 2014. Effects of solvent fractions of *Caylusea abyssiinica* (Fresen.) Fisch. and Mey. on blood glucose levels on normoglycemic, glucose loaded and streptozotocin-induced diabetic rodents. *J. Nat. Remedies* 14: 67–75.

Geetha, B.S., Mathew, B.C., and Augusti, K.T. 1994. Hypoglycemic effects of leucodelphinidin derivative isolated from *Ficus bengalensis* (Linn.). *Indian J. Physiol. Pharmacol.* 38: 220–222.

Geetha, G., Gopinathapillai, P.K., and Sankar, V. 2011. Anti diabetic effect of *Achyranthes rubrofusca* leaf extracts on alloxan induced diabetic rats. *Pak. J. Pharm. Sci.* 24: 193–199.

Geetha, G., Prasanth, K.G., Thavamani, B.S., Prudence, A.R., and Hari Bhaskar, V. 2008. Hypolipidemic activity of *Achyranthes rubrofusca* Linn. whole plant extracts in high fat diet induced hyperlipidemic rats. *Pharmacologyonline* 1: 466–473.

Geetha, R.K. and Vasudevan, D.M. 2004. Inhibition of lipid peroxidation by botanical extracts of *Osimum sanctum*: *In vivo* and *in vitro* studies. *Life Sci.* 76: 21–28.

Geetham, P.K. and Prince, P.S. 2008. Anti-hyperglycemic effect of D-pinitol on streptozotocin-induced diabetic Wistar rats. *J. Biochem. Mol. Toxicol.* 22: 220–224.

Geng, H.W., Zhang, X.L., Wang, G.G., Yang, X.X., Wu, X. et al. 2011. Anti-viral dicaffeoyl derivatives from *Elephantopus scaber*. *J. Asian Nat. Prod. Res.* 13: 665–669.

Genta, S.B., Cabrera, W.M., Mercado, M.I., Grau, A., Catalán, C.A., and Sánchez, S.S. 2010. Hypoglycemic activity of leaf organic extracts from *Smallanthus sonchifolius*: Constituents of the most active fractions. *Chem. Biol. Interact.* 185: 143–152.

George, C., Lochner, A., and Huisamen, B. 2011. The efficacy of *Prosopis glandulosa* as antidiabetic treatment in rat models of diabetes and insulin resistance. *J. Ethnopharmacol.* 137: 298–304.

George, S., Thushar, K.V., Unnikrishnan, K.P., Hashim, K.M., and Balachandran, I. 2008. *Hemidesmus indicus* (L) R.Br.: A review. *J. Plant Sci.* 3: 146–156.

Georgewill, U.O. and Georgewill, O.A. 2009. Effect of extract of *Pseudocedrela kotschyi* on blood glucose concentration of alloxan induced diabetic albino rats. *Eastern J. Med.* 14: 17–19.

Ghannam, N., Kingston, M., Al-Meshaal, I.A., Tariq, M., Parman, N.S., and Woodhouse, N. 1986. Antidiabetic activity of Aloes: Preliminary clinical and experimental observation. *Horm. Res.* 24: 288–294.

Ghazanfar, K., Ganai, B.A., Akbar, S., Mubashir, K., Dar, S.A., Dar, M.Y., and Tantry, M.A. 2014. Antidiabetic activity of *Artimisia amygdalina* Decne in streptozotocin induced diabetic rats. *Biomed. Res. Int.* 2014: Article ID 185676, 10 pages.

Ghisalberti, E.L. 2005. Australian and New Zealand plants with anti-diabetic properties. In: *Traditional Medicines for Modern Times: Antidiabetic Plants*, Soumyanath, A. (ed.). Boca Raton, FL: CRC Press, pp. 241–256.

Gholamhoseinian, A., Fallah, H., and Sharififar, F. 2009. Inhibitory effect of methanol extract of *Rosa damascena* Mill. flowers on alpha-glucosidase activity and postprandial hyperglycemia in normal and diabetic rats. *Phytomedicine* 16: 935–941.

Gholamhoseinian, A., Fallah, H., Sharififar, F., and Mirtajaddini, M. 2008. The inhibitory effect of some Iraian plant extracts on the alpha glucosidase. *Iranian J. Basic Med. Sci.* 11: 1–9.

Gholap, S. and Kar, A. 2005. Regulation of cortisol and glucose concentration by some plant extract in mice. *J. Med. Aromatic Plant Sci.* 27: 478–482.

Ghorbani, A. 2013. Best herbs for managing diabetes: A review of clinical studies. *Braz. J. Pharm. Sci.* 49: 413–418.

Ghosal, S. and Jaiswal, D.K. 1980. Chemical constituents of Gentianaceae XXVIII: Flavonoides of *Enicostemma hyssopifolium* (Willd.) Verd. *J. Pharm. Sci.* 69: 53–56.

Ghosesawar, M.G., Ahamed, R.N., Ahmed, M., and Aladakatti, R.H. 2003. *Azadirachta indica* adversely affects sperm parameters and fructose levels in vas deferens fluid of albino rats. *J. Basic Clin. Physiol. Pharmacol.* 14: 387–395.

Ghosh, D., Bera, T.K., Chatterjee, K., Ali, K.M., and De, D. 2009. Anti-diabetic and antioxidative effects of aqueous extract of seeds of *Psoralea corylifolia* (somraji) and seeds of *Trigonella foenum-graecum* L. (methi) in separate and composite manner in streptozotocin-induced diabetic male albino rats. *Int. J. Pharm. Res. Dev. Online* 7: 1–6.

Ghosh, D., Jana, D., Maiti, R., and Das, V.K. 2004a. Antidiabetic effect of aqueous extract of seed of *Tamarindus indica* in streptozotocin induced diabetic rats. *J. Ethnopharmacol.* 92: 85–91.

Ghosh, P.S., Das, N., and Dinda, B. 2014a. Anti-oxidant flavones glycosides and other constituents from *Premna latifolia* leaves. *Indian J. Chem.* 53: 746–749.

Ghosh, R., Dhande, I., Kakade, V., Vohra, R., Kadam, V., and Mehra. 2008b. Antihyperglycemic activity of *Madhuca longifolia* in alloxan-induced diabetic rats. *Internet J. Pharmacol.* 6: 1–12.

Ghosh, R., Sharathchandra, K.H., Rita, S., and Thokchom, I.S. 2004b. Hypoglycemic activity of *Ficus hispida* (bark) in normal and diabetic albino rats. *Indian J. Pharmacol.* 36: 222–225.

Ghosh, S., Ahire, M., Patil, S., Jabgunde, A., Bhat, M.D., Joshi, B.N. et al. 2012. Antidiabetic activity of *Gnidia glauca* and *Dioscorea bulbifera*: Potent amylase and glucosidase inhibitors. *Evid. Based Complement. Altern. Med.* 2012: 929051.

Ghosh, S., More, P., Derle, A., Patil, A.B., Markad, P., Asok, A. et al. 2014b. Diosgenin from *Dioscorea bulbifera*: Novel hit for treatment of type 2 diabetes mellitus with inhibitory activity against α-amylase and α-glucosidase. *PLoS ONE* 9: e106039. DOI: 10.1371/journal.pone.0106039.

Ghosh, S. and Roy, T. 2013. Evaluation of antidiabetic potential of methanolic extract of *Coccinia indica* leaves in streptozotocin induced diabetic rats. *Inter. J. Pharm. Sci. Res.* 4: 4325–4328.

Ghosh, T., Maity, T.K., Sengupta, P., Dash, D.K., and Bose, A. 2008a. Antidiabetic and *in vivo* antioxidant activity of ethanolic extract of *Bacopa monnieri* Linn. aerial parts: A possible mechanism of action. *Iranian J. Pharm. Res.* 7: 61–68.

Ghosh, T., Maity, T.K., and Singh, J. 2011. Antihyperglycemic activity of bacosine, a triterpene from *Bacopa monnieri*, in alloxan-induced diabetic rats. *Planta Med.* 77: 804–808.

Ghosian-Moghadam, M., Roghani, M., Mohammadnezhad, Z., and Mehrabian, A. 2014. *Fumaria parviflora* Lam effect on serum levels of glucose and lipids in streptozotocin-induced diabetic rats. *J. Basic Clin. Pathophysiol.* 2: 35–42.

Ghoul, J.E. and Ben-Attia, M. 2014. Vasorelaxant effects of aqueous extract of *Zygophyllum album* and antihyperglycemic activities in streptozotocin-induced diabetic mice. *J. Diabetes Metab.* 5: 426. DOI: 10.4172/2155-6156.1000426.

Ghoul, J.E., Smiri, M., Ghrab, S., and Boughattas, N.A. 2013. Antihyperglycemic and antihyperlipidemic activities of ethanolic extract of *Zygophyllum album* in streptozotocin-induced diabetic mice. *Toxicol. Ind. Health* 29: 43–51.

Ghoul, J.E., Smiri, M., Ghrab, S., Boughattas, N.A., and Ben-Attia, M. 2012. Antihyperglycemic, antihyperlipidemic and antioxidant activities of traditional aqueous extract of *Zygophyllum album* in streptozotocin diabetic mice. *Pathophysiology* 19: 35–42.

Giaginis, C., Giagini, A., and Theocharis, S. 2009. Peroxisome proliferator-activated receptor-gamma (PPAR-gamma) ligands as potential therapeutic agents to treat arthritis. *Pharmacol. Res.* 60: 160–169.

Giancarlo, S., Rosa, L.M., Nadjafi, F., and Francesco, M. 2006. Hyroglycaemic activity of two spices extracts: *Rhus coriaria* L. and *Bunium persicum* Boiss. *Nat. Prod. Res.* 20: 882–886.

Gireesh, G., Thomas, S.K., Joseph, B., and Paulose, C.S. 2009. Anti-hyperglycemic and insulin secretory activity of *Costus pictus* leaf extract in streptozotocin induced diabetic rats and in *in vitro* pancreatic islet culture. *J. Ethnopharmacol.* 123: 470–474.

Giribaba, N., Srinivasarao, N., Rekha, S.S., Muniandy, S., and Salleh, N. 2014. *Centella asiatica* attenuates diabetes induced hippocampal changes in experimental diabetic rats. *Evid. Based Complement. Altern. Med.* 2014: 592062, 10 pages.

Girija, K., Lakshman, K., Udaya, C., Sachi, G.S., and Divya, T. 2011. Anti-diabetic and anti-cholesterolemic activity of methanol extracts of three species of *Amaranthus*. *Asian Pac. J. Trop. Biomed.* 1: 133–138.

Gnananath, K., Reddy, K.R., Kumar, G.P., Krishna, B., Reddy, K.S., and Kumar, A.S. 2013. Evaluation of antidiabetic activity of *Corallocarpus epigaeus* rhizomes. *Int. Curr. Pharm. J.* 2: 53–56.

Gnangoran, B.N., Nguessan, B.B., Amoateng, P., Dosso, K., Yapo, A.P., and Ehile, E.E. 2012. Hypoglycaemic activity of ethanolic leaf extract and fractions of *Holarrhena floribunda* (Apocynaceae). *J. Med. Biomed. Sci.* 1: 46–54.

Godhwani, S., Godhwani, J.L., and Vyas, D.S. 1988. *Ocimum sanctum*-a preliminary study evaluating its immunoregulatory profile in albino rats. *J. Ethnopharmacol.* 24: 193–198.

Gohil, T.A., Patel, J.K., Vaghasiya, J.D., and Manick 2008. Anti-hyperglycemic and anti-hyperinsulinema effect of *Aegle marmelos* leaf and *Enicostemma littorale*. *Indian J. Pharm.* 40: 66–91.

Gokce, G. and Haznedaroglu, M.Z. 2008. Evaluation of antidiabetic, antioxidant and vasoprotective effects of *Posidonia oceanica* extract. *J. Ethnopharmacol.* 115: 122–130.

Gomase, P.V. 2012. *Sesbania sesban* Linn. A review on its ethnobotany, phytochemical and pharmacological profile. *Asian J. Biomed. Pharm. Sci.* 2: 1307–1312.

Gomathy, V., Jayakar, B., Kothai, R., and Ramakrishnan, G. 2010. Anti-diabetic activity of *Spinacia oleracea* Linn. in alloxan-induced diabetic rats. *J. Chem. Pharm. Res.* 2: 266–274.

Gomes, A., Vedasiromoni, J.R., Das, M., Sharma, R.M., and Ganguly, D.K. 1995. Anti-hyperglycemic effect of black tea (*Camellia sinensis*) in rat. *J. Ethnopharmacol.* 45: 223–226.

Gomes, J.A.D., Faria, B.G., Silva, V.D., Zangeronimo, M.G., Miranda, J.R., de Lima, A.R. et al. 2013. Influence of integral and decaffeinated coffee brews on metabolic parameters of rats fed with hiperlipidemic diets. *Braz. Arch. Biol. Technol.* 56: 829–836.

Gondi, M., Basha, S.A., Bhaskar, J.J., Salimath, P.V., and Prasada Rao, U.J.S. 2015. Anti-diabetic effect of dietary mango (*Mangifera indica* L.) peel in streptozotocin-induced diabetic rats. *J. Sci. Food Agric.* 95: 991–999.

Gondwe, M., Kamadyaapa, D.R., Tufts, M.A., Chuturgoon, A.A., Ojewole, J.A., and Musabayane, C.T. 2008. Effects of *Persea americana* Mill (Lauraceae) ["Avocado"] ethanolic leaf extract on blood glucose and kidney function in streptozotocin-induced diabetic rats and on kidney cell lines of the proximal (LLCPK1) and distal tubules (MDBK). *Methods Find. Exp. Clin. Pharmacol.* 30: 25–35.

Gong, H., Shen, P., Jin, L., Xing, C., and Tang, F. 2005. Therapeutic effects of *Lycium barbarum* polysaccharide (LBP) on irradiation or chemotherapy-induced myelosuppressive mice. *Cancer Biother. Radiopharm.* 20: 155–162.

Gong, Y., Chen, K., Yu, S.Q., Liu, H.R., and Qi, M.Y. 2012. Protective effect of terpenes from *Fructus corni* on the cardiomyopathy in alloxan-induced diabetic mice. *Zhongguo Ying Yong Sheng Li Xue Za Zhi* 28: 378–380.

Gonzfilez, A.T., Ortiz, G.G., Puebla-Pdrez, A.M., Huizar-Contreras, M.D., Mazariegos, M.R.M., Arreguin, S.M. et al. 1996. A purified extract from prickly pear cactus (*Opuntia fuliginosa*) controls experimentally induced diabetes in rats. *J. Ethnopharmacol.* 55: 27–33.

Gopal, V. and Chauhan, M.G. 1994. *Holarrhena antidysenterica*—A novel herbal anti-diabetes seed drug. *Indian J. Pharm. Sci.* 56: 156–159.

Gopinathan, S. and Naveenraj, D. 2014. Antidiabetic activity of *Clerodendrum phlomidis* Linn. and *Gymnema sylvestre* Linn. in alloxan induced diabetic rats—A comparative preclinical study. *World J. Pharm. Res.* 3045: 1640–1675.

Gorelick, J., Kitron, A., Pen, S., Rosenzweig, T., and Mada, Z. 2011. Anti-diabetic activity of *Chiliadenus iphionoides*. *J. Ethnopharmacol.* 137: 1245–1249.

Govindarajan, R., Asare-Anane, H., Persaud, S., Jones, P., and Houghton, P.J. 2007. Effect of *Desmodium gangeticum* extract on blood glucose in rats and on insulin secretion *in vitro*. *Planta Med.* 73: 427–432.

Govindarajan, R., Vijayakumar, M., Rao, C.V., Pushpangadan, P., Asare-Anane, H., Persuad, S. et al. 2008. Antidiabetic activity of *Croton klozchianus* in rats and direct stimulation of insulin secretion *in vitro*. *J. Pharm. Pharmacol.* 60: 371–376.

Govindasamy, C., Al-Numair, K.S., Alsaif, M.A., and Viswanathan, K.P. 2011. Influence of 3 hydroxymethyl xylitol, a novel antidiabetic compound isolated from *Caseariae sculenta* (Roxb.) root, on glycoprotein components in streptozotocin-diabetic rats. *J. Asian Nat. Prod. Res.* 13: 700–706.

Govorko, D., Logendra, S., Wang, Y., Esposito, D., Komarnytsky, S., Ribnicky, D. et al. 2007. Polyphenolic compounds from *Artemisia dracunculus* L. inhibit PEPCK gene expression and gluconeogenesis in an H4IIE hepatoma cell line. *Am. J. Physiol. Endocrinol. Metab.* 293: E1503–E1510.

Gowri, P.M., Tiwari, A.K., Ali, A.Z., and Rao, J.M. 2007. Inhibition of alpha-glucosidase and amylase by bartogenic acid isolated from *Barringtonia racemosa* Roxb. seeds. *Phytother. Res.* 21: 796–799.

Grace, I.A., Oluwakemi, T.A., Bamidele, V.O.W., and Ayodele, O.S. 2009a. Anti-diabetic properties of the aqueous leaf extract of *Bougainvillea glabra* (glory of the garden) on alloxan-induced diabetic rats. *Rec. Nat. Prod.* 3: 187–192.

Grace, M.H., Ribnicky, D.M., Kuhn, P., Poulev, A., Logendra, S., Yousef, G.G. et al. 2009b. Hypoglycemic activity of a novel anthocyanin-rich formulation from lowbush blueberry, *Vaccinium angustifolium* Aiton. *Phytomedicine* 16: 406–415.

Gray, A.M., Abdel-Wahab, Y.H., and Flatt, P.R. 2000. The traditional plant treatment, *Sambucus nigra* (elder), exhibits insulin-like and insulin-releasing actions *in vitro*. *J. Nutr.* 130: 15–20.

Gray, A.M. and Flatt, P.R. 1997. Pancreatic and extra-pancreatic effects of the traditional anti-diabetic plant, *Medicago sativa* (lucerne). *Br. J. Nutr.* 78: 325–334.

Gray, A.M. and Flatt, P.R. 1998. Antihyperglycemic actions of *Eucalyptus globulus* (eucalyptus) are associated with pancreatic and extra-pancreatic effects in mice. *J. Nutr.* 128: 2319–2323.

Gray, A.M. and Flatt, P.R. 1999. Insulin-releasing and insulin-like activity of the traditional anti-diabetic plant *Coriandrum sativum* (coriander). *Br. J. Nutr.* 81: 203–209.

Green, D.E. 2007. New therapies for diabetes. *Clin. Cornerstone* 8: 58–63.

Grover, I.S. and Bala, S. 1993. Studies on antimutagenic effects of guava (*Psidium guajava*) in *Salmonella typhimurium*. *Mutat. Res.* 300: 1–3.

Grover, J.K., Yadav, S., and Vats, V. 2002. Medicinal plants of India with anti-diabetic potential. *J. Ethnopharmacol.* 81: 81–100.

Gu, M., Zhang, Y., Fan, S., Ding, X., Ji, G., and Huang, C. 2013. Extracts of rhizoma *Polygonati odorati* prevent high-fat diet-induced metabolic disorders in C57BL/6 mice. *PLoS ONE* 8: e81724. DOI: 10.1371/journal.pone.0081724.

Guasch, L., Sala, E., Ojeda, M.J., Valls, C., Blade, C., Mulero, M. et al. 2012. Identification of novel human dipeptidyl peptidase-IV inhibitors of natural origin (Part II): *In silico* prediction in antidiabetic extracts. *PLoS ONE* 7: e44972. DOI: 10.1371/journal.pone.0044972.

Guerrero-Analco, J., Medina-Campos, O., Brindis, F., Bye, R., Pedraza-Chaverri, J., Navarrete, A. et al. 2007. Anti-diabetic properties of selected Mexican copalchis of the Rubiaceae family. *Phytochemistry* 68: 2087–2095.

Guerrero-Analco, J.A., Hersch-Martinez, P., Pedraza-Chaverri, J., Navarrete, A., and Mata, R. 2005. Antihyperglycemic effect of constituents from *Hintonia standleyana* in streptozotocin-induced diabetic rats. *Planta Med.* 71: 1099–1105.

Guerrero-Analco, J.A., Martineau, L., Saleem, A., Madiraju, P., Muhammad, A. et al. 2010. Bioassay-guided isolation of the antidiabetic principle from *Sorbus decora* (Rosaceae) used traditionally by the Eeyou Istchee Cree First Nations. *J. Nat. Prod.* 73: 1519–1523.

Guevara, A.P., Lim-Sylianco, C., Dayrit, F., and Finch, P. 1990. Antimutagens from *Momordica charantia*. *Mutat. Res.* 230: 121–126.

Guillermo, A.A.J., Maria, G.G.A., Felipe, M.R., Tzasna, H.D.C., Maximiliano, I.B., Alfonso, R.V. et al. 2012. Antihyperglycemic effect and genotoxicity of *Psittacanthus calyculatus* extract in streptozotocin-induced diabetic rats. *Bol. Latinoam. Caribe Plant. Med. Aromat.* 11: 345–353.

Gul, M.Z., Bhakshu, L.M., Ahmad, F., Kondapi, A.K., Qureshi, I.A., and Ghazi, I.A. 2011. Evaluation of *Abelmoschus moschatus* extracts for antioxidant, free radical scavenging, antimicrobial and antiproliferative activities using *in vitro* assays. *BMC Complement. Altern. Med.* 11: 64. DOI: 10.1186/1472-6882-11-64.

Gulati, V., Harding, I.H., and Palombo, E.A. 2012. Enzyme inhibitory and antioxidant activities of traditional medicinal plants: Potential application in the management of hyperglycemia. *BMC Complement. Altern. Med.* 12: 77. DOI: 10.1186/1472-6882-12-77.

Gul-e-Rana, Karim, S., Khurhsid, R., Saeed-ul-Hassan, S., Tariq, I., Sultana, M. et al. 2013. Hypoglycemic activity of *Ficus racemosa* bark in combination with oral hypoglycemic drug in diabetic human. *Acta Pol. Pharm.* 70: 1045–1049.

Gulfraz, M., Ahmad, A., Asad, M.J., Sadiq, A., Afzal, U., Imran, M. et al. 2011. Antidiabetic activities of leaves and root extracts of *Justicia adhatoda* Linn. against alloxan-induced diabetes in rats. *Afr. J. Biotechnol.* 10: 6101–6106.

Gulshan, M.D. and Roa, N.R. 2013. An overview of medicinal plants with antidiabetic potential. *Int. J. Phar. Sci. Rev. Res.* 23: 335–342.

Gunawan-Puteri, M.D.P.T. and Kawabata, J. 2012. Novel α-glucosidase inhibitors from *Macaranga tanarius* leaves. *Food Chem.* 123: 384–389.

Gunjan, M., Ravindran, M., and Jana, G.K. 2011. A review on some potential traditional phytomedicine with anti-diabetic properties. *Int. J. Phytomed.* 3: 448–458.

Guo, C., Li, R., Zheng, N., Xu, L., Liang, T., and He, Q. 2013. Anti-diabetic effect of ramulus mori polysaccharides, isolated from *Morus alba* L. on STZ-diabetic mice through blocking inflammatory response and attenuating oxidative stress. *Int. Immunopharmacol.* 16: 93–99.

Guo, J., Liu, T., Han, L., and Liu, Y. 2009. The effects of corn silk on glycaemic metabolism. *Nutr. Metab.* 6: 47–52.

Guo, X.M., Cui, R.J., and Dong, Q. 2005. Effect of herba *Potenillae discoloris* on mRNA expression of insulin-degrading enzyme in rat model of type 2 diabetes mellitus. *Chin. J. Inform. TCM* 12: 40–42.

Gupta, A., Singh, P., Trivedi, N., Jha, K.K., Kumar, S., and Singh, B. 2014. A review on pharmacognostical and pharmacological activities of plant *Nicandra physalodes*. *Pharma Res.* 11: 42–47.

Gupta, B., Nayak, S., and Solanki, S. 2011a. *Abroma augusta* L. f: A review. *Der Pharmacia Sinica* 2: 253–261.

Gupta, J.K., Mishra, P., Rani, A., and Maumder, P.M. 2010. Blood glucose lowering potential of stem bark of *Berberis aristata* DC in alloxan induced diabetic rats. *Iranian J. Pharmacol. Ther.* 9: 21–24.

Gupta, M., Mazumder, U.K., Kumar, R.S., Sivakumar, T., and Vamsi, M.L. 2004a. Anti tumour activity and antioxidant status of *Caesalpinia bonducella* against Ehrlich ascites carcinoma in Swiss albino mice. *J. Pharmacol. Sci.* 94: 177–184.

Gupta, M.P., Solis, N.G., Avella, M.E., and Sanchez, C. 1984. Hypoglycemic activity of *Neurolaena lobata* (L.) R. Br. *J. Ethnopharmacol.* 10: 323–327.

Gupta, P.C. 2012. Biological and pharmacological properties of *Terminalia chebula* R. (haritaki)—A review. *Int. J. Pharm. Pharm. Sci.* 4: 62–68.

Gupta, R. and Saxena, A.M. 2011. Hypoglycemic and anti-hyperglycemic activities of *Syzygium cumini* (Linn.) Skeels whole fruit in normal and streptozotocin-induced diabetic rats. *Asian J. Pharm. Biol. Res.* 267–272.

Gupta, R., Katariya, P., Mathur, M., Bajaj, V.K., Yadav, S., Kamal, R. et al. 2011b. Antidiabetic and renoprotective activity of *Momordica dioica* in diabetic rats. *Diabetol. Croat.* 40: 81–88.

Gupta, R., Sharma, A.K., Dobhal, M.P., Sharma, M.C., and Gupta, R.S. 2011c. Antidiabetic and antioxidant potential of β-sitosterol in streptozotocin induced experimental hyperglycemia. *J. Diabetes* 3: 29–37.

Gupta, R., Sharma, A.K., Sharma, M.C., Dobhal, M.P., and Gupta, R.S. 2012. Evaluation of antidiabetic and antioxidant potential of lupeol in experimental hyperglycaemia. *Nat. Prod. Res.* 26: 1125–1129.

Gupta, R.N., Pareek, A., Suthar, M., Rathore, G.S., Basniwal, P.K., and Jain, D. 2009a. Study of glucose uptake activity of *Helicteres isora* Linn. fruits in L-6 cell lines. *Int. J. Diabetes Dev. Countries* 29: 170–173.

Gupta, R.S. and Sharma, A. 2003. Antifertility effect of *Tinospora cordifolia* (Willd) stem extract in male rats. *Indian J. Exp. Biol.* 41: 885–889.

Gupta, S., Dadheech, N., Singh, A., Soni, S., and Bhonde, R.R. 2010. *Enicostemma littorale*: A new therapeutic target for islet neogenesis. *Int. J. Integr. Biol.* 9: 50–56.

Gupta, S., Kataria, M., Gupta, P.K., Muruganandan, S., and Yashroy, R.C. 2004b. Protective role of extraction of neem seeds in diabetes caused by streptozotocin in rats. *J. Ethnopharmacol.* 90: 185–189.

Gupta, S., Mediratta, P.K., Singh, S., Sharma, K.K., and Shukla, R. 2006. Antidiabetic, antihypercholesterolaemic and antioxidant effect of *Ocimum sanctum* (Linn.) seed oil. *Indian J. Exp. Biol.* 44: 300–304.

Gupta, S., Sharma, S.B., and Prabhu, K.M. 2009a. Ameliorative effect of *Cassia auriculata* L. leaf extract on glycemic control and atherogenic lipid status in alloxan-induced diabetic rabbits. *Indian J. Exp. Biol.* 47: 974–980.

Gupta, S.S., Verma, S.C., Garg, V.P., and Khandelwal P. 1967. Studies on the anti-diabetic effects of *Casearia esculenta*. *Indian J. Med. Res.* 55: 754–763.

Gupta, V., Jadhav, J.K., Masirkar, V.J., and Deshmukh, V.N. 2009b. Antihyperglycemic effect of *Diospyros melanoxylon* (Roxb.) bark against alloxan-induced diabetic rats. *Int. J. PharmTech Res.* 1: 196–200.

Gurrola-Diaz, C.M., Borelli, M.I., Przybyl, A.K., Garcia-Lopez, J.S., de la Mora, P.G., and Garcia-Lopez, P.M. September 14–18, 2008. Insulin secretion effect of 2,17-dioxosparteine, 17-thionosparteine, multiflorine and 17-hydroxylupanine on rat Langerhan's islets. In: *Lupins for Health and Wealth*, Palta, J.A. and Berger, J.D. (eds.), *Proceedings of the 12th International Lupin Conference*, Fremantle, Western Australia, Australia.

Gurudeeban, S., Satyavani, K., Ramanathan, T., and Balasubramanian, T. 2012. Antidiabetic effect of a black mangrove species *Aegiceras corniculatum* in alloxan-induced diabetic rats. *J. Adv. Pharm. Technol. Res.* 3: 52–56.

Gutierrez, R.M.P. 2012. Inhibition of advanced glycation end-product formation by *Origanum majorana* L. *in vitro* and in streptozotocin-induced diabetic rats. *Evid. Based Complement. Altern. Med.* 598638: 1–8.

Gutierrez, R.M.P. and Baez, E.G. 2014. Evaluation of antidiabetic, antioxidant and antiglycating activities of the *Eysenhardtia polystachy*. *Pharmacogn. Mag.* 10(Suppl. 2): S404–S418.

Gutierrez, R.M.P., Juarez, V.A., Sauceda, J.V., and Sosa, I.A. 2014. *In vitro* and *in vivo* antidiabetic and antiglycation properties of *Apium graveolens* in Type 1 and 2 diabetic rat. *Int. J. Pharmacol.* 10: 368–379.

Guzman, A. and Guerrero, R.O. 2005. Inhibition of aldose reductase by herbs extracts and natural substances and their role in prevention of cataracts. *Revista Cubana de Plantas Medicinales* 10: 3–4.

Guzman-Partida, A.M., Jatomea-Fino, O., Robles-Burgueño, M.R., Ortega-Nieblas, M., and Vazquez-Moreno, L. 2007. Characterization of α-amylase inhibitor from Palo fierro seeds. *Plant Physiol. Biochem.* 45: 711–715.

Ha, D.T., Tuan, D.T., Thu, N.B., Nhiem, N.X., Ngoc, T.M., Yim, N. et al. 2009. Palbinone and triterpenes from Moutan Cortex (*Paeonia suffruticosa*, Paeoniaceae) stimulate glucose uptake and glycogen synthesis via activation of AMPK in insulin-resistant human HepG2 cells. *Bioorg. Med. Chem. Lett.* 19: 5556–5559.

Habs, M., Jahn, S.A., and Schmahl, D. 1984. Carcinogenic activity of condensate from coloquint seeds (*Citrullus colocynthis*) after chronic epicutaneous administration to mice. *J. Cancer Res. Clin. Oncol.* 108: 154–156.

Haddad, P.S., Martineau, L.C., Lyoussi, B., and Le, P.M. 2005. Antidiabetic plants of North Africa and the Middle East. In: *Traditional Medicines for Modern Times: Antidiabetic Plants,* Soumyanath, A. (ed.). Boca Raton, FL: CRC Press, pp. 221–224.

Hagiwara, H., Seki, T., and Ariga, T. 2004. The effect of pre-germinated brown rice intake on blood glucose, PAI-1 levels in streptozotocin-induced diabetic rats. *Biosci. Biotechnol. Biochem.* 68: 444–447.

Hahm, S.W., Park, J., and Son, Y.S. 2011. *Opuntia humifusa* stems lower blood glucose and cholesterol levels in streptozotocin-induced diabetic rats. *Nutr. Res.* 31: 479–487.

Haidari, F., Seyed-Sadjadi, N., Taha-Jalali, M., and Mohammed-Shahi, M. 2011. The effect of oral administration of *Carum carvi* on weight, serum glucose, and lipid profile in streptozotocin-induced diabetic rats. *Saudi Med. J.* 32: 695–700.

Hajzadeh, M.A.R., Rajaei, Z., Ghamami, G., and Tamiz, A. 2011. The effect of *Salvia officinalis* leaf extract on blood glucose in streptozotocin-diabetic rats. *Pharmacologyonline* 1: 213–220.

Halder, N., Joshi, S., and Gupta, S.K. 2003. Lens aldose reductase inhibiting potential of some indigenous plants. *J. Ethnopharmacol.* 86: 113–116.

Hamdan, I.I. and Afifi, F.U. 2004. Studies on the *in vitro* and *in vivo* hypoglycemic activities of some medicinal plants used in treatment of diabetes in Jordanian traditional medicine. *J. Ethnopharmacol.* 93: 117–121.

Hamden, K., Mnafgui, K., Amri, Z., Aloulou, A., and Elfeki, A. 2013. Inhibition of key digestive enzymes related to diabetes and hyperlipidemia and protection of liver-kidney functions by trigonelline in diabetic rats. *Sci. Pharm.* 81: 233–246.

Hamid, K., Ullah, M.O., Sultana, S., Howlader, M.A., Basak, D., Nasrin, N. et al. 2011. Evaluation of the leaves of *Ipomoea aquatica* for its hypoglycemic and antioxidant activity. *J. Pharm. Sci. Res.* 3: 1330–1333.

Hamidpour, M., Hamidpour, R., and Shahlari, M. 2014. Chemistry, pharmacology and medicinal property of Sage (Salvia) to prevent and cure illnesses such as obesity, diabetes, depression, dementia, lupus, autism, heart disease and cancer. *J. Tradit. Complement. Med.* 4: 82–88.

Hamza, N., Berke, B., Cheze, C., Agli, A.N., Robinson, P., Gin, H. et al. 2010. Prevention of type 2 diabetes induced by high fat diet in the C57BL/6J mouse by two medicinal plants used in traditional treatment of diabetes in the east of Algeria. *J. Ethnopharmacol.* 128: 513–518.

Hamza, N., Berke, B., Cheze, C., Garrec, R., Lassalle, R., Agli, A.N. et al. 2011. Treatment of high fat diet induced type 2 diabetes in C57BL/6J mice by two medicinal plants used in traditional treatment of diabetes in the east of Algeria. *J. Ethnopharmacol.* 133: 931–933.

Hamzah, R.U., Odetola, A.A., Erukainure, O.L., and Oyagbemi, A.A. 2012. *Peperomia pellucida* in diets modulates hyperglycemia, oxidative stress and dyslipidemia in diabetic rats. *J. Acute Dis.* 2012: 135–140.

Han, L.-K., Zheng, Y.-N., Yoshikawa, M., Okuda, H., and Kimura, Y. 2005. Anti-obesity effects of chikusetsusaponins isolated from *Panax japanicus* rhizomes. *BMC Complement. Altern. Med.* 5: 9–15.

Handral, H.K., Pandith, A., and Shruthi, S.D. 2012. A review on *Murraya koenigii*: Multi potential medicinal plant. *Asian J. Pharm. Clin. Res.* 5: 5–14.

Hanna, E.T., Aniess, W.I.M., Khalil, A.F., Abdulla, E.S., Hassanin, E.A., and Nagib, E.W. 2014. The effect of ginger and thyme on some bio-chemical parameters in diabetic rats. *IOSR J. Pharm. Biol. Sci.* 9: 54–61.

Hao, L.N., Zhang, Y.Q., Shen, Y.H., Wang, S.Y., Wang, Y.H., Zhang, H.F., and He, S.Z. 2010. Effect of puerarin on retinal pigment epithelial cells apoptosis induced partly by peroxynitrite *via* Fas/Fasl pathway. *Int. J. Ophthalmol.* 3: 283–287.

Harborne, J.B. 1967. Comparative biochemistry of flavonoids-VI. Flavonoid patterns in the Bignoniaceae and the Gesneriaceae. *Phytochemistry* 6: 1643–1651.

Hari Kumar, K.B., Sabu, M.C., Lima, P.S., and Kuttan, R. 2004. Modulation of haematopoetic system and antioxidant enzymes by *Emblica officinalis* Gaertn and its protective role against gamma-radiation induced damages in mice. *J. Radiat. Res.* (Tokyo) 45: 549–555.

Hasan, I. and Khatoon, S. 2012. Effect of *Momordica charantia* (bitter gourd) tablets in diabetes mellitus: Type 1 and Type 2. *Prime Res. Med.* 2: 72–74.

Hasanein, P. and Shahidi, S. 2011. Effects of *Hypericum perforatum* extract on diabetes-induced learning and memory impairment in rats. *Phytother. Res.* 25: 544–549.

Hasani-Ranjbar, S., Zahedi, H.S., Abdollahi, M., and Larijani, B. 2013. Trends in publication of evidence-based traditional Iranian medicine in endocrinology and metabolic disorders. *J. Diabetes Metab. Disord.* 12: 49–55.

Hashim, M.A., Yam, M.F., Hor, S.Y., Lim, C.P., Asmawi, M.Z., and Sadikun, A. 2013. Anti-hyperglycemic activity of *Swietenia macrophylla* King (Meliaceae) in normoglycemic rats undergoing glucose tolerance test. *Chinese Med.* 8: 11–17.

Hashmat, I., Azad, H., and Ahmed, A. 2012. Neem (*Azadirachta indica* A. Juss)—A nature's drug store: An overview. *Int. Res. J. Biol. Sci.* 1: 76–79.

Hashmat, M.A. and Hussain, R. 2013. A review on *Acacia catechu* Willd. *Int. J. Contemp. Res. Bus.* 5: 593–600.

Hassan, F., El-Razek, A., and Hassan, A.A. 2012a. Nutritional value and hypoglycemic effect of prickly cactus pear (*Opuntia ficus-indica*) fruit juice in alloxan-induced diabetic rats. *Aust. J. Basic Appl. Sci.* 5: 356–377.

Hassan, S.A., Barthwal, R., Nair, M.S., and Haque, S.S. 2012b. Aqueous bark extract of *Cinnamomum zeylanicum*: A potential therapeutic agent for streptozotocin-induced type 1 diabetes mellitus (T1DM) rats. *Trop. J. Pharm. Res.* 11: 429–435.

Hazra, M., Kundusen, S., Bhattacharya, S., Haldar, P.K., Gupta, M., and Mazumder, U.K. 2011. Evaluation of hypoglycemic and anti-hyperglycemic effects of *Luffa cylindrica* fruit extracts in rats. *J. Adv. Pharm. Edu. Res.* 2: 138–146.

He, K., Li, X., Chen, Y., Ye, X., Huang, J., Jin, Y. et al. 2011. Evaluation of antidiabetic potential of selected traditional Chinese medicines in streptozotocin-induced diabetic mice. *J. Ethnopharmacol.* 137: 1135–1142.

Heacock, P.M., Hertzler, S.R., Williams, J.A., and Wolf, B.W. 2005. Effect of medicinal food containing an herbal α-glucosidase inhibitor on post prandial glycemia and insulinemia in healthy adults. *J. Am. Diet. Assoc.* 105: 65–71.

Hedge, P.L., Rao, H.A., and Rao, P.N. 2014. A review on Insulin plant (*Costus igneus* Nak). *Pharmacogn. Rev.* 8: 67–72.

Heidarian, E. and Soofiniya, Y. 2011. Hypolipidemic and hypoglycemic effects of aerial parts of *Cynara scolymus* in streptozotocin-induced diabetic rats. *J. Med. Plant Res.* 5: 2717–2723.

Helal, E.G.E., Abd-Elwahab, S.M., Atia, T.A., and Mohammad, A.L. 2013. Hypoglycemic effect of the aquous extractof *Lupinus alba*, *Medicago sativa* (seeds) and their mixture on diabetic rats. *Egypt. J. Hosp. Med.* 52: 685–690.

Helen, O.T., Olusola, A.E., Iyare, E.E., and Anyaehie, U.B. 2013. *Dioscorea alata* L. reduces body weight by reducing food intake and fasting blood glucose level. *Br. J. Med. Med. Res.* 3: 1871–1880.

Helmstadter, A. and Schuster, N. 2010. *Vaccinium myrtillus* as an antidiabetic medicinal plant—research through the ages. *Pharmazie* 65: 315–321.

Hemalatha, S., Wahi, A.K., Singh, P.N., and Chansouria, J.P.N. 2010. Evaluation of anti-hypergycemic and free radical scavenging activity of *Melothria maderaspatana* Linn. in streptozotocin-induced diabetic rats. *Afr. J. Pharm. Pharmacol.* 4: 817–822.

Hemamalini, H. and Vijusha, M. 2012. Antidiabetic activity of methanolic extracts of leaves of *Anogeissus acuminata* Roxb. ex Candolle and *Solanum pubescens* Willd. by alloxan induced model in rats. *Der Pharmacia Lett.* 4: 1445–1460.

Hemmati, A.A., Jalali, M.T., Rashidi, I., and Kalantar-Hormozi, T. 2010. Impact of aqueous extract of black mulberry (*Morus nigra*) on liver and kidney function of diabetic mice. *Jundishapur J. Nat. Pharm. Prod.* 5: 18–25.

Heo, S.I., Jin, Y.S., Jung, M.J., and Wang, M.H. 2007. Antidiabetic properties of 2,5-dihydroxy-4,3′-di(beta-D-glucopyranosyloxy)-trans-stilbene from mulberry (*Morus bombycis* Koidzumi) root in streptozotocin-induced diabetic rats. *J. Med. Food* 10: 602–607.

Hepcy-Kalarani, D., Dinakar, A., and Senthilkumar, N. 2011. Hypoglycemic and anti-diabetic activity of *Alangium salvifolium* Wang in alloxan-induced diabetic rats. *Asian J. Pharm. Clin. Res.* 4: 131–133.

Herrera-Arellano, A.J., Aguilar-Santamaria, L., Garcia-Hernansez, B., Nicasio-Terres, P., and Tortoriello, J. 2004. Clinical trial of *Cecropia obtusifolia* and *Marrubium vulgare* leaf extracts on blood glucose and serum lipids in type 2 diabetes. *Phytomedicine* 11: 561–566.

Hetta, M.H., Aly, H.F., and Arafa, A. 2014. Inhibitary effect of *Eruca sativa* Mill. on carbohydrate metabolizing enzymes *in vitro*. *Int. J. Pharm. Sci. Rev. Res.* 26: 205–208.

Hettiarachchi, P., Jiffry, M.T., Jansz, E.R., Wickramasinghe, A.R., and Fernando, D.J. 2001. Glycaemic indices of different varieties of rice grown in Sri Lanka. *Ceylon Med. J.* 46: 11–14.

Higashino, H., Kinoshita, K., Kurita, T., Hamada, H., Etoh, H., and Hukunaga, Y. 2011. Inhibitory effects of shiso (*Perilla frutescens*) extracts on an increase of blood glucose level in rats. *Nippon Shokuhin Kagaku Kogaku Kaishi* 58: 164–169.

Hii, C.S. and Howell, S.L. 1985. Effects of flavonoids on insulin secretion and Ca handling in rat islets of Langerhans. *J. Endocrinol.* 107: 1–8.

Hikawczuk, V.J., Saad, J.R., Guardia, T., Juarez, A.O., and Giordano, O.S. 1998. Anti-inflammatory activity of compounds isolated from *Cecropia pachystachya*. *Anales des la Asociacion Quimica Argentina* 86: 167–170.

Hikino, H., Konno, C., Takahasi, M., Murakami, M., Kato, Y., and Karikura, M. 1986a. Isolation and hypoglycemic activity of dioscorans A, B, C, D, E, and F: Glycans of *Dioscorea japonica* rhizophors. *Planta Med.* 52: 168–171.

Hikino, H., Mizuno, T., Oshima, Y., and Konno, C. 1985. Isolation and hypoglycemic activity of moran A, a glycoprotein of *Morus alba* root barks. *Planta Med.* 51: 159–160.

Hikino, H., Murakami, M., Oshima, Y., and Konno, C. 1986b. Isolation and hypoglycemic activity of oryzarans A, B, C, and D: Glycans of *Oryza sativa* roots. *Planta Med.* 6: 490–492.

Hikino, H., Yoshizawa, M., Suzyki, Y., Oshima, Y., and Konno, C. 1989. Isolation and hypoglycemic activity of trichosans A, B, C, D and E: Glycans of *Trichosanthes kirilowii* roots. *Planta Med.* 55: 349–350.

Hima, B.N., Suvarnalatha Devi, P., Rukmini, K., and Singara Charya, M.A. 2012. Phytochemical screening and antibacterial activity of *Hemionitis arifolia* (Burm.) Moore. *Indian J. Nat. Prod. Resour.* 3: 9–12.

Hnatyszyn, O., Mino, J., Ferraro, G., and Acevedo, C. 2002. The hypoglycemic effect of *Phyllanthus sellowianus* fractions in streptozotocin-induced diabetic mice. *Phytomedicine* 9: 556–559.

Hoa, N.K., Phan, D.V., Thuan, N.D., and Ostenson, C.G. 2004. Insulin secretion is stimulated by ethanol extract of *Anemarrhena asphodeloides* in isolated islet of healthy Wistar and diabetic Goto-Kakizaki Rats. *Exp. Clin. Endocrinol. Diabetes* 112: 520–525.

Hojberg, P.V., Zander, M., and Vilsboll, T. 2008. Near normalisation of blood glucose improves the potentiating effect of GLP-1 on glucose-induced insulin secretion in patients with type 2 diabetes. *Diabetologia* 51: 632–640.

Hong, E.-J., Hong, S.-J., Jung, K.H., Ban, J.Y., Baek, Y.-H., Woo, H.-S. et al. 2008. *Ecommia ulmoides* extract stimulates glucose uptake through PI 3-kinase mediated pathway in L6 rat skeletal muscle cells. *Mol. Cell. Toxicol.* 4: 224–229.

Hong, G., Bhandari, M.R., John-Anurakkun, N., and Kwabata, J. 2008. α-Glucosidase and α-amylase inhibitory activities of Nepalese medicinal herb Pakhanbhed (*Bergenia ciliata*, Haw.). *Food Chem.* 106: 247–252.

Honma, A., Koyama, T., and Yazawa, K. 2010. Antihyperglycemic effects of sugar maple *Acer saccharum* and its constituent acertannin. *Food Chem.* 123: 390–394.

Honore, S.M., Cabrera, W.M., Genta, S.B., and Sanchez, S.S. 2012. Protective effect of yacon leaves decoction against early nephropathy in experimental diabetic rats. *Food Chem. Toxicol.* 50: 1704–1715.

Hooda, M.S., Pal, R., Bhandari, A., and Singh, J. 2014. Antihyperglycemic and antihyperlipidemic effects of *Salvadora persica* in streptozotocin-induced diabetic rats. *Pharm. Biol.* 52: 745–749.

Hoseini, H.F., Gohari, A.R., Saeidnia, S., Majd, N.S., and Hadjiakhoondi, A. 2009. The effect of *Nasturtium officinale* on blood glucose levels in diabetic rats. *Pharmacologyonline* 3: 866–871.

Hossain, M.K., Prodhan, M.K., Hasan-Evan, A.S.M.I., Morshed, H., and Hossain, M.M. 2012a. Anti-inflammatory and antidiabetic activity of ethanolic extracts of *Sterculia villosa* barks on albino Wistar rats. *J. Appl. Pharm. Sci.* 2: 96–100.

Hossain, M.S., Sokeng, S., Shoeb, M., Hasan, K., Mosihuzzaman, Nahar, N. et al. 2012b. Hypoglycemic effect of *Irvinga gabonensis* (Aubry-Lacomate ex. Ororke), Baill in type 2 diabetic Long-Evan's rats. *Dhaka Univ. J. Pharm. Sci.* 11: 19–24.

Hosseini, S., Huseini, H.F., Larijani, B., Mohammad, K., Najmizadeh, A., Nourijelyani, K., and Jamshidi, L. 2014. The hypoglycemic effect of *Juglans regia* leaves aqueous extract in diabetic patients: A first human trial. *J. Pharm. Sci.* 22: 19. http://www.darujps.com/content/22/1/19.

Hossen, S.M.M., Uddin, M., and Sarkar, M.R. 2014. Medicinal potential of *Phyllanthus emblica* L. fruits extracts: Biological and pharmacological activities. *Int. J. Pharmacogn.* 1: 307–316.

Hosseinzadeh, H., Haddadkhodaparast, M.H., and Shokoohizadeh, H. 1998. Antihyperglycemic effect of *Salvia leriifolia* Benth. leaf and seed extract in mice. *Iran J. Med. Sci.* 23: 74–80.

Hosseinzadeh, H., Ramezani, M., and Danaei, A.R. 2002. Anti-hyperglycemic effect and acute toxicity of *Securigera securidaca* L. seed extracts in mice. *Phytother. Res.* 16: 745–747.

Hou, C., Lin, S.J., Cheng, J.T., and Hsu, F.L. 2003. Antidiabetic dimeric guianolides and a lignan glycoside from *Lactuca indica*. *J. Nat. Prod.* 66: 625–629.

Hou, S.Z., Chen, S.X., Huang, S., Jiang, D.X., Zhou, C.J., Chen, C.Q. et al. 2011. The hypoglycemic activity of *Lithocarpus polystachyus* Rehd. leaves in the experimental hyperglycemic rats. *J. Ethnopharmacol.* 138: 142–149.

Hou, W., Li, Y., Zhang, Q., Wei, X., Peng, A., Chen, L., and Wei, Y. 2009. Triterpene acids isolated from *Lagerstroemia speciosa* leaves as α-glucosidase inhibitors. *Phytother. Res.* 23: 614–618.

Hou, Z., Zhang, Z., and Wu, H. 2005. Effects of *Sanguish draxonis* (a Chinese traditional herbal drug) on the formation of insulin resistance in rats. *Diabetes Res. Clin. Pract.* 68: 3–11.

Hozayen, W.G.M., Bastawy, M., and Hamed, M.Z. 2011. Biochemical effects of *Cichorium intybus* and *Sonchus oleraceus* infusions and esculetin on streptozotocin-induced diabetic albino rats. *J. Am. Sci.* 7: 1124–1137.

Hsu, F.-L., Chen, Y.C., and Cheng, J.T. 2000. Caffeic acid as active principle from the fruit of *xanthium strumarium* to lower plasma glucose in diabetic rats. *Planta Med.* 66: 228–230.

Hsu, F.-L., Huang, C.-F., Chen, Y.-W., Yen, Y.-P., Wu, C.-T., Uang, B.-J. et al. 2013. Anti-diabetic effects of pterosin A, a small molecular-weight natural product, on diabetic mouse models. *Diabetes* 62: 628–638.

Hsu, Y.-J., Lee, T.-H., Chang, C.L.-T., Huang, Y.-T., and Yang, W.-C. 2009. Anti-hyperglycemic effects and mechanism of *Bidens pilosa* water extract. *J. Ethnopharmacol.* 122: 379–383.

Hsu, Y.S., Kuo, Y.H., Cheng, H.L., Flatt, P.R., and Liu, H.K. 2012. *Schizandra arisanensis* extract attenuates cytokine-mediated cytotoxicity in insulin-secreting cells. *World J. Gastroenterol.* 18: 6809–6818.

Hu, C.K., Lee, Y.J., Colitz, C.M., Chang, C.J., and Lin, C.T. 2012. The protective effects of *Lycium barbarum* and *Chrysanthemum morifolum* on diabetic retinopathies in rats. *Vet. Ophthalmol.* 15: 65–71.

Hu, C.M., Li, J.S., Cheah, K.P., Lin, C.W., Yu, W.Y., Chang, M.L. et al. 2011a. Effect of Sanguis draconis (a dragon's blood resin) on streptozotocin- and cytokine-induced β-cell damage, in vitro and in vivo. *Diabetes Res. Clin. Pract.* 94: 417–425.

Hu, X., Xing, X., Zhang, Z., Wu, R., Guo, Q., Cui, S.W., and Wang, Q. 2011b. Antioxidant effects of *Artemisia sphaerocephala* Krasch. gum on streptozotocin-induced type 2 diabetic rats. *Food Hydrocolloids* 25: 207–213.

Hua, J., Qi, J., and Yu, B.-Y. 2014. Iridoid and phenylpropanoid glycosides from *Scrophularia ningpoensis* Hemsl and their α-glucosidase inhibitory activities. *Fitoterapia* 93: 67–73.

Huang, C.F., Chen, Y.W., Yang, C.Y., Lin, H.Y., Way, T.D., Chiang, W. et al. 2011. Extract of lotus leaf (*Nelumbo nucifera*) and its active constituent catechin with insulin secretagogue activity. *J. Agric. Food Chem.* 59: 1087–1094.

Huang, J., Huang, K., Lan, T., Xie, X., Shen, X., Liu, P., and Huang, H. 2013. Curcumin ameliorates diabetic nephropathy by inhibiting the activation of the SphK1-S1P signaling pathway. *Mol. Cell. Endocrinol.* 365: 231–240.

Huang, M., Xie, Y., Chen, L., Chu, K., Wu, S., Lu, J. et al. 2012. Antidiabetic effect of the total polyphenolic acids fraction from *Salvia miltiorrhiza* Bunge in diabetic rats. *Phytother. Res.* 26: 944–988.

Huang, T., Peng, G., Li, G., Yamahara, J., Roufogalis, B., and Li, Y. 2006. *Salacia oblonga* root improves postprandial hyperlipidemia and hepatic steatosis in Zucker diabetic fatty rats: Activation of PPAR. *Toxicol. Appl. Pharmacol.* 210: 225–235.

Huang, T.H., He, L., Qin, Q., Yang, Q., Peng, G., Harada, M. et al. 2008. *Salacia oblonga* root decreases cardiac hypertrophy in Zucker diabetic fatty rats: Inhibition of cardiac expression of angiotensin II type 1 receptor. *Diabetes Obes. Metab.* 10: 574–585.

Huang, T.H., Peng, G., Kota, B.P., Li, G.Q., Yamahara, J., Roufogalis, B.D., and Li, Y. 2005. Pomegranate flower improves cardiac lipid metabolism in a diabetic rat model: Role of lowering circulating lipids. *Br. J. Pharmacol.* 145: 767–774.

Hukeri, V.I., Kalyani, G.A., and Kakrani, H.K. 1998. Hypoglycemic activity of flavonoids of *Phyllanthus fraternus* in rats. *Fitoterapia* 59: 68–70.

Hule, A.K., Shah, A.S., Gambhire, M.N., and Juvekar, A.R. 2011. An evaluation of the anti-diabetic effects of *Elaeocarpus ganitrus* in experimental animals. *Indian J. Pharmacol.* 43: 56–59.

Husain, G.M., Singh, P.N., and Kumar, V. 2009. Antidiabetic activity of standardized extract of *Picrorhiza kurroa* in rat model of NIDDM. *Drug Discov. Ther.* 3: 88–92.

Husain, G.M., Singh, P.N., Sing, R.K., and Kumar, V. 2011. Antidiabetic activity of standardized extract of *Quassia amara* in nicotinamide-streptozotocin-induced diabetic rats. *Phytother. Res.* 25: 1806–1812.

Huseini, H.F., Darvishzadeh, F., Heshmat, R., Jafariazar, Z., Raza, M., and Larijani, B. 2009. The clinical investigation of *Citrullus colocynthis* (L.) schrad fruit in treatment of Type II diabetic patients: A randomized, double blind, placebo-controlled clinical trial. *Phytother. Res.* 23: 1186–1189.

Hussain, F., Zin, N.N.M., Zullkefli, M.R., Choon, Y.S., Abdullah, N.A., and Lin, T.S. 2013. *Piper sarmentosum* water extract attenuates diabetic complications in streptozotocin-induced Sprague-Dawley rats. *Sains Malaysiana* 42: 1605–1612.

Hussain, Z., Waheed, A., Qureshi, R.A., Burdi, D.K., Verspohl, E.J., Khan, N. et al. 2004. The effect of medicinal plants of Islamabad and Murree region of Pakistan on insulin secretion from INS-1 cells. *Phytother. Res.* 18: 73–77.

Hussein, M.A. 2008. Antidiabetic and antioxidant activity of *Jasonia montana* extract in streptozotocin-induced diabetic rats. *Saudi Pharm. J.* 16: 214–221.

Hussen, H.M., Sayed, E.M., and Said, A.A. 2006. Antihyperglycemic, antihyperlipidemic and antioxidant effects of *Zizyphus spina* Christi and *Zizyphus jujube* in Alloxan diabetic rats. *Int. J. Pharmacol.* 2: 563–570.

Huyssteen, M., Milne, P.J., Campbell, E.E., and van de Venter, M. 2011. Antidiabetic and cytotoxicity screening of five medicinal plants used by traditional African health practitioners in the Nelson Mandela Metropole, South Africa. *Afr. J. Tradit. Complement. Altern. Med.* 8: 150–158.

Hwang, J.K., Kong, T.W., Baek, N.I., and Pyun, Y.R. 2000. Alpha-glycosidase inhibitory activity of hexagalloylglucose from the galls of *Quercus infectoria*. *Planta Med.* 66: 273–274.

Hwang, J.T., Kim, S.H., Lee, M.S., Kim, S.H., Yang, H.J., Kim, M.J. et al. 2007. Anti-obesity effects of gingenoside Rh2 are associated with the activation of AMPK signaling pathways in 3T3-L1 adipocytes. *Biochem. Biophys. Res. Commun.* 364: 1002–1008.

Hwang, J.T., Kwon, D.Y., and Yoon, S.H. 2009a. AMP-activated protein kinase: A potential target for the diseases prevention by natural occurring polyphenols. *Nat. Biotechnol.* 26: 17–22.

Hwang, J.T., Lee, M.S., Kim, H.J., Sung, M.J., Kim, H.Y., Kim, M.S. et al. 2009b. Anti-obesity effect of gingenoside Rg3 involves the AMPK and PPAR-γ signal pathways. *Phytother. Res.* 23: 262–266.

Iacobellis, N.S., Lo Cantore, P., Capasso, F., and Senatore, F. 2005. Antibacterial activity of *Cuminum cyminum* L. and *Carum carvi* L. essential oils. *J. Agric. Food Chem.* 53: 58–61.

Ibegbulem, C.O. and Chikezie, P. 2013. Hypoglycemic properties of ethanolic extracts of *Gongronema latifolium*, *Aloe perryi*, *Viscum album* and *Allium sativum* administered to alloxan-induced diabetic albino rats (*Rattus norvegicus*). *Pharmacogn. J.* 3: 12–16.

Ibeh, B.O. and Ezeaja, M.I. 2011. Preliminary study of antidiabetic activity of the methanolic leaf extract of *Axonopus compressus* (P. Beauv) in alloxan-induced diabetic rats. *J. Ethnopharmacol.* 138: 713–716.

Ibraheim, Z.Z., Ahmed, A.S., and Douda, Y.G. 2011. Phytochemical and biological studies of *Adiantum capillus-veneris* L. *Saudi Pharm. J.* 19: 65–74.

Ichikawa, H., Yagi, H., Tanaka, T., Cyong, J.C., and Masaki, T. 2010. *Lagerstroemia speciosa* extract inhibit TNF-induced activation of nuclear factor-κB in rat cardiomyocyte H9c2 cells. *J. Ethnopharmacol.* 128: 254–256.

IDF. 2011. *International Diabetes Federation Diabetes Atlas*, 5th edn. Brussels, Belgium: International Diabetes Federation Bulletin.

IDF. 2012. *International Diabetes Federation Diabetes Atlas*, update 2012, 5th edn. Brussels, Belgium: International Diabetes Federation.

Idris, M.H.M., Budin, S.B., Osman, M., and Mohamed, J. 2012. Protective role of *Hibiscus sabdariffa* calyx extract against streptozotocin induced sperm damage in diabetic rats. *EXCLI J.* 11: 659–669.

Ijaola, T.O., Osunkiyesi, A.A., Taiwo, A.A., Oseni, O.A., Lanreiyanda, Y.A., Ajayi, J.O. et al. 2014. Antidiabetic effect of *Ipomoea batatas* in normal and alloxan-induced diabetic rats. *IOSR J. Appl. Chem.* 7: 16–25.

Ikarashi, N., Toda, T., Okaniwa, T., Ito, K., Ochiai, W., and Sugiyama, K. 2011. Anti-obesity and anti-diabetic effects of *Acacia* polyphenol in obese diabetic KKAy mice fed high-fat diet. *Evid. Based Complement. Altern. Med.* 2011: Article ID 952031, 10 pages.

Ikeda, Y., Murakami, A., and Ohigashi, H. 2008. Ursolic acid: An anti- and pro-inflammatory triterpenoid. *Mol. Nutr. Food Res.* 52: 26–42.

Ikeda, Y., Noguchi, M., Kishi, S., Masuda, K., Kusumoto, A., Zedia, M., Abe, K., and Kiso, Y. 2002. Blood glucose controlling effects and safety of single and long-term administration on the extract of banaba leaves. *J. Nutr. Food* 5: 41–53.

Ikewuchi, C.C. 2010. Effect of aqueous extract of *Sansevieria senegambica* Bakeron plasma chemistry, lipid profile and atherogenic indices of alloxan treated rats: Implications for the management of cardiovascular complications in diabetes mellitus. *Pac. J. Sci. Technol.* 11: 524–529.

Ilango, K. and Chitra, V. 2009. Anti-diabetic and anti-oxidant activities of *Limonia acidissima* Linn. in aloxan-induced diabetic rats. *Der Pharmacia Lett.* 1: 117–125.

Ilham, S., Ali, M.S., Hasan, C.M., Kaisar, M.A., and Bachar, S.C. 2012. Antinociceptive and hypoglycemic activity of methanolic extract of *Phlogacanthus thyrsiflorus* Nees. *Asian J. Pharm. Clin. Res.* 5: 15–18.

Indradevi, S., Ilavenil, S., Kaleeswaran, B., Srigopalram, S., and Ravikumar, S. 2012. Ethanolic extract of *Crinum asiaticum* attenuates hyperglycemia-mediated oxidative stress and protects hepatocytes in alloxan-induced experimental diabetic rats. *J. King Saud Univ. Sci.* 24: 171–177.

Inman, W.D., King, S.R., Evans, J.L., and Luo, J. 1998. Furanoeremophilane and eremophilanolide sesquiterpenes for treatment of diabetes. Patent No. US 5747527.

Inman, W.D., Luo, J., Jolad, S.D., King, S.R., and Cooper, R. 1999. Antihyperglycemic sesquiterpenes from *Psacalium decompositum*. *J. Nat. Prod.* 62: 1088–1092.

Iranloye, B.O., Arikawe, A.P., Rotimi, G., and Sogbade, A.O. 2011. Anti-diabetic and anti-oxidant effects of *Zingiber officinale* on alloxan-induced and insulin-resistant diabetic male rats. *Nig. J. Physiol. Sci.* 26: 89–96.

Iriadam, M., Musa, D., Gumushan, H., and Baba, F. 2006. Effects of two Turkish medicinal plants *Artemisia herba-alba* and *Teucrium polium* on blood glucose levels and other biochemical parameters in rabbits. *J. Cell. Mol. Biol.* 5: 19–24.

Irshad, N., Akhtar, M.S., Bashir, S., Hussain, A., Shafiq, M., Iqbal, J. et al. 2014. Hypoglycaemic effects of methanolic extract of *Canscora decussate* (Schult) whole-plant in normal and alloxan-induced diabetic rabbits. *Pak. J. Pharm. Sci.* 28: 167–174.

Irshad, N., Akhtar, M.S., Kamal, Y., Qayyum, M.I., Malik, A., and Hussain, R. 2013. Antihyperlipidemic and renoprotective activities of methanolic extract of *Canscora decussata* extract in alloxan-induced diabetic rabbits. *Bangladesh J. Pharmacol.* 8: 323–327.

Isalm, M.S., Islam, M.R., Rahman, M.M., Ali, M.S., Mosaddik, A., and Sadik, G. 2011. Evaluation of antidiabetic activity of *Peltophorum pterocarpum* on alloxan and glucose induced diabetic mice. *Int. J. Pharm. Sci. Res.* 2: 1543–1547.

Ishiki, M. and Klip, A. 2005. Minireview: Recent developments in the regulation of glucose transporter-4 traffic: New signals, locations, and partners. *Endocrinology* 146: 5071–5078.

Islam, M.S. 2011. Effects of the aqueous extract of white tea (*Camellia sinensis*) in a streptozotocin-induced diabetes model of rats. *Phytomedicine* 19: 25–31.

Islam, M.S. and Choi, H. 2008. Dietary red chilli (*Capsicum frutescens* L.) is insulinotropic rather than hypoglycemic in type 2 diabetes model of rats. *Phytother. Res.* 22: 1025–1029.

Islam, M.S., Choi, H., and Loots, D.T.L. 2008. Effects of dietary onion, *Allium cepa* L. in a high-fat diet streptozotocin-induced diabetes rodent model. *Ann. Nutr. Metab.* 53: 6–12.

Islam, T., Rahman, A., and Islam, A.U. 2012. Effects of extracts of fresh leaves of *Abroma augusta* L. on oral absorption of glucose and metformin hydrochloride in experimental rats. *ISRM Pharm.* 2012: 472586.

Ismail, M.Y.M., Assem, N.M., and Zakriya, M. 2010. Role of spices in diabetes mellitus. *Res. J. Pharm. Biol. Chem. Sci.* 1: 30–34.

Ismail, T.S., Gopalakrishnan, S., Begum, V.H., and Elango, V. 1997. Anti inflammatory activity of *Salacia oblonga* wall. and *Azima tetracantha* Lam. *J. Ethnopharmacol.* 56: 145–152.

Itankar, P.R., Lokhande, S.J., Verma, P.R., Arora, S.K., Sahu, R.A., and Patil, A.T. 2011. Antidiabetic potential of unripe *Carissa carandas* Linn. fruit extract. *J. Ethnopharmacol.* 135: 430–433.

Ivorra, M.D., D'Ocan, M.P., Paya, M., and Villar, A. 1998. Antihyperglycemic and insulin-releasing effects of β-sitosterol 3-beta-D-glucoside and its aglycone, β-sitosterol. *Arch. Int. Pharmacodyn. Ther.* 296: 224–231.

Ivorra, M.D., Paya, M., and Villar, A. 1989. Effect of tormentic acid on insulin secretion in isolated rat islets of langerhans. *Phytother. Res.* 3: 145–147.

Ivorra, M.D., Paya, M., and Villar, A. 1990. Effect of β-sitosterol-3-beta-D-glucoside on insulin secretion *in vivo* in diabetic rats and *in vitro* in isolated rat islets of Langerhans. *Pharmazie* 45: 271–273.

Iweala, E.E.J. and Oludare, F.D. 2011. Hypoglycemic effect, biochemical and histological changes of *Spondias mombin* Linn. and *Parinari polyandra* Benth seeds ethanolic extracts in alloxan-induced diabetic rats. *J. Pharmacol. Toxicol.* 6: 101–102.

Iweala, E.E.J., Uhegbu, F.O., and Adeesanoyel, O.A. 2013. Biochemical effects of leaf extracts of *Gongronema latifolium* and selenium supplementation in alloxan induced diabetic rats. *J. Pharmacognosy Phytother.* 5: 91–97.

Iwu, M.M., Okunji, C.O., Akah, P., Tempesta, M.S., and Corley, D. 1990a. Dioscoretine: The hypoglycemic principle of *Dioscorea dumetorum. Planta Med.* 56: 119–120.

Iwu, M.M., Okunji, C.O., Ohiaeri, G.O., Akah, P., Corley, D., and Tempesta, M.S. 1990b. Hypoglycemic activity of dioscoretine from tubers of *Dioscorea dumetorum* in normal and alloxan diabetic rabbits. *Planta Med.* 56: 264–267.

Iwueke, A.V. and Nwodo, O.F.C. 2008. Antihyperglycaemic effect of aqueous extract of *Daniella oliveri* and *Sarcocephalus latifolius* roots on key carbohydrate metabolic enzymes and glycogen in experimental diabetes. *Biokemistri* 20: 63–70.

Iyer, U.M. and Mani, U.V. 1990. Studies on the effect of curry leaves supplementation (*Murraya koenigi*) on lipid profile, glycated proteins and amino acids in non-insulin-dependent diabetic patients. *Plant Foods Hum. Nutr.* 40: 275–282.

Jafri, M.A., Aslam, M., Javed, K., and Singh, S. 2000. Effect of *Punica granatum* L. (flowers on blood glucose level in normal and alloxan-induced diabetic rats. *J. Ethnopharmacol.* 70: 309–314.

Jagetia, G.C. and Baliga, M.S. 2003. Evaluation of the radioprotective effect of the leaf extract of *Syzygium cumini* (Jamun) in mice exposed to the lethal dose of gamma-irradiation. *Nahrung* 47: 181–185.

Jagetia, G.C. and Baliga, M.S. 2004. The evaluation of nitric oxide scavenging activity of certain Indian medicinal plants *in vitro*: A preliminary study. *J. Med. Food* 7: 343–348.

Jagetia, G.C., Baliga, M.S., and Vankatesh, P. 2005. Influence of seed extract of *Syzygium cumini* (Jamun) on mice exposed to different doses of gamma-radiation. *J. Radiat. Res.* (Tokyo) 46: 59–65.

Jahan, I.A., Nahar, N., Mosihuzzaman, M., Rokeya, B., Ali, L., Azadkhan, A.K. et al. 2009. Hypoglycaemic and antioxidant activities of *Ficus racemosa* Linn. fruits. *Nat. Prod. Res.* 23: 399–408.

Jahromi, M.A. and Ray, A.B. 1993. Antihyperlipidemic effect of flavanoids from *Pterocarpus marsupium. J. Nat. Prod.* 56: 989–994.

Jaiarj, P., Khoohaswan, P., Wongkrajang, Y., Peungvicha, P., Suriyawong, P., Saraya, M.L. et al. 1999. Anticough and antimicrobial activities of *Psidium guajava* L. leaf extract. *J. Ethnopharmacol.* 67: 203–212.

Jaikumar, N. and Loganathan, P. 2010. Hypoglycemic effect of *Spinacea oleracea* in alloxan-induced diabetic rats. *Global J. Biotechnol. Biochem.* 5: 87–91.

Jain, P. and Sharma, H.P. 2013. A potential ethnomedicinal plant: *Semecarpus anacardium* L.—A review. *Int. J. Res. Pharm. Chem.* 3: 564–572.

Jain, P.S., Bari, S.B., and Surana, S.J. 2009. Isolation of stigmasterol and γ-sitosterol from petroleum ether extract of woody stem of *Abelmoschus manihot. Asian J. Biol. Sci.* 2: 112–117.

Jain, R.K., Junaid, M.A., and Rao, S.L. 1998. Receptor interactions of beta-N-oxalyl-L-alpha,beta-diaminopropionic acid, the *Lathyrus sativa* putative excitotoxin, with synoptic membranes. *Neurochem. Res.* 23: 1191–1196.

Jain, S., Bhatia, G., Barik, R., Kumar, P., Jain, A., and Dixit, V.K. 2010. Antidiabetic activity of *Paspalum scrobiculatum* Linn. in alloxan induced diabetic rats. *J. Ethnopharmacol.* 127: 325–328.

Jain, S., Katare, Y., and Patil, U.K. 2011. Effect of n-butanol extract of *Cassia tora* L. seeds on biochemical profile and β-cell regeneration in normal and alloxan-induced diabetic rats. *Der. Pharm. Lett.* 3: 183–193.

Jain, S.R. and Sharma, S.N. 1967a. Studies on anti-diabetes activities of bark of *Pinus roxburghi.* 15: 44–49.

Jain, S.R. and Sharma, S.N. 1967b. Studies on anti-diabetes activities of *Rhus chinensis, Ricinus communis* and *Rosa brunani. Planta Med.* 15: 439–442.

Jain, V., Viswanatha, G.L., Manohar, D., and Shivaprasad, H.N. 2012. Isolation of antidiabetic principle from fruit rinds of *Punica granatum. Evid. Based Complement. Altern. Med.* 2012: 147202.

Jaiswal, D., Kumar Rai, P., Kumar, A., Mehta, S., and Watal, G. 2009. Effect of *Moringa oleifera* Lam. leaves aqueous extract therapy on hyperglycemic rats. *J. Ethnopharmacol.* 123: 392–396.

Jaiswal, D., Rai, P.K., Kumar, A., and Watal, G. 2008. Study of glycemic profile of *Cajanus cajan* leaves in experimental rats. *Indian J. Clin. Biochem.* 23: 167–170.

Jaiswal, N., Bhatia, V., Srivastava, S.P., Srivastava, A.K., and Tamrakar, A.K. 2012. Antidiabetic effect of *Eclipta alba* associated with the inhibition of α-glucosidase and aldose reductase. *Nat. Prod. Res.* 26: 2363–2367.

Jalalpure, S.S., Bamne, S., Patil, M.B., Shah, B., and Salahuddin, M. 2006. Anti-diabetic activity of *Holarrhena antidysenterica* (Linn.) Wall, bark on alloxan induced diabetic rats. *J. Nat. Remedies* 6: 26–30.

Jana, K., Chatterjee, K., Ali, K.M., De, D., Bera, T.K., and Ghosh D. 2012. Antihyperglycemic and antioxidative effects of the hydro-methanolic extract of the seeds of *Caesalpinia bonducella* on streptozotocin-induced diabetes in male albino rats. *Pharmacognosy Res.* 4: 57–62.

Janapati, Y., Ahemad, R., Jayaveera, K., and Reddy, R. 2008c. Anti-diabetic activity of ethanolic extract of *Holostemma ada kodien* Schults in alloxan induced diabetic rats. *Int. J. Endocrinol.* 5: 9.

Janapati, Y.K., Jayaveera, K.N., Ravidrareddy, K., Rupesh, K., and Raghavendra, D. 2008a. Anti-diabetic activity of aqueous extract of *Talinum cuneifolium* Linn. in rats. *Pharmacologyonline* 2: 198–206.

Janapati, Y.K., Rasheed, A., Jayaveera, K.N., Ravidrareddy, K., Srikar, A., and Siddaiah, M. 2008b. Anti-diabetic activity of alcohol extract of *Talinum cuneifolium* in rats. *Pharmacologyonline* 2: 63–73.

Jaouhari, J.T., Lazrek, H.B., and Jana, M. 2000. The hypoglycemic activity of *Zygophyllum gaetulum* extracts in alloxan-induced hyperglycemic rats. *J. Ethnopharmacol.* 69: 17–20.

Jarald, E., Joshi, S.B., and Jain D.C. 2008a. Antidiabetic activity of aquous extract and non polysaccharide fraction of *Cyanodon dactylon. Indian J. Exp. Biol.* 46: 660–667.

Jarald, E., Joshi, S.B., and Jain, D.C. 2008b. Diabetes and herbal medicines. *Iranian J. Pharmacol. Ther.* 7: 97–106.

Jarald, E., Joshi, S.B., and Jain, D.C. 2009a. Antidiabetic activity of extracts and fraction of *Zizyphus mauritiana. Pharm. Biol.* 47: 328–334.

Jarald, E., Joshi, S.B., and Jain, D.C. 2009b. Biochemical study on the hypoglycemic effects of extract and fraction of *Acacia catechu* Willd in alloxan-induced diabetic rats. *Int. J. Diabetes Metab.* 17: 63–69.

Jarald, E., Joshi, S.B., Jain, D.C., and Edwin, S. 2013. Biochemical evaluation of the hypoglycemic effects of extract and fraction of *Cassia fistula* Linn. in alloxan-induced diabetic rats. *Indian J. Pharm. Sci.* 75: 427–434.

Jarvill-Taylor, K.A., Anderson, R.A., and Graves, D.J. 2001. A hydroxychalcone derived from cinnamon functions as a mimetic for insulin in 3T3-L1 adipocytes. *J. Am. Coll. Nutr.* 20: 327–336.

Javed, I., Faisal, I., Zia, U.R., Khan, M.Z., Muhammad F., Aslam, B. et al. 2012. Lipid lowering effect of *Cinnamomum zeylanicum* in hyperlipidaemic albino rabbits. *Pak. J. Pharm. Sci.* 25: 141–147.

Jawla, S., Kumar, Y., and Khan, M.S.A.Y. 2013. Isolation of antidiabetic principle from *Bougainvillea spectabilis* Willd. (Nyctaginaceae) stem bark. *Trop. J. Pharm. Res.* 12: 761–765.

Jawla, S., Kumar, Y., and Khan, M.S.Y. 2011. Hypoglycemic potential of *Bougainvillea spectabilis* rootbark in normal and alloxan-induced diabetic rats. *Pharmacologyonline* 3: 73–87.

Jawla, S., Kumar, Y., and Khan, M.S.Y. 2012. Antimicrobial and anti-hyperglycemic activities of *Musa paradisiaca* flowers. *Asian Pac. J. Trop. Biomed.* S914–S918.

Jay, M.A. and Ren, J. 2007. Peroxisome proliferator-activated receptor (PPAR) in metabolic syndrome and type 2 diabetes mellitus. *Curr. Diabetes Rev.* 3: 33–39.

Jaya, A., Shanthi, P., and Sachdanandam, P. 2010. Hypoglycemic effect of *Semecarpus anacardium* in streptozotocin induced diabetic rats. *Int. J. Pharmacol.* 6: 435–443.

Jaya, A., Shanthi, P., and Sachdanandam, P. 2011. *Semecarpus anacardium* (bhallataka) alters the glucose metabolism and energy production in diabetic rats. *Evid. Based Complement. Alternat. Med.* 2011: Article ID 142978, 9 pages.

Jaya Preethi, P. 2013. Herbal medicine for diabetes mellitus: A review. *Int. J. Phytopharm.* 3: 1–22.

Jaya Preethi, P., Bindu Sree, K., Pavan kumar, K., Rajavelu, R., and Sivakumar, T. 2012. Antidiabetic and antioxidant activity of latex of *Calotropis procera. Int. J. Adv. Pharm. Res.* 3: 975–980.

Jayakar, B., Rajkapoor, B., and Suresh, B. 2004. Effect of *Caralluma attenuata* in normal and alloxan induced diabetic rats. *J. Herb. Phamacother.* 4: 35–40.

Jayakar, B. and Suresh, B. 2003. Antihyperglycemic and hypoglycemic effect of *Aporosa lindleyana* in normal and alloxan induced diabetic rats. *J. Ethnopharmacol.* 84: 247–249.

Jayanthi, M., Sowbala, N., Rajalakshmi, R., Kanagavalli, U., and Sivakumar, V. 2010. Study of anti-hyperglycemic effect of *Catharanthus roseus* in alloxan-induced diabetic rats. *Int. J. Pharm. Pharm. Sci.* 2: 114–116.

Jayasooriya, A.P., Sakono, M., Yukizaki, C., Kawano, M., Yamamoto, K., and Fukuda, N. 2000. Effects of *Momordica charantia* powder on serum glucose levels and various lipid parameters in rats fed with cholesterol free and cholesterol enriched diets. *J. Ethnopharmacol.* 72: 331–336.

Jelastin, K.S.M., Tresina, P.S., and Mohan, V.R. 2012. Antioxidant, antihyperlipidaemic and antidiabetic activity of *Eugenia floccosa* Bedd leaves in alloxan induced diabetic rats. *J. Basic Clin. Pharm.* 3: 235–240.

Jelodar, G., Mohsen, M., and Shahram, S. 2007. Effect of walnut leaf, coriander, and pomegranate on blood glucose and histo-pathology of pancreas of alloxan diabetic rats. *Afr. J. Tradit. CAM* 43: 299–305.

Jenkins, D.J., Kendall, C.W., Josse, A.R., Salvatore, S., Brighenti, F., Augustin, L.S. et al. 2006. Almonds decrease postprandial glycemia, insulinemia, and oxidative damage in healthy individuals. *J. Nutr.* 136: 2987–2992.

Jeong-Lan, K., Cho-Rong, B., and Youn-Soo, C. 2010. *Helianthus tuberosus* extract has anti-diabetes effects in HIT-T15 cells. *J. Korean Soc. Food Sci. Nutr.* 39: 31–35.

Jessen, N. and Goodyear, L.J. 2005. Contraction signaling to glucose transport in skeletal muscle. *J. Appl. Physiol.* 99: 330–337.

Jeyanthi, K.A. 2012. Beneficial effect of *Cassia fistula* (L.) flower extract on antioxidant defense in streptozotocin induced diabetic rats. *Int. J. Pharm. Pharm. Sci.* 4: 274–279.

Ji, Y., Chen, S., Zhang, K., and Wang, W. 2002. Effects of *Hovenia dulcis* Thunb on blood sugar and hepatic glycogen in diabetic mice. *Zhong Yao Cai* 25: 190–191.

Jia, Q., Liu, X., Wu, X., Wang, R., Hu, X., Li, Y. et al. 2009. Hypoglycemic activity of a polyphenolic oligomer-rich extract of *Cinnamomum parthenoxylon* bark in normal and streptozotocin-induced diabetic rats. *Phytomedicine* 16: 744–750.

Jian, X.U. and Ming-Hui, Z. 2009. Molecular insights and therapeutic targets for diabetic endothelial dysfunction. *Circulation* 120: 1266–1286.

Jianfang, F., Yuan, L., Li, W., Gao, B., Zhang, N., and Ji, Q. 2009. Paeoniflorin prevents diabetic nephropathy in rats. *Comp. Med.* 59: 557–566.

Jiang, S., Jiang, Y., Guan, Y.-F., Tu, P.-F., Wang, K.-Y., and Chen, J.-M. 2011. Effects of 95% ethanol extract of *Aquilaria sinensis* leaves on hyperglycemia in diabetic db/db mice. *J. Chin. Pharm. Sci.* 20: 609–614.

Jiang, W.L., Zhang, S.P., Hou, J., and Zhu, H.B. 2012. Effect of loganin on experimental diabetic nephropathy. *Phytomedicine* 19: 217–222.

Jiangsu New Medical College. 1975. *Dictionary of Chinese Material Medica*. Shanghai, China: Shanghai Scientific and Technical Publishers, pp. 2705–2706.

Jiang, C.-S., Liang, L., and Guo, Y. 2012. Natural products possessing protein tyrosine phosphatase 1B (PTP1B) inhibitory activity found in the last decades. *Acta Pharmacol. Sin.* 33: 1217–1245.

Jilka, C., Stifler, B., Fortner, G.W., Hay, E.F., and Takemoto, D.J. 1983. *In vivo* antitumor activity of the bitter melon (*Mommordia charantia*). *Cancer Res.* 43: 5151–5155.

Jimam, N.S., Wannang, N.N., Omale, S., and Gotom, B. 2010. Evaluation of the hypoglycemic activity of *Cucumis metuliferus* (Cucurbitaceae) fruit pulp extract in normoglycemic and alloxan-induced hyperglycemic rats. *J. Young Pharm.* 2: 394–397.

Jimenez, I., Jimenez, J., Gamez, M.J., Gonzalez, M., De Medina, F.S., Zarzuela, A. et al. 1995. Effects of *Salvia lavandulifolia* Vahl. ssp. *oxyodon* extract on pancreatic endocrine tissue in streptozotocindiabetic rats. *Phytother. Res.* 9: 536–537.

Jimenez-Escrig, A., Rincon, M., Pulido, R., and Saura-Calixto, F. 2001. Guava fruit (*Psidium guajava* L.) as a new source of antioxidant dietary fiber. *J. Agric. Food Chem.* 49: 5489–5493.

Jimenez-Estrada, M., Merino-Aguilar, H., Lopez-Fernandez, A., Rojano-Vilchis, N.A., Roman-Ramos, R., and Alarcon-Aguilar, F.J. 2011. Chemical characterization and evaluation of the hypoglycemic effect of fructo-oligosaccharides from *Psacalium decompositum*. *J. Complement. Integr. Med.* 8: 1–10.

Jimoh, A., Tanko, Y., and Mohammed, A. 2013. Antidiabetic effect of methanolic extract of *Cissus cornifolia* on alloxan-induced hyperglycemic Wister rats. *Ann. Biol. Res.* 4: 46–54.

Jin, C.B., Hamid, M., Ismail, A., and Pei, C.P. 2012. The potential of cocoa extract as a hypoglycemic agent. *J. Trop. Med. Plants* 13: 193–197.

Jing, L., Cui, G., and Feng, Q., and Xiao, Y. 2009. Evaluation of hypoglycemic activity of polysaccharides extracted from *Lycium barbarum*. *Arf. J. Tradit. Complement. Alter. Med.* 6: 579–584.

Joglegar, G.V., Chaudhary, N.Y., and Aimam, R. 1959. Studies on the hypoglycaemic effects of *Casearia esculenta*. *Indian J. Physiol. Pharmacol.* 3: 76.

Johnson, S.A. and Arjmandi, B.H. 2013. Evidence for anti-cancer properties of blueberries: A mini-review. *Anticancer Agents Med. Chem.* 13: 1142–1148.

Joost, H.G. and Thorens, B. 2001. The extended GLUT-family of sugar/polyol transport facilitators: Nomenclature, sequence characteristics, and potential function of its novel members. *Mol. Membr. Biol.* 18: 247–256.

Jorge, A.P., Horst, H., de Sousa, E., Pizzolatti, M.G., and Silva, F.R. 2004. Insulinomimetic effects of kaempferitrin on glycaemia and on ^{14}C-glucose uptake in rat soleus muscle. *Chem. Biol. Interact.* 149: 89–96.

Jose, B. and Reddy, L.J. 2010. Analysis of the essential oils of the stems, leaves and rhizomes of the medicinal plant *Costus pictus* from southern India. *Int. J. Pharm. Pharm. Sci.* 2(Suppl. 2): 100–101.

Jose, M.A., Abraham, A., and Narmadha, M.P. 2011. Effect of silymarin in diabetes mellitus patients with liver disease. *J. Pharmacol. Pharmacother.* 2: 287–289.

Joseph, B. and Jini, D. 2011. Insights into the hypoglycemic effect of traditional Indian herbs used in the treatment of diabetes. *Res. J. Med. Plant* 5: 352–376.

Joseph, B. and Jini, D. 2013. Anti-diabetic effects of *Momordica charantia* (bitter melon) and its medicinal potency. *Asian Pac. J. Trop. Dis.* 3: 93–102.

Joshi, B.N., Munot, H., Hardikar, M., and Kulkarni, A.A. 2013. Orally active hypoglycemic protein from *Costus igneus* N. E. Br.: An *in vitro* and *in vivo* study. *Biochem. Biophys. Res. Commun.* 436: 278–282.

Joshi, N.C., Murugananthan, G., Thabah, P., and Nandakumar, K. 2008. Hypoglycemic and antidiabetic activity of *Tephrosia purpurea* (Linn.) root extracts. *Pharmacologyonline* 3: 926–933.

Joshi, R.K., Patil, P.A., Mujawar, M.H.K., Kumar, D., and Kholkute, S.D. 2009. Hypoglycaemic activity of *Bambusa arundinacea* leaf aqueous extract in euglycemic and hyperglycaemic Wistar rats. *Pharmacologyonline* 3: 789–795.

Jothivel, N., Ponnusamy, S.P., Appachi, M., Singaravel, S., Rasilingam, D., Deivasigamani, K. et al. 2007. Anti-diabetic activity of methanol leaf extract of *Costus pictus* D.Don in alloxan-induced diabetic rats. *J. Health Sci.* 53: 655–663.

Jouad, H., Eddouks, M., Lacaille-Dubois, M.A., and Lyoussi, B. 2000. Hypoglycaemic effect of *Spergularia purpurea* in normal and streptozotocin-induced diabetic rats. *J. Ethnopharmacol.* 71: 169–177.

Jouad, H., Maghrani, M., and Eddouks, M. 2002a. Hypoglycemic effect of *Rubus fructicosis* L. and *Globularia alypum* L. in normal and streptozotocin induced diabetic rats. *J. Ethnopharmacol.* 81: 351–356.

Jouad, H., Maghrani, M., and Eddouks, M. 2002b. Hypoglycemic effect of aqueous extract of *Ammi visnaga* in normal and streptozotocin-induced diabetic rats. *J. Herb. Pharmacother.* 2: 19–29.

Joy, K. and Kuttan, R. 1999. Anti-diabetic activity of *Picrorrhiza kurrooa* extract. *J. Ethnopharmacol.* 67: 143–148.

Joy, V., Peter, M.P.J., Raj, J.Y., and Ramesh. 2012. Medicinal values of avaram (*Cassia auriculata* L.): A review. *Int. J. Curr. Pharm. Res.* 4: 1–3.

Juarez-Rojop, I.E., Diaz-Zagoya, J.C., Ble-Castillo, J.L., Miranda-Osorio, P.H., Castell-Rodriguez, A.E., Tovilla-Zarate, C.A. et al. 2012. Hypoglycemic effect of *Carica papaya* leaves in streptozotocin-induced diabetic rats. *BMC Complement. Altern. Med.* 12: 236–240.

Judy, W.V., Hari, S.P., Stogsdill, W.W., Judy, J.S., Naguib, Y.M.A., and Passwater, R. 2003. Antidiabetic activity of a standardized extract (Glucosol) from *Lagerstroemia speciosa* leaves in type 2 diabetes: A dose-dependence study. *J. Ethnopharmacol.* 87: 115–117.

Jung, C.H., Zhou, S., Ding, G.X., Kim, J.H., Hong, M.H., Shin, Y.C. et al. 2006a. Antihyperglycemic activity of herb extracts on streptozotocin-induced diabetic rats. *Biosci. Biotechnol. Biochem.* 70: 2556–2559.

Jung, H.A., Karki, S., Ehom, N.-Y., Yoon, M.-H., Kim, E.J., and Choi, J.S. 2014. Anti-diabetic and anti-inflammatory effects of green and red kohlrabi cultivars (*Brassica oleracea* var. gongylodes). *Prev. Nutr. Food Sci.* 19: 281–290.

Jung, H.A., Kim, J.E., Chung, H.Y., and Choi, J.S. 2003. Anti-oxidant principles of *Nelumbo nucifera* stamens. *Arch. Pharm. Res.* 26: 279–285.

Jung, H.A., Yoon, N.Y., Bae, H.J., Min, B.-S., and Choi, J.S. 2008. Inhibitory activities of the alkaloids from Coptidis Rhizoma against aldose reductase. *Arch. Pharm. Res.* 31: 1405–1412.

Jung, M., Park, M., Lee, H.C., Kang, Y.H., Kang, E.S., and Kim, S.K. 2006b. Antidiabetic agent from medicinal plants. *Curr. Med. Chem.* 13: 1203–1218.

Jung, S.H., Ha, Y.J., Shim, E.K., Choi, S.Y., Jin, J.L., Yun-Choi, H.S. et al. 2007. Insulin-mimetic and insulin-sensitizing activities of a pentacyclic triterpinoid insulin receptor activator. *Biochem. J.* 403: 243–250.

Jung, U.J., Lee, M.K., Park, Y.B., Kang, M.A., and Choi, M.S. 2006c. Effects of citrus flavonoids on lipid metabolizing and glucose regulating enzymes mRNA levels in type 2 diabetic mice. *Int. J. Biochem. Cell Biol.* 38: 1134–1145.

Jurbe, G. 2011. Glycosides fraction extracted from fruit pulp of *Cucumis metuliferus* E. Meyer has anti-hyperglycemic effect in rats with alloxan-induced diabetes. *J. Nat. Pharm.* 2: 48–56.

Jyoti, B.S. 2013. *Chelidonium majus* L.—A review on pharmacological activities and clinical effects. *Global J. Res. Med. Plants Indigen. Med.* 2: 238–245.

Kabeer, F.A. and Prathapan, R. 2014. Phytopharmacological profile of *Elephantopus scaber*. *Pharmacologia* 5: 272–285.

Kabir, A.U., Samad, M.B., D'Costa, N.M., Akhter, F., Ahmed, A., and Hannan, J.M.A. 2014. Anti-hyperglycemic activity of *Centella asiatica* is partly mediated by carbohydrase inhibition and glucose-fiber binding. *BMC Complement. Altern. Med.* 14: 31.

Kabra, M.P., Rachhadiya, R.M., Desai, N.V., and Sharma, S. 2012. Hypoglycemic effect of *Cocos nucifera* flower alcoholic extract and oil in normal and alloxanised hyperglycemic rats. *Int. J. Pharm. Phytopharmacol. Res.* 2: 29–31.

Kade, I.J., Barbosa, N.B.V., Ibukun, E.O., Igbakinc, A.P., Nogueirab, C.W., and Rochab, J.B.T. 2010. Aqueous extracts of *Sphagneticola trilobata* attenuates streptozotocin-induced hyperglycaemia in rat models by modulating oxidative stress parameters. *Biol. Med.* 2: 1–13.

Kaeidi, A., Esmaeili-Mahani, S., Sheibani, V., Abbasnejad, M., Rasoulian, B., Hajializadeh, Z. et al. 2011. Olive (*Olea europaea* L.) leaf extract attenuates early diabetic neuropathic pain through prevention of high glucose-induced apoptosis: *In vitro* and *in vivo* studies. *J. Ethnopharmacol.* 136: 188–196.

Kahn, S.E., Cooper, M.E., and Prato, S.D. 2014. Pathophysiology and treatment of type 2 diabetes: Perspectives on the past, present and future. *Lancet* 383: 1068–1083.

Kaithwas, G. and Majumdar, D.K. 2012. *In vitro* antioxidant and *in vivo* antidiabetic, antihyperlipidemic activity of linseed oil against streptozotocin-induced toxicity in albino rats. *Eur. J. Lipid Sci. Technol.* 144: 1237–1245.

Kako, M., Miura, T., Nishiyama, Y., Ichimaru, M., Moriyasu, M., and Kato, A. 1996. Hypoglycemic effect of the rhizomes of *Polygala senega* in normal and diabetic mice and its main component, the triterpenoid glycoside senegin-II. *Planta Med.* 62: 440–443.

Kako, M., Miura, T., Nishiyama, Y., Ichimaru, M., Moriyasu, M., and Kato, A. 1997. Hypoglycemic activity of some triterpenoid glycosides. *J. Nat. Prod.* 60: 604–605.

Kako, M., Miura, T., Usami, M., Nishiyama, Y., Ichimaru, M., Moriyasu, M., and Kato, A. 1995. Effect of senegin-II on blood glucose in normal and NIDDM mice. *Biol. Pharm. Bull.* 18: 1159–1161.

Kalailingam, P., Sekar, A.D., Samuel, J.S., Gandhirajan, P., Govindaraju, Y., Kesavan, M. et al. 2011. The efficacy of *Costus igneus* rhizome on carbohydrate metabolic, hepatoproductive and antioxidative enzymes in steptozotocin-induced diabetic rats. *J. Health Sci.* 57: 37–46.

Kalpana, K. and Pugalendi, K.V. 2011. Antioxidative and hypolipidemic efficacy of alcoholic seed extract of *Swietenia macrophylla* in streptozotocin diabetic rats. *J. Basic Clin. Physiol. Pharmacol.* 22: 11–21.

Kamboj, V.K. 2000. Herbal medicine. *Curr. Sci.* 75: 35–39.

Kamalakannan, K. and Balakrishnan, V. 2014. Hypoglycemic activity of aqueous leaf extract of *Limonia elephantum* in alloxan-induced diabetic rats. *Bangladesh J. Pharmacol.* 9: 383–388.

Kamalakannan, N. and Prince, P.S.M. 2006. Antihyperglycaemic and antioxidant effect of rutin, a polyphenolic flavonoid, in streptozotocin-induced diabetic Wistar rats. *Basic Clin. Pharmacol. Toxicol.* 98: 97–103.

Kamalakkanan, N., Rajadurai, M., and Prince, P.S. 2003. Effect of *Aegle marmelos* fruits on normal and streptozotocin-diabetic Wistar rats. *J. Med. Food* 6: 93–98.

Kamanyi, A., Njamen, D., and Nkeh, B. 2006. Hypoglycaemic properties of the aqueous root extract of *Morinda lucida* (Benth) (Rubiaceae). Studies in the mouse. *Phytother. Res.* 8: 369–371.

Kamble, S.M., Kamlakar, P.L., Vaidya, S., and Bambole, V.D. 1998. Influence of *Coccinia indica* on certain enzymes in glycolytic and lipolytic pathway in human diabetes. *Indian J. Med. Sci.* 52: 143–146.

Kamboj, J., Sharma, S., and Kumar, S. 2011. *In vivo* anti-diabetic and anti-oxidant potential of *Psoralea corylifolia* seeds in streptozotocin-induced type 2 diabetic rats. *J. Health Sci.* 57: 225–235.

Kambouche, N., Merah, B., Derdour, A., Bellahouel, S., Bouayed, J., Dicko, A. et al. 2009. Hypoglycemic and antihyperglycemic effects of *Anabasis articulata* (Forssk) Moq (Chenopodiaceae), an Algerian medicinal plant. *Afr. J. Biotechnol.* 820: 5589–5594.

Kamel, M.S., Ohtani, K., Kurokawa, T., Assaf, M.H., El-Shanawany, M.A., Ali, A.A. et al. 1991. Studies on *Balanites aegyptiaca* fruits, an antidiabetic Egyptian folk medicine. *Chem. Pharm. Bull.* 39: 1229–1233.

Kameswararao, B., Giri, R., Kesavulu, M.M., and Apparao, C. 2001. Effect of oral administration of bark extracts of *Pterocarpus santalinus* L. on blood glucose level in experimental animals. *J. Ethnopharmacol.* 74: 69–74.

Kameswararao, B., Kesavulu, M.M., and Apparao, C. 2003. Evaluation of anti-diabetic effect of *Momordica cymbalaria* fruit in alloxan-diabetic rats. *Fitoterapia* 74: 7–13.

Kamgang, R., Mboumi, R.Y., Fondjo, A.F., Tagne, M.A., Ndille, G.P., and Yonkeu, J.N. 2008. Antihyperglycaemic potential of the water-ethanol extract of *Kalanchoe crenata* (Crassulaceae). *J. Nat. Med.* 62: 34–40.

Kamon, C., Weerapan, K., and Supeecha, W. 2009. Pharmacokinetic and the effect of capsaicin in *Capsicum frutescens* on decreasing plasma glucose level. *J. Med. Assoc. Thai.* 92: 108–113.

Kamtchouing, P., Kahpui, S.M., Dzeufiet, P.D., Tedong, L., Asongalem, E.A., and Demo, T. 2005. Anti-diabetic activity of methanol/methylene chloride stem bark extracts of *Terminalia superba* and *Canarium schweinfurthii* on streptozotocin-induced diabetic rats. *J. Ethnopharmacol.* 104: 306–309.

Kamtchouing, P., Sokeng, S.D., Moundipa, P.F., Watcho, P., Jatsa, H.B., and Lontsi, D. 1998. Protective role of *Anacardium occidentale* extract against streptozotocin-induced diabetes in rats. *J. Ethnopharmacol.* 62: 95–99.

Kan, C.G., Ko, S.K., Sung, J.H., and Sung, S.H. 2007. Compound K enhances insulin secretion with beneficial metabolic effects in db/db mice. *J. Agric. Food. Chem.* 55: 10641–10648.

Kanaya, Y., Doi, T., Sasaki, H., Fujita, A., Matsuno, S., Okamoto, K. et al. 2004. Rice bran extract prevents the elevation of plasma peroxylipid in KKAy diabetic mice. *Diabetes Res. Clin. Pract.* 66: 157–160.

Kanetkar, P., Singhal, R., and Kamat, M. 2007. *Gymnema sylvestre*: A memoir. *J. Clin. Biochem. Nutr.* 41: 77–81.

Kang, J., Lee, J., Kwon, D., and Song, Y. 2013. Effect of *Opuntia humifusa* supplementation and acute exercise on insulin sensitivity and associations with PPAR-γ and PGC-1α protein expression in skeletal muscle of rats. *Int. J. Mol. Sci.* 14: 7140–7154.

Kang, T.H., Moon, E., Hong, B.N., Choi, S.Z., Son, M., Park, J.-H., and Kim, S.Y. 2011. Diosgenin from *Dioscorea nipponica* ameliorates diabetic neuropathy by inducing nerve growth factor. *Biol. Pharm. Bull.* 34: 1493–1498.

Kang, Y. and Kim, H.Y. 2004. Glucose uptake-stimulatory activity of Amomi Semen in 3T3-L1 adipocytes. *J. Ethnopharmacol.* 92: 103–105.

Kanthalal, S.K., Suresh, V., Arunachalam, G., Royal, F.P., and Kameshwaran, S. 2012. Anti-obesity and hypolipidemic activity of methanol extract of *Tabernaemontana divaricata* on atherogenic diet induced obesity in rats. *Int. Res. J. Pharm.* 3: 157–161.

Kanthlal, S.K., Kumar, B.A., Joseph, J., Aravind, R., and Frank, P.R. 2014. Amelioration of oxidative stress by *Tabernamontana divaricata*on in alloxan-induced diabetic rats. *Anc. Sci. Life* 33: 222–228.

Kapoor, R., Srivastava, S., and Kakkar, P. 2009. *Bacopa monnieri* modulates antioxidant responses in brain and kidney of diabetic rats. *Environ. Toxicol. Pharmacol.* 27: 62–69.

Kappel, V.D., Pereire, D.F., Cazarolli, L.H., Guesser, S.M., da Silva, C.H.B., Schenkel, E.P. et al. 2012. Short and long term effects of *Baccharis articulata* on glucose homeostasis. *Molecules* 17: 6754–6768.

Kar, A., Choudhary, B.K., and Bandyopadhyay, N.G. 2003. Comparative evaluation of hypoglycemic activity of some Indian medicinal plants in alloxan diabetic rats. *J. Ethnopharmacol.* 84: 105–108.

Karan, S.K., Mondal, A., Mishra, S.K., Pal, D., and Rout, K.K. 2013. Antidiabetic effect of *Streblus asper* in streptozotocin-induced diabetic rats. *Pharm. Biol.* 51: 369–375.

Karawya, M.S., Abde, S.M., EI-Olemy, M.M., and Farrag, N.M. 1984. Diphenylamine, an antihyperglycemic agent from onion and tea. *J. Nat. Prod.* 47: 775–780.

Karthikeyan, K., Ravichandran, P., and Govindasamy, S. 1999. Chemopreventive effect of *Ocimum sanctum* on DMBA-induced hamster buccal pouch carcinogenesis. *Oral Oncol.* 35: 112–119.

Karuna, R., Bharathi, V.G., Reddy, S.S., Ramesh, B., and Saralakumari, D. 2011. Protective effects of *Phyllanthus amarus* aqueous extract against renal oxidative stress in Streptozotocin-induced diabetic rats. *Indian J. Pharmacol.* 43: 414–418.

Karunanayake, E.H., Jeevathayaparan, S., and Tennekoon, K.H. 1990. Effect of *Momordica charantia* fruit juice on streptozotocin-induced diabetes in rats. *J. Ethnopharmacol.* 30: 199–204.

Karuppusamy, A. and Thangaraj, P. 2013. Anti-diabetic of *Ficus amplissima* Smith bark extract in streptozotocin-induced diabetic rats. *J. Ethnopharmacol.* 146: 302–310.

Kasabri, V., Abu-Dahab, R., Afifi, F.U., Naffa, R., and Majdalawi, L. 2012a. Modulation of pancreatic MIN6 insulin secretion and proliferation and extrapancreatic glucose absorption with *Achillea santolina*, *Eryngium creticum* and *Pistacia atlantica* extracts: *In vitro* evaluation. *J. Exp. Integr. Med.* 2: 245–254.

Kasabri, V., Abu-Dahab, R., Afifi, F.U., Naffa, R., Majdalawi, L., and Shawash, H. 2012b. *In vitro* modulation of pancreatic MIN6 insulin secretion and proliferation and extrapancreatic glucose absorption by *Paronychia argentea, Rheum ribes* and *Teucrium polium* extracts. *Jordan J. Pharm. Sci.* 5: 203–219.

Kasabri, V., Afifi, F.U., and Hamdan, I. 2011. Evaluation of the acute antihyperglycemic effects of four selected indigenous plants from Jordan used in traditional medicine. *Pharm. Biol.* 49: 687–695.

Kasabri, V., Flatt, P.R., and Abdel-Wahab, Y.H.A. 2010. *Terminalia bellirica* stimulates the secretion and action of insulin and inhibits starch digestion and protein glycation in vitro. *Br. J. Nutr.* 103: 212–217.

Kashiwada, Y., Aoshima, A., Ikeshiro, Y., Chen, Y.P., Furukawa, H., Itoigawa, M. et al. 2005. Anti-HIV benzylisoquinoline alkaloids and flavonoids from the leaves of *Nelumbo nucifera* and structure activity correlations with related alkaloids. *Bioorg. Med. Chem.* 13: 443–448.

Kaskoos, R.A. 2013. *In vitro* α-glucosidase inhibition and anti-oxidant activity of methanolic extract of *Centaurea calcitrapa* from Iraq. *Am. J. Essential Oils Nat. Prod.* 1: 122–125.

Katare, C., Saxena, S., Agrawal, S., and Prasad, G.B.K.S. 2013. Alleviation of diabetes-induced dyslipidemia by *Lageraria siceraria* fruit extract in human type 2 diabetes. *J. Herbal Med.* 3: 1–8.

Kato, A. and Miura, T. 1994. Hypoglycemic action of the rhizome of *Polygonatum officinale* in normal and diabetic mice. *Planta Med.* 60: 201–203.

Kaur, A.K., Jain, S.K., Gupta, A., Gupta, S.K., Bansal, M., and Sharma, P.K. 2011. Anti-diabetic activity of *Bauhinia tomentosa* L. roots extract in alloxan-induced diabetic rats. *Der Pharmacia Lett.* 3: 456–459.

Kaur, D. and Sharma, R. 2012. An update on pharmacological properties of cumin. *Int. J. Res. Pharm. Sci.* 2: 14–24.

Kaur, I., Ahmad, S., and Harikumar, S.L. 2014. Pharmacognosy, phytochemistry and pharmacology of *Cassia occidentalis* Linn. *Int. J. Pharmacogn. Phytochem. Res.* 6: 151–155.

Kaur, N., Rashmi, Tripathy, Y.C., and Upadhyay, L. 2012. Phytochemical and antidiabetic evaluation of *Phaseolus trilobus* roots. *J. Pharm. Res.* 5: 5202–5205.

Kaushik, P., Khokra, S.L., Rana, A.C., and Kaushik, D. 2014. Pharmacophore modeling and molecular docking studies on *Pinus roxburghii* as a target for diabetes mellitus. *Adv. Bioinformatics* 2014: Article ID 903246, 8 pages.

Kavalal, G., Tuncel, H., Goksel, S., and Hatemi, H.H. 2003. Hypoglycemic activity of *Urtica pilulifera* in streptozotocin-diabetic rats. *J. Ethnopharmacol.* 84: 241–245.

Kavimani, S. and Manisenthlkumar, K.T. 2000. Effect of methanol extract of *Enicostemma littorale* on Dalton's ascetic lymphoma. *J. Ethnopharmacol.* 71: 349–352.

Kavipriya, S., Tamilselvan, N., Thirumalai, T., and Arumugam, G. 2013. Anti-diabetic effect of methanolic leaf extract of *Pongamia pinnata* on streptozotocin induced diabetic rats. *J. Coastal Life Med.* 1: 113–117.

Kavishankar, G.B., Lakshmidevi, L., Mahadeva Murthy, S., Prakash, H.S., and Niranjana, S.R. 2011. Diabetes and medicinal plants—A review. *Int. J. Pharm. Biomed. Sci.* 2: 65–80.

Kavitha, L., Kumaravel, B., Sriram, P.G., and Subramanian, S. 2013. Beneficial role of *Areca catechu* nut extract in alloxan-induced diabetic rats. *Res. J. Pharmacogn. Phytochem.* 5: 100–108.

Kavoosi, G. 2011. *Zataria multiflora* essential oil reduces diabetic damages in streptozotocin-induced diabetic rats. *Afr. J. Biotechnol.* 10: 1763–1767.

Kawabata, J., Mizuhata, K., Sato, E., Nishioka, T., Aoyama, Y., and Kasai, T. 2003. 6-Hydroxyflavonoids as α-glucosidase inhibitors from marjoram (*Origanum majorana*) leaves. *Biosci. Biotechnol. Biochem.* 67: 445–447.

Kawano, A., Nakamura, H., Hata, S., Minakawa, M., Miura, Y., and Yagasaki, K. 2009. Hypoglycemic effect of aspalathin, a rooibos tea component from *Aspalathus linearis*, in type 2 diabetic model db/db mice. *Phytomedicine* 16: 437–443.

Kazeem, M.I., Adamson, J.O., and Ogunwande, I.A. 2013a. Modes of Inhibition of α-amylase and α-glucosidase by aqueous extract of *Morinda lucida* Benth Leaf. *Biomed. Res. Int.* 2013: Article ID 527570, 6 pages.

Kazeem, M.I., Oyedapa, B.F., Raimi, O.G., and Adu, O.B. 2013b. Evaluation of *Ficus exasperata* Vahl. leaf extract in the management of diabetes mellitus *in vitro*. *J. Med. Sci.* 13: 269–275.

Kedari, T.S. and Khan, A.A. 2014. Guyabano (*Annona muricata*): A review of its traditional uses. Phytochemistry and pharmacology. *Am. J. Res. Commun.* 2: 247–268.

Keenan, H., Sun, J.K., Levine, J., Doria, A., Aiello, L.P., Eisenbarth, G. et al. 2010. Residual insulin production and pancreatic β-cell turnover after 50 years of diabetes: Joslin Medalist study. *Diabetes* 59: 2846–2853.

Keller, A.C., Ma, J., Kavalier, A., He, K., Brillantes, A.M., and Kennelly, E.J. 2011. Saponins from the traditional medicinal plant *Momordica charantia* stimulate insulin secretion in vitro. *Phytomedicine* 19: 32–37.

Keller, A.C., Vandebroek, I., Liu, Y., Balick, M.J., Kronenberg, F., Kennelly, E.J. et al. 2009. *Costus pictus* tea failed to improve diabetic progression in C57BLKS/J db/db mice, a model of type 2 diabetes mellitus. *J. Ethnopharmacol.* 121: 248–254.

Kelly, M.A., Mijovic, C.H., and Barnett, A.H. 2001. Genetics of type 1 diabetes. *Best Pract. Res. Clin. Endocrinol. Metab.* 15: 279–291.

Kelm, M.A., Nair, M.G., Strasburg, G.M., and DeWitt, D.L. 2000. Anti-oxidant and cyclooxygenase inhibitory phenolic compounds from *Ocimum sanctum* Linn. *Phytomedicine* 7: 7–13.

Kesari, A.N., Gupta, R.K., and Watal, G. 2005. Hypoglycemic effect of *Murraya koenigii* on normal and alloxan diabetic rats. *J. Ethnopharmacol.* 97: 247–251.

Khalaf, G. and Mohamed, A. 2008. Effect of barley (*Hordeum vulgare*) on the liver of diabetic rats: Histological and biochemical study. *Egypt. J. Histol.* 31: 245–255.

Khalil, O.A., Ramadan, K.S., Danial, E.N., Alnahdi, H.S., and Ayaz, N.O. 2012. Antidiabetic activity of *Rosmarinus officinalis* and its relationship with the antioxidant property. *Afr. J. Pharm. Pharmacol.* 6: 1031–1036.

Khan, A.A. 2005. An overview of diabetes. In: *Modern and Alternative Medicine for Diabetes*, Khan, I.A., Khanum, A., and Khan, A.A. (eds.). Hyderabad, India: Ukaaz Publications, pp. 1–42.

Khan, A.K., Akhtar, S., and Mahtab, H. 1979. *Coccinia indica* in the treatment of patients with diabetes mellitus. *Bangladesh Med. Res. Counc. Bull.* 5: 60–66.

Khan, A.K., Akhtar, S., and Mahtab, H. 1980. Treatment of diabetes mellitus with *Coccinia indica*. *Br. Med. J.* 280: 1044–1048.

Khan, B.A., Abraham, A., and Leelamma, S. 1995. Hypoglycemic action of *Murraya koenigii* (curry leaf) and *Brassica junea* (mustard): Mechanism of action. *Indian J. Biochem. Biophys.* 32: 106–108.

Khan, B.A., Abraham, A., and Leelamma, S. 1996. *Murraya koenigii* and *Brassica junea*—Alterations on lipid profile in 1–2 dimethyl hydrazine induced colon carcinogenesis. *Invest. New Drugs* 14: 365–369.

Khan, C.R. and White, M.F. 1998. The insulin receptor and the molecular mechanism of insulin action. *J. Clin. Invest.* 82: 1151–1156.

Khan, H.B., Vinayagam, K.S., Palanivelu, S., and Panchanadham, S. 2012b. Anti-diabetic effect of *Semecarpus anacardium* Linn. nut milk extract in a high fat diet streptozotocin-induced type 2 diabetic rat model. *Comp. Clin. Pathol.* 21: 1395–1399.

Khan, H.B., Vinayagam, K.S., Renny, C.M., Palanivelu, S., and Panchanadham, S. 2013a. Potential antidiabetic effect of the *Semecarpus anacardium* in a type 2 diabetic rat model. *Inflammopharmacology* 21: 47–53.

Khan, H.B., Vinayagam, K.S., Sekar, A., Palanivelu, S., and Panchanadham, S. 2012a. Antidiabetic and antioxidant effect of *Semecarpus anacardium* Linn. nut milk extract in a high-fat diet STZ-induced type 2 diabetic rat model. *J. Diet. Suppl.* 9: 19–33.

Khan, K.H. 2009. Roles of *Emblica officinalis* in medicine—A review. *Bot. Res. Int.* 2(4): 218–228.

Khan, K.Y., Khan, M.A., Ahmad, M., Hussain, I., Mazari, P., Fazal, H. et al. 2011. Hypoglycemic potential of genus *Ficus* L.: A review of ten years of plant based medicine used to cure diabetes (2000–2010). *J. Appl. Pharm. Sci.* 1: 221–227.

Khan, M., Ali, M., Ali, A., and Mir, S.R. 2014. Hypoglycemic and hypolipidemic activities of Arabic and Indian origin *Salvadora persica* root extract on diabetic rats with histopathology of their pancreas. *Int. J. Health Sci.* 8: 45–58.

Khan, M.R.I. 2010. Antidiabetic effects of the different fractions of ethanolic extracts of *Ocimum sanctum* in normal and alloxan-induced diabetic rats. *J. Sci. Res.* 2: 158–168.

Khan, M.S., Habib, M.A., Chowdhury, I., Islam, M.A., and Saha, D. 2013b. Hypoglycemic effect of *Polyalthia longifolia* on glucose tolerance in glucose-induced hyperglycemic mice and analgesic effect of *Polyalthia longifolia* in acetic acid induced writhing model mice. *Int. J. Phytopharm.* 3: 86–89.

Khan, M.Y., Panchal, S., Vyas, N., Butani, A., and Kumar, V. 2007. *Olea europaea*: A phyto-pharmacological review. *Pharmacogn. Rev.* 1: 112–116.

Khan, P.K. and Awasthy, K.S. 2003. Cytogenetic toxicity of neem. *Food Chem. Toxicol.* 41: 1325–1328.

Khan, S.E., Hull, R.L., and Utzschneider, K.M. 2006. Mechanisms linking obesity to insulin resistance and type 2 diabetes. *Nature* 444: 840–846.

Khan, V., Najmi, A.K., Akhtar, M., Aqil, M., Mujeeb, M., and Pillai, K.K. 2012b. A pharmacological appraisal of medicinal plants with antidiabetic potential. *J. Pharm. Bioallied Sci.* 4: 27–42.

Khan, Z.A., Nahar, B., Jakaria, M.A., Rahman, S., Chowdhury, M.H., and Rahmatullah, M. 2010. An evaluation of anti-hyperglycemic and antinociceptive effects of methanol extract of *Cassia fistula* L. (Fabaceae) leaves in Swiss albino mice. *Adv. Nat. Appl. Sci.* 4: 305–310.

Khanavi, M., Taheri, M., Rajabi, A., Fallah-Bonekohal, S., Baeeri, M., Mohammadirad, A. et al. 2012. Stimulation of hepatic glycogenolysis and inhibition of gluconeogenesis are the mechanisms of antidiabetic effect of *Centaurea bruguierana* ssp. *belangerana*. *Asian J. Anim. Vet. Adv.* 7: 1166–1174.

Khanna, P., Jain, S.C., Panagariya, A., and Dixit, V.P. 1981. Hypoglycemic activity of polypeptide-p from a plant source. *J. Nat. Prod.* 44: 648–655.

Khera, N. and Bhargava, S. 2013. Phytochemical and pharmacological evaluation of *Tectona grandis* Linn. *Int. J. Pharm. Pharm. Sci.* 5: 923–927.

Kho, M.C., Lee, Y.J., Ahn, Y.M., Choi, Y.H., Kim, A.Y., Kang, D.G. et al. 2013. Effects of ethanol extract of *Gastrodia elata* Blume on high-fructose induced metabolic syndrome. *FASEB J.* 27 (Meeting Abstract Supplement): 1108.3.

Kho, M.C., Lee, Y.J., Cha, J.D., Choi, K.M., Kang, D.G., and Lee, H.S. 2014. *Gastrodia elata* ameliorates high-fructose diet-induced lipid metabolism and endothelial dysfunction. *Evid. Based Complement. Alternat. Med.* 2014: Article ID 101624, 10 pages.

Khosa, R.L., Pandey, O.B., and Singh, J.P. 1983. Experimental studies on *Zizyphus rugosa* bark. *Indian Drugs* 20: 241–246.

Khosla, P., Bhanwra, S., Singh, J., Seth, S., and Srivastava, R.K. 2000. A study of hypoglycemic effects of *Azadirachta indica* (neem) in normal and alloxan diabetic rabbits. *Indian J. Physiol. Pharmacol.* 44: 69–74.

Khyade, M.S., Kasote, D.M., and Vaikos, N.P. 2014. *Alstonia scholaris* (L.) R. Br. and *Alstonia macrophylla* Wall. ex G. Don: A comparative review on traditional uses, phytochemistry and pharmacology. *J. Ethnopharmacol.* 153: 1–18.

Kianbakht, S., Abasi, B., and Dabaghian, F.H. 2013. Anti-hyperglycemic effect of *Vaccinium arctostaphylos* in type 2 diabetic patients: A randomized controlled trial. *Forsch Komplementmed.* 20: 17–22.

Kianbakht, S. and Hajiaghaee, R. 2013. Anti-hyperglycemic effects of *Vaccinium arctostaphylos* L. fruit and leaf extracts in alloxan-induced diabetic rats. *J. Med. Plants* 12: 43–50.

Kihara, H.W. 2012. Investigation of the hypoglycemic effect of *Erythrina abyssinica*. B Pharm dissertation, College of Health Sciences, University of Nairobi. http://hdl.handle.net/11295/74405.

Kim, D., Park, K., Lee, S.K., and Hwang, J. 2011. *Cornus kousa* F. Buerger ex Miquel increases glucose uptake through activation of peroxisome proliferator-activated receptor γ and insulin sensitization. *J. Ethnopharmacol.* 133: 803–809.

Kim, D.-H., Yu, K.-W., Bae, E.-A., Park, H.-J., and Choi, J.-W. 1998. Metabolism of kalopanaxsaponin B and H by human intestinal bacteria and antidiabetic activity of their metabolites. *Biol. Pharm. Bull.* 21: 360–365.

Kim, E.-K., Kwon, K.-B., Han, M.-J., Song, M.-Y., Lee, J.-H., Lv, N. et al. 2007. Inhibitory effect of *Artemisia capillaris* extract on cytokine-induced nitric oxide formation and cytotoxicity of RINm5F cells. *Int. J. Mol. Med.* 19: 535–540.

Kim, H.J., Kong, M.K., and Kim, Y.C. 2008. Beneficial effects of *Phellodendri cortex* extract on hyperglycemia and diabetic nephropathy in streptozotocin-induced diabetic rats. *BMB Rep.* 41: 710–715.

Kim, H.-L., Sim, J.-E., Choi, H.-M., Choi, I.-Y., Jeong, M.-Y., Park, J. et al. 2014. The AMPK pathway mediates an anti-adipogenic effect of fruits of *Hovenia dulcis* Thumb. *Food Funct.* 5: 2961–2968.

Kim, H.Y., Moon, B.H., Lee, H.J., and Choi, D.H. 2004. Flavonol glycosides from the leaves of *Eucommia ulmoides* with glycation inhibitory potential. *J. Ethnopharmacol.* 93: 227–230.

Kim, J. and Park, Y. 2012. Anti-diabetic effect of sorghum extract on hepatic gluconeogenesis of streptozotocin-induced diabetic rats. *Nutr. Metab.* 9: 106–113.

Kim, J.-S., Yang, J., and Kim, M.-J. 2011a. Alpha glucosidase inhibitory effect, anti-microbial activity and UPLC analysis of *Rhus verniciflua* under various extract conditions. *J. Med. Plant Res.* 5: 77–83.

Kim, K. and Kim, H.Y. 2008. Korean red ginseng stimulates insulin release from isolated rat pancreatic islets. *J. Ethnopharmacol.* 120: 190–195.

Kim, K.-H., Kim, K.-S., Shin, M.H., Jang, E.G., Kim, E.Y., Lee, J.H. et al. 2013a. Aqueous extract of *Anemarrhena asphodeloides* stimulate glucagon-like peptide-1 secretion in enteroendocrine NCI-H716 cells. *BioChip J.* 7: 188–193.

Kim, K.J., Lee, M.S., Jo, K., and Hwang, J.K. 2011b. Piperidine alkaloids from *Piper retrofractum* Vahl. protect against high-fat diet-induced obesity by regulating lipid metabolism and activating AMP-activated protein kinase. *Biochem. Biophys. Res. Commun.* 411: 219–225.

Kim, L. and Park, Y. 2012. Anti-diabetic effect of sorghum extract on hepatic gluconeogenesis of streptozotocin-induced diabetic rats. *Nutr. Metab. (Lond.)* 9: 106–112.

Kim, N., Kim, S.-H., Kim, Y.-J., Kim, J.-K., Nam, M.-K., Rhim, H. et al. 2011c. Neurotrophic activity of DA-9801, a mixture extract of *Dioscorea japonica* Thunb and *Dioscorea nipponica* Makino, *in vitro*. *J. Ethnopharmacol.* 137: 312–319.

Kim, S.H., Hyun, S.H., and Choung, S.Y. 2006. Anti-diabetic effect of cinnamon extract on blood glucose in db/db mice. *J. Ethnopharmacol.* 104: 119–123.

Kim, S.O., Yun, S.-J., and Lee, E.H. 2007. The water extract of adlay seed (*Coix lachrymajobi* var. *mayuen*) exhibits anti-obesity effects through neuroendocrine modulation. *Am. J. Chin. Med.* 35: 297. DOI: 10.1142/S0192415X07004825.

Kim, T.H., Kim, J.K., Kang, Y.H., Lee, J.Y., Kang, I.J., and Lim, S.S. 2013b. Aldose reductase inhibitory activity of compounds from *Zea mays* L. *Biomed. Res. Int.* 2013: 727143.

Kim, Y.M., Kim, S.G, Khil, L.Y., and Moon, C.K. 1995. Brazilin stimulates the glucose transport in 3T3-L1 cells. *Planta Med.* 61: 297–301.

Kimura, M., Waki, I., Chujo, T., Kikuchi, T., Hiyama, C., Yamazaki, K. et al. 1981. Effects of hypoglycemic components in ginseng radix on blood insulin levels in alloxan diabetic mice and on insulin release from perfused rat pancreas. *J. Pharmacobiodyn.* 4: 410–417.

Kingsley, O., Georgina, E.O., Esosa, U.S., and Sunday, J.J. 2010. Evaluation of hypoglycemic and antioxidant properties of *Garcinia kola* seeds in Wister rats. *Curr. Res. J. Biol. Sci.* 3: 326–329.

Kingwell, K. 2014. GIG1 goes long to tackle diabetes. *Nat. Rev. Drug Discov.* 13: 652–653.

Kirana, H. and Srinivasan, B.P. 2008. *Trichosanthes cucumerina* Linn. improves glucose tolerance and tissue glycogen in non insulin dependent diabetes mellitus induced rats. *Indian J. Pharmacol.* 40: 103–106.

Kirk-Ballard, H., Wang, Z.Q., Acharya, P., Zhang, X.H., Yu, Y., Kilroy, G. et al. 2013. An extract of *Artemisia dracunculus* L. inhibits ubiquitin-proteasome activity and preserves skeletal muscle mass in a murine model of diabetes. *PLoS One* 8: e57112. DOI: 10.1371/journal.pone.0057112.

Kirtikar, K.R. and Basu, B.D. 1975. *Indian Medicinal Plants*. Dehrudun, India: Bishen Singh Mahendra Pal Singh. (Reprint of 1935 ed.; Edited, revised, and enlarged by Blatter, E., Caius, J.F., and Mhaskar, K.S.)

Kishi, A., Morikawa, T., Matsuda, H., and Yoshikawa, M. Structures of new friedelane and norfriedelane-type triterpenes and polyacylated eudesmane-type sesquiterpene from *Salacia chinensis* Linn. (*S. prinoides* DC., Hippocrateaceae) and radical scavenging activities of principal constituents. *Chem. Pharm. Bull.* 51: 1051–1055.

Klein, G., Kim, J., Himmeldirk, K., Cao, Y., and Chen, X. 2007. Antidiabetes and anti-obesity activity of *Lagerstroemia speciosa*. *eCAM* 4: 401–407. DOI: 10.1093/ecam/nem013.

Klover, P.J. and Mooney, R.A. 2004. Hepatocytes: Critical for glucose homeostasis. *Int. J. Biochem. Cell Biol.* 36: 753–758.

Ko, B.S., Jang, J.S., Hong, S.M., Sung, S.R., Lee, J.E., Lee, M.Y. et al. 2007. Changes in components, glycyrrhizin and glycyrrhetinic acid, in raw *Glycyrrhiza uralensis* Fisch, modify insulin sensitizing and insulinotropic actions. *Biosci. Biotechnol. Biochem.* 71: 1452–1461.

Ko, B.S., Kim, H.K., and Park, S. 2002. Insulin sensitizing and insulin-like effects of water extracts from *Kalopanax pictus* Nakai in 3T3-L1 adipocyte. *Agric. Chem. Biotechnol.* 45: 42–46.

Kobayashi, K., Ishihara, T., Khono, E., Miyase, T., and Yoshizaki, F. 2006. Constituents of stem bark of *Callistemon rigidus* showing inhibitory effects on mouse alpha-amylase activity. *Biol. Pharm. Bull.* 29: 1275–1277.

Kobayashi, T., Song, Q.H., Hong, T., Kitamura, H., and Cyong, J.C. 2002. Preventative effects of the flowers of *Inula britannica* on autoimmune diabetes in C57BL/KsJ mice induced by multiple low doses of streptozotocin. *Phytother. Res.* 16: 377–382.

Koetter, U. and Biendl, M. 2010. Hops (*Humulus lupulus*): A review of its historic and medicinal uses. *HerbalGram* 87: 44–57.

Koffi, N., Ernest, A.K., Marie-Solange, T., Beugre, K., and Noel, Z.G. 2009. Effect of aqueous extract of *Chrysophyllum cainito* leaves on the glycemia of diabetic rabbits. *Afr. J. Pharm. Pharmacol.* 3: 501–506.

Koffuor, G.A., Amoateng, P., Okai, C.A., and Fiagbe, N.I.Y. 2011. Hypoglycemic effects of whole and fractionated *Azadirachta indica* (neem) seed oils on alloxan-induced diabetes in New Zealand white rabbits. *J. Ghana Sci. Assoc.* 13: 117–127.

Kohli, K.R. and Singh, R.H. 1993. A clinical trial of jambu (*Eugenia jambolana*) in non-insulin dependent diabetes mellitus. *J. Res. Ayurveda Siddha* 14: 89–97.

Kohno, H., Suzuki, R., Yasui, Y., Hosokawa, M., Miyashita, K., and Tanaka, T. 2004. Pomegranate seed oil rich in conjugated linolenic acid suppresses chemically-induced colon carcinogenesis in rats. *Cancer Sci.* 95: 481–486.

Kondeti, V.K., Badri, K.R., Maddirala, D.R., Thur, S.K., Fatima, S.S., Kasetti, R.B. et al. 2010. Effect of *Pterocarpus santalinus* bark, on blood glucose, serum lipids, plasma insulin and hepatic carbohydrate metabolic enzymes in streptozotocin-induced diabetic rats. *Food Chem. Toxicol.* 48: 1281–1287.

Koneri, R.B., Samaddar, S., and Ramaiah, C.T. 2014a. Antidiabetic activity of a triterpenoid saponin isolated from *Momordica cymbalaria* Fenzl. *Indian J. Exp. Biol.* 52: 46–52.

Koneri, R.B., Samaddar, S., Simi, S.M., and Rao, S.T. 2014b. Neuroprotective effect of a triterpenoid saponin isolated from *Momordica cymbalaria* Fenzl in diabetic peripheral neuropathy. *Indian J. Pharmacol.* 46: 76–81.

Kong, W., Wei, J., Abidi, P., Lin, M., Inaba, S., Li, C. et al. 2004. Berberine is a novel cholesterol-lowering drug working through a unique mechanism distinct from statins. *Nat. Med.* 10: 1344–1351.

Konig, G.M., Wright, A.D., Keller, W.J., Judd, R.L., Bates, S., and Day, C. 1998. Hypoglycaemic activity of an HMG-containing flavonoid glucoside, chamaemeloside, from *Chamaemelum nobile*. *Planta Med.* 64: 612–614.

Konno, C., Mizuno, T., and Hikino, H. 1985a. Isolation and hypoglycemic activity of lithospermans A, B and C, glycans of *Lithospermum erythrorhizon* roots. *Planta Med.* 51: 157–158.

Konno, C., Murayama, M., Sugiyama, K., Arai, M., Murakami, M., Takahashi, M. et al. 1985b. Antidiabetes drugs. 5. Isolation and hypoglycemic activity of aconitans A, B, C and D, glycans of *Aconitum carmichaeli* roots. *Planta Med.* 51: 160–161.

Konno, C., Suzuki, Y., Oishi, K., Munakata, E., and Hikino, H. 1985c. Isolation and hypoglycemic activity of atractans A, B and C, glycans of *Atractylodes japonica* rhizomes. *Planta Med.* 51: 102–103.

Koo, H.J., Kwak, J.H., and Kang, S.C. 2014. Anti-diabetic properties of *Daphniphyllum macropodum* fruit and its active compound. *Biosci. Biotechnol. Biochem.* 78: 1392–1401.

Koo, M.W.L. 1994. *Aloe vera*: Anti-ulcer and anti-diabetic effects. *Phytother. Res.* 8: 461–467.

Korec, R., Sensch, K.H., and Zoukas, T. 2000. Effects of the neoflavonoid coutareagenin, one of the antidiabetic active substances of *Hintonia latiflora*, on streptozotocin-induced diabetes mellitus in rats. *Arzneimittel-Forschung* 50: 189–224.

Korecova, M. and Hladikova, M. 2014. Treatment of mild and moderate type-2 diabetes: Open prospective trial with *Hintonia latiflora* extract. *Eur. J. Med. Res.* 19: 16.

Korkmaz, H. and Gurdal, A. 2002. Effect of *Artemisia santonicum* L. on blood glucose in normal and alloxan-induced diabetic rabbits. *Phytother. Res.* 16: 675–676.

Kostava, I. 2001. *Fraxinus ornus* L. *Fitoterapia* 72: 471–480.

Kotani, H., Tanabe, H., Mizukami, H., Amagaya, S., and Inoue, M. 2012. A naturally occurring rexinoid, honokiol, can serve as a regulator of various retinoid X receptor heterodimers. *Biol. Pharm. Bull.* 35: 1–9.

Koti, B.C., Biradar, S.M., Karadi, R.V., Taranalli, A.D., and Benade, V.S. 2009. Effect of *Bauhinia variegata* bark extract on blood glucose level in normal and alloxanised diabetic rats. *J. Nat. Remedies* 9: 27–34.

Koti, B.C., Maddi, V.S., Jamakandi, V.G., and Patil, P.A. 2004. Effect of crude petroleum ether extract of *Brassica oleracea* on euglycemic and alloxan induced hyperglycemic albino rats. *Indian Drugs* 41: 54–57.

Kotowska, D., El-Houri, R.B., Borkowski, K., Petersen, R.K., Frette, X.C., Wolber, G. et al. 2014. Isomeric C_{12}-alkamides from the roots of *Echinacea purpurea* improve basal and insulin-dependent glucose uptake in 3T3-L1 adipocytes. *Planta Med.* 80: 1712–1720.

Kouambou, C.C.N., Dimo, T., Dzeufiet, P.D.D., Ngueguim, F.T., Tchamadeu, M.C., Wembe, E.F. et al. 2007. Antidiabetic and hypolipidemic effects of *Canarium schweinfurthii* hexane bark extract in streptozotocin-diabetic rats. *Pharmacologyonline* 1: 209–219.

Kovacic, S., Soltys, C.-L.M., Barr, A.J., Shiojima, I,, Walsh, K., and Dyck, J.R.B. 2003. Akt activity negatively regulates phosphorylation of AMP-activated protein kinase in the heart. *J. Biol. Chem.* 278: 39422–39427.

Koyagura, N., Kumar, V.H., Jamadar, M.G., Huilgol, S.V., Nayak, N., Yendigeri, S.M. et al. 2013. Anti-diabetic and hepato-protective activities of *Tamarindus indica* fruit pulp in alloxan induced diabetic rats. *Int. J. Pharmacol. Clin. Sci.* 2: 33–40.

Krawinkel, M.B. and Keding, G.B. 2006. Bitter gourd (*Momordica charantia*): A dietary approach to hyperglycemia. *Nutr. Rev.* 64: 331–337.

Krenisky, J.M., Luo, J., Reed, M.J., and Carney, J.R. 1999. Isolation and antihyperglycemic activity of bakuchiol from *Otholobium pubescens* (Fabaceae), a Peruvian medicinal plant used for the treatment of diabetes. *Biol. Pharm. Bull.* 22: 1137–1140.

Krisanapun, C., Peungvicha, P., Temsiririrkkul, R., and Wongkrajang, Y. 2009. Aqueous extract of *Abutilon indicum* Sweet inhibits glucose absorption and stimulates insulin secretion in rodents. *Nutr. Res.* 29: 579–587.

Krisanapun, C., Wongkrajang, Y., Temsiririkkul, R., Phornchirasilp, S., and Peungvicha, P. 2012. *In vitro* evaluation of anti-diabetic potential of *Piper sarmentosum* Roxb. extract. *FASEB J.* 26: 686–687.

Krishnakumar, K., Augusti, K.T., and Vijayammal, P.L. 1999. Hypoglycemic and anti-oxidant activity of *Salacia oblonga* Wall. extract in streptozotocin-induced diabetic rats. *Indian J. Physiol. Pharmacol.* 43: 510–514.

Krishnakumar, K., Augusti, K.T., and Vijayammal, P.L. 2000. Anti-peroxidative and hypoglycaemic activity of *Salacia oblonga* extract in diabetic rats. *Pharm. Biol.* 38: 101–105.

Krishnamurthy, B., Nammi, S., Kota, M.K., Krishnarao, R.V., Koteswararao, N., and Annapurna, A. 2004. Evaluation of hypoglycemic and antihyperglycemic effects of *Datura metel* (Linn.) seeds in normal and alloxan-induced diabetic rats. *J. Ethnopharmacol.* 91: 95–98.

Krishnamurthy, G., Kuruba, L., Nagaraj, P., and Udaya, C.P. 2011. Antihyperglycemic and hypolipidemic activity of methanolic extract of *Amaranthus viridis* leaves in experimental diabetes. *Indian J. Pharmacol.* 43: 450–458.

Krishnan, K., Vijayalakshmi, N.R., and Helen, A. 2011. Beneficial effects of *Costus igneus* and dose response studies in streptozotocin induced diabetic rats. *Int. J. Curr. Pharm. Res.* 3: 42–46.

Ku, S.-K., Kwak, S., Kim, Y., and Bae, J.-S. 2015. Aspalathin and nothofagin from rooibos (*Aspalathus linearis*) inhibits high glucose-induced inflammation *in vitro* and *in vivo*. *Inflammation* 38: 445–455.

Kuang, Q.T., Zhao, J.J., Ye, C.L., Wang, J.R., Ye, K.H., Zhang, X.Q. et al. 2012. Nephro-protective effects of total triterpenoids from *Psidium guajava* leaves on type 2 diabetic rats. *Zhong Yao Cai* 35: 94–97.

Kuate, D., Etoundi, B.C., Ngondi, J.L., Muda, W.A.M.B.W., and Oben, J.E. 2013. Anti-inflammatory, anthropometric and lipomodulatory effects Dyglomera® (aqueous extract of *Dichrostachys glomerata*) in obese patients with metabolic syndrome. *Funct. Foods Health Dis.* 3: 416–427.

Kubacey, T.M., Haggag, E.G., El-Toumy, S.A., Ahmed, A.A., El-Ashmawy, I.M., and Youns, M.M. 2012. Biological activity and flavonoids from *Centaurea alexandrina* leaf extract. *J. Pharm. Res.* 5: 3352–3361.

Kubo, H., Inoue, M., Kamei, J., and Higashiyama, K. 2006. Hypoglycemic effects of multiflorine derivatives in normal mice. *Biol. Pharm. Bull.* 29: 2046–2050.

Kubow, S., Hobson, L., Iskandar, M.M., Sabally, K., Donnelly, D.J., and Agellon, L.B. 2014. An extract of Irish potatoes (*Solanum tuberosum* L.) decreases body weight gain and adiposity and improves glucose control in the mouse model of diet-induced obesity. *Mol. Nutr. Food Res.* 58: 2235–2238.

Kukreja, A. and Maclaren, K. 1999. Autoimmunity and diabetes. *J. Clin. Endocrinol. Metab.* 84: 4371–4378.

Kulkarni, C.R., Joglekar, M.M., Patil, S.B., and Arvindekar, A.U. 2012. Antihyperglycemic and antihyper-lipidemic effect of *Santalum album* in streptozotocin induced diabetic rats. *Pharm. Biol.* 50: 360–365.

Kulkarni, Y.A. and Veeranjaneyulu, A. 2013. Effects of *Gmelina arborea* extract on experimentally induced diabetes. *Asian Pac. J. Trop. Med.* 6: 602–608.

Kumanan, R., Manimaran, S., Saleemulla, K., Dhanabal, S.P., and Nanjan, M.J. 2010. Screening of bark of *Cinnamomum tamala* (Lauraceae) by using α-amylase inhibition assay for anti-diabetic activity. *Int. J. Pham. Biomed. Res.* 1: 69–72.

Kumar, A., Lingadurai, S., Shrivastava, T.P., Bhattacharia, S., and Haldar, P.K. 2011a. Hypoglycemic activity of *Erythrina variegata* leaf in streptozotocin-induced diabetic rats. *Pharm. Biol.* 49: 577–582.

Kumar, A.Y., Nandakumar, K., Handral, M., Talwar, S., and Dhayabarn, D. 2011b. Hypoglycaemic and anti-diabetic activity of stem bark extracts *Erythrina indica* in normal and alloxan-induced diabetic rats. *Saudi Pharm. J.* 19: 35–42.

Kumar, B., Gupta, S.K., Nag, T.C., Srivastava, S., and Saxena, R. 2012a. Green tea prevents hyperglycemia-induced retinal oxidative stress and inflammation in streptozotocin-induced diabetic rats. *Ophthalmic Res.* 47: 103–108.

Kumar, C., Kumar, R., and Nehar, S. 2013a. Hypoglycemic effect of acetone extract of *Terminalia arjuna* Roxb. stem bark on type 2 diabetic albino rats. *Bioscan* 8: 709–712.

Kumar, D., Gaonkar, R.H., Ghosh, R., and Pal, B.C. 2012b. Bio-assay guided isolation of α-glucosidase inhibitory constituents from *Eclipta alba. Nat. Prod. Commun.* 7: 989–990.

Kumar, D., Mitra, A., and Mahadevappa, M. 2010a. Antidiabetic and hypolipidemic effects of Mahanimbine (carbazole alkaloid) from *Murraya koenigii* (Rutaceae). *Int. J. Phytomed.* 2: 22–30.

Kumar, E.K. and Janardhana, G.R. 2011. Antidiabetic activity of alcoholic stem extract of *Nervilia plicata* in streptozotocin-nicotinamide induced type 2 diabetic rats. *J. Ethnopharmacol.* 133: 480–483.

Kumar, G., Banu, G.S., Murugesan, A.G., and Pandian, M.R. 2006. Hypoglycemic effect of *Helicterees isora* bark extract in rats. *J. Ethnopharmacol.* 107: 304–307.

Kumar, G., Banu, G.S., Murugesan, A.G., and Pandian, M.R. 2007. Antihyperglycemic and antiperoxidative effect of *Helicteres isora* L. bark extracts in streptozotocin-induced diabetic rats. *J. Appl. Biomed.* 5: 97–104.

Kumar, G.P., Sudheesh, S., and Vijayalakshmi, N.R. 1993. Hypoglycemic effect of *Coccinia indica*: Mechanism of action. *Planta Med.* 59: 330–332.

Kumar, K.V., Rao, C.A., Babu, K.R., Kumar, M.T.S., Fatima, S.S., and Rajasekhar, M.D. 2009a. Antidiabetic and antihyperlipidemic activity of ethyl acetate: Isopropanol (1:1) fraction of *Vernonia anthelmintica* seeds in streptozotocin induced diabetic rats. *Food Chem. Toxicol.* 48: 495–501.

Kumar, K.V., Sharief, S.D., Rajkumar, R., Ilango, B., and Sukumar, E. 2010b. Anti-diabetic potential of *Lantana aculeata* root extract in alloxan-induced diabetic rats. *Int. J. Phytomed.* 2: 299–303.

Kumar, M., Goutam, M.K., Singh, A., and Goel, R.K. 2013b. Healing effects of *Musa sapientum* var. *paradisiaca* in diabetic rats with co-occurring gastric ulcer: Cytokines and growth factor by PCR amplification. *BMC Complement. Altern. Med.* 2013: 30530–30539.

Kumar, M., Rawat, P., Khan, M.F., Tamarkar, A.K., Srivastava, A.K., and Arya, K.R. 2010c. Phenolic glycosides from *Dodecadenia grandiflora* and their glucose-6-phosphatase inhibitory activity. *Fitoterapia* 81: 475–479.

Kumar, M., Rawat, P., Rahuja, N., Srivastava, A.K., and Maurya, R. 2009b. Antihyperglycemic activity of phenylpropanoyl esters of catechol glycoside and its dimers from *Dodecadenia grandiflora*. *Phytochemistry* 70: 1448–1455.

Kumar, M., Sharma, S., and Vasudeva, N. 2013d. Antihyperglycemic and antioxidant potential of oil from *Arachis hypogaea* L. in streptozotocin-nicotinamide induced diabetic rats. *Afr. J. Pharm. Pharmacol.* 7: 2374–2381.

Kumar, N. and Singh, A.K. 2014. Plant profile, phytochemistry and pharmacology of avartani (*Helicteres isora* Linn): A review. *Asian Pac. J. Trop. Biomed.* 4(Suppl. 1): S22–S26.

Kumar, P., Baraiya, S., Gaidhani, S.N., Gupta, M.D., and Wanjari, M.M. 2012c. Antidiabetic activity of stem bark of *Bauhinia variegata* in alloxan-induced hyperglycemic rats. *J. Pharmacol. Pharmacother.* 3: 64–66.

Kumar, R., Patel, D.K., Prasad, S.K., Laloo, Krishnamurthy, S., and Hemalatha, S. 2012e. Type 2 antidiabetic activity of bergenin from the roots of *Caesalpinia digyna* Rottler. *Fitoterapia* 83: 395–401.

Kumar, R., Patel, D.K., Prasad, S.K., Sairam, K., and Hemalatha, S. 2012d. Antidiabetic activity of alcoholic root extract of *Caesalpinia digyna* in streptozotocin-nicotinamide induced diabetic rats. *Asian Pac. J. Trop. Biomed.* 2012: S934–S940.

Kumar, R.P., Sujatha, D., Saleem, T.S.M., Chetty, C.M., and Ranganayakulu, D. 2010d. Potential anti-diabetic and antioxidant activities of *Morus indica* and *Asystasia gangetica* in alloxan-induced diabetes mellitus. *J. Exp. Pharmacol.* 2: 29–36.

Kumar, R.V. and Augusti, K.T. 1994. Insulin sparing action of a leucocyanidin derivative isolated from *Ficus bengalensis* Linn. *Indian J. Biochem. Biophys.* 31: 71–76.

Kumar, S. 2014. Preclinical evaluation of anti-diabetic and hypolipidemic effects of *Hibiscus tiliaceus*. *World J. Pharm. Res.* 3: 891–900.

Kumar, S., Kumar, V., and Prakash O. 2011c. Antidiabetic, hypolipidemic and histopathological analysis of *Dillenia indica* (L.) leaves extract on alloxan induced diabetic rats. *Asian Pac. J. Trop. Med.* 4: 347–352.

Kumar, S., Kumar, V., and Prakash, O. 2010e. Anti-diabetic and hypolipidemic activities of *Hibiscus tiliaceus* (L.) flowers extract in streptozotocin-induced diabetic rats. *Pharmacologyonline* 2: 1037–1044.

Kumar, S., Kumar, V., and Prakash, O. 2012f. Antidiabetic and hypolipidemic activities of *Kigelia pinnata* flowers extract in streptozotocin induced diabetic rats. *Asian Pac. J. Trop. Biomed.* 2: 543–546.

Kumar, S., Kumar, V., and Prakash, O. 2013c. Enzymes inhibition and antidiabetic effect of isolated constituents from *Dillenia indica*. *BioMed. Res. Int.* 2013: Article ID 382063, 7 pages.

Kumar, S., Kumar, V., and Prakash, O.M. 2011d. Antihyperglycemic, antihyperlipidemic potential and histopathological analysis of ethyl acetate fraction of *Callistemon lanceolatus* leaves extract on alloxan induced diabetic rats. *J. Exp. Integr. Med.* 1: 185–190.

Kumar, S., Pramanik, G., and Haldar, P.K. 2011e. Preclinical evaluation of antihyperglycemic activity of *Clerodendron infortunatum* leaf against streptozotocin-induced diabetic rats. *Diabetes Ther.* 2: 92–100.

Kumar, S., Sharma, S., and Suman, J. 2011f. *In vivo* anti-hyperglycemic and antioxidant potential of *Piper longum* fruit. *J. Pharm. Res.* 4: 471–474.

Kumar, S., Vasudeva, N., and Sharma S. 2012g. GC-MS analysis and screening of antidiabetic, antioxidant and hypolipidemic potential of *Cinnamomum tamala* oil in streptozotocin induced diabetes mellitus in rats. *Cardiovasc. Diabetol.* 11: 95.

Kumar, S., Vasudeva, N., and Sharma, S. 2012h. GC-MS analysis and screening of antidiabetic, antioxidant and hypolipidemic potential of *Cinnamomum tamala* oil in streptozotocin-induced diabetes mellitus in rats. *Cardiovasc. Diabetol.* 10: 1–11.

Kumar, V., Khanna, A.K., Khan, M.M., Singh, R., Singh, S., Chander, R. et al. 2009c. Hypoglycemic, lipid lowering and anti-oxidant activities in root extract of *Anthocephalus indicus* in alloxan-induced diabetic rats. *Indian J. Clin. Biochem.* 24: 65–69.

Kumar, V., Thakur, A.K., Barothia, N.D., and Chatterjee, S.S. 2011g. Therapeutic potentials of *Brassica juncea*: An overview. *Int. J. Genuine Tradit. Med.* 1: 1–17. DOI: http://dx.doi.org./10.5667/tang.2011.0005.

Kumar, V.L. and Padhy, B.M. 2011. Protective effect of aqueous suspension of dried latex of *Calotropis procera* against oxidative stress and renal damage in diabetic rats. *Biocell* 35: 63–69.

Kumarappan, C.T., Rao, T.N., and Mandal, S.C. 2007. Polyphenolic extract of *Ichnocarpus frutescens* modifies hyperlipidemia status in diabetic rats. *J. Cell. Mol. Biol.* 6: 175–187.

Kumarappan, C.T., Thilagam, M., Vijayakumar, M., and Mandal, S.C. 2012. Modulatory effect of polyphenolic extracts of *Ichnocarpus frutescens* on oxidative stress in rats with experimentally induced diabetes. *Indian J. Med. Res.* 136: 815–821.

Kumaraswamy, M., Sudipta, K.M., Lokesh, P., Neeki, A., Rashmi, W., Bhaumik, H. et al. 2012. Phytochemical screening and in vitro antimicrobial activity of *Bougainvillea spectabilis* flower extracts. *Int. J. Phytomed.* 4: 375–379.

Kumaresan, P., Jeyanthi, K.A., and Kalaivani, R. 2014. Biochemical evaluation of anti-diabetic activity of aqueous extract of *Gmelina arborea* in alloxan-induced albino rats. *Int. J. Herbal Med.* 2: 90–94.

Kumari, K. and Augusti, K.T. 2002. Antidiabetic and antioxidant effects of S-methyl cysteine sulfoxide isolated from onions (*Allium cepa* Linn.) as compared to standard drugs in alloxan diabetic rats. *Indian J. Exp. Biol.* 40: 1005–1009.

Kumari, K., Mathew, B.C., and Augusti, K.T. 1995. Antidiabetic and hypolipidemic effects of S-methyl cysteine sulfoxide isolated from *Allium cepa* Linn. *Indian J. Biochem. Biophys.* 32: 49–54.

Kumari, S., Wanjari, M., Kumar, P., and Palani, S. 2012. Antidiabetic activity of *Pandanus fascicularis* Lamk—aerial roots in alloxan-induced hyperglycemic rats. *Int. J. Nutr. Pharmacol. Neurol. Dis.* 2: 105–110.

Kumavat, U.C., Shimpi, S.N., and Jagdale, S.P. 2012. Hypoglycemic activity of *Cassia javanica* Linn. in normal and streptozotocin-induced diabetic rats. *J. Adv. Pharm. Technol. Res.* 3: 47–51.

Kumawat, N.S., Chaudhari, S.P., Wani, N.S., Deshmukh, T.A., and Patil, V.R. 2010. Anti-diabetic activity of ethanol extract of *Colocasia esculenta* leaves in alloxan-induced diabetic rats. *Int. J. Pharm. Technol. Res.* 2: 1246–1249.

Kumudhavalli, M.V. and Jaykar, B. 2012. Pharmacological screening on leaves of the plant of *Hemionitis arifolia* (Burm.) Moore. *Res. J. Pharm. Biol. Chem. Sci.* 3: 72–83.

Kundusen, S., Haldar, P.K., Gupta, M., Mazumder, U.K., Saha, P., Bala, B. et al. 2011. Evaluation of antihyperglycemic activity of *Citrus limetta* fruit peel in streptozotocin-induced diabetic rats. *ISRN Endocrinol.* 2011: Article ID 869273, 6 pages.

Kunyanga, C.N., Vellingiri, V., and Imungi, K.N. 2014. Nutritional quality, phytochemical composition and health protective effects of an under-utilized prickly Cactus fruit (*Opuntia stricta* Haw.) collected from Kenya. *Afr. J. Food Agric. Nutr. Dev.* 14: 55–62.

Kupiecki, F.P., Ogzewalla, C.D., and Schell, F.M. 1974. Isolation and characterization of a hypoglycemic agent from *Xanthium strumarium*. *J. Pharm. Sci.* 63: 1166–1167.

Kurihara, H., Fukami, H., Kusumoto, A., Toyoda, Y., Shibata, H., Matsui, Y. et al. 2003. Hypoglycemic action of *Cyclocarya paliurus* (Batal.) Iljinskaja in normal and diabetic mice. *Biosci. Biotechnol. Biochem.* 67: 877–880.

Kuriyan, R., Rajendran, R., bantwal, G., and Kurpad, A.V. 2008. Effect of supplementation of *Coccinia cordifolia* extract on newly detected diabetic patients. *Diabetes Care* 31: 216–220.

Kuroda, M., Mimaki, Y., Nishiyama, T., Mae, T., Kishida, H., Tsukagawa, M. et al. 2005. Hypoglycemic effect of turmeric (*Curcuma longa* L. rhizomes) on genitically diabetic KK-Ay mice. *Biol. Pharm. Bull.* 28: 937–939.

Kuroda, M., Mimaki, Y., Sashida, Y., Mae, T., Kishida, H., Nishiyama, T. et al. 2003. Phenolics with PPAR-ligand binding activity obtained from licorice (*Glycyrrhiza uralensis* roots) and ameliorative effects of glycyrin on genetically diabetic KK-Ay mice. *Bioorg. Med. Chem. Lett.* 13: 4267–4272.

Kuroda, M., Mimaki, Y., Sashida, Y., Mae, T., Kishida, H., Nishiyama, H. et al. 2004. Phenolics with PPAR-ligand binding activity obtained from licorice (*Glycyrrhiza uralensis* roots) and ameliorative effects of glycyrin on genetically diabetic KK-Ay mice. *Bioorg. Med. Chem. Lett.* 13: 4267–4272.

Kusano, A. and Abe, H. 2000. Antidiabetic activity of white skinned sweet potato (*Ipomoea batatas* L.) in obese Zucker fatty rats. *Biol. Pharm. Bull.* 23: 23–26.

Kushwah, A.S., Patil, B.M., and Thippeswamy, B.S. 2010. Effect of *Phyllanthus fraternus* on fructose induced insulin resistance in rats. *Int. J. Pharmacol.* 210: 1–7.

Kwon, D.Y., Kim, D.S., Yang, H.Y., and Park, S. 2011. The lignan rich fractions of Fructus Schisandrae improve insulin sensitivity via the PPAR-γ pathway in *in vitro* and *in vivo* studies. *J. Ethnopharmacol.* 135: 455–462.

Kwon, S.-U., Im, J.-Y., Jcon, S.-B., Jee, H.-K., Park, Y.-S., Lee, H.Y. et al. 2013. Anti-hyperglycemic effect of fermented *Gastrodia elata* Blume in streptozotocin-induced diabetic mice. *Food Sci. Biotechnol.* 22: 1403–1408.

Labban, L. 2014. Medicinal and pharmacological properties of turmeric (*Curcuma longa*): A review. *Int. J. Pharm. Biomed. Sci.* 5: 17–23.

Labban, L., Mustafa, U.E., and Ibrahim, Y.M. 2014a. The effects of rosemary (*Rosmarinus officinalis*) leaves powder on glucose level, lipid profile and lipid perodoxation. *Int. J. Clin. Med.* 5: 297–304.

Lakshmi, M.S., Sandhya Rani, K.S., and Reddy, U.K.T. 2012. A review on diabetes milletus and the herbal plants used for its treatment. *Asian J. Pharm. Clin. Res.* 5: 15–21.

Lakshmi, T., Roy, A., and Geetha, R.V. 2011. *Acacia catechu* Willd—A gift from Ayurveda to mankind—A review. *Pharma Res.* 5: 273–293.

Lakshmi, V., Agarwal, S.K., Ansari, J.A., Mahdi, A.A., and Srivastava, A.K. 2014. Anti-diabetic potential of *Musa paradisiaca* in streptozotocin-induced diabetic rats. *J. Phytopharmacol.* 3: 77–81.

Lal, V.K., Gupta, P.P., Pandey, A., and Tripathi, P. 2011. Effect of hydro-alcoholic extract of *Cucurbita maxima* fruit juice and glibenclamide on blood glucose in diabetic rats. *Am. J. Pharmacol. Toxicol.* 6: 84–87.

Lamela, M., Cadavid, I., and Calleja, J.M. 1986. Effects of *Lythrum salicaria* extracts on hyperglycemic rats and mice. *J. Ethnopharmacol.* 15: 153–160.

Lammi, N., Karvonen, M., and Tuomilehto, J. 2005. Do microbes have a causal role in type 1 diabetes? *Med. Sci. Monit.* 11: 63–69.

Lanjhiyana, S., Garabadu, D., Ahirwar, D., Bigoniya, P., Rana, A.C., Patra, K.C. et al. 2011. Antidiabetic activity of methanolic extract of stem bark of *Elaeodendron glaucum* Pers. in alloxanized rat model. *Adv. Appl. Sci. Res.* 2: 47–62.

Latha, M. and Pari, L. 2003. Antihyperglycemic effect of *Cassia auriculata* in experimental diabetes and its effects on key metabolic enzymes involved in carbohydrate metabolism. *Clin. Exp. Pharmacol.* 30: 38–43.

Latha, M. and Pari, L. 2003b. Preventive effect of *Cassia auriculata* L. flowers on brain lipid peroxidation in rats treated with streptozotocin. *Mol. Cell. Biochem.* 243: 23–28.

Latha, M. and Pari, L. 2004. Effect of an aqueous extract of *Scoparia dulcis* on blood glucose, plasma insulin and some polyol pathway enzymes in experimental rat diabetes. *Braz. J. Med. Biol. Res.* 37: 577–586.

Latha, M. and Pari, L. 2005. Effect of an aqueous extract of *Scoparia dulcis* on plasma and tissue glycoproteins in streptozotocin induced diabetic rats. *Pharmazie* 60: 151–154.

Latha, M., Pari, L., Ramkumar, K.M., Rajaguru, P., Suresh, T., Dhanabal, T., Sitasawad, S., and Bhonde, R. 2009. Antidiabetic effects of scoparic acid D isolated from *Scoparia dulcis* in rats with streptozotocin induced diabetes. *Nat. Prod. Res.* 23: 1528–1540.

Latha, M., Pari, L., Sitasawad, S., and Bhonde, R. 2004. Insulin secretagogue activity and cytoprotective role of the traditional antidiabetic plant *Scoparia dulcis* (sweet broom weed). *Life Sci.* 75: 2003–2014.

Latha, R.C.R. and Daisy, P. 2010. Influence of *Terminalia bellerica* Roxb. fruit extracts on biochemical parameters in streptozotocin diabetic rats. *Int. J. Pharmacol.* 6: 89–96.

Latha, R.C.R. and Daisy, P. 2011. Insulin-secretagogue, antihyperlipidemic and other protective effects of gallic acid isolated from *Terminalia bellerica* Roxb. in streptozotocin-induced diabetic rats. *Chem. Biol. Interact.* 189: 112–118.

Lau, C.W. and Yao, X.Q. 2001. Cardiovascular actions of berberine. *Cardiovasc. Drug Rev.* 19: 234–244.

Laurenz, J.C., Collier, C.C., and Kut, J.O. 2003. Hypoglycaemic effect of *Opuntia lindheimeri* Engelm. in a diabetic pig model. *Phytother. Res.* 17: 26–29.

Lawag, I.L., Aguinaldo, A.M., Naheeds, S., and Mosihuzzaman, M. 2012. α-Glucosidase inhibitory activity of selected Philippine plants. *J. Ethnopharmacol.* 144: 217–219.

Lawson-Evi, P., Eklu-Gadeg, K., Agbonon, A., and Gbeassor, M. 2011. Anti-diabetic activity of *Phyllanthus amarus* Schum and Thonn on alloxan-induced diabetes in male Wistar rats. *J. Appl. Sci.* 11: 2968–2973.

Lee, B., Jung, K., and Kim, D.H. 2010a. Timosaponin AIII, a saponin isolated from *Anemarrhena asphodeloides*, ameliorates learning and memory deficits in mice. *Pharmacol. Biochem. Behav.* 93: 121–127.

Lee, H.S. 2002. Rat lens aldose reductase inhibitory activities of *Coptis japonica* root derived isoquinoline alkaloids. *J. Agric. Food Chem.* 50: 7013–7016.

Lee, K.T., Sohn, I.C., Kim, D.H., Choi, J.W., Kwon, S.H., and Park, H.J. 2000. Hypoglycemic and hypolipidemic effects of tectorigenin and kaikasaponin III in the streptozotocin-induced diabetic rat and their antioxidant activity *in vitro. Arch. Pharmacol. Res.* 23: 461–466.

Lee, M.K., Kim, M.J., Cho, S.Y., Park, S.A., Park, K.K., Jung, U.J. et al. 2005. Hypoglycemic effect of Du-zhong (*Eucommia ulmoides* Oliv.) leaves in streptozotocin-induced diabetic rats. *Diabetes Res. Clin. Pract.* 67: 22–28.

Lee, S.-C., Xu, W.-X., Lin, L.-Y., Yang, J.J., and Liu, C.-T. 2013. Chemical composition and hypoglycemic and pancreas-protective effect of leaf essential oil from indigenous cinnamon (*Cinnamomum osmophloeum* Kanehira). *J. Agric. Food Chem.* 61: 4905–4913.

Lee, S.-K., Hwang, J.-Y., Song, J.-H., Jo, J.-R., Kim, M.-J., Kim, M.E. et al. 2007. Inhibitory activity of *Euonymus alatus* against alpha-glucosidase *in vitro* and *in vivo. Nutr. Res. Pract.* 1: 184–188.

Lee, S.Y., Eom, S.H., Kim, Y.K., Park, N.I., and Park, S.U. 2009a. Cucurbitane-type triterpenoids in *Momordica charantia* Linn. *J. Med. Plants Res.* 3: 1264–1269.

Lee, W.-C., Wang, C.-J., Chen, Y.-H., Hsu, J.-D., Cheng, S.-Y., Chen, H.-C., and Lee, H.-J. 2009b. Polyphenol extracts from *Hibiscus sabdariffa* Linnaeus attenuate nephropathy in experimental type 1 diabetes. *J. Agric. Food Chem.* 57: 2206–2210.

Lee, W.K., Kao, S.T., Liu, I.M., and Cheng, J.T. 2006a. Increase of insulin secretion by ginsenoside Rh2 to lower plasma glucose in Wistar rats. *Clin. Exp. Pharmacol. Physiol.* 33: 27–32.

Lee, Y.S., Kim, S.H., Jung, S.H., Kim, J.K., Pan, C.H., and Lim, S.S. 2010b. Aldose reductase inhibitory compounds from *Glycyrrhiza uralensis*. *Biol. Pharm. Bull.* 33: 917–921.

Lee, Y.S., Kim, W.S., Kim, K.H., Yoon, M.J., Cho, H.J., Shen, Y. et al. 2006b. Berberine, a natural plant product, activates AMP-activated protein kinase with beneficial metabolic effects in diabetic and insulin-resistant states. *Diabetes* 55: 2256–2264.

Lee-Huang, S., Huang, P.L., Bourinbaiar, A.S., Chen, H.C., and Kung, H.F. 1995. Inhibition of the integrase of human immunodeficiency virus (HIV) type 1 and anti-HIV plant proteins MAP30 and GAP31. *Proc. Natl. Acad. Sci. USA* 92: 8818–8822.

Leeuwen, P.A. 2001. Flavonoids: A review of probable mechanism of action and potential applications. *Am. J. Clin. Nutr.* 74: 418–425.

Lehrke, M. and Lazar, M.A. 2005. The many faces of PPAR gamma. *Cell* 123: 993–999.

Lei, M. 2008. Screening the dipeptidyl peptidase IV inhibitors from the Chinese herb and its pharmacodynamics. Master's Degree thesis, Jiangnan University, Jiangnan, China.

Leite, A.C.R., Araujo, T.G., Carvalho, B.M., Maia, M.B.S., and Lima, V.L.M. 2011. Characterization of the antidiabetic role of *Parkinsonia aculeata* (Caesalpineaceae). *Evid. Based Complement. Altern. Med.* 692378: 1–9.

Leite, A.C.R., Araujo, T.G., Carvalho, B.M., Silva, N.H., Lima, V.L.M., and Maiab, M.B.S. 2007. *Parkinsonia aculeate* aqueous extract fraction: Biochemical studies in alloxan-induced diabetic rats. *J. Ethnopharmacol.* 111: 547–552.

Lekshmi, P.C., Arimboor, R., Indulekha, P.S., and Menon, A.N. 2012. Turmeric (*Curcuma longa* L.) volatile oil inhibits key enzymes linked to type 2 diabetes. *Int. J. Food Sci. Nutr.* 63: 832–834.

Lemhadri, A., Eddouks, M., Sulpice, T., and Burcelin, R. 2007. Anti-hyperglycaemic and anti-obesity effects of *Capparis spinosa* and *Chamaemelum nobile* aqueous extracts in HFD mice. *Am. J. Pharmacol. Toxicol.* 2: 106–110.

Lemhadri, A., Zeggwagh, N.A., Maghrani, M., Jouad, H., Michal, J.B., and Eddouks, M. 2004. Hypoglycemic effect of *Calamintha officinalis* Moench. in normal and streptozotocin induced diabetic rats. *J. Pharm. Pharmacol.* 56: 795–799.

Lemus, I., Garcia, R., Delvillar, E., and Knop, G. 1999. Hypoglycaemic activity of four plants used in Chilean popular medicine. *Phytother. Res.* 13: 91–94.

Leshweni, Jerimia, L., and Shai, J. 2012. The effects of *Clausena anisata* (Wild) Hook leaf extracts on selected diabetic related metabolizing enzymes. *J. Med. Plant Res.* 6: 4200–4027.

Lestari, K., Hwang, J.K., Kariadi, S.H., Wijaya, A., Ahmad, T., Subarnas, A. et al. 2012. Screening for PPAR γ agonist from *Myristica fragrans* Houtt seeds for the treatment of type 2 diabetes by *in vitro* and *in vivo*. *Med. Health Sci. J.* 12: 7–15.

Leung, S.O., Yeung, H.W., and Leung, K.N. 1987. The immunosuppressive activities of two abortifacient proteins isolated from the seeds of bitter melon (*Momordica charantia*). *Immunopharmacology* 13: 159–171.

Leyama, T., Gunawan-Puteri, M.D.P.T., and Kawabata, J. 2011. α-Glucosidase inhibitors from the bulb of *Eleutherine americana*. *Food Chem.* 128: 308–311.

Leyon, P.V. and Kuttan, G. 2004a. Effect of *Tenospora cordifolia* on the cytokine profile of angiogenesis-induced animals. *Int. Immunopharmacol.* 4: 1569–1575.

Leyon, P.V. and Kuttan, G. 2004b. Inhibitory effect of a polysaccharide from *Tenospora cordifolia* on experimental metastasis. *J. Ethnopharmacol.* 90: 233–237.

Li, C.G., Zhang, S.Q., Wang, S.Y., Cui, M.M., Zhou, M.B., and Qin, Q. 2006a. Intervention of litchi nucleus extract fluid in the blood biochemical indexes of model mice with diabetes mellitus induced by alloxan. *J. Zhonguo. Linchuang Kangfu.* 10: 61–63.

Li, H.M., Hwang, S.H., Kang, B.G., Hong, J.S., and Lim, S.S. 2014. Inhibitory effects of *Colocasia esculenta* (L.) Schott constituents on aldose reductase. *Molecules* 19: 13212–13224.

Li, H.M., Kim, J.K., Jang, J.M., Kwon, S.O., Cui, C.B., and Lim, S.S. 2012. The inhibitory effect of *Prunella vulgaris* L. on aldose reductase and protein glycation. *J. Biomed. Biotechnol.* 2012: Article ID 928159, 7 pages.

Li, J.M., Li, Y.C., Kong, L.D., and Hu, Q.H. 2010a. Curcumin inhibits hepatic protein-tyrosine phosphatase 1B and prevents hypertriglyceridemia and hepatic steatosis in fructose-fed rats. *Hepatology* 51: 1555–1566.

Li, M., Qu, W., Wang, Y., Wan, H., and Tian, C. 2002. Hypoglycemic effect of saponin from *Tribulus terrestris*. *Zhong Yao Cai.* 25: 420–422.

Li, Q., Fu, C., Rui, Y., Hu, G., and Cai, T. 2005a. Effects of protein bound polysaccharide isolated from pumpkin on insulin in diabetic rats. *Plant Foods Hum. Nutr.* 60: 13–16.

Li, Q. and Qu, H. 2012. Study on the hypoglycemic activities and metabolism of alcohol extract of Alismatis Rhizoma. *Fitoterapia* 83: 1046–1053.

Li, Q., Xiao, Y., Gong, H., Shen, D., Zhu, F., Wu, Q., Chen, H., and Zhong, H. 2009. Effect of puerarin on the expression of extracellular matrix in rats with streptozotocin-induced diabetic nephropathy. *Nat. Med. J. India* 22: 9–12.

Li, S., An, T.Y., Li, J., Shen, Q., Lou, F.C., and Hu, L.H. 2006b. PTP1B inhibitors from *Saussrurea lappa. J. Asian Nat. Prod. Res.* 8: 281–286.

Li, X., Cui, X., Sun, X., Li, X., Zhu, Q., and Li, W. 2010b. Mangiferin prevents diabetic nephropathy progression in streptozotocin-induced diabetic rats. *Phytother. Res.* 24: 893–899.

Li, X., Cui, X., Wang, J., Yang, J., Sun, X. Li, X. et al. 2013. Rhizome of *Anemarrhena asphodeloides* counteracts diabetic ophthalmopathy progression in streptozotocin-induced diabetic rats. *Phytother. Res.* 27: 1243–1250.

Li, X., Cui, X., Sun, X., Zhu, Q., and Li, W. 2010. Mangiferin prevents diabetic nephropathy progression in streptozotocin-induced diabetic rats. *Phytother. Res.* 24: 893–899.

Li, X.M. 2007. Protective effect of *Lycium barbarum* polysaccharides on streptozotocin-induced oxidative stress in rats. *Int. J. Biol. Macromol.* 40: 461–465.

Li, Y., Peng, G., Li, Q., Wen, S., Huang, T.H., Roufogalis, B.D. et al. 2004. *Salacia oblonga* improves cardiac fibrosis and inhibits postprandial hyperglycemia in obese Zucker rats. *Life Sci.* 75: 1735–1746.

Li, Y., Wen, S., Kota, B.P., Peng, G., Li, G.Q., Yamahara, Y. et al. 2005b. *Punica granatum* flower extract, a potent alpha-glucosidase inhibitor, improves postprandial hyperglycemia in Zucker diabetic fatty rats. *J. Ethnopharmacol.* 99: 239–244.

Li, Z., Jie, Y., Xiao-qing, C., Ke, Z., Xiao-dong, W., Hong, C. et al. 2010c. Antidiabetic and antioxidant effects of extracts from *Potentilla discolor* Bunge on diabetic rats induced by high fat diet and streptozotocin. *J. Ethnopharmacol.* 132: 518–524.

Liang, C.-H., Chan, L.-P., Chou, T.-H., Chiang, F.-Y., Yen, C.-M., Chen, P.-J. et al. 2013. Brazilein from *Caesalpinia sappan* L. antioxidant inhibits adipocyte differentiation and induces apoptosis through caspase-3 activity and anthelmintic activities against *Hymenolepis nana* and *Anisakis simplex. Evid. Based Complement. Altern. Med.* 2013: Article ID 864892, 14 pages.

Lieberman, I., Lentz, D.P., Trucco, G.A., Seow, W.K., and Thong, Y.H. 1992. Prevention by tetrandrine of spontaneous development of diabetes mellitus in BB rats. *Diabetes* 41: 616–619.

Lien, D.N., Phuc, D.V., Lien, P.Q., Trang, N.T., Kien, T.T., Lien, T.T.P. et al. 2011. Effect of sweet potato (*Ipomoea batatas* (L.) Lam) leaf extract on hypoglycaemia, blood insulin secretion, and key carbohydrate metabolic enzymes in experimentally obese and streptozotocin-induced diabetic mice. *VNU J. Sci. Nat. Sci. Technol.* 27: 118–124.

Lim, T.K. 2014. *Edible Medicinal and Non-Medicinal Plants*, Vol. 8: Flowers. Dordrecht, the Netherlands: Springer Sciences and Business, 1038pp.

Lima, C.F., Azevedo, M.F., Araujo, R., Fernandes-Ferreira, M., and Pereira- Wilson, C. 2006. Metformin-like effect of *Salvia officinalis* (common sage): Is it useful in diabetes prevention? *Br. J. Nutr.* 96: 326–333.

Lima, C.R., Vasconcelos, C.F., Costa-Silva, J.H., Maranhao, C.A., Costa, J., Batista, T.M. et al. 2012. Antidiabetic activity of extract from *Persea americana* Mill. leaf via the activation of protein kinase B (PKB/Akt) in streptozotocin-induced diabetic rats. *J. Ethnopharmacol.* 141: 517–525.

Lin, G.-P., Jiang, T., Hu, X.-B., Qiao, X.-H., and Tuo, Q. 2007. Effect of *Siraitia grosvenorii* polysaccharide on glucose and lipid of diabetic rabbits induced by feeding high fat/high sucrose chow. *Exp. Diabetes Res.* 2007: 67435.

Ling, Z.Q., Xie, B.L., and Yang, E.L. 2005. Isolation, characterization and determination of anti-oxidative activity of oligomeric procyanidins from the seed pod of *Nelumbo nucifera* Gaertn. *J. Agric. Food Chem.* 53: 2441–2445.

Liu, C.P., Tsai, W.J., Lin, Y.L., Liao, J.F., Chen, C.F., and Kuo, Y.C. 2004. The extract from *Nelumbo nucifera* suppress cell cycle progression, cytokine genes expression, and cell proliferation in human peripheral blood mononuclear cells. *Life Sci.* 75: 699–716.

Liu, I.M., Chi, T.C., Hsu, F.L., Chen, C.F., and Cheng, J.T. 1999. Isoferulic acid as active principle from the rhizoma of *Cimicifuga dahurica* to lower plasma glucose in diabetic rats. *Planta Med.* 65: 712–714.

Liu, I.M., Hsu, F.-L., and Cheng, J.-T. 2000. Antihyperglycemic action of isoferulic acid in streptozotocin-induced diabetic rats. *Br. J. Pharmacol.* 129: 631–636.

Liu, I.M., Liou, S.S., Lan, T.W., Hsu, F.L., and Cheng, J.T. 2005a. Myricetin as the active principle of *Abelmoschus moschatus* to lower plasma glucose in streptozotocin-induced diabetic rats. *Planta Med.* 71: 617–621.

Liu, I.M., Tzeng, T.F., and Liou, S.S. 2010. *Abelmoschus moschatus* (Malvaceae), an aromatic plant, suitable for medical or food uses to improve insulin sensitivity. *Phytother. Res.* 24: 233–239.

Liu, I.M., Tzeng, T.F., Liou, S.S., and Chang, C.J. 2011. *Angelica acutiloba* root alleviates advanced glycation end-product-mediated renal injury in streptozotocin-diabetic rats. *J. Food Sci.* 76: H165–H174.

Liu, J. 1995. Pharmacology of oleanolic acid and ursolic acid. *J. Ethnopharmacol.* 49: 57–68.

Liu, J.C. and Chan, P.O. 1999. The antihypertensive effect of the berberine derivative: 6-protoberberine in spontaneously hypertensive rats. *Pharmacology* 59: 283–289.

Liu, K.Y., Wu, Y.C., Liu, I.M., Yu, W.C., and Cheng, J.T. 2008. Release of acetylcholine by syringin, an active principle of *Eleutherococcus senticosus,* to raise insulin secretion in Wistar rats. *Neurosci. Lett.* 434: 195–199.

Liu, T.P., Liu, I.M., and Cheng, J.T. 2005b. Improvement of insulin resistance by *Panax ginseng* in fructose-rich chow-fed rats. *Horm. Metab. Res.* 37: 333–339.

Liu, X., Zhu, L., Tan, J., Zhu, X., Xiao, L., Yang, X. et al. 2014. Glucosidase inhibitory activity and antioxidant activity of flavonoid compound and triterpenoid compound from *Agrimonia pilosa* Ledeb. *BMC Complement. Altern. Med.* 14: 12–16.

Liu, Z., Liu, W., Li, X., Zhang, M., Li, C., Zheng, Y. et al. 2013a. Anti-diabetic effects of malonyl ginsenosides from *Panax ginseng* on type 2 diabetic rats. *J. Ethnopharmacol.* 145: 233–240.

Liu, Z., Wan, L., Yue, Y., Xiao, Z., Zhang, Y., Wang, Y. et al. 2013b. Hypoglycemic activity and antioxidative stress of extracts and corymbiferin from *Swertia bimaculata in vitro* and *in vivo*. *Evid. Based Complement. Altern. Med.* 2013: Article ID 125416, 12 pages.

Lodha, S.R., Joshy, S.V., Vyas, B.A., Upadyye, M.C., Kirve, M.S., Salunke, S.S. et al. 2010. Assessent of the anti-diabetic potential of *Cassia grandis* using an *in vivo* model. *J. Adv. Pharm. Technol. Res.* 1: 330–333.

Lokman, F.E., Gu, H.F., Mohamud, W.N.W., Yusoff, M.M., Chia, K.L., and Ostenson, C.-G. 2013. Anti-diabetic effect of oral borapetol B compound, isolated from the plant *Tinospora crispa*, by stimulating insulin release. *Evid. Based Complement. Altern. Med.* 2013: 7.

Loots, D.T., Pieters, M., Islam, M.S., and Botes, L. 2011. Antidiabetic effects of *Aloe ferox* and *Aloe greatheadii* var. *davyana* leaf gel extracts in a low-dose streptozotocin diabetes rat model. *S. Afr. J. Sci.* 107: 1–6.

Lopes, A.A. 2009. End-stage renal disease due to diabetes in racial ethnic minorities and disadvantaged populations. *Ethn. Dis.* 19: S1–47–51.

Lopez, J.A. 1976. Isolation of astilbin and sitosterol from *Hymenaea courbaril*. *Phytochemistry* 5: 2027F.

Lopez, J.L. Jr. 2007. Use of *Opuntia cactus* as a hypoglycemic agent in managing type 2 diabetes mellitus among Mexican American patients. *Nutr. Bytes* 12: 1–6.

Lovati, M.R., Manzoni, C., Castiglioni, S., Parolari, A., Magni, C., and Duranti, M. 2012. Lupin seed γ-conglutin lowers blood glucose in hyperglycaemic rats and increases glucose consumption of HepG2 cells. *Br. J. Nutr.* 107: 67–73.

Lovejoy, J.C, Most, M.M., Lefevre, M., Greenway, F.L., and Rood, J.C. 2002. Effects of diet enriched in almond on insulin action and serum lipids in adults with normal glucose tolerance or type 2 diabetes. *Am. J. Clin. Nutr.* 76: 1000–1006.

Lozoya, X., Reyes-Morales, M.-H., Chavez-Soto, M.A., Martinez-Garcia, M.C., Soto-Gonzalez, Y., and Doubova, S.V. 2002. Intestinal anti-spasmodic effect of a phytodrug of *Psidium guajava* in the treatment of acute diarrheic disease. *J. Ethnopharmacol.* 83: 19–24.

Lozoya-Meckes, M. and Mellado-Campos, V. 1985. Is the *Tecoma stans* infusion an anti-diabetic remedy? *J. Ethnopharmacol.* 14: 1–9.

Lu, X.Y., Chen, X.M., Fu, D.X., Cong, W., and Ouyang, F. 2002. Effect of *Amorphophallus konjac* oligosaccharides on STZ-induced diabetes model of isolated islets. *Life Sci.* 72: 711–719.

Ludvik, B., Neuffer, B., and Pacini, G. 2004. Efficacy of *Ipomoea batatas* (Caiapo) on diabetes control in type 2 diabetic subjects treated with diet. *Diabetes Care* 27: 436–440.

Lukmanul, F., Hakkim, F., Girija, S., Senthil Kumar, R., and Jalaludeen, M.D. 2007. Effect of aqueous and ethanol extracts of *Cassia auriculata* L. flowers on diabetes using alloxan induced diabetic rats. *Int. J. Diabetes Metab.* 15: 100–106.

Luo, J., Chuang, T., Cheung, J., Tsai, J., Sullivan, C., Hector, R.F. et al. 1998. Masoprocol (nordihydroguaiaretic acid): A new antihyperglycemic agent isolated from the creosote bush (*Larrea tridentata*). *Eur. J. Pharmacol.* 346: 77–79.

Luo, Q., Cai, Y., Yan, J., Sun, M., and Corke, H. 2004. Hypoglycemic and hypolipidemic effects and antioxidant activity of fruit extracts from *Lycium barbarum*. *Life Sci.* 76: 137–149.

Lutterodt, G.D. 1992. Inhibition of Microlax induced experimental diarrhea with narcotic like extracts of *Psidium guajava* leaf in rats. *J. Ethnopharmacol.* 37: 151–157.

Lutterodt, G.D. and Maleque, A. 1988. Effects of mice locomotor activity of a narcotic-like principle from *Psidium guajava* leaves. *J. Ethnopharmacol.* 24: 219–231.

Ma, D.Q., Jing, Z.J., Xu, S.Q., Yu, X., Hu, X.M., and Pan, H. 2012. Effects of flavonoids in *Morus indica* on blood lipids and glucose in hyperlipidemia-diabetic rats. *Chinese Herbal Med.* 4: 314–318.

Ma, W., Nomura, M., Takahashi-Nishioka, T., and Kobayashi, S. 2007. Combined effects of fangchinoline from *Stephania tetrandra* Radix and formononetin and calycosin from *Astragalus membranaceus* Radix on hyperglycemia and hypoinsulinemia in streptozotocin-diabetic mice. *Biol. Pharm. Bull.* 30: 2079–2083.

Ma, W., Wang, K.-J., Cheng, C.-H., Yan, G., Lu, W.-L., Ge, J.-F. et al. 2014. Bioactive compounds from *Cornus officinalis* fruits and their effects on diabetic nephropathy. *J. Ethnopharmacol.* 153: 840–845.

Ma, X., Egawa, T., Kimura, H., Karaike, K., Masuda, S., Iwanaka, N. et al. 2010. Berberine-induced activation of 5'-adenosine monophosphate-activated protein kinase and glucose transport in rat skeletal muscles. *Metabolism* 59: 1619–1627.

MacFarlane, A.J., Burghardt, K.M., Kelly, J., Simell, T., Simell, O., Altosaar, I., and Scott, F.W. 2003. A type 1 diabetes related protein from wheat (*Triticum aestivum*). *J. Biol. Chem.* 278: 54–63.

Macharla, S.P. 2011. Antidiabetic activity of *Bambusa arundinacea* seed extract on alloxan induced diabetic rats. *Int. J. Pharm. Res. Dev.* 3: 83–86.

Macharla, S.P., Goli, V., and Sattla, S.R. 2012. Anti-diabetic activity of *Rephanus sativus* L. leaves extracts on alloxan induced diabetic rats. *J. Chem. Pharm. Res.* 4: 1519–1522.

Mackenzie, R.W.A. and Elliott, B.T. 2014. Akt/PKB activation and insulin signaling: A novel insulin signaling pathway in the treatment of type 2 diabetes. *Diabetes Metab. Syndr. Obes. Targets Ther.* 7: 55–64.

Madan, S., Ahmad, S., Singh, J.N., Kohli, K., Kumar, Y., Singh, R. et al. 2010. *Stevia rebaudiana* (Bert.) Bertoni—A review. *Indian J. Nat. Prod. Resour.* 1: 267–286.

Madani, H., Mahmoodabady, N.A., and Vahdati, A. 2005. Effects of hydroalcoholic extract of *Anethum graveollens* (dill) on plasma glucose and lipid levels in diabetes induced rats. *Iranian J. Diabetes Lipid Disord.* 5: 35–42.

Madhavan, V., Nagar, J.C., Murali, A., Mythreyi, R., and Yoganarasimhan, S.N. 2008. Antihyperglycemic activity of alcohol and aqueous extracts of *Pandanus fascicularis* Lam. roots in alloxan induced diabetic rats. *Pharmacologyonline* 3: 529–536.

Maghrani, M., Lemhadri, A., Jouad, H., Michel, J.B., and Eddouks, M. 2003. Effect of the desert plant *Retama raetam* on glycaemia in normal and streptozotocin-induced diabetic rats. *J. Ethnopharmacol.* 87: 21–25.

Maghrani, M., Lemhadri, A., Zeggwagh, N.A., El-Amraoui, A., Haloui, M., Jouad, H. et al. 2004a. Effect of *Retama raetam* on lipid metabolism in normal and recent-onset diabetic rats. *J. Ethnopharmacol.* 90: 323–329.

Maghrani, M., Lemhadri, A., Zeggwagh, N.A., El-Amraoui, M., Haloui, M., Jouad, H. et al. 2004b. Effect of an aqueous extract of *Triticum repens* on lipid metabolism in normal and recent onset diabetic rats. *J. Ethnopharmacol.* 90: 331–337.

Maghrani, M., Michel, J.B., and Eddouks, M. 2005. Hypoglycaemic activity of *Retama raetam* in rats. *Phytother. Res.* 19: 125–128.

Maghrani, M., Zeggwagh, N.A., Lemhadri, A., El-Amraoui, M., Michel, J.B., and Eddoukks, M. 2004c. Study of the hypoglycaemic activity of *Fraxinus excelsior* and *Silybum marianum* in an animal model of type 1 diabetes mellitus. *J. Ethnopharmacol.* 91: 309–316.

Magied, M.M.A., Salama, N.A.R., and Ali, M.R. 2014. Hypoglycemic and hypocholesterolemia effects of intragastric administration of dried red chili pepper (*Capsicum annum*) in alloxan-induced diabetic male albino rats fed with high-fat-diet. *J. Food Nutr. Res.* 2: 850–856.

Magni, C., Sessa, F., Accardo, E., Vanoni, M., Morazzoni, P., Scarafoni, A., and Durant, M. 2004. Conglutin-γ, a lupin seed protein binds insulin *in vitro* and reduces plasma glucose levels of hyperglycemic rats. *J. Nutr. Biochem.* 15: 646–650.

Mahadevan, S. and Park, Y. 2008. Multifaceted therapeutic benefits of *Ginkgo biloba* L.: Chemistry, efficacy, safety, and uses. *J. Food Sci.* 73: 14–19.

Mahadevan, S.N. and Kamboj, P. 2010. Anti-hyperlipidemic effect of hydroalcoholic extract of Kenaf (*Hibiscus cannabinus* L.) leaves in high fat fed rats. *Ann. Biol. Res.* 1: 174–181.

Mahendran, G., Thamotharan, G., Sengottuvelu, S., and Bai, V.N. 2014. Anti-diabetic activity of *Swertia corymbosa* (Griseb.) Wight ex C.B. Clarke aerial parts extract in streptozotocin induced diabetic rats. *J. Ethnopharmacol.* 151: 1175–1183.

Mai, T.T. and Chuyen, N.V. 2007. Anti-hyperglycemic activity of an aqueous extract from flower buds of *Cleistocalyx operculatus* (Roxb.) Merr and Perry. *Biosci. Biotechnol. Biochem.* 71: 69–76.

Mai, T.T., Fumie, N., and Chuyen, N.V. 2009. Antioxidant activities and hypolipidemic effects of an aqueous extract from flower buds of *Cleistocalyx operculatus* (roxb.) Merr. and Perry. *J. Food Biochem.* 33: 780–807.

Maithili, V., Dhanabal, S.P., Mahendran, S., and Vadivelan, R. 2011. Antidiabetic activity of ethanolic extract of tubers of *Dioscorea alata* in alloxan induced diabetic rats. *Indian J. Pharmacol.* 43: 455–459.

Maiti, A., Dewanjee, S., Jana, G., and Mandal, S.C. 2008. Hypoglycemic effect of *Swietenia macrophylla* seeds against type 2 diabetes. *Int. J. Green Pharm.* 2: 224–227.

Maiti, R., Jana, D., Das, U.K., and Ghosh, D. 2004. Anti-diabetic effect of aqueous extract of seed of *Tamarindus indica* in streptozotocin-induced diabetic rats. *J. Ethnopharmacol.* 92: 85–91.

Maity, P., Hansda, D., Bandyopadhyay, U., and Mishra, D.K. 2009. Biological activities of crude extracts and chemical constituents of Bael, *Aegle mammelos* (L.) Corr. *Int. J. Exp. Biol.* 47: 849–861.

Maity, S., Firdous, S.M., and Debnath, R. 2013. Evaluation of antidiabetic activity of ethanolic extract of *Sechium edule* fruits in alloxan-induced diabetic rats. *World J. Pharm. Pharm. Sci.* 5: 3612–3621.

Majeed, H., Bokhari, T.Z., Sherwani, S.K., Younis, U., Shah, M.H.R., and Khaliq, B. 2013. An overview of biological, phytochemical and pharmacological values of *Abes pindrow*. *J. Pharmacogn. Phytochem.* 2: 182–187.

Majumder, P. and Paridhavi, M. 2013. An ethnophytochemical and pharmacological review on novel Indian medicinal plants used in herbal formulations. *Int. J. Pharm. Pharm. Sci.* 5: 74–83.

Majumder, R. and Akter, S. 2014. Anti-oxidant and anti-diabetic activities of the methanolic extract of *Premna integrifolia* bark. *Adv. Biol. Res.* 8: 29–36.

Makare, N., Bodhankar, S., and Rangari, V. 2001. Immunomodulatory activity of alcohol extract of *Mangifera indica* L. in mice. *J. Ethnopharmacol.* 78: 133–137.

Malakul, W., Thirawarapan, S., Ingkaninan, K., and Sawasdee, P. 2011. Effects of *Kaempferia parviflora* Wall. ex Baker on endothelial dysfunction in streptozotocin-induced diabetic rats. *J. Ethnopharmacol.* 133: 371–377.

Malalavidhane, T.S., Wickramasinghe, S.M., and Jansz, E.R. 2000. Oral hypoglycemic activity of *Ipomoea aquatica*. *J. Ethnopharmacol.* 72: 293–298.

Malalavidhane, T.S., Wickramasinghe, S.M., and Jansz, E.R. 2001. An aqeous extract of the green leafy vegetable *Ipomoea aquatic* is as effective as the oral hypoglycemic drug tolbutamide in reducing the blood sugar levels of Wistar rats. *Phytother. Res.* 15: 635–637.

Malalavidhane, T.S., Wickramasinghe, S.M., Perera, M.S., and Jansz, E.R. 2003. Oral hypoglycemic activity of *Ipomoea aquatic* in streptozotocin-induced diabetic Wistar rats and type 2 diabetes. *Phytother. Res.* 17: 1098–1100.

Malarvili, A., Selvaraja, P., Ndyeabura, A.W., Akowuah, G.A., and Okechukwu, P.N. 2011. Antidiabetic activity of crude stem extracts of *Coscinium fenestratum* on streptozotocin-induced type-2 diabetic rats. *Asian J. Pharm. Clin. Res.* 4: 47–51.

Malathi, R., Rajan, S.S., Gopalakrishnan, G., and Suresh, G. 2002. Azadirachtol, a tetranortriterpenoid from neem kernels. *Acta Crystallogr. C* 58: 708–710.

Mali, P.Y., Bigoniya, P., Panchal, S.S., and Muchhandi, I.S. 2013. Anti-obesity activity of chloroform-methanol extract of *Premna integrifolia* in mice fed with cafeteria diet. *J. Pharm. Bioallied Sci.* 5: 229–236.

Malini, T. and Vanithakumari, G. 1987. Estrogenic activity of *Cuminum cyminum* in rats. *Indian J. Exp. Biol.* 5: 442–444.

Mall, G.K., Mishra, P.K., and Prakash, V. 2009. Antidiabetic and hypolipidemic activity of *Gymnema sylvestre* in alloxan induced diabetic rats. *Global J. Biotech. Biochem.* 4: 37–42.

Maloney, K.P., Truong, V., and Allen, J.C. 2014. Susceptibility of sweet potato (*Ipomoea batatas*) peel proteins to digestive enzymes. *Food Sci. Nutr.* 2: 351–360.

Malpani, S.N. and Manjunath, K.P. 2013. *In vitro* α-amylase inhibition and chromatographic isolation of *Cassia fistula* Linn. bark. *Int. J. Adv. Pharm. Biol. Chem.* 2: 16–20.

Malvi, R., Jain, S., Khatri, S., Patel, A., and Mishra, S. 2011. A review on anti-diabetic medicinal plants and marketed herbal formulations. *Int. J. Pharm. Biol. Arch.* 2: 1344–1355.

Manaiyan, M., Ghannadi, A., Movahedian, A. Ramezanlou, P., and Osooli, F.S. 2014. Effect of the hydroalcoholic extract and juice of *Prunus divaricata* fruit on blood glucose and serum lipids of normal and streptozotocin-induced diabetic rats. *Res. Pharm. Sci.* 9: 421–429.

Mandade, S. and Sreenivas, S.A. 2011. Anti-diabetic effects of aqueous ethanolic extract of *Hibiscus rosasinensis* L. on streptozotocin-induced diabetic rats and the possible morphologic changes in the liver and kidney. *Int. J. Pharmacol.* 7: 363–369.

Mandal, S., Barik, B., Mallick, C., De, D., and Ghosh, D. 2008. Therapeutic effect of ferulic acid, an ethereal fraction of ethanolic extract of seed of *Syzygium cumini* against streptozotocin-induced diabetes in male rat. *Methods Find. Exp. Clin. Pharmacol.* 30: 121–128.

Mani, P., Kumar, A.R., Bastin, T.M., Jenifer, S., and Arumugam, M. 2010. Comparative evaluation of extracts of *C. igneus* (or *C. pictus*) for hypoglycemic and hypolipidemic activity in alloxan diabetic rats. *Int. J. Pharm. Tech.* 2: 183–195.

Manickam, M., Ramanathan, M., Jahromi, M.A., Chansouria, J.P., and Ray, A.B. 1997. Antihyperglycemic activity of phenolics from *Pterocarpus marsupium*. *J. Nat. Prod.* 60: 609–610.

Manickam, V.S. and Irudayaraj, V. 1992. *Pteridophytes Flora of the Western Ghats-South India*. New Delhi, India: BI Publications.

Manikandan, R., Thiagarajan, R., Beulaia, S., Sivakumar, M.R., Meiyalagan, V., Sundaram, R. et al. 2011. 1, 2 di-substituted idopyranose from *Vitex negundo* L. protects against streptozotocin-induced diabetes by inhibiting nuclear factor-kappa B and inducible nitric oxide synthase expression. *Microsc. Res. Tech.* 74: 301–307.

Manjula, S. and Ragavan, B. 2007. Hypoglycemic and hypolipidemic effects of *Coccinia indica* Wight & Arn in alloxan-induced diabetic rats. *Anc. Sci. Life* 27: 34–37.

Manohar, D., Viswanatha, G.L., Nagesh, S., Jain, V., and Shivaprasad, H.N. 2012. Ethnopharmacology of *Lepidium sativum* L. (Brassicaceae): A review. *Int. J. Phytother. Res.* 2: 1–7.

Manohar, R., Shaik, M.Z., Keerthi, V., Sravani, E., Srinivasulu, G., Janapati, Y.K. et al. 2011. Hypoglycemic and hypolipidemic effects of ethanolic extract of *Talinum cuneifolium* (Willd) whole plant on streptozotocin induced diabetic rats. *Int. J. Adv. Pharm. Sci.* 2: 184–192.

Manoharan, S. and Kaur, J. 2013. Anticancer, antiviral, antidiabetic, antifungal and phytochemical constituents of medicinal plants. *Am. J. Pharm. Tech. Res.* 3: 152–163.

Manoharan, S., Umadevi, S., Jayanthi, S., and Baskaran, N. 2011. Antihyperglycemic effect of *Coscinium fenestratum* and *Catharanthus roseus* in alloxan-induced diabetic rats. *Int. J. Nutr. Pharmacol. Neurol. Dis.* 1: 189–193.

Manosroi, J., Moses, Z.Z., Manosroi, W., and Manosroi, A. 2011a. Hypoglycemic activity of Thai medicinal plants selected from the Thai/Lanna medicinal recipe database MANOSROI II. *J. Ethnopharmacol.* 138: 92–98.

Manosroi, J., Zaruwa, M.Z., and Manosroi, A. 2011b. Potent hypoglycemic effect of Nigerian anti-diabetic medicinal plants. *J. Complement. Integr. Med.* 8. DOI: 10.2202/1553-3840.1482. PMID: 22754948.

Mansar-Benhamza, L., Djerrou, Z., and Hamdi, P.Y. 2013. Evaluation of anti-hyperglycemic activity and side effects of *Erythraea centaurium* (L.) Pers. in rats. *Afr. J. Biotechnol.* 12: 6980–6985.

Mansour, A.A., Al-Maliky, A.A., Kasem, B., Jabar, A., and Mosbeh, K.A. 2014. Prevalence of diagnosed and undiagnosed diabetes mellitus in adults aged 19 years and older in Basrah, Iraq. *Diabetes Metab. Syndr. Obes. Targets Ther.* 2014: 139–144.

Mansour, H.A. and Newairy, A.A. 2000. Amelioration of impaired renal function associated with diabetes by *Balanites aegyptiaca* fruits in streptozotocin-induced diabetic rats. *J. Med. Res. Inst.* 21: 115–125.

Mansouri, M., Nayebi, N., Keshtkar, A., Hasani-Ranjbar, S., Taheri, E., and Larijani, B. 2012. The effect of 12 weeks *Anethum graveolens* (dill) on metabolic markers in patients with metabolic syndrome; a randomized double blind controlled trial. *Daru* 20: 47.

Mao, X.Q., Yu, F., Wang, N., Wu, Y., Zou, F., Wu, K. et al. 2009. Hypoglycemic effect of polysaccharides enriched extract of *Astragalus membranaceus* in diet induced insulin resistant C57BL/6J mice and its potential mechanism. *Phytomedicine* 16: 416–425.

Mapanga, R.P., Tufts, M.A., Shode, F.O., and Musabayane, C.T. 2009. Renal effects of plant drived oleanoic acid in streptozotocin-induced diabetic rats. *Endocr. Abstr.* 19: 124.

Maquvi, K.J., Ali, M., and Ahamad, J. 2012. Anti-diabetic activity of aqueous extract of *Coriandrum sativam* L. fruits in streptozotocin-induced rats. *Int. J. Pharm. Pharm. Sci.* 4: 239–240.

Mard, S.A., Jalalvand, K., Jafarinejad, M., Balochi, H., and Naseri, M.K. 2010. Evaluation of the antidiabetic and antilipaemic activities of the hydroalcoholic extract of *Phoenix dactylifera* palm leaves and its fractions in alloxan-induced diabetic rats. *Malays. J. Med. Sci.* 17: 4–13.

Mariee, A.D., Abd-Allah, G.M., and El-Yamany, M.F. 2009. Renal oxidative stress and nitric oxide production in streptozotocin-induced diabetic nephropathy in rats: The possible modulatory effects of garlic (*Allium sativum* L.). *Biotechnol. Appl. Biochem.* 5: 227–232.

Marin, N.J.A. 1988. Cardiovascular effects of berberine in patients with severe congestive heart failure. *Clin. Cardiol.* 11: 253–260.

Markey, O., McClean, C.M., Medlow, P., Davison, G.W., Trinick, T.R., Duly, E., and Shafat, A. 2011. Effect of cinnamon on gastric emptying, arterial stiffness, postprandial lipemia, glycemia, and appetite responses to high-fat breakfast. *Cardiovasc. Diabetol.* 10: 78–82.

Marles, R.J. and Farnsworth, N.R. 1995. Anti-diabetic plants and their active constituents. *Phytomedicine* 2: 137–189.

Maroo, J., Vasu, V.T., and Gupta, S. 2003. Dose dependent hypoglycemic effect of aqueous extract of *Enicostemma littorale* Blume in alloxan-induced diabetic rats. *Phytomedicine* 10: 196–199.

Martineau, L.C., Couture, A., Spoor, D., Benhaddou-Andaloussi, A., Harris, C., Meddah, B. et al. 2006. Anti-diabetic properties of the Canadian lowbush blueberry *Vaccinium angustifolium* Ait. *Phytomedicine* 13: 612–623.

Martz, L. and Writer, S. 2014. Critical mass in diabetes. *SciBX* 7: 1–2.

Masih, M., Banerjee, T., Banerjee, B., and Pal, A. 2011. Anti-diabetic activity of *Acalypha indica* L. in normal and alloxan-induced diabetic rats. *Int. J. Pharm. Pharm. Sci.* 3: 51–55.

Masirkar, V.J., Deshmukh, V.N., Jadhav, J.K., and Sakarkar, D.M. 2008. Anti-diabetic activity of ethanol extract of *Pseudarthria viscida* root against alloxan-induced diabetes in rats. *Res. J. Pharm. Technol.* 1: 541–542.

Masjedi, F., Gol, A., and Dabira, S. 2013. Preventive effects of garlic (*Allium sativum* L.) on serum biochemical factors and histopathology of pancreas and liver in streptozotocin induced rats. *Iranian J. Pharm. Res.* 12: 325–338.

Masri, M.A., Mohammad, M.K., and Tahaa, M.O. 2009. Inhibition of dipeptidyl peptidase IV (DPP IV) is one of the mechanisms explaining the hypoglycemic effect of berberine. *J. Enzyme Inhib. Med. Chem.* 24: 1061–1066.

Masso, L.J. and Adzet, T. 1976. Hypoglycaemic activity of *Centaurea aspera* L. *Rev. Esp. Fisiol.* 32: 313–316. [Article in Spanish].

Mastache, J.M.N., Ramrez, M.L.G., Alvarez, L., and Delgado, G. 2006. Antihyperglycemic activity and chemical constituents of *Eysenhardtia platycarpa*. *J. Nat. Prod.* 69: 1687–1691.

Mathew, K.O., Olugbenga, O.S., Olajide, A.O., and Doyin, A.O. 2006a. The effect of *Bridelia ferruginea* and *Senna alata* on plasma glucose concentration in normoglycemic and glucose induced hyperglycemic rats. *Ethnobot. Leaflets* 10: 209–218.

Mathew, O.O., Olusola, L., and Mathew, A. 2012. Preliminary study of hypoglycaemic and hypolipidemic activity of aqueous root extract of *Ricinus communis* in alloxan-induced diabetic rats. *J. Phys. Pharm. Adv.* 2: 354–359.

Mathew, P.T. and Augusti, K.T. 1973. Studies on the effect of allicin (diallyldisulphide-oxide) on alloxan diabetes I. Hypoglycaemic action and enhancement of serum insulin effect and glycogen synthesis. *Indian J. Biochem. Biophys.* 10: 209–212.

Mathew, S., Singh, D., Jaiswal, S., Kumar, M., Jayakar, B., and Bhowmik, D. 2013. Anti-diabetic activity of *Kalanchoe pinnata* (Lam.) Pers. in alloxan induced diabetic rats. *J. Chem. Pharm. Sci.* 6: 1–7.

Mathews, J.N., Flatt, P.R., and Abdel-Wahab, Y.H. 2006b. *Asparagus adscendens* (Shweta musali) stimulates insulin secretion, insulin action and inhibits starch digestion. *Br. J. Nutr.* 95: 576–581.

Matsuda, H., Cai, H., Kubo, M., Tosa, H., and Iinuma, M. 1995. Study on anti-cataract drugs from natural sources. Effects of buddlejae flos on *in vitro* ALR2 activity. *Biol. Pharm. Bull.* 18: 463–466.

Matsuda, H., Morikawa, T., Ueda, H., and Yoshikawa, M. 2001. Medicinal food stuffs. XXVI. Inhibitors of aldose reductase and new triterpene and its oligoglycoside, centellasapogenol A and centellasaponin A, from *Centella asiatica* (Gotu Kola). *Heterocycles* 55: 1499–1504.

Matsuda, H., Murakami, T., Li, Y., Yamahara, J., and Yoshikawa, M. 1998. Mode of action of escins Ia and IIa and E,Z-senegin II on glucose absorption in gastrointestinal tract. *Bioorg. Med. Chem.* 6: 1019–1023.

Matsuda, H., Murakami, T., Yashiro, K., Yamakara, J., and Yoshikawa, M. 1999. Anti-diabetic principles of natural medicines. IV. Aldose reductase and alpha glucosidase inhibitors from the roots of *Salacia oblonga* Wall. (Celastraceae): Structure of a new friedelane-type triterpene, kotalagenin 16-acetate. *Chem. Pharm. Bull.* (*Tokyo*) 47: 1725–1729.

Matsuda, H., Nishida, N., and Yoshikawa, M. 2002. Antidiabetic principles of natural medicines. V. Aldose reductase inhibitors from *Myrcia multiflora* DC. (2): Structures of myrciacitrins III, IV, and V. *Chem. Pharm. Bull.* 50: 429–431.

Matsumura, T., Kasai, M., Hayashi, T., Arisawa, M.Y., Arai, I., Amagaya, S., and Komatsu, Y.A. 2000. Glucosidase inhibitors from Paraguayan natural medicine, Nangapiry, the leaves of *Eugenia uniflora*. *Pharm. Biol.* 38: 302–307.

Matsura, T., Yoshikawa, Y., Masui, H., and Sano, M. 2004. Suppression of glucose absorption by various health teas in rats. *Yakugaku Zasshi* 124: 217–223.

Mawa, S., Husain, K., and Jantan, I. 2013. *Ficus carica* L. (Moraceae): Phytochemistry, traditional uses and biological activities. *Evid. Based Complement. Altern. Med.* 2013: Article ID 974256, 8 pages.

Mazumdar, P.M., Faeswan, M., and Parcha, V. 2009. Hypoglycemic effect of *Ficus arnottiana* Miq. bark extracts on streptozotocin-induced diabetes in rats. *Nat. Prod. Radiance* 8: 478–482.

Mazundar, U.K., Gupta, M., and Rajeshwar, Y. 2005. Anti-hyperglycemic effect and anti-oxidant potential of *Phyllanthus niruri* (Euphorbiaceae) in streptozotocin-induced diabetic rats. *Eur. Bull. Drug Res.* 13: 15–23.

Mbagwu, H.O.C., Jackson, C., Jackson, I., Ekpe, G., Eyaekop, U., and Essien, G. 2011. Evaluation of the hypoglycemic effect of aqueous extract of *Phyllanthus amarus* in alloxan-induced diabetic albino rats. *Int. J. Pharm. Biomed. Res.* 2: 158–160.

Mbaka, G.O., Ogbonnia, S.O., and Banjo, A.E. 2011. Antihyperglycaemic and antihyperlidaemic effects of *Raphia hookeri* root extract on alloxan induced diabetic rats. Abstract from the *59th International Congress and Annual Meeting of the Society for Medicinal Plant and Natural Product Research*, Antalya, Turkey, September 4–9, 2011. *Planta Med.* 77: PH 14.

Mbaka, G.O., Ojewale, A.O., Ogbonnia, S.O., and Ota, D.A. 2014. Anti-hyperglycemic and antihyperlipidaemic activities of aqueous ethanol root extract of *Pseudocedrela kotschyi* on alloxan-induced diabetic rats. *Br. J. Med. Med. Res.* 4: 5839–5852.

Mbouangouere, R.N., Tane, P., Ngamga, D., Khan, S.N., Choudhary, M.I., and Ngadjui, B.T. 2007. A new steroid and α-glucosidase inhibitors from *Anthocleista schweinfurthii*. *Res. J. Med. Plant* 1: 106–111.

McCarty, M.F. 2005. Potential utility of natural polyphenols for reversing fat-induced insulin resistance. *Med. Hypotheses* 64: 628–635.

McIntosh, C.H. 2008. Dipeptidyl peptidase IV inhibitors and diabetes therapy. *Front. Biosci.* 13: 1753–1773.

McIntosh, C.H., Demuth, H.U., Pospisilik, J.A., and Pederson, R. 2005. Dipeptidyl peptidase IV inhibitors: How do they work as new antidiabetic agents? *Regul. Pept.* 128: 159–165.

McKenna, D.J., Jones, K., and Hughes K. 2001. Efficacy, safety, and use of *Ginkgo biloba* in clinical and preclinical applications. *Altern. Ther. Health Med.* 86: 88–90.

Mediratta, P.K., Sharma, K.K., and Singh, S. 2002. Evaluation of immunomodulatory potential of *Ocimum sanctum* seed oil and its possible mechanism of action. *J. Ethnopharmacol.* 80: 15–20.

Meena, M.C., Meena, R.K., and Patni, V. 2014. Ethnobotanical studies of *Citrullus colocynthis* (Linn.) Schrad. an important threatened medicinal herb. *J. Med. Plants Stud.* 2: 15–22.

Meenakshi, P., Bhuvaneshwari, R., Rathi, M.A., Thiruoorthi, L., Guravaiah, D.C., Jiji, M.J. et al. 2010. Antidiabetic activity of *Zaleya decandra* in alloxan-induced diabetic rats. *Appl. Biochem. Biotechnol.* 162: 1153–1159.

Meenu, J., Sunil, S., and Manoj, K. 2011. Evaluation of anti-hyperglycemic activity of *Dodonaea viscosa* leaves in normal and streptozotocin-diabetic rats. *Int. J. Pharm. Pharm. Sci.* 3: 69–74.

Mehranjani, M.S., Shariatzadeh, M.A., Desfulian, A.R., Noori, M., Abnosi, M.H., and Moghadam, Z.H. 2007. Effects of *Medicago sativa* on nephropathy in diabetic rats. *Indian J. Pharm. Sci.* 69: 768–772.

Mehta, R. and Lansky, E.P. 2004. Breast cancer chemopreventive properties of pomegranate (*Punica granatum*) fruit extracts in a mouse mammary organ culture. *Eur. J. Cancer Prev.* 13: 345–348.

Melendez-Camargo, M.E., Castillo-Najera, R., Silva-Torres, R., and Campos-Aldrete, M.E. 2006. Evaluation of the diuretic effect of the aqueous extract of *Costus pictus* D. Don in rat. *Proc. West Pharmacol. Soc.* 49: 72–74.

Meliani, N., Dib, M.E.A., Attali, H., and Tabti, B. 2011. Hypoglycaemic effect of *Berberis vulgaris* L. in normal and streptozotocin-induced diabetic rats. *Asian J. Trop. Biomed.* 1: 468–471.

Memişoğullari, R. and Bakan, E. 2004. Levels of ceruloplasmin, transferin and lipid peroxidation in the serum of patients with type 2 diabetes mellitus. *J. Diabetes Complications* 18: 193–197.

Meng, G., Wang, M., Zhang, K., Guo, Z., and Shi, J. 2014. Research progress on the chemistry and pharmacology of *Prunella vulgaris* species. *Open Access Libr. J.* 1: 1–19.

Meng, L.Y., Zhu, L.X., Zheng, H.H., Gu, C.S., and Cai, X.Y. 2004. Effect of PDB on hyperglycemic animal models. *Chin. Pharmacol. Bull.* 20: 588–590.

Meng, Q.Y., Lv, X.F., and Jin, X.D. 2008. Effect of *Rehmannia glutinosa* Libosch water extraction on gene expression of proinsulin in type 2 diabetes mellitus rats. *Zhong Yao Cai* 31: 397–399.

Menon, P.V. and Kurup, P.A. 1976. Hypolipidemic action of the polysaccharide from *Phaseolus mungo* (blackgram): Effect on lipid metabolism. *Indian J. Biochem. Biophys.* 13: 46–48.

Mentreddy, S.R., Mohamed, A.I., and Rimando, A.M. 2005. Medicinal plants with hypoglycemic/anti-hyperglycemic properties: A review. *Proceedings of the Annual Meeting of the Association for the Advancement of Industrial Crops,* September 17–21, 2005, Murica, Spain, pp. 341–353.

Merghache, S., Zerriouh, M., Merghache, D., Tabti, B., Djaziri, R., and Ghalem, S. 2013. Evaluation of hypoglycaemic and hypolipidemic activities of Globularin isolated from *Globularia alypum* L. in normal and streptozotocin-induced diabetic rats. *J. Appl. Pharm. Sci.* 3: 1–7.

Merina, A.J., Kesavan, D., and Sulochana, D. 2011. Isolation and anti-hyperglycemic activity of flavonoid from flower petals of *Opuntia stricta*. *Pharm. Chem. J.* 45: 317–326.

Mesaik, M.A., Ahmed, A., Khalid, A.S., Jan, S., Siddiqui, A.A., Perveen, S. et al. 2013. Effect of *Grewia asiatica* fruit on glycemic index and phagocytosis tested in healthy human subjects. *Pak. J. Pharm. Sci.* 26: 85–89.

Metwally, N.S., Mohamed, A.M., and Sharabasy, F.S.E.L. 2012. Chemical constituents of the Egyptian plant *Anabasis articulata* (Forssk) Moq and its anti-diabetic effects on rats with streptozotocin-induced diabetic hepatopathy. *J. Appl. Pharm. Sci.* 2: 54–65.

Michael, U.A., David, B.U., Theophine, C.O., Philip, F.U., Ogochukwu, A.M., and Benson, V.A. 2010. Antidiabetic effect of combined aqueous leaf extract of *Vernonia amygdalina* and metformin in rats. *J. Basic Clin. Pharm.* 1: 197–202.

Middha, S.K., Usha, T., and Pande, V. 2013. A review on anti-hyperglycemic and hepatoprotective activity of eco-friendly *Punica granatum* peel waste. *Evid. Based Complement. Altern. Med.* 2013: Article ID 656172, 10 pages.

Minaiyan, M., Ghannadi, A., Movahedian, A., and Hakim-Elahi, I. 2013. Effect of *Hordeum vulgare* L. (barley) on blood glucose levels of normal and streptozotocin-induced diabetic rats. *Res. Pharm. Sci.* 9: 173–178.

Minaiyan, M., Moattar, F., and Vali, A. 2006. Effect of *Securigera securidaca* seeds on blood glucose levels of normal and diabetic rats. *Iranian J. Pharm. Sci.* 2: 151–156.

Minaiyan, M., Zolfaghari, B., and Kamal, A. 2011. Effect of hydroalcoholic and buthanolic extract of *Cucumis sativus* seeds on blood glucose level of normal and streptozotocin-induced diabetic rats. *Iran. J. Basic Med. Sci.* 14: 436–442.

Mingrone, G. and Castagneto-Gissey, L. 2014. A central role of the gut in glucose homeostasis. *Nat. Rev. Endocrinol.* 10: 73–74.

Minich, D.M., Lerman, R.H., Darland, G., Babish, J.G., Pacioretty, M., Bland, J.S., and Tripp, M.L. 2010. Hop and Acacia phytochemicals decreased lipotoxicity in 3T3-L1 cells, db/db mice and individuals with metabolic syndrome. *J. Nutr. Metabol.* 2010: Article ID 467316, 11 pages.

Miranda-Perez, M.E., Escober-Villanueva, M.D., Camarillo, C.O., Sanchez-Munoz, F., Almanza-Perez, J.C., and Alarcon-Aguilar, F.J. 2013. *Cucurbita ficifolia* Bouche fruit acts as an insulin secretagogue in RINm5F cells. *Int. Biotechnol. Colour J.* 3: 8–15.

Mirza, R.H., Chi, N., and Chi, Y. 2013. Therapeutic potential of the natural product mangiferin in metabolic syndrome. *J. Nutr. Ther.* 2: 74–79.

Mishra, N. 2013. Haematological and hypoglycemic potential of *Anethum graveolens* seeds extract in normal and diabetic Swiss albino mice. *Vet. World* 6: 502–507.

Mishra, S., Aeri, V., Gaur, P.K., and Jachak, S.M. 2014. Phytochemical, therapeutic, and ethnopharmacological overview for a traditionally important herb: *Boerhavia diffusa* L. *Biomed. Res. Int.* 2014: Article ID 808302, 19 pages.

Mishra, S.B., Verma, A., Mukerjee, A., and Vijayakumar, M. 2011. Anti-hyperglycemic activity of leaves extract of *Hyptis suaveolens* L. Poit in streptozotocin induced diabetic rats. *Asian Pac. J. Trop. Med.* 4: 689–693.

Mishra, S.B., Verma, A., Mukerjee, A., and Vijayakumar, M. 2012. *Amaranthus spinosus* L. (Amaranthaceae) leaf extract attenuates streptozotocin-nicotinamide induced diabetes and oxidative stress in albino rats: A histopathological analysis. *Asian Pac. J. Trop. Biomed.* 2012: S1647–S1652.

Mishra, S.B., Verma, A., and Vijayakumar, M. 2013. Preclinical valuation of anti-hyperglycemic and antioxidant action of Nirmali (*Strychnos potatorum*) seeds in streptozotocin-nicotinamide-induced diabetic Wistar rats: A histopathological investigation. *Biomarkers Genomic Med.* 5: 157–163.

Mishra, S.B., Vijayakumar, M., Ojha, S.K., and Verma, A. 2010. Antidiabetic effect of *Jatropha curcas* leaves extract in normal and alloxan-induced diabetic rats. *Int. J. Pharm. Sci.* 2: 482–487.

Misra, H., Soni, M., Silawat, N., Mehta, D., Mehta, B.K., and Jain, D.C. 2011. Antidiabetic activity of medium-polar extract from the leaves of *Stevia rebaudiana* Bert. (Bertoni) on alloxan-induced diabetic rats. *J. Pharm. Bioallied Sci.* 3: 242–248.

Mitra, S. and Pandey, A. 2014. Potential antidiabetic plants of eastern India—A review article. *BMR Phytomed.* 1: 1–11.

Miura, T., Ichiki, H., Iwamoto, N., Kato, M., Kubo, M., Sasaki, H., Okada, M., Ishida, T., Seino, Y., and Tanigawa, K. 2001a. Antidiabetic activity of the rhizoma of *Anemarrhena asphodeloides* and active components, mangiferin and its glucoside. *Biol. Pharm. Bull.* 24: 1009–1011.

Miura, T., Itoh, C., Iwamoto, N., Kato, M., Kawai, M., Park, S.R., and Suzuki, I. 2001b. Hypoglycemic activity of the fruit of the *Momordica charantia* in type 2 diabetic mice. *J. Nutr. Sci. Vitaminol.* 47: 340–344.

Miura, T., Nosaka, K., Ishii, H., and Ishida, T. 2005a. Antidiabetic effect of Nitobegiku, the herb *Tithonia diversifolia*, in KK-Ay diabetic mice. *Biol. Pharm. Bull.* 28: 2152–2154.

Miura, T., Takagi, S., and Ishida, T. 2012. Management of diabetes and its complications with banaba (*Lagerstroemia speciosa* L.) and corosolic acid. *Evid. Based Complement. Altern. Med.* 2012: 1–8. DOI: 10.1155/2012/871495.

Miura, Y., Hosono, M., Oyamada, C., Odai, H., Oikawa, S., and Kondo, K. 2005b. Dietary, isohumulones, the bitter components of beer, rise plasma HDL-cholesterol levels and reduce liver cholesterol and triglyceride contents similar to PPAR-α activation in C57BL/6 mice. *Br. J. Nutr.* 93: 559–567.

Mnafgui K., Hamden, K., Salah, H.B., Kchaou, M., Nasri, M., Slama, S. et al. 2012. Inhibitory activities of *Zygophyllum album*: A natural weight-lowering plant on key enzymes in high-fat diet-fed rats. *Evid. Based Complement. Altern. Med.* 2012: 620384, 9 pages.

Mnafgui, K., Kchaou, M., Hamden, K., Derbali, F., Slama, S., Nasri, M. et al. 2014. Inhibition of carbohydrate and lipid digestive enzymes activities by *Zygophyllum album* extracts effect on blood and pancreas inflammatory biomarkers in alloxan-induced diabetic rats. *J. Physiol. Biochem.* 70: 93–106.

Mnonopi, N., Levendal, R.A., Mzilikazi, N., and Frost, C.L. 2012. Murrabiin, a constituent of *Leonotis leonurus*, alleviates diabetic symptoms. *Phytomedicine* 19: 488–493.

Modilal, M.R.D. and Daisy, P. 2011. Hypoglycemic effects of *Elephantopus scaber* in alloxan-induced diabetic rats. *Indian J. Novel Drug Deliv.* 3: 98–103.

Moerman, D.E. 1998. *Native American Ethnobotany.* Portland, OR: Timber Press.

Mohamed, A. 2011b. Antidiabetic, antihyperlipedemic and antioxidant effects of aqueous extract of the roots of *Cynara cornigera* in alloxan-induced experimental diabetes mellitus. *Int. J. Pharmacol.* 7: 782–789.

Mohamed, A.E.H., El-Sayed, M.A., Hegazy, M.E., Helaly, S.E., Esmail, A.M., and Mohamed, N.S. 2010. Chemical constituents and biological activities of *Artemisia herba-alba*. *Rec. Nat. Prod.* 4: 1–25.

Mohamed, E.A. K. 2011a. Antidiabetic, antihypercholestermic and antioxidative effect of *Aloe vera* gel extract in alloxan induced diabetic rats. *Aust. J. Basic Appl. Sci.* 5: 1321–1327.

Mohamed, E.A., Yam, M.F., Ang, L.F., Mohamed, A.J., and Asmawi, M.Z. 2013. Antidiabetic properties and mechanism of action of *Orthosiphon stamineus* Benth bioactive sub-fraction in streptozotocin-induced diabetic rats. *J. Acupunct. Meridian. Stud.* 6: 31–40.

Mohammadi, A., Gholamhoseinian, A., and Fallah, H. 2014. *Zataria multiflora* increases insulin sensitivity and PPARγ gene expression in high fructose fed insulin resistant rats. *Iran. J. Basic Med. Sci.* 17: 263–270.

Mohammadi, J., Delaviz, H., Malekzadeh, J.M., and Roozbehi, A. 2012. The effect of hydroalcoholic extract of *Juglans regia* leaves in streptozotocin-nicotinamide induced diabetic rats. *Pak. J. Pharm. Sci.* 25: 407–411.

Mohammadi, J., Saadipour, K., Delaviz, H., and Mohammadi, B. 2011. Antidiabetic effects of an alcoholic extract of *Juglans regia* in an animal model. *Turk J. Med. Sci.* 41: 685–691.

Mohammadi, S., Montasser, K.S., and Monaver, F.A. 2010. Antidiabetic properties of the ethanolic extract of *Rhus coriaria* fruits in rats. *Daru* 18: 270–275.

Mohammed, A. and Atiku, M.K. 2012. Antihyperglycemic and antihyperlipidemic effects of leaves and stem bark methanol extracts of *Senna siamea* in alloxan induced diabetic rats. *Curr. Res. Cardiovasc. Pharmacol.* 1: 10–17.

Mohammed, A., Ibrahim, M.A., and Islam, M.D.S. 2014. African medicinal plants with antidiabetic potentials: A review. *Planta Med.* 80: 354–377.

Mohammed, F., Mohammed, A., Ghori, S., and Ali, S. 2010. Antidiabetic activity of *Vinca rosea* extracts in alloxan-induced diabetic rats. *Int. J. Endocrinol.* 2010: Article ID 841090.

Mohammed, H., Hassan, A.D., Touseef, H., Mohammed, S.A.Q., and Ali, A.H.H. 2008. Anti-diabetic effect of alcohol extract of *Caralluma sinaica* L. on streptozotocin-induced diabetic rabbits. *J. Ethnopharmacol.* 117: 215–220.

Mohammed, R.K., Ibrahim, S., Atawodi, S.E., Eze, E.D., and Suleiman, J.B. 2012. Anti-diabetic and haematological effects of n-butanol fraction of *Alchornea cordifolia* leaf extract in streptozotocin-induced diabetic Wistar rats. *Global J. Med. Plant Res.* 1: 14–21.

Mohammed, S.A., Yaqub, A.G., Sanda, K.A., Nicholas, A.O., Arastus, W., Muhammad, M. et al. 2013. Review on diabetes, synthetic drugs and glycemic effects of medicinal plants. *J. Med. Plant Res.* 7: 2628–2637.

Mohan, C.G., Viswanatha Shastry, G.L., Savinay, G., Rajendra, C.E., and Halemani, P.D. 2013a. 1,2,3,4,6 penta-O-galloyl-β-D-glucose, a bioactivity guided isolated compound from *M. indica* inhibits 11-β-HSD-1 and ameliorates high fat diet induced diabetes in C57BL/6 mice. *Phytomedicine* 20: 417–426.

Mohan, M.R.M.K. and Mishra, S.H. 2010. Comprehensive review of *Clerodendron phlomidis*: A traditionally used bitter. *J. Chin. Integr. Med.* 8: 510–524.

Mohan, R.S., Shahid, M., Vinoth, K.S., Kumar, C.S., and Subal, D. 2012. Antidiabetic effect of *Luffa acutangula* fruits and histology of organs in streptozotocin induced diabetic in rats. *Res. J. Pharmacogn. Phytochem.* 4: 64–69.

Mohan, V., Sandeep, S., Deepa, R., Shah, B., and Varghese, C. 2007. Epidemiology of type 2 diabetes: Indian scenario. *Indian J. Med. Res.* 125: 217–230.

Mohan, V.R., Thangakrishnakumari, S., Nishanthini, A., and Muthukumarasamy, S. 2013b. Hypoglycemic and hypolipidemic effects of ethanol extract of *Sarcostemma secamone* (L.) Bennett (Asclepliadaceae) in alloxan induced diabetic rats. *Asian J. Pharm. Clin Res.* 16: 65–70.

Mohan, Y., Jesuthankaraj, G.N., and Thangavelu, N.R. 2013c. Anti-diabetic and anti-oxidant properties of *Triticum aestivam* in streptozotocin-induced diabetic rats. *Adv. Pharmacol. Sci.* 2013: 9. http://dx.doi. org/10.1155/2013/716073.

Mohanraj, R. and Sivasankar, S. 2014. Sweet potato (*Ipomoea batatas* (L.) Lam)—A valuable medicinal food: A review. *J. Med. Food* 17: 733–741.

Mohmood, S., Talat, A., Karim, S., Khurshid, R., and Zia, A. 2011. Effect of cinnamon extract on blood glucose level and lipid profile in alloxan induced diabetic rats. *Pak. J. Physiol.* 7: 13–16.

Mojekwu, T.O., Yama, O.E., and Ojokuku, S.A. 2011. Hypoglycaemic effects of aqueous extract of *Aframomum melegueta* leaf on alloxan-induced diabetic male albino rats. *Pac. J. Med. Sci.* 8: 28–36.

Mokhber-Dezfuli, N., Saeidnia, S., Gohari, A.R., and Kurepaz-Mahmoodabadi, M. 2014. Phytochemistry and pharmacology of *Berberis* species. *Pharmacogn. Rev.* 8: 8–15.

Mokhtari, M., Sharifi, S., and Shahamir, T.M. 2011. The effect of hydro-alcoholic seeds extract of *Ceratonia siliqua* on the blood glucose and lipids concentration in diabetic male rats. *Int. Conf. Life Sci. Technol.* 3: 82–86.

Molitch, M.E., DeFronzo, R.A., Franz, M.J., Keane, W.F., Mogensen, C.E., Parving, H.H. et al. 2004. Nephropathy in diabetes. *Diabetes Care* 27: 79–83.

Mollik, A.H., Khan, S.S., Lucky, S.A.S., Jahan, R., and Rahmatullah, M. 2009. Hypoglycemic effect of methanol and chloroform extract of *Cuscuta reflexa* Roxb. in mice. *Planta Med.* 75: 30–35.

Molokovskij, D.S., Davydov, V.V., and Khegay, M.D. 2002. Comparative estimation of antidiabetic activity of different adaptogenic vegetative preparations and extractions from plant material of some officinal medicinal plants. *Rastitelnye Resursy* 38: 15–28.

Moloudizargari, M., Mikaili, P., Aghajanshakeri, S., Asghari, M.H., and Shayegh, J. 2013. Pharmacological and therapeutic effects of *Peganum harmala* and its main alkaloids. *Pharmacogn. Rev.* 7: 199–212.

Momo, C.E.N. and Oben, E.J. 2009. *In vivo* assessment of hypoglycaemic and antioxydant activities of aqueous extract of *Terminalia superba* in alloxan-diabetic rats. *Planta Med.* 75: PH 1.

Momo, C.E.N., Oben, J.E., Tazoo, D., and Dongo, E. 2006a. Antidiabetic and hypolipidaemic effects of a methanol/methylene-chloride extract of *Laportea ovalifolia* (Urticaceae), measured in rats with alloxan-induced diabetes. *Ann. Trop. Med. Parasitol.* 100: 69–74.

Momo, C.E.N., Oben, J.E., Tazoo, D., and Dongo, E. 2006b. Antidiabetic and hypolipidemic effects of *Laportea ovalifolia* (Urticaceae) in alloxan induced diabetic rats. *Afr. J. Tradit. Complement. Altern. Med.* 3: 33–43.

Monago, C.C. and Akhidue, V. 2002. Antidiabetic effect of crude glycoside of *Abrus precatorius* in alloxan diabetic rabbits. *Global J. Pure Appl. Sci.* 9: 35–38.

Monago, C.C. and Alumanah, E.O. 2005. Antidiabetic effect of chloroform-methanol extract of *Abrus precatorius* Linn. seed in alloxan diabetic rabbit. *J. Appl. Sci. Environ. Manage.* 9: 85–88.

Moorthy, R., Prabhu, K.M., and Murthy, P.S. 2010a. Anti-hyperglycemic compound GII from fenugreek (*Trigonella foenum-graceum* Linn.) seeds, its purification and effect in diabetes mellitus. *Ind. J. Exp. Biol.* 48: 1111–1118.

Moorthy, R., Prabhu, K.M., and Murthy, P.S. 2010b. Mechanism of anti-diabetic action, efficacy and safety profile of GII purified from fenugreek (*Trigonella foenum-graceum* Linn.) seeds in diabetic animals. *Ind. J. Exp. Biol.* 48: 1119–1122.

Moqbel, F.S., Naik, P.R., Najma, H.M., and Selvaraj, S. 2011. Antidiabetic properties of *Hibiscus rosa sinensis* L. leaf extract fractions on non-obese diabetic (NOD) mouse. *Indian J. Exp. Biol.* 49: 24–29.

Moradi-Afrapoli, F., Asghari, B., Saeidnia, S., Ajani, Y., Mirjani, M., Malmir, M. et al. 2012. *In vitro* α-glucosidase inhibitory activity of phenolic constituents from aerial parts of *Polygonum hyrcanicum*. *DARU J. Pharm. Sci.* 20: 37–43.

Moraes, E.A., Rempel, C., Périco, E., and Strohschoen, A.A.G. 2010. Avaliação do perfil glicêmico de portadores de diabetes mellitus tipo II em UBSs que utilizam infusão de folhas de *Bauhinia forficata* Link. *Conscientiae Saúde* 9: 569–574.

Morikawa, T., Kishi, A., Ponpiriyadacha, Y., Matsuda, H., and Yoshikawa, M. 2003. Structures of friedelane-type triterpenes and eudesmanetype sesquiterpene and aldose reductase inhibitors from *Salacia*. *Chin. J. Nat. Prod.* 66: 1191–1196.

Mori-Okamota, J., Otawara-Hamamoto, Y., Yamato, H., and Yoshimura, H. 2004. Pomegranate extract improves a depressive state and bone properties in menopausal syndrome model ovariectomized mice. *J. Ethnopharmacol.* 92: 93–101.

Morrison, E.Y., Thompson, H., Pascoe, K., West, M., and Fletcher, C. 1991. Extraction of an hyperglycaemic principle from the annatto (*Bixa orellana*), a medicinal plant in the West Indies. *Trop. Geogr. Med.* 43: 184–188.

Morton, J. 1987. Soursop. In: *Fruits of Warm Climates*, Morton, J.F. (ed.). Published by Julia F. Morton, Miami, FL, pp. 75–80.

Moshi, M.J. and Nagpa, V. 2000. Effect of *Caesalpinia bonducella* seeds on blood glucose in rabbits. *J. Pharm. Biol.* 38: 81–86.

Mosihuzzaman, M., Nahar, N., Ali, L., Rokeya, B., Khan, A.K., Nur-E-Alam, M. et al. 1994. Hypoglycemic effects of three plants from eastern Himalayan belt. *Diabetes* 26: 127–138.

Mossa, J.S., Cassady, J.M., Antoun, M.D., Byrn, S.R., Mckenzie, A.T., Kozlowski, J.F. et al. 1985. Saudin, a hypoglycemic diterpenoid, with a novel 6,7-seco-labdane carbon skeleton, from *Cluytia richardiana*. *J. Org. Chem.* 50: 916–918.

Mostafa, U.E. and Labban, L. 2013. Effect of *Zizyphus jujube* on serum lipid profile and some anthropometric measurements. *Adv. Med. Plant Res.* 1: 49–55.

Mousinho, N.M., vanTonder, J.J., and Steenkampa, V. 2013. *In vitro* anti-diabetic activity of *Sclerocarya birrea* and *Ziziphus mucronata*. *Nat. Prod. Commun.* 6: 1279–1284.

Movahedian, A., Zolfaghari, B., Sajjadi, E., and Moknatjou, R. 2010. Antihyperlipidemic effect of *Peucedanum pastinacifolium* extract in streptozotocin-induced diabetic rats. *Clinics* 65: 629–633.

Mowla, M., Alauddin, M.D., Rahman A., and Ahmed, K. 2009. Antihyperglycemic effect of *Trigonella foenum-graecum* (fenugreek) seed extract in alloxan-induced diabetic rats and its use in diabetes mellitus: A brief qualitative phytochemical and acute toxicity test on the extract. *Afr. J. Tradit. CAM* 6: 255–261.

Mpiana, P.T., Masunda, T.A., Longoma, B.F., Tshibangu, D.S.T., and Ngbolua, K.N. 2013. Anti-hyperglycemic activity of *Raphia gentiliana* De Wild. (Arecaceae). *Eur. J. Med. Plants* 3: 233–240.

Mueller, M., Lukas, B., Novak, J., Simoncini, T., Genazzani, A.R., and Jungbauer, A. 2008. Oregano: A source for peroxisome proliferator-activated receptor gamma antagonists. *J. Agric. Food Chem.* 56: 11621–11630.

Mujeeb, M., Alam, K.S., Momhd, A., and Mall, A. 2009. Antidiabetic activity of the aqueous extract of *Annona squamosa* in streptozotocin induced hyperglycemic rats. *Pharm. Res.* 2: 59–63.

Mukesh, R. and Namita, P. 2013. Medicinal plants with anti-diabetic potential—A review. *American-Eurasian J. Agric. Environ. Sci.* 13: 81–94.

Mukherjee, P.K., Saha, K., Pal, M., and Saha, B.P. 1997. Effect of *Nelumbo nucifera* rhizome extract on blood sugar levels in rats. *J. Ethnopharmacol.* 58: 207–213.

Mukherjee, R., Dash, P.K., and Ram, G.C. 2005. Immunotherapeutic potential of *Ocimum sanctum* in bovine subclinical mastitis. *Res. Vet. Sci.* 79: 37–43.

Mukhtar, H.M., Ansari, S.H., Ali, M., Bhat, Z.A., and Naved, T. 2004a. Effect of aqueous extract of *Cyamopsis tetragonoloba* Linn. beans on blood glucose level in normal and alloxan-induced diabetic rats. *Ind. J. Exp. Biol.* 42: 1212–1215.

Mukhtar, H.M., Ansari, S.H., Ali, M., Naved, T., and Bhat, Z.A. 2004b. Effect of water extract of *Psidium guajava* leaves on alloxan-induced diabetic rats. *Pharmazie* 59: 734–735.

Mukhtar, H.M., Ansari, S.H., Bhat, Z.A., and Naved, T. 2006. Anti-hyperglycemic activity of *Cyamopsis tetragonaloba* Linn. beans on blood glucose levels in alloxan-induced diabetic rats. *Pharm. Biol.* 44: 10–13.

Mukti, M., Taznin, I., Banna, H., Morshed, M.T., Roney, M.S.I., Suvro, K.F.A. et al. 2013. *In vivo* anti-hyperglycemic and antinociceptive effects of methanolic extract of *Tamarindus indica* L. leaves. *Adv. Nat. Appl. Sci.* 7: 186–191.

Muller, C.J.F., Joubert, E., deBeer, D., Sanderson, M., Malherbe, C.J., Fey, S.J. et al. 2012. Acute assessment of an aspalathin-enriched green rooibos (*Aspalathus linearis*) extract with hypoglycemic potential. *Phytomedicine.* 20: 32–39.

Murali, B., Upadhyaya, U.M., and Goyal, P.K. 2002. Effect of chronic treatment with *Enicostemma littorale* in non-insulin-dependent diabetic (NIDDM) rats. *J. Ethnopharmacol.* 81: 199–204.

Murali, Y.K., Anand, P., Tandon, V., Singh, R., Chandra, R., and Murthy, P.S. 2007. Long-term effects of *Terminalia chebula* Retz. on hyperglycemia and associated hyperlipidemia, tissue glycogen content and *in vitro* release of insulin in streptozotocin induced diabetic rats. *Exp. Clin. Endocrinol. Diabetes* 115: 641–646.

Murali, Y.K., Chandra, R., and Murthy, P.S. 2004. Antihyperglycemic effect of water extract of dry fruits of *Terminalia chebula* in experimental diabetes mellitus. *Indian J. Clin. Biochem.* 19: 202–204.

Muralikrishna, K.S., Latha, K.P., Shreedhara, C.S., Vaidya, V.P., and Krupanidhi, A.M. 2008. Effect of *Bauhinia purpurea* Linn. on alloxan-induced diabetic rats and isolated frogs heart. *Int. J. Green Pharm.* 2: 83–86.

Murugan, E. 2009. A novel steroid from *Elephantopus scaber* L. an ethnomedicinal plant with antidiabetic activity. *Phytomedicine* 16: 252–257.

Murugan, M., Uma, C., and Reddy, M. 2009. Hypoglycemic and hypolipidemic activity of leaves of *Mucuna pruriens* in alloxan-induced diabetic rats. *J. Pharm. Sci. Technol.* 1: 69–73.

Muruganandan, S., Pant, S., Srinivasan, K., Chandra, S.K., Lal, J., Tandan, S., and Prakash, R.V. 2002. Inhibitory role of *Syzygium cumini* on autocoid-induced inflammation in rats. *Indian J. Physiol. Pharmacol.* 46: 482–486.

Muruganandan, S., Srinivasan, K., Gupta, S., Gupta, P.K., and Lal, J. 2005. Effect of mangiferin on hyperglycemia and atherogenicity in streptozotocin diabetic rats. *J. Ethnopharmacol.* 97: 497–501.

Musabayane, C.T., Mahlalela, N., Shode, F.O., and Ojewole J.A.O. 2005. Effects of *Sysygium cordatum* Hoscht. Ex. C.Krauss (Mytraceae) leaf extract on plasma glucose and hepatic glygogen in streptozotocin-induced diabetic rats. *J. Ethnopharmacol.* 97: 485–490.

Musabayane, C.T., Tufts, M.A., and Mapanga, R.F. 2010. Synergistic antihyperglycemic effects between plant-derived oleanolic acid and insulin in streptozotocin-induced diabetic rats. *Renal Fail.* 32: 832–839.

Muthukumran, P., Begum, V.H., and Kalaiarasan, P. 2011. Anti-diabetic activity of *Dodonaea viscosa* (L.) leaf extracts. *Int. J. PharmTech Res.* 3: 136–139.

Muthulakshmi, A., Jothibai, M.R., and Mohan, V.R. 2013. Anitidiabetic, antihyperlipidaemic and antioxidant activity of ethanol extract of *Feronia elephantum* Correa leaf and bark in normal and alloxan induced diabetic rats. *Int. Res. J. Pharm.* 4: 190–196.

Muthulingam, M. 2010. Antidiabetic efficacy of leaf extract of *Asteracantha longifolia* (Linn.) Nees on alloxan-induced diabetes in male albino Wistar rats. *Int. J. Pharm. Biomed. Res.* 1: 28–34.

Muthusamy, V.S., Anand, S., Sangeetha, K.N., Sujatha, S., Arun, B., and Lakshami, B.S. 2008. Tannins present in *Cichorium intybus* enhance glucose uptake and inhibit adipogenesis in 3T3-L1 adepocytes through PTP1B inhibition. *Chem. Biol. Interact.* 174: 69–78.

Mwafy, S.N. and Yassin, M.M. 2011. Antidiabetic activity evaluation of glimepiride and *Nerium oleander* extract on insulin, glucose levels and some liver enzymes activities in experimental diabetic rat model. *Pak. J. Biol. Sci.* 14: 984–990.

Nabeel, M.A., Kathiresan, K., and Manivannan, S. 2010. Anti-diabetic activity of the mangrove species *Ceriops decendra* in the alloxan-induced diabetic rats. *J. Diabetes* 2: 97–101.

Nabi, S.A., Kasetti, R.B., Sirasanagandla, S., Tilak, T.K., Kumar, M.V.J., and Rao, C.A. 2013. Antidiabetic and antihyperlipidemic activity of *Piper longum* root aqueous extract in streptozotocin induced diabetic rats. *BMC Complement. Altern. Med.* 13: 37.

Nadeem, M., Anjum, F.M., Khan, M.I., and Tehseen, S. 2013. Nutritional and medicinal aspects of coriander (*Coriandrum sativum* L.): A review. *Br. Food J.* 115: 743–755.

Nagamine, R., Ueno, S., Tsubata, M., Yamaguchi, K., Takagaki, K., Hira, T. et al. 2014. Dietary sweet potato (*Ipomoes batatas* L.) extract attenuates hyperglycemia by enhancing the secretion of glucagon-like peptide-1 (GLP-1). *Food Funct.* 5: 2309–2316.

Nageswara Rao, G., Mahesh Kumar, P., Dhandapani, V.S., Ramakrishna, T., and Hayashi, T. 2000. Constituents of *Cassia auriculata*. *Fitoterapia* 71: 82–83.

Nagy, M.A. and Mohamed, S.A. 2014. Antidiabetic effect of *Cleome droserifolia* (Cleomaceae). *Am. J. Biochem.* 4: 68–75.

Naik, S.R., Barbosa Filho, J.M., Dhuley, J.N., and Deshmukh, V. 1991. Probable mechanism of hypoglycemic activity of bassic acid, a natural product isolated from *Bumelia sartorum*. *J. Ethnopharmacol.* 33: 37–43.

Nain, P., Saini, V., Sharma, S., and Nain J. 2012. Antidiabetic and antioxidant potential of *Emblica officinalis* Gaertn. leaves extract in streptozotocin-induced type-2 diabetes mellitus (T2DM) rats. *J. Ethnopharmacol.* 142: 65–71.

Nair, P.K., Rodriguez, S., Ramachandran, R., Alamo, A., Melnick, S.J., Excalon, E. et al. 2004. Immune stimulating properties of a novel polysaccharide from the medicinal plant *Tinospora cordifolia*. *Int. Immunopharmacol.* 4: 1465–1459.

Nair, R., Shukla, V., and Chanda, S. 2007. Assessment of *Polyalthia longifolia* var. pendula for hypoglycemic and antihyperglycemic activity. *J. Clin. Diagn. Res.* 1: 116–121.

Nair, R.B. and Santhakumari, G. 1986. Anti-diabetic activity of the seed kernel of *Syzygium cumini* L. *Anc. Sci. Life* 6: 80–84.

Najla, O.A., Olfat, A.K., Kholoud, S.R, Enas N.D., and Hanan S.A. 2012. Hypoglycemic and biochemical effects of *Matricaria chamomilla* leave extract in streptozotocin-induced diabetic rats. *J. Health Sci.* 2: 43–48.

Nakamura, S., Takahira, K., Tanabe, G., Morikawa, T., Sakano, M., Ninomiya, K., Yoshikawa, M., Muraoka, O., and Nakanishi, I. 2010. Docking and SAR studies of Salacinol derivatives as alpha-glucosidase inhibitors. *Bioorg. Med. Chem. Lett.* 20: 4420–4423.

Nakashima, N., Kimura, I., and Kimura, M. 1993. Isolation of pseudoprototimosaponin AIII from rhizomes of *Anemarrhena asphodeloides* and its hypoglycemic activity in streptozotocin-induced diabetic mice. *J. Nat. Prod.* 56: 345–350.

Nakhaee, A., Bokaeian, M., Saravani, M., Farhangi, A., and Akbar-zadeh, A. 2009. Attenuation of oxidative stress in streptozotocin-induced diabetic rats by *Eucalyptus globules*. *Indian J. Clin. Biochem.* 24: 419–425.

Nalamolu, R.K., Boini, K.M., and Nammi, S. 2004. Effect of chronic administration of *Boerhaavia diffusa* Linn. leaf extract on experimental diabetes in rats. *Trop. J. Pharm. Res.* 3: 305–309.

Nammi, S., Boini, K.M., Lodagala, S.D., and Behara, R.B.S. 2003. The juice of fresh leaves of *Cathararanthes roseus* Linn. reduces blood glucose in normal and alloxan diabetic rabbits. *BMC Complement. Altern. Med.* 3: 4–9.

Narasimhulu, G., Adam, A., Lavanya, T., Reddy, S.R.S., and Reddy, K.S. 2012. Protective effect of *Pimpinella tirupatiensis* Bal. and Subr. an endangered medicinal plant of Tirumala Hills against streptozotocin generated oxidative stress induced diabetic rats. *Int. Res. J. Pharm.* 3: 308–314.

Narayan, D.S., Patra, V.J., and Dinda, S.C. 2012. Diabetes and Indian traditional medicines: An overview. *Int. J. Pharm. Pharm. Sci.* 4: 45–53.

Narayanan, C.R., Joshi, D.D., and Majumdar, A.M. 1984. Hypoglycemic action of *Bougainvillea spectabilis* leaves. *Curr. Sci.* 53: 579–581.

Narayanan, R., Joshi, D.D., Mujumdar, A.M., and Dhekne, V.V. 1987. Pinitol, a new antidiabetic compound from the leaves of *Bougainvillea spectabilis*. *Curr. Sci.* 56: 138–141.

Nardos, A., Makonnen, E., and Debella, A. 2011. Effects of crude extracts and fractions of *Moringa stenopetala* (Baker f) Cufodontis leaves in normoglycemic and alloxan-induced diabetic mice. *Afr. J. Pharm. Pharmacol.* 5: 2220–2225.

Narender, T., Shweta, S., Tiwari, P., Khaliq, T., Prathipati, P., Puri, A., Srivastava, A.K., Chander, R., Agarwal, S.C., and Raj, K. 2007. Antihyperglycemic and antidyslipidemic agent from *Aegle marmelos*. *Bio-Org. Med. Chem. Lett.* 17: 1808–1811.

Narendhirakannan, R.T. and Subramanian, S. 2010. Biochemical evaluation of the protective effect of *Aegle marmelos* (L.), Corr. leaf extract on tissue antioxidant defense system and histological changes of pancreatic beta-cells in streptozotocin-induced diabetic rats. *Drug Chem. Toxicol.* 33: 120–130.

Narita, Y. and Inouye, K. 2011. Inhibitory effects of chlorogenic acids from green coffee beans and cinnamate derivatives on the activity of porcine pancreas α-amylase isozyme I. *Food Chem.* 127: 1532–1539.

Narkhede, M.B., Ajimire, P.V., Wagh, A.E., Mohan, M., and Shivashanmugam, A.T. 2012. *In vitro* anti-diabetic activity of *Caesalpina digyna* (R.) methanol extract. *Asian J. Plant Sci. Res.* 1: 101–106.

Narvaez-Mastache, J.M., Garduno-Ramirez, M.L., Alvarez, L., and Delgado, G. 2006. Antihyperglycemic activity and chemical constituents of *Eysenhardtia platycarpa*. *J. Nat. Prod.* 69: 1687–1691.

Naseem, M.Z., Patil, S.R., Patil, S.R., and Ravindra-Patil, R.S.1998. Anti-spermatogenic and androgenic activities of *Momordica charantia* (karela) in albino rats. *J. Ethnopharmacol.* 61: 9–16.

Natarajan, A., Ahmed, K.S.Z., Sundaresan, S., Sivaraj, A., and Devi, K. 2012. Effect of aqueous flower extract of *Catharanthus roseus* on alloxan induced diabetes in male albino rats. *Int. J. Pharm. Sci. Drug Res.* 4: 150–153.

Natarajan, V. and Arul Gnana Dhas, A.S. 2013. Effect of active fraction isolated from the leaf extract of *Dregea volubilis* (Linn.) Benth. on plasma glucose concentration and lipid profile in streptozotocin-induced diabetic rats. *SpringerPlus* 2: 394–399.

Nauck, M.A. 2014. Update on developments with SGLT2 inhibitors in the management of type 2 diabetes. *Drug Design Ther.* 8: 1335–1380.

Navarro, M., Coussio, J., Hnatyszyn, O., and Ferraro, G. 2004. Efecto hipoglucemiante del extracto acuoso de *Phyllanthus sellowianus* (sarandí blanco) en ratones C57BL/Ks. *Acta Farm. Bonaerense* 23: 520–523.

Nayak, A., Mandal, S., Banerji, A., and Banerji, J. 2010. Review on chemistry and pharmacology of *Murraya koenigii* Spreng (Rutaceae). *J. Chem. Pharm. Res.* 2: 286–299.

Nayak, B.S., Ellaiah, P., and Dinda, S.C. 2012. Anti-bacterial, anti-oxidant and anti-diabetic activities of *Gmelina arborea* Roxb. fruit extracts. *Int. J. Green Pharm.* 6: 224–230.

Nayak, P.S., Kar, D.M., and Nayak, S.P. 2013. Evaluation of antidiabetic and antioxidant activity of aerial parts of *Hyptis suaveolens* Poit. *Afri. J. Pharm. Pharmacol.* 7: 1–7.

Nazaruk, J. and Kluczyk, M.B. 2014. The role of triterpenes in the management of diabetes mellitus and its complications. *Phytochem. Rev.* 14: 675–690.

Nazni, P., Poongodi, T.V., Alagianambi, P., and Amirthaveni, M. 2006. Hypoglycemic and hypolipidemic effect of *Cynara scolymus* among selected type 2 diabetic individuals. *Pak. J. Nutr.* 5: 147–151.

Nazreem, S., Kaur, G., Alam, M.M., Haider, S., Shafi, S., Hamid, H. et al. 2011. Hypoglycemic activity of *Callistemon lanceolatus* leaf ethanol extract in streptozotocin-induced diabetic rats. *Pharmacologyonline* 1: 799–808.

Nazreen, S., Kaur, G., Alam, M.H., Haider, S., Namid, H., and Alam, M.S. 2011. Nypoglycemic activity of *Bambusa arundinacea* leaf ethanolic extract in streptozotocin induced diabetic rats. *Pharmacologyonline* 1: 964–972.

Nazreen, S., Kaur, G., Alam, M.M., Shafi, S. Hamid, H., Ali, M., and Alam, M.S. 2012. New flavones with anti-diabetic activity from *Callistemon lanceolatus* DC. *Fitoterapia* 83: 1623–1627.

Ndiaye, M., Diatta, W, Sy, A.N., Dieye, A.M., Faye, B., and Bassene, E. 2008. Antidiabetic properties of aqueous barks extract of *Parinari excelsa* in alloxan-induced diabetic rats. *Fitoterapia.* 79: 267–270.

Nelson, R., Knowler, W., Pettitt, D., and Bennett, P. 1995. Kidney diseases in diabetes. In: *Diabetes in America*, Harris, M.I., Cowie, C.C., Stern, M.P., Boyko, E.J., Reiber, G.E. and Bennett, P.H. (eds.) Bethesda, MD: National Institutes of Diabetes and Digestive and Kidney Diseases. NIH Publication No: 95-1468, pp. 349–385.

Nerurkar, P.V., Nishioka, A., Eck, P.O., Johns, L.M., Volper, E., and Nerurkar, V.R. 2012. Regulation of glucose metabolism via hepatic forkhead transcription factor 1 (FoxO1) by *Morinda citrifolia* (noni) in high-fat diet-induced obese mice. *Br. J. Nutr.* 108: 218–228.

Nerurkar, P.V., Pearson, L., Efird, J.T., Adeli, K., Theriault, A.G., and Nerurkar, V.R. 2004. Effect of *Momordica charantia* on lipid profile and oral glucose tolerancein diabetic rats. *Phytother. Res.* 18: 954–956.

Neto, M.C.C., de Vasconcelos, C.F.B., Thijan, V.N., Caldas, G.F.R., Araújo, A.V., Costa-Silva, J.H. et al. 2013. Evaluation of antihyperglycaemic activity of *Calotropis procera* leaves extract on streptozotocin-induced diabetes in Wistar rats. *Rev. Bras. Farmacogn.* 23: 913–919.

Neurath, A.R., Strick, N., Li, Y.Y., and Debnath, A.K. 2004. *Punica granatum* (pomegranate) juice provides HIV-1 entry inhibitor and candidate topical microbicide. *BMC Infect. Dis.* 4: 41–46.

Ng, T.B., Chan, W.Y., and Yeung, H.W. 1992. Proteins with abortifacient, ribosome inactivating, immunomodulatory, anti-tumor and anti-AIDS activities from Cucurbitaceae plants. *Gen. Pharmacol.* 23: 579–590.

Ng, T.B., Li, W.W., and Yeung, H.W. 1987a. Effect of gingenosides, lectins and *Momordica charantia* insulin-like peptide on corticosterone production by isolated rat adrenal cells. *J. Ethnopharmacol.* 21: 21–29.

Ng, T.B., Wong, C.M., Li, W.W., and Yeung, H.W. 1986. Insulin-like molecules in *Momordica charantia* seeds. *J. Ethnopharmacol.* 15: 107–117.

Ng, T.B., Wong, C.M., Li, W.W., and Yeung, H.W. 1987b. Acid ethanol extractable compounds from fruits and seeds of the bitter gourd, *Momordica charantia*: Effect on lipid metabolism in isolated rat adipocytes. *Am. J. Chin. Med.* 15: 31–42.

Ngondi, J.L., Djiotsa, E.J., Fossouo, Z., and Oben, J. 2006. Hypoglycaemic effect of the methanol extract of *Irvingia gabonensis* seeds on streptozotocin diabetic rats. *Afr. J. Tradit. CAM* 3: 74–77.

Ngondi, J.L., Oben, J.E., and Minka, S.R. 2005. The effect of *Irvingia gabonensis* seeds on body weight and blood lipids of obese subjects in Cameroon. *Lipids Health Dis.* 4: 12–16.

Nguelefack, T.B., Dutra, R.C., Paszcuk, A.F., Andrade, E.L., Tapondjou, L.A., and Calixto, J.B. 2010. Antinociceptive activities of the methanol extract of the bulbs of *Dioscorea bulbifera* L. var. sativa in mice is dependent on NO-cGMP-ATP-sensitive -K+channel activation. *J. Ethnopharmacol.* 128: 567–574.

Nguessan, K., Fofie, N.V.Y., and Zirihi, G.N. 2011. Effect of aqueous extract of *Terminalia catappa* leaves on the gucaemia of rabbits. *J. Appl. Pharm. Sci.* 1: 59–64.

Nicasio, P., Anguilar-Santamaria, L., Aranda, E., Ortiz, S., and Gonzalez, M. 2005. Hypoglycemic effect and chlorogenic acid content in two *Cecropia* species. *Phytother. Res.* 9: 661–664.

Nickavar, B. and Amin, G. 2004. Anthocyanins from *Vaccinium arctostaphylos* berries. *Pharm. Biol.* 42: 289–291.

Nickavar, B. and Amin, G. 2010. Bioassay-guided separation of an alpha-amylase inhibiting anthocyanin from *Vaccinium arctostaphylos* berries. *Z. Naturforsch C.* 65: 567–570.

Nickavar, B. and Yousefian, N. 2009. Inhibitory effects of six *Allium* species on α-amylase enzyme activity. *Iran. J. Pharm. Res.* 8: 53–57.

Nidavani, R.B. and Mahalakshmi, A.M. 2014. Pharmacology of *Tectona grandis* Linn.: Short review. *Int. J. Pharmacogn. Phytochem. Res.* 6: 86–90.

Nim, D.K., Shankar, P., Chaurasia, R., Goel, B., Kumar, N., and Dixit, R.K. 2013. Clinical evaluation of antihyperglycemic activity of *Boerhaavia diffusa* and *Ocimum sanctum* extracts in streptozotocin-induced type 2 DM rat models. *Int. J. Pharm. Biomed. Sci.* 4: 30–34.

Nimenibo-Uadia, R. 2003. Effect of aqueous extract of *Canavalia ensiformis* seeds on hyperlipidemia and hyperketonaemia in alloxan-induced diabetic rats. *Biokemistri* 15: 7–15.

Niranjan, P.S., Singh, D., Prajapati, K., and Jain, S.K. 2010. Antidiabetic activity of ethanolic extract of *Dalbergia sissoo* L. leaves in alloxan-induced diabetic rats. *Int. J. Curr. Pharm. Res.* 2: 24–27.

Nirmala, A., Eliza, J., Rajalakshmi, M., Priya, E., and Daisy, P. 2008. Effect of hexane extract of *Cassia fistula* barks on blood glucose and lipid profile in streptozotocin diabetic rats. *Int. J. Pharmacol.* 4: 292–296.

Nishioka, T., Kawabata, J., and Aoyama, Y. 1998. Baicalein, an α-glucosidase inhibitor from *Scutellaria baicalensis*. *J. Nat. Prod.* 61: 1413–1415.

Nishiyama, T., Mae, T., Kishida, H., Tsukagawa, M., Mimaki, Y., Kuroda, M. et al. 2005. Curcuminoids and sesquiterpenoids in turmeric (*Curcuma longa* L.) suppress an increase in blood glucose level in type 2 diabetic KK-Ay mice. *J. Agric Food Chem.* 53: 959–963.

Niu, H.S., Liu, I.M., Cheng, J.T., Lin, C.L., and Hsu, F.L. 2008. Hypoglycemic effect of syringin from *Eleutherococcus senticosus* in streptozotocin-induced diabetic rats. *Planta Med.* 74: 109–113.

Niwa, A., Tajiri, T., and Higashino, H. 2011. *Ipomoea batatas* and *Agarics blazei* ameliorate diabetic disorders with therapeutic antioxidant potential in streptozotocin-induced diabetic rats. *J. Clin. Biochem. Nutr.* 48: 194–202.

Niyonzima, G., Laekeman, G.M., Scharpe, S., Mets, T., and Vlietinck, A.J. 1990. Hypoglycemic activity of *Spathodea campanulata* bark decoctions on streptozotocin-diabetic mice. *Planta Med.* 56: 682–689.

Niyonzima, G., Scharpe, S., Beeck, L.V., Vlietinck, A.J., Laekeman, G.M., and Mets, T. 2006. Hypoglycaemic activity of *Spathodea campanulata* stem bark decoction in mice. *Phytother. Res.* 7: 64–67.

Njagi, J.M., Piero, M.N., Ngeranwa, J.J.N., Njagi, E.N.M, Kibiti, C.M., Njue, W.M. et al. 2012. Assessment of anti-diabetic potential of *Ficus syncomors* on alloxan-induced diabetic mice. *Int. J. Diabetes Res.* 1: 47–51.

Njike, G.N., Watcho, P., Nguelefack, T.B., and Kamanyi, A. 2005. Hypoglycaemic activity of the leaves extracts of *Bersama engleriana* in rats. *Afr. J. Tradit. CAM* 2: 215–221.

Nmila, R., Gross, R., Rchid, H., Roye, M., Manteghetti, M., Petit, P. et al. 2000. Insulinotropic effect of *Citrullus colocynthis* fruit extracts. *Planta Med.* 66: 418–423.

Nofal, S.M., Mahmoud, S.S., Ramadan, A., Soliman, G., and Abdel-Rahman, R.F. 2009. Antidiabetic effect of *Artemisia judaica* extracts. *Res. J. Med. Med. Sci.* 4: 42–48.

Nojima, H., Kimura, I., Chen, F.J., Sugihara, Y., Haruno, M., Kato, A., and Asano, N. 1998. Antihyperglycemic effects of N-containing sugars from *Xanthocercis zambesiaca*, *Morus bombycis*, *Aglaonema treubii*, and *Castanospermum australe* in streptozotocin-diabetic mice. *J. Nat. Prod.* 61: 397–400.

Noor, A., Bansal, V.S., and Vijayalakshmi, M.A. 2013. Current update on anti-diabetic biomolecules from key traditional Indian medicinal plants. *Curr. Sci.* 104: 721–727.

Noor, H.A.A., Wafaa, M.A., and Al, S.H. 2014. Preliminary phytochemical screening and *in vitro* evaluation of anti-oxidant activity of Iraqi species of *Silybum marianum* seeds. *Int. Res. J. Pharm.* 5: 378–383.

Noorshahida, A., Wong, T.W., and Choo, C.Y. 2009. Hypoglycemic effect of quassinoids from *Brucea javanica* (L.) Merr (Simaroubaceae) seeds. *J. Ethnopharmacol.* 124: 586–591.

Noreen, W., Wadood, A., Hidayat, H.K., and Wahid, S.A. 1988. Effect of *Eriobotrya japonica* on blood glucose levels of normal and alloxan diabetic rabbits. *Planta Med.* 54: 196–199.

Noreen, Y., Serrano, G., Perera, P., and Bohlin, L. 1998. Flavan-3-ols isolated from some medicinal plants inhibiting COX-1 and COX-2 catalyzed prostaglandin biosynthesis. *Planta Med.* 64: 520–524.

Nosalova, G., Mokry, J., and Hassan, K.M. 2003. Anti-tussive activity of the fruit extract of *Emblica officinalis* Gaertn. (Euphorbiaceae). *Phytomedicine* 10: 583–589.

Nostro, A., Cellini, L., Bartolomea, S.D., Campli, E.D., Grande, R., Cannatelli, M.A. et al. 2005. Anti-bacterial effect of plant extracts against *Helicobactor pylori. Phytother. Res.* 19: 198–202.

Noumi, E., Snoussi, M., Hajlaoui, H., Valentin, E., and Bakhrouf, A. 2010. Anti-fungal properties of *Salvadora persica* and *Juglans regia* extracts against oral *Candida* strains. *Eur. J. Clin. Microbiol. Infect. Dis.* 29: 81–88.

Novaes, A.P., Rossi, C., Poffo, C., Junior, E.P., Oliveira, A.E., Schlemper, V. et al. 2001. Preliminary evaluation of the hypoglycemic effect of some Brazilian medicinal plants. *Therapie* 56: 427–430.

Nugroho, A.E., Andrie, M., Warditiani, N.K., Siswanto, E., Suwidjivo, P., and Lukitaningsih, E. 2012. Antidiabetic and antihiperlipidemic effect of *Andrographis paniculata* (Burm. f.) Nees and andrographolide in high-fructose-fat-fed rats. *Indian J. Pharmacol.* 44: 377–381.

Nugroho, A.E., Rais, I.R., Setiawan, I., Pratiwi, P.Y., Hadibarata, T., Tegar, M. et al. 2014. Pancreatic effect of andrographolide isolated from *andrographis paniculata* (Burm. F.) Nees. *Pak. J. Biol. Sci.* 17: 22–31.

Nunez-Lopez, M.A., Paredes-Lopez, O., and Reynoso-Camacho, R. 2013. Functional and hypoglycemic properties of nopal cladodes (*O. ficus-indica*) at different maturity stages using *in vitro* and *in vivo* tests. *J. Agric. Food Chem.* 61: 10981–10986.

Nwanjo, H.U. 2006. Studies on the effect of aqueous extract of *Phyllanthus niruri* on the plasma glucose level and some hepatospecific markers in diabetic Wistar rats. *Internet J. Lab. Med.* 2: 2

Nwanjo, H.U. 2008. Hypoglycemic and hypolipidemic effects of aqueous extracts of *Abrus precatorius* Linn. seeds in streptozotocin-induced diabetic Wistar rats. *J. Herbs Spices Med. Plants* 14: 68–76.

Nwozo, S.O, Adaramoye, O.A., and Ajaiyeoba, E.O. 2004. Anti-diabetic and hypolipidemic studies of *Telifairia occidentalis* on alloxan induced diabetic rabbits. *Nig. J. Nat. Prod. Med.* 8: 45–47.

Nyarko, A.K., Ankrah, N.A., Ofosuhene, M., and Sittie, A.A. 1999. Acute and subchronic evaluation of *Indigofera arrecta:* Absence of both toxicity and modulation of selected cytochrome P450 isozymes in ddY mice. *Phytother. Res.* 13: 686–688.

Nyarko, A.K., Asare-Anane, H., Ofosuhene, M., and Addy, M.E. 2002a. Extract of *Ocimum canum* lowers blood glucose and facilitates insulin release by isolated pancreatic beta-islet cells. *Phytomedicine* 9: 346–351.

Nyarko, A.K., Asare-Anane, H., Ofosuhene, M., Addy, M.E., Teye, K., and Addo, P. 2002b. Aqueous extract of *Ocimum canum* decreases levels of fasting blood glucose and free radicals and increases antiathero-genic lipid levels in mice. *Vascul. Pharmacol.* 39: 273–279.

Nyarko, A.K., Sittie, A.A., and Addy, M.E. 1993. The basis for the antihyperglycemic activity of *Indigofera arrecta* in the rat. *Phytother. Res.* 7: 1–4.

Nyunaï, N., Manguelle-Dicoum, A., Njifutié, N., Abdennebi, E.H., and Gérard, C. 2010. Antihyperglycaemic effect of *Ageratum conyzoides* L. fractions in normoglycemic and diabetic male Wistar rats. *Int. J. Biomed. Pharm. Sci.* 4: 38–42.

Nyunai, N., Njikam, N., Abdennebiel, H., Mbafor, J.T., and Lamnaouer, D. 2009. Hypoglycaemic and antihy-perglycaemic activity of *Ageratum conyzoides* L. in rats. *Afr. J. Complement. Altern. Med.* 6: 123–130.

Nyunai, N., Njikam, N., Mounier, C., and Pastoureau, P. 2006. Blood glucose lowering effect of aqueous leaf extracts of *Ageratum conyzoides* in rats. *Afr. J. Tradit. CAM* 3: 76–79.

Obaineh, O.M., Olusola, L., and Oluwole, A.M. 2012. Preliminary study of hypoglycaemic and hypolipidemic activity of aqueous root extract of *Ricinus communis* in alloxan-induced diabetic rats. *J. Phys. Pharm. Adv.* 2: 354–359.

Obanda, D.N., Hernandez, A. Ribnicky, D., Yu, Y., Zhang, X.H., Wang, Z.Q. et al. 2012. Bioactives of *Artemisia dracunculus* L. mitigate the role of ceramide in attenuating insulin signaling in rat skeletal muscle cells. *Diabetes* 61: 597–605.

Obara, K., Mizutani, M., Hitomi, Y., Yajima, H., and Kondo, K. 2009. Isohumulones, the bitter component of beer, improve hyperglycemia and decrease body fat in Japanese subjects with prediabetes. *Clin. Nutr.* 28: 278–284.

Obatomi, D.K., Bikomo, F.O., and Temple, V.J. 1994. Antidiabetic properties of the African mistletoe in streptozotocin-induced diabetic rats. *J. Ethnopharmacol.* 43: 13–17.

Oberg, A.I., Yassin, K., Csikasz, R.I., Dehvari, N., Shabalina, I.G., Hutchinson, D.S., Wilcke, M., Ostenson, C.G., and Bengtsson, T. 2011. Shikonin increases glucose uptake in skeletal muscle cells and improves plasma glucose levels in diabetic Goto-Kakizaki rats. *PLoS One* 6: e22510.

Oboh, G., Ademosun, A.O., Ademiluyi, A.O., Omojokun, O.S., Nwanna, E.E., and Longe, K.O. 2014. *In vitro* studies on the antioxidant property and inhibition of α-amylase, α-glucosidase, and angiotensin I-converting enzyme by polyphenol-rich extracts from cocoa (*Theobroma cacao*) bean. *Pathol. Res. Int.* 2014: Article ID 549287, 6 pages.

Oboh, G., Akinyemi, A.J., and Ademiluvi, A.O. 2012. Inhibition of α-amylase and α-glucosidase activities by ethanolic extract of *Telfairia occidentalis* (fluted pumpkin) leaf. *Asian Pac. J. Trop. Biomed.* 2: 733–738.

Obolskiy, D., Pischel, I., Feistel, B., Glotov, N., and Heinrich, M. 2011. *Artemisia dracunculus* L. (tarragon): A critical review of its traditional use, chemical composition, pharmacology and safety. *J. Agric. Food Chem.* 59: 11367–11384.

Odetola, A.A., Akinloye, O., Egunjobi, C., Adekunle, W.A., and Ayoola, A.Q. 2006. Possible antidiabetic and antihyperlipidaemic effect of fermented *Parkia biglobosa* (JACQ) extract in alloxan-induced diabetic rats. *Clin. Exp. Pharmacol. Physiol.* 33: 818–912.

Odoh, U.E., Ezugwu, C.O., and Ezea, C.S. 2013. Antidiabetic and biochemical effect of the different fractions of methanol extract of *Detarium microcarpum* (Fabaceae) Guill and Perr stem bark on normal and alloxan induced diabetic rats (congress abstract). *Planta Med.* 79: PN 71.

Odoh, U.E., Ndubuokwu, R.I., Inyagha, S.I., and Ezejiofor, M. 2013. Antidiabetic activity and chemical char-acterization of the *Acalypha wilkesiana* (Euphorbiaceae) Mull Arg. roots in alloxan-induced diabetic rats. *Planta Med.* DOI: 10.1055/s-0033-1351836.

Odoh, U.E., Ndubuokwu, R.I., Inya-Agha, S.I., Osadebe, P.O., Philip, U., and Ezejiofor, M. 2014. Antidiabetic activity and phytochemical screening of *Acalypha wilkesiana* (Euphorbiaceae) Mull Arg. roots in alloxan-induced diabetic rats. *Sci. Res. Essays* 9: 204–212.

Oduola, T., Bello, I., Adeosun, G., Ademosun, A.-W., Raheem, G., and Avwioro, G. 2010. Hepatotoxicity and nephrotoxicity evaluation in Wistar albino rats exposed to *Morinda lucida* leaf extract. *N. Am. J. Med. Sci.* 2: 230–233.

Ofusori, D.A., Komolafe, O.A., Adewole, O.S., Obuotor, E.M., Fakunle, J.B., and Ayoka, A.O. 2012. Effect of ethanolic extract of *Croton zambesicus* (Mull. Arg.) on lipid profile in streptozotocin-induced diabetic rats. *Diabetol. Croat.* 41: 68–76.

Ogbonnia, S.O., Adekunle, A., Olagbende-Dada, S.O., Anyika, E.N., Enwuru, V.N., and Orolepe, M. 2008a. Assessing plasma glucose and lipid levels, body weight and acute toxicity following oral administration of an aqueous ethanolic extract of *Parinari curatellifolia* Planch. (Chrysobalanaceae) seeds in alloxan-induced diabetes in rats. *Afr. J. Biotechnol.* 7: 2998–3003.

Ogbonnia, S.O., Mbaka, G.O., Anjika, E.N., Ladiju, O., Igbokwe, H.N., Emordi, J.E. et al. 2011. Evaluation of anti-diabetics and cardiovascular effects of *Parinari curatellifolia* seed extract and *Anthoclista vogelli* root extract individually and combined on postprandial and alloxan-induced diabetic albino rats. *Br. J. Med. Med. Sci.* 1: 146–162.

Ogbonnia, S.O., Odimegwu, J.I., and Enwuru, V.N. 2008b. Evaluation of hypoglycaemic and hypolipidaemic effects of aqueous ethanolic extracts of *Treculia africana* Decne and *Bryophyllum pinnatum* Lam. and their mixture on streptozotocin (STZ)-induced diabetic rats. *Afr. J. Biotechnol.* 7: 2535–2539.

Ogunbolude, Y., Ajayi, M.A., Ajagbawa, T.M., Igbakin, A.P., Rocha, J.B.T., and Kade, I.J. 2009. Ethanolic extracts of seeds of *Parinari curatellifolia* exhibit potent antioxidant properties: A possible mechanism of its antidiabetic action. *J. Pharmacogn. Phytother.* 1: 67–75.

Oh, W.K., Lee, C.H., Lee, M.S., Bae, E.Y., Sohn, C.B., Oh, H. et al. 2005. Anti-diabetic effects of extracts from *Psidium guajava*. *J. Ethnopharmacol.* 96: 411–415.

Ohno, H., Nagai, J., Kurakawa, T., Sonoda, M., Yumoto, R., and Takano, M. 2012. Effects of aqueous extract from the root cortex of *Aralia elata* on intestinal alpha-glucosidases and postprandial glycemic response in mice. *Int. J. Phytomed.* 4: 567–572.

Ojewole, J.A. 2002. Hypoglycaemic effect of *Clausena anisata* (Willd.) Hook methanolic root extract in rats. *J. Ethnopharmacol.* 81: 231–237.

Ojewole, J.A. 2007. Analgesic, anti-inflammatory and hypoglycaemic effects of *Rhus chirindensis* (Baker F.) [Anacardiaceae] stem-bark aqueous extract in mice and rats. *J. Ethnopharmacol.* 113: 338–345.

Ojewole, J.A.O. 2003. Hypoglycemic effect of *Sclerocarya birrea* {(A.Rich.) Hochst} [Anacardiaceae] stem bark aqueous extract in rats. *Phytomedicine* 10: 675–681.

Ojewole, J.A.O. 2005. Antinociceptive, anti-inflammatory and antidiabetic effects of *Bryophyllum pinnatum* (Crassulaceae) leaf aqueous extract. *J. Ethnopharmacol.* 100: 310–313.

Ojewole, J.A.O. 2006. Antinociceptive, anti-inflammatory and anti-diabetic properties of *Hypoxis hemerocallidea* Fisch. & C.A. Mey. (Hypoxidaceae) corm ['African potato'] aqueous extract in mice and rats. *J. Ethnopharmacol.* 103: 126–134.

Ojieh, G.C., Oluba, O.M., Erifeta, G.O., and Eidangbe, G.O. 2009. Effect of aqueous root extract of *Treculia africana* on haemoglobin glycosylation and plasma lipid peroxidation in streptozotocin-induced diabetic rabbits. *Int. J. Plant Physiol. Biochem.* 1: 5–8.

Oki, N., Nonaka, S., and Ozaki, S. 2011. The effects of an arabinogalactan-protein from the white-skinned sweet potato (*Ipomoea batatas* L.) on blood glucose in spontaneous diabetic mice. *Biosci. Biotechnol. Biochem.* 75: 2011–2018.

Okokon, E.J., Antia, B.S., Osuji, L.C., and Udia, P.M. 2007a. Antidiabetic and hypolipidemic effects of *Mammea africana* (Guttiferae) in streptozotocin induced diabetic rats. *J. Pharmacol. Toxicol.* 2: 278–283.

Okokon, J.E. and Antia, B.S. 2007. Hypoglycemic and anti-diabetic effect of *Setaria megaphylla* on normal and alloxan induced diabetic rats. *J. Nat. Remedies* 7: 220–224.

Okokon, J.E., Antia, B.S., and Ita, B.N. 2007b. Antidiabetic effects of *Homalium letestui* (Flacourtiaceae) in streptozotocin induced diabetic rats. *Res. J. Med. Plant* 1: 134–138.

Okokon, J.E., Antia, B.S., and Udobang, J.A. 2012. Antidiabetic activities of ethanolic extract and fraction of *Anthocleista djalonensis*. *Asian Pac. J. Trop. Biomed.* 2: 461–464.

Okokon, J.E., Bassey, A.L., and Nwidu, L.L. 2007c. Antidiabetic and hypolipidaemic effects of ethanolic root extract of *Setaria megaphylla*. *Int. J. Pharmacol.* 3: 91–95.

Okokon, J.E., Bassey, A.L., and Obot, J. 2006. Anti-diabetic activity of ethanolic leaf extract of *Croton zambensicus* Muell. (thunder plant) in alloxan diabetic rats. *Afr. J. Tradit. CAM* 3: 21–26.

Okokon, J.E., Umoh, E.E., Etim, E.L., and Jackson, C.L. 2009. Antiplasmodial and anti-diabetic activities of ethanolic leaf extract of *Heinsia crinata*. *J. Med. Food* 12: 131–136.

Okon, J.E. and Ofeni, A.A. 2013. Anti-diabetic effect of *Dioscorea bulbifera* on alloxan-induced diabetic rats. *CIBTech J. Pharm. Sci.* 2: 14–19. http://www.cibtech.org/cjps.htm.

Okonkwo, P. and Okoye, Z. 2009. Hypoglycaemic effects of the aqueous extract of *Newbouldia laevis* root in rats. *Int. J. Biol. Chem. Sci.* 3: 998–1004.

Okwuosa, C., Azubike, N., and Nebo, I. 2011. Evaluation of the anti-hyperglycemic activity of the crude leaf extracts of *Sida acuta* in normal and diabetic rabbits. *Indian J. Novel Drug Deliv.* 3: 200–213.

Okyar, A., Can, A., Akev, N., Baktir, G., and Sutlupinar, N. 2001. Effects of *Aloe vera* on blood glucose levels in type 1 and type 2 diabetic rat models. *Phytother. Res.* 15: 157–161.

Olajide, O.A., Aderogba, M.A., Adedapo, A.D., and Makinde, J.M. 2004. Effects of *Anacardium occidentale* stem bark extract on *in vivo* inflammatory models. *J. Ethnopharmacol.* 95: 139–142.

Olajide, O.A., Awe, S.O., Makinde, J.M., and Morebise, O. 1999. Evaluation of the anti-diabetic property of *Morinda lucida* leaves in streptozotocin-diabetic rats. *J. Pharm. Pharmacol.* 51: 1321–1324.

Olaokun, O.O., Mcgaw, L.J., Awouafack, M.D., Eloff, J.N., and Naido, V. 2014. The potential role of GLUT4 transporters and insulin receptors in the hypoglycaemic activity of *Ficus lutea* acetone leaf extract. *BMC Complement. Altern. Med.* 14: 269.

Olaokun, O.O., Mcgaw, L.J., Eloff, J.N., and Naido, V. 2013. Evaluation of the inhibition of carbohydrate hydrolyzing enzymes, anti-oxidant activity and polyphenolic content of extracts of ten African *Ficus* species (Moraceae) used traditionally to treat diabetes. *BMC Complement. Altern. Med.* 13: 94–99.

Olatunji, B.P., Suleiman, M., and Moody, J.O. 2014. Hypoglycemic effect of hydroacetone extracts of *Treculia africana* Decne root and stem bark in alloxan-induced diabetic rats. *Afr. J. Biomed. Res.* 17: 23–29.

Olatunji, L.A., Okwusidi, J.I., and Soladoye, A.O. 2005. Antidiabetic effect of *Anacardium occidentale* stem-bark in fructose-diabetic rats. *Pharm. Biol.* 43: 589–593.

Oliveira, A.C., Endringer, D.C., Amorim, L.A., das Gracas, L.B.M., and Coelho, M.M. 2005. Effect of the extracts and fractions of *Baccharis trimera* and *Syzygium cumini* on glycaemia of diabetic and non-diabetic mice. *J. Ethnopharmacol.* 102: 465–469.

Oliveira, D.M., Freitas, H.S., Souza, M.F.F., Arcari, D.P., Ribeiro, M.L., Carvalho, P.O. et al. 2008. Yerba Mate (*Ilex paraguariensis*) aqueous extract decreases intestinal SGLT1 gene expression but does not affect other biochemical parameters in alloxan-diabetic Wistar rats. *J. Agric. Food Chem.* 56: 10527–10532.

Olubomehin, O.O., Abo, K.A., and Ajaiyeoba, E.O. 2013. Alpha-amylase inhibitory activity of two *Anthocleista* species and *in vivo* rat model anti-diabetic activities of *Anthocleista djalonensis* extracts and fractions. *J. Ethnopharmacol.* 146: 811–814.

Olubunmi, A., Ademola, O.G., and Adenike, F.O. 2009. *Blighia sapida*: The plant and its hypoglycins: An overview. *J. Sci. Res.* 39: 15–25.

Omara, E.A., Nada, S.A., Farrag, A.R.H., Sharaf, W.M., and El-Toumy, S.A. 2012. Therapeutic effect of *Acacia nilotica* pods extract on streptozotocin-induced diabetic nephropathy in rat. *Phytomedicine* 19: 1059–1067.

Omer, S.A., Elobeid, M.A., Elamin, M.H., Hassan, Z.K., Virk, P., Daghestani, M.H. et al. 2012. Toxicity of olive leaves (*Olea europaea* L.) in Wistar albino rats. *Asian J. Anim. Vet. Adv.* 7: 1175–1182.

Omigie, I.O. and Agoreyo, F.O. 2014. Effects of watermelon (*Citrullus lanatus*) seed on blood glucose and electrolyte parameters in diabetic Wistar rats. *J. Appl. Sci. Environ. Manag.* 18: 231–233.

Omodamiro, O.D. and Jimoh, M.A. 2014. Hypoglycemic effect of ethanolic leaves extracts of *Anacardium occidentalis* and *Gomphrena globosa* plants on alloxan induced-diabetic rats. *J. Chem. Pharm. Res.* 6: 492–498.

Omoniwa, B.P. and Luka, C.D. 2012. Antidiabetic and toxicity evaluation of aqueous stem extract of *Ficus asperifolia* in normal and alloxan-induced diabetic albino rats. *Asian J. Exp. Biol. Sci.* 3: 726–732.

Omonkhua, A.A. and Onoagbe, I.O. 2012. Effects of long-term oral administration of aqueous extracts of *Irvingia gabonensis* bark on blood glucose and liver profile of normal rabbits. *J. Med. Plant Res.* 6: 2581–2589.

Onakpa, M.M. and Ajagbonna, O.P. 2012. Anti-diabetic potential of *Cassia occidentalis* leaf extract on alloxan-induced diabetic albino rats. *Int. J. PharmTech Res.* 4: 1766–1769.

Ong, K.W., Hsu, A., Song, L., Huang, D., and Tan, B.K. 2011. Polyphenols-rich *Vernonia amygdalina* shows anti-diabetic effects in streptozotocin-induced diabetic rats. *J. Ethnopharmacol.* 133: 598–607.

Ong, K.W., Hsu, A., and Tan, B.K.H. 2012. Chlorogenic acid stimulates glucose transport in skeletal muscle *via* AMPK activation: A contributor to the beneficial effects of coffee on diabetes. *PLoS One* 7: e32718. DOI: 10.1371/journal.pone.0032718.

Ooi, K.L., Muhammad, T.S., Tan, M.L., and Sulaiman, S.F. 2011. Cytotoxic, apoptotic and anti-α-glucosidase activities of 3,4-di-O-caffeoyl quinic acid, an antioxidant isolated from the polyphenolic-rich extract of *Elephantopus mollis* Kunth. *J. Ethnopharmacol.* 135: 685–695.

Orhan, N., Berkkan, A., Deliorman, O.D., Aslam, M., and Ergun, F. 2011. Effects of *Juniperus oxycedrus* ssp. *oxycedrus* on tissue lipid peroxidation, trace elements (Cu, Zn, Fe) and blood glucose levels in experimental diabetes. *J. Ethnopharmacol.* 133: 759–764.

Orhan, N., Aslan, M., Sukuroglu, M. and Deliorman Orhan, D. 2013. *In vivo* and *in vitro* anti-diabetic effect of *Cistus laurifolius* L. and detection of major phenolic compounds by UPLC-TOF-MS analysis. *J. Ethnopharmacol.* 146: 859–865.

Orhan, N., Hocbac, S., Orhan, D.D., Asian, M., and Ergun, F. 2014. Enzyme inhibitory and radical scavenging effects of some antidiabetic plants of Turkey. *Iran. J. Basic Med. Sci.* 17: 426–432.

Ortiz-Andrade, R.R., Garcia-Jimenez, S., Castillo-Espana, P., Ramirez-Avila, G., Villalobos-Molina, R., and Estrada-Soto, S. 2007. Alpha-glucosidase inhibitory activity of the methanolic extract from *Tournefortia hartwegiana*: An anti-hyperglycemic agent. *J. Ethnopharmacol.* 109: 48–53.

Ortiz-Andrade, R.R., Rodriguez-Lopez, V., Garduno-Ramirez, M.L., Castillo-Epsana, P., and Estrada-Soto, S. 2005. Anti-diabetes effects on alloxanized and normoglycemic rats and pharmacological evaluations of *Tournefortia hartwegiana*. *J. Ethnopharmacol.* 101: 37–42.

Oryan, S., Eydi, M., Yazdi, E., Eydi, A., and Soulati, J. 2003. Hypoglycemic effect of alcohol extract of *Morus nigra* L. leaves in normal and diabetic rats. *J. Med. Plants* 2: 27–32.

Osadebe, P.O., Okide, G.B., and Akabogu, I.C. 2004. Studies on antidiabetic activities of crude methanolic extract of *Loranthus micranthus* (Linn.) sourced from five different host trees. *J. Ethnopharmacol.* 95: 133–138.

Osadebe, P.O., Omeje, E.O., Nworu, S.C., Esimone, C.O., Uzor, P.F., and David, E.K. 2010a. Antidiabetic principles of *Loranthus micranthus* Linn. parasitic on *Persea americana*. *Asian Pac. J. Trop. Med.* 3: 619–623.

Osadebe, P.O., Omeje, E.O., Uzor, P.F., David, E.K., and Obiorah, D.C. 2010b. Seasonal variations for the antidiabetic activity of *Loranthus micranthus* methanol extract. *Asian Pac. J. Trop. Med.* 3: 196–199.

Osuagwu, A.N, Ekpo, I.A, Okpako, E.C., Out, P., and Ottoho, E. 2013. The biology, utilization and phytochemical composition of the fruits and leaves of *Gongronema latifolium* Benth. *J. Agrotechnol.* 2: 115. DOI: 10.4172/2168-9881.1000115.

Ou, H.C., Tzang, B.S., Chang, M.H., Liu, C.T., Liu, H.W., Lii, C.K. et al. 2010. Cardiac contractile dysfunction and apoptosis in streptozotocin-induced diabetic rats are ameliorated by garlic oil supplementation. *J. Agric. Food Chem.* 58: 10347–10355.

Ovesna, Z., Vachalkova, A., Horvathova, K., and Tothova, D. 2004. Pentacyclic triterpenoic acids: New chemoprotective compounds. *Neoplasma* 51: 327–333.

Owolabi, O.J., Amaechina, F.C., and Okoro, M. 2011. Effect of ethanol leaf extract of *Newboulda laevis* on blood glucose levels of diabetic rats. *Trop. J. Pharm. Res.* 10: 249–254.

Oyedemi, S.O., Adewusi, E.A., Aiyegoro, O.A., and Akinpelu, D.A. 2011. Antidiabetic and haematological effect of aqueous extract of stem bark of *Afzelia africana* (Smith) on streptozotocin-induced diabetic Wistar rats. *Asian Pac. J. Trop. Med.* 5: 353–358.

Oyelola, O.O., Moody, J.O., Odeniyi, M.A., and Fakeye, T.O. 2007. Hypoglycemic effect of *Treculia africana* Decne root bark in normal and alloxan-induced diabetic rats. *Afr. J. Tradit. Complement. Altern. Med.* 4: 387–391.

Oyinlade, O.C. 2014. Phytochemical and physicochemical analysis of three different types of apples. *Int. J. Sci. Res. Rev. (IJSSR)* 3: 67–78.

Ozaki, S., Oe, H., and Kitamura, S. 2008. Alpha-glucosidase inhibitor from Kothala-himbutu (*Salacia reticulata* Wight). *J. Nat. Prod.* 71: 981–984.

Paarakh, P.M. 2009. *Coriandrum sativum* Linn.—Review. *Pharmacologyonline* 3: 561–573.

Padee, P., Naulkaew, S., Talubmook, C., and Sakuliaitrong, S. 2010. Hypoglycemic effect of leaf extracts of *Pseuderanthemum palatiferum* (Nees) Radlk. in normal and streptozotocin-induced diabetic rats. *J. Ethnopharmacol.* 132: 491–496.

Padmashree, R., Probhu, P.P., and Pandey, S. 2010. Anti-diabetic activity of methanol/methylene chloride extract of *Terminalia superba* leaves on streptozotocin-induced diabetes in rats. *Int. J. PharmTech Res.* 2: 2415–2419.

Paek, J.H., Shin, K.H., Kang, Y.-H., Lee, J.-Y., and Lim, S.S. 2013. Rapid identification of aldose reductase inhibitory compounds from *Perilla frutescens*. *Bio-Med. Res. Int.* 2013: Article ID 679463, 8 pages.

Pal, D., Mishra, P., Sachan, N., and Ghosh, A.K. 2011. Biological activities and medicinal properties of *Cajanus cajan* L. Millsp. *J. Adv. Pharm. Technol. Res.* 2: 207–224.

Palanisamy, U., Ling, L.T., Manaharan, T., and Appleton, D. 2011. Rapid isolation of geraniin from *Nephelium lappaceum* rind waste and its antihyperglycemic activity. *Food Chem.* 127: 21–27.

Palsamy, P. and Malathi, R. 2007. Evaluation of hypoglycemic and hypolipidemic activity of methanolic extract of *Cocculus hirsutus* (L.) Diels leaves in streptozotocin-induced diabetes mellitus in rats. *Int. J. Biol. Chem.* 1: 205–212.

Palsamy, P. and Subramanian, S. 2010. Ameliorative potential of resveratrol on proinflammatory cytokines, hyperglycemia mediated oxidative stress and pancreatic cell dysfunction in streptozotocin-nicotinamide induced diabetic rats. *J. Cell. Physiol.* 224: 423–432.

Pan, J., Liu, G., Liu, H., Qiu, Z., and Chen, L. 1998. Effects of *Artemisia capillaris* on blood glucose and lipid in mice. *Zhong Yao Cai.* 21: 408–411.

Pan, L.-H., Li, X.-F., Wang, M.-N., Zha, X.-Q., Yang, X.-F., Liu, Z.-J. et al. 2014. Comparison of hypoglycemic and anti-oxidant effects of polysaccharides from four different *Dendrobium* species. *Int. J. Biol. Macromol.* 64: 420–427.

Panda, S. 2008. The effect of *Anethum graveolens* L. (dill) on corticosteroid induced diabetes mellitus: Involvement of thyroid hormones. *Phytother. Res.* 22: 1695–1697.

Panda, S. and Kar, A. 1998. *Ocimum sanctum* leaf extract in the regulation of thyroid function in the male mouse. *Pharmacol. Res.* 38: 107–110.

Panda, S. and Kar, A. 2006. Evaluation of the antithyroid, antioxidative and antihyperglycemic activity of scopoletin from *Aegle marmelos* leaves in hyperthyroid rats. *Phytother. Res.* 20: 1103–1105.

Panda, S.P., Haldar, P.K., Bera, S., Adhikary, S., and Kander, C.C. 2010. Antidiabetic and antioxidant activity of *Swietenia mahagoni* in streptozotocin-induced diabetic rats. *Pharm. Biol.* 48: 974–979.

Pandey, A.K., Gupta, P.P., and Lal, V.K. 2013. Preclinical evaluation of hypoglycemic activity of *Ipomoea digitata* tuber in streptozotocin-induced diabetic rats. *Basic Clin. Physiol. Pharmacol.* 24: 35–39.

Pandey, M. and Khan, A. 2002. Hypoglycemic effect of defatted seeds and water soluble fiber from the seeds of *Syzygium cumini* (Linn.) skeels in alloxan diabetic rats. *Indian J. Exp. Biol.* 40: 1178–1182.

Pandey, M.M., Rastogi, S., and Rawat, A.K.S. 2007. *Saussurea costus*: Botanical, chemical and pharmacological review of an Ayurvedic medicinal plant. *J. Ethnopharmacol.* 110: 379–390.

Pandhare, R., Balakrishnan, S., Mohite, P., and Khanage, S. 2012. Antidiabetic and antihyperlipidaemic potential of *Amaranthus viridis* (L.) Merr. in streptozotocin induced diabetic rats. *Asian Pac. J. Trop. Dis.* 2: S180–S185.

Pandhare, R.B., Sangameswaran, B., Mohite, P.B., and Khanage, S.G. 2011. Anti-diabetic activity of aqueous leaves extract of *Sebania sesban* (L.) Merr. in streptozotocin-induced diabetic rats. *Avicenna J. Med. Biotechnol.* 3: 37–43.

Pandikumar, P., Babu, N.P., and Ignacimuthu, S. 2009. Hypoglycemic and antihyperglycemic effect of *Begonia malabarica* Lam. in normal and streptozotocin induced diabetic rats. *J. Ethnopharmacol.* 124: 111–115.

Pandit, R., Phadke, A., and Jagtap, A. 2010. Antidiabetic effect of *Ficus religiosa* extract in streptozotocin-induced diabetic rats. *J. Ethnopharmacol.* 128: 462–466.

Paneda, C., Villar, A.V., Alonso, A., Goni, F.M., Varela, F. Brodbeck, U. et al. 2001. Purification and characterization of insulin-mimetic inositol phosphoglycan-like molecules from grass pea (*Lathyrus sativus*) seeds. *Mol. Med.* 7: 454–460.

Panosian, A.G., Avetisian, G.M., Karagezian, K.G., Vartanian, G.S., and Buniatian, G.K. 1981. Hypoglycemic action of *Bryonia alba* L. unsaturated fatty trihydroxy acids in alloxan diabetes. *Dokl. Akad. Nauk. SSSR* 256: 1267–1269.

Panosian, A.G., Parsadanian, G.K., and Karagezian, K.G. 1984. Effect of trihydroxyoctadecadiene acids from *Bryonia alba* L. on the activity of glycogen metabolism enzymes in alloxan diabetes. *Biull. Eksp. Biol. Med.* 97: 295–297.

Paramesha, B., Kumar, V.P., Bankala, R., Manasa, K., and Tamilanban, T. 2014. Antidiabetic and hypolipidaemic activity of *Ceiba pentandra, Amaranthus viridis* and their combination on dexamethasone induced diabetic Swiss albino rats. *Int. J. Pharm. Pharm. Sci.* 6: 242–246.

Parameshwar, S., Srinivasan, K.K., and Mallikarjuna, R. 2007. Oral antidiabetic activity of *Caesalpinia bonducella* seed kernels. *J. Pharm. Biol.* 40: 10–15.

Parameshwar, S., Srinivasan, K.K., and Rao, C.M. 2002. Oral antidiabetic activities of different extracts of *Caesalpinia bonducella* seed kernels. *Pharm. Biol.* 40: 590–595.

Parasuraman, S., Sankar, N., Chandrasekar, T., Murugesh, K., and Neelaveni, T. 2010. Phytochemical analysis and oral hypoglycemic activity of leaf extract of leaves of *Andrographis stenophylla* C.B. Clarke (Acanthaceae). *IJABPT* 1: 442–448.

Pareek, A., Suthar, M., Godavarthi, A., Goyal, M., and Bansal, V. 2010. Negative regulation of glucose uptake by *Costus pictus* in L6 myotube cell line. *J. Pharm. Negative Results* 1: 24–26.

Pareek, H., Sharma, S., Khajja, B.S., Jain, K., and Jain, G.C. 2009. Evaluation of hypoglycemic and antihyperglycemic potential of *Tridax procumbens* (Linn.). *BMC Complement. Altern. Med.* 9: 48–52.

Pari, L. and Maheswari, J.U. 1999. Hypoglycaemic effect of *Musa sapientum* L. in alloxan-induced diabetic rats. *J. Ethnopharmacol.* 68: 321–325.

Pari, L. and Maheswari, J.U. 2000. Antihyperglycaemic activity of *Musa sapientum* flowers: Effect on lipid peroxidation in alloxan diabetic rats. *Phytother. Res.* 14: 136–138.

Pari, L. and Satheesh, A.M. 2004a. Antidiabetic activity of *Boerhaavia diffusa* L.: Effect on hepatic key enzymes in experimental diabetes. *J. Ethnopharmacol.* 91: 109–113.

Pari, L. and Latha, M. 2002. Effect of *Cassia auriculata* flowers on blood sugar levels, serum and tissue lipids in streptozotocin diabetic rats. *Singapore Med. J.* 43: 617–621.

Pari, L. and Latha, M. 2004. Protective role of *Scoparia dulcis* plant extract on brain antioxidant status and lipid peroxidation in streptozotocin diabetic male Wistar rats. *BMC Complement. Altern. Med.* 4: 16–19.

Pari, L. and Latha, M. 2006. Antihyperlipidemic effect of *Scoparia dulcis* (Sweet broom weed) in streptozotocin diabetic rats. *J. Med. Food.* 9: 102–107.

Pari, L. and Satheesh, A.M. 2004b. Antidiabetic effect of *Boerhavia diffusa*: Effect on serum and tissue lipids in experimental diabetes. *J. Med. Food* 7: 472–476.

Pari, L. and Satheesh, M.A. 2006. Effect of pterostilbene on hepatic key enzymes of glucose metabolism in streptozotocin- and nicotinamide-induced diabetic rats. *Life Sci.* 79: 641–645.

Pari, L. and Venkateswaran, S. 2002. Hypoglycaemic activity of *Scoparia dulcis* L. extract in alloxan induced hyperglycaemic rats. *Phytother. Res.* 16: 662–664.

Pari, L. and Venkateswaran, S. 2003a. Effect of an aqueous extract of *Phaseolus vulgaris* on plasma insulin and hepatic key enzymes of glucose metabolism in experimental diabetes. *Pharmazie* 58: 916–919.

Pari, L. and Venkateswaran, S. 2003b. Protective effect of *Coccinia indica* on changes in the fatty acid composition in streptozotocin-induced diabetic rats. *Pharmazie* 58: 409–412.

Pari, L. and Venkateswaran, S. 2004. Protective role of *Phaseolus vulgaris* on changes in the fatty acid composition in experimental diabetes. *J. Med. Food* 7: 204–209.

Parimala, M. and Shoba, F.G. 2014. Evaluation of anti-diabetic potential of *Nymphaea nouchali* Burm. f. seeds in streptozotocin-induced diabetic rats. *Int. J. Pharm. Pharm. Sci.* 6: 536–541.

Parivash, R., Asgari, S., Madani, H., and Mahzoni, P. 2009. Hypoglycemic effect of hydroalcoholic extract of *Carthamus tinctorius* petal on alloxan-induced diabetic rats. *Knowledge Health* 4: 1–5.

Park, B.H. and Park, J.W. 2001. The protective effect of *Amomum xanthoides* extract against alloxan-induced diabetes through the suppression of NF kappa B activation. *Exp. Mol. Med.* 33: 64–68.

Park, C. and Lee, J.S. 2013. Mini review: Natural ingredients for diabetes which are approved by Korean FDA. *Biomed. Res.* 24: 164–169.

Park, H.G., Bak, E.J., Woo, G.H., Kim, J.M., Quan, Z., Yoon, H.K. et al. 2012a. Licochalcone E has an antidiabetic effect. *J. Nutr. Biochem.* 23: 759–767.

Park, H.J., Kim, D.H., Choi, J.W., Park, J.H., and Han, Y.N. 1998. A potent antidiabetic agent from *Kalopanax pictus*. *Arch. Pharm. Res.* 21: 24–29.

Park, J.H., Lee, S.H., Chung, I.M., and Park, Y. 2012b. Sorghum extract exerts an anti-diabetic effect by improving insulin sensitivity *via* PPAR-γ in mice fed a high-fat diet. *Nutr. Res. Pract.* 6: 322–327.

Park, K.Y., Jung, G.O., Choi, J., Lee, K.T., and Park, H.U. 2002. Potent antimutagenic and their anti-lipid peroxidative effect of kaikasaponin III and tectorigenin from the flower of *Pueraria thunbergiana*. *Arch. Pharm. Res.* 25: 320–324.

Park, M.W., Ha, J., and Chung, S.H. 2008a. 20(S)-gingenoside Rg3 enhances glucose-stimulated insulin secretion and activates AMPK. *Biol. Pharm. Bull.* 31: 748–751.

Park, S., Ahn, I.S., Kwon, D.Y., Ko, B.S., and Jun, W.K. 2008b. Gingenoside Rb1 and Rg1 suppress triglyceride accumulation in 3T3-L1 adipocytes and enhance β-cell insulin secretion and viability in Min6 cells via PKA-dependent pathways. *Biosci. Biotechnol. Biochem.* 72: 2815–2823.

Park, S., Kim, D.S., and Kang, S. 2011. *Gastrodia elata* Blume water extracts improve insulin resistance by decreasing body fat in diet-induced obese rats: Vanillin and 4-hydroxy benzaldehyde are the bioactive candidates. *Eur. J. Nutr.* 50: 107–118.

Park, S.A., Choi, M.-S., Kim, M.-J., Jung, U.J., Kim, H.-J., Park, K.K. et al. 2006. Hypoglycemic and hypolipidemic action of Du-zhong (*Eucommia ulmoides* Oliv.)leaves water extract in C57BL/KsJ-db/db mice. *J. Ethnopharmacol.* 107: 42–47.

Park, S.M., Hong, S.M., Sung, S.R., Lee, J.E., and Kwon, D.Y. 2008c. Extracts of Rehmanniae radix, Ginseng radix and Scutellariae radix improve glucose-stimulated insulin secretion and beta-cell proliferation through IRS2 induction. *Genes Nutr.* 2: 347–351.

Parmar, H.S. and Kar, A. 2007. Antidiabetic potential of *Citrus sinensis* and *Punica granatum* peel extracts in alloxan treated male mice. *BioFactors* 31: 17–24.

Parmar, H.S. and Kar, A. 2008a. Medicinal values of fruit peels from *Citrus sinensis*, *Punica granatum*, and *Musa paradisiaca* with respect to alterations in tissue lipid peroxidation and serum concentration of glucose, insulin, and thyroid hormones. *J. Med. Food* 11: 376–381.

Parmar, H.S. and Kar, A. 2008b. Possible amelioration of atherogenic diet induced dyslipidemia, hypothyroidism and hyperglycemia by the peel extracts of *Mangifera indica*, *Cucumis melo* and *Citrullus vulgaris* fruits in rats. *Biofactors* 33: 13–24.

Parmar, S., Gangwal, A., and Sheth, N. 2010. *Solanum xanthocarpum* (Yellow berried night shade): A review. *Der Pharmicia Lett.* 2: 373–383.

Parthasarathy, R., Ilavarasan, R., and Karrunakaran, C.M. 2009. Antidiabetic activity of *Thespesia populnea* bark and leaf extract against streptozotocin induced diabetic rats. *Int. J. PharmTech Res.* 1: 1069–1072.

Parveen, A., Ifran, M., and Mohammed, F. 2012. Anti-hyperglycemic activity in *Grewia asiatica,* a comparative investigation. *Int. J. Pharm. Pharm. Sci.* 4: 210–213.

Parvin, A., Alam, M.M., Haque, M.A., Bhowmilk, A., Ali, L., and Rokeya, B. 2013. Study of the hypoglycemic effects of *Tamarindus indica* Linn. seeds on non-diabetic and diabetic model rats. *Br. J. Pharm. Res.* 3: 1094–1105.

Pastors, T.G., Blaisdell, P.W., Balm, T.K., Asplin, C.M., and Pohl, S.L. 1991. Phyllium fiber reduces rise in postprandial glucose and insulin concentrations in patients with non-insulin dependent diabetes. *Am. J. Clin. Nutr.* 53: 1431–1435.

Patade, G.R. and Marita, A.R. 2014. Metformin: A journey from countryside to the bedside. *J. Obes. Metab. Res.* 1: 127–130.

Patel, D.K., Prasad, S.K., and Hemalatha, S. 2012a. An overview on antidiabetic medicinal plants having insulin mimetic property. *Asian Pac. J. Trop. Biomed.* 2: 320–330.

Patel, J., Kumar, S., Patel, H., Prasad, A.K., Iyer, S.V., and Vaidya, S.K. 2012b. Hypoglycaemic and hypolipidaemic potential of aerial parts of *Amaranthus viridis* (L.) Merr. in streptozotocin induced diabetic rats. *Int. J. Pharm. Arch.* 1: 1–6.

Patel, M.B. and Mishra, S.M. 2009. Aldose reductace inhibitory activity and anti cataract potential of some traditionally acclaimed antidiabetic medicinal plants. *Oriental Pharm. Exp. Med.* 9: 245–251.

Patel, P., Harde, P., Pillai, J., Darji, N., and Patel, B. 2012. Anti-diabetic herbal drugs: A review. *Pharmacophore* 3: 18–29.

Patel, V.S., Chitra, V., Prasanna, P.L., and Krishnaraju, V. 2008. Hypoglycemic effect of aqueous extract of *Parthenium hysterophorus* L. in normal and alloxan induced diabetic rats. *Indian J. Pharmacol.* 40: 183–185.

Pathania, S., Randhawa, V., and Bagler, G. 2013. Prospecting for novel plant-derived molecules of *Rauvolfia serpentina* as inhibitors of aldose reductase, a potent drug target for diabetes and its complications. *PLOS One* 8: e61327.

Patil, R.A., Jagdale, S.C., and Kasture, S.B. 2006. Anti-hyperglycemic, antistress and nootropic activity of roots of *Rubia cordifolia* L. *Indian J. Exp. Biol.* 44: 987–992.

Patil, R.N., Patil, R.Y., Ahirwar, B., and Ahirwar, D. 2011. Isolation and characterization of anti-diabetic component (bioactivity-guided fractionation) from *Ocimum sanctum* L. (Lamiaceae) aerial part. *Asian Pac. J. Trop. Med.* 4: 278–282.

Patil, S. and Jain, G. 2014. Holistic approach of *Trigonella foenum-graecum* in phytochemistry and pharmacology—A review. *Curr. Trends Technol. Sci.* 3: 34–49.

Patil, V.V. and Patil, V.R. 2010. *Ficus bengalensis* L.—An overview. *Int. J. Pharm. Biol. Sci.* 1: 1–11.

Patil, V.V., Sutar, N.G., Pimprikar, R.B., Patil, A.P. Chaudhari, R.Y., and Patil, V.R. 2010. Anti-hyperglycemic effects of *Ficus racemosa* leaves. *J. Nat. Remedies* 10: 11–16.

Patole, P.P., Agte, V.V., and Phadnes, M.C. 1998. Effect of mucilaginous seeds on in vitro rate of starch hydrolysis and blood glucose levels on NIDDM subjects with special reference to garden cress seeds. *J. Med. Arom. Plant Sci.* 20: 1005–1010.

Patra, A., Jha, S., and Sahu, A.N. 2009. Antidiabetic activity of aqueous extract of *Eucalyptus citriodora* Hook. in alloxan induced diabetic rats. *Pharmacogn. Mag.* 5: 51–54.

Paul, S., Bhattacharyya, S.S., Boujedaini, N., and Khuda-Bukhsh, A.R. 2011. Anticancer potentials of root extract of *Polygala senega* and its PLGA nanoparticles-encapsulated form. *Evid. Based Complement. Altern. Med.* 2011: Article ID 517204, 13 pages.

Paul, T., Das, B., Apte, K.G., Banerjee, S., and Saxena, R.C. 2012. Evaluation of anti-hyperglycemic activity of *Adiantum philippense* Linn., a pteridophyte in alloxan induced diabetic rats. *Diabetes Metab.* 3: 9–17.

Pavana, P., Sethupathy, S., and Manoharan, S. 2007a. Antihyperglycemic and anti-lipidperoxidative effects of *Tephrosia purpurea* seed extract in streptozotocin-induced diabetic rats. *Indian J. Clin. Biochem.* 22: 77–83.

Pavana, P., Manoharan, S., and Sethupathy, S. 2007b. Anti-hyperglycemic and anti-lipid peroxidative effects of *Tephrosia purpurea* leaf extract in streptozotocin-induced diabetic rats. *J. Nat. Remedies* 7: 43–49.

Pavankumar, K., Vidyasagar, G., Ramakrishna, D., Reddy, I.M., Gupta Atyam, V.S.S.S., and Raidu, C.S. 2011. Screening of *Madhuca indica* for antidiabetic activity in streptozotocin and streptozotocin-nicotinamide induced diabetic rats. *Int. J. PharmTech Res.* 3: 1073–1077.

Payahoo, L., Ostadrahimi, A., Mobasseri, M., Bishok, K.Y., Jafarabadi, M.A., Mahdavi, A.B. et al. 2014. *Anethum graveolens* L. supplementation has anti-inflammatory effect in type 2 diabetic patients. *Indian J. Tradit. Knowledge* 13: 461–465.

Pedreschi, R., Campos, D., Noratto, G., Chrinos, R., and Cisneros-Zevallos, L. 2003. Andean Yacon roots (*Smallanthus sonchifolius* Poepp. Endl) frutooligossaccharides as potential novel source of probiotics. *J. Agric. Food Chem.* 15: 213–218.

Pekamwar, S.S., Kalyankar, T.M., and Kokate, S.S. 2013. Pharmacological activities of *Coccinia grandis*: Review. *J. Appl. Pharm. Sci.* 3: 114–119.

Peng, Y., Liu, F., and Ye, J. 2005. Determination of phenolic acids and flavones in *Lonicera japonica* Thumb. by capillary electrophoresis with electrochemical detection. *Electroanalysis* 17: 356–362.

Pengvicha, P., Thirawarapan, S.S., and Watanabe, H. 1996. Hypoglycaemic effect of water extract of the root of *Pandanus odorus* Ridl. *Biol. Pharm. Bull.* 19: 364–366.

Penner, E.A., Buettner, H., and Mittleman, M.A. 2013. The impact of marijuana use on glucose, insulin, and insulin resistance among US adults. *Am. J. Med.* 126: 583–589.

Pepato, M.T., Baviera, A.M., Vendramini, R.C., and Brunetti, I.L. 2004. Evaluation of toxicity after one-month treatment with *Bauhinia forficata* decoction in streptozotocin-induced diabetic rats. *BMC Complement. Altern. Med.* 4: 7

Pepato, M.T., Keller, E.H., Baviera, A.M., Kettelhut, I.C., Vendramini, R.C., and Brunetti, I.L. 2002. Anridiabetic activity of *Bauhinia forficata* decoction in streptozotocin-diabetic rats. *J. Ethnopharmacol.* 81: 191–197.

Pepato, M.T., Oliveira, J.R., Kettelhut, I.C., and Migliorini, R.H. 1993. Assessment of the antidiabetic activity of *Myrcia uniflora* extracts in streptozotocin diabetic rats. *Diabetes Res.* 22: 49–57.

Pereiraa, D.F., Kappela, V.D., Cazarolli, L.H., Boligonc, A.A., Athaydec, M.L., Guessera, S.M. et al. 2012. Influence of the traditional Brazilian drink *Ilex paraguariensis* tea on glucose homeostasis. *Phytomedicine* 19: 868–877.

Perera, P.R.D., Ekanayake, S., and Ranaweera, K.K.D.S. 2013. *In vitro* study on antiglycation activity, anti-oxidant activity and phenolic content of *Osbeckia octandra* L. leaf decoction. *J. Pharmacogn. Phytochem.* 2: 198–201.

Perez, C., Dominguez, E., Canal, J.R., Campillo, J.E., and Torres, M.D. 2000a. Hypoglycaemic activity of an aqueous extract from *Ficus carica* (fig tree) leaves in streptozotocin diabetic rats. *Pharm. Biol.* 38: 181–186.

Perez, G.R.M., Perez, G.S., Zavala, M.A., and Perez, G.S.C. 1996. Effect of *Agarista mexicana* and *Verbesina persicifolia* on blood glucose level of normoglycaemic and alloxan-diabetic mice and rats. *Phytother. Res.* 10: 351–353.

Pérez, G.R.M., Pérez-González, C., Zavala-Sánchez, M.A., and Pérez-Gutiérrez, S. 1998. Hypoglycemic activity of *Bouvardia terniflora*, *Brickellia veronicaefolia*, and *Parmentiera edulis*. *Salud pública de México* 40: 354–358.

Perez, G.R.M., Ramirez, L.E., and Vargas, S.R. 2001. Effect of *Cirsium pascuarense* on blood glucose levels of normoglycaemic and alloxan-diabetic mice. *Phytother. Res.* 15: 1–3.

Perez, G.R.M. and Vargas, S.R. 2002. Triterpenes from *Agarista mexicana* as potential antidiabetic agents. *Phytother. Res.* 16: 55–58.

Perez Gutierrez R.M. Inhibition of advanced glycation end-product formation by *Origanum majorana* L. *in vitro* and in streptozotocin-induced diabetic rats. *J. Evid. Based Complement. Altern. Med.* 2012: 8.

Perez, R.M., Perez, C., Zavala, M.A., Perez, S., Hernandez, H., and Lagunes, F. 2000b. Hypoglycemiceffects of lactucin-8-O-methylacrylate of *Parmentiera edulis* fruit. *J. Ethnopharmacol.* 71: 391–394.

Perez, S., Perez, R.M., Perez, C., Zavala, M.A., and Vargas, R. 1997. Coyolosa, a new hypoglycemic from *Acrocomia mexicana*. *Pharm. Acta Helv.* 72: 105–111.

Perez-Gutierrez, R.M., Vargas, S.R., Garcia, B.E., and Gallardo, N.Y. 2009. Hypoglycemic activity of constituents from *Astianthus viminalis* in normal and streptozotocin-induced diabetic mice. *J. Nat. Med.* 63: 393–401.

Perez da Silva, G.G., Zanoni, J.N., and Buttow, N.C. 2011. Neuroprotective action of *Ginkgo biloba* on the enteric nervous system of diabetic rats. *World J. Gastroenterol.* 17: 898–905.

Perfumi, M., Arnold, N., and Tacconi, R. 1991. Hypoglycemic activity of *Salvia fruticosa* Mill. from Cyprus. *J. Ethnopharmacol.* 34: 135–140.

Perry, R.J., Samuel, V.T., Peterson, K.F., and Shulman, G.I. 2014. The role of hepatic lipids in hepatic insulin resistance and type 2 diabetes. *Nature* 510: 84–91.

Persaud, S.J., Al-Majed, H., Raman, A., and Jones, P.M. 1999. *Gymnema sylvestre* stimulates insulin release *in vitro* by increased membrane permeability. *J. Endocrinol.* 163: 207–212.

Petchi, R.R., Parasuraman, S., Vijaya, C., Darwhekar, G., and Devika, G.S. 2011. Antidiabetic effect of kernel seeds extract of *Mangifera indica* (Anacardiaceae). *Int. J. Pharm. Biol. Sci.* 2: 385–393.

Peter, E.-S. and Obi, N. 2010. Antilipidaemic studies of mistletoe (*Loranthus micranthus*) leaf extracts on diabetic rats. *Int. J. Curr. Res.* 8: 48–55.

Petrus, A.J.A. 2012. *Mukia maderaspatana* (L.) M. Roem.—A review of its global distribution, phytochemical profile and anti-oxidant capacity. *Asian J. Chem.* 24: 2361–2368.

Petrus, A.J.A. 2013. Ethanobotanical and pharmacological profile with propagation strategies of *Mukia maderaspatana* (L.) M. Roem.—A concise overview. *Indian J. Nat. Prod. Resour.* 4: 9–26.

Peungvicha, P., Temsiririrkkul, R., Prasain, J.K., Tezuka, Y., Kadota, S., Thirawarapan, S., and Watanabe, H.S. 1998a. 4-Hydroxybenzoic acid: A hypoglycemic constituent of aqueous extract of *Pandanus odorus* root. *J. Ethnopharmacol.* 62: 79–84.

Peungvicha, P., Thirawarapan, S.S., Temsiririrkkul, R. Watanabe, H., Kumar Prasain, J., and Kodota, S. 1998b. Hypoglycemic effect of the water extract of *Piper sarmentosum* in rats. *J. Ethnopharmacol.* 60: 27–32.

Peungvicha, P., Thirawarapan, S.S., and Watanabe, H. 1998c. Possible mechanism of hypoglycemic effect of 4-Hydroxybenzoic acid, a constituent of *Pandanus odorus* root. *Jpn. J. Pharmacol.* 78: 395–398.

Phan, M.A.T., Wang, J., Tang, J., Lee, Y.Z., and Ng, K. 2013. Evaluation of α-glucosidase inhibition potential of some flavonoids from *Epimedium brevicornum*. *Food Sci. Technol.* 53: 492–498.

Phan, P.V., Sohrabi, A., Polotsky, A., Hungerford, D.S., Lindmark, L., and Frondoza, C.G. 2005. Ginger extract components suppress induction of chemokine expression in human synoviocytes. *J. Altern. Complement. Med.* 11: 149–154.

Piero, N.M., Joan, M.N., Cromwell, K.M., Joseph, N.J., Wilson, N.M. et al. 2011. Hypoglycemic activity of some Kenyan plants traditionally used to manage diabetes mellitus in Eastern Province. *J. Diabetes Metab.* 2: 155. DOI: 10.4172/2155-6156.1000155.

Pierre, W., Gildas, A.J.H., Ulrich, M.C., Modeste, W.-N., Benoit, N.T., and Albert, K. 2012. Hypoglycemic and hypolipidemic effects of *Bersama engleriana* leaves in nicotinamide/streptozotocin-induced type 2 diabetic rats. *BMC Complement. Altern. Med.* 12: 264.

Pietropolo, M. 2001. Pathogenesis of diabetes: Our current understanding. *Clini. Cornerstone Diabetes* 4: 1–21.

Pillai, N.R., Seshadri, C., and Santhakumari, G. 1979. Hypoglycemic activity of the root bark of *Salacia prenoides*. *Indian J. Exp. Biol.* 17: 1279–1270.

Pimple, B.P., Kadam, P.V., and Patil, M.J. 2011. Anti-diabetic and anti-hyperlipidemic activity of *Luffa acutangula* fruit extracts in streptozotocin-induced NIDDM rats. *Asian J. Pharm. Clin. Res.* 4: 156–163.

Pimple, B.P., Kadam, P.V., and Patil, M.J. 2012. Comparative antihyperglycaemic and antihyperlipidemic effect of *Origanum majorana* extracts in NIDDM rats. *Oriental Pharm. Exp. Med.* 12: 41–50.

Platel, K. and Srinivasan, K. 1995. Effect of dietary intake of freeze dried bitter gourd (*Momordica charantia*) in streptozotocin induced diabetic rats. *Nahrung* 39: 262–268.

Polyxeni, A. and Vassilis, J.D. 2010. Medicinal plants used for the treatment of diabetes and its long-term complications. In: *Plants in Traditional and Modern Medicine: Chemistry and Activity*, Kokkalou, E. (ed.). Thessaloniki, Greece: Aristotle University of Thessaloniki, pp. 69–175.

Poonum, T., Prakash, G.P., and Kumar, L.V. 2013. Influence of *Allium sativum* extract on the hypoglycemic activity of glibenclamide: An approach to possible herb-drug interaction. *Drug Metabol. Drug Interact.* 28: 225–230.

Poornima, M., Anita, S., and Kumar, J.V. 2013. *Semecarpus anacardiu*: A review. *Int. Ayurvedic Med. J.* 1: 69–76.

Porchezhian, E. and Ansari, S.H. 2001. Effect of *Securigera securidaca* on blood glucose levels of normal and alloxan-induced diabetic rats. *Pharm. Biol.* 39: 62–64.

Porchezhian, E., Ansari, S.H., and Shreedharan, N.K.K. 2000. Antihyperglycemic activity of *Euphrasia officinale* leaves. *Fitoterapia* 71: 522–526.

Porchezhian, E. and Dobriyal, R.M. 2003. An overview on the advances of *Gymnema sylvestre*: Chemistry, pharmacology and patents. *Pharmazie* 58: 5–12.

Postic, C., Dentin, R., and Girard, J. 2004. Role of the liver in the control of carbohydrate and lipid homeostasis. *Diabetes Metab.* 30: 398–408.

Potterat, O. 2010. Goji (*Lycium barbarum* and *L. chinense*): Phytochemistry, pharmacology and safety in the perspective of traditional uses and recent popularity. *Planta Med.* 76: 7–19.

Powers, A.C. 2008. Diabetes mellitus. In: *Principles of Harrison's Internal Medicine*, Vol. 2, 17th edn., Fauci, A.S., Braunwald, E., Kasper, D.L., Hauser, S.L., Longo, D.L., Jameson, J.L., and Loscalzo, J. (eds.). New York (and worldwide): McGraw Hill, pp. 2275–2304.

Prabhakar, P.K. and Doble, M. 2011. Effect of natural products on commercial oral antidiabetic drugs in enhancing 2-deoxyglucose uptake by 3T3-L1 adipocytes. *Ther. Adv. Endocrinol. Metab.* 2: 103–114.

Prabhakar, P.K. and Doble, M.A. 2008. Target based therapeutic approach towards diabetes mellitus using medicinal plants. *Curr. Diabetes Rev.* 4: 291–308.

Prakasam, A., Sethupathy, S., and Pugalendi, K.V. 2003a. Effect of *Casearia esculenta* root extract on blood glucose and plasma antioxidant status in streptozotocin diabetic rats. *Pol. J. Pharmacol.* 55: 43–49.

Prakasam, A., Sethupathy, S., and Pugalendi, K.V. 2003b. Hypolipidaemic effect of *Casearia esculenta* root extracts in streptozotocin-induced diabetic rats. *Pharmazie* 58: 828–832.

Prakash, D., Kumar, P., and Kumar, N. 2009. Anti-oxidant and hypoglycemic activity of some Indian medicinal plants. *Pharmacologyonline* 3: 513–521.

Pranakhon, R., Pannangpetch, P., and Aromdee, C. 2011. Anti-hyperglycemic activity of agarwood leaf extract in streptozotocin-induced diabetic rats and glucose uptake enhancement activity in rat adipocytes. *Songklanakarin J. Sci. Technol.* 33: 405–410.

Prasad, K. 2001. Secoisolariciresinol diglucoside from flaxseed delays the development of type 2 diabetes in Zucker rat. *J. Lab. Clin. Med.* 138: 32–39.

Prasannaraja, P., Sivakumar, V., and Riyazullah, M.S. 2012. Antidiabetic potential of aqueous and ethanol leaf extracts of *Vitex negundo*. *Int. J. Pharmacogn. Phytochem. Res.* 4: 38–40.

Prashanth, D., Padmaja, R., and Samiulla, D.S. 2001. Effect of certain plant extracts on alpha-amylase activity. *Fitoterapia* 72: 179–181.

Prashanth, S., Kumarb, A.A., Madhub, B., and Kumar, Y.P. 2010. Antihyperglycemic and antioxidant activity of ethanolic extract of *Madhuca longifolia* bark. *Int. J. Pharm. Sci. Rev. Res.* 5: 89–94.

Pravin, B., Krishnkant, L., Shreyas, J., Ajay, K., and Prinka, G. 2013. Recent pharmacological review on *Cinnamomum tamala*. *Res. J. Pharm. Biol. Chem. Sci.* (*RJPBCS*) 4: 916–921.

Preetha, P.P., Girija Devi, V., and Rajamohan, T. 2013. Comparative effects of mature coconut water (*Cocos nucifera*) and glibenclamide on some biochemical parameters in alloxan induced diabetic rats. *Revista Brasileira de Farmacognosia*. 23: 481–487.

Preeti, K. and Tripathi, S. 2014. *Ziziphus jujuba*: A phytopharmacological review. *Int. J. Res. Dev. Life Sci.* 3: 959–966.

Prince, P.S., Kamalakkannan, N., and Menon, V.P. 2004a. Restoration of anti-oxidants by ethanolic extract of *Tinospora cordifolia* in alloxan-induced diabetic Wistar rats. *Acta Pol. Pharm.* 61: 283–287.

Prince, P.S.M., Kamalakkannan, N., and Menon, V.P. 2004b. Antidiabetic and antihyperlipidaemic effect of alcoholic *Syzigium cumini* seeds in alloxan induced diabetic albino rats. *J. Ethnopharmacol.* 91: 209–213.

Prince, P.S.M. and Menon, V.P. 1997. Hypolipidaemic effect of *Syzigium cumini* (Jamun) seeds in alloxan diabetic rats. *Med. Sci. Res.* 25: 819–821.

Prince, P.S., Menon, V.P., and Pari, L. 1998. Hypoglycemic activity of *Syzigium cumini* (Jamun) seeds: Effect on lipid peroxidation in alloxan diabetic rats. *J. Ethnopharmacol.* 61: 1–7.

Prince, P.S.M. and Srinivasan, M. 2005. *Enicostemma littorale* Blume aqueous extract improves the antioxidant status in alloxan-induced diabetic rat tissues. *Acta Pol. Pharm.* 62: 363–367.

Priyadarshini, L., Mazumder, P.B., and Choudhury, M.D. 2014. Acute toxicity and oral glucose tolerance test of ethanol and methanol extracts of anti-hyperglycaemic plant *Cassia alata* Linn. *IOSR—J. Pharm. Biol. Sci.* 9: 43–46.

Priyanka, P.S., Patel, M.M., and Bhavsar, C.J. 2010. Comparative anti-diabetic activity of some herbal plants extracts. *Pharma Sci. Monitor* 1: 12–19.

Proph, T.P. and Onoagbe, I.O. 2012. Effects of extracts of *Triplochiton scleroxylon* (K. Schum) on plasma glucose and lipid peroxidation in normal and streptozotocin-induced diabetic rats. *J. Phys. Pharm. Adv.* 2: 380–388.

Proph, T.P., Onoagbe, I.O., and Arhoghro, M.E. 2014. Effect of aqueous and ethanolic extracts of *Triplochiton scleroxylon* K. Schum. On serum cholesterol, triglyceride and phospholipid levels in normal and streptozotocin-induced diabetic rats. *Am. J. Res. Commun.* 2: 79–90.

Prohp, T.P., Onoagbe, I.O., and Joel, P. 2012. Plasma glucose concentration and lipid profile in streptozotocin-induced diabetic rats treated with extracts of *Triplochiton scleroxylon* K. Schum. *Int. J. Anal. Pharm. Biomed. Sci.* 1: 22–29.

Pund, K.V., Vyawahare, N.S., Gadakh, R.T., and Murkute, V.K. 2012. Antidiabetic evaluation of *Dalbergia sissoo* against alloxan induced diabetes mellitus in Wistar albino rats. *J. Nat. Prod. Plant Resour.* 2: 81–88.

Punitha, R. and Manoharan, S. 2006. Antihyperglycemic and antilipidperoxidative effects of *Pongamia pinnata* (Linn.) Pierre flowers in alloxan-induced diabetic rats. *J. Ethnopharmacol.* 105: 39–46.

Puratchimani, V. and Jha, S. 2004. Standardization of *Gymnema sylvestre* R.Br. with reference to gymnemagenin by high performance thin-layer chromatography. *Phytochem. Anal.* 15: 164–166.

Puri, D., Prabhu, K.M., Dev, G., Agarwal, S., and Murthy, P.S. 2011. Mechanism of antidiabetic action of compound GII purified from fenugreek (*Trigonella foenum graecum*) seeds. *Indian J. Clin. Biochem.* 26: 335–346.

Puri, D., Prabhu, K.M., and Murthy, P.S. 2012. Antidiabetic effect of GII compound purified from fenugreek (*Trigonella foenum graecum* L.) seeds in diabetic rabbits. *Indian J. Clin. Biochem.* 27: 21–27.

Purohit, A. and Sharma, A. 2006. Antidiabetic efficacy of *Bougainvillea spectabilis* bark extract in streptozotocin induced diabetic rats. *J. Cell Tissue Res.* 6: 537–540.

Purohit, A., Vyas, K.B., and Vyas, S.K. 2008. Hypoglycemic activity of *Embelia ribes* berries (50% ethanol) extract in alloxan-induced diabetic rats. *Anc. Sci. Life* 27: 41–44.

Pushparaj, P., Low, H.K., Manikandan, J., Tan, B.K., and Tan, C.H. 2007. Anti-diabetic effects of *Cichorium intybus* in streptozotocin-induced diabetic rats. *J. Ethnopharmacol.* 111: 430–434.

Pushparaj, P., Tan, C.H., and Tan, B.K. 2000. Effects of *Averrhoa bilimbi* leaf extract on blood glucose and lipids in streptozotocin-diabetic rats. *J. Ethnopharmacol.* 72: 69–76.

Pushparaj, P., Tan, C.H., and Tan, B.K. 2001. The mechanism of hypoglycemic action of the semi-purified fractions of *Averrhoa bilimbi* in streptozotocin-induced diabetic rats. *Life Sci.* 21: 535–547.

Qadan, F., Verspohl, E.J., Nahrstedt, A., Petereit, F., and Matalka, K.Z. 2009. Cinchonain Ib isolated from *Eriobotrya japonica* induces insulin secretion *in vitro* and *in vivo*. *J. Ethnopharmacol.* 124: 224–227.

Qazi, N., Khan, R.A., Rizwani, G.H., and Feroz, Z. 2014. Effect of *Carthamus tinctorius* (Safflower) on fasting blood glucose and insulin levels in alloxan induced diabetic rabbits. *Pak. J. Pharm. Sci.* 27: 377–380.

Qi, M.Y., Kai-Chen, Liu, H.R., Su, Y.H., and Yu, S.Q. 2011. Protective effect of icariin on the early stage of experimental diabetic nephropathy induced by streptozotocin *via* modulating transforming growth factor β1 and type IV collagen expression in rats. *J. Ethnopharmacol.* 138: 731–736.

Qi, M.Y., Liu, H.R., Dai, D.Z., Li, N., and Dai, Y. 2008. Total triterpene acids, active ingredients from *Fructus corni*, attenuate diabetic cardiomyopathy by normalizing ET pathway and expression of FKBP12.6 and SERCA2a in streptozotocin-diabetic rats. *J. Pharm. Pharmacol.* 60: 1687–1694.

Qian, H. and Nihorimbere, V. 2004. Antioxidant power of phytochemicals from *Psidium guajava* leaf. *J. Zhejiang Univ. Sci.* 5: 676–683.

Qiao, W., Zhao, C., Qin, N., Zhai, H.Y., and Duan, H.Q. 2011. Identification of trans-tilliroside as active principle with anti-hyperglycemic, anti-hyperlipidemic and anti-oxidant effects from *Potentilla chinensis*. *J. Ethnopharmacol.* 135: 515–521.

Qu, W.H., Li, J.G., and Wang, M.S. 1991. Chemical studies on the *Helicteres isora*. *Zhongguo Yaoke Daxue Xuebao* 22: 203–206.

Quanhong, L.I., Caili, F., Yukui, R., Guanghui, H., and Tongyi, C. 2005. Effect of protein-bound polysaccharide isolated from pumpkin on insulin in diabetic rats. *Plant Food Hum. Nutr.* 60: 13–16.

Qujeq, D., Tatar, M., Feizi, F., Parsian, H., Faraji, A.S., and Halalkhor, S. 2013. Effect of *Urtica dioica* leaf alcoholic and aqueous extracts on the number and the diameter of the islets in diabetic rats. *Int. J. Mol. Cell. Med.* 2: 21–26.

Radha, R. and Amrithaveni, M. 2009. Role of medicinal plant *Salacia reticulata* in the management of type II diabetic subjects. *Anc. Sci. Life* 29: 14–16.

Radhika, R., Krishnakumari, S., and Sudarsanam, D. 2010. Antidiabetic activity of *Rheum emodi* in alloxan induced diabetic rats. *Int. J. Pharm. Sci. Res.* 1: 296–300.

Radhika, S., Senthilkumar, R., and Arumugam, P. 2013. A review on ethnic florae with antihyperglycemic efficacy. *Int. J. Herbal Med.* 1: 55–62.

Ragavan, B. and Krishnakumari, S. 2006. Antidiabetic effect of *Terminalia arjuna* bark extract in alloxan induced diabetic rats. *Indian J. Clin. Biochem.* 21: 123–128.

Raghuram, T.C., Sharma, R.D., Sivakumar, B., and Sahay, B.K. 1994. Effect of fenugreek seeds on intravenous glucose disposition in non-insulin dependent diabetic patients. *Phytother. Res.* 8: 83–86.

Raghuram Reddy, A., Goverdhan Reddy, P., Venkateshwarlu, E., Srinivas, N., Ramya, C.H., and Nirmala, D. 2012. Anti-diabetic and hypo-lipidemic effect of *Acalypha indica* in streptozotocin-nicotinamide induced type-2 diabetic rats. *Int. J. Pharm. Pharm. Sci.* 4: 205–212.

Rahideh, S.T., Shidfar, F., Khandozi, N., Rajab, A., Hosseini, S.P., and Mirtaher, S.M. 2014. The effect of sumac (*Rhus coriaria* L.) powder on insulin resistance, malondialdehyde, high sensitive C-reactive protein and paraoxonase 1 activity in type 2 diabetic patients. *J. Res. Med. Sci.* 19: 933–938.

Rahman, A.U. and Zaman, K. 1989. Medicinal plants with hypoglycemic activity. *J. Ethopharmacol.* 26: 1–55.

Rahman, M., Sayeed, M.A., Biplab, K.P., and Siddique, S.A. 2012. Anti-diabetic and cytotoxic activities of methanol extract of *Tabernaemontana divaricata* (L.) leaves in alloxan-induced mice. *Asian J. Pharm. Clin. Res.* 5: 49–92.

Rahman, N., Noor, K., Hlaing, K., Suhaimi, F., Kutty, M., and Sinor, M. 2010. Piper Sarmentosum influences the oxidative stress involved in experimental diabetic rats. *The Internet J. Herbal Plant Med.* 2: 1. ispub. com/IJHPM/2/1/7637.

Rai, P.K., Gupta, S.K., Srivastava, A.K., and Gupta, R.K. 2013. A scientific validation of anti-hyperglycemic and anti-hyperlipidemic attributes of *Trichosanthes dioica* Roxb. *ISRN Pharmacol.* 2013: Article ID 473059, 7 pages.

Rai, V., Iyer, U., and Mani, U.V. 1997. Effect of Tulasi (*Ocimum sanctum*) leaf powder supplementation on blood sugar levels, serum lipids and tissue lipids in diabetic rats. *Plant Foods Hum. Nutr.* 50: 9–16.

Raj, R.A., Kumar, A.S., and Gandhimathi, R. 2009. Hypoglycemic and hypolipidemic activity of *Wrightia tinctoria* L. in alloxan-induced diabetes in albino Wistar rats. *Pharmacologyonline* 3: 550–559.

Raj, R.A., Kumar, A.S., and Gandhimathi, R. 2010. Anti-diabetic effect of *Wrightia tinctoria* extracts in streptozotocin-induced diabetic rats. *Int. J. Phytopharmacol.* 1: 47–52.

Rajagopal, K. and Sasikala, K. 2008b. Antidiabetic activity of hydro-ethanolic extracts of *Nymphaea stellata* flowers in normal and alloxan induced diabetic rats. *Afr. J. Pharm. Pharmacol.* 2: 173–178.

Rajagopal, K. and Sasikala, K. 2008a. Antihyperglycaemic and antihyperlipidaemic effects of *Nymphaea stellata* in alloxan-induced diabetic rats. *Singapore Med. J.* 49: 137–141.

Rajagopal, K., Sasikala, K., and Ragavan, B. 2008. Hypoglycemic and antihyperglycemic activity of *Nymphaea stellata* flowers in normal and alloxan diabetic rats. *Pharm. Biol.* 46: 654–659.

Rajagopal, P.L., Premaletha, K., Kiron, S.S., and Sreejith, K.R. 2013. Phytochemical and pharmacological review on *Cassia fistula* Linn. The golden shower. *Int. J. Pharm. Chem. Biol. Sci.* 3: 672–679.

Rajak, S., Banerjee, S.K., Sood, S., Dinda, A.K., Gupta, Y.K., Gupta, S.K. et al. 2004. *Emblica officinalis* causes myocardial adaptation and protects against oxidative stress in ischemic-reperfusion injury in rats. *Phytother. Res.* 18: 54–60.

Rajarajeswari, A., Vijayalakshmi, P., and Mohamed Sadiq, A. 2012. Influence of *Trigonella foenum graecum* (fenugreek) in alloxan induced diabetic rats. *Bioscan* 7: 395–400.

Rajaram, K. 2013. Anti-oxidant and anti-diabetic activity of *Tectona grandis* Linn. in alloxan-induced albino rats. *Asian J. Pharm. Clin.* Res. 6: 174–177.

Rajasekaran, S., Sivagnanam, K., and Subramaniam, S. 2005. Modulatory effects of *Aloe vera* leaf gel extract on oxidative stress in rats treated with streptozotocin. *J. Pharm. Pharmacol.* 57: 241–246.

Rajavashisth, T.B., Shaheen, M., Norris, K.C., Pan, D., Sinha, S.K., Ortegal, J., and Friedman, T.C. 2012. Decreased prevalence of diabetes in marijuana users: Cross-sectional data from the National Health and Nutrition Examination Survey (NHANES) III. *BMJ Open* 2: e000494. DOI: 10.1136/bmjopen-2011-000494.

Rajesh, V., Sarthaki, R., Palani, R., and Jayaraman, P. 2014. *In vitro* evaluation of *Memecylon umbellatum* Burm. F for antihyperglycemic activity and phytochemical potential. *Int. J. Pharmacogn. Phytochem. Res.* 6: Article 18.

Rajesh, M., Mukhopadhyay, P., Batkai, S., Patel, V., Saito, K., Matsumoto, S. et al. 2010. Cannabidiol attenuates cardiac dysfunction, oxidative stress, fibrosis, and inflammatory and cell death siganaling pathways in diabetic cardiomyopathy. *J. Am. Coll. Cardiol.* 56: 2015–2025.

Rajesh, M.S., Harish, M.S., Sathyaprakash, R.J., Shetty, A.R., and Shivananda, T.N. 2009. Antihyperglycemic activity of the various extracts of *Costus speciosus* rhizomes. *J. Nat. Remedies* 9: 235–241.

Rajesh, R., Chitra, K., and Paarakh, P.M. 2012. Anti hyperglycemic and antihyperlipidemic activity of aerial parts of *Aerva lanata* Linn Juss in streptozotocin induced diabetic rats. *Asian Pac. J. Trop. Biomed.* 2012: S924–S929.

Rajeshkumar, N.V., Pillai, M.R., and Kuttan, R. 2003. Induction of apoptosis in mouse and human carcinoma cell lines by *Emblica officinalis* piolyphenols and its effect on chemical carcinogenesis. *J. Exp. Clin. Cancer Res.* 22: 201–212.

Rajeswari, J., Kesavan, K., and Jayakar, B. 2012. Antidiabetic activity and chemical characterization of aqueous/ethanol prop roots extracts of *Pandanus fascicularis* Lam in streptozotocin-induced diabetic rats. *Asian Pac. J. Trop. Med.* 2: S170–S174.

Raji, Y., Ogunwande, I.A., Osadebe, C.A., and John, G. 2004. Effect of *Azaderachia indica* extract on gastric ulceration and acid secretion in rats. *J. Ethnopharmacol.* 90: 167–170.

Raji, Y., Udoh, U.S., Mewoyeka, O.O., Ononye, F.C., and Bolarinwa, A.F. 2003. Implication of reproductive malfunction in male antifertility efficacy of *Azadirachta indica* extract in rats. *Afr. J. Med. Med. Sci.* 32: 159–165.

Rakesh, P., Sekar, D.S., and Senthilkumar, K.L. 2011. A comparative study on the antidiabetic effect of *Nelumbo nucifera* and glimepiride in streptozotocin induced diabetic rats. *Int. J. Pharma BioSci.* 2: 63–69.

Ramachandran, S., Rajasekaran, A., and Adhirajan, N. 2013. *In vivo* and *in vitro* antidiabetic activity of *Terminalia paniculata* bark: An evaluation of possible phytoconstituents and mechanisms for blood glucose control in diabetes. *ISRN Pharmacol.* 2013: Article ID 484675, 10 pages.

Ramachandran, S., Rajasekaran, A., and Manisenthil, K.T. 2011. Antidiabetic, antihyperlipidemic and antioxidant potential of methanol extract of *Tectona grandis* flowers in streptozotocin induced diabetic rats. *Asian Pac. J. Trop. Med.* 4: 624–631.

Ramachandran, S., Rajasekaran, A., and Manisenthil, K.T. 2012. Investigation of hypoglycemic, hypolipidemic and antioxidant activities of *Terminalia paniculata* bark in diabetic rats. *Asian Pac. J. Trop. Biomed.* 2: 262–268.

Ramachandrana, V., Saravanana, R., and Senthilraja, P. 2014. Anti-diabetic and anti-hyperlipidemic activity of asiatic acid in diabetic rats, role of HMG CoA: *In vivo* and *in silico* approaches. *Phytomedicine* 21: 225–232.

Ramakrishnan, M., Bhuvaneshwari, R., Duraipandiyan, R., and Dhandapani, R. 2011. Hypoglycemic activity of *Coccinia india* Wight & Arn. fruits in alloxan-induced diabetic rats. *Indian J. Nat. Prod. Resour.* 2: 350–353.

Raman, B.V., Krishna, A.N.V., Rao, B.N., Sarathi, M.P., and and Rao, M.V.B. 2012. Plants with antidiabetic activities and their medicinal values. *Int. Res. J. Pharm.* 3: 11–15.

Ramesh, A., Kumar, P., Silvia, N., Narendra, R.J., and Prasad, K. 2014. Spasmolytic and hypoglycemic effects of different fractions isolated from methanolic extract of whole plant of *Hygrophila auriculata* in Wistar albino rats. *Int. J. Pharm. Pharm. Sci.* 6: 436–439.

Ramesh, B. and Ragavan, B. 2007. Hypoglycemic activity of *Asteracantha longifolia* Nees in alloxan-induced diabetic rats. *Asian J. Microbiol. Biotechnol. Environ. Sci.* 9: 825–828.

Ramful, D., Tarnus, E., Rondeau, P., Da Dilva, C.R., Bahorun, T., and Bourdon, E. 2010. Citrus fruit extracts reduce advanced glycation end products (AGEs)- and H_2O_2-induced oxidative stress in human adipocytes. *J. Agric. Food Chem.* 58: 11119–11129.

Ramirez, P. 1992. Cultivation harvesting and storage of sweet potato products. In: *Roots, Tubers, Plantains and Bananas in Animal Feeding*, Machin, D. and Nyvold, S. (eds.). *Proceedings of the FAO Expert Consultation* held in CIAT, Cali, Colombia, January 21–25, 1991, FAO Animal Production and Health Paper -95.

Ramirez, R.O. and Rao, C.J. Jr. 2003. The gastroprotective effects of tannins extracted from duhat (*Syzygium cumini* Skeels) bark on HCL/ethanol induced gastric mucosal injury I Sprague-Dawley rats. *Clin. Hemorheol. Microcirc.* 29: 253–261.

Ramirez-Espinosa, J.J., Garcia-Jimenez, S., Rios, M.Y., Medina-Franco, J.L., Lopez-Vallejo, F., Webster, S.P. et al. 2013. Antihyperglycemic and sub-chronic antidiabetic actions of morolic and moronic acids, *in vitro* and *in silico* inhibition of 11β-HSD 1. *Phytomedicine* 20: 571–576.

Ramkumar, K.M., Vanitha, P., Uma, C., Suganya, N., Bhakkiyalakshmi, E., and Sujatha, J. 2011. Antidiabetic activity of alcoholic stem extract of *Gymnema montanum* in streptozotocin-induced diabetic rats. *Food Chem. Toxicol.* 49: 3390–3394.

Ramkumar, K.M., Lee, A.S., Krishnamurthi, K., Devi, S.S., Chakrabarti, T., Kang, K.P. et al. 2009a. *Gymnema montanum* Hook protects against alloxan-induced oxidative stress and apoptosis in pancreatic β-cells. *Cell Physiol. Biochem.* 24: 429–440.

Ramkumar, K.M., Ponmanickam, P., Velayuthaprabhu, S., Archunan, G., and Rajaguru, P. 2009b. Protective effect of *Gymnema montanum* against renal damage in experimental diabetic rats. *Food Chem. Toxicol.* 47: 2516–2521.

Ramsewak, R.S., Nair, M.G., Strasburg, G.M., DeWitt, D.L., and Nitiss, J.L. 1999. Biologically active carbozole alkaloids from *Murraya koenigii*. *J. Agric. Food. Chem.* 47: 444–447.

Ramudu, S.K., Korivi, M., Kesireddy, N., Lee, L.C., Cheng, I.S., Kuo, C.H., and Kesireddy, S.R. 2011. Nephroprotective effects of a ginger extract on cytosolic and mitochondrial enzymes against streptozotocin (STZ)-induced diabetic complications in rats. *Chin. J. Physiol.* 54: 79–86.

Ramya, S.S., Vijayanand, N., and Rathinavel, S. 2014. Antidiabetic activity of *Cynodon dactylon* (L.) Pers extracts in alloxan-induced rats. *Int. J. Pharm. Pharm. Sci.* 6: 348–352.

Ramyashree, M., Krishna Ram, H., and Shivabasavaiah. 2013. Reproductive toxicity of *Opuntia* fruit extract in male Swiss albino mice. *Int. J. Adv. Biol. Res.* 3: 464–469.

Ranasinghe, P., Jayawardana, R., Galappaththy, P., Constantine, G.R., de Vas Gunawardana, N., and Katulanda, P. 2012. Efficacy and safety of 'true' cinnamon (*Cinnamomum zeylanicum*) as a pharmaceutical agent in diabetes: A systematic review and meta-analysis. *Diabet. Med.* 29: 1480–1492.

Ranasinghe, P., Pigera, S., Premakumara, G.A.S., Galappaththy, P., Constantine, G.R., and Katulada, P. 2013. Medicinal properties of 'true' cinnamon (*Cinnamomum zeylanicum*): A systemic review. *BMC Complement. Altern. Med.* 13: 275–284.

Rani, S., Mandave, P., Khadke, S., Jagtap, S., Patil, S., and Kuvalekar, A. 2013. Antiglycation, antioxidant and antidiabetic activity of traditional medicinal plant: *Rubia cordifolia* Linn. for management of hyperglycemia. *Int. J. Plant Anim. Environ. Sci.* 3: 42–48.

Ranjan, V., Vats, M., Gupta, N., and Sardana, S. 2014. Antidiabetic potential of whole plant of *Adiantum capillus-veneris* Linn. in streptozotocin induced diabetic rats. *Int. J. Pharm. Clin. Res.* 6: 341–347.

Ranjbar-Heidari, A., Khaiatzadeh, J., Madhavi-Shahri, N., and Tehranipoor, M. 2012. The effect of fruit pod powder and aquatic extracts of *Prosopis farcta* on healing cutaneous wounds in diabetic rat. *Zahedan J. Res. Med. Sci.* 14: 16–20.

Rao, A.R., Veeresham, C., and Asres, K. 2013. *In vitro* and *in vivo* inhibitory activities of four Indian medicinal plant extracts and their major components on rat aldose reductase and generation of advanced glycation end products. *Phytother. Res.* 27: 753–760.

Rao, A.V., Madhuri, V.R.S., and Prasad, Y.R. 2012. Evaluation of the *in vivo* hypoglycemic effect of neem (*Azadirachta indica* A.Juss) fruit aqueous extract in normoglycemic rabbits. *RJPBCS* 3: 799–806.

Rao, B.K. and Rao, C.H. 2001. Hypoglycemic and antihyperglycemic activity of *Syzygium alternifolium* (Wt.) Walp. seed extracts in normal and diabetic rats. *Phytomedicine* 8: 88–93.

Rao, H.J. and Lakshmi. 2012. Therapeutic application of almonds (*Prunus amygdalus* L.): A review. *J. Clin. Diagn. Res.* 6: 130–135.

Rao, K.B., Renuka, S.P., Rajasekhar, M.D., Nagaraju, N., and Rao, A.C. 2003. Antidiabetic activity of *Terminalia pallida* fruit in alloxan induced diabetic rats. *J. Ethnopharmacol.* 85: 169–172.

Rao, M.U., Sreenivasulu, M., Chengaiah, B., Jaganmohan Reddy, K., and Madhusudhans Chetty, C. 2010. Herbal medicines for diabetes mellitus: A review. *Int. J. PharmTech Res.* 2: 1883–1892.

Rao, N.V., Benoy, K., Hemamalini, K., Shanta Kumar, S.M., and Satyanarayana, S. 2007. Anti-diabetic activity of root extracts of *Tragia involucrate*. *Pharmacologyonline* 2: 203–217.

Rao, N.K. and Nammi, S. 2006. Antidiabetic and renoprotective effects of the chloroform extract of *Terminalia chebula* Retz. seeds in streptozotocin-induced diabetic rats. *BMC Complement. Altern. Med.* 6: 17.

Rao, N.K., Bethala, K., Sisinthy, S.P., and Rajeswari, K.S. 2014. Antidiabetic activity of *Orthosiphon stamineus* Benth roots in streptozotocin induced type 2 diabetic rats. *Asian J. Pharm. Clin. Res.* 7: 149–153.

Rao, P.S. and Krishnamohan, G. 2014. Evaluation of anti-hyperglycaemic and anti-hyperlipidaemic effect of chloroform extract of *Hydnocarpus laurifolia* seed on streptozotocin induced diabetic rats. *Pharmanest* 5: 2351–2355.

Rao, P.V. and Gan, S.H. 2014. Cinnamon: A multifaceted medicinal plant. *Evid. Based Complement. Altern. Med.* 2014: Article ID 642942, 12 pages.

Rao, P.V. and Naidu, M.D. 2010. Anti-diabetic effect of *Rhinacanthus nasutus* leaf extract in streptozotocin induced diabetic rats. *Libyan Agric. Res. Center J. Int.* 1: 310–312.

Rao, P.V., Madhavi, K., Naidu, M.D., and Gan, S.H. 2013a. *Rhinacanthus nasutus* improves the levels of liver carbohydrate, protein, glycogen, and liver markers in streptozotocin-induced diabetic. *Evid. Based Complement. Altern. Med.* 2013: Article ID 102901, 7 pages.

Rao, P.V., Madhavi, K., Naidu, M.D., and Gan, S.H. 2013b. *Rhinacanthus nasutus* ameliorates cytosolic and mitochondrial enzyme levels in streptozotocin-induced diabetic rats. *Evid. Based Complement. Altern. Med.* 2013: Article ID 486047, 6 pages.

Rao, S., Mohan, K.G., and Srinivas, P. 2014. Evaluation of anti-diabetic activity of *Hydnocarpus laurifolia* in streptozotocin induced diabetic rats. *Asian J. Pharm. Clin. Res.* 7: 62–64.

Rao, S.A., Srinivas, P.V., Tiwari, A.K., Vanka, U.M., Rao, R.V., Dasari, K.R. et al. 2007. Isolation, characterization and chemobiological quantification of alpha-glucosidase enzyme inhibitory and free radical scavenging constituents from *Derris scandens* Benth. *J. Chromatogr. B: Analyt. Technol. Biomed. Life Sci.* 855: 166–172.

Raphael, K.R., Sabu, M.C., and Kuttan, R. 2002. Hypoglycemic effect of methanol extract of *Phyllanthus amarus* Schum & Thonn on alloxan induced diabetes mellitus, in rats and its relation with antioxidant potential. *Indian J. Exp. Biol.* 40: 905–909.

Rasal, V., Shetty, B., Sinnathambi, A., Yeshmaina, S., and Ashok, P. 2005. Antihyperglycaemic and antioxidant activity of *Brassica oleracea* in streptozotocin diabetic rats. *Int. J. Pharmacol.* 4: 1–6.

Rashid, F., Sharif, N., Ali, I., Sharif, S., Nisa, F.U., and Naz, S. 2013. Phytochemical analysis and inhibitory activity of ornamental plant (*Bougainvillea spectabilis*). *Asian J. Plant Sci. Res.* 3: 1–5.

Rashidi, A.A. and Noureddini, M. 2011. Hypoglycemic effect of the aromatic water of leaves of *Ficus carica* in normal and streptozotocin induced diabetic rats. *Pharmacologyonline* 1: 372–379.

Rasineni, K., Bellamkonda, R., Singareddy, S.R., and Desireddy, S. 2010. Antihyperglycemic activity of *Catharanthus roseus* leaf powder in streptozotocin-induced diabetic rats. *Pharmacogn. Res.* 2: 195–201.

Rastogi, R.P. and Mehrotra, B.N. 1999. *Compendium of Indian Medicinal Plants*. New Delhi, India: Central Drug Research Institute, Lucknow & National Institute of Science Communication.

Rathee, S., Mogla, O.P., Sardana, S., Vats, M., and Rathee, P. 2010. Antidiabetic activity of *Capparis decidua* Forsk Edgew. *J. Pharm. Res.* 3: 231–238.

Rathi, S.S., Grover, J.K., and Vats, V. 2002b. The effect of *Momordica charantia* and *Mucuna pruriens* in experimental diabetes and their effect on key metabolic enzymes involved in carbohydrate metabolism. *Phytother. Res.* 16: 236–243.

Rathi, S.S., Grover, J.K., Vikrant, V., and Biswas, N.R. 2002a. Prevention of experimental diabetic cataract by Indian Ayurvedic plant extracts. *Phytother. Res.* 16: 774–777.

Rathinamoorthy, R. and Thilagavathi, G. 2014. *Terminalia chebula*—review on pharmacological and biochemical studies. *Int. J. PharmTech Res.* 6: 97–116.

Rathod, N.R., Chitme, H.R., Irchhaiya, R., and Chandr, R. 2011. Hypoglycemic effect of *Calotropis gigantea* Linn. leaves and flowers in streptozotocin-induced diabetic rats. *Oman Med. J.* 26: 104–108.

Rau, O., Wurglics, M., Dingermann, T., Abdel-Tawab, M., and Schubert-Zsilavecz, M. 2006. Screening of herbal extracts for activation of the human peroxisome proliferator-activated receptor. *Pharmazie* 61: 952–956.

Raut, N. 2012. Antidiabetic potential of fractions of hydro-ethanol extract of *Cyperus rotundus* L. (Cyperaceae). *Res. J. Pharm. Biol. Chem. Sci.* 2012. 3: 1014–1018.

Raut, N. and Gaikwad, N.J. 2006. Antidiabetic activity of hydro-ethanolic extract of *Cyperus rotundus* in alloxan induced diabetes in rats. *Fitoterapia* 77: 585–588.

Rauter, A.P., Martins, A., Lopes, R., Ferreira, J., Serralheiro, L.M., Araujo, M.E. et al. 2009. Bioactivity studies and chemical profile of the anti-diabetic plant *Genista tenera. J. Ethnopharmacol.* 122: 384–393.

Ravanshad, S.H., Naserollahzadeh, J., Sovaid, M., and Setoudehmaram, E. 2006. Effect of sour orange (*Citrus aurantium* L.) juice consumption on blood glucose and lipid profiles in diabetic patients with dyslipidemia. *J. Guilan Univ. Med. Sci.* 15: 48–53.

Ravichandiran, R., Ram, P.S., Saritha, C., and Shankaraiah, P. 2014. Anti diabetic and anti dyslipidemia activities of *Cleome gynandra* in alloxan induced diabetic rats. *J. Pharmacol. Toxicol.* 9: 55–61.

Ravichandran, V., Duraisankar, M., and Nirmala, S. 2014. Anti-diabetic activity of ethanol extract from the roots of *Cephlandra indica* on alloxan-induced diabetic rats. *Int. J. Front. Sci. Technol.* 2: 82–96.

Raza, H., Ahmed, I., John, A., and Sharma, A.K. 2000. Modulation of xenobiotic metabolism and oxidative stress in chronic streptozotocin induced diabetic rats fed with *Momordica charantia* fruit extract. *J. Biochem. Mol. Toxicol.* 14: 131–139.

Reddy, G.T., Kumar, B.R., and Mohan, G.K. 2005. Antihyperglycemic activity of *Momordica dioica* fruits in alloxan-induced diabetic rats. *Nig. J. Nat. Prod. Med.* 9: 33–34.

Reddy, J.K., Rao, B.S., Reddy, T.S., and Priyanka, B. 2013. Anti-diabetic activity of ethanolic extract of *Hydnocarpus wightiana* Blume using streptozotocin induced diabetes in rats. *IOSR J. Pharm.* 3: 29–40.

Reddy, N.M. and Reddy, R.N. 2014. *Solanum xanthocarpum* chemical constituents and medicinal properties: A review. *Scholars Acad. J. Pharm.* 3: 146–149.

Reddy, S.S., Ramatholisamma, P., Karuna, R., and Saralakumari D. 2009. Preventive effect of *Tinospora cordifolia* against high-fructose diet-induced insulin resistance and oxidative stress in male Wistar rats. *Food Chem. Toxicol.* 47: 2224–2229.

Reddy, S.V., Tiwari, A.K., Kumar, U.S., Rao, R.J., and Rao, J.M. 2005. Free radical scavenging, enzyme inhibitory constituents from antidiabetic Ayurvedic medicinal plant *Hydnocarpus wightiana* Blume. *Phytother. Res.* 19: 277–281.

Reed, M.J., Meszaros, K., Entes, L.J., Claypool, M.D., Pinkett, J.G., Gadbois, T.M., and Reaven, G.M. 2000. A new rat model of type 2 diabetes, the fat fed streptozotocin treated rats. *Metabolism* 49: 1390–1394.

Reeds, D.N., Patterson, B.W., Okunade, A., Holloszy, J.O., Polonsky, K.S., and Klein, S. 2011. Ginseng and ginsenoside Re do not improve β-cell function or insulin sensitivity in overweight and obese subjects with impaired glucose tolerance or diabetes. *Diabetes Care* 34: 1071–1076.

Ren, D., Zhao, Y., Nie, Y., Yang, N., and Yang, X. 2014. Hypoglycemic and hepatoprotective effects of polysaccharides from *Artemisia sphaerocephala* Krasch seed. *Int. J. Biol. Macromol.* 69: 296–306.

Renata, A., Sancho, S., and Pastore, G.M. 2012. Evaluation of the effects of anthocyanins in type 2 diabetes. *Food Res. Int.* 46: 378–386.

Renjith, R.S., Chikku, A.M., and Rajamohan, T. 2013. Cytoprotective, antihyperglycemic and phytochemical properties of *Cocos nucifera* (L.) inflorescence. *Asian Pac. J. Trop. Med.* 6: 804–810.

Resurreccion-Magno, M., Villasenor, I., Harada, N., and Monde, K. 2005. Antihyperglycaemic flavonoids from *Syzygium samarangense* (Blume) Merr. and Perry. *Phytother. Res.* 19: 246–251.

Retna, A.M. and Ethalsha, P. 2013. A review of the taxonomy, ethnobotany and pharmacology of *Catharanthus rosus* (Apocyanaceae). *Int. J. Eng. Res. Technol.* 2: 3899–3912.

Revilla, M.C., Andrade-Cetto, A., Islas, S., and Wiedenfeld, H. 2002. Hypoglycemic effect of *Equisetum myriochaetum* areal parts on type 2 diabetic patients. *J. Ethnopharmacol.* 81: 117–120.

Revilla-Monsalve, M.C., Andrade-Cetto, A., Palomino-Garibay, M.A., Wiedenfeld, H., and Islas-Andrade, S. 2007. Hypoglycemic effect of *Cecropia obtusifolia* Bertol aqueous extracts on type 2 diabetic patients. *J. Ethnopharmacol.* 111: 336–340.

Reza, H., Haq, W.M., Das, A.K., Rahman, S., Jahan, R., and Rahmatullah, M. 2011. Anti-hyperglycemic and antinociceptive activity of methanol leaf and stem extract of *Nypa fruticus* Wurmb. *Pak. J. Pharm. Sci.* 24: 485–488.

Rhodes, C.J. 2005. Type 2 diabetes–a matter of beta-cell life and death? *Science* 307: 380–384.

Riaz, A., Khan, R.A., and Ahmed, M. 2013. Glycemic response of *Citrus limon*, pomegranate and their combinations in alloxan-induced diabetic rats. *Aust. J. Basic Appl. Sci.* 7: 215–219.

Riaz, H., Raza, S.A., Hussain, S., Mahmood, S., and Malik, F. 2014. An overview of ethnopharmacological properties of *Boerhaavia diffusa*. *Afr. J. Pharm. Pharm.* 8: 49–58.

Ribnicky, D.M., Kuhn, P., Poulev, A., Logendra, S., Zuberi, A., Cefalu, W.T., and Raskin, I. 2009. Improved absorption and bioactivity of active compounds from an anti-diabetic extract of *Artemisia dracunculus* L. *Int. J. Pharm.* 370: 87–92.

Richter, B., Bandeira-Echtler, E., Bergerhoff, K., and Lerch, C.L. 2008. Dipeptidyl peptidase-4 (DPP-4) inhibitors for type 2 diabetes mellitus. *Cochrane Database Syst. Rev.* 16: CD 006739.

Rodrigues, G.R., Nasa, F.C.D., Porawski, M., Marcolin, E., Kretzmann, N.A., Ferraz, A.D.F. et al. 2012. Treatment with aqueous extract from *Croton cajucara* Benth reduces hepatic oxidative stress in streptozotocin-diabetic rats. *J. Biomed. Biotechnol.* 2012: Article ID 902351, 7 pages.

Rodriguez-Mateos, A., Cifuents-Gomez, T., Tabatabaee, S., Lecras, C., and Spencer, J.P. 2012. Procyanidin, anthocyanin, and chlorogenic acid contents of highbush and lowbush blueberries. *J. Agric. Food. Chem.* 60: 5772–5778.

Rodriguez-Mateos, A., Ishisaka, A., Mawatari, K., Vidal-Diez, A., Spencer, J.P., and Terao, J. 2013. Blueberry intervention improves vascular reactivity and lowers blood pressure in high-fat-, high-cholesterol-fed rats. *Br. J. Nutr.* 109: 1746–1754.

Roghani, M., Baluchnejadmojarad, T., Amin, A., and Amirtouri, R. 2007. The effect of administration of A. *graveolens* aqueous extract on the serum levels of glucose and lipids of diabetic rats. *Indian J. Endocrinol. Metabol.* 9: 177–181.

Roman-Ramos, R., Alarcon-Aguilar, F., Lara-Lemus, A., and Flores-Saenz, J.L. 1992. Hypoglycemic effect of plants used in Mexico as antidiabetics. *Arch. Med. Res.* 23: 59–64.

Roman-Ramos, R., Flores-Saenz, J.L., Partida-Hernandez, G., Lara-Lemus, A., and Alarcon-Aguilar, F. 1991. Experimental study of the hypoglycemic effect of some antidiabetic plants. *Arch. Invest. Med. (Mex.)* 22: 87–93.

Roman-Ramos, R., Flores-Saenz, J.L., and Alarcon-Aguilar, F.J. 1995. Anti-hyperglycemic effect of some edible plants. *J. Ethnopharmacol.* 48: 25–32.

Romila, Y., Mazumder, P.B., and Choudhury, M.D.D. 2010. A review on antidiabetic plants used by the people of Manipur characterized by hypoglycemic activity. *J. Sci. Tech.* 6: 167–175.

Rosemary, H., Rosidah, Y., and Haro, G. 2014. Anti-diabetic effect of roselle calyces extract (*Hibiscus sabdariffa* L.) in streptozotocin induced mice. *Int. J. PharmTech Res.* 6: 1703–1711.

Rosen, E.D. and Spiegelman, B.M. 2006. Adipocytes as regulators of energy balance and glucose homeostasis. *Nature* 444: 847–853.

Rosenzweig, T., Abitbol, G., and Taler, D. 2007. Evaluating the anti-diabetic effects of *Sarcopoterium spinosum* extracts *in vitro*. *Israel J. Plant Sci.* 55: 103–109.

Rouibi, A., Khali, M., and Saidi, F. 2013. Antispasmodic and hypoglycemic effects of aqueous extract of *Ajuga iva* L. in induced diabetic rats. *Bull. UASVM Anim. Sci. Biotechnol.* 70: 339–343.

Roy, M.G., Rahman, S., Rehana, F., Munmun, M., Sharmin, N., Hasan, Z. et al. 2010. Evaluation of anti-hyperglycemic potential of methanolic extract of *Tamarindus indica* L. (Fabaceae) fruits and seeds in glucose-induced hyperglycemic rats. *Adv. Nat. Appl. Sci.* 4: 159–162.

Royhan, A. Susilowati. R., and Sunarti. 2009. Effect of white-skinned sweet potato (*Ipomoea batatas* L.) pancreatic beta cells and insulin expression in streptozotocin induced diabetic rats. *Majalah Kesehatan Pharma Medika* 1: 45–49.

Ruela, H.S., Sabino, K.C., Leal, I.C., Landeira-Fernandez, A.M., De Almeida, M.R., Rocha, T.S. et al. 2013. Hypoglycemic effect *of Bumelia sartorum* polyphenolic rich extracts. *Nat. Prod. Commun.* 28: 207–210.

Rungprom, W., Siripornvisal, S., Keawsawai, S., and Songtrai, M. 2010. α-Glucosidase inhibitors from medicinal plants for diabetes therapy. *Agric. Sci. J.* 41: 30–34.

Russell, K.R., Morrison, E.Y., and Ragoobirsingh, D. 2005. The effect of annatto on insulin binding properties in the dog. *Phytother. Res.* 19: 433–436.

Russell, K.R., Omoruvi, F.O., Pascoe, K.O., and Morrison, E.Y. 2008. Hypoglycaemic activity of *Bixa orellana* extract in the dog. *Methods Find. Exp. Clin. Pharmacol.* 30: 301–305.

Russo, E.M., Reichelt, A.A., De-Sa, J.R., Furlanetto, R.P., Moises, R.C.S., Kasamatsu, T.S. et al. 1990. Clinical trial of *Myrcia uniflora* and *Bauhinia forficata* leaf extracts in normal and diabetic patients. *Braz. J. Med. Biol. Res.* 23: 11–20.

Saba, A.B., Oyagbemi, A.A., and Azeez, O.I. 2010. Antidiabetic and haematinic effects of *Parquetina nigrescens* on alloxan induced type-1 diabetes and normocytic normochromic anaemia in Wistar rats. *Afr. Health Sci.* 10: 276–282.

Sabah, B. 2010. Anti-oxidant properties of *Tamus communis* L., *Carthamus caerulens* L. and *Ajuga iva* L. extracts. Doctorat of Sciences thesis, University of Ferhat Abbas Setif, Setif, Algeria.

Sabitha Rani, S., Sulaakshana, G., and Patnaik, S. 2012. *Costus speciosus*, an antidiabetic plant—Review. *FS J. Pharm. Res.* 1: 52–53.

Saboo, S.S., Chavan, R.W., Tapadia, G.G., and Khadabadi S.S. 2014. An important ethnomedicinal plant *Balanite aegyptiaca* Del. *Am. J. Ethnomed.* 1: 122–128.

Sabu, M.C. and Kuttan, R. 2002. Anti-diabetic activity of medicinal plants and its relationship with their antioxidant property. *J. Ethnopharmacol.* 81: 155–160.

Sabu, M.C. and Kuttan. R. 2009. Antidiabetic and antioxidant activity of *Terminalia belerica* Roxb. *Indian J. Exp. Biol.* 47: 270–275.

Sabu, M.C. and Kuttan, R. 2004. Antidiabetic activity of *Aegle marmelos* and its relationship with its antioxidant properties. *Indian J. Physiol. Pharmacol.* 48: 81–88.

Sabu, M.C. and Subburaju, T. 2002. Effect of *Cassia auriculata* Linn. on serum glucose level and glucose utilization by isolated rat hemidiaphram. *J. Ethnopharmacol.* 80: 203–206.

Sachan, N.K., Kumar, Y., Pushkar, S., Thakur, R.N., Gangwar, S.S., and Kalaichelvan, V.K. 2009. Antidiabetic potential of alcohol and aqueous extracts of *Ficus racemosa* Linn. bark in normal and alloxan-induced diabetic rats. *Int. J. Pharm. Sci. Drug Res.* 1: 24–27.

Sachdewa, A. and Khemani, L.D. 2003. Effect of *Hibiscus rosa-sinensis* L. ethanol extract on blood glucose and lipid profile in streptozotocin-induced diabetic rats. *J. Ethnopharmacol.* 89: 61–66.

Sadiq, S., Nagi, A.H., Shahzad, M., and Zia, A. 2010.The reno-protective effect of aqueous extract of *Carum carvi* (Black Zeera) seeds in streptozotocin-induced diabetic nephropathy in rodents. *Saudi J. Kidney Dis. Transplant* 21: 1058–1065.

Sadique, J., Chandra, T., Thenmozhi, V., and Elango, V. 1987. The anti-inflammatory activity of *Enicostemma littorale* and *Mollugo cerviana*. *Biochem. Med. Metab. Biol.* 37: 167–176.

Sah, S.P., Sah, M.L., Juyal, V., and Pandey, S. 2010. Hypoglycemic activity of *Urtica parviflora* Roxb. in normoglycemic rats. *Int. J. Phtomed.* 2: 47–51.

Saha, B.K., Bhuiyan, M.N.H., Mazumder, K., and Haque, K.M.F. 2009. Hypoglycemic activity of *Lagerstroemia speciosa* L. extract on streptozotocin-induced diabetic rat: Underlying mechanism of action. *Bangladesh J. Pharmacol.* 4: 79–83.

Saha, P., Bala, A., Kar, B., Naskar, S., Mazumder, U.K., Haldar, P.K. et al. 2011a. Anti-diabetic activity of *Cucurbita maxima* aerial parts. *Res. J. Med. Plants* 5: 577–586.

Saha, P., Mazumdar, U.K., Haldar, P.K., Sen, S.K., and Naskar, S. 2011b. Antihyperglycemic activity of *Lagenaria siceraria* aerial parts on streptozotocin induced diabetes in rats. *Diabetol. Croat.* 40: 49–60.

Sahu, G. and Gupta, P.K. 2012. A review on *Bauhonia veriegata* L. *Int. Res. J. Pharm.* 3: 48–51.

Sahu, P.K., Giri, D.D., Singh, R., Pandey, P., Gupta, S., Shrivastava, A.K. et al. 2013. Therapeutic and medicinal uses of *Aloe vera*: A review. *Pharmacol. Pharm.* 4: 599–610.

Saidu, A.N., Mann, A., and Onuegbu, C.D. 2012. Phytochemical screening and hypoglycemic effect of aqueous *Blighia sapida* root bark extract on normoglycemic albino rats. *Br. J. Pharm. Res.* 2: 89–97.

Saidu, A.N., Oibiokpa, F.I., and Olukotom, I.O. 2014. Phytochemical screening and hypoglycemic effect of methanolic fruit pulp extract of *Cucumis sativus* in alloxan induced diabetic rats. *J. Med. Plant Res.* 8: 1173–1178.

Saikia, R., Choudhury, M.D., Talukdar, D.A., and Chetia, P. 2011. Antidiabetic activity of ethno medicinal plant *Scoparia dulcis* L. (Family: Scrophulariaceae): A review. *Assam Univ. J. Sci. Technol: Biol. Environ. Sci.* 7: 173–180.

Saini, S. and Sharma, S. 2013. Anti-diabetic effect of *Helianyhus annuus* L. seeds ethanolic extract in streptozotin-nicotinamide induced type 2 diabetic mellitus. *Int. J. Pharm. Pharm. Sci.* 5: 382–387.

Salahuddin, M. and Jalalpure, S.S. 2010. Antidiabetic activity of aqueous fruit extract of *Cucumis trigonus* Roxb. in streptozotocin-induced-diabetic rats. *J. Ethnopharmacol.* 127: 565–567.

Saleem, R., Ahmad, M., Hussain, S.A., Qazi, A.M., Ahmad, S.I., Qazi, M.H. et al. 1999. Hypotensive, hypoglycaemic and toxicological studies on the flavonol C-glycoside shamimin from *Bombax ceiba*. *Planta Med.* 65: 331–334.

Salgado, J.M., Mansi, D.N., and Gagliardi, A. 2009. *Cissus sicyoides*: Analysis of glycemic control in diabetic rats through biomarkers. *J. Med. Food* 12: 72207.

Salib, J.Y., Michael, H.N., and Eskande, E.F. 2013. Anti-diabetic properties of flavonoid compounds isolated from *Hyphaene thebaica* epicarp on alloxan induced diabetic rats. *Pharmacogn. Res.* 5: 22–29.

Saltiel, A.R. and Pessin, J.E. 2003. Insulin signaling in microdomains of the plasma membrane. *Traffic* 4: 711–716.

Samarghandian, S., Borji, J., and Tabasi, S.H. 2013. Effects of *Cichorium intybus* Linn. on blood glucose, lipid constituents and selected oxidative stress parameters in streptozotocin-induced diabetic rats. *Cardiovasc. Hematol. Disord. Drug Targets* 13: 231–236.

Samy, R.P. and Ignacimuthu, S. 2000. Anti-bacterial activity of some folklore medicinal plants used by tribals in Western Ghats of India. *J. Ethnopharmacol.* 69: 63–71.

Sanadhya, I., Lobo, V., Bhot, M., Varghese, J., and Chandra, N. 2013. Anti-diabetic activity of *Anthocephalus indicus* A.Rich fruits in alloxan-induced diabetic rats. *Int. J. Pharm. Pharm. Sci.* 5: 519–523.

Sanadhya, I.U., Bhot, M., Varghese, J., and Chandra, N. 2012. Anti-diabetic activity of leaves of *Anthocephalus indicus* A.Rich in alloxan-induced diabetic rats. *Int. J. Phytomed.* 4: 511–518.

Sancheti, S., Sancheti, S., Bafna, M., and Seo, S. 2010. Antihyperglycemic, antihyperlipidemic, and antioxidant effects of *Chaenomeles sinensis* fruit extract in streptozotocin-induced diabetic rats. *Eur. Food Res. Technol.* 231: 415–421.

Sancheti, S., Sancheti, S., and Seo, S.Y. 2013. Antidiabetic and antiacetylcholinesterase effects of ethyl acetate fraction of *Chaenomeles sinensis* (Thouin) Koehne fruits in streptozotocin-induced diabetic rats. *Exp. Toxicol. Pathol.* 65: 55–60.

Sanchez, G.M., Rodriguez, H.M.A., Giuliani, A., Nunez Selles, A.J., Rodriguez, N.P., Leon Fernandez, O.S. et al. 2003. Protective effects of *Mangifera indica* L. extract (Vimang) on the injury associated with hepatic ischaemia reperfusion. *Phytother. Res.* 17: 197–201.

Sancho, R.A.S. and Pastore, G.M. 2012. Evaluation of the effects of anthocyanins in type 2 diabetes. *Food Res. Int.* 46: 378–386.

Sanchooli, N. 2011. Antidiabetic properties of *Physalis alkekengi* extract in alloxan-induced diabetic rats. *Res. J. Pharm. Biol. Chem. Sci.* 2: 168–173.

Sanda, K.A., Sandabe, U.K., Auwal, M.S., Bulama, I., Bashir, T.M., Sanda, F.A. et al. 2013. Hypoglycemic and anti-diabetic profile of the aqueous root extracts of *Leptadenia hastata* in albino rats. *Pak. J. Biol. Sci.* 16: 190–194.

Sandhyarani, M., Pippalla, R.S., Mohan, G.K., and Gangaraju, M. 2012. Anti-diabetic activity of methanolic and ethyl acetate extracts of *Wrightia tinctoria* R.Br. fruit. *Int. J. Pharm. Sci. Res.* 3: 3861–3866.

Sang, H., Gu, J., Yuan, J., Zhang, M., Jia, X., and Feng, L. 2014. The protective effect of Smilax glabra extract on advanced glycation end products-induced endothelial dysfunction in HUVECs via RAGE-ERK1/2-NF-κB pathway. *J. Ethnopharmacol.* 155: 785–795.

Sangameswaran, B. and Jayakar, B. 2007. Anti-diabetic and spermatogenic activity of *Cocculus hirsutus* (L.) Diels. *Afr. J. Biotechnol.* 6: 1212–1216.

Sangameswaran, B. and Jayakar, B. 2008. Anti-diabetic, anti-hyperlipidemic and spermatogenic effects of *Amaranthus spinosus* Linn. on streptozotocin-induced diabetic rats. *J. Nat. Med.* 62: 79–82.

Sangeswaran, B., Ilango, K., Chaurey, M., and Bhaskar, V.H. 2010. Anti-hyperglycemic and anti-hyperlipidemic effects of extracts of *Ipomoea reniformis* Chois on alloxan-induced diabetic rats. *Ann. Biol. Res.* 1: 157–163.

Sang-Mi, K., Min, L.Y., Mi-ju, K., Song-Yee, N., Sung-Hee, K., and Hwan-Hee, J. 2013. Effects of *Agrimonia pilosa* Ledeb. water extract on α-glucosidase inhibition and glucose uptake in C2C12 skeletal muscle cells. *Korean J. Food Nutr.* 26: 806–813.

Sangwan, S., Rao, D.V., and Sharma, R.A. 2010. A review on *Pongamia pinnata* L. Pierre: A great versatile leguminous plant. *Nat. Sci.* 8: 130–139.

Sani, U.M. 2014. phytochemical screening and anti-diabetic activity of extracts of *Citrullus lanatus* rind in alloxan-induced diabetic albino mice. *Int. J. Chem. Pharm. Sci.* 2: 1211–1215.

Sanjeeva, A., Kumar, Gnananath, K., Kiran, D., Reddy, A.M., and Ch, R. 2011. Anti diabetic activity of ethanolic extract of *Cynodon dactylon* root stalks in streptozotocin induced diabetic rats. *Int. J. Adv. Pharm. Res.* 2: 418–422.

Sanjita, D. and Uttam, S. 2014. Antidiabetic and antioxidant potential of the fruits of *Callistemon lanceolatus* DC. *J. Pharm. Res.* 3: 92–97.

Sankaran, M. and Vadivel, A. 2011. Antioxidant and antidiabetic effect of *Hibiscus rosasinensis* flower extract on streptozotocin induced experimental rats-a dose response study. *Not Sci. Biol.* 3: 13–21.

Sankaranarayanan, C. and Pari, L. 2011. Thymoquinone ameliorates chemical-induced oxidative stress and β-cell damage in experimental hyperglycemic rats. *Chem. Biol. Interact.* 190: 148–154.

Sankhala, L.N., Saini, R.K., and Saini, B.S. 2012. A review on chemical and biological properties of *Tinospora cordifolia*. *Int. J. Med. Arom. Plants* 2: 340–344.

Santhakumari, P., Prakasam, A., and Pugalendi, K.V. 2006. Antihyperglycemic activity of *Piper betle* leaf on streptozotocin-induced diabetic rats. *J. Med. Food* 9: 108–112.

Santhosh, G., Prakash, T., Kotresha, D., Roopa, K., Surendra, V., and Divakar, G. 2010. Anti-diabetic activity of *Celosia argentea* root in streptozotocin-induced rats. *Int. J. Green Pharm.* 4: 206–211.

Santos, F.A., Frota, J.T., Arruda, B.R., de Melo, T.S., da Silva, A.A.C.A., Brito, G.A.C. et al. 2012. Antihyperglycemic and hypolipidemic effects of α, β-amyrin, a triterpenoid mixture from *Protium heptaphyllum* in mice. *Lipids Health Dis.* 11: 98–116.

Saral, O., Olmez, Z., and Sahins, H. 2014. Comparison of antioxidant properties of wild blueberries (*Vaccinium arctostaphylos* L. and *Vaccinium myrtillus* L.) with cultivated blueberry varieties (*Vaccinium corymbosum* L.) in Artvin region of Turkey. *Turkish J. Agric. Food Sci. Technol.* 3: 40–44.

Saranya, R., Thirumalai, T., Hemalatha, M., Balaji, R., and David, E. 2013. Pharmocognosy of *Enicostemma littorale:* A review. *Asian Pac. J. Trop. Biomed.* 3: 79–84.

Saranya, S., Pradeepa, S.S., and Subramanian, S. 2014. Biochemical evaluation of antidiabetic activity of *Cocos nucifera* flowers in streptozotocin-induced diabetic rats. *Int. J. Pharm. Sci. Rev. Res.* 26: 67–72.

Sarathchandran, I., Vrushabendraswamy, B.M., Pavenkumar, B., Narayanan, V., Vinothkumar, S., Muralidharan, P. et al. 2008. Hypoglycaemic and hypolipidemic activity of methanolic root extract of *Caesalpinia digyna* (Rottler) in streptozotocin-induced diabetic rats. *Pharmacologyonline* 2: 781–789.

Saravanamuttu, S. and Sudarsanan, D. 2012. Antidiabetic plants and their active ingredients: A review. *Int. J. Pharm. Sci. Res.* 3: 3639–3650.

Saravanan, D., Aparnalakshmi, I., Govinath, M., Girieshkumar, B., Priya, S., Syamala, E. et al. 2011. Potential antioxidant, hypoglycemic and hypolipidemic effect of leaves of *Hibiscus platanifolius* Linn. *Int. J. Pharm. Sci. Drug Res.* 3: 236–240.

Saravanan, G. and Leelavinothan, P. 2006. Effects of *Syzygium cumini* bark on blood glucose, plasma insulin and C-peptide in streptozotocin-induced diabetic rats. *Int. J. Endocrinol. Metab.* 4: 96–105.

Saravanan, G. and Pari, L. 2008. Hypoglycaemic and antihyperglycaemic effect of *Syzygium cumini* bark in streptozotocin-induced diabetic rats. *J. Pharmacol. Toxicol.* 3: 1–10.

Saravanan, G. and Ponmurugan, P. 2011. Ameliorative potential of S-allylcysteine on oxidative stress in STZ-induced diabetic rats. *Chem. Biol. Interact.* 189: 100–106.

Sarkar, A., Lavania, S.C., Pandey, D.N., and Pant, M.C. 1994. Changes in the lipid profile after administration of *Ocimum sanctum* (tulsi) leaves in the normal albino rabbits. *Indian J. Physiol. Pharmacol.* 38: 311–312.

Sarkar, M., Biswas, P., and Samanta, A. 2013. Study of hypoglycemic activity of aqueous extract of *Leucas indica* Linn. aerial parts on streptozotocin induced diabetic rats. *Int. J. Pharm. Sci. Drug Res.* 5: 50–55.

Sarkar, P., Mahmud, A.K., and Mohanty, J.P. 2011. Antidiabetic activity of ethanolic extract of *Mirabilis jalapa* roots. *Int. J. Pharm. Technol.* 3: 1470–1479.

Sarkar, S., Pranava, M., and Marita, R. 1996. Demonstration of the hypoglycemic action of *Momordica charantia* in a validated animal model of diabetes. *Pharmacol. Res.* 33: 1–4.

Sarkodie, J.A., Fleischer, T.C., Edoh, D.A., Dickson, R.A., Mensah, M., and Annan, K. 2013. Antihyperglycaemic activity of ethanolic extract of the stem of *Adenia lobata* Engl (Passifloraceae). *IJPSR* 4: 1370–1377.

Sasaki, T., Lee, W., Morimura, H., Li, S., Li, Q., and Asada, Y. 2011. Chemical constituents from *Sambucus adnata* and their protein–tyrosine phosphatase 1B inhibitory activities. *Chem. Pharm. Bull.* 59: 1396–1399.

Satdive, R.K., Abhilash, P., and Fulzele, D.P. 2003. Anti-microbial activity of *Gymnema sylvestre* leaf extract. *Fitoterapia* 74: 699–701.

Satheesh, A.M. and Pari, L. 2004. Antioxidant effect of *Boerhavia diffusa* L. in tissues of alloxan induced diabetic rats. *Indian J. Exp. Biol.* 42: 989–992.

Sathisekar, D.S., Sivagnanam, K., and Subramanian, S. 2005a. Anti-diabetic activity of *Momordica charantia* seeds on streptozotocin induced diabetic rats. *Pharmazie* 60: 383–387.

Sathishsekar, D. and Subramanian, S. 2005a. Beneficial effects of *Momordica charantia* seeds in the treatment of streptozotocin-induced diabetes in experimental rats. *Biol. Pharm. Bull.* 28: 978–983.

Sathishsekar, D. and Subramanian, S. 2005b. Anti-oxidant properties of *Momordica charantia* (bitter gourd) seeds on streptozotocin-induced diabetic rats. *Asia Pac. J. Clin. Nutr.* 14: 153–158.

SatishKumar, B.N. 2011. Phytochemistry and pharmacological studies on *Pongamia pinnata* (L. Pierre). *Int. J. Pharm. Sci. Rev. Res.* 9: 12–19.

Satyaprakash, R.J., Rajesh, M.S., Bhanumathy, M., Harish, M.S., Shivananda, T.N., Shivaprasad, H.N. et al. 2013. Hypoglycemic and antihyperglycemic effect of *Ceiba pentandra* L. Gaertn in normal and streptozotocin-induced diabetic rats. *Ghana Med. J.* 47: 121–127.

Saumya, S.M. and Basha, P.M. 2011. Antioxidant effect of *Lagerstroemia speciosa* Pers (Banaba) leaf extract in streptozotocin-induced diabetic mice. *Indian J. Exp. Biol.* 49: 125–131.

Sauvaire, Y., Petit, P., Broca, C., Manteghetti, M., Baissac, Y., and Fernandez Alvarez, J. 1998. 4-hydroxyisoleucine: A novel amino acid potentiator of insulin secretion. *Diabetes* 47: 206–210.

Savitha, K., Wanjari, M., and Kumar, P. 2012. Anti-diabetic activity of *Pandanus fascicularis* aerial parts in alloxan-induced hyperglycemic rats. *IJNPND* 2: 105–110.

Savitha, R.C., Padmavathy, S., and Sundhararajan, A. 2011. *In vitro* antioxidant activities on leaf extracts of *Syzygium malaccense* (L.) Merr and Perry. *Anc. Sci. Life* 30: 110–113.

Sawant, T.P. and Gogle, D.P. 2014. A brief review on recent advances in clinical research of *Annona muricata*. *Int. J. Universal Pharm. Bio Sci.* 3: 267–304.

Saxena, A.M., Bajpai, M.B., Murthy, P.S., and Mukherjee, S.K. 1993. Mechanism of blood sugar lowering by a swerchirin-containing hexane fraction (SWI) of *Swertia chirayita*. *Indian J. Exp. Biol.* 31: 178–181.

Saxena, A.M., Murthy, P.S., and Mukherjee, S.K. 1996. Mode of action of three structurally different hypoglycemic agents: A comparative study. *Indian J. Exp. Biol.* 34: 351–355.

Saxena, M., Bajpai, M.B., and Mukherjee, S.K. 1991. Swerchirin induced blood sugar lowering of streptozotocin treated hyperglycemic rats. *Indian J. Exp. Biol.* 29: 674–675.

Schade, D.S. and Eaton, R.P. 1983. Diabetic ketoacidosis—pathogenesis, prevention and therapy. *Clin. Endocrinol. Metab.* 12: 321–338.

Schafer, A. and Hogger, P. 2007. Oligomeric procyanidins of French maritime pine bark extract (Pycnogenol®) effectively inhibit α-glucosidae. *Diabetes Res. Clin. Pract.* 77: 41–46.

Schwartz, I.F., Hershkovitz, R., Iaina, A., Gnessin, E., Wollman, Y., Chernichowski, T. et al. 2002. Garlic attenuates nitric oxide production in rat cardiac myocytes through inhibition of induceable nitric oxide synthase and the arginine transporter CAT-2 (cationic amino acid transporter-2). *Clin. Sci. (Lond)* 102: 487–493.

Seaforth, C.E. 2005. Antidiabetic plants in the Caribbean. In: *Traditional Medicines for Modern Times: Antidiabetic Plants,* Soumyanath, A. (ed.). Boca Raton, FL: CRC Press, pp. 221–224.

Sebai, L., Selmi, S., Rtibi, K., Souli, A., Gharbi, N., and Sakly, M. 2013. Lavender (*Lavandula stoechas* L.) essential oils attenuate hyperglycemia and protect against oxidative stress in alloxan-induced diabetic rats. *Lipids Health Dis.* 12: 189–194.

Sedaghat, R., Roghani, M., Ahmadi, M., and Ahmadi, F. 2011. Antihyperglycemic and antihyperlipidemic effect of *Rumex patientia* seed preparation in streptozotocin-diabetic rats. *Pathophysiology* 18: 111–115.

Sedigheh, A., Jamal, M.S., Mahbubeh, S., Somayeh, K., Mahmoud, R., Azadeh, A. et al. 2011. Hypoglycemic and hypolipidemic effects of pumpkin (*Cucurbita pepo* L.) on alloxan-induced diabetic rats. *Afr. J. Pharm. Pharmacol.* 5: 2620–2626.

Seetharam, Y.N., Chalageri, G., Setty, S.R., and Bheemachar. 2002. Hypoglycemic activity of *Abutilon indicum* leaf extracts in rats. *Fitoterapia* 73: 156–159.

Sefi, M., Fetoui, H., Soudani, N., Chtourou, Y., Makni, M., and Zeghal, N. 2012. *Artemisia campestris* leaf extract alleviates early diabetic nephropathy in rats by inhibiting protein oxidation and nitricoxide end products. *Pathol. Res. Pract.* 208: 157–162.

Seifarth, C., Littmann, L., Resheg, Y., Rossner, S., Goldwich, A., Pangratz, N. et al. 2011. MCS-18, a novel natural plant product prevents autoimmune diabetes. *Immunol. Lett.* 139: 58–67.

Sekiya, K., Ohtani, A., and Kusano, S. 2004. Enhancement of insulin sensitivity in adipocytes by ginger. *Biofactors* 22: 153–156.

Sellamuthu, P.S., Arulselvan, P., Fakurazi, S., and Kandasamy, M. 2014. Beneficial effects of mangiferin isolated from *Salacia chinensis* on biochemical and hematological parameters in rats with streptozotocin-induced diabetes. *Pak. J. Pharm. Sci.* 27: 161–167.

Selvamani, P., Latha, S., Elayaraja, K., Babu, P.S., Gupta, J.K., Pal, T.K. et al. 2008. Antidiabetic activity of the ethanol extract of *Capparis sepiaria* L. leaves. *Indian J. Pharm. Sci.* 70: 378–380.

Selvaraj, G., Kaliamurthi, S., and Thirugnanasambandam, R. 2014. Molecular docking studies on potential PPAR-γ agonist from *Rhizophora apiculata. Bangladesh J. Pharmacol.* 9: 298–302.

Sembulingam, K., Sembulingam, P., and Namasivayam, A. 1997. Effect of *Ocimum sanctum* Linn. on noise induced changes in plasma corticosterol level. *Indian J. Physiol. Pharmacol.* 41: 139–143.

Semwal, B.C., Gupta, J., Singh, S., Kumar, Y., and Giri, M. 2009. Anti hyperglycemic activity of *Berberis aristata* DC. in the alloxan induced diabetic rats. *Int. J. Green Pharm.* 3: 259–262.

Sen, S., Chen, S., Feng, B., Wu, Y., Lui, E., and Chakrabarti, S. 2012. Preventive effects of North American ginseng (*Panax quinquefolium*) on diabetic nephropathy. *Phytomedicine* 19: 494–505.

Sen, S., Chen, S., Wu, Y., Feng, B., Lui, E.K., and Chakrabarti, S. 2013. Preventive effects of North American ginseng (*Panax quinquefolius*) on diabetic retinopathy and cardiomyopathy. *Phytother. Res.* 27: 290–298.

Sen, S.K., Gupta, M., Mazumder, U.K., Haldar, P.K., Saha, P., Bhattacharya, S. et al. 2011. Antihyperglycemic effect and antioxidant property of *Citrus maxima* leaf in streptozotocin-induced diabetic rats. *Diabetol. Croat.* 40: 113–120.

Sendogdu, N., Aslan, M., Orhan, D.D., Ergun, F., and Yesilada, E. 2006. Anti-diabetic and antioxidant effects of *Vitis vinifera* L. leaves in streptozotocin-diabetic rats. *Pharm. Sci.* 3: 7–18.

Sener, G., Sacan, O., Yanardaq, R., and Ayanoqlu- Dulger, G. 2002. Effects of chard (*Beta vulgaris* L. var. cicla) extract on oxidative injury in the aorta and heart of streptozotocin-diabetic rats. *J. Med. Food.* 5: 37–42.

Sener, G., Sacan, O., Yanardag, R., and Ayanoglu-Dulger, G. 2003. Effects of parsley (*Petroselinum crispum*) on the aorta and heart of streptozotocin- induced diabetic rat. *Plant Foods Hum. Nutr.* 58: 1–7.

Senthilkumar, G.P. 2007. Evaluation of antioxidant potential of *Terminalia chebula* fruits studied in streptozotocin-induced diabetic rats. *Pharm. Biol.* 45: 511–518.

Senthilkumar, G.P., Arulselvan, P., Kumar, D.S., and Subramanian, S.P. 2006. Anti-diabetic activity of fruits of *Terminalia chebula* on streptozotocin induced diabetic rats. *J. Health Sci.* 52: 283–291.

Senthilkumar, G.P. and Subramanian, S.P. 2008. Biochemical studies on the effect of *Terminalia chebula* on the levels of glycoproteins in streptozotocin-induced experimental diabetes in rats. *J. Appl. Biomed.* 6: 105–115.

Senthilkumar, M.K., Sivakumar, P., Changanakkattil, F., Rajesh, V., and Perumal, P. 2011. Evaluation of anti-diabetic activity of *Bambusa vulgaris* leaves in streptozotocin induced diabetic rats. *Int. J. Pharm. Sci. Drug Res.* 3: 208–210.

Seo, E., Lee, E.-K., Lee, C.S., Chun, K.H., Lee, M-Y., and Jun, H.-S. 2014. *Psoralea corylifolia* L. seed extract ameliorates streptozotocin-induced diabetes in mice by inhibition of oxidative stress. *Oxid. Med. Cell. Longiv.* 2014: 9. http://dx.doi.org/10.1155/2014/897296.

Sepehri, G., Khazaee, M., Khaksari, M., and Gholamhoseinian, A. 2014. The effect of water extract of *Zataria multiflora* on microvascular permeability in streptozocin induced diabetic rats. *Annu. Res. Rev. Biol.* 4: 3119–3127.

Seshagiri, M., Gaikwad, R.D., Paramjyothi, S., Jyothi, K.S., and Ramchandra, S. 2007. Antiinflammatory, anti-ulcer and hypoglycemic activities of ethanolic and crude alkaloid extract of *Madhuca indica* (Koenig) Gmelin seed cake. *Oriental Pharm. Exp. Med.* 7: 141–149.

Sethumathi, P.P., Nandhakumar, J., Sengottuvelu, S., Duraisam, R., Karthikeyan, D., Ravikumar, V.R. et al. 2009. Antidiabetic and antioxidant activity of methanolic leaf extracts of *Costus pictus* D.Don in alloxan induced diabetic rats. *Pharmacologyonline* 1: 1200–1213.

Sezik, E., Aslan, M., Yesilada, E., and Ito, S. 2005. Hypoglycemic activity of *Gentiana olivieri* and isolation of the active constituent through bio-assay-directed fractionation techniques. *Life Sci.* 76: 1223–1238.

Shabana, M.M., Mirhom, Y.W., Genenah, A.A., Aboutabl, E.A., and Amer, H.A. 1990. Study into wild Egyptian plants of potential medicinal activity. Ninth communication: Hypoglycaemic activity of some selected plants in normal fasting and alloxanised rats. *Arch. Exp. Veterinarmed.* 44: 389–394.

Shabeer, J., Srivastava, R.S., and Singh, S.K. 2009. Antidiabetic and antioxidant effect of various fractions of *Phyllanthus simplex* in alloxan-diabetic rats. *J. Ethnopharmacol.* 124: 34–38.

Shaffer, J.E. 1985. Inotropic and chronotropic activity of berberine on isolated guinea pig atria. *J. Cardiovasc. Pharmacol.* 92: 307–315.

Shafi, P.M., Rosamma, M.K., Jamil, K., and Reddy, P.S. 2002. Anti-bacterial activity of *Syzygium cumini* and *Syzygium travancoricum* leaf essential oils. *Fitoterapia* 73: 414–416.

Shah, G., Shri, R., Panchal, V., Sharma, N., Singh, B., and Mann, A.S. 2011. Scientific basis for the therapeutic use of *Cymbopogon citratus* stapf (Lemon grass). *J. Adv. Pharm. Technol. Res.* 2: 3–8.

Shah, J.G., Patel, B.G., Patel, S.B., and Patel, R.K. 2012. Protective effect of *Hordeum vulgare* Linn. seeds against renal oxidative stress in streptozotocin-induced diabetic rats. *J. Pharm. Res.* 5: 3577–3581.

Shah, S.A.W., Wadood, A., and Wadood, N. 1988. Effects of *Eugenia jambolana* and *Zizyphus sativa* on blood glucose levels of normal and alloxan diabetic rats. *J. Postgrad. Med. Inst.* 3: 18–30.

Shah, T.I., Sharma, E., and Ahmad, G. 2014. *Juglans regia* L.: A phytopharmacological review. *World J. Pharm. Sci.* 2: 364–373.

Shaheen, H.M., Ali, B.H., Alqarawi, A.A., and Bashir, A.K. 2000. Effect of *Psidium guajava* leaves on some aspects of central nervous system in mice. *Phytother. Res.* 14: 107–111.

Shahida, A.N., Choo, C.Y., and Wong, T.W. 2011. Acute/subchronic oral toxicity of *Brucea javanica* seeds with hypoglycemic activity. *J. Nat. Remedies* 11: 60–68.

Shahraki, A. and Shahraki, M. 2013. The comparisons of eucalyptus aqueous extract and insulin on blood sugar and liver enzymes in diabetic male rats. *Zahedan J. Res. Med. Sci.* 15: 25–28.

Shahwar, D., Ullah, S., Ahmad, M., Ullah, S., Ahmad, N., and Khan, M.A. 2011. Hypoglycemic activity of *Ruellia tuberosa* Linn. (Acanthaceae) in normal and alloxan-induced diabetic rabbits. *Iran. J. Pharm. Sci.* 7: 107–115.

Shajeela, P.S., Kalpanadevi, V., and Mohan, V.R. 2012. Potential anti-diabetic, hypolipidemic and anti-oxidant effects of *Nymphaea pubescens* extract in alloxan induced diabetic rats. *J. Appl. Pharm. Sci.* 2: 83–88.

Shalaby, N.M.M., Abd-Alla, H.I., Aly, H.F., Albalawy, M.A., Shaker, K.H., and Bouajila. J. 2014. Preliminary *in vitro* evaluation of antidiabetic activity of *Ducrosia anethifolia* Boiss. and its linear furanocoumarins. *BioMed. Res. Int.* 2014: Article ID 480545, 13 pages.

Shamim A., Qureshi, Asad, W., and Sultana, V. 2009. The effect of *Phyllantus emblica* Linn. on Type-II diabetes, triglycerides and liver-specific enzyme. *Pak. J. Nutr.* 8: 125–128.

Shamin, T., Ahmad, M., and Mukhtar, M. 2014. Antihyperglycemic and hypolipidemic effects of methanolic extract of *Euphorbia prostrata* on alloxan induced diabetic rabbits. *Eur. Sci. J.* (Special edition) 458–465.

Shan, J.J. and Tian, G.Y. 2003. Studies on physico-chemical properties and hypoglycemic activity of complex polysaccharide AMP-B from *Atractylodes macrocephala* Koidz. *Yao Xue Xue Bao* 38: 438–441.

Shang, W., Yang, Y., Zhou, L., Jiang, B., Jin, H., and Chen, M. 2008. Gingenoside Rb1 stimulates glucose uptake through insulin-like signaling pathways in 3T3-L1 adipocytes. *J. Endocrinol.* 198: 561–569.

Shankar Murthy, K. and Kiran, B.R. 2012. Medicinal plants used as an anti-diabetic drug in pharmaceutical industry and their conservation: An overview. *Int. Res. J. Pharm.* 3: 65–72.

Shanmugasundaram, E.R.B., Rajeswari, G., Baskaran, K., Kumar, B.R.R., Shanmugasundaram, K.R., and Ahmath, B.K. 1990. Use of *Gymnema sylvestre* leaf extract in the control of blood glucose in insulin-dependent diabetes mellitus. *J. Ethnopharmacol.* 30: 281–294.

Sharma, A. and Sharma, V. 2013. A brief review of medicinal properties of *Asparagus racemosus* (Shatawari). *Int. J. Pure Appl. Biosci.* 1: 48–52.

Sharma, A.K., Agarwal, V., Kumar, R., Balasubramoniam, A., Mishra, A., and Gupta, R. 2011a. Pharmacological studies on seeds of *Alangium salvifolium* Linn. *Acta Pol. Pharm.* 68: 897–904.

Sharma, B., Salunke, R., Srivastava, S., Majumder, C., and Roy, P. 2009. Effects of guggulsterone isolated from *Commiphora mukul* in high fat diet induced diabetic rats. *Food Chem. Toxicol.* 47: 2631–2639.

Sharma, M., Fernandes, J., Ahirwar, D., and Jain, R. 2008. Hypoglycemic and hypolipidimic activity of alcoholic extract of *Citrus aurantium* in normal and alloxan-induced diabetic rats. *Pharmacologyonline* 3: 161–171.

Sharma, M.K., Kumar, M., and Kumar, A. 2002. *Ocimum sanctum* aqueous leaf extract provides protection against mercury induced toxicity in Swiss albino mice. *Indian J. Exp. Biol.* 40: 1079–1082.

Sharma, N. and Garg, V. 2009. Antidiabetic and antioxidant potential of ethanolic extract of *Butea monosperma* leaves in alloxan-induced diabetic mice. *Indian J. Biochem. Biophys.* 46: 99–105.

Sharma, P., Dubey, G., and Kaushik, S. 2011b. Chemical and medico-biological profile of *Cyamopsios tetragonoloba* (L.) Taub: An overview. *J. Appl. Pharm. Sci.* 1: 32–37.

Sharma, P., Singh, R., and Bhardwaj, P. 2013a. Anti-diabetic activity of *Cassia sophera* in streptozotocin-induced diabetic rats and its effect on insulin secretion from isolated pancreatic islats. *Int. J. Phytomed.* 5: 314–319.

Sharma, P.V. and Samanta, K.C. 2011. Hypoglycemic activity of methanolic extract of *Tectona grandis* Linn. root in alloxan induced diabetic rats. *J. Appl. Pharm. Sci.* 1: 106–109.

Sharma, R. and Arya, V. 2011. A review on fruits having antidiabetic potential. *J. Chem. Pharm. Res.* 3: 204–212.

Sharma, R., Mehan, S., Kalra, S., and Khanna, D. 2013b. *Tephrosia purpurea*—A magical herb with blessings in human biological system. *Int. J. Recent Adv. Pharm. Res.* 3: 12–22.

Sharma, R.A. and Singh, R. 2013. A review on *Phyla nodiflora* L.: A wild wetland medicinal herb. *Int. J. Pharm. Sci. Rev. Res.* 20: 57–63.

Sharma, S. and Agarwal, N. 2011. Nourishing and healing prowess of garden cress (*Lepidium sativum* Linn.)—A review. *Indian J. Nat. Prod. Resour.* 2: 292–297.

Sharma, S., Choudhary, M., Bhardwaj, S., Choudhary, N., and Rana, A.C. 2014. Hypoglycemic potential of alcoholic root extract of *Cassia occidentalis* Linn. in streptozotocin induced diabetes in albino mice. *Bull. Faculty Pharm. Cairo Univ.* 52: 211–217.

Sharma, S., Mehta, B.K., Mehta, D., Nagar, H., and Mishra, A. 2012. A review of pharmacological activity of *Syzygium cumini* extracts using different solvents and their effective doses. *Int. Res. J. Pharm.* 3: 54–58.

Sharma, S., Sharma, A.K., Chand, T., Khardiya, M., and Yadav, K.C. 2013c. Antidiabetic and antihyperlipidemic activity of *Cucurbita maxima* Duchense (pumpkin) seeds on streptozotocin induced diabetic rats. *J. Pharmacogn. Phytochem.* 1: 108–116.

Sharma, S.B., Gupta, S., Ac, R., Singh, U.R. Rajpoot, R., and Shukla, S.K. 2010a. Antidiabetogenic action of *Morus rubra* L. leaf extract in streptozotocin-induced diabetic rats. *J. Pharm. Pharmacol.* 62: 247–255.

Sharma, S.B., Nasir, A., Prabhu, K.M., and Murthy, P.S. 2006. Antihyperglycemic effect of the fruit-pulp of *Eugenia jambolana* in experimental diabetes mellitus. *J. Ethnopharmacol.* 104: 367–373.

Sharma, S.B., Nasir, A., Prabhu, K.M., Murthy, P.S., and Dev, G. 2003. Hypoglycaemic and hypolipidemic effect of ethanolic extract of seeds of *Eugenia jambolana* in alloxan-induced diabetic rabbits. *J. Ethnopharmacol.* 85: 201–206.

Sharma, S.B., Tanwar, R.S., Rini, A.C., Singh, U.R., Gupta, S., and Shukla, S.K. 2010b. Protective effect of *Morus rubra* L. leaf extract on diet-induced atherosclerosis in diabetic rats. *Indian J. Biochem. Biophys.* 47: 26–31.

Sharma, S.R., Dwivedi, S.K., and Swarup, D. 1997. Hypoglycemic, anti-hyperglycemic and hypolipidemic activities of *Caesalpinia bonducella* seeds in rats. *J. Ethnopharmacol.* 58: 39–44.

Sharma, U., Sahu, R., Roy, A., and Golwala, D. 2010c. *In vivo* antidiabetic and antioxidant potential of *Stephania hernandifolia* in streptozotocin-induced diabetic rats. *J. Young Pharm.* 2: 255–260.

Sharma, V. and Agrawal, R.C. 2013. *Glycyrrhiza glabra*—A plant for the future. *Mintage J. Pharm. Med. Sci.* 2: 15–20.

Sharma, V., Pooja, and Marwaha, A. 2011c. Hypoglycemic activity of methanolic extracts of *Nyctanthes arbor-tristis* Linn. root in alloxan induced diabetic rats. *Int. J. Pharm. Pharm. Sci.* 3: 210–212.

Sharma, V.J. and Shah, U.D. 2010. Antihyperglycemic activity of flavonoids from methanolic extract of aerial parts of *Scoparia dulcis* in streptozotocin induced diabetic rats. *Int. J. Chem. Tech. Res.* 2: 214–218.

Sharma, V.K., Kumar, S., and Patel, H.J. 2010d. Hypoglycemic activity of *Ficus glomerata* in alloxan induced diabetic rats. *Int. J. Pharm. Sci. Rev. Res.* 1: 18–22.

Sharmin, R., Khan, M.R.I., Akhter, M.A., Alim, A., Islam, M.A., Anisuzzaman, A.S.M. et al. 2013. Hypoglycemic and hypolipidemic effect of cucumber, white pumpkin and ridge gourd in alloxan induced diabetic rats. *J. Sci. Res.* 5: 161–170.

Shaw, J.E., Sicree, R.A., and Zimmet, P.Z. 2010. Global estimates of the prevalence of diabetes for 2010 and 2030. *Diabetes Res. Clin. Pract.* DOI: 10.1016/j.diabres. 2009.10.007.

Sheehan, E.W., Zemaitis, M.A., Slatkin Jr., D.J., and Schiff, P.L. 1983. A constituent of *Pterocarpus marsupium*, (–)-epicatechin, as a potential antidiabetic agent. *J. Nat. Prod.* 46: 232–234.

Shen, P., Liu, M.H., Ng, T.Y., Chan, Y.H., and Yong, E.L. 2006. Differential effects of isoflavones, from *Astragalus membranaceus* and *Pueraria thomsonii*, on the activation of PPARalpha, PPARgamma, and adipocyte differentiation *in vitro*. *J. Nutr.* 136: 899–905.

Shen, S.C. and Chang, W.C. 2013. Hypotriglyceridemic and hypoglycemic effects of vescalagin from Pink wax apple [*Syzygium samarangense* (Blume) Merrill and Perry cv. Pink] in high-fructose diet-induced diabetic rats. *Food Chem.* 136: 858–863.

Shen, S.-C., Chang, W.-C., and Chang, C.-L. 2012. Fraction from wax apple [*Syzygium samarangense* (Blume) Merrill and Perry] fruit extract ameliorates insulin resistance *via* modulating insulin signaling and inflammation pathway in tumor necrosis factor α-treated FL83B mouse hepatocytes. *Int. J. Mol. Sci.* 13: 8562–8577.

Shen, S.C., Cheng, F.C., and Wu, N.J. 2008. Effect of guava (*Psidium guajava* Linn.) leaf soluble solids on glucose metabolism in type 2 diabetic rats. *Phytother. Res.* 22: 1458–1464.

Shen, Z.F., Chen, Q.M., Liu, H.F., and Xie, M.Z. 1989. The hypoglycemic effect of clausenacoumarine. *Yao. Xue. Xue. Bao.* 24: 391–392.

Shen, Z.F. and Xie, M.Z. 1985. The anti-hyperglycemic effect of kakonein and aspirin. *Acta Pharmaceutica Sinica* 20: 863–865.

Shende, V.S., Sawant, V.A., Turuskar, A.O., Chatap, V.K., and Vijaya, C. 2009. Evaluation of hypoglycemic and antihyperglycemic effects of alcoholic extract of *Chonemorpha fragrans* root in normal and alloxan induced diabetic rats. *Pharmacogn. Mag.* 5: 36–41.

Sheng, Q., Yao, H., Xu, H., Ling, X., and He, T. 2004. Isolation of plant insulin from *Momordica charantia* seeds by gel filtration and RP-HPLC. *Zhong Yao Cai* 27: 414–416.

Sher, H. and Alyemeni, M. 2011. Evaluation of anti-diabetic activity and toxic potential of *Lycium shawii* in animal models. *J. Med. Plants Res.* 5: 3387–3395.

Shetty, A.J., Choudhury, D., Rejeesh, V.N., Nair, V., Kuruvilla, M., and Kotian, S. 2010a. Effect of the insulin plant (*Costus igneus*) leaves on dexamethasone-induced hyperglycemia. *Int. J. Ayurveda Res.* 1: 100–102.

Shetty, A.J., Parampalli, S.M., Bhandarkar, R., and Kotian, S. 2010b. Effect of the insulin plant (*Costus igneus*) leaves on blood glucose levels in diabetic patients: A cross sectional study. *J. Clin. Diagn. Res.* 4: 2617–2621.

Shi, L., Zhang, W., Zhou, Y.Y., Zhang, Y.N., Li, J.Y., Hu, L.H., and Li, J. et al. 2008. Corosolic acid stimulates glucose uptake *via* enhancing insulin receptor phosphorylation. *Eur. J. Pharmacol.* 584: 21–29.

Shibano, M., Tsukamoto, D., Masuda, A., Tanaka, Y., and Kusano, G. 2001. Two new pyrrolidine alkaloids, radicamines A and B as inhibitors of alpha-glucosidase from *Lobelia chinensis* Lour. *Chem. Pharm. Bull.* 49: 1362–1365.

Shigematsu, N., Asano, R., Shimosaka, M., and okazaki, M. 2001. Effect of administration with the extract of *Gymnema sylvestre* R. Br. leaves on lipid metabolism in rats. *Biol. Pharm. Bull.* 24: 713–717.

Shih, C.-C., Wu, Y.-W., and Lin, W.-C. 2002. Antihyperglycaemic and anti-oxidant properties of *Anoectochilus formosanus* in diabetic rats. *Clin. Exp. Pharmacol. Physiol.* 29: 684–688.

Shim, K.S., Lee, S.U., Ryu, S.Y., Min, Y.K., and Kim, S.H. 2009. Corosolic acid stimulates osteoblast differentiation by activating transcription factors and MAP kinases. *Phytother. Res.* 23: 1754–1758.

Shimura, M., Hasumi, A., Minato, T., Hosono, M., Miura, Y., Mizutani, S. et al. 2005. Isohumulones modulate blood lipid status through the activation of PPAR-α. *Biochem. Biophys. Acta* 1736: 51–60.

Shin, D.M., Choi, K.M., Lee, Y.S., Kim, W., Shin, K.O., Oh, S. et al. 2014. *Echinacea purpurea* root extract enhances the adipocyte differentiation of 3T3-L1 cells. *Arch. Pharm. Res.* 37: 803–812.

Shinde, U.A., Sharma, G., and Goyal, R.K. 2004. *In vitro* insulin mimicking action of Bis (maltolato) oxovanadium (IV). *Indian J. Pharm. Sci.* 66: 392–395.

Shiny, C.T., Saxena, A., and Gupta, S.P. 2013. Phytochemical and hypoglycaemic activity investigation of *Costus pictus* plants from Kerala and Tamilnadu. *Int. J. Pharm. Sci. Invent.* 2: 11–18.

Shirdel, Z., Madani, H., and Mirbadalzadeh, R. 2009. Investigation into the hypoglycemic effect of hydroalcoholic extract of *Ziziphus jujuba* leaves on blood glucose and lipids in alloxan-induced diabetes in rats. *Iran. J. Diabetes Lipid Disord.* 8: 13–19.

Shirdel, Z. and Mirbadalzadeh, R. 2011. Improvement of hyperglycemia in diabetic rats by ethanolic extract of red date leaves. *Elixir Hor. Sig.* 3: 1957–1959.

Shirwaikar, A., Rajendran, K., and Barik, R. 2006. Effect of aqueous bark extract of *Garuga pinnata* Roxb. in streptozotocin-nicotinamide induced type-II diabetes mellitus. *J. Ethnopharmacol.* 107: 285–290.

Shirwaikar, A., Rajendran, K., and Punitha, I.S.R. 2005. Antidiabetic activity of alcoholic stem extract of *Coscinium fenestratum* in streptozotocin-nicotinamide induced type 2 diabetic rats. *J. Ethnopharmacol.* 97: 369–374.

Shirwaikar, A., Rajendran, K., and Punitha, I.S.R. 2008. Antihyperglycemic activity of the aqueous stem extract of *Coscinium fenestratum* in non-insulin dependent diabetic rats. *Pharm. Biol.* 43: 707–712.

Shishtar, E., Jovanovski, E., Jenkins, A., and Vuksan, V. 2014. Effects of Korean white ginseng (Panax ginseng C.A. Meyer) on vascular and glycemic health in type 2 diabetes: Results of a randomized, double blind, placebo-controlled, multiple-crossover, acute dose escalation trial. *Clin. Nutr. Res.* 3: 89–97.

Shivkumar, H., Sahu, A., Rao, R.N., Prakash, T., Swamy, B.H.M.J., Shivkaur, B.S.V.K., and Govil, J.N. 2008. Anti-diabetic and hypoglycemic activities of flower extracts of *Thespesia populnea* (Linn.) Soland ex Correa in rats. In: *Phytopharmacology and Therapeutic*, Vol. III, Singh, V.K. and Govil, J.N. (eds.). Houston, Texas: Studium Press LLC, pp. 483–492.

Shobana, S., Harsha, M.R., Platel, K., Srinivasan, K., and Malleshi, N.G. 2010. Amelioration of hyperglycaemia and its associated complications by finger millet (*Eleusine coracana* L.) seed coat matter in streptozotocin-induced diabetic rats. *Br. J. Nutr.* 104: 1787–1795.

Shokeen, P., Anand, P., Murali, Y.K., and Tandon, V. 2008. Antidiabetic activity of 50% ethanolic extract of *Ricinus communis* and its purified fractions. *Food Chem. Toxicol.* 46: 3458–3466.

Shokeen, P., Ray, K., Bala, M., and Tandon, V. 2005. Preliminary studies on activities of *Ocimum sanctum*, *Drynaria quercifolia*, and *Annona squamosa* against *Neisseria gonorrhoeae*. *Sex. Transm. Dis.* 32: 106–111.

Shrivastava, A., Longia, G.S., Singh, S.P., and Joshi, L.D. 1987. Hypoglycemic and hypolipidemic effect of *Cyamopsis tetragonoloba* (guar) in normal and diabetic rats. *Ind. J. Physiol. Pharmacol.* 31: 77–83.

Shruthi, A., Latha, K.P., Vagdevi, H.M., Pushpa, B., and Shwetha, C. 2012. Anti-diabetic activity of the leaves extracts of *Wrightia tinctoria* on alloxan induced diabetic rats. *J. Chem. Pharm. Res.* 4: 3125–3129.

Shu, X.S., Lv, J.H., Tao, J., Li, G.M., Li, H.D., and Ma, N. 2009. Anti-hyperglycemic effects of total flavonoids from *Polygonatum odoratum* in streptozotocin and alloxan-induced diabetic rats. *J. Ethnopharmacol.* 124: 539–543.

Shukla, A., Bigoniya, P., and Srivastava, B. 2012a. Hypoglycemic activity of *Lepidium sativum* Linn. seed total alkaloid on alloxan induced diabetic rats. *Res. J. Med. Plant* 2012: 1–10.

Shukla, K., Dikshit, P., Shukla, R., and Gambhir, J.K. 2012b. The aqueous extract of *Withania coagulans* fruit partially reverses nicotinamide/streptozotocin-induced diabetes mellitus in rats. *J. Med. Food* 15: 718–725.

Shukla, K., Narain, J.P., Puri, P., Gupta, A., Bijlani, R.L., Mahapatra, S.C. et al. 1991. Glycaemic response to maize, bajra and barley. *Indian J. Physiol. Pharmacol.* 35: 249–254.

Shukla, R., Gupta, S., Gambhir, J.K., Prabhu, K.M., and Murthy, P.S. 2004. Anti-oxidant effect of aqueous extract of the bark of *Ficus bengalensis* in hypercholesterolaemic rabbits. *J. Ethnopharmacol.* 92: 47–51.

Shukla, S., Chatterji, S., Mehta, S., Rai, P.K., Singh, R.K., Yadav, D.K. et al. 2011. Antidiabetic effect of *Raphanus sativus* root juice. *Pharm. Biol.* 49: 32–37.

Siddiqui, B.S., Afshan, F., Gulzar, T., Sultana, R., Naqvi, S.N., and Tariq, R.M. 2003. Tetracyclic triterpenoids from the leaves of *Azadirachta indica* and their insecticidal activities. *Chem. Pharm. Bull.* (*Tokyo*) 51: 415–417.

Siddiqui, B.S., Hasan, M., Mairai, F., Mehmood, I., Hafizur, R.M., Hameed, A. et al. 2014. Two new compounds from the aerial parts of *Bergenia himalaica* Boriss and their anti-hyperglycemic effect in streptozotocin-nicotinamide induced diabetic rats. *J. Ethnopharmacol.* 152: 561–567.

Sidhu, M.C. and Sharma, T. 2013. Medicinal plants from twelve families having anti-diabetic activity: A review. *Am. J. PharmTech Res.* 3: 36–52.

Sievenpiper, J.L., Arnason, J.T., Leiter, L.A., and Vuksan, V. 2003. Variable effects of American ginseng: A batch of American ginseng (*Panax quinquefolis* L.) with a depressed ginsenoside profile does not affect postprandial glycemia. *Eur. J. Clin. Nutr.* 57: 243–248.

Sikarwar, M.S. and Patil, M.B. 2010. Antidiabetic activity of *Pongamia pinnata* leaf extracts in alloxan-induced diabetic rats. *Int. J. Ayurveda Res.* 1: 199–204.

Silawat, N., Jarald, E.E., Jain, N., Yadav, A., and Deshmukh, P.T. 2009. The mechanism of hypoglycemic and anti-diabetic action of hydro-alcoholic extract of *Cassia fistula* Linn. in rats. *Pharma Res.* 1: 82–92.

Silva, F.R., Szpoganicz, B., Pizzolatti, M.G., Willrich, M.A., and de Sousa, E. 2002. Acute effect of *Bauhinia forficata* on serum glucose levels in normal and alloxan-induced diabetic rats. *J. Ethnopharmacol.* 83: 33–37.

Silva, M., Valencia, A., Grossi-de-Sa, M.F., Rocha, T.L., Freire, É., de Paula, J.E., and Espindola, L.S. 2009. Inhibitory action of Cerrado plants against mammalian and insect α-amylases. *Pesticide Biochem. Physiol.* 95: 141–146.

Silva, R.M., Santos, F.A., Rao, V.S.N., Maciel, M.A., and Pinto, A.C. 2001. Blood glucose- and triglyceride-lowering effect of *trans*-dehydrocrotonin, a diterpene from *Croton cajucara* Benth., in rats. *Diabetes Obes. Metab.* 3: 452–456.

Simmonds, M.S.J. and Howes, M.-J.R. 2005. Plants used in the treatment of diabetes. In: *Traditional Medicine for Modern Times: Antidiabetic plants*, Soumyanath, A. (ed.), CRC Press, Boca Raton, FL, 2005, pp. 19–82.

Simoes, C.M.O., Schenkel, E.P., Bauer, L., and Langeloh, A. 1988. Pharmacological investigations on *Achyrocline satureioides* (Lam.) DC., Compositae. *J. Ethnopharmacol.* 22: 281–293.

Singab, A.N., El-Beshbishy, H.A., Yonekawa, M., and Fukai, T. 2005. Hypoglycemic effect of Egyptian *Morus alba* root bark extract: Effect on diabetes and lipid peroxidation of streptozotocin-induced diabetic rats. *J. Ethnopharmacol.* 100: 333–338.

Singab, A.N., Youssef, F.S., and Ashour, M.L. 2014. Medicinal plants with potential anti-diabetic activity and their assessment. *Med. Aromat. Plants* 3: 151. http://dx.doi.org/10.4172/2167-0412.1000151.

Singh, A.B., Chaturvedi, J.P., Narender, T., and Srivastava, A.K. 2008. Preliminary studies on the hypoglycemic effect of *Peganum harmala* L. seeds ethanol extract on normal and streptozotocin induced diabetic rats. *Indian J. Clin. Biochem.* 23: 391–393.

Singh, D. and Kakkar, P. 2009. Antihyperglycemic and anti-oxidant effect of *Berberis aristata* root extract and its role in regulating carbohydrate metabolism in diabetic rats. *J. Ethinopharmacol.* 123: 22–26.

Singh, D., Yadav, K., Maurya, R., and Srivastava, A.K. 2009b. Antihyperglycaemic activity of alpha-amyrin acetate in rats and db/db mice. *Nat. Prod. Res.* 23: 876–882.

Singh, J. and Kakar, P. 2009. Anti hyperglycemic and anti oxidant effect of *Berberis aristata* root extract and its role in regulating carbohydrate metabolism in diabetic rats. *J. Ethnopharmacol.* 123: 22–26.

Singh, K.L., Srivastava, P., Kumar, S., Singh, D.K., and Singh, V.K. 2014. *Mimusops elengi* Linn. (Maulsari): A potential medicinal plant. *Arch. Biomed. Sci.* 2: 18–29.

Singh, K.N. and Chandra, V. 1977. Hypoglycaemic and hypocholesterolaemic effects of proteins of *Acacia milanoxylon* and *Bauhinia retusa* wild leguminous seeds in young albino rats. *J. Indian Med. Assoc.* 68: 201–203.

Singh, N., Sing, S.P., Vrat, S., Misra, N., Dixit, K.S., and Kohli, R.P. 1985. A study on the antidiabetic activity of *Coccinia indica* in dogs. *Indian J. Med. Sci.* 39: 27–29.

Singh, N., Singh, S.M., and Shrivastava, P. 2004. Immunomodulatory and anti-tumour actions of medicinal plant *Tinospora cordifolia* are mediated through activation of tumour associated macrophages. *Immunopharmacol. Immunotoxicol.* 26: 145–162.

Singh, P.K., Baxi, D., Doshi, A., and Ramachandran, A.V. 2011b. Antihyperglycaemic and renoprotective effect of *Boerhaavia diffusa* L. in experimental diabetic rats. *J. Complement. Integr. Med.* 8.

Singh, P.S., Salwan, C., and Mann, A.S. 2011c. Evaluation of anti-diabetic activity of leaves of *Cassia occidentalis*. *Int. J. Res. Pharm. Chem.* 1: 904–913.

Singh, R., Kaur, N., Kishore, L., and Gupta, G.K. 2013. Management of diabetic complications: A chemical constituents based approach. *J. Ethnopharmacol.* 150: 51–70.

Singh, R., Rajesree, P.H., and Sankar, C. 2012. Screening for anti-diabetic activity of the ethanolic extract of *Bryonia alba* roots. *Int. J. Pharm. BioSci.* 2: 210–215.

Singh, R., Seherawat, A., and Sharma, P. 2011a. Hypoglycemic, antidiabetic and toxicological evaluation of *Momordica dioica* fruit extracts in alloxan induced diabetic rats. *J. Pharmacol. Toxicol.* 6: 454–467.

Singh, R.K., Mehta, S., Jaiswal, D., Rai, P.K., and Watal, G. 2009a. Antidiabetic effect of *Ficus bengalensis* aerial roots in experimental animals. *J. Ethnopharmacol.* 123: 110–114.

Singh, S. and Majumdar, D.K. 1999. Evaluation of the gastric anti-ulcer activity of fixed oil of *Ocimum sanctum* (holy basil). *J. Ethnopharmacol.* 65: 13–19.

Singh, S.K., Kesari, A.N., Rai, P.K., and Watal, G. 2007. Assessment of glycemic potential of *Musa paradisiaca* stem juice. *Indian J. Clin. Biochem.* 22: 48–52.

Singh, S.N., Vats, P., Suri, S., Shyam, R., Kumria, M.M., Ranganathan, S. et al. 2001a. Effect of an antidiabetic extract of *Catharanthus roseus* on enzyme activities in streptozotocin-induced diabetic rats. *J. Ethnopharmacol.* 79: 269–277.

Singh, V. and Radhav, P.K. 2012. Review on pharmacological properties of *Caesalpinia bonduc* L. *Int. J. Med. Arom. Plants* 2: 514–530.

Singha, A.K., Bhattacharjee, B., Das, N., Dinda, B., and Maiti, D. 2013. *Ichnocarpus frutescens* (Linn.)—A plant with different biological activities. *Asian J. Pharm. Clin. Res.* 6: 74–77.

Sinha, S., Mukherjee, P.K., Mukherjee, K., Pal, M., Mandal, S.C., and Saha, B.P. 2000. Evaluation of anti-pyretic potential of *Nelumbo nucifera* stalk extract. *Phytother. Res.* 14: 272–274.

Sitasawad, S.L., Shewade, Y., and Bhonde, R. 2000. Role of bitter gourd fruit juice in streptozotocin-induced diabetic state *in vivo* and *in vitro*. *J. Ethnopharmacol.* 73: 71–79.

Sivashanmugan, A.T. and Chatterjee, T.K. 2013. *In vitro* and *in vivo* anti-diabetic activity of *Polyalthia longifolia* (Sonner.) Thw. leaves. *Orient. Pharm. Exp. Med.* 13: 289–300.

Smirin, P., Taler, D., Abitbol, G., Brutman-Barazani, T., Kerem, Z., Sampson, S.R. et al. 2010. *Sarcopoterium spinosum* extract as an antidiabetic agent: *In vitro* and *in vivo* study. *J. Ethnopharmacol.* 129: 10–17.

Smith, Y.R.A., Adanlawo, I.G., and Oni, O.S. 2012. Hypoglycemic effect of saponin from the root of *Garcinia kola* (bitter kola) on alloxan-induced diabetic rats. *J. Drug Deliv. Ther.* 2: 9–12.

Soares de Moura, R., da Costa, G.F., Moreira, A.S., Queiroz, E.F., Garcia-Souza, E.P., Resende, A.C. et al. 2012. *Vitis vinifera* L. grape skin extract activates the insulin-signalling cascade and reduces hyperglycaemia in alloxan-induced diabetic mice. *J. Pharm. Pharmacol.* 64: 268–276.

Soetikno, V., Sari, F.R., Sukumaran, V., Lakshmanan, A.P., Mito, S., and Harima, M. 2012. Curcumin prevents diabetic cardiomyopathy in streptozotocin-induced diabetic rats: Possible involvement of PKC–MAPK signaling pathway. *Eur. J. Pharm. Sci.* 47: 604–614.

Sohn, D.H., Kim, Y.C., Oh, S.H., Park, E.J., Li, X., and Lee, B.H. 2003. Hepatoprotective and free radical scavenging effects of *Nelumbo nucifera*. *Phytomedicine* 10: 165–169.

Sokeng, S.D., Lontsi, D., Moundipa, P.F., Jatsa, H.B., Watcho, P., and Kamtchouing, P. 2007a. Hypoglycemic effect of *Anacardium occidentale* L. methanol extract and fractions on streptozotocin-induced diabetic rats. *Global J. Pharmacol.* 1: 1–5.

Sokeng, S.D., Rokeya, B., Hannan, J.M., Junaida, K., Zitech, P., Ali, L. et al. 2007b. Inhibitory effect of *Ipomoea aquatica* extracts on glucose absorption using a perfused rat intestinal preparation. *Fitoterapia* 78: 526–529.

Soltani, R., Hakimi, M., Asgary, S., Ghanadian, S.M., Keshvari, M., and Sarrafzadegan, N. 2014. Evaluation of the effects of *Vaccinium arctostaphylos* L. fruit extract on serum lipids and hs-CRP levels and oxidative stress in adult patients with hyperlipidemia: A randomized, double-blind, placebo-controlled clinical trial. *Evid. Based Complement. Altern. Med.* 2014: Article ID 217451, 6 pages.

Soman, S., Rajamanickam, C., Rauf, A.A., and Indira, M. 2013. Beneficial effects of *Psidium guajava* leaf extract on diabetic myocardium. *Exp. Toxicol. Pathol.* 65: 91–95.

Soman, S., Rauf, A.A., Indira, M., and Rajamanickam, C. 2010. Antioxidant and anti-glycative potential of ethyl acetate fraction of *Psidium guajava* leaf extract in streptozotocin-induced diabetic rats. *Plant Foods Hum. Nutr.* 65: 386–391.

Somani, R., Kasture, S., and Singhai, A.K. 2006. Anti-diabetic potential of *Butea monosperma* in rats. *Fitoterapia* 77: 86–90.

Somani, R.S. and Singhai, A.K. 2008. Hypoglycaemic and antidiabetic activities of seeds of *Myristica fragrans* in normoglycaemic and alloxan-induced diabetic rat. *Asian J. Exp. Sci.* 22: 95–102.

Somani, R.S., Singhai, A.K., Shivgunde, P., and Jain, D. 2012. *Asparagus racemosus* Willd (Liliaceae) ameliorates early diabetic nephropathy in STZ-induced diabetic rats. *Indian J. Exp. Biol.* 50: 469–475.

Song, C., Huang, L., Rong, L., Zhou, Z., Peng, X., Yu, S. et al. 2012a. Anti-hyperglycemic effect of *Potentilla discolor* decoction on obese-diabetic (Ob-db) mice and its chemical composition. *Fitoterapia* 83: 1474–1483.

Song, M.K., Roufogalis, B.D., and Huang, T.H. 2012b. Reversal of the caspase-dependent apoptotic cytotoxicity pathway by taurine from *Lycium barbarum* (Goji Berry) in human retinal pigment epithelial cells: Potential benefit in diabetic retinopathy. *Evid. Based Complement. Altern. Med.* 2012: 323784.

Song, Y., Zhang, Y., Zhou, T., Zhang, H., Hu, X., and Li, Q. 2012c. A preliminary study of monosaccharide composition and α-glucosidase inhibitory effect of polysaccharides from pumpkin (*Cucurbita moschata*) fruit. *Int. J. Food Sci. Technol.* 47: 357–361.

Soni, R., Metha, N.M., Trivedi, R., and Srivastava, D.N. 2013. Effect of aqueous extract of *Bougainvillea glabra* Choisy on cutaneous wound healing in diabetic rats. *Asian J. Complement. Altern. Med.* 1: 12–15.

Sood, S., Narang, D., Dinda, A.K., and Maulik, S.K. 2005. Chronic oral administration of *Ocimum sanctum* Linn. augments cardiac endogenous anti-oxidants and prevents isoproterenol-induced myocardial necrosis in rats. *J. Pharm. Pharmacol.* 57: 127–133.

Sood, S., Narang, D., Thomas, M.K., Gupta, Y.K., and Maulik, S.K. 2006. Effect of *Ocimum sanctum* Linn. on cardiac changes in rats subjected to chronic restraint stress. *J. Ethnopharmacol.* 108: 423–427.

Sophia, D. and Manoharan, S. 2007. Hypolipidemic activities of *Ficus racemosa* Linn. bark in alloxan induced diabetic rats. *Afr. J. Tradit. CAM* 4: 279–288.

Sotaniemi, E.A., Haapakoski, E., and Rautio, A. 1995. Ginseng therapy in non-insulin dependent diabetic patients. *Diabetes Care* 18: 1373–1375.

Souza, A., Mbatchi, B., and Herchuelz, A. 2011. Induction of insulin secretion by an aqueous extract of *Tabernanhte jboga* Bail. (Apocynaceae) in rat pancreatic islets of Langerhans. *J. Ethnopharmacol.* 133: 1015–1020.

Souza, C.R.F., Georgetti, S.R., Salvador, M.J., Fonseca, M.J.V., and Oliveira, W.P. 2009. Antioxidant activity and physical-chemical properties of spray and spouted bed dried extracts of *Bauhinia forficate. Braz. J. Pharm. Sci.* 45: 209–218.

Spadiene, A., Savickiene, N., Jurgeviciene, N., Zalinkevicius, R., Norkus, A., Ostrauskas, R. et al. 2013. Effect of gingo extract on eye microcirculation in patients with diabetes. *Cent. Eur. J. Med.* 8: 736–741.

Sreenathkumar, S. and Arcot, S. 2010. Anti-diabetic activity of *Nymphaea pubescens* Willd—A plant drug of aquatic flora. *J. Pharm. Res.* 3: 3067–3069.

Sridevi, P., Bhagavan, R.M., Harikumar, V.V., and Rangarao, P. 2014. Synthesis of derivatives of 4-hydroxy isoleucine from fenugreek and evaluation of their anti-diabetic activity. *Int. J. Phytopharm.* 4: 6–10.

Srilatha, B.R. and Ananda, S. 2014. Antidiabetic effects of *Mukia maderaspatana* and its phenolics: An *in vitro* study on gluconeogenesis and glucose uptake in rat tissues. *Pharm. Biol.* 52: 597–602.

Srinivasa, U., Venkateshwara, J., Krupanidhi, A.M., and Divakar, K. 2008. Antidiabetic activity of *Justicia beddomei* leaves in alloxan induced diabetic rats. *J. Res. Educ. Indian Med.* 26: 45–47.

Srinivasan, K., Patole, P.S., Kaul, C.L., and Ramarao, P. 2004. Reversal of glucose intolerance by pioglitazone in high-fat diet fed rats. *Methods Find. Exp. Clin. Pharmacol.* 26: 327–333.

Srinivasan, K., Viswanad, B., Asrat, L., Kaul, C.L., and Ramarao, P. 2005. Combination of high-fat diet-fed and low-dose streptozotocin-treated rat: A model for type 2 diabetes and pharmacological screening. *Pharmacol. Res.* 52: 313–320.

Srinivasan, M., Padmanabhan, M., and Prince, P.S. 2005. Effect of *Enicostemma littorale* Blume extract on key carbohydrate metabolic enzymes, lipid peroxides and antioxidants in alloxan-induced diabetic rats. *J. Pharm. Pharmacol.* 57: 497–503.

Srivastava, A. and Joshi, L.D. 1990. Effect of feeding black gram (*Phaseolus mungo*) on serum lipids of normal and diabetic guinea pigs. *Indian J. Med. Res.* 92: 383–386.

Srivastava, A., Longia, G.S., Singh, S.P., and Joshi, L.D. 1987. Hypoglycemic and hypolipidemic effects of *Cyamopsis tetragonoloba* (guar) in normal and diabetic guinea pigs. *Indian J. Physiol. Pharmacol.* 31: 77–83.

Srivastava, K.C. 1989. Extracts from two frequently consumed spices—Cumin (*Cuminum cyminum*) and turmeric (*Curcuma longa*)—inhibit platelet aggregation and alter ecosanoid biosynthesis in human blood platelets. *Prostaglandins Leukot. Essent. Fatty Acids* 37: 57–64.

Srivastava, R., Verma, A., Mukerjee, A., and Soni, N. 2014. Phytochemical, pharmacological and pharmacognostical profile of *Vernonia anthelmintica*: An overview. *J. Pharmacogn. Phytochem.* 2: 22–28.

Srivastava, S., Singh, P., Mishra, G., Jha, K.K., and Khosa, R.L. 2011. *Costus speciosus* (Keukand): A review. *Phelagia Res. Libr.* 2: 118–128.

Srivastava, Y., Venkatakrishna-Bhatt, H., and Verma, Y. 1988. Effect of *Momordica charantia* Linn. aqueous extract on cataractogenesis in murine alloxan diabetics. *Pharmacol. Res. Commun.* 20: 201–209.

Stalin, C., Dineshkumar, P., and Nithiyananthan, K. 2012. Evaluation of anti-diabetic activity of methanolic leaf extract of *Ficus carica* in alloxan-induced diabetic rats. *Asian J. Pharm. Clin. Res.* 5: 85–87.

Stanely, M.P.P., Kamalakkannan, N., and Menon, V.P. 2003. *Syzygium cumini* seed extract reduces tissue damage in diabetic rat brain. *J. Ethnopharmacol.* 84: 205–209.

Stanely, M.P.P. and Menon, V.P. 2001. Anti-oxidant action of *Tinospora cordifolia* root extract in alloxan diabetic rats. *Phytother. Res.* 15: 213–218.

Stanely, M.P.P. and Menon, V.P. 2003. Hypoglycemic and hypolipidemic action of alcohol extract of *Tinospora cordifolia* roots in chemical-induced diabetes in rats. *Phytother. Res.* 17: 410–413.

Stefkov, G., Kulevanova, S., Miova, B., Dinevska-Kjovkarovska, S., Molgaard, P., Jager, A.K.. et al. 2011. Effects of *Teucrium polium* spp. capitatum flavonoids on the lipid and carbohydrate metabolism in rats. *Pharm. Biol.* 49: 885–892.

Stéphane, D.K., Koffi, N., and Bernard, A.C. 2013. Effect of aqueous extract of *Ageratum conyzoides* leaves on the glycaemia of rabbits. *Pharma Innovat. J.* 2: 1–8.

Stephen, I.S., Sunil, C., Duraipandiyan, V., and Ignacimuthu, S. 2012. Antidiabetic and antioxidant activities of *Toddalia asiatica* (L.) Lam. leaves in streptozotocin induced diabetic rats. *J. Ethnopharmacol.* 143: 515–523.

Stohs, S.J., Miller, H., and Kaata, G.R. 2012. A review on the efficacy and safety of banaba (*Lagerstroemia speciosa* L.) and corosolic acid. *Phytother. Res.* 26: 317–324.

Strippoli, G.F., Craig, M., Deeks, J.J., Schena, F.P., and Craig, J.C. 2004. Effects of angiotensin converting enzyme inhibitors and angiotensin II receptor antagonists on mortality and renal outcomes in diabetic nephropathy: Systematic review. *Br. Med. J.* 329: 828.

Stull, A.J., Cash, K.C., Johnson, W.D., Champagne, C.M., and Cefalu, W.T. 2010. Bioactives in blue berries improve insulin sensitivity in obese, insulin-resistant men and women. *J. Nutr.* 140: 1764–1768.

Su, C.F., Cheng, J.T., and Li, I.M.2007. Increase of acetylcholine release by *Panax ginseng* root enhances insulin secretion in Wistar rats. *Neurosci. Lett.* 412: 101–104.

Suba, V., Murugesan, T., Rao, R.B., Ghosh, L., Pal, M., Mandal, S.C., and Saha, B.P. 2004. Antidiabetic potential of *Barleria lupulina* extract in rats. *Fitoterapia* 75: 1–4.

Subapriya, R., Bhuvaneswari, V., Ramesh, V., and Nagini, S. 2004. Ethanolic extract of neem (*Azadirachta indica*) inhibits vocal pouch carcinogenesis in hamsters. *Cell. Biochem. Funct.* 22: 10–14.

Subapriya, R. and Nagini, S. 2003. Ethanolic neem leaf extract protects against N-methyl, N-nitro-N-nitrosoguanidine-induced gastric carcinogenesis in Wistar rats. *Asian Pac. J. Cancer Prev.* 4: 215–223.

Subashbabu, P., Ignacimuthu, S., Agastian, P., and Varghese, B. 2009. Partial regeneration of β-cells in the islets of Langerhans by nymphayol, a sterol isolated from *Nymphaea stellata* (Willd.) flowers. *Bioorgan. Med. Chem.* 17: 2864–2870.

Subrahmanyam, G.V., Sushma, M., Alekya, A., Neeraja, C., Harsha, H.S.S., and Ravindra, J. 2011. Antidiabetic activity of *Abelmoschus esculentus* fruit extract. *Int. J. Res. Pharm. Chem.* 1: 17–20.

Subramanian, M., Chintalwar, G.J., and Chattopadhyay, S. 2002. Anti-oxidant properties of a *Tinospora cordifolia* polysaccharide against iron mediated lipid damage and gamma-ray induced protein damage. *Redox. Rep.* 7: 137–143.

Subramanian, P.M. and Misra, G.S. 1978. Chemical constituents of *Ficus bengalensis* (Part II). *Pol. J. Pharmacol. Pharm.* 30: 559–562.

Subramanian, S., Rajeswari, S., and Prasath, G.S. 2011. Antidiabetic, antilipidemic and antioxidant nature of *Tridax procumbens* studied in alloxan-induced experimental diabetes in rats: A biochemical approach. *Asian J. Res. Chem.* 4: 1732

Subramoniam, A. 2001. Pharmacological evaluation of ecotypes of medicinal plants. *Renaissance* 2: 9–11.

Subramoniam, A. 2003. Pharmacological evaluation in herbal drug development. In: *GMP for Botanicals*, Verpoorte, R. and Mukherjee, P.K. (eds.). New Delhi, India: Business Horizons, pp. 321–330.

Subramoniam, A. and Babu, V. 2003. Standardized phytomedicines for diabetes. In: *Role of Biotechnology in Medicinal and Aromatic Plants*, Vol. 8, Khan, I.A. and Khanum, A. (eds.). Hyderabad, India: Ukaaz Publications, pp. 46–69.

Subramoniam, A., Evans, D.A., Rajasekharan, S., and Sreekandan Nair, G. 2001. Effect of *Hemigraphis colorata* (Blume) H.G. Hallier leaf on wound healing and inflammation in mice. *Indian J. Pharmacol.* 33: 283–286.

Subramoniam, A. and Pushpangadan, P. 1999. Development of phytomedicines for liver diseases. *Indian J. Pharmacol.* 31: 166–175.

Subramoniam, A., Pushpangadan, P., Rajasekharan, S., Evans, D.A., Latha, P.G., and Valsaraj, R. 1996. Effects of *Artemisia pallens* Wall. on blood glucose levels in normal and alloxan-induced diabetic rats. *J. Ethnopharmacol.* 50: 13–17.

Subramoniam, A. and Satyanarayana, M.N. 1989. Influence of certain plant constituents on platelet aggregation. *J. Food Saf.* 9: 201–204.

Subramoniam, A., Shylesh, B.S., and Ajikumaran, N.S. 2004. Medicinal plants for the prevention and cure of viral diseases. In: *Role of Biotechnology in Medicinal and Aromatic Plants*, Vol. 6, Khan, I.A. and Khanum, A. (eds.), Ukaaz Publications, Hyderabad, India, pp. 225–256.

Suganya, S., Narmadha, R., Gopalakrishnan, V.K., and Devaki, K. 2012. Hypoglycemic effect of *Costus pictus* D. Don on alloxan induced type 2 diabetes mellitus in albino rats. *Asian Pac. J. Trop. Dis.* 2: 117–123.

Sugihara, Y., Nojima, H., Matsuda, H., Murakami, T., Yoshikawa, M., and Kimura, I. 2000. Antihyperglycemic effects of gymnemic acid IV, a compound derived from *Gymnema sylvestre* leaves in streptozotocin diabetic mice. *J. Asian Nat. Prod. Res.* 2: 321–327.

Suh, J.M., Jonker, J.W., Ahmadian, M., Goetz, R., Lackey, D., Osborn, O. et al. 2014. Endocrinization of FGF1 produces a neomorphic and potent insulin sensitizer. *Nature.* 513: 436–439.

Suhashini, R., Sindhu, S., and Sagadevan, E. 2014. In vitro evaluation of anti-diabetic potential and phytochemical profile of *Psoralea corylifolia* seeds. *Int. J. Pharmacogn. Phytochem. Res.* 6: 414–419.

Sui, D.Y., Lu, Z.Z., Li, S.H., and Cai, Y. 1994. Hypoglycemic effect of saponin isolated from leaves of *Acanthoganax senticosus* (Rupr. et Maxin.) Harms. *Zhongguo Zhong Yao Za Zhi* 19: 683–685.

Sultana, M.C., Alam, M.B., Asadujjaman, M., and Akter, M.A. 2012. Antihyperglycemic and antihyperlipidemic effects of *Stephania japonica* (thunb.) Miers. Tenril in alloxan induced diabetic mice. *Int. J. Pharm. Sci. Res.* 3: 2726–2730.

Sultana, N., Choudhary, M.I., and Khan, A. 2009. Protein glycation inhibitory activities of *Lawsonia inermis* and its active principles. *J. Enzyme Inhib. Med. Chem.* 24: 257–261.

Sultana, S., Ahmed, S., Sharma, S., and Jahangir, T. 2004. *Embilica officinalis* reverses thioacetamide-induced oxidative stress and early promotional events of primary hepatocarcinogeneis. *J. Pharm. Pharmacol.* 56: 1573–1579.

Sultana, Z., Jami, S., Ali, E., Begum, M., and Haque, M. 2014. Investigation of antidiabetic effect of ethanolic extract of *Phyllanthus emblica* Linn. fruits in experimental animal models. *Pharmacol. Pharm.* 5: 11–18. DOI: 10.4236/pp.2014.51003.

Sumbul, S., Ahmad, M.A., Asif, M., and Akhtar, M. 2011. *Myrtus communis* Linn. a review. *Indian J. Nat. Prod. Resour.* 2: 395–402.

Sun, L.H., Xing, D.M., Sun, H., Li, M., Jin, W., and Du, L.J. 2002. Effect of pueraria flavonoid on diabetes in mice complicated by hyperlipidemia. *Tsinghua Sci. Technol.* 7: 369–373.

Sundaram, R., Naresh, R., Shanthi, P., and Sachdanandam, P. 2012. Antihyperglycemic effect of iridoid glucoside, isolated from the leaves of *Vitex negundo* in streptozotocin-induced diabetic rats with special reference to glycoprotein components. *Phytomedicine* 19: 211–216.

Sundarrajan, T., Rajkumar, T., Udhayakumar, E., Sekar, M., and Kumar, M.K.S. 2011. Anti-diabetic activity of methanolic extract of *Hibiscus cannabinus* in streptozotocin-induced diabetic rats. *Int. J. Pharma BioSci.* 2: 125–130.

Sunday, E.A., Ify, E., and Uzoma, O.K. 2014. Hypoglycemic, hypolipidemic and body weight effects of unripe pulp of *Carica papaya* using albino rat model. *J. Pharmacogn. Phytochem.* 2: 109–114.

Sunday, O., Graeme, B., and Anthony, A. 2012. Anti-diabetic activities of aqueous stem bark extract of *Strychnos henningsii* Gilg in streptozotocin-nicotinamide type 2 diabetic rats. *Indian J. Pharm. Res.* 11: 221–227.

Sundeepkumar, H.K., Raju, M.B.V., Dinda, S.C., and Sahu, S.K. 2012. Antihyperglycemic activity of *Bambusa arundinacea. Rasayan J. Chem.* 5: 112–116.

Sunil, C., Ignacimuthu, S., and Agastian, P. 2011. Antidiabetic effect of *Symplocos cochinchinensis* (Lour.) S. Moore. in type 2 diabetic rats. *J. Ethnopharmacol.* 134: 298–304.

Supkamonseni, N., Thinkratok, A., Meksuriyen, D., and Srisawat, R. 2014. Hypolipidemic and hypoglycemic effects of *Centella asiatica* (L.) extract *in vitro* and *in vivo. Indian J. Exp. Biol.* 52: 965–971.

Sur, T.K., Seal, T., Pandit, S., and Bhattacharya, D. 2004. Hypoglycemic activities of a mangrove plant *Rhizophora apiculata* Blume. *Nat. Prod. Sci.* 10: 11–15.

Suresh, B., Rajkapoor, B., and Jayakar, B. 2004. Effect of *Caralluma attenuata* in normal and alloxan induced diabetic rats. *J. Herbal Pharmacother.* 4: 35–40.

Suresh, J., Rajan, D., and Nagamani. 2014. Anti-diabetic activity of aerial parts of *Xanthium strumarium* Linn. *World J. Pharm. Pharm. Sci.* 3: 2185–2200.

Suresh Babu, K., Tiwari, A.K., Srinivas, P.V., Ali, A.Z., Chinaraju, B., and Rao, J.M. 2004. Yeast and mammalian alpha-glucosidase inhibitory constituents from Himalayan rhubarb *Rheum emodi* Wall. ex Meisson. *Bioorg. Med. Chem. Lett.* 14: 3841–3845.

Suryanarayana, P., Kumar, P.A., Saraswat, M., Petrash, J.M., and Reddy, G.B. 2004. Inhibition of aldose reductase by tannoid principles of *Embilica officinalis*: Implications for the prevention of sugar cataract. *Mol. Vis.* 10: 148–154.

Suryanarayana, P., Saraswat, M., Petrash, J.M., and Reddy, G.B. 2007. *Emblica officinalis* and its enriched tannoids delay streptozotocin-induced diabetic cataract in rats. *Mol. Vis.* 24(13): 1291–1297.

Susheela, T., Balaravi, P., Theophilus, J., Reddy, T.N., and Reddy, P.U.N. 2008. Evaluation of hypoglycemic and anti-diabetic effect of *Melia dubia* Cav fruits in mice. *Curr. Sci.* 94: 1191–1195.

Sutar, N.G., Sutar, U.G., and Behera, B.C. 2009. Antidiabetic activity of the leaves of *Mimosa pudica* Linn. in albino rats. *J. Herbal Med. Toxicol.* 3: 123–126.

Suthar, M., Rathore, G.S., and Pareek, A. 2009. Antioxidant and antidiabetic activity of *Helicteres isora* (L.) fruits. *Indian J. Pharm. Sci.* 71: 695–699.

Swanston-Flatt, S.K., Day, C., Bailey, C.J., and Flatt, P.R. 1990. Traditional plant treatments for diabetes. Studies in normal and streptozotocin diabetic mice. *Diabetologia* 33: 462–464.

Swanston-Flatt, S.K., Day, C., Flatt, P.R., Gould, B.J., and Bailey, C.J. 1989. Glycaemic effects of traditional European plant treatments for diabetes. Studies in normal and streptozotocin diabetic mice. *Diabetes Res.* 10: 69–73.

Swargiary, A., Boro, H., Brahma, B.K., and Rahman, S. 2013. Ethno-botanical study of anti-diabetic medicinal plants used by the local people of Kokrajhar district of bodoland territorial council, India. *J. Med. Plants Stud.* 1: 51–58.

Sy, G.Y.A., Cisse, A., Nongonierma, R.B., Sarr, M., Mbodja, N.A., and Fayea, B. 2005. Hypoglycaemic and antidiabetic activity of acetonic extract of *Vernonia colorata* leaves in normoglycaemic and alloxan-induced diabetic rats. *J. Ethnopharmacol.* 98: 171–175.

Syed, M.A., Vrushabendraswamy, B.M., Gopakumar, P., Dhanapal, R., and Chandrashekara, V.M. 2005. Anti-diabetic activity of *Terminalia catappa* Linn. leaf extracts in alloxan-induced diabetic rats. *Iran. J. Pharmacol. Ther.* 4: 36–39.

Syiem, D., Ayngai, G., Khup, P.Z., Khongwir, B.S., Kharbuli, B., and Kayang, H. 2002. Hypoglycemic effects of *Potentilla fulgens* L. in normal and alloxan-induced diabetic mice. *J. Ethnopharmacol.* 83: 55–61.

Syiem, D., Khup, P.Z., and Syiem, A.B. 2009. Effects of *Potentilla fulgens* Linn. on carbohydrate and lipid profiles in diabetic mice. *Pharmacologyonline* 2: 787–795.

Syiem, D. and Majaw, S. 2011. Effect of different solvent extracts of *Potentilla fulgens* L. on aldose reductase and sorbitol dehydrogenase in normoglycemic and diabetic mice. *Pharmacologyonline* 3: 63–72.

Syiem, D., Monsang, S.W., and Sharma, R. 2010. Hypoglycemic and anti-hyperglycemic activities of *Curcuma amada* Roxb. in normal and alloxan-induced diabetic mice. *Pharmacologyonline* 3: 364–372.

Tachibana, Y., Kikuzaki, H., Lajis, N.H., and Nakatani, N. 2001. Anti-oxidant activity of carbazoles from *Murraya koenigii* leaves. *J. Agric. Food Chem.* 49: 5589–5594.

Tahrani, A.A., Piya, M.K., Kennedy, A., and Barnett, A.H. 2010. Glycaemic control in type 2 diabetes: Targets and new therapies. *Pharmacol. Ther.* 125: 328–361.

Taiwo, I.A., Adewumi, O.O., and Odeigah, P.G.C. 2012. Assessment of *Bridelia ferruginea* Benth for its therapeutic potential in pregnancy-induced impaired glucose tolerance in rats. *Int. J. Med. Biomed. Res.* 1: 49–55.

Takagi, S., Miura, T., Ishibashi, C., Kawata, T., Ishihara, E., Gu, Y. et al. 2008. Effect of corosolic acid on the hydrolysis of disaccharides. *J. Nutr. Sci. Vitaminol.* 54: 266–268.

Takagi, S., Miura, T., Ishihara, E., Ishida, T., and Chinzei, Y. 2010. Effect of corosolic acid on dietary hypercholesterolemia and hepatic steatosis in KK-Ay diabetic mice. *Biomed. Res.* 31: 213–218.

Takahashi, M., Konno, C., and Hikino, H. 1986. Isolation and hypoglycemic activity of coixans A, B, and C, glycans of *Coix lachrymal-jobi* var. mayuen seeds. *Planta Med.* 1: 64–65.

Takeuchi, H., Mooi, L.Y., Inagaki, Y., and He, P. 2001. Hypoglycemic effect of a hot-water extract from defatted sesame (*Sesamum indicum* L.) seed on the blood glucose level in genetically diabetic KK-Ay mice. *Biosci. Biotechnol. Biochem.* 65: 2318–2321.

Takii, H., Kometani, T., Nishimura, T., Nakae, T., Okada S., and Fushiki, T. 2000. Anti-diabetic effect of glycyrrhizin in genetically diabetic KK-Ay mice. *Biol. Pharm. Bull.* 24: 484–487.

Takin, M., Ahokpe, M., Zohoun, L., Assou, E., Aivodji, N., Agossou, E. et al. 2014. Effect of total *Khaya senegalensis* (Meliaceae) barks extracts on hepatic liberation of glucose. *Nat. J. Physiol. Pharm. Pharmacol.* 4: 112–117.

Talaviya, P.A., Rao, S.K., Vyas, B.M., Indoria, S.P., Suman, R.K., and Suvagiya, V.P. 2014. Review on potential antidiabetic herbal medicines. *Int. J. Pharm. Sci. Res.* 5: 302–319.

Talubmook, C. and Buddhakala, N. 2012. Anti-oxidant and anti-diabtic activities of flower extract from *Butea monosperma* (Lam.) Taub. *GSTF Int. J. BioSci.* 2: 7–11.

Talukder, M.J. and Nessa, J. 1998. Effect of *Nelumbo nucifera* rhizome extract on the gastrointestinal tract of rat. *Bangladesh Med. Res. Counc. Bull.* 24: 6–9.

Tambekar, D.H., Khante, B.S., Panzade, B.K., Dahikar, S., and Banginwar, Y. 2008. Evaluation of phytochemical and antibacterial potential of *Helicteres isora* L. fruits against enteric bacterial pathogens. *Afr. J. Tradit. Complement. Altern. Med.* 5: 290–293.

Tamilselvi, S., Balasubramani, S.P., Venkatasubramanian, P., and Vasanthi, N.S. 2014. A review on the pharmacognosy and pharmacology of the herbals traded as 'Daruharidra'. *Int. J. Pharm. BioSci.* 5: 556–570.

Tamiru, W., Engidawork, E., and Asres, K. 2012. Evaluation of the effects of 80% methanolic leaf extract of *Caylusea abyssinica* (Fresen.) Fisch. & Mey. on glucose handling in normal, glucose loaded and diabetic rodents. *BMC Complement. Altern. Med.* 12: 151–157.

Tamrakar, A.K., Yadav, P.P., Tiwari, P., Maurya, R., and Srivastava, A.K. 2008. Identification of pongamol and karanjin as lead compounds with antihyperglycemic activity from *Pongamia pinnata* fruits. *J. Ethnopharmacol.* 118: 435–439.

Tan, M.J., Ye, J.M., Turner, N., Hohnen-Behrens, C., Ke, C.Q., Tang, C.P. et al. 2008. Antidiabetic activities of triterpenoids isolated from bitter melon associated with activation of the AMPK pathway. *Chem. Biol.* 15: 263–273.

Tanabe, K., Nakamura, S., Omagari, K., and Oku, T. 2011. Repeated ingestion of the leaf extract from *Morus alba* reduces insulin resistance in KK-Ay mice. *Nutr. Res.* 31: 848–854.

Tanayen, J.K., Ajayi, A.M., Ezeonwumelu, J.O.C., Oloro, J., Tanayen, G.G., Adzu, B. et al. 2014. Antidiabetic properties of an aqueous-methanolic stem bark extract of *Spathodea campanulata* (Bignoniaceae) P. Beauv. *Br. J. Pharmacol. Toxicol.* 5: 163–168.

Taniguchi, H., Kobayashi-Hattori, K., Tenmyo, C., Kamei, T., Uda, Y., Sugita-Konishi, Y. et al. 2006. Effect of Japanese radish (*Raphanus sativus*) sprout (Kaiwara-daikon) on carbohydrate and lipidmetabolism in normal and streptozotocin-induced diabetic rats. *Phytother. Res.* 20: 274–278.

Tanira, M.O.M., Ali, B.H., Bashir, A.K., and Chandernath, I. 1996. Some pharmacological and toxicological studies of *Rhazya stricta* Decne. in rats, mice, and rabbits. *Gen. Pharmacol.* 27: 1261–1267.

Tanko, Y., Abdulaziz, M.M., Adelaiye, A.B., Faithu, M.Y., and Musa, K.Y. 2009a. Effects of hydromethanolic leaves extract of *Indigofera pulchra* on blood glucose levels of normoglycemic and alloxan-induced diabetic Wistar rats. *Int. J. Appl. Res. Nat. Prod.* 1: 13–18.

Tanko, Y., Abdulaziz, M.M., Adelaiye, A.B., Faithu, M.Y., and Musa, K.Y. 2009b. Effects of ethyl acetate portion of *Indigofera pulchra* leaves extract on blood glucose levels of alloxan-induced diabetic and normoglycemic Wistar rats. *Asian J. Med. Sci.* 1: 10–14.

Tanko, Y., Mabrouk, M.A., Adelaiye, A.B., Faithu, M.Y., and Musa, K.Y. 2011. Anti-diabetic and some haematological effects of ethylacetate and n-butanol fractions of *Indigofera pulchra* extract on alloxan-induced diabetic Wistar rats. *J. Diabetes Endocrinol.* 2: 1–7.

Tanko, Y., Muhammad, A., Elaigwu, J., Mohammed, K.A., Jimoh, A., Sada, N.M. et al. 2013. Antidiabetic effects of ethyl acetate and n-butanol fractions of*Acacia nilotica* methanol extract on alloxan-induced diabetic Wistar rats. *J. Appl. Pharm. Sci.* 3: 89–93.

Tanko, Y., Okasha, M.A., Saleh, M.I.A., Mohammed, A., Yerima, M., Yaro, A.H. et al. 2008a. Anti-diabetic effect of ethanolic flower extracts of *Newbouldia laevis* (Bignoniaceae) on blood glucose levels of streptozotocin-induced diabetic Wistar rats. *Res. J. Med. Sci.* 2: 62–65.

Tanko, Y., Yaro, A.H., Isa, A.I., Yerima, M., Saleh, M.I.A., and Mohammed, A. 2007. Toxicological and hypoglycemic studies on the leaves of *Cissampelos mucronata* (Menispermaceae) on blood glucose levels of streptozotocin-induced diabetic Wistar rats. *J. Med. Plants Res.* 1: 113–116.

Tanko, Y., Yerima, M., Mahdi, M.A., Yaro, A.H., Musa, K.Y., and Mohammed, A. 2008b. Hypoglycemic activity of methanolic stem bark of *Adansonnia digitata* extract on blood glucose levels of streptozotocin-induced diabetic Wistar rats. *Int. J. Appl. Res. Nat. Prod.* 1: 32–36.

Tanquilut, N.C., Tanquilut, M.R.C., Estacio, M.A.C., Torres, E.B., Rosario, J.C., and Reyes, B.A.S. 2009. Hypoglycemic effect of *Lagerstroemia speciosa* (L.) Pers. on alloxan-induced diabetic mice. *J. Med. Plant Res.* 3: 1066–1071.

Tanwar, R.S., Sharma, S.B., Singh, U.R., and Prabhu, K.M. 2010. Attenuation of renal dysfunction by anti-hyperglycemic compound isolated from fruit pulp of *Eugenia jambolana* in streptozotocin-induced diabetic rats. *Indian J. Biochem. Biophys.* 47: 83–89.

Taskinen, M.R. 2002. Diabetic dyslipidemia. *Atherosclerosis Suppl.* 3: 47–51.

Tastekin, D., Atasever, M., Adiguzel, G., Keles, M., and Tastekin, A. 2006. Hypoglycemic effect of *Artemisia herba-alba* in experimental hyperglycemic rats. *Bull. Vet. Inst. Pulawy* 50: 235–238.

Tatiya, A.U., Kulkarni, A.S., Surana, S.J., and Bari, N.D. 2010. Antioxidant and hypolipidemic effect of *Caralluma adscendens* Roxb. in alloxanized diabetic rats. *Int. J. Pharmacol.* 6: 400–406.

Tayade, P.M., Jagtap, S.A., Borde, S., Chandrasekar, N., and Joshi, A. 2012. Effect of *Psoralea corylifolia* on dexamethasone-induced insulin resistance in mice. *J. King Saud Univ. Sci.* 24: 251–255.

Tayyab, F., Lal, S.S., Mishra, M., and Kumar, U. 2012. A review: Medicinal plants and its impact on diabetes. *World J. Pharm. Res.* 1: 1019–1046.

Tchamadeu, M.C., Dzeufiet, P.D., Nouga, C.C., Azebaze, A.G., Allard, J., Girolami, J.P. et al. 2010. Hypoglycaemic effects of *Mammea africana* (Guttiferae) in diabetic rats. *J. Ethanopharmacol.* 127: 368–372.

Tedong, L., Madiraju, P., Martineau, L.C., Vallerand, D., Arnason, J.T., Desire, D.D., Lavoie, L., Kamtchouing, P., and Haddad, P.S. 2010. Hydroethanolic extract of cashew tree (*Anacardium occidentale*) nut and its principal compound, anacardic acid, stimulate glucose uptake in C2C12 muscle cells. *Mol. Nutr. Food Res.* 54: 1753–1762.

Teixeira, C.C., Weinert, L.S., Barbosa, D.C., Ricken, C., Esteves, J.F., and Fuchs, F.D. 2004. *Syzygium cumini* (L.) Skeels in the treatment of type 2 diabetes: Results of a randomized, double blind, double dummy, controlled trial. *Diabetes Care* 27: 3019–3020.

Telang, M., Dhulap, S., Mandhare, A., and Hirwani, R. 2013. Therapeutic and cosmetic applications of mangiferin: A patent review. 13543776.2013.

Teotia, S. and Singh, M. 1997. Hypoglycemic effect of *Prunus amygdalus* seeds in albino rabbits. *Indian J. Exp. Biol.* 35: 295–296.

Teotia, S., Singh, M., and Pant, M.C. 1997. Effect of *Prunus amygdalus* seeds on lipid profile. *Indian J. Physiol. Pharmacol.* 41: 383–389.

Teponno, R.B., Tapondjou, A.L., Abou-Mansour, E., Stoeckli-Evans, H., Tane, P., Luciano, B. et al. 2008. Further clerodane diterpenoids from *Dioscorea bulbifera* L. var sativa and revised structure of Bafoudiosbulbin B. *Phytochemistry* 69: 2374–2379.

Terashima, S., Shimizu, M., Horie, S., and Morita, N. 1991. Studies on aldose reductase inhibitors from natural products. IV. Constituents and aldose reductase inhibitory effect of *Chrysanthemum morifolium*, *Bixa orellana* and *Ipomoea batatas*. *Chem. Pharm. Bull.* (Tokyo) 39: 3346–3347.

Terruzzi, I., Senesi, P., Magni, C., Montesano, A., Scarafoni, A., Luzi, L., and Durant, M. 2011. Insulin mimetic action of conglutin-γ, a lupin seed protein in mouse myoblasts. *Nutr. Metab. Cardiovasc. Dis.* 21: 197–205.

Teugwa, C.M., Mejiato, P.C., Zofou, D., Tchinda, B.T., and Boyom, F.F. 2013. Anti-oxidant and anti-diabetic profiles of two African medicinal plants: *Picralima nitida* (Apocynaceae) and *Sonchus oleraceus* (Asteraceae). *BMC Complement. Altern. Med.* 13: 175–180.

Thamotharang, G., Revathy, P., Haja, S.S., Alavala, V., Vijayakumar, K., Sengottuvelu, S. et al. 2013. Evaluation of hypoglycemic and hypolipidemic studies in ethanol leaf extract of *Ficus pumila* Linn. on streptozotocin-induced diabetic rats. *Int. J. Pharm. Pharm. Sci.* 5: 766–769.

The Diabetes Control and Complications Trial Research Group. 1993. The effect of intensive treatment of diabetes on the development and progression of long-term complications in insulin-dependent diabetes mellitus. *N. Engl. J. Med.* 329: 977–986.

The Wealth of India. A dictionary of Indian raw materials and industrial products. 1948–1992, Vols. 1–11 (including revised editions). Publication and Information Directorate, CSIR, New Delhi, India.

Thirumalai T., Therasa, S.V., Elumalai, E.K., and David, E. 2011. Hypoglycemic effect of *Brassica juncea* (seeds) on streptozotocin induced diabetic male albino rat. *Asian Pac. J. Trop. Biomed.* 1: 323–325.

Thiruvenkatasubramoniam, R. and Jayakar, B. 2010. Anti-hyperglycemic and anti-hyperlipidemic activities of *Premna corymbosa* (Burm.F.) Rottl on streptozotocin induced diabetic rats. *Der Pharmacia Lettre* 2: 505–509.

Thiruvenkatasubramoniam, R. and Jayakar, B. 2010. Anti-hyperglycemic and anti-hyperlipidemic activities of *Bauhinia variegata* L. on streptozotozin-induced diabetic rats. *Der Pharmacia Lettre* 2: 330–334.

Thuppia, A., Rabintossaporn, P., Saenthaweesuk, S., Ingkaninan, K., and Sireeratawong, S. 2009. The hypoglycemic effect of water extract from leaves of *Lagerstroemia speciosa* L. in streptozotocin-induced diabetic rats. *Songklanakarin J. Sci. Technol.* 31: 133–137.

Tian, L.Y., Bai, X., Chen, X.H., Fang, J.B., Liu, S.H., and Chen, J.C. 2010. Antidiabetic effect of methylswertianin and bellidifolin from *Swertia punicea* Hemsl. and its potential mechanism. *Phytomedicine* 17: 533–539.

Timothy, K.O., Adewumi, A.M., Emmanuel, A.O., and Olayemi, A.M. 2013. Ethanolic extract of leaves of *Newbouldia laevis* attenuates glycosylation of hemoglobin and lipid peroxidation in diabetic rats. *Am. J. Pharmacol. Toxicol.* 8: 176–186.

Tiwari, A.K., Rao, R.R., Madhusudhana, K., Rao, V.R.S., and Ali, A.Z. 2009. New labdane diterpenes as intestinal alpha-glucosidase inhibitor from antihyperglycemic extract of *Hedychium spicatum* (Ham. Ex Smith) rhizomes. *Bioorg. Med. Chem. Lett.* 19: 2562–2565.

Tiwari, A.K., Reddy, K.S., Radhakrishnan, J., Kumar, D.A., Zehra, A., and Agawane, S.B. 2011. Influence of antioxidant rich fresh vegetable juices on starch induced postprandial hyperglycemia in rats. *Food Funct.* 2: 521–528.

Tiwari, P., Mishra, B.N., and Sangwan, N.S. 2014. Phytochemical and pharmacological properties of *Gymnema sylvestre*: An important medicinal plant. *BioMed. Res. Int.* 214: Article ID 830285, 18 pages.

Toi, M., Bando, H., Ramachandran, C., Melnick, S.J., Imai, A., Fife, R.S. et al. 2003. Preliminary studies on the anti-angiogenic potential of pomegranate fractions *in vitro* and *in vivo. Angiogenesis* 6: 121–128.

Toma, A., Makonnen, E., Mekonnen, Y., Debella, A., and Addisakwattana, S. 2014. Intestinal α-glucosidase and some pancreatic enzymes inhibitory effect of hydroalcholic extract of *Moringa stenopetala* leaves. *BMC Complement. Altern. Med.* 14: 180–184.

Tomar, R.S. and Sisodia, S.S. 2013. Assessment of anti-diabetic activity of *Bougainvillea glabra* Choisy in streptozotocin-induced diabetic rats. *Asian J. Biochem. Pharm. Res.* 3: 79–91.

Tomczyk, M. and Latte, K.P. 2009. Potentilla—A review of its phytochemical and pharmacological profile. *J. Ethnopharmacol.* 122: 184–204.

Tominaga, M., Kimura, M., Sugiyama, K., Abe, T., Igarashi, K., Igarashi, M. et al. 1995. Effects of seishin-renshi-in and *Gymnema sylvestre* on insulin resistance in streptozotocin-induced diabetic rats. *Diabetes Res. Clin. Pract.* 29: 11–17.

Tomoda, M., Shimizu, N., and Oshima, Y. 1987. Hypoglycemic activity of twenty plant mucilages and three modified products. *Planta Med.* 53: 8–12.

Tong, H., Liang, Z., and Wang, G. 2008. Structural characterization and hypoglycemic activity of a polysaccharide isolated from the fruit of *Physalis alkekengi* L. *Carbohydr. Polym.* 71: 316–323.

Trejo-Gonzalez, A., Gabriel-Ortiz, G., Puebla-Perez, A.M., Huizar-Contreras, M.D., Munguia-Mazariegos, M.R., Mella-Arreguin, S. et al. 1996. A purified extract from prickly pear cactus (*Opuntia fuliginosa*) controls experimentally induced diabetes in rats. *J. Ethnopharmacol.* 55: 27–33.

Tripathi, N., Kumar, S., Singh, R., Singh, C.J., Singh, P., and Varshney, V.K. 2013. Isolation and identification of γ-sitosterol by GC-MS from the leaves of *Girardinia heterophylla* Decne. *Open Bioactive Compounds J.* 4: 25–27.

Tripathi, V. and Verma, J. 2014. Current updates of Indian anti-diabetic medicinal plants. *Int. J. Res. Pharm. Chem.* 4: 114–118.

Tripathi, J. and Joshi, T. 1988. Phytochemical investigation of roots of *Pterocarpus marsupium*. Isolation and structural studies of two new flavanone glycosides. *Z. Naturforsch.* [C] 43: 184–186.

Tripathi, Y.G. and Chaturvedi, P. 1995. Assessment of endocrine response of *Inula racemosa* in relation to glucose homeostasis in rats. *Indian J. Exp. Biol.* 33: 686–689.

Tsutsumi, T., Kobayashi, S., Liu, Y.Y., and Kontani, H. 2003. Anti-hyperglycemic effect of fangchinoline isolated from *Stephania tetrandra* Radix in streptozotocin-diabetic mice. *Biol. Pharm. Bull.* 26: 313–317.

Tungtrakanpoung, N. and Rhienpanish, K. 1992. The toxicity of *Mimosa invisa* Mart. var. *inermis* Adelbert to buffaloes. *Buffalo Bull.* 11: 230–231.

Turner, N., Li, J.Y., Gosby, A., To, S.W., Cheng, Z., Miyoshi, H. et al. 2008. Berberine and its more biologically available derivative, dihydroberberine, inhibit mitochondrial respiratory complex I: A mechanism for the action of berberine to activate AMP-activated protein kinase and improve insulin action. *Diabetes* 57: 1414–1418.

Tzeng, T.F., Liou, S.S., Chang, C.J., and Liu, I.M. 2014. The ethanol extract of *Lonicera japonica* (Japanese honeysuckle) attenuates diabetic nephropathy by inhibiting p-38 MAPK activity in streptozotocin-induced diabetic rats. *Planta Med.* 80: 121–129.

Tzeng, Y.M., Chen, K., Rao, Y.K., and Lee, M.J. 2009. Kaempferitrin activates the insulin signaling pathway and stimulates secretion of adiponectin in 3T3-L1 adipocytes. *Eur. J. Pharmacol.* 607: 27–34.

Ubaid, R.S., Anantrao, K.M., Jaju, J.B., and Mateenuddin, M. 2003. Effect of *Ocimum sanctum* leaf extract on hepatotoxicity induced by anti-tubercular drugs in rats. *Indian J. Physiol. Pharmacol.* 47: 465–470.

Ubaka, C.M. and Ukwe, C.V. 2010. Anti-diabetic effect of the methanolic seed extract of *Sphenostylis stenocarpa* in rats. *J. Pharm. Res.* 3: 2192–2194.

Ubillas, R.P., Mendez, C.D., Jolad, S.D., Luo, J., King, S.R., Carlson, T.J. et al. 2000. Antihyperglycemic acetylenic glucosides from *Bidens pilosa. Planta Med.* 66: 82–83.

Uchenna, E.O., Emeruwa, O., Ezugwa, C.O., Tchimene, M., and Ezejiofor, M. 2014. Antidiabetic activity of the methanol extract of *Detarium microcarpum* Guill and Perr (Fabaceae*). Planta Med.* 80: PD 122.

Udayabhanu, J., Kaminidevi, S., and Thayumanavan, T.T. 2014. Evaluation of phytochemical and antioxidant contents of *Mesua ferrea*, *Hemionitis arifolia* and *Pimenta dioica*. *Int. J. Adv. Pharm. Biol. Chem.* 3: 273–276.

Udayakumar, R., Kasthurirengan, S., Mariashibu, T.S., Rajesh, M., Anbazhagan, V.R., Kim, S.C. et al. 2009. Hypoglycaemic and hypolipidaemic effects of *Withania somnifera* root and leaf extracts on alloxan-induced diabetic rats. *Int. J. Mol. Sci.* 10: 2367–2382.

Udenta, E.A., Obizoba, I.C., and Oguntibeju, O.O. 2014. Anti-diabetes effects of Nigerian indigenous plant foods/diets. In: *Anti-Oxidant and Anti-Diabetic Agents and Human Health*, Oguntibeju, O. (ed.) Croatia: InTech.

Ueda, H., Kuroiwa, E., Tachibana, Y., Kawanishi, K., Ayala, F., and Moriyasu, M. 2004. Aldose reductase inhibitors from the leaves of *Myrciaria dubia* (H. B. & K.) McVaugh. *Phytomedicine* 11: 652–656.

Ugochukwu, N.H. and Babady, N.E. 2002. Antihyperglycemic effect of aqueous and ethanolic extracts of *Gongronema latifolium* leaves on glucose and glycogen metabolism in livers of normal and streptozotocin-induced diabetic rats. *Life Sci.* 73: 1925–1938.

Ugochukwu, N.H. and Cobourne, M.K. 2003. Modification of renal oxidative stress and lipid peroxidation in streptozotocin-induced diabetic rats treated with extracts from *Gongronema latifolium* leaves. *Clin. Chim. Acta* 336: 73–81.

Ulbricht, C., Windsor, R.C., Brigham, A., Bryan, J.K., Conquer, L., Costa, D. et al. 2012. An evidence-based systematic review of annatto (*Bixa orellana* L.) by the natural standard. *J. Dietary Suppl.* 9: 57–77.

Umadevi, P. 2001. Radioprotective, anti-carcinogenic and anti-oxidant properties of the Indian holy basil, *Ocimum sanctum* (tulasi). *Indian J. Exp. Biol.* 39: 185–190.

Umadevi, P., Selvi, S., Suja, S., Selvam, K., and Chinnaswamy, P. 2006. Antidiabetic and hypolipidemic effect of *Cassia auriculata* in alloxan induced diabetic rats. *Int. J. Pharmacol.* 2: 601–607.

Umar, A., Ahmed, Q.U., Muhammad, B.Y., Dogaraj, B.B., and Soad, S.Z. 2010. Anti-hyperglycemic activity of the leaves of *Tetracera scandens* Linn. Merr. (Dilleniaceae) in alloxan induced diabetic rats. *J. Ethnopharmacol.* 131: 140–145.

Umar, I.A., Mohammed, A., Ndidi, U.S., Abdulazeez, A.B., Olica, W.C., and Adam, M. 2014. Anti-hyperglycemic and anti-hyperlipidemic effects of aqueous stem bark extract of *Acacia albida* Delile. in alloxan-induced diabetic rats. *Asian J. Biochem.* 9: 170–178.

Umasankar, K., Nambikkairaj, B., and Manley Backyavathy, D. 2013. Wound healing activity of topical *Mentha piperita* and *Cymbopogan citratus* essential oil on streptozotocin induced diabetic rats. *Asian J. Pharm. Clin. Res.* 6: 180–183.

Une, H., Ghodke, M.S., Mubashir, M., and Naik, J.B. 2013. Effects of *Zizyphus mauritiana* Lam. leaves in alloxan induced diabetic and its secondary complications in rats. *Der Pharmacia Sinica* 4: 92–97.

Upadhyay, O.P., Singh, R.H., and Dutta, S.K. 1996. Studies on antidiabetic medicinal plants used in Indian folklore. *Aryavaidyan* 9: 159–167.

Upadhyay, U.M. and Goyal, R.K. 2004. Efficacy of *Enicostemma littorale* in type 2 diabetic patients. *Phytother. Res.* 18: 233–235.

Uraku, A.J., Ajah, P.M., Okaka, A.N.C., Ibiam, U.A., and Onu, P.N. 2010. Effects of crude extracts of *Abelmoschus esculentus* on albumin and total bilirubin of diabetic albino rats. *Int. J. Sci. Nat.* 1: 38–41.

Urias-Silvas, J.E., Cani, P.D., Delmee, E., Neyrinck, A., Lopez, M.G., and Delzenne, M. 2007. Physiological effects of dietary fructans extracted from *Agave tequilana* Gto and *Dasylirion* spp. *Br. J. Nutr.* 13: 1–8.

Usharani, P., Fatima, N., and Muralidhar, N. 2013. Effects of *Phyllanthus emblica* extract on endothelial dysfunction and biomarkers of oxidative stress in patients with type 2 diabetes mellitus: A randomized, double-blind, controlled study. *Diabetes Metab. Syndr. Obes.: Targets Ther.* 6: 275–284.

Utsunomiya, H., Yamakawa, T., Kamei, J., Kadonosono, K., and Tanaka, S. 2005. Anti hyperglycemic effects of plum in a rat model of obesity and type 2 diabetes, Wistar fatty rat. *Biomed Res.* 26: 193–200.

Uzayisenga, R., Ayeka, P.A., and Wang, Y. 2014. Anti-diabetic potential of *Panax notoginseng* saponins (PNS): A review. *Phytother. Res.* 28: 510–516.

Vadivelan, R., Dhanabal, S.P., Mohan, P., Shanish, A., Elango, K., and Suresh, B. 2010. Anti-diabetic activity of *Mukia maderaspatana* (L.) Roem in alloxan-induced diabetic rats. *Res. J. Pharmacol. Pharmacodynamics* 2: 78–80.

Vadivelan, R., Dhanabal, S.P., Wadhawani, A., and Elanko, K. 2012. α-Glucosidasae and α-amylase inhibitory activities of *Mukia madraspatana* (L.) Roem. *J. Intercult. Ethnopharmacol.* 1: 97–100.

Valcheva-Kuzmanova, S., Kuzmanov, K., Tancheva, S., and Belcheva, A. 2007. Hypoglycemic and hypolipid-emic effects of *Aronia melanocarpa*, fruit juice in streptozotocin-induced diabetic rats. *Methods Find. Exp. Clin. Pharmacol.* 29: 1–5.

Valentina, P., Ilango, K., Kiruthiga, B., and Parimala, M.J. 2013. Preliminary phytochemical analysis and biological screening of extracts of leaves of *Melia dubia* Cav. *Int. J. Res. Ayurveda Pharm.* 4: 417–419.

Valentova, K., Cvak, L., Muck, A., Ulrichova, J., and Simancj, V. 2003. Anti oxidant effect of extracts from the leaves of *Smallanthus sonchifolius*. *Eur. Nutr. J.* 47: 61–65.

Valentova, K., Moncion, A., Wazier, I., and Ulrichova, J. 2004. The effect of *Smallanthus sonchifolius* leaf extract on rat hepatic metabolism. *Cell Biol. Toxicol.* 20: 109–120.

Vamsikrishna, A.N., Ramgopal, M., Venkataraman, B., and Balagi, M. 2012. Anti diabetic efficacy of etha-nolic extract of *Phragmites vallatoria* on streptozotocin-induced diabetic rats. *Int. J. Pharm. Pharm. Sci.* 4: 118–120.

van Dam, R.M. and Hu, F.B. 2005. Coffee consumption and risk of type 2 diabetes: A systematic review. *JAMMA* 294: 97–104.

van de Venter, M., Roux, S., Bungu, L.C., Louw, J., Crouch, N.R., Grace, O.M. et al. 2008. Anti-diabetic screening and scoring of 11 plants traditionally used in South Africa. *J. Ethnopharmacol.* 119: 81–86.

van Wyk, B-E., van Oudtshoorn, B., and Gericke, N. 2009. *Medicinal Plants of South Africa*, 2nd edn. Pretoria, South Africa: Briza.

Vandanmagsar, B., Havnie, K.R., Wicks, S.E., Bermudez, E.M., Mendoza, T.M., Ribnicky, D. et al. 2014. *Artemisia dracunculus* L. extract ameliorates insulin sensitivity by attenuating inflammatory signaling in human skeletal muscle culture. *Diabetes Obes. Metab.* 16: 728–738.

Vangapelli, S., Guntha, S., Naveen, P., Santosh, P., and Srinivas B. 2014. Evaluation of anti-diabetic activity of leaves of *Diospyros melanoxylon* (Roxb.). *Indo Am. J. Pharm. Res.* 4: 969–981.

VanWyk, B.E. and Albrecht, C. 2008. A review of the taxonomy, ethnobotany, chemistry and pharmacology of *Sutherlandia frutescens* (Fabaceae). *J. Ethnopharmacol.* 119: 620–629.

Vareda, P.M.P., Saldanha, L.L., Camaforte, N.A.D., Violata, N.M., Dokkedal, A.L., and Bosqueiro, J.R. 2014. *Myrcia bella* leaf extract presents hypoglycemic activity via PI3K/Akt insulin signaling pathway. *Evid. Based Complement. Altern. Med.* 2014: 10. http://dx.doi.org/10.1155/2014/543606.

Varghese, G.K., Bose, L.V., and Habtemariam, S. 2013b. Antidiabetic components of *Cassia alata* leaves: Identification through α-glucosidase inhibition studies. *Pharm. Biol.* 51: 345–349.

Varghese, S., Narmadha, R., Gomathy, D., Kalaiselvi, M., and Devaki, K. 2013a. Evaluation of hypoglycemic effect of ethanolic seed extracts of *Citrullus lanatus*. *J. Phytopharmacol.* 2: 31–40.

Varma, S.D., Mikuni, I., and Kinoshita, J.H. 1975. Flavonoids as inhibitors of lens aldose reductase. *Science* 188: 1215–1216.

Varma, S.V. and Jaybhaye, D.L. 2010. Antihyperglycemic activity of *Tectona grandis* Linn. bark extract on alloxan induced diabetes in rats. *J. Ayurveda Res.* 1: 163–166.

Vasconcelosa, C.F.B., Maranhaoa, H.M.L., Batista, T.M., Carneirob, E.M., Ferreirac, F., Costad, J. et al. 2011. Hypoglycaemic activity and molecular mechanisms of *Caesalpinia ferrea* Martius bark extract on streptozotocin-induced diabetes in Wistar rats. *J. Ethnopharmacol.* 137: 1553–1541.

Vats, V., Grover, J.K., and Rathi, S.S. 2002. Evaluation of anti-hyperglycemic and hypoglycemic effect of *Trigonella foenum-graecum* Linn., *Ocimum sanctum* Linn. and *Pterocarpus marsupium* Linn. in nor-mal and alloxan induced diabetic rats. *J. Ethnopharmacol.* 79: 95–100.

Vats, V., Yadav, S.P., and Grover, J.K. 2004. Ethanol extract of *Ocimum sanctum* leaves partially attenu-ates streptozotocin-induced alterations in glycogen content and carbohydrate metabolism in rats. *J. Ethnopharmacol.* 90: 155–160.

Veerapur, V.P., Prabhakar, K.R., Kandadi, M.R., Srinivasan, K.K., and Unnikrishnan, M.. 2010b. Antidiabetic effect of Dodonaea viscosa aerial parts in high fat diet and low dose streptozotocin-induced type 2 dia-betic rats: A mechanistic approach. *Pharm. Biol.* 48: 1137–1148.

Veerapur, V.P., Prabhakar, K.R., Parihar, V.K., and Bansal, P. 2010a. Antidiabetic, hypolipidemic, and anti-oxidant activities of *Dodonaea viscosa* aerial parts in streptozotocin-induced diabetic rats. *Int. J. Phytomed.* 2: 59–70.

Veerapur, V.P., Prabhakar, K.R., Thippeswamy, B.S., Bansal, P., Srinivasan, K.K., and Unnikrishnan, M. 2010c. Anti-diabetic effect of *Dodonaea viscosa* (L.) Lacq aerial parts in high fructose fed insulin resis-tant rats: A mechanism based study. *Indian J. Exp. Biol.* 48: 800–810.

Vega-Avila, E., Cano-Velasco, J.L., Alarcon-Aguilar, F.J., Ortiz, M.D.F., Almanza-Perez, J.C., and Román-Ramos, R. 2012. Hypoglycemic activity of aqueous extracts from *Catharanthus roseus*. *Evid. Based Complement. Altern. Med.* 2012: 934258. http://dx.doi.org/10.1155/2012/934258.

Velmurugan, C. and Bhargava, A. 2014. Anti-diabetic activity of *Gossypium herbaceum* by alloxan induced model in rats. *Pharma Tutor* 2: 126–132.

Velmurugan, C., Sundaram, T., Kumar, R.S., Vivek, B., Sheshadri, S.D., and Kumar, B.S.A. 2011. Antidiabetic and hypolipidemic activity of bark of ethanolic extract of *Ougeinia oojeinensis* (Roxb). *Med. J. Malaysia* 66: 22–26.

Venkata, N., Rao, R., Babu, A.N., Kumar, M.S., Sharmila, P., Vutla, V. et al. 2013. Evaluation of anti-diabetic potential of leaves of *Nelumbo nucifera* in streptozotocin-induced diabetic rats. *Int. J. Allied Med. Sci. Clin. Res.* 1: 18–24.

Venkatesan, T. and Pillai, S.S. 2012. Antidiabetic activity of gossypin, a pentahydroxyflavone glucoside, in streptozotocin-induced experimental diabetes in rats. *J. Diabetes* 4: 41–46.

Venkatesh, S., Dayanand Reddy, G., Reddy, Y.S., Sathyavathy, D., and Madhava Reddy, B. 2004. Effect of *Helicteres isora* root extracts on glucose tolerance in glucose-induced hyperglycemic rats. *Fitoterapia* 75: 364–367.

Venkatesh, S., Reddy, G.D., Reddy, B.M., Ramesh, M., and Rao, A.V. 2003. Anti-hyperglycemic activity of *Caralluma attenuata*. *Fitoterapia* 74: 274–279.

Venkateswaran, S. and Pari, L. 2002. Antioxidant effect of *Phaseolus vulgaris* in streptozotocin-induced diabetic rats. *Asia Pac. J. Clin. Nutr.* 11: 206–209.

Venkateswaran, S., Pari, L., and Saravanan, G. 2002. Effect of *Phaseolus vulgaris* on circulatory anti-oxidants and lipids in rats with streptozotocin-induced diabetes. *J. Med. Food* 5: 93–103.

Venkateswarlu, V., Kokate, C.K., Rambhau, D., and Veeresham, C. 1993. Antidiabetic activity of roots of *Salacia macrosperma*. *Planta Med.* 59: 391–393.

Verma, M., Khatri, A., Kaushik, B., Patil, U.K., and Pawar, R.S. 2010a. Antidiabetic activity of *Cassia occidentalis* (Linn.) in normal and alloxan-induced diabetic rats. *Indian J. Pharmacol.* 42: 224–228.

Verma, P.R., Itankar, P.R., and Arora, S.K. 2013. Evaluation of antidiabetic, antihyperlipidemic and pancreatic regeneration potential of aerial parts of *Clitoria ternatea*. *Rev. Bras. Farmacogn.* 23: 819–829.

Verma, N., Amresh, G., Sahu, P.K., Rao, Ch, V., and Singh, A.P. 2012a. Antihyperglycemic activity of *Woodfordia fruticosa* (Kurz) flowers extracts in glucose metabolism and lipid peroxidation in streptozotocin-induced diabetic rats. *Indian J. Exp. Biol.* 50: 351–358.

Verma, S., Kumar, D., Chaudhary, A.K., Singh, R., and Tyagi, S. 2012b. Anti-diabetic potential of methanolic bark extract of *Ficus religiosa* L. in streptozotocin-induced diabetic rats. *Int. J. Biopharm. Toxicol. Res.* 2: 261–265.

Verma, S., Sharma, H., and Garg, M. 2014. *Phyllanthus amarus*: A review. *J. Pharmacognosy Phytochem.* 3: 18–22.

Verma, S.M., Suresh, K.B., and Verma A. 2010b. Antidiabetic activity of leaves of *Indigofera tinctoria* Linn. (Fabaceae). *Int. J. Toxicol. Pharmacol. Res.* 1: 42–43.

Vessal, G., Akmali, M., Najafi, P., Moein, M.R., and Sagheb, M.M. 2010. Silymarin and milk thistle extract may prevent the progression of diabetic nephropathy in streptozotocin-induced diabetic rats. *Renal Failure* 32: 733–739.

Vessal, M., Hemmati, M., and Vasei, M. 2003. 2003. Antidiabetic effects of quercetin in streptozocin-induced diabetic rats. *Comp. Biochem. Physiol. C: Toxicol. Pharmacol.* 135C: 357–364.

Vetere, A., Choudhary, A., Burns, S.M., and Wagner, B.K. 2014. Targeting the pancreatic β-cells to treat diabetes. *Nat. Rev. Drug Discov.* 13: 278–289.

Vetrichelvan, T. and Jegadeesan, M. 2002. Anti-diabetic activity of alcoholic extract of *Aerva lanata* (L.) Juss. ex Schultes in rats. *J. Ethnopharmacol.* 80: 103–107.

Vetrichelvan, T., Jegadeesan, M., and Devi, B.A. 2002. Anti-diabetic activity of alcohol extract of *Celosia argentea* Linn. seeds in rats. *Biol. Pharm. Bull.* 25: 526–528.

Viana, G.S.B., Lino, C., Diogenes, J.P.L.D., Pereira, B.A., Faria, R.A.P.G., Neto, M.A. et al. 2004a. Antidiabetic activity of *Bouhinia forficata* extracts in alloxan diabetic rats. *Biol. Pharm. Bull.* 27: 125–127.

Viana, G.S.B., Medeiros, A.C.C., Lacerda, A.M.R., Leal, L.K.A.M., Vale, T.G., and deAbreu, M.F.J. 2004b. Hypoglycemic and anti-lipidemic effects of the aqueous extract from *Cissus sicyoides*. *BMC Pharmacol.* 4: 1–7.

Vidhya, R., Rajiv Gandhi, G., Jothi, G., Radhika, J., and Brindha, P. 2012. Evaluation of antidiabetic potential of *Achyranthus aspera* Linn. on alloxan-induced diabetic animals. *Int. J. Pharm. Pharm. Sci.*
4: 577–580.

Vijayakumar, P., Thirumurugan, V., Bharathi, K., Surya, S., Kavitha, M., Murugananthan, G. et al. 2014. An
evaluation of anti-diabetic potential and physico-phytochemical properties of *Tephrosia purpurea* (Linn.)
Pers in streptozotocin-induced diabetes model in albino rats. *Int. J. Pharm. Sci. Rev. Res.* 24: 56–60.

Vijayalakshmi, N., Anbazhagan, M., and Arumugam, K. 2014. Medicinal plants for diabetes used by the
people of Thirumoorthy Hills region of Western Ghats, Tamil Nadu, India. *Int. J. Curr. Microbiol.
Appl. Sci.* 3: 405–410.

Vijayvargia, R., Kumar, M., and Gupta, S. 2000. Hypoglycemic effect of aqueous extract of *Enicostemma
littorale* Blume (chhota chirayata) on alloxan induced diabetes mellitus in rats. *Indian J. Exp. Biol.* 38:
781–784.

Villasenor, I.M. and Lamadrid, M.R. 2006. Comparative antihyperglycemic potential of medicinal plants.
J. Ethnopharmacol. 104: 129–131.

Vinagre, A.S., Ronnau, A.D.S.R., Pereire, S.F., da Silveira, L.U., Wiilland, E.D., and Suyenara, E.S. 2010.
Anti-diabetic effects of *Campomanesia xanthocarpa* (Berg) leaf decoction. *Braz. J. Pharm. Sci.*
46: 169–177.

Vinuthan, M.K., Girish Kumar, V., Ravindra, J.P., Jayaprakash, and Narayana, K. 2004. Effect of extracts of
Murraya koenigii leaves on the levels of blood glucose and plasma insulin in alloxan-induced diabetic
rats. *Indian J. Physiol. Pharmacol.* 48: 384–352.

Virdi, J., Sivakami, S., Shahani, S., Suthar, A.C., Banavalikar, M.M., and Biyani, M.K. 2003. Antihyperglycemic effects of three extracts from *Momordica charantia*. *J. Ethnopharmacol.* 88(1): 107–111.

Visen, P., Saraswat, B., Visen, A., Roller, M., Bily, A., Mermet, C. et al. 2009. Acute effects of *Fraxinus
excelsior* L. seed extract on postprandial glycemia and insulin secretion on healthy volunteers.
J. Ethnopharmacol. 126: 226–232.

Vishnusithan, K.S., Jeyakumar, J.J., Kamaraj, M., and Ramachandran, B. 2014. Anti-diabetic and anti-oxidant
property of *Wattakaka volubilis*. *Int. J. Pharma Res. Rev.* 3: 12–15.

Vishwakarma, S.L., Sonawane, R.D., Rajani, M., and Goyal, R.K. 2010. Evaluation of effect of aqueous
extract of *Enicostemma littorale* Blume in streptozotocin-induced type 1 diabetic rats. *Indian J. Exp.
Biol.* 48: 26–30.

Vispute, S. and Khopade, A. 2011. *Glycyrrhiza glabra* Linn.—Klitaka: A review. *Int. J. Pharm. BioSci.*
2: 42–51.

Viswanathan, R., Sekar, V., Velpandian, V., Sivasaravanan, K.S., and Ayyasamy, S. 2013. Anti-diabetic activity of thottal vadi choornam (*Mimosa pudica*) in alloxan-induced diabetic rats. *Int. J. Nat. Prod. Sci.*
3: 13–20.

Viswanathaswamy, A.H.M., Koti, B.C., Gore, A., Thippeswamy, A.H., and Kulkarni, R.V. 2011a.
Antihyperglycemic and antihyperlipidemic activity of *Plectranthus amboinicus* on normal and alloxan-
induced diabetic rats. *Indian J. Pharm. Sci.* 73: 139–145.

Viswanathaswamy, A.H.M., Koti, B.C., Singh, A.K., and Thippeswam, A.H.M. 2011b. Antihyperglycemic
and antihyperlipidemic effect of *Rubia cordifolia* leaf extract on alloxan-induced diabetes. *RGUHS J.
Pharm. Sci.* 1: 49–54.

Vogl, S., Atanasov, A.G., Binder, M., Bulusu, M., Zehl, M., Fakhrudin, N. et al. 2013. The herbal drug
Melampyrum pratense L. (Koch): Isolation and identification of its bioactive compounds targeting
mediators of inflammation. *Evid. Based Complement. Altern. Med.* 2013: 395316.

Volpato, G.T., Calderon, I.M.P., Sinzato, S., Campos, K.E., Rudge, M.V.C., and Damasceno, D.C. 2011. Effect
of *Moris nigra* aqueous extract treatment on the maternal-fetal outcome, oxidative stress status and lipid
profile of streptozotocin-induced diabetic rats. *J. Ethnopharmacol.* 138: 691–696.

Vosough-Ghanbari, S., Rahimi, R., Kharabaf, S., Zeinali, S., Mohammadirad, A., Amini, S. et al. 2010.
Effects of *Satureja khuzestanica* on serum glucose, lipid, and markers of oxidative stress in patients
with type 2 diabetes mellitus: A double blind randomized controlled trial. *Evid. Based Complement.
Altern. Med.* 7: 465–470.

Vuksan, V., Jenkins, D.J., Spadafora, P., Sievenpiper, J.L., Owen, R., Vidgen, E. et al. 1999. Konjac-mannan
(glucomannan) improves glycemia and other associated risk factors for coronary heart disease in type 2
diabetes. A randomized controlled metabolic trial. *Diabetes Care* 22: 913–919.

Vuksan, V., Sievenpiper, J.L., Vernon, Y.Y.K., Francis, T., Beljan-Zdravkovic, U., Xu, Z. et al. 2000a. American ginseng (*Panax quinquefolis* L.) reduces postprandial glycemia in nondiabetic subjects and subjects with type 2 diabetes mellitus. *Arch. Intern. Med.* 160: 1009–1013.

Vuksan, V., Stavro, M.P., Sievenpiper, J.L., Beljan-Zdravkovic, U., Leiter, L.A., Josse, R.G., and Xu, Z. 2000c. Similar postprandial glycemic reductions with escalation of dose and administration time of American ginseng in type 2 diabetes. *Diabetes Care* 23: 1221–1226.

Vuksan, V., Stavro, M.P., Sievenpiper, J.L., Koo, V.Y., Wong, E., Beljan-Zdravkovic, U. et al. 2000b. American ginseng improves glycemia in individuals with normal glucose tolerance: Effect of dose and time escalation. *J. Am. Coll. Nutr.* 19: 738–744.

Vuksan, V., Sung, M.K., Sievenpiper, J.L., Starvo, P.M., Jenkins, A.L., Di Buono, M. et al. 2008. Korean red ginseng (*Panax ginseng*) improves glucose and insulin regulation in well-controlled study of efficacy and safety. *Nutr. Metab. Cardiovasc. Dis.* 18: 46–56.

Wada, J. and Makino, H. 2009. Historical chronology of basic and clinical research in diabetic nephropathy and contributions of Japanese scientists. *Clin. Exp. Nephrol.* 13: 405–414.

Wadood, N., Wadood, A., and Shah, S.A. 1992. Effect of *Tinospora cordifolia* on blood glucose and total lipid levels of normal and alloxan-diabetic rabbits. *Planta Med.* 58: 131–136.

Waheed, A., Miana, G.A., Ahmad, S.I., and Khan, M.A. 2006. Clinical investigation of hypoglycemic effect of *Coriandrum sativum* in type 2 (NIDDM) diabetic patients. *Pak. J. Pharmacol.* 23: 7–11.

Wang, B.L., Zhang, W., Zhao, S., Wang, F., Fan, L., and Hu, Z. 2011a. Gene cloning and expression of a novel hypoglycemic peptide from *Momordica charantia*. *J. Sci. Food Agric.* 91: 2443–2448.

Wang, H.X. and Ng, T.B. 1999. Natural products with hypoglycemic, hypotensive, hypocholesterolemic, anti-atherosclerotic and antithrombic activities. *Life Sci.* 65: 2663.

Wang, J., Ji, L., Liu, H., and Wang, Z. 2010d. Study of the hepatotoxicity induced by *Dioscorea bulbifera* L. rhizome in mice. *Biosci. Trends* 4: 79–85.

Wang, J.F., Luo, H., Miyoshi, M., Imoto, T., Hiji, Y., and Sasaki, T. 1998. Inhibitory effect of gymnemic acid on intestinal absorption of oleic acid in rats. *Can. J. Physiol. Pharmacol.* 76: 1017–1023.

Wang, J.L., Nong, Y., Xia, G.J., Yao, W.X., and Jiang, M.X. 1993. Effects of linesinine on slow action potentials in myocardium and slow inward current in canine cardiac Purkinje fibers. *Yao Xue Xue Bao* 28: 812–816.

Wang, K., Cao, P., Shui, W., Yang, Q., Tang, Z., and Zhang, Y. 2015. *Angelia sinensis* polysaccharide regulates glucose and lipid metabolism disorder in prediabetic and streptozotocin-induced diabetic mice through the elevation of glycogen levels and reduction of inflammatory factors. *Food Funct.* 6: 902–909.

Wang, L., Waltenberger, B., Pferschy-Wenzing, E.-M., Blunder, M., Liu, X., Malainer, C. et al. 2014. Natural product agonists of peroxisome proliferator-activated receptor gamma (PPARγ): A review. *Biochem. Pharmacol.* 92: 73–89

Wang, L., Zhang, X.T., Zhang, H.Y., Yao, H.Y., and Zhang, H. 2010c. Effect of *Vaccinium bracteatum* Thunb. leaves extract on blood glucose and plasma lipid levels in streptozotocin-induced diabetic mice. *J. Ethnopharmacol.* 130: 465–469.

Wang, L., Zhang, Y., Xu, M., Wang, Y., Cheng, S., Liebrecht, A. et al. 2013. Anti-diabetic activity of *Vaccinium bracteatum* Thunb. leaves' polysaccharide in streptozotocin-induced diabetic mice. *Int. J. Biol. Macromol.* 61: 317–321.

Wang, M., Zhang, W.B., Zhu, J.H., Fu, G.S., and Zhou, B.Q. 2009a. Breviscapine ameliorates hypertrophy of cardiomyocytes induced by high glucose in diabetic rats *via* the PKC signaling pathway. *Acta Pharmacol. Sin.* 30: 1081–1091.

Wang, M., Zhang, W.B., Zhu, J.H., Fu, G.S., and Zhou, B.Q. 2010b. Breviscapine ameliorates cardiac dysfunction and regulates the myocardial Ca(2+)-cycling proteins in streptozotocin-induced diabetic rats. *Acta Diabetol.* 47: 209–218.

Wang, P.H., Tsai, M.J., Hsu, C.Y., Wang, C.Y., Hsu, H.K., and Weng, C.F. 2008b. *Toona sinensis* Roem (Meliaceae) leaf extract alleviates hyperglycemia *via* altering adipose glucose transporter 4. *Food Chem. Toxicol.* 46: 2554–2560.

Wang, S., Fang, M., Ma, Y.L., and Zhang, Y.Q. 2014. Preparation of the branch bark ethanol extract in mulberry *Morus alba*, its antioxidation and antihyperglycemic activity *in vivo*. *Evid. Based Complement. Altern. Med.* 2014: Article ID 569652, 7 pages.

Wang, S., Zheng, Z., Weng, Y., Yu, Y., Zhang, D., Fan, W. et al. Angiogenesis and anti-angiogenesis activity of Chinese medicinal herbal extracts. *Life Sci.* 74: 2467–2478.

Wang, S.C., Lee, S.F., Wang, C.J., Lee, C.H., Lee, W.C., and Lee, H.J. 2011b. Aqueous extract from *Hibiscus sabdariffa* Linnaeus ameliorates diabetic nephropathy *via* regulating oxidative status and Akt/Bad/14–3–3γ in an experimental animal model. *Evid. Based Complement. Altern. Med.* 2011: Article ID 938126, 9 pages.

Wang, Y., Xin, X., Jin, Z., Hu, Y., Li, X., Wu, J. et al. 2011c. Anti-diabetes effects of pentamethylquercetin in neonatally streptozotocin-induced diabetic rats. *Eur. J. Pharmacol.* 668: 347–353.

Wang, Y.N., Shi, G.L., Zhao, L.L., Liu, S.Q., Yu, T.O., Clarke, S.R., and Sun, J.H. 2007. Acaricidal activity of *Juglans regia* leaf extracts on *Tetranychus viennensis* and *Tetranychus cinnabarinus* (Acari: Tetranychidae). *J. Econ. Entomol.* 100: 1298–1303.

Wang, Z.Q., Ribnicky, D.M., Zhang, X.H., Zuberi, A., Raskin, I., Yu, U. et al. 2010a. An extract of *Artemisia dracunculus* L. enhances insulin receptor signaling and modulates gene expression in skeletal muscle in KKay mice. *J. Nutr. Biochem.* 22: 71–78.

Wang, Y.N., Wang, H.X., Shen, Z.J., Zhao, L.L., Clarke, S.R., Sun, J.H. et al. 2009b. Methyl palmitate, an acaricidal compound occurring in green walnut husks. *J. Econ. Entomol.* 102: 196–202.

Wang, Z.Q., Zhang, X., Ribnicky, D.M., Raskin, I., and Cefalu, W.T. 2008a. Bioactives of *Artemisia dracunculus* L. enhance cellular insulin signaling in primary human skeletal muscle culture. *Metabolism* 57(Suppl. 1): S58–S64.

Wang, Z.W., Huang, Z.S., Yang, A.P., Li, C.Y., Huang, H., Lin, X. et al. 2005. Radio-protective effect of *Aloe* polysaccharides on three non-tumor cell lines. *Ai Zheng* 24: 38–42.

Wani, V.K., Dubey, R.D., Verma, S., Sengottuvelu, S., and Sivakumar, T. 2011. Anti-diabetic activity of methanol extract of *Mukia maderaspatana* in alloxan-induced diabetic rats. *Int. J. Pharm. Technol. Res.* 3: 214–220.

Wargent, E.T., Zaibi, M.S., Silvestri, C., Hislop, D.C., Stocker, C.J., Stott, C.G. et al. 2013. The cannabinoid Δ(9)-tetrahydrocannabivarin (THCV) ameliorates insulin sensitivity in two mouse models of obesity. *Nutr. Diabetes* 3: e68. DOI: 10.1038/nutd.2013.9.

Wasfi, I.A., Bashir, A.K., Amiri, M.H., and Abdalla, A.A. 1994. The effect of *Rhazya stricta* on glucose homeostasis in normal and streptozotocin diabetic rats. *J. Ethnopharmacol.* 43: 141–147.

Watal, G., Gupta, R.K., Kesari, A.N., Murthy, P.S., Chandra, R., and Tandon, V. 2005. Hypoglycemic and antidiabetic effect of ethanolic extract of leaves of *Annona squamosa* L. in experimental animals. *J. Ethnopharmacol.* 99: 75–81.

Watanabe, M., Kato, M., and Ayugase, J. 2012. Anti-diabetic effects of adlay protein in type 2 diabetic db/db mice. *Food. Sci. Technol. Res.* 18: 383–390.

Watcho, P., Stavniichuk, R., Ribnicky, D.M., Raskin, H., and Obrosova, I.G. 2010. High-fat diet-induced neuropathy of prediabetes and obesity: Effect of PMI-5011, an ethanolic extract of *Artemisia dracunculus* L. *Mediat. Inflamm.* 2010: 268547.

Wee, J.J., Mee Park, K., and Chung, A.S. 2011. Biological activities of ginseng and its application to human health. In: *Herbal Medicine: Biomolecular and Clinical Aspects*, 2nd edn., Benzie, I.F.F. and Wachtel-Galor, S. (eds.). Boca Raton, FL: CRC Press; Chapter 8. http://www.ncbi.nlm.nih.gov/books/NBK92776/.

Wei, H., Yan, L.H., Feng, W.H., Ma, G.X., Peng, Y., Wang, Z.M., and Xiao, P.G. 2014. Research progress on active ingredients and pharmacologic properties of *Saussurea lappa*. *Curr. Opin. Complement. Altern. Med.* 1: e00005. DOI: 10.7178/cocam.00005.

Wei, L., Li, Z., and Chen, B. 2000. Clinical study on treatment of infantile rotaviral enteritis with *Psidium guajava* L. *Zhongguo Zhong Xi Yi Jie He Za Zhi* 20: 893–895.

Wei, X., Chen, D., Yi, Y., Qi, H., Gao, X., Fang, H. et al. 2012. Syringic acid extracted from *Herba Dendrobii* prevents diabetic cataract pathogenesis by inhibiting aldose reductase activity. *Evid. Based Complement. Altern. Med.* 2012: Article ID 426537, 13 pages.

Weidner, C., de Groot, J.C., Prasad, A., Freiwald, A., Quedenau, C., Kliem, M. et al. 2012. Amorfrutins are potent antidiabetic dietary natural products. *Proc. Natl. Acad. Sci. USA* 109: 7257–7262.

Weinoehrl, S., Feistel, B., Pischel, I., Kopp, B., and Butterweck, V. 2012. Comparative evaluation of two different *Artemisia dracunculus* L. cultivars for blood sugar lowering effects in rats. *Phytother. Res.* 26: 625–629.

Weiss, L., Zeira, M., Reich, S., Har-Noy, M., Mechoulam, R., Slavin, S., and Gallily, R. 2006. Cannabidiol lowers incidence of diabetes in non-obese mice. *Autoimmunity* 39: 143–151.

Welt, K., Fitzl, G., and Schepper, A. 2001. Experimental hypoxia of streptozotocin-diabetic rat myocardium and protective effects of *Ginkgo biloba* extract. II. Ultrastructural investigation of microvascular endothelium. *Exp. Toxicol. Pathol.* 52: 503–512.

Weng, Y., Yu, L., Cui, J., Zhu, Y.-R., Guo, C., Wei, G. et al. 2014. Antihyperglycemic, hypolipidemic and antioxidant activities of total saponins extracted from *Aralia taibaiensis* in experimental type 2 diabetic rats. *J. Ethnopharmacol.* 152: 553–560.

WHO, 1991. Guidelines for the assessment of herbal medicines. World Health Organization, Geneva, Document no. WHO/TRM/91.4.

WHO. 1999a. *Definitions, Diagnosis and Classification of Diabetes Mellitus and Its Complications.* Geneva, Switzerland: Department of Non-Communicable Diseases Sarveillane. Part 1 Diagnosis and Classification of Diabetes Mellitus: 2.

WHO. 1999b. *WHO Monographs on Selected Medicinal Plants.* Geneva, Switzerland: WHO Press.

WHO. 2002. *WHO Monographs on Selected Medicinal Plants*, Vol. 2. Geneva, Switzerland: WHO Press.

WHO. 2007. *WHO Monographs on Selected Medicinal Plants*, Vol. 3. Geneva, Switzerland: WHO Press.

WHO. 2009. *WHO Monographs on Selected Medicinal Plants*, Vol. 4. Geneva, Switzerland: WHO Press.

Wickenberg, J., Ingemansson, S.L., and Hlebowicz, J. 2010. Effects of *Curcuma longa* (turmeric) on postprandial plasma glucose and insulin in healthy subjects. *Nutr. J.* 9: 43–48.

Wien, M.A., Sabate, J.M., Ikle, D.N., Cole, S.E., and Kandeel, F.R. 2003. Almonds vs complex carbohydrates in a weight reduction program. *Int. J. Obes. Relat. Metab. Disord.* 27: 1365–1372.

Wolf, B.W. and Weisbrode, S.E. 2003. Safety evaluation of an extract from *Salacia oblonga*. *Food Chem. Toxicol.* 41: 867–874.

Wolfram, R.M., Kritz, H., Efthimiou, Y., Stomatopoulos, J., and Sinzinger, H. 2002. Effect of prickly pear (*Opuntia robusta*) on glucose- and lipid-metabolism in non-diabetics with hyperlipidemia—A pilot study. *Wien Klin Wochenschr.* 114: 840–846.

Wong, C.M., Yeung, H.W., and Ng, T.B. 1985. Screening of *Trichosanthes kirilowii, Momordica charantia* and *Cucurbita maxima* (family Cucurbitaceae) for compounds with anti-lipolytic activity. *J. Ethnopharmacol.* 13: 313–321.

Wood, I.S. and Trayhurn, P. 2003. Glucose transporters (GLUT and SGLT): Expanded families of sugar transport proteins. *Br. J. Nutr.* 89: 3–9.

Wu, C., Li, Y., Chen, Y., Lao, X., Sheng, L., Dai, R. et al. 2011. Hypoglycemic effect of *Belamcanda chinensis* leaf extract in normal and streptozotocin-induced diabetic rats and its potential active fraction. *Phytomedicine* 18: 292–297.

Wu, H., Guo, H., and Zhao, R. 2006. Effect of *Lycium barbarum* polysaccharide on the improvement of antioxidant ability and DNA damage in NIDD rats. *Yakugaku Zasshi* 126: 365–371.

Wu, K., Liang, T., Duan, X., Xu, L., Zhang, K., and Li, R. 2013. Anti-diabetic effects of puerarin, isolated from *Pueraria lobata* (Willd.), on streptozotocin-diabetogenic mice through promoting insulin expression and ameliorating metabolic function. *Food Chem. Toxicol.* 60: 341–347.

Wu, M.J., Wang, L., Weng, C.Y., and Yen, J.H. 2003. Anti-oxidant activity of the methanol extract of the lotus leaf (*Nulumbo nucifera* Gertn). *Am. J. Chin. Med.* 31: 687–698.

Wu, T., Zhou, X., Deng, Y., Jing, Q., Li, M., and Yuang, L. 2011. In vitro studies of *Gynura divaricata* (L.) DC extracts as inhibitors of key enzymes relevant for type 2 diabetes and hypertension. *J. Ethnopharmacol.* 136: 305–308.

Xi, S., Zhou, G., Zhang, X., Zhang, W., Cai, L., and Zhao, C. 2009. Protective effect of total araloses of *Aralia elata* (Miq) Seem (TASAES) against diabetic cardiomyopathy in rats during the early stage, and possible mechanisms. *Exp. Mol. Med.* 41: 538–547.

Xia, T. and Wang, Q. 2006a. Antihyperglycemic effect of *Cucurbita ficifolia* fruit extract in streptozotocin-induced diabetic rats. *Fitoterapia* 77: 530–533.

Xia, T. and Wang, Q. 2006b. D-chiro-inositol found in *Cucurbita ficifolia* (Cucurbitaceae) fruit extracts plays the hypoglycaemic role in streptozocin-diabetic rats. *J. Pharm. Pharmacol.* 58: 1527–1532.

Xia, X., Yan, J., Shen, Y., Tang, K., Yin, J., Zhang, Y. et al. 2011. Berberine improves glucose metabolism in diabetic rats by inhibition of hepatic gluconeogenesis. *PLoS One* 6: e16556. DOI: 10.1371/journal.pone.0016556.

Xiang-Yang, Q., Wei-Jun, C., Li-Qin, Z., and Bi-Jun, X. 2008. Mogrosides extract from *Siraitia grosvenori* scavenges free radicals *in vitro* and lowers oxidative stress, serum glucose, and lipid levels in alloxan-induced diabetic mice. *Nutr. Res.* 28: 278–284.

Xiao, J.H., Zhang, J.H., Chen, H.L., Feng, X.L., and Wang, J.L. 2005. Inhibitory effect of isoliensinine on bleomycin-induced pulmonary fibrosis in mice. *Planta Med.* 71: 225–230.

Xie, C., Xu, L.Z., Li, X.M., Li, K.M., Zhao, B.H., and Yang, S.L. 2001. Studies on chemical constituents in fruit of *Lycium barbarum* L. *Zhongguo Zhong Yao Za Zhi* 26: 323–324.

Xie, J.T., Wang, C.Z., Li, X.L., Ni, M., Fishbein, A., and Yuan, C.S. 2009. Anti-diabetic effect of American ginseng may not be linked to antioxidant activity: Comparison between American ginseng and *Scutellaria baicalensis* using an ob/ob mice model. *Fitoterapia* 80: 306–311.

Xie, J.T., Wang, A., Mehendale, S., Wu, J., Aung, H.H., Dey, L. et al. 2003. Anti-diabetic effects of *Gymnema yunnanense* extract. *Pharmacol. Re*s. 47: 323–329.

Xin, H., Zhou, F., Liu, T., Li, G.-Y., Liu, J., Gao, Z.-Z. et al. 2012. Icariin ameliorates streptozotocin-induced diabetic retinopathy *in vitro* and *in vivo*. *Int. J. Mol. Sci.* 13: 866–878.

Xing, X., Zhang, Z., Hu, X., Wu, R., and Xu, C. 2009. Antidiabetic effects of *Artimisia sphaerocephala* Krasch gum, a novel food additive in China, on streptozotocin-induced type 2 diabetic rats. *J. Ethnopharmacol.* 125: 410–416.

Xiong, W.T., Gu, L., Wang, C., Sun, H.X., and Lu, X. 2013. Anti-hyperglycemic and hypolipidemic effects of *Cistanche tubulosa* in type 2 diabetic db/db mice. *J. Ethnopharmacol.* 150: 935–945.

Xiu, L.M., Miura, A.B., Yamamoto, K., Kobayashi, T., Song, Q.H., Kitamura, H. et al. 2001. Pancreatic islet regeneration by ephedrine in mice with streptozotocin-induced diabetes. *Am. J. Chin. Med.* 29: 493–500.

Xu, J. and Zou, M.-H. 2009. Molecular insights and the therapeutic targets for diabetic endothelial dysfunction. *Circulation* 120: 1266–1286.

Xu, M., Zhang, H., and Wang, Y. 2002. The protective effects of *Lycium barbarum* polysaccharide on alloxan-induced isolated islet cells damage in rats. *Zhong Yao Cai* 25: 649–651.

Xu, X.X., Zhang, W., Zhang, P., Qi, X.M., Wu, Y.G., and Shen, J.J. 2013. Superior renoprotective effects of the combination of breviscapine with enalapril and its mechanism in diabetic rats. *Phytomedicine* 20: 820–827.

Xu, Z., Wang, X., Zhou, M., Ma, L., Deng, Y., Zhang, H. et al. 2008. The antidiabetic activity of total lignan from *Fructus Arctii* against alloxan-induced diabetes in mice and rats. *Phytother. Res.* 22: 97–101.

Xue, J., Ding, W., and Liu, Y. 2010. Anti-diabetic effects of emodin involved in the activation of PPARgamma on high-fat diet-fed and low dose of streptozotocin-induced diabetic mice. *Fitoterapia* 81: 173–177.

Xue, P.F., Zhao, Y.Y., Wang, B., and Liang, H. 2006. Secondary metabolites from *Potentilla discolor* Bunge (Rosaceae). *Biochem. Syst. Ecol.* 34: 825–828.

Yadav, J.P., Saini, S., Kalia, A.N., and Dangi, A.S. 2008. Hypoglycemic and hypolipidemic activity of ethanolic extract of *Salvadora oleoides* in normal and alloxan-induced diabetic rats. *Indian J. Pharmacol.* 40: 23–27.

Yadav, M., Khan, K.K., and Beg, M.Z. 2012. Medicinal plants used for the treatment of diabetes by the Baiga tribe living in Rewa district M.P. *Indian J. L. Sci.* 2: 99–102.

Yadav, P., Sarkar, S., and Bhatnagar, D. 1997. Action of *Capparis deciduas* against alloxan-induced oxidative stress and diabetes in rat tissues. *Pharmacol. Res.* 36: 221–228.

Yadav, P.P., Ahmad, G., and Maurya, R. 2004b. Furanoflavonoids from *Pongamia pinnata* fruits. *Phytochemistry* 65: 439–441

Yadav, S. and Mukundan, U. 2011. *In vitro* anti-oxidant properties of *Salvia coccinea* Buc'hoz ex. et L. and *Salvia officinalis* L. *Indian J. Fundam. Appl. Life Sci.* 1: 232–238.

Yadav, S.P., Vata, V., Ammini, A.C., and Grover, J.K. 2004a. *Brassica juncea* (rai) significantly prevented the development of insulin resistance in rats fed fructose-enriched diet. *J. Ethnopharmacol.* 93: 113–116.

Yadav, S.P., Vats, V., Dhunnoo, Y., and Grover, J.K. 2004c. Hypoglycemic and anti-hyperglycemic activity of *Murraya koenigii* leaves in diabetic rats. *J. Ethnopharmacol.* 82: 111–116.

Yagi, A., Hegazy, S., Kabbash, A., and Wahab, E.A. 2009. Possible hypoglycemic effect of *Aloe vera* L. high molecular weight fractions on type 2 diabetic patients. *Saudi Pharm. J.* 17: 209–215.

Yajima, H., Ikeshima, E., Shiraki, M., Kanaya, T., Fujiwara, D., Odai, H. et al. 2004. Isohumulones, bitter acids derived from hops, activate both peroxisome proliferator-activated receptor alpha and gamma and reduce insulin resistance. *J. Biol. Chem.* 279: 33456–33462.

Yamada, K., Hogokowa, M., and Fujimoto, S. 2008. Effect of corosolic acid on gluconeogenesis in rat liver. *Diabetes Res. Clin. Pract.* 80: 48–55.

Yamada, K., Hosokawa, M., Yamada, C., Watanabe, R., Fujimoto, S., Fujiwara, H. et al. 2008. Dietary coro-solic acid ameliorates obesity and hepatic steatosis in KK-Ay mice. *Biol. Pharm. Bull.* 31: 651–655.

Yamashiro, S., Noguchi, K., Matsuzaki, T., Miyagi, K., Nakasone, J., Sakanashi, M. et al. 2003. Cardioprotective effects of extracts from *Psidium guajava* L. and *Limonium wrightii*, Okinawan medicinal plants, against ischemia-reperfusion injury in perfused rat hearts. *Pharmacology* 67: 128–135.

Yanardag, R., Bolkents, S., Tabakoglu-Oguz, A., and Ozsoy-Sacan, O. 2003. Effects of *Petroselinum crispum* extracts on pancreatic β-cells and blood glucose of streptozotocin-induced diabetic rats. *Biol. Pharm. Bull.* 26: 1206–1210.

Yang, C.Y., Wang, J., Zhao, Y., Shen, L., Jiang, X., Xie, Z.G. et al. 2010a. Anti-diabetic effects of *Panax notoginseng* saponins and its major anti-hyperglycemic components. *J. Ethnopharmacol.* 130: 231–236.

Yang, H., Ding, C., Duan, Y., and Liu, J. 2005. Variation of active constituents of an important Tibet folk medicine *Swertia mussotii* Franch. (Gentianaceae) between artificially cultivated and naturally distrib-uted. *J. Ethnopharmacol.* 98: 31–35.

Yang, J., Chen, H., Zhang, L., Wang, Q., and Lai, M.X. 2010b. Anti-diabetic effect of standardized extract of *Potentilla discolor* Bunge and identification of its active components. *Drug Dev. Res.* 71: 127–132.

Yang, M.H., Vasquez, Y., Ali, Z., Khan, I.A., and Khan, S.I. 2013. Constituents from *Terminalia* species increase PPARalpha and PPARgamma levels and stimulate glucose uptake without enhancing adipo-cyte differentiation. *J. Ethnopharmacol.* 149: 490–498.

Yang, Q.-H., Liang, Y., Xu, Q., Zhang, Y., Xiao, L., and Si, L.-Y. 2011. Protective effect of tetramethylpyrazine isolated from *Ligusticum chuanxiong* on nephropathy in rats with streptozotocin-induced diabetes. *Phytomedicine* 18: 1148–1152.

Yang, S., Na, M.K., Jang, J.P., Kim, K.A., Kim, B.Y., Sung, N.J. et al. 2006. Inhibition of protein tyrosine phosphatase 1B by lignans from *Myristica fragrans*. *Phytother. Res.* 20: 680–682.

Yang, W.-C. 2014. Botanical, pharmacological, phytochemical, and toxicological aspects of the antidiabetic plant *Bidens pilosa* L. *Evid. Based Complement. Altern. Med.* 2014: Article ID 698617, 14 pages.

Yang, Y.C., Hsu, H.K., Hwang, J.H., and Hong, S.J. 2003. Enhancement of glucose uptake in 3T3-L1 adipo-cytes by *Toona sinensis* leaf extract. *Kaohsiung J. Med. Sci.* 19: 327–333.

Yankuzo, H., Ahmed, Q.U., Santosa, R.I., Akter, S.F.U., and Talib, N.A. 2011. Beneficial effect of the leaves of *Murraya koenigii* (Linn.) Spreng (Rutaceae) on diabetes-induced renal damage *in vivo*. *J. Ethnopharmacol.* 135: 88–94.

Yan-Lin, S., Lin-De, L., and Soon-Kwan, H. 2011. *Eleutherococcus senticosus* as a crude medicine: Review of biological and pharmacological effects. *J. Med. Plants Res.* 5: 5946–5952.

Yanpallewar, S.U., Rai, S., Kumar, M., and Acharya, S.B. 2004. Evaluation of anti-oxidant and neuroprotec-tive effect of *Ocimum sanctum* on transient cerebral ischemia and long-term cerebral hypoperfusion. *Pharmacol. Biochem. Behav.* 79: 155–164.

Yanpallewar, S.U., Rai, S., Kumar, M., Chauhan, S., and Acharya, S.B. 2005. Neuroprotective effect of *Azadirachta indica* on cerebral post-ischemic reperfusion and hypoperfusion in rats. *Life Sci.* 76: 1325–1338.

Yanpallewar, S.U., Sen, S., Tapas, S., Kumar, M., Raju, S.S., and Acharya, S.B. 2003. Effect of *Azadirachta indica* on paracetamol-induced hepatic damage in albino rats. *Phytomedicine* 10: 391–396.

Yarapa, L.R., Rai, S.P., and Basavaraj, T.K. 2012. Evaluation of antidiabetic activity of *Vitis vinifera* stem bark. *J. Pharm. Res.* 5: 5239–5244.

Yarnell, E. 2009. Chapter 11: Integrating botanicals into diabetes treatment protocol. In: *Clinical Botanical Medicine*, 2nd edn. New Rochelle, NY: Mary Ann Liebert Inc.

Yasir, M., Jain, P., Debajyoti, D., and Kharya, M.D. 2010. Hypoglycemic and antihyperglycemic effect of different extracts of *Acacia arabica* Lamk bark in normal and alloxan induced diabetic rats. *Int. J. Phytomed.* 2: 133–138.

Yasuda, K., Kizu, H., Yamashita, T., Kameda, Y., Kato, A., Nash, R.J. et al. 2002. New sugar-mimic alkaloids from the pods of *Angylocalyx pynaertii*. *J. Nat. Prod.* 65: 198–202.

Yazdanparast, R., Ardestani, A., and Jamshidi, S. 2007. Experimental diabetes treated with *Achillea santo-lina*: Effect on pancreatic oxidative parameters. *J. Ethnopharmacol.* 112: 13–18.

Yeh, G.Y., Eisenberg, D.M., Kaptchuk, T.J., and Phillips, R.S. 2003. Systematic review of herbs and dietary supplements for glycemic control in diabetes. *Diabetes Care* 26: 1277–1294.

Yeo, J., Kang, Y.-J., Jeon, S.-M., Jung, U.N., Lee, M.K., Song, H. et al. 2008. Potential hypoglycemic effect of an ethanol extract of *Gynostemma pentaphyllum* in C57BL/KsJ-db/db mice. *J. Med. Food* 11: 709–716.

Yerima, M., Anuka, J.A., Salawu, O.A., and Abdu-Aguve, I. 2014. Antihyperglycaemic activity of the stem-bark extracts of *Tamarindus indica* L. on experimentally induced hyperglycaemic and normoglycaemic Wistar rats. *Pak. J. Biol. Sci.* 17: 414–418.

Yessoufou, A., Ghenou, J., Grissa, O., Hichami, A., Simonin, A., Tabka, Z. et al. 2013. Anti-hyperglycemic effects of three medicinal plants in diabetic pregnancy: Modulation of T cell proliferation. *BMC Complement. Altern. Med.* 13: 77–84.

Ying, H., Siu-Chun, H., and Tak-Yuen, C. 2002. Effect of berberine on regression of pressure-overload induced cardiac hypertrophy in rats. *Am. J. Chin. Med.* 30: 589–599.

Yogisha, S. and Raveesha, K.A. 2010. Dipeptidyl peptidase IV inhibitary activity of *Mangifera indica*. *J. Nat. Prod.* 3: 76–79.

Yokoyama, H., Hiai, S., Oura, H., and Hayashi, T. 1982. Effects of total saponins extracted from several crude drugs on rat adrenocortical hormone secretion. *Yakugaku Zasshi* 102: 555–559.

Yokozawa, T., Kim, H.Y., Cho, E.J., Choi, J.S., and Chung, H.Y. 2002. Anti-oxidant effects of isorhamnetin 3,7-di-O-beta-D-glucopyranoside isolated from mustard leaf (*Brassica juncea*) in rats with streptozoto-cin-induced diabetes. *J. Agric. Food Chem.* 50: 5490–5495.

Yokozawa, T., Kim, H.Y., Cho, E.J., Yamabi, N., and Choi, J.S. 2003. Protective effects of mustard leaf (*Brassica juncea*) against diabetic oxidative stress. *J. Nutr. Sci. Vitaminol.* 49: 87–93.

Yoo, N.H., Jang, D.S., Yoo, J.L., Lee, Y.M, Kim, Y.S., Cho, J.H., and Kim, J.S. 2008. Erigeroflavanone, a flavanone derivative from the flowers of *Erigeron annuus* with protein glycation and aldose reductase inhibitory activity. *J. Nat. Prod.* 71: 713–715.

Yoshikawa, K., Amimoto, K., Arihara, S., and Matsuura, K. 1989. Structure studies of new antisweet constitu-ents from *Gymnema sylvestre*. *Tetr. Lett.* 30: 1103–1106.

Yoshikawa, M., Harada, E., Murakami, T., Matsuda, H., Wariishi, N., Yamahara, J. et al. 1994a. Escins-Ia, Ib, IIa, IIb, and IIIa, bioactive triterpene oligoglycosides from the seeds of *Aesculus hippocastanum* L.: Their inhibitory effects on ethanol absorption and hypoglycemic activity on glucose tolerance test. *Chem. Pharm. Bull.* 42: 1357–1359.

Yoshikawa, M., Matsuda, H., Harada, E., Murakami, T., Wariishi, N., Jamahara, J. et al. 1994b. Elatoside E, a new hypoglycemic principle from the root cortex of *Aralia elata* Seem: Structure-related hypoglycemic activity of oleanolic acid glycosides. *Chem. Pharm. Bull.* 42: 1354–1356.

Yoshikawa, M., Murakami, T., and Matsuda, H. 1997a. Medicinal foodstuffs IX. The inhibitors of glucose absorption from the leaves of *Gymnema sylvestre* R. BR. (Asclepiadaceae): Structures of gymnemo-sides a and b. *Chem. Pharm. Bull (Tokyo)* 45: 1671–1676.

Yoshikawa, M., Murakami, T., Kadoya, M., Matsuda, H., Yamahara, J., Muraoka, O., and Murakami, N. 1996a. Medicinal foodstuff. III. Sugar beet. Hypoglycemic oleanolic acid oligoglycosides, betavulgaro-sides I, II, III, and IV, from the root of *Beta vulgaris* L. (Chenopodiaceae). *Chem. Pharm. Bull. (Tokyo)* 44: 1212–1217.

Yoshikawa, M., Murakami, T., Matsuda, H., Ueno, T., Kadoya, M., Yamahara, J. et al. 1996b. Bioactive sapo-nins and glycosides. II. Senegae radix. (2): Chemical structures, hypoglycemic activity, and ethanol absorption-inhibitory effect of *E*-senegasaponin C, *Z*-senegasaponin C, and *Z*-senegins II, III and IV. *Chem. Pharm. Bull. (Tokyo)* 44: 1305–1313.

Yoshikawa, M., Murakami, T., Matsuda, H., Yamahara, J., Murakami, N., and Kitagawa, I. 1996c. Bioactive saponins and glycosides. III. Horse chestnut. (1): The structures, inhibitory effects on ethanol absorp-tion, and hypoglycemic activity of escins Ia, Ib, IIa, IIb, and IIIa from the seeds of *Aesculus hippocas-tanum* L. *Chem. Pharm. Bull.* 44: 1454–1464.

Yoshikawa, M., Pongpiriyadacha, Y., Kishi, A., Kageura, T., Wang, T., Morikawa, T. et al. 2003. Biological activities of *Salacia chinensis* originating in Thailand: The quality evaluation guided by alpha-glucosi-dase inhibitory activity. *Yakugaku Zasshi*. 123: 871–880.

Yoshikawa, M., Shimada, H., Morikawa, T., Yoshizumi, S., Matsumura, N., Murakami, T., Matsuda, H., Hori, K., and Yamahara, J. 1997b. Medicinal food stuffs. VII. On the saponin constituents with glucose and alcohol absorption-inhibitory activity from a food garnish "Tonburi", the fruit of Japanese *Kochia scoparia* (L.) Schrad.: Structures of scoparianosides A, B, and C. *Chem. Pharm. Bull.* 45: 1300–1305.

Yoshikawa, M., Shimada, H., Nishida, N., Li, Y., Toguchida, I., Yamahara, J. et al. 1998. Antidiabetic prin-ciples of natural medicines. II. Aldose reductase and alpha-glucosidase inhibitors from Brazilian natu-ral medicine, the leaves of *Myrcia multiflora* DC (Myrtaceae): Structures of myrciacitrins I and II and myrciaphenones A and B. *Chem. Pharm. Bull.* 46: 113–119.

Yoshikawa, M., Yoshizumi, S., Ueno, T., Matsuda, H., Murakami, T., Yamahara, J., and Murakami, N. 1995. Medicinal foodstuffs. I. Hypoglycemic constituents from a garnish foodstuff "taranome," the young shoot of *Aralia elata* Seem.: Elatosides G, H, I, J and K. *Chem. Pharm. Bull.* 43: 1878–1882.

Yoshioka, K., Yoshida, T., Kamanaru, K., Hiraoka, N., and Kondo, M. 1990. Caffeine activates brown adipose tissue thermogenesis and metabolic rate in mice. *J. Nutr. Sci. Vitaminol.* 36: 173–178.

Youn, J.Y., Park, H.Y., and Cho, K.H. 2004. Anti-hyperglycemic activity of *Commelina communis* L.: Inhibition of alpha-glucosidase. *Diabetes Res. Clin. Pract.* 66: S149–S155.

Youn, L., Tu, D., Ye, X., and Wu, J. 2006. Hypoglycemic and hypocholesterolemic effects of *Coptis chinensis* franch inflorescence. *Plant Foods Hum. Nutr.* 61: 139–144.

Yu, Z.L., Liu, X.R., McCulloch, M., and Gao, J. 2004. Anticancer effects of various fractions extracted from *Dioscorea bulbifera* on mice bearing HepA. *Zhongguo Zhong Yao Za Zhi.* 29: 563–567.

Yu, B.C., Chang, C.K., Su, C.F., and Cheng, J.T. 2008. Mediation of β-endorphin in andrographolide-induced plasma glucose-lowering action in type I diabetes-like animals. *Naunyn Schmiedebergs Arch. Pharmacol.* 377: 529–540.

Yu, B.C., Hung, C.R., Chen, W.C., and Cheng, J.T. 2003. Antihyperglycemic effect of andrographolide in streptozotocin-induced diabetic rats. *Planta Med.* 69: 1075–1079.

Yu, J.G. and Hu, W.S. 1997. Effects of neferine on platlet aggregation in rabbits. *You Xue Xue Bao* 32: 1–4.

Yu, Y., Liu, L., Wang, X., Liu, X., Xie, L., and Wang, G. 2010. Modulation of glucagon-like peptide-1 release by berberine: *In vivo* and *in vitro* studies. *Biochem. Pharmacol.* 79: 1000–1006.

Yuan, Y., Hartland, K., Boskovic, Z., Wang, Y., Walpita, D., Lysy, P.A. et al. 2013. A small molecule inducer of PDX1 expression identified by high-throughput screening. *Chem. Biol.* 20: 1513–1522.

Yuan-Sung, K., Chien, H.-F., and Lu, W. 2012. *Plectranthus amboinicus* and *Centella asiatica* cream for the treatment of diabetic foot ulcers. *Evid. Based Complement. Altern. Med.* 2012: Article ID 418679, 9 pages.

Yun, T. 2001. Brief introduction of *Panax ginseng* C.A. Meyer. *J Korean Med. Sci.* 16(Suppl.): S3–S5.

Yunlong, C., Guoging, H., Ming, Z., and Huijun, L. 2003. Hypoglycemic effect of the polysaccharide from *Dendrobium moniliforme* (L.) Sw. *J. Zhejiang Univ.* 30: 693–696.

Yusof, R.M. and Said, M. 2004. Effect of high fibre (Guva, *Psidium guajava* L.) on the serum glucose level in induced diabetic mice. *Asia Pac. J. Clin. Nutr.* 13: 135–139.

Zangiabadi, N., Asadi-Shekaari, M., Sheibani, V., Jafari, M., Shabani, M., Asadi, A.R. et al. 2011. Date fruit extract is a neuroprotective agent in diabetic peripheral neuropathy in streptozotocin-induced diabetic rats: A multimodal analysis. *Oxi. Med. Cell. Longev.* 2011: Article ID 976948, 9 pages.

Zapata-Bustos, R., Alonsa-Castro, A.J., Gomez-Sanchez, M., and Salazar-Pliva, L.A. 2014. *Ibervillea sonorae* (Cucurbitaceae) induces the glucose uptake in human adipocytes by activating a PI3K-independent pathway. *J. Ethnopharmacol.* 152: 546–552.

Zapata-Bustos, R., Alonso-Castro, A.J., Romo-Yañez, J., and Salazar-Olivo, L.A. 2009. *Teucrium cubense* induces glucose-uptake in insulin-sensitive and insulin-resistant murine and human adipocytes. *Planta Med.* 75: PJ 137.

Zar Kalai, F., Han, J., Ksouri, R., Abdelly, C., and Isoda, H. 2014. Oral administration of *Nitraria retusa* ethanol extract enhances hepatic lipid metabolism in db/db mice model 'BKS.Cg-Dock7(m)+/+ Lepr(db/)J' through the modulation of lipogenesis-lipolysis balance. *Food. Chem. Toxicol.* 72: 247–256.

Zaruwa, M.Z., Manosroi, J., Akihisa, T., Manosroi, W., and Manosroi, A. 2012. Anti-diabetic activity of *Anogeissus acuminate* a medicinal plant selected from the Thai medicinal plant recipe database MANOSROI II. *Wudpecker J. Med. Plants* 1: 8–15.

Zaruwa, M.Z., Manosroi, A., Akihisa, T., Manosroi, W., Rangdaeng, S., and Manosroi, J. 2013. Hypoglycemic activity of the *Anisopus mannii* N. E. Br. methanolic leaf extract in normal and alloxan-induced diabetic mice. *J. Complement. Integr. Med.* 10: 37–46.

Zarzuelo, A., Risco, S., Gamez, M.J., Jimenez, J.J., Camara, M., and Martinez, M.A. 1990. Hypoglycemic action of *Salvia lavandulifolia* Vahl. spp. oxyodon: A contribution to studies on the mechanism of action. *Life Sci.* 47: 909–915.

Zeggwagh, N.A. and Ouahidi, M.L. 2006. Study of hypoglycaemic and hypolipidemic effects of *Inula viscosa* L. aqueous extract in normal and diabetic rats. *J. Ethnopharmacol.* 108: 233–237.

Zeggwagh, N.A., Sulpice, T., and Eddouks, M. 2007. Anti-hyperglycemic and hypolipidemic effects of *Ocimum bacilicum* aqueous extract in diabetic rats. *Am. J. Pharmacol. Toxicol.* 2: 123–129.

Zeng, X.H., Zeng, X.J., and Li, Y.Y. 2003. Efficacy and safety of berberine for congestive heart failure secondary to ischemic or idiopathic dilated cardiomyopathy. *Am. J. Cardiol.* 92: 173–176.

Zhang, D., Zheng, S., Yang, J., Gu, Y., Xie, X., and Luo, M. 2003. The rat model of type 2 diabetes mellitus and its glycometabolism characters. *Exp. Anim.* 52: 401–407.

Zhang, F., Huang, Y., Hou, T., and Wang, Y. 2006. Hypoglycemic effect of *Artemisia sphaerpcephala* Krasch seed polysaccharide in alloxan-induced diabetic rats. *Swiss Med. Wkly.* 136: 529–532.

Zhang, F., Ye, C., Li, G., Ding, W., Zhou, W., Zhu, H. et al. 2003. The rat model of type 2 diabetic mellitus and its glycometabolism characters. *Exp. Anim.* 52: 401–407.

Zhang, H., Matsuda, H., Yamashita, C., Nakamura, S., and Yoshikawa, M. 2009a. Hydrangeic acid from the processed leaves of *Hydrangea macrophylla* var. *thunbergii* as a new type of anti-diabetic compound. *Eur. J. Pharmacol.* 606: 255–261.

Zhang, J., Xie, X., Li, C., and Fu, P. 2009b. Systematic review of the renal protective effect of *Astragalus membranaceus* (root) on diabetic nephropathy in animal models. *J. Ethnopharmacol.* 126: 189–196.

Zhang, L., He, H., and Balschi, J.A. 2007. Metformin and phenformin activate AMP-activated protein kinase in the heart by increasing cytosolic AMP concentration. *Am. J. Physiol. Heart Circ. Physiol.* 293: 457–466.

Zhang, L., Hogan, S., Li, J.R., Sun, S., Canning, C., Jian Zheng, S. et al. 2011. Grape skin extract inhibits mammalian intestinal α-glucosidase activity and suppresses postprandial glycemic response in streptozocin-treated mice. *Food Chem.* 126: 466–471.

Zhang, L., Mao, W., Guo, X., Wu, Y., Li, C., Lu, Z. et al. 2013. *Ginkgo biloba* extract for patients with early diabetic nephropathy: A systematic review. *Evid. Based Complement. Altern. Med.* 2013: 689142.

Zhang, L., Yang, J., Chen, X.Q., Zan, K., Wen, X.D., Chen, H. et al. 2010a. Antidiabetic and antioxidant effects of extracts from *Potentilla discolor* Bunge on diabetic rats induced by high fat diet and streptozotocin. *J. Ethnopharmacol.* 132: 518–524.

Zhang, M., Chen, M., Zhang, H.Q., Sun, S., Xia, B., and Wu, F.H. 2009c. *In vivo* hypoglycemic effects of phenolics from the root bark of *Morus alba. Fitoterapia* 80: 475–477.

Zhang, R., Zhou, J., Jia, Z., Zhang, Y., and Gu, G. 2004. Hypoglycemic effect of *Rehmannia glutinosa* oligosaccharide in hypoglycemic and alloxan induced diabetic rats and its mechanism. *J. Ethnopharmacol.* 90: 39–43.

Zhang, S., Yang, J., Li, H., Li, Y., Liu, Y., Zhang, D. et al. 2012. Skimmin, a coumarin, suppresses the streptozotocin-induced diabetic nephropathy in Wistar rats. *Eur. J. Pharmacol.* 692: 78–83.

Zhang, S.-Y., Zheng, C.-G., Yan, X.-Y., and Tian, W.-X. 2008. Low concentrations of condensed tannins from catechu significantly inhibit fatty acid synthase and growth of MCF-7 cells. *Biochem. Biophys. Res. Commun.* 371: 654–658.

Zhang, W., Zhao, J., Wang, J., Pang, X., Zhuang, X., Zhu, X., and Qu, W. 2010. Hypoglycemic effect of aqueous extract of seabuckthorn (*Hippophae rhamnoides* L.) seed residues in streptozotocin-induced diabetic rats. *Phytother. Res.* 24: 228–232.

Zhang, Y., Cai, J., Ruan, H., Pi, H., and Wu, Y. 2007. Antihyperglycemic activity of kinsenoside, a high yielding constituent from *Anoectochilus roxburghii* in streptozotocin-diabetic rats. *J. Ethnopharmacol.* 141–145.

Zhang, Y., Chen, C., and Huang, C. 2010. Berberine inhibits PTP1B activity and mimics insulin action. *Biochem. Biophys. Res. Commun.* 397: 543–547.

Zhang, Z., Chen, J., Jiang, X., Wang, J., Yan, X., Zheng, Y. et al. 2014. The *Magnolia* bioactive constituent, 4-O-methylhonokiol protects against high-fat diet-induced obesity and systemic insulin resistance in mice. *Oxi. Med. Cell Longev.* 2014: Article ID 965954, 10 pages.

Zhang, Z., Luo, A., Zhong, K., Huang, Y., Gao, Y., Zhang, J. et al. 2013. α-Glucosidase inhibitory activity by the flower buds of *Lonicera japonica* Thunb. *J. Funct. Food* 5: 1253–1259.

Zhao, G., Li, X., Chen, W., Xi, Z., and Sun, L. 2012b. Three new sesquiterpenes from *Tithonia diversifolia* and their anti-hyperglycemic activity. *Fitoterapia* 83: 1590–1597.

Zhao, R., Li, Q., and Xiao, B. 2005. Effect of *Lycium barbarum* polysaccharide on the improvement of insulin resistance in NIDD rats. *Yakugaku Zasshi* 125: 981–988.

Zhao, L.Y., Lan, Q.J., Huang, Z.C., Ouyang, L.J., and Zeng, F.H. 2011a. Anti-diabetic effect of a newly identified component of *Opuntia dillenii* polysaccharides. *Phytomed. Int. J. Phytother. Phytopharmacol.* 18: 1–13.

Zhao, L.Y., Lan, Q.J., Huang, Z.C., Ouyang, L.J., and Zeng, F.H. 2011b. Antidiabetic effect of a newly identified component of *Opuntia dillenii* polysaccharides. *Phytomedicine* 18: 661–668.

Zhao, R., Li, Q., Long, L., Li, J., Yang, R., and Gao, D. 2007. Antidiabetic activity of flavone from *Ipomoea Batatas* leaf in non-insulin dependent diabetic rats. *Int. J. Food Sci. Technol.* 42: 80–85.

Zhao, R., Li, Q.W., Li, J., and Zhang, T. 2009. Protective effect of *Lycium barbarum* polysaccharide 4 on kidneys in streptozotocin-induced diabetic rats. *Can. J. Physiol. Pharmacol.* 87: 711–719.

Zhao, W., Yin, Y., Yu, Z., Liu, J., and Chen, F. 2012a. Comparison of anti-diabetic effects of polysaccharides from corn silk on normal and hyperglycemia rats. *Int. J. Biol. Macromol.* 50: 1133–1137.

Zheng, J., He, J., Ji, B., Li, Y., and Zhang, X. 2007. Anti-hyperglycemic activity of *Prunella vulgaris* L. in streptozotocin-induced diabetic mice. *Asia Pac. J. Clin. Nutr.* 16: 427–431.

Zheng, X.K., Li, Y.J., Zhang, L., Feng, W.S., and Zhang, X. 2011a. Antihyperglycemic activity of *Selaginella tamariscina* (Beauv.) Spring. *J. Ethnopharmacol.* 133: 531–537.

Zheng, X.K., Wang, W.W., Zhang, L., Su, C.F., Wu, Y.Y., Ke, Y.Y. et al. 2013. Antihyperlipidaemic and antioxidant effect of the total flavonoids in *Selaginella tamariscina* (Beauv.) Spring in diabetic mice. *J. Pharm. Pharmacol.* 65: 757–766.

Zheng, X.K., Xhang, L., Wang, W.W., Wu, Y.Y., Zhang, Q.B., and Feng, W.S. et al. 2011b. Anti-diabetic activity and potential mechanism of total flavonoids of *Selaginella tamariscina* (Beauv.) Spring in rats induced by high fat diet and low dose streptozotocin. *J. Ethnopharmacol.* 137: 662–668.

Zhigila, D.A., Rahaman, A.A.A., Kolawole, O.S., and Oladele, F.A. 2014. Fruit morphology as taxonomic features in five varieties of *Capsicum annuum* L. Solanaceae. *J. Bot.* 2014: Article ID 540868, 6 pages.

Zhno, Y., Son, Y.-O., Kim, S.-S., Jang, Y.-S., and Lee, J.-C. 2007. Anti-oxidant and anti-hyperglycemic activity of polysaccharide isolated from *Dendrobium chrysotoxum* Lindl. *J. Biochem. Mol. Biol.* 40: 670–677.

Zhou, C.X., Qiao, D., Yan, Y.Y., Wu, H.S., Mo, J.X., and Gan, L.S. 2012. A new anti-diabetic sesquiterpenoid from *Acorus calamus*. *Chin. Chem. Lett.* 23: 1165–1168.

Zhou, Q.X., Liu, F., Zhang, J.S., Lu, J.G., Gu, Z.L., and Gu, G.X. 2013. Effects of triterpenic acid from *Prunella vulgaris* L. on glycemia and pancreas in rat model of streptozotozin diabetes. *Chin. Med. J.* 126: 1647–1653.

Zhu, B.H., Guan, Y.Y., He, H., and Lin, M.J. 1999. *Erigeron breviscapus* prevents defective endothelium-dependent relaxation in diabetic rat aorta. *Life Sci.* 65: 1553–1559.

Ziai, S.A., Lariijani, B. Akhoondzadeh, S., Fakhrzadeh, H., Dastpak, A., and Bandarian, F. 2005. Psyllium decreased serum glucose and glycosylated hemoglobin significantly in diabetic outpatients. *J. Ethnopharmacol.* 102: 202–207.

Zia-ul-haq, M., Cavar, S., Qayum, M., Imran, I., and deFeo, V. 2011. Compositional studies: Antioxidant and antidiabetic activities of *Capparis decidua* (Forsk.) Edgew. *Int. J. Mol. Sci.* 12: 8846–8861.

Zia-ul-haq, M., Riaz, M., Feo, V.D., Jaafar, H.Z.E., and Moga, M. 2014. *Rubus fruticosus* L.: Constituents, biological activities and health related uses. *Molecules* 19: 10998–11029.

Zottich, U., Da Cunha, M., Carvalho, A.O., Dias, G.B., Silva, N.C.M., Santos, I.S. et al. 2011. Purification, biochemical characterization and antifungal activity of a new lipid transfer protein (LTP) from *Coffea canephora* seeds with α-amylase inhibitor properties. *Biochim. Biophys. Acta* 1810: 375–383.

Zulet, M.A., Navas-Carretero, S., Sanchez, L.D., Abete, I., Flanagan, J., Issaly, N. et al. 2014. A *Fraxinus excelsior* L. seeds/fruits extract benefits glucose homeostasis and adiposity related markers in elderly overweight/obese subjects: A longitudinal, randomized, crossover, double-blind, placebo-controlled nutritional intervention study. *Phytomedicine* 21: 1162–1169.

Барнаулов, О.Д. and Поспелова, М.Л. 2005. Anti-diabetic properties of flores *Filipendula ulmaria* infusion. *Psychopharmacol. Boil. Narcol.* 5: 1044–1120

Index

Milton Keynes UK
Ingram Content Group UK Ltd.
UKHW052028071024
449327UK00027B/2479